D1546605

MICROWAVE SOLID STATE CIRCUIT DESIGN

MICROWAVE SOLID STATE CIRCUIT DESIGN

INDER BAHL
Gallium Arsenide Technology Center
ITT Corporation
Roanoke, Virginia

PRAKASH BHARTIA
Department of National Defence
National Defence Headquarters
Ottawa, Ontario, Canada

WILEY

A Wiley-Interscience Publication

JOHN WILEY & SONS

New York • Chichester • Brisbane • Toronto • Singapore

Library of Congress Cataloging in Publication Data:

Microwave solid state circuit design.

"A Wiley-Interscience publication."
Includes bibliographies and index.
1. Microwave circuits—Design and construction.
I. Bahl, I. J. II. Bhartia, P.

TK7876.M536 1988 621.381'32 87-31720
ISBN 0-471-83189-1

Printed in the United States of America

10 9 8 7 6 5 4 3 2 1

CONTRIBUTORS

I. J. Bahl, Gallium Arsenide Technology Center, ITT Corporation, Roanoke, VA 24019

P. Bhartia, Department of National Defence, National Defence Headquarters, Ottawa, Ontario, Canada K1A 0K2

E. L. Griffin, Gallium Arsenide Technology Center, ITT Corporation, Roanoke, VA 24019

K. C. Gupta, Electrical Engineering Department, University of Colorado, Boulder, CO 80309

R. G. Harrison, Department of Electronics, Carleton University, Ottawa, Ontario, Canada K1S 5B6

J. Irvine, ITT-DCD, Nutley, NJ 07110

A. P. S. Khanna, Avantek, Inc., Santa Clara, CA 95051

M. Kumar, Microwave Semiconductor Corporation, Somerset, NJ 08873

P. Pramanick, Satellite Communication Department, COMDEV Ltd. Cambridge, Ontario, Canada N1R 7H6

A. K. Sharma, Advanced Technology Department, TRW Electronic Systems Group, Redondo Beach, CA 90278

R. J. Trew, Electrical Engineering Department, North Carolina State University, Raleigh, NC 27650

P. Wahi, The Analytical Sciences Corporation, Mclean, VA 22102

PREFACE

Over the past decade, microwave circuit technology has shifted from the conventional waveguide and coaxial line components and systems to the use of planar circuits. These circuits are characterized by their ability to be etched on a suitable dielectric substrate and are generically referred to as microwave integrated circuits (MIC). In the hybrid form (HMIC), active and passive discrete components (in packaged or chip form), such as transistors, diodes, capacitors, resistors, etc., are attached externally to the etched circuit. However, monolithic microwave integrated circuits (MMICs), wherein all components are fabricated using deposition and etching processes, thereby eliminating discrete components, are fast becoming commonplace.

At present, there is no textbook that provides a comprehensive treatment of passive circuit design, solid state devices, and microwave solid state circuits. Textbooks that are currently in use address only a specific area, such as active or passive devices or semiconductor devices, etc., have very little, if any, design information, little information on state-of-the-art technology and circuits, and do not mention computer-assisted design techniques or MICs. In addition they do not really provide the student or practicing engineer with an adequate background for being productive in industrial positions. This design ability is only developed through hands-on experience in industry, and it is usually many years before a suitable level of competence is developed. This text is designed primarily to fill this gap.

Unlike other texts, which tend to deal more with analysis techniques and leave out synthesis, this text presents a balance that allows a student or designer both to analyze or synthesize a circuit and carry it through to fabrication. While system design is not dealt with due to the diversity of possible systems, their configurations, and the fact that system design depends primarily on the creativity of the designer, the book covers the design procedure for most microwave components.

The individual chapters were written by university professors and field engineers who have demonstrated a special competence and ability in their respective topics and are able to bring out specific issues related to the design of the specific components. With proper coordination and care in manuscript preparation, the negative aspects of edited volumes, such as notational difficulties, have been avoided. The result is a book that is well suited for use in the senior level program in an electrical engineering curriculum or at the first year graduate level. Besides this, the book is ideal for use by designers and engineers in industry, and an extensive bibliography makes it very attractive to the researcher.

The book is divided into seventeen chapters, with the material treated precisely and thoroughly in each chapter. Thus each chapter in the book is essentially self-contained. For material of less interest, the reader is referred to appropriate, easily accessible references. References for future research are also included, and potential future developments in the topic are discussed. Design procedures and examples are provided in each chapter. The step-by-step procedures help to eliminate any doubts and help the student sharpen his or her design skills. In addition, technical information and remarks on various components, devices, and circuits update the reader on the most widely used microwave techniques. Finally, most chapters have a set of suggested exercises.

Chapter 1 provides an introduction to the text and the general subject area. In Chapter 2, the fundamentals of transmission lines and lumped elements are considered, including their characteristics. Since the book is primarily devoted to planar circuits, planar transmission media such as strip lines, microstrip lines, etc., are given greater coverage. Discontinuities and coupling aspects in these media are also covered. The design of lumped elements such as capacitors, inductors, and resistors is described at the end of the chapter.

Chapter 3 deals with resonators. Fundamental parameters are defined, together with circuit models. Emphasis is placed on planar media resonators such as microstrip, slot line, dielectric, etc. YIG resonators are also discussed together with methods for coupling resonators to transmission lines and for electrical and mechanical tuning.

In Chapter 4, impedance-matching networks that are fundamental to any microwave circuit or system are presented. Impedance-matching circuits for narrow band and wide band matching and their design techniques are discussed together with methods for synthesis of such structures. Finally, practical realization aspects are considered.

All of which leads naturally into Chapter 5 on hybrids and couplers. The fundamental differences between the devices are presented together with the methods for design and analysis for both hybrids and couplers. Power dividers and combiners are a form of coupling circuits, and these are also covered in this chapter. For assistance in fabrication consideration, losses, and broadbanding aspects are also dealt with.

Filters and multiplexers, which are again very commonly used components in microwave systems, are discussed extensively in Chapter 6. Analysis and synthesis procedures are described together with design techniques using lumped or distributed elements. This important topic is backed up with a number of design examples, and special fabrication considerations are presented. Finally, design procedures for multiplexers are presented and suitably illustrated by sample designs.

Chapters 7 and 8 deal with active and passive solid state devices, respectively. Basic device principles, characteristics of devices and their applications and limitations are delved into. Three-terminal devices, such as the GaAs FETs, which are fast becoming the most commonly used solid state device, are treated in depth, especially from the point of view of their use in microwave circuits such as oscillators, amplifiers, etc. In Chapter 8, passive devices, such as Schottky-barrier diodes, *pin* diodes, varactor diodes, etc., are covered. Again the treatment is limited to emphasize characteristics that will be of interest to the design engineer.

Using the material in Chapter 7 as background, Chapter 9 deals with oscillator design. This critical component is treated in depth, starting with a section on theory and including aspects such as tuning by various techniques. Both Si bipolar and GaAs transistor oscillators are considered and their design techniques are presented. Dielectric resonator oscillators (DROs), which are gaining widespread acceptance and application, are discussed. Design examples are provided and important factors such as the frequency stability for the device with temperature are discussed. Bandwidth for these devices is always an important consideration, and methods for tuning oscillators, using varactors, optical, and other tuning techniques are presented.

Other critical component analysis, design, and fabrication procedures are dealt with as well. Amplifiers are covered extensively in Chapter 10. Analysis, design procedures, and typical design examples and circuits are presented for GaAs FET amplifiers for both low noise and power amplications. The material presented is adequate to allow the reader to design any of these amplifier types with little prior experience. Similarly, two other components, namely detectors and mixers, are dealt with in the next chapter. Again for each component, the basic theory is presented, with a discussion of the types of mixers and detectors and the methods of design. Design examples and special considerations will provide the designer and student with the insight needed to avoid errors and produce suitable designs the first time around.

A large number of microwave circuit components that can essentially be classified as "microwave control circuits" are covered in Chapter 12, which has the same title. These include switches, phase shifters, attenuators, and modulators. As in previous chapters, the emphasis is on the design of these components, and thus an exposure to the different circuit possibilities for each component, design considerations, limitations of design, etc., are presented. Again suitable design examples for each of these components help to make the material easily understood.

The final type of component design discussed is that for frequency multipliers and dividers in Chapter 13. These circuits are seeing widespread use, particularly in electronic warfare, countermeasures, and support systems. Multiplier and divider circuit types are discussed together with design techniques, realization aspects, and special design considerations. Examples of circuits developed are presented, although the material is not very conductive to illustration of a step-by-step design procedure.

With the common usage of computers, a good discipline in computer-aided design (CAD) of microwave circuits has developed. A number of texts are available on the topic. The fundamentals, however, are presented in Chapter 14, which includes topics such as modeling, analysis and optimization, computer-aided microwave circuit analysis by use of S-matrix or two-dimensional analysis, and compensation of discontinuity reactance effects. Optimization procedures using pattern search method, computations of gradients, and the gradient method are covered. Illustrative examples are included for the different methods, and the material provides the reader with an adequate background in CAD procedures. This allows the student to understand and use many of the commercially available packages for CAD of microwave circuits or to develop his or her own.

Using the material presented in the previous chapters, it is naturally hoped that the reader will embark on the fabrication of microwave integrated circuits (MICs). While mastering the complexity of the manufacture of these circuits can only come from years of practical experience, Chapter 15 exposes one to the types of MICs, their design considerations, fabrication procedures, design criteria, etc. Adequate material is presented to allow the designer to make a good choice of substrates and materials for design and fabrication, for both hybrid MICs and monolithic MICs. Examples of MICs fabricated or in current production are illustrated.

Chapter 16 covers the fast evolution of optical, acoustic, and magnetostatic devices and particularly integrated optics on monolithic substrates. The intension of this chapter is to stimulate thought and encourage consideration of the integration of other technologies with microwaves, should it provide for some advantages. Chapter 17, which is the final chapter, looks broadly along similar lines to identify where microwave/millimeter-wave technology is heading. Even at the time of writing, many of these programs are already taking shape in government or university laboratories, sponsored by funding mainly from the military. The principal challenge lies in converting a larger part of this technology into products that can find a commercial market.

The book contains enough material for a one year course at the senior or graduate level. With judicious selection of specific topics, one can use the book for a one-semester, a one-quarter, two semester, or a two-quarter course. One possible breakup is to divide the material into active and passive circuits and components. The following table recommends a possible organization of topics for typical courses of different duration and levels. No matter what scheme is followed, essential material with design examples can be covered, with the remaining sections being left for self-study to provide the student with a broad overview of the topic. Problems

Course Length	Level	Chapters and Sections Covered	Remarks
One semester 14 Weeks 3 hr/week	Senior	Chapters 1–6 Chapters 8, 11, 12 Selected material Chapters 13–17	Microwave passive circuits course
One semester 15 Weeks 3 hr/week	Senior	Chapter 1 and 2 Chapters 7, 10, 14–17 Selected material Chapter 13	Microwave active circuits course
Two semesters 30 Weeks 3 hr/week	Senior	The whole book	Microwave circuits design course
One semester Graduate 15 Weeks 3 hr/week	Graduate	Quick review of Chapters 1–6, Lectures 7–13 Selected topics 14–17	Microwave circuits design course
Two semesters 30 Weeks 3 hr/week	Graduate	Complete book with greater emphasis on design assignments and system design term paper	Microwave circuits design course
Two semesters 30 Weeks 3 hr/week	Graduate	Chapters 1–6, 12	Microwave passive components
One quarter 10 Weeks 3 hr/week	Senior or graduate	Review Chapter 1, 2, and 7, Lecture 8–13	Microwave active components

are given at the end of most chapters. They have been tested to ensure that their level of difficulty and complexity is suitable for the student.

As with most edited texts, this book has required the cooperation and coordination of the many authors who have contributed to this text. Their patience—and the understanding of their families—in preparing, correcting, and rewriting their manuscripts to result in a comprehensive text is appreciated. Although some errors may remain in the text, special precautions have been taken to ensure that the equations are correct. Suggestions for corrections from the readers will be greatly appreciated.

Many other individuals have been involved in the background in bringing this book to publication. Though it is not possible to mention each of them individually here, we would like to express our sincere appreciation to our managements for their constant encouragement and support.

Dr. I. J. Bahl is indebted to Dr. Dennis Fisher, Director, and Mr Ed Donoho, Senior Manager, for extending facilities at ITT/GTC. Most of the manuscript was keyed-in expertly by Ms. Gradytene M. Pitzer, and Glenn Brookman devoted many hours to the figures. Their dedication and that of other staff members who helped is much appreciated. Finally, we appreciate the patience of our own families for their dedication and support through this exercise.

I. J. Bahl
P. Bhartia

Roanoke, Virginia
Ottawa, Canada
1987

CONTENTS

1 INTRODUCTION 1

P. Bhartia

1.1 Characteristics of Microwaves/Millimeter Waves, 2
1.2 History of Microwave Planar Circuits, 3
1.3 Applications of Microwave Planar Circuits, 4
References, 6

2 TRANSMISSION LINES AND LUMPED ELEMENTS 7

I. J. Bahl

2.1 Transmission Lines, 7
- 2.1.1 Basics of Transmission Lines, 7
- 2.1.2 Characteristics of Conventional Transmission Structures, 10
- 2.1.3 Characteristics of Planar Transmission Lines, 13
- 2.1.4 Comparison of Various MIC Transmission Media, 24

2.2 Coupled Lines, 26
2.3 Discontinuities, 33
- 2.3.1 Coaxial-Line Discontinuities, 34
- 2.3.2 Rectangular Waveguide Discontinuities, 37
- 2.3.3 Strip-line Discontinuities, 38
- 2.3.4 Microstrip Discontinuities, 39
- 2.3.5 Discontinuity Compensation, 45

2.4 Lumped Elements, 45
- 2.4.1 Design of Lumped Elements, 46
- 2.4.2 Design of Inductors, 46
- 2.4.3 Design of Capacitors, 52
- 2.4.4 Design of Resistors, 55

References, 57
Problems, 62

3 RESONATORS 65

A. K. Sharma and A. P. S. Khanna

3.1 Introduction, 65
3.2 Resonator Parameters, 65

3.2.1 Resonant Frequency, 65
3.2.2 Quality Factor, 66
3.2.3 Fractional Bandwidth, 67
3.2.4 Loaded Quality Factor, 67
3.2.5 Damping Factor, 67
3.2.6 Coupling, 68
3.3 Cavity Resonators, 69
3.3.1 Coaxial Resonators, 69
3.3.2 Reentrant Coaxial Resonators, 71
3.3.3 Rectangular Waveguide Resonators, 72
3.3.4 Circular Waveguide Resonators, 73
3.3.5 Elliptic Waveguide Resonators, 76
3.4 Planar Microstrip Resonant Structures, 77
3.4.1 Rectangular Microstrip Resonators, 79
3.4.2 Circular Microstrip Disk Resonators, 79
3.4.3 Circular Microstrip Ring Resonators, 82
3.4.4 Triangular Microstrip Resonators, 84
3.4.5 Hexagonal Microstrip Resonators, 89
3.4.6 Elliptic Microstrip Disk Resonators, 90
3.4.7 Interacting Resonant Structures, 92
3.5 Resonant Structures in Fin Lines, 94
3.6 Dielectric Resonators, 98
3.6.1 Material, 99
3.6.2 Resonant Frequency, 100
3.6.3 Coupling of a Dielectric Resonator in MIC Configuration, 103
3.6.4 Spurious Modes, 107
3.6.5 Frequency Tuning, 108
3.7 YIG Resonators, 109
3.7.1 Resonant Frequency, 111
3.7.2 Frequency of Operation and Quality Factor, 112
3.7.3 Equivalent Circuit, 115
3.7.4 Spurious Modes, 118
3.7.5 Magnetic Tuning Circuit, 119
3.8 Resonator Measurements, 120
3.8.1 Single-Port Resonator, 121
3.8.2 Two-Port Resonator, 124
References, 126
Problems, 129

4 IMPEDANCE-MATCHING NETWORKS **131**
P. Wahi

4.1 Introduction, 131
4.1.1 Importance and Applications, 131

4.1.2 One-Port and Two-Port Networks, 132
4.1.3 Transmission-Line Matching Schemes, 133
4.2 Lossless Matching Networks, 138
4.2.1 Transfer Function, 138
4.2.2 Network Theory, 141
4.2.3 Ladder Networks, 145
4.2.4 Approximate Solution, 150
4.3 Impedance-Matching Circuits, 153
4.3.1 Impedance Transformations: Lumped or Distributed, 153
4.3.2 Gain-Bandwidth Limitations, 159
4.3.3 Real Frequency Approach, 162
4.4 Network Synthesis and Optimization, 163
4.4.1 Insertion-Loss Synthesis, 164
4.4.2 Topology Selection/Parasitic Element Inclusion, 166
4.4.3 Microwave Realization, 169
4.5 CAD Tools, 170
References, 171
Problems, 172

5 HYBRIDS AND COUPLERS **173**
P. Bhartia and P. Pramanick

5.1 Introduction, 173
5.1.1 Basics of Hybrids and Couplers, 173
5.1.2 Types of Hybrids and Couplers, 175
5.1.3 Applications, 177
5.2 Design of Hybrids, 178
5.2.1 90° Hybrid, 178
5.2.2 Ring Form of Branch-Line Hybrid, 181
5.2.3 Matched Hybrid T (Rat-Race Hybrid), 181
5.3 Coupled-Line Directional Couplers, 185
5.3.1 Directional Couplers Using Aperture-Coupled Lines, 187
5.3.2 TEM Line Directional Couplers, 198
5.3.3 Multiconductor Couplers, 209
5.3.4 Distributed-Type Couplers, 211
5.3.5 Wilkinson Couplers, Power Dividers, and Combiners, 214
5.3.6 Other Couplers, 222
5.4 Design Considerations, 231
5.4.1 Losses in Hybrids, 231
5.4.2 Directivity Improvement, 231
References, 233
Problems, 235

6 FILTERS AND MULTIPLEXERS **237**

E. L. Griffin and I. J. Bahl

6.1 Introduction, 237
  6.1.1 Filter Parameter Definition, 239
  6.1.2 Basic Types, 243
  6.1.3 Applications, 243
6.2 Filter Measurements, 245
  6.2.1 Insertion Loss and Return Loss, 245
  6.2.2 *S*-Parameters, 246
6.3 Filter Synthesis, 246
  6.3.1 Filter Design from Low-Pass Filter Synthesis, 246
  6.3.2 Special Response Filter Synthesis, 254
  6.3.3 Filter Transformations, 256
  6.3.4 Impedance and Admittance Inverters, 262
6.4 Experimental Method of Designing Filters, 265
6.5 Filter Modeling, 271
  6.5.1 Narrow-Band Approximation, 271
  6.5.2 Filter Analysis, 271
6.6 Numerical Techniques, 275
6.7 Filter Realizations, 276
  6.7.1 Printed Circuit Filters, 277
  6.7.2 Dielectric Resonator Filters, 284
6.8 Practical Considerations, 287
  6.8.1 Size, Weight, and Cost, 287
  6.8.2 Finite *Q*, 288
  6.8.3 Power-Handling Capability, 290
  6.8.4 Temperature Effects, 290
  6.8.5 Group Delay, 293
  6.8.6 Electrically Tuned Filters, 293
6.9 Multiplexers, 293
  6.9.1 Multiplexing Techniques, 294
  6.9.2 Diplexer Design, 298
  6.9.3 Multiplexer Realization, 299
References, 299
Problems, 302

7 ACTIVE DEVICES **304**

R. J. Trew

7.1 Introduction, 304
7.2 Basic Semiconductor Device Equations, 304
7.3 Material Parameters, 307
7.4 Bipolar Transistors, 311

7.4.1 Basic Transistor Operation, 312
7.4.2 Current Gain, 316
7.4.3 Limitations and Second-Order Effects, 318
7.4.4 Microwave Transistor, 319
7.4.5 Equivalent Circuit, 323
7.4.6 Noise Figure Analysis, 326
7.5 Field-Effect Transistors, 329
7.5.1 Basic Operation Principles, 330
7.5.2 MESFET Model, 337
7.5.3 Small-Signal Model, 341
7.5.4 Equivalent Circuit and Figures of Merit, 346
7.5.5 Noise Figure Analysis, 350
7.5.6 Arbitrary Doping Profile Model and Deep Levels, 356
7.5.7 Power FETs, 358
7.6 Comparison of Bipolar Transistor and MESFET Noise Figures, 366
References, 367
Problems, 369

8 PASSIVE DEVICES **373**
R. J. Trew

8.1 Introduction, 373
8.2 *pn* Junctions, 374
8.2.1 The Ideal Diode Equation, 378
8.2.2 Deviations from the Ideal Diode Equation, 381
8.2.3 Junction Capacitance, 384
8.3 Schottky–Barrier Junctions, 386
8.3.1 Surface Effects, 388
8.3.2 Image Force Lowering, 389
8.3.3 Schottky Model, 392
8.3.4 Junction Capacitance, 396
8.3.5 Rectifying Contact Materials, 397
8.3.6 Series Resistance, 398
8.3.7 Equivalent Circuit, 400
8.3.8 Figure of Merit, 400
8.4 Varactor Diodes, 401
8.4.1 Equivalent Circuit, 402
8.4.2 Figures of Merit, 405
8.5 Varistors, 408
8.6 *pin* Diodes, 409
8.6.1 Basic Device Physics, 409
8.6.2 Switching Speed, 412
8.6.3 Equivalent Circuit, 412

8.6.4 Figure of Merit, 413
8.7 Step Recovery Diodes, 416
8.7.1 Basic Device Physics, 416
8.7.2 Frequency Limits, 421
8.7.3 Equivalent Circuit, 421
References, 422
Problems, 422

9 OSCILLATORS **426**
A. P. S. Khanna

9.1 Introduction, 426
9.2 Concept of Negative Resistance, 428
9.3 Three-Port *S*-Parameter Characterization of Transistors, 429
9.4 Oscillation and Stability Conditions [7], 431
9.5 Fixed-Frequency Oscillators, 435
9.5.1 Design of Oscillators, 436
9.5.2 Temperature Stability of DROs, 449
9.5.3 Tuning of Transistor DROs, 452
9.6 Wide-Band Tunable Oscillators, 456
9.6.1 YIG-Tuned Oscillators (YTO), 457
9.6.2 Voltage-Controlled Oscillators, 462
9.7 Oscillator Measurements, 467
9.7.1 Measurements Using Network Analyzer, 467
9.7.2 Measurement of Pulling Figure, 472
9.7.3 Measurement of FM Noise, 474
References, 479
Problems, 481

10 AMPLIFIERS **483**
I. J. Bahl and E. L. Griffin

10.1 Introduction, 483
10.2 Amplifier Characterization, 485
10.2.1 Power Gain, 485
10.2.2 Noise Characterization, 487
10.2.3 Stability, 492
10.2.4 Nonlinear Behavior, 496
10.2.5 Dynamic Range, 498
10.3 Biasing Networks, 499
10.4 Linear Amplifier Design, 502
10.4.1 FET Selection, 504
10.4.2 Narrow-Band Low-Noise Design, 504
10.4.3 Maximum Gain Amplifier Design, 508

10.4.4 Broadband Amplifiers, 509
10.5 Power Amplifiers, 514
10.5.1 Selection of Power FET, 515
10.5.2 Large-Signal Characterization, 519
10.5.3 Power Amplifier Design, 523
10.5.4 Design of Internally Matched Power FET Amplifiers, 525
10.5.5 Power-Combining Techniques, 526
References, 534
Problems, 536

11 DETECTORS AND MIXERS **540**

J. Irvine

11.1 Introduction, 540
11.1.1 Basics of Detection: Video and Heterodyne, 542
11.1.2 Applications, 544
11.2 Detectors, 546
11.2.1 Basic Theory, 546
11.2.2 Types of Detectors, 560
11.2.3 Detector Devices, 562
11.2.4 Design Considerations, 562
11.2.5 Detector Design Examples, 564
11.3 Mixers, 569
11.3.1 Basic Theory, 570
11.3.2 Types of Mixers, 576
11.3.3 Analysis Techniques, 586
11.3.4 Design Considerations, 591
11.3.5 Mixer Design Examples, 592
References, 596
Problems, 598

12 MICROWAVE CONTROL CIRCUITS **601**

K. C. Gupta

12.1 Devices for Microwave Control Circuits, 601
12.1.1 *pin* Diodes, 601
12.1.2 GaAs MESFETs, 602
12.2 Design of Switches, 604
12.2.1 Basic Configurations, 604
12.2.2 Insertion Loss and Isolation, 606
12.2.3 Compensation of Device Reactances, 609
12.2.4 Single-Pole Double-Throw Switches, 611
12.2.5 Series-Shunt Switching Configurations, 613

12.2.6 Switching Speed Considerations, 621
12.3 Design of Phase Shifters, 626
12.3.1 General, 626
12.3.2 Switched-Line Phase Shifters, 628
12.3.3 Loaded-Line Phase Shifters, 631
12.3.4 Reflection-Type Phase Shifters, 637
12.3.5 Switched-Network Phase Shifters, 648
12.3.6 Amplifier-Type Phase Shifters, 652
12.4 Design of Limiter Circuits, 659
12.4.1 Various Phenomena Used for Limiting, 659
12.4.2 *pin* Diode Limiters, 663
12.4.3 Limiters in Microstrip Configuration, 664
12.5 Design of Variable Attenuators, 666
12.5.1 *pin* Diode Attenuators, 667
12.5.2 MESFET Attenuators, 669
References, 671
Problems, 672

13 FREQUENCY MULTIPLIERS AND DIVIDERS, 676
R. G. Harrison

13.1 Introduction, 676
13.1.1 Basics of Frequency Multiplication and Division, 676
13.1.2 Applications, 677
13.2 Frequency Multiplication, 687
13.2.1 Types of Multipliers, 687
13.2.2 Passive Multipliers Using Diodes, 687
13.2.3 Parametric Multipliers Using Varactors, 691
13.2.4 Active Multipliers Using GaAs FETs, 699
13.3 Frequency Division, 717
13.3.1 Types of Dividers, 717
13.3.2 Parametric Frequency Dividers Using Varactors, 717
13.3.3 Mixer-with-Feedback Dividers, 727
13.3.4 Digital Frequency Dividers, 736
References, 744
Problems, 751

14 COMPUTER-AIDED DESIGN 754
K. C. Gupta

14.1 Basic Aspects of CAD, 754
14.2 Modeling of Circuit Components, 756

14.3 Computer-Aided Analysis Techniques, 759
14.3.1 General Scattering-Matrix Analysis, 760
14.4 Circuit Optimization, 765
14.4.1 Pattern Search Optimization Method, 767
14.4.2 Conjugate Gradient Method, 768
14.5 CAD of Nonlinear Circuits, 773
14.5.1 Linear and Nonlinear Subnetworks, 774
14.5.2 Harmonic Balance Method, 775
14.5.3 Optimization of Nonlinear Circuits, 776
14.6 Use of Supercomputers, 777
References, 778

15 MICROWAVE INTEGRATED CIRCUITS **781**
M. Kumar and I. J. Bahl

15.1 Introduction, 781
15.2 Materials, 782
15.2.1 Substrate Materials, 782
15.2.2 Conductor Materials, 784
15.2.3 Dielectric Materials, 785
15.2.4 Resistive Films, 785
15.3 Mask Layouts and Mask Fabrication, 786
15.3.1 Mask Layouts, 786
15.3.2 Mask Fabrication, 790
15.4 Hybrid Microwave Integrated Circuits, 792
15.4.1 Hybrid Microwave Integrated Circuits, 792
15.4.2 Miniature Hybrid Microwave Integrated Circuits, 796
15.5 Monolithic Microwave Integrated Circuits, 804
15.5.1 Brief History, 805
15.5.2 Why GaAs for MMICs, 805
15.5.3 Design Considerations, 807
15.5.4 Design Procedure, 808
15.5.5 MMIC Fabrication, 810
15.5.6 Examples of MMICs, 817
15.6 Hybrid versus Monolithic MICs, 827
15.6.1 Cost, 829
15.6.2 Size and Weight, 830
15.6.3 Design Flexibility, 831
15.6.4 Circuit Tweaking, 831
15.6.5 Broadband Performance, 831
15.6.6 Reproducibility, 832
15.6.7 Reliability, 832
References, 833
Problems, 835

16 MICROWAVE OPTIC, ACOUSTIC, AND MAGNETOSTATIC CIRCUITS **836**
P. Wahi

16.1 Introduction, 836
16.2 Microwave Modulation of Optical Sources, 838
16.2.1 Direct Modulation, 838
16.2.2 Indirect Modulation, 843
16.3 Fiber-Optic RF Links, 848
16.4 RF/Optical Interaction, 855
16.5 Optical Control of Microwave Devices, 867
16.5.1 Switching Applications, 869
16.5.2 Oscillator Tuning, 872
16.5.3 Injection Locking, 873
16.6 Optical Techniques for Millimeter-Wave Circuits, 875
References, 879
Problems, 881

17 FUTURE TRENDS IN MICROWAVE CIRCUITS **883**
I. J. Bahl

17.1 Monolithic Microwave Integrated Circuits, 883
17.1.1 MMIC Systems, 884
17.1.2 Millimeter-Wave Monolithic ICs, 884
17.2 Optics for Microwave Applications, 885
17.2.1 Microwave Fiber Optics, 885
17.2.2 Optical Control of Microwave Devices, 886
17.3 Microwave Acoustic Technology, 886
17.4 Magnetostatic-Wave Technology, 887

APPENDIX A. UNITS AND SYMBOLS **888**

APPENDIX B. PHYSICAL CONSTANTS AND OTHER DATA **891**

APPENDIX C. *ABCD* AND *S*-PARAMETERS **892**

APPENDIX D. TRANSFER FUNCTION RESPONSES **897**

INDEX **899**

MICROWAVE SOLID STATE CIRCUIT DESIGN

1 INTRODUCTION

The history of the development of microwave circuits has in many ways followed that of the lower frequency electronics circuits. There have been constant pressures in both these disciplines to move from tubes to solid state devices, from large components to small, and to the development of integrated circuits, devices, and systems. However, unlike the electronics field, where a large public impact was possible and strong consumer demand generated through ownership of radios, digital watches, calculators, television sets, video cassette recorders, etc., the microwave oven seems to be the single piece of equipment that comes to the layman's mind when the word microwaves is mentioned. The greatest impact and perhaps use of microwave circuits and systems has been in areas such as communications, radar, electronic warfare, navigation, surveillance, and weapon guidance systems, which are largely military in nature and have been supported strongly by the defense community. Although there is also a lucrative market for defense-products-oriented industries, the profits are often limited by small volume, limited customers due to export restrictions, and hence the drive for development in the area of microwave circuits seems traditionally to have lagged behind that of electronic circuits. For example, electronic integrated circuits (ICs) on chips were available commercially long before microwave integrated circuits (MICs) were. The former technology has matured to such a degree as to allow manufacturers to offer digital watches, calculators, etc., for a few dollars, whereas, yield and design problems still plague the MIC manufacturer with circuits still costing in the hundreds of dollars or more. Under Department of Defense sponsorship a major program was launched a few years ago to develop very high speed integrated circuits (VHSIC). This program was meant to correct the problem of semiconductor manufacturers not having any incentives to design complex-function chips needed for unique military applications. The investment could not be justified at the advanced development stage of a system due to lack of assurances that the product would go into full production. But if the system was approved for production, usually there was little or no time to develop these chips. The microwave circuit market being even more defense oriented was a prime candidate for a similar effort and only recently has the Department of Defense initiated a program for the microwave/millimeter-wave circuits with similar goals as the VHSIC program. The program called MIMIC, an acronym for Microwave/Millimeter-wave Monolithic Integrated Circuits, will mimic the VHSIC effort, and the name reflects the program similarities to the VHSIC program. It is hoped that spin-offs to the commercial world will perhaps broaden the meaning of the word microwaves to the public by making available low cost commercial consumer systems such as direct broadcast satellite (DBS) receivers and cellular mobile communication radios (CMCR).

1.1 CHARACTERISTICS OF MICROWAVES/MILLIMETER WAVES

The term microwave/millimeter wave generally refers to the frequency range where wavelengths are of the order of centimeters down to 1 mm. Some authors have suggested that microwave/millimeter-wave spectrum corresponds to the frequency range 1 GHz to 300 GHz (wavelength 30 cm to 1 mm), but the generally accepted convention is that the frequency range 300 MHz to 300 GHz (wavelength 100 cm to 1 mm) is more appropriate. Figure 1.1 depicts the position of the microwave and millimeter-wave part of the spectrum in relation to the rest of the electromagnetic spectrum.

In the microwave/millimeter-wave region, for convenience, a letter band designation has also been used to indicate the portion of the spectrum being referred to. Over the years, the radar community, etc., have created their own band designations, leading to general confusion. For example, *K*-band designates the frequency ranges 18–26.5 GHz, 19–27 GHz, 10.9–36 GHz, and 20–40 GHz in the radar users terminology, the U.K. frequency designation, the old U.S. military, and the new U.S. military designations, respectively. For the sake of clarity and uniformity, only the new U.S. military designation should be used. Figure 1.2 depicts the frequency band breakdown for the microwave/millimeter-wave region of interest, for the purposes of this test.

Some other features of microwave/millimeter waves are worth noting; for example, the decrease in wavelength as frequency increases corresponds to a reduction in component size, resulting in more compact systems and narrow beamwidths, which allow for greater resolution and precision in target tracking, etc. Many advantages and disadvantages of microwave via-à-vis millimeter wave are discussed in Reference 1.

Another interesting and important characteristic is the free-space propagation attenuation over the frequency range. As shown in Fig. 1.3, atmospheric absorption increases with frequency, but in addition, there exist high absorption windows in the millimeter-wave band due to atmospheric

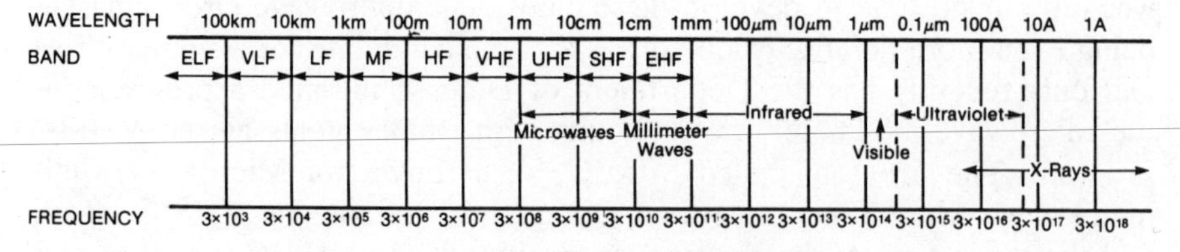

Figure 1.1 The electromagnetic spectrum.

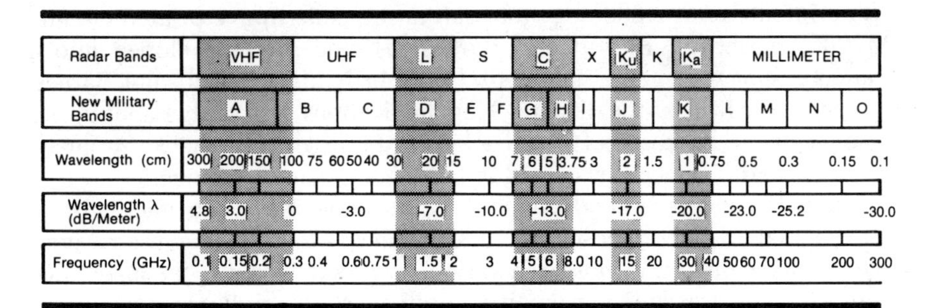

Figure 1.2 Band designation chart.

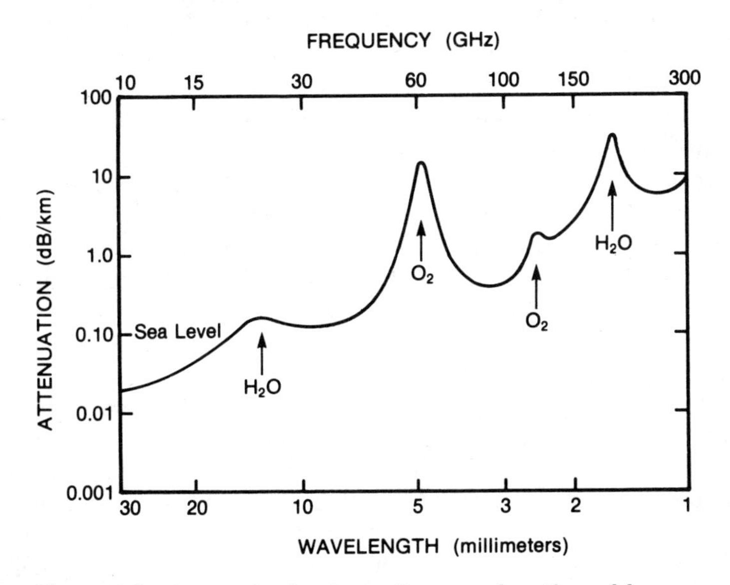

Figure 1.3 Atmospheric absorption as a function of frequency.

water vapor and oxygen. While this characteristic is not of significant interest for the purposes of this book in microwave circuits, it should be realized that a similar behavior occurs with transmission media, and one must choose the appropriate low-loss media for the frequency of operation of one's circuit.

1.2 HISTORY OF MICROWAVE PLANAR CIRCUITS

The first evolution of the conventional waveguide and coaxial line was a flat-strip coaxial transmission line used by Rumsey and Jamieson—as mentioned, for example, by Barrett [2] for producing an antenna system and power division network during World War II. This was gradually integrated with printed-circuit technology to result in the "microwave printed circuit" (MPC) reported by Barrett and Barnes [3], and soon developed into a printed-circuit waveguide handbook [4]. Shortly after Barrett and Barnes' [3] report on MPCs, the Federal Communications Research Laboratories announced the microstrip [5] and King [6] reported the "dielectric image line." The MPC concept was used in fabricating many components, such as directional couplers, filters, attenuators, and antennas. The microstrip is fundamental to integrated-circuit design, and is today being used extensively in its many forms in a large variety of circuits. Finally, the image line is an excellent transmission medium, particularly for millimeter-wave circuits.

Following the preceding developments, the microwave integrated circuits area grew rapidly in the 1960s. Between 1964 and 1968, the largest quantity production for a MIC was perhaps the 600 transmit/receive (T/R) modules produced by Texas Instruments for the MERA radar [7]. Many other significant developments also occurred over the 1960–1980 period, and it was evident that the field of microwave integrated circuits was fast maturing. Over these years, the idea of monolithic microwave integrated circuits (MMICs) also evolved, where all microwave functions of analog circuits, as well as new digital applications, could be incorporated on a single chip [8]. In the earlier MICs and MMICs, high-resistivity *p*-type (boron) silicon was used as the microwave substrate and host for the devices. However, two factors have been primarily responsible for the emergence of gallium arsenide (GaAs) in the development of MMICs. First, the semi-insulating substrate is almost an ideal dielectric medium for microstrip transmission; second is the GaAs field-effect transistor (FET), which is the workhorse of all analog ICs. This latter has benefited from the application of silicon processing technology. Thus, GaAs technology promises a new breed of components that will allow designers greater flexibility and lower cost approaches to achieve their design goals. Finally, the MIMIC program discussed before will undoubtedly considerably enhance the growth pace in MMICs, and will increase MMIC popularity by allowing for increased levels of integration. This will lead to lower costs, higher yields, and better performance, reproducibility, and reliability.

1.3 APPLICATIONS OF MICROWAVE PLANAR CIRCUITS

Microwave planar circuits can be applied to and substituted for the conventional form of microwave circuitry in virtually every application in the fields of communications, electronic warfare, radar, and weapon systems. In general, the limitations are few, one fundamental one being the power-handling capability. However, realization of good matching circuitry has allowed high power levels to be achieved in some laboratory MMICs, but to achieve high powers while maintaining high yield still requires hybrid MIC as well as conventional waveguide and coaxial line techniques. In many cases, requiring high power, such as space-based radar, the use of thousands of solid state transmitter-receivers for an active aperture phased array, allows for power distribution and hence use of MMICs or MICs. Hopefully, the extensive number of T/R modules required will allow for a substantial cost reduction for MMICs over hybrid MICs or conventional circuitry.

Another area that should see high microwave planar circuit usage is electronic warfare. In particular, expendable systems such as expendable jammers and smart munitions using microwave/millimeter-wave guidance systems, both passive radiometer type and active radar, should benefit from the cost reductions and lower weight properties of these circuits. Naturally, satellite systems, where weight is an all-important issue due to high cost per kilogram of launched payload, also stand to benefit significantly from the use of this technology.

Similarly, receivers and transmitters for communications, electronic support measures, electronic countermeasures, and systems for electronic communication or signal intelligence (ELINT, COMINT, SIGINT) all make extensive use of microwave/millimeter-wave circuits and devices. As systems become more complex due to the nature of the electronic threat, hardware complexity increases, resulting in larger and heavier systems. Use of planar technology, and in particular MMICs, helps to achieve significant reduction in both these factors, thus making it possible to deploy these systems easily on aircraft, where space and power are driving constraints.

Most radars being built currently use hybrid circuitry that needs tweaking for optimum performance. In these cases, microwave planar circuits offer the advantages of smaller size, lighter weight, potentially lower cost, high reliability, broad bandwidth capability, and function reproducibility. These advantages allow the development of active element phased radars with significant beam agility, multifunction capability and reliability. In addition, in most applications the redundancy of the T/R modules that can be built in, allows for the desirable feature of graceful degradation in case of failure. A typical active element configuration for one of these is shown in Fig. 1.4.

While the military and space applications stand to gain the most from use of planar microwave/millimeter-wave circuits, other unconventional applications, such as highway-traffic control using microwave systems, microwave sensor systems used in microwave heating and drying will likewise see cost reductions and smaller sizes upon the availability of multifunction monolithic circuits.

The advances in microwave/millimeter-wave planar circuits coupled with advances that are currently occurring in electrooptics, magnetooptics, microwave-optics, and microwave acoustics point to exciting decades ahead for engineering and the sciences. In particular, these fields will have strong impact on consumer electronics and gadgets, while at the same time

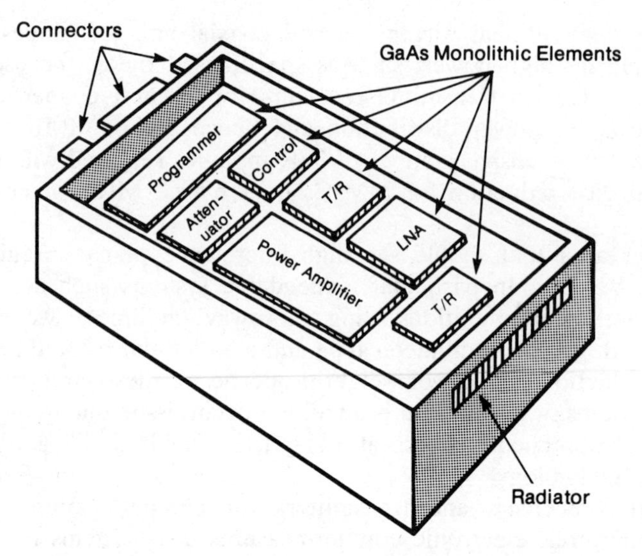

Figure 1.4 A T/R module using GaAs MMIC chips.

contributing in a large way to fields such as robotics, smart weapons, satellite technology and capabilities, smart-skins for aircraft, photonics, and smart built-in testing. Toward this objective, this text should serve as a reference for the fundamental microwave/millimeter-wave planar circuit active and passive components.

REFERENCES

1. Bhartia, P., and I. J. Bahl, *Millimeter Wave Engineering and Applications*, Wiley, New York, 1984.
2. Barrett, R. M., "Microwave Printed Circuits—the Early Years," *IEEE Trans. Microwave Theory Tech.*, Vol. MTT-32, Sept. 1984, pp. 983–990.
3. Barrett, R. M., and M. H. Barnes, "Microwave Printed Circuits," *IRE National Conf. on Airborne Electronics*, Dayton, Ohio, May 23–25, 1951.
4. Sanders Associates Inc., *Handbook of Tri-Plate Microwave Components*, Nashua, New Hampshire, 1956.
5. Greig, D. D., and H. F. Engelmann, "Microstrip—a New Transmission Technique for the Kilomegacycle L Range," *IRE Proc.*, Vol. 40, Dec. 1952, pp. 1644–1650.
6. King, D. D., "Properties of Dielectric Image Lines," *IRE Trans. Microwave Theory Tech.*, Vol. MTT-3, Mar. 1955, pp. 75–78.
7. Howe, H., "Microwave Integrated Circuits—an Historical Perspective," *IEEE Trans. Microwave Theory Tech.*, Vol. MTT-32, Sept. 1984, pp. 991–996.
8. Mcquiddy, Jr., D. N., et al., "Monolithic Microwave Integrated Circuits: an Historical Perspective," *IEEE Trans. Microwave Theory Tech.*, Vol. MTT-32, Sept. 1984, pp. 997–1008.

2 TRANSMISSION LINES AND LUMPED ELEMENTS

Transmission lines in microwave circuits are normally used to carry information or energy from one point to the other and as circuit elements for passive circuits like filters, impedance transformers, couplers, delay lines, and baluns. Passive elements in conventional microwave circuits are mostly distributed and employ sections of transmission lines and waveguides. This is because the sizes of discrete lumped elements (resistors, inductors, and capacitors) used in electronic circuits at lower frequencies become comparable to the wavelength at microwave frequencies. However, when the sizes of lumped elements are reduced to dimensions much smaller than the wavelength, they are also used at microwave frequencies.

This chapter is intended to provide accurate and simple closed-form expressions for characteristics of various types of lines, their discontinuities, and coupled structures. Here no attempt has been made to derive these formulas, but adequate references are provided for their sources as well as for more complicated and accurate expressions. Brief descriptions of conventional waveguide and coaxial lines are included to help the reader in designing packages and good transitions between these media and planar transmission lines. Design information for lumped elements (inductors, capacitors, and resistors) is also included in the last section.

2.1 TRANSMISSION LINES

This section describes the basic parameters of transmission media, followed by various transmission structures and their performance characteristics.

2.1.1 Basics of Transmission Lines

Multiconductor structures that support transverse electromagnetic (TEM) or non-TEM modes of propagation are commonly referred to as "transmission lines." Waveguide (single conductor) or dielectric rod (nonconductor) or their derivatives support the non-TEM mode of propagation. The TEM transmission lines are characterized by four basic parameters, the characteristic impedance Z_0, the phase velocity v_p, the attenuation constant

α, and peak power-handling capability P_{max}, in terms of physical parameters (like the geometrical cross section) and properties of the dielectric and the conductor materials used.

In a TEM line (perfectly terminated), the ratio of the voltage to the current at any point along the line is constant, having the units of resistance for a lossless medium. This ratio is defined as the characteristic impedance. The propagation constant for a lossy transmission structure is a complex quantity, comprising a real part known as the attenuation constant (which contains information about dissipation due to conductor and dielectric losses) and an imaginary part known as the phase constant (which contains information about phase velocity). The attenuation constant is defined as

$$\alpha = \frac{\text{average power lost per unit length}}{2 \times \text{power transmitted}} \tag{2.1}$$

Consider a uniform transmission line with series resistance ($\bar{R}$), series inductance ($\bar{L}$), shunt conductance ($\bar{G}$), and shunt capacitance ($\bar{C}$), all defined per unit length of the line as shown in Fig. 2.1. Important transmission line expressions are summarized in Table 2.1.

An extensive variety of transmission and waveguide structures are used at microwave frequencies. Figure 2.2 shows cross-sectional views of commonly used structures. Half wavelength, quarter wavelength, or smaller sections of these lines form the basic building blocks in most microwave circuits.

The power-handling capability of a transmission line is limited by dielectric breakdown and by heating due to attenuation. The electrical breakdown limits the peak power, while the increase in temperature due to conductor and dielectric losses limits the average power.

At normal temperature and pressure, the breakdown electric field of dry air is 2.9×10^6 V/m. Using this and by calculating the maximum field strength, the peak power-handling capability of a transmission line is readily determined. The average power-handling capability of a transmission is determined by the temperature rise of the line in an air environment. The parameters that play major roles in the calculation of average power capacity are (1) attenuation constant, (2) surface area of the line, (3)

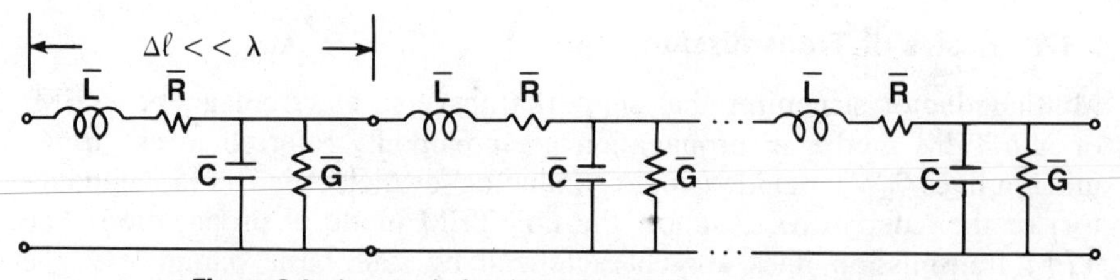

Figure 2.1 Lumped circuit representation of a transmission line.

TABLE 2.1 Summary of General Formulas for Transmission Lines

Quantity	General Line Expressions	Ideal Line Expressions	Approximate Results for Low-Loss Lines
Propagation constant $\gamma = \alpha + j\beta$	$\sqrt{(\bar{R}+j\omega\bar{L})(\bar{G}+j\omega\bar{C})}$	$j\omega\sqrt{\bar{L}\bar{C}}$	(See α and β below)
Phase constant β	$\mathrm{Im}(\gamma)$	$\omega\sqrt{\bar{L}\bar{C}} = \frac{\omega}{v_p} = \frac{2\pi}{\lambda}$	$\omega\sqrt{\bar{L}\bar{C}}\left[1 - \frac{\bar{R}\bar{G}}{4\omega^2\bar{L}\bar{C}} + \frac{\bar{G}^2}{8\omega^2\bar{C}^2} + \frac{\bar{R}^2}{8\omega^2\bar{L}^2}\right]$
Attenuation constant α	$\mathrm{Re}(\gamma)$	0	$\frac{\bar{R}}{2Z_0} + \frac{\bar{G}Z_0}{2}$
Characteristic impedance Z_0	$\sqrt{\frac{\bar{R}+j\omega\bar{L}}{\bar{G}+j\omega\bar{C}}}$	$\sqrt{\frac{\bar{L}}{\bar{C}}}$	$\sqrt{\frac{\bar{L}}{\bar{C}}}\left[1 + j\left(\frac{\bar{G}}{2\omega\bar{C}} - \frac{\bar{R}}{2\omega\bar{L}}\right)\right]$
Input impedance Z_i	$Z_0\left[\frac{Z_L \cosh \gamma l + Z_0 \sinh \gamma l}{Z_0 \cosh \gamma l + Z_L \sinh \gamma l}\right]$	$Z_0\left[\frac{Z_L \cos \beta l + jZ_0 \sin \beta l}{Z_0 \cos \beta l + jZ_L \sin \beta l}\right]$	
Impedance of shorted line ($Z_L = 0$)	$Z_0 \tanh \gamma l$	$jZ_0 \tan \beta l$	$Z_0\left[\frac{\alpha l \cos \beta l + j \sin \beta l}{\cos \beta l + j\alpha l \sin \beta l}\right]$
Impedance of open line ($Z_L = \infty$)	$Z_0 \coth \gamma l$	$-jZ_0 \cot \beta l$	$Z_0\left[\frac{\cos \beta l + j\alpha l \sin \beta l}{\alpha l \cos \beta l + j \sin \beta l}\right]$
Impedance of quarter-wave line	$Z_0\left[\frac{Z_L \sinh \alpha l + Z_0 \cosh \alpha l}{Z_0 \sinh \alpha l + Z_L \cosh \alpha l}\right]$	$\frac{Z_0^2}{Z_L}$	$Z_0\left[\frac{Z_0 + Z_L\alpha l}{Z_L + Z_0\alpha l}\right]$
Impedance of half-wave line	$Z_0\left[\frac{Z_L \cosh \alpha l + Z_0 \sinh \alpha l}{Z_0 \cosh \alpha l + Z_L \sinh \alpha l}\right]$	Z_L	$Z_0\left[\frac{Z_L + Z_0\alpha l}{Z_0 + Z_L\alpha l}\right]$
Reflection coefficient ρ (at the load location)	$\frac{Z_L - Z_0}{Z_L + Z_0}$	$\frac{Z_L - Z_0}{Z_L + Z_0}$	
Standing-wave ratio	$\frac{1+\lvert\rho\rvert}{1-\lvert\rho\rvert}$	$\frac{1+\lvert\rho\rvert}{1-\lvert\rho\rvert}$	

λ: wavelength measured along line; ω: angular frequency.

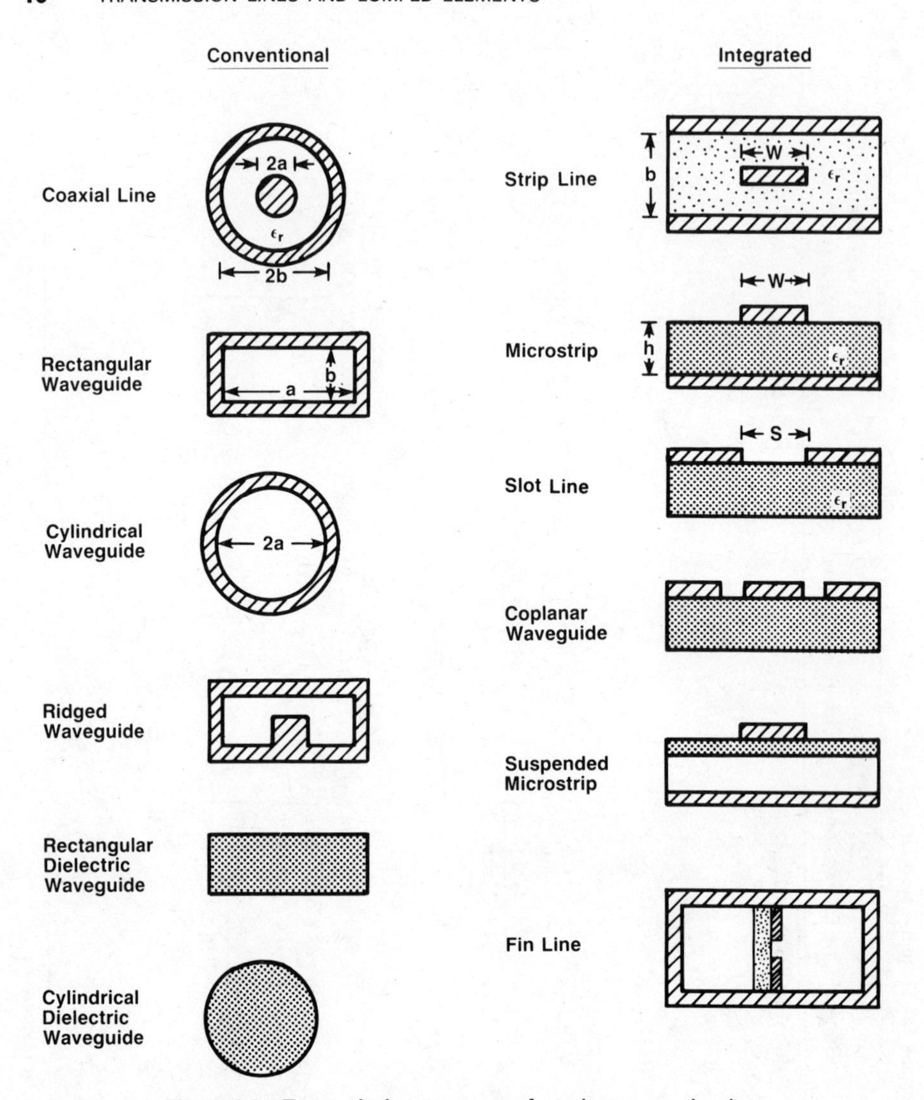

Figure 2.2 Transmission structures for microwave circuits.

maximum tolerable temperature rise, and (4) ambient temperature (i.e., the temperature of the medium surrounding the transmission structure. Thus the average power rating can be raised, if desired, by choosing a higher temperature limit and using forced cooling or cooling fins.

2.1.2 Characteristics of Conventional Transmission Structures

Coaxial lines and waveguides are frequently used in microwave circuits. These are briefly discussed in this section.

TABLE 2.2 Coaxial-Line Characteristics

Parameter	Expression	Units
Capacitance	$C = \frac{55.556\epsilon_r}{\ln(b/a)}$	pF/m
Inductance	$L = 200 \ln \frac{b}{a}$	nH/m
Characteristic impedance	$Z_0 = \frac{60}{\sqrt{\epsilon_r}} \ln \frac{b}{a}$	Ω
Phase velocity	$v_p = \frac{3 \times 10^8}{\sqrt{\epsilon_r}}$	m/s
Delay	$\tau_d = 3.33\sqrt{\epsilon_r}$	ns/m
Dielectric attenuation constant	$\alpha_d = 27.3\sqrt{\epsilon_r} \frac{\tan \delta}{\lambda_0}$	dB/unit length
Conductor attenuation constant (for copper at 20°C)	$\alpha_c = \frac{9.5 \times 10^{-5}\sqrt{f}(a+b)\sqrt{\epsilon_r}}{ab \ln(b/a)}$	dB/unit length
Cutoff wavelength for higher order modes	$\lambda_c = \pi\sqrt{\epsilon_r}(a+b)$	unit of a or b
Maximum peak power	$P_{max} = 44\|E_{max}\|^2 a^2 \sqrt{\epsilon_r} \ln \frac{b}{a}$	kW

λ_0 = free-space wavelength; $\tan \delta$ = loss tangent of the dielectric; f = operating frequency in GHz; E_{max} = breakdown electric field.

Coaxial Line. The dominant mode of propagation in a coaxial line is the TEM, and the characteristics obtained from static field analysis [1, 2] are given in Table 2.2. As shown in Fig. 2.2, a and b are the radii of the inner and outer conductors, respectively. The effect of conductor temperature and roughness on the attenuation is very well described in Reference 3. To compute attenuation for temperatures other than 20°C, multiply the attenuation expression α_c in Table 2.2 by $[1 + 0.0039(T - 20)]^{1/2}$, where the temperature T is expressed in °C.

Waveguide. Waveguides of rectangular and circular cross-sections (Fig. 2.2) find frequent applications in high-power and low-loss systems. The properties of waveguide structures can be determined by solving the wave equation with appropriate boundary conditions [1–3]. The structure supports transverse electric (TE) and transverse magnetic (TM) waves, and the properties of these modes in a rectangular waveguide are summarized in Table 2.3. The quantities η_0, k_0, and c are defined as follows

$$\eta_0 = \sqrt{\frac{\mu_0}{\epsilon_0}}, \qquad k_0 = \omega\sqrt{\mu_0\epsilon_0}, \qquad \text{and} \qquad c = \frac{1}{\sqrt{\mu_0\epsilon_0}} \tag{2.2}$$

TABLE 2.3 Properties of Waves in Empty Rectangular Waveguides

Property	TE_{mn} Modes	TM_{mn} Modes
Cutoff wave number k_c	$\sqrt{\left(\frac{m\pi}{a}\right)^2+\left(\frac{n\pi}{b}\right)^2}$	$\sqrt{\left(\frac{m\pi}{a}\right)^2+\left(\frac{n\pi}{b}\right)^2}$
Propagation constant γ_{mn}	$\sqrt{k_c^2-k_0^2}$	$\sqrt{k_c^2-k_0^2}$
Guide wavelength λ_g	$\frac{\lambda_0}{\sqrt{1-(k_c/k_0)^2}}$	$\frac{\lambda_0}{\sqrt{1-(k_c/k_0)^2}}$
Group velocity v_g	$c\frac{\lambda_0}{\lambda_g}$	$c\frac{\lambda_0}{\lambda_g}$
Phase velocity v_p	$c\frac{\lambda_g}{\lambda_0}$	$c\frac{\lambda_g}{\lambda_0}$
Wave impedance Z	$\frac{jk_0\eta_0}{\gamma_{mn}}$	$\frac{-j\gamma_{mn}\eta_0}{k_0}$
Longitudinal magnetic field H_z	$k_c^2\cos\left(\frac{m\pi x}{a}\right)\cos\left(\frac{n\pi y}{b}\right)$	0
Longitudinal electric field E_z	0	$k_c^2\sin\left(\frac{m\pi x}{a}\right)\sin\left(\frac{n\pi y}{b}\right)$
Transverse magnetic field		
H_x	$\frac{\gamma_{mn}m\pi}{a}\sin\left(\frac{m\pi x}{a}\right)\cos\left(\frac{n\pi y}{b}\right)$	$\frac{k_0 n\pi}{b\eta_0}\sin\left(\frac{m\pi x}{a}\right)\cos\left(\frac{n\pi y}{b}\right)$
H_y	$\frac{\gamma_{mn}n\pi}{b}\cos\left(\frac{m\pi x}{a}\right)\sin\left(\frac{n\pi y}{b}\right)$	$-\frac{k_0 m\pi}{a\eta_0}\cos\left(\frac{m\pi x}{a}\right)\sin\left(\frac{n\pi y}{b}\right)$
Transverse electric field		
E_x	$\frac{jk_0\eta_0 n\pi}{b}\cos\left(\frac{m\pi x}{a}\right)\sin\left(\frac{n\pi y}{b}\right)$	$-\frac{\gamma_{mn}m\pi}{a}\cos\left(\frac{m\pi x}{a}\right)\sin\left(\frac{n\pi y}{b}\right)$
E_y	$-\frac{jk_0\eta_0 m\pi}{a}\sin\left(\frac{m\pi x}{a}\right)\cos\left(\frac{n\pi y}{b}\right)$	$-\frac{\gamma_{mn}n\pi}{b}\sin\left(\frac{m\pi x}{a}\right)\cos\left(\frac{n\pi y}{b}\right)$

where ϵ_0 and μ_0 are the free-space permittivity and permeability, respectively. The lowest order propagating mode is the TE_{10} ($m=1$, $n=0$), and is also called the dominant mode of the rectangular waveguide. Since the modes of propagation in waveguides are non-TEM, the characteristic impedance cannot be defined uniquely [4].

For these structures, Z_0 can be defined either in terms of the voltage–current ratio or in terms of the power transmitted for a given voltage or a given current; that is,

$$Z_0(v,i)=\frac{v}{i}, \quad \text{or} \quad Z_0(W,i)=\frac{2W}{ii^*}, \quad \text{or} \quad Z_0(W,v)=\frac{vv^*}{2W} \tag{2.3}$$

TABLE 2.4 Rectangular Waveguide Characteristics

Parameter	Expression	Units
Cutoff wavelength	$\lambda_c = 2a$	m
Guide wavelength	$\lambda_g = \dfrac{\lambda_0}{\sqrt{1-(\lambda_0/\lambda_c)^2}}$	m
Phase velocity	$v_p = \dfrac{c\lambda_g}{\lambda_0}$	m
Wave impedance	$Z = \dfrac{\eta_0}{\sqrt{1-(\lambda_0/\lambda_c)^2}}$	Ω
Characteristic impedance (v, i)	$Z_0 = \dfrac{\pi b}{2a} Z$	Ω
Attenuation constant (copper)	$\alpha = \dfrac{7.14\times 10^{-2}\sqrt{f}}{b\eta_0\sqrt{1-(\lambda_0/\lambda_c)^2}}\left[1+\dfrac{2b}{a}\left\{\dfrac{\lambda_0}{\lambda_c}\right\}^2\right]$	dB/unit length
Maximum peak power	$P_{max} = 6.63\times 10^{-7}ab\lvert E_{max}\rvert^2\sqrt{1-\left(\dfrac{\lambda_0}{\lambda_c}\right)^2}$	kW

$c = 3\times 10^8$ m/s, f in GHz.

For TEM-mode transmission structures, these definitions are identical, but for waveguides they lead to three different values of Z_0. All of these three different characteristic impedances may be expressed in terms of the wave impedance Z, which is defined as the ratio of transverse components of electric and magnetic fields. For the dominant TE_{10} mode the characteristics of a rectangular waveguide are summarized in Table 2.4.

2.1.3 Characteristics of Planar Transmission Lines

For a transmission structure to be suitable as a circuit element in microwave integrated circuits, one of the principal requirements is that the structure should be "planar" in configuration. A planar geometry implies that the characteristics of the element can be determined from the dimensions in a single plane. As shown in Fig. 2.2, various forms of planar transmission lines have been developed for use in microwave integrated circuits (MICs). The strip line [5–11], microstrip line [12–14], inverted microstrip line [15], slot line [14, 16], coplanar waveguide [14, 17], and coplanar strip line [14, 17] are representative planar transmission lines. The circuits realized using any one of the aforementioned transmission lines or combinations of them have distinct advantages, such as light weight, small size, improved performance, better reliability and reproducibility, and low cost, as compared to conventional microwave circuits. They are also compatible with solid state chip devices. Integrated circuits employing these structures at

TABLE 2.5 Strip Line Characteristics

Parameter	Expressions	Remarks
Characteristic impedance (Ω) for $t=0$	$Z_0=\frac{30\pi}{\sqrt{\epsilon_r}}\frac{K'(k)}{K(k)}, \quad k=\tanh\frac{\pi W}{2b}$	Virtually exact [5]
Characteristic impedance (Ω) for $t\neq 0$	$Z_0=\frac{30\ln}{\sqrt{\epsilon_r}}\left\{1+\frac{4}{\pi}\frac{b-t}{W'}\left[\frac{8}{\pi}\frac{b-t}{W'}+\sqrt{\left(\frac{8}{\pi}\frac{b-t}{W'}\right)^2+6.27}\right]\right\}$	For $W'/(b-t)<10$ accuracy within 0.5 percent [11]
	$\frac{W'}{b-t}=\frac{W}{b-t}+\frac{\Delta W}{b-t}$	
	$\frac{\Delta W}{b-t}=\frac{x}{\pi(1-x)}\left\{1-\frac{1}{2}\ln\left[\left(\frac{x}{2-x}\right)^2+\left(\frac{0.0796x}{W/b+1.1x}\right)^m\right]\right\}$	
	$m=2\left[1+\frac{2}{3}\frac{x}{1-x}\right]^{-1}$ and $x=\frac{t}{b}$	
Attenuation constant (dB/unit length)	$\alpha_c=\frac{0.0231R_s\sqrt{\epsilon_r}}{Z_0}\frac{\partial Z_0}{\partial W'}\left\{1+\frac{2W'}{b-t}-\frac{1}{\pi}\left[\frac{3x}{2-x}+\ln\frac{x}{2-x}\right]\right\}$	[4]
	$\frac{\partial Z_0}{\partial W'}=\frac{30e^{-A}}{W'\sqrt{\epsilon_r}}\left[\frac{3.135}{Q}-\left(\frac{8}{\pi}\frac{b-t}{W'}\right)^2(1+Q)\right]$	
	$A=\frac{Z_0\sqrt{\epsilon_r}}{30\pi}, \quad Q=\sqrt{1+6.27\left(\frac{\pi}{8}\frac{W'}{b-t}\right)^2}$	
	$\alpha_d=27.3\sqrt{\epsilon_r}\frac{\tan\delta}{\lambda_0}$	
Cutoff for higher order mode (GHz)	$f_c=\frac{15}{b\sqrt{\epsilon_r}}\frac{1}{(W/b+\pi/4)}$, W and b in cm	TE is the lowest order mode [4]

microwave frequencies have been widely discussed in the literature [9, 14, 18, 19].

Strip line. The strip line is one of the most commonly used transmission lines for passive MICs. The dominant mode in a strip line is TEM, and the design data can be obtained completely by electrostatic analysis such as conformal mapping. Expressions for Z_0, $\alpha_T = \alpha_c + \alpha_d$ and f_c are given in Table 2.5, where K represents a complete elliptic function of the first kind with K' its complementary function. An approximate expression for K/K' (accurate to 8 ppm) is given by

$$\frac{K(k)}{K'(k)} = \begin{cases} \left[\dfrac{1}{\pi}\ln\left(2\dfrac{1+\sqrt{k'}}{1-\sqrt{k'}}\right)\right]^{-1} & \text{for } 0 \le k \le 0.7 \\ \dfrac{1}{\pi}\ln\left(2\dfrac{1+\sqrt{k}}{1-\sqrt{k}}\right) & \text{for } 0.7 \le k \le 1 \end{cases} \tag{2.4}$$

where $K'(k) = K(k')$; $k' = \sqrt{1-k^2}$. The characteristic impedance versus W/b for various values of t/b is shown in Fig. 2.3.

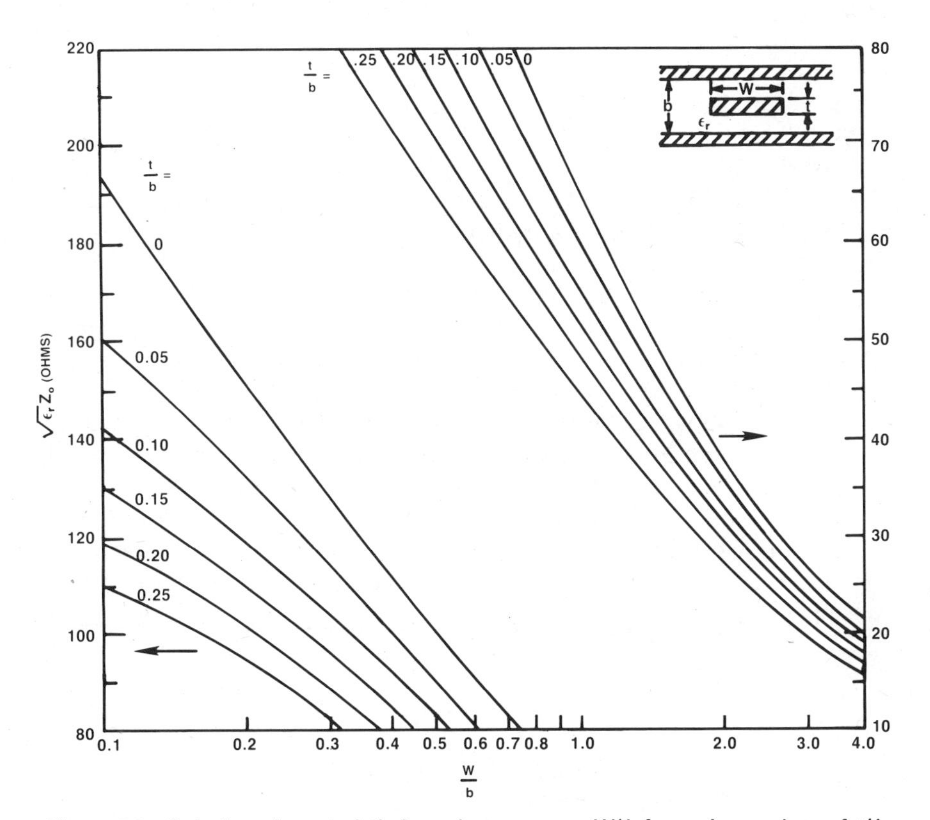

Figure 2.3 Strip-line characteristic impedance versus W/b for various values of t/b.

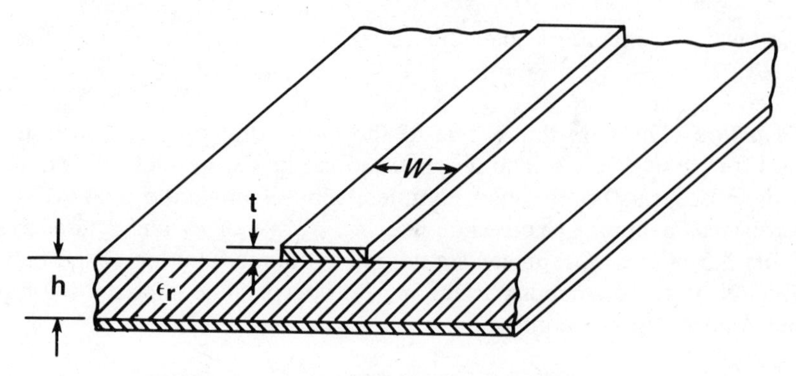

Figure 2.4 A microstrip-line configuration.

The total loss can be used to determine the quality factor Q of a $\lambda/2$ resonator as

$$Q = \frac{8.68\pi\sqrt{\epsilon_r}}{\lambda_0 \alpha_T} \tag{2.5}$$

Microstrip. Unlike the strip line, the microstrip line (shown in Fig. 2.4) is an inhomogeneous transmission line, since the field lines between the strip and the ground plane are not contained entirely in the substrate. Therefore, the mode propagating along the microstrip is not purely TEM but quasi-TEM.

Extensive literature dealing with the analytical and numerical solutions of this medium [14] exists. Of these solutions, the quasi-static approach is perhaps the simplest but has a limited range of validity, while the full-wave approach is complete and rigorous. In the quasi-static method, the nature of the mode of propagation is considered to be pure TEM, and the transmission-line characteristics are calculated from the electrostatic capacitances of the structure. It is found that this analysis is adequate for designing circuits when the strip width and the substrate thickness are much smaller than the wavelength in the dielectric material. In the quasi-static approach, the transmission characteristics are calculated from the values of two capacitances: one is C_a, for a unit length of the microstrip configuration with the dielectric materials replaced by air, and the other is C for a unit length of the microstrip with the dielectric present. The characteristic impedance Z_0 and the phase constant β can be written in terms of these capacitances as follows

$$Z_0 = \frac{1}{c\sqrt{CC_a}} \tag{2.6}$$

$$\beta = k_0 \left(\frac{C}{C_a}\right)^{1/2} = k_0\sqrt{\epsilon_e} \tag{2.7}$$

where

$$\epsilon_e = \left(\frac{\lambda_0}{\lambda_g}\right)^2 = \frac{C}{C_a} \tag{2.8}$$

λ_0 is the free space wavelength and λ_g is the guide wavelength. The effective dielectric constant ϵ_e takes into account the fields in the air regions. Numerical methods for the characterization of microstrip lines involve extensive computations. Closed-form expressions are necessary for optimization and computer-aided design of microstrip circuits. Closed-form expressions for Z_0 and ϵ_e have been reported by Wheeler [13, 20], Schneider [21], and Hammerstad [22, 23]. Both Wheeler and Hammerstad have also given synthesis expressions for Z_0. Closed-form expressions for microstrip characteristics are summarized in Table 2.6. The characteristic impedance and effective dielectric constant versus W/h are shown in Fig. 2.5.

The maximum frequency of operation with a microstrip is limited due to several factors such as excitation of spurious modes, higher losses, tight

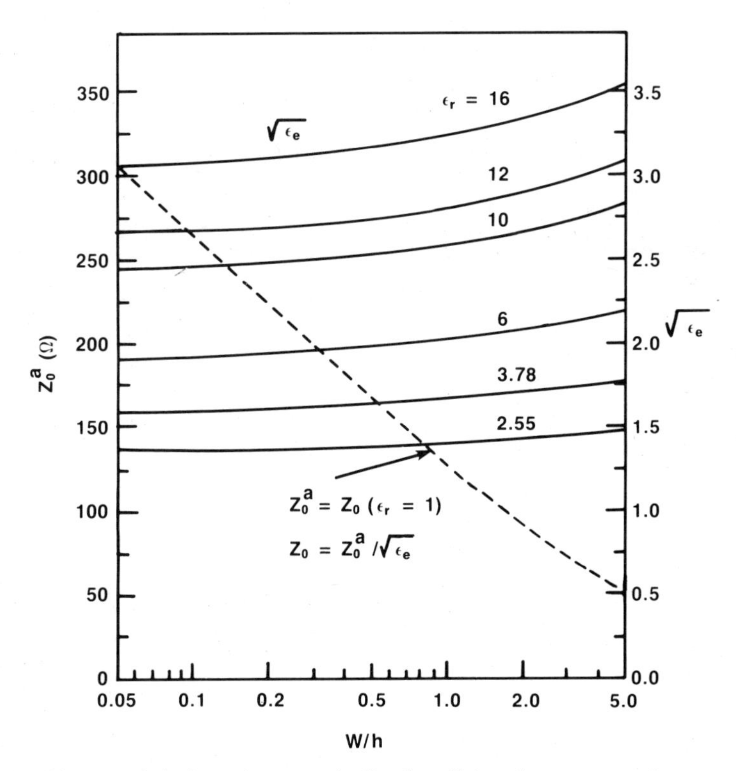

Figure 2.5 Characteristic impedance and effective dielectric constant of open microstrip lines.

TABLE 2.6 Microstrip-Line Characteristics

Parameters	Expressions
Characteristic impedance (Ω)	$Z_0 = \begin{cases} \dfrac{\eta_0}{2\pi\sqrt{\epsilon_e}} \ln\left\{\dfrac{8h}{W'} + 0.25\dfrac{W'}{h}\right\}, & \dfrac{W}{h} \leq 1 \\ \dfrac{\eta_0}{\sqrt{\epsilon_e}}\left\{\dfrac{W'}{h} + 1.393 + 0.667\ln\left(\dfrac{W'}{h} + 1.444\right)\right\}^{-1}, & \dfrac{W}{h} \geq 1 \end{cases}$
	$\eta_0 = 120\pi\,\Omega$
	$\dfrac{W'}{h} = \dfrac{W}{h} + \dfrac{1.25}{\pi}\dfrac{t}{h}\left(1 + \ln\dfrac{4\pi W}{t}\right), \quad \dfrac{W}{h} \leq \dfrac{1}{2\pi}$
	$\dfrac{W'}{h} = \dfrac{W}{h} + \dfrac{1.25}{\pi}\dfrac{t}{h}\left(1 + \ln\dfrac{2h}{t}\right), \quad \dfrac{W}{h} \geq \dfrac{1}{2\pi}$
Effective dielectric constant	$\epsilon_e = \dfrac{\epsilon_r + 1}{2} + \dfrac{\epsilon_r - 1}{2} F\left(\dfrac{W}{h}\right) - \dfrac{\epsilon_r - 1}{4.6}\dfrac{t/h}{\sqrt{W/h}}$
	$F\left(\dfrac{W}{h}\right) = \begin{cases} \left(1 + 12\dfrac{h}{W}\right)^{-1/2} + 0.04\left(1 - \dfrac{W}{h}\right)^2, & \dfrac{W}{h} \leq 1 \\ \left(1 + 12\dfrac{h}{W}\right)^{-1/2} & \dfrac{W}{h} \geq 1 \end{cases}$
Attenuation constant (dB/unit length)	$\alpha_c = \begin{cases} 1.38\dfrac{R_s}{hZ_0}\dfrac{32 - (W'/h)^2}{32 + (W'/h)^2}\Lambda, & \dfrac{W}{h} \leq 1 \\ 6.1 \times 10^{-5}\dfrac{R_s Z_0 \epsilon_e}{h}\left(\dfrac{W'}{h} + \dfrac{0.667\,W'/h}{W'/h + 1.444}\right)\Lambda, & \dfrac{W}{h} \geq 1 \end{cases}$
	$\Lambda = \begin{cases} 1 + \dfrac{h}{W'}\left(1 + \dfrac{1.25t}{\pi W} + \dfrac{1.25}{\pi}\ln\dfrac{4\pi W}{t}\right), & \dfrac{W}{h} \leq \dfrac{1}{2\pi} \\ 1 + \dfrac{h}{W'}\left(1 - \dfrac{1.25t}{\pi h} + \dfrac{1.25}{\pi}\ln\dfrac{2h}{t}\right), & \dfrac{W}{h} \geq \dfrac{1}{2\pi} \end{cases}$
	$\alpha_d = 27.3\dfrac{\epsilon_r}{\sqrt{\epsilon_e}}\left(\dfrac{\epsilon_e - 1}{\epsilon_r - 1}\right)\dfrac{\tan\delta}{\lambda_0}$
Dispersion	$\epsilon_e(f) = \left(\dfrac{\sqrt{\epsilon_r} - \sqrt{\epsilon_e}}{1 + 4F^{-1.5}} + \sqrt{\epsilon_e}\right)^2$
	$Z_0(f) = Z_0\dfrac{\epsilon_e(f) - 1}{\epsilon_e - 1}\sqrt{\dfrac{\epsilon_e}{\epsilon_e(f)}}$
	$F = \dfrac{4h\sqrt{\epsilon_r - 1}}{\lambda_0}\left\{0.5 + \left[1 + 2\log\left(1 + \dfrac{W}{h}\right)\right]^2\right\}$
Quality factor	$\dfrac{1}{Q} = \dfrac{1}{Q_0} + \dfrac{1}{Q_r}, \quad Q_0 = \dfrac{8.68\pi\sqrt{\epsilon_e(f)}}{\lambda_0(\alpha_c + \alpha_d)}, \quad Q_r = \dfrac{Z_0(f)}{480\pi\left(\dfrac{h}{\lambda_0}\right)^2 R}$
	$R = \dfrac{\epsilon_e(f) + 1}{\epsilon_e(f)} - \dfrac{[\epsilon_e(f) - 1]^2}{2[\epsilon_e(f)]^{3/2}}\ln\left[\dfrac{\sqrt{\epsilon_e(f)} + 1}{\sqrt{\epsilon_e(f)} - 1}\right]$

fabrication tolerances, handling fragility, pronounced discontinuity effects, low Q due to radiation from discontinuities, and, of course, technological processes. The frequency at which significant coupling occurs between the quasi-TEM mode and the lowest order surface-wave spurious mode is given in Reference 24.

$$f_T = \frac{150}{\pi h} \sqrt{\frac{2}{\epsilon_r - 1}} \tan^{-1}(\epsilon_r) \tag{2.9}$$

where f_T is in gigahertz and h is in millimeters. Thus the maximum thickness of the GaAs substrate ($\epsilon_r \simeq 12.9$) for microstrip circuits designed at 100 GHz is less than 0.3 mm. The maximum thickness is also restricted by the high radiation losses incurred in microstrip discontinuities, such as open ends, gaps, slits, steps in widths, and bends. For a $\lambda/2$ resonator, the radiation Q factor may be approximated as

$$Q_r = \frac{3\epsilon_r Z_0 \lambda_0^2}{32 \eta_0 h^2} \tag{2.10}$$

Thus for thicker substrates, where $Q \simeq Q_r$, the variation of Q is proportional to $1/(fh)^2$. For example, a 50-Ω resonator on a GaAs substrate has a $Q_r \simeq 1.44 \times 10^4/(fh)^2$ where f is in gigahertz and h is in millimeters. At 100 GHz, the substrate thickness is less than 0.125 mm for Q_r greater than 100. A substrate thickness of this order results in attenuation of the order of 1 dB/cm. Fabrication tolerances and technological processes such as photoetching limit the minimum strip width and the spacing between two adjacent strips in the case of coupled lines. High-impedance lines of about 100 Ω require strip widths of the order of 0.01 mm on a 0.125-mm-thick GaAs substrate, thereby also setting a limit on the frequency of operation in microstrip lines because of low radiation Q_r.

Suspended and Inverted Microstrip Lines. Suspended and inverted microstrip lines (shown in Fig. 2.6) provide a higher Q (500–1500) than microstrip. The wide range of impedance values achievable makes these media particularly suitable for filters. Expressions for the characteristic impedance and effective dielectric constant for $t/h \ll 1$ are given as follows [25]

$$Z_0 = \frac{60}{\sqrt{\epsilon_e}} \ln\left[\frac{f(u)}{u} + \sqrt{1 + \left(\frac{2}{u}\right)^2}\right] \tag{2.11}$$

where

$$f(u) = 6 + (2\pi - 6) \exp\left[-\left(\frac{30.666}{u}\right)^{0.7528}\right]$$

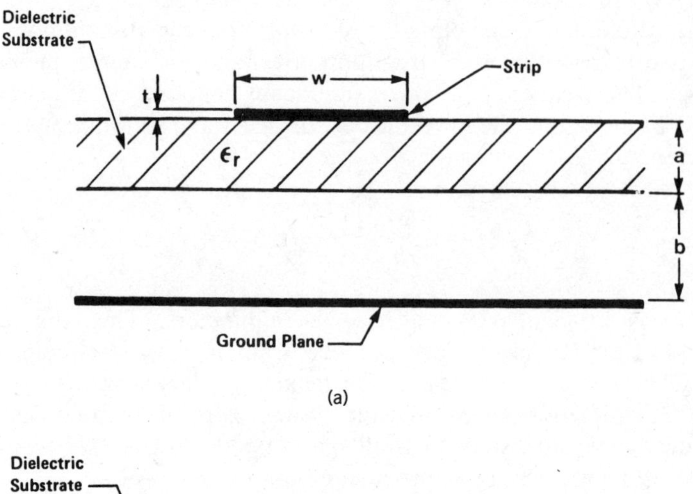

(a)

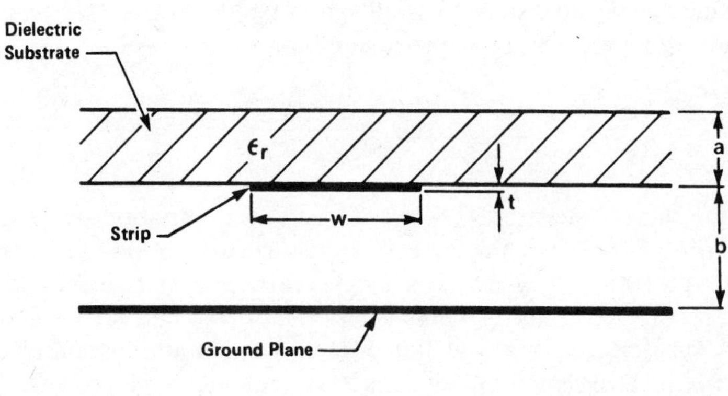

(b)

Figure 2.6 Suspended-substrate microstrip-line configurations. (*a*) Suspended. (*b*) Inverted.

For suspended microstrip $u = w/(a+b)$ and for inverted microstrip $u = w/b$, where all the variables are defined in Fig. 2.6.

For suspended microstrip the effective dielectric constance ϵ_e is obtained from

$$\sqrt{\epsilon_e} = \left[1 + \frac{a}{b}\left(a_1 - b_1 \ln \frac{w}{b}\right)\left(\frac{1}{\sqrt{\epsilon_r}} - 1\right)\right]^{-1} \tag{2.12}$$

where

$$a_1 = \left(0.8621 - 0.1251 \ln \frac{a}{b}\right)^4$$

$$b_1 = \left(0.4986 - 0.1397 \ln \frac{a}{b}\right)^4$$

and for inverted microstrip the effective dielectric constant is obtained from

$$\sqrt{\epsilon_e} = 1 + \frac{a}{b}\left(\bar{a}_1 - \bar{b}_1 \ln \frac{w}{b}\right)(\sqrt{\epsilon_r} - 1) \tag{2.13}$$

where

$$\bar{a}_1 = \left(0.5173 - 0.1515 \ln \frac{a}{b}\right)^2$$

$$\bar{b}_1 = \left(0.3092 - 0.1047 \ln \frac{a}{b}\right)^2$$

The accuracy of (2.12) and (2.13) is within $\pm 1\%$ for $1 < w/b \leq 8$, $0.2 \leq a/b \leq 1$ and $\epsilon_r \leq 6$. For $\epsilon_r \simeq 10$, the error is less than $\pm 2\%$.

Slot Line. The slot-line configuration (shown in Fig. 2.7) is useful in circuits requiring high-impedance lines, series stubs, short circuits, and in hybrid combinations with microstrip circuits in microwave integrated circuits. The mode of propagation is non-TEM and almost transverse electric (TE) in nature. Various methods of analysis discussed in the literature do not lead to closed-form expressions for slot-line wavelength and impedance. This becomes a serious handicap for circuit analysis and design, especially when computer-aided design techniques are used. However, closed-form expressions for characteristic impedance and slot-line wavelength have been obtained [26] by curve fitting the numerically computed results based

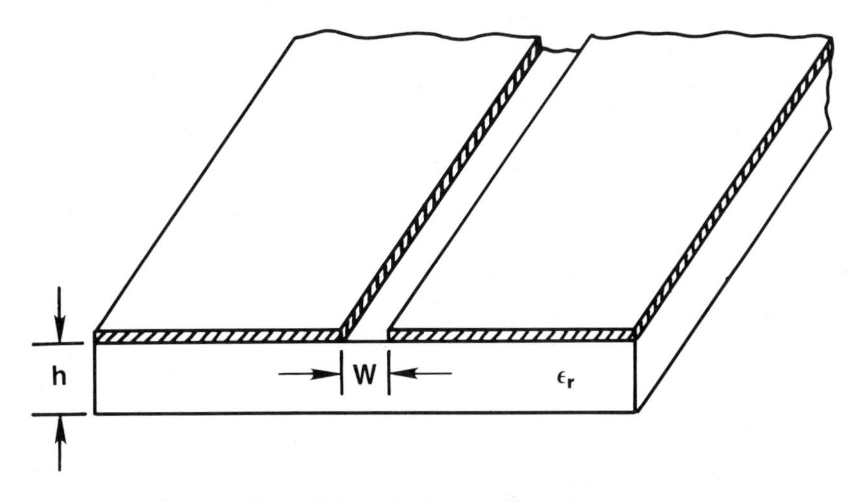

Figure 2.7 A slot-line configuration.

on Reference 27. These expressions have an accuracy of about 2% for the following sets of parameters

$$9.7 \le \epsilon_r \le 20$$

$$0.02 \le \frac{W}{h} \le 1.0$$

and

$$0.01 \le \frac{h}{\lambda_0} \le \left(\frac{h}{\lambda_0}\right)_c$$

where $(h/\lambda_0)_c$ is the cutoff value for the TE_{10} surface-wave mode on slot line, and is given by

$$\left(\frac{h}{\lambda_0}\right)_c = 0.25/\sqrt{\epsilon_r - 1} \tag{2.14}$$

1. For $0.02 \le W/h \le 0.2$

$$\frac{\lambda_g}{\lambda_0} = 0.923 - 0.195 \ln \epsilon_r + 0.2 \frac{W}{h} - \left(0.126 \frac{W}{h} + 0.02\right) \ln\left(\frac{h}{\lambda_0} \times 10^2\right) \tag{2.15}$$

$$Z_0 = 72.62 - 15.283 \ln \epsilon_r + 50 \frac{(W/h - 0.02)(W/h - 0.1)}{W/h} + \ln\left(\frac{W}{h} \times 10^2\right)(19.23 - 3.693 \ln \epsilon_r) - \left[0.139 \ln \epsilon_r - 0.11 + \frac{W}{h}(0.465 \ln \epsilon_r + 1.44)\right] \times \left(11.4 - 2.636 \ln \epsilon_r - \frac{h}{\lambda_0} \times 10^2\right)^2 \tag{2.16}$$

2. For $0.2 \le W/h \le 1.0$

$$\frac{\lambda_g}{\lambda_0} = 0.987 - 0.21 \ln \epsilon_r + \frac{W}{h}(0.111 - 0.0022\epsilon_r) - \left(0.053 + 0.041 \frac{W}{h} - 0.0014\epsilon_r\right) \ln\left(\frac{h}{\lambda_0} \times 10^2\right) \tag{2.17}$$

$$Z_0 = 113.19 - 23.257 \ln \epsilon_r + 1.25 \frac{W}{h}(114.59 - 22.531 \ln \epsilon_r)$$
$$+ 20\left(\frac{W}{h} - 0.2\right)\left(1 - \frac{W}{h}\right)$$
$$- \left[0.15 + 0.1 \ln \epsilon_r + \frac{W}{h}(-0.79 + 0.899 \ln \epsilon_r)\right]$$
$$\times \left[10.25 - 2.171 \ln \epsilon_r + \frac{W}{h}(2.1 - 0.617 \ln \epsilon_r) - \frac{h}{\lambda_0} \times 10^2\right]^2 \qquad (2.18)$$

More accurate expressions for slotline wavelength for $\epsilon_r = 9.7$ and 20 are also available [4, 14].

Coplanar Lines. Coplanar waveguides are finding extensive applications in microwave integrated circuits. Inclusion of coplanar waveguides in microwave circuits adds to the flexibility of circuit design and improves the performance for some circuit functions. The configuration of a coplanar waveguide (CPW) is shown in Fig. 2.8(a). Another promising configuration

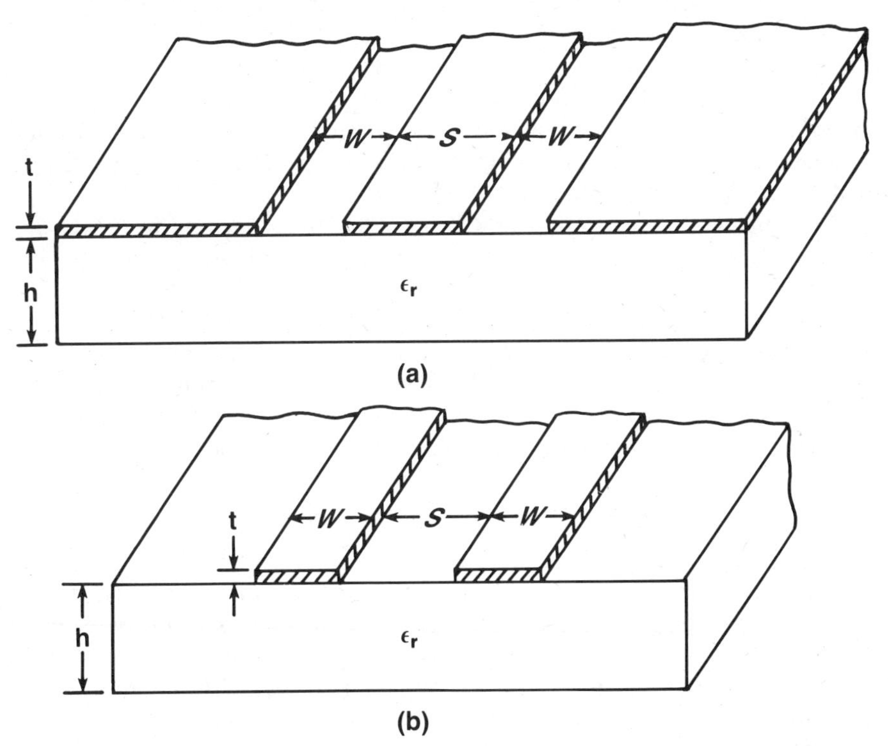

Figure 2.8 (a) A coplanar waveguide (CPW). (b) A "coplanar strips" (CPS) transmission line.

TABLE 2.7 Expressions for Coplanar-Line Characteristic Impedance[a]

Structure	Characteristic Impedance (Ω)	Effective Dielectric Constant
Coplanar waveguide	$Z_0 = \dfrac{30\pi}{\sqrt{\epsilon_e}} \dfrac{K(k')}{K(k)}$	$\epsilon_e = 1 + \dfrac{\epsilon_r - 1}{2} \dfrac{K(k')K(k_1)}{K(k)K(k_1')}$
Coplanar strips	$Z_0 = \dfrac{120\pi}{\sqrt{\epsilon_e}} \dfrac{K(k)}{K(k')}$	$\epsilon_e = 1 + \dfrac{\epsilon_r - 1}{2} \dfrac{K(k')K(k_1)}{K(k)K(k_1')}$

$$k = \frac{a}{b}, \quad a = \frac{S}{2}, \quad b = \frac{S}{2} + W$$

$$k_1 = \frac{\sinh(\pi a/2h)}{\sinh(\pi b/2h)}$$

[a]Ghione, G., and C. Naldi, "Analytical Formulas for Coplanar Lines in Hybrid and Monolithic MICs," *Electron. Lett.*, Vol. 20, 1984, pp. 179–181.

that is complementary to CPW is known as coplanar strips (CPS) and is shown in Fig. 2.8(*b*). Both of these configurations belong to the category of "coplanar lines," wherein all the conductors are in the same plane (i.e., on the top surface of the dielectric substrate). A distinct advantage of these two lines lies in the fact that mounting lumped (active or passive) components in shunt or series configuration is much easier. Drilling holes or slots through the substrate is not needed.

Coplanar waveguides and coplanar strips support quasi-TEM modes and have been analyzed using quasi-static as well as full-wave methods [14]. Expressions for Z_0 and ϵ_e of CPW and CPS are given in Table 2.7. Expressions for $K(k')/K(k)$ are given in (2.4). Approximate expressions for attenuation for these lines are reported in Reference 14. The variation of total loss $(\alpha_c + \alpha_d)$ for coplanar lines on alumina substrate of thickness 0.63 mm as a function of aspect ratio is plotted in Fig. 2.9. It is observed that loss decreases with decreasing impedance or increasing strip width.

2.1.4 Comparison of Various MIC Transmission Media

For hybrid MIC applications, microstrip, slot-line, coplanar waveguide and coplanar strips have been used; whereas for monolithic MICs, microstrip has been used extensively, although there is an interest in CPW. Several other parameters of the four types of lines are compared qualitatively in Table 2.8. It can be generally seen that CPW and CPS combine some advantageous features of microstrip lines and slot lines. Their power-handling capabilities, radiation losses, Q factors, and dispersion behavior lie in between the corresponding values for microstrip and slot line. Perhaps

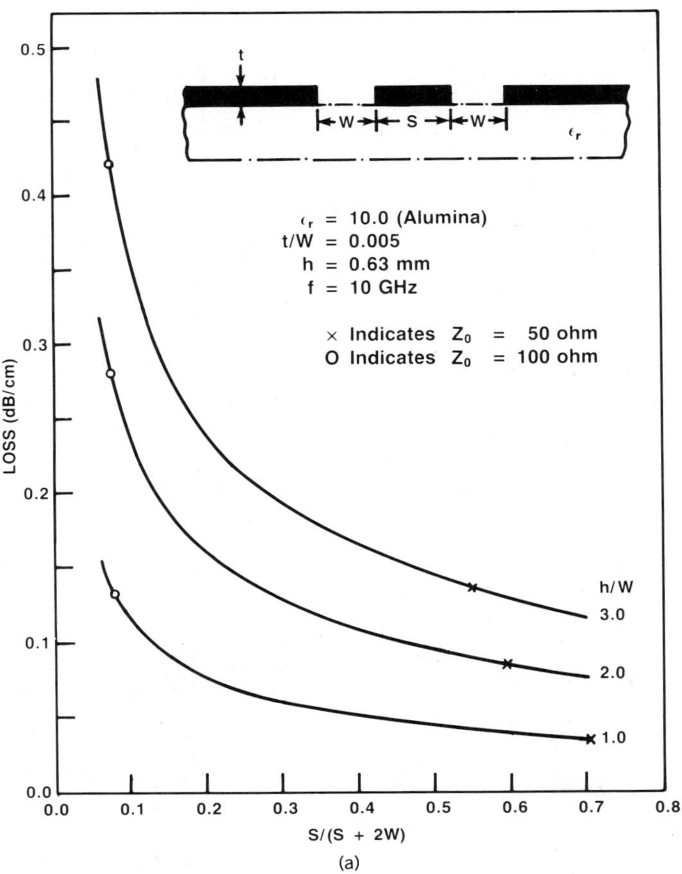

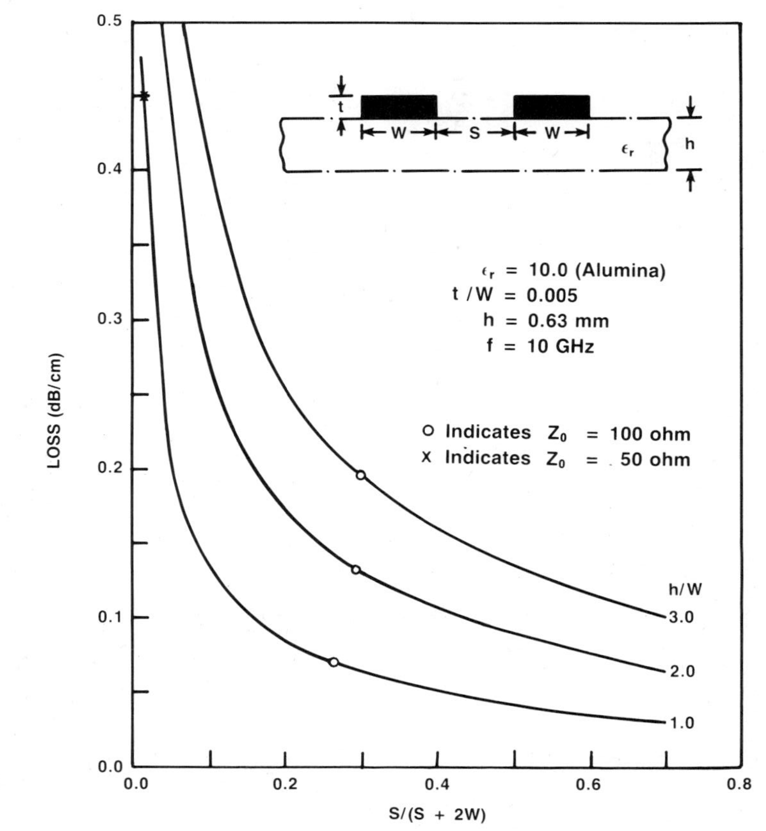

Figure 2.9 (*a*) Attenuation constant for CPW on alumina substrate. (*b*) Attenuation constant for CPS on alumina substrate.

TABLE 2.8 Qualitative Comparison of Various MIC Lines

Characteristic	Microstrip	Slot Line	Coplanar Waveguide	Coplanar Strips
Impedance range	20–110	55–300	25–155	45–280
Effective dielectric constant ($\epsilon_r = 10$ and $h = 0.025$ in.)	~6.5	~4.5	~5	~5
Power-handling capability	High	Low	Medium	Medium
Radiation loss	Low	High	Medium	Medium
Unloaded Q	High	Low	Medium	Low (lower impedances) High (higher impedance)
Dispersion	Small	Large	Medium	Medium
Mounting of components:				
in shunt configuration	Difficult	Easy	Easy	Easy
in series configuration	Easy	Difficult	Easy	Easy
Technological difficulties	Ceramic holes Edge plating	Double side etching	—	—
Elliptically polarized magnetic field configuration	Not available	Available	Available	Available
Enclosure dimensions	Small	Large	Large	Large

the best feature of the two coplanar lines is the ease of mounting components in series and shunt configurations, whereas microstrip lines are convenient only for series mounting, and slot lines can accommodate only shunt-mounted components.

2.2 COUPLED LINES

A "coupled line" configuration consists of two transmission lines placed parallel to each other and in close proximity. In such a case, there is a continuous coupling between the electromagnetic fields of two lines. Coupled lines are utilized extensively as basic elements for directional couplers, filters, phase shifters, baluns, matching networks, and a variety of other useful circuits.

Data such as characteristic impedance, phase velocity, and insertion loss are needed for the design of circuit components. Because of the coupling of electromagnetic fields, a pair of coupled lines can support two different modes of propagation. These modes have different characteristic impedances. The velocity of propagation of these two modes is equal when the lines are imbedded in a homogeneous dielectric medium. This is a desirable property for the design of circuits such as directional couplers. However, for transmission lines such as coupled microstrip lines the dielectric medium is not homogeneous. A part of the field extends into the air above the substrate. This fraction of total field is different for the two modes of coupled lines. Consequently, the effective dielectric constants (and the phase velocities) are not equal for the two modes. This nonsynchronous feature deteriorates the performance of circuits using these types of coupled lines. For the sake of simplicity only symmetrical configuration (both identical conductors) is considered here. Figure 2.10 shows coupled strip lines, microstrip lines, and broadside-coupled strip lines, and their characteristics are given in Table 2.9.

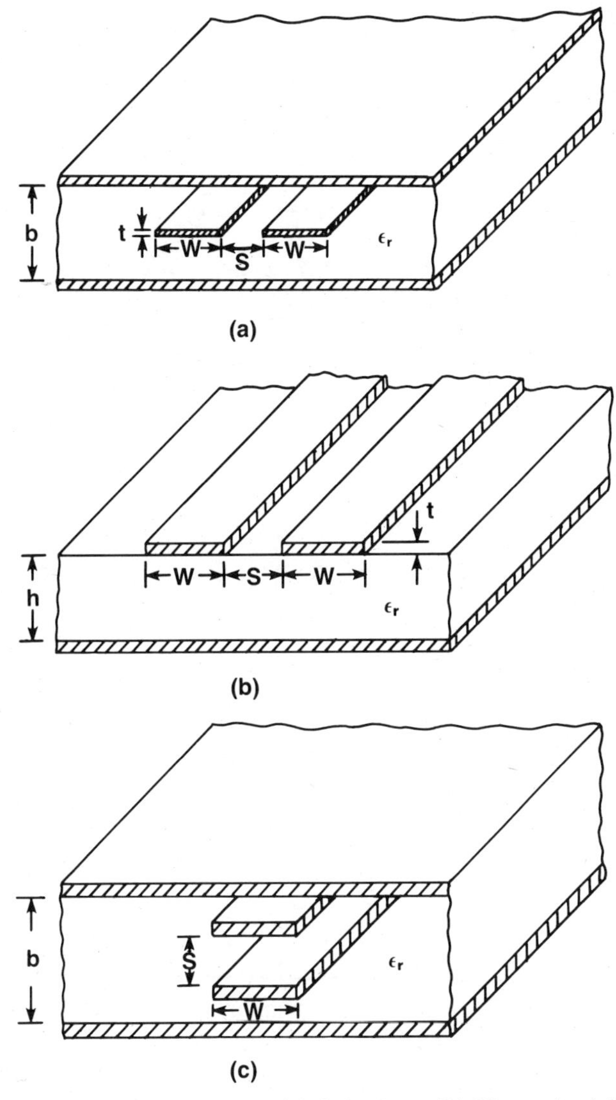

Figure 2.10 Coupled transmission lines. (*a*) Strip lines. (*b*) Microstrip. (*c*) Broadside strip lines.

TABLE 2.9 Characteristics for Coupled Structures

Structure	Characteristic Impedances (Ω)	Attenuation Constants (dB/unit length)	Remarks
Strip lines	$Z_{0e} = \dfrac{30\pi}{\sqrt{\epsilon_r}} \dfrac{K(k_e')}{K(k_e)}$ $Z_{0o} = \dfrac{30\pi}{\sqrt{\epsilon_r}} \dfrac{K(k_o')}{K(k_o)}$ $k_e = \tanh\left(\dfrac{\pi}{2}\dfrac{W}{b}\right) \tanh\left(\dfrac{\pi}{2}\dfrac{W+S}{b}\right)$ $k_o = \tanh\left(\dfrac{\pi}{2}\dfrac{W}{b}\right) \coth\left(\dfrac{\pi}{2}\dfrac{W+S}{b}\right)$		$t=0$, Virtually Exact [28]
	$Z_{0e} = \dfrac{30\pi(b-t)}{\sqrt{\epsilon_r}\left\{W + \dfrac{bC_f}{2\pi} A_e\right\}}$ $Z_{0o} = \dfrac{30\pi(b-t)}{\sqrt{\epsilon_r}\left\{W + \dfrac{bC_f}{2\pi} A_o\right\}}$ $A_e = 1 + \dfrac{\ln(1+\tanh\theta)}{\ln 2}$ $A_o = 1 + \dfrac{\ln(1+\coth\theta)}{\ln 2}$ $\theta = \dfrac{\pi S}{2b}$ $C_f = 2\ln\left(\dfrac{2b-t}{b-t}\right) - \dfrac{t}{b}\ln\left[\dfrac{t(2b-t)}{(b-t)^2}\right]$	$\alpha_c^e = \dfrac{0.0231 R_s\sqrt{\epsilon_r}}{30\pi(b-t)}\left\{60\pi + Z_{0e}\sqrt{\epsilon_r}\left[1 - \dfrac{A_e}{\pi}\left(\ln\dfrac{2b-t}{b-t} + \dfrac{1}{2}\ln\dfrac{t(2b-t)}{(b-t)^2}\right) + C_f \dfrac{(1+S/b)}{4\ln 2}\dfrac{\text{sech}^2\,\theta}{1+\tanh\theta}\right]\right\}$ $\alpha_c^o = \dfrac{0.0231 R_s\sqrt{\epsilon_r}}{30\pi(b-t)}\left\{60\pi + Z_{0o}\sqrt{\epsilon_r}\left[1 - \dfrac{A_o}{\pi}\left(\ln\dfrac{2b-t}{b-t} + \dfrac{1}{2}\dfrac{t(2b-t)}{(b-t)^2}\right) - C_f \dfrac{(1+S/b)}{4\ln 2}\dfrac{\text{cosech}^2\,\theta}{1+\coth\theta}\right]\right\}$ $\alpha_d^e = \alpha_d^o = 27.3\sqrt{\epsilon_r}\,\dfrac{\tan\delta}{\lambda_0}$	$\dfrac{t}{b} < 0.1$ $\dfrac{W}{b} \geq 0.35$ [28]

Microstrip lines	$Z_{0i}=[c\sqrt{C_i C_i^a}]^{-1}, \qquad i=e \text{ or } o$ $C_e=C_p+C_f+C_f'$ $C_o=C_p+C_f+C_{ga}+C_{gd}$ $C_p=\epsilon_0\epsilon_r\dfrac{W}{h}$ $2C_f=\sqrt{\epsilon_{re}}/(cZ_0)-C_p, \qquad c=3\times10^8 \quad \text{m/s}$ $C_f'=\dfrac{C_f\sqrt{\epsilon_r/\epsilon_e}}{1+\exp[-0.1\exp(2.33-2.53W/h)](h/s)\tanh(10\,s/h)}$ $C_{ga}=\epsilon_0\dfrac{K(k')}{K(k)}; \qquad k=\dfrac{s/h}{s/h+2W/h}$ $C_{gd}=\dfrac{\epsilon_0\epsilon_r}{\pi}\ln\left[\coth\left(\dfrac{\pi S}{4h}\right)\right]+0.65C_f\left[\dfrac{0.02}{s/h}\sqrt{\epsilon_r}+1-\epsilon_r^{-2}\right]$ $\epsilon_e^i=\dfrac{C_i}{C_i^a}$	$\alpha_c^i=\dfrac{8.686R_s}{240\pi Z_{0i}}\cdot\dfrac{2}{h}\cdot\dfrac{1}{c(C_i^{at})^2}$ $\cdot\left[\dfrac{\partial C_f^{at}}{\partial(W/h)}\left(1+2\dfrac{W}{2h}\right)-\dfrac{\partial C_i^{at}}{\partial(s/h)}\left(1-2\dfrac{s}{2h}\right)\right.$ $\left.+\dfrac{\partial C_i^{at}}{\partial(t/h)}\left(1+2\dfrac{t}{2h}\right)\right]$ $\alpha_d^i=27.3\dfrac{\epsilon_r}{\sqrt{\epsilon_e^i}}\dfrac{\epsilon_e^i-1}{\epsilon_r-1}\dfrac{\tan\delta}{\lambda_0}$ $\dfrac{W_t^e}{h}=\dfrac{W}{h}+\dfrac{\Delta W}{h}\left(1-0.5\exp\dfrac{-0.69\Delta W}{\Delta t}\right)$ $\dfrac{W_t^o}{h}=\dfrac{W_t^e}{h}+\dfrac{\Delta t}{h}, \qquad \dfrac{\Delta t}{h}=\dfrac{1}{\epsilon_r}\dfrac{t/h}{s/h}$	Accuracy better then 3% for $0.2\le\dfrac{W}{h}\le2$ $0.05\le\dfrac{S}{h}\le2$
Broadside-coupled strip lines	$Z_{0o}=\dfrac{60\pi/\sqrt{\epsilon_r}}{\dfrac{W}{b-s}+C_{f0}+2\{(1+t/s)\ln(1+t/s)-(t/s)\ln(t/s)\}/\pi}$ $Z_{0e}=\dfrac{60\pi/\sqrt{\epsilon_r}}{\dfrac{W}{b-s}+0.443+\left[\ln\dfrac{b+2t}{b-s}+\dfrac{S+2t}{b-s}\ln\{(b+2t)/(S+2t)\}\right]/\pi}$ $C_{f0}=\dfrac{b}{s\pi}\left[\ln\dfrac{1}{1-S/b}+\dfrac{S/b}{1-S/b}\ln\dfrac{b}{S}\right]$	$\alpha_c^i=\dfrac{0.0231R_s\sqrt{\epsilon_r}}{Z_{0i}}\left[\dfrac{\partial Z_{0i}}{\partial b}+\dfrac{\partial Z_{0i}}{\partial S}-\dfrac{\partial Z_{0i}}{\partial W}-\dfrac{\partial Z_{0i}}{\partial t}\right]$ $\alpha_d=27.3\sqrt{\epsilon_r}\dfrac{\tan\delta}{\lambda_0}$	[31–33] $\dfrac{W/b'}{1-s/b}\ge0.35$ $\dfrac{W}{s}\ge0.35$

Superscripts a and t denote air as dielectric and strip thickness effect, respectively.

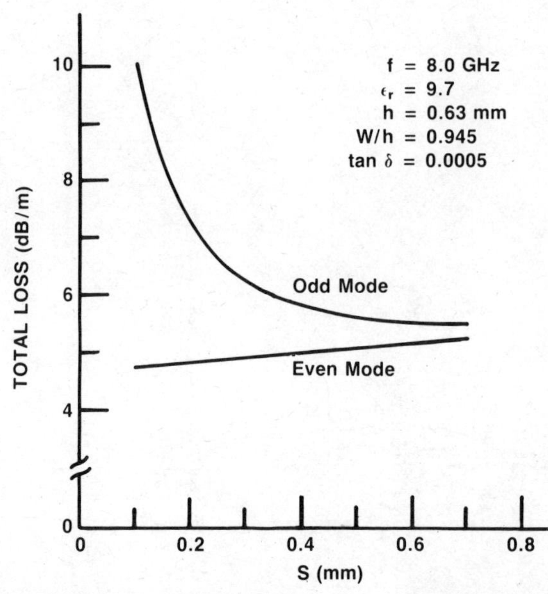

Figure 2.11 Even- and odd-mode losses in coupled microstrip lines.

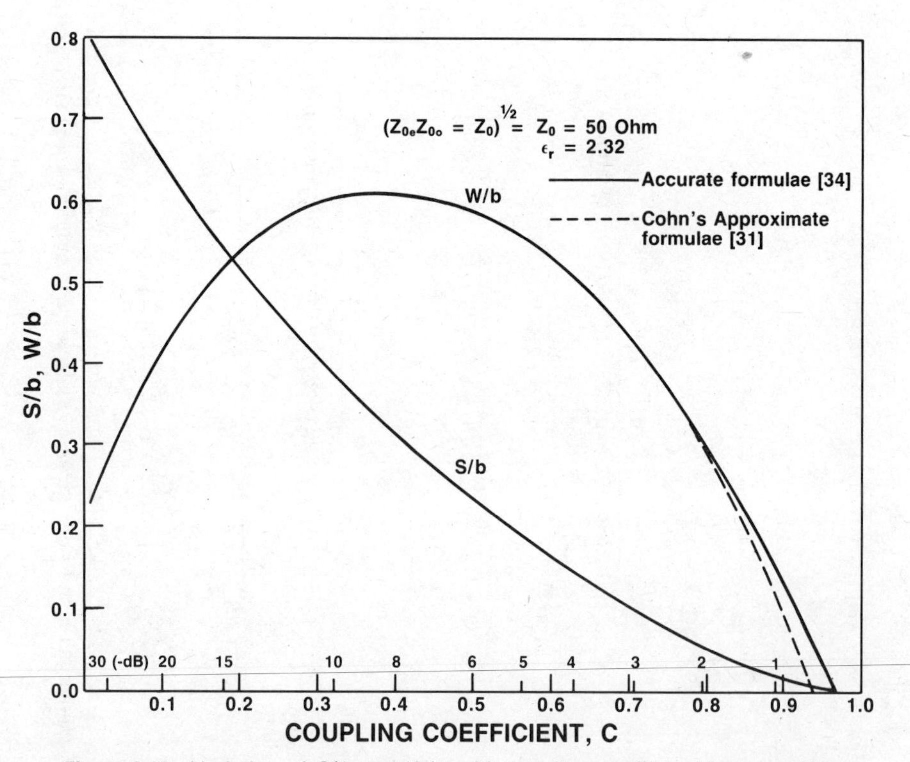

Figure 2.12 Variation of S/b and W/b with coupling coefficient for $\epsilon_r = 2.32$.

In coupled microstrip lines, the effect of thickness of conductors on capacitances can be evaluated by using the concept of effective width W_e [29]. The expressions given in Table 2.9 are valid for $S \geq 2t$, where $\Delta W = W_e - W$ defined in Table 2.5. Variation of even- and odd-mode total loss as a function of gap between the two conductors is shown in Fig. 2.11 [30].

Broadside-coupled strip lines have been widely used for realizing tight couplings (e.g., 3-dB hybrids) because for greater than -8 dB coupling, the spacing between the strips in the case of parallel coupled transmission lines (i.e., strip lines, microstrip lines, etc.) becomes prohibitively small. Expressions for characteristic impedances and attenuation constants are given in Table 2.9 [31–33]. Figure 2.12 shows the variation of W/b and S/b as a

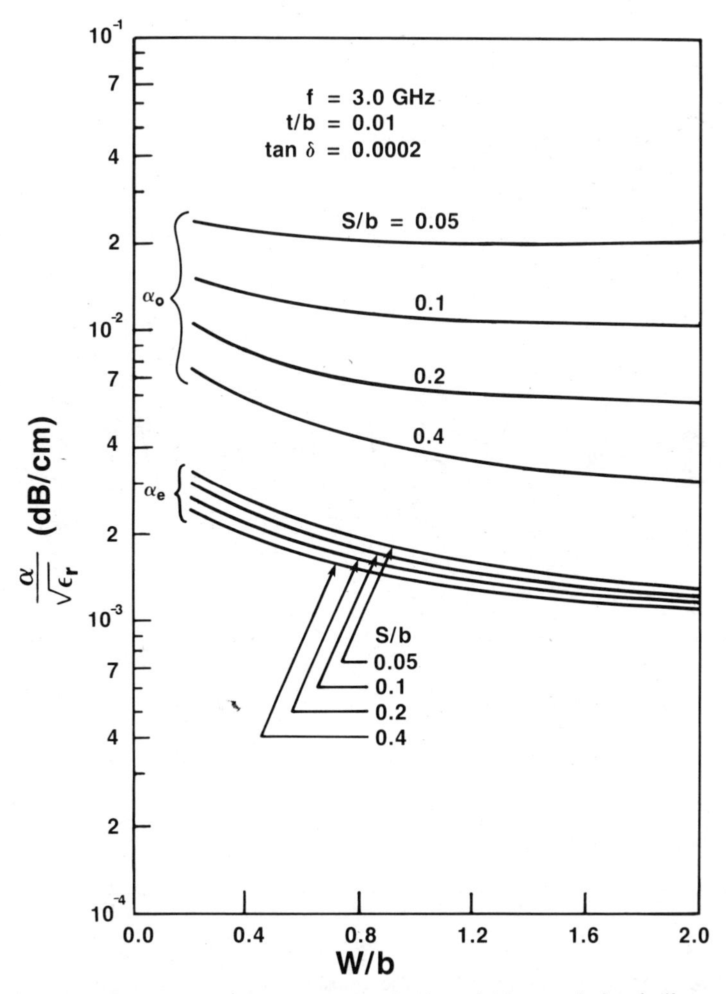

Figure 2.13 Attenuation constants for broadside-coupled strip lines.

function of coupling coefficient for $\epsilon_r = 2.32$ and $t = 0$. More accurate results [34] are also included for comparison. The total loss, $\alpha = \alpha_c + \alpha_d$, in broadside-coupled strip lines is plotted in Fig. 2.13 as a function of W/b for various values of S/b and $f = 3.0$ GHz. It may be observed that the odd-mode attenuation constant is always higher than the even-mode values.

Broadside-coupled strip lines in inhomogeneous media [35, 36] have also been used extensively for designing directional couplers and filters. However, to design high-directivity directional couplers, phase velocities for the even and odd modes must be the same. Multilayered broadside-coupled strip lines have been analyzed [36] to obtain equal phase velocities for the two modes.

Design curves for coupling coefficient are shown in Figs. 2.14 and 2.15. Definitions of coupling coefficient and Z_0 are given in Section 5.4.

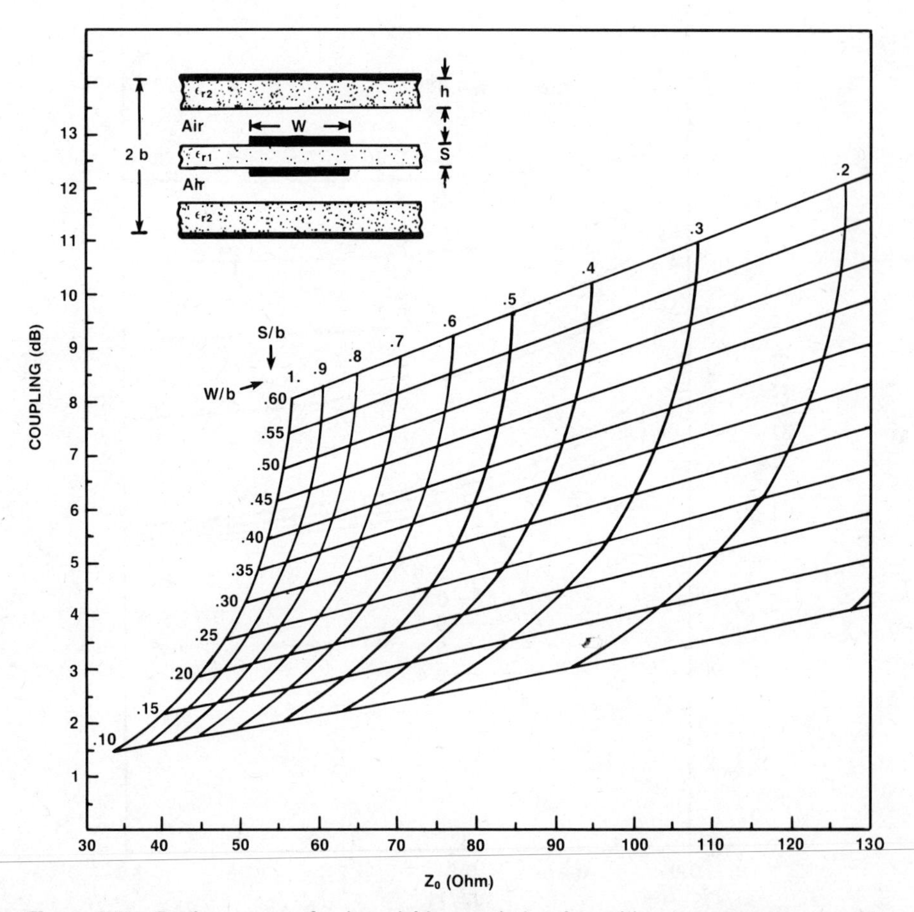

Figure 2.14 Design curves for broadside-coupled strips with two substrates $\epsilon_{r1} = 2.5$ and $\epsilon_{r2} = 10.0$. (After M. Horno and F. Medina [36] 1986 IEEE, reprinted with permission.)

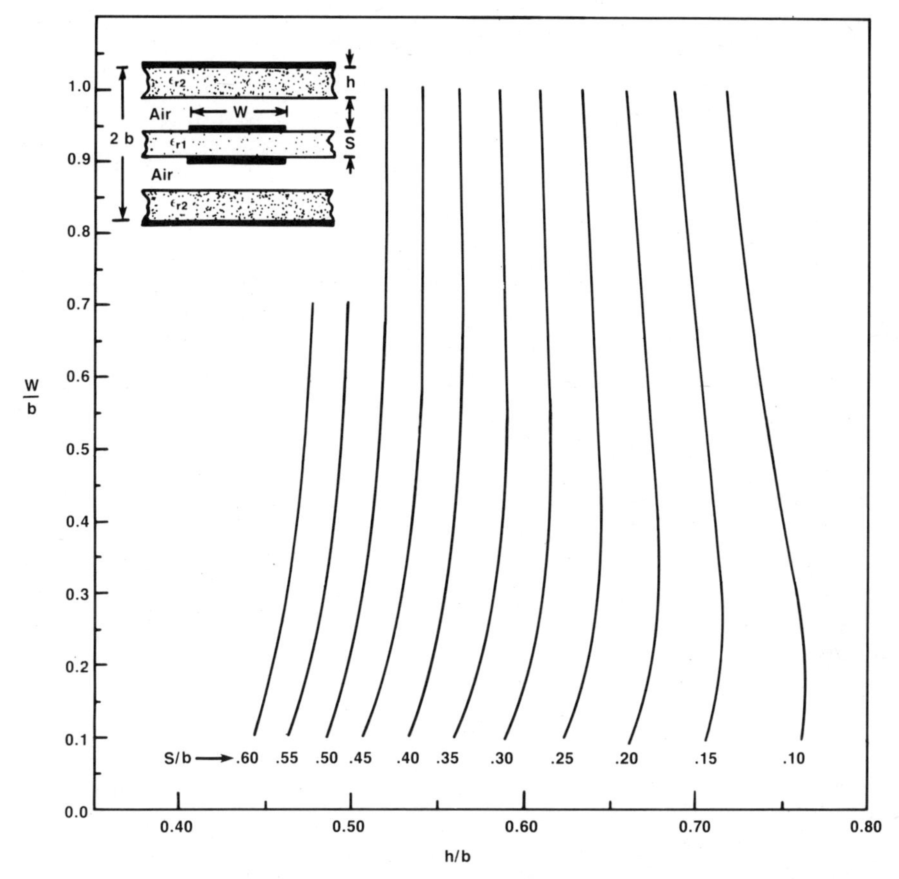

Figure 2.15 Optimum h/b ratio for broadside-coupled strips with two substrates $\epsilon_{r1} = 2.5$ and $\epsilon_{r2} = = 10.0$. (After M. Horno and F. Medina [36] 1986 IEEE, reprinted with permission.)

2.3 DISCONTINUITIES

In microwave circuits, discontinuities between distributed elements, between lumped elements, and between distributed and lumped circuit elements always exist. Typical discontinuities are (1) open circuits and short circuits, (2) bends (right-angled and others), (3) step change in dimensions (introduced for a change in impedance level), and (4) T and cross-junctions. The reactances associated with these discontinuities may be called parasitic, as they are not introduced intentionally. Some of the effects of discontinuities on circuit performances are

1. frequency shift in narrow-band circuits,
2. degradation in input and output VSWRs,

3. higher ripple in gain flatness of broadband ICs,
4. interfacing problem in multifunction circuits,
5. lower circuit yield due to degradation in circuit performance,
6. surface wave and radiation couplings that may cause oscillations in high-gain amplifiers.

The effect of discontinuities becomes more critical at higher frequencies. The discontinuities should be either taken into account or compensated for at the final stage of design. In most cases discontinuities are basically undesirable circuit reactances, and in a good circuit design, efforts are made to reduce or compensate for these reactances. A complete understanding of the design of microwave circuits requires characterization of the discontinuities present in these circuits. Since it is impossible to do tuning on GaAs monolithic microwave integrated circuits (MMICs), an accurate and comprehensive modeling of each device and circuit element is required to save expensive and time-consuming iteration of mask and wafer fabrication and evaluation. As the yield of MMICs depends on the size (the smaller the chip, the higher the yield) and acceptable circuits' electrical performance, discontinuities play an important part in the development of MMICs.

Since the discontinuity dimensions are usually much smaller than a wavelength, the discontinuities are represented by lumped-element equivalent circuits. In many cases when the longitudinal dimension of a discontinuity is very short, the equivalent circuit consists of a single shunt or series-connected reactance located at the point of the discontinuity. However, when the discontinuity has a larger longitudinal extent, the equivalent circuit is usually a π or a T network.

Various general methods are used to determine discontinuity reactances. The most commonly used techniques are variational method, mode matching, and spectral domain. A more complete characterization involves determination of the frequency-dependent scattering matrix coefficients associated with the discontinuity. Such analyses are available for several types of discontinuities [14].

2.3.1 Coaxial-Line Discontinuities

Various types of coaxial-line discontinuities have been described in the literature [4, 37, 38]. These include capacitive gaps, steps, and T junctions. The configurations and the equivalent circuits for these discontinuities are shown in Fig. 2.16. For an open gap in the center conductor, the values of capacitances C_1 and C_2, are given in Table 2.10. If an open-circuit plane is located at AA, C_2 gives the open-end capacitance of the coaxial line, and the capacitance with the shorting plane at a distance $S/2$ from the inner conductor is C, given by $(2C_1 + C_2)$. A closed-form expression for the

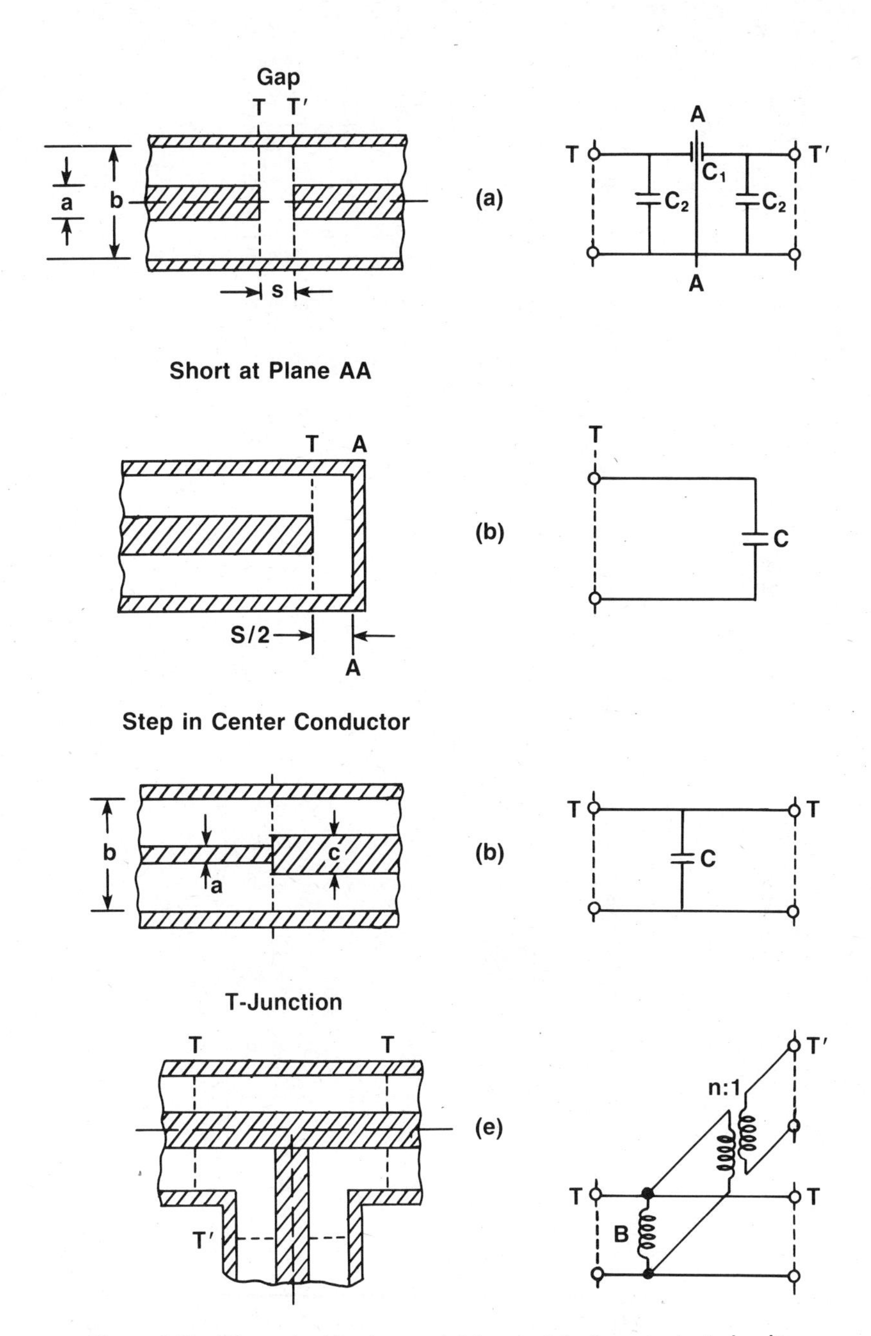

Figure 2.16 Discontinuities in coaxial line and their equivalent circuits.

TABLE 2.10 Series (C_1) and Shunt (C_2) Capacitances of Equivalent Circuits for Gaps in Coaxial Lines (pF/$2\pi b$, b in cm) [38]

Gap Ratio (s/b)	Diameter ratio ($b:a$)											
	10:9		4:3		5:3		2:1		3:1		5:1	
	C_1	C_2	C_1	C_2	C_1	C_2	C_1	C_2	C_1	C_2	C_1	C_2
0.05	0.367	0.0354	0.275	0.0143	0.188	0.0082	0.138	0.0061	0.0702	0.0039	0.0316	0.0026
0.075	0.238	0.0486	0.183	0.0206	0.127	0.0120	0.0946	0.0089	0.0498	0.0057	0.0231	0.0038
0.10	0.173	0.0598	0.136	0.0265	0.0960	0.0156	0.0719	0.0116	0.0384	0.0074	0.0183	0.0050
0.15	0.106	0.0767	0.0858	0.0366	0.0623	0.0221	0.0474	0.0166	0.0259	0.0105	0.0127	0.0067
0.20	0.0718	0.0890	0.0598	0.0450	0.0443	0.0277	0.0340	0.0210	0.0188	0.0133	0.0093	0.0087
0.25	0.0516	0.0985	0.0436	0.0520	0.0328	0.0327	0.0254	0.0248	0.0143	0.0157	0.0070	0.0102
0.30	0.0383	0.106	0.0328	0.0579	0.0249	0.0369	0.0194	0.0281	0.0109	0.0178	0.0054	0.0113

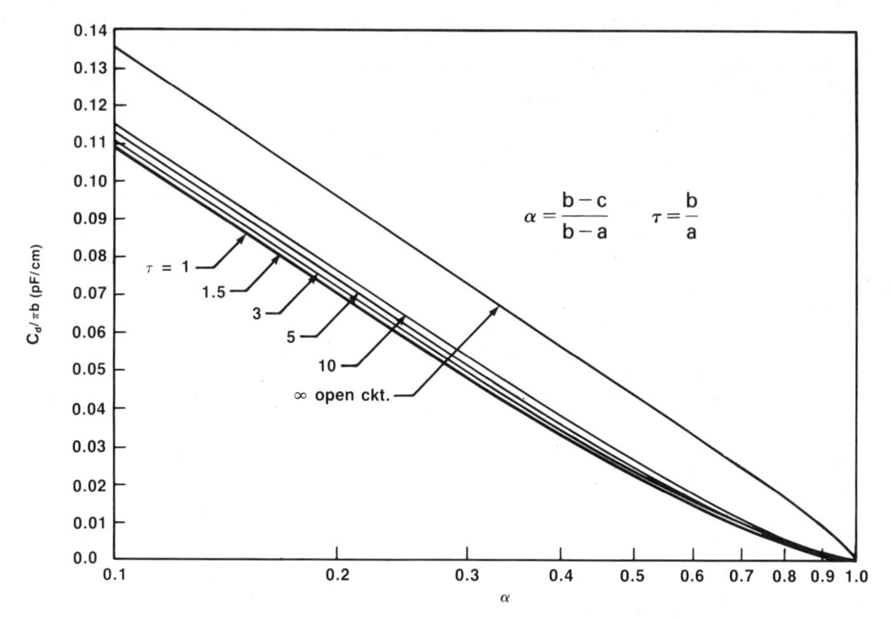

Figure 2.17 Discontinuity capacitance of a coaxial line having a step in inner conductor.

capacitance C is written as [2]

$$C = \frac{\pi a^2 \epsilon_0 \epsilon_r}{2S} + 2a\epsilon_0\epsilon_r \ln \frac{b-a}{S} \tag{2.19}$$

Equation 2.19 is valid under the following restrictions

$$\lambda_0 \gg (b-a) \gg S$$

It is seen that (2.19) is accurate to within 5% for small gaps ($S/a < 0.1$). For a 50-Ω line the error is less than 1.5%. Open-circuited coaxial lines terminated in circular waveguides below cutoff are used as standard open-circuit terminations. The capacitance offered by such a termination has been calculated by Bianco et al. [39] using the variational method.

The discontinuity capacitance for the abrupt step in the inner conductor is plotted in Fig. 2.17, while for T junction a specific set of parameters is available for $f = 3$ GHz in Reference [40].

2.3.2 Rectangular Waveguide Discontinuities

The characterizations of various types of discontinuities in rectangular waveguides have been reported in the literature. These discontinuities include posts, strips, diaphragms, windows, steps, bends, T junctions, and apertures. Due to the limited scope of this book, these discontinuities are not treated here, but an excellent description of them can be found in [4, 37].

2.3.3 Strip-line Discontintuities

Various types of discontinuities that occur in the conductor of planar transmission lines, such as strip lines or microstrips, are shown in Fig. 2.18. Examples of circuits and circuit elements, wherein these discontinuities are frequently encountered, are also shown in this figure.

The discontinuities in the center conductor of strip lines have been studied comprehensively by Oliner [41], and Altschuler and Oliner [42].

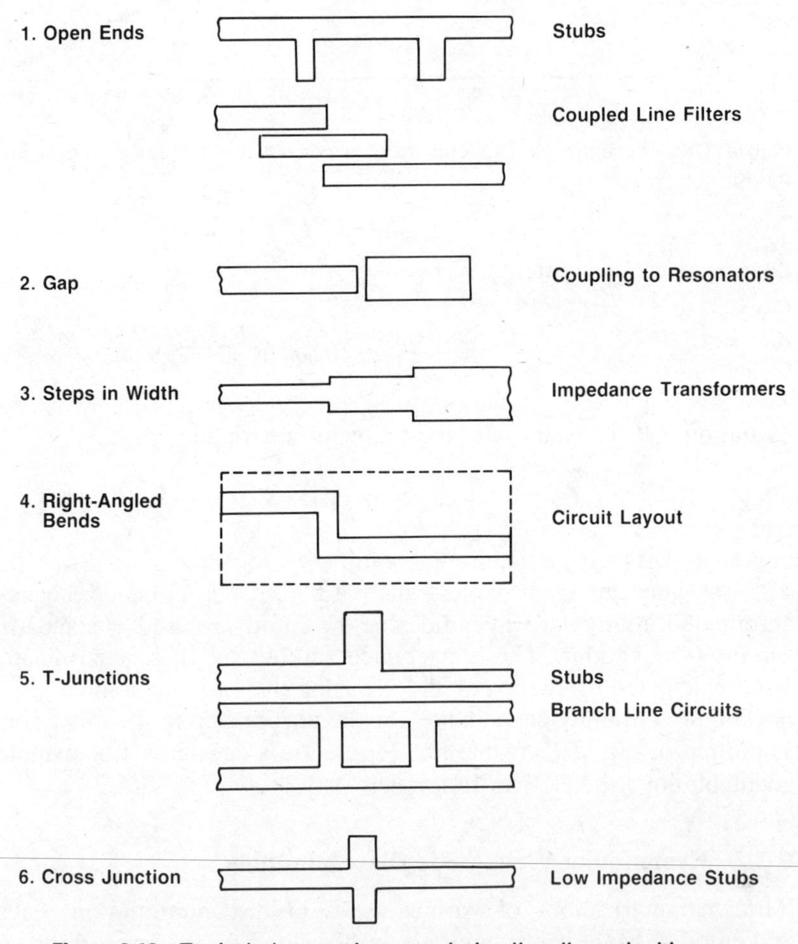

Figure 2.18 Typical planar strip transmission-line discontinuities.

Some of the important discontinuities described here are an open end, a step, a round hole, a gap, a right-angled bend, and a T junction. These configurations, their equivalent circuits and approximate closed-form expressions [4, 41–43] are given in Table 2.11. In these expressions, a bar over the reactances and susceptances denotes normalization with respect to the characteristic impedance and characteristic admittance, respectively. For the analysis of discontinuities, the strip width W of the center conductor is replaced by an equivalent strip width D, which is given by

$$D = \begin{cases} b\dfrac{K(k)}{K(k')} + \dfrac{t}{\pi}\left(1 - \ln\dfrac{2t}{b}\right) & \text{for } \dfrac{W}{b} \leq 0.5 \quad (2.20) \\[2ex] W + \dfrac{2b}{\pi}\ln 2 + \dfrac{t}{\pi}\left(1 - \ln\dfrac{2t}{b}\right) & \text{for } \dfrac{W}{b} > 0.5 \quad (2.21) \end{cases}$$

where $k = \tanh(\pi W/2b)$, W is the width, t is the thickness of the conductor, and b is the spacing between the two ground planes. The elliptic functions $K(k)$ and $K(k')$ are defined in (2.4).

The equivalent circuit of an open-ended strip line (open-ended shunt stub) represented by an excess capacitance can be expressed as an equivalent length of a transmission line Δl. This means that an effective open circuit is located at a distance Δl from the physical open end. A gap discontinuity is generally introduced intentionally to provide a dominant series capacitance C_{12} (susceptance B_B). The shunt capacitance C_1 (susceptance B_A) is a result of the disturbance in the electric field distribution at the edge of the gap. For very large gap ($S \rightarrow \infty$), C_{12} decreases to zero and C_1 tends to the value of end-capacitance for the open-ended line. Holes are also introduced purposely to provide inductive reactive elements for filters and other circuits.

Step in width discontinuity is formed by the junction of two lines having different widths, while bend discontinuity is introduced in the conductors to make the size of a circuit compact. The T junction is perhaps the most important discontinuity and is found in most microwave circuits.

2.3.4 Microstrip Discontinuities

A complete understanding of the design of microwave integrated circuits requires characterizations of the discontinuities present in these circuits. These discontinuities have been studied extensively and a detailed review is given in [14]. A collection of closed-form expressions for various discontinuities is available in [44]. The capacitive components of the discontinuities have been obtained through analyses by Benedek and Silvester [45], Silvester and Benedek [46, 47], Farrar and Adams [48, 49], Maeda [50], Horton [51], and Itoh et al. [52]. The inductive components have been determined by Thomson and Gopinath [53], and Gopinath et al. [54].

TABLE 2.11 Expressions for Strip Line Discontinuities

Discontinuity	Equivalent Circuit	Expressions
Open End 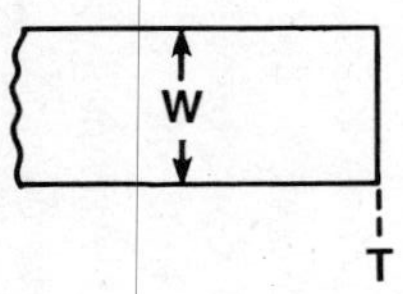 W, T	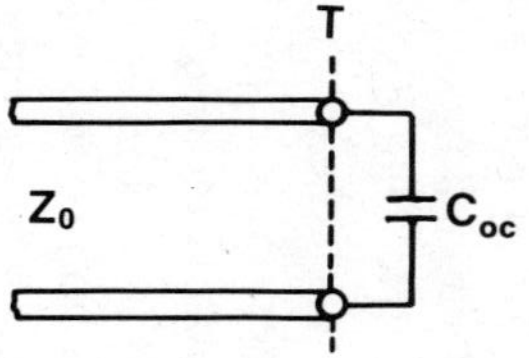 T, Z_0, C_{oc}	$C_{oc} = \dfrac{\beta \Delta l}{\omega Z_0}, \quad \beta = \dfrac{2\pi}{\lambda}, \quad \lambda = \dfrac{\lambda_0}{\sqrt{\epsilon_r}}$ $\beta \Delta l = \tan^{-1}\left[\dfrac{\delta + 2W}{4\delta + 2W}\tan(\beta\delta)\right], \quad \delta = \dfrac{b \ln 2}{\pi}$
Gap 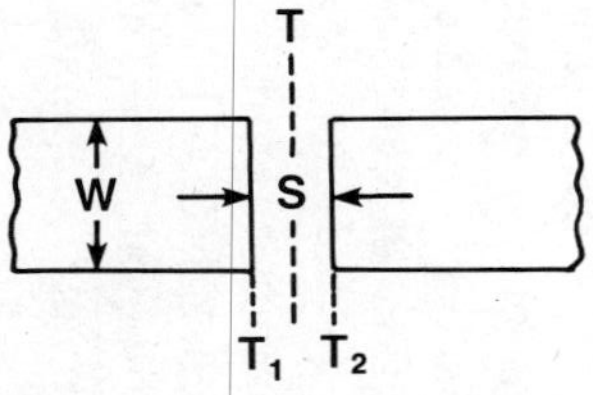 T, W, S, T_1, T_2	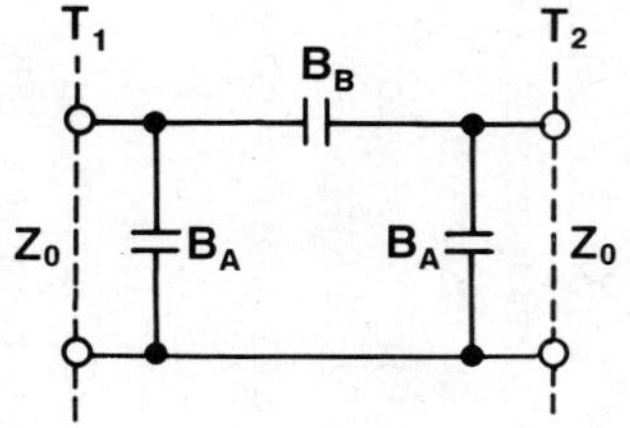 T_1, T_2, B_B, Z_0, B_A, B_A, Z_0	$\bar{B}_A = \dfrac{1 + \bar{B}_a \cot(\beta S/2)}{\cot(\beta S/2) - \bar{B}_a} = \dfrac{\omega C_1}{Y_0}$ $2\bar{B}_B = \dfrac{1 + (2\bar{B}_b + \bar{B}_a)\cot(\beta S/2)}{\cot(\beta S/2) - (2\bar{B}_b + \bar{B}_a)} - \bar{B}_A = \dfrac{2\omega C_{12}}{Y_0}$ $\lambda \bar{B}_a = -2b \ln \cosh \dfrac{\pi S}{2b}, \quad \lambda \bar{B}_b = b \ln \coth \dfrac{\pi S}{2b}$
Round Hole 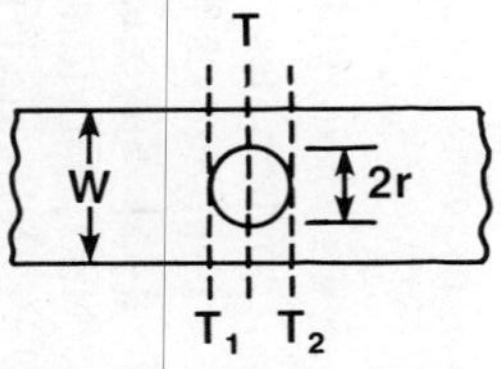 T, W, 2r, T_1, T_2	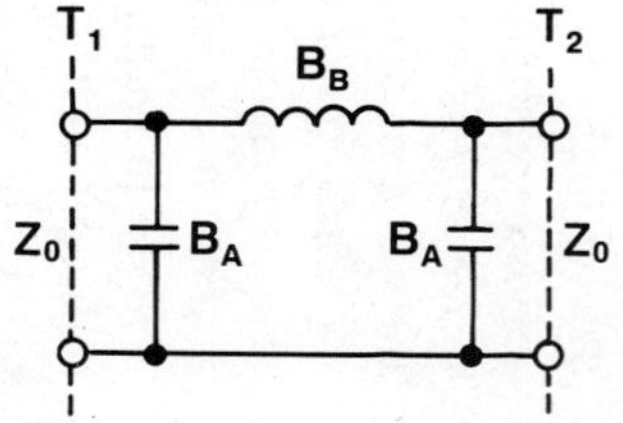 T_1, T_2, B_B, Z_0, B_A, B_A, Z_0	$\bar{B}_A = \dfrac{1 + \bar{B}_a \cot(\beta r)}{\cot(\beta r) - \bar{B}_a}$ $2\bar{B}_B = \dfrac{1 + 2\bar{B}_b \cot(\beta r)}{\cot(\beta r) - 2\bar{B}_b} - \bar{B}_A$ $\bar{B}_b = -\dfrac{3}{16\beta}\dfrac{bD}{r^3}, \quad \bar{B}_a = \dfrac{1}{4\bar{B}_b}$

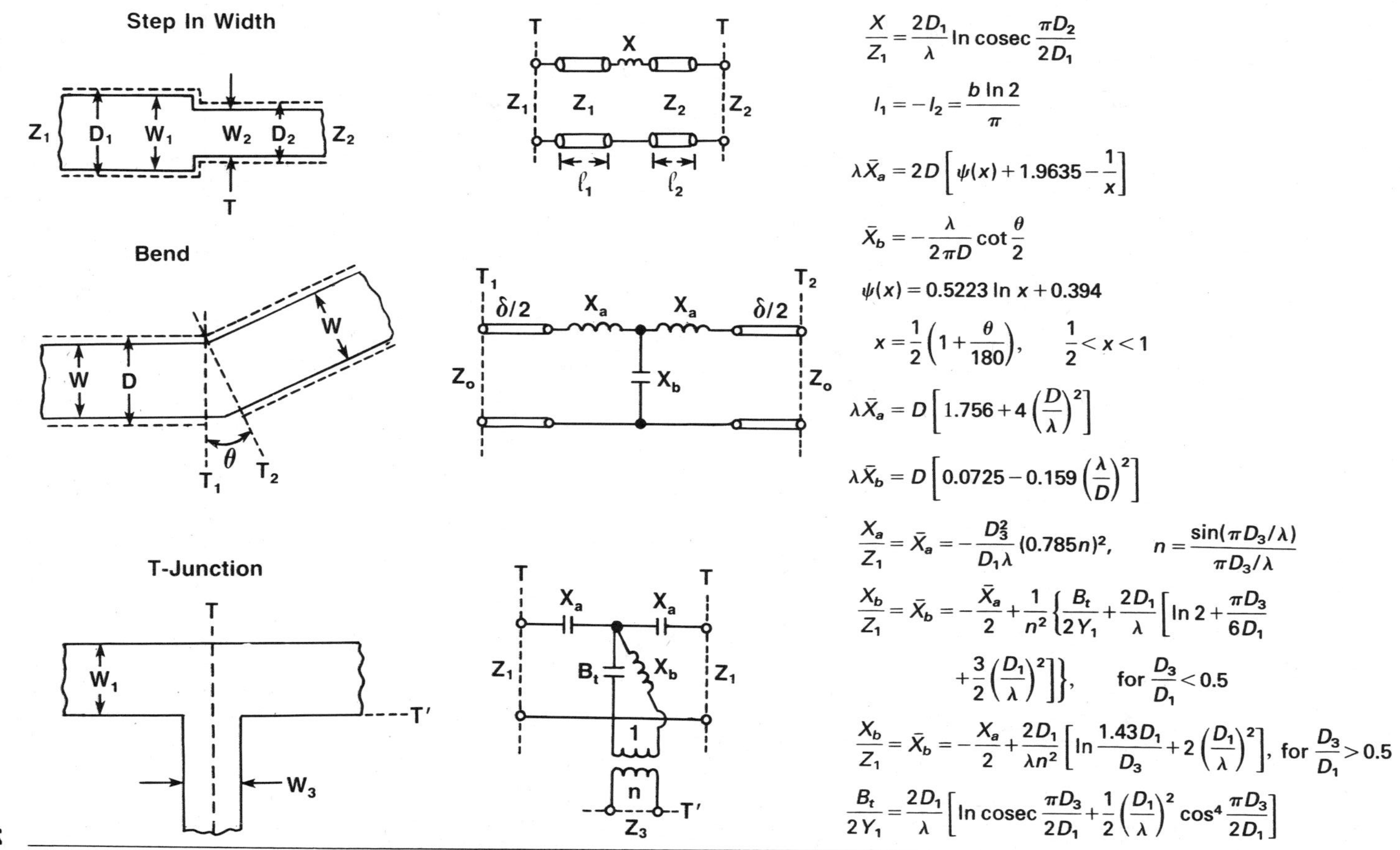

$$\frac{X}{Z_1} = \frac{2D_1}{\lambda} \ln \operatorname{cosec} \frac{\pi D_2}{2D_1}$$

$$l_1 = -l_2 = \frac{b \ln 2}{\pi}$$

$$\lambda \bar{X}_a = 2D \left[\psi(x) + 1.9635 - \frac{1}{x} \right]$$

$$\bar{X}_b = -\frac{\lambda}{2\pi D} \cot \frac{\theta}{2}$$

$$\psi(x) = 0.5223 \ln x + 0.394$$

$$x = \frac{1}{2}\left(1 + \frac{\theta}{180}\right), \qquad \frac{1}{2} < x < 1$$

$$\lambda \bar{X}_a = D \left[1.756 + 4 \left(\frac{D}{\lambda}\right)^2 \right]$$

$$\lambda \bar{X}_b = D \left[0.0725 - 0.159 \left(\frac{\lambda}{D}\right)^2 \right]$$

$$\frac{X_a}{Z_1} = \bar{X}_a = -\frac{D_3^2}{D_1 \lambda}(0.785n)^2, \qquad n = \frac{\sin(\pi D_3/\lambda)}{\pi D_3/\lambda}$$

$$\frac{X_b}{Z_1} = \bar{X}_b = -\frac{\bar{X}_a}{2} + \frac{1}{n^2}\left\{ \frac{B_t}{2Y_1} + \frac{2D_1}{\lambda}\left[\ln 2 + \frac{\pi D_3}{6D_1} + \frac{3}{2}\left(\frac{D_1}{\lambda}\right)^2 \right] \right\}, \qquad \text{for } \frac{D_3}{D_1} < 0.5$$

$$\frac{X_b}{Z_1} = \bar{X}_b = -\frac{X_a}{2} + \frac{2D_1}{\lambda n^2}\left[\ln \frac{1.43 D_1}{D_3} + 2\left(\frac{D_1}{\lambda}\right)^2 \right], \quad \text{for } \frac{D_3}{D_1} > 0.5$$

$$\frac{B_t}{2Y_1} = \frac{2D_1}{\lambda}\left[\ln \operatorname{cosec} \frac{\pi D_3}{2D_1} + \frac{1}{2}\left(\frac{D_1}{\lambda}\right)^2 \cos^4 \frac{\pi D_3}{2D_1} \right]$$

TABLE 2.12 Expressions for Microstrip Discontinuities

Discontinuity	Equivalent Circuit	Expressions
Open End	(W; T)	(T; $\Delta\ell$; Z_0)	$\frac{\Delta l}{h} = 0.412 \frac{\epsilon_e + 0.3}{\epsilon_e - 0.258} \left[\frac{W/h + 0.264}{W/h + 0.8}\right]$
	(T; Z_0; C_{oc})	$C_{0c} = \frac{\Delta l \sqrt{\epsilon_e}}{cZ_0}$
Gap 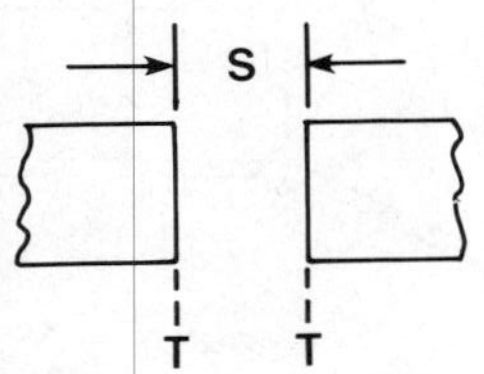 (S; T; T)	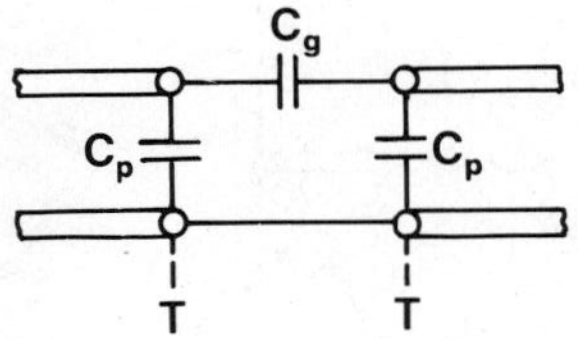 (C_g; C_p; C_p; T; T)	$C_p = \frac{C_{even}}{2}, \quad C_g = \frac{C_{odd} - C_p}{2}$ $C_{even} = \left(\frac{\epsilon_r}{9.6}\right)^{0.9} C_e, \quad C_{odd} = \left(\frac{\epsilon_r}{9.6}\right)^{0.8} C_o$ $\frac{C_e}{W}(\text{pF/m}) = \left(\frac{S}{W}\right)^{m_e} \exp(k_e), \quad \frac{C_0}{W}(\text{pF/m}) = \left(\frac{S}{W}\right)^{m_o} \exp(k_o)$ $m_e = 0.8675, \quad k_e = 2.043 \left(\frac{W}{h}\right)^{0.12}, \quad 0.1 \le \frac{S}{W} \le 0.3$ $m_e = \frac{1.565}{(W/h)^{0.16}} - 1, \quad k_e = 1.97 - \frac{0.03}{W/h}, \quad 0.3 \le \frac{S}{W} \le 1.0$ $\left.\begin{aligned} m_o &= \frac{W}{h}\left(0.619 \log \frac{W}{h} - 0.3853\right), \\ k_o &= 4.26 - 1.453 \log \frac{W}{h} \end{aligned}\right\} \quad 0.1 \le \frac{S}{W} \le 1.0$
Notch 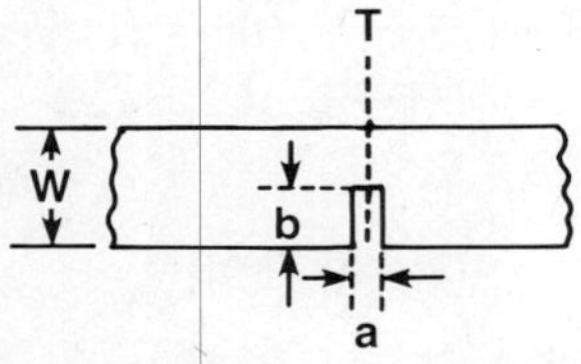 (T; W; b; a)	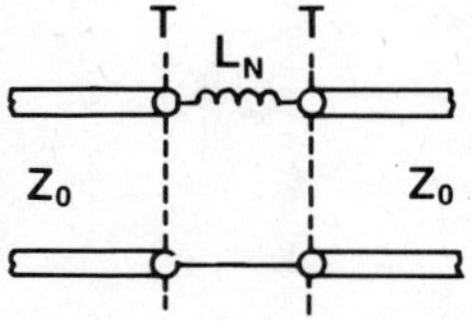 (T; T; L_N; Z_0; Z_0)	$\frac{L_N}{h}(\mu\text{H/m}) = 2\left[1 - \frac{Z_0}{Z_0'} \sqrt{\frac{\epsilon_e}{\epsilon_e'}}\right]^2$

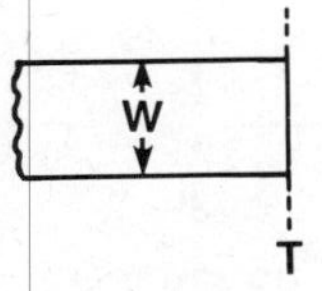

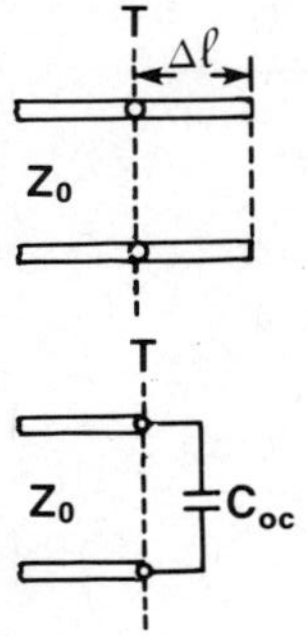

Step In Width

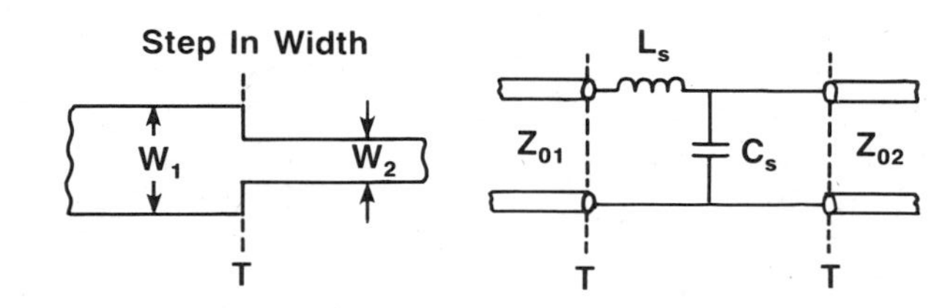

$$\frac{C_S}{h}\,(\text{pF/m}) = 1370\,\frac{\sqrt{\epsilon_{e1}}}{Z_{01}}\left(1-\frac{W_2}{W_1}\right)\left[\frac{\epsilon_{e1}+0.3}{\epsilon_{e1}-0.258}\right]\left[\frac{W_1/h+0.264}{W_1/h+0.8}\right]$$

$$\frac{L_S}{h}\,(\mu\text{H/m}) = \left[1-\frac{Z_{01}}{Z_{02}}\sqrt{\frac{\epsilon_{e1}}{\epsilon_{e2}}}\right]^2$$

Chamferred Bend

$$C_b = \frac{l_b\sqrt{\epsilon_e(b)}}{cZ_{0B}}, \qquad L_b = \frac{l_b Z_{0B}\sqrt{\epsilon_e(b)}}{2c}$$

$$\Delta l = \left[\frac{a}{2\cos(\phi/2)} - W\right]\tan\frac{\phi}{2}$$

$$W_b = W - \Delta l \tan\frac{\phi}{2}, \qquad b = W_b \cos\frac{\phi}{2}$$

$$l_b = W_b \sin\frac{\phi}{2}, \qquad \epsilon_e(b) = \epsilon_e \text{ of } b$$

$$Z_{0A} = Z_0 \text{ of } \frac{W+W_b}{2}, \qquad Z_{0B} = Z_0 \text{ of } b$$

T-Junction

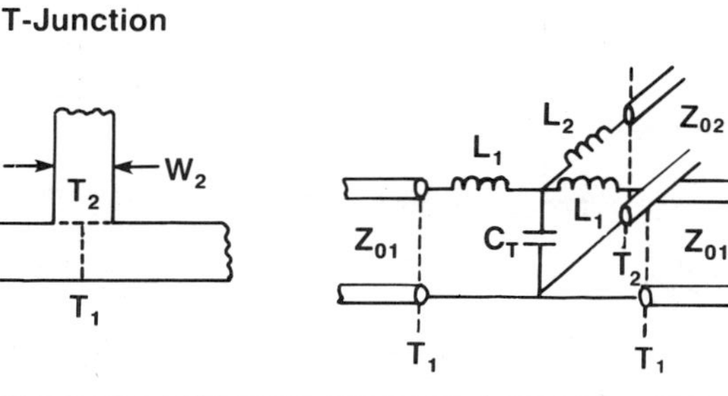

$$\frac{C_T}{W_1}\,(\text{pF/m}) = \frac{100}{\tanh(0.0072 Z_{02})} + 0.64 Z_{02} - 261 \qquad (25 \le Z_{02} \le 100)$$

$$\frac{L_1}{h}\,(\text{nH/m}) = -\frac{W_2}{h}\left\{\frac{W_2}{h}\left(-0.016\,\frac{W_1}{h}+0.064\right)+\frac{0.016}{W_1/h}\right\} L_{w1},\ 0.5 \le \left(\frac{W_1}{h}, \frac{W_2}{h}\right) \le 2.0$$

$$\frac{L_2}{h}\,(\text{nH/m}) = \left\{\left(0.12\,\frac{W_1}{h}-0.47\right)\frac{W_2}{h}+0.195\,\frac{W_1}{h}-0.357 + 0.0283\sin\left(\pi\frac{W_1}{h}-0.75\pi\right)\right\} L_{w2},\ 1 \le \frac{W_1}{h} \le 2.0,\ 0.5 \le \frac{W_2}{h} \le 2$$

$$L_w = \frac{Z_0\sqrt{\epsilon_e}}{c}\,(\text{H/m}), \qquad c = 3\times 10^8 \text{ m/s}$$

Several experimental measurements on the characterization of discontinuities have been reported. Some of the prominent ones are Napoli and Hughes [55], Easter [56], Stephens and Easter [57], and Groll and Weidmann [58]. The dynamic behavior of discontinuities has been studied by Wolff and Menzel [59], and Mehran [60, 61]. More recent methods for full-wave analysis of microstrip discontinuities include spectral domain technique and its variations [14, 62] and moment method using Green's function for a grounded dielectric slab [63, 64].

Closed-form expressions for various discontinuity elements are given in Table 2.12. For open-end discontinuity, the results are within 4% of the numerical values [46] for $W/h \geq 0.2$ and $2 \leq \epsilon_r \leq 50$. The expressions for the gap discontinuity give results that are within 7% of numerical values

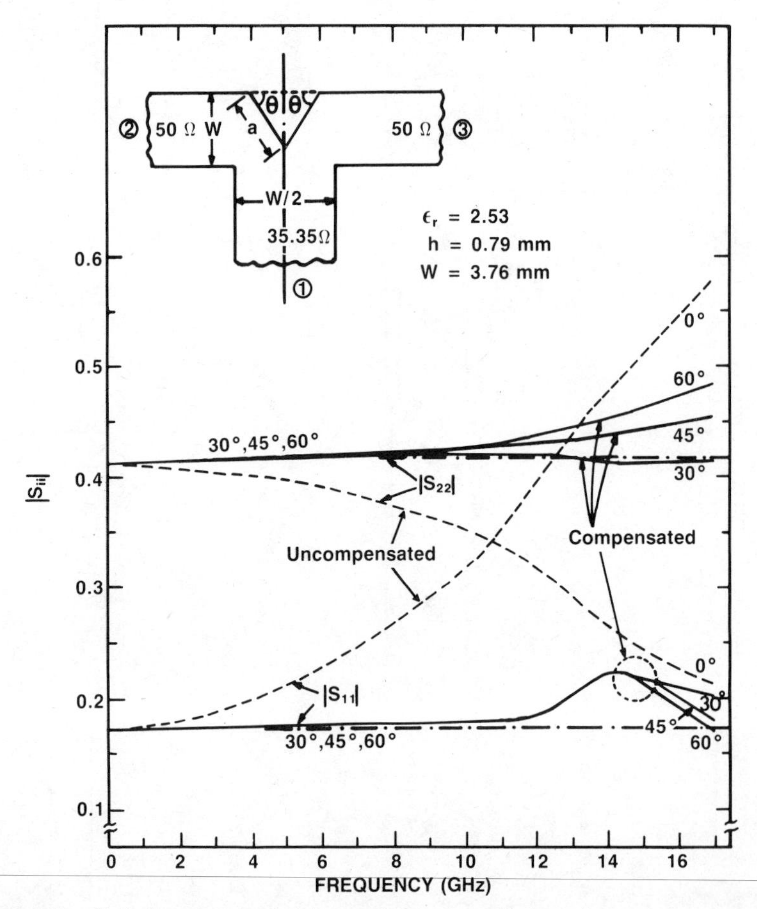

Figure 2.19 Main-line and branch-line reflection coefficients for uncompensated and compensated T junctions-impedance ratio $1/\sqrt{2}:1:1$). (After R. Chadha and K. C. Gupta [71] 1982 IEEE reprinted with permission.)

[45] for $0.5 \leq W/h \leq 2$ and $2.5 \leq \epsilon_r \leq 15$. A notch or a narrow slit in the strip conductor can be used to realize a pure series inductance. In the expressions, ϵ_e and ϵ_e' are the effective dielectric constants for microstrip lines with widths W and $(W-b)$, respectively, and Z_0 and Z_0' are the corresponding impedances. The expression for L_N is valid for $0 \leq b/W \leq 0.9$ and $a \leq h$ [65]. A simplified model for step in width consisting of only two components, L_s and C_s [66, 67], has been included here. In MICs, bends are usually chamfered. For a right-angled bend ($\phi = 90°$), the optimum chamfered length $a = 1.8$. For T junctions, very limited data are available in the literature. The expressions for the equivalent circuit parameters given in Table 2.12 are for a junction with main-line impedance of 50 Ω and for $\epsilon_r = 9.9$. The results are within 5 percent of the numerical values [47]. A more general formulation of a T junction discontinuity has been described in [14, 68]. The closed-form expressions for cross-junction discontinuities also have been derived [14, 44] by curve fitting the available numerical results for capacitances [47] and inductances [54] for $\epsilon_r = 9.9$.

2.3.5 Discontinuity Compensation

Compensation of discontinuity reactances in planar transmission lines can be obtained by removing appropriate portions from the discontinuity configurations, such as chamfering the bend. Compensation of step in width and T junctions have been treated in the literature [69–72]. The step-in-width discontinuity can be easily compensated by gradually narrowing the wider strip. Discontinuity reactances associated with a T junction can be compensated by removing a triangular portion from the junction, as shown in the Fig. 2.19 inset. It can be seen that by chopping off an isosceles triangle with $\theta = 30°$ the effects of discontinuity reactances are minimized.

2.4 LUMPED ELEMENTS

The size of a lumped-circuit element, by definition, is very much smaller than the operating wavelength, and therefore exhibits negligible phase shift. Lumped elements for use at microwave frequencies are also designed on the basis of this consideration.

Lumped-element circuits that have lower Q than distributed circuits have the advantage of smaller size, lower cost, and wide-band characteristics. These are especially suitable for monolithic MICs and for broadband hybrid MICs where real-estate requirements are of prime importance. Impedance transformations of the order of 20:1 can be easily accomplished using the lumped-element approach. Therefore, high-power devices that have very low impedance values can be easily tuned with large impedance transformers using lumped elements. Consequently, lumped elements find applications in high-power oscillators, power amplifiers, and broadband circuits.

With the advent of new photolithographic techniques, the fabrication of lumped elements that was limited to X-band frequencies can now be extended to about 60 GHz. The three basic building blocks for circuit design—inductors, capacitors and resistors—are available in lumped form. Computer-aided design of circuits using lumped elements requires a complete and accurate characterization of lumped elements at microwave frequencies. This necessitates the development of comprehensive mathematical models that take into account the presence of ground planes, proximity effect, fringing fields, parasitics, etc. In this section we discuss the available design and characterization details of lumped elements [73–82].

2.4.1 Design of Lumped Elements

Design of lumped elements at microwave and the lower end of millimeter-wave frequencies may be arrived at by considering them as very short sections of (smaller than $\lambda/10$) TEM lines. The driving point or input impedance of the line of length l given in Table 2.1 (when $\gamma l \ll 1$) may be written as

$$Z_{in} = Z_0 \frac{Z_L + Z_0 \gamma l}{Z_0 + Z_L \gamma l} \tag{2.22}$$

where Z_L is the terminating or load impedance. The characteristic impedance Z_0 and the propagation constant γ for the TEM line shown in Fig. 2.1 are given by

$$Z_0 = \sqrt{\frac{\bar{R} + j\omega\bar{L}}{\bar{G} + j\omega\bar{C}}} \tag{2.23a}$$

$$\gamma = \sqrt{(\bar{R} + j\omega\bar{L})(\bar{G} + j\omega\bar{C})} \tag{2.23b}$$

Equations 2.22 and 2.23 are used to derive equivalent circuits for L, C, and R elements.

2.4.2 Design of Inductors

The input impedance of a very small length of a short-circuited ($Z_L = 0$) line is expressed as

$$Z_{in} = Z_0 \gamma l = (\bar{R} + j\omega\bar{L})l = R + j\omega L \tag{2.24}$$

where R is the resistance and L is the inductance of the line of length l. When the thickness and width of conductors is much greater than the skin depth, for first-order approximation, conductors may be assumed lossless ($R = 0$).

However, in order to realize an inductor or a resistor, it is not necessary to have both the conductors of a transmission line. A lumped inductor may be realized using a metallic strip or a wire either in the form of straight section or a circular or square spiral.

The important characteristics of an inductor are the inductance value and its parasitics, which determine its Q factor and resonant frequency.

Inductance. Straight sections of ribbons and wires are used for low inductance values typically up to 2 to 3 nH. Spiral inductors (circular or rectangular) have higher Q and can provide higher inductance values. These inductors are commonly used for high-density circuits. The presence of a ground plane also affects the inductance value, which decreases as the ground plane is brought nearer [83]. This decrease can be taken into account by means of a correction factor K_g. With this correction, the effective inductance L may be written as

$$L = K_g L_0 \tag{2.25}$$

where L_0 is the free-space inductance value. A closed-form expression for K_g of a ribbon is given by [4, 84]

$$K_g = 0.57 - 0.145 \ln \frac{W}{h}, \quad \frac{W}{h} > 0.05 \tag{2.26}$$

where W is the width and h is the substrate thickness. To first-order approximation, the above expression can also be used with other types of inductors. Typical inductance values for MMICs fall in the range from 0.2 to 10 nH.

Resistance and Unloaded Q. In order to find Q, it is necessary to determine R, the high-frequency resistance of the conductors. The expression for R is given by

$$R = KR_s l/\text{perimeter} \tag{2.27}$$

where R_s is the sheet resistance in ohms per square, l is the length of the conductor, and K is a correction factor that takes into account the crowding of the current at the corners of the inductor conductors. For

TABLE 2.13 Expressions for Lumped Inductors

Inductors Strip		Equivalent Circuit	Expressions
Strip	W, ℓ	R, L, C, C	$L\,(\text{nH}) = 2\times 10^{-4} l \left[\ln\left(\frac{l}{W+t}\right) + 1.193 + 0.2235\,\frac{W+t}{l}\right]\cdot K_g$ $R\,(\Omega) = \frac{KR_s l}{2(W+t)}$
Loop	W, a, S′, W′	R, L, C, C	$L\,(\text{nH}) = 1.257\times 10^{-3} a \left[\ln\left(\frac{a}{W+t}\right) + 0.078\right]\cdot K_g$ $R\,(\Omega) = \frac{KR_s}{W+t}\,\pi a$
Spiral	D_i, s, W, D_o	C_3, R, L, C_1, C_2	$L\,(\text{nH}) = 0.03937\,\frac{a^2 n^2}{8a+11c}\cdot K_g$ $a = \frac{D_o + D_i}{4}, \quad c = \frac{D_o - D_i}{2}$ $R(\Omega) = \frac{k\pi a n R_s}{W}$ $C_3\,(\text{pF}) = 3.5\times 10^{-5} D_o + 0.06$

various structures, expressions for K are given as follows

$$K = \begin{cases} 1 & \text{(Wire)} \quad (2.28a) \\ 1.4 + 0.217 \ln\left(\dfrac{W}{5t}\right), \quad 5 < \dfrac{W}{t} < 100 & \text{(Ribbon)} \quad (2.28b) \\ 1 + 0.333\left(1 + \dfrac{S}{W}\right)^{-1.7} & \text{(Spiral)} \quad (2.28c) \end{cases}$$

where t is the thickness of the conductors. The unloaded Q of an inductor then may be calculated from

$$Q = \frac{\omega L}{R} \tag{2.29}$$

Table 2.13 gives expressions for inductances and resistances of various types of inductors. In the case of spirals n is the number of turns and S is the spacing between the conductors. Figure 2.20 shows the variation of L and Q as a function of number of turns of a spiral. Figure 2.21 [85] shows the typical variation of self-resonant frequency of a planar spiral inductor. It

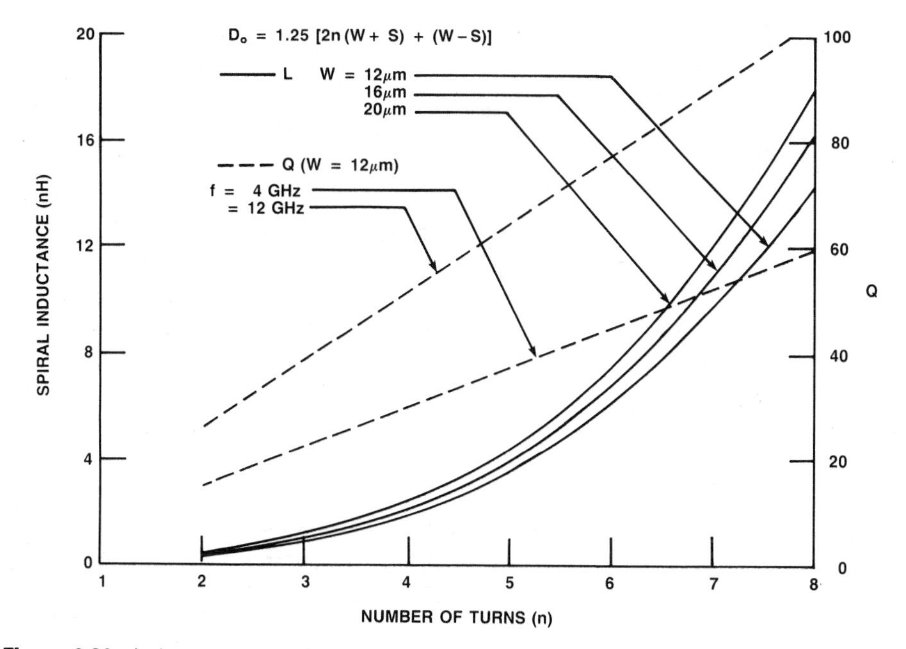

Figure 2.20 Inductance and Q vs. number of turns for a spiral inductor ($h = 200\ \mu$m and $S = 10\ \mu$m).

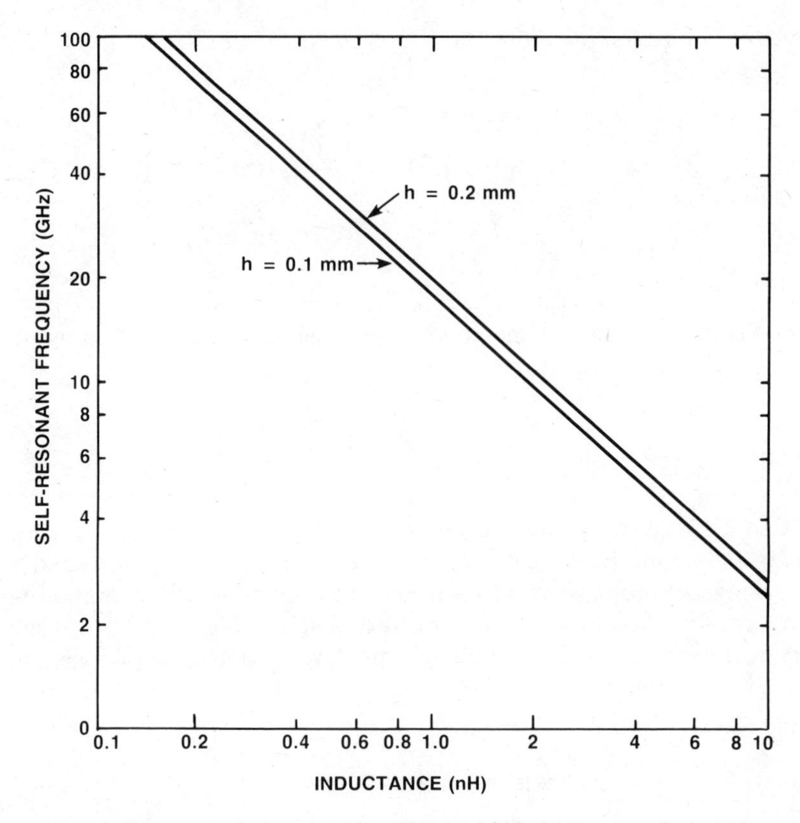

Figure 2.21 Self-resonant frequency of a planar spiral inductor on a GaAs substrate as a function of inductance and substrate thickness.

is recommended that for operating frequencies above one-third of the self-resonant frequency of the inductors, the parasitic capacitances associated with the inductors be included in the design. The values of shunt capacitors for strip and loop inductors can be easily obtained as described in the next section. However, there is no easy way to calculate shunt capacitances for a spiral. Measured inductance, parasitic capacitance and Q values for a few typical spiral inductors are given in Table 2.14 [82]. The substrate was 0.63-mm-thick alumina ($\epsilon_r = 9.6$).

In practice, inductor Q factors on the order of 50 are observed at X band, with higher values at higher frequencies. The maximum value is about 100. There appears to be no way to improve the Q factor significantly, because of the highly unfavorable ratio of metal surface area to dielectric volume. Some salient considerations useful in the design of

TABLE 2.14 Typical Spiral Inductor Parameters

W (mm)	S (mm)	n	D_0 (mm)	L (theoretical) (nH)	L (measured) (nH)	C_3 (pF)	C_1 (pF)	C_2 (pF)	Q (at 4 GHz)
0.10	0.10	1.5	1.30	3.7	4.0	0.13	0.10	0.07	87
0.05	0.05	2.5	1.50	14.1	14.3	0.11	0.09	0.05	84
0.10	0.10	2.5	1.80	9.2	9.0	0.14	0.18	0.08	81
0.10	0.10	3.5	2.00	18.7	19.0	0.15	0.17	0.09	91

these inductors are summarized as follows.

1. In the spiral, $W \gg S$, that is, the separation S between the turns should be as small as possible. However, in monolithic circuits processing limits require $W > 5\ \mu m$ and $S > 5\ \mu m$, and in hybrid MICs $W > 10\ \mu m$ and $S > 10\ \mu m$.
2. There should be some space at the center of the spiral to allow the flux lines to pass through, which increases the stored energy. It has been found that $D_o/D_i = 5$ yields an optimum value of Q, but not the maximum inductance value.
3. It is observed that for the same value of D_o, the Q of a circular spiral is higher than that of a square spiral by about 10%.
4. Multiturn inductors have a high Q due to higher inductance per unit area; also, they have lower self-resonance frequencies due to interturn capacitance.
5. The maximum diameter of the inductor should be less than $\lambda/30$ in order to avoid distributed effects.
6. Spiral inductors require air bridges.

Wire Inductor. In hybrid MICs, bonding wire connections are used to connect active and passive circuit components, and in MMICs bonding wire connections are used to connect the MMIC chip to the real world. The free-space inductance L (in nanohenrys) of a wire of diameter d and length l (in microns) is given by [4, 67, 73]

$$L = 2 \times 10^{-4} l \left(\ln \frac{4l}{d} + 0.5 \frac{d}{l} - 0.75 \right) \tag{2.30}$$

The resistance R (in ohms) obtained from (2.27) is

$$R = \frac{R_s l}{\pi d} \tag{2.31}$$

The effect of the ground plane on the inductance value of a wire has also been considered [4, 74]. If the wire is at a distance h above the ground plane,

$$L = 2\times 10^{-4} l \left[\ln\frac{4h}{d} + \ln\left(\frac{l+\sqrt{l^2+d^2/4}}{l+\sqrt{l^2+4h^2}}\right) + \sqrt{1+\frac{4h^2}{l^2}} - \sqrt{1+\frac{d^2}{4l^2}} - 2\frac{h}{l} + \frac{d}{2l}\right] \tag{2.32}$$

2.4.3 Design of Capacitors

A lumped capacitor may be visualized as a small length of an open-circuited line. For $\gamma l \ll 1$ and $\bar{G} \ll \omega\bar{C}$,

$$Z_{\text{in}} = \frac{Z_0}{\gamma l} = \frac{\bar{G}}{(\omega\bar{C})^2 l} + \frac{1}{j\omega\bar{C}l} + \frac{\bar{R}l}{3} \tag{2.33}$$

Its equivalent circuit will thus be a resistance $[\bar{R}l/3 + \bar{G}/(\omega\bar{C})^2 l]$ in series with the capacitor $\bar{C}l$.

There are three types of passive capacitors generally used in microwave circuits: microstrip patch, interdigital, and metal-insulator-metal (MIM). Microstrip patch capacitors are comprised of a conductor patch on a dielectric substrate having the ground plane on the other side. These capacitors can only be used for low capacitance values (<0.2 pF) due to practical considerations such as capacitance per unit area. In order to have a large value of capacitance per unit area, it becomes necessary to use the interdigital structure or to decrease the distance between the two conductors (top and bottom) of the transmission-line section. The choice between the interdigital and MIM capacitors depends on the capacitance value to be realized and the processing technology available. Usually for values less than 1 pF, interdigital capacitors can be used, while for values greater than 1 pF, MIM techniques are generally used to minimize the overall size. Four types of capacitors used in MMICs are compared in Table 2.15; Schottky capacitors are described in Chapter 8.

TABLE 2.15 Comparison of Various MMIC Capacitors

Capacitor Type	Capacitance Range	Tolerance
Microstrip	0–0.1 pF (shunt only)	2%
Interdigital	0.05–0.5 pF	±10%
MIM	0.1–25 pF	±10%
Schottky	0.5–100 pF	Voltage dependent

TABLE 2.16 Expressions for Lumped Capacitors

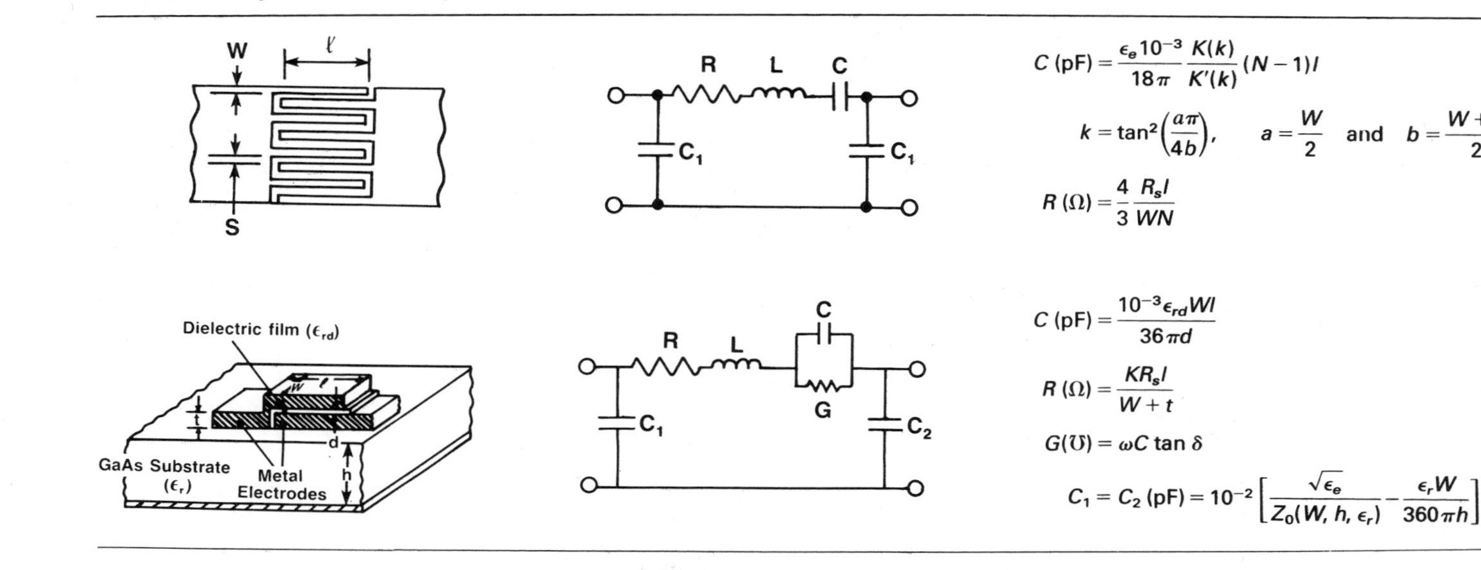

$$C\,(\mathrm{pF}) = \frac{\epsilon_e 10^{-3}}{18\pi}\frac{K(k)}{K'(k)}(N-1)l$$

$$k = \tan^2\left(\frac{a\pi}{4b}\right), \qquad a = \frac{W}{2} \quad \text{and} \quad b = \frac{W+S}{2}$$

$$R\,(\Omega) = \frac{4}{3}\frac{R_s l}{WN}$$

$$C\,(\mathrm{pF}) = \frac{10^{-3}\epsilon_{rd} W l}{36\pi d}$$

$$R\,(\Omega) = \frac{K R_s l}{W+t}$$

$$G(\mho) = \omega C \tan\delta$$

$$C_1 = C_2\,(\mathrm{pF}) = 10^{-2}\left[\frac{\sqrt{\epsilon_e}}{Z_0(W, h, \epsilon_r)} - \frac{\epsilon_r W}{360\pi h}\right] l$$

Capacitor configurations, equivalent circuits, and closed-form expressions for circuit elements are summarized in Table 2.16. Analysis of interdigital capacitors has been reported by Alley [76]. These capacitors can be fabricated employing an interdigital microstrip conductor by the technique used in the fabrication of MICs and do not require any additional processing steps.

On the other hand, MIM capacitors can be used in chip form or can be fabricated by sandwiching a thin layer of dielectric between two parallel-plate conductors during the MIC process. In later stages the dielectric layer completely covers the bottom conductor pad with some overlap to prevent shorts. In MMICs, the dielectric layer normally used is silicon dioxide or

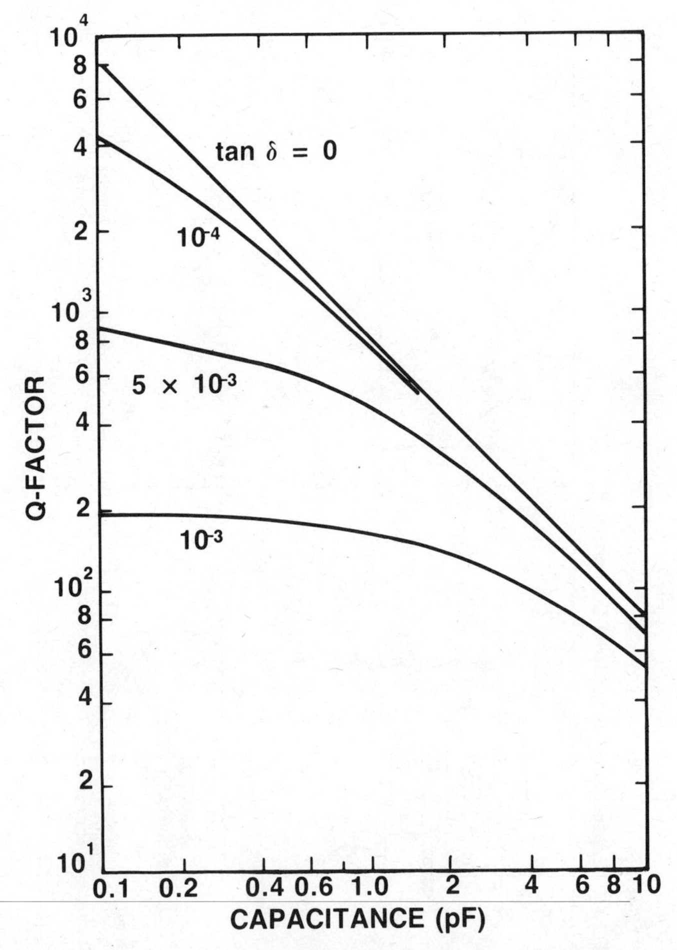

Figure 2.22 Quality factor of a square thin-film capacitor as a function of capacitance and dielectric loss tangent for $f = 10$ GHz.

silicon nitride and is 0.05 to 0.5 μm thick. The largest dimension is less than 0.1λ in dielectric film. An accurate calculation for MIM capacitors should take into account the effect of fringing field [86].

For expressions in Table 2.16, N is the number of fingers, ϵ_e is the effective dielectric constant of microstrip line of width W, Z_0 is the characteristic impedance, h is the substrate thickness, ϵ_r is the dielectric constant of the substrate, ϵ_{rd} is the dielectric constant of the dielectric film, and $\tan\delta$ is its loss tangent. All dimensions are in microns. The elliptic functions $K(k)$ and $K'(k)$ are defined in (2.4).

When the capacitor is used in a circuit, the parasitic inductance L caused by the connection to the capacitor plates must be accounted for. The effective capacitance C_e is then given by

$$C_e = C\left(1 + \frac{\omega^2}{\omega_0^2}\right) \tag{2.34}$$

where $\omega_0^2 = 1/LC$, and ω is the operating angular frequency.

The Q of MIM capacitors is given by

$$Q = \frac{Q_d Q_c}{Q_d + Q_c} \tag{2.35}$$

where $Q_c = 1/(\omega CR)$ and $Q_d = 1/\tan\delta$. Quality factors of a square MIM capacitor as a function of capacitance for various values of the loss tangent are shown in Fig. 2.22.

2.4.4 Design of Resistors

Planar resistors can be realized either by depositing thin films of lossy metal on a dielectric base or by employing semiconductor films on a semi-insulating substrate. Nichrome and tantalum nitride are the most popular and useful film materials for thin-film resistors (thickness 0.05–0.2 μm). Resistors based on semiconductor (e.g., GaAs or Si) films can be fabricated by forming an isolated land of semiconductor conducting layer (thickness 0.05–0.5 μm).

These techniques are illustrated in Fig. 2.23 from the transmission-line analog. It can easily be shown that for a shorted line (neglecting L and G),

$$Z_{\text{in}} = \frac{R}{1 + j\omega CR/3} \tag{2.36}$$

where

$$R = R_s \frac{l}{W} \tag{2.37a}$$

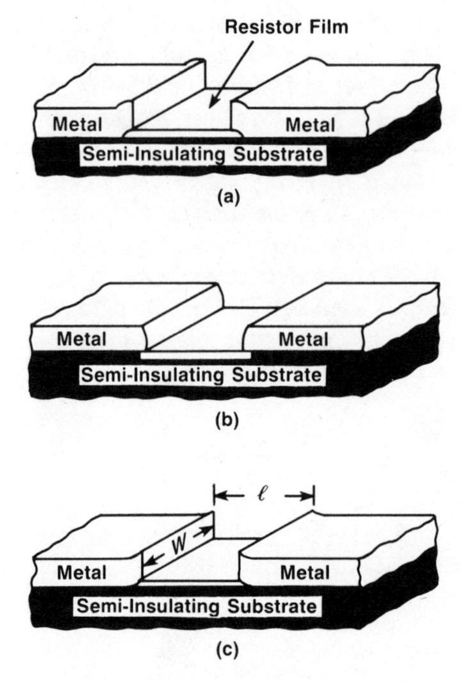

Figure 2.23 Planar resistors. (*a*) Thin-film. (*b*) Mesa. (*c*) Implanted.

or

$$R = \rho_s \frac{l}{dW} \tag{2.37b}$$

Here R_s is the surface resistance (Ω/square) and ρ_s is the specific resistivity (Ω-m) of the resistor film. The thickness d, width W, and length l of the film are measured in meters. The capacitance can be determined from the microstrip-line considerations. When film thickness $d \geq 1\ \mu$m, the formula containing R_s should be used. However, for very thin films, $d \leq 1\ \mu$m, the formula with ρ_s should be used. Desirable characteristics of film resistors are

- good stable-resistance value, which should not change with time,
- low temperature coefficient of resistance (TCR),
- adequate dissipation capability,
- sheet resistivities in the range of 10 to 1000 Ω/square, so that parasitics can be minimized,
- maximum resistor length less than 0.1λ if transmission line effects are to be ignored.

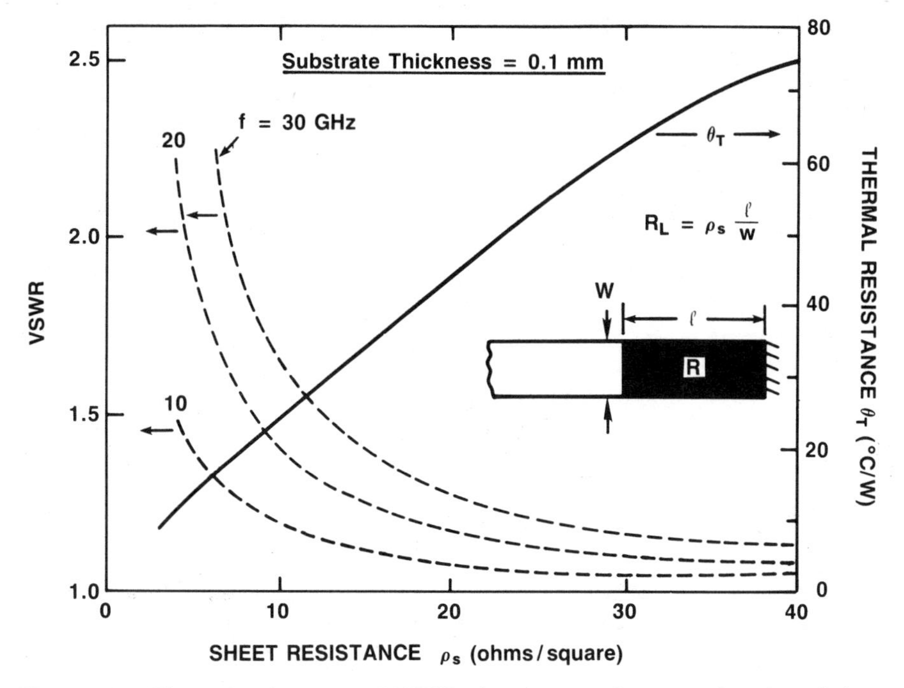

Figure 2.24 Thermal resistance and VSWR of a planar resistor as a function of sheet resistance and frequency.

A problem common to all planar resistors is the parasitic capacitance attributable to the underlying dielectric region and the distributed inductance, which makes such resistors exhibit a frequency dependence at high frequencies. If the substrate has a ground plane, one may determine the frequency dependence by treating the resistor as a very lossy microstrip line. Figure 2.24 shows how the VSWR increases dramatically at low values of ρ_s because the length of the resistor becomes too large. Also shown is the thermal resistance. Clearly, a trade-off is necessary between VSWR and thermal resistance.

REFERENCES

1. Coaxial lines and waveguides have been discussed in several texts. For example, see: Ramo, S., J. R. Whinnery, and T. Van Duzer, *Fields and Waves in Communication Electronics*, 2nd ed., Wiley, New York, 1984.
2. Montgomery, C. G., R. H. Dicke, and E. M. Purcell (Eds.), *Principles of Microwave Circuits*, McGraw-Hill, New York, 1948.

3. Bhartia, P., and I. J. Bahl, *Millimeter Wave Engineering and Applications*, Wiley, New York, 1984.
4. Gupta, K. C., R. Garg, and R. Chadha, *Computer-Aided Design of Microwave Circuits*, Artech House, Dedham, Mass., 1981.
5. Cohn, S. B., "Characteristic Impedance of Shielded Strip Transmission Line," *IRE Trans. Microwave Theory Tech.*, Vol. MTT-2, July 1954, pp. 52–55.
6. Cohn, S. B., "Problems in Strip Transmission Lines," *IRE Trans. Microwave Theory Tech.*, Vol. MTT-3, Mar. 1955, pp. 119–126.
7. Vendelin, G. D., "Limitations on Stripline Q," *Microwave J.*, Vol. 13, May 1970, pp. 63–69.
8. Gunston, M. A. R., *Microwave Transmission-Line Impedance Data*, Van Nostrand–Reinhold, London, 1972, pp. 38–39.
9. Howe, H., Jr., *Stripline Circuit Design*, Artech House, Dedham, Mass., 1974.
10. Bahl, I. J., and Ramesh Garg, "Designer's Guide to Stripline Circuits," *Microwaves*, Vol. 17, Jan. 1978, pp. 90–96.
11. Wheeler, H. A., "Transmission Line Properties of a Stripline Between Parallel Planes", *IEEE Trans. Microwave Theory Tech.*, Vol. MTT-26, Nov. 1978, pp. 866–876.
12. Barrett, R. M., "Microwave Printed-Circuit—Historical Survey," *IRE Trans. Microwave Theory Tech.*, Vol. MTT-3, Mar. 1955, pp. 1–9.
13. Wheeler, H. A., "Transmission Line Properties of Parallel Strips Separated by a Dielectric Sheet," *IEEE Trans. Microwave Theory Tech.*, Vol. MTT-13, Mar. 1965, pp. 172–185.
14. Gupta, K. C., R. Garg, and I. J. Bahl, *Microstrip Lines and Slotlines*, Artech House, Dedham, Mass., 1979.
15. Schneider, M V., B. Glance, and W. F. Bodtmann, "Microwave and Millimeter Wave Integrated Circuits for Radio Systems," *Bell Syst. Tech. J.*, Vol. 48, July–Aug. 1969, pp. 1703–1726.
16. Cohn, S. B., "Slot Line on a Dielectric Substrate," *IEEE Trans. Microwave Theory Tech.*, Vol. MTT-17, Oct. 1969, pp. 768–778.
17. Wen, C. P., "Coplanar Waveguide: A Surface Strip Transmission Line Suitable for Non-Reciprocal Gyromagnetic Device," *IEEE Trans. Microwave Theory Tech.*, Vol. MTT-17, Dec. 1969, pp. 1087–1090.
18. Gupta, K. C., and A. Singh (Eds.), *Microwave Integrated Circuits*, Wiley Eastern, New Delhi, India, 1974.
19. Young, L., and H. Sobol (Eds.), *Advances in Microwaves*, Vol. 8, Academic Press, New York, 1974.
20. Wheeler, H. A., "Transmission Line Properties of a Strip on a Dielectric Sheet on a Plane," *IEEE Trans. Microwave Theory Tech.*, Vol. MTT-25, Aug. 1977, pp. 631–647.
21. Schneider, M. V., "Microstrip Lines for Microwave Integrated Circuits," *Bell Syst. Tech. J.*, Vol. 48, 1969, pp. 1422–1444.
22. Hammerstad, E. O., "Equations for Microstrip Circuit Design," *Proc. European Microwave Conf.* Microwave Exhibitors & Publishers Ltd., Kent, U.K., 1975, pp. 268–272.

23. Hammerstad, E. O., "Accurate Models for Microstrip Computer-aided Design," *Digest of the IEEE Int. Microwave Symp.*, 1980, pp. 407–409.
24. Bahl, I. J., and D. K. Trivedi, "A Designer's Guide to Microstrip Line," *Microwaves*, Vol. 16, May 1977, pp. 174–182.
25. Pramanick, P. and P. Bhartia, "CAD Models for Millimeter-wave Finlines and Suspended-Substrate Microstrip Lines", *IEEE Trans. Microwave Theory Tech.*, Vol. MTT-33, Dec. 1985, pp. 1429–1435.
26. Garg, R., and K. C. Gupta, "Expression for Wavelength and Impedance of Slotline," *IEEE Trans. Microwave Theory Tech.*, Vol. MTT-24, Aug. 1976, p. 532.
27. Cohn, S. B., "Slotline on a Dielectric Substrate," *IEEE Trans. Microwave Theory Tech.*, Vol. MTT-17, Oct. 1969, pp. 768–778.
28. Cohn, S. B., "Shielded Coupled-Strip Transmission Line," *IRE Trans. Microwave Theory Tech*, Vol. MTT-3, Oct. 1955, pp. 29–38.
29. Jansen, R. H., "High-speed Computation of Single and Coupled Microstrip Parameters Including Dispersion, High Order Modes, Loss and Finite Strip Thickness," *IEEE Trans. Microwave Theory Tech.*, Vol. MTT-26, Feb. 1978, pp. 75–82.
30. Garg, R., and I. J. Bahl, "Characteristics of Coupled Microstriplines," *IEEE Trans. Microwave Theory Tech.*, Vol. MTT-27, July 1979, pp. 700–705; also see correction in *IEEE Trans., Microwave Theory, Tech.*, Vol. MTT-28, Mar. 1980, p. 272.
31. Cohn, S. B., "Characteristic Impedances of Broadside-coupled Strip Transmission Lines," *IRE Trans. Microwave Theory Tech.*, Vol. MTT-8, Nov. 1960, pp. 633–637.
32. Howe, H., Jr., *Stripline Circuit Design*, Artech House, Dedham, Mass., 1974, p. 125.
33. Shelton, J. P., Jr., "Impedances of Offset Parallel-coupled Strip Transmission Lines," *IEEE Trans. Microwave Theory Tech.*, Vol. MTT-14, Jan. 1966, pp. 7–15.
34. Bahl, I. J., and P. Bhartia, "The Design of Broadside-Coupled Stripline Circuits," *IEEE Trans. Microwave Theory Tech.*, Vol. MTT 29, Feb. 1981, pp. 165–168.
35. Bahl, I. J., and P. Bhartia, "Characteristics of Inhomogeneous Broadside-coupled Striplines," *IEEE Trans. Microwave Theory Tech.*, Vol. MTT-28, June 1980, pp. 529–535.
36. Horno, M., and F. Medina, "Multilayer Planar Structures for High-directivity Directional Coupler Design," *IEEE Trans. Microwave Theory Tech.*, Vol. MTT-34, Dec. 1986, pp. 1442–1449.
37. Marcuvitz, N. (Ed.), *Waveguide Handbook*, McGraw-Hill, New York, 1951.
38. Green, H. E., "The Numerical Solution of Transmission Line Problems," in *Advances in Microwaves*, Vol. 2, Academic Press, New York, 1967, pp. 327–393.
39. Bianco, B., et al., "Open-circuited Coaxial Lines as Standards for Microwave Measurements," *Electron. Lett.*, Vol. 16, 1980, pp. 373–374.

40. Montgomery, C. G., R. H. Dicke, and E. M. Purcell (Eds.), *Principles of Microwave Circuits*, McGraw-Hill, New York, 1948, p. 295.
41. Oliner, A. A., "Equivalent Circuits for Discontinuities in Balanced Strip Transmission Line," *IRE Trans. Microwave Theory Tech.*, Vol. MTT-3, Mar. 1955, pp. 134–143.
42. Altschuler, H. M., and A. A. Oliner, "Discontinuities in the Center Conductor of Symmetric Strip Transmission Line," *IRE Trans. Microwave Theory Tech.*, Vol. MTT-8, May 1960, pp. 328–339.
43. Bahl, I. J., and R. Garg, "A Designer's Guide to Stripline Circuits," *Microwaves*, Vol. 17, Jan. 1978, pp. 90–96.
44. Garg, R., and I. J. Bahl, "Microstrip Discontinuities," *Int. J. Electron.*, Vol. 45, 1978, pp. 81–87.
45. Benedek, P., and P. Silvester, "Equivalent Capacitance for Microstrip Gaps and Steps," *IEEE Trans. Microwave Theory Tech.*, Vol. MTT-20, Nov. 1972, pp. 729–733.
46. Silvester, P., and P. Benedek, "Equivalent Capacitances of Microstrip Open Circuits," *IEEE Trans. Microwave Theory Tech.*, Vol. MTT-20, Aug. 1972, pp. 511–516.
47. Silvester, P., and P. Benedek, "Microstrip Discontinuity Capacitances for Right-Angled Bends, T-junctions and Crossings," *IEEE Trans. Microwave Theory Tech.*, Vol. MTT-21, May 1973, pp. 341–346. (Also, Correction, Vol. MTT-23, May 1975, p 456).
48. Farrar, A., and A. T. Adams, "Matrix Methods for Microstrip Three-Dimensional Problems," *IEEE Trans. Microwave Theory Tech.*, Vol. MTT-20, Aug. 1972, pp. 497–504.
49. Farrar, A., and A. T. Adams, "Computation of Lumped Microstrip Capacities by Matrix Methods—Rectangular Sections and End Effect," *IEEE Trans. Microwave Theory Tech.*, Vol. MTT-19, May 1971, pp. 495–497.
50. Maeda, M., "Analysis of Gap in Microstrip Transmission Lines," *IEEE Trans. Microwave Theory Tech.*, Vol. MTT-20, June 1972, pp. 390–396.
51. Horton, R., "The Electrical Characterization of a Right-angled Bend in Microstrip Line," *IEEE Trans. Microwave Theory Tech.*, Vol. MTT-21, June 1973, pp. 427–429.
52. Itoh, T., et al., "A Method for Computing Edge Capacitance of Finite and Semi-infinite Microstrip Lines," *IEEE Trans. Microwave Theory Tech.*, Vol. MTT-20, Dec. 1972, pp. 847–849.
53. Thomson, A. F., and A. Gopinath, "Calculation of Microstrip Discontinuity Inductances," *IEEE Trans. Microwave Theory Tech.*, Vol. MTT-23, Aug. 1975, pp. 648–655.
54. Gopinath, A., et al., "Equivalent Circuit Parameters of Microstrip Change in Width and Cross-Junctions," *IEEE Trans. Microwave Theory Tech.*, Vol. MTT-24, Mar. 1976, pp. 142–144.
55. Napoli, L. S., and J J. Hughes, "Foreshortening of Microstrip Open Circuits on Alumina Substrates," *IEEE Trans. Microwave Theory Tech.*, Vol. MTT-19, June 1971, pp. 559–561.
56. Easter, B., "The Equivalent Circuit of Some Microstrip Discontinuities," *IEEE Trans. Microwave Theory Tech.*, Vol. MTT-23, Aug. 1975, pp. 655–660.

57. Stephens, I. M., and B. Easter, "Resonant Techniques for Establishing the Equivalent Circuits for Small Discontinuities in Microstrip," *Electron Lett.*, Vol. 7, Sept. 23, 1971, pp. 582–584.
58. Groll, H., and W. Weidmann, "Measurement of Equivalent Circuit Elements of Microstrip Discontinuities by a Resonant Method," *NTZ*, Vol. 28, 1975, p. 74.
59. Wolff, I., and W. Menzel, "A Universal Method to Calculate the Dynamical Properties of Microstrip Discontinuities," *Proc. 5th European Microwave Conf.*, Hamburg, 1975, pp. 263–267.
60. Mehran, R., "The Frequency-dependent Scattering Matrix of Microstrip Right-angle Bends, T-Junctions and Crossings," *AEU*, Vol. 29, 1975, pp. 454–460.
61. Mehran, R., "Frequency-dependent Equivalent Circuits for Microstrip Right-angle Bends, T-junctions and Crossings," *AEU*, Vol. 30, 1975, pp. 80–82.
62. Jansen, R. H., "The Spectral Domain Approach for Microwave Integrated Circuits," *IEEE Trans. Microwave Theory Tech.*, Vol. MTT-33, Oct. 1985, pp. 1043–1056.
63. Katehi, P. B., and N. G. Alexopolous, "Frequency-dependent Characteristics of Microstrip Discontinuities in Millimeter Wave Integrated Circuits," *IEEE Trans. Microwave Theory Tech.*, MTT-33, Oct. 1985, pp. 1029–1035.
64. Jackson, R. W., and D. M. Pozar, "Full-wave Analysis of Microstrip Open-end and Gap Discontinuities," *IEEE Trans. Microwave Theory Tech.*, Vol. MTT-33, Oct. 1985, pp. 1036–1042.
65. Hoefer, W. J. R., "Equivalent Series Inductivity of a Narrow Transverse Slit in Microstrip," *IEEE Trans. Microwave Theory Tech.*, Vol. MTT-25, Oct. 1977, pp. 822–824.
66. Jansen, R. H., "Problems of Design and Measuring Techniques of Planar Circuits," *NTZ*, Vol. 34, July 1981, pp. 412–417.
67. Besser, L., "Microwave Circuit Design Lecture Notes " UCLA Extension Course, University of California, Los Angeles.
68. Edwards, T. C., *Foundations for Microstrip Circuit Design*. Wiley, New York, 1981, Chap. 5.
69. Malherbe, J. A. G., and A. F. Steyn, "The Compensation of Step Discontinuities in TEM-mode Transmission Lines," *IEEE Trans. Microwave Theory Tech.*, Vol. MTT-26, Nov. 1978, pp. 883–885.
70. Larson, M. A., "Compensation of Step Discontinuities in Stripline," *Proc. 10th European Microwave Conf.*, Warszawa, Sept. 8–11, 1980, pp. 367–371.
71. Chadha, R., and K. C. Gupta, "Compensation of Discontinuities in Planar Transmission Lines," *IEEE Trans. Microwave Theory Tech.*, Vol. MTT-30, Dec. 1982, pp. 2151–2155.
72. Hoefer, W. J. R., "A Contour Formula for Compensated Microstrip Steps and Open Ends," IEEE Int. Microwave Symp. Digest, 1983, pp. 524–526.
73. Wheeler, H. A., "Simple Inductance Formulas for Radio Coils," *Proc. IRE*, Vol. 16, Oct. 1928, pp. 1398–1400.
74. Terman, F. E., *Radio Engineer Handbook*, McGraw-Hill, New York, 1943, p. 51.
75. Grover, F. W., *Inductance Calculations*, Van Nostrand, Princeton, N.J., 1946.
76. Alley, G. D., "Interdigital Capacitors and Their Applications to Lumped-

element Microwave Integrated Circuits," *IEEE Trans. Microwave Theory Tech.*, Vol. MTT-18, Dec. 1970, pp. 1028–1033.
77. Daly, D. A., et al., "Lumped Elements in Microwave Integrated Circuits," *IEEE Trans. Microwave Theory Tech.*, Vol. MTT-15, Dec. 1967, pp. 713–721.
78. Aitchison, C. S , "Lumped Components for Microwave Frequencies," *Philips Tech. Rev.*, Vol. 32, No. 9/10/11/12, 1971, pp. 305–314.
79. Aitchison, C. S., et al , "Lumped Microwave Circuits: Part III: Filters and Tunnel-Diode Amplifiers," *Design Electron.*, Vol. 9, Nov. 1971, pp. 42–51.
80. Aitchison, C. S., et al., "Lumped-Circuit Elements at Microwave Frequencies," *IEEE Trans. Microwave Theory Tech.*, Vol. MTT-19, Dec 1971, pp. 928–937.
81. Caulton, M., "Lumped Elements in Microwave Integrated Circuits," in *Adances in Microwaves*, Vol. 8, L. Young and H. Sobol (Eds.), Academic Press, New York, 1974.
82. Pengelly, R. S., and D. C. Rickard, "Design, Measurement and Application of Lumped Elements up to *J*-Band," *Proc. 7th European Microwave Conf.*, Copenhagen, 1977, pp. 460–464.
83. Gopinath, A., and P. Silvester, "Calculation of Inductance of Finite-length Strips and Its Variations with Frequency," *IEEE Trans. Microwave Theory Tech.*, Vol. MTT-21, June 1973, pp. 380–386.
84. Chaddock, R. E., "The Application of Lumped Element Techniques to High Frequency Hybrid Integrated Circuits," *Radio Electron. Eng.*, Vol. 44, 1974, pp. 414–420.
85. Ferry, D. K. (Ed.) *Gallium Arsenide Technology*, Howard Sams, Indianapolis, Ind., 1985, Chap. 6.
86. Wolff, I. and N. Knoppik, "Rectangular and Circular Microstrip Disk Capacitors and Resonators," *IEEE Trans. Microwave Theory Tech.*, Vol. MTT-22, Oct. 1974, pp. 857–864.

PROBLEMS

2.1 Calculate the magnitude of the characteristic impedance and the propagation constant for a lossy coaxial line at 2 GHz. Assume $b = 3a = 0.5$ cm and $\epsilon = (2.56 - j0.005)\epsilon_0$.

2.2 For a lossless transmission line terminated with a load, we have the characteristic impedance Z_0 equal to 100 Ω. The value of VSWR measured at the input is equal to 4.0 and first voltage maximum is $\lambda/8$ away from load. What is the load impedance? (Use the Smith chart.)

2.3 A rectangular waveguide is to be designed for operation in the frequency range 7.5–10 GHz. Calculate the inside dimensions so that the following design criteria are satisfied: (i) There is only one mode of propagation, (ii) the lowest usable frequency is 10% above the cut-off, and (iii) highest usable frequency is 5% below the frequency where next higher mode can propagate.

2.4 A rectangular waveguide with inside dimensions 2.3 cm by 1.0 cm is used for operation at 10 GHz. Waveguide walls are made of copper (surface resistivity $R_s = 2.6 \times 10^{-7}\sqrt{f}\,\Omega$). Calculate the attenuation constant in nepers/meter and decibels/meter.

2.5 A rectangular waveguide with inside dimensions 3.1 mm by 1.55 mm is used for operation at 75 GHz. The waveguide walls are made of copper with a surface roughness of 1 μm rms. Calculate the attenuation constant in decibels/centimeter for the waveguide.

2.6 A strip line of characteristic impedance 20 is used at a frequency of 10 GHz with a load that is a microwave diode with conductance 0.05 S in parallel with a 1-pF capacitor. The line is one-eighth wavelengths long at the design frequency. Find reflection coefficient at the load and the input admittance. If the dielectric constant of the substrate is 2.5, what is the physical length of the line?

2.7 Show that a pure TEM mode cannot be supported by a microstrip line. Calculate various parameters such as Z_0, β, and α for a microstrip with $\epsilon_r = 10$, $W = 1$ mm and $h = 0.6$ mm at 10 GHz. Here gold conductors are 6 μm thick. Also calculate Q for a $\lambda/2$ resonator.

2.8 A microstrip line of characteristic impedance Z_{01} is connected to another microstrip of higher characteristic impedance Z_{02}. If the step discontinuity reactance is represented by a series inductance (jX), determine the S-matrix for the junction.

2.9 A microstrip line having a characteristic impedance of 50 Ω has a width $W = 1$ mm. The dielectric constant of the substrate is $\epsilon_r = 10$. What is the limiting frequency at which radiation starts? Is it necessary to correct for dispersion when operating at that frequency?

2.10 Since modes on microstrip lines are only quasi-TEM, derive from the basic theory of a lossless line that the inductance L and capacitance C per unit length of a microstrip line are given by

$$L = \frac{Z_0}{v} = \frac{Z_0\sqrt{\epsilon_e}}{c}$$

$$C = \frac{1}{Z_0 v} = \frac{\sqrt{\epsilon_e}}{Z_0 c}$$

where

Z_0 = characteristic impedance of the microstrip line,
v = wave velocity in the microstrip line,
$c = 3 \times 10^8$ m/s is the velocity of light in vacuum,
ϵ_e = effective dielectric constant.

Also determine L, C, and time delay per unit meter for the microstrip described in Problem 2.9.

2.11 A $\lambda/10$ section of an open-circuited transmission line, with parameters given by $\bar{R} = 0.1\ \Omega$/cm, $\bar{L} = 0.1$ nH/cm, $\bar{G} = 0.2$ mΩ/cm, and $\bar{C} = 1$ pF/cm, is to be used as a lumped capacitor at 10 GHz. Find the value of capacitance, its equivalent circuit and Q factor.

2.12 Show that for a lumped-element series-mounted in the middle of a resonator, the value of reactance is given in terms of resonance frequency f as

$$X = 2Z_0 \cot \frac{\pi f}{2f_n}$$

where f_n is the resonant frequency for the line length l without reactance X being present and Z_0 is the characteristic impedance of the line. Stray reactances at open ends may be ignored.

3 RESONATORS

3.1 INTRODUCTION

Resonant structures are extensively used network elements in the realization of various microwave components [1]. At low frequencies, resonant structures are invariably composed of the lumped elements. As the frequency of operation increases, lumped elements in general cannot be used. Microwave resonant circuits can be realized by various forms of transmission lines. However, microwave resonant structures are almost invariably understood as the cavity resonator. Conventional resonators consist of a bounded electromagnetic field in a volume enclosed by metallic walls. The electric and magnetic energies are stored in the electric and magnetic fields, respectively, of the electromagnetic field inside the cavity and the equivalent lumped inductance and capacitance of the structure can be determined from the respective stored energy. Some energy is dissipated due to finite conductivity of the metallic walls and the equivalent resistance can therefore be determined from the currents flowing on the walls of the cavity resonator [2, 3].

In this chapter, a brief description of the resonant structures most commonly employed in various microwave components is presented. As far as possible, simple expressions have been provided for design applications. Basic parameters of microwave resonators are discussed in Section 3.2. TE and TM cavity resonators are described in Section 3.3, and microstrip resonant structures are described in Section 3.4. Fin-line resonators are included in Section 3.5. Finally dielectric resonators and YIG resonators are treated in Sections 3.6 and 3.7, respectively. Resonator measurements are discussed in Section 3.8.

3.2. RESONATOR PARAMETERS

3.2.1 Resonant Frequency

The parameters of a resonator at microwave frequencies are essentially similar to those of a lumped-element resonator circuit at low frequencies. They can easily be described utilizing an *RLC* series or parallel network. Consider, for instance, an *RLC* parallel network as shown in Fig. 3.1. The input impedance of such a network as a function of frequency has both real and imaginary parts. At resonance, the input impedance is real and is equal

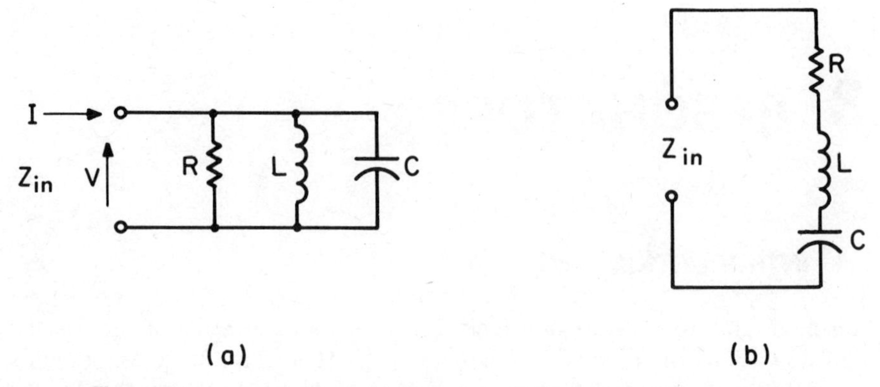

Figure 3.1 Lumped element (*a*) parallel and (*b*) series resonant circuits.

to the resistance of the circuit. The electric and magnetic stored energies are also equal, leading to the expression for the resonant frequency as

$$\omega_0 = \frac{1}{\sqrt{LC}} \tag{3.1}$$

3.2.2 Quality Factor

The performance of a resonant circuit is described in terms of the quality factor Q, and various features such as frequency selectivity, bandwidth, and damping factors can be deduced from this. The quality factor is defined as

$$Q = \omega \cdot \frac{\text{time averaged stored energy}}{\text{energy lost per second}} \tag{3.2}$$

For the lumped resonant circuits

$$Q = \omega RC = \frac{R}{\omega L} \tag{3.3a}$$

for the parallel network [Fig. 3.1(*a*)], and

$$Q = \frac{1}{\omega CR} = \frac{\omega L}{R} \tag{3.3b}$$

for the series network [Fig. 3.1(*b*)].

3.2.3 Fractional Bandwidth

The input impedance of the parallel resonant circuit of Fig. 3.1 is given by

$$Z_{\text{in}} = \left[\frac{1}{R} + \frac{1}{j\omega L} + j\omega C\right]^{-1} \tag{3.4}$$

At a frequency $\omega_0 \pm \Delta\omega$ in the vicinity of the resonant frequency, (3.4) reduces to

$$Z_{\text{in}} = \frac{R}{1 \pm j2Q\left(\dfrac{\Delta\omega}{\omega_0}\right)} \tag{3.5}$$

From (3.5) it is clear that at $\omega = \omega_0$ the input impedance is only resistive. However, when

$$\Delta\omega = \frac{\omega_0}{2Q} \tag{3.6}$$

the magnitude of the input impedance decreases to $R\sqrt{2}$ of its maximum value R, and the phase angle is $\pi/4$ for $\omega < \omega_0$ and $-\pi/4$ for $\omega > \omega_0$. From (3.6) the fractional bandwidth BW is defined as,

$$\text{BW} = \frac{2\Delta\omega}{\omega_0} = \frac{1}{Q} \tag{3.7}$$

3.2.4 Loaded Quality Factor

In practical situations, the resonant circuit is coupled to an external load R_L that also dissipates power, and the loaded quality factor Q_L is given by

$$\frac{1}{Q_L} = \frac{1}{Q} + \frac{1}{Q_e} \tag{3.8}$$

where Q_e is the external quality factor for a lossless resonator in the presence of the load.

3.2.5 Damping Factor

Another important parameter associated with a resonant circuit is the damping factor δ_d. It is a measure of the rate of decay of the oscillations in the absence of an exciting source. For high Q resonant circuits, the rate at which the stored energy decays is proportional to the average energy stored. Consequently, the stored energy as a function of time is given by

$$W = W_0 e^{-2\delta_d t} = W_0 e^{-\omega_0 t/Q} \tag{3.9}$$

which implies that

$$\delta_d = \frac{\omega_0}{2Q} \tag{3.10}$$

Thus, we see that the damping factor is inversely proportional to the Q of the resonant circuit. In the presence of an external load, the Q should be replaced by Q_L.

Alternately, the input impedance in the vicinity of resonance Z_{in} given by (3.5) can be rewritten to take into account the effect of losses in terms of the complex resonant frequency

$$\omega_c = \omega_0 + j\delta_d = \omega_0 \left(1 + j\frac{1}{2Q}\right) \tag{3.11}$$

so that

$$Z_{\text{in}} = \frac{\omega_0 R/2Q}{j(\omega - \omega_c)} \tag{3.12}$$

In (3.12) the parameter R/Q is called the figure of merit and describes the effect of the cavity on the gain bandwidth product. In terms of the lumped elements of the resonant circuit,

$$\frac{R}{Q} = \sqrt{\frac{L}{C}} \tag{3.13}$$

3.2.6 Coupling

Coupling structures provide a means of coupling energy into and/or out of the cavity. The excitation of the cavity can be accomplished by

1. *Current Loops*. The plane of the loop is perpendicular to the direction of the magnetic field in the cavity [Fig. 3.2(*a*)].
2. *Electric Probes*. The direction of probe is parallel to the direction of the electric field in the cavity [Fig. 3.2(*b*)].
3. *Apertures*. The aperture is located between the cavity and input waveguide so that a field component in the cavity mode has the same direction to the one in the input waveguide [Fig. 3.2(*c*)].

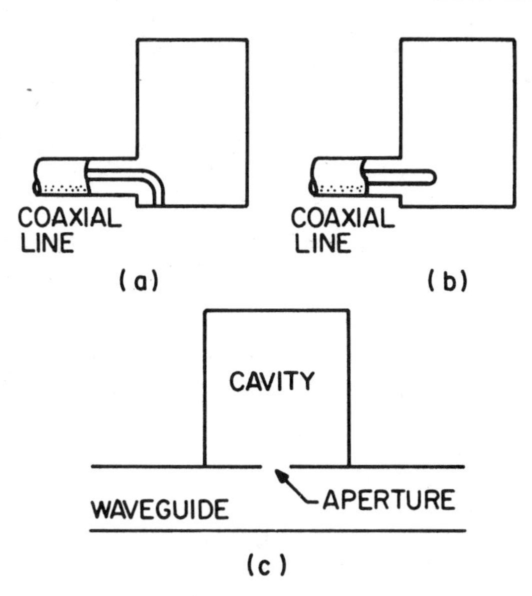

Figure 3.2 Cavity excitation using (*a*) loop coupling, (*b*) electric probe coupling, and (*c*) aperture coupling.

3.3 CAVITY RESONATORS

At microwave frequencies, the dimensions of lumped resonator circuits become comparable to the wavelength and this may cause energy loss by radiation. Therefore, resonant circuits at these frequencies are shielded to prevent radiation. Perfectly conducting enclosures, or cavities, provide a means of confining energy. Usually, cavities with the largest possible surface area for the current path are preferred for low-loss operation and the energy coupled to them by the methods previously described.

3.3.1 Coaxial Resonators

A coaxial cavity resonator (Fig. 3.3) supporting TEM waves can easily be formed by a shorted section of coaxial line. Resonances appear whenever the length $2d$ of the cavity is an integral number of wavelengths. The resonance modes occur at

$$f = \frac{nc}{2d}, \qquad n = 1, 2, \ldots \tag{3.14}$$

where c is the speed of light. The lowest resonant frequency corresponds to

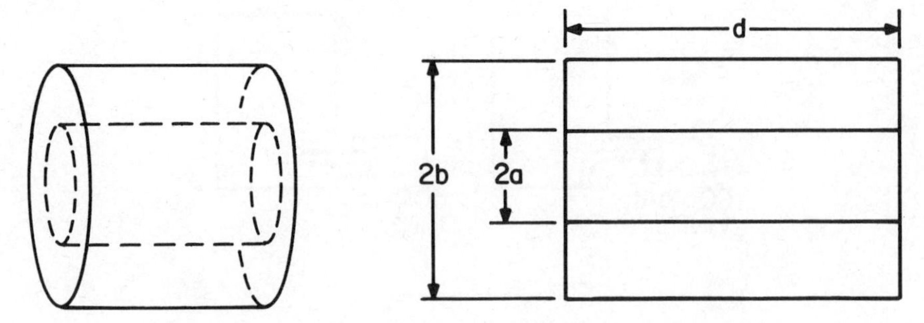

Figure 3.3 Coaxial cavity resonator and its cross section.

$n = 1$ and the Q of the cavity for this mode is given by [4]

$$Q\frac{\delta}{\lambda_0} = \frac{1}{4 + 2(d/b)(1 + b/a)/\ln(b/a)} \qquad (3.15)$$

where δ is the skin depth, and a and b are inner and outer radii, respectively.

It is also possible to have higher order resonance modes, depending on the structural parameters of the coaxial line. The first higher order mode appears when the average circumference is equal to the wavelength in the dielectric medium of the line. The cutoff frequency of this mode is

$$f_c = \frac{c}{\pi\sqrt{\epsilon_r}(a + b)} \qquad (3.16)$$

where ϵ_r is the dielectric constant of the medium. Other higher order modes correspond to TE and TM waves that exist in a circular waveguide with the radius of the center conductor approaching zero. The resonance condition is

$$k_{nml} = \left[p_{nm}^2 + \left(\frac{l\pi}{2d}\right)^2\right]^{1/2} \qquad (3.17)$$

where $k_{nml} = 2\pi f_{nml}/c$, and p_{nm} is the cutoff wavenumber that is obtained as the mth root of the transcendental equations,

$$J_n'(ka)N_n'(kb) - J_n'(kb)N_n'(ka) = 0 \qquad (3.18)$$

for TE modes, and

$$J_n(ka)N_n(kb) - J_n(kb)N_n(ka) = 0 \qquad (3.19)$$

for TM modes. Here J_n and N_n are the nth-order Bessel functions of the first and second kind, respectively, and the prime denotes their derivatives with respect to their arguments.

3.3.2 Reentrant Coaxial Resonators

Another coaxial cavity configuration consists of a short section of coaxial line with a gap in the center conductor. Figure 3.4(a) shows a capacitively loaded coaxial cavity. Radial cavity as shown in Fig. 3.4(b) is another possible variation. They are also referred to as reentrant coaxial cavities, since the metallic boundaries extend into the interior of the cavity. They are widely used in microwave tubes. The resonant frequency of such a structure can be evaluated from the solution of the transcendental equation [5],

$$\tan \beta l = \frac{dc}{\omega a^2 \ln(b/a)} \tag{3.20}$$

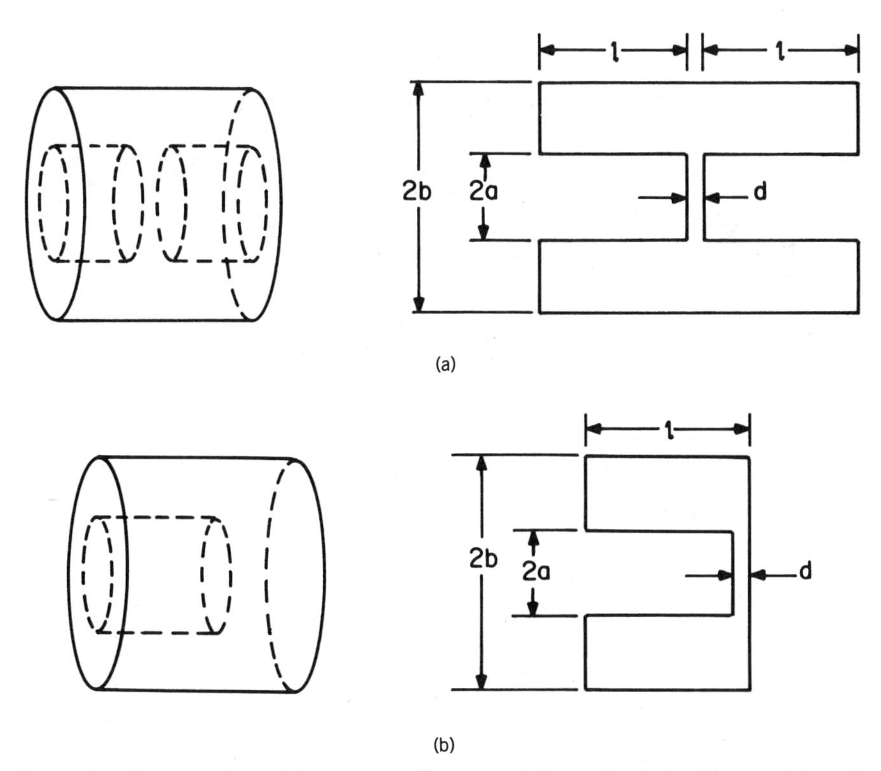

Figure 3.4 Reentrant coaxial cavity resonators. (a) Capacitively loaded coaxial cavity resonator. (b) Radial-cavity resonators.

where d is the gap in center conductor, and $2l+d$ is the length of the cavity. From (3.20) it is obvious that the capacitively loaded coaxial cavity can have an infinite number of modes.

For the radial reentrant cavity of Fig. 3.4(b), the resonant frequency can be evaluated by calculating the inductance and capacitance of the structure. The expression for the resonant frequency is

$$f = \frac{c}{2\pi\sqrt{\epsilon_r}}\left[al\left(\frac{a}{2d} - \frac{2}{l}\ln\frac{0.765}{\sqrt{l^2+(b-a)^2}}\right)\ln\frac{b}{a}\right]^{-1/2} \tag{3.21}$$

An approximate expression for the Q of the cavity is

$$Q\frac{\delta}{\lambda} = \frac{2l}{\lambda}\frac{\ln(b/a)}{2\ln(b/a) + l[(1/a)+(1/b)]} \tag{3.22}$$

For a tunable reentrant cavity, d is large, and $(l-d)$ is also large compared with b. The resonances occur whenever the length of the center conductor is approximately a quarter wavelength.

3.3.3 Rectangular Waveguide Resonators

Rectangular resonant cavities are formed by a section of rectangular waveguide of length d. This cavity can also support an infinite number of modes. The field configuration of the standing wave pattern for the incident and reflected wave is not unique, that is, it depends on the assumed direction of propagation of the wave. In order to be consistent, we shall assume that wave propagation is in the positive z direction. The standing wave pattern is then formed by the incident and reflected waves traveling in $+z$ and $-z$ directions, respectively.

The cutoff wavenumber k_{cmn} is given by

$$k_{cmn}^2 = \left(\frac{m\pi}{a}\right)^2 + \left(\frac{n\pi}{b}\right)^2, \qquad m = 0, 1, 2, \ldots, \quad n = 0, 1, 2, \ldots \tag{3.23}$$

where a and b are waveguide dimensions. The resonant wavenumber is then expressed as

$$k_{mnp} = \left[\left(\frac{m\pi}{a}\right)^2 + \left(\frac{n\pi}{b}\right)^2 + \left(\frac{p\pi}{d}\right)^2\right]^{1/2} \tag{3.24}$$

and the resonant frequency is defined as

$$f_{mnp} = \frac{k_{mnp}c}{2\pi} \tag{3.25}$$

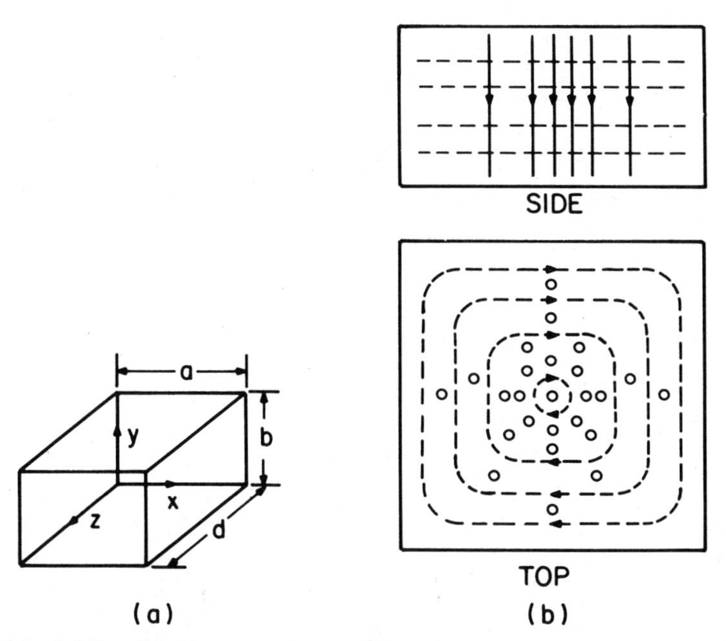

Figure 3.5 (*a*) Rectangular waveguide cavity resonator. (*b*) Field configuration of the dominant TE_{101} mode.

From the preceding discussion, we see that the resonant frequency is the same for TE and TM modes. Therefore, they are referred to as degenerate modes. The field configuration of the dominant TE_{101} mode is shown in Fig. 3.5(*b*). The quality factor, Q, of the dominant TE_{101} mode in the rectangular resonant cavity having surface resistance R_s can be evaluated using the expression,

$$Q = \frac{120\pi^2}{4R_s}\left[\frac{2b(a^2+d^2)^{3/2}}{ad(a^2+d^2)+2b(a^3+b^3)}\right] \tag{3.26}$$

In rectangular cavities, the resonant frequency increases for higher order modes, as does the Q at a given frequency. Higher order mode cavity or "echo boxes" are useful in applications where a slow rate of decay of the energy stored in the cavity after it has been excited is required.

3.3.4 Circular Waveguide Resonators

Circular waveguide cavities are most useful in various microwave applications. Most commonly, they are used in wavemeters to measure frequency and have a high Q factor and provide greater resolution. It consists of a section of a circular waveguide of radius a and length d [Fig. 3.6(*a*)], with end plates to provide short circuit.

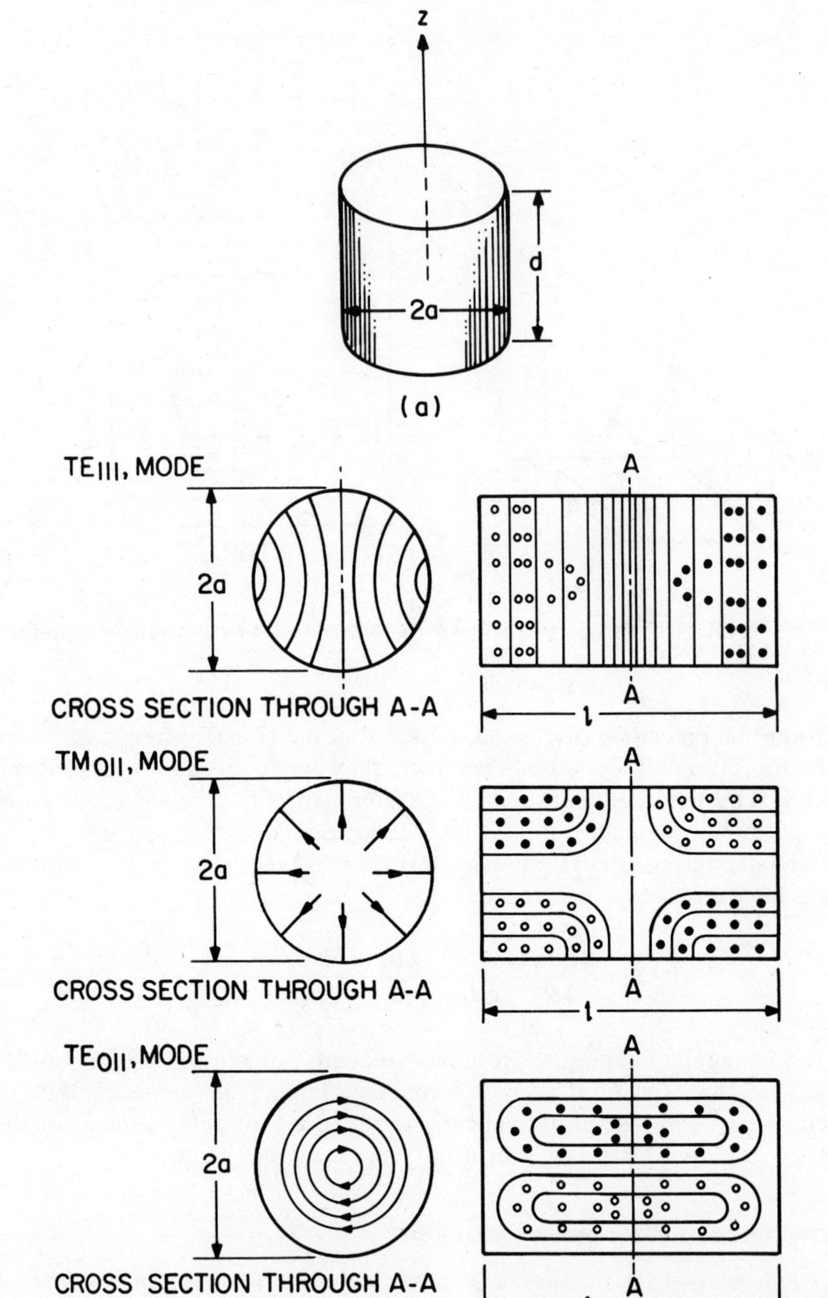

Figure 3.6 (*a*) Circular cylindrical waveguide cavity resonator. (*b*) Field configurations for TE_{111}, TM_{011}, and TE_{011} modes in cylindrical cavities.

TABLE 3.1 Roots of the Transcedental Equation $J'_n(ka) = 0$

Modes		
n	m	p'_{nm}
0	1	0
1	1	1.841
2	1	3.054
0	2	3.832
3	1	4.201
4	1	5.318

The resonance wavenumber of the circular waveguide cavity is given by

$$k_{nml} = \left[\left(\frac{x_{nm}}{a}\right)^2 + \left(\frac{l\pi}{d}\right)^2\right]^{1/2}, \qquad l = 0, 1, 2, \ldots \tag{3.27}$$

where

$$x_{nm} = \begin{cases} p'_{nm} & \text{for TE modes} \\ p_{nm} & \text{for TM modes} \end{cases}$$

Values for p'_{nm} for various modes are given in Table 3.1. Field lines for TE_{111}, TM_{011}, and TE_{011} modes are shown in Fig. 3.6(*b*). Simplifying (3.27) yields

$$(2af_{nml})^2 = \left(\frac{cx_{nm}}{\pi}\right)^2 + \left(\frac{cl}{2}\right)^2\left(\frac{2a}{d}\right)^2 \tag{3.28}$$

The Q of the circular cavity for TE_{nml} modes can be evaluated from

$$Q\frac{\delta}{\lambda_0} = \frac{[1-(n/p'_{nm})^2][(p'_{nm})^2 + (l\pi a/d)^2]^{3/2}}{2\pi[(p'^2_{nm} + 2a/d(l\pi a/d)^2 + (1-2a/d)(nl\pi a/p'_{nm}d)^2]} \tag{3.29}$$

and for the dominant TE_{111} mode, Q can be obtained by substituting $n = m = l = 1$ in the preceding equation.

Using (3.28), plots of $(2af)^2$ versus $(2a/d)^2$ can be used to construct mode charts, as shown in Fig. 3.7. From this it can be seen that for the TE_{011} mode operation, the safe value of $(2a/d)^2$ is between 2 and 3.

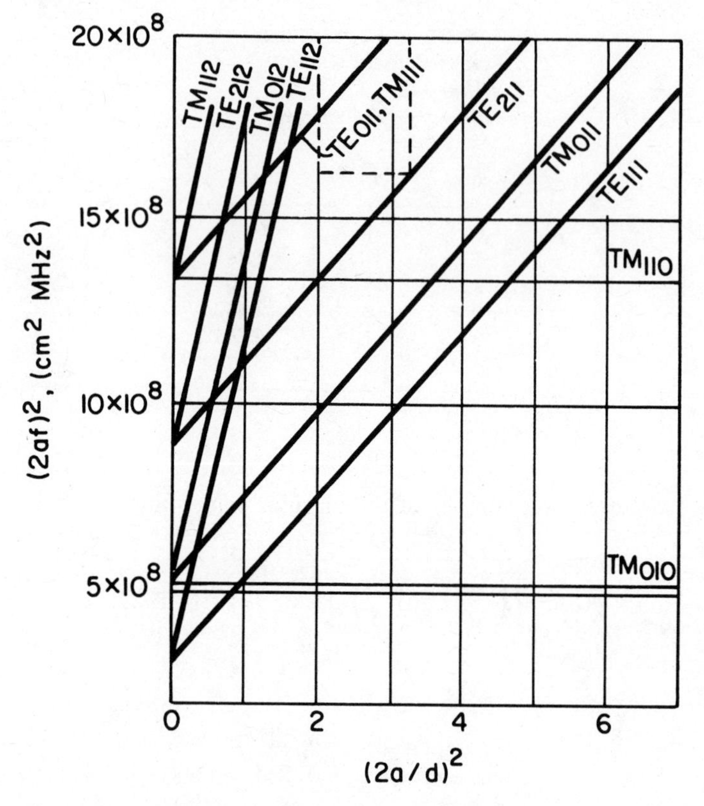

Figure 3.7 Mode chart of a circular cylindrical cavity resonator.

For TM modes, the Q is given by

$$Q\frac{\delta}{\lambda_0}=\frac{[p_{nm}^2+(l\pi a/d)^2]^{1/2}}{2\pi(1+2a/d)}, \qquad \text{for } l>0$$
$$\simeq\frac{p_{nm}}{2\pi(1+a/d)}, \qquad \text{for } l=0 \tag{3.30}$$

As with rectangular cavity resonators, the Q is higher for higher order modes.

3.3.5 Elliptic Waveguide Resonators

Elliptic resonant cavities formed using a section of an elliptic waveguide offer several advantages. There is no mode splitting due to slight deformations in the cavity surface, and the electric field configuration in the transverse plane is fixed with respect to its axes. Also, the longitudinal electric field of the TM_{111} mode in an elliptic cavity with semimajor axis a is always greater than the circular cavity with radius a. This feature may be useful in the dielectric material characterization utilizing perturbation techniques [6]. The resonance wavenumbers of an elliptic resonant cavity are given by Kretzschmar [7]. However, since elliptical waveguide cavities are not widely used, further details are not included and the reader is referred to References [6–8].

3.4 PLANAR MICROSTRIP RESONANT STRUCTURES

Particularly in the last decade, there has been a spate of technical publications on various planar transmission lines, resonant structures, and circuits. Of the various planar transmission lines, the microstrip lines, slot line, and fin line have evolved as the most popular transmission lines in practice, and have received the maximum attention from the theoretical point of view. This interest has stemmed from their applications in a wide variety of microwave integrated circuit (MIC) components, such as directional couplers, filters, and oscillators. They are also used in the measurement of dielectric constant and of dispersion in a transmission line. The resonant structures have been analyzed using various analytical and numerical techniques. The main objective has been to evolve suitable design criteria so that complex microwave circuits can be designed to perform their prescribed functions as accurately as possible without resorting to the "cut-and-try" approach.

Microstrip lines are commonly employed in microwave integrated circuits. A microstrip resonator is a planar conductor patch on a dielectric substrate, generally of a regular shape. The electromagnetic energy is confined to the dielectric region between the top conductor patch and the bottom ground plane, surrounded by a perfect magnetic wall on its contour. The dielectric materials frequently used are alumina, sapphire, and glass or ceramic-reinforced Teflon. The properties of the dielectric material greatly influence the characteristics of the resonant structures. Both the dielectric and conductor contribute to the losses. In the open configuration, radiation from microstrip resonators depends mainly on the dielectric constant (ϵ_r) of the substrate material and its thickness (h). Microstrip resonators are normally coupled through microstrip lines that excite transverse magnetic (TM) modes.

Of the various microstrip resonant structures, rectangular, circular disk, and ring geometries find extensive applications in oscillators, filters, and circulators. Microstrip resonant structures of complex geometrical shapes can, in general, be fabricated to provide better performance and greater flexibility in the design. This has been demonstrated for the equilateral triangle, regular hexagon, elliptic disk, and ring resonators.

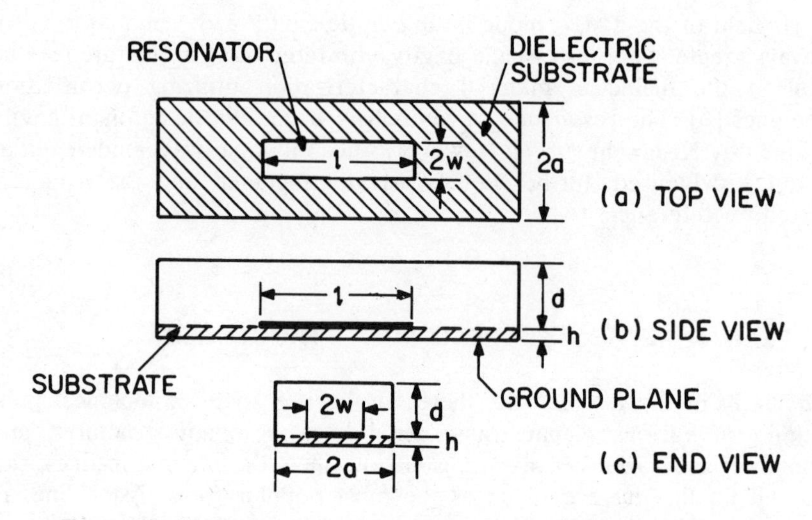

Figure 3.8 Rectangular microstrip resonator in a shielding waveguide.

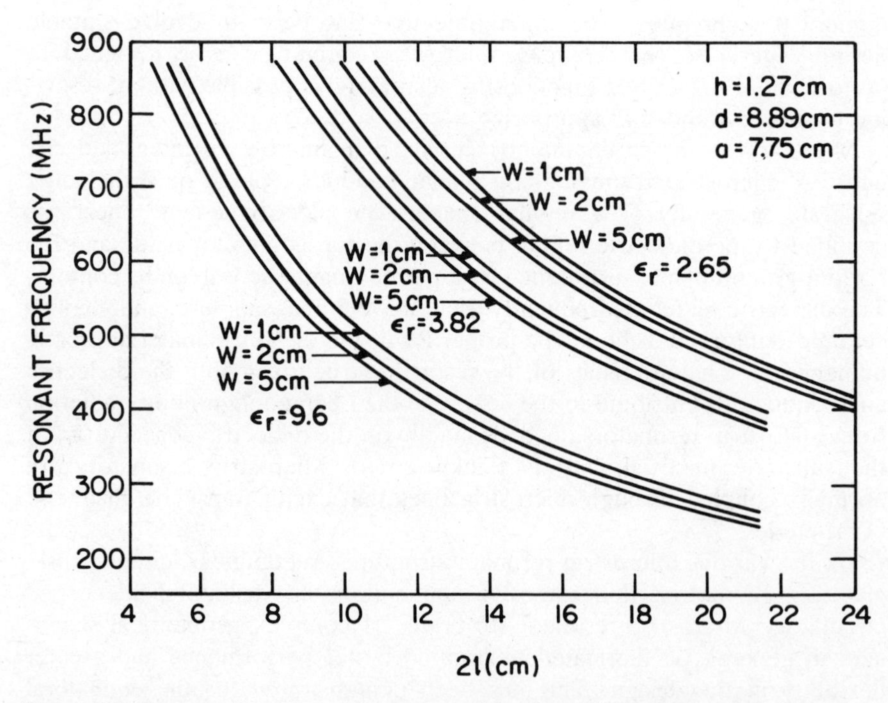

Figure 3.9 Resonant frequency of a rectangular microstrip resonator as a function of resonator length. (After T. Itoh [13], 1974 IEEE, reprinted with permission.)

3.4.1 Rectangular Microstrip Resonators

Rectangular microstrip resonators are extensively used in filters, mixers, and other microwave circuit components. As shown in Fig. 3.8, the rectangular microstrip resonator consists of a rectangle of length l and width $2W$ on a dielectric substrate of thickness h and relative dielectric constant ϵ_r.

To a first approximation, it can be viewed as an open-ended transmission line of length l. The transmission line is resonant whenever the length of the line is an integer multiple of the half guide wavelength. The open end in a microstrip is not truly an open circuit due to the fringing fields [9].

The fringing capacitance can be modeled as an hypothetical extension Δl of the microstrip line at each end [10, 11], as given in Section 2.3. The resonant frequency of a rectangular microstrip resonator under quasi-static approximation is then given by [12]

$$f_r = \frac{c}{2\sqrt{\epsilon_e}(l + 2\Delta l)} \tag{3.31}$$

where ϵ_e is the effective dielectric constant of the microstrip, as given in Section 2.2.

An accurate description of the resonant frequency can be obtained from a solution of the three-dimensional electromagnetic boundary-value problem. Itoh [13] formulated the rectangular microstrip resonator problem utilizing hybrid modes in a rectangular waveguide shielding enclosure. The associated boundary-value problem was solved utilzing the spectral domain technique. The resonant frequencies computed using this technique are shown in Fig. 3.9.

3.4.2 Circular Microstrip Disk Resonators

Circular microstrip disk resonators have applications similar to the rectangular microstrip resonator. The resonant structure consists of a circular disk of radius a on a dielectric substrate, as shown in Fig. 3.10(a). For a substrate thickness much less than the free-space wavelength ($h \ll \lambda_0$), the microstrip disk can be modeled as a cylindrical cavity with magnetic walls. The field inside the dielectric region corresponds to those of the TM_{nml} modes with $l = 0$. The resonant frequency of the cavity is given by [14]

$$f_{nm0} = \frac{p'_{nm} c}{2\pi a_e \sqrt{\epsilon_r}} \tag{3.32}$$

where p'_{nm} is mth root of the equation

$$J'_n(ka) = 0 \tag{3.33}$$

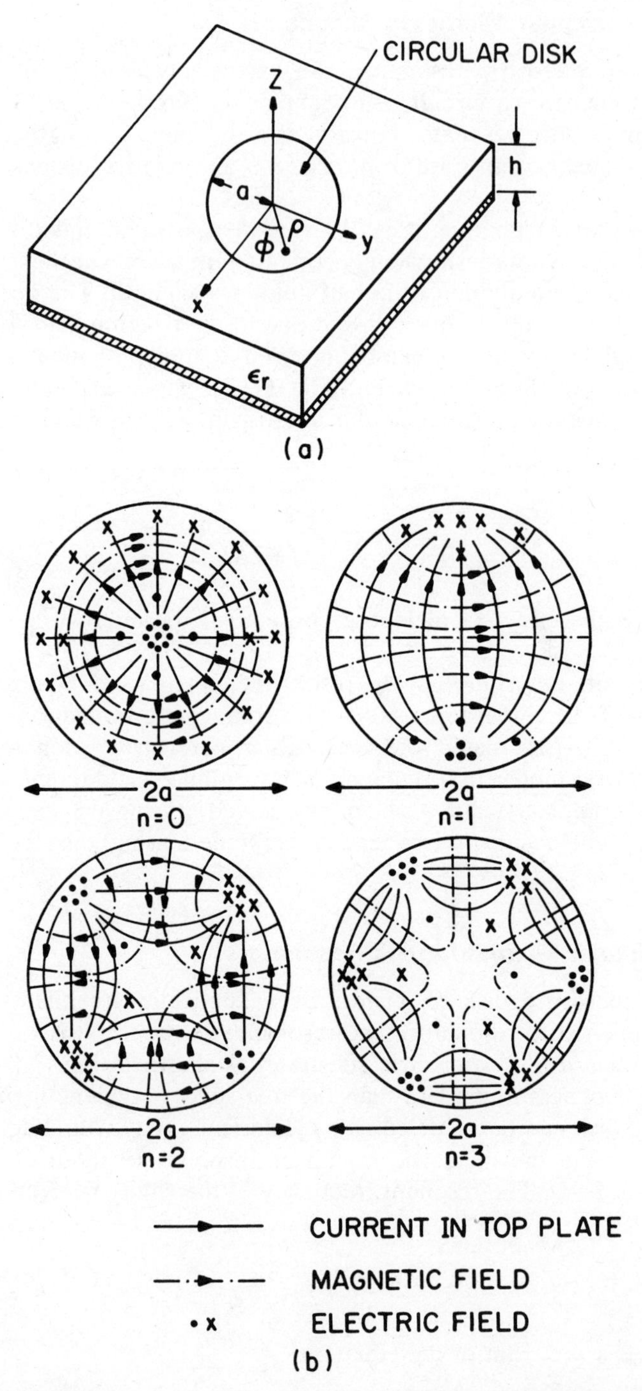

Figure 3.10 (*a*) Circular microstrip resonator. (*b*) Field configurations in circular microstrip disk resonator.

where J'_n is the derivative with respect to the argument of the Bessel function of the first kind and order n. For various values of n and m, the root of the (3.33) can easily be found. Table 3.1 gives the first few roots. The effective radius a_e is given by [15–17]

$$a_e = a\left[1 + \frac{2h}{\pi a \epsilon_r}\left(\ln\frac{\pi a}{2h} + 1.7726\right)\right]^{1/2} \tag{3.34}$$

The field configurations of the first few resonance modes are shown in Fig. 3.10(*b*).

The theoretical results calculated using a quasi-static formulation of the spectral domain technique [14, 15] yields results that are within 2.5% of the experimental values. To simplify the design of microstrip disk resonators

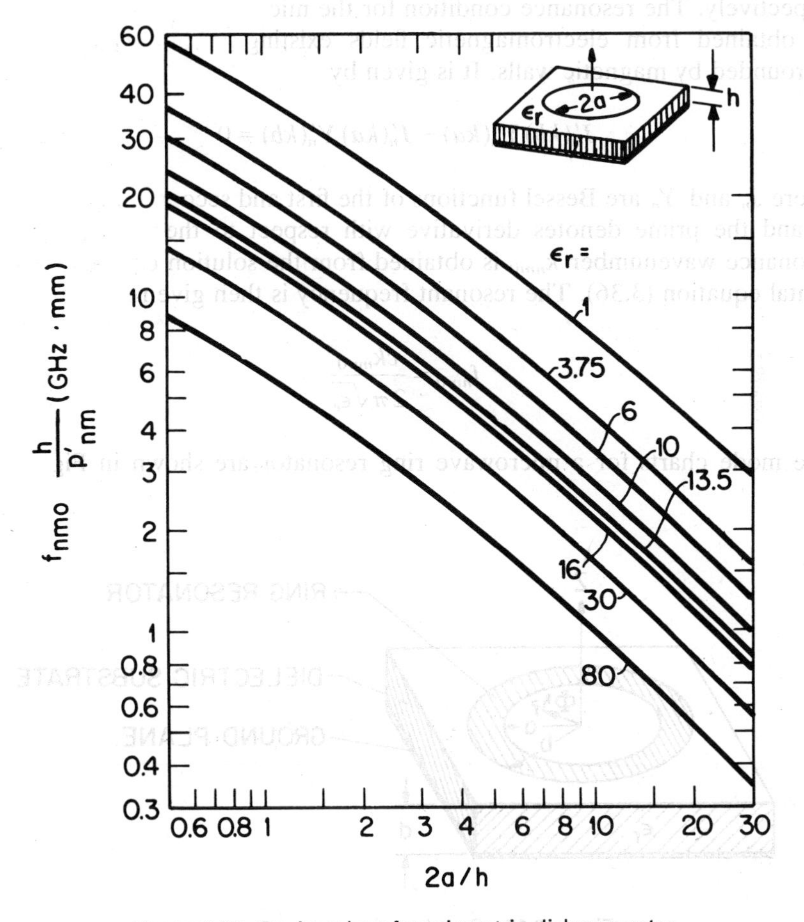

Figure 3.11 Design chart for microstrip disk resonator.

[18], rearranging (3.32) produces

$$f_{nm0} \frac{h}{p'_{nm}} = F\left(\frac{h}{a}; \epsilon_r\right) \tag{3.35}$$

The function on the left side is plotted in Fig. 3.11 as a function of $2a/h$ for various values of ϵ_r.

3.4.3 Circular Microstrip Ring Resonators

The microstrip ring resonator is another extensively used microwave integrated-circuit component for the measurement of dispersion characteristics of microstrip lines [19–21]. The geometry of the microstrip ring resonator is shown in Fig. 3.12, with the inner and outer radii a and b, respectively. The resonance condition for the microstrip ring resonator can be obtained from electromagnetic fields existing in an annular cavity surrounded by magnetic walls. It is given by

$$J'_n(kb)\,Y'_n(ka) - J'_n(ka)\,Y'_n(kb) = 0 \tag{3.36}$$

where J_n and Y_n are Bessel functions of the first and second kind and order n, and the prime denotes derivative with respect to the argument. The resonance wavenumber k_{nm0} is obtained from the solution of the transcendental equation (3.36). The resonant frequency is then given by

$$f_{nm0} = \frac{ck_{mn0}}{2\pi\sqrt{\epsilon_e}} \tag{3.37}$$

The mode charts for a microwave ring resonator are shown in Fig. 3.13.

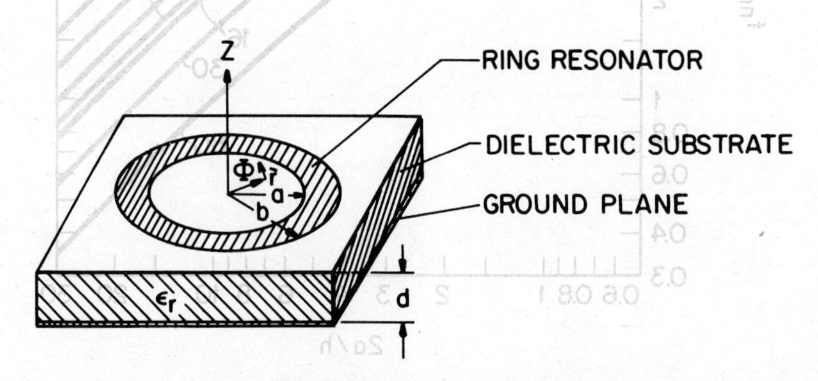

Figure 3.12 Circular microstrip ring resonator.

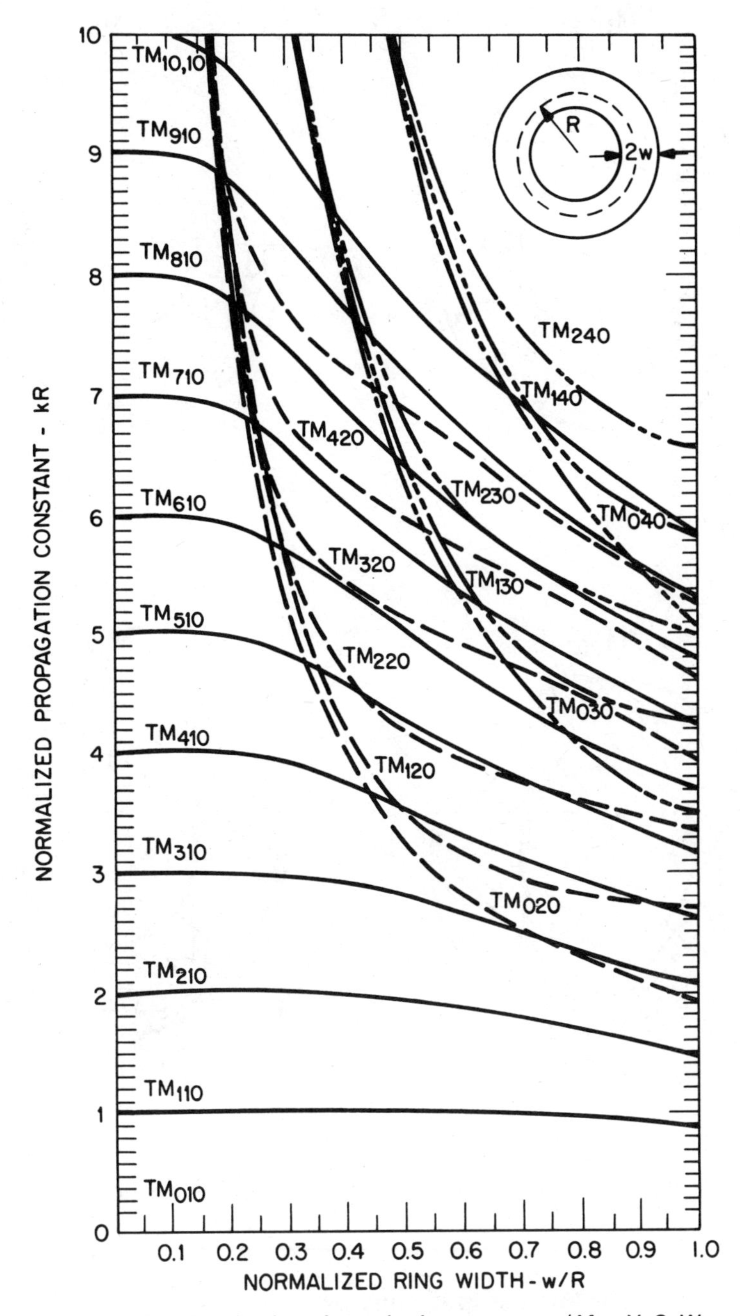

Figure 3.13 Mode chart for circular microstrip ring resonator. (After Y. S. Wu and F. J. Rosenbaum [22], 1973 IEEE reprinted with permission.)

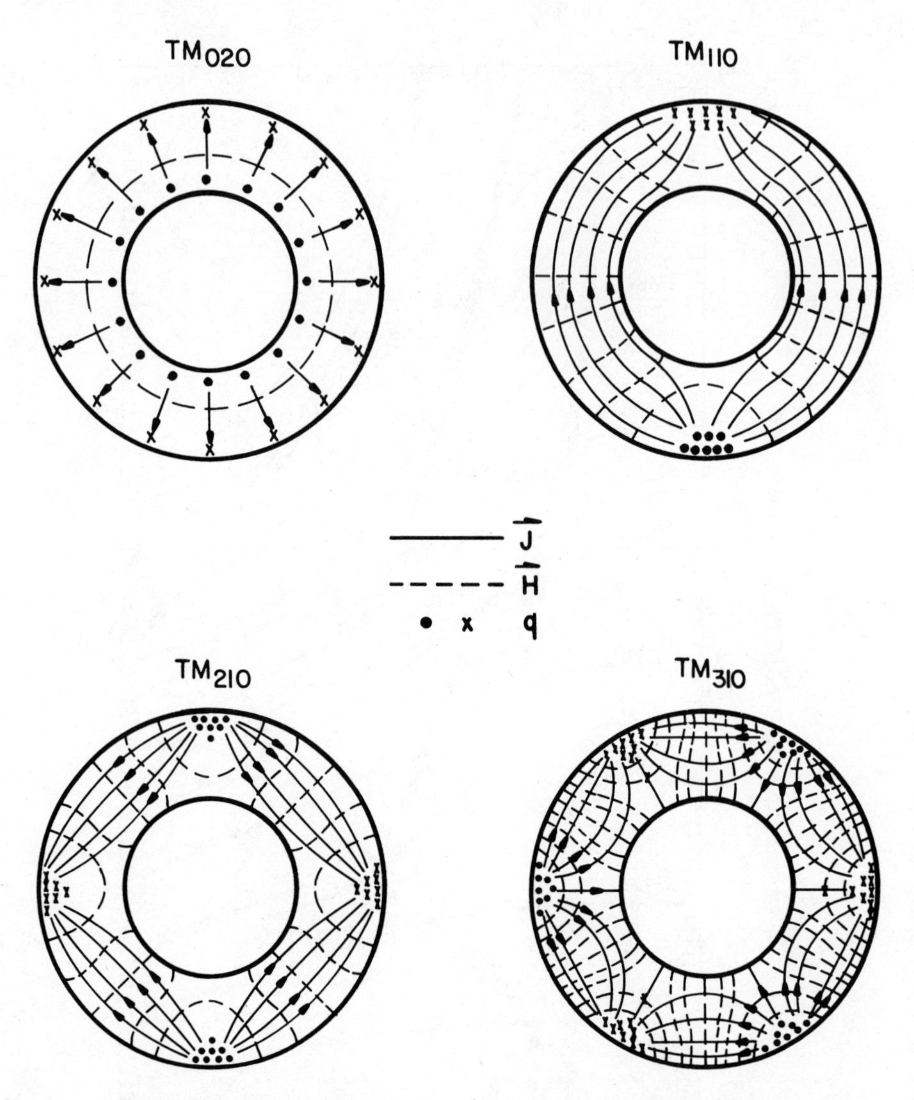

Figure 3.14 Field configurations for circular microstrip ring resonators.

The field patterns of some lower order resonance modes are shown in Fig. 3.14 [22].

3.4.4 Triangular Microstrip Resonators

Another important geometry for a wide variety of applications is the equilateral triangular microstrip resonator, as shown in Fig. 3.15. The 120° symmetry property of this element was utilized in an articulate design of

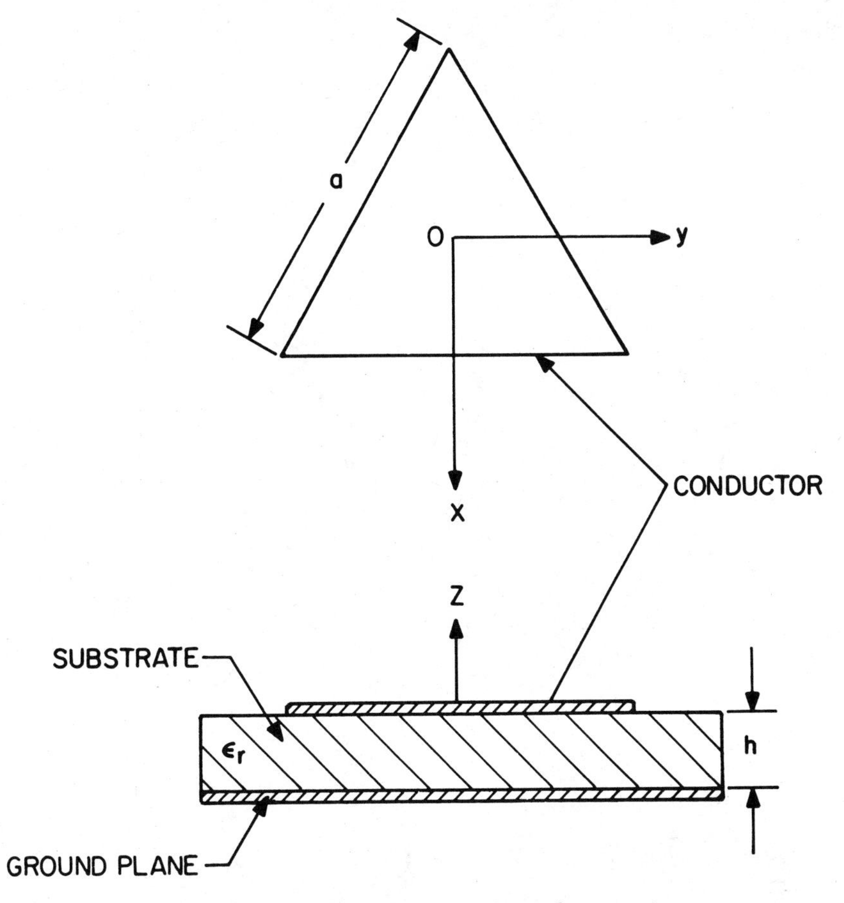

Figure 3.15 Equilateral triangular microstrip resonator.

circulators [23, 24]. Cuhaci and James [25] showed that, as a resonator, this element exhibits a slightly higher radiation Q factor (Q_r) than the corresponding circular microstrip disk resonator. This is a significant advantage in the design of low-loss microwave integrated circuits.

The resonance wavenumber for equilateral triangular resonator is given [23]

$$k_{nmp} = \frac{4\pi}{3a}\sqrt{m^2 + mn + n^2} \tag{3.38}$$

where a is the triangle side. The integers m, n, and p are such that

$$m + n + p = 0 \tag{3.39}$$

The dominant mode is achieved from one of the sets of following integer values:

$$\begin{aligned} &\text{(i)} \quad m=1,\ n=0,\ p=-1 \\ &\text{(ii)} \quad m=0,\ n=1,\ p=-1 \\ &\text{(iii)} \quad m=1,\ n=-1,\ p=0 \end{aligned} \tag{3.40}$$

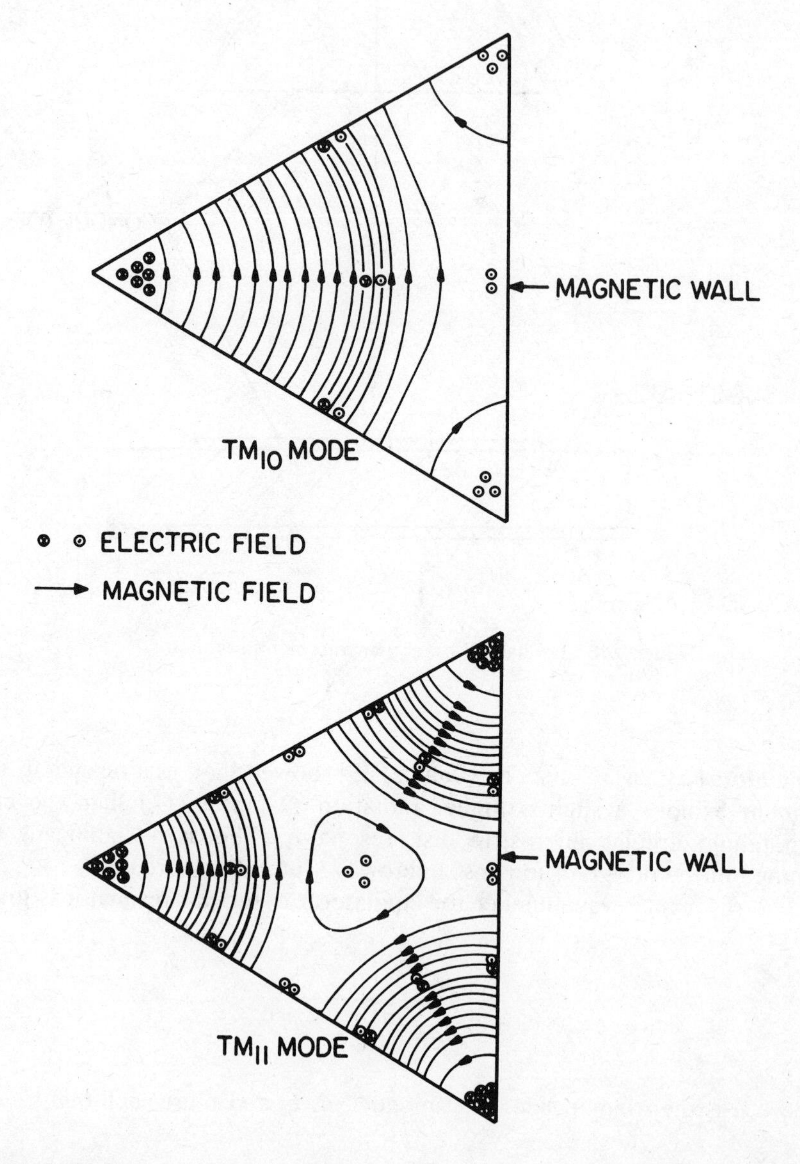

Figure 3.16 Field patterns for TM_{10} and TM_{11} modes in an equilateral triangular microstrip resonator.

Thus the dominant wavenumber is given by

$$k = \frac{4\pi}{3a} \tag{3.41}$$

The field patterns for the TM_{10} and TM_{11} modes in a triangular resonator are shown in Fig. 3.16. Another triangular shape that has not received much attention is the isosceles right-angled triangular microstrip resonator. The resonance condition, in this case, is given by [26]

$$k_{mn} = \left(\frac{\pi}{a}\right)\sqrt{m^2 + mn + 2n^2} \tag{3.42}$$

where a is the length of the isosceles triangle side. The integers m and n determine the mode of resonance, and the dominant mode is obtained when $m = 0$, and $n = 1$. The resonance wavenumber is then given by [27]

$$k_{01} = \sqrt{2}\,\frac{\pi}{a} \tag{3.43}$$

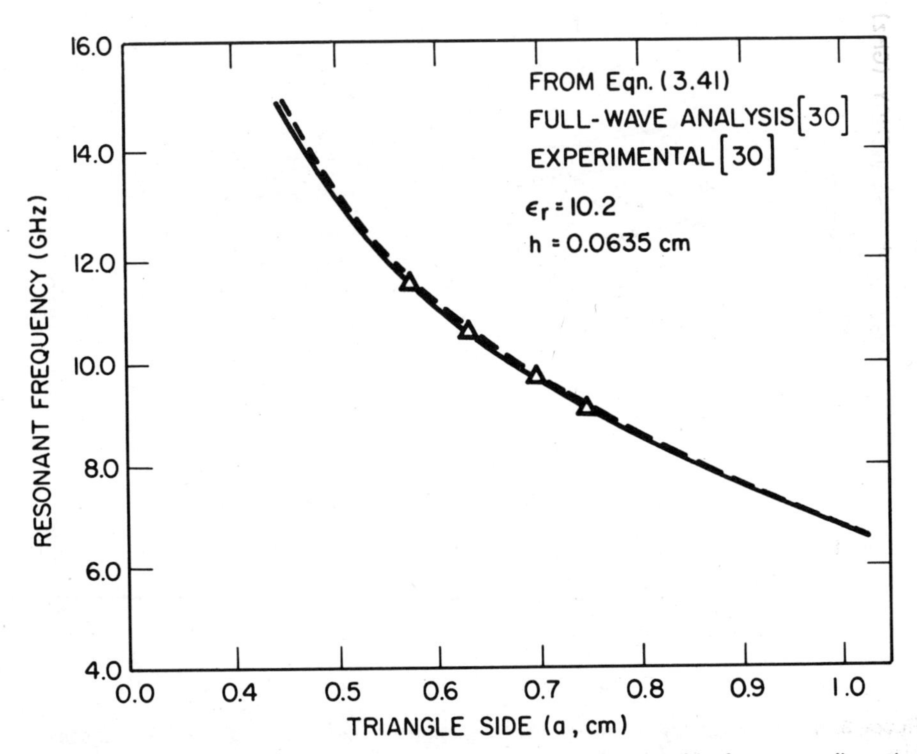

Figure 3.17 Resonant frequency as a function of a triangle side for an equilateral triangular microstrip resonator.

As such, no convenient exact method exists for solving the general triangular microstrip resonator except for equilateral and isosceles right-angled shapes [26]. A numerical method for the eigenvalues of the TE and TM modes propagating in a triangular waveguide of arbitrary dimension do exist [28]. Also, the general isosceles triangular microstrip resonator has been analyzed using the spectral domain technique [29, 30]. The resonant frequency for an equilateral triangular microstrip resonator as a function of triangle side is shown in Fig. 3.17. Those for an isosceles triangle as a function of apex angle (2α) for various values of triangle height are plotted in Fig. 3.18.

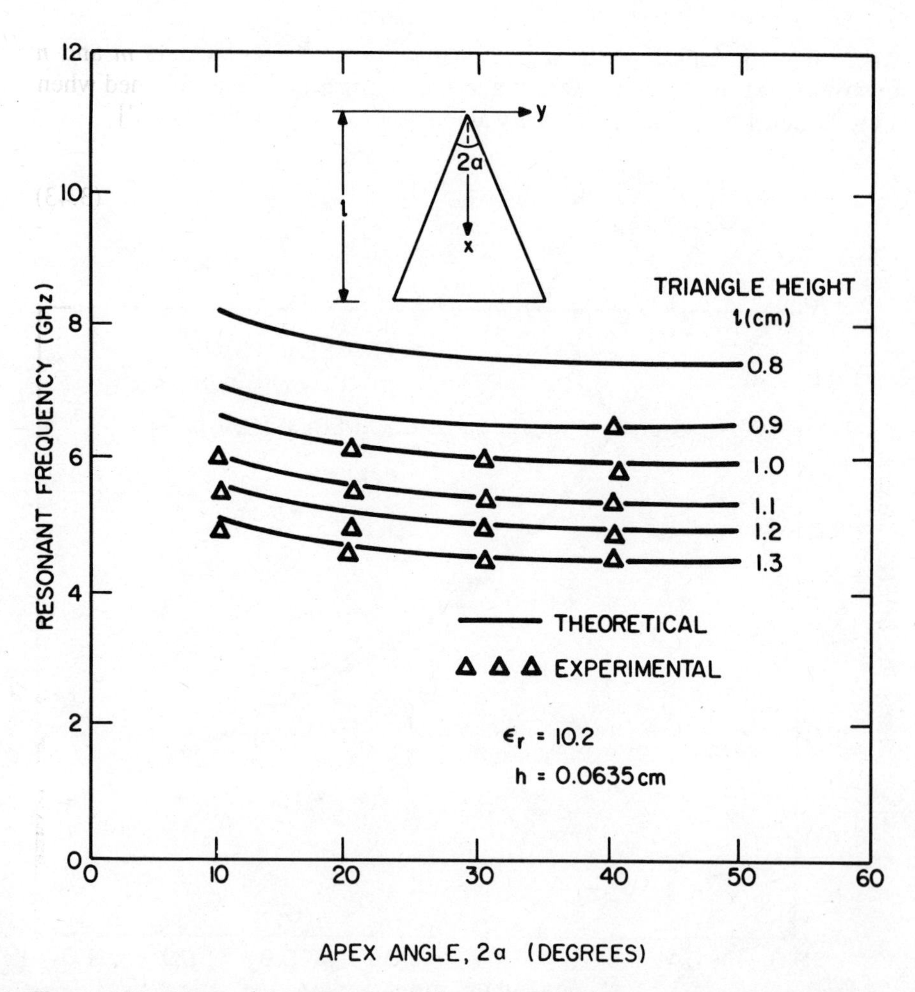

Figure 3.18 Resonant frequency as a function of the apex angle for the isosceles triangular microstrip resonator.

3.4.5 Hexagonal Microstrip Resonators

A regular hexagonal microstrip resonator has been used in the past as a junction resonator and as a circulator element [31]. The resonance condition for the lowest order mode can be calculated from [32]

$$ka = 2.011 \tag{3.44}$$

where a is the side of the hexagon shown in Fig. 3.19 (inset). The resonant frequency as a function of the hexagon side is shown in Fig. 3.19. Also shown there are the results obtained using the quasi-static formulation of the spectral domain technique [33].

The resonant frequencies of the circular and hexagonal resonators of identical dimensions a are related by the following empirical expression

$$\frac{f_{\text{hexagonal}}}{f_{\text{circular}}} = 1.05 \tag{3.45}$$

provided that the ϵ_r is high ($\epsilon_r > 6$) and the ratio $h/a \ll 1$.

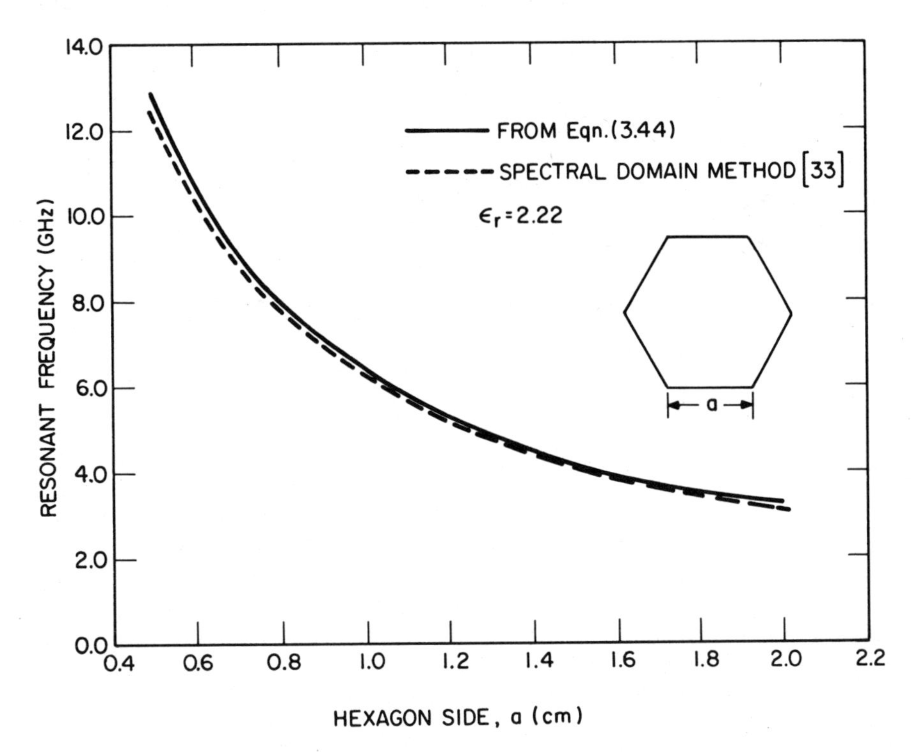

Figure 3.19 Resonant frequency as a function of the hexagon side.

3.4.6 Elliptic Microstrip Disk Resonators

For applications in harmonic multipliers and parametric amplifiers, the circular microstrip disk resonator is rendered unsuitable due to the non-harmonious relationship of mode frequencies. However, several different modes in an elliptic disk resonator are harmonically related for the appropriate choice of eccentricity. Thus, eccentricity as a design parameter provides additional flexibility and enhances the usefulness of this structure.

Extensive work by Kretzschmar [34, 35], McLachlan [36], and Sharma [37, 38] on elliptic disk and ring resonators provide excellent design information. The resonant frequency of the elliptic microstrip disk resonator, shown in Fig. 3.20, is given by

$$ka = 2\frac{\sqrt{q}}{e} \tag{3.46}$$

where $q = \bar{q}_{cmn}$ ($\bar{q}_{smn}$) is the nth parametric zero of even (odd) modified Mathieu function and e is the eccentricity. The TM mode resonance condition is obtained from duality with cutoff wavenumbers of TE mode in a hollow conducting elliptic waveguide having semimajor axis a and semiminor axis b. Under resonance condition, $\bar{q}_{cmn}$ is obtained for even

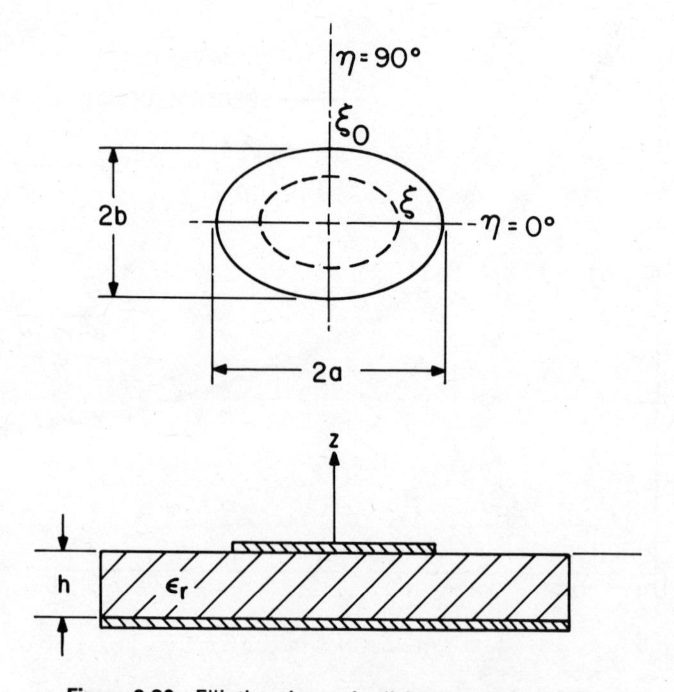

Figure 3.20 Elliptic microstrip disk resonator.

TM_{cmn} modes from the equation

$$Ce'(\xi_0, q) = 0 \tag{3.47a}$$

and $\bar{q}_{smn}$ is obtained for odd TM_{smn} mode from the solution of the equation

$$Se'(\xi_0, q) = 0 \tag{3.47b}$$

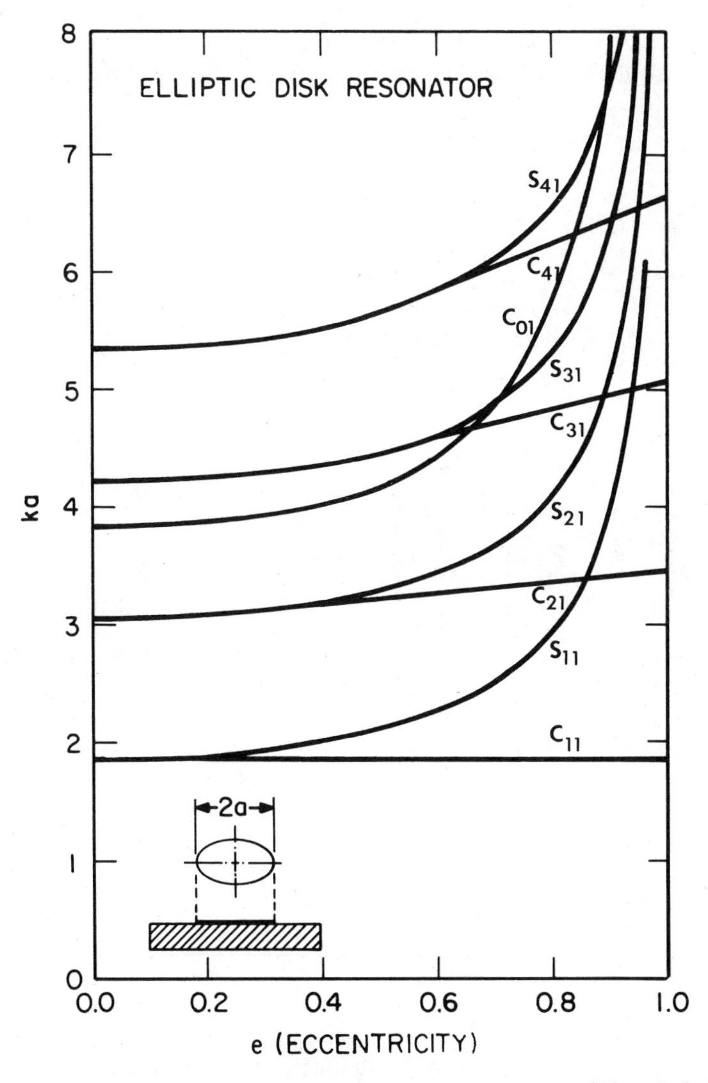

Figure 3.21 Mode chart for an elliptic microstrip disk resonator. (After J. G. Kretzschmar [34], 1972 IEEE reprinted with permission.)

In Fig. 3.21, *ka* values for the elliptic disk resonator are presented for various modes [37].

3.4.7 Interacting Resonant Structures

Among various types of interacting resonant structures in microwave integrated-circuit applications, the half-wave coupled and the quarter-wave coupled rectangular microstrip resonators are extensively used as network elements. In such structures, the propagation of waves is described in terms of the even- and odd-modes.

The interacting rectangular microstrip resonant structures in a shielded waveguide configuration are shown in Fig. 3.22. The basic building block in

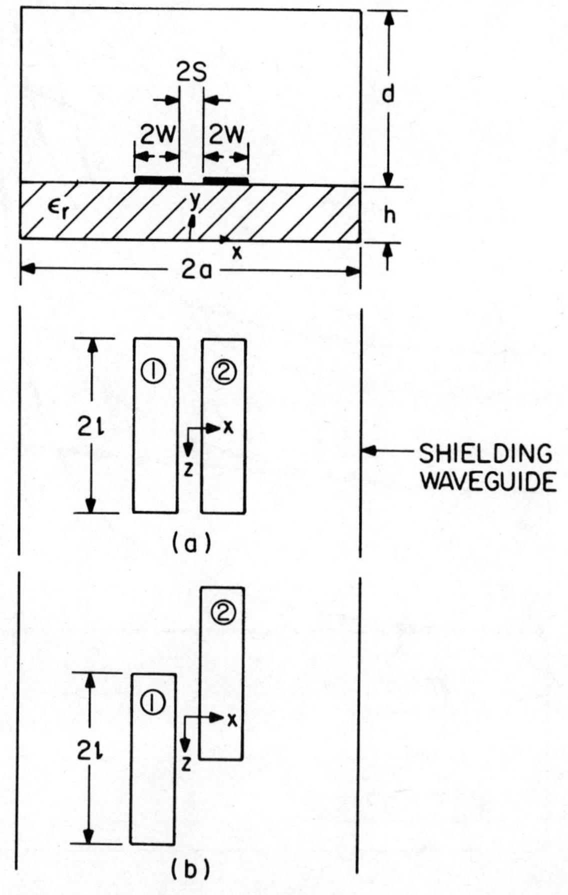

Figure 3.22 Interacting rectangular microstrip resonators. (*a*) Half-wave coupled parallel rectangular microstrip resonator. (*b*) Quarter-wave coupled parallel rectangular microstrip resonator.

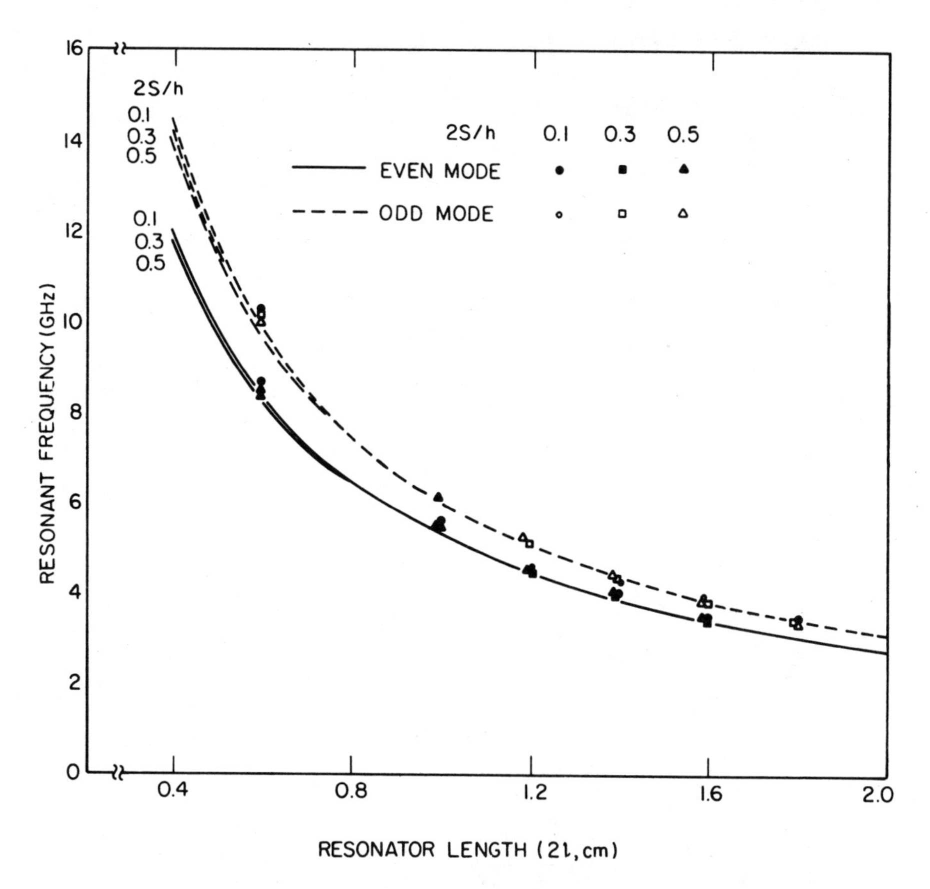

Figure 3.23 Resonant frequencies in the even and odd modes for half-wave coupled rectangular microstrip resonators as a function of $2l$ for $2W/h = 1$ and various values of $2S/h$: $\epsilon_r = 10.2$, $h = 0.0635$ cm, $h + d = 1.27$ cm, and $2a = 1.27$ cm.

each case is a rectangular microstrip resonator of length $2l$ and width $2W$. The shielding waveguide has dimensions of $2a$ and $h + d$.

The resonant frequencies of the half-wave coupled and the quarter-wave coupled rectangular microstrip resonators in the even- and odd-resonance modes have been evaluated with the full-wave formulation of the spectral domain technique [39–42]. The numerical results have been obtained for resonators with various normalized widths ($2W/h$), normalized gaps ($2S/h$), and length ($2l$). In Fig. 3.23, the effect of varying $2S/h$ is plotted for various resonator lengths while keeping the other parameters fixed. The experimental verification of the resonant frequencies in the even- and odd-modes is also provided in Fig. 3.23. Similar study on the quarter-wave coupled rectangular resonators has also been performed [39–42].

3.5 RESONANT STRUCTURES IN FIN LINES

Among various transmission media suitable for millimeter-wave integrated circuits, fin lines have been established as a potential transmission line. Various components and subsystems have been realized using them [43]. Resonant structures in fin lines are useful network elements in the realization of various millimeter-wave integrated-circuit components such as filters and oscillators. An accurate description of a rectangular slot resonator in a unilateral or bilateral fin line can be obtained with the full-wave formulation of the spectral domain technique [44]. In this section, we present resonance characteristics of various isolated and coupled unilateral and bilateral fin resonators [45] as shown in Fig. 3.24. The building block in each case is a rectangular fin resonator of length $2l$ and width $2W$ on a

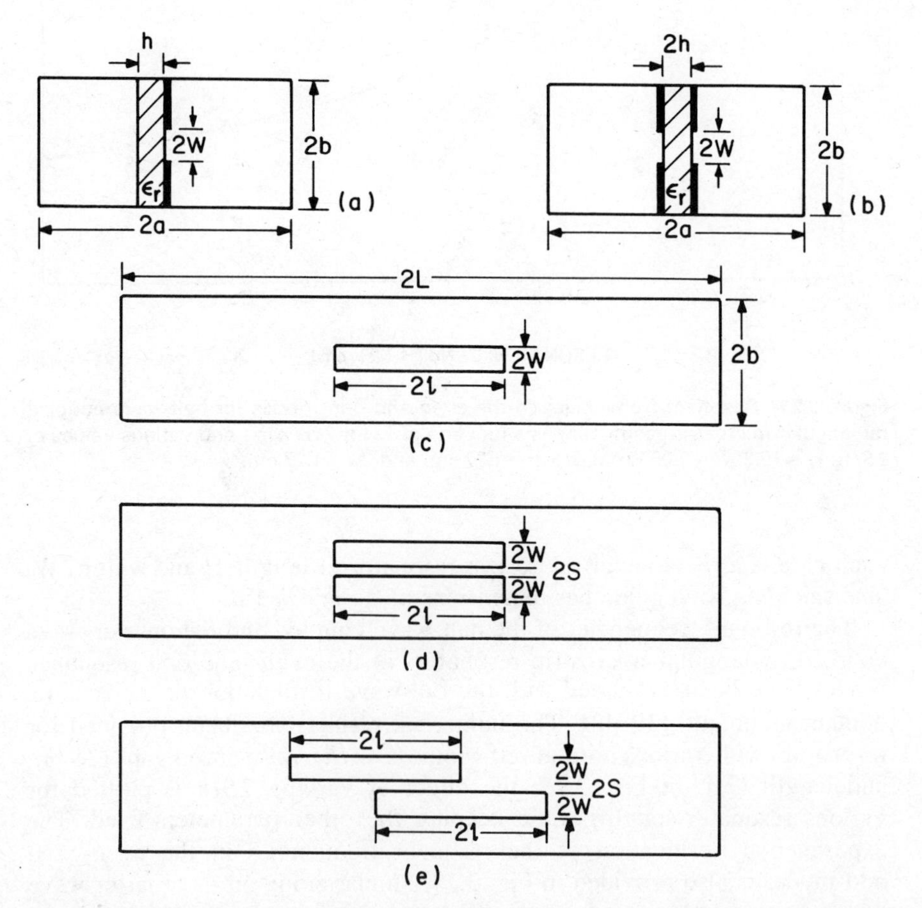

Figure 3.24 Cross-sections of (*a*) unilateral and (*b*) bilateral fin lines. Side views of (*c*) isolated, (*d*) and (*e*) interacting unilateral and bilateral fin resonators.

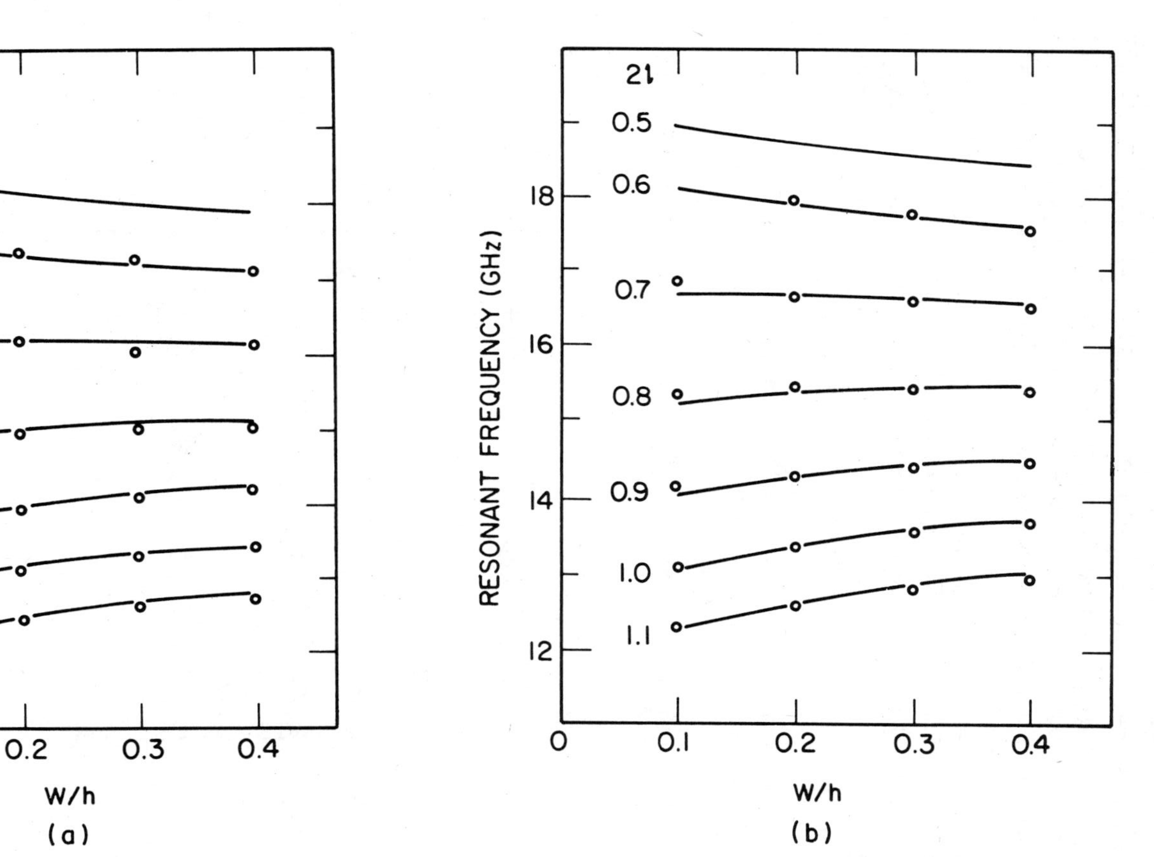

Figure 3.25 Resonant frequency as a function of W/b of (a) unilateral fin resonator with $d/2a = 0.02455$, (b) bilateral fin resonator with $d/a = 0.2455$, $a = 2b = 0.78994$ cm, and $\epsilon_r = 2.22$. The length l is in cm. — calculated, ∘∘∘ measured.

dielectric substrate of thickness h or $2h$ and relative permittivity ϵ_r. The waveguide enclosure has dimensions $2a$ and $2b$. The resonant frequency of a resonator is computed by the full-wave formulation of the spectral domain technique. These are shown in Fig. 3.25 for unilateral and bilateral fin resonators, along with experimental results.

The fin-line shorting septum is not a perfectly reflecting termination. The equivalent reactance of shorting septum in a unilateral fin line can be found by determining the guide wavelength and then the shortest resonant length of the structure at that same frequency. The reactance is then calculated from the difference between the resonant length and $\lambda/2$. The design curves [46] are shown in Fig. 3.26.

The even- and odd-mode resonant frequencies for the half-wave interactively coupled unilateral and bilateral fin resonators are shown in Fig. 3.27. They are plotted for resonators with various normalized gaps ($2S/h$ or S/h), and lengths ($2l$) for a given value of normalized substrate thickness ($h/2a$ or h/a).

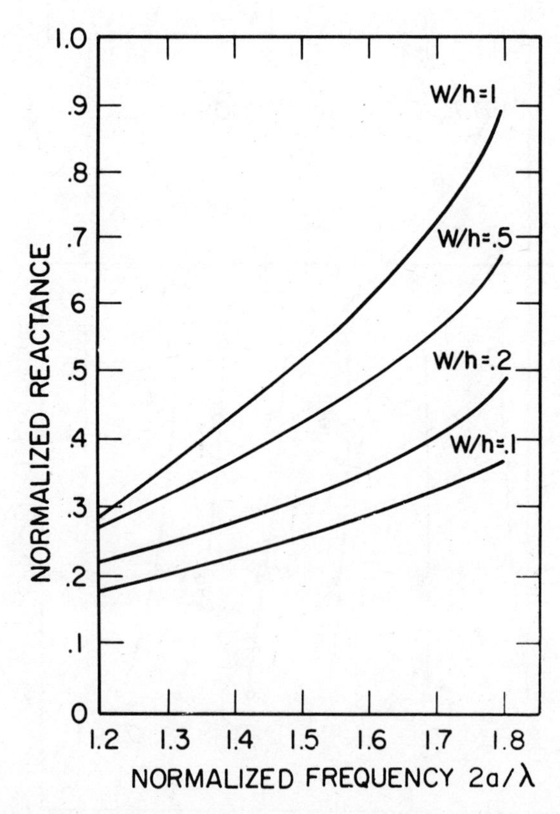

Figure 3.26 Normalized septum reactance vs. normalized frequency for a fin line with $\epsilon_r = 2.2$, $b/a = 0.5$, $h/2a = 0.05$, and $2L \rightarrow \infty$. (After J. B. Knorr [46], 1981 IEEE reprinted with permission.)

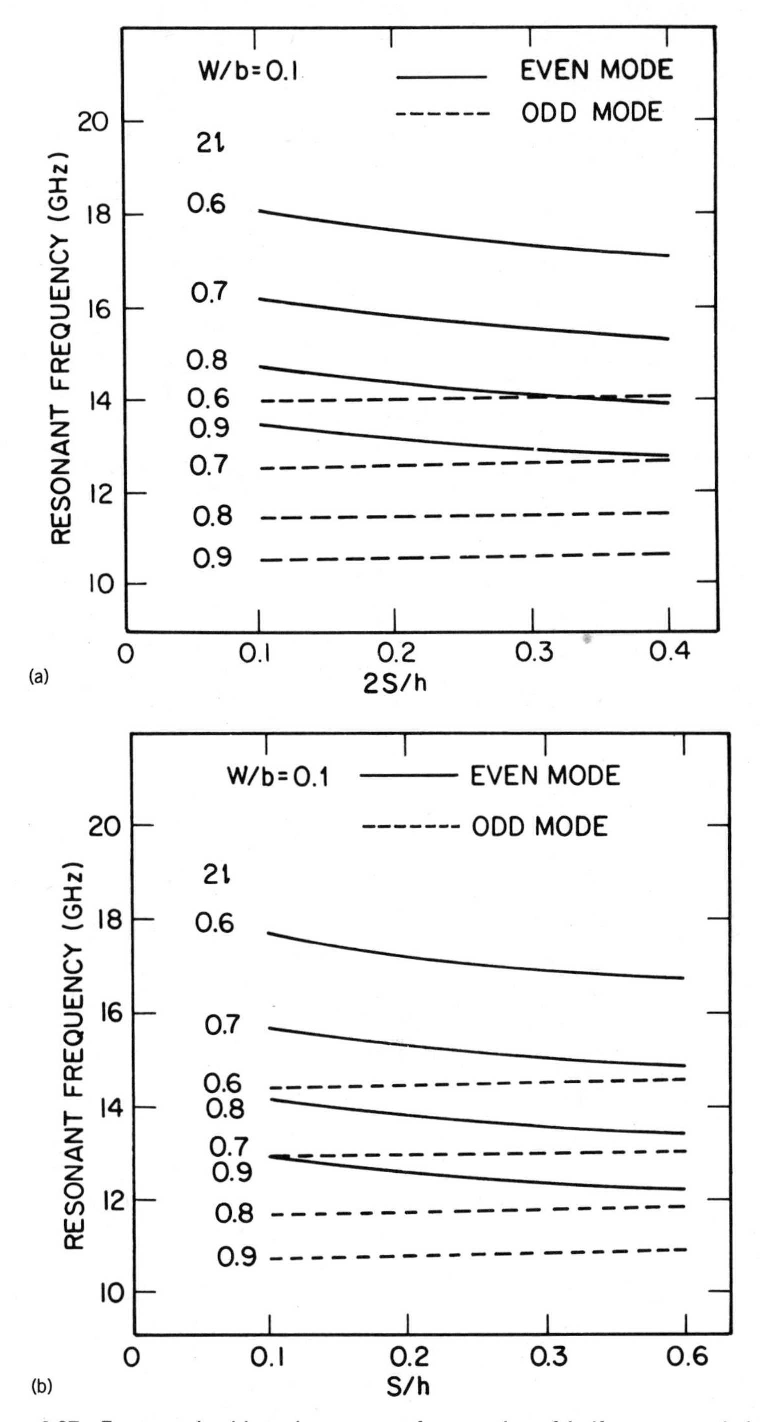

Figure 3.27 Even- and odd-mode resonant frequencies of half-wave coupled (*a*) unilateral and (*b*) bilateral fin resonators as a function of normalized width: $\epsilon_r = 2.2$, $b/a = 0.5$, and $h/2a = 0.02455$. The length l is in cm.

3.6 DIELECTRIC RESONATORS

The dielectric resonator is made of a low-loss, temperature-stable, high-permittivity, and high-Q ceramic material in a regular geometrical form. It resonates in various modes at frequencies determined by its dimensions and shielding conditions. Because of its small size, low price, and excellent integrability in MICs, its application in active and passive microwave circuits has been rapidly increasing [47].

The dimensions of a dielectric resonator are considerably smaller than those of an empty metallic cavity resonant at the same frequency, by a factor of approximately $1/\sqrt{\epsilon_r}$. If ϵ_r is high, the electric and magnetic fields are confined in and near the resonator, thus having small radiation losses. The unloaded quality factor Q_u is thus limited by the losses in the dielectric resonator. To a first approximation, a dielectric resonator is the dual of a metallic cavity. The radiation losses of the dielectric resonators with the

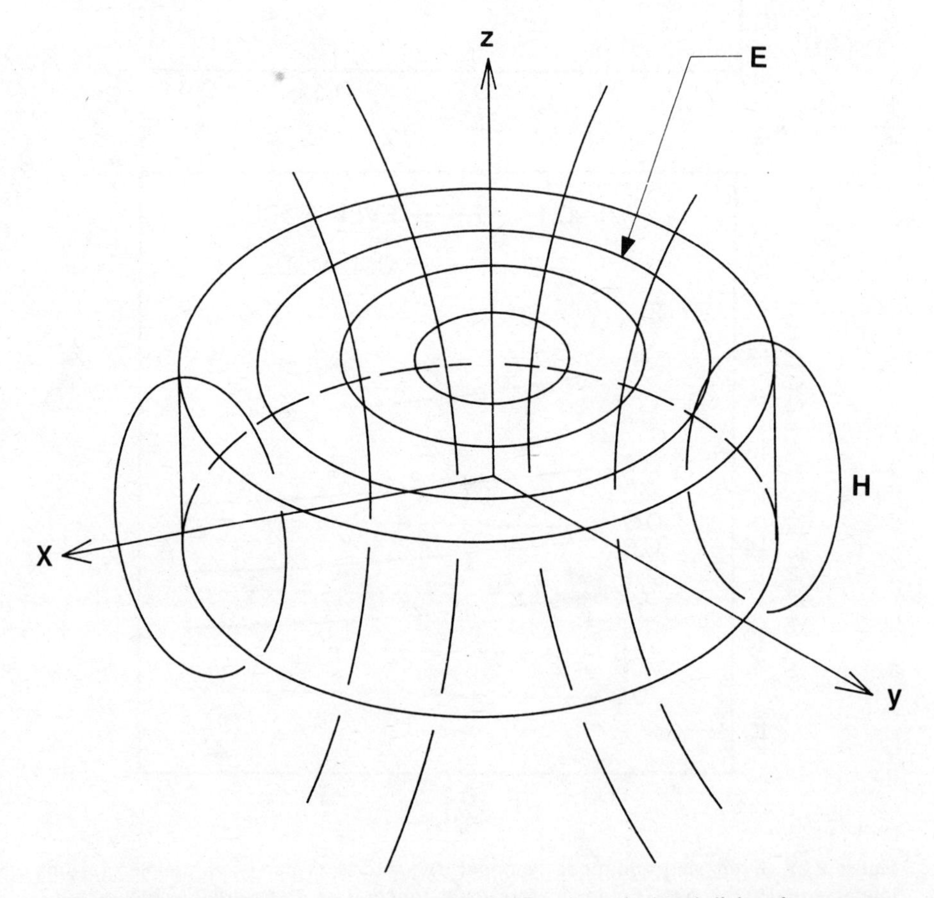

Figure 3.28 Field lines of the resonant mode $TE_{01\delta}$ in an isolated dielectric resonator.

commonly used permittivities, however, are generally much greater than the energy losses in the metallic cavities, which makes proper shielding of the dielectric resonator a necessity.

The shape of a dielectric resonator is usually a solid cylinder, but one can also find tubular, spherical, and parallelopiped shapes. A commonly used resonant mode in cylindrical resonators is denoted by $TE_{01\delta}$. The magnetic field lines are contained in the meridian plane, while the electric field lines are concentric circles around the z axis, as shown in Fig. 3.28. For a distant observer, this mode appears as a magnetic dipole and for this reason sometimes this mode is referred to as a "magnetic dipole mode." When the relative dielectric constant is around 40, more than 95% of the stored electric energy of the $TE_{01\delta}$ mode, as well as more than 60% of the stored magnetic energy are located within the cylinder. The remaining energy is distributed in the air around the resonator, decaying rapidly with distance away from the resonator surface.

3.6.1 Material

The important properties of the ceramic material to be used for dielectric resonators are

- the Q factor, which is equal to the inverse of the loss tangent,
- the temperature coefficient of the resonant frequency T_f, which includes the combined effects of the temperature coefficient of the dielectric constant and the thermal expansion of the dielectric,
- the dielectric constant ϵ_r.

The Q, T_f, and ϵ_r values required for various applications differ, and in general proper combination can be achieved by choosing an appropriate material and composition. Until several years ago, the lack of suitable materials possessing Q, T_f, and ϵ_r all of acceptable values severely limited dielectric resonator applications. Materials such as TiO_2 (rutile phase), which has $Q \simeq 10{,}000$ at 4 GHz and $\epsilon_r \simeq 100$, were most often used for exploratory work. But TiO_2 has a T_f value of 400 ppm/°C, which makes it impractical for most of the applications.

A number of material compositions have been explored in attempts to develop suitable dielectric materials. These include ceramic mixtures containing TiO_2 [48], various titanates and zirconates [49, 50], glass ceramic [51], and alumina-based ceramics. Development of temperature-stable dielectric resonators dates back to less than a decade. At present, several ceramic compositions have been developed offering excellent dielectric properties. Table 3.2 compares the important properties of different materials developed commercially.

It is not yet established if any of the dielectric compositions shown in

TABLE 3.2 Properties of Dielectric Resonators

Composition	Dielectric Constant	Q @ 10 GHz	Temperature Coefficient of Frequency	Frequency Range	Manufacturer
$Ba_2Ti_9O_{20}$	40	4,000	+2	1 to 50 GHz	Bell Labs
$BaTi_4O_9$	39	3,500	+4	1 to 50 GHz	Raytheon, Trans Tech
$(Zr\text{-}Sn)TiO_4$	38	4,000	−4 to −10[a]	1 to 60 GHz	Trans Tech Thomson-CSF, Murata
$Ba(Zn_{1/3}Ta_{2/3})O_2$	29	10,000	0 to 10[a]	5 to 60 GHz	Murata
$(Ba, Pb)Nd_2Ti_5O_{14}$	90	5,000 @ 1 GHz	0 to 6[a]	0.8 to 4 GHz	Murata, Trans Tech

[a] Adjustable with composition.

Table 3.2 has overall superiority over the others, since many factors, such as ease of ceramic processing and ability to hold tolerances on the dielectric properties, must be considered. The performance limitations, if any, of the lower dielectric constant material remain to be determined, since most component work reported thus far has used dielectric resonators possessing ϵ_r in the range of 37–100. The lower dielectric constant material performance is likely to be more sensitive to shielding due to increase in fields outside the isolated resonator.

The temperature coefficient T_f of the resonator can be controlled in some materials, by modifying the composition, to be anywhere within +9 to −9 ppm/°C. Circuit effects also shift T_f by a few parts per million, depending upon the circuit configuration. An initial resonator T_f of +1 to +4 ppm/°C often results in effective temperature compensation in transistor oscillators. Thus, the limit to achievable temperature compensation results from ceramic tolerances, and the usual need for frequency tuning affecting the temperature effects.

The quality factor Q of the dielectric resonator decreases with the increase in frequency. Typically the product f_0 (GHz) $\times$ Q_0 is constant. Some degradation of the Q is usually incurred in component applications. Losses due to housing walls, dielectrics, and adhesives used to support the resonators, and other effects typically reduce the Q by 10%–20% as described later. Variations in the Q's of different materials being significant (Table 3.2), it is possible to select the right material for different applications. A higher Q material is preferred for lower noise oscillations as well as for sharp tuned filters, and lower Q for frequency tunable wider band components.

3.6.2 Resonant Frequency

The resonant frequency of a resonator is determined by its dimensions and surroundings. Although the geometrical form of a dielectric resonator is extremely simple, an exact solution of the Maxwell equations is considerably more difficult than for the hollow metallic cavity. For this reason, the exact resonant frequency of a certain resonant mode, such as $TE_{01\delta}$, can only be computed by rigorous numerical procedures. A number of theories on the subject are available in the literature that have an accuracy of ±1% for the given configuration. Unfortunately these methods call for the use of high-level computers. Kajfez [52] has presented an approximate solution of the equations involved, both for the isolated case and for the more commonly used MIC configuration. These equations, given below, can be easily programmed on a desktop computer and give accuracies of better than ±2%.

1. *Isolated Dielectric Resonator*. The resonant frequency is given in gigahertz by

$$f_r = \frac{34}{a\epsilon_r}\left[\frac{a}{H} + 3.45\right] \tag{3.48}$$

where a represents the radius of the resonator in millimeters and H the height. This relation is accurate to about 2% in the range

$$0.5 < \frac{a}{H} < 2 \qquad \text{and } 30 < \epsilon_r < 50 \tag{3.49}$$

2. *Dielectric Resonator in MIC Configuration*. Figure 3.29 (inset) resents a dielectric resonator in a MIC configuration. The various steps to determine the size of the dielectric resonator for a given frequency are as follows.

(a) The diameter ($D = 2a$) of the resonator is selected to be

$$\frac{5.4}{k_0\sqrt{\epsilon_s}} > 2a > \frac{5.4}{k_0\sqrt{\epsilon_r}} \tag{3.50}$$

where ϵ_s and ϵ_r are the relative permittivities of the substrate and the resonator, respectively, and k_0 is the free-space propagation constant.

(b) Calculate k' from

$$k' = \frac{2.405}{a} + \frac{Y_0}{2.405a[1 + (2.43/Y_0) + 0.291Y_0]} \tag{3.51}$$

where

$$Y_0 = \sqrt{(k_0a)^2(\epsilon_s - 1) - 2.405^2} \tag{3.52}$$

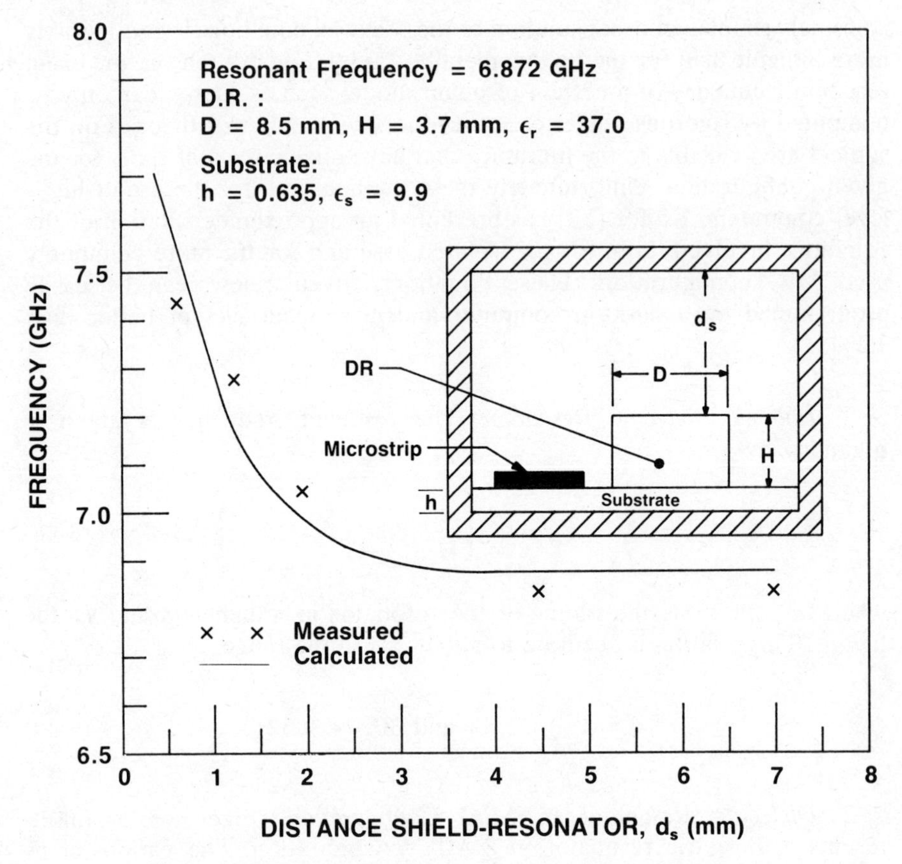

Figure 3.29 Resonant frequency as a function of cover height.

(c) Calculate propagation constant β for the $TE_{01\delta}$ mode as

$$\beta = \sqrt{k_0^2 \epsilon_r - h^2} \tag{3.53}$$

(d) Evaluate the attenuation constants α_1 and α_2

$$\begin{aligned} \alpha_1 &= \sqrt{k'^2 - k_0^2 \epsilon_s} \\ \alpha_2 &= \sqrt{k'^2 - k_0^2} \end{aligned} \tag{3.54}$$

(c) Find the resonator height H from

$$H = \frac{1}{\beta}\left[\tan^{-1}\left[\frac{\alpha_1}{\beta}\coth(\alpha_1 h)\right] + \tan^{-1}\left[\frac{\alpha_2}{\beta}\coth(\alpha_2 d_s)\right]\right] \tag{3.55}$$

The preceding relations can also be used to determine the frequency of a dielectric resonator with known dimensions. Figure 3.29 presents the variation of resonant frequency with the cover height. The measured points show that the accuracy of this method is within $\pm 2\%$.

3.6.3 Coupling of a Dielectric Resonator in MIC Configuration

The dielectric resonator is used in a number of different configurations depending upon the application. In order to effectively use dielectric resonators in microwave circuits, it is necessary to have an accurate knowledge of the coupling between the resonator and different transmission lines. The $TE_{01\delta}$ mode of the cylindrical resonator can be easily coupled to microstrip-line, fin-line, magnetic-loop, metallic, and dielectric waveguides [52]. In this section, we discuss the most commonly used configuration of the dielectric resonator, that is, $TE_{01\delta}$ mode coupling with a microstrip line.

Figure 3.30 shows the magnetic coupling between a dielectric resonator and microstrip. The resonator is placed on the top of the microstrip substrate. The lateral distance between the resonator and the microstrip conductor primarily determines the amount of coupling between the resonator and the microstrip transmission line. Proper metallic shielding, required to minimize the radiation losses (hence to increase Q) also affects the resonant frequency of the $TE_{01\delta}$ mode. The reason for the modification of the resonant frequency can be explained by the cavity perturbation theory. Namely, when a metal wall of a resonant cavity is moved inwards, the resonant frequency will decrease if the stored energy of the displaced field is predominately electric. Otherwise, when the stored energy close to

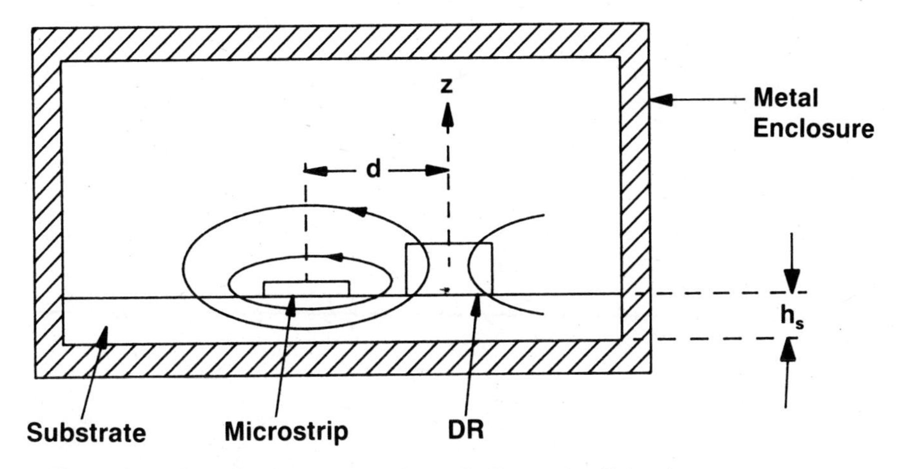

Figure 3.30 Coupling between a microstrip line and a dielectric resonator.

the metal wall is mostly magnetic, as in the case of the shielded $TE_{01\delta}$ dielectric resonator considered here, the resonant frequency will increase when the wall moves inwards.

The $TE_{01\delta}$ mode in a dielectric resonator can be approximated by a magnetic dipole of moment M. The coupling between the line and the resonator is accomplished by orienting the magnetic moment of the resonator perpendicular to the microstrip plane so that the magnetic lines of the resonator link with those of the microstrip line, as shown in Fig. 3.30. The dielectric resonator placed adjacent to the microstrip line operates like a reaction cavity that reflects the RF energy at the resonant frequency [53]. The equivalent circuit of the resonator coupled to a microstrip line is shown in Fig. 3.31. In this figure L_r, C_r, and R_r are the equivalent parameters of the dielectric resonator, L_1, C_1, and R_1 are the equivalent parameters of the microstrip line, and L_m characterizes the magnetic coupling. The transformed resonator impedance Z in series with the transmission line is easily determined to be

$$Z = j\omega L_1 + \frac{\omega^2 L_m^2}{R_r + j\omega(L_r - 1/\omega^2 C_r)} \tag{3.56}$$

Around the center frequency, ωL_1 can be neglected and Z becomes

$$Z = \omega \cdot Q_u \frac{L_m^2}{L_r} \cdot \frac{1}{1 + jX} \tag{3.57}$$

where $X = 2Q_u(\Delta\omega/\omega)$, and unloaded Q and the resonant frequency of the resonator are given by

$$Q_u = \frac{\omega_0 L_r}{R_r} \tag{3.58a}$$

$$\omega_0 = \frac{1}{\sqrt{L_r C_r}} \tag{3.58b}$$

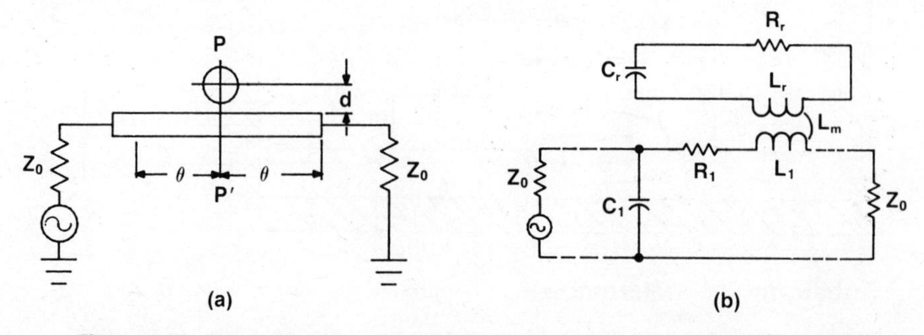

Figure 3.31 Equivalent circuit of the dielectric resonator coupled with a line.

At the resonance frequency, $X = 0$ and

$$Z = R = \omega_0 Q_u \frac{L_m^2}{L_r} \tag{3.59}$$

Equation 3.59 indicates that the circuit shown in Fig. 3.31 can be represented by the simple parallel tuned circuit as shown in Fig. 3.32, where L, R, C satisfy the following equations.

$$L = \frac{L_m^2}{L_r}$$

$$C = \frac{L_r}{\omega_0^2 L_m^2} \tag{3.60}$$

and

$$R = \omega_0 Q_u \frac{L_m^2}{L_r}$$

The coupling coefficient β at the resonant frequency ω_0 is defined by

$$\beta = \frac{R}{R_{\text{ext}}} = \frac{R}{2Z_0} = \frac{\omega_0 Q_u}{2Z_0} \cdot \frac{L_m^2}{L_r} \tag{3.61}$$

If S_{110} and S_{210} are defined as the reflection and transmission coefficients of the resonance frequency of the resonator coupled to the microstrip, β can be shown to be given by [54]

$$\beta = \frac{S_{110}}{1 - S_{110}} = \frac{1 - S_{210}}{S_{210}} = \frac{S_{110}}{S_{210}} \tag{3.62}$$

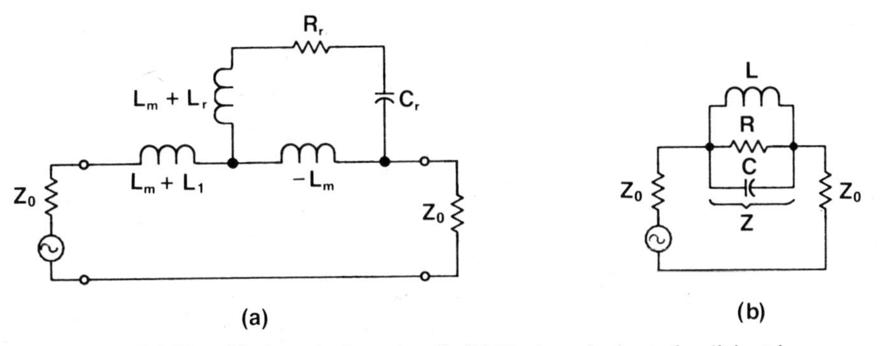

Figure 3.32 (*a*) Simplified equivalent circuit. (*b*) Final equivalent of a dielectric resonator coupled with a microstrip line.

This relation can be used to determine the coupling coefficient from the directly measurable reflection and transmission coefficients. The value of β can also be accurately calculated from a knowledge of the circuit configuration. The quantity L_m^2/L_r in (3.61) is a strong function of the distance between the resonator and the microstrip line for given shielding conditions and substrate thickness and permittivity. The analysis of β involves the use of known electromagnetic concepts and finite-element techniques.

The relation between different quality factors is well known and given by

$$Q_u = Q_L(1+\beta) = Q_e\beta \tag{3.63}$$

The external quality factor Q_e $(= Q_u/\beta)$ is generally used to characterize the coupling. Figure 3.33 shows an example of the variation of Q_e with the distance between the resonator and the line.

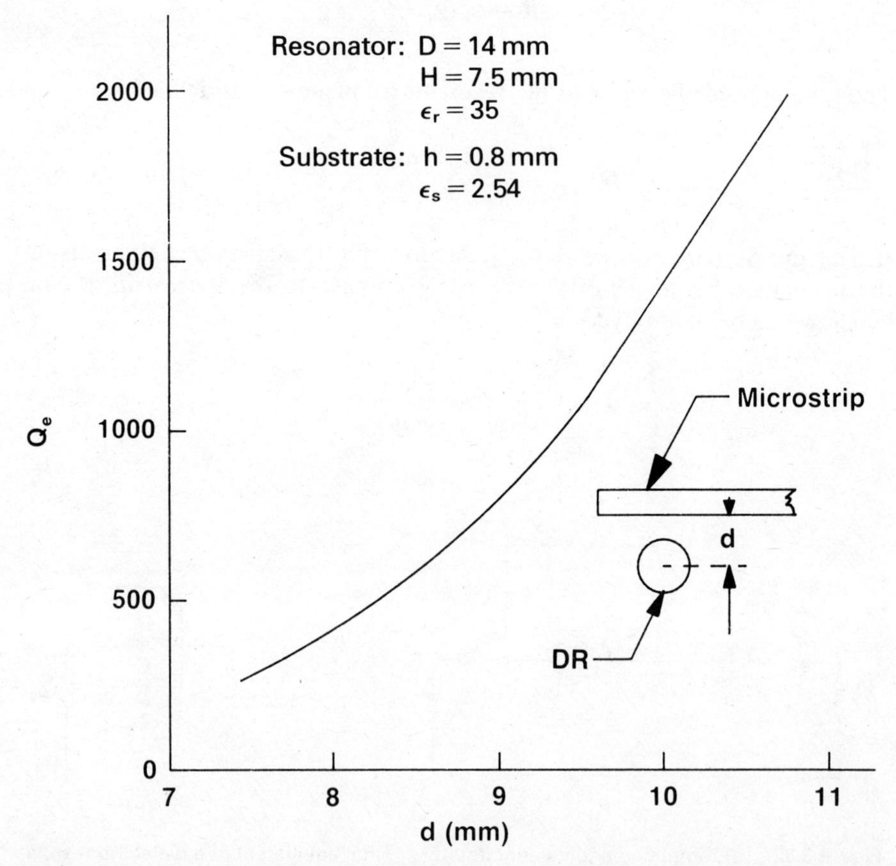

Figure 3.33 External Q factor as a function of the distance between the line and the resonator. (After P. Guillon [52], reprinted with permission of Artech House.)

The S-parameters of the dielectric resonator coupled to a microstrip with the lengths of transmission lines on input and output, as shown in Fig. 3.31, can be determined from the relations previously presented, and are given by [52]

$$S = \begin{bmatrix} \frac{\beta}{\beta + 1 + jQ_u \Delta\omega/\omega_0} e^{-2j\theta} & \frac{1 + jQ_u \Delta\omega/\omega_0}{\beta + 1 + jQ_u \Delta\omega/\omega_0} e^{-2j\theta} \\ \frac{1 + jQ_u \Delta\omega/\omega_0}{\beta + 1 + jQ_u \Delta\omega/\omega_0} e^{-2j\theta} & \frac{\beta}{\beta + 1 + jQ_u \Delta\omega/\omega_0} e^{-2j\theta} \end{bmatrix} \quad (3.64)$$

where 2θ is the electrical line length between the input and output planes.

3.6.4 Spurious Modes

As in the case of all resonant cavities, there are many possible resonant modes that can be excited in dielectric resonators. These modes can be divided into three families: transverse electric (TE), transverse magnetic (TM), and hybrid electromagnetic (HEM) modes. Each of the three families have a large number of individual modes, so that one encounters a dilemma in choosing which mode is best suited for a particular application. The $TE_{01\delta}$ is the principal or main mode traditionally used, but for certain applications, such as for a dual-mode filter, the $HEM_{11\delta}$ mode has definite advantages.

The resonant frequency of the principal and spurious modes is determined by the physical dimensions and the dielectric constant of the resonator for fixed shielding conditions. The resonant frequency of some spurious modes, like $TM_{01\delta}$ and $HE_{11\delta}$, may be close to the resonant frequency of the principal $TE_{01\delta}$ mode. Figure 3.34 shows a mode chart of the dielectric resonator in the shown shielded configuration [55]. The aspect ratio D/H of the dielectric resonator can be used as a design parameter to place the resonance of spurious modes outside the operating frequency band of the principal $TE_{01\delta}$ mode, thus minimizing undesirable interference. As is clear from Fig. 3.34, an aspect ratio between 2 and 2.5 results in the best separation of the spurious modes.

The $TM_{01\delta}$ and $HE_{mn\delta}$ modes are far more sensitive to frequency-tuning screws than the principal mode. For example, the same amount of tuning that will tune the principal mode of a 6-GHz resonator by 25 MHz, can move the resonant frequency of the $TM_{01\delta}$ or $HE_{01\delta}$ in the opposite direction across the entire 500-MHz radio band. Frequency tuning must be limited to avoid a significant reduction of mode separation, which can bring the spurious response into the frequency band of interest.

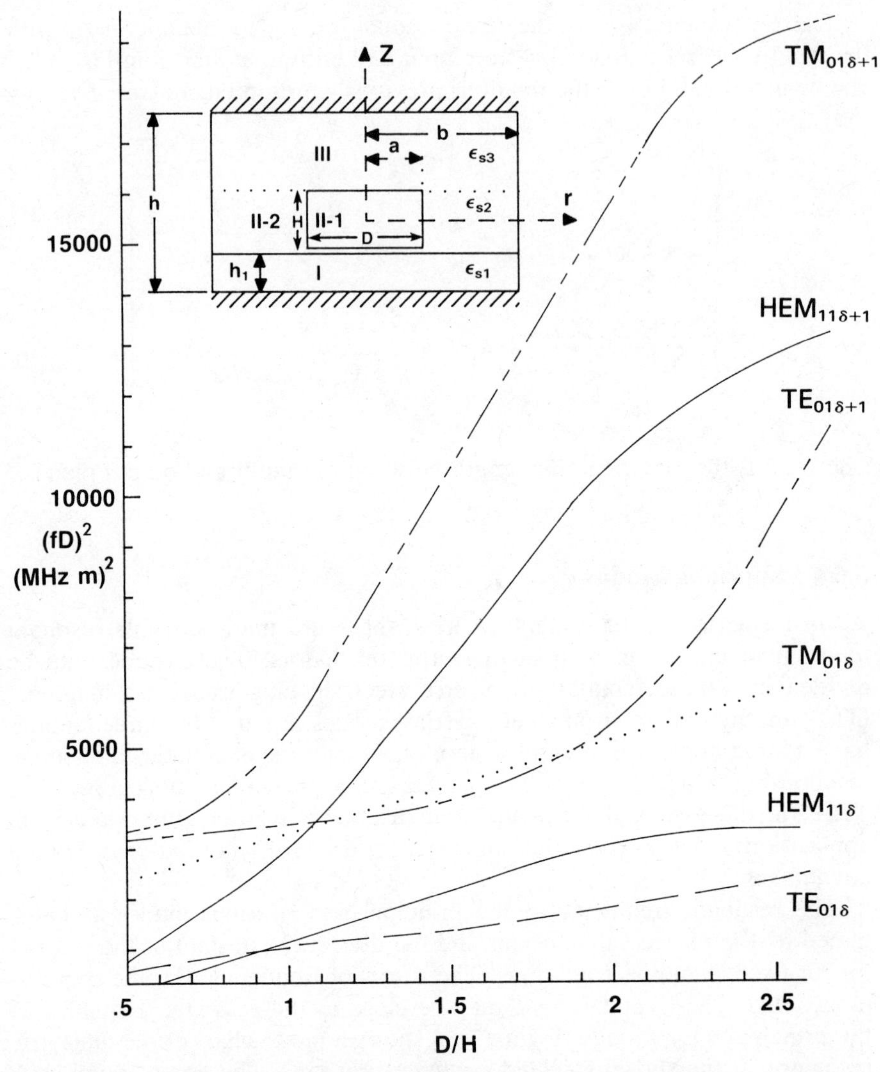

Figure 3.34 Mode chart of a dielectric resonator in a cavity. $\epsilon_r = 37.2$, $\epsilon_{s1} = \epsilon_{s2} = \epsilon_{s3} = 1.0$, $b = 16$ mm, $h = 21$ mm, $H = 8$ mm and $h_1 = 1$ mm.

3.6.5 Frequency Tuning

Most of the applications of the dielectric resonator demand frequency tunability over a narrow band. The resonant frequency of the dielectric resonator can be tuned with the help of a tuning screw placed directly above the resonator, in the metallic shield, as shown in Fig. 3.35. The resonant frequency being sensitive to the distance between the resonator and the shield, as explained earlier, the resonant frequency increases with

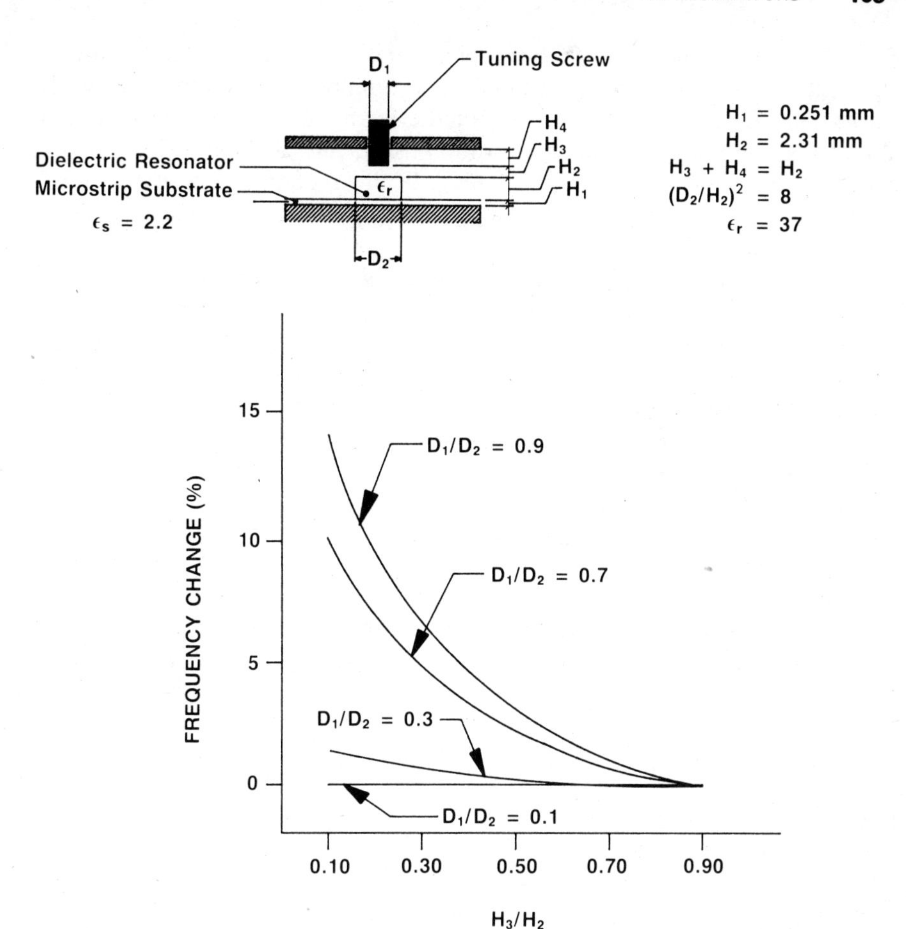

Figure 3.35 Frequency change of a dielectric resonator for several tuning-screw diameters. (After F. H. Gil et al. [56], 1985 IEEE reprinted with permission.)

the tuning-screw depth. This effect can be calculated using the finite-element method [56]. Figure 3.35 also shows the possible frequency tuning for several tuning-screw diameters. The curves presented in this figure allow one to choose an optimum screw diameter to obtain the desired tuning margin.

3.7 YIG RESONATORS

The YIG Resonator is a high-Q, ferrite resonator that can be tuned over a wide band by varying the biasing dc magnetic field. Its high performance and convenient size for applications in microwave integrated circuits makes it an excellent choice in a large number of commercial and military applications, such as filters, multipliers, discriminators, limiters, and oscillators. A YIG resonator makes use of the ferrimagnetic resonance, which depending on the material composition, size, and applied field, can be achieved from 500 MHz to 50 GHz [57]. In this section, material selection, theory of operation, important properties, equivalent circuit, and coupling to transmission line are discussed.

Single-crystal yittrium iron garnet (YIG) and gallium-doped YIG are part of a family of ferrites that resonate at microwave frequencies when immersed in a magnetic field. This resonance is directly proportional to the applied magnetic field and therefore linear tuning is achieved by changing the magnetic field with an electric current. The resonator consists of a YIG sphere, an electromagnet, and a coupling loop. A typical YIG resonator in a MIC configuration is shown in Fig. 3.36. YIG tuning magnets are electromagnets with a single air gap. The dc current through the two series-connected main coils provides the required dc magnetic field across the gap. The YIG sphere, FM coil, and MIC substrate are placed between the poles of the electromagnet.

Theoretically, an electronically tunable microwave resonant element can

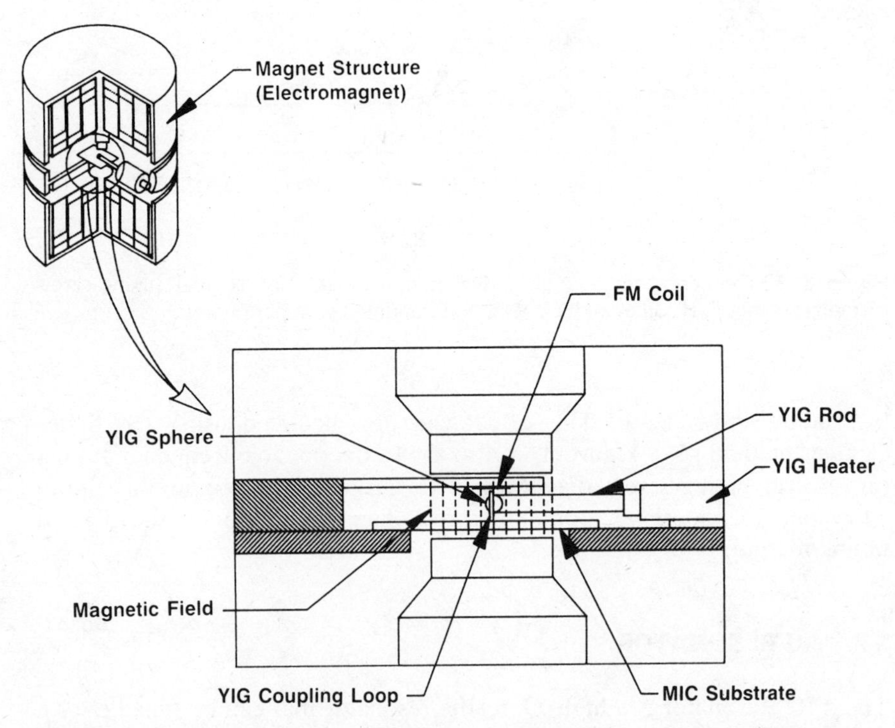

Figure 3.36 YIG oscillator elements. (After N. K. Osbrink [57], reprinted with permission of MSN & Communications Technology.)

be formed from any ferrimagnetic material operated at resonance. YIG is preferred over other ferrite materials due to

- lower losses,
- smaller anisotropy (directional dependence of magnetic properties leading to preferred directions of magnetization),
- no magnetic disorders in the crystal structure,
- more reasonable value of saturation magnetization (magnetic moment per unit-volume) $4\pi Ms$.

The most practical geometry for a YIG resonator is a sphere, because it is easily oriented in the magnetic field and the resonant frequency is not strongly dependent upon its orientation. The sphere is also the easiest to prepare with precision.

3.7.1 Resonant Frequency

The basic ferrimagnetic resonance phenomenon can be explained in terms of spinning electrons that create a new magnetic moment in each molecule of a YIG crystal [Fig. 3.37(*a*)]. Application of an external magnetic biasing field causes these magnetic dipoles to align themselves in the direction of the field, thus producing a strong net magnetization M [58]. Any magnetic force at right angles to the field results in precession of the dipoles around the biasing field at a rate that depends on the strength of the magnetic biasing field, and to some extent on the basic properties of the material [59].

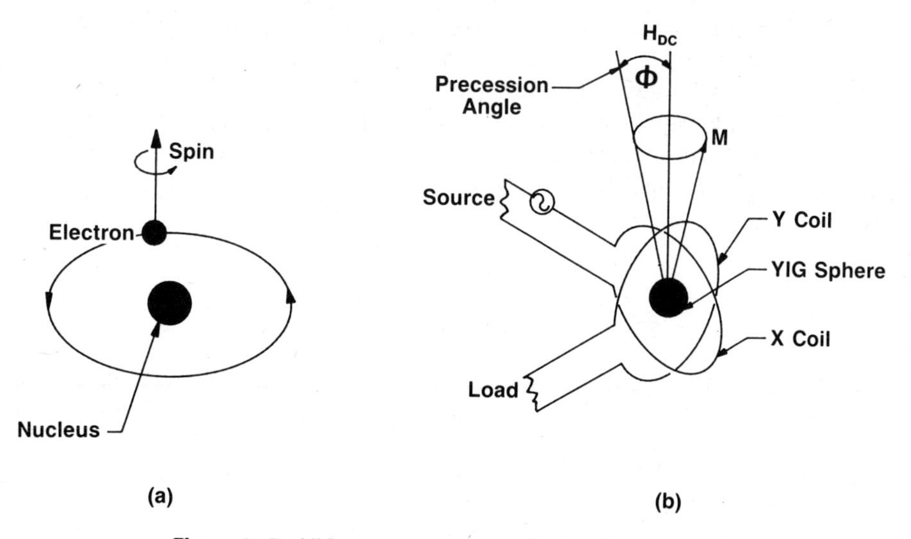

Figure 3.37 YIG resonator spin motion and loop coupling.

Such lateral forces may result from RF magnetic fields orthogonal to the biasing magnetic field. If the frequency of these RF fields coincides with the natural precessional frequency of the material, a strong interaction occurs, resulting in resonance.

For spherical resonators, this resonant frequency is given by

$$f_0 = \gamma(H_0 \pm H_a) \tag{3.65}$$

where H_0 is the applied biasing field measured in oersteds (Oe), H_a is an internal field within the crystal known as the anisotropy field, and γ is the charge–mass ratio of an electron and has a value of 2.8 MHz/Oe. By proper orientation of the crystalline axis of the spherical resonator, the temperature dependence of the anisotropy field may be eliminated. RF coupling to the YIG is achieved by concentrating the RF magnetic field in the vicinity of the sphere with a coupling loop. Orthogonality between the biasing field and the RF magnetic field has to be maintained for proper coupling (Fig. 3.37(b)).

3.7.2 Frequency of Operation and Quality Factor

The upper frequency of YIG resonance is limited by the available magnetic field, and the lower frequency limit of operation is directly proportional to the value of its saturation magnetization ($4\pi Ms$), which is a measure for net density of precessing electron spins in the material. The externally applied tuning field must be sufficient to produce alignment of all magnetic dipoles within the crystal. A pure YIG material has a $4\pi Ms$ value of 1780 gauss (G) at room temperature. A YIG of substantially lower saturation magnetization values (typically 250 G) can be grown by doping the crystal with gallium. However, doping increases YIG resonator losses, and this can be expressed by the linewidth parameter (ΔH).

The linewidth, expressed in oersteds, is a direct measure of the unloaded Q_u of the YIG resonator and can be compared to the 3-dB bandwidth of an unloaded cavity. Linewidth ΔH is related to Q_u and Ms by the following relation [60]

$$Q_u = \frac{H_0 - \frac{1}{3} \cdot 4\pi Ms}{\Delta H} \tag{3.66}$$

where H_0 is the dc biasing field strength at resonance. Typical values of linewidths are less than 0.05 Oe for pure YIG and 1.5 Oe for doped 400-G YIG. Using (3.65) we have

$$Q_u = \frac{f_0 - \frac{1}{3} \cdot \gamma \cdot 4\pi Ms}{\gamma \Delta H} \tag{3.67}$$

This expression demonstrates an interesting phenomenon in YIG, that is, the unloaded Q of the resonator decreases with the decrease in frequency and the cutoff frequency f_c, where $Q_u = 0$ is given by

$$f_c = \frac{\gamma 4\pi Ms}{3} \tag{3.68}$$

For a pure YIG

$$f_c = 1670 \text{ MHz}$$

Figure 3.38 shows typical measured results for $1/Q_u$ of a YIG resonator from 6 to 12 GHz [61].

The behavior of the single crystal previously described is due to the uniform precession made by spinning electrons. If the amplitude of the UHF field (or the RF incident power) increases beyond a certain threshold, the energy of the uniform modes is transferred to the unlimited subharmonic mode through the excited spin waves. The theoretical analysis can demonstrate that there exist two zones of frequencies.

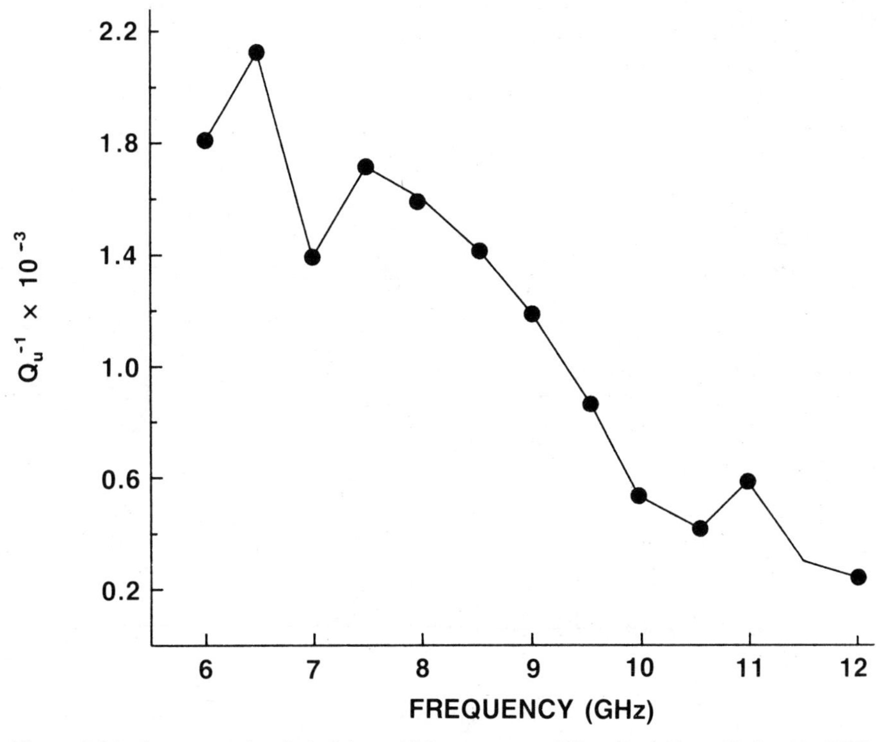

Figure 3.38 Inverse unloaded Q for a YIG resonator. (After R. J. Trew [61], 1979 IEEE, reprinted with permission.)

In the first frequency range—$f_c < f < 2f_c$—the threshold RF power is very low. The normal limiting levels for the low-level limiting are −25 to −15 dBm, and the YIG sphere is no longer usable as a high Q element. In this region the YIG can, however, be used as a limiter. The effective lower frequency limit of the pure YIG resonator is thus 3.3 GHz. Because the limiting frequency is dependent on the saturation frequency of the material (which is temperature variable), the frequency at which this type of operation may occur is also temperature dependent.

Figure 3.39 shows the limiting level and useful frequency range of operation as a function of the amount of gallium doping (saturation magnetization) of the material [62].

In the second frequency range—$f > 2f_c$, that is, $f > \frac{2}{3}\gamma \cdot 4\pi Ms$, which corresponds to $f > 3.3$ GHz for a pure YIG—the theoretical threshold power is more than 100 mW. For all practical purposes, to effectively use YIG as a resonator, this condition should be satisfied.

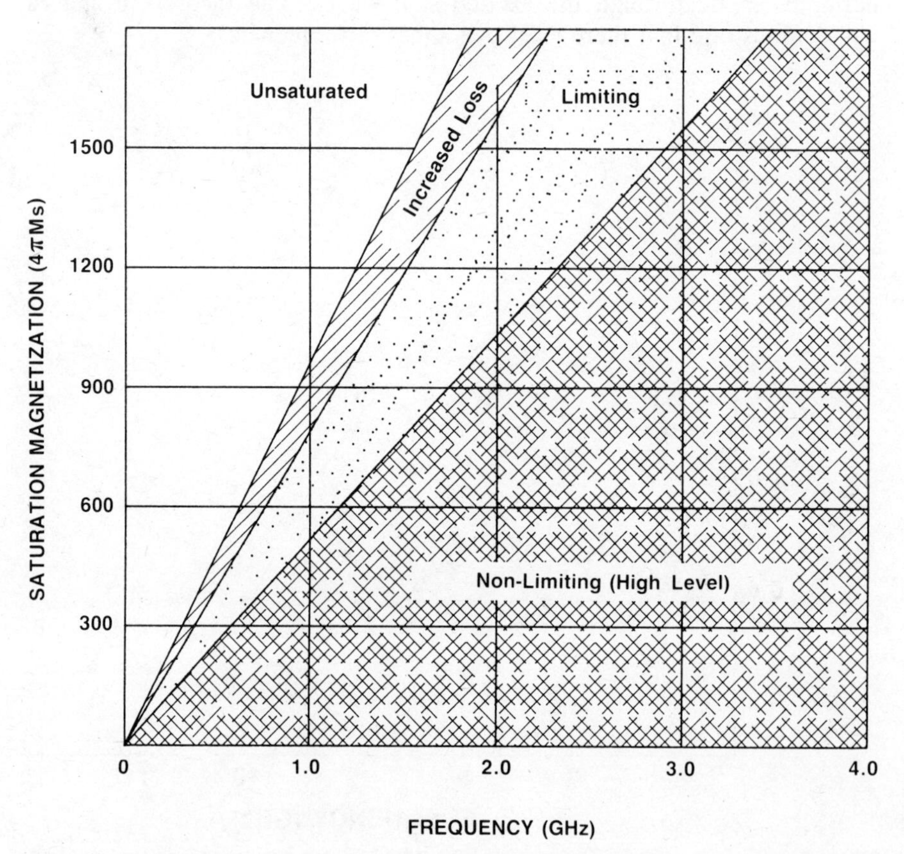

Figure 3.39 Frequency of operation vs. gallium doping. (In [64], reprinted with permission of Watkins and Johnson.)

3.7.3 Equivalent Circuit

In order to make the discussion of the YIG crystal properties more meaningful, it is worthwhile to examine the equivalent circuit of the YIG and its relation to the functioning of a microwave resonator. Figure 3.40(*a*) illustrates the important parameters of the loop-coupled YIG, and Fig. 3.40(*b*) shows the equivalent circuit. In Fig. 3.40(*a*), the YIG is located at the center of a loop of wire. The loop angle may be a full 360 degrees or, as suggested in Fig. 3.40(*a*), less. It can also be greater than 360 degrees in cases where especially tight coupling is required. Figure 3.40(*b*) is the equivalent circuit corresponding to Fig. 3.40(*a*). The parallel resonant circuit is induced in series with the loop impedance by the coupling of the YIG; L_L and R_L are the inductance and resistance, respectively, of the coupling loop.

The input impedance of this circuit is given by

$$Z_y = R_L + j\omega L_L + \frac{j\omega^3 L_y}{\omega_0^2 - \omega^2 + j(\omega \cdot \omega_0 / Q_u)} \tag{3.69}$$

where ω_0 is the resonant frequency of YIG and

$$Q_u = \frac{R_y}{\omega_0 L_y} = \frac{H_0}{\Delta H}$$

The equivalent circuit parameters are related to the basic YIG resonator and loop parameters as follows

$$L_y = \mu_0 V_y K^2 \omega_m \tag{3.70}$$

$$R_y = \omega_0 \omega_m \mu_0 V_y K^2 Q_u = \omega_0 L_y Q_u \tag{3.71}$$

$$C_y = \frac{1}{\omega_0^2 L_y} \tag{3.72}$$

where μ_0 is the permeability of free space, K is the coupling coefficient between the YIG sphere and the coupling loop, $\omega_m = \gamma \cdot 4\pi Ms$, and V_y is the volume of a YIG sphere of diameter D_y.

The coupling coefficient K is a geometric coupling factor and is related to the loop angle, θ, and the loop diameter, D_L, as follows

$$K = \frac{1}{D_L} \frac{\theta}{360} \tag{3.73}$$

The tightness of coupling between the loop and the YIG can also be expressed in terms of the external Q, Q_e of the resonator. If an external

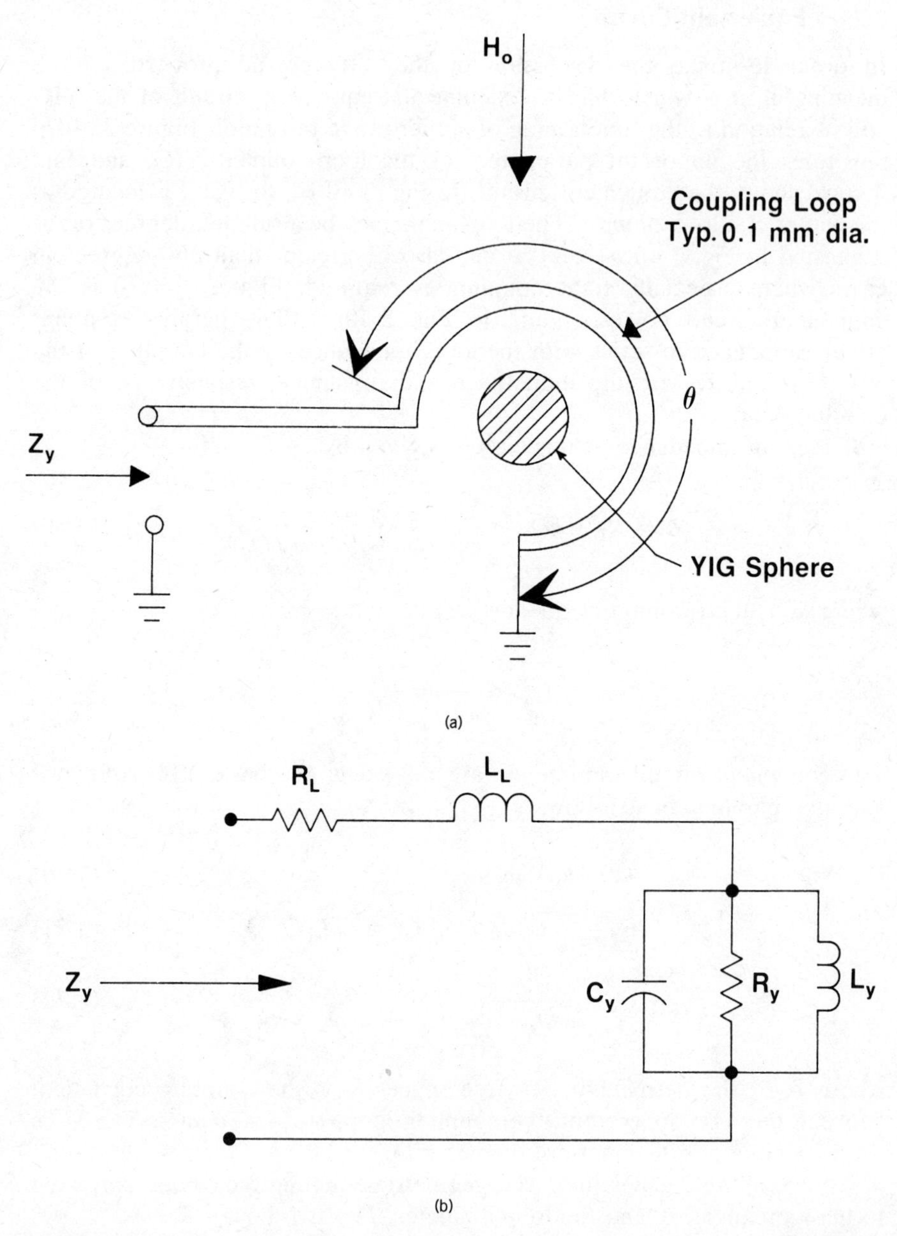

Figure 3.40 (*a*) Loop coupling circuit. (*b*) Equivalent circuit of loop-coupled YIG resonator.

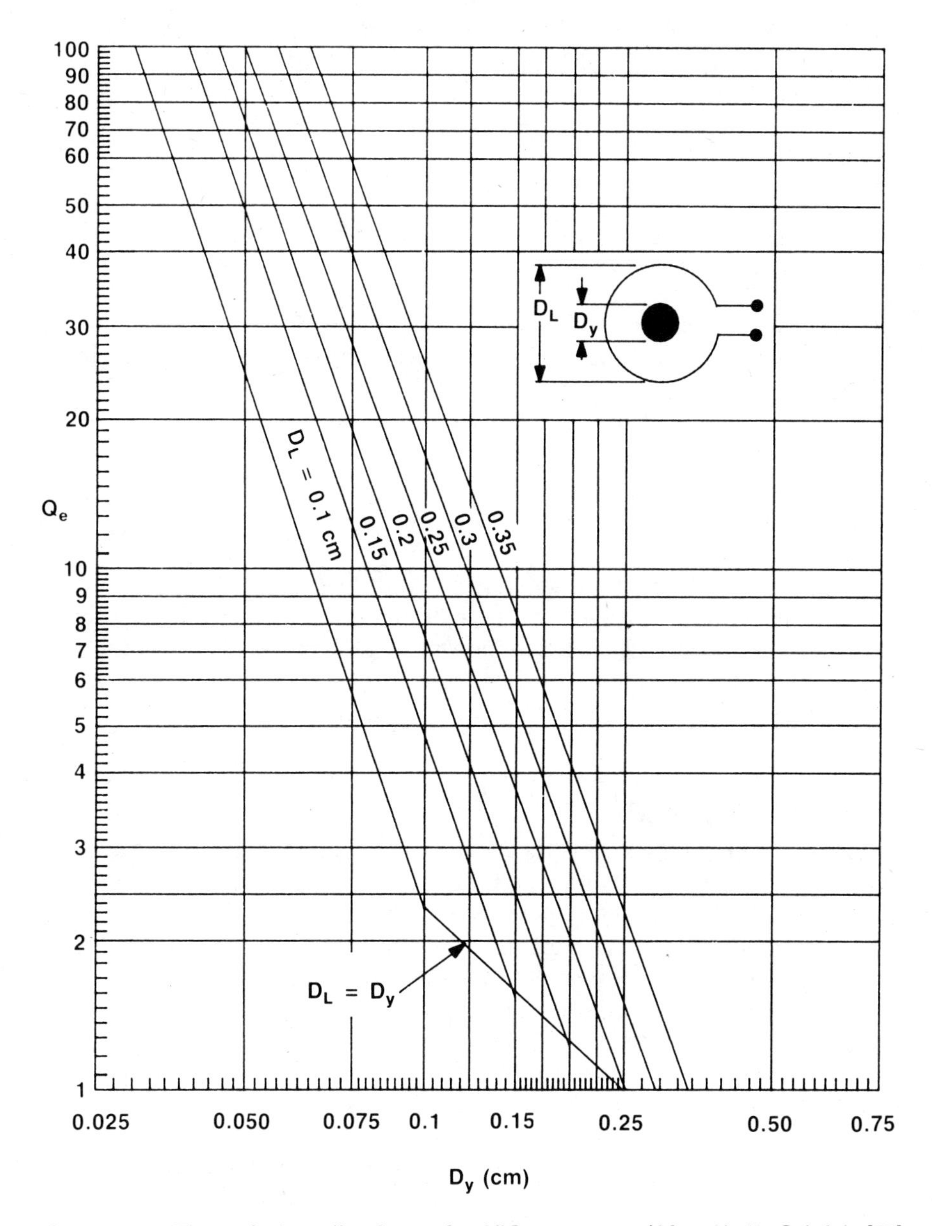

Figure 3.41 Theoretical quality factor for YIG resonator. (After N. K. Osbrink [57] reprinted with permission of MSN & Communications Technology.)

resistance R_0 is connected to the terminals of the loop, then Q_e is defined as

$$Q_e = \frac{R_0}{\omega_0 L_y} \tag{3.74}$$

and

$$R_y = \omega_0 L_y Q_u = \frac{R_0 Q_u}{Q_e} \tag{3.75}$$

Thus, the impedance level R_y of the parallel resonant circuit at resonance is proportional to two quantities that are easily obtained from a single bandwidth measurement.

Similarly, the inductance L_y is related to Q_e as follows

$$L_y = \frac{R_0}{\omega_0 Q_e} \tag{3.76}$$

YIG equivalent circuit parameters can thus be estimated quite easily and quickly from a knowledge of the external Q, Q_e. Figure 3.41 shows theoretical values of Q_e as a function of a YIG sphere diameter with the loop diameter as a parameter for the 360 degree loop case, and for a pure YIG ($4\pi Ms = 1780$ G) sphere.

3.7.4 Spurious Modes

One important problem in YIG device manufacture is the elimination of unwanted spurious oscillations and spectral line broadening effects due to the YIG. All YIG resonators pure or doped exhibit multiple resonance at a given magnetic bias. It is generally accepted that most of these spurious effects are due to the presence of magnetostatic modes [63].

The YIG resonance commonly used is the free precession or 110 mode called the "main" mode. The RF magnetization associated with this mode is uniform throughout the sphere. The "main" mode is only one of the many modes that tune linearily with the applied magnetic field strength, but the other linearly tuning modes are separated too far from the main mode to interfere with the frequency band of interest for all practical purposes. The higher order magnetostatic modes, tuning at rates that are different from the main mode, may have a resonant frequency, at the same applied magnetic field, within the normal useful range of the resonator. The RF magnetization associated with these modes is nonuniform throughout the sphere. Nonuniformities in the applied RF magnetic field and the magnetization interaction excite these modes and cause spurious resonances that couple and distort the main mode. The number of such modes in a given band is dependent upon the doping level of the material. For example, a pure YIG has many magnetostatic mode coincidences in the *C*-band range. It is therefore desirable to use a lower saturation magnetization GaYIG to minimize spurious resonances. In addition to spurious resonances, magnetostatic modes can cause detuning effects. When the higher order magnetostatic resonance is close to the main mode resonance, it causes the main resonant frequency to be pushed upward if the higher order mode frequency is the lower frequency and vice versa.

3.7.5 Magnetic Tuning Circuit

Design of the magnetic tuning circuit forms an important aspect of any YIG device design due to the fact that it is the value of the magnetic field that determines the resonance frequency of the YIG resonator. For the frequency to be maintained within 0.1% requires that the magnetic field be maintained within 0.1%. Likewise, a 0.1% tuning linearity requires that the magnetic field be linearly related to a control input to a maximum deviation of 0.1%.

The magnetic circuit most commonly used is shown in Fig. 3.42 [64].

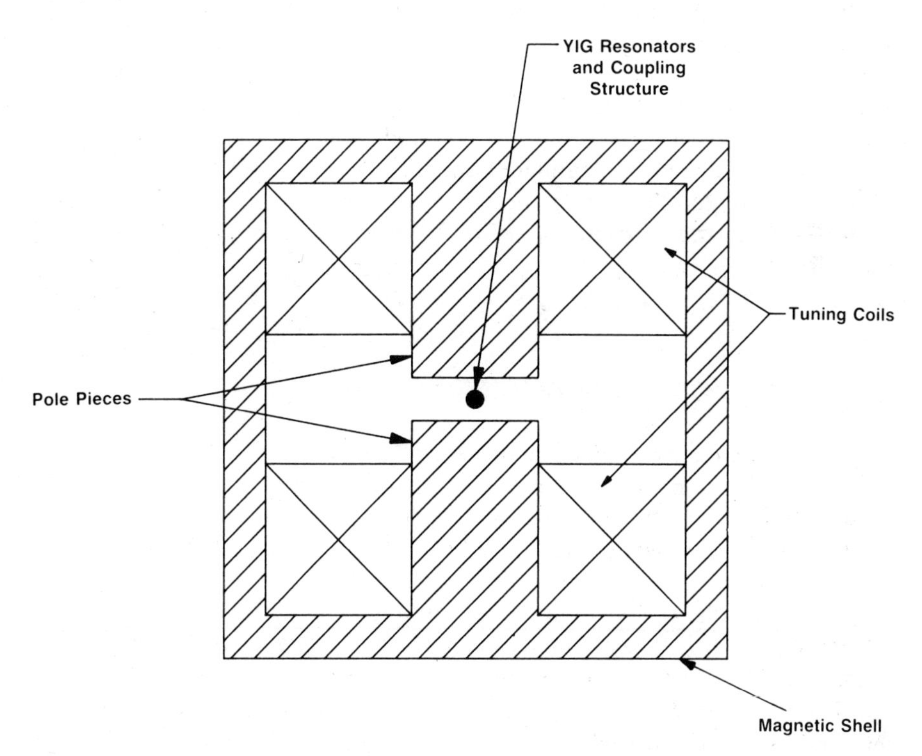

Figure 3.42 Reentrant magnetic structure. (In [64], reprinted with permission of Watkins and Johnson.)

This structure is known as the reentrant self-shielding magnetic circuit. It is made of soft magnetic material (nickel iron alloy, for example), with a high initial permeability and is fully annealed after machining to provide for minimum hysteresis. The pole tips are tapered in order to improve the magnetic design and to allow the placement of a larger heat sink close to the active device. Because of the high permeability of the material, the reluctance of the magnetic circuit is primarily in the air gap. Thus, temperature-dependent changes in permeability of the magnetic material have little effect on the magnetic field.

YIG magnetic tuning dissipates large dc power. Typically, it is about 5 W at 12 GHz, and the tuning sensitivity lies between 15 to 25 MHz/mA. The latter implies that the current supply must have very low noise and ripple to minimize FM noise. For comparison, if the inherent spectral linewidth is expected to be better than 10 KHz, the magnet power ripple should not exceed 0.5 μA.

In addition to the standard magnetic structure, previously described, two complementary types of structures are possible, permanent magnet and laminated magnet circuit. The permanent magnet type is attractive because it reduces the power required to tune the resonator to the low end or middle of its tuning range, and the electronic sweep adds or subtracts a magnetic field sufficient to tune the entire range of the resonator. A permanent magnet design also provides considerably faster switching speed, and its low drive coil dissipation, especially for less than one octave coverage, makes it particularly applicable for temperature stabilization in a thermal enclosure.

The third version, a laminated magnetic circuit has an even faster switching configuration. This design is particularly suitable for digital control where the current is switched from zero for each change in frequency.

3.8 RESONATOR MEASUREMENTS

Accurate characterization of microwave resonators is essential for their effective use. The important parameters that are required to fully describe a resonator for a given mode are the resonant frequency f_0,, the coupling coefficient, and the quality factors Q_u (unloaded Q), Q_L (loaded Q), and Q_e (external Q).

Figure 3.43 shows an experimental setup for the measurement of resonator parameters using a commonly available network analyzer. The network analyzer displays the magnitude and phase of the reflection and transmission coefficients as required, for the single-port or two-port resonators. Many methods for the Q measurement are possible, but we will describe here only one simple technique using Q loci on the Smith chart.

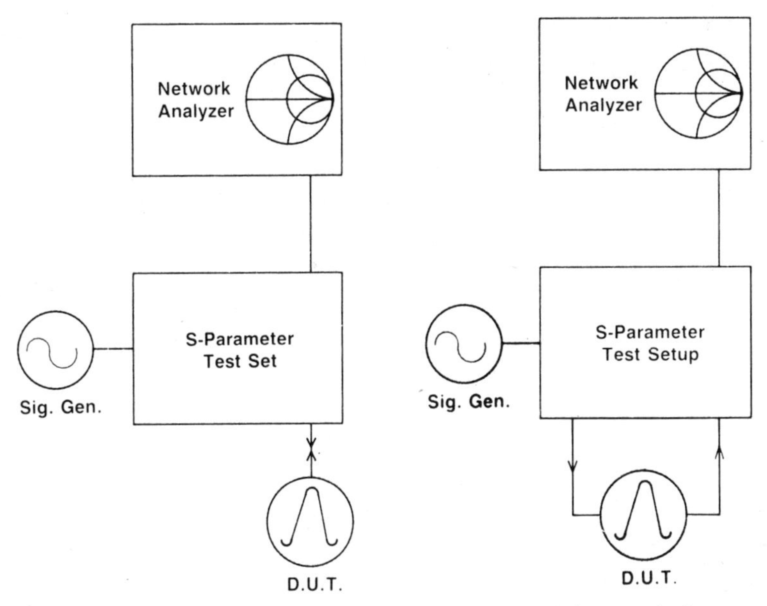

Figure 3.43 Measurement setups for (*a*) reflection, and (*b*) transmission resonators.

3.8.1 Single-Port Resonator

The single-ended resonator is the most commonly used configuration among microwave resonant circuits. The equivalent circuits of the two possible configurations are shown in Fig. 3.1, where R, L, and C are the equivalent lumped resistance, inductance, and capacitance. The parallel tuned circuit of Fig. 3.1(a) is known as the detuned short configuration, and the series-tuned circuit of Fig. 3.1(b) is known as the detuned open configuration. As shown, either configuration can be converted to the other by displacing the reference plane by quarter wavelength. The important parameters of these resonant circuits are defined by the following table:

Parameter	Series Tuned	Parallel Tuned
f_0	$\frac{1}{\sqrt{LC}}$	$\frac{1}{\sqrt{LC}}$
Q_u	$\frac{\omega L}{R}$	$\frac{R}{\omega L}$
β	$\frac{Z_0}{R}$	$\frac{R}{Z_0}$
Q_L	$\frac{Q_u}{1+\beta}$	$\frac{Q_u}{1+\beta}$

Further analysis of the resonant circuits being similar for the two configurations, we restrict our discussion to only the parallel tuned circuit. The input impedance of the circuit in Fig. 3.1(a) can be written as

$$\frac{1}{Z_{\text{in}}} = \frac{1}{R} + \frac{1}{j\omega L} + j\omega C \tag{3.77}$$

or

$$Z_{\text{in}} = \frac{R}{1 + 2jQ_u\delta} \tag{3.78}$$

where $\delta = (\omega - \omega_0)/\omega_0$ represents the frequency detuning parameter.

The locus of the impedance, using (3.78) can be drawn by varying δ [65]. As impedance is a linear function of frequency, a circular locus will be produced when plotted on the Smith chart as illustrated by circles A, B, and C in Fig. 3.44. Circle A for which $R = Z_0$ passes through the origin is called the condition of critical coupling ($\beta = 1$ from the preceding table), since it provides a perfect match to the transmission line at resonance. Circle C with $R > Z_0$ is said to be overcoupled ($\beta > 1$ from the preceding table), and circle B with $R < Z_0$ is undercoupled. The coupled coefficient for any given impedance locus can be easily determined by measuring the reflection coefficient S_{110} at resonance.

For the undercoupled case

$$\beta = \frac{1 - S_{110}}{1 + S_{110}} \tag{3.79}$$

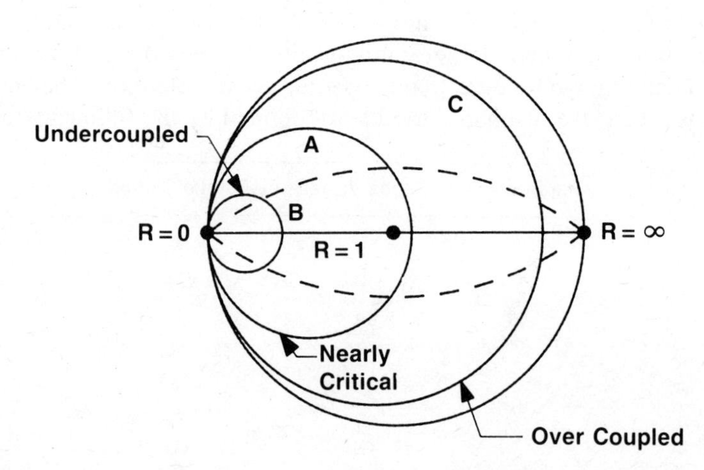

Figure 3.44 Input impedance of a resonant cavity referred to the detuned-short position plotted on the Smith chart for three degrees of coupling.

and for the overcoupled case

$$\beta = \frac{1 + S_{110}}{1 - S_{110}} \tag{3.80}$$

The evaluation of β locates the intersection of the impedance circle with the real axis, as shown in Fig. 3.45.

In order to measure various quality factors, (3.78) can be written as

$$\bar{Z}_{\text{in}} = \frac{Z_{\text{in}}}{Z_0} = \frac{\beta}{1 + 2jQ_u\delta} = \frac{\beta}{1 + 2jQ_L(1+\beta)} = \frac{\beta}{1 + 2jQ_e\beta} \tag{3.81}$$

where Q_u, Q_L, and Q_e are interrelated by the well-known relation

$$Q_u = Q_L(1+\beta) = Q_e\beta \tag{3.82}$$

The normalized frequency deviations corresponding to various quality factors are given by

$$\delta_u = \pm\frac{1}{2Q_u}, \qquad \delta_L = \pm\frac{1}{2Q_L}, \qquad \text{and} \qquad \delta_e = \pm\frac{1}{2Q_e} \tag{3.83}$$

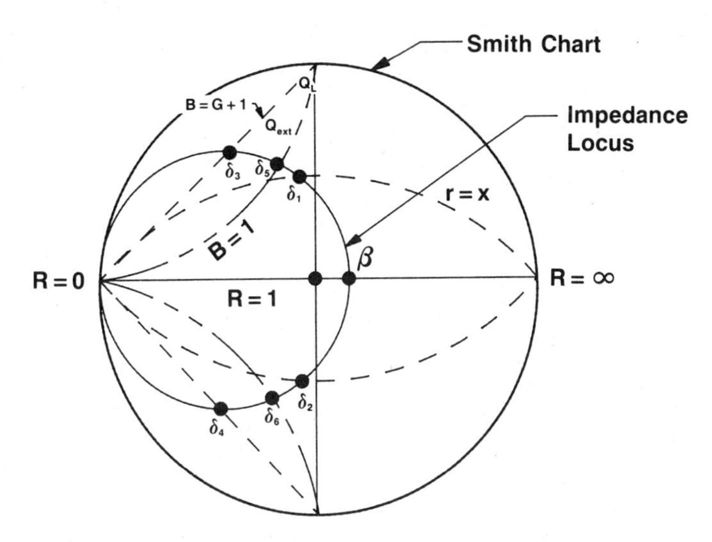

Figure 3.45 Identification of the half power points from the Smith chart. Q_0 locus is given by $X = R(B = G)$; Q_L by $X = R + 1$; Q'_{ext} by $X = 1$.

The impedance locus of Q_u, for example, can be determined by using (3.83) in (3.81) and is given by

$$(Z_{\text{in}})_u = \frac{\beta}{1 \pm j} \tag{3.84}$$

Equation 3.84 represents the points on the impedance locus where the real and imaginary parts of the impedance are the same. Figure 3.45 represents the locus of these points (corresponding to $R = X$) for all possible values of β. This locus is an arc whose center is at $Z = 0 \pm j$ and the radius is the distance to the point $0 \pm j0$. The intersection of this arc with the impedance locus determines the Q_u measurement points

$$Q_u = \frac{f_0}{f_1 - f_2} \tag{3.85}$$

The frequencies f_1 and f_2 are called half-power points, because these points correspond to $R = X$ on the impedance locus.

The loaded and external Q values can be determined in a similar way. From (3.81) and (3.83), the impedances corresponding to Q_e and Q_L are given by

$$(Z_{\text{in}})_e = \frac{\beta}{1 \pm j\beta} \tag{3.86}$$

and

$$(Z_{\text{in}})_L = \frac{\beta}{1 \pm j(1 + \beta)} \tag{3.87}$$

Using (3.86) and (3.87), the Q_e and Q_L loci can be easily determined. These loci are shown in Fig. 3.45.

3.8.2 Two-Port Resonator

Two-port resonant circuits are commonly used as transmission components in a number of applications. The equivalent circuit of a commonly used two-port resonator is shown in Fig. 3.46(a). The input and output coupling coefficients are represented by β_1 and β_2 in

$$\beta_1 = \frac{Y_{01}}{n_1^2 G} \quad \text{and} \quad \beta_2 = \frac{Y_{02}}{n_2^2 G} \tag{3.88}$$

where Y_{01} and Y_{02} are the input and output transformed admittances. The coupling coefficients can be directly determined by measuring the VSWR at

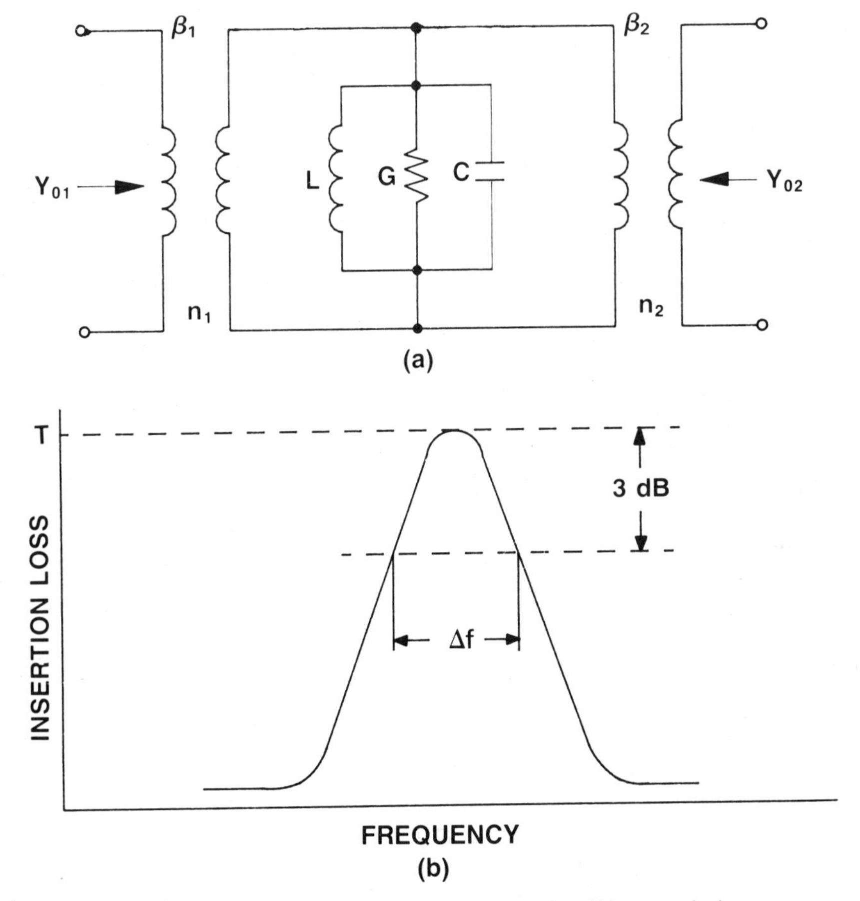

Figure 3.46 (*a*) Equivalent circuit of a two-port resonator, (*b*) transmission response of a two-port resonator.

the input and output port with the other port open circuited. The transmission response of such a resonant circuit measured using the setup of Fig. 3.44(*b*), is shown in Fig. 3.46(*b*). The coupling coefficients and the quality factors can be determined from the measurement of the insertion loss T at resonant frequency and the 3-dB bandwidth Δf using the following well-known relations [66]

$$T = \frac{2\sqrt{\beta_1 \beta_2}}{1 + \beta_1 + \beta_2} \tag{3.89}$$

$$Q_L = \frac{f_0}{\Delta f} \tag{3.90}$$

$$Q_u = Q_L(1 + \beta_1 + \beta_2) \tag{3.91}$$

REFERENCES

1. Collin, R. E., *Foundations of Microwave Engineering*, McGraw-Hill, New York, 1966, Chap. 7.
2. Ramo, S., J. Whinnery, and T. VanDuzer, *Fields and Waves in Communication Electronics*, Wiley, New York, 1965, Chap. 10.
3. Ghose, R. N., *Microwave Circuit Theory and Analysis*, McGraw-Hill, 1963, Chap. 8.
4. Moreno, T., *Microwave Transmission Design Data*, Dover, New York, 1958, Chap. 18.
5. Liao, S. Y., *Microwave Devices and Circuits*, Prentice-Hall, Englewood Cliffs, N.J., 1980, Chap. 4.
6. Kretzschmar, J. G., "Wave Propagation in Hollow Conducting Elliptic Waveguide," *IEEE Trans. Microwave Theory Tech.*, Vol. MTT-18, Sept. 1970, pp. 547–554.
7. Kretzschmar, J. G., "Mode Chart for Elliptical Resonant Cavities," *Electron. Lett.*, Vol. 6, July 1970, pp. 432–433.
8. Rengarajan, S. R., and J. E. Lewis, "Quality Factor of Elliptical Cylindrical Resonant Cavities," *J. Microwave Power*, Vol. 15, Jan. 1980, pp. 53–57.
9. Itoh, T., R. Mittra, and R. D. Wards, "A Method for Computing Edge Capacitance of Finite and Semi-Finite Microstrip Lines," *IEEE Trans. Microwave Theory Tech.*, Vol. MTT-20, Dec. 1972, pp. 847–849.
10. James, D. S., and S. H. Tse, "Microstrip End Effects," *Electron. Lett.*, Vol. 8, Jan. 1972, pp. 46–47.
11. Garg, R., and I. J. Bahl, "Microstrip Discontinuities," *Int. J. Electron.*, Vol. 45, Jan. 1978, pp. 81–87.
12. Carlile, D., T. Itoh, and R. Mittra, "A Study of Rectangular Microstrip Resonators," *AEU*, Vol. 30, Jan. 1976, pp. 38–41.
13. Itoh, T. "Analysis of Microstrip Resonators," *IEEE Trans. Microwave Theory Tech*, Vol. MTT-22, Nov. 1974, pp. 946–952.
14. Itoh, T., and R. Mittra, "Analysis of Microstrip Disk Resonators," *AEU*, Vol. 27, Nov. 1973, pp. 456–458.
15. Sharma, A. K., and B. Bhat, "Influence of Shielding on the Capacitance of Microstrip Disk and Ring Structures," *AEU*, Vol. 34, Jan. 1980, pp. 41–44.
16. Wolff, I., and N. Knoppik, "Rectangular and Circular Microstrip Disk Capacitors and Resonators," *IEEE Trans. Microwave Theory Tech.*, Vol. MTT-22, Oct. 1974, pp. 857–864.
17. Watkins, J., "Circulator Resonator Structures in Microstrip," *Electron. Lett.*, Vol. 5, Oct. 1969, pp. 524–525.
18. Entschladen, H., "Simple Design of Microstrip Disk Resonators," *AEU*, Vol. 29, 1975, pp. 184–185.
19. Troughton, P., "Measurement Techniques in Microstrip," *Electron. Lett.*, Vol. 5, Jan. 1969, pp. 25–26.
20. Wolff, I., and N. Knoppik, "Microstrip Ring Resonator and Dispersion Measurement on Microstrip Line," *Electron. Lett.*, Vol. 7, 1971, pp. 779–781.

21. Sharma, A. K., and B. Bhat, "Spectral Domain Analysis of Microstrip Ring Resonators," *AEU*, Vol. 33, Mar. 1979, pp. 130–132.
22. Wu, Y. S., and F. J. Rosenbaum, "Mode Chart for Microstrip Ring Resonators," *IEEE Trans. Microwave Theory Tech.*, Vol. MTT-21, July 1973, pp. 487–489.
23. Helszajn, J., and D. S. James, "Planar Triangular Resonators with Magnetic Walls," *IEEE Trans. Microwave Theory Tech.*, Vol. MTT-26, Feb. 1978, pp. 95–100.
24. Helszajn, J., D. S. James, and W. T. Nisbet, "Circulators Using Planar Triangular Resonators," *IEEE Trans. Microwave Theory Tech.*, Vol. MTT-27, Feb. 1979, pp. 188–193.
25. Cuhaci, M., and D. S. James, "Radiation from Triangular and Circular Resonators in Microstrip," presented at the IEEE Int. Microwave Symp., June 1977, Paper D 5.2.
26. Schelkunoff, S. A., *Electromagnetic Waves*, Van Nostrand, New York, 1943, p. 393.
27. Ng, F. L., "Tabulation of Method for the Numerical Solution of the Hollow Waveguide Problem," *IEEE Trans. Microwave Theory Tech.*, Vol. MTT-22, Mar. 1974, pp. 322–329.
28. Thomas, D. T., "Functional Approximation for Solving Boundary Value Problems by Computer," *IEEE Trans. Microwave Theory Tech.*, Vol. MTT-17, Aug. 1969, pp. 447–454.
29. Sharma, A. K., "Spectral Domain Analysis of Microstrip Resonant Structures," Ph.D. Thesis, Indian Institute of Technology, Delhi, Dec. 1979.
30. Sharma, A. K., and B. Bhat, "Analysis of Triangular Microstrip Resonators," *IEEE Trans. Microwave Theory Tech.*, Vol. MTT-30, Nov. 1982, pp. 2029–2031.
31. Nisbet, W. T., and J. Helszajn, "Mode Chart for Microstrip Resonators on Dielectric and Magnetic Substrates Using a Transverse Resonance Method," *Microwaves, Opt., Acoust.*, Vol. 3, Mar. 1979, pp. 69–77.
32. Hsu, J. P., O. Kondo, T. Anada, and H. Makino, "Measurement and Calculation of Eigen-Values of Various Triplate-Type Microwave Planar Circuits," Rec. of Professional Groups, IECEJ, Paper MW 73-117, Feb. 23, 1974.
33. Sharma, A. K., and W. J. R. Hoefer, "Spectral Domain Analysis of a Hexagonal Microstrip Resonator," *IEEE Trans. Microwave Theory Tech.*, Vol. MTT-30, May 1982, pp. 825–828.
34. Kretzschmar, J. G., "Theoretical Results for the Elliptic Microstrip Resonator," *IEEE Trans. Microwave Theory Tech.*, Vol. MTT-20, May 1972, pp. 342–343.
35. Kretzschmar, J. G., "Theory of the Elliptic Disk and Ring Resonator," *Proc. 5th Coll. Microwave Commun.*, Budapest, June 1974.
36. McLachlan, N. W., *Theory and Application of Matthieu Functions*, Oxford Univ. Press, London, 1951.
37. Sharma, A. K., and B. Bhat, "Spectral Domain Analysis of Elliptic Microstrip Disk Resonators," *IEEE Trans. Microwave Theory Tech.*, Vol. MTT-28, June 1980, pp. 573–576.

38. Sharma, A. K., "Spectral Domain Analysis of an Elliptic Microstrip Ring Resonator," *IEEE Trans. Microwave Theory Tech.*, Vol. MTT-32, Feb. 1984, pp. 212–218.
39. Sharma, A. K., and B. Bhat, "Spectral Domain Analysis of Discontinuity Microstrip Structures," presented at the 1981 National Radio Science Meeting, Los Angeles, Calif., June 15–19, 1981.
40. Sharma, A. K., and B. Bhat, "Analysis of Interacting Rectangular Microstrip Resonators," presented at the Int. Electrical, Electronics Conf. and Exposition, Toronto, Canada, Oct. 5–7, 1981.
41. Sharma, A. K., and B. Bhat, "Analysis of Microstrip Resonant Structures," presented at the Int. Symp. Microwaves and Communication, Kharagpur, India, Dec. 29–30, 1981, Paper MN 3.7.
42. Sharma, A. K., and B. Bhat, "Spectral Domain Analysis of Interacting Microstrip Resonant Structures," *IEEE Trans. Microwave Theory Tech.*, Vol. MTT-31, Aug. 1983, pp. 681–685.
43. Solbach, K., "The Status of Printed Millimeter-Wave E-Plane Circuits," *IEEE Trans. Microwave Theory Tech.*, Vol. MTT-31, Feb. 1983, pp. 107–121.
44. Sharma, A. K., and W. J. R. Hoefer, "Evaluation of Resonant Frequency of a Rectangular Slot Resonator in Fin Line," *Proc. 1982 IEEE Int. Antenna and Propagation Symp.*, 644–657.
45. Sharma, A. K., "Analysis of Interacting Resonant Structure in Fin Lines," *Proc. 1984 IEEE Int. Antenna and Propagation Symp.*, 376–379.
46. Knorr, J. B., "Equivalent Reactance of a Shorting Septum in a Fin-Line: Theory and Experiments," *IEEE Trans. Microwave Theory and Tech.*, Vol. MTT-29, Nov. 1981, pp. 1196–1202.
47. Plourde, J. K., and C. L. Ren, "Application of Dielectric Resonators in Microwave Components," *IEEE Trans. Microwave Theory Tech.*, Vol. MTT-29, Aug. 1981, pp. 754–770.
48. Ready, D. W., et al., "Microwave High Dielectric Constant Materials," Quarterly Repts. 1–4, Final Rep., Contract DAAB07-69-C-0455, Raytheon Company, Waltham, Mass., Nov. 1969–Feb. 1971.
49. Masse, D. J., et al., "A New Low-Loss High-K Temperature Compensated Dielectric for Microwave Applications," *Proc. IEEE*, Vol. 59, Nov. 1971, pp. 1628–1629.
50. Simonet, W., and J. C. Mage, "(Zr, Sn) TiO_4 Dielectric Materials with Adjustable Temperature Coefficients and Low Microwave Loss," presented at the American Ceramic Society Meet., Chicago, ITT., Apr. 29, 1980.
51. Montvala, A. J., "Microwave Dielectric Constant Materials," Quarterly Reports 1–3, Contract DAAB07-69-C-0402, Ill Research Institute, Chicago, Ill., July 1969–Jan. 1970.
52. Kajfez, D., and P. Guillon, *Dielectric Resonators*, Artech House, Dedham, Mass., 1986.
53. Guillon, P., et al., "Microstrip Bandstop Filter Using a Dielectric Resonator," *Proc. IEE*, Vol. 128, Number 3, June 1981, pp. 151–154.
54. Khanna, A. P. S., and Y. Garault, "Determination of Loaded, Unloaded and External Quality Factors of a Dielectric Resonator Coupled to a Microstrip

Line," *IEEE Trans. Microwave Theory Tech.*, Vol. MTT-31, Mar. 1983, pp. 261–264.

55. Guillon, P., private communication.
56. Gil, F. H., and J. P. Martinez, "Analysis of Dielectric Resonators with Tuning Screw and Supporting Structure," *IEEE Trans. Microwave Theory Tech.*, Vol. MIT-33, Dec. 1985, pp. 1453–1457.
57. Osbrink, N. K., "YIG-Tuned Oscillator Fundamentals," *Microwave Systems News*, Vol. 13, Nov. 1983, pp. 207–225.
58. Matthaei, G. L., L. Young, and E. M. T. Jones, *Microwave Filters, Impedance Matching Networks and Coupling Structures*, McGraw-Hill, New York, 1964.
59. Lax, B., and K. J. Button, *Microwave Ferrites and Ferromagnetics*, McGraw-Hill, New York, 1962, pp. 145–196.
60. Ollivier, P. M., "Microwave YIG-Tuned Transistor Oscillator Amplifier Design Application to C-Band," *IEEE J. Solid-State Circuits*, Vol. SC-7, Feb. 1972, pp. 54–60.
61. Trew, R. J., "Design Theory for Broad-Band YIG-Tuned FET Oscillators," IEEE Trans. *Microwave Theory Tech.*, Vol. MTT-27, Jan. 1979, pp. 8–14.
62. Aldecoa, J. A., and F. J. Bell, "Application Note YIG RF Components," YIG-TEK Corporation, Santa Clara, Cal., Sept. 1972.
63. Fletcher, P. C., and R. O. Bell, "Ferromagnetic Resonance Modes in Sphere," *J. Appl. Phys.*, Vol. 30, May 1959, pp. 687–698.
64. Watkins Johnson Technotes, "YIG-Tuned Integrated Devices," Vol. 4, No. 5 Sept./Oct. 1977, and "Advanced Product Technical Description Tunable YIG Discriminators," Feb. 1969, Watkins Johnson, Palo Alto, Cal.
65. Ginzton, E. L., *Microwave Measurements*, McGraw-Hill, New York, 1957, Chap. 9.
66. Sucher, M., and J. Fox, *Handbook of Microwave Measurements*, 3rd ed., Vol. 2, Wiley, New York, 1963, Chap. 8.

PROBLEMS

3.1 Identify the degenerate modes of a rectangular waveguide resonator when

(a) all sides are unequal,
(b) two sides are equal,
(c) all sides are equal.

3.2 Determine the lowest frequency at which degenerate modes exist in a cylindrical waveguide cavity.

3.3 Calculate the resonant frequency of a radial reentrant coaxial cavity having inner and outer radii of 0.5 and 1.2 cm and length of 10 cm when the center post is 5 cm from the end wall. Determine the tuning range when d is varied from 2 to 6 cm.

3.4 Find the resonant frequency and Q for a TE_{111} mode cylindrical cavity (copper) of radius 2 cm and length 10 cm. Repeat this for the case when the cavity is filled with material of dielectric constant 4.0.

3.5 Calculate the lowest order resonant frequency of a microstrip resonator having $\epsilon_r = 10$, $h = 0.6$ mm, $t = 6$ μm, $W = 2$ mm, and $L = 1$ cm.

3.6 Derive expressions for Q factors of rectangular and circular patch microstrip resonators.

3.7 Calculate and compare the quasi-static resonant frequencies of circular, triangle, and hexagonal microstrip resonator with identical dimension a and substrate thickness h.

3.8 Design a $TE_{01\delta}$ mode cylindrical dielectric resonator at 35 GHz when placed in the shielded MIC package shown in Fig. 3.29, using following parameters.

Substrate: $\epsilon_s = 9.9$, height $h = 0.25$ mm,
D.R.: $\epsilon_r = 36$, $D/H = 2.5 \pm 0.1$,
Distance: d_s: 1 mm.

3.9 In Problem 3.8, find the change in the resonant frequency if the dielectric constant of the dielectric resonator is changed to 38 from 36.

3.10 The equivalent circuit of a dielectric resonator coupled to a microstrip line is shown in Figs. 3.31 and 3.32. Find the relations, in terms of S_{110} and S_{210}, to draw the loci of the unloaded, loaded and external quality factors determination points (Q circles) in the S_{11} and S_{21} planes at PP'. Draw the Q circles on a Smith chart and a Polar chart from the S_{11} and S_{21} planes respectively.

3.11 In the case of a parallel-tuned resonant circuit, 40% of the incident power is reflected at the resonant frequency. The real part (R) equals imaginary part (X) on the impedance locus at 9230 and 9240 MHz. Calculate the resonant frequency, the coupling coefficient, and the loaded, unloaded, and external quality factors for an undercoupled case and an overcoupled case.

4 IMPEDANCE-MATCHING NETWORKS

4.1 INTRODUCTION

An impedance-matching network is perhaps the most extensively used circuit element in designing a microwave component or a subsystem. For optimum power transfer efficiency in the passband it is desirable to match impedances of sections connected together in a passive or an active circuit.

Almost all impedance-matching networks can be grouped under two categories, impedance transformers and filters. Some networks are, however, synthesized using the reverse and forward transfer coefficient characteristics of the device. For example, a feedback amplifier matching network is designed between the output and input terminals. This network utilizes inherent reverse transmission coefficient of the field-effect transistor.

Section 4.1 discusses the single-frequency parameters for two-port networks and use of transmission lines as matching elements. Ladder networks and their approximate solutions are described in Section 4.2. Impedance transformations are used to adjust the terminating impedance values and to obtain realizable elements. These are discussed in Section 4.3 along with analytical limitations imposed by the gain-bandwidth theory. Section 4.4 describes the synthesis of a matching network and selection of topologies for a network.

4.1.1 Importance and Applications

Passive as well as active microwave circuits require impedance matching of complex loads with reactive constraints. For example, a microwave system may require broadband matching of an antenna, while a microwave module containing amplifiers, mixers, oscillators, switches, etc., will require a semiconductor device to be matched to the system impedance level. Single-frequency or narrow-band (less than 10% typical) impedance matching is simply achieved using a single fixed tuned network or, at most, a two-element network depending upon the Q of the structure. Broadband matching network design, particularly for bandwidths greater than 50%, is a difficult and challenging task due to prescribed reactive constraints, broadband impedance transformations of large impedance ratios, and prescribed tapered magnitude characteristics. The latter is particularly true in wide-band amplifier design where it is necessary to compensate for the inherent transistor gain rolloff with frequency.

4.1.2 One-Port and Two-Port Networks

The usual impedance transformation problem is considered to be one-port in nature where a complex or a real load is to be matched to a real source impedance as shown in Fig. 4.1. Fano's results [1], which are discussed in detail in Section 4.3.2, show that an ideal impedance-matching network for a complex load would be a bandpass filter that cuts off sharply at the band edges. Alternatively, a low-pass or a high-pass filter design can be used to match a load impedance. Low-pass filter structures can provide a relatively broadband impedance match for microwave loads that can be approximated by an inductance and resistance in series, or by a capacitance and conductance in parallel. Similarly, high-pass networks may be used for loads that can be approximated by a capacitance and resistance in series or by an inductance and conductance in parallel. But both low-pass and high-pass networks are suitable for limited applications compared with bandpass networks because of some inherent disadvantages. For example, very few circuits really need a good match over a band extending from dc to microwave frequencies. Additionally, the choice of driving-point impedance in low-pass (high-pass) filter design is limited by R–L (R–C) or G–C (L–G) in the load circuit.

A more generalized matching network synthesis is based on the network theory. Here the device is represented by a two-port (input and output) Y-, Z-, or scattering coefficient (S) matrix. The increased complexity of matching network design based on active two-port is a necessary evil in several cases because of the coupling of the input and output circuits through the device. For example, in an amplifier design the input-matching network is dependent on the output-matching network (Fig. 4.2), and vice versa. The device parameters may, however, be measured under actual operating conditions for a meaningful matching network design. Active

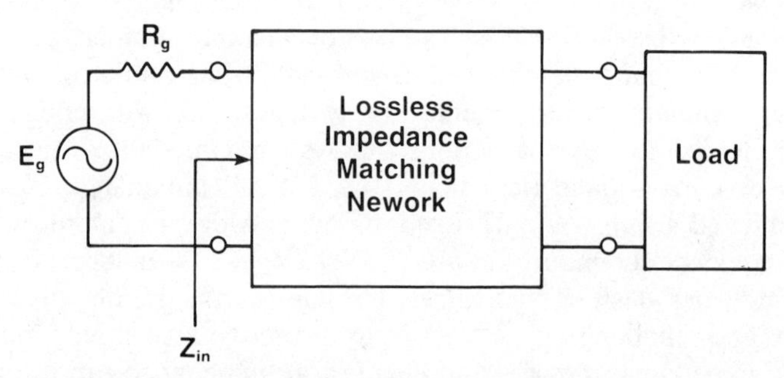

Figure 4.1 A one-port impedance-matching network problem.

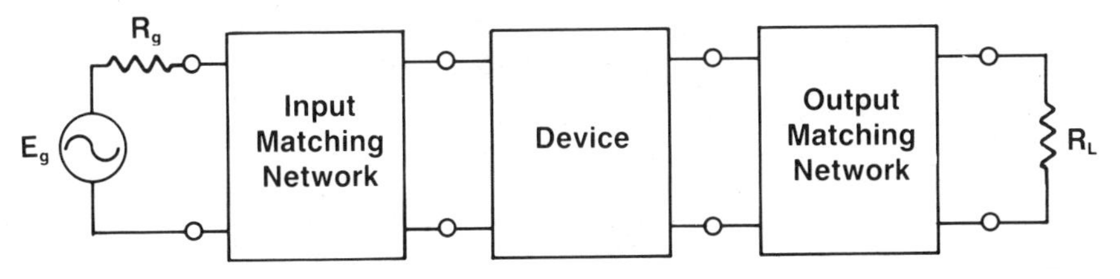

Figure 4.2 Two-port-matching networks.

devices or microwave networks, requiring impedance match, can be completely characterized in terms of parameters measured at the network terminals without regard to the contents of network. Although a particular device may have any number of ports, the two-port representation is applicable only between the input and the output ports, while all the other ports are terminated in known impedances. For example, a two-port representation of a three-terminal FET may characterize the device behavior between the gate as input and drain as the output port, while the source is assumed to be RF grounded.

4.1.3 Transmission-Line Matching Schemes

This section reviews the commonly used impedance transforming properties of transmission lines for designing narrow- as well as broadband matching networks. A standard transmission line, such as a microstrip, can be used as a series transmission line, as an open-circuited or a short-circuited stub, and as a quarter-wavelength transformer section for impedance matching.

Series Single Section. Simplest of all matching networks is a transmission line of electrical length θ and characteristic impedance Z_0 used to match a complex load $(R_L + jX_L)$ to real resistance R. The required transmission-line parameters are given as [2]

$$Z_0 = \sqrt{\frac{RR_L - (R_L^2 + X_L^2)}{(1 - R_L/R)}} \tag{4.1}$$

and

$$\tan\theta = \frac{\sqrt{(1 - R_L/R)(RR_L - [R_L^2 + X_L^2])}}{X_L} \tag{4.2}$$

It is a narrow-band matching technique with limited use, since only those impedances can be matched that result in a real value of Z_0 in (4.1).

In a practical design, the characteristic impedance of the transformer is limited by the type of transmission line used. For example, the impedance values lie between 20 and 100 Ω when microstrip or striplines are used.

Using (4.1) and (4.2) one can show that with these restrictions on the impedance values, tuning can be done over approximately 22% of the Smith chart. This range can be enhanced by using two transmission lines in series [3]. One section brings the impedance into the matchable area and the other one matches it to the source impedance. The two-line section transformer with characteristic impedances limited within 20 and 100 Ω can match loads over 75% of the Smith chart.

Expressions for the design of transmission-line transformers between complex loads and complex source impedances, for maximum power transfer or a conjugate match, have also been given [4]. These are

$$Z_0 = \left[\frac{R_S \cdot |Z_L|^2 - R_L \cdot |Z_S|^2}{R_L - R_S}\right]^2 \tag{4.3}$$

and

$$\tan\theta = \frac{Z_0 \cdot (R_L - R_S)}{R_L \cdot X_S - R_S \cdot X_L} \tag{4.4}$$

where $Z_L = R_L + jX_L$ is the complex load impedance and $Z_S = R_S + jX_S$ is the complex source impedance.

Stub Matching. Alternatively, single-stub or multiple-stub tuners are used to match complex impedances, such as in a feeder connection between a transmitter and an antenna or between the generator and the input of an active device or output of the device and the load. Double-stub matching is needed in cases where it is impractical to place a single stub physically in the ideal location. Location and length of open-circuited or short-circuited stubs is easily calculated using graphical techniques on a Smith chart [5]. The stub-tuned matching circuits can, of course, be designed into a microstrip or strip-line circuit. Once again, the desired impedance matching of a complex load is achieved over a narrow bandwidth.

Single-Stub Matching. Since the stub-matching problems involve parallel connections on the transmission line, it is easier to design the circuit using admittance values rather than impedances. In a single-stub-matching network, the total admittance of the terminated line and the stub is matched with the admittance of the source for maximum power transfer. For example, in Fig. 4.3, the stub is located at AA in such a way that

$$Y_{AA} = Y_0 = Y_S + Y_D \tag{4.5}$$

where Y_S is the admittance of the stub (short circuited or open circuited) of length l, and Y_D is the admittance of the load transformed at AA location.

Double-Stub Matching. A double-stub-matching network consists of two stubs (short-circuited sections preferred, because they are easier to obtain

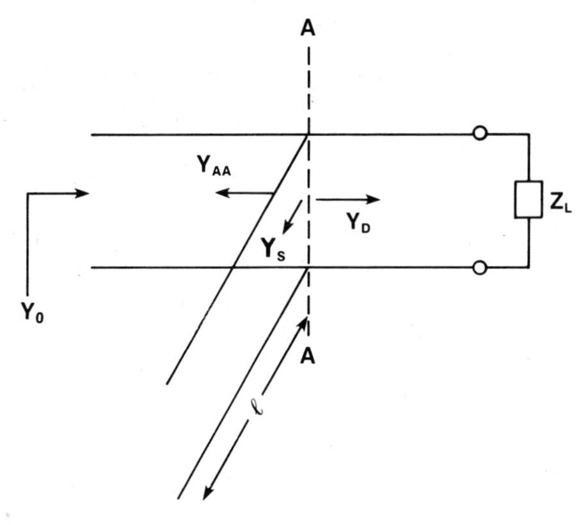

Figure 4.3 Single-stub matching.

than a good open circuit) connected in parallel with a fixed length between them (Fig. 4.4). Usually the length of the transmission line between the stubs is $\frac{1}{8}$, $\frac{3}{8}$, or $\frac{5}{8}$ of a wavelength. Lengths and impedances of both stubs and the location of the stub closest to the load can be adjusted to get a perfect match. In Fig. 4.4,

$$Y_{BB} = Y_{S2} + Y_{D2} = Y_0 \tag{4.6}$$

where Y_{BB} is total admittance at the left of the second stub, Y_{S2} is the admittance of the second stub, Y_{D2} is the admittance Y_{AA} $(= Y_{S1} + Y_{D1})$ at

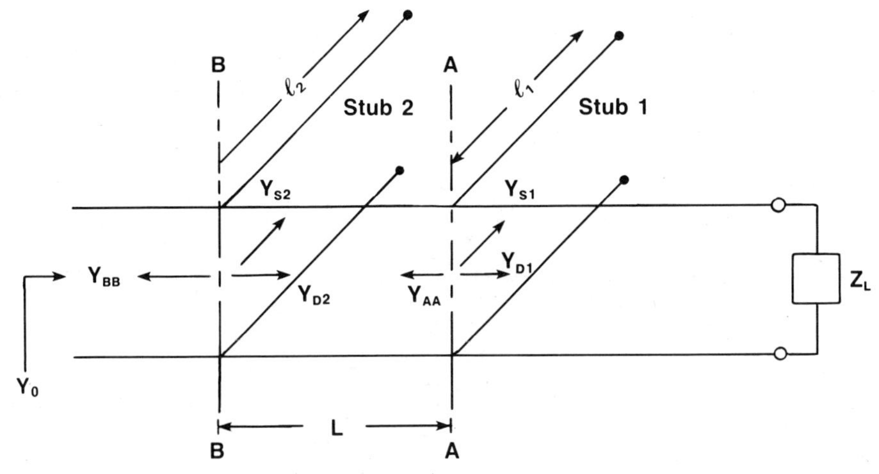

Figure 4.4 Double-stub matching.

location AA transformed to location BB by the line section between the two stubs, Y_{S1} is the admittance of stub one, and Y_{D1} is the load admittance transformed to the location of the first stub. There is no definite procedure for solving a double-stub-matching problem. One can go through an iterative design, starting with an arbitrary location for the first stub (closest to the load).

Quarter-Wavelength Transformer. Quarter-wavelength multisection impedance transformers are commonly used to match real impedances over wide bandwidths. The reflection coefficient at the input of the matching network is given by a sum of the series of multiple partial reflections arising at each of the impedance discontinuities. For small reflections, it can be approximated to be [6]:

$$\Gamma = |\Gamma_1| + |\Gamma_2|e^{-j2\theta} + |\Gamma_3|e^{-j4\theta} \cdots |\Gamma_{N+1}|e^{-j2N\theta} \tag{4.7}$$

where Γ_i is the first-order reflection at the ith discontinuity in an N-section transformer as shown in Fig. 4.5. For a transformer designed with symmetrical steps ($|\Gamma_i| = |\Gamma_{N+2-i}|$), the expression in (4.7) is simplified to be

$$\Gamma = 2e^{-jN\theta}[\rho_1 \cos N\theta + \rho_2 \cos(N-2)\theta + \cdots + A] \tag{4.8}$$

where

$$\rho_i = |\Gamma_i|$$

and

$$A = \begin{cases} \rho_{(N+1)/2} \cos \theta & N \text{ odd} \\ \frac{1}{2}\rho_{(N/2)+1} & N \text{ even} \end{cases}$$

A minimum passband VSWR for a given input–output impedance ratio and bandwidth is possible when the reflection coefficients, ρ_i in (4.8), are chosen to have ratios corresponding to the like terms in a Nth-order Chebyshev

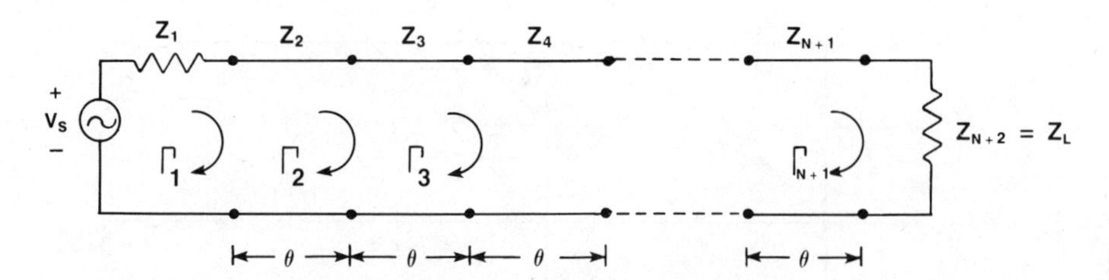

Figure 4.5 *N*-section impedance transformer.

TABLE 4.1 Proportionality Constants a_i for the Chebyshev Transformer

N	a_1	a_2	a_3	a_4
1	x_0			
2	x_0^2	$2x_0^2-2$		
3	x_0^3	$3x_0^3-3x_0$		
4	x_0^4	$4x_0^4-4x_0^2$	$6x_0^4-8x_0^2+2$	
5	x_0^5	$5x_0^5-5x_0^3$	$10x_0^5-15x_0^3+5x_0$	
6	x_0^6	$6x_0^6-6x_0^4$	$15x_0^6-24x_0^4+9x_0^2$	$20x_0^6-36x_0^4+18x_0^2-2$

$x_0 = \sec\theta_1$

$\theta_1 = \frac{\pi}{2} - \frac{w\pi}{4}$

w = fractional bandwidth

polynomial [6]. Approximating

$$\rho_m = \frac{Z_{m+1} - Z_m}{Z_{m+1} + Z_m} \simeq \frac{1}{2} \ln \frac{Z_{m+1}}{Z_m} \tag{4.9}$$

and retaining only the first term in a Taylor series expansion of logarithm, the value for the $(m+1)$th section impedance in the transformer is given in terms of mth section impedance to be

$$\ln \frac{Z_{m+1}}{Z_m} = \frac{a_m \ln(Z_{N+2}/Z_1)}{\sum_{i=1}^{N+1} a_i} \tag{4.10}$$

where the a_i are the proportionality constants related to the Chebyshev polynomial, Z_{N+2} is the load impedance, and Z_1 is the source impedance as shown in Fig. 4.5. Table 4.1 gives the value of these constants up to $N=6$ from a simple recursion formula given by Cohn [6]. The passband response is applicable between θ_1 and $(\pi - \theta_1)$, as shown in Fig. 4.6. Nomograms

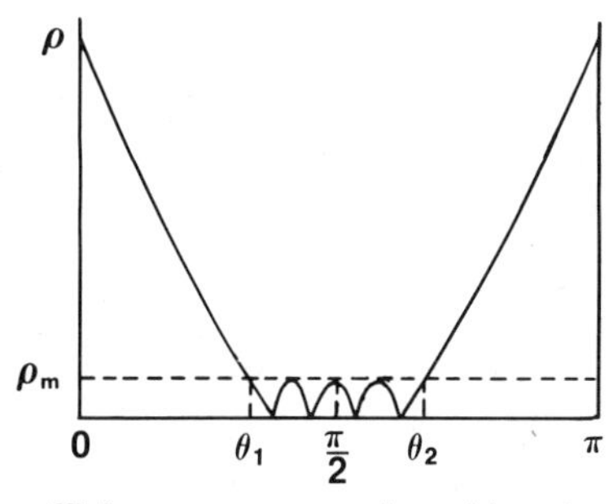

Figure 4.6 Reflection-coefficient response of multisection Chebyshev impedance transformer.

have been developed to speed up the design of quarter-wave transformers [7].

Tapered Transmission Lines. These represent a practical and effective solution to impedance matching. Some of the tapers that have been used for impedance matching are of the linear, the exponential, and the Chebyshev designs [8, 9].

The synthesis of a tapered-line transformer is done at the lowest frequency of interest because the derivative of the reflection coefficient decreases rapidly with frequency. The length of the exponential taper is given as [9]

$$L = \frac{\lambda_g}{\rho_e} \frac{\ln(\bar{Z}_L)}{4\pi} \tag{4.11}$$

where ρ_e is the maximum reflection coefficient, $\bar{Z}_L$ is the normalized [with respect to source impedance] value of the load impedance to be matched, and λ_g is the guide wavelength at the lowest frequency of interest. The normalized impedance profile as a function of length parameter l is given as

$$\bar{Z}(l) = e^{\ln[\bar{Z}_L \cdot l/L]} \tag{4.12}$$

For example, to design an exponential TEM line taper for matching a 50-Ω source to a 100-Ω load when maximum reflection coefficient is $\rho_e = 0.1$ and frequency of interest is 4 GHz, one gets

$$L = \frac{\lambda_g}{0.1} \cdot \frac{\ln(2)}{4\pi} = 0.55\lambda_g$$

and $\bar{Z}(l) = e^{0.3l/\lambda_g}$. Assuming an effective dielectric constant of 6.25

$$\lambda_g = \frac{30}{4} \times \frac{1}{\sqrt{6.25}} = 3\text{ cm}$$

$$L = 1.65\text{ cm}$$

The design of linear and Chebyshev tapered-line transformers has been described in Reference 8.

4.2 LOSSLESS MATCHING NETWORKS

4.2.1 Transfer Function

A function that algebraically expresses a fixed relationship between the input and output of a circuit such that the output can be determined for any

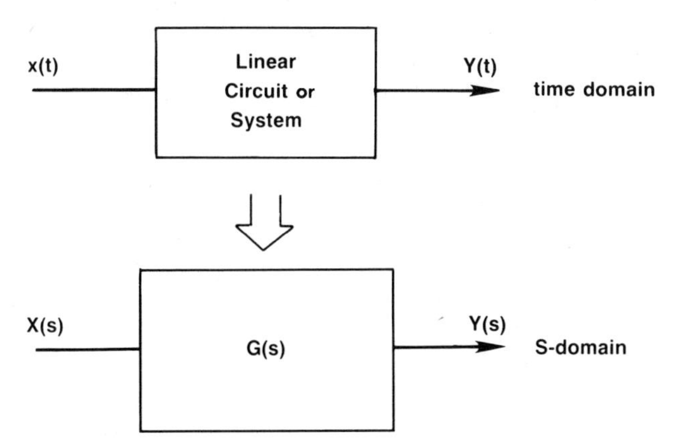

Figure 4.7 Transfer-function representation of an input–output relationship of a linear circuit or a system.

input is known as a transfer function. In the S-domain, for a linear system:

$$Y(s) = G(s)X(s) \tag{4.13}$$

where $X(s)$ represents an input $X(t)$ in the S-domain, and $Y(s)$ is the output in the S-domain (Fig. 4.7). Equation 4.13 gives,

$$G(s) = \frac{Y(s)}{X(s)} \tag{4.14}$$

where $G(s)$ is called the transfer function of the circuit. It is a generalization of the more common gain (loss) parameter used to characterize the input–output relationship of a circuit. It is fixed by the nature of the circuit, and it is independent of the type of excitation. Many different circuit configurations could all possess identical transfer functions. For example, in Fig. 4.8 both circuits have the same transfer function. Thus, when viewed from outside the enclosed boxes, both circuits look similar in the input–output relationship.

Other circuit parameters can be used to identify the separate circuits, but the transfer function should be considered as a function of significance rather than the circuit details.

In addition to the finite poles (roots of the denominator polynomial) and finite zeros (roots of the numerator polynomial) of a function, we would like to introduce the concept of "poles and zeros at infinity." If the transfer function $G(s)$ can be represented as the ratio of two polynomials such that

$$G(s) = \frac{N(s)}{D(s)} = \frac{a_0 + a_1 s + \cdots + a_n s^n}{b_0 + b_1 s + \cdots + b_n s^m} \tag{4.15}$$

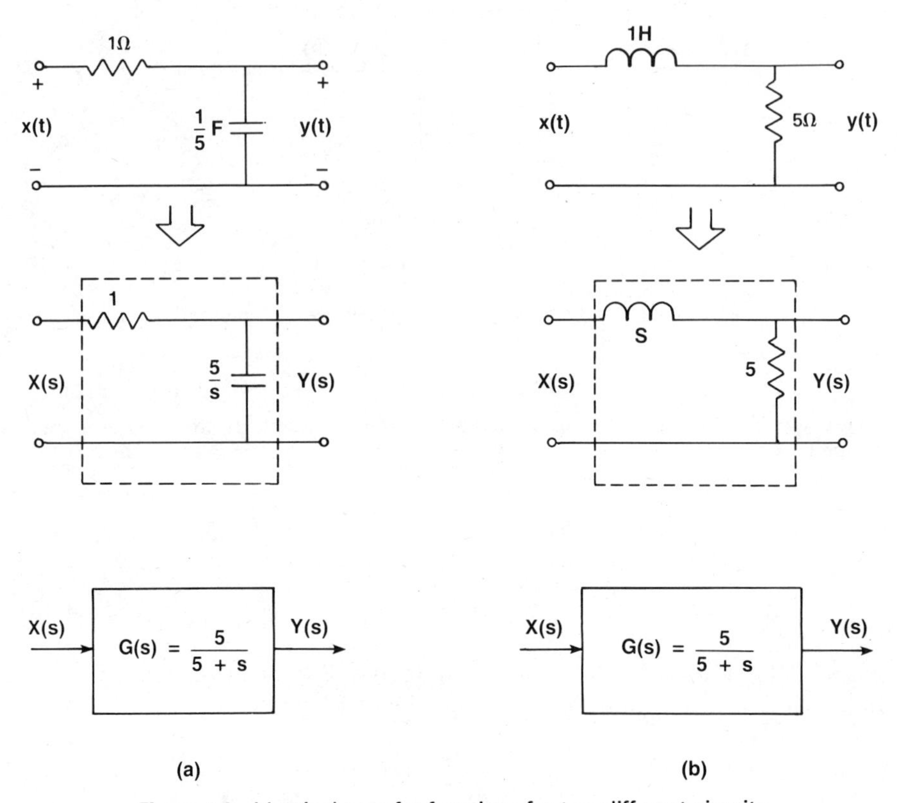

Figure 4.8 Identical transfer functions for two different circuits.

where a_i and b_j are the coefficients of polynomials, then the order of the transfer function is the larger of the two integers m and n. If $m > n$, that is, the degree of the denominator polynomial is greater than the degree of the numerator polynomial, then the transfer function is said to have $(m - n)$ zeros at infinity. Alternately, if $n > m$, then the transfer function is said to have $(n - m)$ poles at infinity.

Poles of a transfer function produce the natural response of a circuit or they display the natural time constants and frequencies of the circuit. Each pole (or a pair of poles for the complex response) may be considered as producing one of the terms in the associated output function $y(t)$. For further details, the reader should see the standard textbooks on network analysis [10]. For example, in the circuit given in Fig. 4.8, the transfer function has a pole at $s = -5$ that will display a natural response function with the exponential term e^{-5t}. In addition to the natural response, the output function $y(t)$ includes the forced response due to the poles associated with the input function $X(s)$.

4.2.2 Network Theory

The purpose of this section is to describe circuit theory concepts that are useful in analyzing lossless matching networks, that is, a network that contains no resistive elements and has only inductors or capacitors for the ideal transformers. The impedance Z of the network may be expressed as a ratio of two polynomials as follows:

$$Z(s) = \frac{N(s)}{D(s)} = \frac{A_N(s - s_{01})(s - s_{02}) \cdots (s - s_{0N})}{B_M(s - s_{p1})(s - s_{p2}) \cdots (s - s_{pM})} \tag{4.16}$$

where all the zeros ($s_{01}, s_{02}, \ldots, s_{0N}$) and poles ($s_{p1}, s_{p2}, \ldots, s_{pM}$) lie on the $j\omega$-axis of the complex frequency plane (s), a condition for lossless networks where the transients neither grow nor decay. Additionally, there are no double roots either in the numerator polynomial, $N(s)$, or in the denominator polynomial, $D(s)$, because double roots correspond to growing transients with time. A circuit with a finite number of lumped reactive elements will have a finite number of poles and zeros. Distributed network impedance functions, however, have an infinite number of zeros and poles, because each distributed element can be represented as an infinite number of infinitesimal lumped elements. Thus, circuits containing microwave transmission lines are easily represented by transcendental impedance functions. For example, the input impedance of a quarter-wavelength, short-circuited transmission line is given as

$$Z(s) = Z_0 \tanh\left(\frac{\pi s}{2\omega_0}\right) \tag{4.17a}$$

$$= Z_0 \frac{\pi s}{2\omega_0} \prod_{n=1}^{\infty} \left(\frac{2n-1}{2n}\right)^2 \frac{(s + j2n\omega_0)(s - j2n\omega_0)}{[s + j(2n-1)\omega_0][s - j(2n-1)\omega_0]} \tag{4.17b}$$

where Z_0 is the characteristic impedance of the line and ω_0 is the radial frequency at which the physical line length is $\lambda/4$. Thus, a short-circuited transmission line has poles at $\pm j(2n-1)\omega_0$ and zeros at $s = 0$ and $\pm j2n\omega_0$ where $n = 1, 2, 3, \ldots, \infty$.

In a reactive network, shown in Fig. 4.9, the terminal impedance at zero frequency and at infinite frequency must be either zero or infinity because capacitive or inductive branches are represented by either opens or shorts in the network. The transient response of the network contains a superposition of complex frequency waveforms of various natural modes of vibration of the circuit. Natural vibrations of the circuit can continue even after all driving signals are tuned off. For an open-circuited terminal impedance, $Z(s)$, natural vibrations can be observed only at frequencies $s_{p1}, s_{p2}, \ldots, s_{pn}$, as given in (4.16). Alternatively, if $Z(s)$ is short circuited,

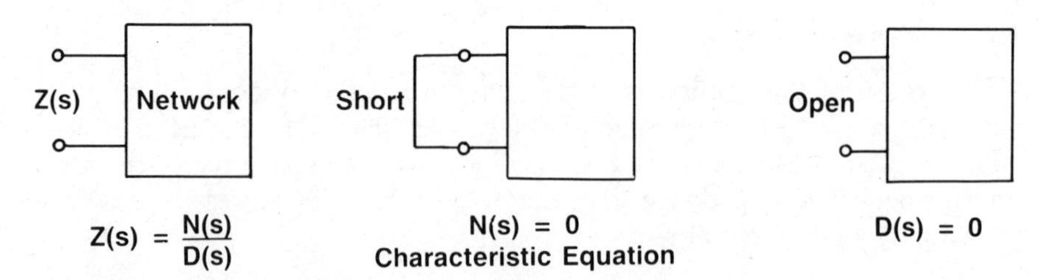

Figure 4.9 A lossless matching network with open-circuit and short-circuit terminations.

the natural vibration frequencies correspond to the zeros of $Z(s)$. The terminal impedance, $Z(s)$ in (4.16), can be rewritten in the factored form to show conjugate pairs of zeros and poles or as a sum of partial fractions, each fraction corresponding to a pole in the function. That is,

$$\begin{aligned} Z(s) &= \frac{A_N(s-s_{01})(s-s_{01}^*)\cdots(s-s_{0N}^*)}{B_M(s-s_{p1})(s-s_{p1}^*)\cdots(s-s_{pM}^*)}\left\{s \quad \text{or} \quad \frac{1}{s}\right\} \\ &= \frac{R_1}{(s-s_{p1})}+\frac{R_1}{(s-s_{p1}^*)}\cdots\frac{R_M}{(s-s_{pM}^*)}+\left\{0 \quad \text{or} \quad \frac{R_0}{s}\right\} \end{aligned} \tag{4.18}$$

where the R's are the residues of the function in its poles. From standard network theory, it is possible to show that all residues are real and positive. Partial fraction representation of the network impedance demonstrates the dominance of that component of the expression whose pole is in the vicinity of the frequency under consideration. In addition, each part of the partial fraction expansion represents a realizable network function. This becomes apparent when terms containing conjugate poles are grouped together as given below:

$$Z(s) = \frac{R_0}{s} + \sum_i \frac{2R_i s}{s^2+\omega_i^2} + Ks \tag{4.19}$$

Figure 4.10 identifies partial fraction components with lossless network elements. It is to be pointed out here that the reactance versus frequency slopes for the elements in a reactance network are positive and so is the slope of the total reactance function.

To synthesize a network from the reactance function (also true for a susceptance function when component networks are connected in parallel) one removes, step by step, partial fraction components by continued fraction expansion. If the partial fraction component that is removed is identified with a pole at infinity, then the remaining reactance function is left with a zero at infinity. The next partial fraction component can be removed, then, from the network susceptance function by inverting the

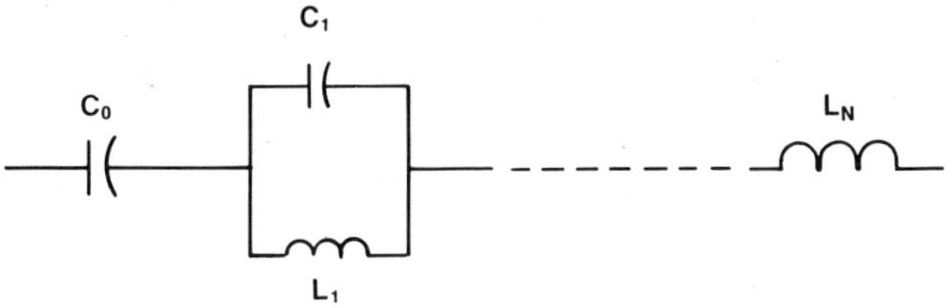

$$\frac{1}{C_0 s}+\frac{\frac{1}{C_1}s}{s^2+\frac{1}{L_1 C_1}}+\cdots+L_N s$$

$$\frac{R_0}{s}+\frac{2R_1 s}{s^2+\omega_1^2}+\cdots+Ks$$

Thus, $C_0 = 1/R_0$, $C_1 = 1/2R, \ldots, L_N = K$, $L_1 = 1/C_1\omega_1^2$

Figure 4.10 Identification of partial fraction components in a series connection of lossless elements.

remainder, as shown by example in Table 4.2. Alternatively, the whole process of continued fraction expansion can be applied to removing poles of reactance function at zero frequency. As shown in Table 4.2, once the first partial fraction component is removed (identified with a pole at zero frequency), the remainder function can be inverted to remove another pole at zero frequency from the susceptance function. The procedure continues until the last pole at zero frequency is extracted. The example given in Table 4.2 has the reactance function:

$$Z(s)=\frac{(s^2+1)(s^2+6)}{s(s^2+2)}=\frac{s^4+7s^2+6}{s^3+2s} \tag{4.20}$$

Table 4.2 shows only the development of canonic networks using the continued fraction expansion approach. The detailed network synthesis procedure is discussed in Section 4.4. Table 4.2 gives two different networks, one of which is obtained when poles at infinity are removed and the other is based on poles at zero frequency.

As given in (4.20), $Z(s)$ as a ratio of even and odd polynomials is a general property of reactance functions. It can be shown that a necessary condition for a stable system is that the sum of even and odd polynomials in a reactance function is a Hurwitz polynomial [11]. A Hurwitz polynomial has all of its zeros in the left half of the complex frequency plane. The relationship between a Hurwitz polynomial and the reactance function is exploited in the next section for selecting a network terminated by a single resistance such that the system has its natural frequencies in the left half of the complex plane.

TABLE 4.2 Cauer Development of Reactance Function

Removing Poles at ∞	Removing Poles at Zero Frequency
Step 1: First Element	
$Z(s) = s + \frac{5s^2+6}{s^3+2s}$	$Z(s) = \frac{3}{5} + \frac{4s^2+s^4}{s^3+2s}$
$= s + Z_1(s)$	$= \frac{3}{s} + Z_1'(s)$
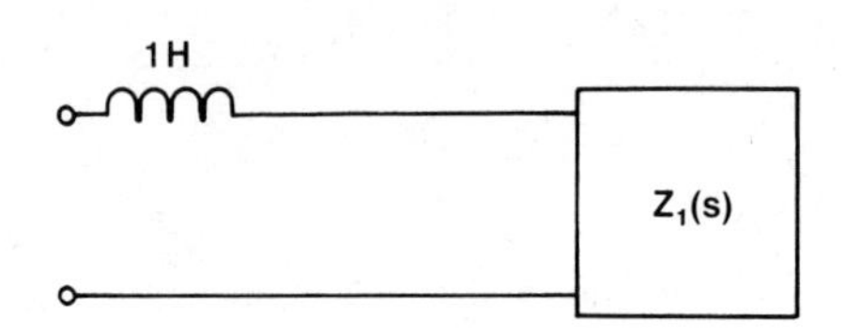	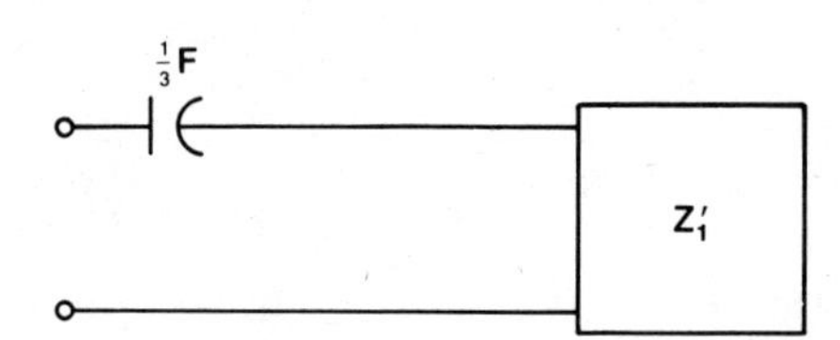
Step 2	
$Y_1(s) = \frac{1}{Z_1(s)}$	$Y_1'(s) = \frac{1}{Z_1'(s)}$
Step 3: Second Element	
$Y_1(s) = \frac{s}{5} + \frac{\frac{4}{5}s}{5s^2+6}$	$Y_1'(s) = \frac{1}{2s} + \frac{s^3/2}{4s^2+s^4}$
$= \frac{s}{5} + Y_2(s)$	$= \frac{1}{2s} + Y_2'(s)$
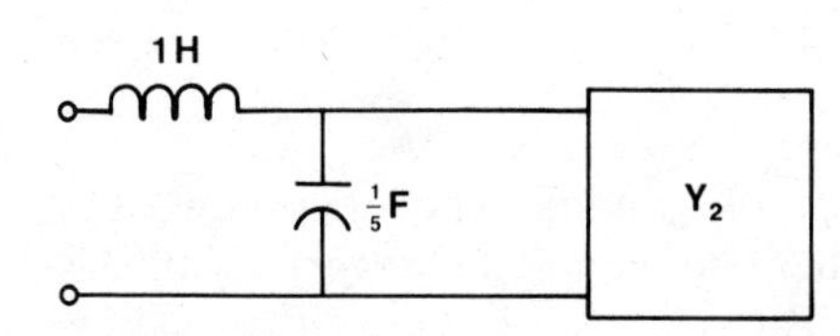	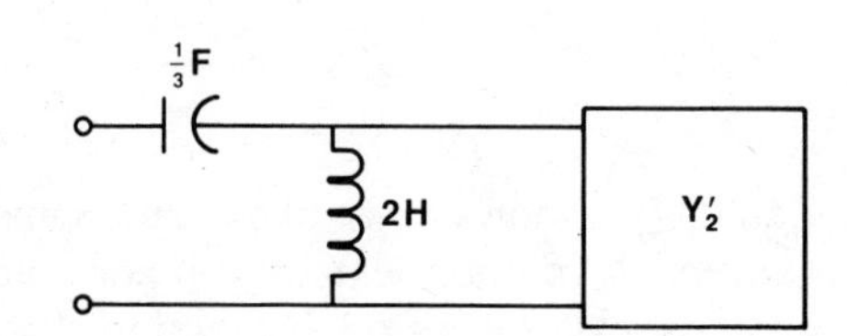
Step 4	
$Z_2(s) = \frac{1}{Y_2(s)}$	$Z_2'(s) = \frac{1}{Y_2'(s)}$
Step 5: Third Element	
$Z_2(s) = \frac{25}{4}s + \frac{6}{\frac{4}{5}s}$	$Z_2'(s) = \frac{8}{s} + \frac{s^4}{s^3/2}$
$= \frac{25}{4}s + Z_3(s)$	$= \frac{8}{s} + Z_3'(s)$

Table 4.2 (*Continued*)

Removing Poles at ∞	Removing Poles at Zero Frequency

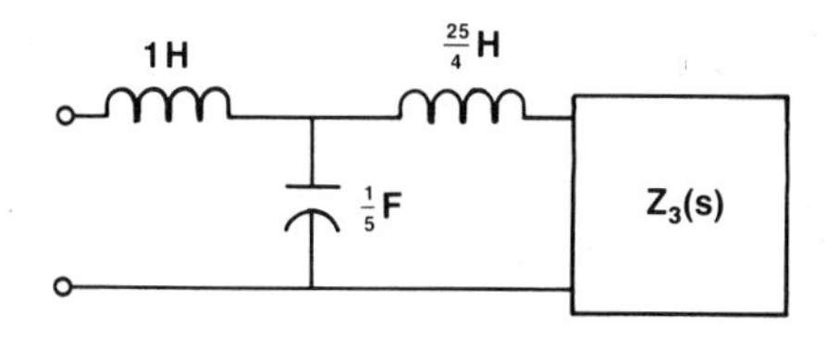

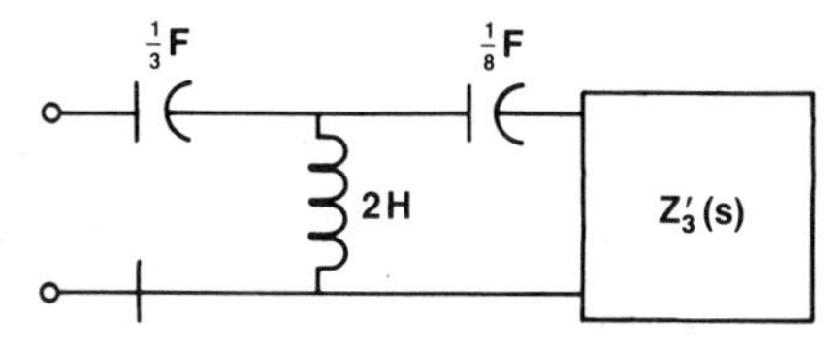

Step 6

$$Y_3(s) = \frac{1}{Z_3(s)} \qquad Y'_3(s) = \frac{1}{Z'_3(s)}$$

Step 7: Fourth Element (Final circuit)

$$Z_3(s) = \frac{2}{15}s \qquad Z'_3(s) = \frac{1}{2}\cdot\frac{1}{s}$$

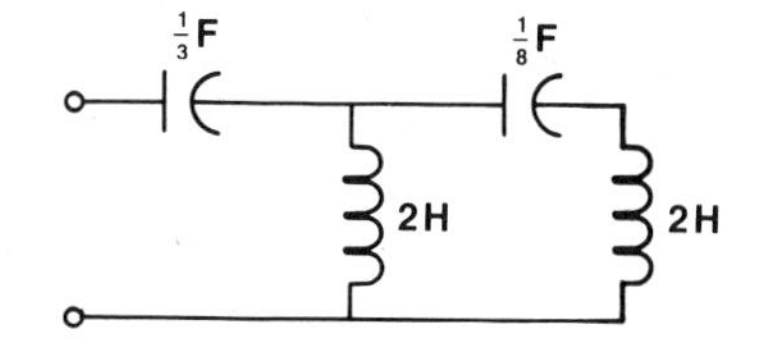

4.2.3 Ladder Networks

Table 4.2 gives the driving point reactance of a lossless network whose poles and zeros are placed on the $j\omega$-axis in the complex frequency plane. Terminating the network with a resistor (a lossy element) results in a movement of the natural frequencies of the system to the left half-plane. For example, if $Z(s) = N(s)/D(s)$ is the reactance function of a lossless network, then the characteristic equation of the lossy network, terminated by a 1-Ω resistor, is given by

$$1 + Z(s) = 1 + \frac{N(s)}{D(s)} = \frac{N(s) + D(s)}{D(s)} = 0 \tag{4.21}$$

or $N(s) + D(s) = 0$, which is a Hurwitz polynomial. Conversely, it is possible to find a lossless reactance network to be connected to a 1-Ω resistor from the natural frequencies in the left plane. The complex natural frequency points should appear in conjugate pairs, a condition true for a lossless reactance function.

Most lossless networks are connected to resistances at both ends, source and the load, as shown in Fig. 4.1. Here, the network input impedance, Z_i, is given as a ratio of two Hurwitz polynomials, $N(s)$ and $D(s)$. That is,

$$Z_i(s) = \frac{N(s)}{D(s)} \tag{4.22}$$

where $N(s)$ and $D(s)$ represent the characteristic equations when the left-hand terminal in the network is shorted and opened, respectively. Additionally, Z_i, a rational function with real coefficients, has a real part that is always positive at real frequencies, because the network draws power instead of providing any power. Synthesis of resistance-terminated networks is discussed in the next section.

The transfer function of a lossless ladder network, from a current source in parallel with any two nodes or a voltage source in series with any branch of the network to a remote resistive termination (Fig. 4.11), has the same natural frequencies with and without the source. The zeros of the transmission in transfer function of the network represent the opening of series elements or shorting of shunt elements. The denominator polynomial gives all the natural frequencies of the network. Changing the elements in a lossless network does not change its natural frequencies, but it does change the transfer characteristic. For example, terminating the network in Table 4.2 with a 1-Ω resistor and attaching a current source across the right hand capacitor, it is possible to write the transfer function, E_o/I_s, as shown in Fig. 4.12. The network has zeros of transmission at only infinite frequency,

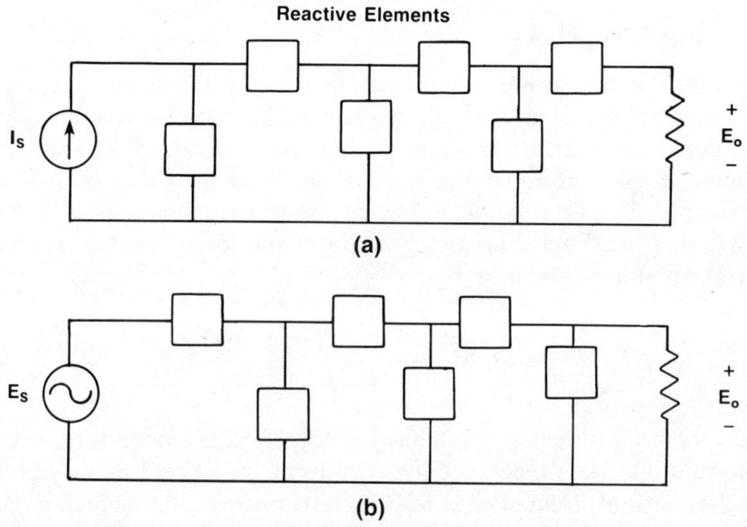

Figure 4.11 Resistance-terminated ladder networks. (*a*) Current driven. (*b*) Voltage driven.

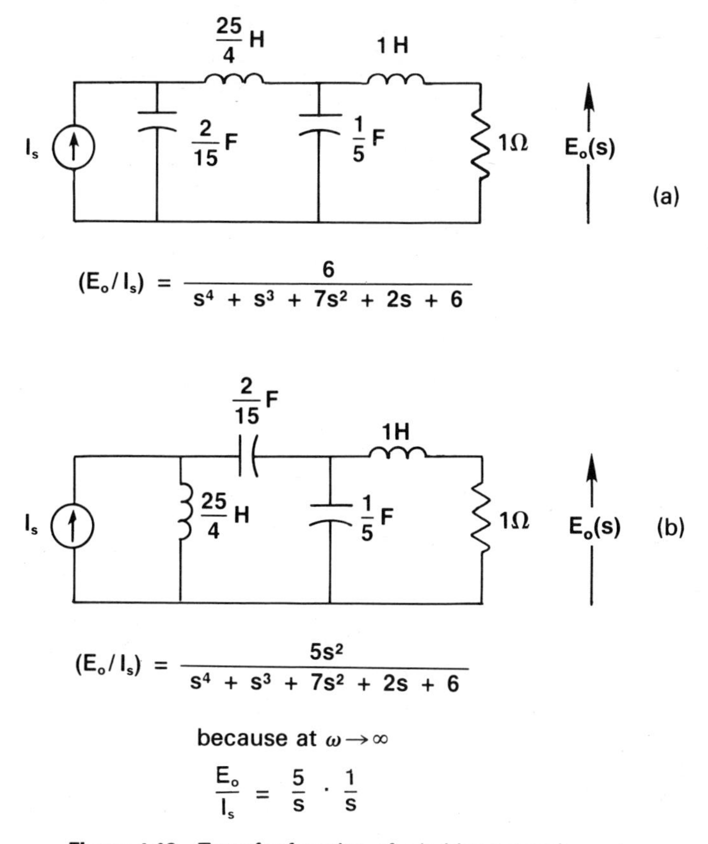

Figure 4.12 Transfer function of a ladder network.

which determines that the transfer function has a constant value in the numerator. Since the denominator is the characteristic equation $[N(s) + D(s)]$, the value of the constant is determined by evaluating the circuit behavior at zero frequency when capacitors act like open circuits and inductors as shorts. Thus, the constant value in the numerator is equal to 6, as shown in Fig. 412(a).

Now exchanging capacitance and the inductance on the left side in Fig. 4.12(a) a new circuit is built, as shown in Fig. 4.12(b). It has the same natural frequencies, but two zeros of transmission are at zero frequency, while the other two are still at infinity. Thus, the new transfer function becomes a constant times s^2 in the numerator with the same characteristic equation in the denominator. It is possible to calculate the constant by evaluating the circuit at very high frequencies. As $\omega \to \infty$, the current source provides a voltage that is determined primarily by the $\frac{1}{5}$-F capacitor and the output current is determined by that voltage divided by the reactance of the 1-H inductor.

The transfer function of a lossless network can be expressed in terms of driving-point impedance or admittance of a terminated network, as shown in Fig. 4.13. A property of rational functions, that is,

$$Z(-j\omega) = Z^*(j\omega) \tag{4.23}$$

gives two important equations, useful for evaluating networks. These are

$$\tfrac{1}{2}[Z(s) + Z(-s)] = \text{Re}[Z(s)] \tag{4.24}$$

$$Z(s) \cdot Z(-s) = |Z^2| \tag{4.25}$$

where $s = j\omega$. If the rational function is a ratio of two polynomials given by

$$Z(s) = \frac{A(s)}{B(s)} = \frac{a_0 + a_1 s + a_2 s^2 + \cdots}{b_0 + b_1 s + b_2 s^2 + \cdots} \tag{4.26}$$

$$= \frac{A_e + A_o}{B_e + B_o}$$

where A_e and B_e are the polynomials containing even powers of s, and polynomials A_o and B_o contain odd powers of s, then it is easy to show that the real part of $Z(s)$ and the magnitude of $Z(s)$, from (4.24), are always functions of s^2. Thus, their critical frequencies, zeros, and poles occur in pairs and are negative. Combining the properties of the rational functions just given and a general characteristic of passive networks that requires all poles to be in the left half-plane, one can synthesize a network if the real part of a driving-point impedance function (a function of s^2) is known. For

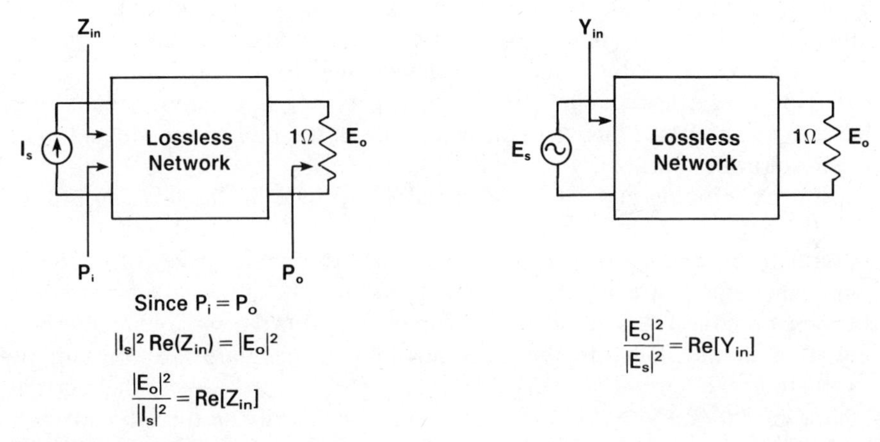

Figure 4.13 Relationship of transfer function and driving-point impedance (admittance) of a lossless matching network.

example, if the transfer function of a terminated network is given as

$$\frac{E_o}{E_s} = \frac{5}{(s+2)(s+3)}, \qquad s = j\omega$$

then

$$\left|\frac{E_o}{E_s}\right|^2 = \frac{25}{(4-s^2)(9-s^2)} = \text{Re}[Y_{\text{in}}] = \frac{1}{2}[Y_{\text{in}}(s) + Y_{\text{in}}(-s)]$$

Expanding Re[Y_{in}] into partial fractions

$$\frac{25}{(4-s^2)(9-s^2)} = \frac{r_1}{2+s} + \frac{r_2}{3+s} + \frac{r_3}{s-2} + \frac{r_4}{s-3}$$

Residues r_1 and r_2 are found to be $\frac{5}{4}$ and $-\frac{5}{6}$, respectively. Thus,

$$Y_{\text{in}}(s) = \frac{2r_1}{s+2} + \frac{2r_2}{s+3} = \frac{\frac{5}{2}}{s+2} + \frac{-\frac{5}{3}}{s+3}$$

$$= \frac{\frac{5}{6}s + \frac{25}{6}}{(s+2)(s+3)} = \frac{5}{6}\frac{s+5}{(s+2)(s+3)}$$

In a doubly terminated lossless network one can calculate the natural frequencies of the network from poles of the reflection coefficient functions at the input or at the output. The reflection coefficients at the input (ρ_i) and at the output (ρ_o) can be expressed in terms of the Z-parameters of the network. That is

$$\rho_i = \frac{\Delta + Z_{11}R_L - Z_{22}R_S - R_LR_S}{\Delta + Z_{11}R_L + Z_{22}R_S + R_LR_S} \tag{4.27}$$

and

$$\rho_o = \frac{\Delta + Z_{22}R_S - Z_{11}R_L - R_LR_S}{\Delta + Z_{11}R_L + Z_{22}R_S + R_LR_S} \tag{4.28}$$

where $\Delta = Z_{22}Z_{11} + Z_{12}Z_{21}$, R_S and R_L are source and load resistances, respectively. Coefficients ρ_i and ρ_o are calculated using the expression for the input and output impedance in terms of the Z-parameters of the network. That is,

$$Z_i = Z_{11} + \frac{Z_{12}Z_{21}}{Z_{22} + R_L} \tag{4.29}$$

and

$$Z_o = Z_{22} + \frac{Z_{21}Z_{12}}{Z_{11} + R_S} \tag{4.30}$$

The denominator in both (4.27) and (4.28) can be shown to be equivalent to the characteristic equation of the network. It can also be shown that zeros of ρ_i are negative of the zeros of ρ_o and are equal at all points on the imaginary axis.

The starting point for network synthesis can be the desired passband performance of the network or its transducer gain. The transducer gain, G_T, is the ratio of power transferred to the load, P_L, and power available from the source, P_S. It is related to the magnitude of the reflection coefficient. That is,

$$G_T = \frac{P_L}{P_S} = 1 - |\rho_i|^2 = 1 - |\rho_o|^2 \tag{4.31}$$

Since the magnitudes of ρ_i and ρ_o are same, the transducer gain is also a function of s^2.

The synthesis procedure is described in Section 4.4, but the following section describes approximate solutions for insertion loss ($1/G_T$) functions.

4.2.4 Approximate Solution

The desired ideal response of a matching network having no loss in the passband and with vertical skirts, as shown in Fig. 4.14, is approximated by polynomial functions for network synthesis. The approximated insertion loss function has to be bounded, be consistent with the desired topology, and it must behave in a specific manner over the band of frequencies of interest. The classical approximations include low-pass, high-pass and bandpass networks where the element values (based on the coefficients of insertion-loss function) provide maximally flat or Chebyshev response. Several network topologies, however, cannot be synthesized from classical approximations. For example, a network topology shown in Fig. 4.15 is commonly used in amplifier matching circuits, but it cannot be synthesized

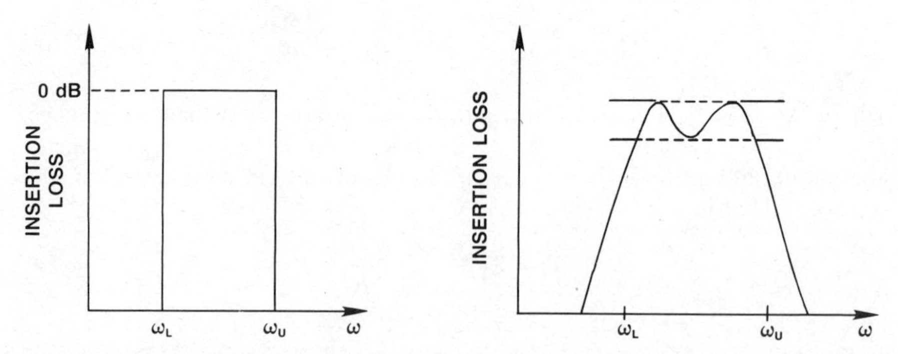

Figure 4.14 (*a*) Desired and (*b*) approximated frequency response of a matching network.

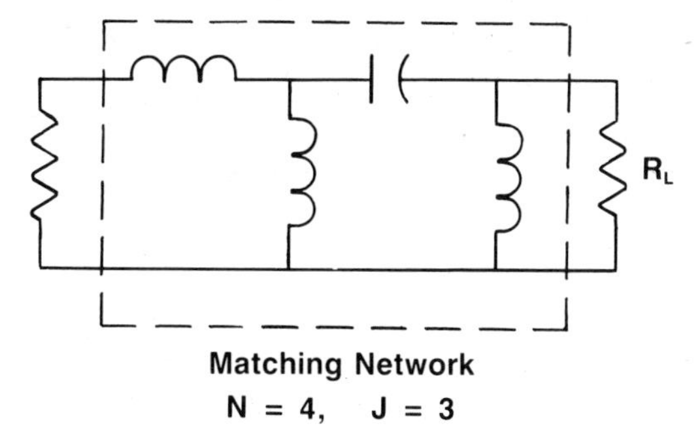

Figure 4.15 A desirable matching network that cannot be synthesized using classical solutions.

using the low-pass design. High-pass and bandpass responses can be derived from low-pass functions or designed with appropriate transformations.

The insertion-loss polynomial form for a maximally flat low-pass network is given by

$$\mathrm{IL}_{\mathrm{MF}} = a_0 + a_c\left(\frac{\omega}{\omega_c}\right)^{2N} \tag{4.32}$$

where the passband extends from dc to cutoff frequency ω_c and a_0 is the insertion loss at dc, while $a_0 + a_c$ represents the loss at $\omega = \omega_c$, as shown in Fig. 4.16.

Alternately, the ideal loss response can be approximated by a function that provides an insertion-loss ripple in the frequency band of interest. The insertion-loss function is given by

$$\mathrm{IL}_{\mathrm{CH}} = a_0 + a_c\left(C_N\frac{\omega}{\omega_c}\right)^2 \tag{4.33}$$

where C_N is the Nth-order Chebyshev polynomial.

Table 4.1 gives Chebyshev polynomials up to $N = 6$. The insertion-loss response ripples between a_0 and $a_0 + a_c$ in the passband (Fig. 4.16). Low-pass to bandpass and low-pass to high-pass transformations are given in Chapter 6. Typically, a low-pass network of prescribed response is transformed to a bandpass network by replacing every series inductor with a series-resonant circuit and every shunt capacitor with a shunt-resonant circuit.

From Table 4.1 for Chebyshev polynomials and using low-pass and high-pass transformation, the bandpass insertion-loss function for the $N = 2$ case is given by

$$\mathrm{IL}_{\mathrm{CH}} = 1 + \frac{\omega^4 - 4\omega^2 + 4}{\omega^2} \tag{4.34}$$

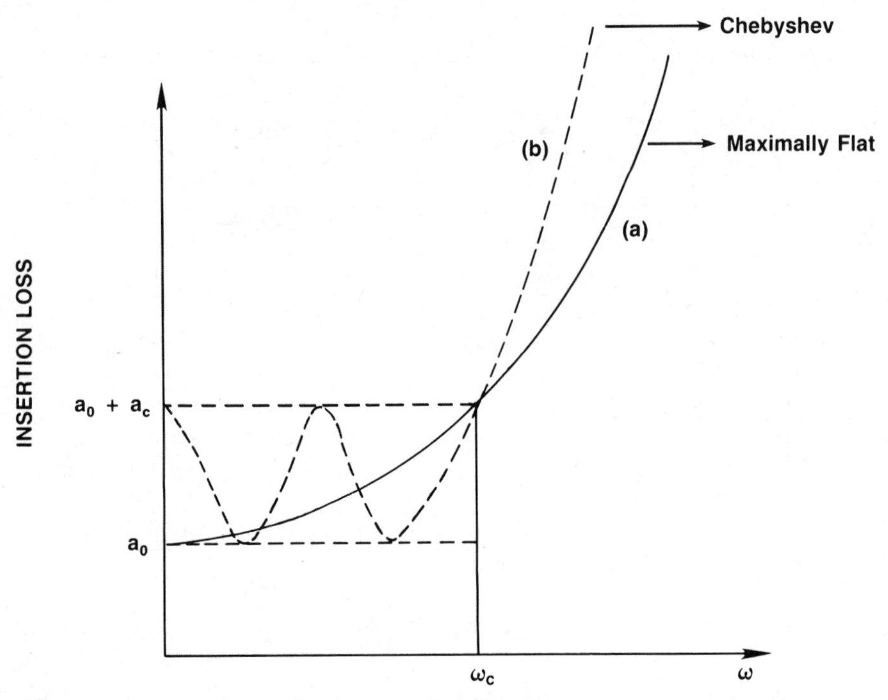

Figure 4.16 A low-pass insertion-loss response of a network. (*a*) Maximally flat. (*b*) Chebyshev.

Network functions for sloped insertion loss characteristics can be approximated by appropriate flat response functions. Integral multiples of 6 dB/octave slopes in the passband insertion loss characteristics are easily obtained by dividing the flat response function by ω^{2I}, where I is the desired integral multiple number. Nonintegral sloped approximations are calculated by combining linearly two integral slope responses, one having an integral slope just less than and another having an integral slope just greater than the desired slope. For example, to obtain a passband loss slope of 4 dB/octave it is possible to combine two functions, one corresponding to an integral slope approximation of 6 dB/octave and another response function for 0 dB/octave slope. That is,

$$\mathrm{IL}_4 = A_1 \mathrm{IL}_0 + A_2 \mathrm{IL}_6$$

The coefficients A_1 and A_2 are given as

$$A_1 = \frac{\omega_L^{2(L_2 - L)} - 1}{\omega_L^2 - 1}$$

$$A_2 = 1 - A_1$$

where ω_L is normalized (with respect to the upper frequency) lower frequency. L is the noninteger value corresponding to the desired slope (the integer value of 1 corresponds to 6 dB/octave) and L_2 is the next higher integer value. For 4 dB/octave slope, $L=0.67$ and $L_2=1.0$. If $\omega_L=0.5$, one gets $A_1=0.49$ and $A_2=0.51$.

4.3 IMPEDANCE-MATCHING CIRCUITS

4.3.1 Impedance Transformations: Lumped or Distributed

Impedance transformation in a network makes it possible to adjust the terminating impedance that, in general, is determined by the frequency response function and the topology during synthesis. Inductors and capacitors connected in L-shaped configuration can be transformed into a T or π network with an impedance transformation as shown in Fig. 4.17 [12]. The range of impedance transformation depends upon the value of original inductors or capacitors in the L-configuration.

The addition of one more element to the simple L-section matching circuit gives the designer much greater control over bandwidth [13] and also permits the use of more practical circuit elements. The circuit configurations and circuit element values are given in Fig. 4.18, where

$$M=\frac{R_2}{R_1}>1 \qquad \text{and} \qquad N>M \tag{4.35}$$

By properly selecting N, a compromise is obtained in terms of bandwidth and realizable circuit element values.

In distributed matching networks, which are either synthesized directly or obtained by converting lumped-element networks into the distributed form using Richard's transformation described in Section 4.4, one can use Kuroda's identities to adjust the terminating impedances. The four Kuroda's identities are listed in Fig. 4.19. These identities can be verified by showing that the overall two-port behavior of each circuit, in the identity, is equal. For example, the first identity left-hand circuit is described by the chain matrix (in the s-plane) as

$$\begin{aligned}\begin{bmatrix} A & B \\ C & D \end{bmatrix}_{\text{LHS}} &= \begin{bmatrix} 1 & 0 \\ \dfrac{s}{Z_2} & 1 \end{bmatrix} \frac{1}{\sqrt{1-s^2}} \begin{bmatrix} 1 & sZ_1 \\ \dfrac{s}{Z_1} & 1 \end{bmatrix} \\ &= \frac{1}{\sqrt{1-s^2}} \begin{bmatrix} 1 & sZ_1 \\ s\left(\dfrac{1}{Z_1}+\dfrac{1}{Z_2}\right) & s^2\dfrac{Z_1}{Z_2}+1 \end{bmatrix}\end{aligned} \tag{4.36}$$

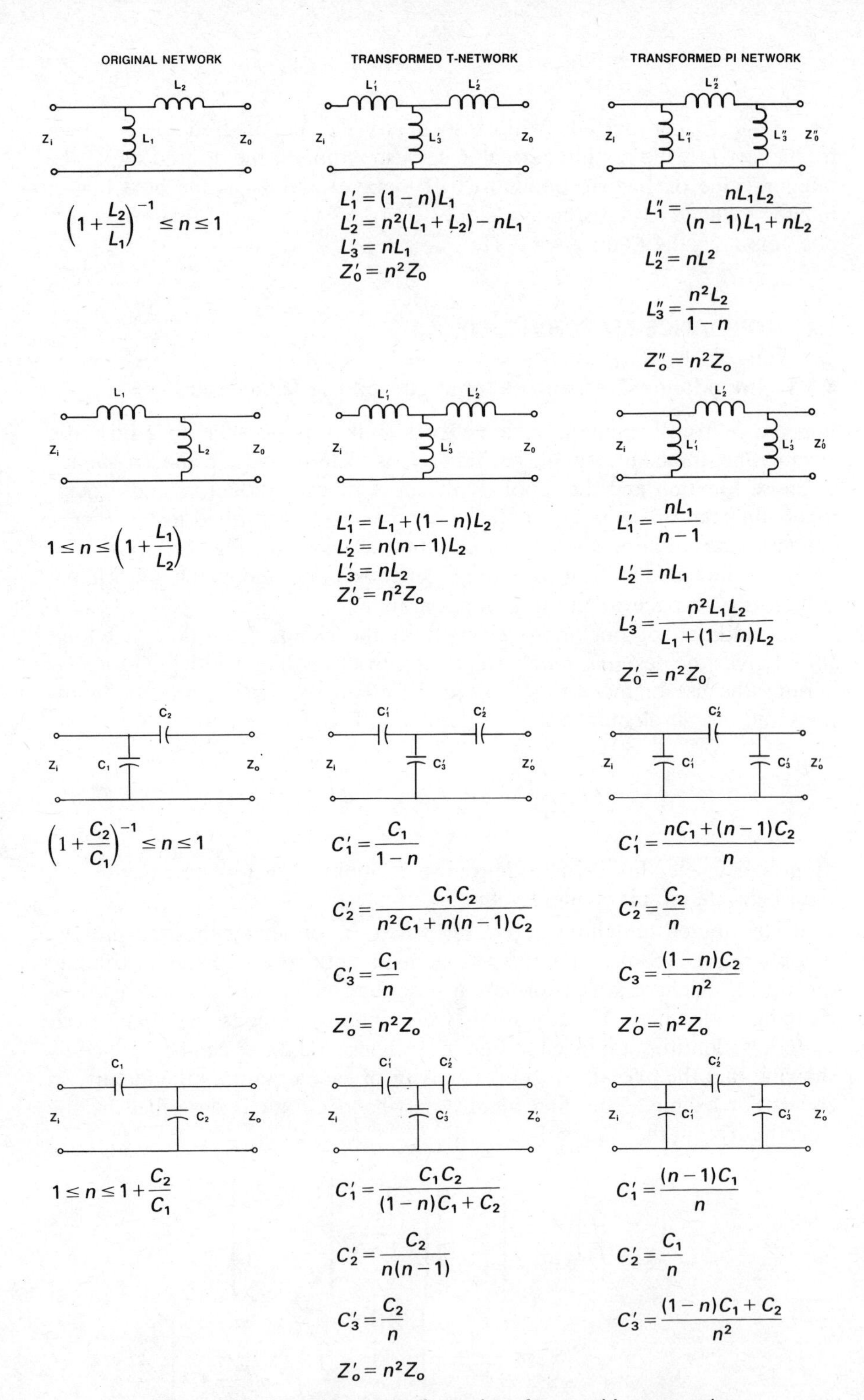

Figure 4.17 Impedance transformations for matching networks.

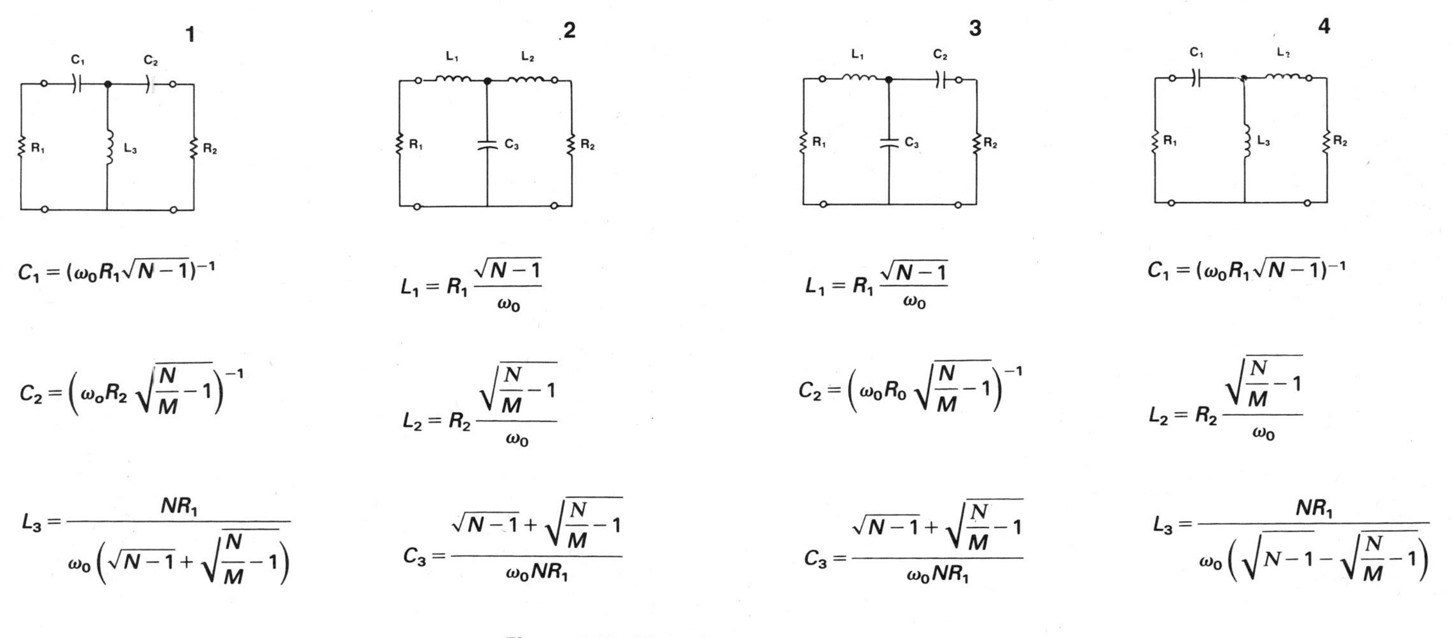
1
2
3
4
C₁
C₂
L₁
L₂
L₃
C₃
R₁
R₂
$C_1 = (\omega_0 R_1 \sqrt{N-1})^{-1}$
$C_2 = \left(\omega_o R_2 \sqrt{\frac{N}{M} - 1}\right)^{-1}$
$L_3 = \frac{NR_1}{\omega_0\left(\sqrt{N-1} + \sqrt{\frac{N}{M} - 1}\right)}$
$L_1 = R_1 \frac{\sqrt{N-1}}{\omega_0}$
$L_2 = R_2 \frac{\sqrt{\frac{N}{M} - 1}}{\omega_0}$
$C_3 = \frac{\sqrt{N-1} + \sqrt{\frac{N}{M} - 1}}{\omega_0 N R_1}$
$L_1 = R_1 \frac{\sqrt{N-1}}{\omega_0}$
$C_2 = \left(\omega_0 R_0 \sqrt{\frac{N}{M} - 1}\right)^{-1}$
$C_3 = \frac{\sqrt{N-1} + \sqrt{\frac{N}{M} - 1}}{\omega_0 N R_1}$
$C_1 = (\omega_0 R_1 \sqrt{N-1})^{-1}$
$L_2 = R_2 \frac{\sqrt{\frac{N}{M} - 1}}{\omega_0}$
$L_3 = \frac{NR_1}{\omega_0\left(\sqrt{N-1} - \sqrt{\frac{N}{M} - 1}\right)}$

Figure 4.18 T-section matching configurations.

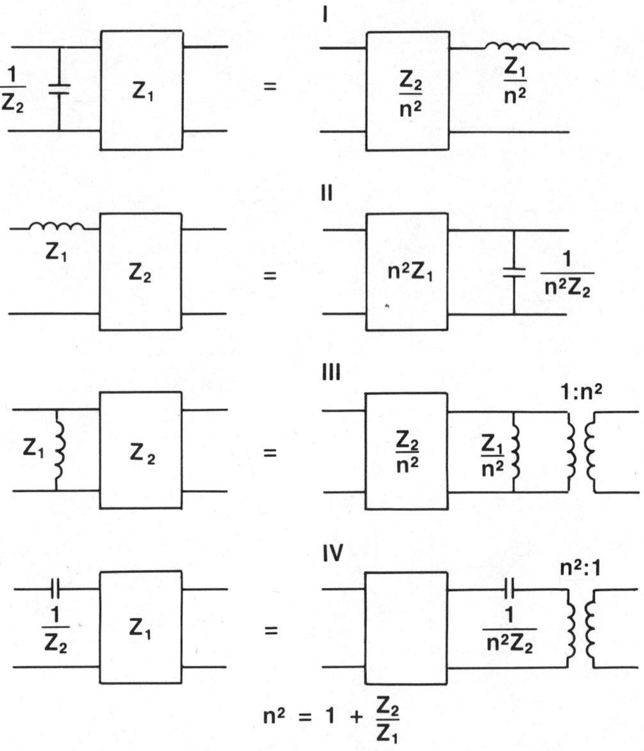

Figure 4.19 Four Kuroda's identities for impedance transformation in distributed networks using unit elements.

and, the right-hand circuit is given by

$$\begin{bmatrix} A & B \\ C & D \end{bmatrix}_{\text{RHS}} = \frac{1}{\sqrt{1-s^2}} \begin{bmatrix} 1 & \dfrac{sZ_2}{n^2} \\ \dfrac{sn^2}{Z_2} & 1 \end{bmatrix} \begin{bmatrix} 1 & \dfrac{sZ_1}{n^2} \\ 0 & 1 \end{bmatrix}$$

$$= \frac{1}{\sqrt{1-s^2}} \begin{bmatrix} 1 & s\left(\dfrac{Z_1}{n^2} + \dfrac{Z_2}{n^2}\right) \\ s\dfrac{n^2}{Z_2} & s^2\dfrac{Z_1}{Z_2} + 1 \end{bmatrix} \tag{4.37}$$

Thus, corresponding matrix elements in the right-hand and left-hand circuit chain matrices are equal if

$$n^2 = 1 + \frac{Z_2}{Z_1} \tag{4.38}$$

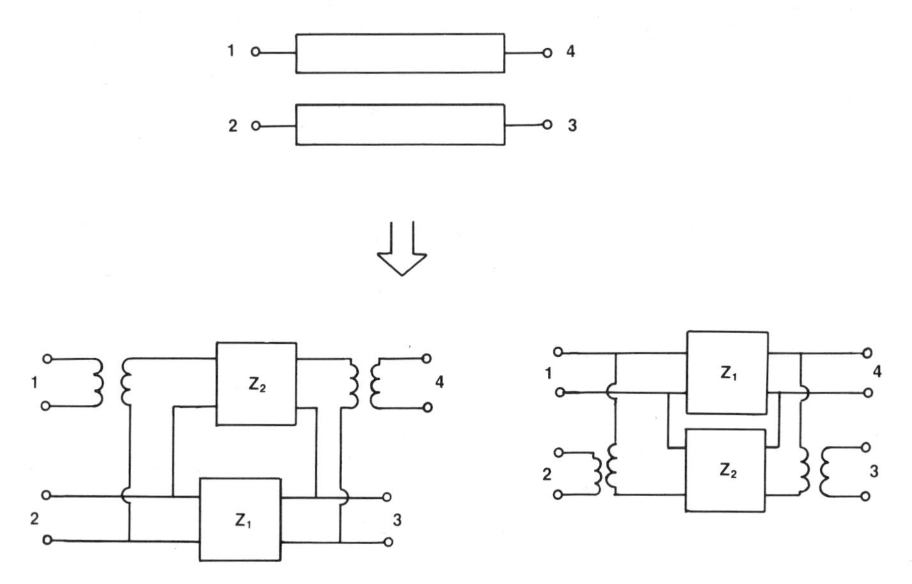

Figure 4.20 Equivalent circuits of a pair of coupled lines.

Alternatively, one can derive these transformations by simplifying network representation of coupled lines.

Figure 4.20 shows two equivalent circuits of a pair of coupled lines. The characteristic impedances of unit elements in the equivalent circuits are denoted by Z_1 and Z_2. For coupled lines having equal widths (impedances)

$$\frac{1}{Z_1} - \frac{n-1}{Z_2} = \frac{n(n-1)}{Z_2}$$

or

$$n^2 = 1 + \frac{Z_2}{Z_1} \tag{4.39}$$

Now consider a pair of coupled lines having open and short circuits at two ports (Ports 1 and 4, respectively) as shown in Fig. 4.21. Both circuits in this figure are identical under all circumstances. This constitutes one of the Kuroda's identities shown in Fig. 4.19. Kuroda's identities become particularly important when unit elements are used in a matching network.

A unit element is a series transmission line of equal length in a commensurate distributed network. An arbitrary number of unit elements of characteristic impedance $Z_0 = R$ can be inserted between a network and the termination, R_L, without altering the response of the network. Kuroda identities, however, have to be applied to have a practical physical realization. For example, Fig. 4.22 shows the use of unit elements in transforming

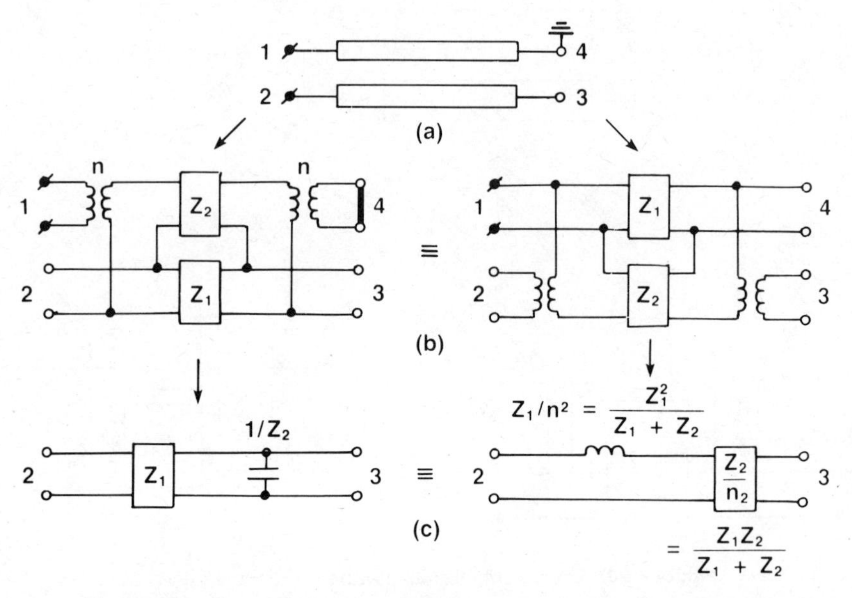

Figure 4.21 Kuroda's first transform by means of the network model.

a 2-to-1 bandwidth lumped-element bandstop filter into a realizable distributed filter using shunt stubs (open-circuited) or series high-impedance transmission lines (for inductors).

Figure 4.23 shows two other impedance transformations, known as Norton transformations, that are useful in distributed matching networks.

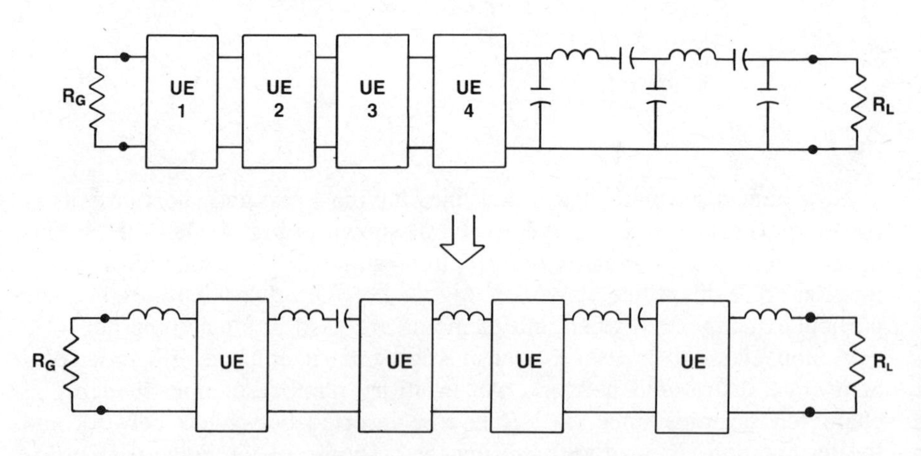

Figure 4.22 Example of inserting unit elements using Kuroda's identities in the design of a distributed matching network.

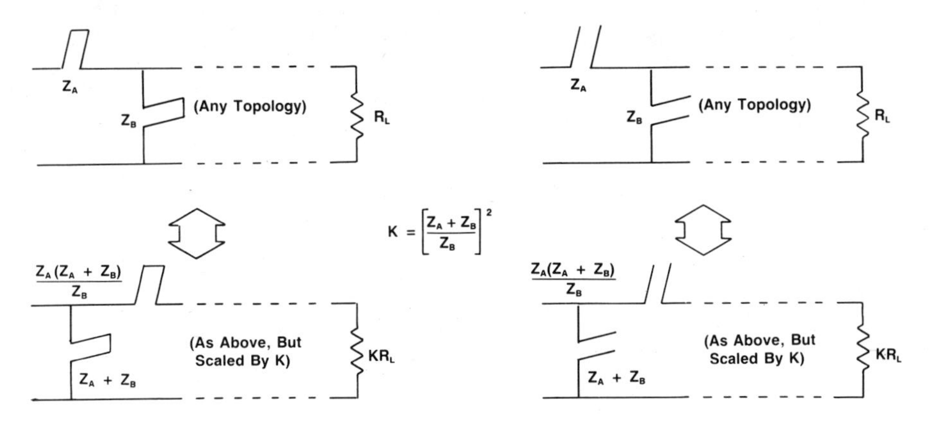

Figure 4.23 Norton transformations.

4.3.2 Gain-Bandwidth Limitations

In designing wide-band matching networks it is important to know the greatest possible bandwidth that can be obtained from the component. In several designs, the input and output impedances of active devices are approximated by lumped elements. For example, input and output impedances of bipolar transistors and FETs are approximated by four general types of equivalent circuits shown in Fig. 4.24. Only circuits in Fig. 4.24(*a*) and (*b*) are fundamentally different, because circuits in Fig. 4.24(*c*) and 4.24(*d*) are duals of the circuits in Fig. 4.24(*a*) and 4.24(*b*), respectively.

It is desirable to optimize a matching network to match out the capacitance over the frequency band of interest, of course, within a given

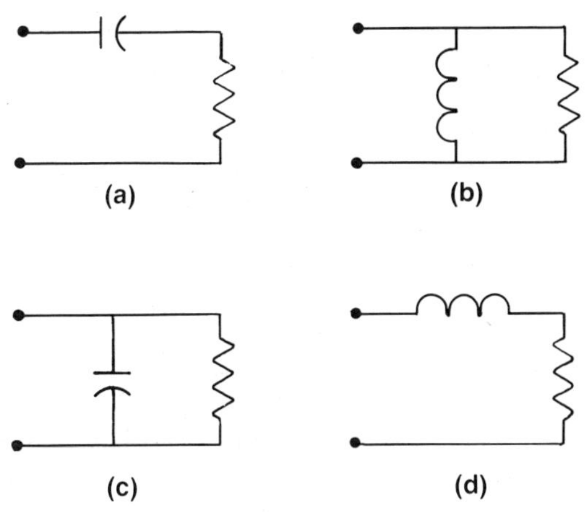

Figure 4.24 Approximate equivalent-circuit models of transistor input/output impedances.

tolerance. For the simple case shown in Fig. 4.24, Bode [14] was the first to derive the fundamental limitation of the matching network. The Bode gain-bandwidth integral restriction is given as

$$\int_0^\infty \ln\left|\frac{1}{\Gamma}\right| d\omega \le \frac{\pi}{RC} \tag{4.40}$$

where Γ is the input reflection coefficient of the matching network. If Γ is constant over the desired band (ω_a to ω_b) and unity everywhere else, then the best reflection coefficient that can be achieved is given by the expression

$$\int_{\omega_a}^{\omega_b} \ln\left|\frac{1}{\Gamma}\right| d\omega = \frac{\pi}{RC}(\omega_b - \omega_a)\ln\left|\frac{1}{\Gamma}\right| \tag{4.41}$$

It becomes obvious that matching is difficult when RC is large, particularly when ordinary filter techniques are used that tend to provide a perfect match (small Γ) at a number of frequencies in the passband. Ideally, a matching network should have a square-shaped response, a maximum $|\Gamma|_{\max}$ in the passband, and $|\Gamma| = 1$ outside the passband. Such a matching network, however, would require an infinite number of matching elements. Fano [1] showed that the value of $|\Gamma|_{\max}$ in the passband converges rapidly to the theoretical minimum ($|\Gamma|_{\max}$ for the $n = \infty$ case) as the number of matching elements increases. Fano's integral restriction for arbitrary load networks (specified only by poles and zeros of the load impedance) is given as

$$\int_0^\infty \ln\left|\frac{1}{\Gamma}\right| d\omega = \frac{\pi}{RC} - \pi\sum_i s_{pi} \tag{4.42}$$

where s_{pi} are defined as the poles of the reciprocal reflection coefficient function ($1/\Gamma$) in the right half complex frequency plane. Since s_{pi} have positive real parts and occur in complex-conjugate pairs, the term $\pi\sum s_{pi}$ reduces the available gain bandwidth.

If instead of flat gain characteristics in the passband, a matching network has to be designed for a tapered gain response given by

$$\begin{aligned} G_M(\omega) &= K\left(\frac{\omega}{\omega_b}\right)^k \qquad && \omega_a \le \omega \le \omega_b \\ &= 0 && \omega_a > \omega > \omega_b \end{aligned} \tag{4.43}$$

where K is the gain scale factor that must be less than or equal to unity, k is the coefficient given as

$$k = \frac{X}{10\log 2} \simeq \frac{X}{3}$$

and X is the desired slope in decibels per octave, then the fundamental integral restriction for the "low-pass" equivalent circuit in Fig. 4.24 becomes

$$\int_{\omega_a}^{\omega_b} \ln\left(\frac{1}{1-K\left(\frac{\omega}{\omega_b}\right)^k}\right) d\omega \le \frac{2\pi}{RC} \tag{4.44}$$

Normalizing, the upper frequency in the passband, $\omega_b = 1$ rad/sec (4.44) can be rewritten as

$$\int_{\omega_a}^{1} \ln \frac{1}{(1-K\omega^k)}\, d\omega \le \frac{2\pi}{\omega_b RC} \tag{4.45}$$

For a flat gain response in the passband ($k = 0$) the optimum gain-bandwidth limitation is given as

$$\left[\ln\left(\frac{1}{1-K}\right)\right] \cdot (1-\omega_a) = \frac{2\pi}{\omega_b RC}$$

or

$$K = 1 - \exp\left\{-\frac{2\pi}{(1-\omega_a)\omega_b RC}\right\} \tag{4.46}$$

Thus, for a specific bandwidth, the limiting value of $K = 1$ cannot be attained. Figure 4.25 gives the optimum gain-bandwidth limitations for

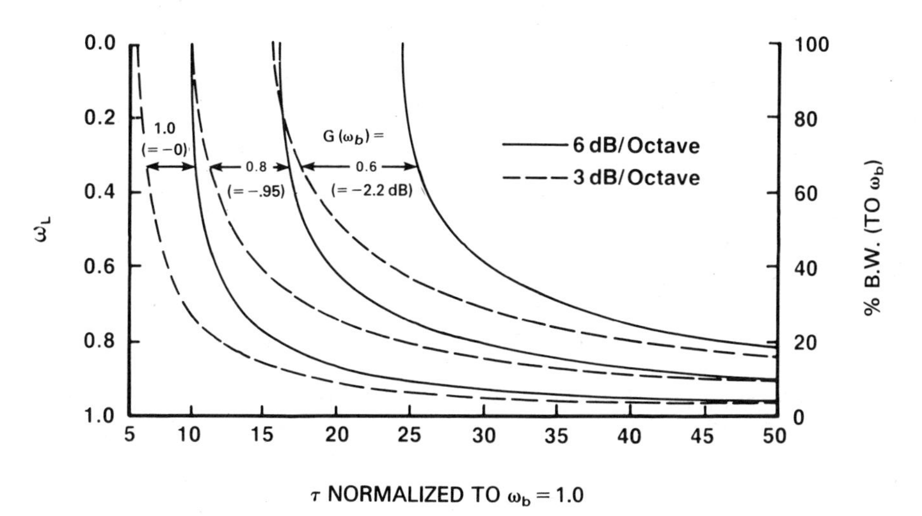

Figure 4.25 Optimum gain-bandwidth limitations for tapered magnitude gain characteristics for equivalent low-pass reactive constraint. (After Ku and Petersen [15], 1975, IEEE, reprinted with permission.)

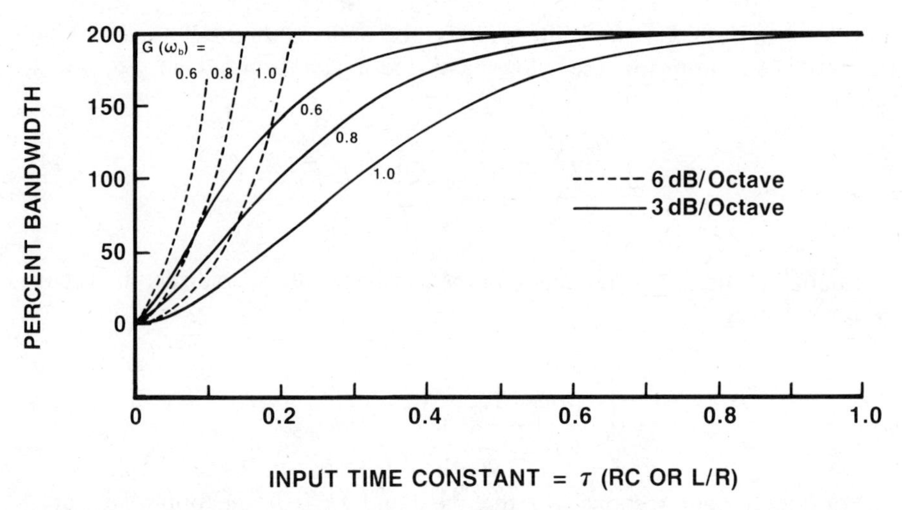

Figure 4.26 Optimum gain-bandwidth limitations for tapered magnitude gain characteristics for equivalent high-pass reactive constraint. (After Ku and Petersen [15], 1975 IEEE, reprinted with permission.)

6-dB/octave and 3-dB/octave gain taper characteristics for an equivalent low-pass reactive constraint. Similar results can be derived for "high-pass" reactive constraint. Figure 4.26 gives the optimum gain-bandwidth limitations for tapered magnitude gain characteristics in a high-pass reactive constraint network [15].

4.3.3 Real Frequency Approach

An alternate approach to design matching networks has been introduced by Carlin and Komiak [16]. This is a real frequency technique that utilizes the measured (real frequency) data, bypassing the analytic gain-bandwidth theory. Carlin and Komiak's results for single matching problems (pure resistance generator and complex load) have been extended to more general cases where both generator and load impedances are complex (double matching problem) [17]. The real frequency approach does not have to assume the topology or analytic form of system transfer function in advance. There is no need to determine an equivalent circuit for the devices.

The heart of the real frequency approach is in the generation of a positive real input impedance, $Z_i(\omega) = R_i(\omega) + jX_i(\omega)$, looking into a lossless matching network with a resistive termination. The transducer power gain of a network is given by

$$G_T(\omega) = \frac{4R_i(\omega)R_L(\omega)}{(R_i + R_L)^2 + (X_i + X_L)^2} \tag{4.47}$$

where $Z_L(\omega) = R_L(\omega) + jX_L(\omega)$ is the load impedance and it is assumed that the unknown input impedance is a minimum reactance function, X_i, related to R_i through Hilbert transformation [18]. The matching network design is divided into two steps. First, $R_i(\omega)$ is described as a set of linear combinations of unknown line segments (X_i becomes a function of the same line segments for a minimum reactance function, Z_i) that are computed in such a way that the transducer gain, G_T, is optimized; for example, to make G_T as flat and as high as possible over a prescribed band. Secondly, Z_i is approximated by a rational function that, in turn, is used to synthesize a lossless network with resistive termination. Linear programming optimization techniques may be employed to compute appropriate line segments and, again, to approximate line segments by rational functions [19].

4.4 NETWORK SYNTHESIS AND OPTIMIZATION

Design of impedance-matching networks, filters, equalizers, and interstage matching networks is best obtained through the use of synthesis, analysis, and optimization steps. For example, in an amplifier design, simple assumptions, such as unilateral devices, make it possible to synthesize and approximate networks. The circuit so obtained is then analyzed without assumptions based upon the actual device parameters. At least, the optimization adjusts the circuit elements of the synthesized network, improves the overall design, and provides room for component variations. While each of the circuit design problems places differing requirement on the synthesis processes, they all can follow basic insertion-loss synthesis techniques. A most general matching network design problem requires the network to operate between complex and unequal impedances and provide a gain slope. Insertion-loss synthesis of passive networks follows the following steps:

Step 1: Formulate the desired insertion-loss function.

Step 2: Approximate the desired response with a rational function of frequency using standard techniques [20].

Step 3: Develop a network that has a frequency response equal to that of the specifying rational approximation. Overall matching network design may find certain network topologies preferable to others, all of which may be consistent with a given synthesis specification. For example, large gain-bandwidth product requirements can only be met with a minimum amount of shunt capacitance at the input. Also, the network design should be able to incorporate parasitic reactances of the active device, at both ends, as shown in Fig. 4.27. Other important considerations include element value realizability and impedance transformation available from the network.

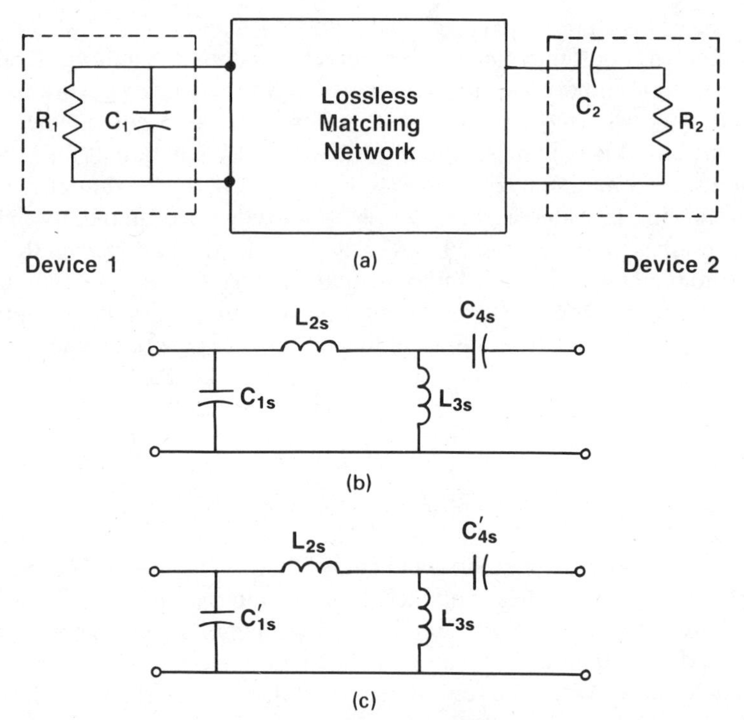

Figure 4.27 Incorporation of device parasitics in the matching network. (*a*) Equivalent circuit of two devices to be matched. (*b*) The synthesized network with appropriate element values, that is, $C_{1s} \geq C_1$ and $C_{4s} \leq C_2$. (*c*) Matching network realization $C'_{1s} = C_{1s} - C_1$ and

$$C'_{4s} = \frac{1}{(1/C_{4s}) - (1/C_2)}$$

Step 4: Adjust the networks that satisfy the frequency response for the required load resistance using impedance transformation as discussed in Section 4.3.1.

Step 5: Lastly, approximate the synthesized network consisting of lumped elements by microwave transmission-line networks.

4.4.1 Insertion-Loss Synthesis

Figure 4.28 shows the schematic of a lossless ladder matching network where the insertion function is given by

$$\text{IL} = \frac{\text{power available from } E_s,\ R_s}{\text{power delivered to load, } R_L} = \frac{1}{G_T} \tag{4.48}$$

The insertion loss synthesis theorem states that a lossless ladder network coupling two resistors having a prescribed insertion loss can be found, provided the insertion-loss function is bounded between 1 and ∞ for all real

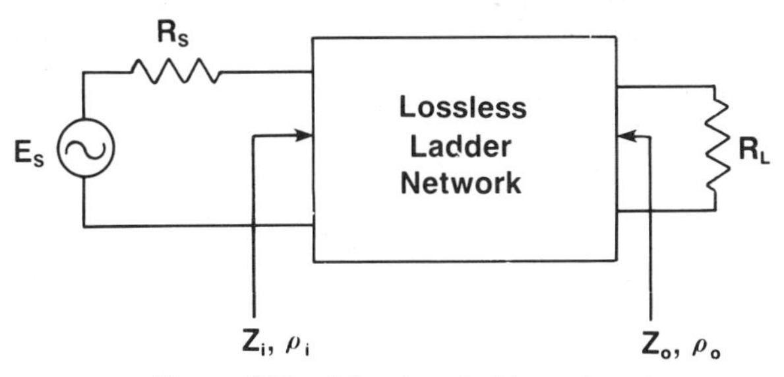

Figure 4.28 A lossless ladder network.

frequencies. That is, the power delivered to the load is always less than the power available from the source. The bounding condition, as the preceding statement is known, forces Z_i to a positive real function. Thus,

$$1 < \mathrm{IL} < \infty \quad \text{for} \quad 0 < \omega < \infty$$

This, in turn, imposes a bound on the reflection coefficient

$$0 < |\rho_i| = |\rho_o| < 1 \tag{4.49}$$

for a lossless ladder network.

The number of transmission zeros at $\omega = 0$ and $\omega = \infty$ and input/output impedance of the network at these frequencies determine the allowable topologies. For example, given an insertion-loss function of the form

$$\mathrm{IL} = \frac{a_0 + a_1\omega^2 + a_2\omega^4 \cdots a_N\omega^{2N}}{\omega^{2J}} \tag{4.50}$$

it is obvious that it has J transmission zeros at dc and another $(N - J)$ transmission zero at $\omega = \infty$. Since the order of the network is equal to N, the network must have J high-pass elements and $(N - J)$ low-pass elements.

From the insertion-loss function it is easy to compute another function that is a ratio of two polynomials

$$|\rho_i|^2 = 1 - \frac{1}{\mathrm{IL}} \tag{4.51}$$

Selecting only the poles of $|\rho_i|^2$ that are in the left half of the complex frequency plane and zeros of $|\rho_i|^2$ in complex pair fashion, the reflection coefficient function $\rho_i(s)$ is constructed. That is,

$$\rho_i(s) = \frac{\pm \prod^{N} (s - s_{0i})}{\prod^{N} (s - s_{pi})} \tag{4.52}$$

The driving-point input impedance function is calculated from (4.52) by substituting

$$Z_i(s) = R_s \frac{1 + \rho_i(s)}{1 - \rho_i(s)} \tag{4.53}$$

where R_s is the source resistance. The actual circuit elements are calculated by performing a partial fraction expansion on (4.53). The following example describes this method. Two alternative solutions are given, depending upon the selection of the zero, while the poles of the function remain the same.

Example. If a network has an insertion-loss function given by a polynomial $\text{IL} = (4 + 5\omega^2 + \omega^4)/4$ and connected to source and load impedance of $1\ \Omega$, determine its circuit values.

Using (4.53)

$$|\rho_i|^2 = \frac{\omega^4 + 5\omega^2}{\omega^4 + 5\omega^2 + 4}$$

$$= \frac{s^4 - 5s^2}{s^4 - 5s^2 + 4} = \frac{s(s + \sqrt{5})(s - \sqrt{5})s}{(s+1)(s+2)(1-s)(2-s)}$$

Thus,

$$\rho_i = \frac{s(s + \sqrt{5})}{(s+1)(s+2)}$$

and

$$Z_i = \frac{2s^2 + 5.236 + 2}{0.764s + 2}$$

for a 1-Ω terminating resistor and zeros of the reflection coefficient at $s = 0$ and at $s = -\sqrt{5}$. Using partial expansion, the network can be described as shown in Fig. 4.29(a).

Selecting the zeros of the reflection coefficient at $s = 0$ and at $s = \sqrt{5}$, the new impedance is given by

$$Z_i = \frac{2s^2 + 0.764s + 2}{5.24s + 2}$$

and the network elements are, once again, obtained using partial fraction expansion and shown in Fig. 4.29(b). Other combination of elements are possible when zero pairs are selected differently.

4.4.2 Topology Selection/Parasitic Element Inclusion

Simply specifying the order of the network and the number of transmission zeros at dc and ∞ can allow several alternate topologies for matching

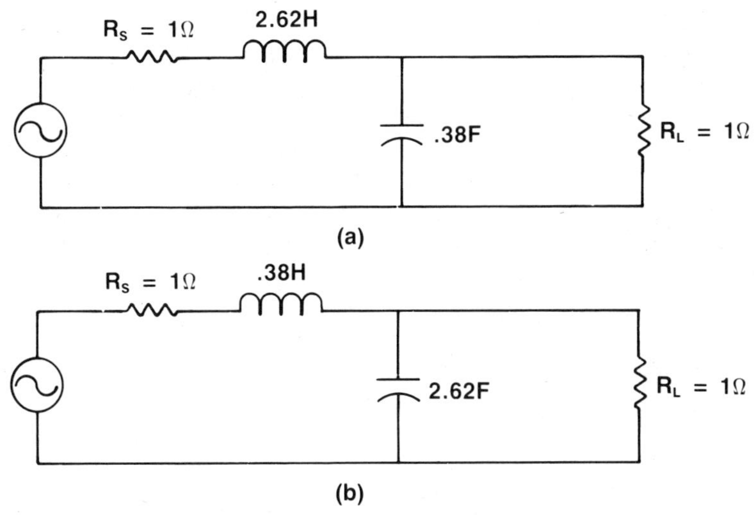

Figure 4.29 Example of insertion-loss synthesis.

networks. For example, if a third-order network has two transmission zeros at dc and another one at ∞, a generalized expression for the loss is given as

$$\mathrm{IL} = \frac{a_0 + a_2\omega^2 + a_4\omega^4 + a_6\omega^6}{\omega^4} \tag{4.54}$$

The network should have two elements that are either shunt inductors or series capacitors and one element that may be a shunt capacitor or a series inductor. Actual circuit elements are determined from the real zeros of ρ_i as described in the previous section. Several possible topologies that have the same frequency response are shown in Fig. 4.30. Each topology is evaluaated based on reliability, convenience, and complexity. Selection of the appropriate topology is based upon its consistency with parasitic elements. For example, for an interstage matching network in an amplifier the first element, at the input side, should be a shunt capacitor and the last element, nearest to the output end, should be a series capacitor to match with the

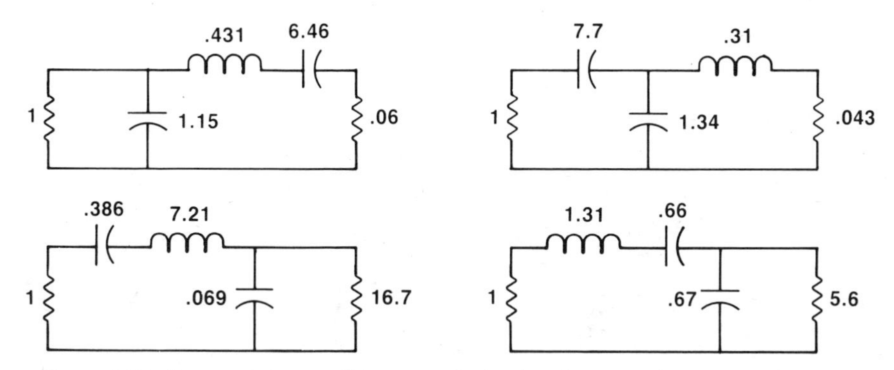

Figure 4.30 Some of the possible networks having the same frequency response

input equivalent circuit of the device (FET). The presence of parasitic elements limits the gain-bandwidth product of the matching network, and thus it is important to incorporate inherent parasitic elements into the synthesized network. Thus, the shunt capacitor value, at the input end of the network, should be larger than the capacitance associated with the device output equivalent circuit. In this way, the device capacitance can be absorbed and it becomes a part of the equivalent circuit as shown in Fig. 4.27. Similarly the series capacitance value, at the other network port, should be smaller than the capacitance in the series R–C equivalent circuit for the device input. Fano's theorem which was discussed in Section 4.3.2, can be used to derive the relationship between the frequency response of the network and the parasitic element absorption. Table 4.3 gives the constraint relations for four lossless elements at the input of a network. It becomes obvious that to maximize parasitic element absorption, the input reflection coefficient zeros should lie in the left half of the complex frequency plane. Since the zeros of ρ_2 (output reflection coefficient) are negative of zeros of ρ_1, in a lossless network it is possible to trade off element absorbability between input and output ends by placing appropriate zeros of ρ_1 and ρ_2 in the right-hand plane. This property is very useful for an optimal network design (maximum gain-bandwidth product). Parasitic absorption is

TABLE 4.3 Fano's Constraint Relations for Four Lossless Elements at the Input of a Matching Network

Parasitic Element	Fano's Constraint on Gain Bandwidth
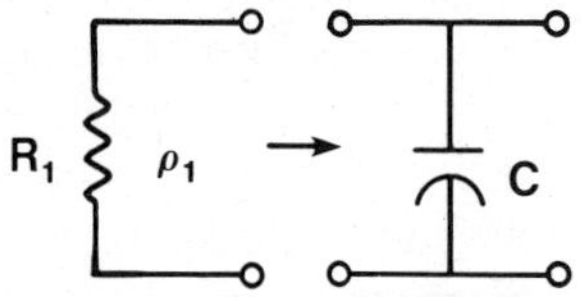 R_1, ρ_1 → shunt C	$\int_0^\infty \ln\left\|\frac{1}{\rho_1}\right\| d\omega \le \pi\left(\frac{1}{R_1C} - \sum Z_{\rho 1 \mathrm{RHP}}\right)$
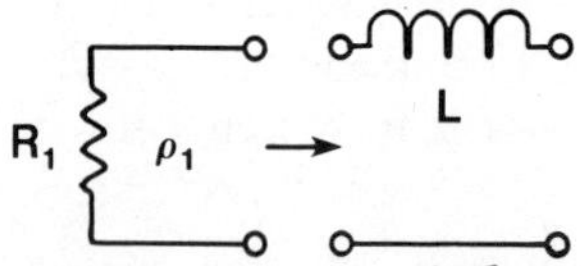 R_1, ρ_1 → series L	$\int_0^\infty \ln\left\|\frac{1}{\rho_1}\right\| d\omega \le \pi\left(\frac{R_1}{L} - \sum Z_{\rho 1 \mathrm{RHP}}\right)$
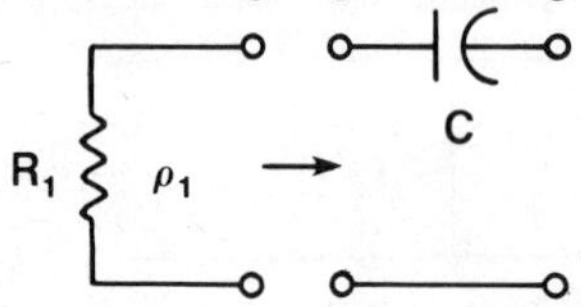 R_1, ρ_1 → series C	$\int_0^\infty \omega^{-2} \ln\left\|\frac{1}{\rho_1}\right\| d\omega \le \pi\left(R_1C - \sum \frac{1}{Z_{\rho 1 \mathrm{RHP}}}\right)$
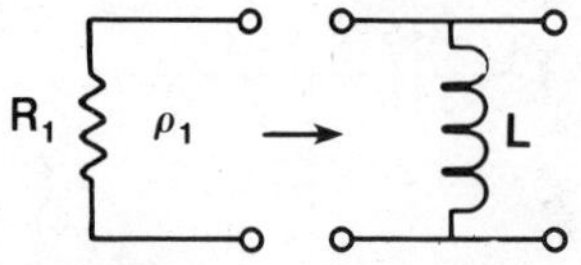 R_1, ρ_1 → shunt L	$\int_0^\infty \omega^{-2} \ln\left\|\frac{1}{\rho_1}\right\| d\omega \le \pi\left(\frac{L}{R_1} - \sum \frac{1}{Z_{\rho 1 \mathrm{RHP}}}\right)$

indirectly affected by all parameters that affect the frequency response of the network. These parameters include the bandwidth (ω_u and ω_L), minimum insertion loss, and the passband slope characteristics. Minimum insertion loss in a network can be chosen based upon the value of the parasitic element to be absorbed, desired bandwidth, and passband insertion-loss slope using the equations given in Section 4.2.3. A slightly larger reduction in gain than previously calculated should be chosen to allow for nonideal insertion loss in an actual circuit. The insertion-loss function is modified to take into account minimum insertion loss (MIL) requirement as given in Section 4.2.3.

4.4.3 Microwave Realization

Exact analysis and synthesis methods developed so far provide a unified approach to the design of networks that is valid at all frequencies. At low RF frequencies (in the MHz region) networks are based on lumped-element circuit models. At microwave frequencies, however, it is necessary to use distributed elements such as transmission lines and stubs. Richards [21] first showed that networks composed of lumped-element resistors and commensurate TEM transmission lines behaved in a manner analogous to the lumped-element network under a transformation known as Richard's transform. This opened the way to new exact methods for the design of microwave networks. Richards's transformation is given as

$$s = j \tan \frac{\pi\omega}{2\omega_0} = j\Omega \tag{4.55}$$

where ω is the usual frequency of the network and ω_0 is the frequency of the network at which all line lengths are a quarter wavelength long. Thus, synthesis using the lumped-element variable $\Omega = j\omega$ can be correlated with synthesis procedures in the distributed frequency variable s. Using the transformation, it can be shown that open-circuited transmission lines map into capacitors and short-circuited transmission lines correspond to lumped inductors. The frequency response of the lumped-element network repeats after every $2\omega_0$ intervals when a lumped-element network is transformed into a distributed equivalent using Richard's transform (Fig. 4.31). This happens because Ω goes from zero to infinity as ω goes from 0 to ω_0. When ω exceeds ω_0, the sign of Ω changes and its value goes from negative infinity to zero at $\omega = 2\omega_0$. This process repeats between $2\omega_0$ and $4\omega_0$, $4\omega_0$ and $6\omega_0$, and so on.

While the preceding approach is valid in theory, it is limited in its use because an actual distributed network designed this way is physically impossible to implement. For example, series-connected stubs cannot be constructed because microwave elements require isolation from each other

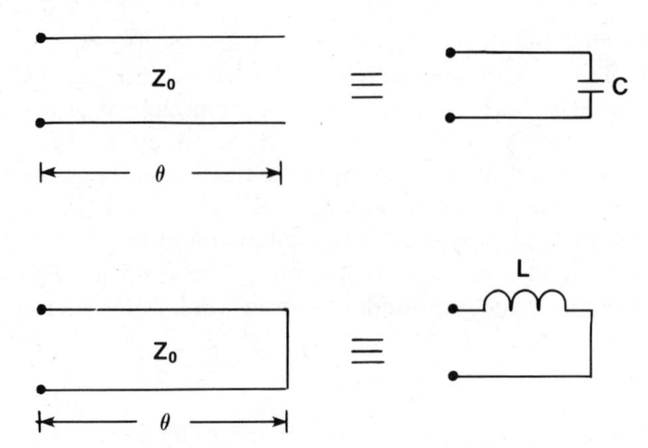

Figure 4.31 Richard's transform to convert a lumped-element network to a distributed network.

to obtain predicted performance. This problem is overcome using a distributed element without equivalent under the inverse Richard's transform, a unit element consisting of a transmission line of length $\lambda_0/4$, and the application of a set of network equivalences given in Section 4.3.1. A unit element is described in the chain matrix form as

$$\begin{bmatrix} A & B \\ C & D \end{bmatrix} = \frac{1}{\sqrt{1-s^2}} \begin{bmatrix} 1 & Z_0 s \\ \dfrac{s}{Z_0} & 1 \end{bmatrix} \tag{4.56}$$

An example using unit elements to obtain a reliable microwave network was given in Section 4.3.1.

4.5 CAD TOOLS

Several commercially available and special-purpose user-oriented interactive programs for impedance-matching networks of desired frequency response have changed the design philosophy of microwave circuits. Passive network synthesis has long been used for the design of lumped-element filters (typically real and equal termination and flat frequency response in the passband). Powerful network synthesis methods have made it possible to optimize filterlike structures where both terminations are complex and unequal (due to parasitic elements associated with active devices) and where the passband response requires a gain rolloff. The synthesis process becomes a powerful and efficient design tool when implemented in a computer-aided design (CAD) program. Chapter 14 discusses computer-aided design of matching networks and other microwave circuits.

REFERENCES

1. Fano, R. M., "Theoretical Limitations on the Broadband Matching of Arbitrary Impedances," *J. Franklin Institute*, Vol. 249, Jan./Feb. 1950, pp. 57–83, 139–154.
2. Arnold, R. M., "Transmission Line Impedance Matching Using the Smith Chart," *IEEE Trans. Microwave Theory Tech.*, Vol. MTT-22, Nov. 1974, pp. 977–978.
3. Lepoff, J. M., "Matching: When Are Two Lines Better than One?," *Microwaves*, Vol. 20, Mar. 1981, pp. 74–78.
4. Jasik, H., *Antenna Engineering Handbook*, McGraw-Hill, New York, 1961, Chap. 31.
5. Smith, P. H., "Transmission Line Calculator," *Electronics*, Vol. 12, Jan. 1939, pp. 29–31; also, Smith, P. H., "An Improved Transmission Line Calculator," *Electronics*, Vol. 17, Jan. 1944, pp. 130–133, 318, 320, 322, 324–25.
6. Cohn, S. B., "Optimum Design of Stepped Transmission Line Transformers," *IEEE Trans. Microwave Theory Tech.*, Vol. MTT-3, April 1955, pp. 16–21.
7. Guccione, S., "Nomograms Speed Design of $\lambda/4$ Transformers," *Microwaves*, Vol. 14, Nov. 1974, pp. 50–52.
8. Khilla, A. M., "Optimum Continuous Microstrip Tapers Are Amenable to Computer-aided Design," *Microwave J.*, Vol. 26, May 1983, pp. 221–224.
9. Grunau, W. C. and C. R. Mason, "A Simplified Solution for Tapered Transformer Lines," *Microwaves*, Vol. 20, Oct. 1981, pp. 82–83.
10. Stanley, W. D., *Network Analysis with Applications*, Reston, New York, 1985.
11. Guillemin, E. A., *Synthesis of Passive Networks*, Wiley, New York, 1957.
12. Vincent, G. A., "Impedance Transformation without a Transformer," *Frequency, Technol.*, Vol. 7, Sept. 1969, pp. 15–21.
13. Chakmanlan, J. S., "Control VSWR Bandwidth in T-section Networks," *Microwaves*, Vol. 20, July 1981, pp. 87–94.
14. Bode, H. W., *Network Analysis and Feedback Amplifier Design*, Van Nostrand, Princeton, N.J., 1945, pp. 365–366.
15. Ku, W. H., and W. C. Petersen, "Optimum Gain-Bandwidth Limitations of Transistor Amplifiers as Reactively Constrained Active Two-Port Networks," *IEEE Trans. Circuits Syst.*, Vol. CAS-22, June 1975, pp. 523–533.
16. Carlin, H. J., and J. J. Komiak, "A New Method of Broadband Equalization Applied to Microwave Amplifiers," *IEEE Trans. Microwave Theory Tech.*, Vol. MTT-27, Feb. 1979, pp. 93–99.
17. Yarman, B. S., "A Simplified Real Frequency Technique for Broadband Matching of Complex Generator to a Complex Load," *RCA Rev.*, Vol. 43, Sept. 1982, pp. 529–541.
18. Balabanian, N., *Network Synthesis*, Prentice-Hall, Englewood Cliffs, N.J., 1958.
19. Yarman, B. S., "Real Frequency Broadband Matching Using Linear Programming," *RCA Rev.*, Vol. 43, Dec. 1982, pp. 626–653.
20. Mellor, D. J., and J. G. Linvill, "Synthesis of Interstage Networks of Prescribed Gain versus Frequency Slopes," *IEEE Trans. Microwave Theory Tech.*, Vol. MTT-23, Dec. 1975, pp. 1013–1020.

21. Richards, P. I., "Resistor-Transmission-line Circuits," *Proc. IRE*, Vol. 36, Feb. 1948, pp. 217–220.

PROBLEMS

4.1 A $(70 + j50)$-Ω impedance of a solid state device has to be matched to a 50-Ω system. Calculate the characteristic impedance and length of a single line matching section at 9 GHz. What is the bandwidth if the maximum acceptable reflection coefficient is 0.1?

4.2 What will be the impedance and length of the matching section if the impedance in Problem 4.1 has to be matched to a complex load of $45 + j45$ Ω?

4.3 Calculate the length of a short-circuited stub to match a $(70 + j50)$-Ω load to a 50-Ω system at 9 GHz. Assume a characteristic impedance of 50 Ω for all transmission lines.

4.4 Design a three-section quarter-wave transformer to match a 50-Ω source to a 125-Ω load at 9 GHz with a reflection coefficient less than 0.1. What is the bandwidth of this transformer?

4.5 If the load in Problem 4.4 has to be matched to the source (50 Ω) using an exponential tapered line, design the tapered section for maximum reflection coefficient $\rho_e = 0.1$.

4.6 Develop a canonic network using continued fraction approach from the reactance function

$$Z(s) = \frac{(s^2 + 3)(s^2 + 4)}{s(s^2 + 1)}$$

4.7 The input impedance of a device is represented by a shunt RC network where $R = 10$ Ω and $C = 5$ pF. The device is to be matched to a 50-Ω source at 9 GHz over a 20% bandwidth using a third-order network. Select an appropriate network, and then calculate element values and the gain bandwidth limitation of the network.

4.8 Recalculate the element values in Problem 4.7 for the case that the matching network has to have an insertion loss slope of 3 dB/octave.

5 HYBRIDS AND COUPLERS

5.1 INTRODUCTION

Hybrids and couplers form an indispensable component group in modern microwave integrated-circuit technology. With the inventions of new planar transmission lines like strip line, microstrip line, fin line, dielectric image line, and their derivatives, hybrid and coupler technology has undergone a substantial change over the past decade, due to the rapidly growing applications of MICs in the electronic-warfare, communications, and radar industries.

Despite the fact that the basic philosophy behind the operation of such couplers remains the same as in couplers designed using conventional two-wire transmission line, their analyses and syntheses are quite involved. This is because most of these lines support hybrid modes due to inhomogeneity in configuration. However, the present-day analysis and synthesis techniques for such hybrids and couplers are believed to have gained maturity.

This chapter describes the design aspects of planar hybrids and couplers in as self-contained a presentation as possible within a limited space. In what follows in this section, we present the basics of hybrids and couplers, and discuss different types of hybrids and couplers and their applications. The next section describes the design of matched hybrid Ts, hybrid rings, and 90° hybrids. This is followed by a section on coupled line couplers, both the TEM and the distributed types, as well as other miscellaneous types of couplers. The final section includes various aspects of coupler design, such as losses and improvement of directivity.

5.1.1 Basics of Hybrids and Couplers

A hybrid or a directional coupler can in principle be represented as a multiport network. In such a network the port into which the electrical power is fed is called the incident port. The ports through which the desired amounts of coupled power are extracted are called coupled ports, while the rest of the ports are called isolated ports. Although hybrids and couplers having up to six ports find applications in many systems, we will mostly restrict ourselves to the discussion of four-port networks without loss of generality.

Consider the four-port network shown in Fig. 5.1. If P_1 is the power fed into Port 1 (which is matched to the generator impedance) and P_2, P_3, and

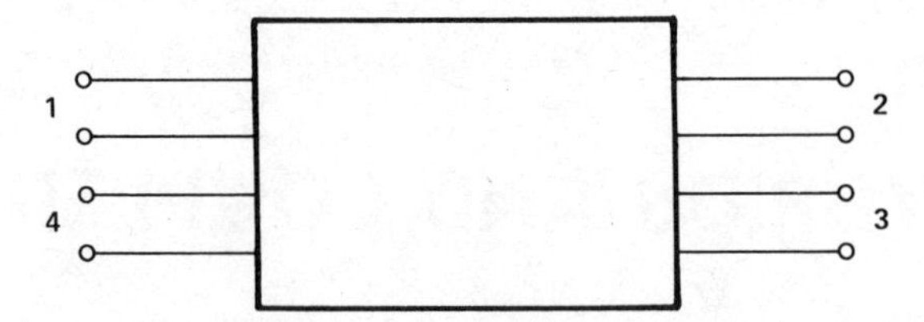

Figure 5.1 Four-port network.

P_4 are the powers available at the Ports 2, 3, and 4, respectively (while each of the ports is terminated by its image impedance), the coupling coefficient is defined as

$$C = -10 \log \left| \frac{P_n}{P_1} \right|, \qquad n = 2, 3, 4 \tag{5.1}$$

If Port 3 happens to be the desired coupled port, the coupling coefficient is given by

$$C = -10 \log \left| \frac{P_3}{P_1} \right| \quad \text{dB} \tag{5.2}$$

If Port 4 is the desired uncoupled port, the desired isolation is given by

$$I = -10 \log \left| \frac{P_4}{P_1} \right| \quad \text{dB} \tag{5.3}$$

The transmission to the primary Port 2 is given by

$$T = 10 \log \left| \frac{P_2}{P_1} \right| \quad \text{dB} \tag{5.4}$$

The measure of directivity between the coupled and the uncoupled ports is given by

$$D = I - C \tag{5.5}$$

As a general practice, the performance of a hybrid or a directional coupler is specified in terms of its coupling, directivity, and the characteristic impedance at the center frequency of its band of operation. These data enable the circuit designer to calculate the structural parameters of the coupler.

5.1.2 Types of Hybrids and Couplers

Hybrids are a direct coupled type of circuit. Couplers can be direct coupled, parallel coupled or aperture coupled.

Figure 5.2(*a*) shows the simplest direct-coupled hybrid. This is a branch-line coupler, consisting of two main lines coupled by two $\lambda/4$ line sections spaced $\lambda/4$ apart, where λ is the wavelength. Such a branch-line coupler can also be in circular form as shown in Fig. 5.2(*b*). In either case, the total length of all the lines is one wavelength.

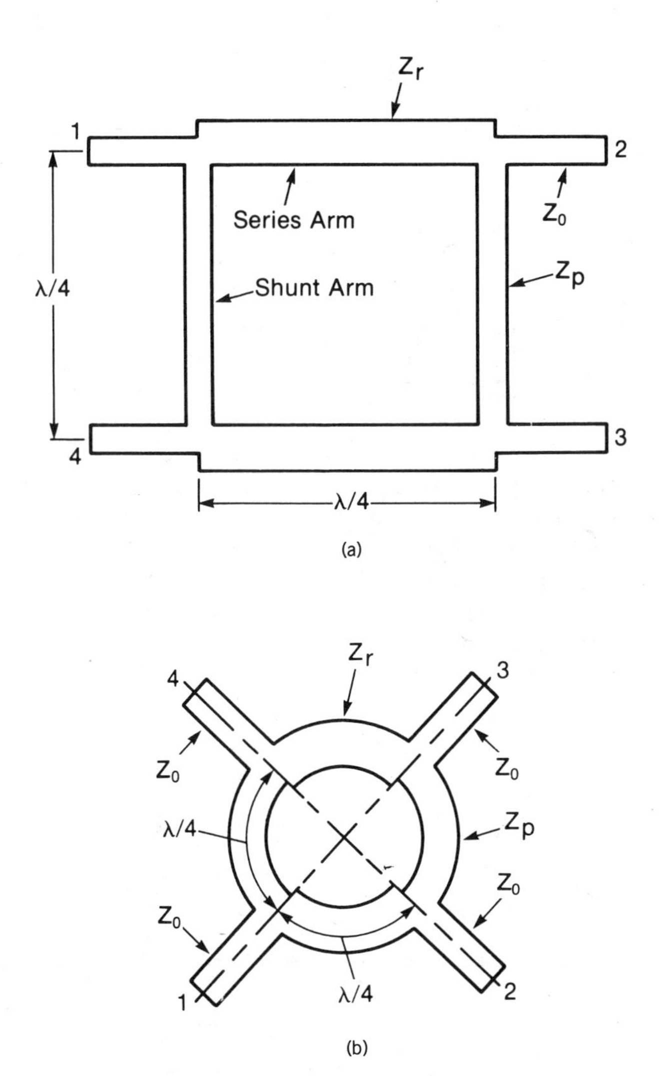

Figure 5.2 (*a*) Single-section branch-line coupler. (*b*) Clrcular form of branch-line coupler.

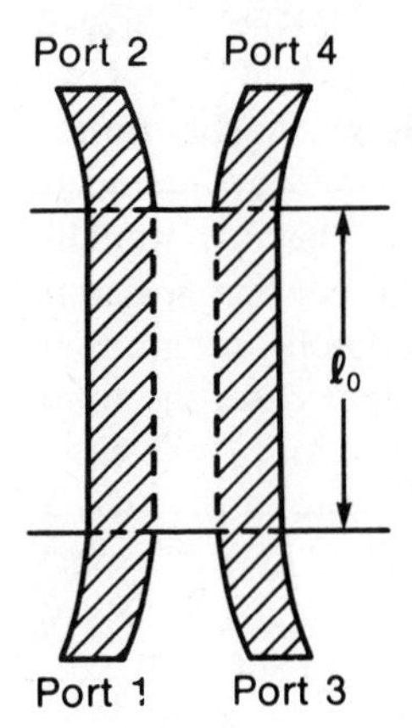

Figure 5.3 Direct-coupled coupler.

Figure 5.3 shows a direct-coupled coupler. It consists of two directly coupled transmission lines. The lines can have any form depending upon the application.

A parallel-coupled line coupler (Fig. 5.4) can be of two types, that is, the TEM type or the distributed type. In the former, the coupled transmission lines support a pure TEM or a quasi-TEM mode. In the latter, the mode supported by the coupled lines are non-TEM in nature. While all of these types of couplers are realized using planar integrated-circuit technology, there can be another type of coupler using aperture coupling, through some common ground plane (as used in conventional aperture-coupled waveguide couplers as shown in Fig. 5.5) with two similar or dissimilar types of planar transmission lines.

In order to achieve tighter coupling over a wider bandwidth multisections of the previously mentioned couplers can be cascaded in tandem (to be discussed in a later section).

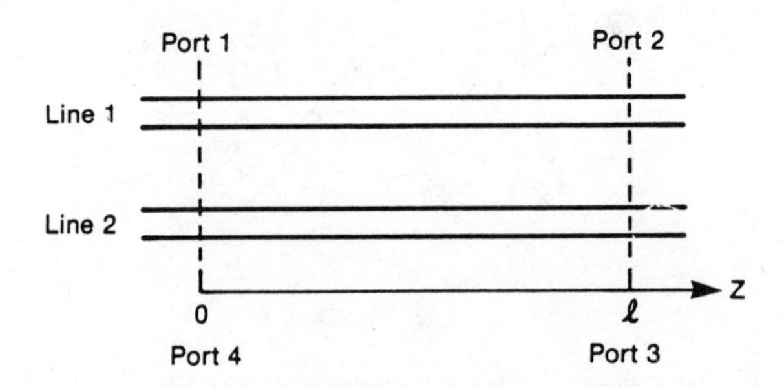

Figure 5.4 Parallel-coupled line coupler.

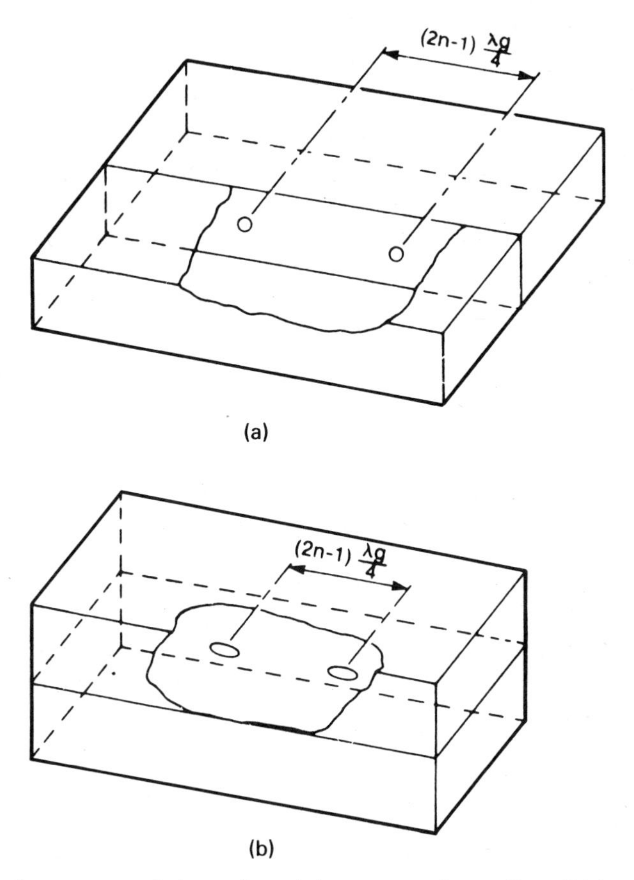

Figure 5.5 Aperture-coupled couplers; (*a*) narrow wall coupling, (*b*) broad wall coupling.

5.1.3 Applications

Virtually all kinds of microwave circuits use hybrids or couplers in one form or the other. In general, the areas of application can be divided into two parts.

1. *Passive*. Tuners, delay lines, filters, and matching networks. Sometimes an array of couplers may be needed for a desired performance of the network.
2. *Active Circuits*. Mainly used as directional couplers in balanced amplifiers, mixers, attenuators, modulators, discriminators and phase shifters.

5.2 DESIGN OF HYBRIDS

5.2.1 90° Hybrid

The simplest 90° hybrid is the branch-line coupler shown in Figure 5.2(*a*). For a certain input power at Port 1, under matched condition, there will be 90° phase difference between the waves appearing at Ports 2 and 3, at the center frequency at which each arm is exactly a quarter wavelength long. This 90° phase difference varies over ±5° for a 10% change in frequency around the center frequency. The coupling bandwidth is 20%, but its usable bandwidth is limited to 10% due to an unacceptable change in the isolation over a bandwidth exceeding 10%. Ideally such couplers can be designed for 3–9 dB coupling.

There are three main arm losses in branch-line couplers. They are losses due to the portion of power coupled to the secondary arm, and the power dissipated due to dielectric and copper losses.

Branch-line couplers can be realized using virtually all kinds of planar transmission lines, for example, strip line, microstrip line, slot line, fin line, and image line. However, the basic design principle is extremely simple and the same in all cases.

The coupling factor is determined by the ratio of the impedance of the shunt and series arms and is optimized to meet the proper match over the required bandwidth.

For 90° hybrids the following conditions hold good [see Fig. 5.2(*a*)].

$$\frac{P_2}{P_3} = \left(\frac{Z_0}{Z_p}\right)^2 \tag{5.6}$$

$$\left(\frac{Z_0}{Z_r}\right)^2 = \left(\frac{Z_0}{Z_p}\right)^2 + 1 \tag{5.7}$$

Although the two-branch coupler is the most fundamental structure, it has a very narrow bandwidth. This disadvantage is overcome by using multisection couplers. But in most cases planar transmission lines require too wide a range of impedances for this purpose. These may sometimes be difficult to realize physically. On the other hand, very wide linewidths may require unreasonable aspect ratios at higher frequencies due to shorter wavelengths. The physically unrealizable high-impedance line can be avoided by using a modified hybrid ring [1].

Complete analytical design techniques for such hybrids using Chebyshev and Zolotarev [2, 3] functions are available. However, such techniques may sometimes be unsuitable for the design of planar circuits because of the wide impedance range problem, as previously mentioned. Muraguchi et al. [4] presented a computer-aided design technique which is most suitable for an optimum design. The method is as follows.

Let us consider the lossless reciprocal four-port branch-line hybrid that has a twofold symmetry around the x, y-planes, as shown in Fig. 5.6(a). The S-matrix of the hybrid can be written as

$$[S]=\begin{bmatrix} S_{11} & S_{21} & S_{31} & S_{41} \\ S_{21} & S_{11} & S_{41} & S_{31} \\ S_{31} & S_{41} & S_{11} & S_{21} \\ S_{41} & S_{31} & S_{21} & S_{11} \end{bmatrix} \tag{5.8}$$

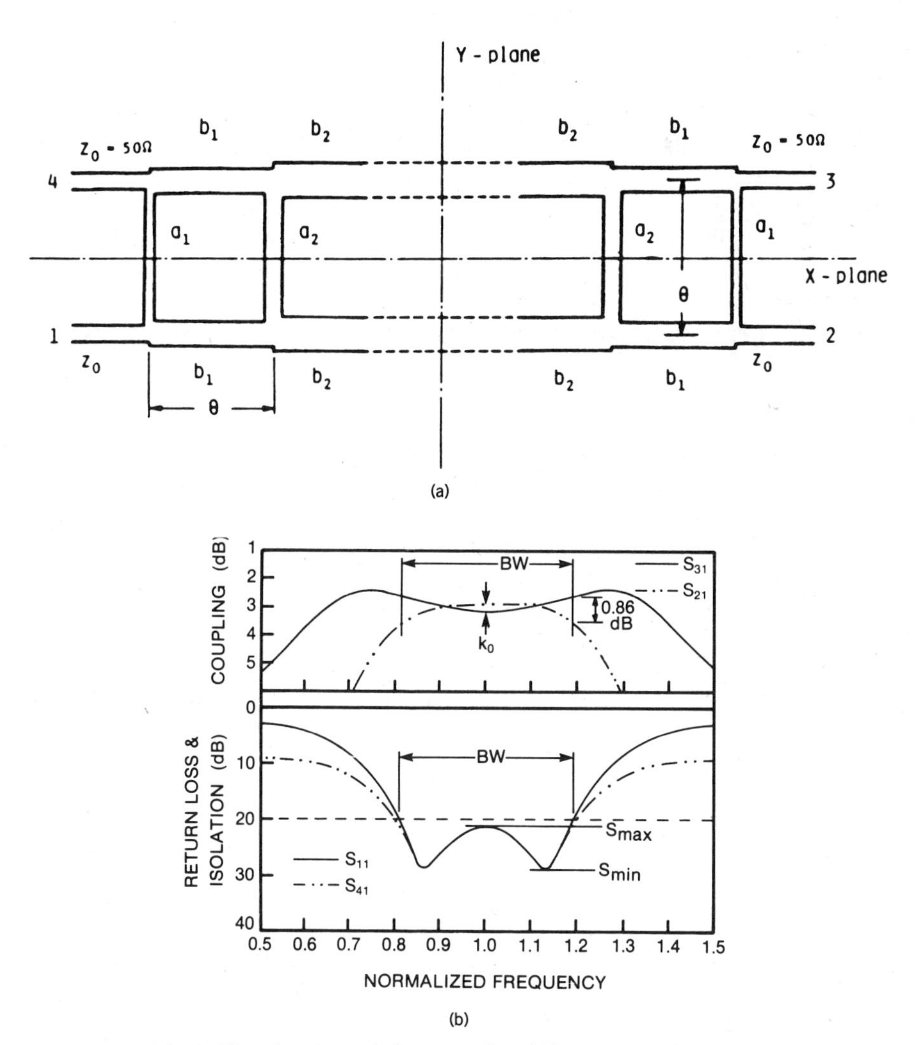

Figure 5.6 (a) Multisection branch-line coupler. (b) Frequency response of a three-branch optimized branch-line coupler. (O. Muraguchi et al. [4]. 1983 IEEE, reprinted with permission.)

together with the unitary relationship

$$|S_{11}|^2 + |S_{21}|^2 + |S_{31}|^2 + |S_{41}|^2 = 1 \tag{5.9}$$

Other conditions for the return loss, the power-coupling coefficient C, and the isolation are

$$|S_{11}|^2 = 0$$
$$|S_{21}|^2 = C$$
$$|S_{31}|^2 = 1 - C$$
$$|S_{41}|^2 = 0$$

The previous quantities are required to be within certain tolerance limits over a broad frequency band. Such tolerance limits depend upon the design requirements. For instance, for a two-branch 3-dB microstrip hybrid reported by Muraguchi et al. [4], assuming less than ±0.43-dB variation in coupling and better than −20-dB isolation, a bandwidth of 30% was realized about the 1-GHz center frequency.

Having defined the tolerance limits, the following penalty function is formed

$$F(a_1, \ldots, a_n, b_1, \ldots, b_m) = \sum_{j=1}^{4} g_j \tag{5.10}$$

$$g_1 = \sum_{i=1}^{N} |S_{11}(f_i)|^2$$

$$g_2 = \sum_{i=1}^{N} [|S_{21}(f_i)|^2 - C]$$

$$g_3 = \sum_{i=1}^{N} [|S_{31}(f_i)|^2 - (1 - C)]$$

and

$$g_4 = \sum_{i=1}^{N} |S_{41}(f_i)|^2$$

where

$$f_i = f_0\left(1 + \frac{i-1}{D}\right), \qquad i = 1, \ldots, N \tag{5.11}$$

and N is the number of sampling points. The f_i are the corresponding sampling frequencies, and f_0/D is the sampling interval. All four scattering parameters S_{11}, S_{21}, S_{31}, and S_{41} are considered despite the fact that they are not independent of each other for a lossless hybrid. The parameters a_1 through a_n and b_1 through b_m are obtained numerically in order that the penalty function F may be minimized by a suitable search method. The optimization process has the following steps.

Step 1: The first computation is performed without any restrictions on the line impedances by changing the sampling interval $1/D$ only.

Step 2: If some undesirably low and/or high impedance is encountered in the result of the first computation, the second computation is performed after one of their impedance values is changed to an appropriate fixed value.

Step 3: If there still happens to be undesirably low- and/or high-impedance lines in the result of the second computation, the third computation is performed after two or three impedance values are held constant.

The computation process is repeated until the prescribed tolerance limits are exceeded.

Because of the two-dimensional nature of a planar transmission line, it is impossible to define the circuit uniquely. Therefore a transmission-line or one-dimensional approach is used. For hybrids with large impedance steps, junction reactance effects are added after the optimization has been carried out. A final experimental correction of the electrical length is also needed. Typical frequency characteristics of an optimized three-branch coupler are shown in Fig. 5.6(*b*).

5.2.2 Ring Form of Branch-line Hybrid

This is a circular ring version of the square 90° hybrid described in the previous section. Therefore all the discussion of the previous section is equally valid for hybrid rings. The configuration is shown in Fig. 5.2(*b*) and is particularly advantageous in the realization of microstrip phase detectors and balanced mixers, with all ports matched.

5.2.3 Matched Hybrid T (Rat-Race Hybrid)

A matched hybrid T is a special kind of ring form of the branch-line coupler in which the circumference is an odd multiple of $(3/2)\lambda$. As a result the phase response is 0°/180°. The simplest version of a matched hybrid T is shown in Fig. 5.7. The Ports A–B, B–C and C–D are separated by 90°, and Port A and Port D are three quarter-wavelengths away from each other.

Because of the impedance and the phase relationships shown in the structure, any power fed into Port C splits equally into two parts that add up in phase at Ports B and D, and out of phase at Port A. As a result Port A is

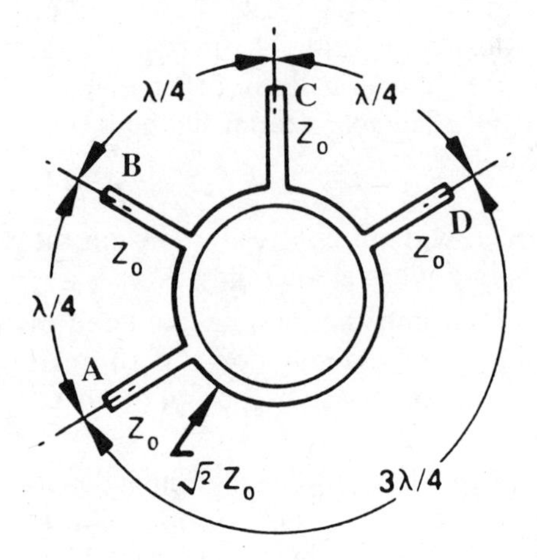

Figure 5.7 Hybrid ring.

isolated from the input. Similarly, power fed at Port A divides equally between Ports B and D with 180° phase difference, and Port C remains isolated.

The frequency response of a typical (3/2) wavelength hybrid T is shown in Figure 5.8(*a*) and 5.8(*b*). It is observed that it offers around a 20% bandwidth in terms of matching, split, and isolation. Moreover, the 180° phase relationship is much more frequency sensitive than the 0° phase relationship. Still, such a response is quite adequate for applications in mixers, single-sideband (SSB) generators, etc.

For high-frequency applications, a circumferential length of (3/2) wavelengths may pose fabricational difficulties due to unrealizable aspect ratios of the transmission lines. This problem is overcome by having a ring of circumferential length of two wavelengths, where Port A and Port D are located 5/4 λ away on the ring. The bandwidth of the ring length is around 20%.

The idealized S-matrix of a matched hybrid T is given by

$$S = \frac{1}{\sqrt{2}} \begin{bmatrix} 0 & 1 & 1 & 0 \\ 1 & 0 & 0 & 1 \\ 1 & 0 & 0 & -1 \\ 0 & 1 & -1 & 0 \end{bmatrix} \tag{5.12}$$

The design of a hybrid amounts to realizing the required transmission-line sections with proper phase velocities and characteristic impedances. Direct synthesis equations can be used in cases of strip line [5], microstrip [6], suspended microstrip, and inverted microstrip [7], which support pure TEM or quasi-TEM modes of propagation.

Realizations of the hybrids using non-TEM transmission lines like slot line, fin line, or image line use iterative techniques with the help of accurate

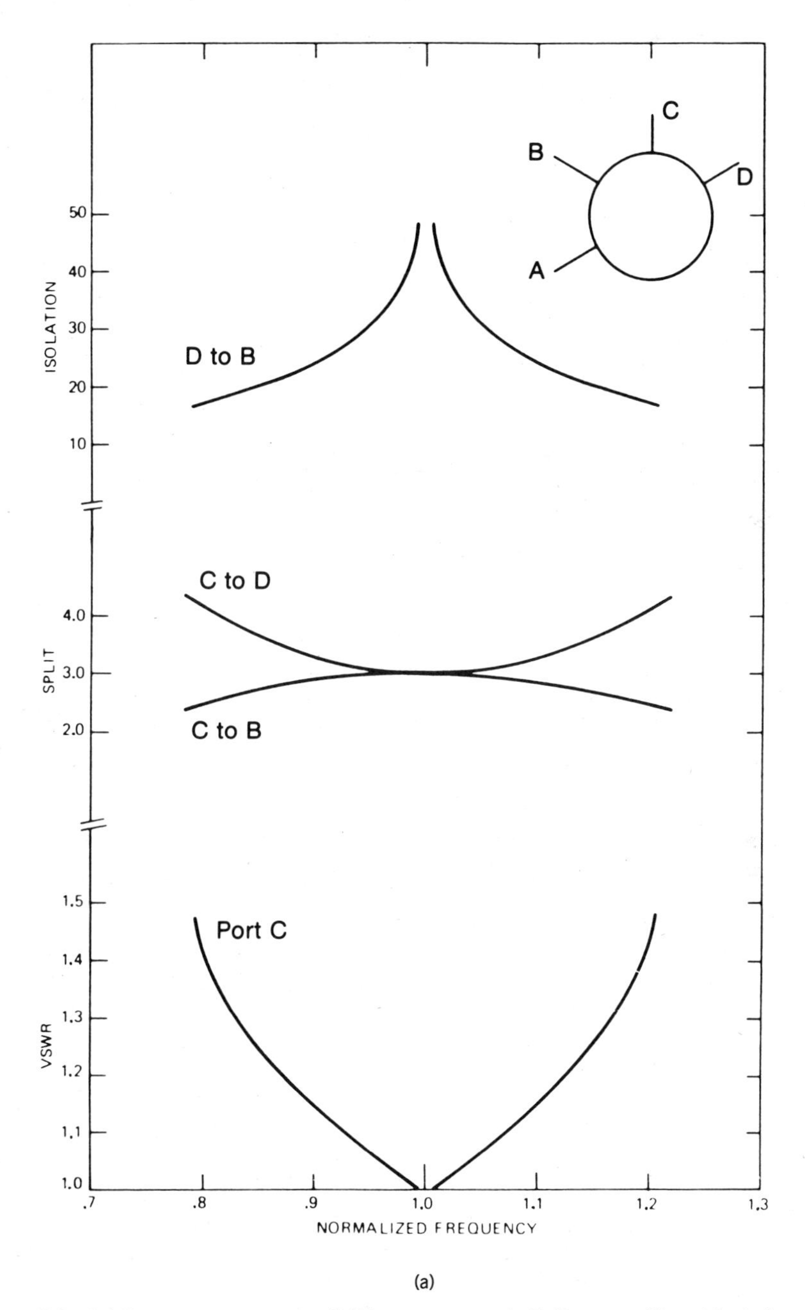

(a)

Figure 5.8 (*a*) Response curves for (3/2) rat-race magic T. Power split and isolation are expressed in decibels.

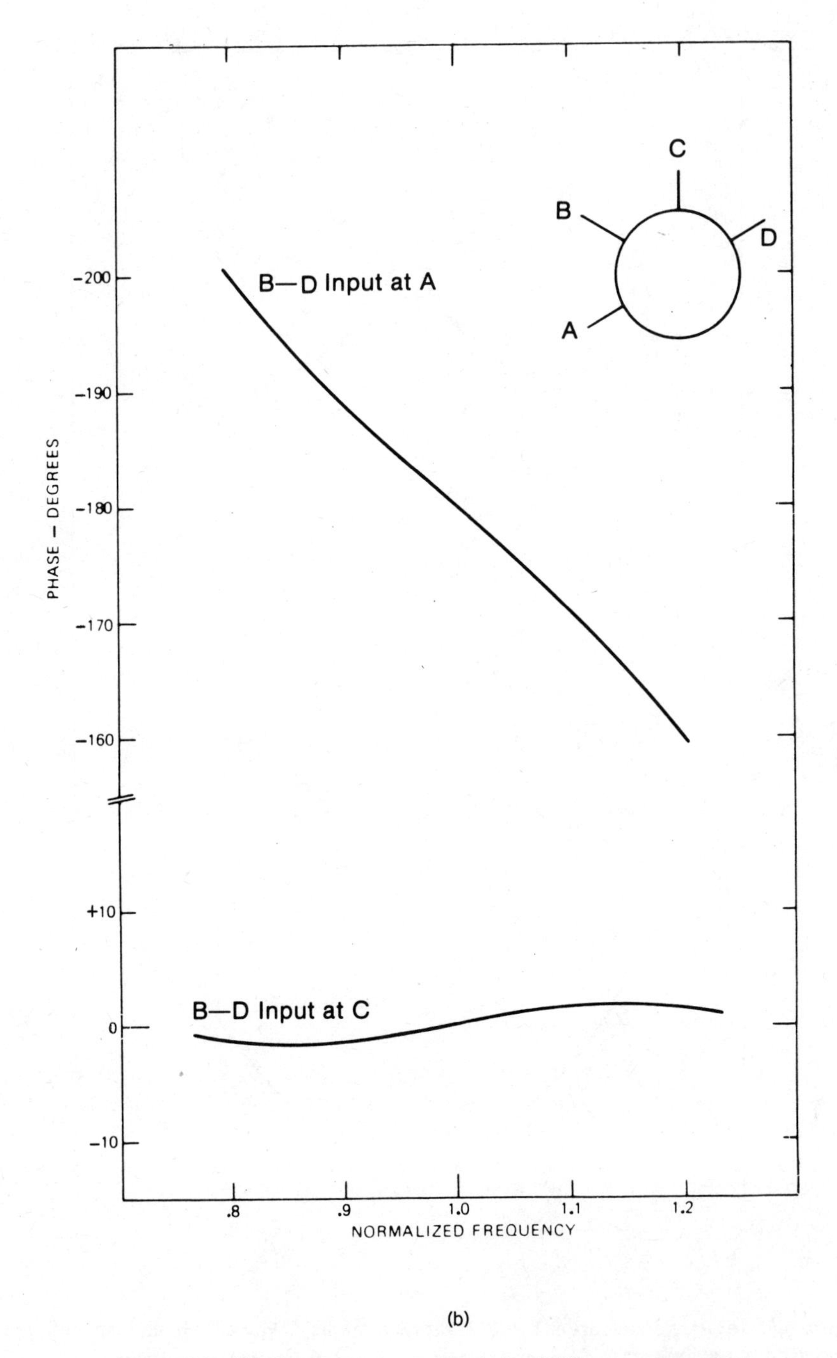

Figure 5.8 (*b*) Phase response curves for (3/2) rat-race magic T.

analysis equations [7–9] depending upon the mode of propagation. However, closed-form design equations are also available.

Microstrip and strip-line hybrids have been realized and successfully used in commercially available microwave balanced mixers and other circuit components. Development of fin-line and image-line hybrids are still at the experimental stage [10]. Figure 5.9(*a*) shows a unilateral fin-line matched hybrid T, and Fig. 5.9(*b*) shows its frequency response in the *Ka*-band [11].

5.3 COUPLED-LINE DIRECTIONAL COUPLERS

The couplers described in the previous sections are inherently of narrow bandwidth. Broadbanding of microwave couplers is achieved in a number of different ways, depending upon the application.

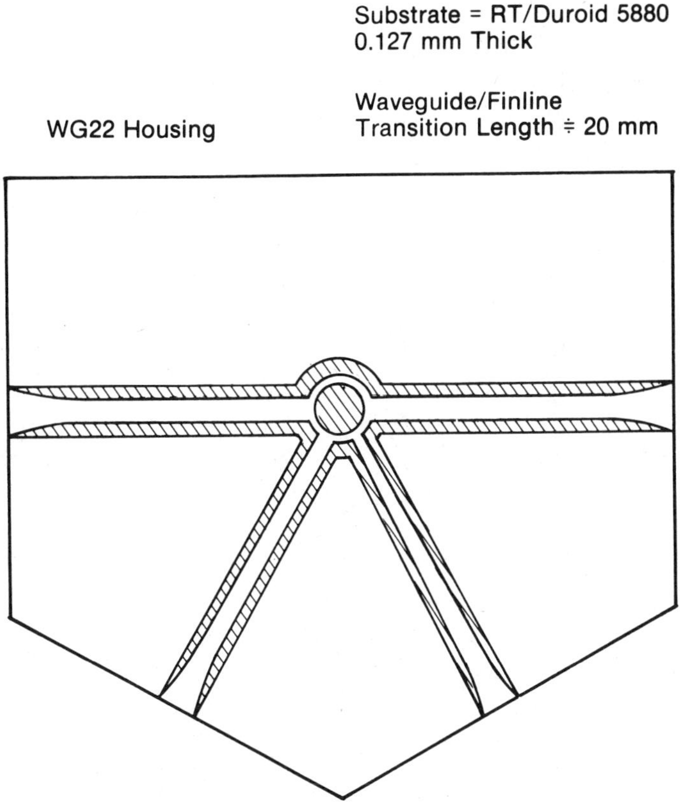

Figure 5.9 (*a*) Fin-line hybrid ring.

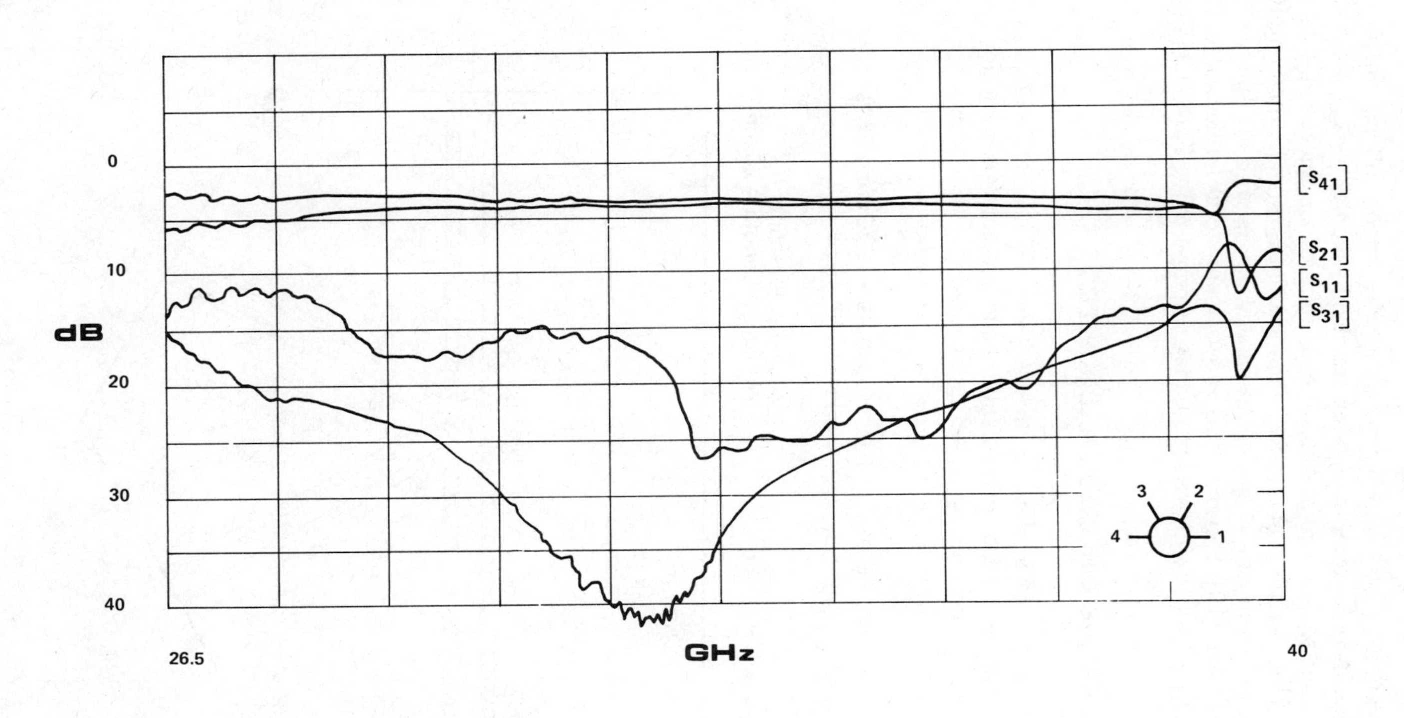

(b)

Figure 5.9 (*b*) Frequency response of fin-line hybrid ring. (C. M. D. Rycroft[11], reprinted with permission of Marconi Research Center.)

Broadband couplers are either aperture coupled or parallel coupled. Aperture coupling is used very successfully in conventional waveguide techniques for realizing directional couplers with large bandwidths. Although aperture coupling is a convenient way of realizing directional couplers using two different planar transmission lines, directional couplers using the same kind of planar transmission lines are realized more efficiently using parallel coupling.

5.3.1 Directional Couplers Using Aperture-Coupled Lines

Figure 5.10 shows a simple two-hole directional coupler. Electromagnetic energy is coupled from the primary guide to the secondary guide due to the field radiated by the excited electric and magnetic dipoles generated at the holes by the propagating electromagnetic wave in the primary guide. The holes are spaced such that the round-trip phase shift of a wave through them should be 180°. Therefore the backward-traveling waves in the secondary guide will be completely out of phase to cancel each other at Port 3. If the coupled lines have the same propagation constant, then the forward-traveling waves in the secondary guide will be of the same phase regardless of hole spacing and are added at Port 4.

Such aperture-coupled directional couplers can be realized using various kinds of planar transmission-line combinations, for example (1) microstrip–microstrip, (2) microstrip–image line, (3) image line–image line, (4) image line–trough guide, or (5) trough guide–trough guide. Some possible combinations are shown in Fig. 5.11.

Expressions for Single-Aperture Coupling. The design of aperture-coupled directional couplers requires an evaluation of the coupling and the directivity of a simple single-aperture coupler. This is accomplished in the following way.

Let α_e and α_m be the electric and magnetic polarizabilities of the aperture, respectively. Then the fields radiated into the secondary guide by

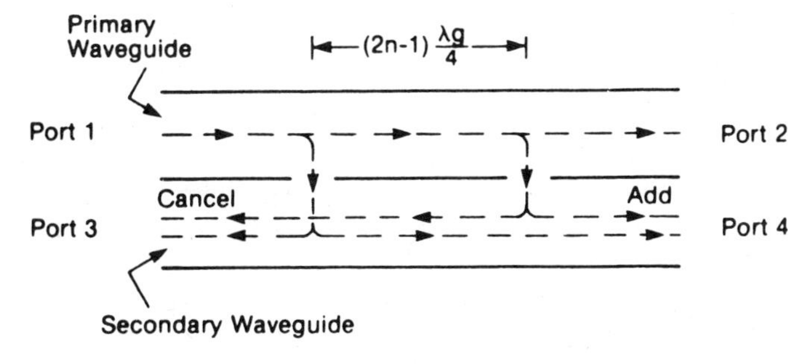

Figure 5.10 Schematic of a two-aperture directional coupler.

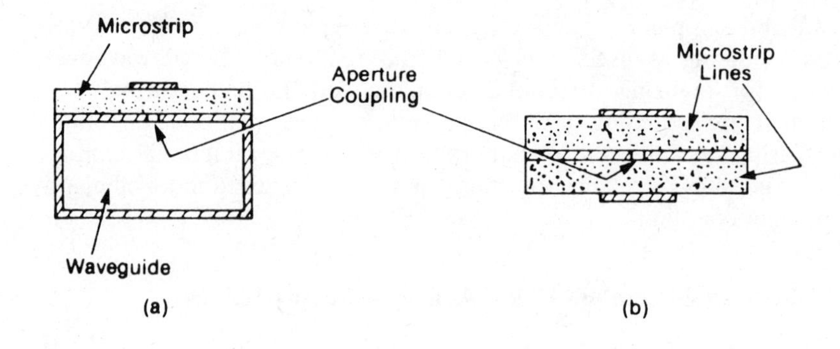

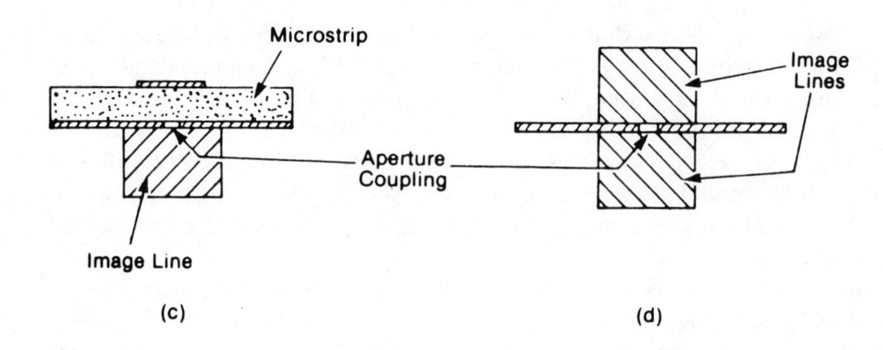

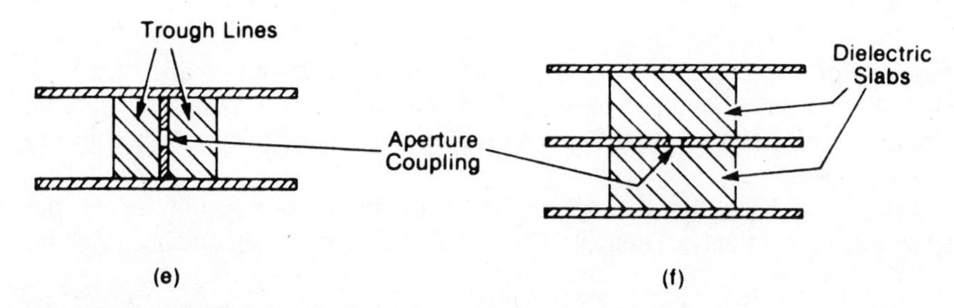

Figure 5.11 Possible combinations of various aperture-coupled lines. (*a*) microstrip–waveguide coupling, (*b*) microstrip–microstrip line, (*c*) microstrip–image line, (*d*) image line–image line, (*e*) trough line–trough line, and (*f*) dielectric guide–dielectric guide.

the electric dipole may be expressed as

$$\mathbf{E}_e = \begin{cases} A_1\mathbf{E}^+, & z \geq 0 \\ A_2\mathbf{E}^-, & z \leq 0 \end{cases} \tag{5.13}$$

$$\mathbf{H}_e = \begin{cases} A_1\mathbf{H}^+, & z \geq 0 \\ A_2\mathbf{H}^-, & z \leq 0 \end{cases} \tag{5.14}$$

whereas those radiated by the magnetic dipole are

$$\mathbf{E}_m = \begin{cases} B_1\mathbf{E}^+, & z \geq 0 \\ B_2\mathbf{E}^-, & z \leq 0 \end{cases} \tag{5.15}$$

$$\mathbf{H}_m = \begin{cases} B_1\mathbf{H}^+, & z \geq 0 \\ B_2\mathbf{H}^-, & z \leq 0 \end{cases} \tag{5.16}$$

where the superscript + denotes propagation along the positive z direction and the superscript − indicates propagation along the negative z direction. The electric and magnetic fields in the secondary guide are represented by $\mathbf{E}^+$, $\mathbf{E}^-$, $\mathbf{H}^+$, and $\mathbf{H}^-$. The subscripts e and m denote the electric and the magnetic components, respectively.

The expressions for the amplitude coefficients can be written as

$$A_1 = A_2 = -\frac{j\omega}{P_n}\mathbf{E}\cdot\mathbf{P} \tag{5.17}$$

$$B_1 = -B_2 = \frac{j\omega\mu_0}{P_n}\mathbf{H}\cdot\mathbf{M}$$

where ω is the angular frequency, μ_0 is the permeability of free space, and P_n is defined at $z = 0$ as

$$P_n = 2\int_S\int \mathbf{E}^+ \times \mathbf{H}^+ \cdot \hat{z}\, ds \tag{5.18}$$

The electric and magnetic dipole moments **P** and **M**, respectively, are given by

$$\mathbf{P} = -\epsilon_0\overline{\epsilon(f)}\alpha_e\mathbf{E} \tag{5.19}$$

$$\mathbf{M} = -\alpha_m\mathbf{H} \tag{5.20}$$

where **E** and **H** are the electric and magnetic fields, respectively, in the primary guide and $\overline{\epsilon(f)}$ is the effective dielectric constant at the aperture. An expression for the effective dielectric constant is given as

$$\overline{\epsilon(f)} = \frac{\epsilon_{ep}(f)\epsilon_{es}(f)}{\epsilon_{ep}(f) + \epsilon_{es}(f)} \tag{5.21}$$

where $\epsilon_{ep}(f)$ and $\epsilon_{es}(f)$ are the frequency-dependent effective dielectric constants in the primary and the secondary guides, respectively.

The total field radiated into the secondary guide is given by

$$\mathbf{E} = \begin{cases} (A_1 + B_1)\mathbf{E}^+, & z \geq 0 \\ (A_2 + B_2)\mathbf{E}^-, & z \leq 0 \end{cases} \tag{5.22}$$

$$\mathbf{H} = \begin{cases} (A_1 + B_1)\mathbf{H}^+, & z \geq 0 \\ (A_2 + B_2)\mathbf{H}^-, & z \leq 0 \end{cases} \tag{5.23}$$

For all cases just described, it can be shown that $A_1 + B_1 \neq 0$. This implies that for the single-aperture coupler the forward coupled wave is never equal to zero, that is, there is always radiation in the forward direction or toward Port 4. For no radiation in the backward direction, that is, to Port 3, the following condition needs to be satisfied [12, 13].

$$\frac{\overline{\epsilon(f)}}{\sqrt{\epsilon_{ep}(f)\epsilon_{es}(f)}} = \frac{\alpha_m}{\alpha_e} \cos \theta \tag{5.24}$$

where θ is the angle between the longitudinal axes of the two guides.

For a circular aperture

$$\alpha_e = -\tfrac{2}{3}r^3 \tag{5.25}$$

$$\alpha_m = \tfrac{4}{3}r^3 \tag{5.26}$$

where r is the radius of the aperture.

For rectangular slots

$$\bar{\alpha}_e = \frac{\pi d^2 l}{12}\, \hat{y} \tag{5.27}$$

$$\bar{\alpha}_m = (0.233\, dl^2 + 0.044\, l^3)\hat{x} + \frac{\pi d^2 l}{16}\, \hat{z} \tag{5.28}$$

where $d \times l$ are the dimensions of the slot.

The forward coupling coefficient is given by

$$C_f = -20 \log|(A_1 + B_1 \cos \theta)| \tag{5.29}$$

and the backward coupling coefficient is given by

$$C_b = -20 \log|(A_2 + B_2 \cos \theta)| \tag{5.30}$$

Using the preceding analysis, expressions for the coupling coefficients for various couplers are shown in Table 5.1.

TABLE 5.1 Expressions for the Coupling Coefficients for Couplers

For Image Guide to Image Guide

$$C_f = -20 \log|(A_1 + B_1 \cos\theta)|$$
$$= -20 \log \frac{\frac{4}{3}\omega r^3 \cos^2(k_x x_0)[\frac{1}{2}\epsilon_0\epsilon_r\eta_0 + (\mu_0 \cos\theta/\eta_0)]}{\{a + [\sin(2k_x a)/2k_x]\}\{b + [\sin(2k_y b)/2k_y]\}}$$

$$C_b = -20 \log|(A_2 + B_2 \cos\theta)|$$
$$= -20 \log \frac{\frac{4}{3}\omega r^3 \cos^2(k_x x_0)[\frac{1}{2}\epsilon_0\epsilon_r\eta_0 - (\mu_0 \cos\theta/\eta_0)]}{\{a + [\sin(2k_x a)/2k_x]\}\{b + [\sin(2k_y b)/2k_y]\}}$$

For definition of parameters, see [21]

For Microstrip Line to Image Line

$$C_f = -20 \log \left| \frac{\frac{4}{3} r^3 \omega [\frac{1}{2}\epsilon_0\epsilon_r\eta_0 + (\mu_0\epsilon_e(f)/\eta_0)]}{a_1 b_1 \{1 + [\sin(2k_x a_1)/2k_x a_1]\}\{1 + [\sin(2k_y b_1)/2k_y b_1]\}} \right|$$

$$C_b = -20 \log \left| \frac{\frac{4}{3} r^3 \omega [\frac{1}{2}\epsilon_0\epsilon_r\eta_0 - (\mu_0\epsilon_e(f)/\eta_0)]}{a_1 b_1 \{1 + [\sin(2k_x a_1)/2k_x a_1]\}\{1 + [\sin(2k_y b_1)/2k_y b_1]\}} \right|$$

$$C_f = -20 \log \left| \frac{\frac{4}{3} r^3 \omega [\frac{1}{2}\epsilon_0\epsilon_r\eta_0 + (\mu_0\epsilon_e(f)/\eta_0)]}{h W_e(f)\sqrt{\epsilon_e(f)}} \right|$$

$$C_b = -20 \log \left| \frac{\frac{4}{3} r^3 \omega [\frac{1}{2}\epsilon_0\epsilon_r\eta_0 - (\mu_0\epsilon_e(f)/\eta_0]}{h W_e(f)\sqrt{\epsilon_e(f)}} \right|$$

For definition of parameters, see [13]

For Waveguide to Image Line

$$C_f = -20 \log \left| \frac{\frac{4}{3} r^3 \omega \cos(k_x d) \cos(\pi d/2a)[(\omega/2)\epsilon_0\epsilon_r\eta_0 + \beta \cos\theta]}{\{a_1 + [\sin(2k_x a_1)/2k_x]\}\{b_1 + [\sin(2k_y b_1)/2k_y]\}} \right|,$$

$$C_b = -20 \log \left| \frac{\frac{4}{3} r^3 \omega \cos(k_x d) \cos(\pi d/2a)[(\omega/2)\epsilon_0\epsilon_r\eta_0 - \beta \cos\theta]}{\{a_1 + [\sin(2k_x a_1)/2k_x]\}\{b_1 + [\sin(2k_y b_1)/2k_y]\}} \right|;$$

For Image Line to Waveguide

$$C_f = -20 \log \left| \frac{\frac{4}{3} r^3 \omega \cos(k_x d) \cos(\pi d/2a)[(\epsilon_0\epsilon_r/2) + \mu_0 Y_w \cos\theta/\eta_0]}{2abY_w} \right|$$

$$C_b = -20 \log \left| \frac{\frac{4}{3} r^3 \omega \cos(k_x d) \cos(\pi d/2a)[(\epsilon_0\epsilon_r/2) - \mu_0 Y_w \cos\theta/\eta_0]}{2abY_w} \right|.$$

For definition of parameters, see [15]

Effects of Large-Size Aperture and Finite Conductor Thickness. The preceding expressions for coupling ignored the effects of the finite common ground plane thickness and large aperture size. It is assumed that the aperture's major dimension is considerably smaller than a quarter

TABLE 5.2 Cutoff Wavelengths for Circular and Rectangular Apertures

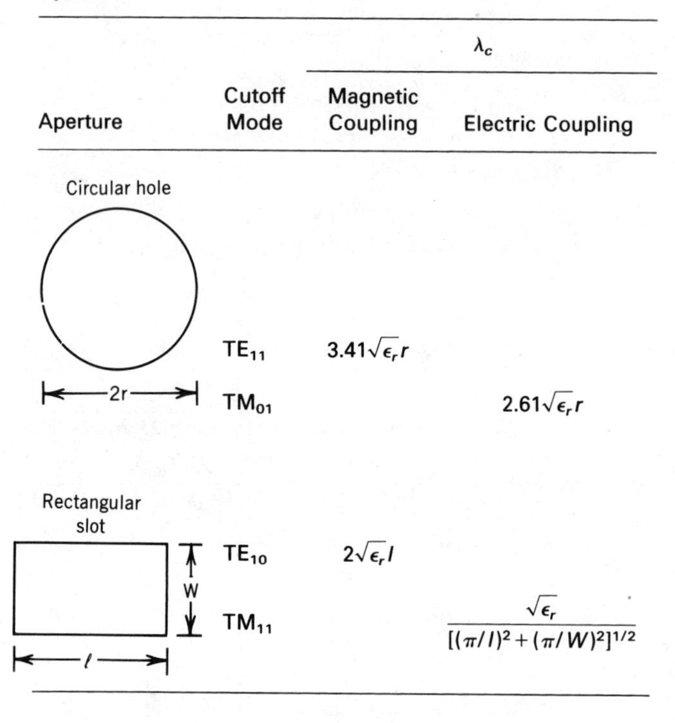

Aperture	Cutoff Mode	λ_c Magnetic Coupling	λ_c Electric Coupling
Circular hole (diameter $2r$)	TE_{11}	$3.41\sqrt{\epsilon_r}\,r$	
	TM_{01}		$2.61\sqrt{\epsilon_r}\,r$
Rectangular slot (length l, width W)	TE_{10}	$2\sqrt{\epsilon_r}\,l$	
	TM_{11}		$\dfrac{\sqrt{\epsilon_r}}{[(\pi/l)^2+(\pi/W)^2]^{1/2}}$

wavelength. The correction factor for both these coupling factors can be expressed as [12, 14].

$$C_F = \frac{\exp\{(-2\pi t A/\lambda_c)\sqrt{[1-(\lambda_c/\lambda_0)^2]}\}}{1-(\lambda_c/\lambda_0)^2} \tag{5.31}$$

where t is the thickness of the ground plane, λ_c is the cutoff wavelength of the aperture for the appropriate mode of operation excited, λ_0 is the operating wavelength, and the factor A accounts for the interactions of local fields on either side of the aperture. The factor A has been determined empirically to be 3 for a narrow slot and unity for a circular aperture. Cutoff wavelengths λ_c for rectangular and circular apertures are given in Table 5.2. The proper values for the coupling factors with the thickness of ground plane and larger aperture effect taken into account are obtained by multiplying A_1, B_1, A_2, and B_2 by C_F with λ_c related to the type of coupling involved.

The coupling characteristics of single-hole couplers using image line–image line reported by Bahl and Bhartia [14] are shown in Figure 5.12(*a*).

Design of Directional Coupler. The design of a directional coupler requires the use of an array of apertures to obtain good directivity and coupling. For $N+1$ apertures the forward and backward coupled waves are given by [15]

$$C_f = Ae^{-jN\beta_1 d} \sum_{n=0}^{N} C_n e^{-jn(\beta_2-\beta_1)d} \tag{5.32}$$

$$C_b = A \sum_{n=0}^{N} D_n e^{-jn(\beta_1+\beta_2)d} \tag{5.33}$$

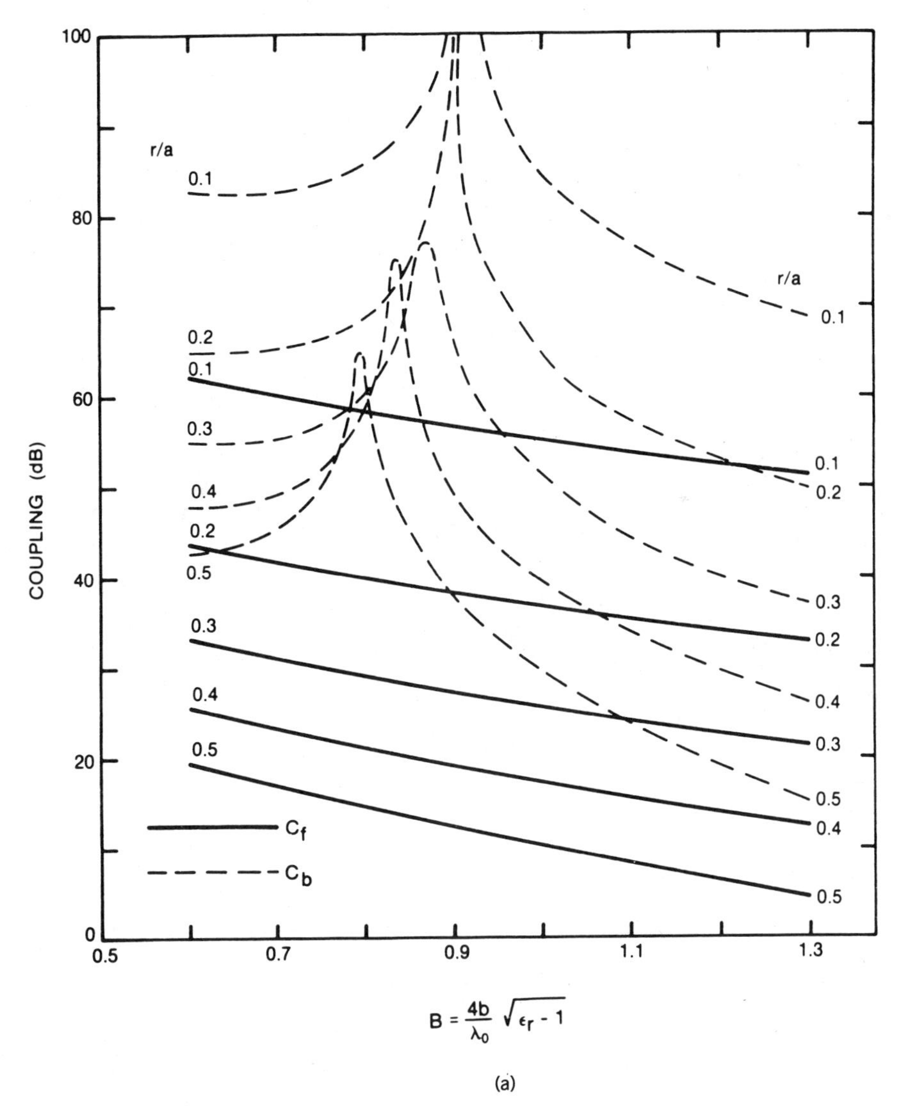

(a)

Figure 5.12 (*a*) Characteristics of single-aperture image-guide coupler.

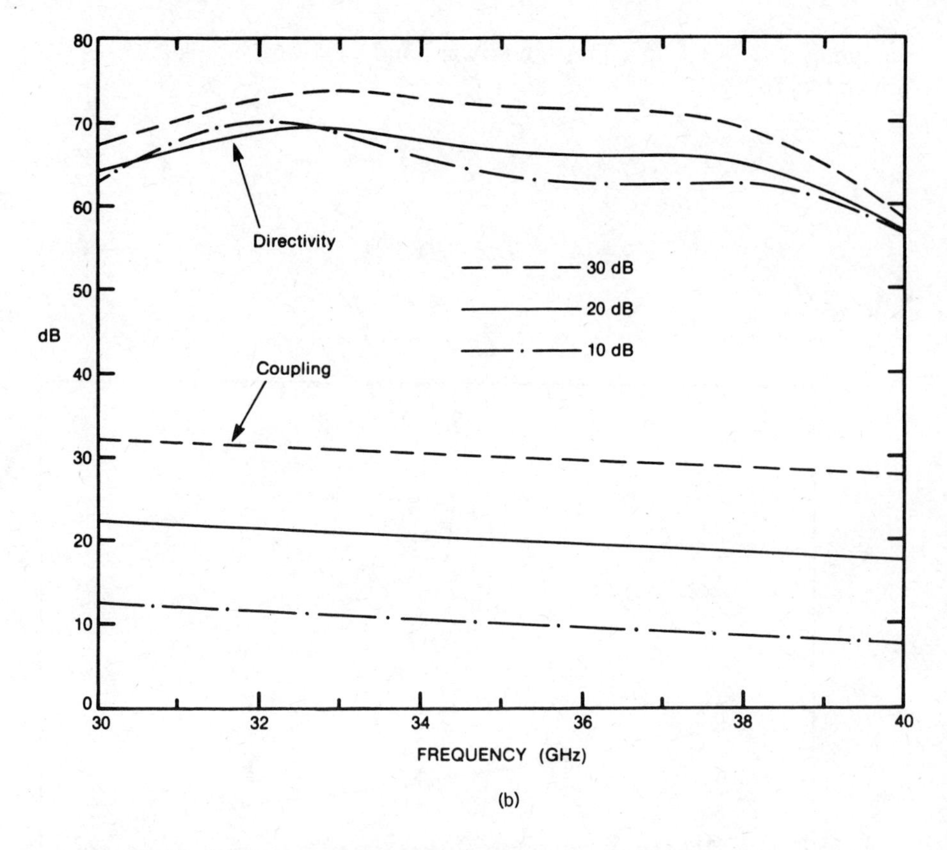

Figure 5.12 (*b*) Frequency response of aperture-coupled image-guide coupler.

where $\beta_1 = \sqrt{\epsilon_{ep}(f)}k_0$, $\beta_2 = \sqrt{\epsilon_{es}(f)}k_0$ ($k_0 = 2\pi/\lambda_0$, and β_1 and β_2 are the propagation constants in the primary and the secondary guide, respectively), and A is a constant.

For a symmetrical array of apertures $r_i = r_{N-i}$ and expressing $C_n = d_n\tau_f$ and $D_n = d_n\tau_b$ where d_n is a frequency-independent term, the forward and backward coupling coefficients are given by

$$\mathrm{CF} = -20\log|\tau_f| - 20\log\left|\sum_{n=0}^{M} 2d_n \cos(N-2n)\cdot(\theta-\beta_1 d)\right| \tag{5.34}$$

$$\mathrm{CB} = -20\log|\tau_b| - 20\log\left|\sum_{n=0}^{M} 2d_n \cos(N-2n)\theta\right| \tag{5.35}$$

where

$$\theta = \tfrac{1}{2}(\beta_1+\beta_2)d \tag{5.36}$$

Hence, the directivity is given by

$$D = \mathrm{CB} - \mathrm{CF} = -20 \log \left| \frac{\tau_b}{2\tau_f} \right| \\ -20 \log \left| \frac{\sum_{n=0}^{M} 2d_n \cos(N-2n)\theta}{\sum_{n=0}^{M} 2d_n \cos(N-2n)(\theta - \beta_1 d)} \right| \tag{5.37}$$

CB, the total backward coupled wave, is zero for $\theta = \pi/2$ at the center frequency, and hence $M = (N-1)/2$ for N odd and $M = N/2$ for N even, and the aperture spacing

$$d = \pi(\beta_1 + \beta_2)^{-1} \tag{5.38}$$

An examination of (5.37) shows that unless β_1 and β_2 are widely different, the numerator of the right-hand side has a much faster variation than the denominator. Hence an approximate broadband coupler design is possible by equating the numerator to an appropriate polynomial according to the desired frequency response of the directivity. We illustrate the method for an approximate Chebyshev-type response. Only the design steps are given. For a detailed derivation of the method, the reader is referred to Reference 15.

We write from (5.37)

$$F \approx \left| \sum_{n=0}^{M} 2d_n \cos(N-2n)\theta \right| = K|T_N(\sec\theta_m \cos\theta)| \tag{5.39}$$

The minimum value of directivity D_m in the passband as contributed by the array factor F is given by

$$D_m = 20 \log|T_N(\sec\theta_m)| \tag{5.40}$$

The Chebyshev polynomial, $T_N(x)$, is of order N with an argument x, and θ_m is the value at the upper and lower edges of the passband. According to (5.40), therefore, if D_m is specified, then θ_m automatically gets fixed and vice versa. Or in other words, the bandwidth and the minimum passband directivity fix each other.

The constant K is chosen to give the desired value of coupling C at the center frequency. This is done in the following way. The coupling at the center frequency is given by

$$C = -20 \log[K|\tau_f||T_N(\sec\theta_m)|] \tag{5.41}$$

Polynomial τ_f is obtained by dividing C_f by the frequency-independent term in C_f. For example, for a circular aperture, τ_f is obtained from

$$\tau_f = \frac{10^{-C_f/20}}{r^3} \tag{5.42}$$

Knowing C and τ_f, K is obtained from (5.41). Having obtained K, the aperture dimensions are obtained from (5.39) by equating the coefficients of $\cos n\theta$ $(n = 0, \ldots, M)$.

Using the preceding procedure, an image guide–image guide broadband coupler was designed for operation in the *Ka*-band. The computed response is shown in Fig. 5.12(*b*) for 10-dB, 20-dB, and 30-dB couplers.

A conceptual design drawing and the characteristics are shown in Figs. 5.13(*a*) and 5.13(*b*), respectively, for a *Ka*-band microstrip–image guide aperture-coupled coupler.

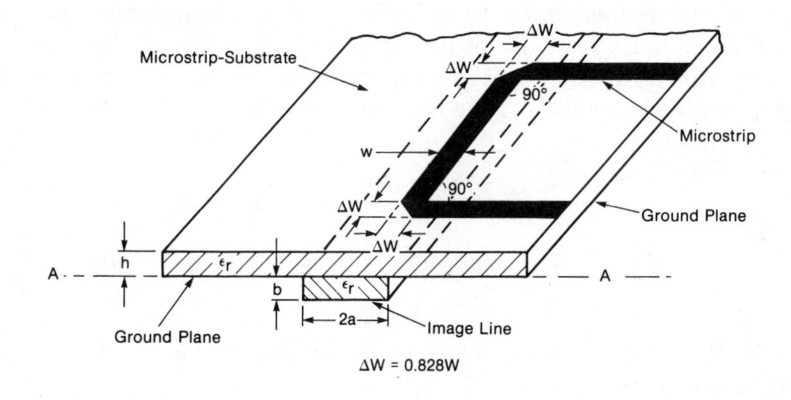

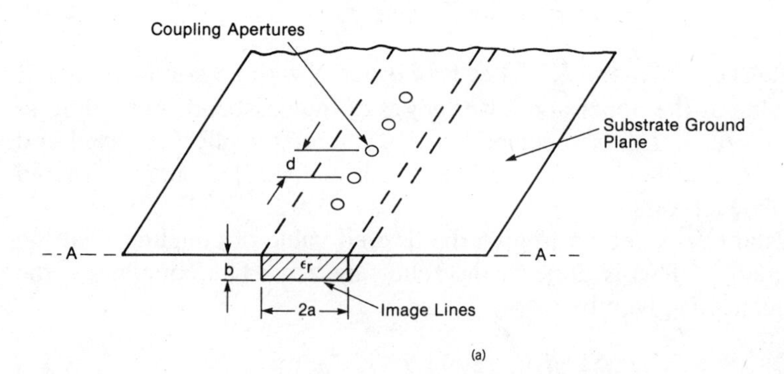

(a)

Figure 5.13 (*a*) Conceptual design drawings of microstrip-image-guide coupler.

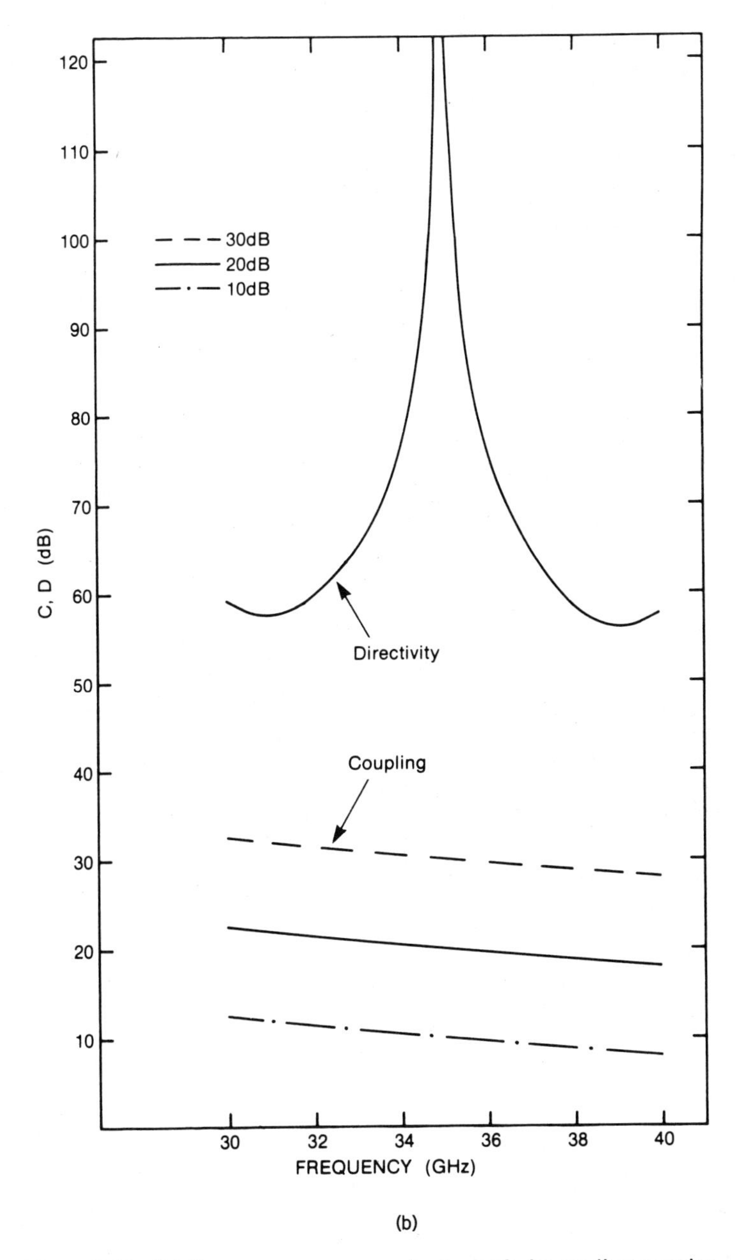

(b)

Figure 5.13 (*b*) Frequency response of microstrip-image-line coupler.

5.3.2 TEM Line Directional Couplers

When the center conductors of two coaxial lines supporting a pure TEM mode of propagation are brought in close proximity of each other, electromagnetic energy is coupled from one line to the other. This property has given rise to a class of broadband planar directional couplers. Most such couplers use strip-line or microstrip transmission lines that support pure TEM or quasi-TEM modes.

Planer TEM line directional couplers can be either edge coupled or broadside coupled as shown in Fig. 5.14(*a*) and 5.14(*b*).

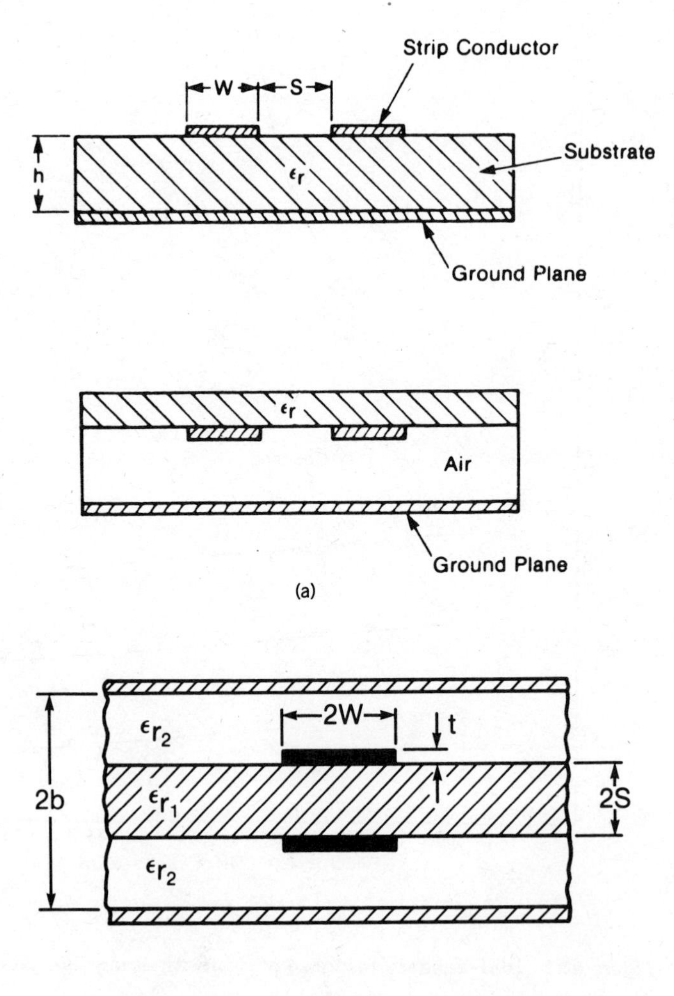

Figure 5.14 (*a*) Edge-coupled microstrip lines. (*b*) Broadside-coupled microstrip line.

In general, such coupled TEM lines support two modes that interact to give rise to the coupling. These are the even and odd modes. Properties of the coupled lines can be evaluated by suitable linear combinations of even and odd modes.

TEM line couplers can be reduced to a four-port network as shown in Fig. 5.15(*a*). There is a plane of symmetry that becomes a perfect magnetic wall for the incident signals of equal amplitude and same phase at Ports 1 and 4.

The plane of symmetry becomes a perfect electric wall for the incident signals of equal amplitude but of exactly opposite phases at the same ports. Therefore each mode corresponds to a two-port network as shown in Fig.

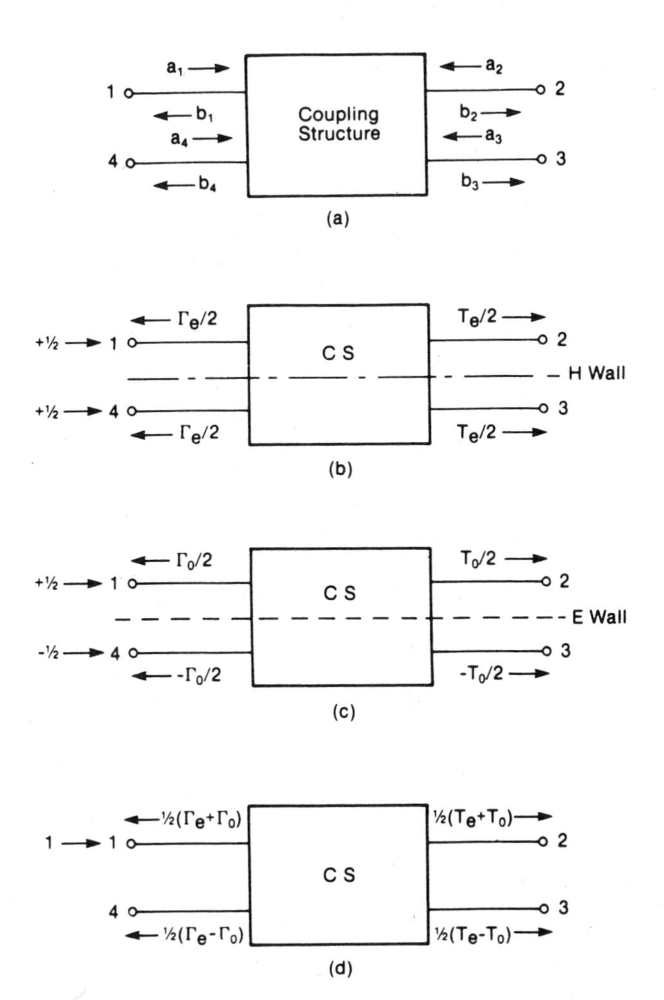

Figure 5.15 Schematic of directional coupler. (*a*) Wave parameters. (*b*) Even-mode excitation. (*c*) Odd-mode excitation. (*d*) Input excitation.

5.15(b) and 5.15(c) for the even and the odd modes, respectively. Analysis of the directional coupler is accomplished by analyzing these two networks and superimposing the responses as shown in Fig. 5.15(d).

The S-matrix equation of a symmetrical and reciprocal network can be written as

$$\begin{bmatrix} b_1 \\ b_2 \\ b_3 \\ b_4 \end{bmatrix} = \begin{bmatrix} S_{11} & S_{12} & S_{13} & S_{14} \\ S_{12} & S_{11} & S_{14} & S_{13} \\ S_{13} & S_{14} & S_{11} & S_{12} \\ S_{14} & S_{13} & S_{12} & S_{11} \end{bmatrix} \begin{bmatrix} a_1 \\ a_2 \\ a_3 \\ a_4 \end{bmatrix} \tag{5.43}$$

For the even-mode excitation $a_1 = a_4 = \frac{1}{2}$. The corresponding reflection coefficient Γ_e and the transmission coefficient T_e are given by

$$\Gamma_e = \frac{b_1}{a_1} = \frac{b_4}{a_4} = S_{11} + S_{14} \tag{5.44}$$

$$T_e = \frac{b_2}{a_1} = \frac{b_3}{a_4} = S_{12} + S_{13} \tag{5.45}$$

For the odd-mode excitation $a_4 = -a_1 = -\frac{1}{2}$. The reflection and the transmission coefficients may be written as

$$\Gamma_o = \frac{b_1}{a_1} = \frac{b_4}{a_4} = S_{11} - S_{14} \tag{5.46}$$

$$T_o = \frac{b_2}{a_1} = \frac{b_3}{a_4} = S_{12} - S_{13} \tag{5.47}$$

Superposition of the preceding solution gives

$$S_{11} = \tfrac{1}{2}(\Gamma_e + \Gamma_o) \tag{5.48a}$$

$$S_{12} = \tfrac{1}{2}(T_e + T_o) \tag{5.48b}$$

$$S_{13} = \tfrac{1}{2}(T_e - T_o) \tag{5.48c}$$

$$S_{14} = \tfrac{1}{2}(\Gamma_e - \Gamma_o) \tag{5.48d}$$

Evaluation of the even- and odd-mode reflection and transmission coefficients is done from the corresponding effective dielectric constants and the characteristic impedances.

$$\Gamma_i = \frac{A_i + \dfrac{B_i}{Z_0} - C_i Z_0 - D_i}{A_i + \dfrac{B_i}{Z_0} + C_i Z_0 + D_i} \tag{5.49}$$

$$T_i = \frac{2}{A_i + \dfrac{B_i}{Z_0} + C_i Z_0 + D_i} \tag{5.50}$$

where the transmission matrix

$$\begin{bmatrix} A_i & B_i \\ C_i & D_i \end{bmatrix} = \begin{bmatrix} \cos\theta_i & jZ_{0i}\sin\theta_i \\ \dfrac{j\sin\theta_i}{Z_{0i}} & \cos\theta_i \end{bmatrix} \qquad i = e \quad \text{or} \quad o \tag{5.51}$$

Knowing Z_{0e}, θ_e, Z_{0o}, θ_o, and Z_0, and the system characteristic equation, the performance of the directional coupler can be calculated by using (5.48) to (5.50). For the computation of the even- and the odd-mode characteristic impedance and phase velocity in terms of the physical parameters, the reader is referred to Section 2.2.

Coupled TEM Line. For coupling between the purely TEM lines shown in Fig. 5.16, we have the following special case

$$\theta_e = \theta_o = \theta \tag{5.52}$$

which means equal, even-, and odd-mode phase velocities, which gives

$$Z_0 = \sqrt{Z_{0e} Z_{0o}} \tag{5.53}$$

$$\Gamma_e = -\Gamma_o = \frac{j\{(Z_{0e}/Z_{0o})^{1/2} - (Z_{0o}/Z_{0e})^{1/2}\}\sin\theta}{\Sigma} \tag{5.54}$$

where

$$\Sigma = 2\cos\theta + j\left\{\left(\frac{Z_{0e}}{Z_{0o}}\right)^{1/2} + \left(\frac{Z_{0o}}{Z_{0e}}\right)^{1/2}\right\}\sin\theta \tag{5.55}$$

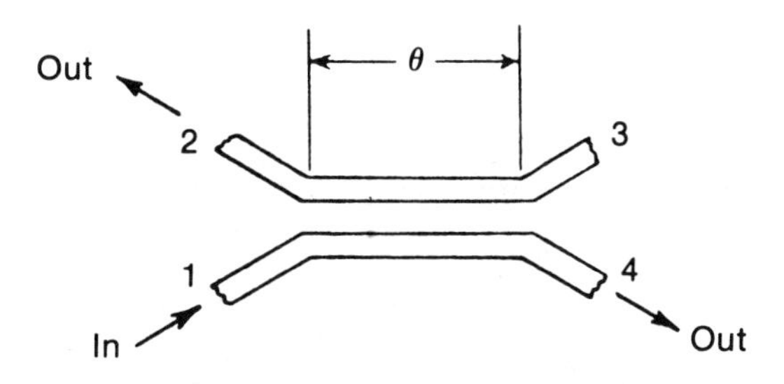

Figure 5.16 Coupling between two TEM lines.

Substitution of (5.54) and (5.55) into (5.48) gives

$$S_{11} = 0 \tag{5.56a}$$

$$S_{12} = T_e \tag{5.56b}$$

$$S_{13} = 0 \tag{5.56c}$$

$$S_{14} = T_e \tag{5.56d}$$

Such couplers are known as backward wave couplers, and are in general a quarter wavelength long at the center frequency or $\theta = \pi/2$.

From (5.53) and (5.54) the coupling coefficient is given by

$$C = -20 \log|S_{14}| = -20 \log\left|\frac{Z_{0e} - Z_{0o}}{Z_{0e} + Z_{0o}}\right| \quad \text{dB} \tag{5.57}$$

Moreover, for a system impedance Z_0 of 50 Ω, the matching condition (condition for a perfect input match) gives

$$Z_0^2 = Z_{0e} Z_{0o} \tag{5.58}$$

Therefore for a specified coupling C the design equations become

$$Z_{0e} = Z_0 \left[\frac{1 + 10^{-C/20}}{1 - 10^{-C/20}}\right]^{1/2} \tag{5.59}$$

$$Z_{0o} = Z_0 \left[\frac{1 - 10^{-C/20}}{1 + 10^{-C/20}}\right]^{1/2} \tag{5.60}$$

Once Z_{0e} and Z_{0o} are known the physical dimensions of the coupler can be obtained using the equations for a coupled TEM transmission lines.

Coupled Quasi-TEM Line. In coupled quasi-TEM lines, for example, in microstrip, the odd-mode phase velocity is different from the even-mode phase velocity. Therefore the condition given by (5.52) does not hold good. However, for weak couplings, (5.52) can be assumed to be approximately true. Hence the designer may go ahead and determine the initial design using (5.57) through (5.60). However, as the coupling gets tighter, the previous equations tend to be less valid. In such a case, the condition for input matching becomes

$$Z_0 = \left(\frac{Z_{0e} \sin \theta_e + Z_{0o} \sin \theta_o}{Z_{0e} \sin \theta_o + Z_{0o} \sin \theta_e}\right)^{1/2} \sqrt{Z_{0o} Z_{0e}} \tag{5.61a}$$

and the electrical length at the center frequency is

$$\theta = \frac{1}{2}(\theta_e + \theta_o) = \frac{2\pi}{\lambda_0} \frac{\sqrt{\epsilon_{ee}} + \sqrt{\epsilon_{eo}}}{2} l = 90° \tag{5.62}$$

where l is the physical length of the coupler.

Single-section quarter-wave parallel-coupled line couplers are used extensively in many applications. They are usually of narrow bandwidth of approximately one octave. In order to obtain the desired coupling at the band edges the coupler has to be designed for over coupling at the center frequency.

Frequency Response of a Single-Section Coupler. Using the analysis equations (5.48) through (5.57), it can be shown that the frequency response of the coupling coefficient is given by

$$C(\theta) = \frac{jC \sin \theta}{\sqrt{1 - C^2} \cos \theta + j \sin \theta} \tag{5.63a}$$

where C is the midband coupling for a matched, loosely coupled coupler. The approximate frequency response of a quasi-TEM coupler is shown in Fig. 5.17. The response is exact for a TEM coupler.

The general expression for the directivity or the undesired coupling is given by [16].

$$D = \left[\frac{\pi\Delta(1 - |\xi|^2)}{4|\xi|}\right]^2 \tag{5.63b}$$

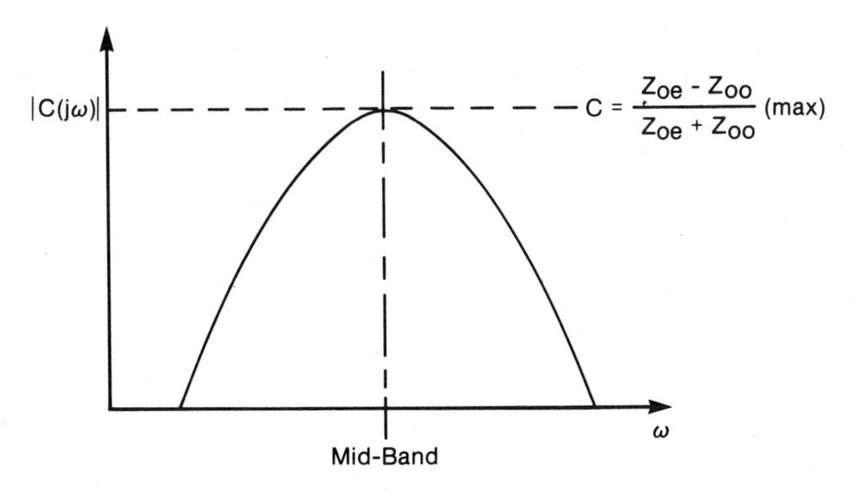

Figure 5.17 Approximate frequency response of a quasi-TEM line.

where

$$\Delta = \frac{\beta_e - \beta_o}{\beta_o}$$

$$\xi = \left(\frac{\rho_e}{1+\rho_e^2}\right) - \left(\frac{\rho_o}{1+\rho_o^2}\right)$$

$$\rho_e = \frac{Z_{0e} - Z_0}{Z_{0e} + Z_0}$$

$$\rho_o = \frac{Z_{0o} - Z_0}{Z_{0o} + Z_0}$$

and β_e and β_o are the even- and odd-mode propagation constants, respectively, $D = 0$ for $\beta_e = \beta_o$, that is, for a TEM coupler.

Multisection Couplers for Wider Bandwidth. For many applications, the single-section coupler proves to be of inadequate bandwidth. Therefore the designer should have recourse to a multisection design. A multisection

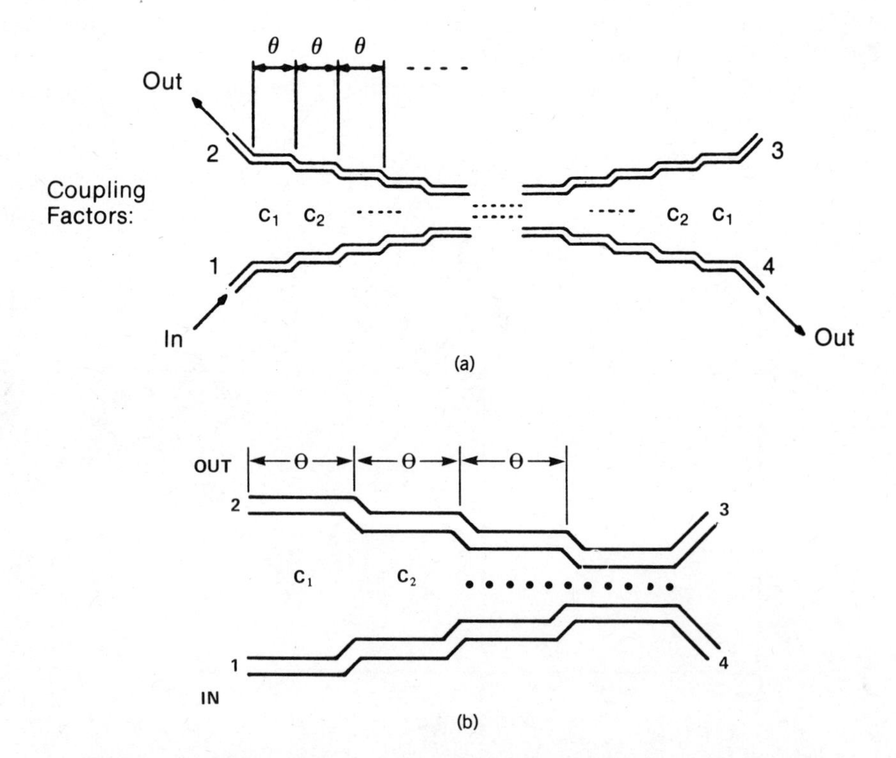

Figure 5.18 Multisection (*a*) symmetrical and (*b*) asymmetrical couplers.

coupler is a cascaded combination of more than one single-section coupler, each being a quarter wavelength long at the center frequency of the band. The number of sections to be used depends upon the tolerable insertion loss, bandwidth, and the available physical space.

Multisection couplers can be either symmetric or asymmetric around the center section, as shown in Fig. 5.18.

Symmetric Coupler. The symmetric coupler gives 90° phase difference between the direct and the coupled output ports under matched conditions. In what follows we present a direct synthesis technique.

Figure 5.18(a) shows an n-section symmetric coupler. The coupling factor for such an n-section symmetric coupler with weak coupling can be written as [17]

$$\begin{aligned} C(\theta) &= \left|\frac{V_2}{V_1}\right| \\ &= C_1 \sin(n\theta) + (C_2 - C_1)\sin[(n-2)\theta] + \cdots \\ &\quad + (C_i - C_{i-1})\sin[(n-2i+2)\theta] \\ &\quad + \{C_{[(n+1)/2]} - C_{[(n-1)/2]}\}\sin\theta \end{aligned} \tag{5.64}$$

If the desired coupling response is maximally flat, then C_i must satisfy a set of $(n-1)/2$ linear equations obtained from

$$\left[\frac{d^r C(\theta)}{d\theta^r}\right]_{\theta=\pi/2} = 0, \qquad r = 2, 4, 6, \ldots, (n-1) \tag{5.65}$$

Note that n is always an odd integer.

The design concept is based on the fact that the backward coupled wave of a TEM coupler corresponds to the reflected wave of a quarter-wave filter. Therefore the designer of a TEM wave coupler has to synthesize only a two-port in place of a four-port with the reflection coefficient response the same as the desired coupling coefficient response of the four-port directional coupler.

In terms of the midband VSWR R of the quarter-wave filter, the coupling coefficient is given by [17]

$$C_0 = \frac{R-1}{R+1} \tag{5.66}$$

Design of Three-Section Maximally Flat Coupler. From (5.66) we obtain

$$R = \frac{1 + C_0}{1 - C_0} \tag{5.67}$$

The step VSWR's, V_1 and V_2, are calculated from

$$V_1 = 1.1592 - 0.01666C_0 + 0.000474C_0^2 \tag{5.68}$$

where C_0 is in decibels and $V_2 = V_1\sqrt{R}$.

The step impedances of the quarter-wave filter are determined as

$$Z_1 = V_1 \tag{5.69a}$$

$$Z_2 = V_1 V_2 \tag{5.69b}$$

Couplings of the individual sections are given by

$$C_1 = \frac{Z_1^2 - 1}{Z_1^2 + 1} \tag{5.70a}$$

$$C_2 = \frac{Z_2^2 - 1}{Z_2^2 + 1} \tag{5.70b}$$

The odd- and even-mode impedances are obtained from

$$(Z_{0e})_i = Z_0\left[\frac{1 + C_i}{1 - C_i}\right]^{1/2} \tag{5.71a}$$

$$(Z_{0o})_i = Z_0\left[\frac{1 - C_i}{1 + C_i}\right]^{1/2}, \qquad i = 1, 2 \tag{5.71b}$$

Once $(Z_{0e})_i$ and $(Z_{0o})_i$ are known, the physical dimensions can be obtained depending upon the type of transmission line. The preceding design method holds exactly for purely TEM lines. However, a multisection symmetric microstrip coupler can be designed approximately with the help of the previous method. Then the design can be improved using optimization techniques.

Asymmetric Coupler. The asymmetric coupler provides a phase difference of 0° or 180° between the coupled and the direct output ports over a wide frequency band. In such couplers the two transmission lines become more lightly coupled gradually or in steps until they reach the center of the coupler where they become uncoupled, as shown in Fig. 5.19. The scattering matrix for the asymmetric coupler is obtained in the following way.

If the Ports 2 and 3 are excited by a pair of even- and odd-mode generators, then analogous to (5.48), we obtain the scattering parameters at Ports 2 and 3 as

$$S_{22} = S_{33} = \tfrac{1}{2}(\Gamma_{2e} + \Gamma_{2o}) \tag{5.72a}$$

$$S_{23} = S_{32} = \tfrac{1}{2}(\Gamma_{2e} - \Gamma_{2o}) \tag{5.72b}$$

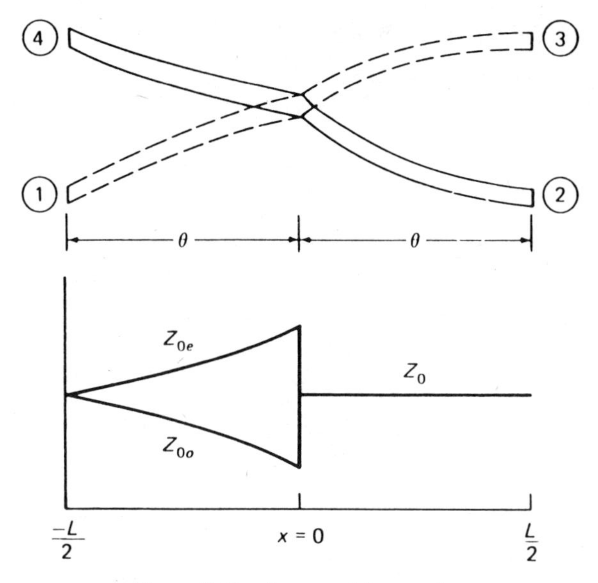

Figure 5.19 Asymmetric coupler.

Now if one assumes that the coupler is matched, so that

$$C = \frac{Z_0}{Z_{0e}(0)} = \frac{Z_{0o}(0)}{Z_0} \tag{5.73}$$

then the maximum coupling is at $x = 0$, where there is a discontinuity due to an abrupt change from either Z_{0e} or Z_{0o} to Z_0. Therefore the even- and the odd-mode reflection coefficients at the input Port 1 are given by [18]

$$\Gamma_{1e} = \frac{C-1}{C+1} e^{-j2\theta} \tag{5.74a}$$

$$\Gamma_{1o} = -\Gamma_{1e} = \frac{1-C}{1+C} e^{-j2\theta} \tag{5.74b}$$

Looking from Port 2, a transmission line of characteristic impedance Z_0 is terminated either by Z_{0e} or Z_{0o} at $x = 0$. Therefore the even-mode reflection coefficient at Port 2 is given by

$$\Gamma_{2e} = -\Gamma_{2o} = \frac{1-C}{1+C} e^{-j2\theta} \tag{5.74c}$$

For the energy conservation principle and the electrical length of the coupler, the transmission coefficients are given by

$$T = \sqrt{1 - |\Gamma|^2}\, e^{-j2\theta} \tag{5.75a}$$

$$T_{1e} = T_{1o} = \frac{2\sqrt{C}}{1 + C}\, e^{-j2\theta} \tag{5.75b}$$

Therefore the scattering matrix is given by

$$[S] = \begin{bmatrix} 0 & p & 0 & -q \\ p & 0 & q & 0 \\ 0 & q & 0 & p \\ -q & 0 & p & 0 \end{bmatrix} \tag{5.76}$$

where

$$p = \frac{2\sqrt{C}}{1 + C}\, e^{-j2\theta} \tag{5.77a}$$

$$q = \frac{1 - C}{1 + C}\, e^{-j2\theta} \tag{5.77b}$$

From (5.76) the relative amplitudes of the waves in the coupled and uncoupled ports are given by coupling response C_∞ at infinite frequency.

The design of the tapered asymmetric coupler can be obtained using optimization programs [19]. But the Klopfenstein [20] taper seems to be the easiest to realize for the best performance.

Three main parameters of an asymmetric coupler are the cutoff frequency f_0, the sidelobe level in the passband, and the physical length of the coupler. Out of these, any two can be chosen by the designer.

The even-mode impedance distribution is obtained from

$$\ln Z(x) = \frac{1}{2} \ln(Z_1 Z_2) + \frac{A^2 \ln(Z_2/Z_1)}{2 \cosh(A)}\, \Phi\left(\frac{2x}{L}, A\right) \tag{5.78}$$

where Z_1 and Z_2 are the even-mode impedances at the ends of the asymmetric coupler, and

$$\Phi(x, A) = \sum_{n=0}^{\infty} a_n b_n \tag{5.79}$$

$$a_0 = 1$$

$$b_0 = \frac{x}{2}$$

$$a_n = \frac{A^2}{4n(n+1)}\, a_{n-1} \tag{5.80a}$$

$$b_n = \frac{(x/2)(1-x^2)^n + 2nb_{n-1}}{2n+1} \tag{5.80b}$$

Usually $Z_1 = 1$ and Z_2 is obtained from the desired coupling response C_∞ at infinite frequency.

$$Z_2 = \frac{Z_{0e}(0)}{Z_0} = \frac{1+|C_\infty|}{1-|C_\infty|} \tag{5.81}$$

Once Z_{0e} has been obtained, Z_{0o} is obtained from (5.58). Suitable taper distribution is used for optimum performance [19, 20]. For each set of Z_{0e} and Z_{0o} the required physical dimensions of the coupler are obtained using a suitable synthesis technique depending upon the type of coupled transmission lines.

5.3.3 Multiconductor Couplers

The interdigital coupler, or multiconductor coupler, invented by Lange[21], has always been a popular component in planar circuits. Figure 5.20 shows a four-element interdigital coupler, although in certain applications the number of elements may be greater than four. The coupler is usually designed for 3-dB coupling and the output phases are in quadrature. Obviously, the best realization is in the microstrip form.

An interdigital coupler has advantages because of its small size and relatively large line separation when compared with the two-coupled line device and has a much larger bandwidth when compared with branch-line couplers.

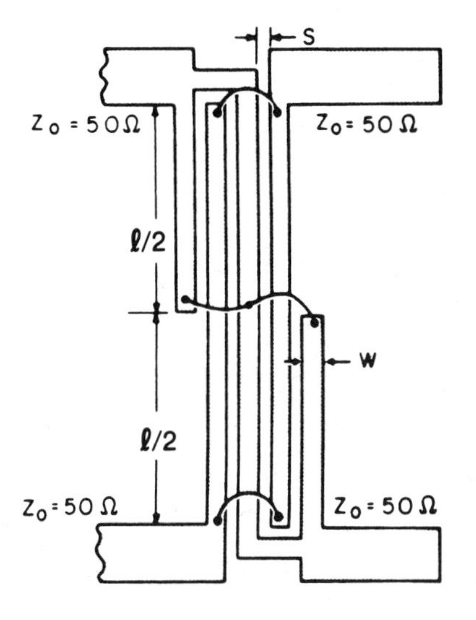

Figure 5.20 Lange coupler.

Interdigital couplers are used for balanced MIC amplifiers, balanced mixers, and binary power divider trees.

Design of Interdigital Couplers. Kajfez et al. [22] have described a simplified design technique for the interdigital coupler. The proposed technique serves many practical purposes but seems to be inadequate for an accurate design. So far the method due to Presser[23] has been found to be the most accurate and simplest. Moreover, it has the provision for finite strip-thickness correction.

Consider the layout shown in Fig. 5.20. The designer is supplied with the desired coupling coefficient, C, and the system characteristic impedance Z_0. The length of the coupled region l has to be a quarter wavelength at the center of the band.

The main design equations for an N element (N even) coupler are written as

$$R = \frac{Z_{0o}}{Z_{0e}} \tag{5.82a}$$

$$C = \frac{(N-1)(1-R^2)}{(N-1)(1+R^2)+2R} \tag{5.82b}$$

and

$$Z = \frac{Z_{0o}}{Z_0} \frac{\sqrt{R[(N-1)+R][(N-1)R+1]}}{(1+R)} \tag{5.82c}$$

Figure 5.21 shows the impedance ratio R as a function of the coupling

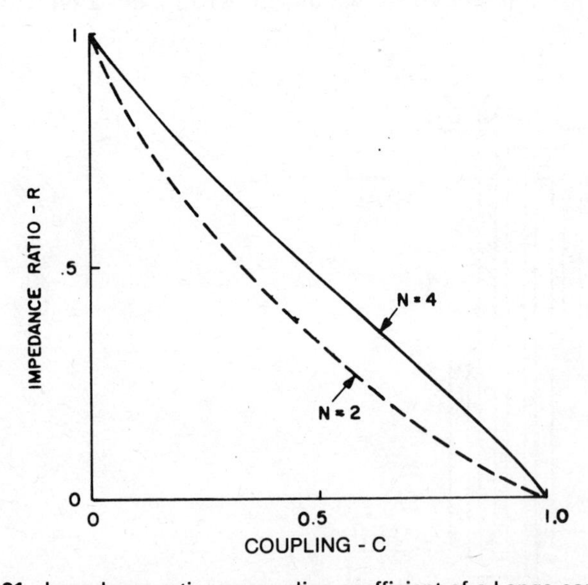

Figure 5.21 Impedance ratio vs. coupling coefficient of a Lange coupler.

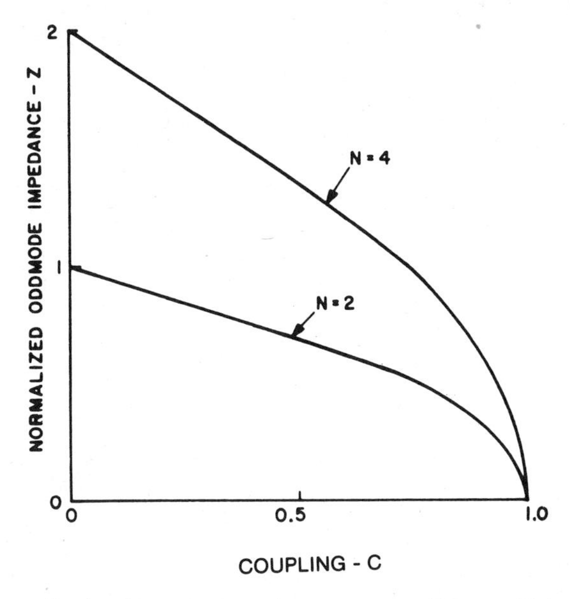

Figure 5.22 Normalized odd-mode impedance vs. coupling coefficient of a Lange coupler.

coefficient, while Fig. 5.22 shows the normalized odd-mode impedance as a function of the coupling coefficient. Equations 5.82(a) through 5.82(c) can be solved for Z_{0o} and Z_{0e}. Knowing Z_{0o} and Z_{0e}, the physical shape ratios can be obtained using the design equations of Garg and Bahl[24].

When the finite thickness of conductors is neglected, the designed coupler shows an overcoupled characteristic. This is because the conductor thickness considerably influences the coupling between the lines. The overcoupling can be corrected by adding an extra gap ΔS to S expressed as

$$\frac{\Delta S}{h} = \frac{t/h}{\pi\sqrt{\epsilon_{eo}}}\left\{1 + \ln\frac{4\pi W/h}{t/h}\right\} \tag{5.83}$$

where t/h is the actual normalized thickness of the metallization and ϵ_{eo} is the odd-mode effective dielectric constant of the coupled lines.

5.3.4 Distributed-type Couplers

Distributed-type coupling takes place between two adjacent transmission lines supporting purely non-TEM modes. For example, distributed-type couplers can be realized using two open dielectric waveguides (or image guides) or fin lines.

In general, two distributed-type coupled lines can be represented as shown in Fig. 5.23(a). Under the assumption that all four ports are matched

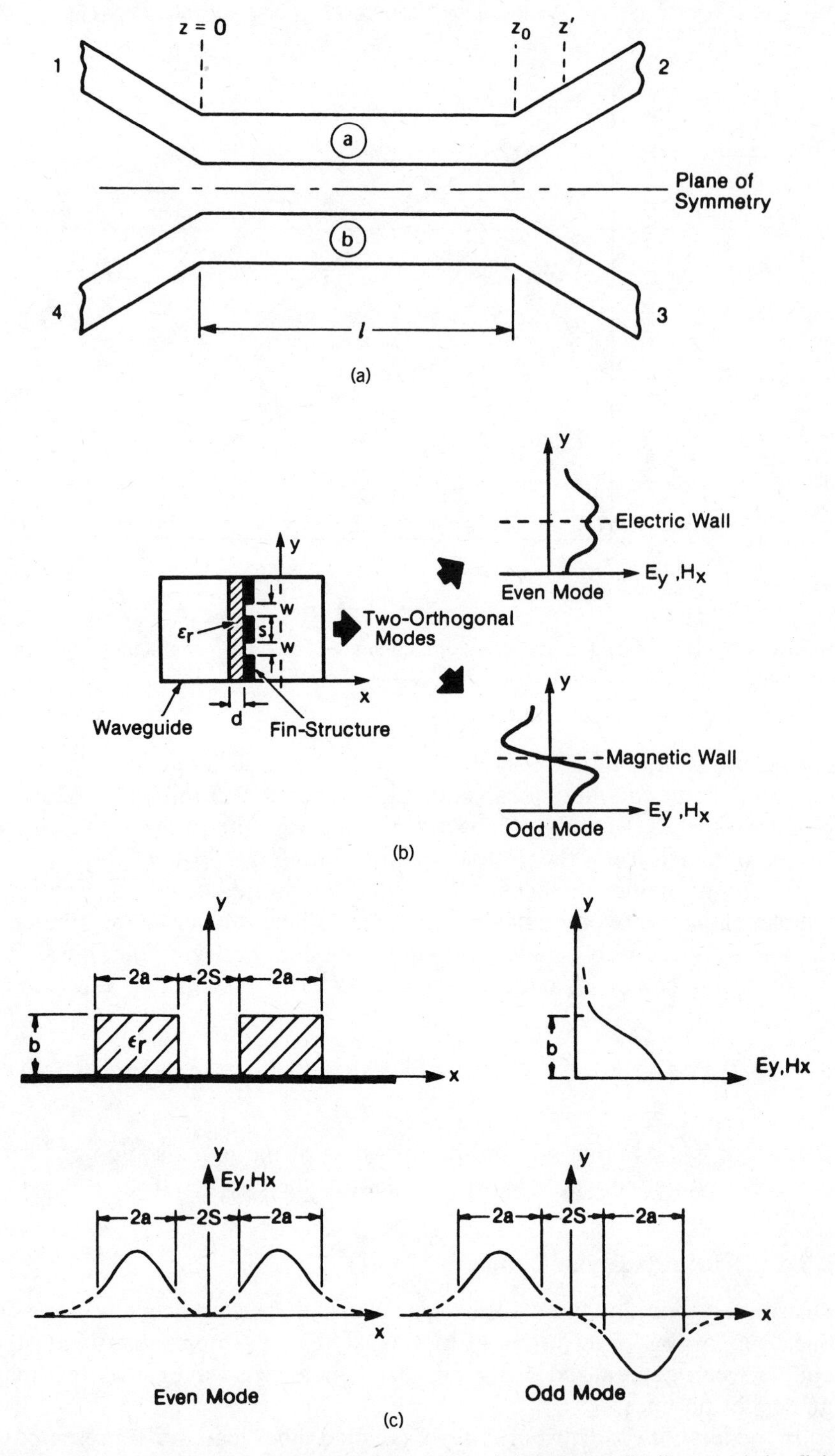

Figure 5.23 (*a*) Schematic of distributed coupler. (*b*) Coupled fin lines and the field distributions. (*c*) Coupled image lines and the field distributions.

and the coupling structure has the length l, the ratio of the fields in the two lines can be shown to be [12]

$$\frac{E_b(l)}{E_a(l)} = \tan\left\{(\beta_e - \beta_o)\frac{l}{2}\right\} \tag{5.84}$$

where β_e and β_o are the even- and the odd-mode phase constants, respectively.

Therefore for complete transfer of power from line a to line b requires

$$(\beta_e - \beta_o)\frac{L}{2} = \frac{\pi}{2} \tag{5.85}$$

or,

$$L = \frac{\pi}{\beta_e - \beta_o} \tag{5.86}$$

The scattering coefficients of the structure can be written as a function of l normalized by L as

$$|S_{12}| = \left|\cos\left(\frac{\pi}{2}\frac{l}{L}\right)\right| \tag{5.87a}$$

$$|S_{13}| = \left|\sin\left(\frac{\pi}{2}\frac{l}{L}\right)\right| \tag{5.87b}$$

Equation 5.87(b) shows that the required length for 3-dB coupling is one half of L.

The preceding equations are based on the assumptions that the bent portion of the guides have no effect on coupling. Branch junction effects near the bends are negligible, but the assumed uncoupled lines $z \leq 0$ and $z \geq z_0$ do get coupled. This extra coupling can be taken into consideration by defining the effective length of the coupler as

$$l_{\text{eff}} = l + \frac{2L}{\pi}\int_{z_0}^{z'} \{\beta_e(z) - \beta_o(z)\}\, dz \tag{5.88}$$

The integration limit z' is chosen to be the point at which the coupling is practically negligible.

From the preceding discussions it appears that the design of any distributed coupler requires a precise knowledge of the even- and the odd-mode phase constants of the coupled lines. Figures 5.23(b) and 5.23(c) show the cross sections of a coupled fin lines and image lines, respectively, as examples of two commonly used distributed couplers.

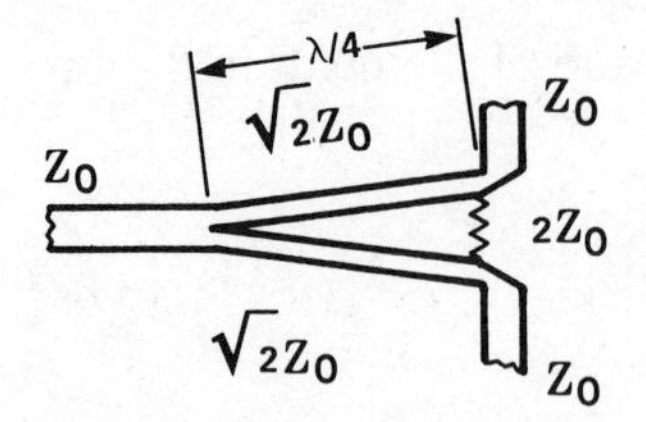

Figure 5.24 Single-section Wilkinson coupler.

5.3.5 Wilkinson Couplers, Power Dividers, and Combiners

A Wilkinson coupler [25, 26] offers broad bandwidth and equal phase characteristics at each of its output ports. Figure 5.24 shows the schematic diagram of a Wilkinson coupler. The output port isolation is obtained by series terminating the output port. Each of the quarter-wave lines has the characteristic impedance of $\sqrt{2}Z_0$ and the output is terminated by a resistor of $2Z_0\,\Omega$, Z_0 being the system impedance.

A Wilkinson power divider offers a bandwidth of about one octave. The typical frequency response is shown in Fig. 5.25. An adequately flat response is obtained over more than one octave band. But at the band edges the isolation is affected by the load impedance.

The performance of a Wilkinson coupler can be further improved, depending upon the availability of space, by the addition of a $\lambda/4$ trans-

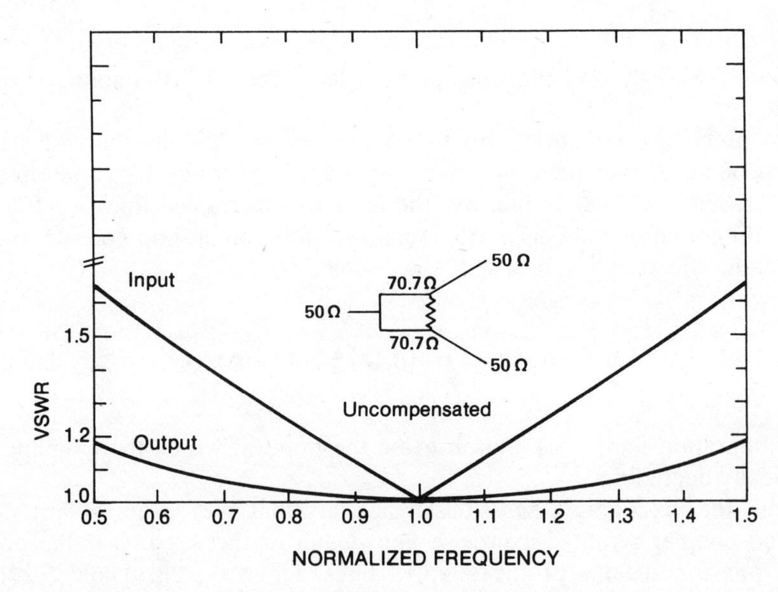

Figure 5.25 Frequency response of Wilkinson coupler. (H. Howe [5], reprinted with permission of Artech House.)

former in front of the power-division step. The input VSWR of the uncompensated coupler is better than the input VSWR of the compensated circuit.

Multisection Wilkinson Coupler. The octave bandwidth of a single-section coupler proves to be inadequate in many applications. Therefore Cohn [26] proposed the use of multisections for bandwidth expansion. The use of multisections makes it possible to obtain a decade bandwidth.

A multisection Wilkinson power-divider–coupler consists of a number of quarter-wave sections with resistive terminations at the end of every section, as shown schematically in Fig. 5.26. Larger bandwidth and greater isolation are obtained when a larger number of sections are used.

The characteristic impedances of the sections are obtained from the normalized impedances for $\lambda/4$ transformer sections for a 2:1 transformer. This can be done with the help of Fig. 5.27, which presents design curves up to four sections. Similar figures are also available for higher numbers of sections. Having obtained the impedance of each section, values of the terminating resistors for each section can be obtained.

For a two-section divider the values of the terminating resistors are given by

$$R_2 = \frac{2Z_1Z_2}{[(Z_1+Z_2)(Z_2 - Z_1\cot^2\phi)]^{1/2}} \tag{5.89a}$$

$$R_1 = \frac{2R_2(Z_1+Z_2)}{R_2(Z_1+Z_2) - 2Z_2} \tag{5.89b}$$

where

$$\phi = \frac{\pi}{2}\left[1 - 0.707\left(\frac{f_2 - f_1}{f_2 + f_1}\right)\right] \tag{5.89c}$$

and f_1, f_2 are the upper and lower band edge frequencies of operation.

Although the design of a two section coupler is straightforward, that of more than two sections is cumbersome. The characteristic admittance of each section is chosen as it is in the case of a two-section coupler. The

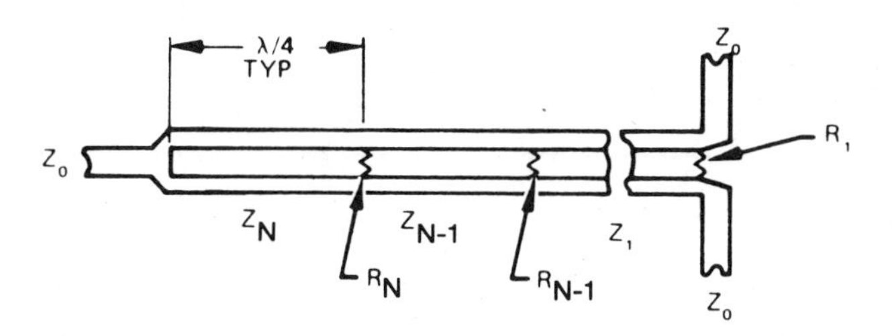

Figure 5.26 Multisection Wilkinson coupler.

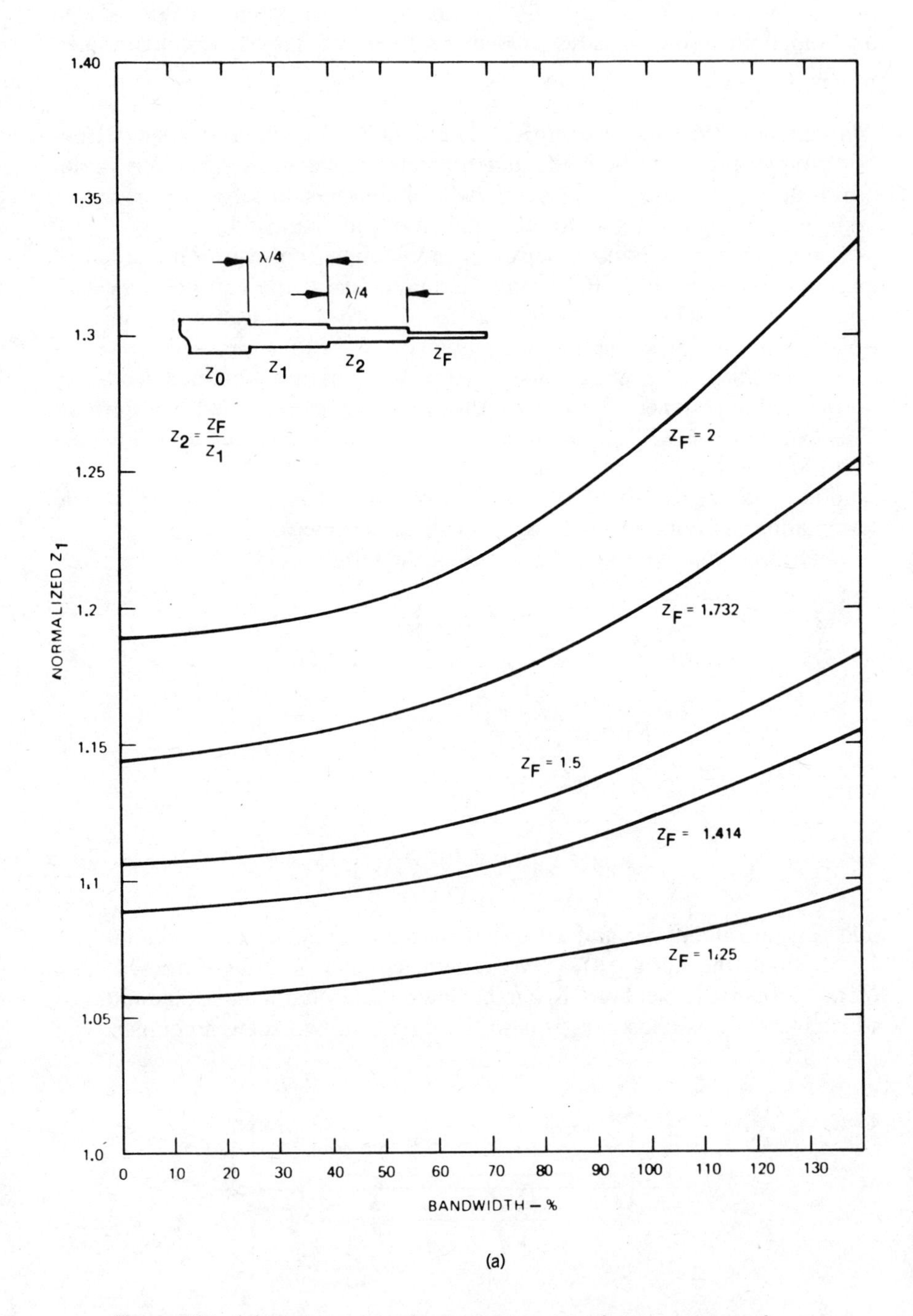

(a)

Figure 5.27 (*a*) Design curves for impedance transformer—two-section.

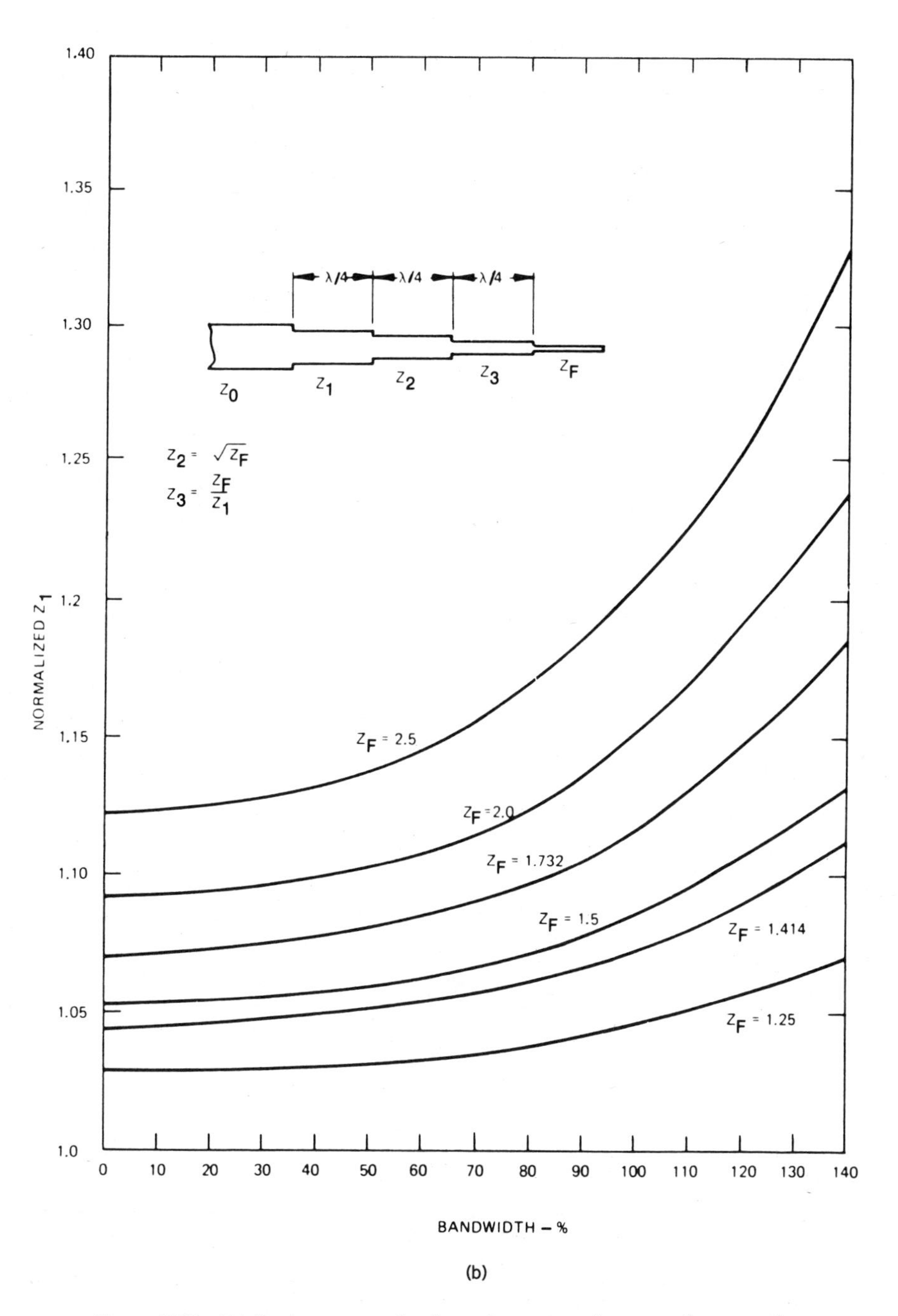

Figure 5.27 (*b*) Design curves for impedance transformer—three-section.

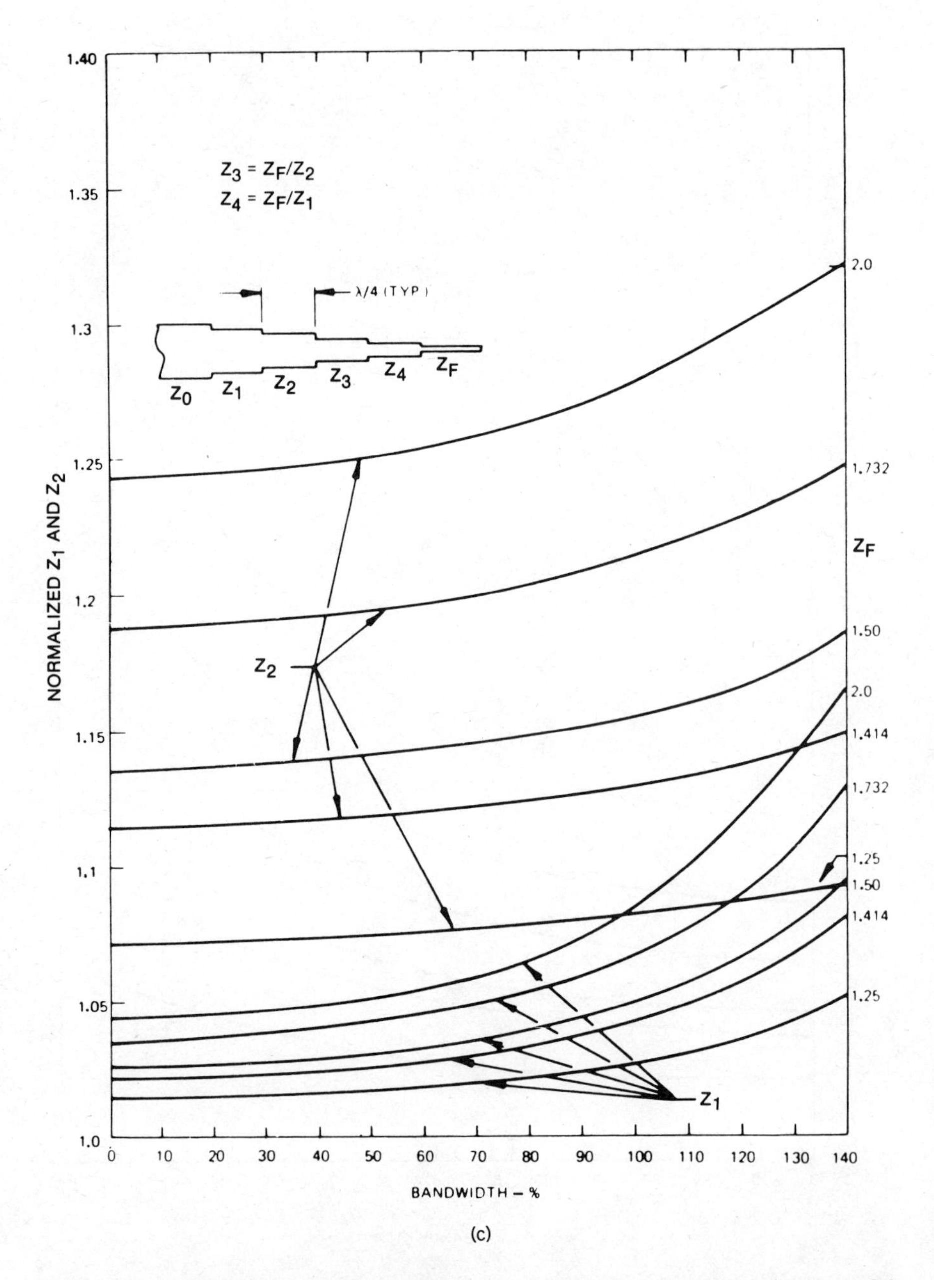

(c)

Figure 5.27 (*c*) Design curves for impedance transformer—four-section. (H. Howe [5], reprinted with permission of Artech House.)

output resistor is obtained from

$$\frac{1}{R_1} = 1 - Y_1 \tag{5.90a}$$

The intermediate resistor is obtained from

$$\frac{1}{R_K} = G_K = \frac{Y_{K-1} - Y_K}{Y_{K-1} T_1 T_2 \cdots T_{K-1}}, \quad \text{for} \quad K = 2 \quad \text{to} \quad N-1 \tag{5.90b}$$

$$T_K = \frac{4 Y_{K-1} Y_K}{(Y_{K-1} + Y_K + 2G_K)^2}, \quad \text{for} \quad K = 1 \quad \text{to} \quad N \tag{5.90c}$$

$$\frac{1}{R_n} = G_n = \cfrac{0.5 Y_{n-1}^2}{-2G_{n-1} + \cfrac{Y_{n-2}^2}{-2G_{n-2} + \cfrac{Y_{n-3}^2}{\ddots \, \cfrac{}{-2G_2 + \cfrac{Y_1^2}{-2G_1 + 1 + 0.7(S_{e90} - 1)}}}}} \tag{5.90d}$$

whose

$$\begin{aligned} S_{e90} &= 1 && \text{for } N \text{ odd} \\ &= S_{em} && \text{for } N \text{ even} \end{aligned}$$

Use of (5.90d) requires the knowledge of the maximum ripple VSWR for the prototype transformer sections chosen for the design. These data are obtained from Fig. 5.28.

The minimum isolation in decibels is computed from

$$I \cong 20 \log\left(\frac{2.35}{S_{em} - 1}\right) \tag{5.91}$$

where S_{em} is the maximum VSWR ripple.

Theoretically speaking, the performance of a multisection Wilkinson power-divider or combiner is the same as that of a single section except for the equiripple VSWR characteristics. In reality, much depends upon the

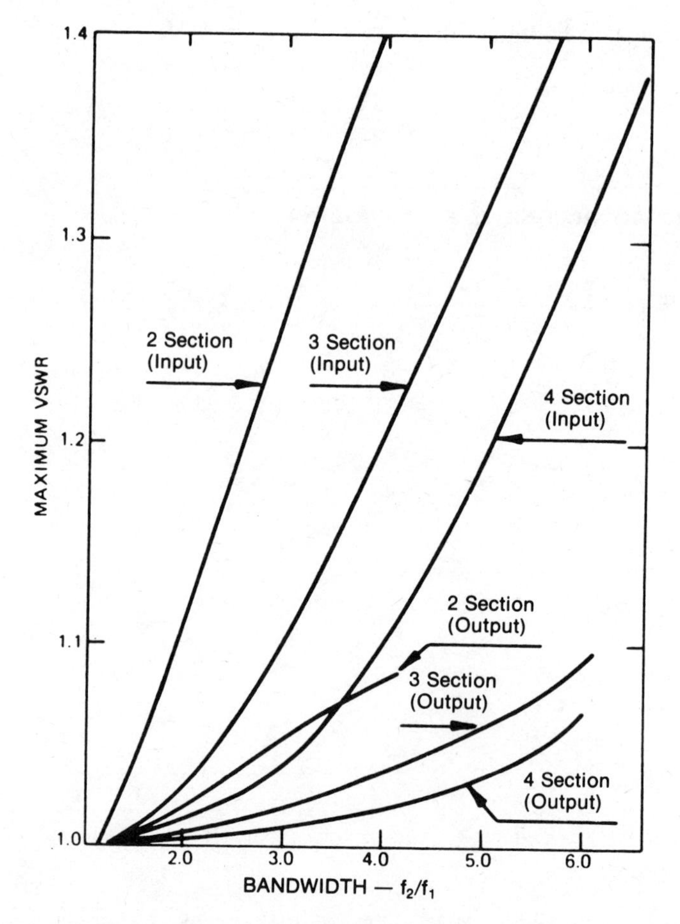

Figure 5.28 Maximum input–output VSWR vs. bandwidth for a multisection Wilkinson coupler. (H. Howe [5], reprinted with permission of Artech House.)

fabrication and the tolerance in the values of the terminating resistors. Smaller degradation in performance is observed over *L*- through *S*-band. However, in *X*- or *Ka*-band the degradation may be severe, lowering the minimum isolation by even 40%. Table 5.3 gives the values of *R*'s and *Z*'s for $N = 3$, 4, and 7 [26].

Unequal Split Wilkinson Power Dividers. Figure 5.29 shows the schematic of the unequal split Wilkinson divider [27]. As can be seen from the figure, output impedance transformers are also required, in contrast to the equal power split case.

TABLE 5.3 Parameters of Power Divider and Common Design Parameters for Both Wilkinson and Improved Version Power Dividers

Number of Expected Section = 3, *Epsilon* = 2.150, *Thickness of Sub* = 1.600 *mm*

Number	*R*	*Z*	Width (mm)
1	400.000	57.485	2.506
2	211.460	70.710	1.729
3	107.180	86.980	1.096

Expected Section = 4, *Epsilon* = 2.150, *Thickness of Sub* = 1.600 *mm*

Number	*R*	*Z*	Width (mm)
1	482.160	55.785	2.547
2	291.630	64.785	1.973
3	172.620	77.175	1.401
4	103.165	89.630	0.985

Expected Section = 7, *Epsilon* = 2.150, *Thickness of Sub* = 1.600 *mm*

Number	*R*	*Z*	Width (mm)
1	442.480	56.370	2.415
2	616.145	60.255	2.161
3	446.230	65.085	1.887
4	319.900	70.710	1.615
5	217.580	76.820	1.365
6	129.620	82.985	1.150
7	248.260	88.700	0.978

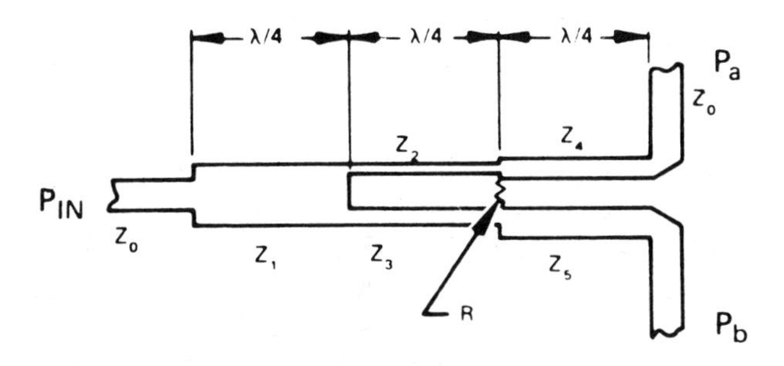

Figure 5.29 Schematic of unequal power split Wilkinson coupler.

The design equations for the compensated case are as follows:

$$K^2 = \frac{P_b}{P_a} \tag{5.92}$$

$$Z_1 = Z_0\left[\frac{K}{1+K^2}\right]^{1/4}$$

$$Z_2 = Z_0\{K^{3/4}(1+K^2)^{1/4}\}$$

$$Z_3 = Z_0\left\{\frac{(1+K^2)^{1/4}}{K^{5/4}}\right\}$$

$$Z_4 = Z_0\sqrt{K}$$

$$Z_5 = \frac{Z_0}{\sqrt{K}}$$

$$R = Z_0\left\{\frac{1+K^2}{K}\right\}$$

while those for the uncompensated case are available in Reference 28.

Example. If one wishes to design a split-T power combiner, with one output port having three times the power of the other, and if the input–output line impedances are 50 Ω, then $K^2 = 3$, and substituting into the preceding equations,

$$Z_1 = 40.56, \qquad Z_2 = 106.77, \qquad Z_3 = 35.59$$

$$Z_4 = 66.0, \qquad Z_5 = 37.88, \qquad R = 115.47$$

5.3.6 Other Couplers

Tandem Coupler. The design of a multisection coupler with tight coupling over a broad bandwidth requires some of its sections to have tighter coupling than the overall coupling. This invariably leads to physically unrealizable spacings between the two conductors or severely reduced directivity due to significant mechanical discontinuities in the sections. To solve this problem, in a restricted physical space, various combinations of symmetric and asymmetric couplers are tandemed [5, 18]. Since in the majority of applications, the tightest coupling may be 3 dB, two couplers may be connected in tandem to achieve the goal. Figure 5.30(*a*) shows the symmetric tandem of two 8.34-dB couplers, while Fig. 5.30(*b*) shows the symmetric tandem of two asymmetric couplers. Each of the couplers of 8.34 dB when tandemed gives an overall coupling of 3 dB. This configuration offers high power-handling capability and often represents a good

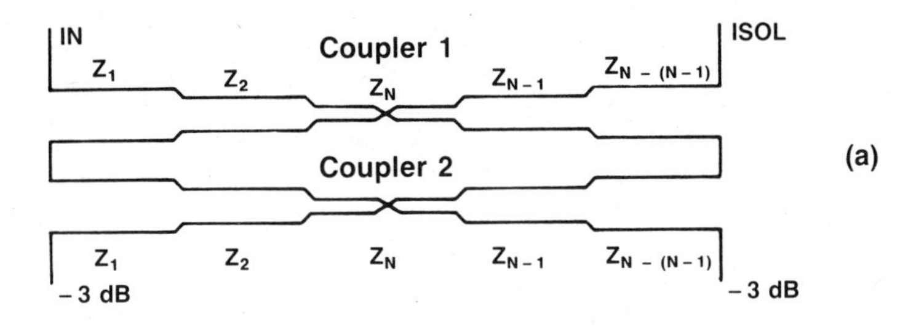

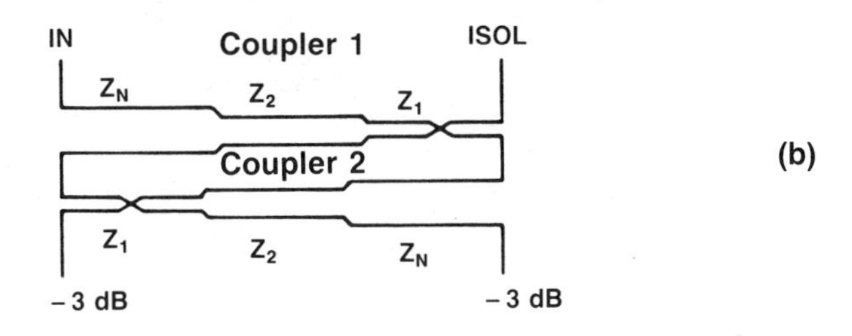

Figure 5.30 Tandem coupler. (*a*) Symmetric tandem of symmetric couplers. (*b*) Symmetric tandem of asymmetric couplers.

choice, provided the particular application does not require maximum bandwidth with very low loss.

Figure 5.31 shows the configuration of an asymmetric tandem of symmetric couplers. As in the symmetric tandem of symmetric couplers, the quadrature phase shift is maintained between the outputs.

As is apparent from the interconnections, the 90° phase relationship is maintained in the symmetric case, but in the asymmetric case, the phase relation depends upon the number of sections.

De Ronde Coupler. The De Ronde coupler is suitable for realization of 3-dB hybrids in MICs. The structure, shown in Fig. 5.32, was first proposed

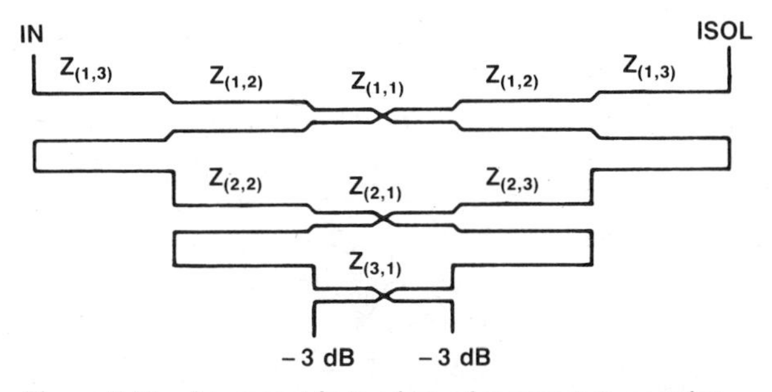

Figure 5.31 Asymmetric tandem of symmetric coupler.

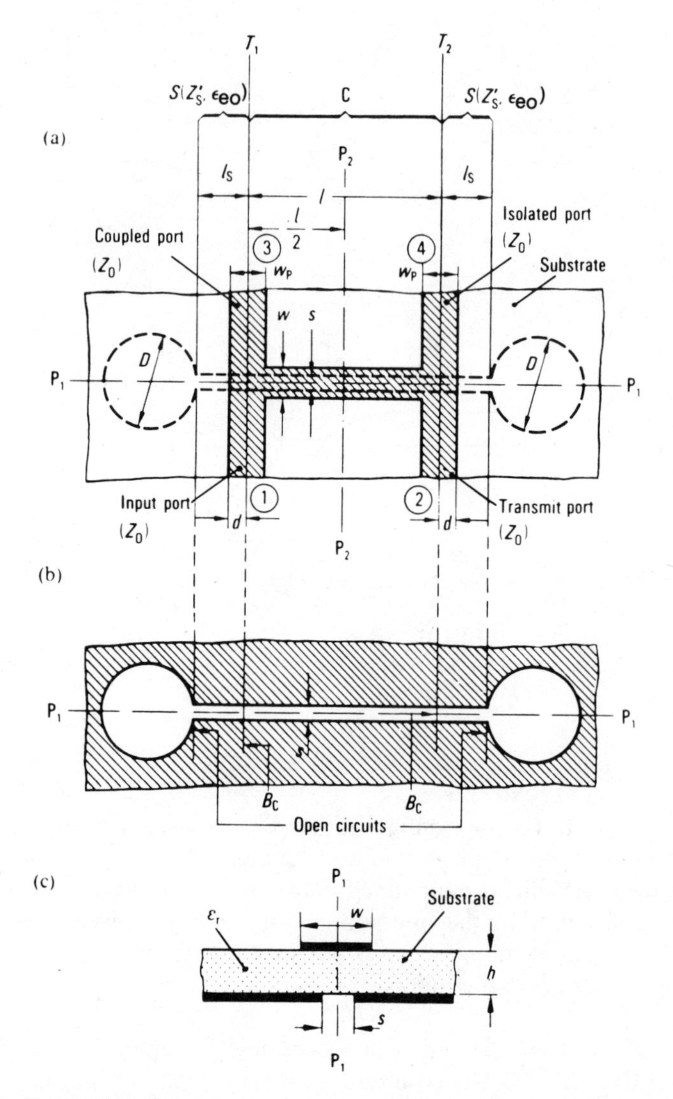

Figure 5.32 De Ronde microstrip slot coupler. (*a*) Upperside of substrate, (*b*) bottom side of substrate, (*c*) cross section.

by De Ronde [29] and has been named after him. Figure 5.32 shows that the hybrid consists of a coupled section of a microstrip and a slot line connected by four microstrip lines of width W_p. The effective length l of the coupled section is defined by the reference planes T_1 and T_2. The slot line of length $l + 2l_s$ is terminated on either side by a disc-shaped slot, D, forming an open circuit. The effects of the parasitics due to the field perturbation occurring at the junctions at each end are nullified by proper choice of the distance d between the two reference planes T_1 and T_2.

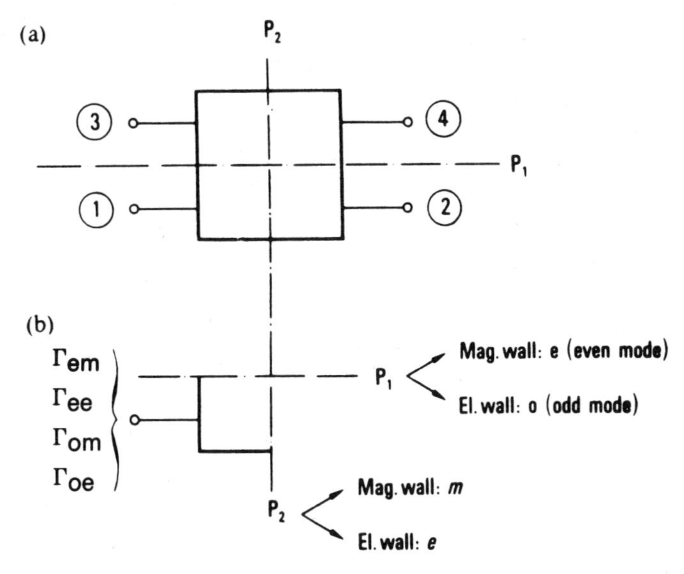

Figure 5.33 Common four-port network with double symmetry.

As can be seen in Fig. 5.32, the hybrid has a double symmetry with respect to the planes P_1 and P_2, as shown in Fig. 5.33, and it is a reciprocal four-port network. The scattering parameters of the network can be written in terms of the even- and the odd-mode reflection coefficients as in [30].

$$S_{11} = \tfrac{1}{4}(\Gamma_{em} + \Gamma_{ee} + \Gamma_{om} + \Gamma_{oe}) \tag{5.93a}$$

$$S_{21} = \tfrac{1}{4}(\Gamma_{em} - \Gamma_{ee} + \Gamma_{om} - \Gamma_{oe}) \tag{5.93b}$$

$$S_{31} = \tfrac{1}{4}(\Gamma_{em} + \Gamma_{ee} - \Gamma_{om} - \Gamma_{oe}) \tag{5.93c}$$

$$S_{41} = \tfrac{1}{4}(\Gamma_{em} - \Gamma_{ee} - \Gamma_{om} + \Gamma_{oe}) \tag{5.93d}$$

The reflection coefficients Γ_{em}, Γ_{ee}, Γ_{om}, and Γ_{oe} are referenced to Z_0 (the characteristic impedance of the connecting section) appearing at each of the ports on application of certain combinations of perfect electric (pec) and perfect magnetic (pmc) planes at the planes of symmetry P_1 and P_2.

A pmc at P_1 [Fig. 5.34(*a*)] corresponds to an even-mode excitation with terminal voltages of equal amplitude and the same phase. The corresponding quasi microstrip field pattern is shown in Fig. 5.34(*b*). This pattern divides the network shown in Fig. 5.34(*b*) into two identical transmission lines of characteristic impedance Z_e

$$Z_e = 2Z_M \qquad \text{and} \qquad \epsilon_{re} = \epsilon_{ee} \tag{5.94}$$

where Z_M is the characteristic impedance of the microstrip and ϵ_{ee}, ϵ_{re} are the effective dielectric constants of the even mode of the coupled lines and the microstrip, respectively. The corresponding relation for the odd mode (pec at P_1) are given by [Fig. 5.34(*c*)]

$$Z_0 = \tfrac{1}{2} Z_s, \qquad \epsilon_{ro} = \epsilon_{eo} \tag{5.95}$$

where Z_s and ϵ_{eo} are the characteristic impedance and the effective

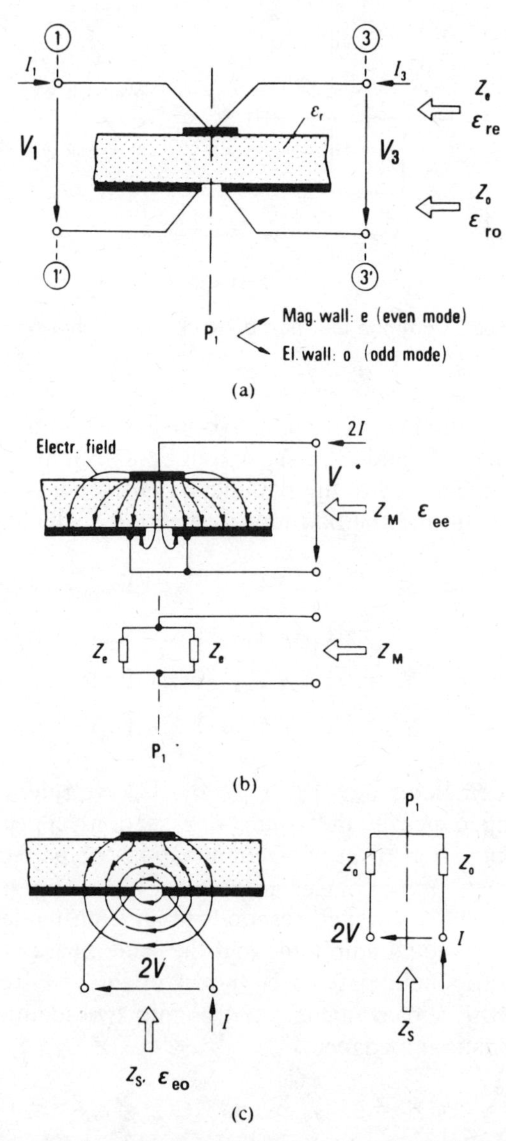

Figure 5.34 (*a*) Even-mode excitation of De Ronde coupler. (*b*) Quasi microstrip field pattern corresponding to even-mode excitation. (*c*) Odd-mode excitation.

dielectric constant of the quasi slot mode due to the odd mode excitation.

In terms of the preceding parameters, the reflection coefficients are given by

$$\Gamma_{em} = \exp\left\{-j2 \arctan\left[\frac{Z_0}{2Z_M} \tan\left(\frac{\theta_e}{2}\right)\right]\right\} \tag{5.96a}$$

$$\Gamma_{ee} = \exp\left\{j2 \arctan\left[\frac{Z_0}{2Z_M} \cot\left(\frac{\theta_o}{2}\right)\right]\right\} \tag{5.96b}$$

$$\Gamma_{om} = \exp\left\{-j2 \arctan\left[\frac{2Z_0}{Z_s} \tan\left(\frac{\theta_e}{2}\right) + 2B_cZ_0\right]\right\} \tag{5.96c}$$

$$\Gamma_{oe} = \exp\left\{-j2 \arctan\left[-\frac{2Z_0}{Z_s} \cot\left(\frac{\theta_o}{2}\right) + 2B_cZ_0\right]\right\} \tag{5.96d}$$

where

$$\theta_e = \beta_e l \tag{5.97a}$$

$$\theta_o = \beta_o l \tag{5.97b}$$

and $\beta_e = (2\pi/\lambda_0)\sqrt{\epsilon_{ee}}$ and $\beta_o = (2\pi/\lambda_0)\sqrt{\epsilon_{eo}}$ are the even- and odd-mode propagation constants, respectively.

The term $2B_cZ_0$ accounts for the slot compensation length l_s (see Fig. 5.35).

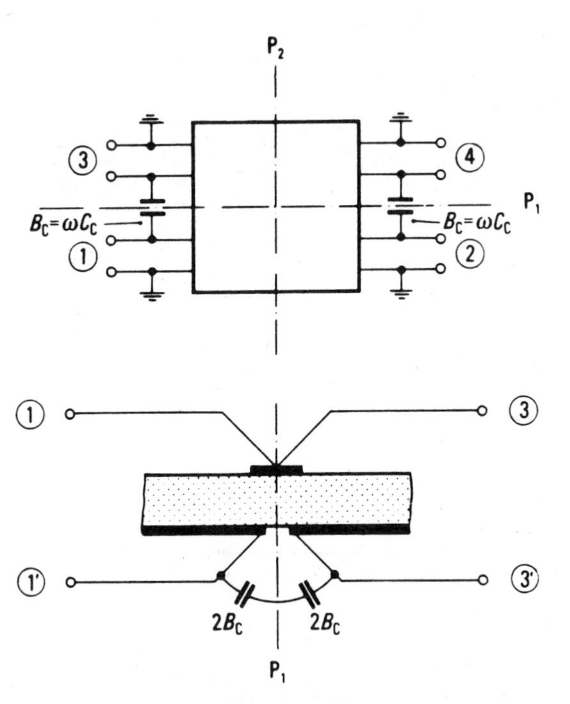

Figure 5.35 Equivalent circuit of compensated De Ronde Coupler.

$$B_c = \frac{1}{Z'_s} \tan \theta'_s = \omega C_c \tag{5.98}$$

$$\theta'_s = \beta_s l'_s \qquad \beta_s = \frac{2\pi}{\lambda_0} \sqrt{\epsilon'_{eo}}$$

Z'_s and ϵ'_{eo} are the characteristic impedance and the effective dielectric constant, respectively, of the slot s. The compensation length l_s does not affect the even-mode excitation. The capacitance C_c effectively increases the odd-mode transmission phase between Terminals 1 and 3, and 2 and 4. Therefore the compensation in terms of C_c leads to equal phase velocities and consequently infinite directivity.

Special Cases of the De Ronde's Coupler

Ideal TEM Case. Although the modes supported by the slot line and the microstrip are truly non-TEM in nature, one can think of an ideal coupler where either supports a purely TEM mode. Such a coupler is physically impossible to realize. However, it can be analyzed to derive the design equations for an ideal coupler.

For an ideal coupler that is matched, $S_{11} = S_{41} = 0$ over the entire useful band of frequencies. This gives from (5.93)

$$\Gamma_{ee} = -\Gamma_{om} \qquad \text{and} \qquad \Gamma_{em} = -\Gamma_{oo}$$

which requires equal, even- and odd-mode effective dielectric constants, or

$$\epsilon_{re} = \epsilon_{ro} = \epsilon_e \qquad \text{and} \qquad \epsilon_{ee} = \epsilon_{eo} = \epsilon_e$$

and

$$Z_0 = \sqrt{Z_e Z_o} = \sqrt{Z_M Z_s} \tag{5.99}$$

With the preceding conditions, the coupling and transmission coefficients are given by

$$S_{21}(f) = \frac{[1 - C^2]^{1/2}}{[1 - C^2]^{1/2} \cos \theta + j \sin \theta} \tag{5.100a}$$

$$S_{31}(f) = \frac{jC \sin \theta}{[1 - C^2]^{1/2} \cos \theta + j \sin \theta} \tag{5.100b}$$

where

$$\theta = \frac{2\pi}{\lambda_0} \sqrt{\epsilon_e}\, l$$

At the center frequency $\theta = \pi/2$. These are the parameters of an ideal TEM coupler as shown in Section 5.3.2. Since

$$C = \frac{Z_{0e} - Z_{0o}}{Z_{0e} + Z_{0o}} \tag{5.101}$$

which gives

$$C = \frac{4Z_M - Z_s}{4Z_M + Z_s} \tag{5.102}$$

Combining (5.99) and (5.102) gives

$$Z_M = \frac{Z_0}{2}\sqrt{\frac{1+C}{1-C}} \tag{5.103a}$$

$$Z_s = 2Z_0\sqrt{\frac{1-C}{1+C}} \tag{5.103b}$$

For a 3-dB coupler in a 50-Ω system, one obtains, using the previous equations,

$$Z_M = 60.35\ \Omega \qquad \text{and} \qquad Z_s = 41.4\ \Omega$$

Real Uncompensated Coupler. In a real uncompensated coupler $l_s = 0$. For all couplers of conventional dimensions, $\epsilon_{eo} \leq \epsilon_{ee}$. This gives the odd-mode phase velocity greater than the even-mode phase velocity. However, the following condition is always satisfied

$$2\left\{\frac{\epsilon_{ee} - \epsilon_{eo}}{\epsilon_{ee} + \epsilon_{eo}}\right\} \leq 1 \tag{5.104}$$

Equation 5.104 implies that for feeding at Port 1, the direct Port 2, the coupled Port 3, and the isolated Port 4 remain unchanged.

In the preceding case certainly the directivity $\log D \leq \infty$ and $S_{11} \geq 0$. However, if the synthesis equation (5.103) is used for designing the coupler, the following conditions hold

$$S_{11}, S_{41} \ll 1$$

and a first-order approximation yields

$$S_{11}(f_0) = -j\frac{\pi}{4}\left\{\frac{\epsilon_{ee} - \epsilon_{eo}}{\epsilon_{ee} + \epsilon_{eo}}\right\} C\sqrt{1 - C^2} \tag{5.105a}$$

$$D(f_0) = -20 \log \left| \frac{\pi}{4} \left\{ \frac{\epsilon_{ee} - \epsilon_{eo}}{\epsilon_{ee} + \epsilon_{eo}} \right\} \frac{1 - C^2}{C} \right\} \right| \tag{5.105b}$$

where f_0 is the center frequency.

Real Compensated Coupler. For compensation, the slot length is increased by an amount l_s and Z_s is changed to Z_s^*. Under the assumption

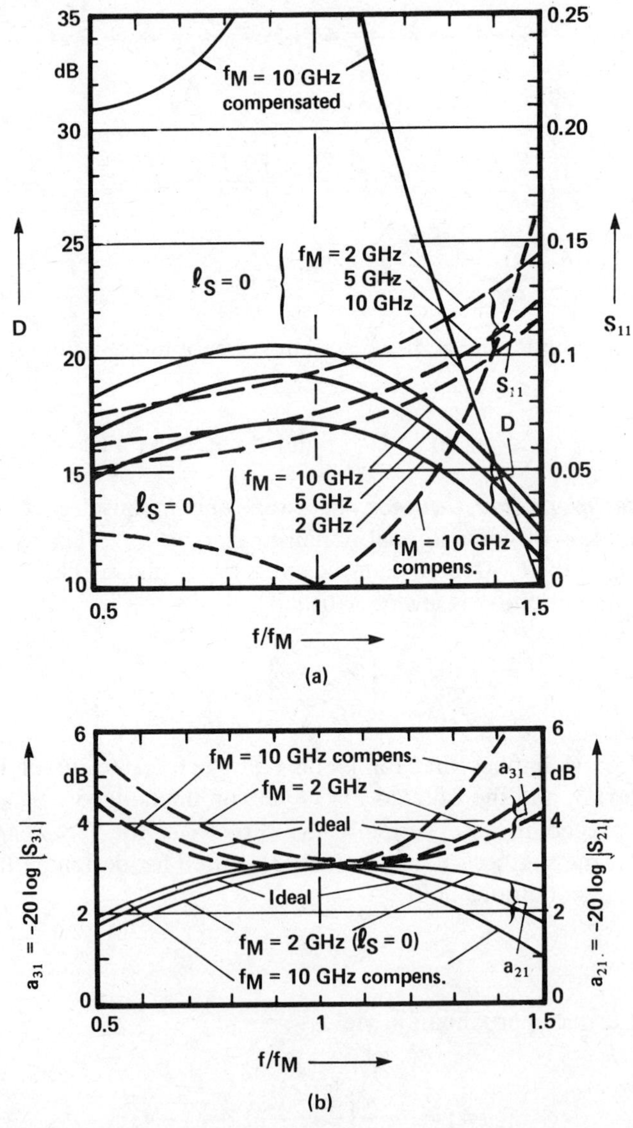

Figure 5.36 Performance of uncompensated coupler. (R. K. Hoffmann and J. Siegel [30], 1982 IEEE, reprinted with permission.)

that the compensation is realized at the center frequency referenced to the microstrip line and $Z_s^* \approx Z_s$, $\epsilon_{eo}^* \approx \epsilon_{eo}$, the required compensation parameters are given by

$$Z_s^* = \frac{Z_s}{2}\left\{\cot\left(\frac{\pi}{4}\sqrt{\frac{\epsilon_{eo}}{\epsilon_{ee}}}\right) + \tan\left(\frac{\pi}{4}\sqrt{\frac{\epsilon_{eo}}{\epsilon_{ee}}}\right)\right\} \tag{5.106a}$$

and

$$l_s = l\sqrt{\frac{\epsilon_{ee}}{\epsilon_{eo}}}\frac{2}{\pi}\tan^{-1}\left\{\frac{Z_s^*}{Z_s}\left[1 - \sin^2\left(\frac{\pi}{4}\sqrt{\frac{\epsilon_{eo}}{\epsilon_{ee}}}\right)\right]\right\} \tag{5.106b}$$

Computation of even- and odd-mode parameters. It is obvious that the coupled section of the device is not purely microstrip nor purely slot line either. An exact analysis of the structure therefore requires the help of numerical techniques. Such techniques are often difficult to carry out. On the other hand, accurate models for analysis of microstrip and slot line are available [6]. These models can be used to approximately realize the design parameters, and the approximations have been found to be valid for all practical purposes as long as the coupling coefficient is less than 6 dB over the 2 to 18 GHz range. The typical performance of an uncompensated coupler is shown in Fig. 5.36(a) and 5.36(b), where a_{21} and a_{31} are the transmission loss and coupling loss, respectively.

5.4 DESIGN CONSIDERATIONS

5.4.1 Losses in Hybrids

The total loss in direct-coupled hybrids can be estimated from the combined dielectric and conductor losses in the individual lines. These are obtained using closed-form equations for microstrip- and strip-line-type transmissions lines [24]. For other types of transmission lines, numerical techniques are used [31]. The loss in coupled lines was dealt with in Chapter 2.

The attenuation due to the even mode α_c^e is always less than that due to the odd mode α_c^o. In coupled lines the loss is given by the average of the losses due to even and odd modes. In almost all planar coupled lines the conductor loss greatly exceeds the dielectric loss [24].

As in strip lines and microstrip lines, the primary contributors to losses in dielectric-based planar waveguide complers are the dielectric loss and the metallic loss. Such losses are always computed numerically from the associated field equations [32].

5.4.2 Directivity Improvement

Because of the inhomogeneity in dielectric structure the directivity of microstrip couplers offers a poor bandwidth resulting from different odd-

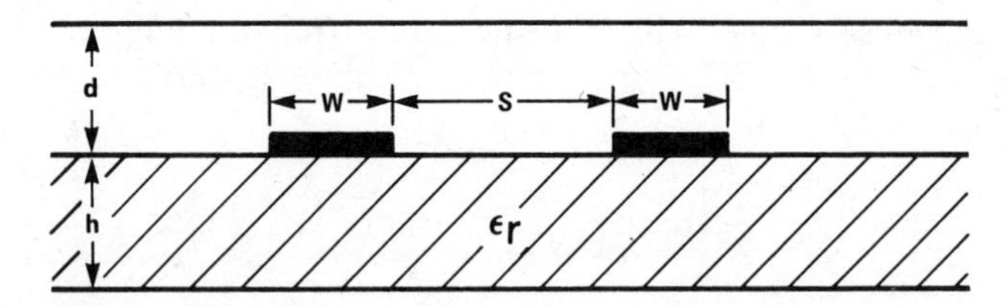

Figure 5.37 Parallel coupler microstrip with grounded shield.

and even-mode phase velocities. There are several ways to equalize the phase velocities, which are as follows.

Use of a Shield. These structures nearly equalize even- and odd-mode phase velocities. The cross section of the structure is shown in Fig. 5.37. This configuration essentially redistributes the field with a substantial amount of the field in the air dielectric medium above the coupled strips. For $d = h$ the phase velocities are exactly equal and each is equal to

$$v_e = v_o = \frac{c}{\sqrt{\frac{\epsilon_r + 1}{2}}} \tag{5.107}$$

Strictly speaking improvement in directivity is obtained at the cost of manufacturing difficulties in this case.

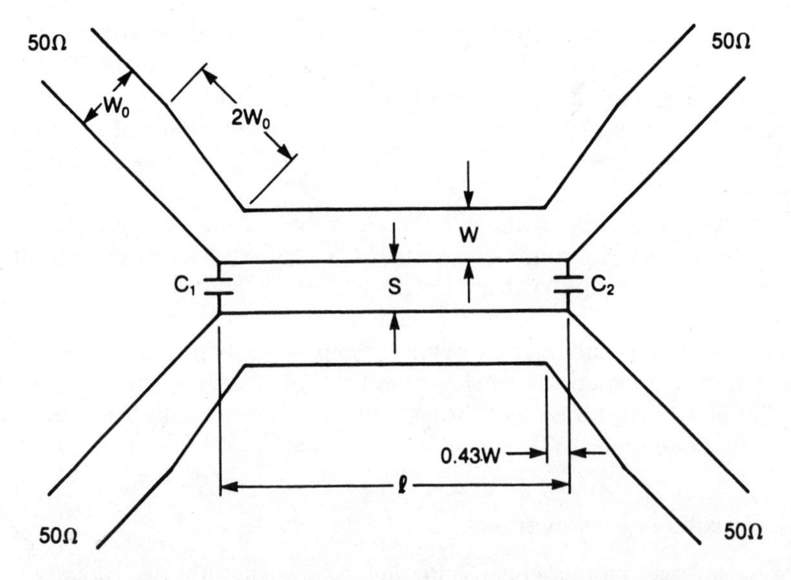

Figure 5.38 Lumped capacitor compensation of microstrip coupler.

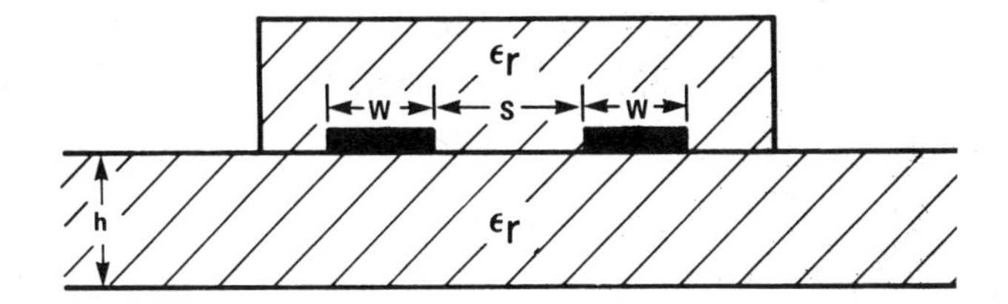

Figure 5.39 Parallel coupled microstrip with overlay compensation.

Use of Lumped Capacitors. Using lumped capacitors at the ends of the coupled section is the simplest way of achieving equal phase velocities. The structure is shown in Fig. 5.38. This effectively increases the odd-mode phase angle by [12]

$$\Delta\theta_o = 2\pi f_0(C_1 + C_2)Z_{0o} \tag{5.108}$$

where f_0 is the center frequency of the coupled line.

Use of a Dielectric Overlay. The presence of another dielectric layer of the same permittivity as that of the substrate reduces the odd-mode phase velocity to a large extent without considerably affecting the even-mode phase velocity [33]. Thus controlling the thickness and width of the overlay, the even- and the odd-mode phase velocities can be equalized within 1% over quite a broad band. As shown in Fig. 5.39, the overlay covers the two strips where the main coupling takes place. The overlay is bonded with the help of some kind of epoxy. However, this method is practically cumbersome due to its poor repeatability. In order to overcome this problem a multilayer structure using different dielectric constants materials has been recently proposed [33].

REFERENCES

1. Agrawal, A. K., and G. F. Mikucki, "A Printed Circuit Hybrid Ring Directional Coupler for Arbitrary Power Divisions," *IEEE Trans. Microwave Theory Tech.*, Vol. MTT-34, Dec. 1986, pp. 1401–1407.
2. Levy, R., "Zolotarev Branch-Guide Couplers," *IEEE Trans. Microwave Theory Tech.* Vol. MTT-21, Feb. 1973, pp. 95–99.
3. Levy, R., and J. Helszajn, "Specific Equation for One or Two Section Quarter-wave Matching Networks for Stub Resistor Loads," *IEEE Trans. Microwave Theory Tech.* Vol. MTT-30, Jan. 1982, pp. 55–62.
4. Muraguchi, O., M. Y. Takeshi, and Y. Naito, "Optimum Design of 3-dB Branch Line Couplers Using Microstrip Lines," *IEEE Trans. Microwave Theory Tech.*, Vol. MTT-31, Aug 1983. pp. 674–678.

5. Howe, H., *Stripline Circuit Design*, Artech House, Dedham, Mass., 1974.
6. Gupta, K. C., R. Garg, and I. J. Bahl, *Microstrip Lines and Slotlines*, Artech House, Dedham, Mass., 1979.
7. Pramanick, P., and P. Bhartia, "Analysis and Synthesis Equations for Suspended and Inverted Microstriplines," *Arch. Elek. Übertragung.*, Vol. 39, Sept. 1985, pp. 323–326.
8. Pramanick, P., and P. Bhartia, "Computer Aided Design Models for Millimeter Wave Finlines and Suspended Substrate Microstrip Lines," *IEEE Trans. Microwave Theory Tech.*, Vol. MTT-33, Dec. 1985, pp. 1429–1435.
9. Bhartia, P., and I. J. Bahl, *Millimeter Wave Engineering and Applications*, Wiley, New York, 1984, pp. 300–346.
10. Koul, S. K., Center for Advanced Research in Electronics, Indian Institute of Technology, New Delhi, India, private communication, 1985.
11. Rycroft, C. M. D., "Finline Rat Race Hybrid Couplers," Interim Report ITM 84, 84/159, Chemsford, Essex, U.K., 1984.
12. Bhartia, P., and I. J. Bahl, *Millimeter Wave Engineering and Applications*, Wiley, New York, 1984, pp. 358–382.
13. Pramanick, P., and P. Bhartia, "Design of Imageguide Microstrip Line Couplers," *Microwave J.*, Vol. 27, May 1984, pp. 247–255.
14. Bahl, I. J., and P. Bhartia, "Aperture Coupling Between Dielectric Image Lines," *IEEE Trans. Microwave Theory Tech.*, Vol. MTT-29, Sept. 1981, pp. 591–596.
15. Pramanick, P., and P. Bhartia, "Waveguide Imageline Directional Coupler," *Arch. Elek. Übertragung.*, Vol. 38, Jan. 1984, pp. 46–50.
16. Edwards, T. C., *Foundations for Microstrip Circuit Design*, Wiley, New York, 1981, pp. 147–148.
17. Matthaei, G. L., L. Young, and E. M. T. Jones, *Microwave Filters Impedance Matching Networks and Coupling Structures* (Reprinted), Artech House, Dedham, Mass., 1980.
18. Davis, W. A., *Microwave Semiconductor Circuit Design*, Van Nostrand, New York, 1983.
19. Hoffmann, R. K., and G. Morler, "Directional Coupler Synthesis—Computer Program Discription," *IEEE Trans. Microwave Theory Tech.*, Vol. MTT-22, Jan. 1974, p. 77.
20. Klopfenstein, R. W., "A Transmission Line Taper of Improved Design," *Proc. IRE*, Vol. 44, Jan. 1956, pp. 31–35.
21. Lange, J., "Interdigited Stripline Quadrature Hybrid," in *IEEE Trans. Microwave Theory Tech.*, Vol. MTT-17, Dec. 1969, pp. 1150–1151.
22. Kajfez, D., Paunovic Z., and Paulin, S., "Simplified Design of Lange Coupler," *IEEE Trans. Microwave Theory Tech.*, Vol. MTT-26, Oct. 1978, pp. 806–808.
23. Presser, A., "Interdigited Microstrip Coupler Design," *IEEE Trans. Microwave Theory Tech.*, Vol. MTT-26, Oct. 1978, pp. 801–805.
24. Garg, R., and I. J. Bahl, "Characteristics of Coupled Microstrip," *IEEE Trans. Microwave Theory Tech.*, Vol. MTT-27, July 1979, pp. 700–705.

25. Wilkinson, E., "An N-Way Hybrid Power Divider," *IEEE Trans. Microwave Theory Tech.*, Vol. MTT-8, Jan. 1960, pp. 116–118.
26. Cohn, S. B., "A Class of Broadband 3-Port TEM Hybrids," *IEEE Trans. Microwave Theory Tech.*, Vol. MIT-16, Feb. 1968, pp. 110–118.
27. Howe, H., *Stripline Circuit Design*, Artech House, Dedham, Mass., 1974, pp. 97–98.
28. Howe, H., *Stripline Circuit Design*, Artech House, Dedham, Mass., 1974, pp. 164–172.
29. De Ronde, F. C., "A New Class of Microstrip Directional Couplers," in *IEEE Int. Microwave Symp. Digest*, 1974, pp. 184–186.
30. Hoffmann, R. K., and J. Siegel, "Microstrip-Slot Coupler Design—Parts I & II," *IEEE Trans. Microwave Theory Tech.*, Vol. MTT-30, Aug. 1982, pp. 1205–1216.
31. Trinh, T. N. and R. Mittra, "Suspended H-waveguide and its mm-Wave Applications," *IEEE Int. Microwave Symp. Digest*, 1983, pp. 305–308.
32. Shelad, B., and Spielman, B. E., "Broadband (7–18 GHz) 10-dB Overlay Coupler for MIC Applications," *Electron Lett.*, Vol. 11, April 17, 1975, pp. 175–176.
33. Horno, M., and F. Medina, "Multilayer Planar Structures for High-Directivity Directional Coupler Design," *IEEE Trans. Microwave Theory Tech.*, Vol. MTT-34, Dec. 1986, pp. 1442–1449.

PROBLEMS

5.1 Determine the aperture radii of five-section Chebyshev rectangular waveguide to image-line and image-line to rectangular waveguide directional couplers for 20 dB coupling, with minimum directivity of 40 dB, a bandwidth of 14%, and operational frequency of 35 GHz. Choose suitable guide dimensions for the fundamental mode of operation. Assume that the image line has a dielectric constant of 2.54.

5.2 Determine the aperture radii of five-section Chebyshev image-line–microstrip and microstrip–image-line couplers for a required coupling of 10 dB, directivity of 50 dB, bandwidth of 16%, ϵ_r of 2.54, and center frequency 35 GHz.

5.3 Design a single-section microstrip directional coupler with the following specifications: Coupling 10 dB, substrate $\epsilon_r = 9.0$, substrate thickness = 0.635 mm, system center frequency = 4.0 GHz, impedance = 50 Ω. Neglect dispersion.

5.4 (a) Find the impedances of a hybrid ring directional coupler for the

following power split ratios:
1. 0-dB,
2. 3-dB,
3. 9-dB.

(b) For a strip line type coupler, find the corresponding strip width dimensions for a substrate $\epsilon_r = 3.8$ and ground plane spacing of 2.5 mm.

5.5 Design an image-line edge-coupled directional coupler with 0-dB coupling at 33 GHz, return loss ≤ 25 dB, and directivity ≤ 30 dB over the 30–40 GHz frequency band. Assume $\epsilon_r = 2.22$.

5.6 A TEM-mode asymmetric directional coupler has $C_\infty = 3.01$ dB. Draw the even- and odd-mode impedance profile as a function of the electrical length of the coupled section, which is required to be of $\lambda/2$ length. Repeat the plot for $C_\infty = 10$ dB.

5.7 Design an edge-coupled 10-dB microstrip coupler on 0.25-mm alumina substrate for operation over the 30–40 GHz band. The required return loss and the directivity are 30 dB and 45 dB, respectively. Use suitable capacitances for phase velocity compensation.

5.8 Design a 3-dB Lange coupler on a 0.38-mm fused quartz microstrip substrate with isolation and return loss ≥ 25 dB over the 8–12 GHz range. Redesign the circuit for operation over 11–15 GHz.

5.9 Design a four-way symmetric power divider in microstrip configuration with the following specifications:

Center frequency = 4 GHz,
Power in the outermost arms = 20%,
Power in the innermost arms = 30%,
Input and output impedance = 50 Ω.

The microstrip parameters are $\epsilon_r = 9.9$, $h = 0.63$ mm, and $t = 5$ μm.

5.10 Design a three-branch symmetric branch-line coupler in strip line having 3 dB coupling at 4 GHz. Compare its 3 ± 0.5 dB coupling bandwidth with a two-branch coupler. The strip-line parameters are $b = 1.2$ mm, $t = 0$, $\epsilon_r = 2.32$, and discontinuity effects may be ignored.

6 FILTERS AND MULTIPLEXERS

6.1 INTRODUCTION

Most microwave systems consist of many active and passive components that are difficult to design and manufacture with precise frequency characteristics. In contrast, microwave passive filters can be designed and manufactured with remarkably predictable performance. As a result, microwave systems are usually designed so that all of the troublesome components are relatively wide in frequency response to minimize their effect on the overall system. Filters are then incorporated to very precisely set the system frequency response. Since the filters are the narrowest bandwidth components in the system, it is usually the filters that limit such system parameters as gain and group delay flatness over frequency.

A passive microwave filter is a circuit component consisting of lumped elements (inductors, capacitors, and resistors) only or distributed elements (waveguide sections or microstrip or fin line or any other medium) or both, arranged in a particular configuration (topology), so that desired signal frequencies are allowed to pass with minimum possible attenuation while undesired frequencies are attenuated. In the broadest sense, all microwave components can be considered as filters, since each will exhibit some band-limiting behavior when used in a system. Sometimes filters are also used for impedance matching as described in Chapter 4.

Immense work has been done on filters, and it is not possible to include every aspect of filters in this book due to its limited scope. However, the intent of this chapter is to touch upon the important topics dealing with design of printed circuit filters and provide references to facilitate more in-depth study.

The microwave filters that are described in this chapter are two-port reciprocal, doubly terminated, passive, linear, reflective, and lossy. A filter may be viewed conceptually as shown in Fig. 6.1. *Two port* simply refers to the fact that the device has two electrical terminals: the input is driven by a signal generator while the output is connected to a load. A *reciprocal* device is one for which there is no preferred signal propagation direction between ports. This does not mean that a reciprocal device can have input and output interchanged without changing electrical performance, because the impedance levels of the two ports may be different. If the generator and load impedances are identical, a reciprocal device will have identical

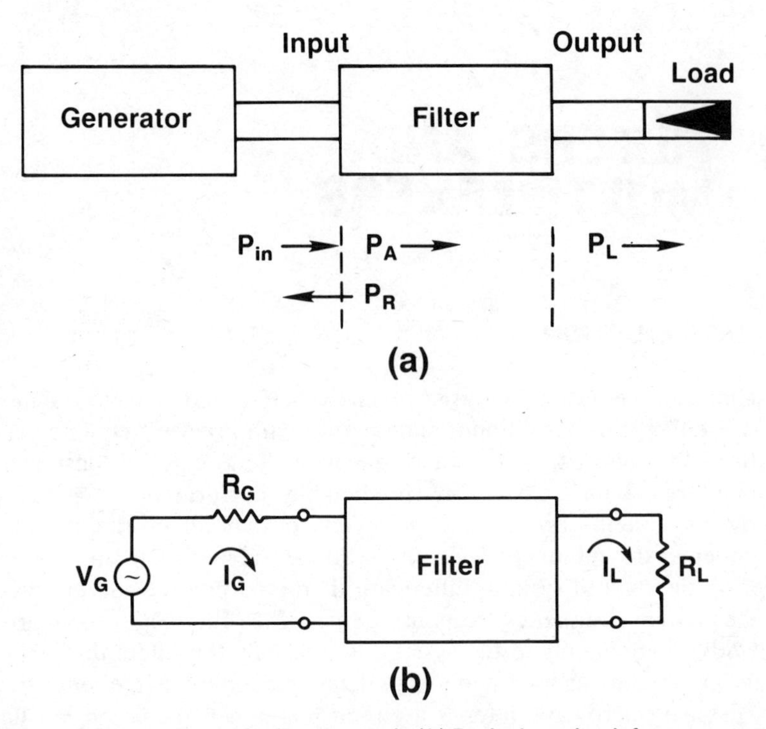

Figure 6.1 (*a*) General form of a filter network. (*b*) Equivalent circuit for power transfer calculations.

transmission characteristics when hooked up in either direction, but the reflection characteristics may be different. Practical filters are never precisely identical in reflection characteristics at the input and output due to manufacturing tolerances. Most microwave systems are designed to operate with resistive 50-Ω source and load impedances. A filter designed to operate with resistive source and load impedances (not necessarily 50 Ω) is called a *doubly terminated* filter.

A *passive* device is one that has no internal energy sources. Incident signal power is reflected, absorbed, or transmitted, but not amplified. A *linear* device is one for which all power levels (reflected, absorbed, transmitted) are proportional to the incident power. A linear device generates no new frequency components, no intermodulation products, and does not compress the signal. Since there is no interaction between signals, problems with multiple transmission can be treated by superposition. For example, independent signals could be propagating both ways simultaneously through the filter.

No device is truly linear for all input powers, since at some power level, the device will either arc out or melt. Practical filters operating below their maximum rated power levels are generally well treated as linear devices.

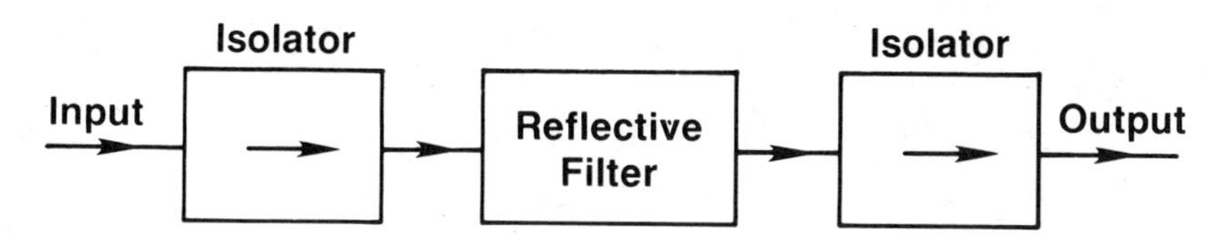

Figure 6.2 A dissipative filter.

One exception to this rule occurs in sensitive systems with high power levels where passive low-level (typically 100 dB below the signal level) intermodulation generation can be a problem, particularly at joints and connections between dissimilar metals.

A *reflective* filter is one that achieves rejection by reflecting the incident power. Reflective filters can be thought of as transformers that impedance match the source and load at frequencies where minimum loss is desired, and mismatch the source and load at frequencies where rejection is desired. Filters using capacitor and inductor lumped elements and low-loss distributed elements are of the reflective type. Filters that dissipate the rejected signal internally are called *dissipative* or *absorbtive filters*. A dissipative filter can also be realized using a reflective filter and two isolators, as shown in Fig. 6.2. A *lossy* filter is one for which the incident power is greater than the sum of the reflected and transmitted power; some power is always attenuated within the filter. All practical filters are lossy to varying degrees. Filters made using superconductors can be nearly lossless. Recent advances in high temperature superconductors may cause much greater exploitation of superconducting elements in filters.

6.1.1 Filter Parameter Definition

We now go on to define some of the commonly used terms in the design of filters. When one is designing a filter, the important specifications one looks into are: frequency range, bandwidth, insertion loss, stopband attenuation and frequencies, input and output impedance levels, VSWR, group delay, phase linearity, temperature range, and transient response.

As shown in Fig. 6.1(a), the input is driven by a signal generator with the output passing to a load. At the input plane of the filter, the power may be broken into three components: P_{in}, the incident power from the generator; P_R, the power reflected back toward the generator; and P_A,, the power absorbed by the filter. The power passed on to the load is P_L.

Conservation of energy demands that

$$\begin{aligned} P_{\text{in}} &= P_R + P_A \\ P_L &\leq P_A \end{aligned} \tag{6.1}$$

where $P_L = P_A$ if the filter is lossless, and $P_L = P_{in}$ if the filter is lossless and there are no reflections.

The incident power is the same as the maximum available power from the generator. Then

$$P_{in} = \frac{V_G^2}{4R_G} \tag{6.2a}$$

$$P_L = |I_L|^2 R_L \tag{6.2b}$$

$$P_A = \text{Re}[I_G(V_G - I_G R_G)] \tag{6.2c}$$

$$P_R = \frac{V_G^2}{4R_G} - \text{Re}[I_G(V_G - I_G R_G)] \tag{6.2d}$$

where Re denotes the real part of the argument and other quantities have their usual meanings. Complex notation has been introduced because filters will usually impose phase shifts on the signals passed.

Insertion Loss. The insertion loss IL (in decibels) at a particular frequency is defined as

$$\text{IL} = -10 \log \frac{P_L}{P_{in}} \tag{6.3}$$

Thus if IL = 3 dB, only 50% of the available or incident power is delivered to the load. The filter rejection RJ (in decibels) at a given frequency is defined as

$$\text{RJ} = -10 \log \frac{P_L}{P_{in}} - \text{IL}_m \tag{6.4}$$

where IL_m is the midband, or sometimes the minimum insertion loss. For example, a filter that has 2-dB minimum insertion loss must reach 22-dB insertion loss to obtain 20 dB of rejection.

Return Loss. Three related parameters, the return loss (RL), voltage standing-wave ratio (VSWR), and reflection coefficient (ρ) are commonly used to characterize filter reflections. Return loss is used with filters because it is a sensitive parameter describing filter performance. The return loss is the ratio of the input to reflected power:

$$\text{RL} = -10 \log \frac{P_R}{P_{in}} = -10 \log \left(\frac{\text{VSWR} - 1}{\text{VSWR} + 1} \right)^2 \tag{6.5}$$

$$= -10 \log(|\rho|^2)$$

TABLE 6.1 Return Loss, Insertion Loss, and Related Parameters

RL (dB)	ρ (Voltage)	VSWR	IL (dB)	Transmitted Power (%)
0	1	∞	∞	0
0.5	0.9441	34.75	9.64	10.87
1	0.8913	17.39	6.87	20.6
2	0.7943	8.72	4.33	36.9
3	0.7079	5.85	3.02	49.9
4	0.6310	4.42	2.20	60.2
6	0.5012	3.01	1.26	74.9
8	0.3981	2.32	0.75	84.2
10	0.3162	1.93	0.46	90.0
12	0.2512	1.67	0.28	93.7
15	0.1778	1.43	0.14	96.8
20	0.1000	1.22	0.04	99.0
25	0.0560	1.12	0.01	99.7
30	0.0316	1.07	<0.01	99.9

Some typical values for RL, ρ, VSWR, IL, and percentage of transmitted power are given in Table 6.1.

In addition to the insertion loss and return loss, phase characteristics of a filter are often very important. The transmission phase, ϕ_T (in radians) is given by

$$\phi_T = \arg(I_L) \tag{6.6a}$$

and the group delay, τ_D (in seconds), is defined as

$$\tau_D = \frac{d\phi_T}{d\omega} = \frac{1}{2\pi}\frac{d\phi_T}{df} \tag{6.6b}$$

where ω is in radians per second.

Group delay is important in several ways. It is a measure of how long a signal takes to propagate through the filter. Deviation from constant group delay over a given frequency band will cause FM signals to become distorted. With constant group delay, components of multifrequency signals will travel at the same velocity through the device with the result that there will be no frequency dispersion; sharp pulses will remain sharp, etc.

Typically, signal delay can be compensated for in a system, but one has to live with dispersion. There are two ways commonly used to define the limits of frequency dispersion. Most common is to specify maximum permissible group delay variation over frequency. Alternatively, the maximum deviation from linear phase (DLP) may be specified. The DLP over a given frequency range is the maximum deviation between the device phase and a linear phase, that is,

$$\text{DLP} = \max[\phi_T - K\omega t] \tag{6.7}$$

where constant K has been chosen to minimize the deviation from linear phase. Note that if the group delay is constant (namely for an idealized transmission line), DLP is zero.

There are some pitfalls associated with equating the device group delay with the signal time delay through the device. When the group delay is calculated by most analytic techniques, the actual physical extent of the filter is ignored, and the transient response is incompletely considered. The group delay will be a reasonable estimate of how long the main body of the signal takes to go through the filter, but the situation is more complicated.

Every filter has both a "steady state" and a "transient" response that may be quite a bit different. That means, for the pulse-modulated signals whose outputs are complex functions of time, the rejection of a filter may be seriously altered from the steady state for pulsed signals with pulse length on the same order as or shorter than the group delay. In a channelized receiver, accounting for a transient phenomenon through filters is one of the most challenging tasks in the design, especially if the receiver is designed to measure the most accurate frequency information on short pulses.

If a single-frequency signal within the passband of a filter is abruptly turned on at time $t = 0$, the envelope of the output signal will look something like that shown in Fig. 6.3. Signal packets that arrive before the

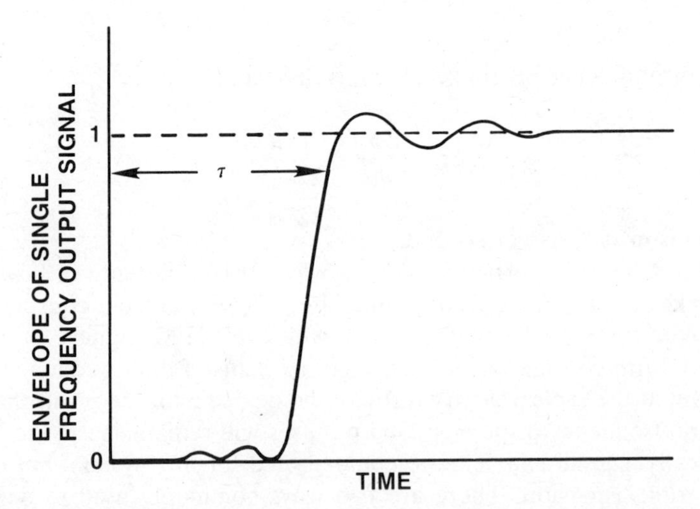

Figure 6.3 Typical output of a filter when a single-frequency input signal is abruptly switched on at $t = 0$.

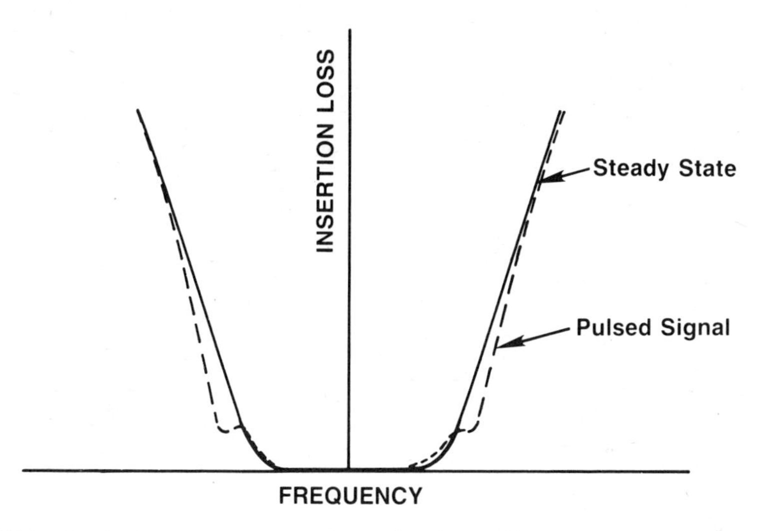

Figure 6.4 Qualitative curves comparing a bandpass filter response to pulsed and steady state signals.

main body of the signal are called *precursors*. Note that the signal will typically overshoot and then follow a damped oscillation about its final value.

A different manifestation of the same phenomenon is shown in Fig. 6.4, which compares the steady state and rapidly pulsed on and off insertion-loss response versus frequency. The two dips in the curve due to the transient response of the filter are called rabbit ears. In most applications, transient effects can be ignored if pulsewidths are longer than the group delays.

6.1.2 Basic Types

Basically there are four types of filters: low pass, bandpass, bandstop (also called band reject or notch) and high pass. Their frequency response is shown in Fig. 6.5. An ideal filter would have zero insertion loss and constant group delay across the desired frequency passbands, and infinite rejection everywhere else. Practical filters deviate *substantially* from the ideal, however. In particular, no filter can operate over an unlimited frequency range. All filters exhibit spurious responses where they have rejection in the passband or regions of low loss where the rejection should be high. The best that can be accomplished is to have the filter perform well over frequencies of interest.

6.1.3 Applications

Almost all microwave receivers, transmitters, and test setups require filter action. The main filter functions are to reject undesirable signal frequencies

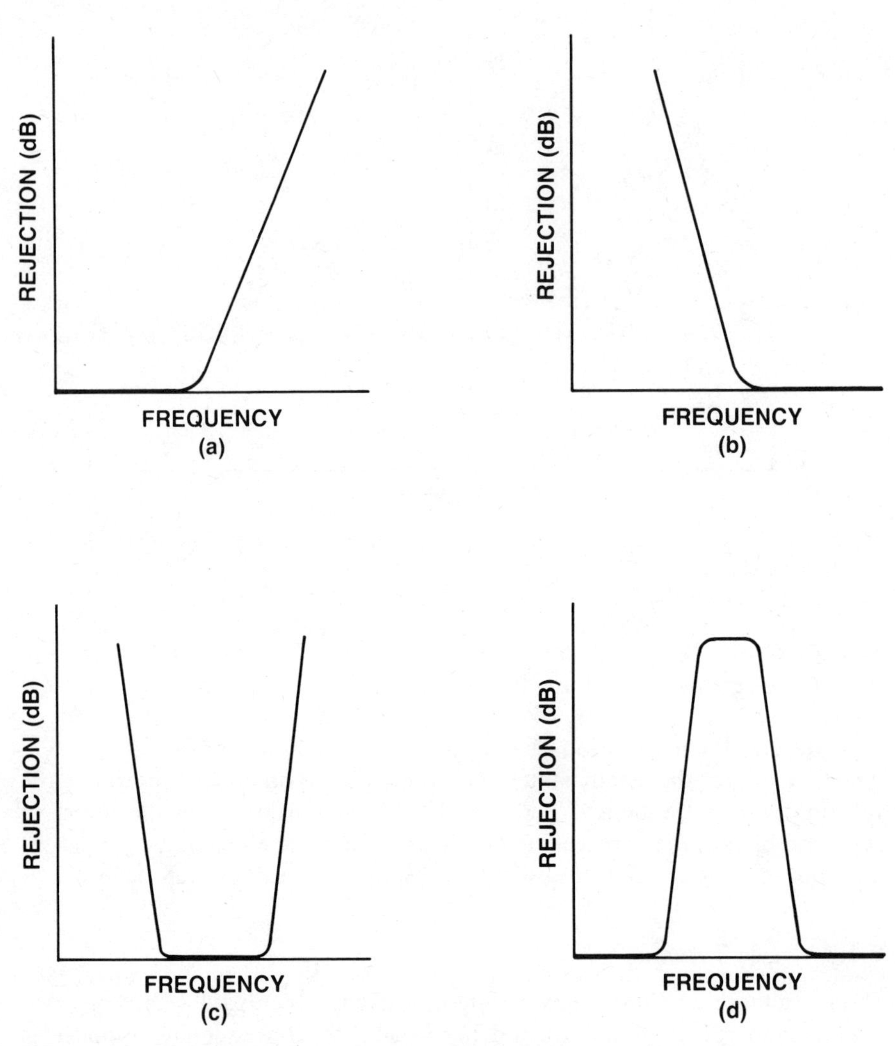

Figure 6.5 Basic types of filters. (*a*) Low pass. (*b*) High pass. (*c*) Bandpass. (*d*) Bandstop.

outside the filter passband and to channelize or combine different frequency signals. Good examples for the former applications are mixers and multipliers. In mixers at the IF port, a low-pass filter is required to pass intermediate frequencies only, while attenuating RF local oscillator, image, and other spurious frequencies. In multipliers, a passband filter is required to pass the desired harmonic while attenuating all other harmonics plus the fundamental. A well-described example for the later application is the channelized receiver in which a bank of filters is used to separate input signals.

Specific applications include ESM receivers, satellite communications, mobile communications, direct broadcast satellite systems, PCM communications, and microwave FM multiplexers.

6.2 FILTER MEASUREMENTS

Insertion loss and return loss of filters can be measured by measuring signal magnitude only, while transmission phase and group delay measurements require a vector network. Filter measurements are briefly described in the following two sections.

6.2.1 Insertion Loss and Return Loss

A typical microwave setup for measuring insertion loss and return loss of a filter is shown in Fig. 6.6. A sweep oscillator is used as the signal source. An isolator minimizes reflections for signals flowing toward the generator. A precision attenuator is used to calibrate the detector. Directional couplers with detectors are used to monitor reflected and transmitted power. A dual-trace oscilloscope or scalar network analyzer with horizontal sweep synchronized to the sweep generator frequency ramp is used to simultaneously display the reflected and transmitted power. When designing a test setup for a particular filter, care must be taken to get the same impedance levels for the source and load as will be used in the filter operation, since filter performance is a strong function of source and load impedances.

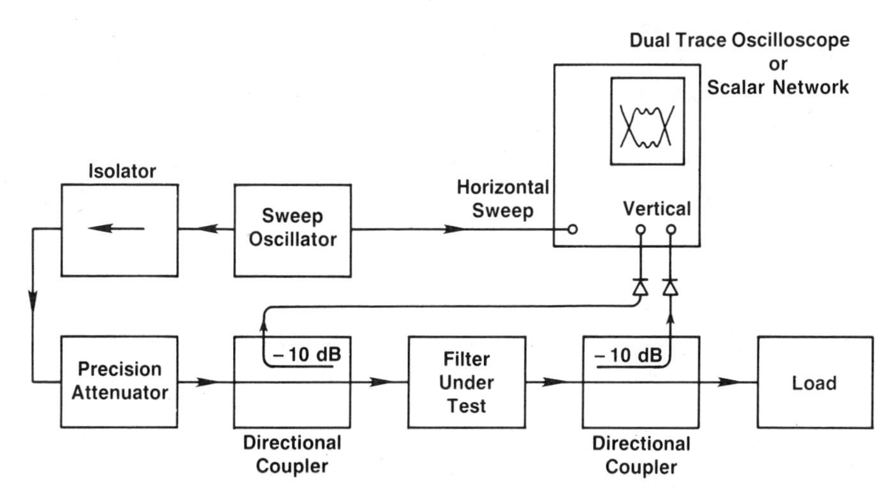

Figure 6.6 Measurement setup for insertion loss and return loss.

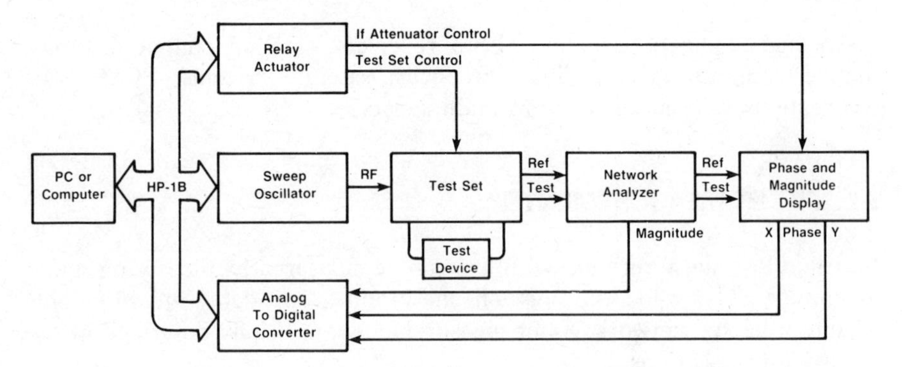

Figure 6.7 Automated *S*-parameter measurement setup.

6.2.2 *S*-Parameters

Filters can be fully characterized by measuring *S*-parameters by a setup using an automatic vector network analyzer, as shown in Fig. 6.7. Measurement of group delay on the HP 8510 is accomplished using the phase–slope difference method. Here adjacent data buckets for S_{21} are manipulated to construct the group delay at each frequency increment, and then (6.6b) is used to determine group delay.

6.3 FILTER SYNTHESIS

Several methods are available for designing filters. Of these, the low-pass prototype filter synthesis and numerical method have been the most successful. Both methods depend critically on the judgment of the designer to choose between many possible solutions. The traditional design approach (low-pass prototype), which dates back to before computers, is to focus on special cases that allow the use of analytic solutions under highly idealized conditions. Despite drawbacks, this technique has been very successful and has been the basis for the vast majority of filter designs [1]. Furthermore, traditional design is the starting point for the second type of design using numerical methods. Therefore, we start with a discussion of the traditional approach [1–5].

6.3.1 Filter Design from Low-Pass Filter Synthesis

This method consists of the following steps:

1. design of a prototype low-pass filter with the desired passband characteristics,
2. transformation of this prototype network to the required type (low

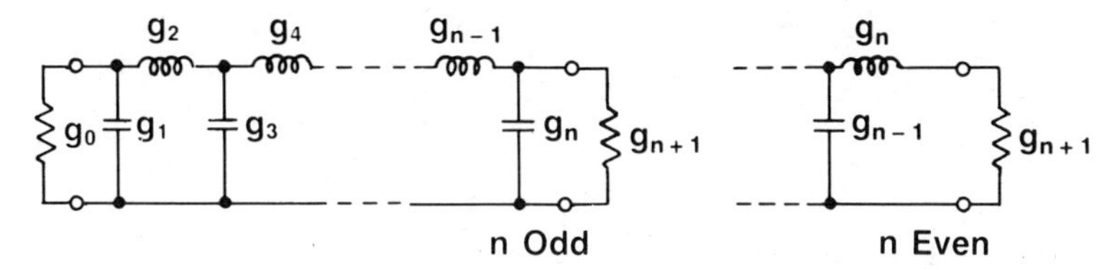

Figure 6.8 Prototype low-pass filter.

pass, high pass, bandpass, or bandstop) filter with the specified center and/or band-edge frequencies,

3. realization of the network in terms of lumped and/or distributed circuit elements.

Low-pass prototype filter design using the insertion-loss method is used quite extensively. In this method, the design of the filter starts with specifying in the desired frequency band the insertion loss as defined by (6.3), or the return loss for a lossless network as defined by (6.5). After specifying the magnitude of the insertion loss as a function of frequency in the passband, the network that will give the desired insertion loss is then synthesized.

The combination of inductors and capacitors shown in Fig. 6.8 is obviously a low-pass circuit, since at high frequencies the series inductors open up and the shunt capacitor shorts out the signal, while at low frequencies, the series inductors short, the shunt capacitors open, and the input is connected without loss to the output. The immediate problem is how to choose element values.

There is virtually an unlimited number of different solutions to the low-pass prototype design. Here we focus on the two that have been applied the most: maximally flat (Butterworth) response, as shown in Fig. 6.9(a),

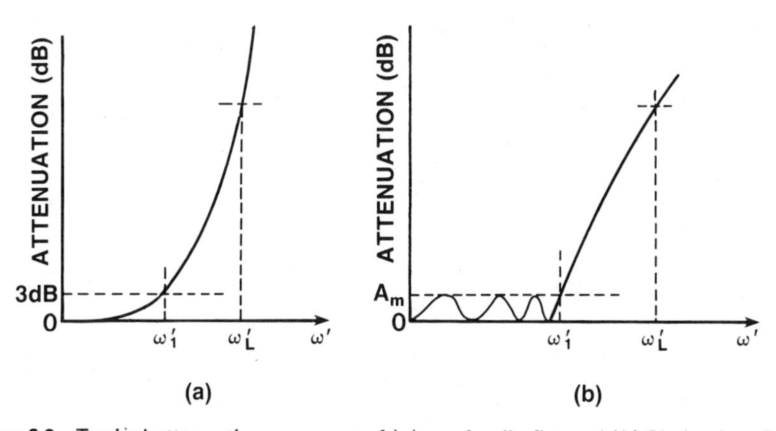

Figure 6.9 Typical attenuation response of (a) maximally flat and (b) Chebyshev filters.

and equal-ripple (Chebyshev also often spelled Tchebycheff), as shown in Fig. 6.9(*b*). Here ω_1' and ω_L' are the passband edge and stopband edge frequencies, respectively, and A_m is the allowed attenuation in the passband.

Maximally Flat Response. In a Butterworth low-pass prototype we want the low-pass insertion loss to be as flat as possible at zero frequency, and then rise monotonically as fast as possible with increasing frequency. The attenuation loss (in decibels) for this response for $\omega_1' = 1$ may be expressed

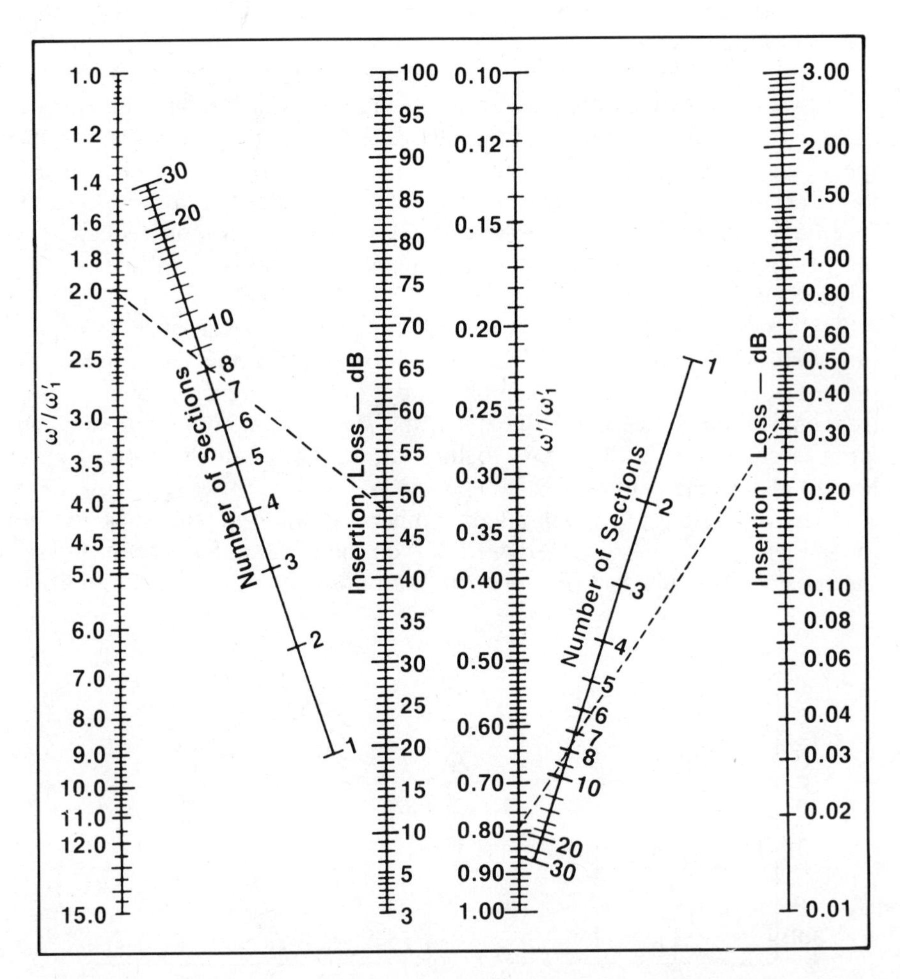

Figure 6.10 Nomograph for selecting number of sections of maximally flat filter for given insertion loss in stop band. This chart is separated into stopband (left-hand side) and passband regions (right-band side). (After T. Milligan[6], 1985 *Microwave & RF*, reprinted with persmission.)

TABLE 6.2 Element Values for the Maximally Flat Low-pass Prototype Filter Having $g_0 = 1$, $\omega_1' = 1$, and $n = 1$ to 10

Value of n	g_1	g_2	g_3	g_4	g_5	g_6	g_7	g_8	g_9	g_{10}	g_{11}
1	2.000	1.000									
2	1.414	1.414	1.000								
3	1.000	2.000	1.000	1.000							
4	0.7654	1.848	1.848	0.7654	1.000						
5	0.6180	1.618	2.000	1.618	0.6180	1.000					
6	0.5176	1.414	1.932	1.932	1.414	0.5176	1.000				
7	0.4450	1.247	1.802	2.000	1.802	1.247	0.4450	1.000			
8	0.3902	1.111	1.663	1.962	1.962	1.663	1.111	0.3902	1.000		
9	0.3473	1.000	1.532	1.879	2.000	1.879	1.532	1.000	0.3473	1.000	
10	0.3129	0.9080	1.414	1.782	1.975	1.975	1.782	1.414	0.9080	0.3129	1.000

as

$$\mathrm{IL} = 10 \log(1 + \omega'^{2n}) \tag{6.8}$$

where n (order of the filter) is the number of reactive elements required to obtain the desired response. The characteristics such as stopband attenuation versus number of sections for the required bandwidth of maximally flat filters can be determined with the help of nomographs like those shown in Fig. 6.10 [6]. In most cases ω_1' is defined as the 3-dB band-edge point. The left-hand-side nomograph is used for $\omega'/\omega_1' \geq 1$ (stopband), and the right-hand-side nomograph for $\omega'/\omega_1' \leq 1$ (passband). For example, $\omega'/\omega_1' = 2.0$ for eight sections ($n = 8$), gives an attenuation of about 48 dB in the stopband, while for $\omega'/\omega_1' = 0.8$, the attenuation is about 0.35 dB in the passband.

For a Butterworth low-pass prototype, the element values g_k (shown in Fig. 6.8) may be calculated from the following equations:

$$g_0 = 1 \tag{6.9a}$$

$$g_k = 2 \sin\left[\frac{(2k-1)\pi}{2n}\right], \qquad k = 1, 2, \ldots, n \tag{6.9b}$$

$$g_{n+1} = 1, \qquad \text{for all } n \tag{6.9c}$$

Table 6.2 gives element values for such filters having $n = 1$ to 10 reactive elements.

The maximally flat filter design is optimum in the sense that it provides an insertion-loss response with a maximum number of frequency derivatives equal to zero at zero frequency with the rejection monotonically increasing as rapidly as possible with frequency. At microwave frequencies, maximally flat design is not as popular as the Chebyshev design.

Chebyshev Response. The Chebyshev solution addresses a slightly different problem: the insertion loss remains less than a specified level, A_m up to a specified frequency ω_1', and then rises monotonically with frequency as fast as possible. For Chebyshev response, the attenuation loss (in decibels) for $\omega_1' = 1$ and $\omega' \leq 1$, may be expressed as

$$A = 10 \log[1 + (10^{A_m/10} - 1)\cos^2(n \cos^{-1} \omega')] \tag{6.10}$$

where n is the order of the filter, A_m is the ripple magnitude in decibels, and ω_1' is bandwidth over which the insertion loss has maximum ripple A_m. The Chebyshev filter nomograph [6] is shown in Fig. 6.11 illustrating (left to right) four variables: normalized frequency ω'/ω_1', number of sections n,

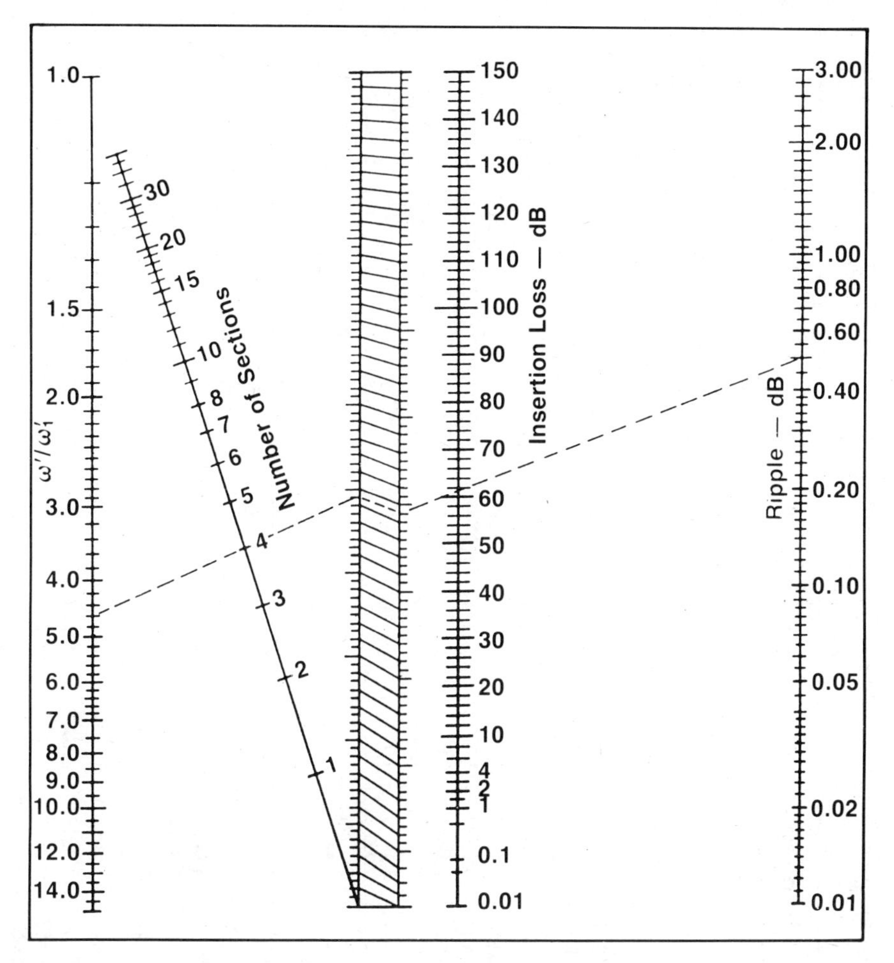

Figure 6.11 Nomograph for selecting number of sections of Chebyshev filter for given ripple and insertion loss in stopband. (After T. Milligan [6], 1985 *Microwave & RF*, reprinted with permission.)

stopband insertion loss, and passband ripple. For example, for passband ripple of 0.5 dB and $\omega'/\omega_1' = 4.6$, the insertion loss in the stopband for four sections is about 61 dB.

The g_k values for Chebyshev response may be calculated from the following equations:

$$g_0 = 1 \tag{6.11a}$$

$$g_1 = \frac{2a_1}{\gamma} \tag{6.11b}$$

TABLE 6.3 Element Values for a Chebyshev Low-pass Prototype Filter Having $g_0 = 1$, $\omega_1' = 1$, and $n = 1$ to 10 with Various Ripple Values

Value of n	g_1	g_2	g_3	g_4	g_5	g_6	g_7	g_8	g_9	g_{10}	g_{11}
					0.01-dB Ripple						
1	0.0960	1.0000									
2	0.4488	0.4077	1.1007								
3	0.6291	0.9702	0.6291	1.0000							
4	0.7128	1.2003	1.3212	0.6476	1.1007						
5	0.7563	1.3049	1.5773	1.3049	0.7563	1.0000					
6	0.7813	1.3600	1.6896	1.5350	1.4970	0.7098	1.1007				
7	0.7969	1.3924	1.7481	1.6331	1.7481	1.3924	0.7969	1.0000			
8	0.8072	1.4130	1.7824	1.6833	1.8529	1.6193	1.5554	0.7333	1.1007		
9	0.8144	1.4270	1.8043	1.7125	1.9057	1.7125	1.8043	1.4270	0.8144	1.0000	
10	0.8196	1.4369	1.8192	1.7311	1.9362	1.7590	1.9055	1.6527	1.5817	0.7446	1.1007
					0.1-dB Ripple						
1	0.3052	1.0000									
2	0.8430	0.6220	1.3554								
3	1.0315	1.1474	1.0315	1.0000							
4	1.1088	1.3061	1.7703	0.8180	1.3554						
5	1.1468	1.3712	1.9750	1.3712	1.1468	1.0000					
6	1.1681	1.4039	2.0562	1.5170	1.9029	0.8618	1.3554				
7	1.1811	1.4228	2.0966	1.5733	2.0966	1.4228	1.1811	1.0000			
8	1.1897	1.4346	2.1199	1.6010	2.1699	1.5640	1.9444	0.8778	1.3554		
9	1.1956	1.4425	2.1345	1.6167	2.2053	1.6167	2.1345	1.4425	1.1956	1.0000	
10	1.1999	1.4481	2.1444	1.6265	2.2253	1.6418	2.2046	1.5821	1.9628	0.8853	1.3554

0.2-dB Ripple

1	0.4342	1.0000									
2	1.0378	0.6745	1.5386								
3	1.2275	1.1525	1.2275	1.0000							
4	1.3028	1.2844	1.9761	0.8468	1.5386						
5	1.3394	1.3370	2.1660	1.3370	1.3394	1.0000					
6	1.3598	1.3632	2.2394	1.4555	2.0974	0.8838	1.5386				
7	1.3722	1.3781	2.2756	1.5001	2.2756	1.3781	1.3722	1.0000			
8	1.3804	1.3875	2.2963	1.5217	2.3413	1.4925	2.1349	0.8972	1.5386		
9	1.3860	1.3938	2.3093	1.5340	2.3728	1.5340	2.3093	1.3938	1.3860	1.0000	
10	1.3901	1.3983	2.3181	1.5417	2.3904	1.5536	2.3720	1.5066	2.1514	0.9034	1.5386

0.5-dB Ripple

1	0.6986	1.0000									
2	1.4029	0.7071	1.9841								
3	1.5963	1.0967	1.5963	1.0000							
4	1.6703	1.1926	2.3661	0.8419	1.9841						
5	1.7058	1.2296	2.5408	1.2296	1.7058	1.0000					
6	1.7254	1.2479	2.6064	1.3137	2.4758	0.8696	1.9841				
7	1.7372	1.2583	2.6381	1.3444	2.6381	1.2583	1.7372	1.0000			
8	1.7451	1.2647	2.6564	1.3590	2.6964	1.3389	2.5093	0.8796	1.9841		
9	1.7504	1.2690	2.6678	1.3673	2.7239	1.3673	2.6678	1.2690	1.7504	1.0000	
10	1.7543	1.2721	2.6754	1.3725	2.7392	1.3806	2.7231	1.3485	2.5239	0.8842	1.9841

$$g_k = \frac{4a_{k-1}a_k}{b_{k-1}g_{k-1}}, \qquad k = 2, 3, \ldots, n \tag{6.11c}$$

$$g_{n+1} = 1, \qquad n \text{ odd} \tag{6.11d}$$

$$g_{n+1} = \tanh^2 \frac{\beta}{4}, \qquad n \text{ even} \tag{6.11e}$$

where

$$a_k = \sin \frac{(2k-1)\pi}{2n}, \qquad k = 1, 2, \ldots, n \tag{6.12a}$$

$$b_k = \gamma^2 + \sin^2 \frac{k\pi}{n}, \qquad k = 1, 2, \ldots, n \tag{6.12b}$$

$$\beta = \ln\left(\coth \frac{A_m}{17.37}\right) \tag{6.12c}$$

$$\gamma = \sinh \frac{\beta}{2n} \tag{6.12d}$$

Table 6.3 gives element values for Chebyshev low-pass prototype filters having $g_0 = 1$, $\omega_1' = 1$, and $n = 1$ to 10 with various ripple values. Most microwave Chebyshev designs use a ripple value in the 0.01 dB (VSWR = 1:1.1) to 0.2 dB (VSWR = 1:1.54) range. Higher ripple values give rise to excessive ripples caused by interaction between cascaded components.

The Chebyshev filter design is optimum in the sense that for any n, and all possible element value choices, it provides the maximum possible stopband monotonic insertion loss for a specified maximum passband insertion-loss ripple.

6.3.2 Special Response Filter Synthesis

The elliptic function response and generalized Chebyshev response are also frequently used for microwave filters. Bessel and Gaussian and many other similar responses realizable with a ladder prototype are seldom used at microwave frequencies because of beating effects caused by long phase lengths between cascaded components and high passband VSWR of the preceding type of designs. Synthesis of elliptic and generalized Chebyshev filters is briefly described below.

Elliptic Function Response. The elliptic filter stopband response has a series of peaks, corresponding to the number of sections, and a minimum attenuation level L_m [Fig. 6.12(a)] rather than monotonically increasing attenuation value provided by Butterworth and Chebyshev filters. The attenuation loss is expressible in several forms using elliptic functions or

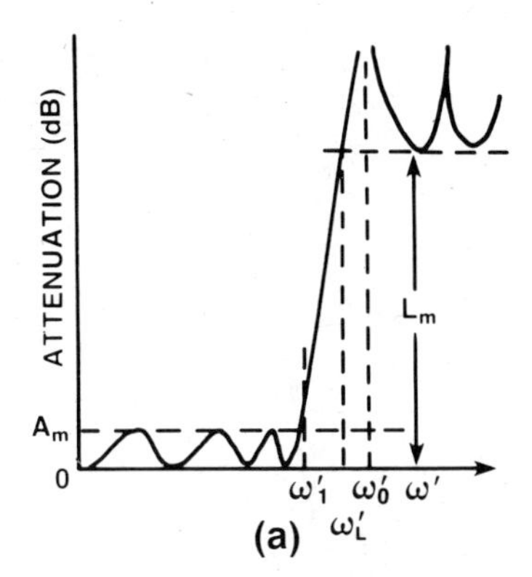

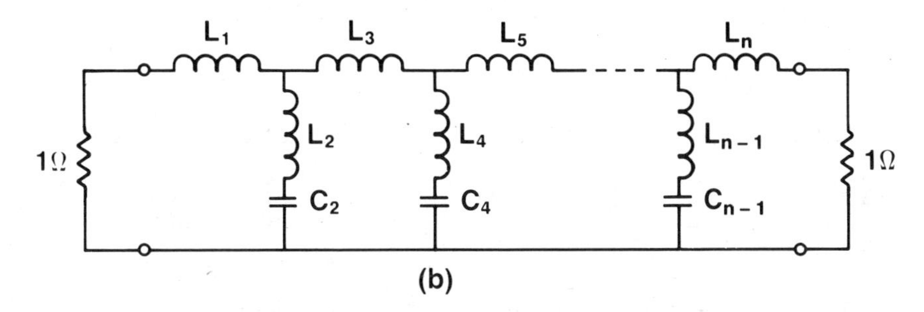

Figure 6.12 (*a*) Elliptic function response. (*b*) Prototype low-pass elliptic filter (*n* is odd integer).

Chebyshev rational functions. However, these forms are not suitable for a simple calculation of attenuation loss. A complete discussion on elliptic filters is given in References [3–5]. Unlike the simple method of calculating g values for the Butterworth and Chebyshev filters previously described, the normalized L- and C-values of this class of filter as shown in Fig. 6.12(*b*) have been derived by synthesis [2–5]. Element values of some of the more useful cases for VSWRs up to 1.5 and $n = 3$, 5, 7, and 9 sections is given in [7]. In each case, the minimum attenuation L_m and the frequency ω'_L at which this attenuation occurs are also given.

The elliptic filter design is optimum in the sense that it provides a much steeper stopband skirt for a given value of n, and passband insertion loss and stopband insertion loss as compared to maximally flat and Chebyshev filters.

Generalized Chebyshev Response. Generalized Chebyshev response filters have been used [8, 9] to improve selectivity and physical realization in MICs. The odd-degree generalized Chebyshev response having an equiripple passband, three transmission zeros at infinity, and an even

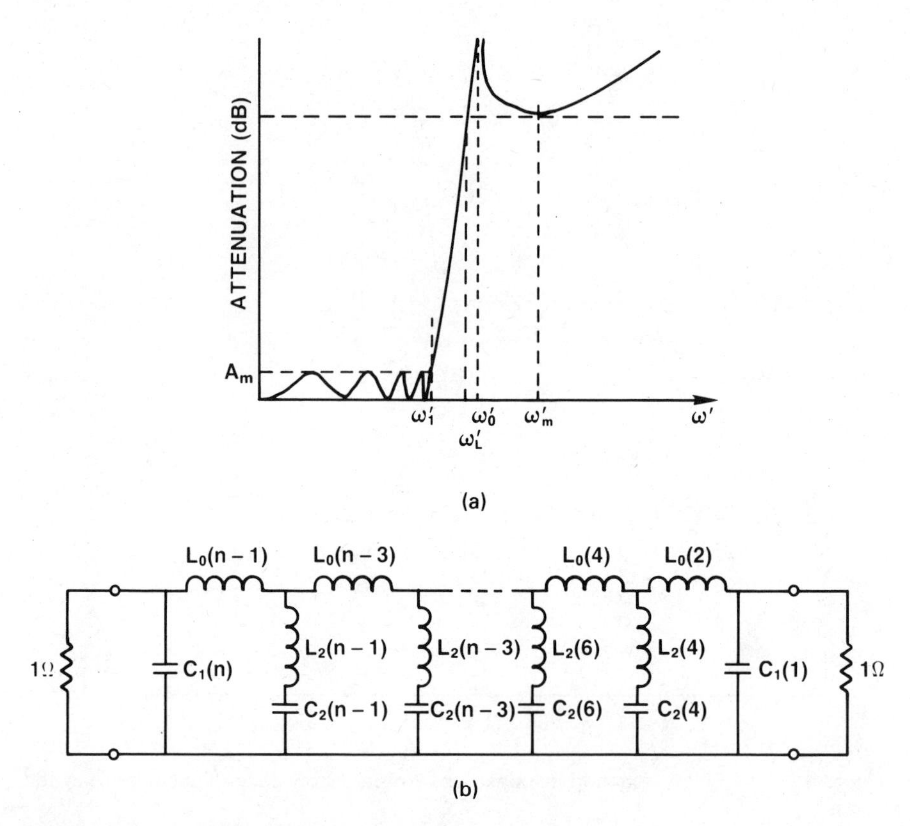

Figure 6.13 (*a*) The generalized Chebyshev insertion-loss response. (*b*) Generalized Chebyshev low-pass prototype.

multiple of transmission zeros close to the band edge has a selectivity nearly as good as the same-degree elliptic function prototype. Such filters are easier to realize physically in MICs, as impedance variation is typically less than 2:1. However, in elliptic filters, for normally required specifications, the impedance variation is as large as 10:1.

A typical insertion-loss response for a generalized Chebyshev filter is shown in Fig. 6.13(a), where ω_1', ω'_L, ω_0', and ω'_m are the passband edge, stopband edge, transmission zeros, and minimum insertion-loss frequencies, respectively. Element values for the doubly terminated low-pass prototype network shown in Fig. 6.13(b) are given in [8].

6.3.3 Filter Transformations

In principle, one could build these designs exactly, having the source and load resistance at 1 Ω and the upper edge of the passband at 1 rad/sec. However, these prototypes can be scaled to any frequency and can be transformed into any of the four filter types: low pass, high pass, bandpass, and bandstop.

The transformations of low-pass prototype filters with cutoff at $\omega_1' = 1$ and terminated in a 1-Ω source and load impedance into low-pass, high-pass, bandpass and bandstop filters with arbitrary source and load impedance, Z_0, are described in this section.

Low Pass. For low-pass filters, transformation of the low-pass prototype to the desired frequency band and impedance level is accomplished by multiplying g_k as follows:

$$L_k = g_k \left(\frac{Z_0}{\omega_{LP}} \right) \qquad \text{(For Series Inductors)} \qquad (6.13a)$$

$$C_k = g_k \left(\frac{1}{\omega_{LP} Z_0} \right) \qquad \text{(For Shunt Capacitors)} \qquad (6.13b)$$

where ω_{LP} is the required low-pass bandwidth. The following example describes step by step the design of a low-pass filter.

Example. Design a low-pass filter with a ripple factor of 0.2 dB having bandwidth of 2 GHz and attenuation at 2.4 GHz of 30 dB.

Step 1: Calculate the number of sections from Fig. 6.11.

$$\frac{\omega'}{\omega_1'} = \frac{2 \times \pi \times 2.4}{2 \times \pi \times 2.0} = 1.2$$

For $\omega'/\omega_1' = 1.2$, ripple $= 0.2$ dB, and insertion loss $= 30$ dB; $n = 9$.

Step 2: Find the prototype element values from Table 6.3 for $n = 9$ and ripple $= 0.2$ dB.

$$g_1 = g_9 = 1.3860, \qquad g_2 = g_8 = 1.3938, \qquad g_3 = g_7 = 2.3093,$$

$$g_4 = g_6 = 1.5340, \qquad g_5 = 2.3728$$

Step 3: Determine the lumped-element values from (6.13). For $Z_0 = 50\ \Omega$ and $\omega_{LP} = 2\pi \times 2 \times 10^9 = 12.57 \times 10^9$ rad/sec.

$$L_1 = L_9 = 1.3860 \times \frac{50}{12.57 \times 10^9} = 5.513 \text{ nH}$$

$$C_2 = C_8 = 1.3938 \times \frac{1}{50 \times 12.57 \times 10^9} = 2.2 \text{ pF}$$

$$L_3 = L_7 = 9.1858 \text{ nH}$$

$$C_4 = C_6 = 2.4 \text{ pF}$$

$$L_5 = 9.4383 \text{ nH}$$

High Pass. The low-pass prototype network is transformed into a high-pass filter by transforming series inductances into series capacitances and shunt capacitances into shunt inductances. Using the frequency transformation gives

$$\frac{\omega'}{\omega'_1} = -\frac{\omega_{HP}}{\omega} \tag{6.14}$$

where ω_{HP} and ω are the band-edge and variable angular frequencies of the high-pass filter. The element values are obtained from

$$C_k = \frac{1}{g_k \omega_{HP} Z_0} \quad \text{(For Series Capacitors)} \tag{6.15a}$$

$$L_k = \frac{Z_0}{g_k \omega_{HP}} \quad \text{(For Shunt Inductors)} \tag{6.15b}$$

In order to illustrate the previously described transformation, we consider an example of a lumped-element high-pass filter with a passband ripple of 0.1 dB, band-edge frequency at 3 GHz, and 30-dB attenuation at 2 GHz.

Step 1: Calculate the number of sections from Fig. 6.11.

$$\left|\frac{\omega'}{\omega'_1}\right| = \frac{\omega_{HP}}{\omega} = \frac{2 \times \pi \times 3}{2 \times \pi \times 2} = 1.5$$

For $\omega'/\omega'_1 = 1.5$, ripple = 0.1 dB, and insertion loss = 25 dB; $n = 6$.

Step 2: Find the prototype element values from Table 6.3 for $n = 6$ and ripple = 0.1 dB.

$$g_1 = 1.1681, \quad g_2 = 1.4039, \quad g_3 = 2.0562, \quad g_4 = 1.5170$$

$$g_5 = 1.9029, \quad g_6 = 0.8618, \quad g_7 = 1.3554$$

Step 3: Determine the lumped-element values from (6.15). For $Z_0 = 50\ \Omega$ and $\omega_{HP} = 2\pi \times 3 \times 10^9 = 18.855 \times 10^9$ rad/sec:

$$C_1 = \frac{1}{1.1681 \times 18.855 \times 10^9 \times 50} = 0.90834 \text{ pF}$$

$$L_2 = \frac{50}{1.4039 \times 18.855 \times 10^9} = 1.8894 \text{ nH}$$

$$C_3 = 0.51602 \text{ pF}, \quad L_4 = 1.7485 \text{ nH}, \quad C_5 = 0.5576 \text{ pF}, \quad L_6 = 3.0779 \text{ nH}$$

$$R_7 = R_L = 50 \times 1.3554 = 67.77 \ \Omega$$

For n even, $R_L \neq 50 \ \Omega$, that is input and output impedances are different. This can be avoided by selecting odd numbers of filter sections.

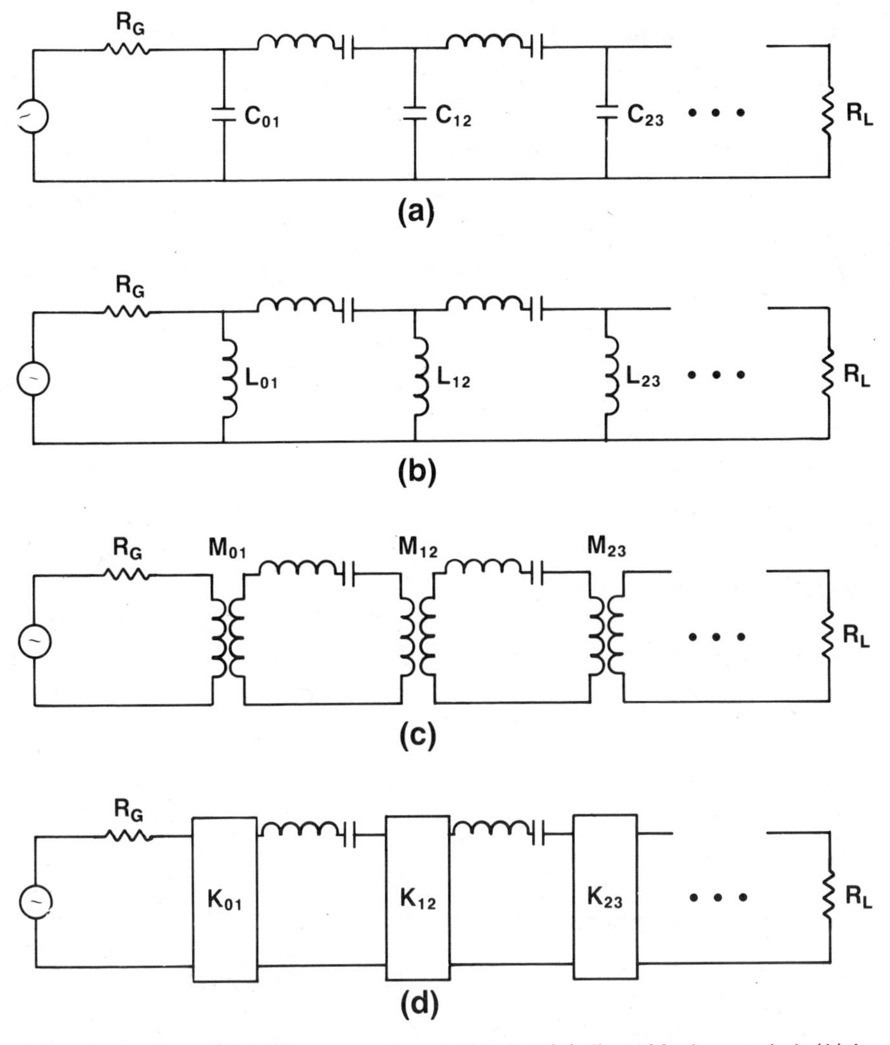

Figure 6.14 Several bandbass prototype networks. (*a*) Capacitively coupled. (*b*) Inductively coupled. (*c*) Transformer coupled. (*d*) Impedance inverted coupled.

Bandpass. To map the low-pass prototype to a bandpass filter, the following transformation is used:

$$\frac{\omega'}{\omega_1'} = \frac{f_0}{\mathrm{BW}}\left(\frac{f}{f_0} - \frac{f_0}{f}\right), \qquad f_0 = \sqrt{f_1 f_2}, \qquad \mathrm{BW} = f_2 - f_1 \tag{6.16}$$

where f_0, f, and BW are the center frequency ($\omega_0/2\pi$), variable frequency, and bandwidth, respectively, and f_1 and f_2 are the frequency band limits. The transformation does not result in a unique bandpass prototype. Several bandpass prototype networks are shown in Fig. 6.14. Note that the element values may be different in each circuit.

Applying the frequency transformation to series inductances and shunt capacitances of the low-pass prototype gives,

$$L_k = g_k \frac{Z_0}{2\pi \mathrm{BW}}; \qquad C_k = \frac{2\pi \mathrm{BW}}{g_k Z_0 \omega_0^2} \qquad \text{(Series-Tuned Series Elements)} \tag{6.17a}$$

$$L_k = \frac{2\pi \mathrm{BW} Z_0}{g_k \omega_0^2}; \qquad C_k = \frac{g_k}{2\pi \mathrm{BW} Z_0} \qquad \text{(Shunt-Tuned Shunt Elements)} \tag{6.17b}$$

where

$$\omega_0^2 = \frac{1}{L_k C_k}$$

Example. Design a Chebyshev bandpass filter using LCs with 0.5-dB ripple and 5% bandwidth centered at 6 GHz. The minimum desirable rejection at 6 ± 1 GHz is 45 dB.

Step 1: Calculate the number of resonators from Fig. 6.11.

$$\frac{\omega'}{\omega_1'} = \frac{6}{0.3}\left(\frac{7}{6} - \frac{6}{7}\right) = 6.19$$

For $\omega'/\omega_1' = 6.19$, ripple = 0.5 dB, and insertion loss = 45 dB; $n = 3$.

Step 2: Find the prototype element values from Table 6.3 for $n = 3$ and ripple = 0.5 dB.

$$g_0 = g_4 = 1.0, \qquad g_1 = g_3 = 1.5963, \qquad g_2 = 1.0967$$

Step 3: Determine the lumped element values from (6.17). For $Z_0 = 50\ \Omega$, $2\pi \mathrm{BW} = 2\pi \times 0.3 \times 10^9 = 1.8850 \times 10^9$ rad/sec, and $\omega_0 = 2\pi \times 6 \times 10^9 =$

37.7×10^9 rad/sec:

$$\left.\begin{aligned} L_1 = L_3 &= \frac{2\pi \mathrm{BW} Z_0}{g_1 \omega_0^2} = 0.0415 \text{ nH} \\ C_1 = C_3 &= \frac{g_1}{2\pi \mathrm{BW} Z_0} = 16.94 \text{ pF} \end{aligned}\right\} \text{(Shunt-Tuned Shunt Elements)}$$

$$\left.\begin{aligned} L_2 &= g_2 \frac{Z_0}{2\pi \mathrm{BW}} = 29.09 \text{ nH} \\ C_2 &= \frac{2\pi \mathrm{BW}}{g_2 Z_0 \omega_0^2} = 0.0242 \text{ pF} \end{aligned}\right\} \text{(Series-Tuned Series Elements)}$$

Bandstop. The transformation from low-pass prototype to bandstop is given by

$$\frac{\omega_1'}{\omega'} = \frac{f_0}{\mathrm{BW}}\left(\frac{f}{f_0} - \frac{f_0}{f}\right) \tag{6.18}$$

where all the quantities used in (6.18) were defined before. Here series inductance is mapped into a shunt-tuned circuit with element values

$$\omega_0 C_k = \frac{1}{\omega_0 L_k} = \frac{\omega_0}{2\pi \mathrm{BW} Z_0 g_k} \tag{6.19a}$$

and shunt capacitance into a series-tuned circuit with element values

$$\omega_0 L_k = \frac{1}{\omega_0 C_k} = \frac{\omega_0 Z_0}{2\pi \mathrm{BW}\, g_k} \tag{6.19b}$$

Example. Design a stopband filter with the following specifications:

Frequency of infinite attenuation, $f_0 = 6$ GHz,
Bandwidth, BW = 300 MHz,
Passband ripple = 0.5 dB,
Minimum attenuation over 2% stop band = 20 dB.

Step 1: Calculate the number of resonators from Fig. 6.11.

$$\frac{\omega_1'}{\omega'} = \frac{6}{0.3}\left(\frac{6.06}{6} - \frac{6}{6.06}\right) = 0.4$$

or

$$\frac{\omega'}{\omega'_1} = 2.5$$

For $\omega'/\omega'_1 = 2.5$, ripple = 0.5 dB, and insertion loss = 20 dB; $n = 3$.

Step 2: Find the prototype element values from Table 6.3 for $n = 3$ and ripple = 0.5 dB.

$$g_0 = g_4 = 1.0, \qquad g_1 = g_3 = 1.5963, \qquad g_2 = 1.0967$$

Step 3: Determine the lumped-element values from (6.19). For $Z_0 = 50\ \Omega$, $2\pi \mathrm{BW} = 2\pi \times 0.3 \times 10^9 = 1.8850 \times 10^9$ rad/sec, and $\omega_0 = 2\pi \times 6 \times 10^9 = 37.7 \times 10^9$ rad/sec:

$$\left.\begin{aligned} L_1 = L_3 &= \frac{2\pi \mathrm{BW} Z_0 g_{1,3}}{\omega_0^2} = 0.106 \text{ nH} \\ C_1 = C_3 &= \frac{1}{2\pi \mathrm{BW} Z_0 g_{1,3}} = 6.65 \text{ pF} \end{aligned}\right\} \quad \text{(Shunt-Tuned Series Elements)}$$

$$\left.\begin{aligned} L_2 &= \frac{Z_0}{2\pi \mathrm{BW} g_2} = 24.19 \text{ nH} \\ C_2 &= \frac{2\pi \mathrm{BW} g_2}{\omega_0^2 Z_0} = 0.029 \text{ pF} \end{aligned}\right\} \quad \text{(Series-Tuned Shunt Elements)}$$

6.3.4 Impedance and Admittance Inverters

Impedance and admittance inverters play a very important role in classical filter design. Because of the inverting action, a series inductance (or shunt capacitance) with an inverter on each side looks like a shunt capacitance (or series inductance) from its external terminals. Making use of this property, low-pass filters can be realized with only one kind of reactance. Similarly, bandpass filters may be realized by series inductance–capacitance (LC), series resonant circuits separated by impedance inverters or shunt LC parallel resonant circuits separated by admittance inverters. Under ideal conditions, both networks; prototypes and circuits with inverters, will have identical transmission characteristics.

The impedance inverter K and admittance inverter J are defined in Fig. 6.15, and their simple realizations are a quarter-wavelength transmission line of characteristic impedance K and characteristic admittance J, respectively. Since in the design equations these inverters operate like quarter-wavelength lines at all frequencies, they are useful in filter designs having

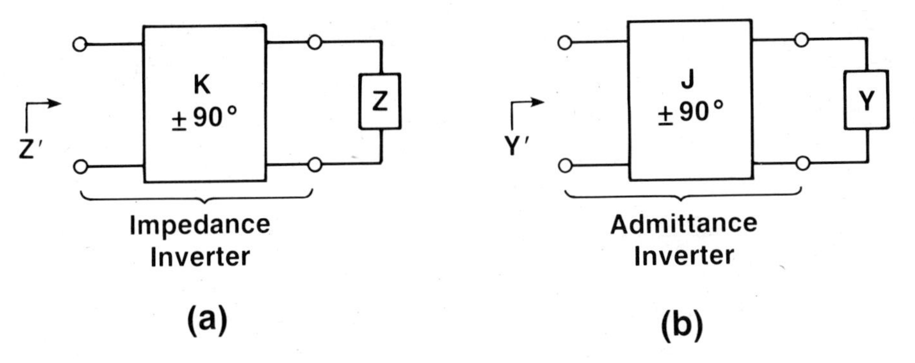

Figure 6.15 Definition of inverters. (*a*) Impedance (*K*). (*b*) Admittance (*J*).

bandwidths up to 20%. This bandwidth limitation may be increased to about 40% if K and J inverters are used alternately.

The inverter derives its name from the impedance seen looking into a reactively terminated inverter:

$$Z' = \frac{K^2}{Z} \qquad (K\text{-Inverter}) \tag{6.20a}$$

$$Y' = \frac{J^2}{Y} \qquad (J\text{-Inverter}) \tag{6.20b}$$

The calculation of element values for bandpass filters using distributed resonators proceeds as follows:

$$\frac{K_{01}}{Z_0} = \frac{J_{01}}{Y_0} = \sqrt{\frac{\pi\nu}{2}\frac{\mathrm{BW}}{f_0}\frac{1}{\sqrt{g_0 g_1}}} \tag{6.21a}$$

$$\frac{K_{r,r+1}}{Z_0} = \frac{J_{r,r+1}}{Y_0} = \frac{\pi\nu}{2}\frac{\mathrm{BW}}{f_0}\frac{1}{\sqrt{g_r g_{r+1}}} \tag{6.21b}$$

$$\frac{K_{n,n+1}}{Z_0} = \frac{J_{n,n+1}}{Y_0} = \sqrt{\frac{\pi\nu}{2}\frac{\mathrm{BW}}{f_0}\frac{1}{\sqrt{g_n g_{n+1}}}} \tag{6.21c}$$

where $g_0, g_1, \ldots, g_{n+1}$ are low-pass prototype values, $\nu = 1$ for half-wavelength-long resonators, and $\nu = \frac{1}{2}$ for quarter-wavelength-long resonators.

There are numerous other circuit elements that have good inverting properties over a much wider bandwidth than does a quarter-wavelength line. Figure 6.16 shows four inverting circuits for use as K-inverters (i.e., inverters to be used with series resonators). They are particularly useful in circuits where the negative L or C can be absorbed into adjacent positive

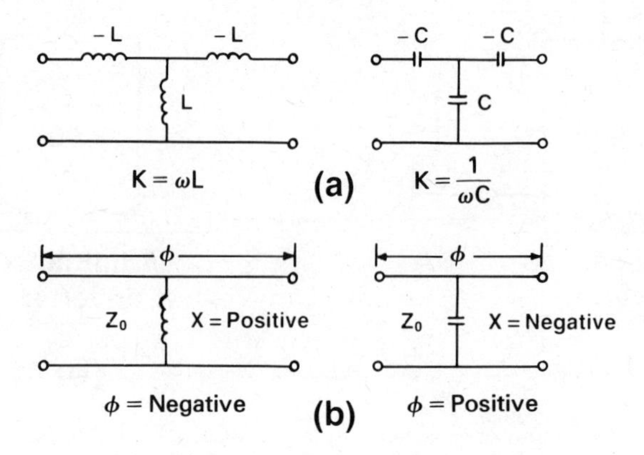

Figure 6.16 *K*-inverter circuit elements.

series elements. For Fig. 6.16(*b*)

$$K = Z_0 \tan\left|\frac{\phi}{2}\right| \qquad (\Omega) \tag{6.22}$$

$$\phi = -\tan 2\frac{X}{Z_0} \qquad \text{(rad)} \tag{6.23}$$

$$\left|\frac{X}{Z_0}\right| = \frac{K/Z_0}{1-(K/Z_0)^2} \tag{6.24}$$

Figure 6.17 shows four inverting circuits that are useful as *J*-inverters. In

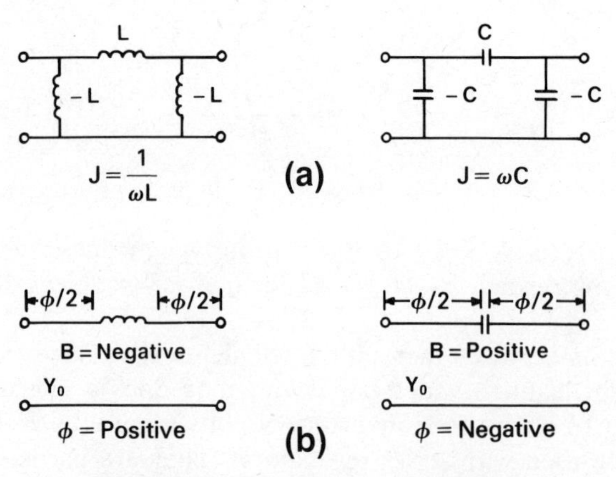

Figure 6.17 J-inverter circuit elements.

this case, for Fig. 6.17(*b*)

$$J = Y_0 \tan\left|\frac{\phi}{2}\right| \qquad \text{(mho)} \tag{6.25}$$

$$\phi = -\tan\frac{2B}{Y_0} \qquad \text{(rad)} \tag{6.26}$$

$$\frac{B}{Y_0} = \frac{J/Y_0}{1-(J/Y_0)^2} \tag{6.27}$$

An open-circuited coupled line functioning as a K-inverter is another example of a commonly used inverter circuit component in bandpass filters.

6.4 EXPERIMENTAL METHOD OF DESIGNING FILTERS

Bandpass filters which are physically symmetric can be realized as a simple combination of resonators. Narrow- to moderate-bandwidth bandpass filters using resonators can be designed by measuring the coupling coefficient between resonators and the external quality factor of the input and output resonators [10–13]. These measured values are then related to a normalized low-pass prototype value and can be used to realize all possible response shapes. This procedure is the most practical design method when the filter structure is complex or its equivalent circuit model is not readily available. The method can be used with microstrip, waveguide, or any other medium. For cases where accurate and relatively simple equivalent circuits are available, the experimental method can be used to test the validity of the circuit models.

The experimental approach is very suitable for microstrip bandpass filters such as interdigital, combline, hairpin-line, and parallel coupled. Tapped-line versions of these filters (Fig. 6.18) have more flexibility in terms of parallel coupling for the end sections and they offer space advantage. Since exact designs of tapped-line microstrip filters are not available [12, 13], experimental procedure and design curves for the sample filters are described in the following.

The first step in the filter design is to determine the coupling coefficients as a function of resonator spacing. This requires an assembly, as shown in Fig. 6.19(*a*), for interdigital structure. Nonresonant probes are attached to a test fixture and both at the input and output the probes are loosely coupled to the electric field of the resonators. The widths of all resonators are identical with dimensions that give good Q and sufficient freedom from spurious responses [12]. When a source is connected to the first resonator and a detector to the second resonator, the pair of resonators become a double-tuned circuit. The pair is tuned so that two peaks are of equal height, and the valley is at least 10 dB down, as shown in Fig. 6.19(*b*). The

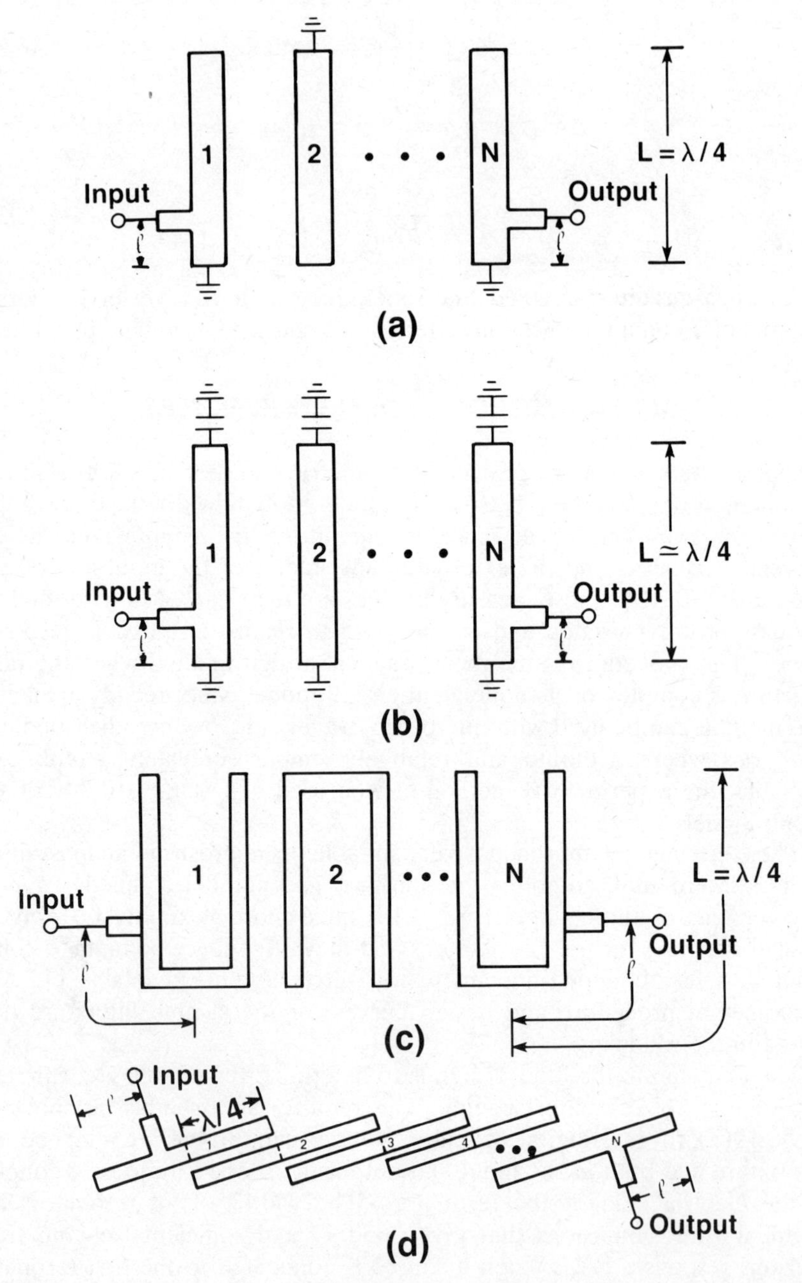

Figure 6.18 Tapped-line filter configurations. (*a*) Interdigital. (*b*) Combline. (*c*) Hairpin-line. (*d*) Parallel coupled.

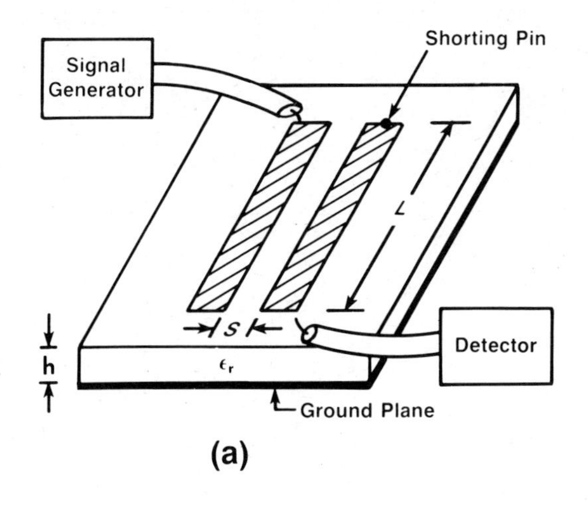

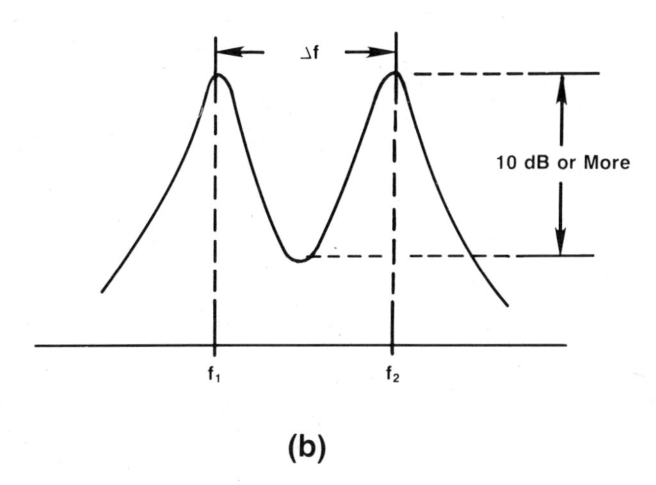

Figure 6.19 (*a*) Measurement of coupling coefficients. (*b*) Frequency response of a double-tuned resonator pair.

coupling coefficient K of the pair of resonators is related as

$$K = \frac{f_2 - f_1}{f_0} = \frac{\Delta f}{f_0} \tag{6.28}$$

where $f_0 = (f_1 + f_2)/2$.

Thus, by selecting a number of pairs of resonators with physically realizable gaps between each pair (ranging from very narrow to very wide), a curve of K versus S/h can be experimentally obtained. Figure 6.20(a) shows measured coupling coefficient as a function of S/h for $\epsilon_r = 2.22$ and

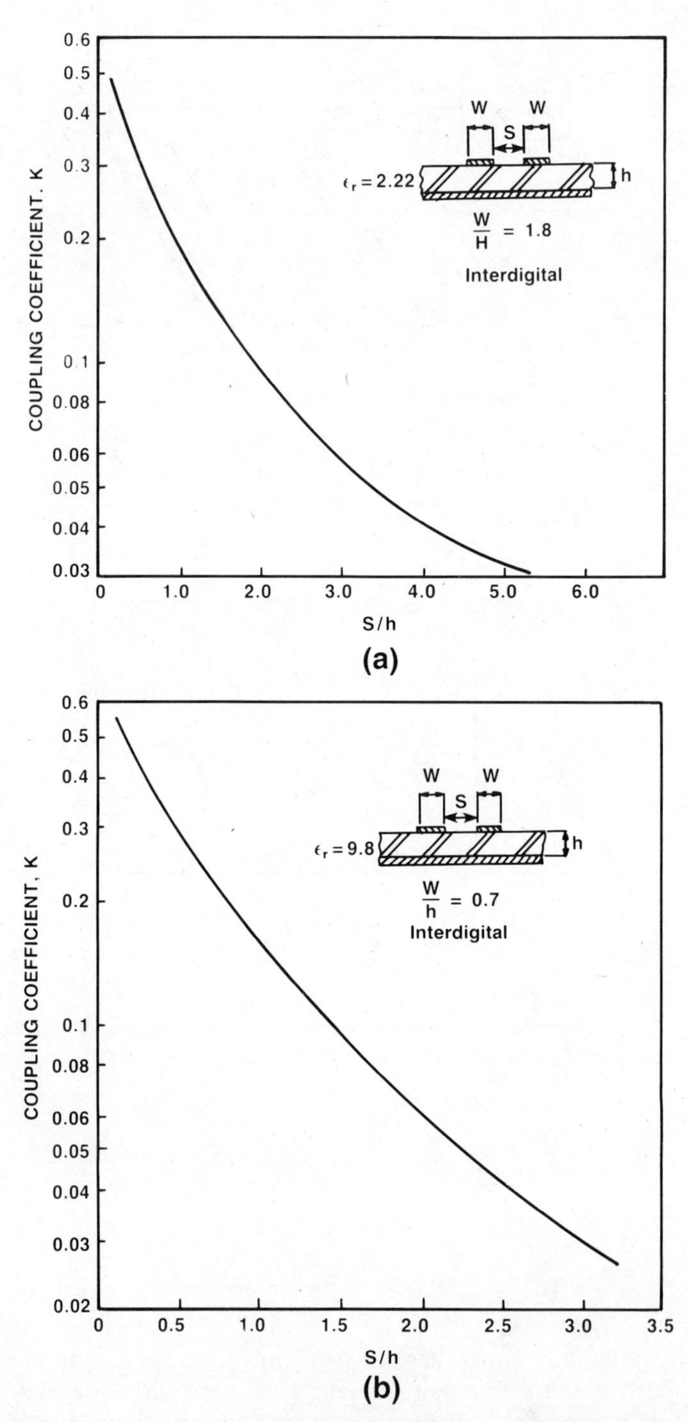

Figure 6.20 Measured combline coupling coefficient versus *S*/*h* for (*a*) RT/duroid and (*b*) alumina. (After J. S. Wong [12], 1979 IEEE, reprinted with permission.)

$W/h = 1.8$, which corresponds to a single-strip impedance of approximately 70 Ω. A similar curve can be obtained for other linewidths and dielectric substrates. Figure 6.20(*b*) shows measured coupling coefficient as a function of S/h for alumina substrate ($\epsilon_r = 9.8$) and $W/h = 0.7$, which corresponds to single-strip impedance of approximately 58 Ω.

The next step in the design procedure is finding the necessary normalized coupling coefficients in terms of low-pass prototype element values and design frequencies as follows:

$$K_{n,n+1} = \frac{\text{BW}}{f_0\sqrt{g_n g_{n+1}}} \tag{6.29}$$

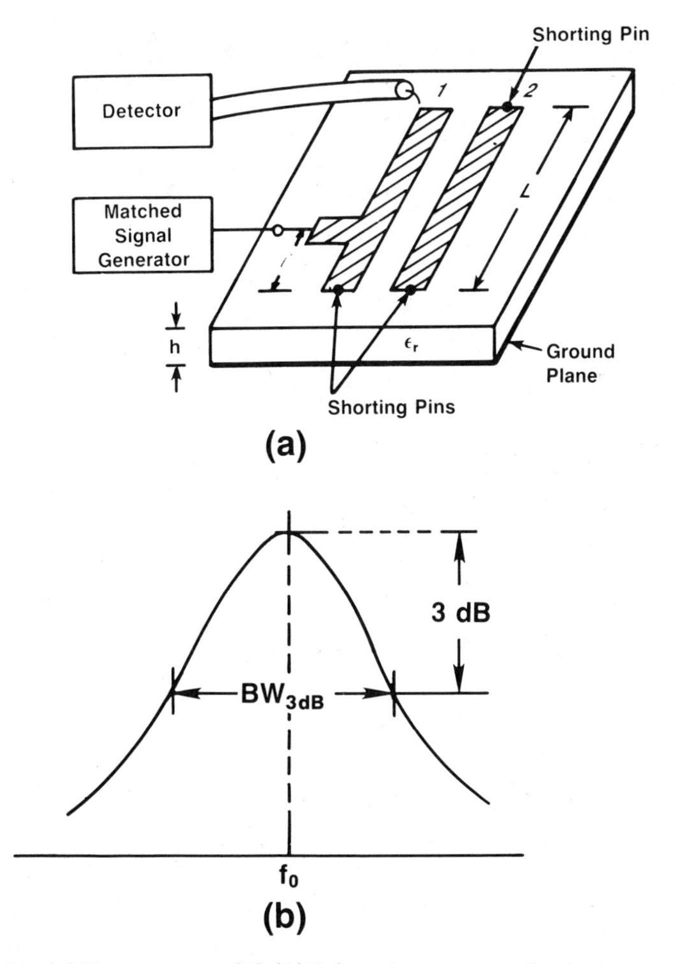

Figure 6.21 (*a*) Measurement of *Q*. (*b*) Frequency response of a singly tuned resonator.

where BW is the equal ripple bandwidth for Chebyshev response or 3-dB bandwidth for maximally flat response, f_0 is the center frequency of the proposed filter, and g_n are the low-pass prototype element values normalized to $\omega_1' = 1$ and $Z_0 = 1$.

The final step in the tapped filter is to measure the loaded Q of the first and the last resonator in order to obtain tap-point locations. With a simple test fixture as shown in Fig. 6.21(a), where the signal generator is well matched and the nonresonant detector is loosely coupled, frequency response of the first resonator is measured for various tap locations. In this case both ends of resonator 2 are short-circuited, and resonators are alone is a single-tuned circuit whose frequency response is shown in Fig. 6.21(b). The loaded Q is then calculated using

$$Q_L = \frac{f_0}{\mathrm{BW}_{3\mathrm{dB}}} \tag{6.30}$$

Variation of singly loaded Q for a tapped interdigital resonator on RT/duroid as described previously is plotted in Fig. 6.22 as a function of

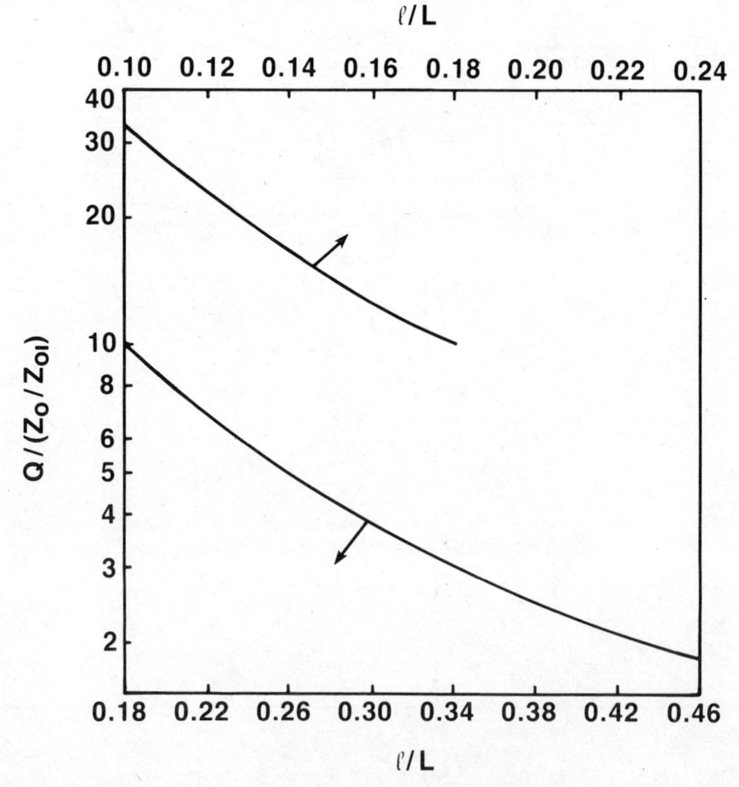

Figure 6.22 Singly loaded Q for tapped interdigital resonator. (After J. S. Wong [12], 1979 IEEE, reprinted with permission.)

l/L for $Z_0 = 50\,\Omega$ and $Z_{0I} = 70\,\Omega$, where Z_{0I} is the filter internal impedance.

In a filter design, the singly loaded Q is calculated from

$$Q_L = \frac{f_0}{\text{BW}}\, g_1 = \frac{f_0}{\text{BW}}\, g_{n+1} \tag{6.31}$$

where all these variables were defined previously. Once the singly loaded Q_L is known, the tap point l/L of Fig. 6.21(a) can be calculated from Fig. 6.22 or from the following equation [12]

$$\frac{Q_L}{(Z_0/Z_{0I})} = \frac{\pi}{4\sin^2(\pi l/2L)} \tag{6.32}$$

The procedure just described is applicable to all types of coupled resonator filters whether realized in microstrip or in any other medium.

6.5 FILTER MODELING

6.5.1 Narrow-band Approximation

In developing the prototype bandpass filter just presented, it has been assumed that couplings between resonators were frequency independent, and that the resonator impedances were antisymmetric in frequency about the center frequency (f_0). These two assumptions make up the *narrow-band approximation* and result in the insertion loss, and group delay being symmetric in frequency about f_0. This "approximation term" derives its name from the fact that the result will be accurate for any type of resonator or coupling provided that the percentage frequency range of interest is small enough. The narrow-band approximation typically works reasonably well for percentage bandwidths of up to 1–2% for waveguide filters and 10–15% for MIC filters.

In all cases the narrow-band approximation is a good design starting point for bandpass filters. Your intuition will at least tell you in what direction you are erring. For example, series capacitively coupled filters will have tighter coupling above f_0 than below, with the result that there will be relatively less rejection above f_0 than below. Series inductively coupled resonators are skewed the other way, however. To calculate the response more exactly, one can apply the techniques to be presented next.

6.5.2 Filter Analysis

Analytical formulations for the insertion loss and return loss for Butterworth filters, Chebyshev filters, etc., were described in Section 6.3, but phase characteristics have not been calculated yet. Also at microwave frequencies, all lumped elements have distributed effects [each element has to be characterized as a suitable combination of LRC, see Chapter 2] and resonators have finite Q. One can calculate phase information along with actual performance of the filter structure by calculating the electrical response directly from the circuits instead of using the analytic formulation. The objective is to calculate I_G and I_L as shown in Fig. 6.1(b), since once these currents are known we have already shown in Section 6.2 how to calculate the parameters we are interested in. There are two techniques commonly used to solve filter circuits: $ABCD$ matrix and Kirchhoff's equations. These methods are briefly described below.

***ABCD*-Matrix Method.** The definition and properties of the $ABCD$ matrix are given in Appendix C. From Fig. 6.1 and Fig. C.1

$$I_G = I_1$$

$$I_L = I_2 = \frac{V_G}{AR_L + B + R_G(CR_L + D)} \tag{6.33}$$

Using (6.33) in (6.3)

$$\text{IL} = -10 \log \frac{4R_G R_L}{[AR_L + B + R_G(CR_L + D)]^2} \tag{6.34}$$

Similarly return loss and other electrical parameters of a filter can be expressed in terms of A, B, C, and D. If two-port devices are connected together, the overall $ABCD$ matrix is just the matrix product of the individual matrices. One must remember that matrices are not multiplicatively transitive; one has to multiply the matrices in the right order. Thus the problem at hand is reduced to finding the $ABCD$ matrix of the overall circuit.

A circuit that can be constructed as a series connection of two-port devices is called a *ladder network*. If one knows the $ABCD$ matrix for each two-port circuit element in the ladder network, one can multiply the matrices together two by two to eventually get the overall $ABCD$ matrix. This is an efficient way of doing the calculation that lends itself well to computer simulation. The $ABCD$ matrices for a variety of elements are given in Table C.1 (in Appendix C). It is a simple matter to write a computer program to calculate the resultant response once the individual $ABCD$ matrices are known.

The main limitation of the $ABCD$-matrix technique is that it cannot handle reentrant combinations (nonladder networks) of two ports that play a very important role in high-performance bandpass filters. This limitation can be overcome by solving Kirchhoff's equations or using nodal analysis.

Kirchhoff's-Equations Method. To illustrate this method, an example of a prototype bandpass filter of order n is chosen. A typical two-resonator bandpass filter with all possible couplings included is shown in Fig. 6.23. The resultant equations for the n resonator network are given in Table 6.4. In this case, each resonator is coupled to every circuit node in the filter. The resonators that lie next to each other in the ladder network (Z_1, Z_2 or Z_7, Z_8) are called adjacent resonators and couplings between them are called *mainline couplings*. Couplings between nonadjacent resonators are called *bridge couplings*. If all the bridge couplings are zero, the network reduces to the original ladder network.

The usual technique to solve a set of equations such as given in Table 6.4 is to perform Gaussian ellimination with pivoting followed by back substitution as is described in most linear algebra texts. The calculation speed will likely not be an issue if all that is desired is to calculate the response, so a generalized linear equation solving subroutine should work fine. If speed is an issue, as may be the case in numerical design, the following comments may be useful:

1. If triangularization begins with the first term and proceeds down the main diagonal, good accuracy can usually be obtained without pivoting.

2. The matrix has symmetry about the main diagonal. About a factor of 2 speed up can be obtained by making use of this symmetry.

3. Finally, many designs, including Butterworth and Chebyshev designs, result in matrix symmetry about the minor axis as well as the major axis. These symmetries can be used to obtain a factor of 4 computer speed up. One way to obtain this is to use an even–odd solution as shown in Fig. 6.24.

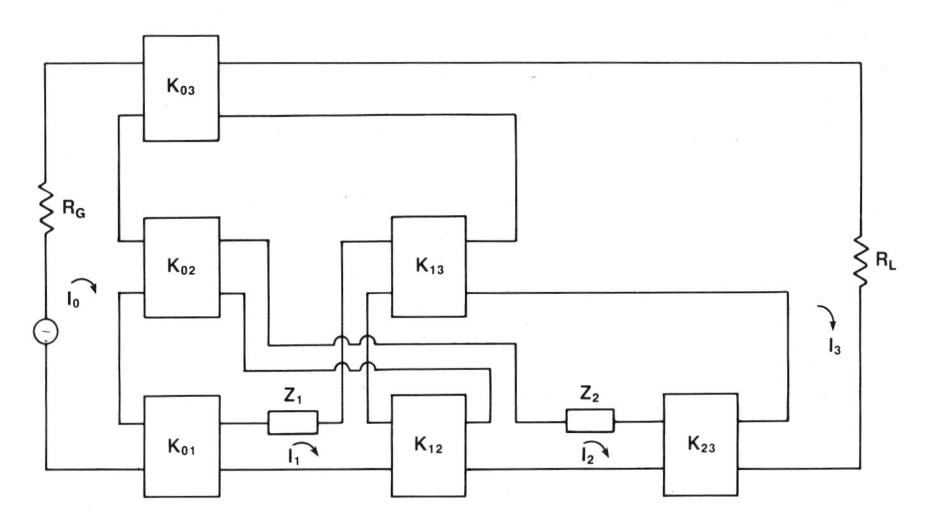

Figure 6.23 A two-resonator bandpass filter with all bridge couplings included.

TABLE 6.4 The Resultant Set of $n+2$ Linear Equations for a Resonator Bandpass Filter with all Possible Bridge Couplings

$$
\begin{bmatrix}
R_G & jK_{01} & jK_{02} & jK_{03} & \cdots & & jK_{0,n+1} \\
jK_{01} & Z_1 & jK_{13} & jK_{13} & & & jK_{1,n+1} \\
jK_{02} & jK_{12} & Z_2 & jK_{23} & & & jK_{2,n+1} \\
jK_{03} & jK_{13} & jK_{23} & Z_3 & & & jK_{3,n+1} \\
\vdots & \vdots & \vdots & \vdots & & & \vdots \\
 & & & & & Z_n & jK_{n,n+1} \\
jK_{0,n+1} & jK_{1,n+1} & jK_{2,n+1} & jK_{3,n+1} & & jK_{n,n+1} & RL
\end{bmatrix}
\begin{bmatrix}
I_0 \\ I_1 \\ I_2 \\ I_3 \\ \vdots \\ \\ I_{n+1}
\end{bmatrix}
=
\begin{bmatrix}
V_G \\ 0 \\ 0 \\ 0 \\ \vdots \\ \\ 0
\end{bmatrix}
$$

When the symmetrical filter is driven at both ends with in phase signals (even symmetry), $I_0 = I_{n+1}$, $I_1 = I_n$, etc., by symmetry, so only $(n+3)/2$ equations need to be solved. Similarly, when both ends are driven with out-of-phase signals (odd symmetry), $I_0 = -I_{n+1}$, $I_1 = -I_n$, etc., and again $(n+3)/2$ equations must be solved. Since the time to solve n equations is roughly proportional to n^3, solving half as many equations twice results in a factor of 4 savings.

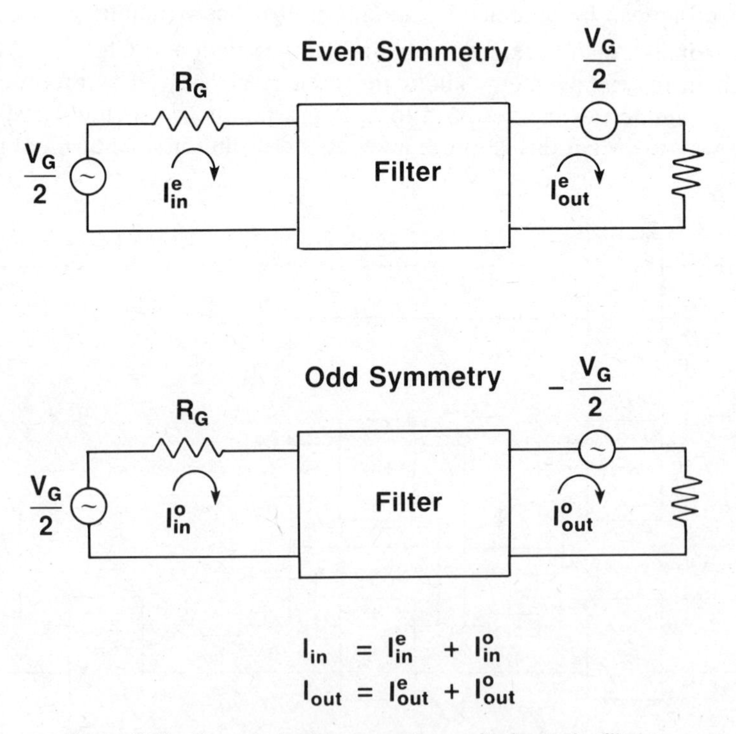

Figure 6.24 Even and odd symmetry excitations of a filter.

6.6 NUMERICAL TECHNIQUES

Microwave filters have traditionally been designed using the filter synthesis approach described in Section 6.3, in which the design is based on a previously derived transfer function. This technique is simple and accurate when applied to narrow-band and high-Q filters where nonideal effects can be ignored. Problems that arise when nonideal effects such as low Q and coupling dispersion become important, are often manageable by designing with enough margin so that, although the filter response differs from the transfer function, the response is still acceptable. A fundamental concern remains. One would really prefer a response that is the best obtainable as limited by physical constraints, not a response limited by how well the filter approximates a transfer function that itself was an approximate solution to the design problem.

In this section, we introduce numerical methods that can be applied when the highest performance and accuracy are needed. These techniques are applied most frequently to high-performance waveguide filters that are designed in minute detail; traditional design will normally suffice for MIC filters. However, it is important to be aware that more advanced techniques are available should the need arise.

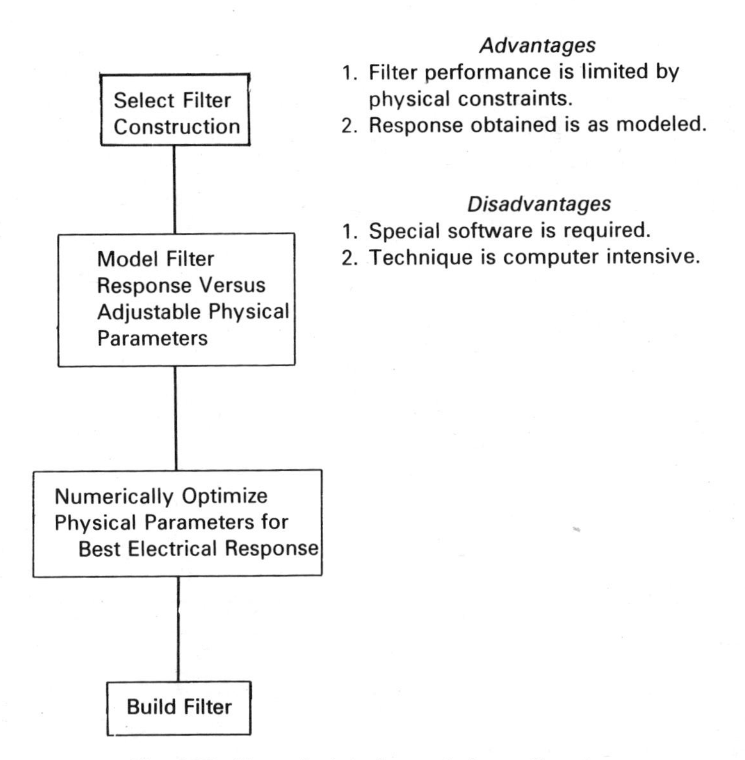

Fig. 6.25 Numerical design techniques flowchart.

The numerical design method, as summarized in Fig. 6.25, has been successfully applied to design filters that are close to truly optimal performance in the sense that the response is limited by physical constraints. The design procedure begins by selecting the construction to be used to meet the filter specifications. Next, one must model the filter response versus the physical parameters of the filter. At this point, accurate models (measured or analytical) for filter structure must be used. Once the response can be calculated, one must optimize the filter parameters to fit the application. This is done by first defining an error function that measures the difference between the calculated and the desired response (for details, see Chapter 14). For filters, the error function chosen might be

$$\text{Error} = (\text{RL}_C - \text{RL}_O)^2 + (\text{RJ}_C - \text{RJ}_O)^2 + \cdots \tag{6.35}$$

where RL and RJ are return loss and rejection defined over the required bandwidth, and subscripts C and O denote calculated and optimum, respectively. Next, a numerical optimization routine is used to minimize the error function. Many numerical optimization schemes are described in Chapter 14.

6.7 FILTER REALIZATIONS

A historical account of microwave filters is given by Levy and Cohn [14] in their recent paper. Filter developments include low pass, bandpass, and high pass in a variety of media such as waveguide, coaxial line, microstrip, and strip line, as well as dielectric resonators. Filters can be realized in any medium depending on the desired application. The maximum useful frequency limits of various circuit elements are given in Table 6.5. Since

TABLE 6.5 Characteristics of Various Structures for Filters

Structure	Frequency Range (GHz)	Useful Bandwidth (Percent)	Q^a
Waveguide	1–100	0.1–20	≃5000
Coaxial	0.1–40	1–30	≃2000
Strip line	0.1–20	5-Octave	≃150
Microstrip	0.1–100	5-Octave	≃200
Suspended Microstrip	1–200	2–20	≃1000
Fin line	20–100	2–50	≃500
Dielectric resonator	1–40	0.2–20	≃10,000
Lumped elements	0.01–10 (Hybrid)	20-Octave	≃200
	0.1–60 (Monolithic)	20-Octave	≃100
SAW	0.01–5		

a: At X-band frequencies.

the emphasis in this book is on integrated-type circuits, filter examples are restricted to microstrip-type and microstrip-coupled dielectric-resonator-type structures.

6.7.1 Printed Circuit Filters

Satellite, airborne communications, and EW systems have requirements for small size, light weight, and low-cost filters. Microstrip and strip-line filters are very suitable for wide-band applications and where the demand on selectivity is not severe. Various kinds of filters, as shown in Figs. 6.18 and 6.26, can be realized using microstrip type structures (Chapter 2 or Reference 15).

The suspended microstrip provides a higher Q than microstrip or strip line, as most of the energy is propagating in the air. This results in lower loss filters with sharper band edges. The wide range of impedance values achievable makes this medium particularly suitable for low-pass and broadband bandpass filters. An account of the development of many different types of filters in this medium was given in Reference 16.

Filters can be realized using lumped elements described in Chapter 2 or by employing microstrip sections. It is difficult to simulate accurately a series inductor in microstrip line. Kuroda's identities [17, 18], also described in Chapter 4, allow one to realize low-pass structures using shunt elements with the identical response. Richards' transformation [18, 19], which establishes a simple relationship between lumped and distributed circuit elements, enables one to design filters using distributed circuits.

In many microwave filter designs, a length of transmission line terminated in either an open circuit or a short circuit is often used as a resonator. Figure 6.27 illustrates four such resonators with their equivalent LRC networks, which were determined by equating slope parameters for both of these configurations at resonance $\omega = \omega_0$.

Low-pass Filters. Low-pass filters in waveguide, coaxial, and strip-line form are very important components in microwave systems. A very good account of the early development is given in the classic book of Matthaei, Young and Jones [1]. TEM structures such as coaxial lines, strip line, and microstrip are ideal for low-pass filters, and the design is approximated as nearly as possible to an idealized lumped-element circuit.

Consider an example for a microstrip low-pass filter with the following specifications:

Cutoff frequency	2 GHz
Response	Chebyshev with 0.2-dB ripple
30-dB attenuation frequency	3.5 GHz
Substrate material	Alumina, $\epsilon_r = 9.9$, $h = 0.63$ mm, $t = 6$ μm

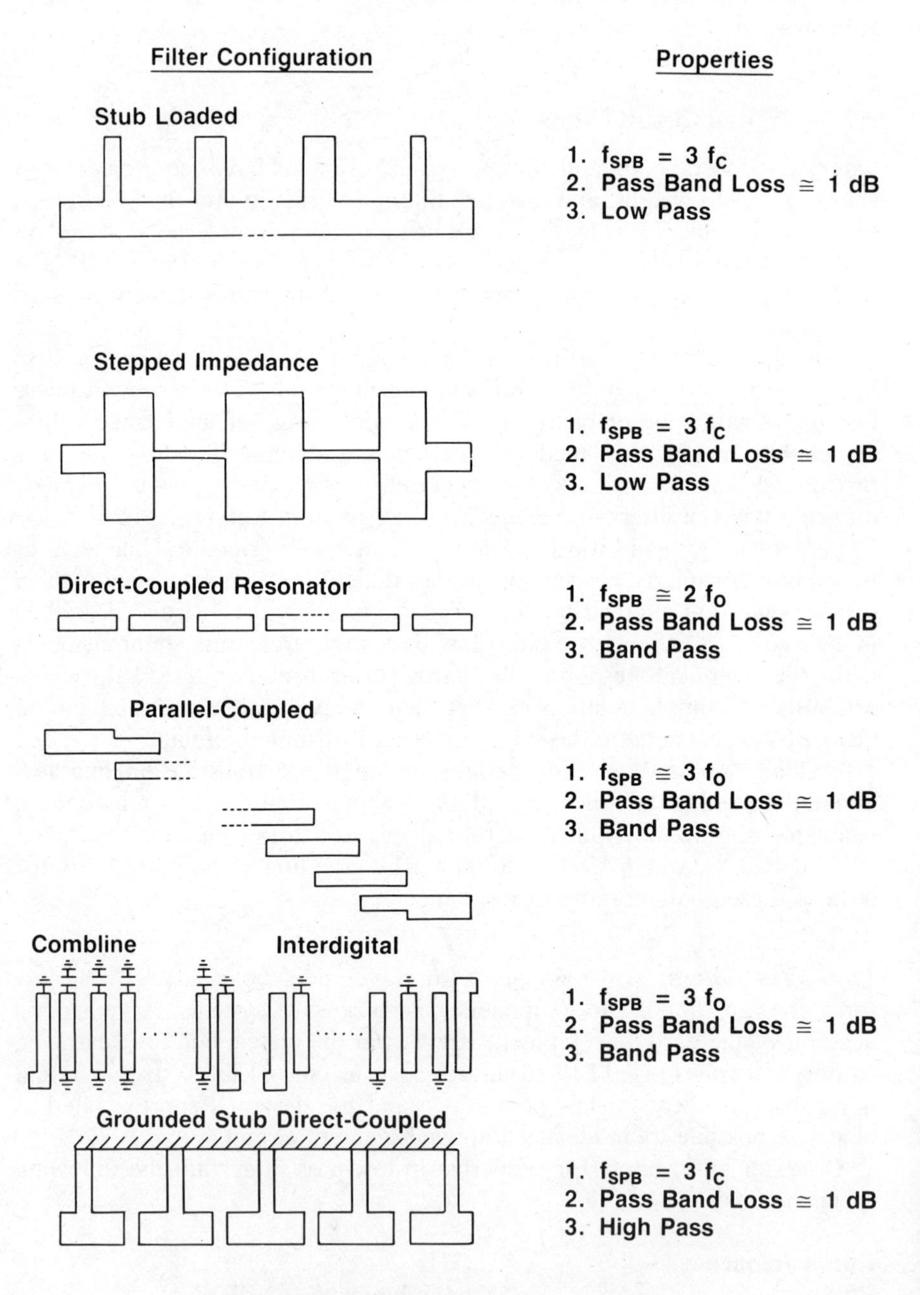

Figure 6.26 Microstrip-type filter configurations.

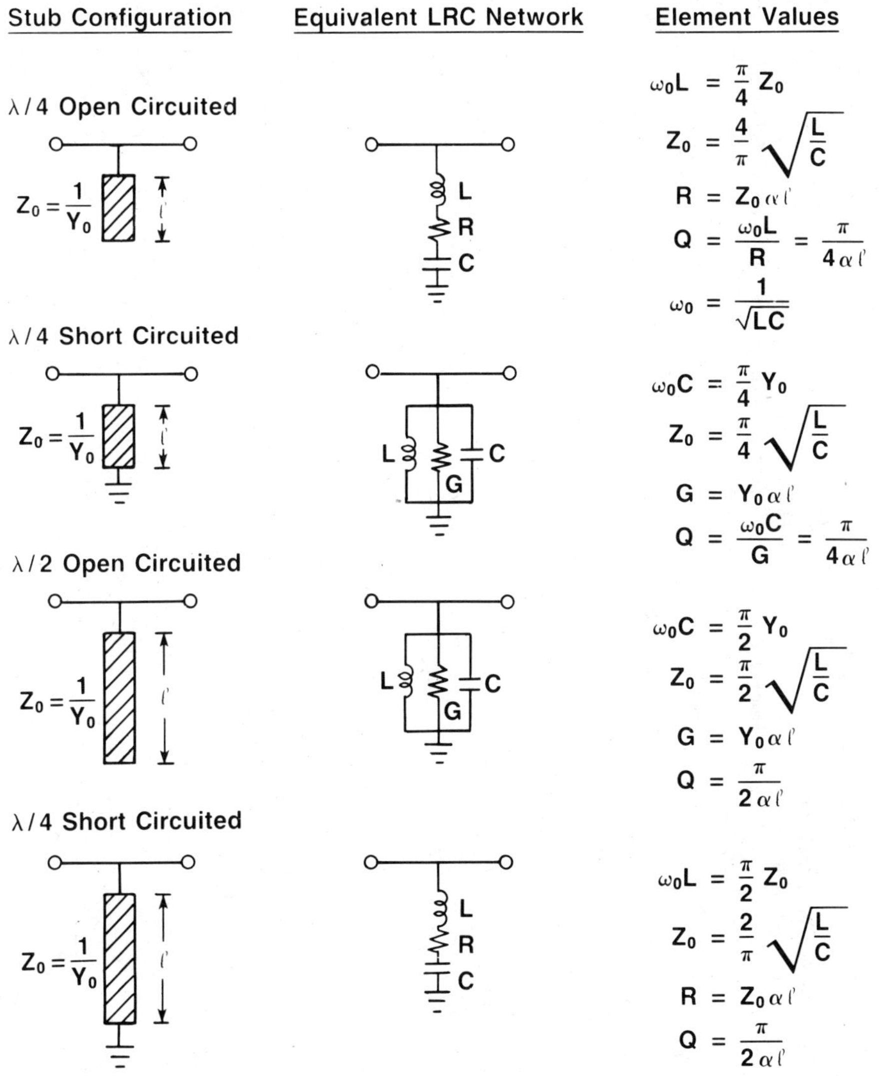

Stub Configuration	Equivalent LRC Network	Element Values
$\lambda/4$ Open Circuited, $Z_0 = \frac{1}{Y_0}$, ℓ	L, R, C	$\omega_0 L = \frac{\pi}{4} Z_0$; $Z_0 = \frac{4}{\pi}\sqrt{\frac{L}{C}}$; $R = Z_0 \alpha \ell$; $Q = \frac{\omega_0 L}{R} = \frac{\pi}{4\alpha\ell}$; $\omega_0 = \frac{1}{\sqrt{LC}}$
$\lambda/4$ Short Circuited, $Z_0 = \frac{1}{Y_0}$, ℓ	L, G, C	$\omega_0 C = \frac{\pi}{4} Y_0$; $Z_0 = \frac{\pi}{4}\sqrt{\frac{L}{C}}$; $G = Y_0 \alpha \ell$; $Q = \frac{\omega_0 C}{G} = \frac{\pi}{4\alpha\ell}$
$\lambda/2$ Open Circuited, $Z_0 = \frac{1}{Y_0}$, ℓ	L, G, C	$\omega_0 C = \frac{\pi}{2} Y_0$; $Z_0 = \frac{\pi}{2}\sqrt{\frac{L}{C}}$; $G = Y_0 \alpha \ell$; $Q = \frac{\pi}{2\alpha\ell}$
$\lambda/4$ Short Circuited, $Z_0 = \frac{1}{Y_0}$, ℓ	L, R, C	$\omega_0 L = \frac{\pi}{2} Z_0$; $Z_0 = \frac{2}{\pi}\sqrt{\frac{L}{C}}$; $R = Z_0 \alpha \ell$; $Q = \frac{\pi}{2\alpha\ell}$

Figure 6.27 Equivalent circuits for TEM strip-line resonators.

The number of resonators required is 5. The prototype values are

$$g_0 = 1.0, \qquad g_1 = g_5 = 1.3394, \qquad g_2 = g_4 = 1.337, \qquad g_3 = 2.166$$

The lumped-element values are

$$C_1 = C_5 = 2.13\ pF, \qquad L_2 = L_4 = 5.318\ \text{nH}, \qquad C_3 = 3.446\ \text{pF}$$

TABLE 6.6 Distributed Inductors and Capacitors

Distributed Element	Equivalent Circuit	Expressions	Approximate Expressions
Mainly Inductive 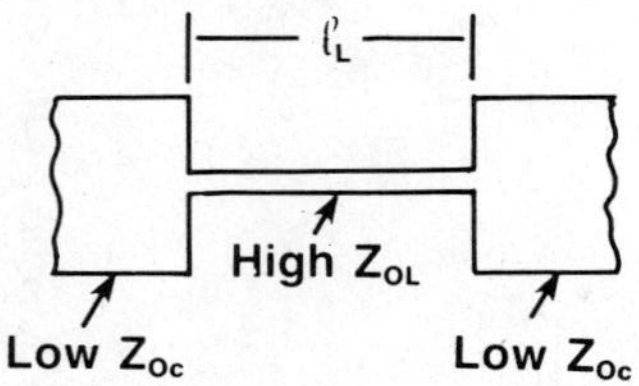 ℓ_L; High Z_{OL}; Low Z_{Oc}; Low Z_{Oc}	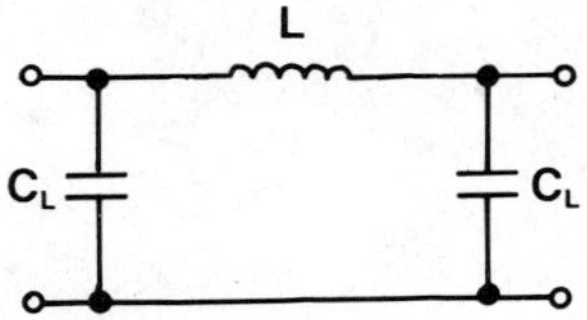 L; C_L; C_L	$L = \frac{Z_{OL}}{\omega} \sin\left(\frac{2\pi l_L}{\lambda_{gL}}\right)$ $C_L = \frac{1}{\omega Z_{OL}} \tan\left(\frac{\pi l_L}{\lambda_{gL}}\right)$	$l_L \simeq \frac{f\lambda_{gL} L}{Z_{OL}}$
Mainly Capacitive 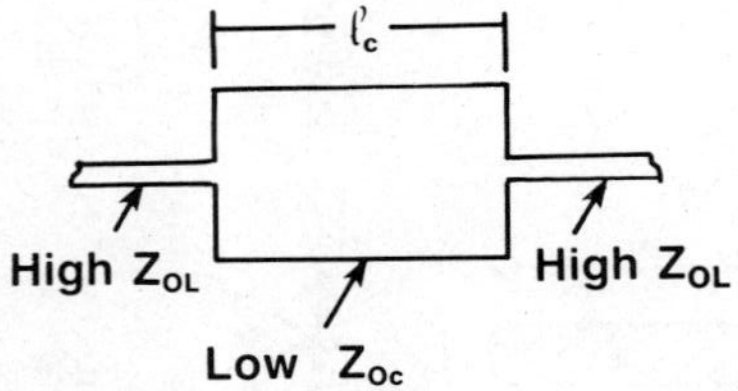 ℓ_c; High Z_{OL}; Low Z_{Oc}; High Z_{OL}	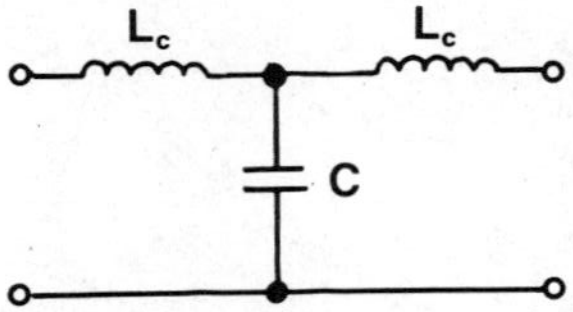 L_c; L_c; C	$C = \frac{1}{\omega Z_{Oc}} \sin\left(\frac{2\pi l_c}{\lambda_{gc}}\right)$ $L_c = \frac{Z_{Oc}}{\omega} \tan\left(\frac{\pi l_c}{\lambda_{gc}}\right)$	$l_c = f\lambda_{gc} Z_{Oc} C$

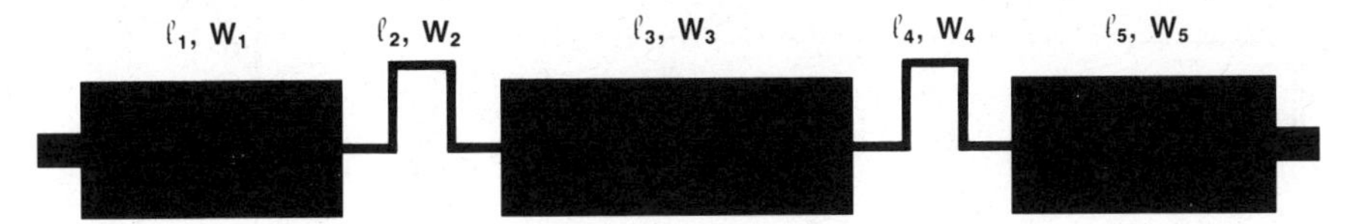

Figure 6.28 Layout of a 5-section low-pass filter.

The filter was realized using microstrip sections as described in Table 6.6. The microstrip layout of the filter is shown in Fig. 6.28. Various dimensions for microstrip sections are given in Table 6.7. Figure 6.29 shows insertion loss and return loss for this filter for ideal design, based on Table 6.6, and when optimized with discontinuities. No conductor and dielectric losses are included so far. Figure 6.30 shows a comparison between the measured and calculated responses (with losses).

Bandpass Filters. Popular configurations for printed circuit filters (Figs. 6.18 and 6.26) are direct coupled, parallel coupled, interdigital, combline and hairpin line. Direct-coupled resonator filters [1] have excessive length. The dimensions can be reduced by a factor of 2 with the introduction of parallel-coupled lines. Parallel coupling is much stronger than end coupling, so that realizable bandwidths could be much greater [17, 20]. In this configuration the first spurious response occurs at three times the center

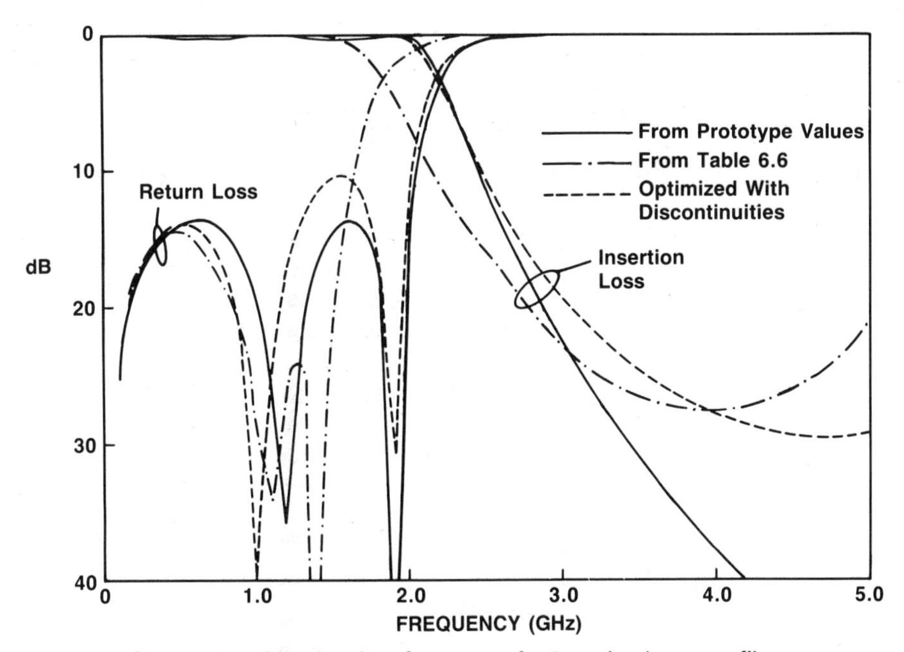

Figure 6.29 Calculated performance of a 5-section low-pass filter.

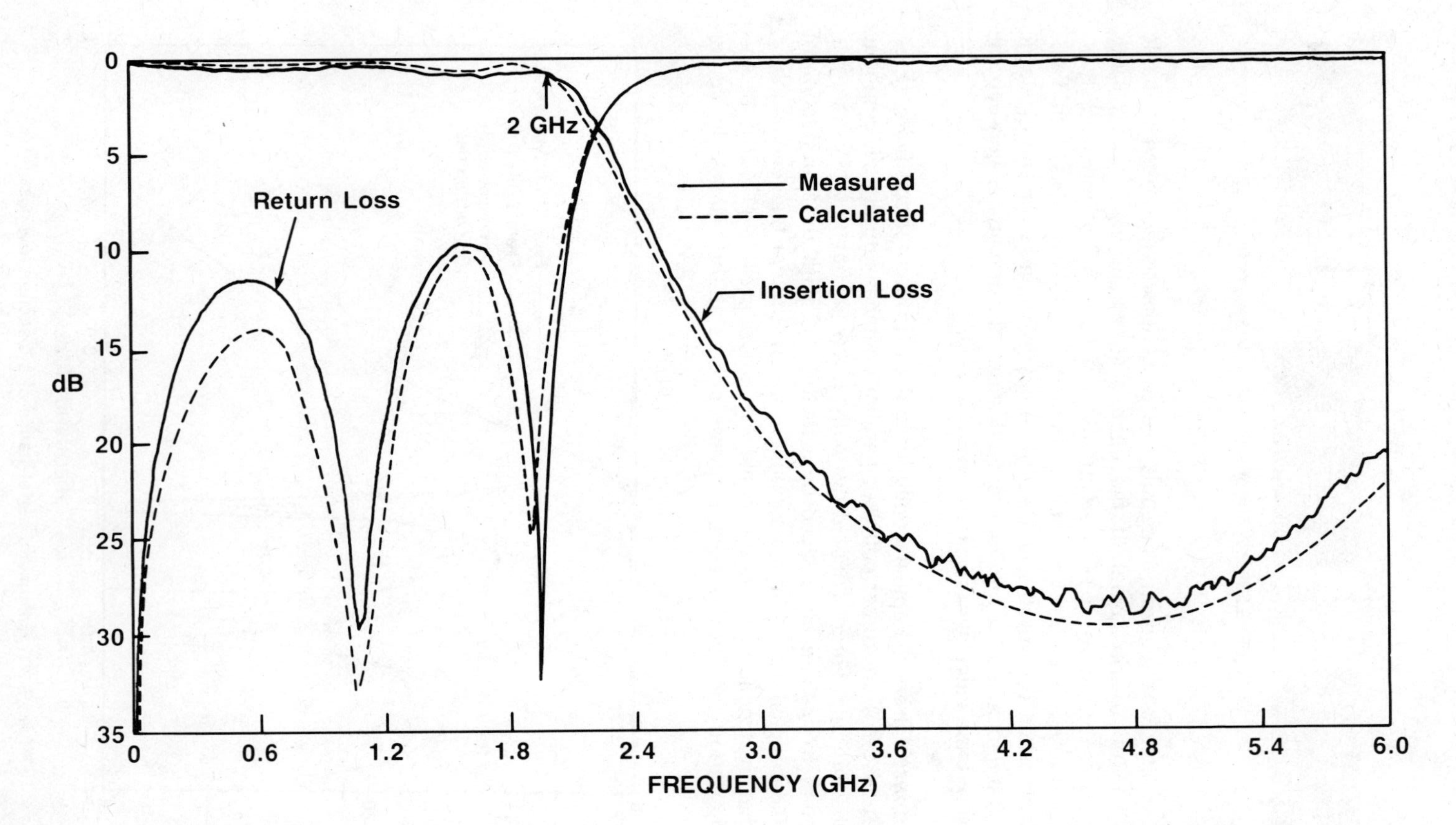

Figure 6.30 Comparison of calculated and measured responses of a five-section low-pass filter realized using microstrip.

TABLE 6.7 Design Parameters for a Low-Pass Filter

	From Table 6.6		Optimized with Discontinuities
$l_{c1} = l_{c5}$ (mm)	4.85		5.15
$l_{L2} = l_{L4}$ (mm)	7.25		6.07
l_{c3} (mm)	8.99		6.92
$Z_0 = 50\ \Omega$	$W = 0.6$ mm	$\epsilon_e = 6.58$	$\lambda_g = 5.85$ cm at 2 GHz
$Z_{0c} = 20\ \Omega$	$W = 2.64$ mm	$\epsilon_e = 7.74$	$\lambda_{gc} = 5.39$ cm at 2 GHz
$Z_{0L} = 100\ \Omega$	$W = 0.075$ mm	$\epsilon_e = 5.82$	$\lambda_{gL} = 6.22$ cm at 2 GHz

frequency and a much larger gap is permitted between parallel adjacent strips. The gap tolerance is also reduced, permitting a broader bandwidth for a given tolerance. Filters can be designed with reasonable accuracy using design information in [1, 20–25]. In order to fabricate compact filters, resonators are placed side by side. Interdigital, combline, and hairpin-line filters are realized using this concept. Accurate design analysis of these filters, interdigital [1, 26, 27], combline [1, 28–30], and hairpin-line [31, 32], are also available.

To illustrate a design example of an interdigital bandpass filter the following specifications have been chosen:

Center frequency f_0	4 GHz
Response	Chebyshev with 0.2-dB ripple
Bandwidth	0.4 GHz
35-dB attenuation points	4 ± 0.4 GHz
Substrate	$\epsilon_r = 9.8$ and $h = 1.27$ mm

From Fig. 6.11, the number of resonators is 5. The prototype values are the same calculated for the low-pass filter example. Here $Z_0 = 50\ \Omega$ and $Z_{0I} = 58\ \Omega$ are selected for the tapped interdigital filter [12]. The coupling parameters from (6.29) and (6.31) are

$$Q_L = \frac{f_0}{\text{BW}} g_1 = 13.4$$

$$K_{12} = K_{45} = 0.0747$$

$$K_{23} = K_{34} = 0.0588$$

The dimensions of this filter, shown in Fig. 6.31, are determined from Figs.

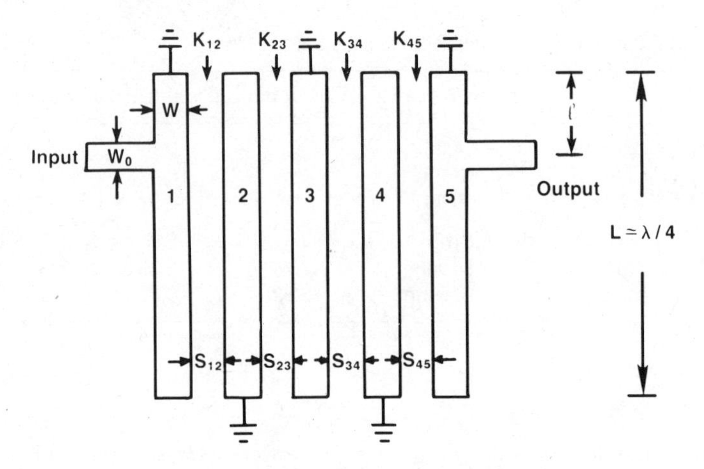

Figure 6.31 Five-section interdigital filter.

6.20 and 6.22 and are given below

$$W = 0.889 \text{ mm}$$

$$L = 7.43 \text{ mm}$$

$$S_{12} = S_{45} = 2.2 \text{ mm}$$

$$S_{23} = S_{34} = 2.52 \text{ mm}$$

$$\frac{l}{L} = 0.145, \qquad l = 1.077 \text{ mm}$$

6.7.2 Dielectric Resonator Filters

MIC structures have suffered from a lack of high-Q miniature elements, which are required to construct high-performance, highly stable, narrow-band filters. Filters such as bandpass and bandstop are frequently realized using high-quality dielectric resonators. The dielectric constant is around 40, and resonator Q ($\simeq 1/\tan\delta$) lies between 5000 and 10,000 in the 2–7-GHz frequency range [33, 34]. The commercially available dielectric resonators have temperature stability as good as Invar these days. Filters may be constructed in all the common transmission media ranging from waveguide to microstrip. Such filters are small in size, lightweight, and low cost.

Most of the early work was carried on in the early 1960s as summarized in [35]. The basic bandpass filter topology consists of an evanescent-mode waveguide section (waveguide below cutoff) in which the dielectric resona-

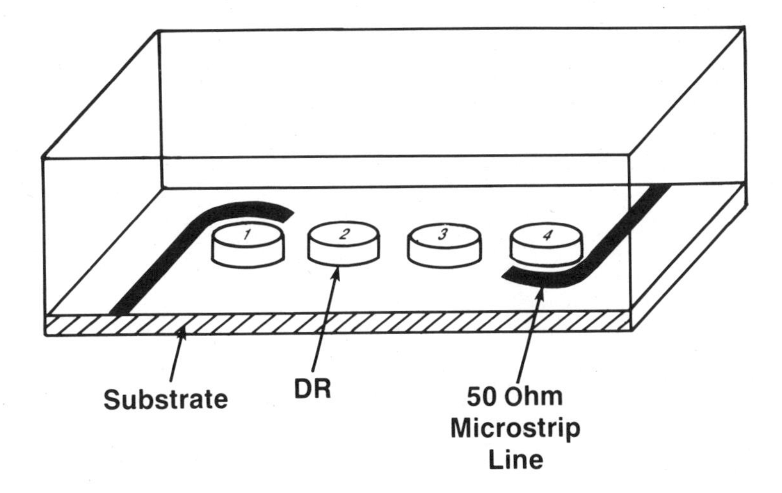

Figure 6.32 Microstrip-coupled dielectric resonator bandpass filter.

tors are housed. The most commonly used mode in the resonators is the $TE_{01\delta}$. The dual-mode or hybrid-mode $HE_{01\delta}$ resonators find applications in sophisticated elliptic function dielectric filters. Various kinds of dielectric resonator filters have been treated in the literature [36–40].

The bandpass dielectric resonator filter can be designed by the experimental method described in Section 6.4. A four-section filter configuration is shown in Fig. 6.32. The coupling to the external circuit is obtained by using 50-Ω microstrip lines. Typical coupling coefficients expressed in terms of loaded quality factor Q_L (required to design filters) are plotted in Fig. 6.33 for resonators having resonant frequency of 4 GHz, and $D = 14$ mm, $H = 7.5$ mm, $\epsilon_r = 35$, $h_1 = 18.5$ mm, $\epsilon_s = 2.54$, and $h = 0.8$ mm [41]. Here the generator, load, and the characteristic impedance of the microstrip are 50 Ω. Coupling between resonators as a function of separation is shown in Fig. 6.34 [40].

Consider an example of a dielectric bandpass filter with the following specifications:

Center frequency	4 GHz
Response	Chebyshev with 0.2-dB ripple
Bandwidth	40 MHz
30-dB attenuation points	4 ± 0.05 GHz
Resonator material	$\epsilon_r = 35$
Substrate	$\epsilon_s = 2.55$ and $h = 0.63$ mm

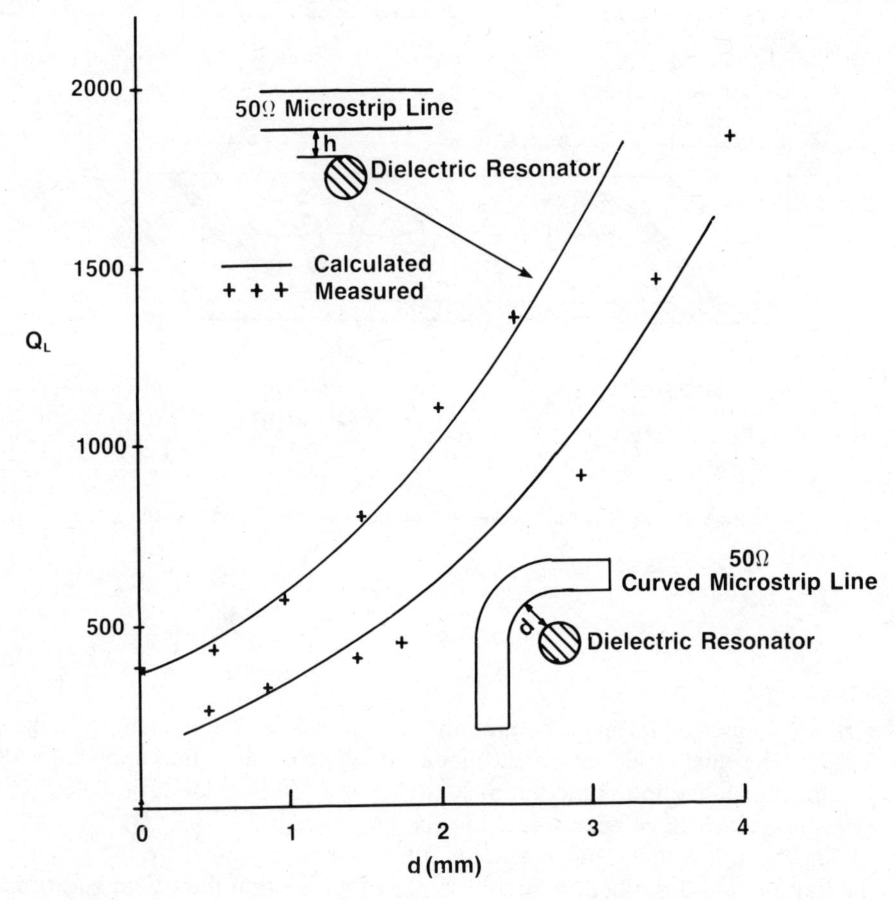

Figure 6.33 Variation of Q_L for a dielectric resonator coupled to straight and curved microstrip lines.

The number of resonators required is 4. The prototype values are

$g_0 = 1$, $g_1 = 1.3028$, $g_2 = 1.2844$, $g_3 = 1.9761$, $g_4 = 0.864$, $g_5 = 1.5386$

The coupling parameters from (6.29) and (6.31) are

$$Q_L = \frac{f_0}{\mathrm{BW}} g_1 = 130$$

$$K_{12} = K_{34} = 7.7 \times 10^{-3}$$

$$K_{23} = 6.3 \times 10^{-3}$$

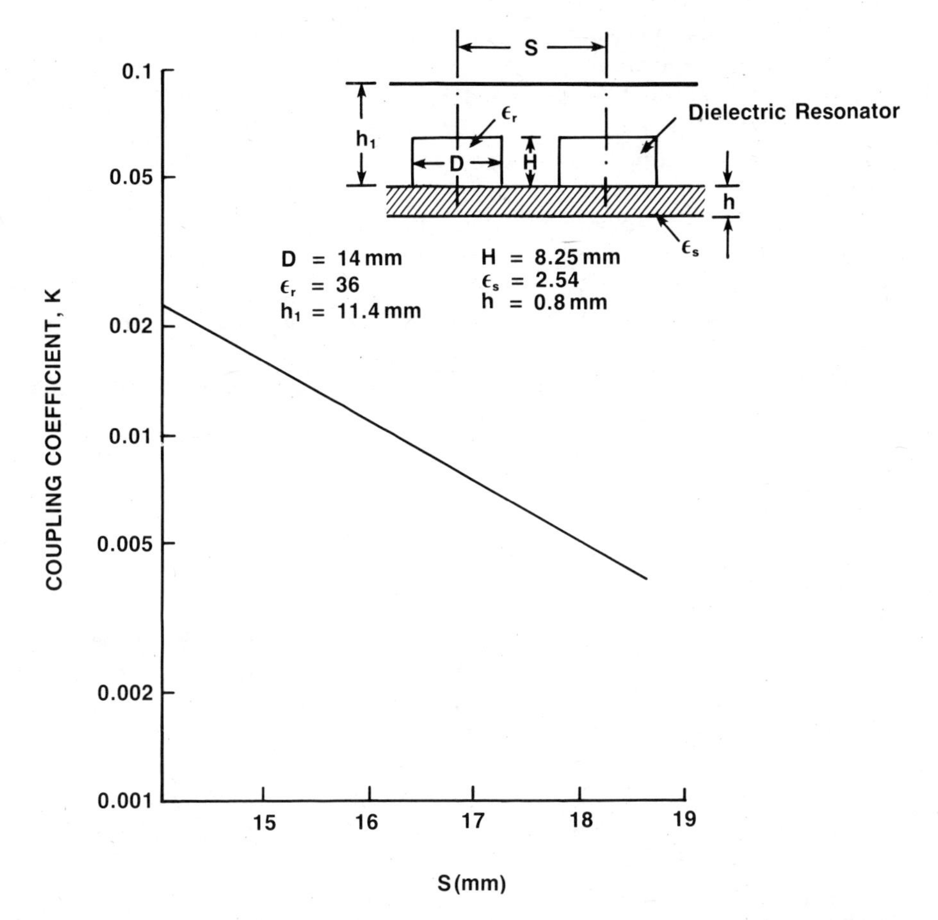

Figure 6.34 Coupling between two adjacent dielectric resonators operating in the $TM_{01\delta}$ mode.

6.8 PRACTICAL CONSIDERATIONS

In the preceding sections, various design parameters of filters were described. The important design considerations are size, weight, cost, effect of finite Q, group delay, temperature effect, power-handling capability, and tuning of the filters.

6.8.1 Size, Weight, and Cost

Applications such as ECM, airborne, satellite, and mobile communication demand small size, light weight, and less expensive filters. Filters constructed using strip line, microstrip, and fin line are limited to relatively wide-band applications where the selectivity is not severe. Furthermore, for highly selective applications, these structures exhibit temperature instability and tuning difficulties. The suspended-substrate techniques provide a higher Q (500 to 1500) than similar microstrip techniques. This results in lower loss filters with sharper band edges and with an excellent temperature stability. In certain cases, filters are available with temperature stabilities of $\pm 0.1\%$ over $-54°C$ to $+125°C$. Dielectric resonator filters have shown potential and are being developed from UHF to millimeter-wave frequencies. Size can also be reduced using a high dielectric constant substrate, such as barium tetratitanate ($BaTi_4O_9$) or other compounds of barium.

Up to a couple of gigahertz SAW filters are commonly used to build channelized receivers that can include several hundred filters. SAW filters are very attractive in terms of size, weight, and cost at lower microwave frequencies.

6.8.2 Finite *Q*

So far no attempt has been made to consider the effect of losses of the filter circuits on their performance. In other words, what happens if the resonators have a finite Q? When the lossy line sections are used, the insertion loss in the passband increases. In the case of the equal-ripple response filter, the losses result in a suppression of the ripples, as shown in Fig. 6.35 for different unloaded Q values of the filter. It is assumed that all the five sections of the filters have the same unloaded Q and resonant frequency.

In practice the bandwidth of a filter is set by the loaded Q of the resonator. The loaded Q, Q_L of a resonator depends on its losses and the external circuit connected to it and is given by

$$\frac{1}{Q_L} = \frac{1}{Q_u} + \frac{1}{Q_e} \tag{6.36a}$$

where Q_u is the unloaded resonator quality factor and Q_e is the external Q which accounts for the energy coupled into the external circuitry. When $Q_e \ll Q_u$ (or $\text{BW} \gg f_0/Q_u$), the bandwidth of the filter (BW) is almost independent of the unloaded Q values, but as Q_e approaches Q_u, the circuit gets very lossy. Thus unloaded Q sets a lower limit for the bandwidth of the filter. As Q_u approaches Q_e, it has three effects on the performance of filters: it broadens the bandwidth, introduces extra insertion loss, and reduces rejection in the stopband.

The insertion loss at the center of the passband of small-ripple Che-

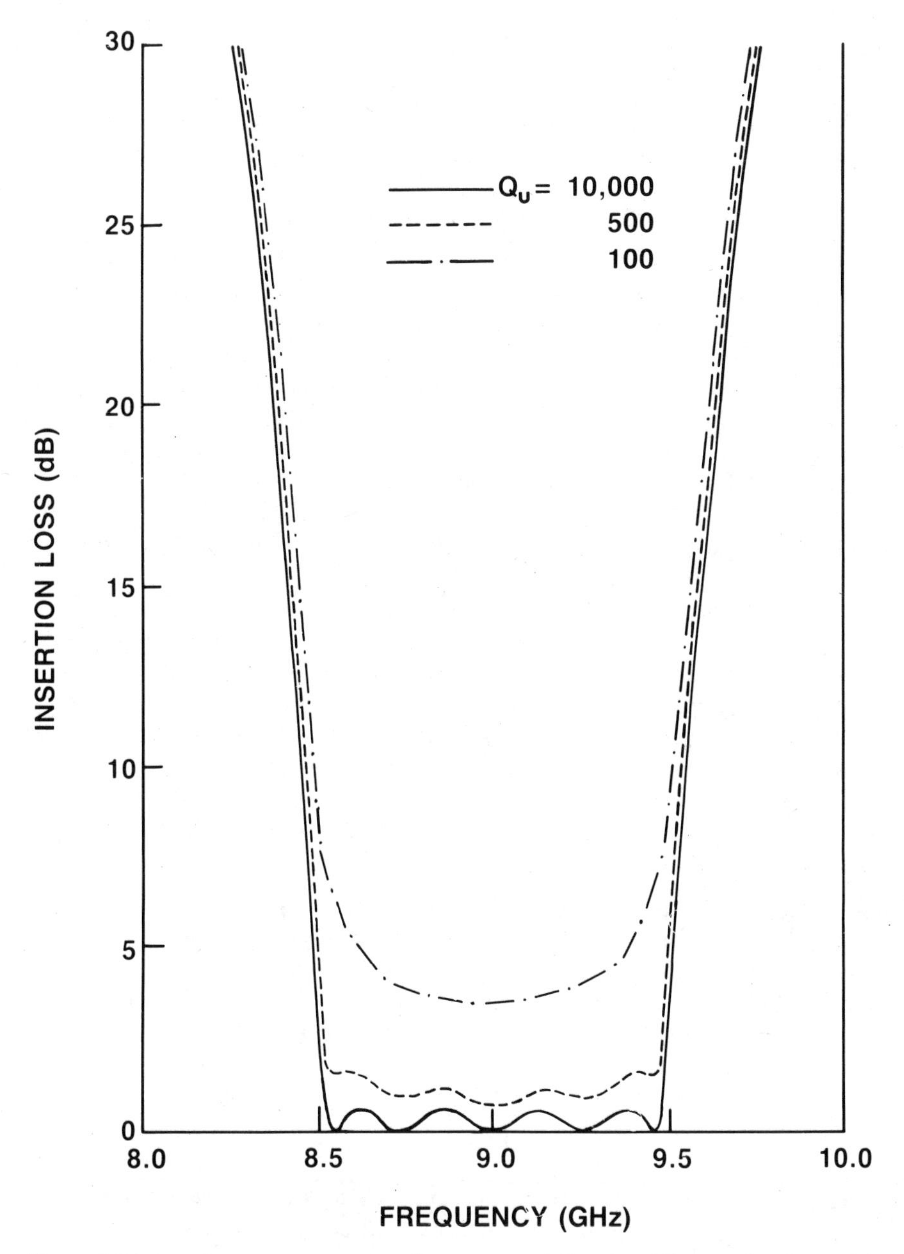

Figure 6.35 Bandpass response of a direct-coupled filter for different Q values of the filter: $f_0 = 9$ GHz, BW = 900 MHz, and $A_m = 0.5$ dB.

byshev bandpass filters is approximately given by

$$\mathrm{IL(dB)} = \frac{4.343 f_0}{\mathrm{BW} Q_u} \sum_{i=1}^{N} g_i \tag{6.36b}$$

where BW/f_0 is the equal-ripple frequency bandwidth.

6.8.3 Power-Handling Capability

The amount of peak power or average power that a filter can handle without voltage breakdown or excessive heating depends on bandwidth, filter size, filter type, and operating environment and altitude. Printed circuit filters can handle up to a few hundred watts, whereas for higher power applications, conventional waveguide and coaxial structures have to be used.

The power-handling capabilities of filters can be determined by knowing the power-handling capability of the transmission media [1, 12, 13] and filter structure geometries. Since it is not in the scope of this book to go into detail, readers are referred to the references. The larger gaps in the filter structure permit higher filter power rating.

For high-power applications (>10 kW) pressurized force-cooled waveguide filters are often the only choice. Coaxial filters are next, while strip-line, microstrip, lumped-element, and SAW filters are usually used for low-power applications. However, printed circuit filters are capable of handling pulse powers on the order of a few kilowatts peak and up to 50-W average.

6.8.4 Temperature Effects

When the temperature of the environment surrounding the filter changes, the physical dimensions of the structure also change. Therefore, the electrical characteristics of the filter are modified. Since most solids expand linearly with an increase in temperature, a physical dimension l will expand by Δl when the temperature increases by ΔT, that is,

$$\frac{\Delta l}{l} = \alpha \, \Delta T \tag{6.37}$$

where α is the linear expansion coefficient. For commonly used materials its values are given in Table 6.8. For bandpass filters using $\lambda/2$ resonators,

$$\frac{\Delta f}{f} = -\alpha \, \Delta T \tag{6.38}$$

TABLE 6.8 Linear Expansion Coefficients for Metals

Metal	α (per °C) × 10^{-6}	Dielectric	α (per °C) × 10^{-6}
Aluminum	23	Teflon	90
Brass	19	Glass	9
Copper	17	Alumina	7
Gold	15	RT/Duroid	6
Steel	11	Glass (Pyrex)	3
Invar	0.7	Fused quartz	0.6

Thus, the temperature coefficient of TEM resonators made of copper is −17 ppm/°C, while Invar resonators has −0.7 ppm/°C. The second effect of temperature on filters is due to the temperature dependence of the relative dielectric constant ϵ_r of dielectric materials. As a first approximation, the change is linear with temperature, that is,

$$\frac{\Delta \epsilon_r}{\epsilon_r} = \alpha_\epsilon \, \Delta T \tag{6.39}$$

where α_ϵ is the linear temperature coefficient of the dielectric constant. If the resonant frequency of resonators is inversely proportional to the square root of ϵ_r, then net change in resonant frequency is given by

$$\frac{\Delta f}{f} = -\alpha \, \Delta T - 0.5 \alpha_\epsilon \, \Delta T \tag{6.40}$$

The previous equation suggests that by properly selecting α and α_ϵ, Δf can be made zero. Table 6.9 gives α and α_ϵ for commonly used materials in

TABLE 6.9 Properties of Dielectric Resonators

Material	Q	ϵ_r	α (ppm/°C)	α_ϵ (ppm/°C)	Company
$(Zr, Sn)TiO_4$ Type C	15,000 (4 GHz)	37.3 ± 0.5	6.5	−13 ± 2	Murata
$Ba(Zr, Zn, Ta)O_3$ Type C	10,000 (10 GHz)	28.6 ± 0.5	10.2	−20.4 ± 4	Murata
Zr/Sn Titanate D-8515 Type	10,000 (4 GHz)	36.0 ± 0.5	5.6	−6.9	Trans Tech
Barium Tetratitante D-8512 Type	10,000 (4 GHz)	38.6 ± 0.6	9.5	−10.4	Trans Tech
$(Zr, Zn)TiO_4$ E-2036 Type	4000 (10 GHz)	37.0 ± 0.4	5	−10 ± 1	Thomson-CSF

dielectric resonators. It is evident from this table that the temperature stability of most dielectric resonators is excellent.

Temperature coefficients of other filter components is given in Table 6.10.

Consider an example of combline filters, whose required performance over operating temperatures $-50°C$ to $125°C$ is as follows

f_0	2 GHz
3-dB points	1.8 GHz
	2.2 GHz
40-dB points	1.5 GHz
	2.5 GHz

If 25°C is the room temperature,

$$\Delta T\ (\text{Cold}) = -75°\text{C}$$

$$\Delta T\ (\text{Hot}) = 100°\text{C}$$

$$\Delta f\ (\text{Cold}) = -75 \times 2000 \times 25 \times 10^{-6} = -3.75\ \text{MHz}$$

$$\Delta f\ (\text{Hot}) = 100 \times 2000 \times 25 \times 10^{-6} = 5\ \text{MHz}$$

Specifications adjusted for temperature variation

	$f_0 = 2$ GHz
3-dB points	$1.8 + (-0.00375) = 1.79625$ GHz
	$2.2 + (0.005) = 2.25$ GHz
40-dB points	$1.5 + (0.005) = 1.505$ GHz
	$2.5 + (-0.00375) = 2.49625$ GHz

In other words, the filter must be designed with about 9 MHz larger 3-dB bandwidth and about 9 MHz shorter 40-dB bandwidth to meet the specifications over the required temperature range.

TABLE 6.10 Temperature Coefficient of Filter Elements

Description	Center Frequency (GHz)	Temperature Coefficients (ppm/°C)
Standard lumped elements	>0.1	75
Stable lumped elements	0.1–16	10
Microstrip resonators (combline, interdigital, etc.)	0.5–18	25

6.8.5 Group Delay

The group delay of a filter is dependent on both the selected prototype and the number of sections used. Band-center group delay and group delay variation both increase with increasing number of sections. Many applications require constant group delay over the desired bandwidth.

The design of constant group delay filters has been described in the literature [1, 2, 42]. A general approach to design a constant group delay in the passband is to specify a function that describes the desired phase response such as [42]

$$\psi_T(\omega) = A\omega[1 + B(\omega/\omega_c)^p] \tag{6.41}$$

and then to force the filter response to assume values that satisfies the above equation at a number of discrete frequencies, while maintaining acceptable amplitude properties. Then the group delay τ_D is calculated using (6.6b).

6.8.6 Electrically Tuned Filters

Electrically tunable filters have many applications in EW and ESM systems. System requirements necessitate that the filters exhibit high-Q and broad tuning bandwidths with minimum degradation in passband performance. YIG spheres [43] and varactor diodes [44–46] are commonly used as tuning devices. Varactor Q factors are considerably lower than those of YIG spheres and thus varactor-tuned filters are not generally used above 10 GHz. On the other hand, the tuning speed of YIG filters is severely limited by magnetic hysteresis effects, whereas in the case of varactor-tuned filters, it is limited by the time constant of the varactor bias circuitry.

6.9 MULTIPLEXERS

To this point, we have been considering filters as individual elements, a simplifying but unrealistic assumption, since filters are never used alone but always in conjunction with other components. The filters that have been described so far are reflective filters whose performance can be specified only for known source and load impedances and thus can vary drastically in mismatched systems. A classic example to illustrate this effect is when two identical filters are used in series to increase rejection, as shown in Fig. 6.36. The filter impedances beat against one another with the result that the combination will often have less attenuation than either filter by itself. It is perhaps a bit paradoxical that putting an isolator with 15-dB return loss and 20-dB isolation between two filters can increase the rejection by 100 dB or more.

When filters are combined to separate a wide frequency band into a number of narrower channels or to combine frequency channels at a

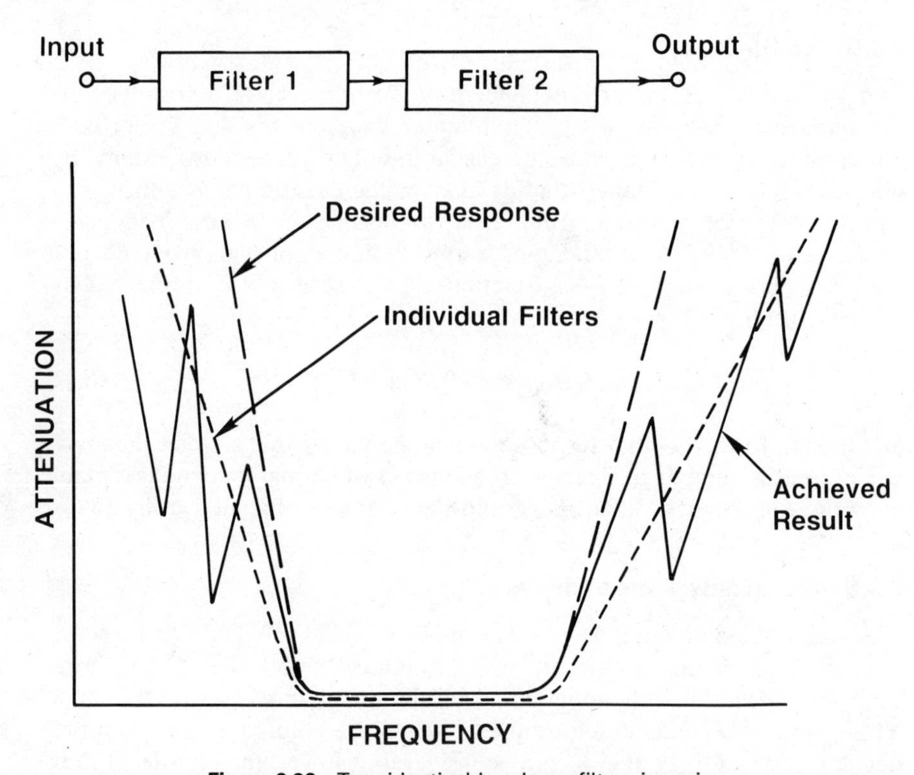

Figure 6.36 Two identical bandpass filters in series.

common port, the resulting component is called a multiplexer. The most common multiplexers that have one common port and multiple frequency selective ports are named according to how many ways the signal is split or combined; diplexer (two), triplexer (three), quadraplexer (four), etc., and can be divided into two types: contiguous and noncontiguous. In a contiguous multiplexer, the two adjacent filters share a common crossover frequency and have a minimum of 3-dB loss at the crossover frequencies, whereas in a noncontiguous type, the passbands are well separated in frequency.

6.9.1 Multiplexing Techniques

Since filters are often severely mismatched outside their bands, combining several of these into a multiplexer requires careful considerations so that undesirable mutual interactions are eliminated. Additionally, in most applications, the overall size of the assembly must be minimized. The channel-dropping (summing) multiplexer shown in Fig. 6.37 is a commonly used configuration that simplifies the handling of the interactions by using circulators. The multiplexer shown in the figure is not reciprocal but can be

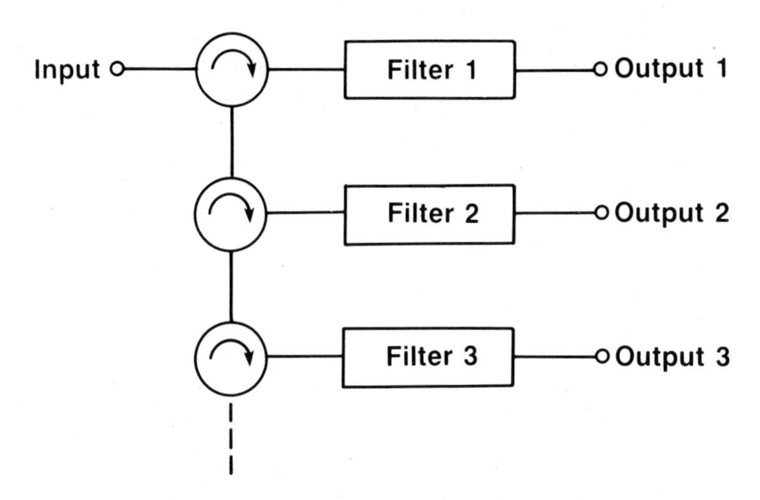

Figure 6.37 Channel-dropping multiplexers.

used to sum the channels instead of separating them by turning around the circulators. This multiplexer works by successively presenting what is left of the signal to each filter. Since each filter can be terminated in a matched load at the input, the response of each channel is simply the algebraic sum of the reflections from the preceding channels plus the loss due to the multiple circulator passes. If the passbands of each of the channels are separated enough in frequency so that the return loss of each filter is small in the passbands, the succeeding filter will replicate the ideal response of each filter quite well, although, if many channels are used, the loss due to multiple circulator passes can be large. If a preceding channel is expected to provide a notching effect, the circulator isolation will limit the maximum notch attenuation. If loss is not important, a power-splitting multiplexer, such as that shown in Fig. 6.38, can be used by combining the properties of a filter and a hybrid. The response will be just that of the individual filters plus the power divider loss. For this multiplexer, channels can overlap. The filters are independent within the bandwidth of the power divider.

The situation becomes much more complex if the multiplexer must be designed for minimum loss. In that case, the filter interactions must be explicitly included in the design. The two standard multiplexing techniques used are shown in Fig. 6.39.

In the simplest implementation of the common junction multiplexer, each filter includes an additional line length (θ_i) to present an open circuit across the passband frequency of all other channels and a match at its own passband frequencies. The common junction noncontiguous multiplexer is usually fairly straightforward when the overall bandwidth is not too wide (less than one octave). In these multiplexers, the filter designs are doubly terminated as described in the preceding sections.

The shunted manifold multiplexer is similar in operation to the common

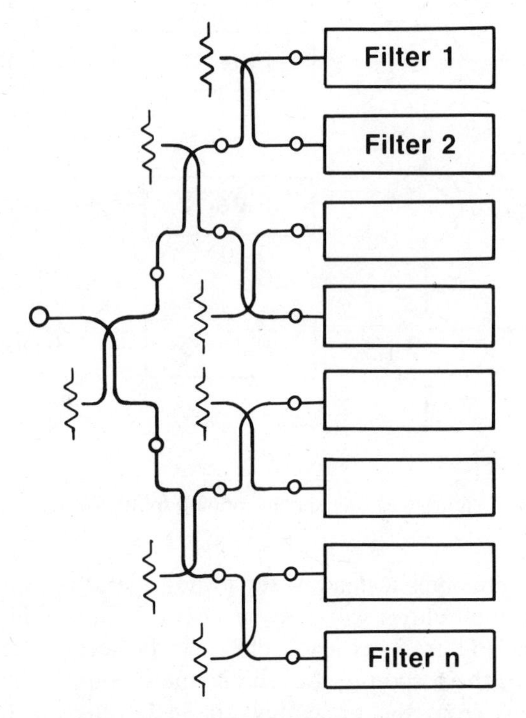

Figure 6.38 Power-splitting multiplexer.

junction multiplexer. In that case, the filters are usually spaced at quarter wavelengths apart along the manifold to cancel reflections from each filter. The manifold multiplexer is harder to make broadband than the common junction type because as the number of channels goes up, the filters end up with long line lengths between them. However, the manifold multiplexer is often mechanically simpler because of the difficulty of bringing many filters physically close together at the common junction point of the common junction multiplexer.

As the filter passbands get closer together, as in the case of contiguous multiplexers, it becomes impossible to maintain a reasonable approximation of an open circuit from each channel across the operating frequencies of the others. The filters interact appreciably with each other, so the designs must deviate from doubly terminated design.

Recent work has shown that contiguous as well as noncontiguous multiplexers may be designed by modifying the first few elements of standard doubly terminated filters. The design of multiplexers with interacting channel filters is beyond the scope of this chapter, but a few comments on the subject follows.

1. Single-ended filters have considerably more attenuation for a given bandwidth than doubly terminated filters. Thus, for a given rejection

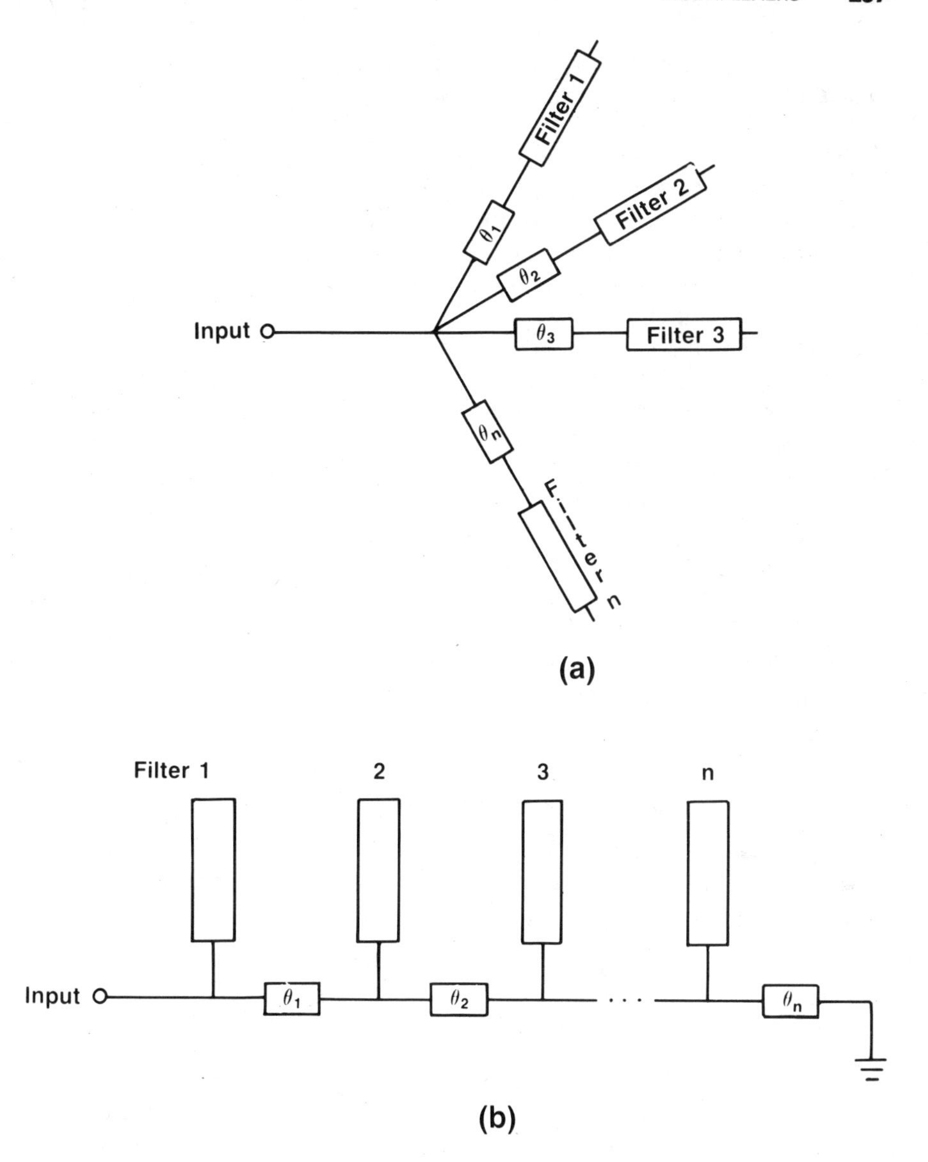

Figure 6.39 Two standard multiplexers. (*a*) Common junction. (*b*) Shorted manifold.

specification, a single-ended filter will have fewer sections with lower passband insertion loss.

2. The output manifold coupling is considerably smaller than for the doubly terminated filter. The couplings gradually approach doubly terminated values as one looks from the common junction to the output.

3. The filters are synchronous, but look wildly detuned when viewed individually in a matched system. This makes it hard to tune individual

channel filters before putting together the multiplexer. Multiplexer tuning is difficult because of the interaction between filters.

In practice, one never has an infinite number of channels. The single-ended designs work reasonably well for channels with at least one neighboring channel on either side of the passband. The end-frequency channels will work poorly, however. Two approaches are used to overcome this problem: one can put on annulling networks that simulate the impedance of the missing channels (one way is to go ahead and put terminated extra channels in the multiplexer), or one can design asynchronous asymmetrical end filters that have the equivalent of internal annulling. The diplexer design described in the next subsection is an example of this type of filter.

Throughout this past discussion, no mention of the line lengths and deviation from the narrow-band approximation has been made. These effects can be handled by the extension of the numerical design techniques described previously.

6.9.2 Diplexer Design

In many mixers, reactive terminations of LO, RF, and other harmonics using reflective low-pass filters is not desirable. In such situations, a diplexer

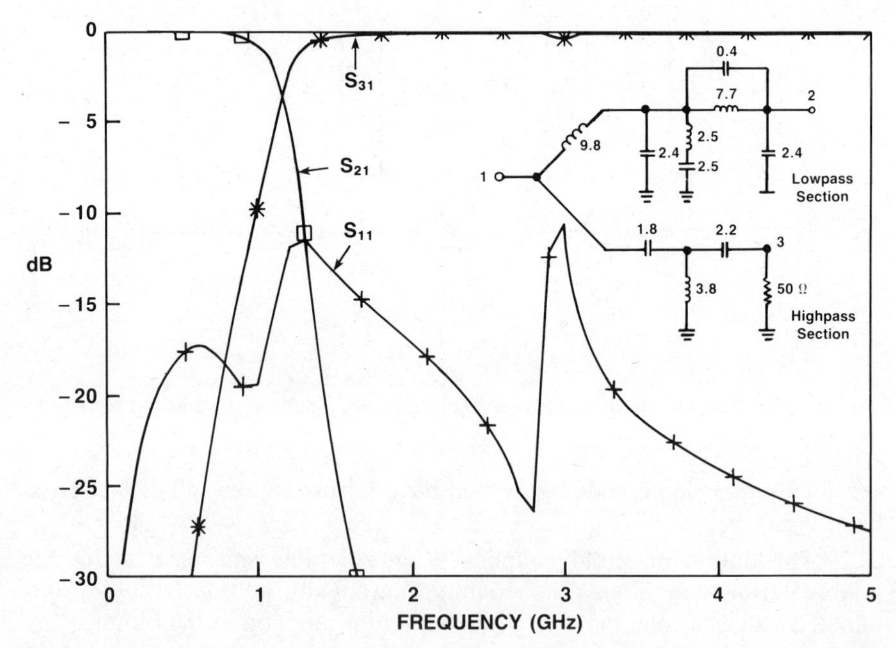

Figure 6.40 Calculated performance of a diplexer. Inductors are in nanohenries and capacitors in picofarads.

is required to provide a good termination load for LO, RF, and other harmonics along with good isolation between IF and LO/RF.

A diplexer consisting of lumped-element components is shown in Fig. 6.40 (inset). The diplexer has a contiguous passband and consists of a singly terminated low-pass filter and a high-pass filter. The low-pass filter is a quasi-elliptic filter. The diplexer was optimized to have a crossover frequency of 1.1 GHz and have 20-dB suppression of signals whose frequency is greater than 1.5 GHz. The input VSWR is 2.0:1 maximum over the 0.01- to 5-GHz frequency range. Figure 6.40 shows calculated S_{11}, S_{21}, and S_{31} for the diplexer.

6.9.3 Multiplexer Realization

Printed-circuit multiplexers using microstrip, suspended microstrip, fin line, etc., have low cost, have fewer machined parts, and are simpler to tune compared with conventional coaxial and waveguide multiplexers. The disadvantages of such multiplexers include higher insertion loss and lower power-handling capability. Construction of multioctave-band contiguous multiplexers in suspended substrate strip line has been described by Rhodes [47, 48].

REFERENCES

1. Matthaei, G. L., L. Young, and E. M. T. Jones, *Microwave Filters, Impedance-Matching Networks and Coupling Structures*, McGraw-Hill, New York, 1964.
2. Rhodes, J. D., *Theory of Electrical Filters*, Wiley, New York, 1976.
3. Saal, R., *The Design of Filters using the Catalogue of Normalized Low-Pass Filters* (in German), Telefunken (GMBH), Backnang, W. Germany, 1961.
4. Skwirzynski, J. K., *Design Theory and Data for Electrical Filters*, Van Nostrand, London, 1965.
5. Zverev, A. I., *Handbook of Filter Synthesis*, Wiley, New York, 1967.
6. Milligan, T., "Nomographs Aid and Filter Designer," *Microwave & RF*, Vol. 24, Oct. 1985, pp. 103–107.
7. Howe, H., Jr., *Stripline Circuit Design*, Artech House, Dedham, Mass., 1974.
8. Alseyab, S. A., "A Novel Class of Generalized Chebyshev Low-pass Prototype for Suspended Substrate Stripline Filters," *IEEE Trans. Microwave Theory Tech.*, Vol. MTT-30, Sept. 1982, pp. 1341–1347.
9. Mobbs, C. I., and J. D. Rhodes, "A Generalized Chebyshev Suspended Substrate Stripline Bandpass Filter," *IEEE Trans. Microwave Theory Tech.*, Vol. MTT-31, May 1983, pp. 397–402.
10. Dishal, M., "Alignment and Adjustment of Synchronously Tuned Multiple-Resonant-Circuit Filters," *Proc. IRE*, Vol. 39, Nov. 1951, pp. 1448–1455.
11. Dishal, M., "A Simple Design Procedure for Small Percentage Bandwidth

Round-Rod Interdigital Filters," *IEEE Trans. Microwave Theory Tech.*, Vol. MTT-13, Sept. 1965, pp. 696–698.

12. Wong, J. S., "Microstrip Tapped-line Filter Design," *IEEE Trans. Microwave Theory Tech.*, Vol. MTT-27, Jan. 1979, pp. 44–50.
13. Wenzel, R. J., "Lecture Notes—Microwave Circuit Design," UCLA Extension Course, University of California, Los Angeles.
14. Levy, R., and S. B. Cohn, "A Brief History of Microwave Filter Research, Design, and Development," *IEEE Trans. Microwave Theory Tech.* (*Special Centennial Issue*), Vol. MTT-32, Sept. 1984, pp. 1055–1067.
15. Gupta, K. C., R. Garg, and I. J. Bahl, *Microstrip Lines and Slotlines*, Artech House, Dedham, Mass., 1979.
16. Rooney, J. P., and L. M. Underkofler, "Printed Circuit Integration of MW Filters," *Microwave J.*, Vol. 21, Sept. 1978, pp. 68–73.
17. Ozaki, H., and J. Ishii, "Synthesis of a Class of Stripline Filters," *IRE Trans. Circuit Theory*, Vol. CT-5, June 1958, pp. 104–109.
18. Davis, W. A., *Microwave Semiconductor Circuit Design*, Van Nostrand–Reinhold, Chap. 3, 1984.
19. Richards, P. I., "Resistor-Transmission-line Circuits," *Proc. IRE*, Vol. 36, Feb. 1948, pp. 217–220.
20. Cohn, S. B., "Parallel-Coupled Transmission-line Resonator Filters," *IRE Trans. Microwave Theory Tech.*, Vol. MTT-6, Apr. 1958, pp. 223–231.
21. Makimoto, M., and S. Yamashita, "Bandpass Filters Using Parallel Coupled Stripline Stepped Impedance Resonators," *IEEE Trans. Microwave Theory Tech.*, Vol. MTT-28, Dec. 1980, pp. 1413–1417.
22. Minnis, B. J., "Printed Circuit Coupled-line Filters for Bandwidths Up To and Greater Than an Octave," *IEEE Trans. Microwave Theory Tech.*, Vol. MTT-29, Mar. 1981, pp. 215–222.
23. Wenzel, R. J., and W. G. Erlinger, "Problems in Microstrip Filter Design," *IEEE Int. Microwave Symp. Digest*, 1981, pp. 203–205.
24. Ho, C. Y., and J. H. Weidman, "Improved Design of Parallel Coupled Line Filters with Tapped Input/Output," *Microwave J.*, Vol. 26, Oct. 1983, pp. 127–130.
25. Bhartia, P., and I. J. Bahl, *Millimeter Wave Engineering and Applications*, Wiley, New York, 1984.
26. Matthaei, G. L., "Interdigital Band-Pass Filters," *IRE Trans. Microwave Theory Tech.*, Vol. MTT-10, Nov. 1962, pp. 479–491.
27. Wenzel, R. J., "Exact Theory of Interdigital Bandpass Filters and Related Coupled Structures," *IEEE Trans. Microwave Theory Tech.*, Vol. MTT-13, Sept. 1965, pp. 558–575.
28. Matthaei, G. L., "Comb-line Band-Pass Filters of Narrow or Moderate Bandwidth," *Microwave J.*, Vol. 6, Aug. 1963, pp. 82–91.
29. Wenzel, R. J., "Synthesis of Combline and Capacitely Loaded Interdigital Bandpass Filters of Arbitrary Bandwidth," *IEEE Trans. Microwave Theory Tech.*, Vol. MTT-19, Aug. 1971, pp. 678–686.
30. Vincze, A. D., "Practical Design Approach to Microstrip Combline-type Fil-

ters," *IEEE Trans. Microwave Theory Tech.*, Vol. MTT-22, Dec. 1974, pp. 1171–1181.

31. Cristal, E. G., and S. Frankel, "Hairpin Line/Half-wave Parallel-Coupled-line Filters," *IEEE Trans. Microwave Theory Tech.*, Vol. MTT-20, Nov. 1972, pp. 719–728.
32. Gysel, U. H., "New Theory and Design for Hairpin-Line Filters," *IEEE Trans. Microwave Theory Tech.*, Vol. MTT-22, May 1974, pp. 523–531.
33. Plourde, J. K., and D. F. Linn, "Microwave Dielectric Resonator Filters Using $Ba_2Ti_9O_{20}$ Ceramics," *IEEE Int. Microwave Symp. Digest*, 1977, pp. 290–293.
34. Ren, C. L., "Waveguide Bandstop Filter Utilizing $Ba_2Ti_9O_{20}$ Resonators," *IEEE Int. Microwave Symp. Digest*, 1978, pp. 227–229.
35. Cohn, S. B., "Microwave Bandpass Filters Containing High-Q Dielectric Resonators," *IEEE Trans. Microwave Theory Tech.*, Vol. MTT-16, Apr. 1968, pp. 218–227.
36. Iveland, T. D., "Dielectric Resonator Filters for Application in Microwave Integrated Circuits," *IEEE Trans. Microwave Theory Tech.*, Vol. MTT-19, July 1971, pp. 643–652.
37. Mekerta, S., and P. Guillon, "Dielectric Resonator Bandstop Filters," *IEEE Int. Microwave Symp. Digest*, 1981, pp. 170–172.
38. Atia, A. E., and R. R. Bonetti, "Generalized Dielectric Resonator Filters," *Comsat Tech. Rev.*, Vol. 11, Fall 1981, pp. 321–343.
39. Fiedziuszko, S. J., "Dual Mode Dielectric Resonator Loaded Cavity Filters," *IEEE Trans. Microwave Theory Tech.*, Vol. MTT-30, Sept. 1982, pp. 1311–1316.
40. Kajfez, D., and P. Guillon (Eds.), *Dielectric Resonators*, Artech House, Dedham, Mass., 1986.
41. Guillon, P., B. Byzery, and M. Chaubet, "Coupling Parameters Between a Dielectric Resonator and a Microstrip Line," *IEEE Trans. Microwave Theory Tech.*, Vol. MTT-33, Mar. 1985, pp. 222–226.
42. Malherbe, J. A. G., *Microwave Transmission Line Filters*, Artech House, Dedham, Mass., 1979.
43. Keane, W. J., "Narrow-Band YIG Filters," *Microwaves*, Vol. 17, Sept. 1978, pp. 52–56.
44. Hunter, I. C., and J. D. Rhodes, "Varactor Tuned Microwave Bandpass Filters," *IEEE Trans. Microwave Theory Tech.*, Vol. MTT-30, Sept. 1982, pp. 1354–1360.
45. Hunter, I. C., and J. D. Rhodes, "Varactor Tuned Microwave Bandstop Filters," *IEEE Trans. Microwave Theory Tech.*, Vol. MTT-30, Sept. 1982, pp. 1361–1367.
46. Makimoto, M., and M. Sagawa, "Varactor Tuned Bandpass Filters Using Microstrip-line Ring Resonators," *IEEE Int. Microwave Symp. Digest*, 1986, pp. 411–414.
47. Rhodes, J. D., "Suspended Substrates Provide Alternative to Coax," *Microwave Syst. News*, Vol. 9, Aug. 1979, pp. 134–143.

48. Rhodes, J. D., and J. E. Dean, "MIC Broadband Filters and Contiguous Multiplexer," *9th European Microwave Conf. Proc.* (Microwave Exhibitions and Publishers Ltd., Sevenoaks, Kent, England, 1979), pp. 407–411.

PROBLEMS

6.1 Calculate low-pass prototype element values (g's) for maximally flat and Chebyshev response (0.1-dB ripple) when the number of sections is 3. If the edge frequency is 1 GHz, determine *LC* values for low-pass and high-pass filters.

6.2 Determine *LC* values for bandpass and bandstop filters when the center frequency is 2 GHz in Problem 6.1.

6.3 Design a low-pass strip-line filter having a ripple of 0.1 dB and stopband attenuation of 30 dB at 3 GHz. The filter may be realized using high impedance and low impedance sections. The input and output impedances are 50 Ω.

6.4 Develop a low-pass filter using lumped elements with the specifications given in Problem 6.1.

6.5 Develop design equations for a parallel-coupled microstrip-line bandpass filter. Design a six-section bandpass filter at 4 GHz. For the microstrip, the parameters to be used are $\epsilon_r = 3.8$ and $h = 0.5$ mm. The input and output impedances are 50 Ω.

6.6 Develop design equations for end-coupled microstrip-line bandpass filter. Describe limitations of this configuration with respect to parallel-coupled microstrip line filter. Suggest how these limitations can be overcome.

6.7 Suppose that a filter is desired with a 0.5-dB ripple Chebyshev passband from 3 to 3.2 GHz and that 30-dB attenuation is required at 2.5 and 3.5 GHz. Determine the physical dimensions for end-coupled strip line filter. The input and output impedances are 50 Ω.

6.8 Design an 8-resonator, 0.1-dB Chebyshev ripple, and 25% bandwidth tapped-line interdigital bandpass filter on a 5880 RT/Duroid substrate using Figs. 6.20 and 6.22. The center frequency is 1 GHz. The input and output impedances are 50 Ω.

6.9 Design an *LC* high-pass filter with an f_c of 4 GHz and a minimum attenuation of 30 dB at 3 GHz by assuming that a 0.1-dB passband ripple is tolerable. Calculate and compare performance using ideal LC components and monolithic LC components. The monolithic substrate is 200-μm thick GaAs ($\epsilon_r = 12.9$) and MIM capacitance is 300 pF/mm^2. The input and output impedances are 50 Ω.

7 ACTIVE DEVICES

7.1 INTRODUCTION

The development of microwave solid state circuits is directly dependent upon the availability of suitable active devices. Active devices are required as gain blocks in circuit applications and to allow a circuit to increase the RF energy at a desired frequency. For microwave circuit applications the microwave bipolar transistor and the GaAs MESFET are the two most important and commonly used active elements. The basic operation principles of these two devices are presented in this chapter. The emphasis upon basic physical phenomena and the development of physically based equivalent circuits should provide the information required as background for device design as well as analysis.

The chapter begins with a presentation of the basic equations that describe the operation of all semiconductor devices. Material parameters and their effect upon the operation of the devices are discussed. This is followed by sections discussing the two types of transistors.

7.2 BASIC SEMICONDUCTOR DEVICE EQUATIONS

The set of equations generally referred to as the *semiconductor equations* serves as the starting point for most device investigations. These equations consist of the current-density equations, the continuity equations, Poisson's equation, and Faraday's law. In general, these equations must be solved simultaneously with the appropriate boundary conditions in order to obtain an accurate representation of the device operation. For devices such as bipolar transistors where both electrons and holes are important to the operation of the device, two sets of equations are required, one for each type of charge carrier. These equations form the drift-diffusion approximation that is applicable to conditions where the mobile charge carriers can be considered to be in thermal equilibrium with the crystal lattice. These conditions are generally valid when the device dimensions are large compared to the wavelength of the operation frequency or when the RF period is long compared to the charge carrier relaxation time. When these conditions are not applicable, nonequilibrium (hot-electron) effects can be significant and the drift-diffusion approximation is in general not valid. Under these conditions additional terms must be added to the semiconductor equations.

The current-density equations for electrons and holes are given in Equations 7.1 and 7.2, respectively, where

$$\mathbf{J}_n = q\mu_n n\mathbf{E} + qD_n \nabla n \tag{7.1}$$

and

$$\mathbf{J}_p = q\mu_p p\mathbf{E} - qD_p \nabla_p \tag{7.2}$$

The p and n are charge densities, q is the electronic charge, and μ_p and D_p are material transport parameters. The first term in these equations shows that the mobile carriers drift in response to an applied electric field, whereas the second term shows that a diffusion current also flows due to a gradient in the mobile charge density. These equations, as written, apply to relatively low electric-field regions where linear charge transport is dominant. In this region the charge-carrier velocity is linearly dependent upon the magnitude of the electric field, as shown in the expression,

$$v = \mu E \tag{7.3}$$

The proportionality constant relating the velocity and electric field is called the mobility (units of cm^2/V-sec) and is a characteristic of the particular semiconductor. In the linear region the drift terms can be written in terms of the semiconductor conductivity where,

$$\sigma = \sigma_n + \sigma_p = q(\mu_n n + \mu_p p) \tag{7.4}$$

The total semiconductor current density is composed of both the electron and hole current densities and is given by the expression

$$\mathbf{J} = \mathbf{J}_n + \mathbf{J}_p \tag{7.5}$$

For nondegenerate semiconductors the diffusion coefficient (units of cm^2/sec) can be determined from the mobility with use of the Einstein relation

$$D = \mu \frac{kT}{q} \tag{7.6}$$

Equations 7.1, 7.2, 7.3, and 7.4 apply to ohmic operation where the carrier mobility and diffusion coefficient are electric-field independent. Many devices, however, are operated under high-field conditions. At high electric fields the carrier velocity saturates and the diffusion coefficient becomes field dependent. The current-density equations must be modified. For high-field operation the equations can be modified by substituting the carrier velocity into the drift term and replacing the diffusion constant with a field-dependent equivalent according to the expressions,

$$\mu E \rightarrow v \tag{7.7}$$

and

$$D \rightarrow D(E) \tag{7.8}$$

This substitution, of course, requires knowledge of the velocity–field and diffusion coefficient-field characteristics.

The continuity equations for electrons and holes are given in (7.9) and (7.10), respectively, as

$$\frac{\partial n}{\partial t} = G_n - U + \frac{1}{q} \nabla \cdot \mathbf{J}_n \tag{7.9}$$

and

$$\frac{\partial p}{\partial t} = G_p - U - \frac{1}{q} \nabla \cdot \mathbf{J}_p \tag{7.10}$$

These equations simply state that the time rate of change of charge within a volume is equal to the rate of flow of charge out of the volume. Charge within the volume may be generated by some mechanism, such as optical excitation or avalanche ionization, extinguished by recombination, or forced out of the volume by divergence. The recombination rate, represented by the U term, is given on a net basis and represents the decay of free charge in excess of thermal equilibrium within the semiconductor. It can be represented, in simple cases, as

$$U = \frac{\delta n}{\tau_n} \quad \text{or} \quad \frac{\delta p}{\tau_p} \tag{7.11}$$

where τ_n and τ_p are the carrier lifetimes and δn and δp are the mobile charge densities in excess of thermal equilibrium. They can be written in the form

$$\delta n \triangleq n - n_0 \quad \text{and} \quad \delta p \triangleq p - p_0 \tag{7.12}$$

where p_0 and n_0 are the thermal equilibrium charge densities. The G_n and G_p terms in the continuity equations represent charge generation mechanisms other than thermal generation. Possible mechanisms include avalanche multiplication, tunneling, or generation by radiation sources such as light and X-rays.

Poisson's equation and Faraday's law complete the basic set of equations required for device analyses. Poisson's equation is

$$\nabla \cdot \mathbf{E} = \frac{q}{\epsilon}(p - n + N_d^+ - N_a^-) \tag{7.13}$$

where N_d^+ and N_a^- represent the density of ionized donor and acceptor atoms, and Faraday's Law is

$$\nabla \times \mathbf{E} = -\frac{\partial \mathbf{B}}{\partial t} \tag{7.14}$$

Generally, the semiconductor equations must be solved simultaneously in a self-consistent manner to accurately describe device performance.

7.3 MATERIAL PARAMETERS

The properties of semiconductor crystals are very important to the operation of semiconductor devices. Normally, it is the particular properties of certain types of crystals that permit a device to be fabricated. Ultimate performance limitations of the device are quite often determined by the properties of the crystal.

All semiconductor devices are only possible because of the unique property that the conductivity of certain crystals can be controlled by the addition of very small amounts of certain impurities. The conductivity is determined by the amount of free charge in the crystal and the transport characteristics of the charge. The operation of the device is determined by the ability to move, generate, and remove this free charge in a controlled manner.

The material characteristics vary widely in various semiconductors and only certain materials are suitable for use in device fabrication. The most common materials in current use are Si and GaAs, although InP and various ternary and quaternary III–V compounds (e.g., GaInAs, GaInAsP, and AlGaAs) are being developed for device applications.

The requirement that the conductivity must be controllable with small impurity levels is probably the single most important material consideration. This implies that the intrinsic background concentration at the device operating temperature must be low, typically in the range of 10^{14}–10^{15} cm^{-3}. Since the intrinsic density in GaAs is less than that in Si, higher resistivity substrates are available in GaAs. High-resistivity substrates are desirable since they allow the active region of a device (e.g., the conducting channel in a GaAs MESFET) to be isolated. Also, for MMIC applications device isolation can be achieved without the use of current-blocking techniques, such as back-biased junctions. The availability of high-resistivity GaAs substrates is one reason that this material has found such widespread use for MESFET devices and for MMICs.

The ability to move charge is determined by the transport characteristics of the material. This information is most often presented in the charge-carrier-velocity/electric-field characteristic. For example, the velocity–field characteristics for some commonly used semiconductors are shown in Fig. 7.1.

For low values of electric field the carrier velocity is linearly related to the electric-field strength. The proportionality constant is the mobility, and this parameter is important in determining the low-field operation of a device. Parasitic resistances in a device are directly dependent upon the low-field mobility. The low-field mobility is also important in determining the RF noise characteristics of a semiconductor device. Generally, a high value of mobility is desired for optimum device performance. Since the mobility of electrons in GaAs ($\sim 6500\ \text{cm}^2/\text{V-sec}$) is about six times that of Si ($\sim 1200\ \text{cm}^2/\text{V-sec}$), GaAs is a more attractive material for high-frequency RF and high-speed digital applications. This mobility advantage along with the availability of high-resistivity substrates makes GaAs the preferred and most widely used semiconductor material for these applications.

As the magnitude of the electric field increases, the carrier velocity saturates. Saturated velocity values for GaAs that have been reported in the literature for equilibrium conditions range from about $2(10)^6$ cm/sec to $8(10)^6$ cm/sec. For maximum high-frequency performance it is desirable to operate the device in a transit-time mode at maximum velocity, which requires high fields. Of the semiconductors Si has a $v_s \sim 10^7$ cm/sec and would appear to be the material of choice. In fact, Si generally demonstrates higher operation frequencies than GaAs or other compound semiconductors only when certain devices are fabricated and compared. For example, fundamental-mode Si IMPATT oscillators have been operated as high as 341 GHz [1], whereas a GaAs device has produced RF power at about 248 GHz [2]. For three-terminal devices the compound semiconductors produce the highest frequency operation. These devices, however, have been operated to only about 110 GHz [3].

The velocity–field transport characteristics are functions of various physical and operation parameters. An increase in doping produces a reduction in the low-field mobility and a decrease in the carrier velocity due to an increase in impurity scattering. The transport characteristics vary with temperature and in the normal operating temperature range the mobility usually decreases with temperature. The saturated velocity also decreases with temperature, demonstrating a $v_s \sim 1/T$ relationship.

The behavior of the charge-carrier diffusion coefficient with electric field is very important in the operation of certain solid state devices. Basically, the diffusion process will be important to a device that operates by the injection of a charge pulse into a high-field drift region and the extraction

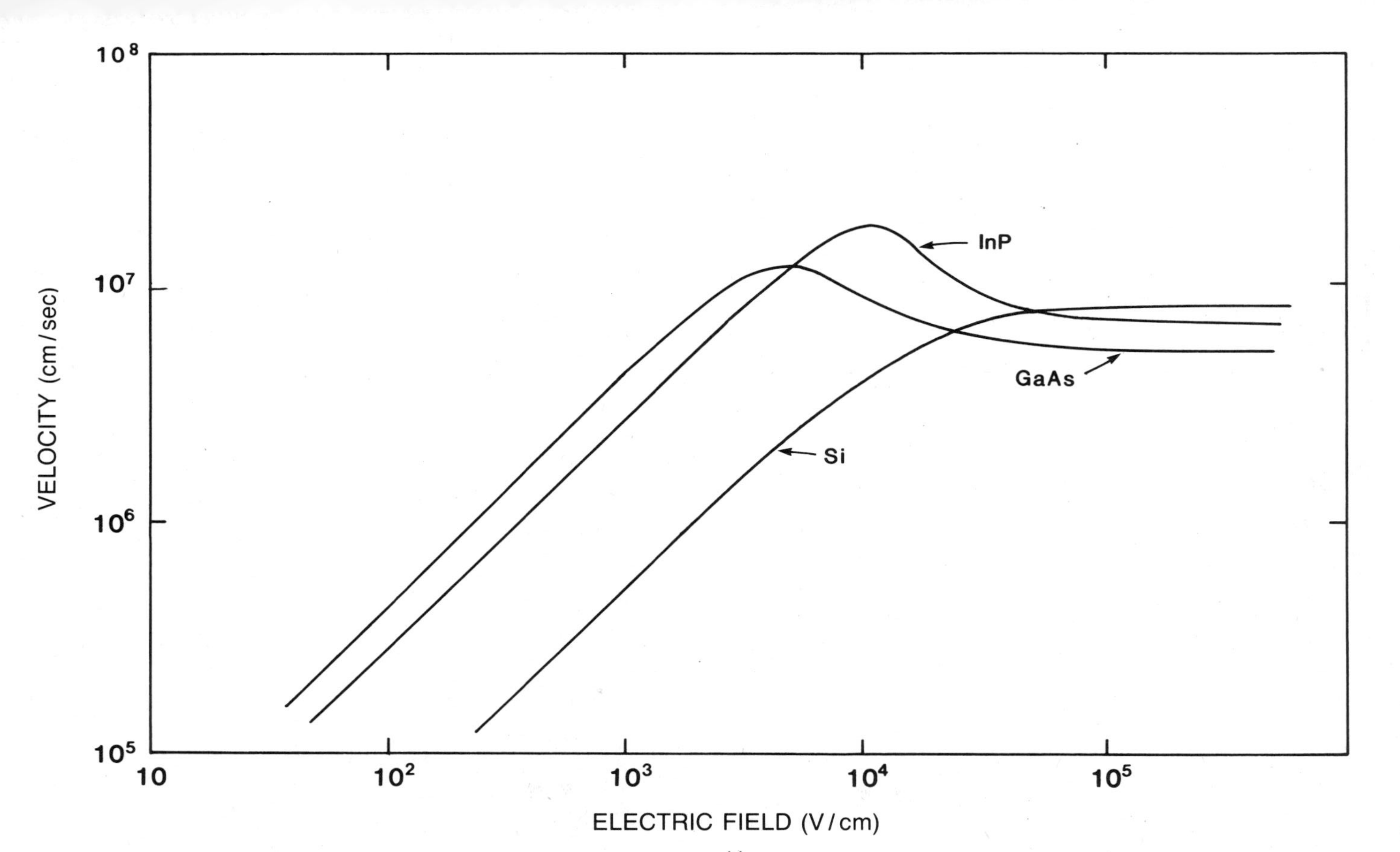

Figure 7.1 (*a*) Electron velocity–field characteristic for some common semiconductors at 200°C.

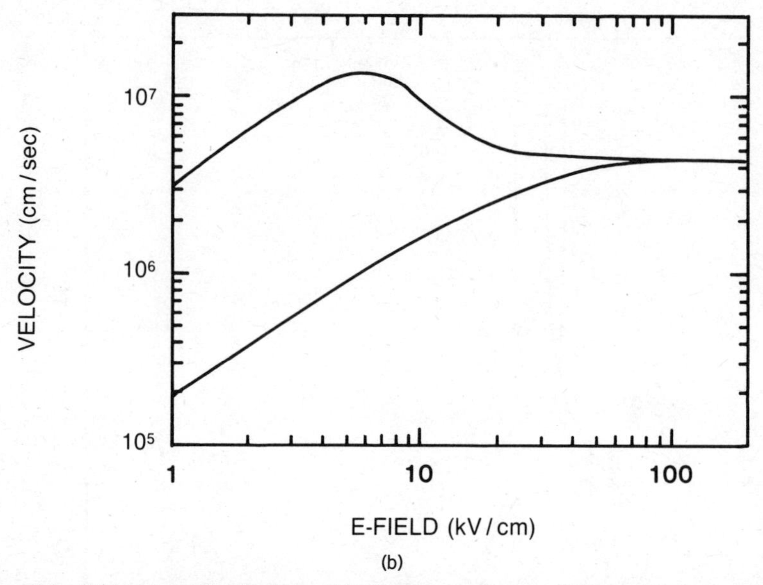

Figure 7.1 (*b*) Velocity-field relationship for *n*-type GaAs at 10^{17} cm^{-3}. The upper curve is for electrons and the lower curve is for holes.

of the pulse from the opposite end. A high magnitude for the diffusion coefficient will result in a spreading out of the pulse and a degradation of information transferring capability. The behavior of the diffusion coefficient with electric field for GaAs is shown in Fig. 7.2.

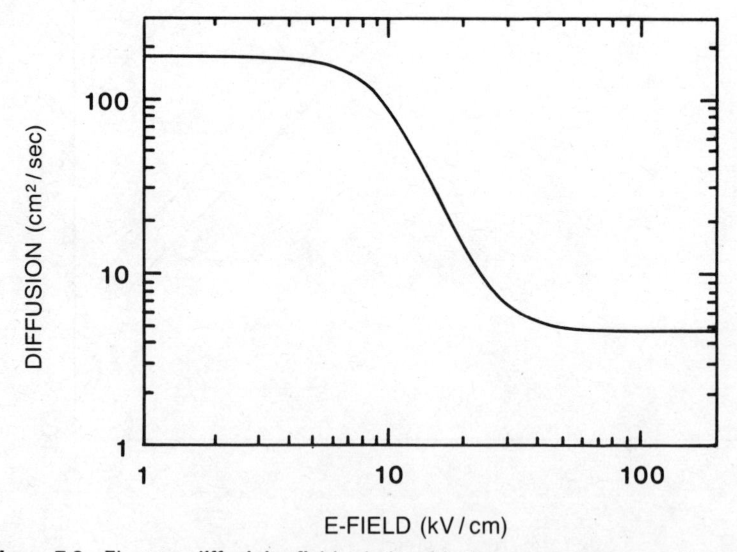

Figure 7.2 Electron diffusivity-field relationship for *n*-type GaAs at 10^{17} cm^{-3}.

Thermal conductivity is important to devices intended for operation at higher power levels or in low noise applications. Power dissipation requires that the device be adequately heat sinked. The thermal conductivity of compound semiconductors is generally much worse than for Si. For this reason Si transistors are often preferred for power applications. When compound semiconductors are used, careful heat sinking is required. Most heat sinks are fabricated from gold-plated copper. However, for high power applications the use of heat sink materials such as Type II diamond can improve the thermal conductivity over that obtainable from copper by about a factor of 6 at room temperature and a factor of 2 at 500 K.

7.4 BIPOLAR TRANSISTORS

The bipolar transistor was invented by Shockley, Bardeen, and Brittain in 1948. Since that time it has undergone continuous development and is currently one of the most widely used semiconductor devices. For microwave applications, it dominates for frequencies ranging from UHF (i.e., hundreds of megahertz) to about *S*-band (i.e., 2–4 GHz). As the technology improves, the upper frequency limit for these devices is continuously being extended, and state-of-the-art devices now are capable of producing useful power through 22 GHz [4]. Predictions of *Ka*-band operation have been made. The majority of bipolar transistors of current interest are fabricated from Si, although GaAs devices offer prospects for improvements in operation frequency, high temperature operation, and radiation hardness. The Si bipolar transistor is inexpensive, durable, integratable, and offers gains much higher than available with competing field-effect devices. It has moderate noise figure in RF applications and $1/f$ noise characteristics that are about 10–20 dB superior to GaAs MESFETs. For these reasons, it dominates in amplifier applications for the lower microwave bands and is often the device of choice for local oscillator applications.

The bipolar transistor is a *pn* junction device and is formed from back-to-back junctions as shown in Fig. 7.3. Since it is a three-terminal device it can be either *pnp* or *npn*. For high-frequency applications, the *npn* structure is preferred because the operation of the device is dependent upon the ability of minority carriers to diffuse across the base region. Since electrons generally have superior transport characteristics compared to holes, the *npn* structure is indicated.

Although there are many ways of fabricating a transistor, diffusion and ion implantation are generally used. For example, the structure in Fig. 7.3(*a*) would typically start with a lightly doped *n*-type epitaxial layer as the collector. The base region would be formed by counter doping the base region *p*-type by diffusion. The emitter would be formed by a shallow heavily doped *n*-type diffusion or ion implantation. The emitter and base

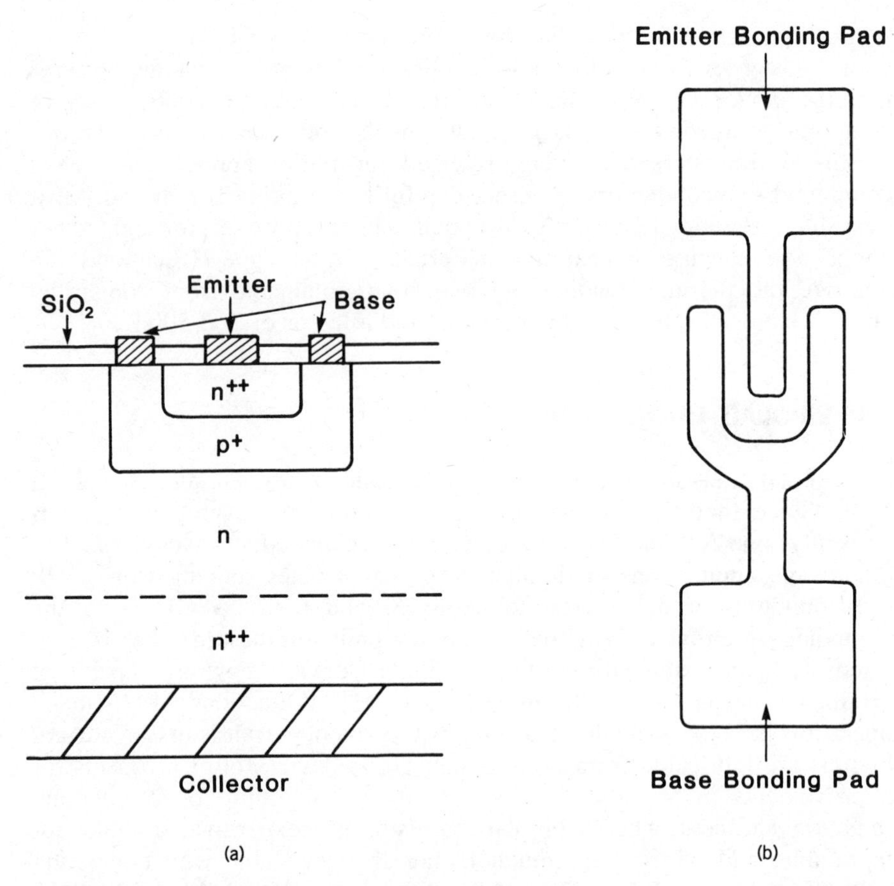

Figure 7.3 (*a*) Cross section for a Si *npn* bipolar transistor. (*b*) Top view for the Si *npn* bipolar transistor showing the interdigitated planar geometry.

contacts are generally located on the semiconductor surface in an interdigitated, planar arrangement as shown in Fig. 7.3(*b*). The interdigitated geometry always provides for $n + 1$ base fingers, where n is the number of emitter fingers. The number of fingers varies with the application, with more fingers required as the output power capability of the transistor increases. Additional fingers, however, increase the device parasitics and degrade the noise and upper frequency capability of the device. An excellent review of microwave bipolar transistors is presented in [5].

7.4.1 Basic Transistor Operation

The basic physics responsible for the operation of bipolar transistors can be revealed through analysis of the common-base device shown in Fig. 7.4. For this analysis a *pnp* transistor is considered, although the arguments are

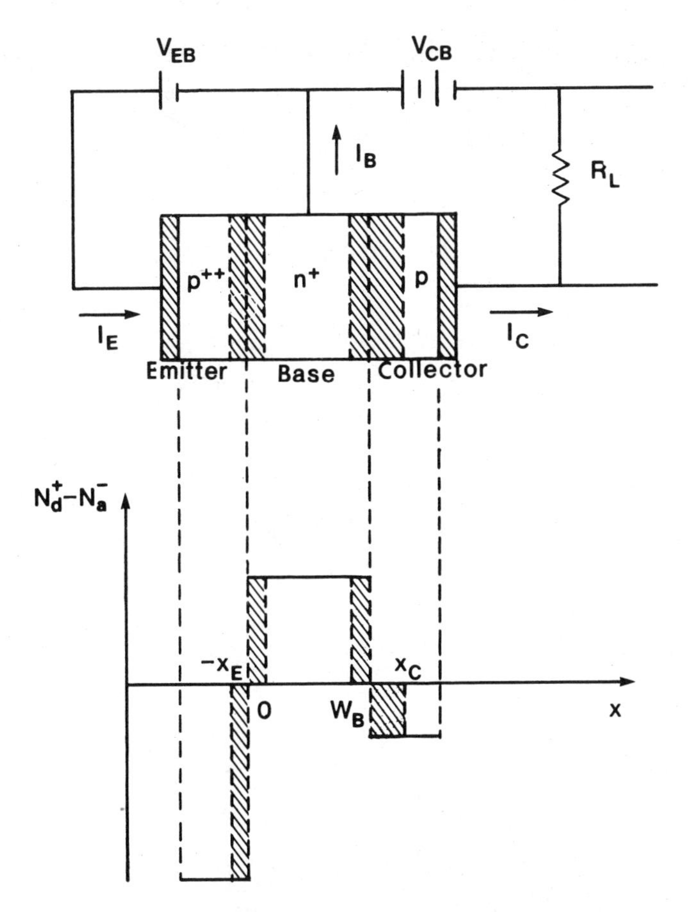

Figure 7.4 A *pnp* bipolar transistor connected in the common-base configuration. Diagram defines terminology used in the formulation given in the text.

analogous for an *npn* device. Under normal bias the emitter–base junction is forward biased and the collector–base junction is reverse biased. Holes are injected across the emitter–base junction, travel through the base region, and are extracted at the collector–base junction. The active characteristics of the device derive from the ability of a small potential applied across the base–emitter junction to control, in an efficient manner, a large emitter–collector current. This mechanism is the source for the name "transistor," which is a unification of the words "transfer resistor."

To derive the device I–V characteristics, the densities of free charges in each region of the device must be determined. From this information the device terminal currents can be calculated. The excess minority density in the base region ($p_B(x)$) can be determined from a one-dimensional solution of the continuity equation. Using the geometry shown in Fig. 7.4, and the injected charge densities for a biased *pn* junction, where

$$\delta p_B(0) = p_{BO}\left[\exp\left(\frac{qV_{EB}}{kT}\right) - 1\right] \tag{7.15}$$

$$\delta p_B(W_B) = p_{BO}\left[\exp\left(\frac{qV_{CB}}{kT}\right) - 1\right] \tag{7.16}$$

and p_{BO} is the thermal equilibrium density of minority charge carriers in the base region, the excess minority density at the edges of the base region can be written for normal biasing (i.e., a forward-biased emitter–base junction and a reverse-biased collector–base junction) as,

$$\delta p_B(0) = p_{BO}\left[\exp\left(\frac{qV_{EB}}{kT}\right) - 1\right] \cong p_{BO}\exp\left(\frac{qV_{EB}}{kT}\right) \tag{7.17}$$

and

$$\delta p_B(W_B) \cong -p_{BO}$$

When these boundary conditions are applied to the solution the result shown in Fig. 7.5 is obtained. The minority charge is injected into the base from the emitter, and moves across the base region to be extracted at the collector. The operation of the transistor is therefore dependent upon the ability of the minority charge to travel the length of the base region. Effects that decrease the minority charge in the base will result in a degradation in transistor operation. For example, if some of the charge recombines before it reaches the collector, the gain capability of the transistor is degraded. For

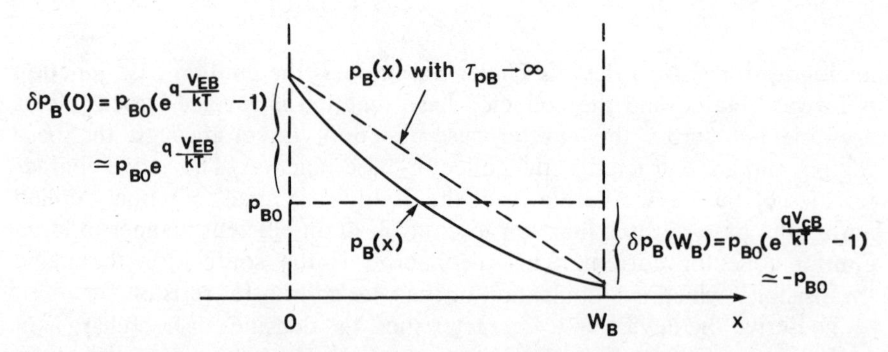

Figure 7.5 Minority charge distribution in the base region of a bipolar transistor operated with normal biasing. The boundary conditions at the emitter and collector sides of the base region are indicated. The dotted line shows the situation for no minority-charge recombination.

no minority-charge recombination in the base, the minority lifetime is infinite and the excess charge has a linear distribution. The minority-charge recombination in the base constitutes the base current, and high-performance transistors require a low base current.

The identical procedure can be applied to determine the minority-charge densities in the emitter and collector regions. The solutions applicable to these regions are

$$\delta n_E(x) = n_E(x) - n_{E0} = n_{E0}\left[\exp\left(\frac{qV_{EB}}{kT}\right) - 1\right]\exp\left(\frac{x + x_E}{L_E}\right) \qquad (7.18)$$

and

$$\delta n_C(x) = n_C(x) - n_{C0} = n_{C0}\left[\exp\left(\frac{qV_{CB}}{kT}\right) - 1\right]\exp\left(\frac{-(x - x_C)}{L_C}\right) \qquad (7.19)$$

where $L_E = \sqrt{D_{nE}\tau_E}$ and $L_C = \sqrt{D_{nC}\tau_C}$ are the minority carrier diffusion lengths in the emitter and collector regions. The majority densities in each of the transistor regions are essentially equal to the background impurity densities.

Expressions for the emitter and collector currents as a function of the applied potentials can now be derived. Following basic *pn* junction theory it is customary to assume that the electron and hole currents are constant across the depletion regions. This assumption allows the depletion regions to be considered as a single node. The minority carrier currents can be summed on each side of the depletion region to get the total current. If, in addition, the electric field in the bulk semiconductor regions is assumed to be very small, the total current can be determined from only diffusion currents. The emitter current density is

$$J_E = J_{pB}(0) + J_{nE}(-x_E)$$

or

$$J_E = -qD_{pB}\left.\frac{\partial p_B(x)}{\partial x}\right|_0 + qD_{nE}\left.\frac{\partial n_E(x)}{\partial x}\right|_{-x_E} \qquad (7.20)$$

which, when applied to a one-sided, abrupt p^+n junction yields

$$J_E \cong \frac{qD_{pB}p_{B0}}{L_B}\coth\left(\frac{W_B}{L_B}\right)\exp\left(\frac{qV_{EB}}{kT}\right) \qquad (7.21)$$

In a similar manner the collector current density is

$$J_C \cong \frac{qD_{pB}p_{B0}}{L_B}\operatorname{csch}\left(\frac{W_B}{L_B}\right)\exp\left(\frac{qV_{EB}}{kT}\right) \qquad (7.22)$$

The base current can be determined as the difference between the emitter and collector currents as

$$I_B = I_E - I_C \tag{7.23}$$

where

$$I_E \triangleq J_E A_{EB} \qquad \text{and} \qquad I_C \triangleq J_C A_{CB}$$

where the term A is the area through which the current flows. Note that, in general, $A_{EB} \neq A_{CB}$.

If it is assumed, for simplicity, that the base–emitter and base–collector junctions have the same area A, the base current can be written as

$$I_B \cong qA \frac{W_B}{2\tau_{pB}} p_{B0} \exp\left(\frac{qV_{EB}}{kT}\right) \tag{7.24}$$

Note that the base current is directly proportional to W_B and inversely proportional to τ_{pB}. For no recombination in the base region $\tau_{pB} \to \infty$ and $I_B \to 0$. Also, for wide base regions, the minority charge has more opportunity to recombine and the base current increases.

7.4.2 Current Gain

The static common-base current gain is called alpha and is defined as

$$\alpha_0 \triangleq \frac{\Delta I_C}{\Delta I_E} \tag{7.25}$$

In the common-emitter configuration, the static current gain is called beta and is

$$\beta_0 \triangleq \frac{\Delta I_C}{\Delta I_B} \tag{7.26}$$

Since the base current is defined as the difference between the emitter and collector currents according to (7.23) it follows that the alpha and beta are related according to the expression

$$\beta_0 = \frac{\alpha_0}{1 - \alpha_0} \tag{7.27}$$

The alpha current gain can be defined as the product of two additional parameters as

$$\alpha_0 \triangleq B\gamma \tag{7.28}$$

where B is a base transport factor and γ is the emitter injection efficiency.

The emitter injection efficiency is defined for a *pnp* transistor as

$$\gamma \triangleq \frac{\text{the hole current injected from the emitter into the base}}{\text{the total emitter current (hole and electron)}} \tag{7.29}$$

This can be written as

$$\gamma = \frac{1}{1 + J_{nE}(-x_E)/J_{pB}(0)} \tag{7.30}$$

and if the expressions previously derived are substituted, the emitter injection efficiency can be written as

$$\gamma \cong \frac{1}{1 + \dfrac{D_{nE} L_B N_B}{D_{pB} L_E N_E} \tanh\left(\dfrac{W_B}{L_B}\right)} \tag{7.31}$$

where N_B and N_E are the base and emitter region dopings, respectively. For high injection efficiency the second term in the denominator must be minimized. This requires that the emitter doping be much greater than that in the base and also that the base region length be much less than the minority carrier diffusion length in the base.

The base transport factor is defined as

$$B \triangleq \frac{\text{the hole current reaching the collector}}{\text{the total hole current injected into the base}} \tag{7.32}$$

Substituting these expressions into the current gain expression, it follows that

$$B = \frac{1}{\cosh\left(\dfrac{W_B}{L_B}\right)} \tag{7.33}$$

The alpha current gain can then be written in the form

$$\alpha_0 = B\gamma = \frac{1}{\cosh\left(\dfrac{W_B}{L_B}\right) + \dfrac{D_{nE} L_B N_B}{D_{pB} L_E N_E} \sinh\left(\dfrac{W_B}{L_B}\right)} \tag{7.34}$$

The current gain expressed in (7.34) is the static or dc current gain. Under RF conditions the current gain will decrease as shown in Fig. 7.6. From this figure the commonly used cutoff frequencies can be defined and

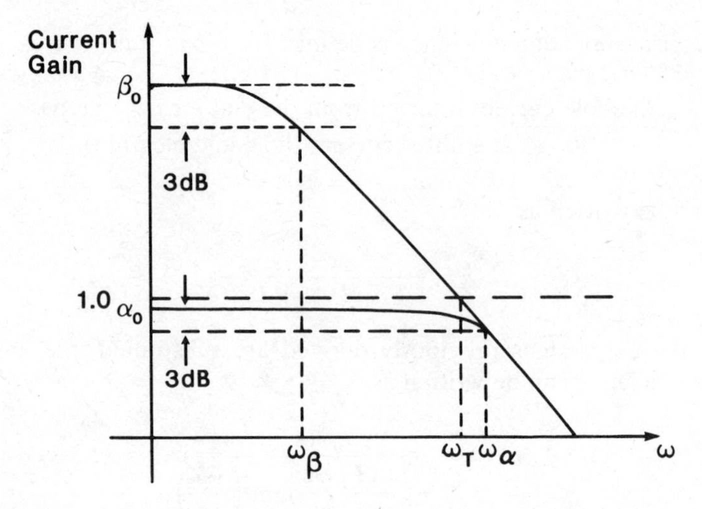

Figure 7.6 Current gains versus frequency, indicating the definitions for the various cutoff frequencies.

are as follows:

1. ω_α—the alpha cutoff frequency is the frequency at which the common-base current gain is reduced to $0.707\alpha_0$ (3-dB down).
2. ω_β—the beta cutoff frequency is the frequency at which the common-emitter current gain is reduced to $0.707\beta_0$.
3. ω_T—the gain-bandwidth product is the frequency at which the common-emitter current gain is reduced to unity.

The frequency dependence of the current gain can be determined by repeating in the time domain the procedure used to determine the dc current gain. For microwave transistors the current gain becomes

$$\alpha(\omega) \cong \operatorname{sech}\left(\frac{W_B}{L'_B}\right) \tag{7.35}$$

where L'_B is a frequency-dependent diffusion length. A simple one-pole model for α results as described by the expression

$$\alpha = \frac{\alpha_0}{1 + j(\omega/\omega_\alpha)}. \tag{7.36}$$

7.4.3 Limitations and Second-order Effects

There are several effects that limit the operation of bipolar transistors to restricted ranges of applied bias. These effects are examined in this section.

Collector Bias Variation (Early Effect). A changing collector-base bias causes a variation in the space-charge layer width at the base–collector junction and, consequently, in the width of the bulk base region. Base width modulation by the collector-base voltage is called the Early effect [6]. An increase in V_{CB} reduces the base width, thereby increasing the gradient of the minority charge. A decrease in total minority charge in the base and an increase in collector current result. The base current is also decreased, since base current is proportional to base-charge storage.

Collector High-Injection Effects (Kirk Effect). Free carriers entering the base–collector depletion region modify the background charge in that region and thereby affect the electric field. For a constant collector-base voltage the width of the depletion region changes to accommodate changes in the electric field. Under high injection the width of the depletion region tends to decrease, thereby increasing the effective neutral base region. This phenomenon is called the Kirk effect [7]. As the current increases, the neutral base region increases and the current gain is reduced.

The Kirk effect can have a significant effect upon device performance, since the neutral base region can increase by a large factor under high-injection conditions. The base transit-time can be significantly increased under high current injection due to the added base length. Under these conditions, the high-frequency performance of the transistor will be degraded.

Base-Region High-Injection Effects. Under high current-injection conditions, it is possible for the injected minority density in the base region to be on the same order of magnitude as the doping density. The large density of injected minority carriers causes an increase in majority carriers in order to satisfy charge neutrality conditions. The increase in majority carriers causes an effective increase in the base region charge and a reduction in the alpha current gain. The reduction in α with I_C is known as the Webster effect [8].

7.4.4 Microwave Transistor

The state of the art in microwave bipolar transistors has developed rapidly, and these devices are now capable of operation approaching *Ka*-band. Much of the improvement in device performance is due to improvements in fine-line lithography where submicron emitter–base spacing and finger widths are now possible. Improvements in material growth and doping techniques have also contributed to the excellent performance now available from these devices. The Si bipolar transistor is the most widely used device in analog circuit design at microwave frequencies ranging from UHF to approximately *S*-band. In the *S*- through *C*-bands, the Si bipolar transistor and the GaAs MESFET are direct competitors.

The frequency performance of the bipolar transistor can be examined by consideration of the delay times encountered by a signal traveling through the device. For this analysis, it is convenient to consider the common-base configuration and normal biasing.

Four principal delay regions can be identified. The total emitter-to-collector delay time can be written as

$$\tau_{ec} = \tau_e + \tau_b + \tau_c + \tau_c' \tag{7.37}$$

where

τ_{ec} is the total emitter-to-collector delay time,
τ_e is the emitter–base junction capacitance charging time,
τ_b is the base region transit time,
τ_c is the collector-region depletion-layer transit time, and
τ_c' is the collector capacitance charging time.

Since a forward-biased emitter–base junction is assumed, the transit time associated with this depletion region can be ignored.

These delay times can be expressed as characteristic frequencies with the relationship

$$\omega = \frac{1}{\tau}$$

Therefore,

$$\frac{1}{\omega_T} = \frac{1}{\omega_e} + \frac{1}{\omega_b} + \frac{1}{\omega_c} + \frac{1}{\omega_c'} \tag{7.38}$$

where

$$\omega_T = 2\pi f_T$$

and f_T is the gain-bandwidth product.

Expressions for each of these delay terms can be determined in the following manner.

Base–Emitter Junction. Since the base–emitter junction is forward biased, an equivalent circuit is as shown in Fig. 7.7. The emitter current i_e entering the junction is divided into two paths. Due to the capacitance, a time delay results. The current that flows through r_e represents the charge carriers that are injected into the base region. This current is

$$i_e' = i_e \frac{1}{1 + j\omega r_e C_e} \tag{7.39}$$

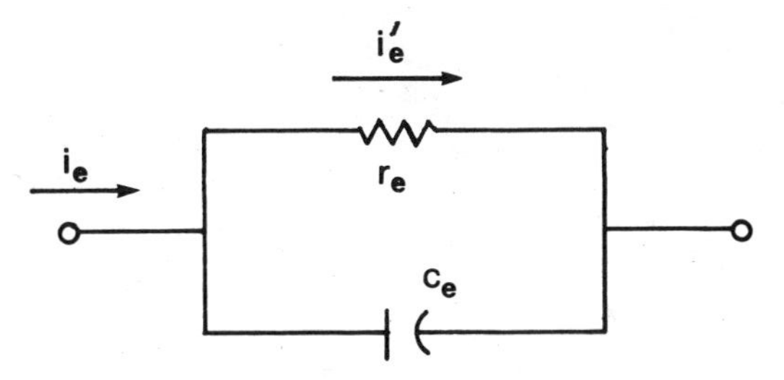

Figure 7.7 Equivalent circuit for the forward-biased base–emitter junction of a bipolar transistor.

By normalizing to the terminal current, a high-frequency emitter efficiency can be defined as

$$\frac{|\gamma_{hf}|}{\gamma_0} \triangleq \frac{\sqrt{2}}{2} \tag{7.40}$$

and it follows that

$$\tau_e = \frac{1}{\omega_e} \cong r_e C_e \tag{7.41}$$

For microwave devices this expression must be modified due to the increased capacitance from the base–collector junction and other external capacitance connected to the base.

The expression is written as

$$\frac{1}{\omega_e} = r_e(C_e + C_c + C_p) \tag{7.42}$$

where C_c is the base–collector junction capacitance and C_p is any other capacitance connected to the base.

An expression can be derived for r_e from the relationship

$$r_e = \frac{dv}{di} = \frac{1}{di/dv} \tag{7.43}$$

where $i = I_E$, the current through the base–emitter junction. It follows that

$$r_e \cong \frac{kT}{qI_E} \tag{7.44}$$

and

$$\tau_e = \frac{1}{\omega_e} \cong \frac{kT}{qI_E}(C_e + C_c + C_p) \tag{7.45}$$

Base Region. The transit time across this region is

$$\tau_b = \frac{W_B^2}{\eta D_{pB}} \tag{7.46}$$

where η is a factor that includes the effects of a graded doping profile in the base region. For uniform doping $\eta = 2$.

Base–Collector Junction. Since the base–collector junction is reverse biased, it has a significant width. The transit time of carriers across the depletion region must be considered. The transit time across the depletion region is

$$\tau_c \triangleq \frac{\text{depletion width}}{\text{carrier velocity}} = \frac{X_C - W_B}{2v_s} \tag{7.47}$$

For a uniformly doped transistor and using the one-sided, abrupt junction approximation it follows that

$$X_C - W_B \cong \left[\frac{2\epsilon(V_{CB} + \varphi_{bi})}{qN_C}\right]^{1/2} \tag{7.48}$$

where φ_{bi} is the base–collector junction built-in potential and N_c is the collector region doping. Therefore,

$$\tau_c \cong \frac{1}{2v_s}\left[\frac{2\epsilon(V_{CB} + \varphi_{bi})}{qN_c}\right]^{1/2} \tag{7.49}$$

Low doping and high V_{CB} cause increases in τ_c.

The base–collector capacitance charging time is calculated in a manner analogous to that for the emitter–base junction. The charging time is

$$\tau'_c = \frac{1}{\omega'_c} = r_c C_c \tag{7.50}$$

where r_c is the collector series resistance.

For typical microwave transistors the total emitter–collector delay time divides in the following manner:

$$\tau_e \sim 40\%\ \tau_{ec}, \qquad \tau_b \sim 10\%\ \tau_{ec}, \qquad \tau_c \sim 45\%\ \tau_{ec}, \qquad \text{and} \qquad \tau'_c \sim 5\%\ \tau_{ec}$$

The emitter–base charging time and collector depletion region transit time dominate (i.e., limit) the transistor frequency response.

The transistor gain-bandwidth product, f_T, is expressed as

$$f_T = \frac{1}{2\pi\tau_{ec}} \cong \left\{2\pi\left[\frac{kT(C_e + C_c + C_p)}{qI_C} + \frac{W_B^2}{\eta D_{pB}} + \frac{X_C - W_B}{2v_s}\right]\right\}^{-1} \tag{7.51}$$

assuming that $I_C = I_E$. For maximum frequency response the transistor should be operated at high collector current and low collector-base potential.

7.4.5 Equivalent Circuit

The Ebers–Moll model [9] is generally considered to be the basic model for the bipolar transistor. It is a device physics model and is shown in both the common-base and common-emitter configurations in Fig. 7.8. The basic model consists of back-to-back diodes in shunt with current sources.

A useful measure of high-frequency performance is the unilateral gain (Mason's gain [10]), which can be defined from the y-parameter expression

$$U \triangleq \frac{|y_{21} - y_{12}|^2}{4 \text{ Re } y_{11} \text{ Re } y_{22} - 4 \text{ Re } y_{12}y_{21}} \tag{7.52}$$

where

$$y_{lm} = g_{lm} + jb_{lm}$$

and g and b are the device terminal conductance and susceptance, respectively. The unilateral gain is of interest because it excludes any effect of package parasitic reactances. It provides a useful figure of merit for the intrinsic device.

By applying the U definition to the common-base equivalent circuit the transistor gain can be determined to be

$$U \cong \frac{\alpha_0}{16\pi^2 r_b C_c f^2 \left(\tau_{ec} + \dfrac{r_e C_c}{\alpha_0}\right)} \tag{7.53}$$

The unilateral gain is therefore proportional to $1/f^2$, $1/r_b$, and $1/\tau_{ec}$. The $1/f^2$ dependence results in the 6-dB/octave gain rolloff, as shown in Fig. 7.9. The maximum frequency of oscillation, $f_{\max}$, is defined as the frequency at which U goes to unity. From the U expression it follows that

$$f_{\max} = \frac{1}{4\pi}\left[\frac{\alpha_0}{r_b C_c \left(\tau_{ec} + \dfrac{r_e C_c}{\alpha_0}\right)}\right]^{1/2} \tag{7.54}$$

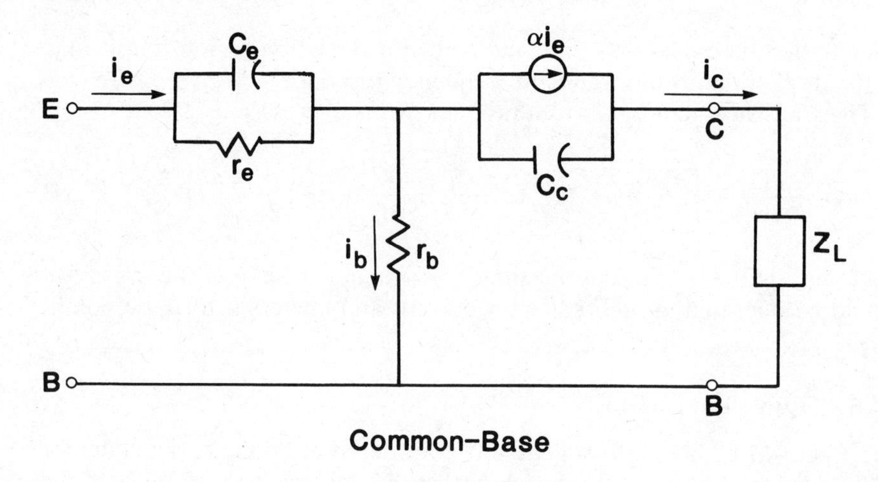

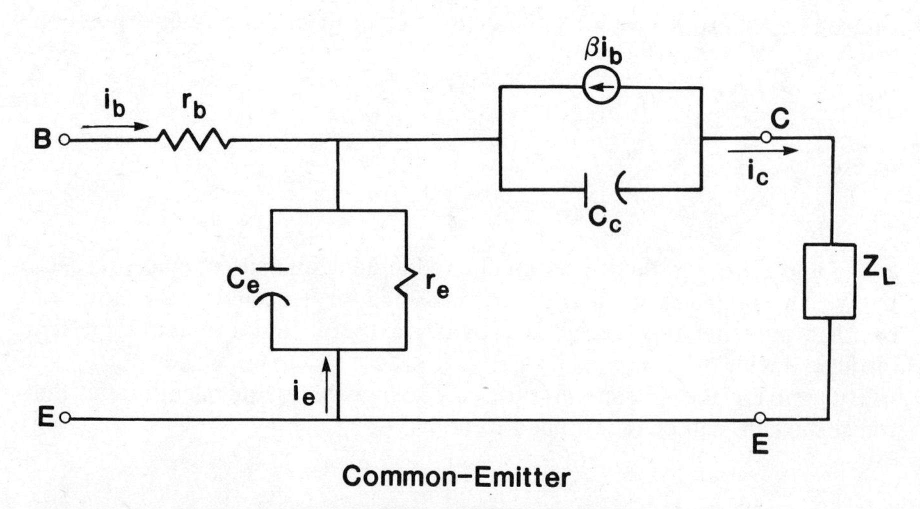

Figure 7.8 Ebers–Moll high-frequency equivalent circuits for a bipolar transistor connected in the common-base and common-emitter configurations.

Maximum oscillation frequency f_{max} can be expressed as a function of f_T as

$$f_{max} = \left[\frac{f_T}{8\pi r_b C_c} \right]^{1/2} \tag{7.55}$$

In general, $f_{max} \gg f_T$ for Si transistors. That is, the transistor is capable of producing power gain frequencies above that at which the current gain is

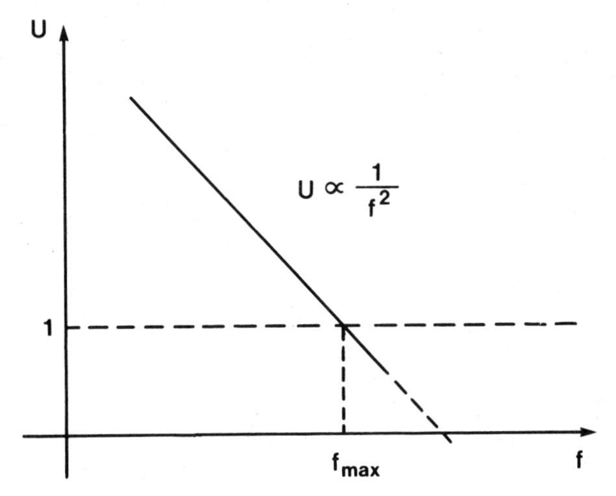

Figure 7.9 Unilateral gain vs. frequency for the single-pole equivalent circuit for a transistor.

unity. In GaAs transistors, it is possible for f_T to be greater than $f_{\max}$ due to the effect the high mobility of GaAs has upon the relative sizes of the terms $(r_b C_c \tau_{ec})$ and τ_{ec} in a particular transistor structure.

Maximum oscillation frequency $f_{\max}$ is a useful figure of merit, since it

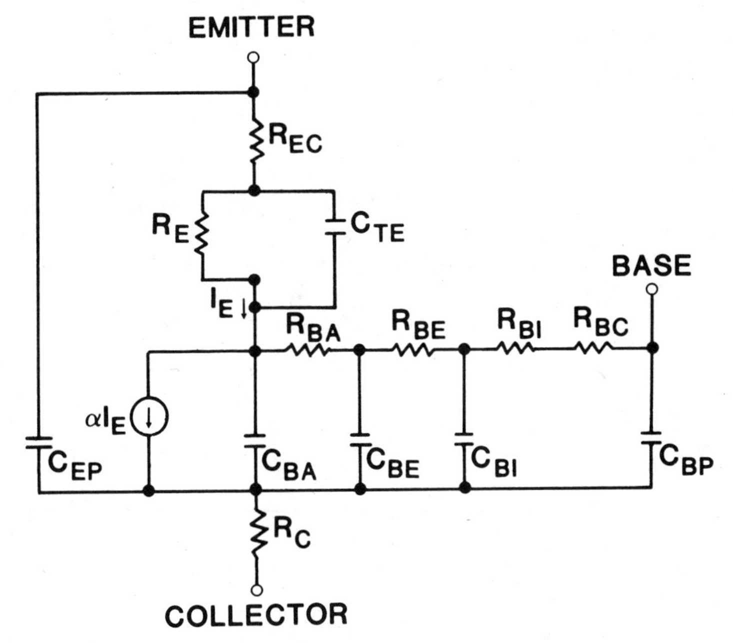

Figure 7.10 High-frequency equivalent circuit for a bipolar transistor including distributed base parameters.

TABLE 7.1 HP 505 Equivalent-Circuit Parameters

Parameter	Symbol	Value	How Obtained
Active base resistance	R_{BA}	12.5 Ω	Calculated
External base resistance	R_{BE}	2.7 Ω	Calculated
Base insert resistance	R_{BI}	0.2 Ω	Calculated
Base contact resistance	R_{BC}	<0.2 Ω	Calculated
Emitter contact resistance	R_{EC}	<1 Ω	Measured
Collector resistance	R_C	5 Ω	Measured
Emitter–base dynamic resistance	r_e	13 Ω	kT/qI_E
Active base C-B junction capacitance	C_{BA}	0.0040 pF	Calculated
External base C-B junction capacitance	C_{BE}	0.0032 pF	Calculated
Base insert C-B junction capacitance	C_{BI}	0.050 pF	Calculated
Base bonding pad capacitance	C_{BP}	0.05 pF	Calculated
Emitter bonding pad capacitance	C_{EP}	0.05 pF	Calculated
Emitter–base junction capacitance	C_{TE}	0.7 pF	τ_e/r_e
Emitter–base junction charging time	τ_e	9 psec	$\tau_T=(\tau_b+\tau_e+\tau_d)$
Neutral base delay time	τ_b	7 psec	From τ_T vs. I/I_e
Charging time of C_{BC} through R_C	τ_c	0.5 psec	R_cC_{BC}
Collector depletion region delay time	τ_d	10 psec	$W_d/2v_s$ ($v_s=0.5\times10^6$ cm/sec)
Low-frequency common-base gain	α_0	0.990	$n_{FE}/(1+n_{FE})$
Total base resistance	R_B	15.6	$R_{BA}+R_{BE}+R_{BI}+R_{EC}$
Total base-collector capacitance	C_{BC}	0.107 pF	$C_{BA}+C_{BE}+C_{BI}+C_{BP}$
Total delay time	τ_T	26.5 psec	$(\tau_e+\tau_b+\tau_c+\tau_d)=1/2\pi f_T$
Emitter-base junction cutoff frequency	f_e	17.7 GHz	$1/2\pi\tau_e$
Base cutoff frequency	f_b	22.7 Ghz	$1/2\pi\tau_b$
Common-emitter Gain-bandwidth product	f_T	6 GHz	Measured

gives an upper limit to the frequency at which a given device should be capable of useful gain. In practice, parasitics and second-order effects will limit operation to frequencies below f_{max}.

For the analysis and design of practical microwave transistors the T-equivalent circuit shown in Fig. 7.10 is useful [11]. This circuit is essentially the Ebers–Moll circuit with a distributed RC network in the base to account for the capacitances and lateral base resistance elements associated with the planar transistor geometry. This circuit, when applied to the HP 505 microwave bipolar transistor, yields the element values shown in Table 7.1.

7.4.6 Noise Figure Analysis

Noise figure is a measure of the amount of noise added to a signal by a lossy device through which the signal passes. It is defined as the ratio of the input to output signal-to-noise ratios and is generally expressed in decibels according to the expression

$$F = 10 \log \frac{(S/N)_{\text{in}}}{(S/N)_{\text{out}}} \tag{7.56}$$

Note that a noise-free device would have equal input and output signal-to-noise ratios and the noise figure is therefore zero. Another parameter sometimes used to describe the noise of an amplifier is noise measure. The noise measure is defined in terms of noise figure and the gain of the device as

$$M = \frac{F' - 1}{1 - (1/G')} \tag{7.57}$$

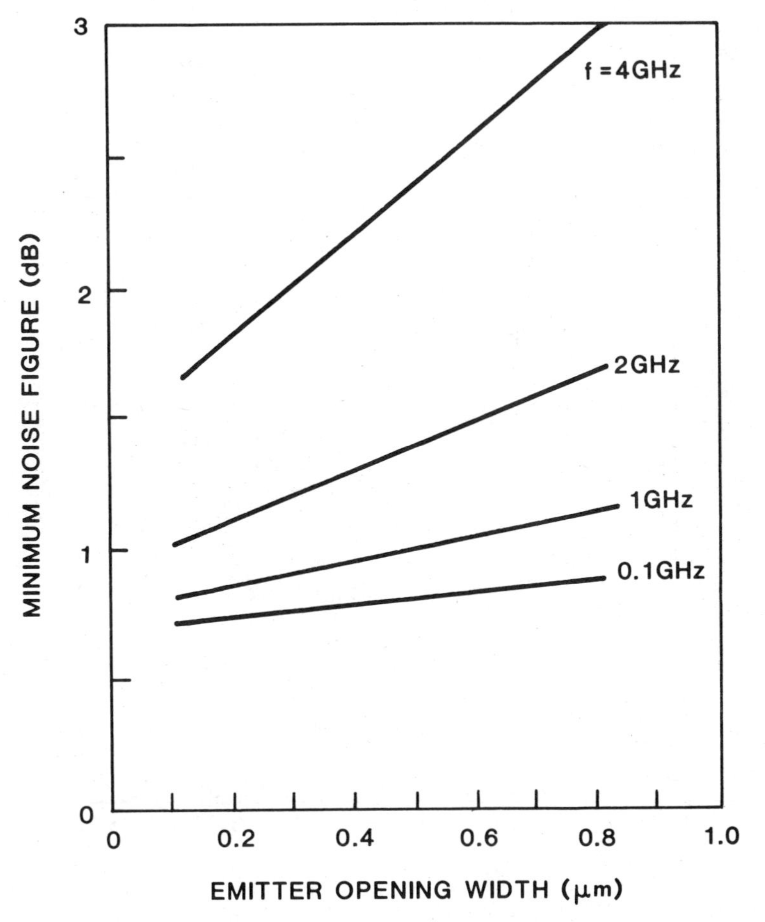

Figure 7.11 Minimum noise figure at 0.1, 1, 2, and 4 GHz vs. emitter opening width for a HP 505 transistor [14].

where the primes indicate non-decibel magnitudes. Noise measure is also generally specified in decibels, and is of interest because it combines the effects of noise figure and gain into a single parameter. In low noise design, especially for front-end stages of receivers, it is often desirable to design for minimum noise measure, rather than simply minimum noise figure.

In terms of a two-port device, the noise figure can be calculated by considering a noise-free two-port to which noise generators are added. The noise added by the two-port can be represented as shunt current generators at the input and output ports, or as a series noise voltage generator and shunt noise current generator at the two-port input [12]. In circuit design the amplifier noise figure calculations are generally performed with all noise sources referenced to the input.

The minimum noise figure for the bipolar transistor can be expressed in

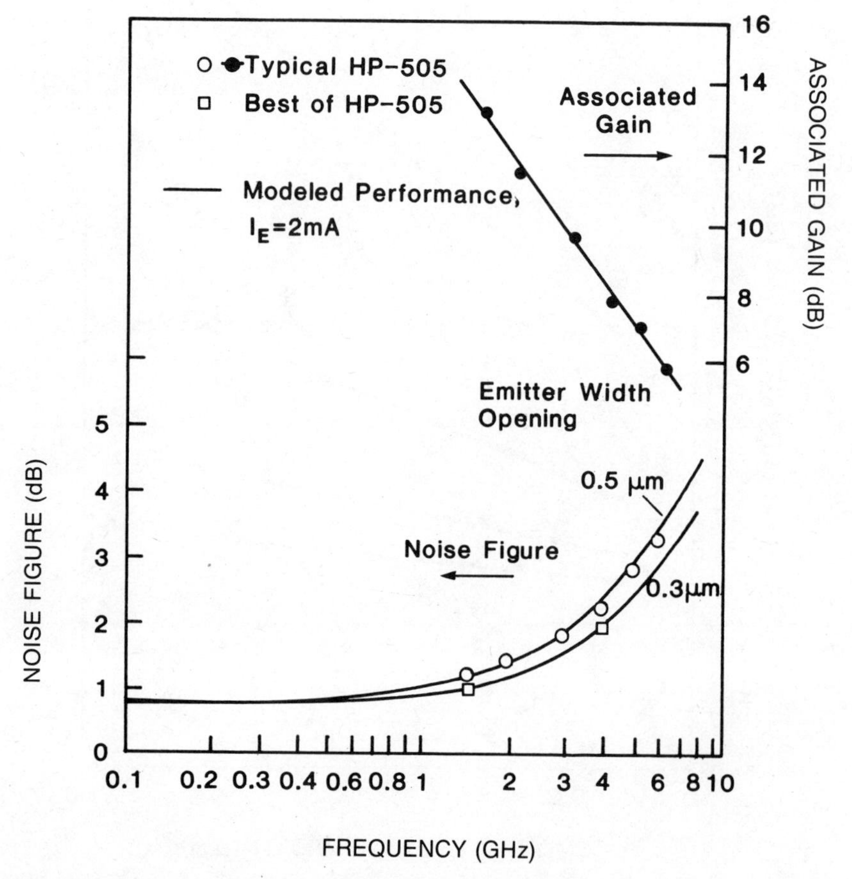

Figure 7.12 Measured and calculated noise figure for the HP 505 as a function of frequency (T. H. Hsu and C. P. Snapp [14], 1978 IEEE, reprinted with permission.)

the form [13]

$$F_{\min} = a\frac{r_{b'} + R_{\text{opt}}}{r_e} + \left(1 + \frac{f^2}{f_b^2}\right)\frac{1}{\alpha_0} \tag{7.58}$$

where the optimum source resistance is [14]

$$R_{\text{opt}} = \left\{r_b^2 - X_{\text{opt}}^2 + \left(1 + \frac{f^2}{f_b^2}\right)\frac{r_e(2r_b + r_e)}{\alpha_0 a}\right\}^{1/2} \tag{7.59}$$

and the optimum source reactance is

$$X_{\text{opt}} = \left(1 + \frac{f^2}{f_b^2}\right)\frac{2\pi f C_{TE} r_e^2}{\alpha_0 a} \tag{7.60}$$

where

$$a = \left\{\left(1 + \frac{f^2}{f_b^2}\right)\left(1 + \frac{f^2}{f_e^2}\right) - \alpha_0\right\}\frac{1}{\alpha_0} \tag{7.61}$$

Here f_b and f_e are the base and emitter-base junction cutoff frequencies, respectively. These expressions were derived neglecting the parasitic bonding pad capacitances and the collector–base junction capacitance. When the parameters listed in Table 7.1 for the HP 505 transistor are used in the noise-figure equations, the noise-figure performance shown in Fig. 7.11 is obtained. The measured and calculated gain and noise-figure performance for this transistor are compared in Fig. 7.12.

7.5 FIELD-EFFECT TRANSISTORS

The basic idea for the field-effect transistor (FET) goes back at least as far as Lilienfeld's patents in 1930 and 1933. The concept for the field-effect transistor predates the invention of the bipolar transistor that was presented in 1948. Development work on FETs was hindered in the early years by poor semiconductor quality, large feature size available with current lithography, and difficulty in obtaining suitable contacts. Most development work was temporarily interrupted by the invention of the bipolar transistor, as this device came under intense investigation and development.

The modern field-effect transistor derives from the work of Stuetzer [15] in 1950 and Shockley [16] in 1952. Most of the early field-effect transistor work was directed toward producing a high-frequency, high-power solid state analog to the vacuum triode. The basic structure was (and still is) similar to a triode in that two conducting electrodes are separated by some distance with a control electrode (grid) in between, as shown in Fig. 7.13.

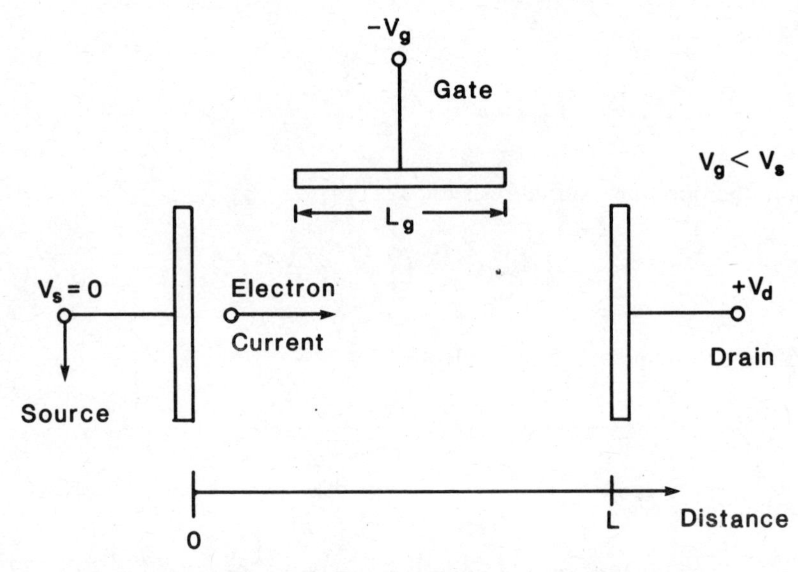

Figure 7.13 Basic electrode placement for a field-effect transistor.

The control grid (called a gate) is biased negative with respect to the source so that a retarding potential barrier is presented to electrons attempting to flow from the source to the drain. The field-effect transistor depends upon only one type of charge-carrier flow and is therefore called "unipolar."

7.5.1 Basic Operation Principles

The GaAs metal semiconductor field-effect transistor (MESFET) is one of the most commonly used and important active devices for use in microwave analog and high-speed digital integrated circuits. The transistor consists of a highly conducting layer of high-quality semiconductor grown epitaxially upon a semi-insulating (i.e., high-resistivity) substrate as shown in Fig. 7.14. The conducting channel is interfaced with external circuitry through two ohmic contacts (called the source and drain), separated by a distance and placed upon the semiconductor. The gate electrode is constructed by placing a rectifying (Schottky) contact between the two ohmic contacts. The conducting channel is very thin, typically on the order of 0.1–0.3 μm, so that the depletion region that forms under the Schottky contact (gate) can efficiently control the flow of current in the thin layer. The device therefore behaves as a voltage-controlled switch, capable of very high modulation rates.

The operation of the device can be understood by first considering a device without the gate, as shown in Fig. 7.15. As the drain–source potential is increased a current will flow, as indicated by region *A*. The

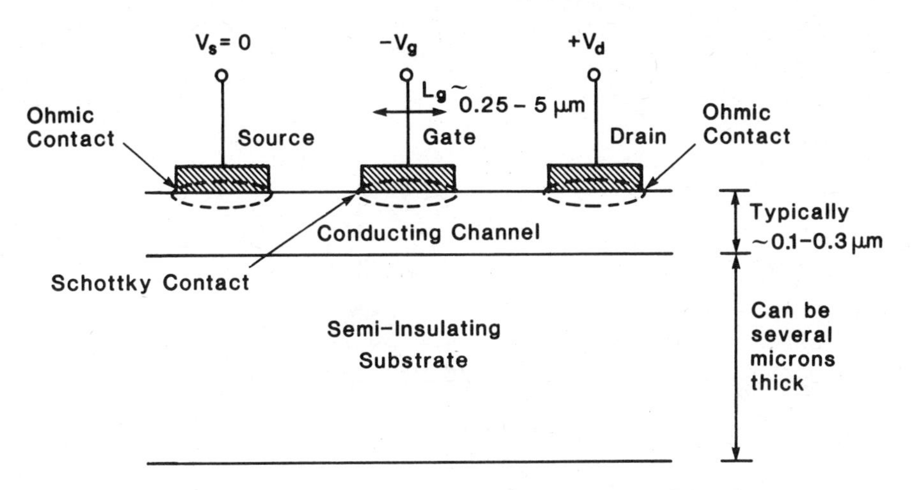

Figure 7.14 Typical cross-sectional structure of a microwave field-effect transistor.

current–voltage behavior is linear and directly follows the velocity–field characteristic for the semiconductor from which the device is fabricated. As the drain–source voltage is increased to the magnitude at which the internal electric field achieves the saturation condition, deviations from linear performance occur (region B). The device is simply a resistor and the slope of the I–V characteristic defines the sum of the channel and contact resistances. Note in Fig. 7.15 that the parameter I_{dss} is defined for such a structure without a gate. The I_{dss} parameter is important in MESFET analyses and appears in the discussion that follows.

If a gate electrode is located between the source and drain contacts as shown in Fig. 7.16 a depletion region is formed under the gate metal. The depletion region affects the device I–V characteristic by constricting the cross-sectional area through which current can flow. This results in an increase in channel resistance [i.e., $R_{\text{ch}}(V_g = 0) > R_{\text{ch}}(\text{no gate})$]. The gate depletion region also causes the electric field under the gate to increase with distance toward the drain because of the potential drop along the channel. This, in turn, results in the saturation field being achieved first at the drain side of the gate. The drain voltage at which the maximum field in the channel achieves E_s is defined as V_{dsat}. As the gate voltage is increased in the negative direction, the depletion region is moved deeper into the conducting channel, thereby reducing the cross-sectional area through which current can flow. The channel resistance increases and the terminal voltage at which saturation occurs is reduced.

Consider a biased condition and the region under the gate when the peak electric field exceeds the saturation field E_s. The current at all cross-sectional planes must be constant to satisfy current continuity requirements. The cross-sectional area under the gate, however, is reduced in the direc-

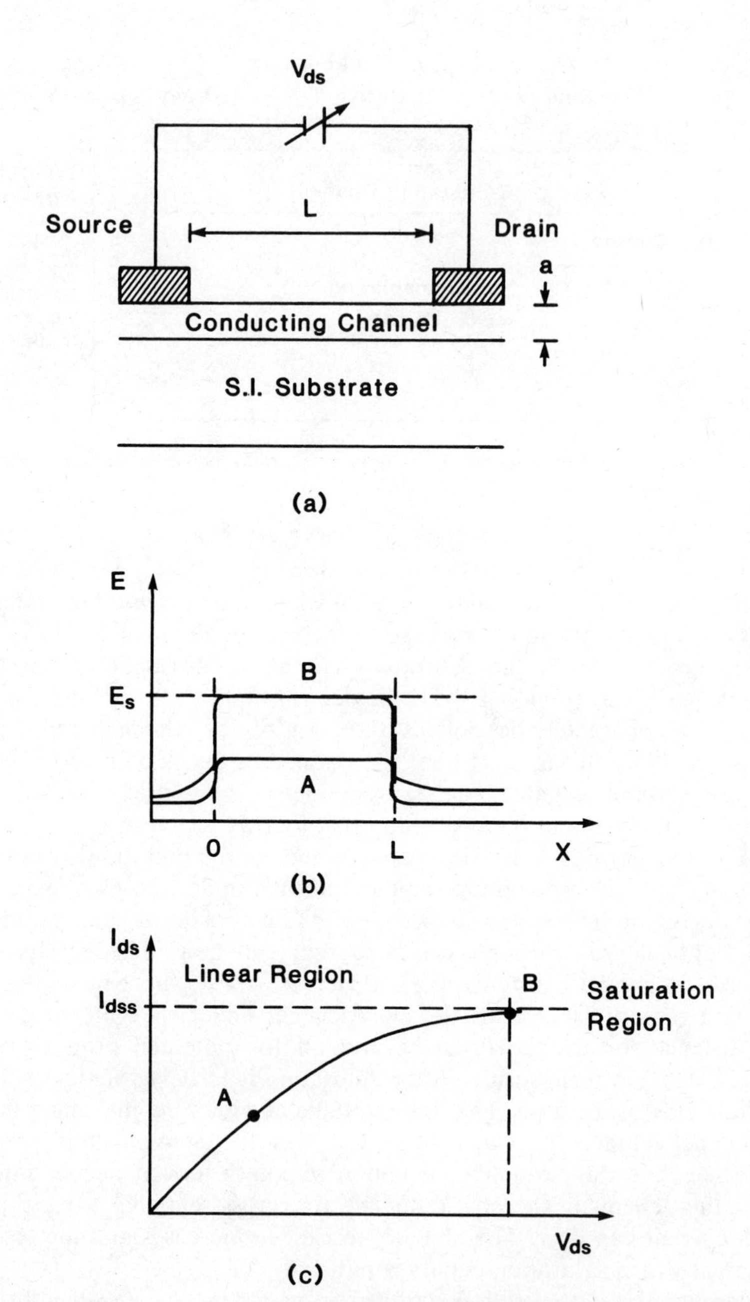

Figure 7.15 (*a*) Cross-sectional view of a gateless FET. (*b*) Internal electric field for the structure biased in the linear (*A*) and saturation (*B*) regions. (*c*) The *I–V* characteristic for the device.

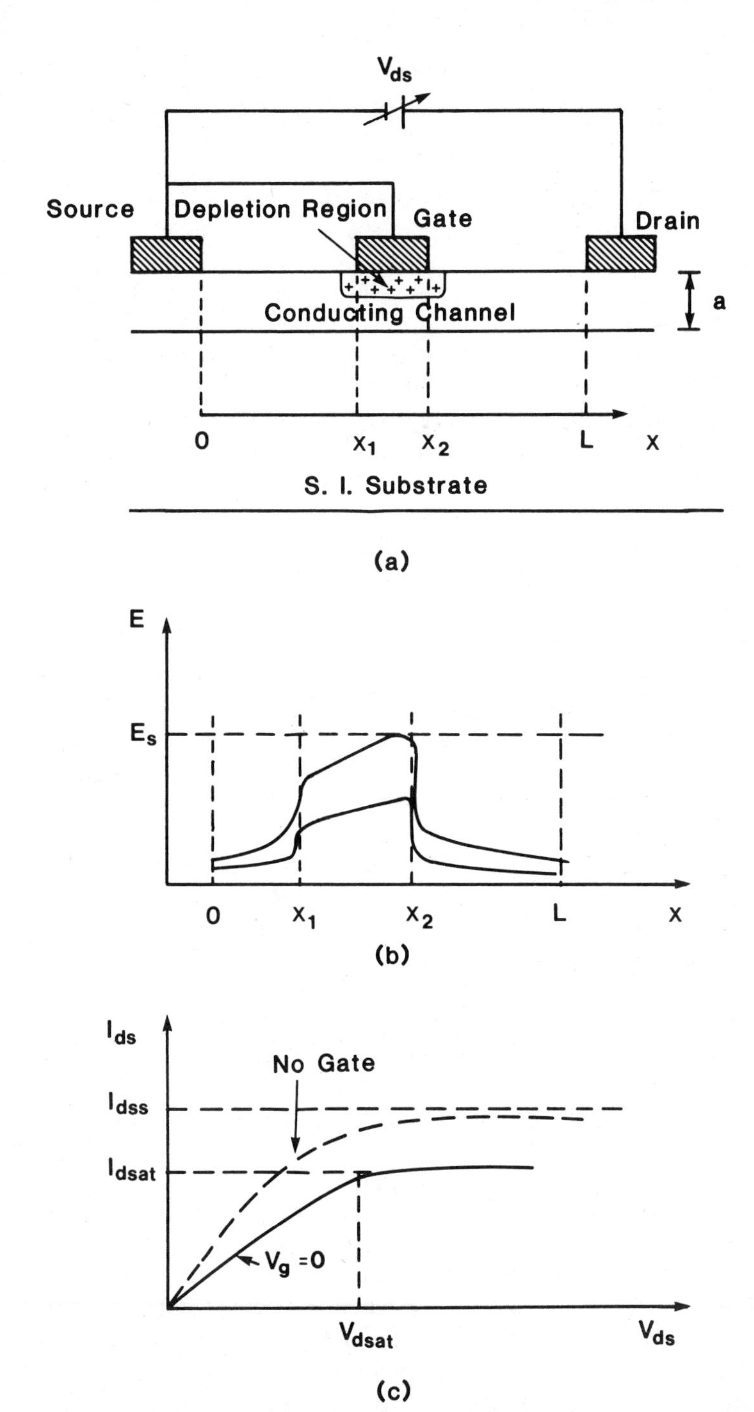

Figure 7.16 (*a*) Cross-sectional view of a biased FET. (*b*) Internal electric field for the device biased in the linear and saturation regions. (*c*) The *I–V* characteristic for the device. The dashed line shows the gateless device for comparison.

tion of current flow. The channel current can be expressed as

$$I_{ds} = Zqn(x)v(x)b(x) \tag{7.62}$$

where $A = Zb(x)$, the channel cross-sectional area, Z is the channel width, and $b(x)$ is the channel opening. In the linear region $n(x) \cong N_d \neq f(x)$ for uniform doping. The reduction in $b(x)$ is compensated for by an increase in $E(x)$ and the corresponding increase in carrier velocity $v(x)$. The result is that I_{ds} is constant.

When $E = E_s$, however, $v(x)$ can no longer increase. The reduction in $b(x)$ can only be compensated for by an increase in $n(x)$. It follows that $n(x) > N_d$ and an accumulation region is created. In the region between the gate and drain $b(x)$ reaches a minimum and then starts to increase again as the depletion region is passed. Since the channel is still in velocity saturation the increase in $b(x)$ is compensated for by a decrease in $n(x)$. A depletion of electrons (i.e., $n(x) < N_d$) is created, resulting in an electric-field dipole in the gate-drain region [17]. This situation is illustrated in Fig. 7.17(a).

As V_{ds} is increased beyond V_{dsat} the depletion region moves toward the drain. The point x_1 [Fig. 7.17(a)] moves toward the source and the potential at x_1 decreases. The channel opening at x_1 increases and more current is injected into the velocity saturation region. Therefore, an increasing V_{ds} produces an increase in I_{ds} and a positive slope to the $I_{ds}-V_{ds}$ characteristic is obtained in the saturation region. The channel has a finite positive resistance.

Note that the dipole formation is material independent, since it occurs because of geometry considerations in conjunction with velocity saturation. Dipoles can, therefore, form in Si as well as III–V materials. The negative differential mobility characteristics of materials such as GaAs, however, enhance the process and much larger dipoles are formed. This situation is illustrated in Fig. 7.17(b). The existence of the dipole has significant influence upon the operation of field-effect-type devices, since it creates a feedback path from the output to the input. This feedback path affects the device gain and frequency performance. It is shown in the equivalent circuit analysis presented in Section 7.5.4 that the dipole results in a −12-dB/octave rolloff in the unilateral gain. The dipole also affects the output impedance of the device and is especially important in understanding the operation of devices designed for power and large-signal applications.

The operation of a MESFET is very sensitive to the channel thickness. Maximum performance is obtained when the device is designed so that the gate depletion region exercises optimum control over the current flow. This generally occurs for $(L/a) \cong 3$, where L is the gate length and a is the conducting channel thickness.

Since the operation of the field-effect transistor is determined by the gating action of a reverse-biased rectifying junction, two modes of opera-

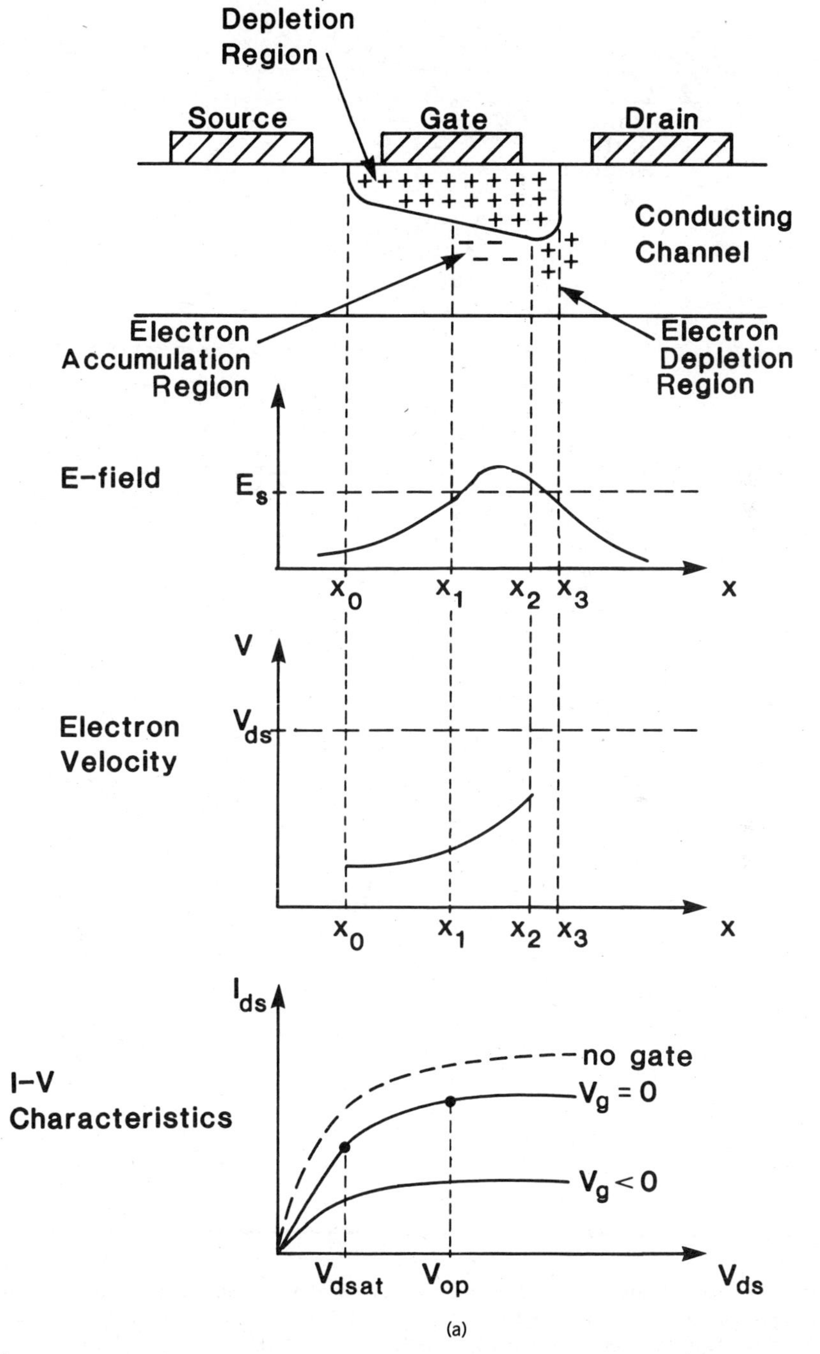

(a)

Figure 7.17 (*a*) Internal device physics for a FET showing the formation of a gate-drain region electric dipole.

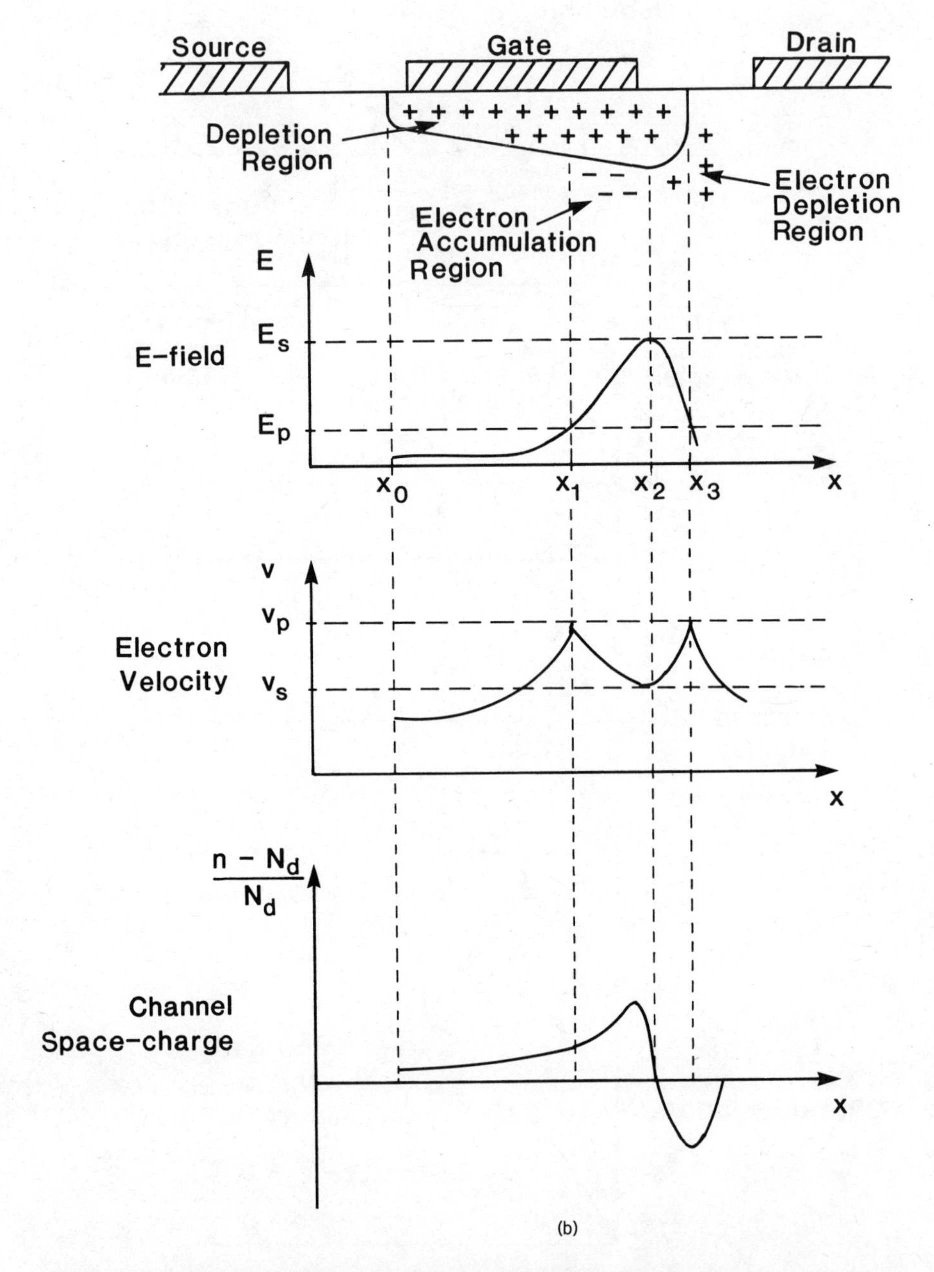

Figure 7.17 (*b*) Internal device physics for a GaAs FET showing the formation of an enhanced gate-drain region electric dipole due to the negative differential mobility region in the GaAs velocity-field characteristic (see Fig. 7.1).

tion are possible. The device can be designed so that with zero gate bias, a channel current flows. Such operation occurs with relatively thick channels and is termed "normally on." The device can also be designed with a relatively narrow channel so that the zero bias depletion region is sufficient to pinch the channel off, resulting in no channel current. This situation is termed "normally off." In this mode of operation the gate must be forward biased to obtain current flow. The normally off mode is particularly interesting for high-speed logic applications, since very low power consumption occurs. The device will only dissipate energy when the channel conducts, and this only occurs for a fraction of an operation cycle. The low power consumption is very attractive for high-density integrated circuits, where large numbers of devices are located in a very small area. Heat dissipation is a fundamental limit to the level of integration that can be achieved.

7.5.2 MESFET Model

In a short-gate GaAs MESFET, part of the channel region under the gate will be dominated by linear current flow and part will be dominated by saturation. That is, electrons will be accelerated from a low velocity at the source end of the channel to their saturation velocity at the drain end of the gate region. The electric fields near the drain end of the gate region almost always exceed the saturation field in short-gate-length GaAs MESFETs. A two-region model that can satisfactorily account for these effects has been presented [18].

The channel region under the gate can be divided into linear and saturation parts, as shown in Fig. 7.18. In Region I, the current is assumed dominated by linear behavior, and the electron velocity is linearly dependent upon the strength of the electric field. In Region II, the carrier velocity is assumed constant and independent of field. The assumed velocity–field characteristic is shown in Fig. 7.19. Note that the saturation velocity is an "effective" parameter, and not the true saturation velocity. An effective velocity is required since the two-region model is not capable of describing the negative differential mobility region of the GaAs velocity–field characteristic. The effective saturation velocity has values between the peak and true saturation values. An averaging method for determining an appropriate value for the effective saturation velocity has been presented [19].

The model is derived with the use of the depletion approximation and Shockley's gradual channel approximation. A two-sided device geometry is assumed, and only the region under the gate is simulated. The other regions of the device are considered as parasitic elements in the complete device equivalent circuit. The two-sided channel region eliminates the substrate from the analysis, thereby simplifying the derivation of the model. The chief advantage in considering only the region under the gate is that the fringing fields that are associated with the edge regions of the gate do not need to be considered. Although this simplification allows solutions to be obtained in a

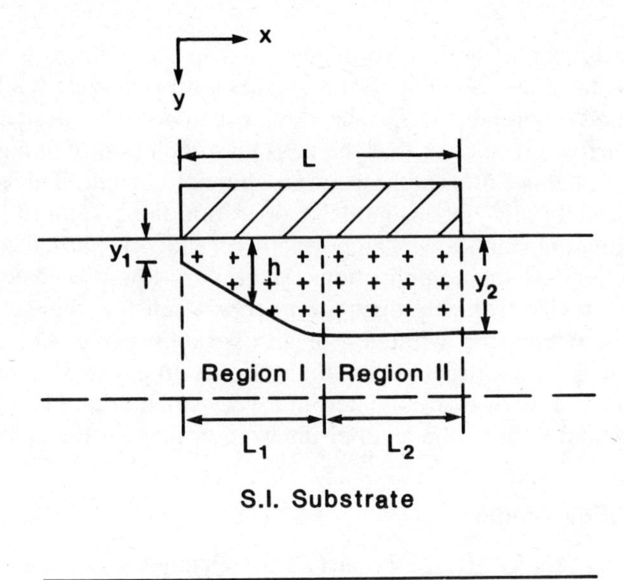

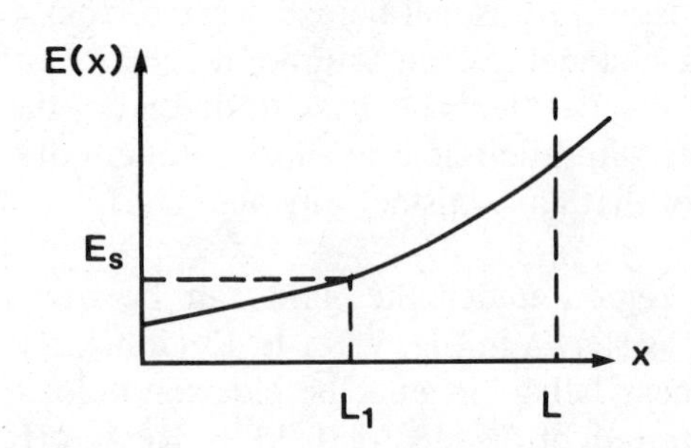

Figure 7.18 Field-effect geometry used in the formulation of the two-region model.

direct manner, the simplification establishes limits to the model's validity. In fact, this model does not work well for submicron gate-length devices where the fringing areas occupy a significant proportion of the gate depletion region. The model works well for 1 μm gate-length devices and produces acceptable results, in certain instances, for gate lengths down to $\frac{1}{2}$ μm. It does not produce accurate results for shorter gate lengths.

The model is derived by writing a potential drop equation along the channel. The channel potential is written as

$$W(x) = V_{gs} + \varphi_{bi} - V(x) \tag{7.63}$$

where V_{gs} is the gate voltage, $V(x)$ is the potential across the depletion region, and φ_{bi} is the built-in potential. At the source end of the channel the

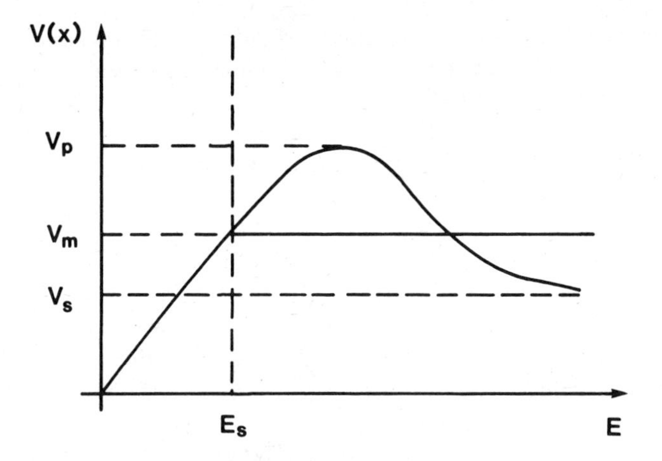

Figure 7.19 Piecewise linear two-region velocity-field characteristic used in the MESFET model.

potential is

$$W_s \triangleq V_{gs} + \varphi_{\text{bi}} \tag{7.64}$$

At the boundary between the linear and saturation regions the potential is

$$W_p \triangleq V_{gs} + \varphi_{\text{bi}} - V_p \tag{7.65}$$

and at the drain end of the channel the potential is

$$W_d \triangleq V_{gs} + \varphi_{\text{bi}} - V_{sd} \tag{7.66}$$

It is convenient to introduce normalized potentials defined as

$$s \triangleq \left(\frac{W_s}{W_{00}}\right)^{1/2} \tag{7.67a}$$

$$p \triangleq \left(\frac{W_p}{W_{00}}\right)^{1/2} \tag{7.67b}$$

$$d \triangleq \left(\frac{W_d}{W_{00}}\right)^{1/2} \tag{7.67c}$$

and

$$w(x) = \left[\frac{W(x)}{W_{00}}\right]^{1/2} \tag{7.68}$$

where $W_{00} \triangleq (qN_d/2\epsilon)a^2$, the channel pinchoff potential. The model is derived by considering the two channel regions separately, and then invoking current continuity at the interface to obtain an overall channel current expression.

Region I. The channel opening is calculated by solving Poisson's equation in the depletion region to get

$$W(x) = W_{00}\left[1 - \frac{b(x)}{a}\right]^2 \tag{7.69}$$

where $b(x)$ is the channel opening under the depletion region. The drain current is calculated to be

$$I_d = J_d A = \sigma E_x A = \sigma\left(\frac{dW(x)}{dx}\right) Z[2b(x)] \tag{7.70}$$

where σ is the conductivity of the conducting channel. After some manipulation this can be written in terms of the normalized potentials as

$$I_d = \frac{g_0 Z W_{00}}{L_1}\left[p^2 - s^2 - \frac{2}{3}(p^3 - s^3)\right] \tag{7.71}$$

where $g_0 = 2a\sigma$, the channel conductance and Z = the gate width.

Region II. In the saturation region the channel current is

$$I_d = J_d A = \sigma E_s (2b_p) Z \tag{7.72}$$

where $2b_p$ is the conducting channel opening at pinchoff. It can be shown that

$$b_p = a(1 - p) \tag{7.73}$$

and

$$I_d = g_0 Z E_s (1 - p) = I_s (1 - p) \tag{7.74}$$

where I_s is the saturation current (i.e., the maximum current that could flow if the channel were fully undepleted and the charge carriers were moving at the saturation velocity v_s).

By equating drain-current expressions of Regions I and II, the boundary L_1 between the linear and saturation regions can be determined. It follows that

$$L_1 = \frac{g_0 Z W_{00}}{I_s(1 - p)}\left[p^2 - s^2 - \frac{2}{3}(p^3 - s^3)\right] \tag{7.75}$$

To define the drain current in terms of the applied potentials, the normalized potentials must be related to the applied bias potentials. This can be accomplished with the relationship

$$V_{sd} = -\int_0^L E_x \, dx \tag{7.76}$$

and

$$V_{ds} = -\underbrace{\int_0^{L_1} E_x \, dx}_{\text{Region I}} - \underbrace{\int_{L_1}^{L} E_x \, dx}_{\text{Region II}} \tag{7.77}$$

For Region I, the integral can be written as

$$V_I = -[W(L_1) - W(0)] \tag{7.78}$$

and

$$V_\mathrm{I} = -(W_p - W_s) = -W_{00}(p^2 - s^2) \tag{7.79}$$

For Region II, the potential is inherently two dimensional and, in general, is very difficult to determine in an analytic fashion. An approximation for the potential can be determined by using the superposition principle and an approach presented by Grebene and Ghandhi [20]. This approach results in an expression for the source–drain potential drop that has the form,

$$V_{ds} = -W_{00}(p^2 - s^2) - \frac{2aE_s}{\pi} \sinh \frac{\pi L_2}{2a} \tag{7.80}$$

where L_2 is the length of the channel region in saturation. This equation, along with those for L_1 and I_d, allow the I–V characteristics for the device to be determined. A small-signal model can then be determined by taking various partial derivatives to define the critical equivalent-circuit elements.

7.5.3 Small-Signal Model

An equivalent circuit for the MESFET is superimposed upon the FET structure in the sketch in Fig. 7.20. The equivalent-circuit elements consist of both passive parasitic parameters and the parameters responsible for the active characteristics of the device. The sketch in Fig. 7.20 shows the physical origin of the various elements. The circuit is generally shown as a T-equivalent circuit as depicted in Fig. 7.21. In this section expressions for some of the most important elements in the circuit are derived.

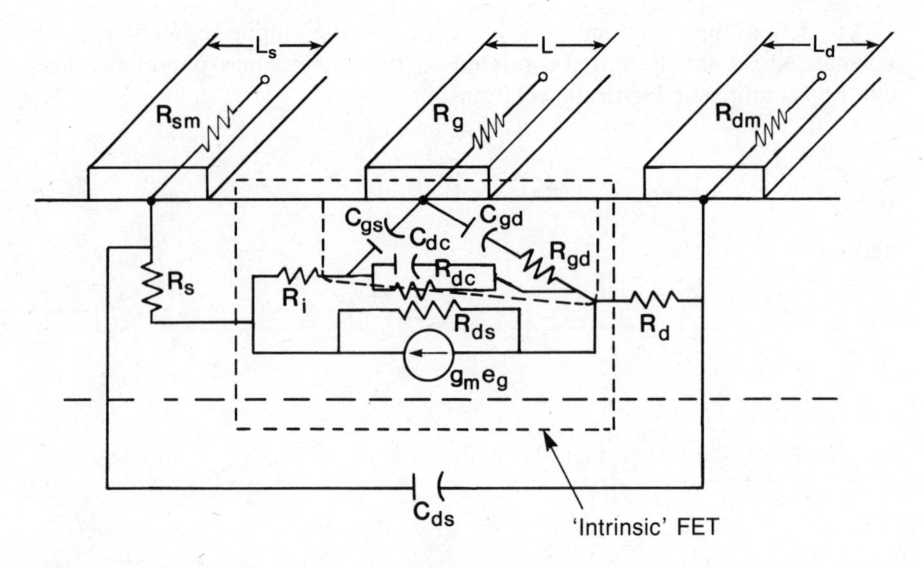

Figure 7.20 Equivalent circuit for the MESFET superimposed upon the device geometry to show the physical basis for the various elements.

Transconductance. The critical elements can be determined by perturbation of the dc I–V characteristics. The device transconductance g_m is defined as

$$g_m = -\left.\frac{\partial I_d}{\partial V_{gs}}\right|_{V_{ds}} = \frac{g_0 E_s Z}{2sW_{00}} \frac{\partial p}{\partial s} \tag{7.81}$$

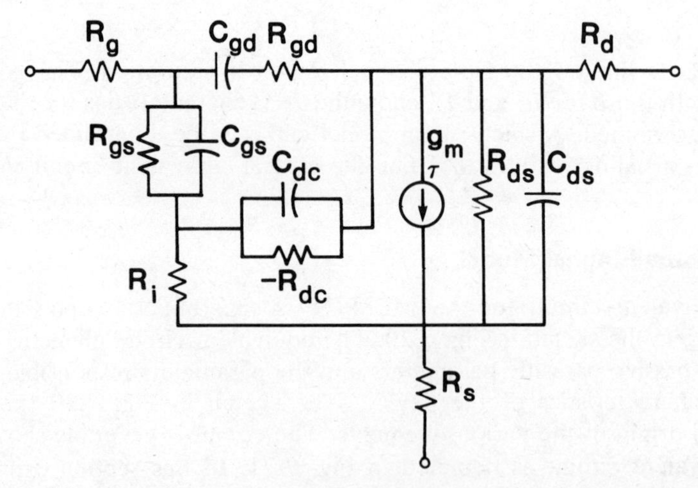

Figure 7.21 T-equivalent circuit for the MESFET shown in standard form.

When this definition is applied to the expressions previously derived the transconductance is written for short-gate-length devices (i.e., $L \cong L_2$, $L_1 \to 0$, and $s \cong p$) as

$$g_m \cong \frac{I_s}{W_{00}(2p)} = \frac{I_s}{2W_{00}}\left[\frac{1}{1 - I_d/I_s}\right] \tag{7.82}$$

This approximation is quite accurate when $I_d/I_s > 0.1$. Note that the transconductance is proportional to the I_d/I_s ratio and also to the value of I_s.

Channel Resistance. The channel resistance R_{ds} (also commonly designated by R_o) is determined from

$$R_{ds} = -\left.\frac{\partial V_{ds}}{\partial I_d}\right|_{V_{gs}} \tag{7.83}$$

which, when applied to short-gate-length devices, results in the expression

$$R_{ds} \cong \frac{|V_{ds}|\pi W_{00} p}{aI_s E_s} = \pi \frac{W_{00}}{aE_s}\frac{|V_{ds}|}{I_s}\left(1 - \frac{I_d}{I_s}\right) \tag{7.84}$$

Note that R_{ds} is independent of gate length and is linearly dependent upon V_{ds}. Also, R_{ds} is linearly dependent upon I_d.

Gate–Source Capacitance. The gate-source capacitance, C_{gs}, is the rate of change of the free charge on the gate electrode with respect to the gate bias voltage when the drain potential is held constant. This definition actually includes the gate–source and gate–drain capacitances. If $C_{gd} \ll C_{gs}$ (typically $C_{gd} \cong 0.1 C_{gs}$ for microwave FETs) the definition for C_{gs} can be used. Therefore,

$$C_{gs} \cong \left.\frac{\partial Q_g}{\partial V_{gs}}\right|_{V_{ds}} \tag{7.85}$$

The gate charge Q_g can be determined from Poisson's equation by integrating the normal component of the electric field under the gate over the gate area. In Region I, the $w(x)$ expression is integrated. In Region II, the contributions from the drain and depletion regions must be considered. The normal components of the electric field are: Region I

$$E_{yI}(x, a) = \frac{2W_{00}}{a}\left[1 - \frac{b(x)}{a}\right] \tag{7.86}$$

Region II

$$E_{yII}(x, a) = \left(\frac{2W_{00}}{a}\right) p + E_s \sinh\left[\frac{\pi(x - L_1)}{2a}\right] \tag{7.87}$$

where

$$p = 1 - \frac{b_p}{a}$$

The gate charge is determined from

$$Q_g = 2\epsilon Z\left[\int_0^{L_1} E_{yI}(x, a)\, dx + \int_{L_1}^{L} E_{yII}(x, a)\, dx\right] \tag{7.88}$$

Integrating and substituting into the capacitance definition, the gate–source capacitance for short-gate-length FETs and bias voltages not close to W_{00} results in the expression,

$$C_{gs} \cong 2\epsilon Z\left[\frac{L}{ap} + 1.56\right] = 2\epsilon Z\left[\frac{L}{a}\left(\frac{1}{1 - I_d/I_s}\right) + 1.56\right] \tag{7.89}$$

The 1.56 term in the expression accounts for fringing area capacitance [21]. The gate depletion region appears as a parallel-plate capacitor with the addition of the fringing area factor.

Drain–Gate and Source–Drain Capacitances. The capacitances C_{gd} and C_{ds} are considered parasitic parameters to the first order in this model. They are not intrinsic to the gain mechanism of the FET and are due to fringing fields of the contact electrodes. Expressions for these elements can be obtained from an analysis of strip conductors on a dielectric medium. Expressions for these capacitances have been determined [18] to be

$$C_{gd}, C_{ds} = (\epsilon_r + 1)\epsilon_0 Z \frac{K(1 - k^2)^{1/2}}{K(k)} \tag{7.90}$$

where $K(k)$ is the complete elliptic integral of the first kind and

$$k_{gd} = \left[\frac{L_{gd}}{L_{gd} + L}\right]^{1/2} \tag{7.91}$$

and

$$k_{ds} = \left[\frac{(2L_s + L_{ds})L_{ds}}{(L_s + L_{ds})^2}\right]^{1/2} \tag{7.92}$$

where L_{gd} and L_{ds} are the separation between gate-drain and drain-source electrodes, respectively. In these expressions it is assumed that $L_d \gg L$, $L_s = L_d$, and the contact areas are large compared to the electrode spacings. Note that C_{gd} and C_{ds} are constant and not functions of potential. This is, of course, an approximation and only accurate in certain applications. This approximation is not generally valid when the switching characteristics of these devices are considered.

Gate–Source Capacitance Charging Resistor. The charging resistor R_i is difficult to relate to bulk material parameters in the device. It represents the time required to charge the gate-depletion region capacitance. A suitable expression can be determined from time-constant arguments. For example, the gate–source capacitance charging time can be expressed as

$$\tau = R_i C_{gs} \tag{7.93}$$

For short-gate-length devices it is known that

$$C_{gs} = \epsilon Z \frac{L}{ap} \tag{7.94}$$

The time constant τ can be determined from the time required for the electrons to travel the length of the gate region. Therefore,

$$\tau = \int_0^L \frac{dx}{v(x)} = \int_0^{L_1} \frac{dx}{\mu E(x)} + \int_{L_1}^L \frac{dx}{v_s} \tag{7.95}$$

For short-gate devices at moderate to high drain potentials the first integral becomes small, since most of the gate region is in velocity saturation. For a first approximation, we can neglect the first integral. Therefore,

$$\tau \cong \frac{L - L_1}{v_s} \cong \frac{L}{v_s} \tag{7.96}$$

and

$$R_i = K' \frac{ap}{\epsilon Z v_s} \tag{7.97}$$

where K' is a proportionality constant chosen to match actual devices. For typical 1-μm gate-length FETs

$$K' \sim 1.65(10)^{10} \ \Omega\text{-m/sec}$$

7.5.4 Equivalent Circuit and Figures of Merit

The GaAs MESFET equivalent circuit is shown in Fig. 7.21. This circuit can be analyzed to determine the performance of the device. For purposes of analysis it is common to reduce the equivalent circuit to a simplified form, retaining only the most significant elements. The circuit reduces as shown in Fig. 7.22. Note that all elements that provide coupling from the output loop to the input loop have been removed. This simplification produces a circuit that is easily analyzed, but does not necessarily accurately predict device performance, especially at microwave frequencies. The simplified circuit produces the gain-bandwidth product f_T, and the maximum frequency of oscillation $f_{\max}$, which are commonly used as figures of merit. These parameters will be derived and then the complete circuit will be analyzed to show the significance of the feedback elements.

The short-circuit current gain is calculated by placing a short circuit on the output and calculating the current gain from the expression,

$$h_{21} = \frac{I_{\text{out}}}{I_{\text{in}}} = \frac{g_m}{j\omega C_{\text{gs}}} \tag{7.98}$$

The frequency at which the magnitude of h_{21} is reduced to unity is defined as the gain-bandwidth product,

$$f_T = \frac{g_m}{2\pi C_{\text{gs}}} \tag{7.99}$$

For short-gate-length devices, the transconductance and gate–source capacitance can be written as

$$g_m = \frac{I_s}{W_{00}(2p)} = \frac{2\epsilon Z v_s}{ap} \tag{7.100}$$

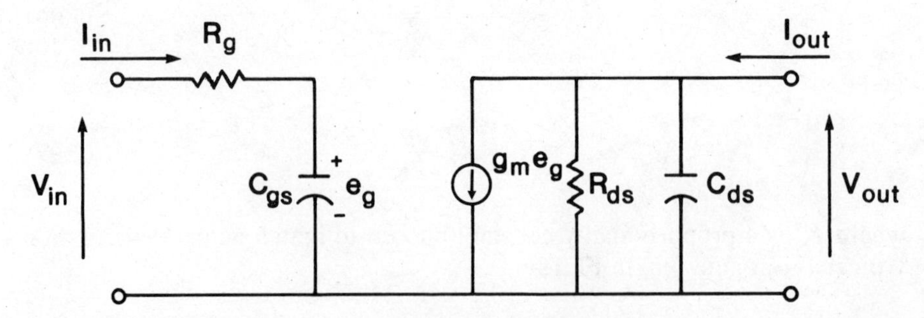

Figure 7.22 Simplified equivalent circuit for the MESFET. Only elements of first-order importance are included.

and

$$C_{gs} = 2\epsilon Z\left[\frac{L}{ap} + 1.56\right] \tag{7.101}$$

If the fringing-field term is neglected, these expressions can be substituted into the f_T expression to get

$$f_T = \frac{v_s}{2\pi L} \tag{7.102}$$

The f_T of the device is directly proportional to the saturated velocity of the charge carriers under the gate and inversely proportional to the gate length. Since high f_T is desirable for high-frequency performance, short-gate lengths and high-velocity semiconductors are required.

Power Gain and f_{max}. The power gain of the device can be calculated from the admittance matrix for the device,

$$\begin{bmatrix} I_{in} \\ I_{out} \end{bmatrix} = \begin{bmatrix} y_{11} & y_{12} \\ y_{21} & y_{22} \end{bmatrix} \begin{bmatrix} V_{in} \\ V_{out} \end{bmatrix} \tag{7.103}$$

The device circuit couples energy from an external source to an external load. The unilateral power gain is defined as the ratio between the power delivered to the load over the power available from the source when the feedback network is tuned so that no feedback occurs and the input and output are simultaneously and conjugately matched.

The unilateral power-gain expression can be applied to the simplified equivalent circuit in Fig. 7.22. It follows that the admittance matrix is

$$\begin{aligned} y_{11} &= \frac{1}{R_g + \dfrac{1}{j\omega C_{gs}}} \\ y_{21} &= \frac{g_m}{1 + j\omega C_{gs} R_g} \\ y_{22} &= \frac{1}{R_{ds}} + j\omega C_{ds} \end{aligned} \tag{7.104}$$

and

$$y_{12} = 0$$

Substituting into the unilateral gain definition, it follows that

$$U = \frac{1}{4}\left(\frac{f_T}{f}\right)^2 \frac{R_{ds}}{R_g} \tag{7.105}$$

The maximum frequency of oscillation, $f_{\max}$, is defined as the frequency at which the unilateral power gain is reduced to unity and is given as

$$f_{\max} = \frac{f_T}{2}\sqrt{\frac{R_{ds}}{R_g}} \tag{7.106}$$

The $f_{\max}$ frequency separates the active and passive regions for the network. Above $f_{\max}$ the network is passive and incapable of amplifying RF energy. The unilateral gain can be written in terms of $f_{\max}$ as

$$U = \left(\frac{f_{\max}}{f}\right)^2 \tag{7.107}$$

and, to increase the gain it is necessary to maximize $f_{\max}$. Also, from the U expression the importance of a large channel resistance R_{ds}, and a low input resistance R_g, for obtaining optimum gain is seen.

This single-pole model predicts that the unilateral gain will decrease with frequency at a rate of 6 dB/octave or 20 dB/decade. At low frequencies (i.e., $f \ll f_{\max}$) this approximation is quite accurate. At higher frequencies, the other elements in the equivalent circuit become important and add complexity to the calculation. For example, when the complete equivalent circuit is analyzed it is found that the unilateral gain decreases at a 6 dB/octave rate at low frequencies, as predicted from the simple analysis, but the rolloff rate increases to about 12 dB/octave at higher frequencies. It has been noted by Das [22] that parasitic common lead elements produce the increased unilateral gain rolloff. Steer and Trew [23] and Trew and Steer [24] have shown that internal feedback, caused by the charge dipole domain that forms in the channels of field-effect-type transistors, produces a complex pole in the unilateral gain, and above the pole frequency the gain decreases at a 12 dB/octave rate. If the complete equivalent circuit of Fig. 7.21 is analyzed, an approximation for the unilateral gain is found to be

$$U \cong \left[\frac{g_m^2 R_{ds}}{4C_{gs}R_i(C_{gs} - C_{dc}g_m R_{ds})}\right]\left[\frac{1}{\omega^2(1 - p^2\omega^2)}\right] \tag{7.108}$$

where

$$p^2 = \frac{(R_i^2 C_{gs})(C_{dc} + C_{gs})^2 + C_{dc}g_m R_{ds}\dfrac{\tau^2}{2}}{(C_{dc}g_m R_{ds} - C_{gs})} \tag{7.109}$$

This calculation includes the elements listed in Table 7.2 and indicates the significance of the domain capacitance (C_{dc}) in establishing the additional pole.

TABLE 7.2 Equivalent Circuit Parameters for a Half-Micron Gate-Length Millimeter-Wave GaAs MESFET

Element	Value
R_s	4.55 Ω
R_g	1.46 Ω
R_d	6.7 Ω
R_i	2.69 Ω
R_{ds}	556 Ω
C_{gs}	0.071 pF
C_{gd}	0.001 pF
C_{ds}	0.025 pF
C_{dc}	0.011 pF
g_m	15.2 mS
τ	1.25 ps

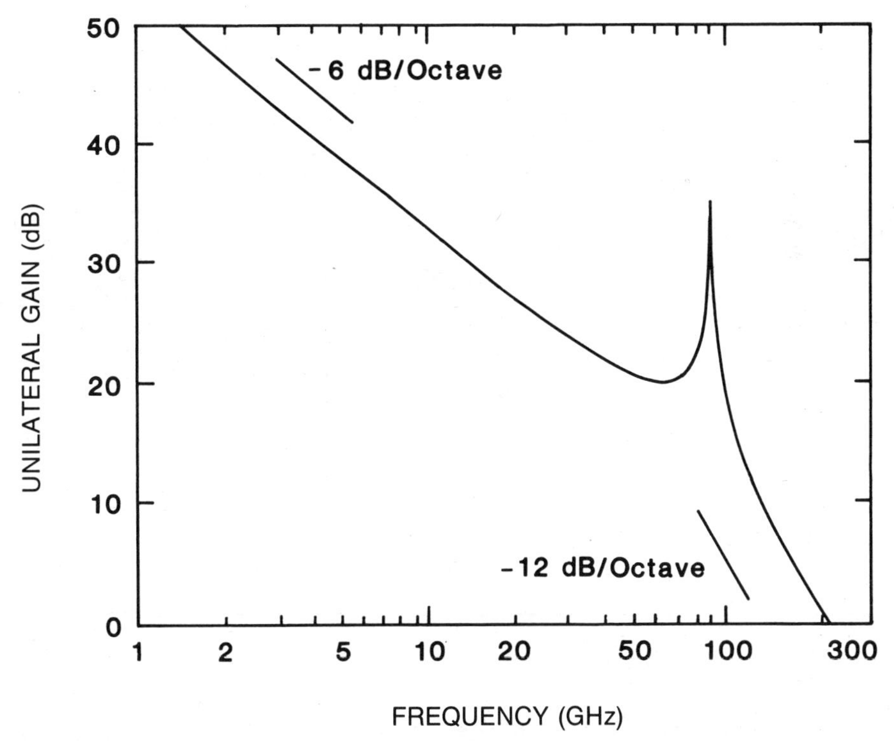

Figure 7.23 Unilateral gain calculated from the complete equivalent circuit shown in Fig. 7.21.

Using the parameter values listed in Table 7.2 for a mm-wave GaAs MESFET, the unilateral gain performance shown in Fig. 7.23 results. Note the 12 dB/octave rolloff at the higher frequencies. If the domain capacitance is removed from the full circuit, the unilateral gain decreases at a 6 dB/octave rate throughout the frequency range. It should be noted that this analysis does not consider distributed effects and these effects will result in even greater rolloff rates. When these effects and the effects of lead parasitics are included, rolloff rates greater than 12 dB/octave can occur.

7.5.5 Noise Figure Analysis

The noise figure performance of GaAs MESFETs is most readily calculated using the equivalent circuit shown in Fig. 7.24, where shunt current noise generators are located at the input and output ports. This representation is particularly meaningful for GaAs MESFETs, since the shunt generators represent noise sources at the gate, source, and drain electrodes. The noise source at the output represents the short-circuit drain-source channel noise. The drain current generator is defined in mean-square terms as [25]

$$\overline{i_{\text{nd}}^2} = 4kT_0\,\Delta f g_m P \tag{7.110}$$

where k is Boltzmann's constant, T_0 is the lattice temperature, Δf is a bandwidth, g_m is the device transconductance, and P is a dimensionless drain noise factor that depends upon device geometry and bias conditions.

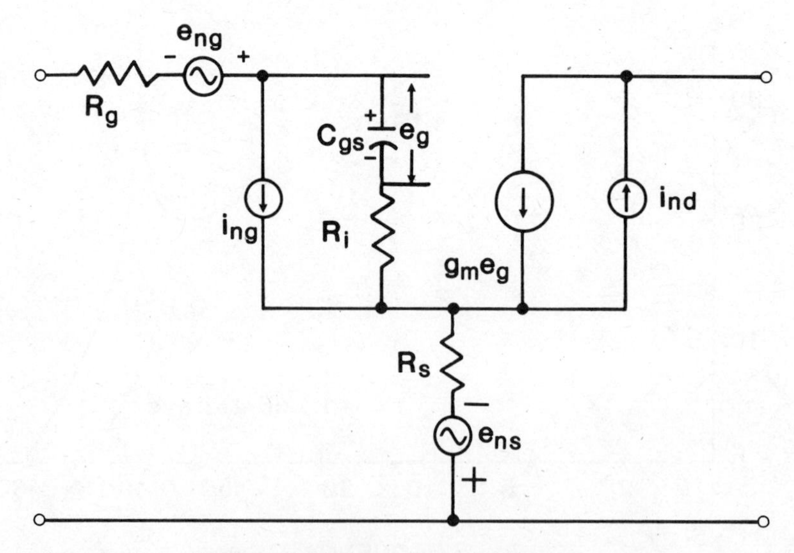

Figure 7.24 Noise equivalent circuit for a GaAs MESFET. The noise sources i_{ng}, i_{nd}, e_{ng}, and e_{ns} represent the induced gate noise, drain circuit noise, and the thermal noise of the gate and source resistances.

For zero drain–source voltage $P = [R_{ds} g_m]^{-1}$ and P usually has a numerical value between 1 and 3. The gate noise generator is defined as [26]

$$\overline{i_{ng}^2} = 4kT_0 \Delta f \frac{\omega^2 C_{gs}^2 R}{g_m} \tag{7.111}$$

where R is a dimensionless gate noise factor that depends upon bias. For zero drain–source voltage, $R \sim g_m R_i$.

The gate and drain current generators are capacitively coupled and, therefore, they are partially correlated. The degree of correlation can be expressed by a factor C [26] where

$$jC = \frac{\overline{i_{ng}^* \cdot i_{nd}}}{\sqrt{\overline{i_{ng}^2} \cdot \overline{i_{nd}^2}}} \tag{7.112}$$

The star stands for the complex conjugate and j is the imaginary representation. The magnitude of C is generally about 0.8 or smaller.

The minimum noise figure for the intrinsic MESFET can be expressed [18] as

$$F_{min} = 1 + 2\sqrt{PR(1 - C^2)} \frac{f}{f_T} + 2 g_m R_i P \left[1 - C \sqrt{\frac{P}{R}}\right] \left(\frac{f}{f_T}\right)^2 \tag{7.113}$$

This expression can be applied to the equivalent circuit in Fig. 7.24 to obtain

$$F_{min} = 1 + 2\left(2\pi f \frac{C_{gs}}{g_m}\right) [K_g(K_r + g_m(R_g + R_s))]^{1/2}$$
$$+ 2\left(2\pi f \frac{C_{gs}}{g_m}\right) [K_g g_m(R_g + R_s + K_c R_i)] \tag{7.114}$$

where

$$K_g = P\left[\left(1 - C\sqrt{\frac{R}{P}}\right)^2 + (1 - C^2)\frac{R}{P}\right]$$

$$K_r = \frac{R(1 - C^2)}{(1 - C\sqrt{R/P})^2 + (1 - C^2)R/P}$$

and

$$K_c = \frac{1 - C\sqrt{R/P}}{(1 - C\sqrt{R/P})^2 + (1 - C^2)R/P}$$

This expression has been simplified by Fukui [27], and for devices operating at frequencies below f_T the minimum noise figure can be approximated as

$$F_{\min} \cong 10 \log[1 + KfL\sqrt{g_m(R_g + R_s)}] \quad (7.115)$$

where L is the gate length and K is a fitting factor selected to match experimental data. Generally, for 1-μm gate-length devices K varies from 0.25 to 0.3 and a value of 0.27 is found to be generally valid. This expression indicates that low minimum noise figures are obtained with short-gate-length devices that have low parasitic resistance. This expression agrees well with experimental data as shown in Fig. 7.25, where the noise measure and noise figure are shown for devices of various gate length [28].

The minimum noise figure can be defined in terms of practical device parameters. The device transconductance can be expressed as [29]

$$g_m \cong KZ\left(\frac{N}{aL}\right)^{1/3} \quad (7.116)$$

where Z is the gate width in millimeters, a is the channel thickness in micrometers, N is the channel doping in $10^{16}\,\text{cm}^{-3}$, and K is the fitting factor previously described. The gate resistance can be written as

$$R_g \cong 17\frac{z^2}{ZhL} \quad (7.117)$$

where z is a unit gate width in millimeters, Z is the actual gate width in millimeters, h is the gate metallization height in micrometers, and L is the gate length in micrometers.

For a recessed gate device the source resistance can be expressed as a sum of three component resistances as

$$R_s = R_1 + R_2 + R_3 \quad (7.118)$$

where R_1 represents the source contact resistance, and R_2 and R_3 are resistances of the semiconductor between the source and gate electrodes. The resistance R_2 is calculated for the region outside the recessed area, and R_3 is calculated for the region under the recess. These resistances are given by the expressions,

$$R_1 \cong \frac{2.1}{Za_1^{0.5}N_1^{0.66}} \quad (7.119a)$$

$$R_2 \cong \frac{1.1L_2}{Za_2N_2^{0.82}} \quad (7.119b)$$

and

$$R_3 \cong \frac{1.1L_3}{Za_3N_3^{0.82}} \quad (7.119c)$$

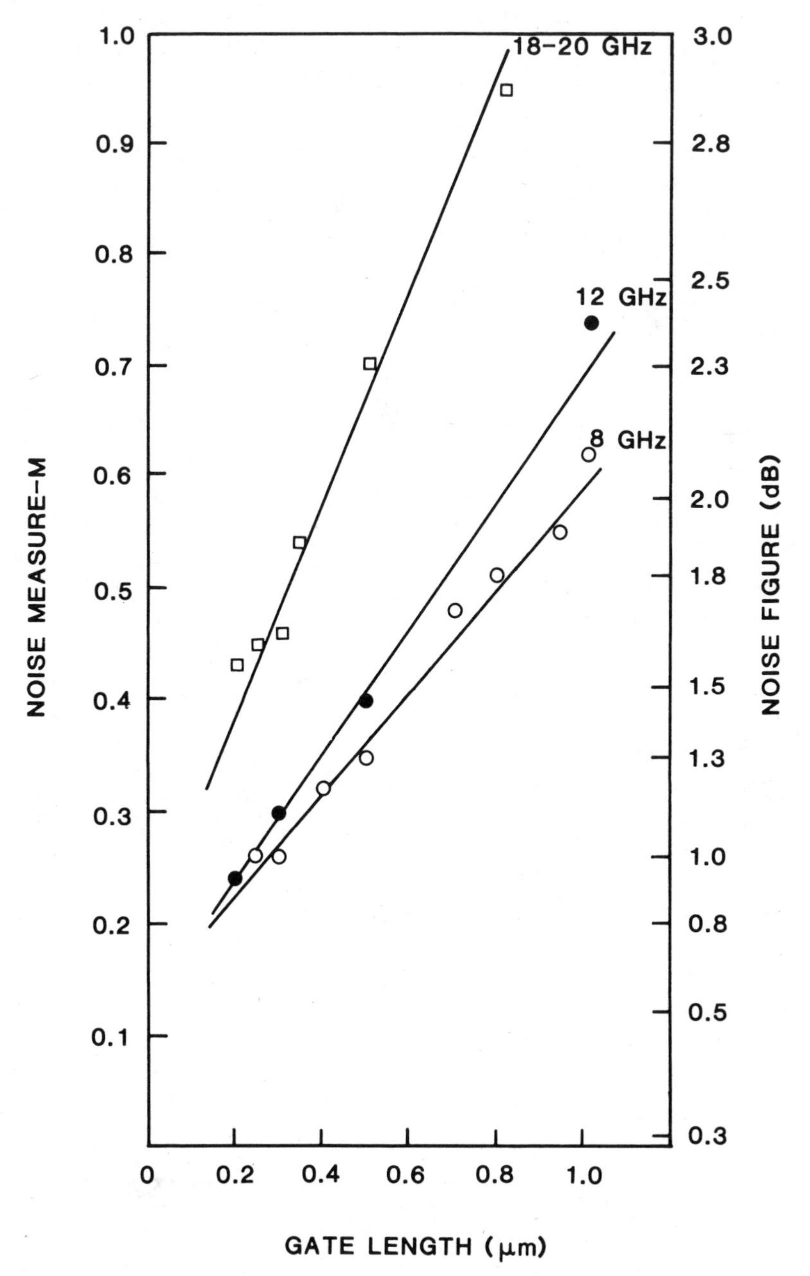

Figure 7.25 Noise measure and noise figure measured for GaAs MESFETs as a function of gate length (H. Goronkin and V. Nair[28], 1985 IEEE, reprinted with permission.)

where a_1 is the channel thickness under the source electrode in micrometers, N_1 is the doping density under the source electrode in $10^{16}\,\text{cm}^{-3}$, L_2 and L_3 are the lengths of the respective fractions of the channel between the source and gate electrodes in micrometers, a_2 and a_3 are the thicknesses of their respective regions in micrometers, and N_2 and N_3 are the doping densities of their respective regions in $10^{16}\,\text{cm}^{-3}$.

By defining the maximum unit gate width as the limit when R_g exceeds R_s, an expression for the minimum noise figure in a form suitable for practical device design is obtained. The expression is

$$F_{\min} = 1 + fK\left[\frac{NL^5}{a}\right]^{1/6}\left[\frac{17z^2}{hL}(1+s) + \frac{2.1}{a_1^{0.5}N_1^{0.66}} + \frac{1.1L_2}{a_2N_2^{0.82}} + \frac{1.1L_3}{a_3N_3^{0.82}}\right]^{1/2} \tag{7.120}$$

where $s = 0.08\sqrt{fhL}$ is a factor that accounts for the skin effect on the gate metal. In this expression a noise coefficient factor of $K = 0.040$ is found to give good results.

From the circuit perspective it is useful to treat the device as a noisy two-port with the noise sources referenced to the input. Using this formulation, the minimum noise figure can be written in the form [30]

$$F = F_{\min} + \frac{R_n}{R_s}\left[\frac{(R_s - R_{\text{op}})^2 + (X_s - X_{\text{op}})^2}{R_{\text{op}}^2 + X_{\text{op}}^2}\right] \tag{7.121}$$

where R_n is an equivalent noise resistance, R_s and X_s are the signal source impedance, and R_{op} and X_{op} are the signal source impedance that produce the lowest noise figure. Note that when the source impedance is equal to the optimum value, the noise figure is equal to its minimum value. This equation is very useful for circuit design, since it shows the effect of input circuit mismatch upon noise figure.

The minimum noise figure is presented in (7.120) in a form useful for practical design. Fukui [31] has presented expressions for the other terms in a similar form. The expressions are

$$R_n = \frac{40}{Z}\left[\frac{aL}{N}\right]^{1/3} \tag{7.122}$$

$$R_{\text{op}} = 2.2Z^{-1}\left[12.5\left(\frac{aL}{N}\right)^{1/3} + \frac{17z^2}{hL} + \frac{2.1}{a^{0.5}N^{0.66}} + \frac{1.1L_{\text{sg}}}{(a - a_s)N^{0.82}}\right] \tag{7.123}$$

and

$$X_{\text{op}} = \frac{450}{fZ}\left[\frac{a}{NL^2}\right]^{1/3} \tag{7.124}$$

TABLE 7.3 Comparison of the Predicted Value of Optimal Noise Figure from the Geometrical and Material Parameters, with the Directly Measured Value for Sample GaAs MESFETs

	Parameter		Device				
	Symbol	Units	*A*	*B*	*C*	*D*	*E*
Predicted by (7.120)	F_{min}	dB	1.72	1.80	2.12	1.56	1.70
Measured directly	F_{min}	dB	1.75	1.76	2.22	1.51	1.74

where L_{sg} is the distance between the source and gate electrodes in micrometers. The a_s term is the depletion depth in micrometers at the surface in the source gate region.

As indicated by (7.121), the device noise figure is a sensitive function of input impedance matching. In order to decrease this sensitivity it is necessary to decrease the magnitude of the equivalent noise resistance R_n. This is accomplished by designing the device with a short-gate length and high channel doping. A comparison of five low-noise designs based upon these calculations is presented in Tables 7.3 and 7.4. The device dimensions are indicated in Fig. 7.26.

TABLE 7.4 Design Parameters of Five Representative GaAs MESFETs Used for Calculation of the Optimal Noise Figure as a Function of Frequency, as Shown in Fig. 7.26

Parameter		Device				
Symbol	Units	*a*	*b*	*c*	*d*	*e*
L	μm	0.9	0.9	0.5	0.5	0.25
L_g	μm	0.9	1.2	0.8	0.8	0.4
L_2	μm	1.0	0.75	0.75	0.75	0.4
L_3	μm	0	0.4	0.3	0.3	0.2
h	μm	0.5	1.0	0.65	0.65	0.4
N	10^{16} cm^{-3}	7	4	8	8	18
N_1	10^{16} cm^{-3}	7	200	200	200	200
N_2	10^{16} cm^{-3}	7	200	200	200	200
N_3	10^{16} cm^{-3}	—	4	8	8	18
a	μm	0.3	0.27	0.15	0.15	0.1
a_1	μm	0.3	0.15	0.15	0.15	0.15
a_2	μm	0.17	0.12	0.12	0.12	0.12
a_3	μm	—	0.27	0.15	0.15	0.1
z	mm	0.25	0.25	0.25	0.1	0.065
z_m	mm	0.24	0.23	0.14	0.14	0.065

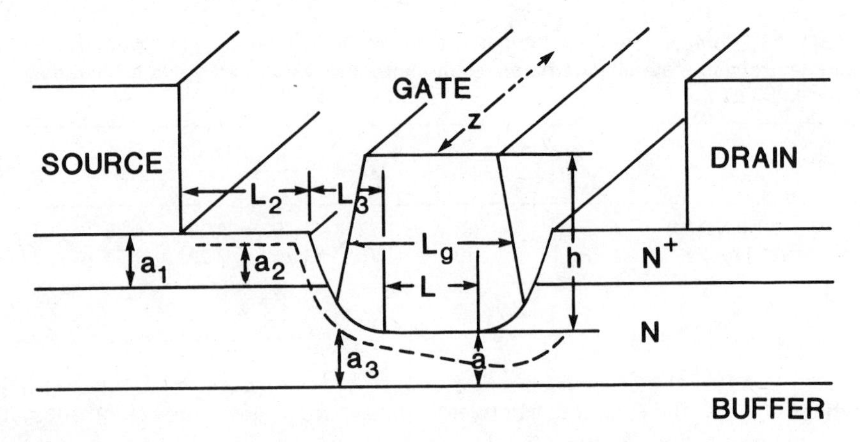

Figure 7.26 Device geometry for a recessed-gate GaAs MESFET showing device dimensions used in the noise figure calculations in the text.

7.5.6 Arbitrary Doping Profile Model and Deep Levels

The two-region model discussed in Section 7.5.2 has been extended by Golio and Trew [32], so that arbitrary doping profile devices and the effects of deep levels in the conducting channel can be investigated. The noise analysis technique presented by Pucel [18] has been applied to the general profile model by Trew et al. [33]. An investigation of recessed-gate, ion-implanted devices shows that an optimum recess depth exists for a given channel doping. For example, the device transconductance and gate–source capacitance are shown in Fig. 7.27 for devices fabricated with various implants and recess depths. Each implanted profile has the same peak doping density. The maximum transconductance is independent of implant energy, but is obtained at a critical recess depth. The optimum recess depth occurs when the gate depletion region is able to exercise maximum control over the channel current, and this occurs with a gate length to channel thickness ratio of about 3. Thicker channels decrease the modulation efficiency, and thinner channels decrease the channel current.

Although deep-levels generally charge and discharge with time constants too long to directly affect the RF performance of microwave devices, they influence the gain and noise performance of MESFETs through an indirect process. They degrade the charge-carrier transport characteristics by introducing additional scattering centers. This effect is illustrated in Fig. 7.28 where the electron mobility for an ion-implanted channel is shown for various deep-level densities. The degree of degradation increases rapidly as the deep-level density approaches the channel donor density.

The gain and noise figure performance of GaAs MESFETs are degraded by deep levels as shown in Fig. 7.29. This figure presents the associated gain and minimum noise figure calculated for devices fabricated with various channel implants. Each device has the same peak doping density and is recessed so that all devices have the same I_{dss}. The data are shown

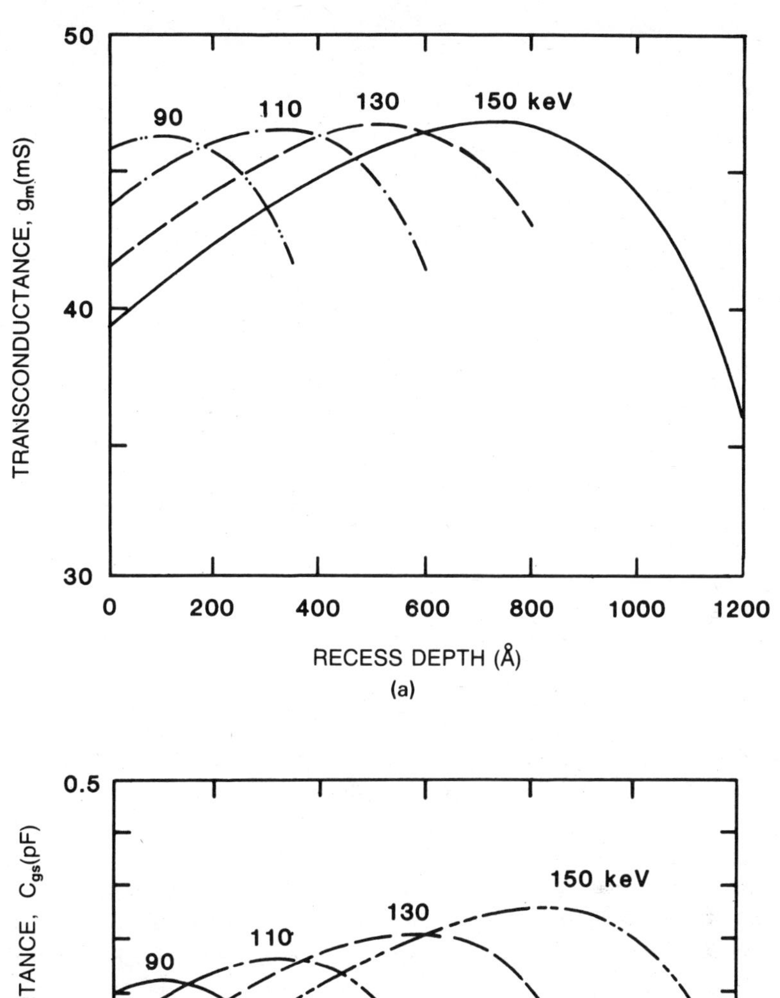

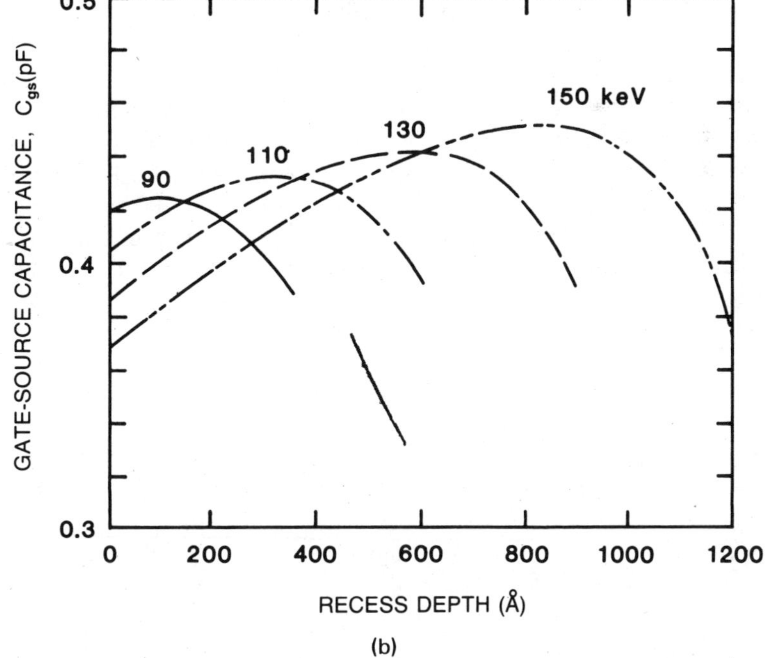

Figure 7.27 (*a*) Device transconductance vs. recess depth for devices fabricated with various implant energies. (*b*) Gate–source capacitance vs. recess depth for GaAs MESFETs fabricated with various implant energies.

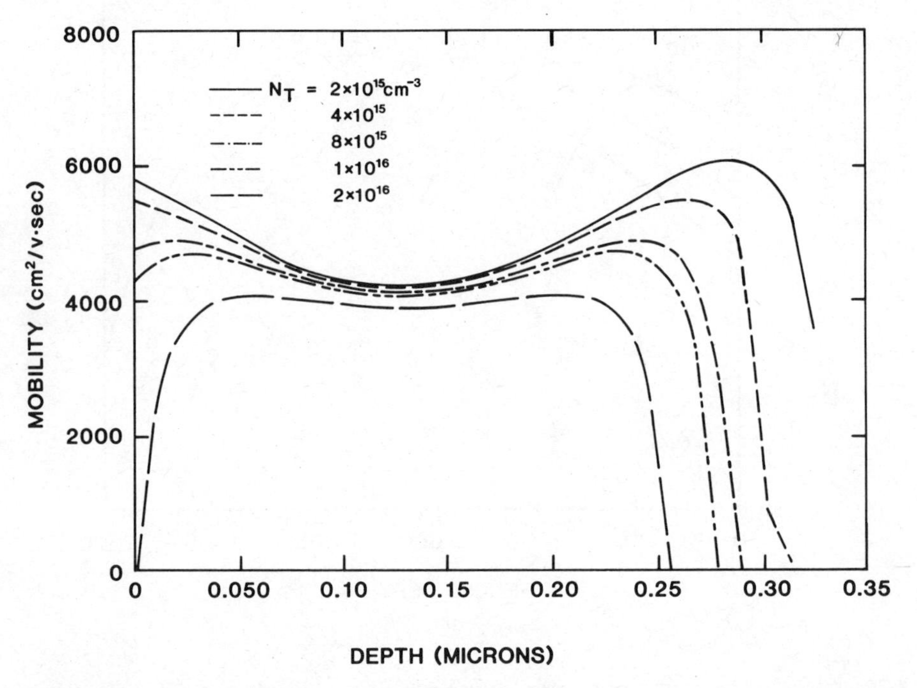

Figure 7.28 Channel electron mobility as a function of depth and deep-level density. The depth is measured from the channel–gate metal interface into the conducting channel.

for a frequency of 8 GHz. This figure shows the associated gain to be independent of implant energy, but to be reduced by the deep-level density. The minimum noise figure decreases with implant energy for low deep-level density, but increases with implant energy for high deep-level density. This indicates that the lowest noise figures will be obtained from devices fabricated from high-quality material (i.e., low deep-level density) and relatively high implant energy. Deep recess depths are indicated.

7.5.7 Power FETs

The GaAs MESFET is useful as a power device. Transistors can be designed for maximum output power or maximum power-added efficiency, although it is difficult to obtain both conditions simultaneously. For example, at X-band commercial devices are available that produce over 6-W RF power with over 40% efficiency, while other designs produce about 10-W RF power with about 20% efficiency. The gain for these devices is usually in the range of 5–10 dB. Large RF powers are generally only possible over relatively narrow bandwidths.

In order to obtain high output RF power it is necessary to operate the device at high drain–source voltage and current. This requirement has

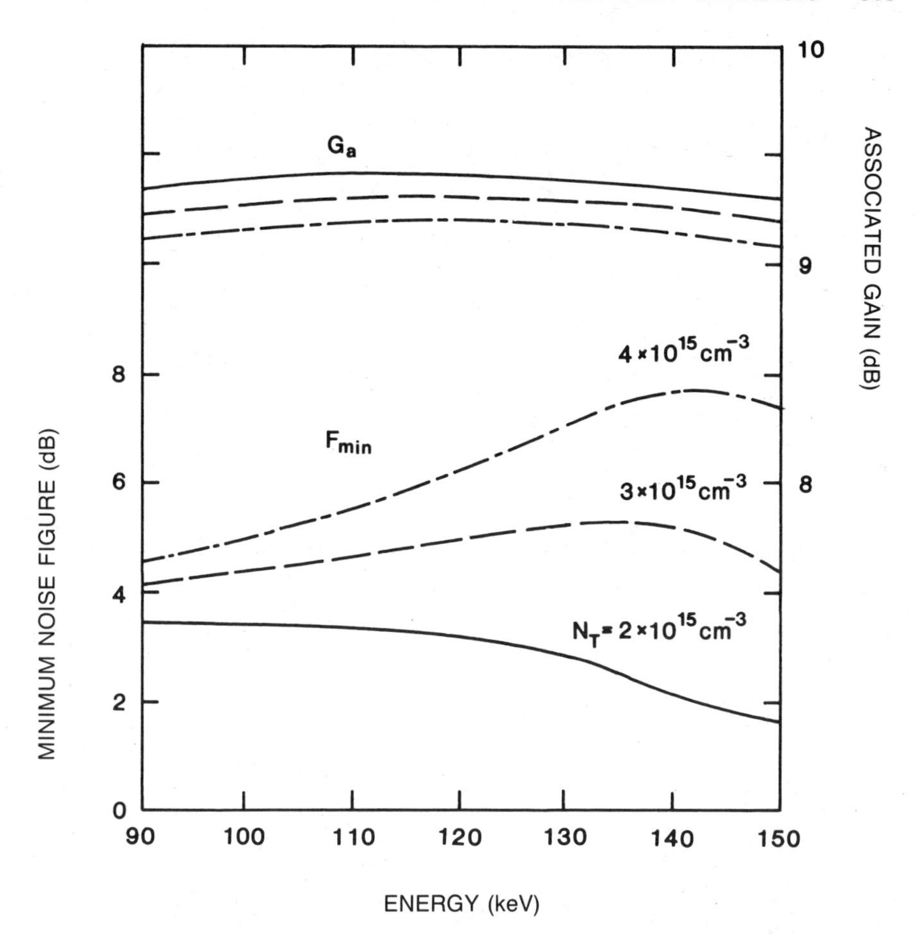

Figure 7.29 Gain and noise figure vs. implant energy. The peak doping for all devices is 2×10^{17} cm^{-3}, the devices are biased at the optimum low-noise bias, $I_{dss} = 30$ mA, and the frequency is 8 GHz.

implications for the design of power devices. In order to support high drain bias voltage it is necessary that the structure have a high drain–gate breakdown voltage. This suggests low channel doping density and large gate–drain spacing. Low channel doping density, however, is inconsistent with high channel current, and large drain–gate spacing increases parasitics and limits operation frequency. The spacing criterion is addressed with the use of recessed-gate structures. For example, the basic power FET structures are shown in Fig. 7.30. The flat channel device has a low breakdown voltage. A simple recess of the gate into the channel increases the gate–drain spacing and results in an increase in breakdown voltage [34]. Frensley [35] has presented a good analysis of the power-limiting effects of breakdown.

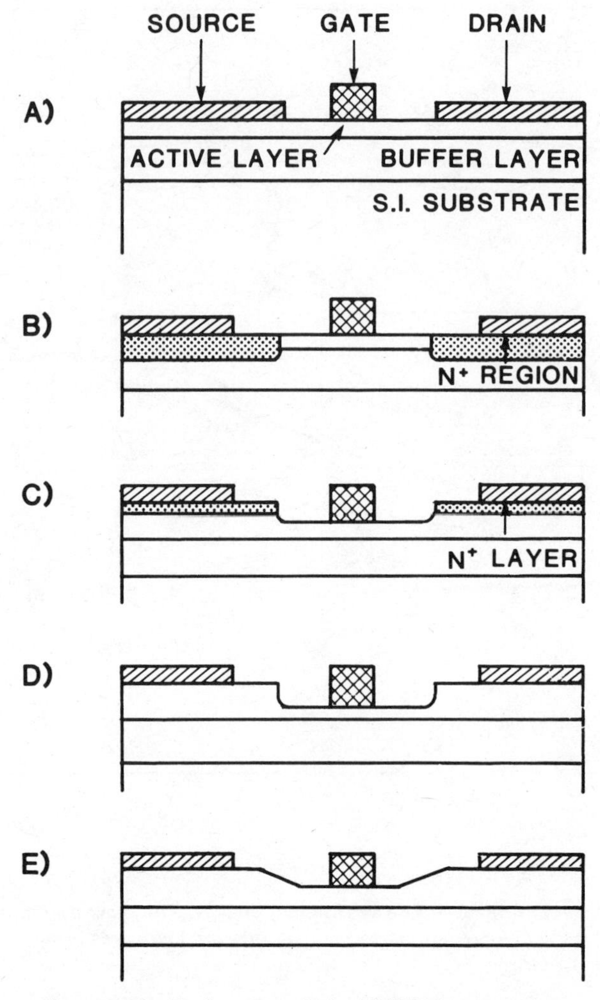

Figure 7.30 Basic power GaAs MESFET geometries.

It is known that breakdown occurs in the high field domain that forms in the gate–drain region of the channel. A smooth electric-field transition in this region results in an increase in the breakdown voltage, and this situation can be achieved with the graded recess structure. Breakdown voltages in excess of 20 V can be obtained with this structure. Excellent performance has also been obtained with an n^+ contact ledge structure with a double recess. An investigation of this structure [36] shows that the gate recess should be as narrow as possible. The wide recess should be 0.5 to 0.7 μm wider than the gate and the gate-recess depth should be about 0.05 to 0.09 μm. Devices of this type produce more than 1 W/mm of gate width at X-band with high gain ($\sim$10 dB) and efficiency ($\sim$40%).

The breakdown condition limits the extent to which doping can be increased to achieve high channel current. Other ways to increase the current are required. One technique is to increase the thickness of the channel. This method, however, decreases the modulation efficiency of the gate and the gain of the device. Typical channel thickness is limited to the range of 0.2 to 0.4 μm at X-band, and the thickness must be scaled inversely with frequency. The most effective method for increasing channel current is to increase the gatewidth. At X-band, gatewidths in the range of 0.4 to 4 mm are typically used, and gatewidths in the range of 20 to 40 mm are possible. Gatewidths of this magnitude are long relative to a wavelength and, in order to avoid distributed effects, multiple gate feeds and power-combining techniques are used. Generally, the gatewidth that an individual gate feeds should be less than a tenth of a wavelength, and at X-band this limits the gatewidth cell to the range of 150 to 300 μm. Higher frequency devices, of course, require reduced gate width per gate feed.

Various power-combining techniques have been used. The most commonly used technique involves the direct paralleling of multiple gate feeds, or fingers. This technique requires multiple bond wires or crossover structures, as shown in Fig. 7.31. The crossover is achieved with a metal that connects the various gate fingers. The metal must be isolated from the source contacts, and this is achieved by depositing the metal over a dielectric or air dielectric. The air bridge crossover is fabricated by depositing the metal over a photoresist layer that has been deposited over the source contact and channel region, excluding the gate contact. The air bridge results when the photoresist is dissolved and removed. The air bridge is preferred to the dielectric overlay due to reduced parasitic capacitance. From a circuit perspective, wide gatewidths produce low input and output impedances and this makes it difficult to design broadband power amplifiers.

Heat dissipation is important in power FET design, and adequate heat sinking must be provided. Various techniques have been investigated for reducing the thermal resistance of the device and mounting structure so that the temperature rise of the channel can be limited. The heat-dissipation problem in power FETs is increased by the poor thermal conductivity of GaAs. This fundamental problem can be addressed by thinning the semi-

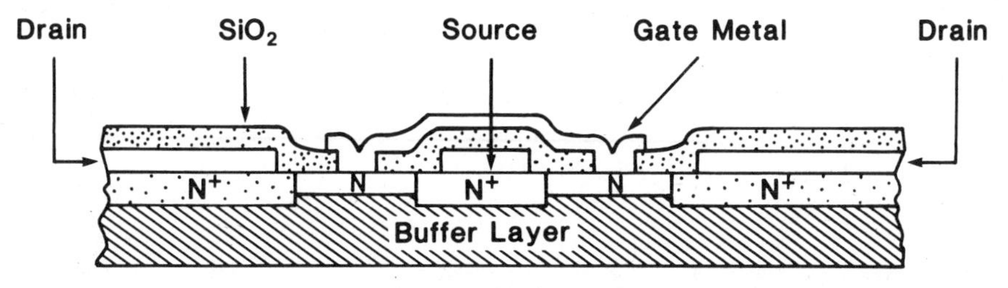

Figure 7.31 A power GaAs MESFET structure using overlays to connect the gate fingers.

insulating substrate as much as possible. The device can also be flip-chip mounted, so that the main heat flow path does not pass through the substrate. This technique is difficult to apply in many situations. The most effective technique for heat sinking involves the use of via holes with thin substrate chips. The via holes are fabricated by etching holes through the substrate under the source contacts and then filling the holes with a metal. When used with plated heat sink technology this technique produces

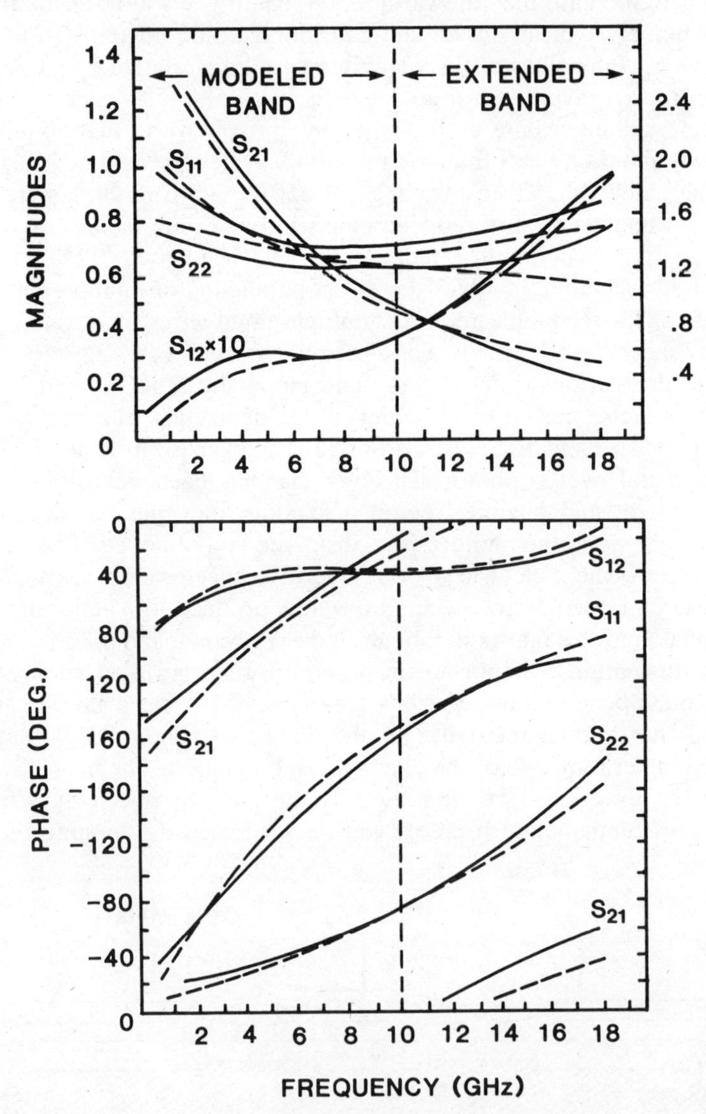

Figure 7.32 Measured and modeled *S*-parameters for a power GaAs MESFET. Measured ——; modeled ----. (H. A. Willing et al. [37], 1978 IEEE, reprinted with permission.)

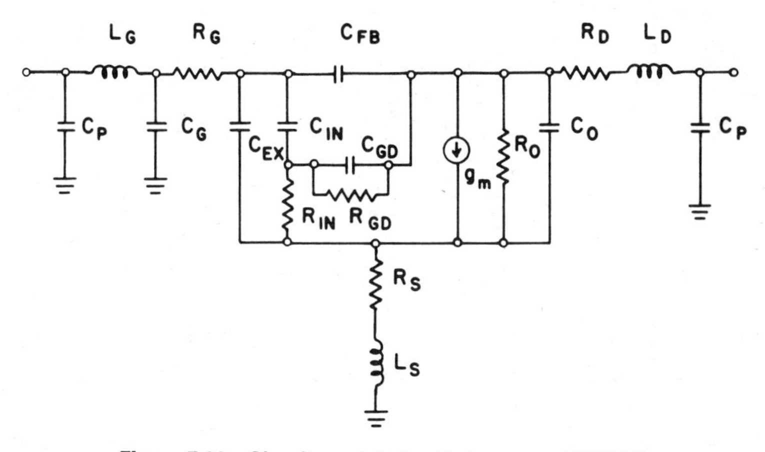

Figure 7.33 Circuit model of a GaAs power MESFET.

devices with low thermal resistance. It also produces low parasitic source resistance and inductance.

In order to analyze the large-signal operation of GaAs MESFETs, a nonlinear model is necessary, and such models are difficult to develop. Most work in this area has been directed toward the development of equivalent-circuit models based upon the characterization of actual devices. This

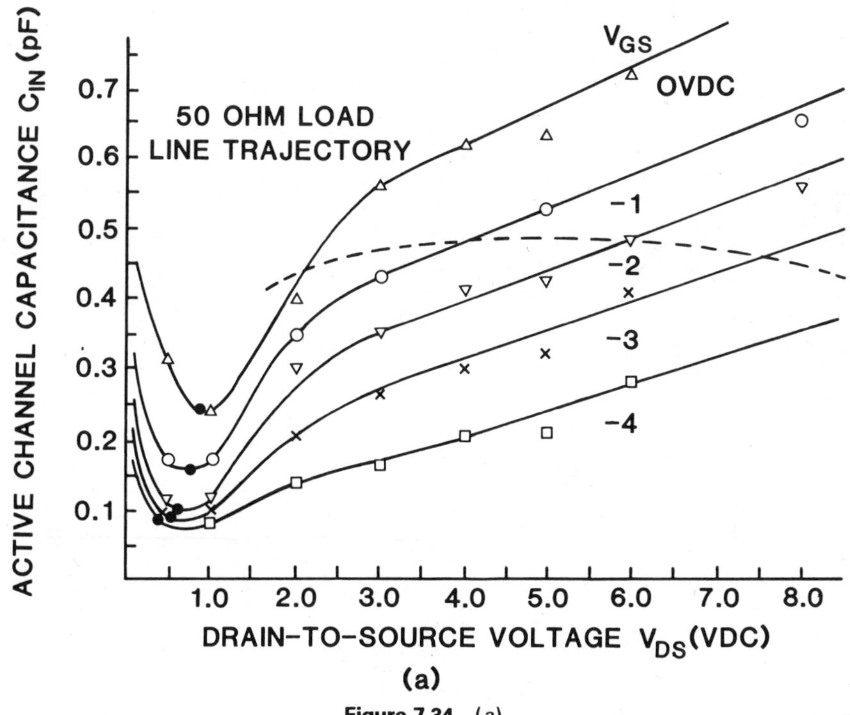

Figure 7.34 (*a*)

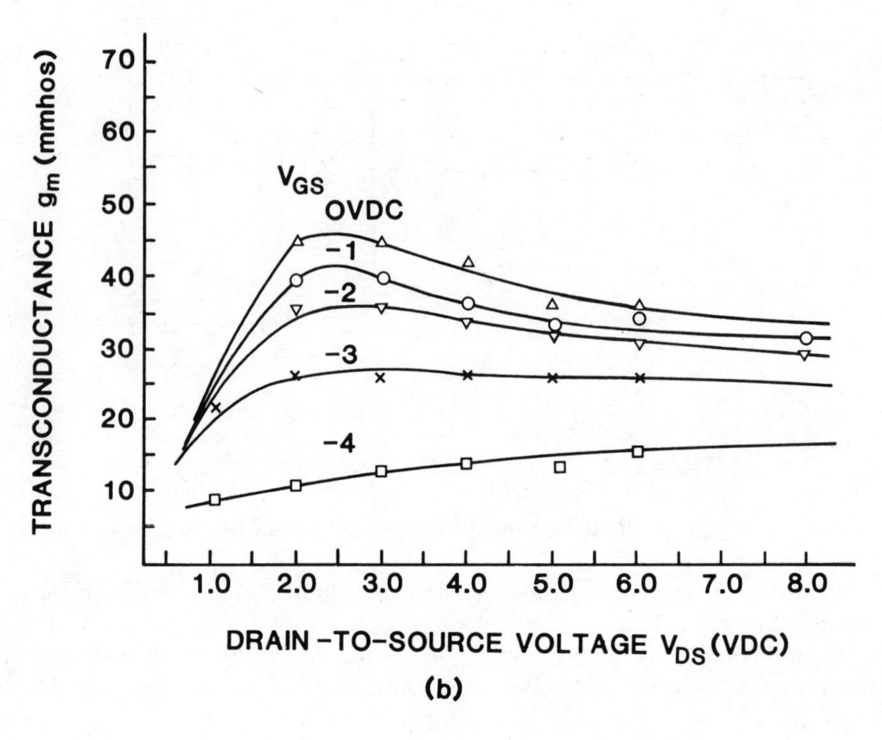

(b)

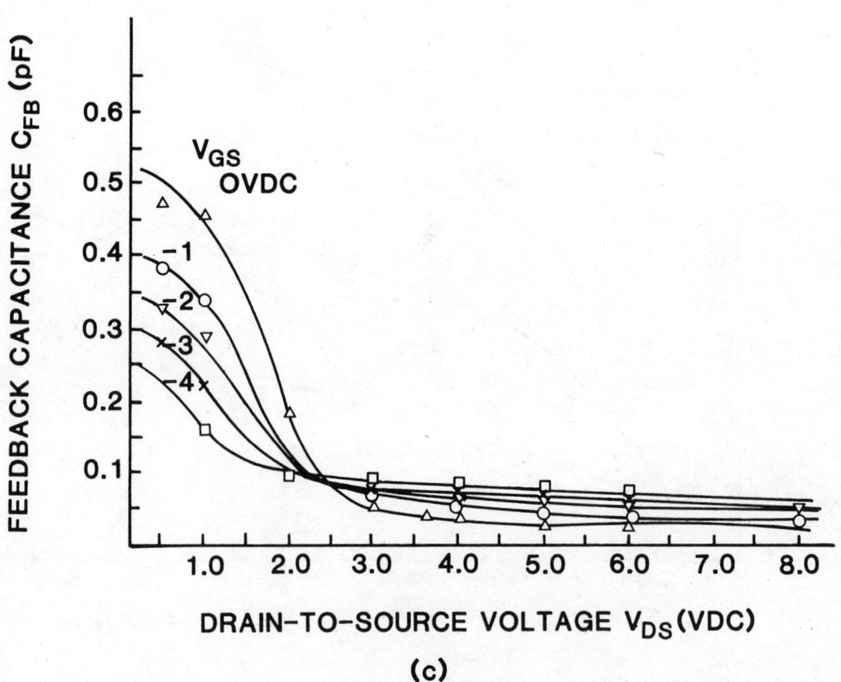

(c)

Figure 7.34 (*b, c*)

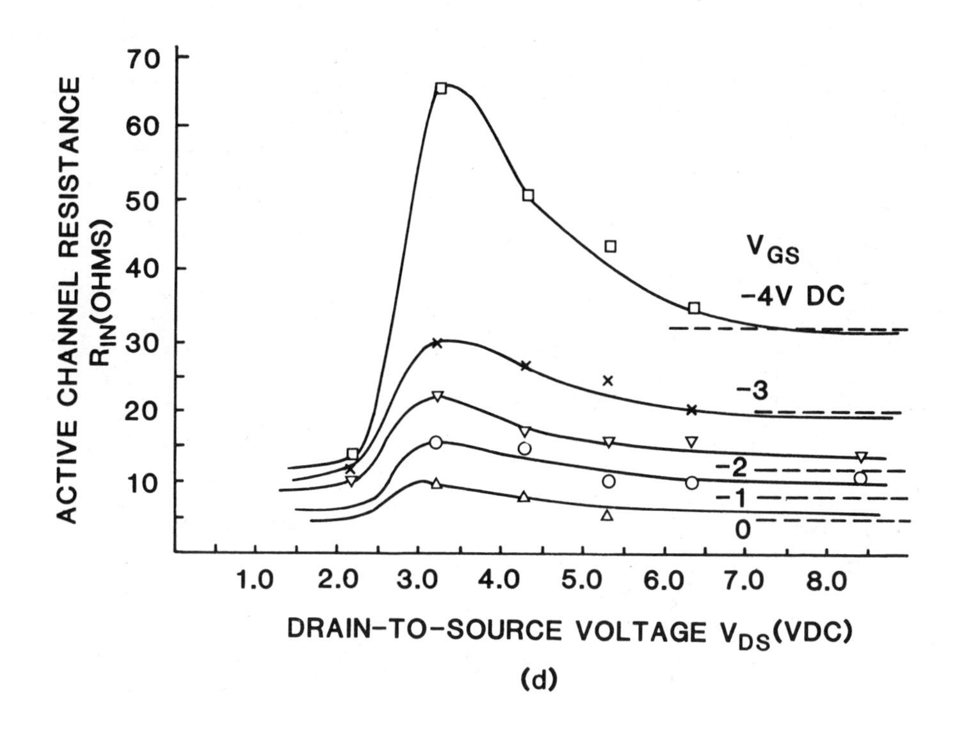

(d)

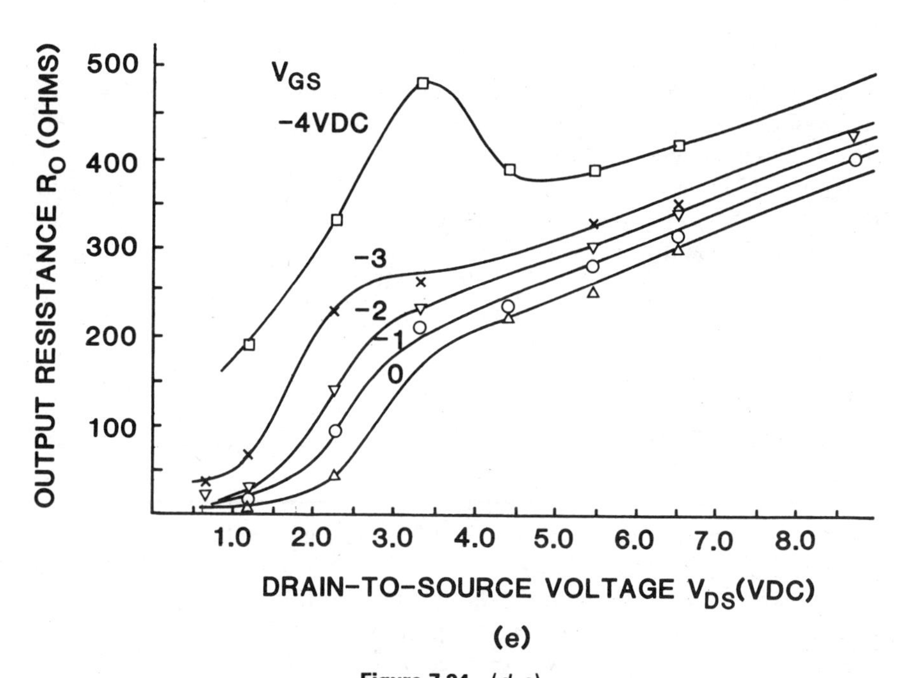

(e)

Figure 7.34 (*d, e*)

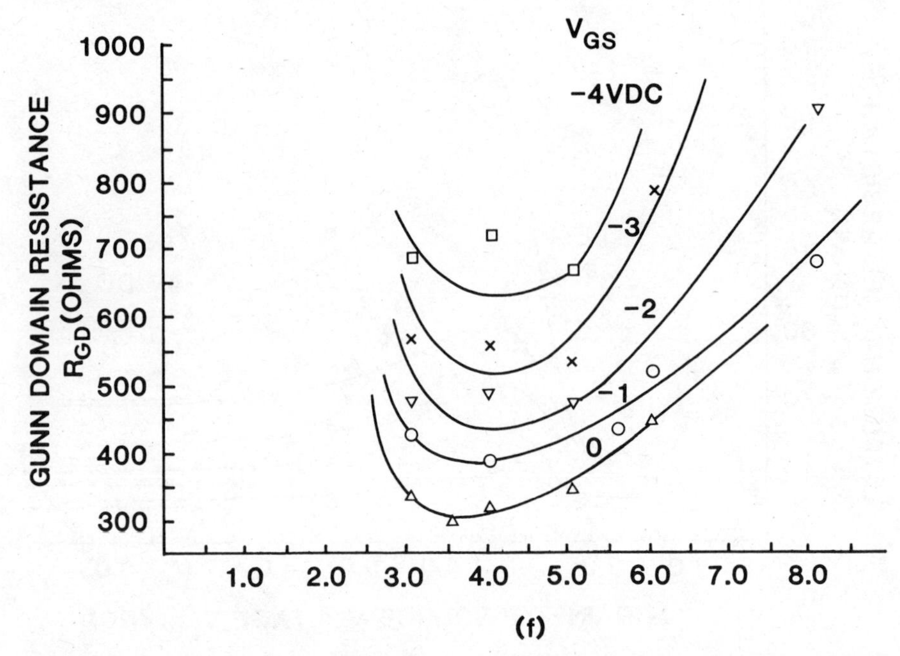

Figure 7.34 Measured bias dependence of the GaAs MESFET equivalent-circuit elements showing characteristics under large-signal operation. (H. A. Willing et al. [37], 1978 IEEE, reprinted with permission.)

semiempirical technique produces useful working models, but has the disadvantage that the device must be fabricated before it can be analyzed. Most large-signal equivalent circuits are derived in this manner.

Willing [37] has presented the large-signal performance of a GaAs MESFET. The FET was characterized with S-parameters for a complete range of drain–source voltages. This information was then used to extract equivalent circuit element values. The S-parameters determined are shown in Fig. 7.32, and the bias-dependent equivalent-circuit elements (Fig. 7.33) are shown in Fig. 7.34.

7.6 COMPARISON OF BIPOLAR TRANSISTOR AND MESFET NOISE FIGURES

In low-noise applications GaAs MESFETs are generally preferred to Si bipolar transistors. The MESFET demonstrates a lower noise figure than the bipolar transistor throughout the microwave frequency range, and the advantage increases with frequency. This advantage is demonstrated by the comparison of the expressions for the minimum noise figure for the two devices. The bipolar transistor has a minimum noise figure of the form

$$F_{\min} \cong 1 + bf^2 \left[1 + \sqrt{1 + \frac{2}{bf^2}} \right] \tag{7.125}$$

where

$$b = \frac{40 I_c r_b}{f_T^2}$$

where r_b is the parasitic base resistance and I_c is the collector current. The minimum noise figure for the MESFET is written in the form

$$F_{\min} \cong 1 + mf \tag{7.126}$$

where

$$m = \frac{2.5}{f_T} \sqrt{g_m (R_g + R_s)}$$

Comparing these expressions shows that the minimum noise figure increases with frequency quadratically for bipolar transistors and linearly for MESFETs. Therefore, the MESFET demonstrates increasingly superior noise figure performance as compared to Si bipolar transistors as the operating frequency is increased.

REFERENCES

1. Ino, M., T. Ishibashi, and M. Ohmori, "Submillimeter Wave Si p^+pn^+ IMPATT Diodes," *Japan. J. Appl. Phys.*, Vol. 16, Supplement 16-1, 1977, pp. 89–92.
2. Nishizawa, J. I., K. Motoya, and Y. Okuno, "GaAs TUNNETT Diodes," *IEEE Trans. Microwave Theory Tech.*, Vol. MTT-26, Dec. 1978, pp. 1029–1035.
3. Tserny, H. Q., and B. Kim, "110 GHz GaAs FET Oscillator," *Electron. Lett.*, Vol. 21, Feb. 1985, pp. 178–179.
4. Leuny, C. C., C. P. Snapp, and V. Grande, "A 0.5 μm Silicon Bipolar Transistor for Low Phase Noise Oscillator Applications Up To 20 GHz," *IEEE Int. Microwave Symp. Digest, 1985*, pp. 383–386.
5. Cooke, H. F., "Microwave Transistors: Theory and Design," *Proc. IEEE*, Vol. 59, Aug. 1971, pp. 1163–1181.
6. Early, J. M., "Effects of Space-Charge Layer Widening in Junction Transistors," *Proc. IRE.*, Vol. 40, Nov. 1952, pp. 1401–1406.
7. Kirk, C. T., "A Theory of Transistor Cutoff (f_T) Fall-Off at High Current Density," *IRE Trans. Electron Devices*, Vol. ED-9, Mar. 1954, pp. 164–174.
8. Webster, W. M., "On the Variation of Junction Transistor Current Amplification with Emitter Current," *Proc. IRE*, Vol. 42, June 1954, pp. 914–920.

9. Ebers, J. J., and J. L. Moll, "Large Signal Behavior of Junction Transistors," *Proc. IRE*, Vol. 42, Dec. 1954, pp. 1761–1772.
10. Mason, S. J., "Power Gain in Feedback Amplifiers," *IRE Trans. Circuit Theory*, Vol. CT-1, June 1954, pp. 20–25.
11. Kronquist, R. L., J. Y. Forrier, J. P. Pestil, and M. E. Brilman, "Determination of a Microwave Transistor Model Based on an Experimental Study of its Internal Structure," *Solid-State Electronics*, Vol. 18, Nov. 1975, pp. 949–963.
12. Netzer, Y., "The Design of Low-Noise Amplifiers," *Proc. IEEE*, Vol. 69, June 1981, pp. 728–741.
13. Hawkins, R. J., "Limitations of Nielsen's and Related Noise Equations Applied to Microwave Bipolar Transistors, and a New Expression for the Frequency and Current Dependent Noise Figure," *Solid-State Electron.*, Vol. 20, Mar. 1977, pp. 191–196.
14. Hsu, T. H., and C. P. Snapp, "Low-Noise Microwave Bipolar Transistor with Sub-Half-Micrometer Emitter Width," *IEEE Trans. Electron Devices*, Vol. ED-25, June 1978, pp. 723–730.
15. Stuetzer, O. M., "A Crystal Amplifier with High Input Impedance," *Proc. IRE.*, Vol. 38, Aug. 1950, p. 868.
16. Shockley, W., "Unipolar 'Field Effect' Transistor," *Proc. IRE.*, Vol. 40, Nov. 1952, p. 1365.
17. Liechti, C. A., "Microwave Field-Effect Transistors—1976," *IEEE Trans. Microwave Theory Tech.*, Vol. MTT-24, June 1976, pp. 279–300.
18. Pucel, R. A., H. Statz, and H. A. Haus, "Signal and Noise Properties of Gallium Arsenide Microwave Field-Effect Transistors," *Advances in Electronics and Electron Physics*, Vol. 38, Academic Press, New York, 1975, pp. 195–265.
19. Golio, J. M., and R. J. Trew, "Compound Semiconductors for Low-Noise Microwave MESFET Applications," *IEEE Trans. Electron Devices*, Vol. ED-27, July 1980, pp. 1256–1262.
20. Grebene, A. B., and S. K. Ghandhi, "General Theory for Pinched Operation of the Junction Gate FET," *Solid-State Electron.*, Vol. 12, July 1969, pp. 573–589.
21. Wasserstron, G., and J. McKenna, "The Potential Due to a Charged Metallic Strip on a Semiconductor Surface," *Bell Sys. Tech. J.*, Vol. 49, May–June 1970, pp. 853–877.
22. Das, M. B., and P. Schmidt, "High-Frequency Limitations of Abrupt Junction FETs," *IEEE Trans. Electron Devices*, Vol. ED-20, Sept. 1973, pp. 779–792.
23. Steer, M. B., and R. J. Trew, "High-Frequency Limits of Millimeter Wave Transistors," *IEEE Electron Devices Lett.*, Vol. EDL-7, Nov. 1986, pp. 640–642.
24. Trew, R. J., and M. B. Steer, "Millimeter-Wave Performance of State-of-the-Art MESFET and PBT Transistors," *Electron. Lett.*, Vol. 23, Feb. 1987, pp. 149–151.
25. A. Vander Ziel, "Thermal Noise in Field-Effect Transistors," *Proc. IRE.*, Vol. 50, Aug. 1962, pp. 1808–1812.
26. A. Vander Ziel, "Gate Noise in Field-Effect Transistors at Moderately High Frequencies," *Proc. IEEE.*, Vol. 51, Mar. 1963, pp. 461–467.

27. Fukui, H., "Determination of the Basic Device Parameters of a GaAs MESFET," *Bell Sys. Tech. J.*, Vol. 58, Mar. 1979, pp. 771–797.
28. Goronkin, H., and V. Nair, "Comparison of GaAs MESFET Noise Figures," *IEEE Electron Devices Lett.*, Vol. EDL-6, Jan. 1985, pp. 47–49.
29. Fukui, H., "Optimal Noise Figure of Microwave GaAs MESFETs," *IEEE Trans. Electron Devices*, Vol. Ed-26, July 1979, pp. 1032–1037.
30. Rothe, H., and W. Dahlke, "Theory of Noisy Fourpoles," *Proc. IRE.*, Vol. 44, June 1956, pp. 811–818.
31. Fukui, H., "Design of Microwave GaAs MESFETs for Broad-Band Low-Noise Amplifiers," *IEEE Trans. Microwave Theory Tech.*, Vol. MTT-27, July 1979, pp. 643–650.
32. Golio, J. M., and R. J. Trew, "Profile Studies of Ion-Implanted MESFETs," *IEEE Trans. Microwave Theory Tech.*, Vol. MTT-31, Dec. 1983, pp. 1066–1071.
33. Trew, R. J., M. A. Khatibzadeh, and N. A. Masnari, "Deep-Level and Profile Effects Upon Low-Noise Ion-Implanted GaAs MESFETs," *IEEE Trans. Electron Devices*, Vol. ED-32, May 1985, pp. 877–882.
34. Furutsuka, T., T. Tsuji, and F. Hasagawa, "Improvement of the Drain Breakdown Voltage of GaAs Power MESFETs by a Simple Recess Structure," *IEEE Trans. Electron Devices*, Vol. ED-25, June 1978, pp. 563–567.
35. Frensley, W. R., "Power Limiting Breakdown Effects in GaAs MESFETs," *IEEE Trans. Electron Devices*, Vol. ED-28, Aug. 1981, pp. 1019–1024.
36. Macksey, H. M., "Optimization of the n^+ Ledge Channel Structure for GaAs Power FETs," *IEEE Trans. Electron Devices*, Vol. ED-33, Nov. 1986, pp. 1818–1824.
37. Willing, H. A., C. Raushcher, and P. deSantis, "A Technique for Predicting Large-Signal Performance of a GaAs MESFET," *IEEE Trans. Microwave Theory Tech.*, Vol. MTT-26, Dec. 1978, pp. 1017–1023.

PROBLEMS

7.1 The one-dimensional time dependent continuity equation for free holes in the base of a uniformly doped *pnp* transistor is

$$D_{pB} \frac{\partial^2 (p_B - p_{Bo})}{\partial x^2} - \frac{p_B - p_{Bo}}{\tau_{pB}} = \frac{\partial (p_B - p_{Bo})}{\partial t}$$

Assume a sinusoidal time dependence so that

$$p_B(x, t) - p_{Bo} = U(x) e^{j\omega t}$$

(a) Derive a differential equation for $U(x)$.

(b) Determine a solution for $U(x)$ and write the solution in the form of a traveling wave (i.e., exponential form with real and imaginary arguments).

(c) Use the $U(x)$ solution to derive an expression for a frequency dependent diffusion length L'_B.

(d) Show that for high frequency (i.e., $\omega\tau_{pB} \gg 1$) $L'_B \simeq 2D_{pB}/\omega$ and for low frequency (i.e., $\omega\tau_{pB} \ll 1$) $L'_B = L_B$.

7.2 Show that

$$\coth x - \operatorname{csch} x = \tanh \frac{x}{2}$$

This relationship was used to arrive at the expression we derived for the base current of a bipolar transistor.

7.3 Discuss the unity gain cutoff frequency (f_T) and the maximum frequency of oscillation ($f_{\max}$) for Si bipolar transistors and GaAs MESFETs. Compare these figures-of-merit for the two devices explaining their physical significance. Estimate (using calculations if possible) their upper frequency limits.

7.4 For a *pnp* transistor the collector current can be written as

$$I_C \simeq \frac{qD_{pB}n_i^2 A \exp(qV_{EB}/kT)}{\int_0^{W_B} n(x)\,dx}$$

Show that the change in collector current as a function of collector-base voltage can be written as

$$\frac{\partial I_C}{\partial V_{CB}} = -\frac{I_C}{V_A}$$

where V_A is the Early voltage. Write an expression for the Early voltage in terms of the Gummel number Q_B.

7.5 When considering the RF operation of a bipolar transistor we derived an expression for the common-base current gain that had the form

$$\alpha(\omega) \simeq \operatorname{sech} \frac{W_B}{L'_B}$$

where W_B is the width of the undepleted base region and L'_B is the effective diffusion length for minority carriers in the base region. Show that this expression leads to a single pole model for α and that α can be written as

$$\alpha(\omega) \simeq \frac{1}{1 + j(\omega/\omega_\alpha)}$$

where

$$\omega_\alpha \triangleq \frac{2D_{pB}}{W_B^2}$$

7.6 Consider a *pnp* bipolar transistor with exponential doping in the base region as shown in the sketch. Let $N_B(x) = N_B(0)\exp(-ax/W_B)$ where $a \triangleq \ln[N_B(0)/N_B(W_B)]$. Assume one dimension, no recombination in the base region, and steady state conditions.

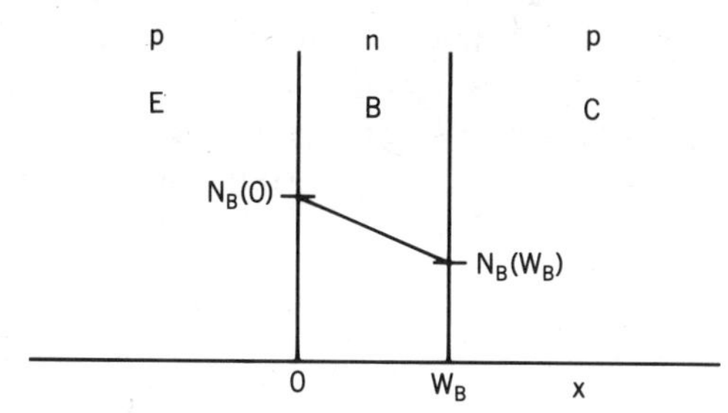

(a) Derive an expression for the electric field that exists in the base region.

(b) Derive an expression for the minority charge density in the base, $p_B(x)$. Use the boundary conditions that at $x = 0$, $p_B = p_B(0)$ and at $x = W_B$, $p_B \simeq 0$.

(c) Derive an expression for the base current, J_{pB}.

(d) Derive an expression for the minority carrier transit time through the base region, τ_t.

(e) Show that for uniformly doped devices $\tau_t = W_B^2/2D_{pB}$ and that $\tau_t \to 0$ as $a \to \infty$.

7.7 A semiconductor switch is fabricated by placing a Schottky contact on a bar of uniformly doped *n*-type semiconductor as shown in the sketch.

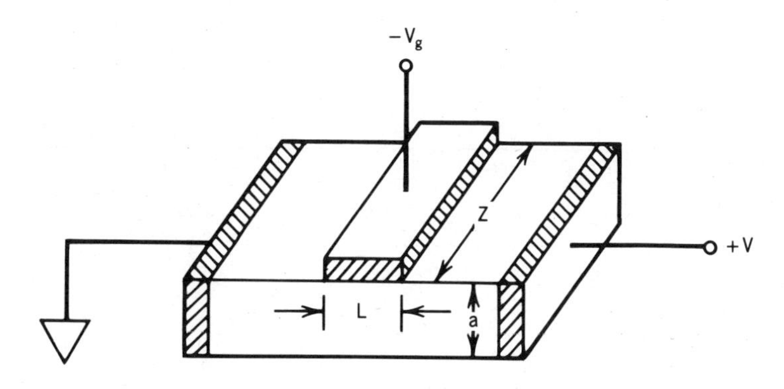

If $L \gg a$, derive an expression for the control electrode voltage V_g required to turn the switch off. State any assumptions or approximations you make.

7.8 A Si *npn* transistor is fabricated with uniform doping concentrations and it is known that the doping densities in the emitter, base, and collector regions are $N_e = 10^{18}\ \text{cm}^{-3}$, $N_b = 10^{16}\ \text{cm}^{-3}$, and $N_c = 10^{14}\ \text{cm}^{-3}$, respectively. The base-emitter and collector-base capacitances were measured at zero bias and were found to be $C_{be} = 5.16$ pF and $C_{cb} = 4.36$ pF. The transistor was then operated in the active mode with a base-emitter voltage of $V_{be} = -0.7$ V and a collector-base voltage of $V_{cb} = 6.0$ V. The collector current was $I_c = 50$ mA. Under these conditions the cutoff frequency for the transistor was determined to be $f_t = 2$ GHz.

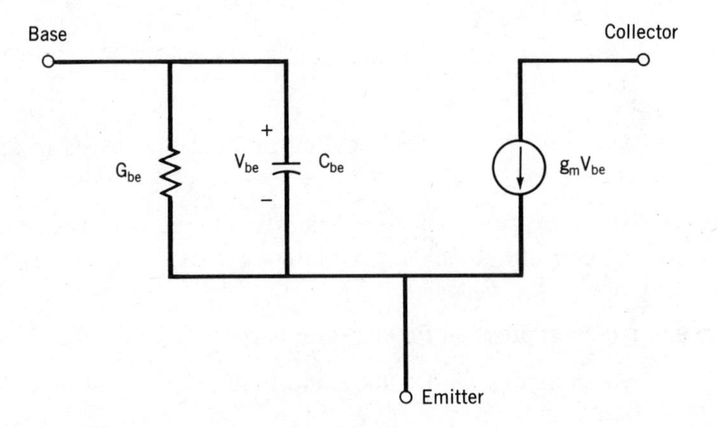

(a) Determine the alpha cutoff frequency (f_α) for the transistor.

(b) A first-order, low-frequency equivalent circuit for the device is shown in the sketch. Determine values for the equivalent circuit elements. Use the material parameters for Si listed below.

Material Parameters for Silicon

$T = 300$ K
$\epsilon_r = 11.9$
$\epsilon_0 = 8.854(10)^{-14}$ F/cm
$n_i = 1.45(10)^{10}\ \text{cm}^{-3}$
$\tau_n = \tau_p = 10^{-6}$ sec (minority carrier lifetime)

Doping (cm^{-3})	μ_n (cm^2/V-sec)	μ_p (cm^2/V-sec)
10^{14}	1450	450
10^{16}	1200	410
10^{18}	270	160

8 PASSIVE DEVICES

8.1 INTRODUCTION

The efficient control of electrical signals in RF circuits requires the use of nonlinear devices. These devices can be either active or passive. In hybrid circuit applications active devices (e.g., transistors or FETs) are generally used for signal generation and amplification, whereas passive devices (e.g., diodes) are often used for signal detection, frequency shifting (i.e., mixing), and control. Active devices are often used in place of passive devices in monolithic integrated-circuit applications where, due to the fabrication process, it is not difficult to obtain large numbers of well-matched devices. In hybrid circuits it is generally preferrable to use passive devices where possible due to their low-loss and optimum performance characteristics.

The rectifying characteristics of certain semiconducting materials have been known for a long time. For example, in 1874 Braun* reported work on the asymmetric conduction properties of metal dots placed upon lead sulphide crystals. Point-contact diodes were probably the first semiconductor devices to make use of this phenomenon and be extensively used in electronic equipment. These devices found widespread use as detectors in early radios, and are still used for a variety of applications (e.g., millimeter-wave detectors).

Rectifying junctions are created by the establishment of an electrostatic barrier in the path of current flow. In order for the junction to conduct electricity, charge carriers must have enough energy to overcome the barrier. Although, under certain conditions tunneling through the barrier is possible, the most common mechanism for charge conduction is thermionic emission over the barrier. By controlling the barrier with external bias, current control is achieved.

In this chapter the basic operating principles for various types of diode elements are discussed. The chapter begins with a description of the physics responsible for the operation of *pn* and Schottky barrier junctions. The remainder of the chapter is concerned with the utilization of the nonlinear characteristics of these rectifying junctions for producing circuit elements useful for controlling RF signals.

*Starting in 1874 Karl Ferdinand Braun (1850–1918) reported observations in which he noted the dependence of crystal electrical resistance on the polarity of the applied bias and on the detailed surface conditions. An excellent review of the historical record of rectifying contacts is presented in Henisch's book [1].

8.2 *pn* JUNCTIONS

pn junctions are formed in semiconductor crystals when the conductivity type changes from *n*-type to *p*-type. Free electrons and holes diffuse across the interface region where they recombine when they reach the opposite conductivity type material. The fixed donor and acceptor atoms that remain on each side of the junction create an electric dipole that results in an electrostatic potential barrier centered at the metallurgical junction between the *n*- and *p*-type material. The energy band diagrams for the formation of a *pn* junction are shown in Fig. 8.1.

The contact potential V_0 represents the barrier that must be overcome for charge to transfer from one side of the junction to the other. In terms of the semiconductor conduction band, the contact potential can be expressed as

$$qV_0 = \mathscr{E}_{cp} - \mathscr{E}_{cn} \tag{8.1}$$

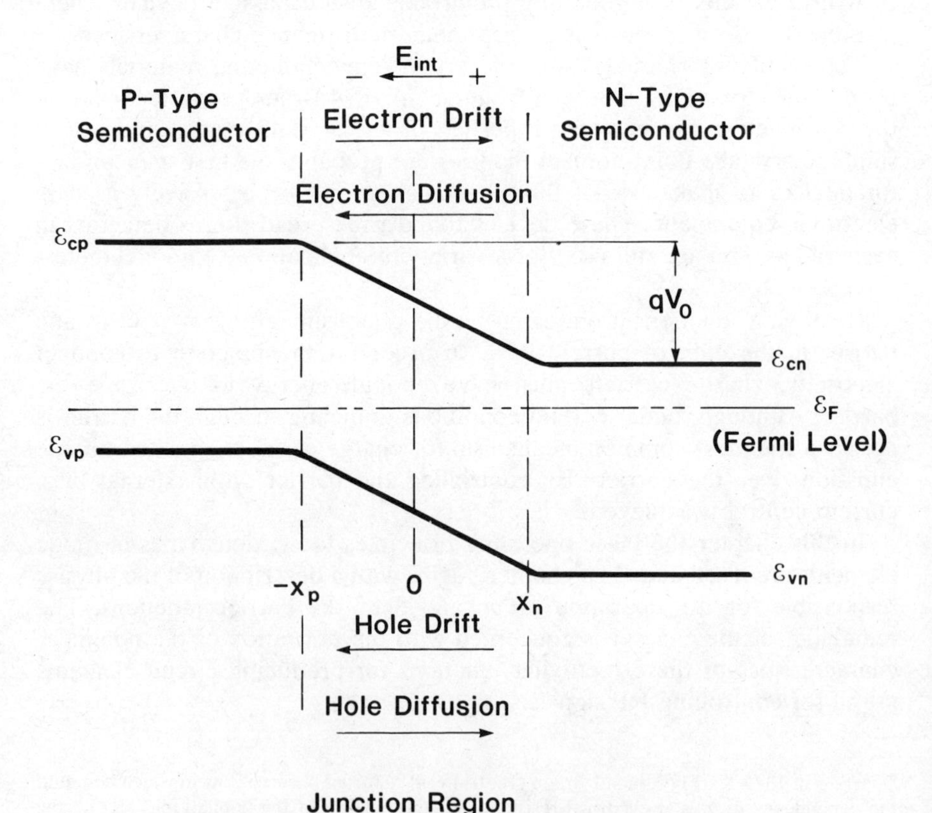

Figure 8.1 Energy band diagram for a *pn* junction diode.

where $\mathscr{E}_{cp}$ and $\mathscr{E}_{cn}$ are the *p*- and *n*-type conduction-band energies, and q is the electron charge. The contact potential depends upon the doping concentrations upon each side of the junction according to the expression

$$V_0 = \frac{kT}{q} \ln \frac{p_{p0} n_{n0}}{n_i^2} \cong \frac{kT}{q} \ln \frac{N_a N_d}{n_i^2} \tag{8.2}$$

where p_{p0} and n_{n0} are the thermal equilibrium densities of majority carriers, and N_a and N_d are the fixed acceptor and donor impurity densities. Note that the contact potential is also a function of the semiconductor material through the intrinsic density term n_i. Generally, wide bandgap semiconductors have lower intrinsic concentrations and, consequently, higher contact potentials.

The contact potential expression given in (8.2) can be manipulated to obtain the minority carrier densities on each side of the junction in terms of the majority (doping) concentrations on the other. For uniform doped junctions the majority charge density on each side of the junction is essentially equal to that of the impurity doping. The resulting expression is sometimes called the "law of the junction" and is expressed as

$$\frac{p_p(-x_p)}{p_n(x_n)} = \frac{n_n(x_n)}{n_p(-x_p)} = e^{q(V_0 \pm V)/kT} \tag{8.3}$$

Equation 8.3 is written for bias situations and the contact potential V_0 has been replaced with the total junction voltage consisting of the contact potential and the applied bias ($V_0 \pm V$). The negative sign is used for forward bias and the positive sign is used for reverse bias.

In order to obtain expressions for the electric field and potential within the device, Poisson's equation can be solved. Poisson's equation is written as

$$\frac{dE}{dx} = -\frac{d^2 \mathscr{V}}{dx^2} = \frac{q}{\epsilon}[p - n + N_d(x) - N_a(x)] \tag{8.4}$$

where p and n represent the free charge densities. The terms N_d and N_a are the fixed (doping) charge densities and ϵ is the permittivity of the material. To solve this equation, it is generally assumed that the semiconductor can be divided into quasi-neutral and space-charge regions. Applying the equation to the one-dimensional junction charge distribution shown in Fig. 8.2 yields for the region $-x_p \leq x \leq 0$ a potential distribution of the form

$$\mathscr{V} = \frac{qN_a}{2\epsilon}[(x + x_p)^2 - x_p^2] \tag{8.5}$$

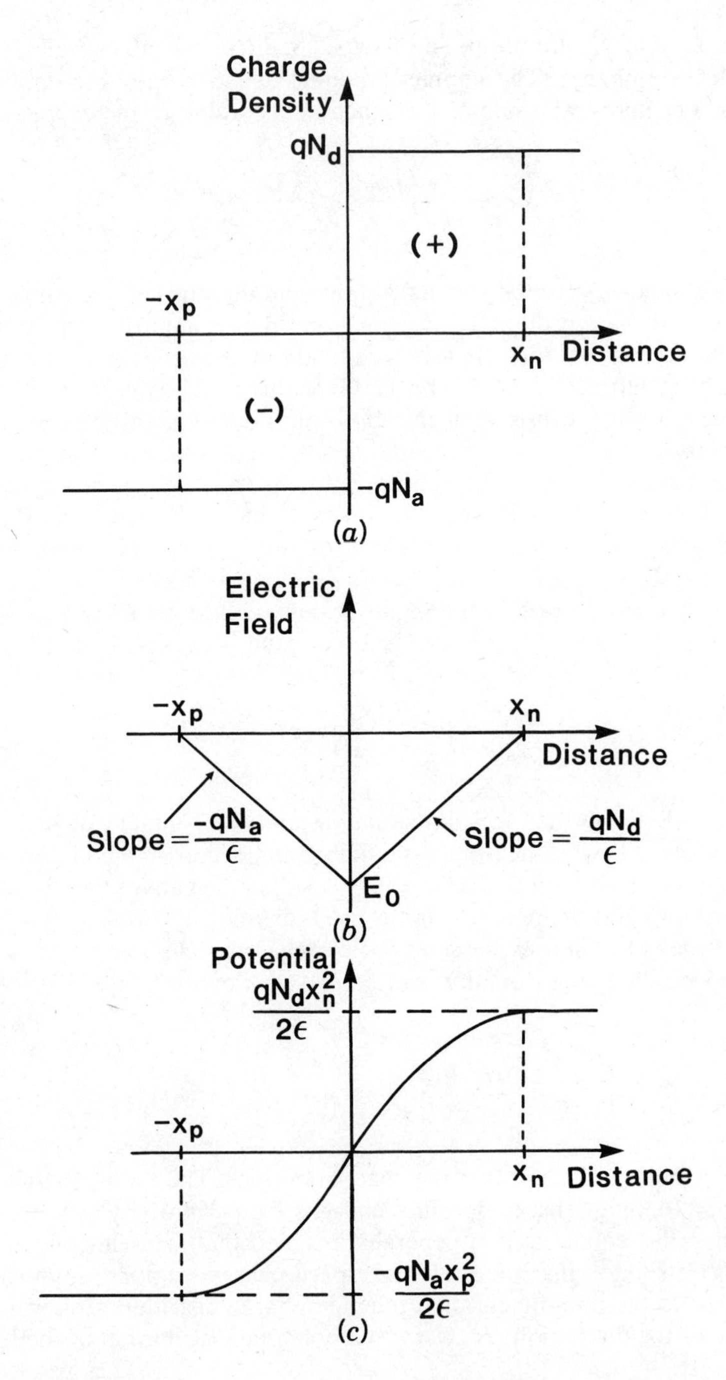

Figure 8.2 (*a*) Charge density, (*b*) electric field, and (*c*) potential distribution for a uniformly doped *pn* junction.

For the region $0 \leq x \leq x_n$, the potential is

$$\mathcal{V} = -\frac{qN_d}{2\epsilon}[(x - x_n)^2 - x_n^2] \tag{8.6}$$

Neglecting the finite resistivity of the neutral semiconductor regions, the potential in these regions will be constant with magnitudes: for $x \geq x_n$,

$$\mathcal{V} = \mathcal{V}(x_n) = \frac{qN_d x_n^2}{2\epsilon} \tag{8.7}$$

and for $x \leq -x_p$,

$$\mathcal{V} = \mathcal{V}(-x_p) = -\frac{qN_a x_p^2}{2\epsilon} \tag{8.8}$$

The electric field and potential throughout the junction area are shown in Fig. 8.2(*b*) and 8.2(*c*).

For single-crystal semiconductors with no trapping states at the interface between the *p* and *n* regions, the electric field must be continuous. The electric field at the interface can be written as

$$E(0) = E_0 = -\frac{d\mathcal{V}(0)}{dx} = -\frac{qN_d x_n}{\epsilon} = -\frac{qN_a x_p}{\epsilon} \tag{8.9}$$

This equation yields the charge balance condition, which is expressed as

$$N_d x_n = N_a x_p \tag{8.10}$$

At thermal equilibrium with no applied bias voltage the total potential drop across the semiconductor is the contact potential and can be obtained from the potential expressions presented in (8.5) and (8.6). Therefore,

$$V_0 = \mathcal{V}(x_n) - \mathcal{V}(-x_p) = \frac{q}{2\epsilon}(N_d x_n^2 + N_a x_p^2) \tag{8.11}$$

This expression can be combined with the charge balance condition to obtain the dimensions of the depletion region. The depletion depths on the *n* and *p* side are

$$x_n = \left[\frac{2\epsilon V_0 N_a}{qN_d}\left(\frac{1}{N_a + N_d}\right)\right]^{1/2} \tag{8.12a}$$

and

$$x_p = \left[\frac{2\epsilon V_0 N_d}{qN_a}\left(\frac{1}{N_a + N_d}\right)\right]^{1/2} \tag{8.12b}$$

The total depletion region width is

$$W = x_n + x_p = \left[\frac{2\epsilon V_0}{q}\left(\frac{N_a + N_d}{N_a N_d}\right)\right]^{1/2} \tag{8.13}$$

Note that the depletion region penetrates most deeply into the less heavily doped side of the junction.

The maximum value for the electric field occurs at the position $x = 0$ and can be expressed as

$$E_{\max} = F_0 = -\left[\frac{2\, V_0 q N_a N_d}{\epsilon(N_a + N_d)}\right]^{1/2} = -\frac{q}{\epsilon}\frac{N_a N_d}{(N_a + N_d)}\,W \tag{8.14}$$

As the doping on both sides of the junction is increased, the maximum electric field increases. For very high doping densities it is possible for the electric field to reach and exceed the critical field for tunnel emission ($E_c \sim 10^6$ V/cm for Si and GaAs) and tunnel breakdown will occur. For lower doping levels the junction will break down by avalanche ionization, and this occurs for electric fields on the order of 500 kV/cm.

8.2.1 The Ideal Diode Equation

When a potential is applied across the *pn* junction, a current will flow. The applied potential modifies the junction potential barrier, making it either easier or more difficult for mobile charge carriers to travel across the junction. The energy-band diagrams for forward and reverse bias are shown in Fig. 8.3.

The voltage–current characteristic for the junction can be derived from the continuity equation, which can be written for the region $x \le -x_p$ in the simplified one-dimensional form,

$$\frac{d^2(n_p - n_{p0})}{dx^2} - \frac{(n_p - n_{p0})}{L_n^2} = 0 \tag{8.15}$$

where $L_n = \sqrt{D_p \tau_p}$ is the electron diffusion length in the *p*-type material, and D_p and τ_p are the hole diffusion coefficient and lifetime, respectively. In writing (8.15) electric-field effects have been neglected. Boundary conditions are supplied by (8.3) and the minority carrier density is

$$n_p(-x_p) = n_{p0} e^{qV/kT} \tag{8.16}$$

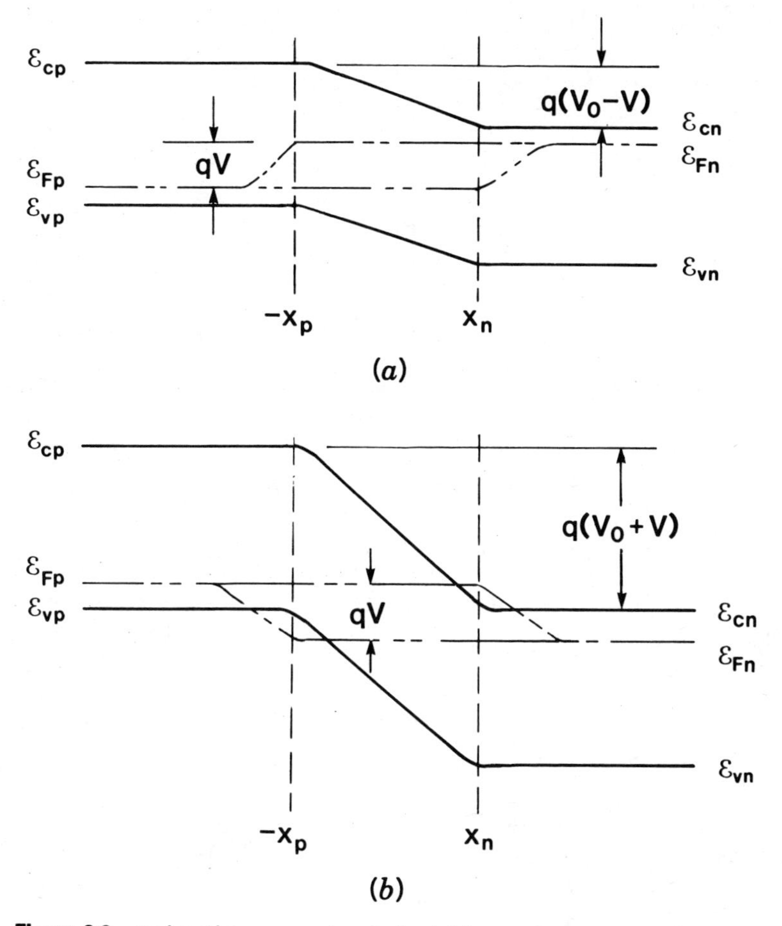

Figure 8.3 *pn* junction energy bands for (*a*) forward and (*b*) reverse bias.

A similar formulation applies to the n-type side of the junction in the region $x \geq x_n$. The continuity equation yields expressions for the minority charge densities in these regions on each side of the junction. The solutions are

$$n_p - n_{p0} = n_{p0}(e^{qV/kT} - 1)e^{(x+x_p)/L_n} \tag{8.17}$$

and

$$p_n - p_{n0} = p_{n0}(e^{qV/kT} - 1)e^{-(x-x_n)/L_p} \tag{8.18}$$

Considering only minority carrier diffusion in the bulk regions, the minority carrier currents are

$$J_n = qD_n \left.\frac{dn_p}{dx}\right|_{-x_p} = \frac{qn_{p0}D_n}{L_n}(e^{qV/kT} - 1) \tag{8.19}$$

and

$$J_p = -qD_p \left. \frac{dp_n}{dx} \right|_{x_n} = \frac{qp_{n0}D_p}{L_p}(e^{qV/kT} - 1) \tag{8.20}$$

The total current density is the sum of (8.19) and (8.20) and is written as

$$J = J_n|_{-x_p} + J_p|_{x_n} = \left(\frac{qp_{n0}D_p}{L_p} + \frac{qn_{p0}D_n}{L_n} \right)(e^{qV/kT} - 1) \tag{8.21}$$

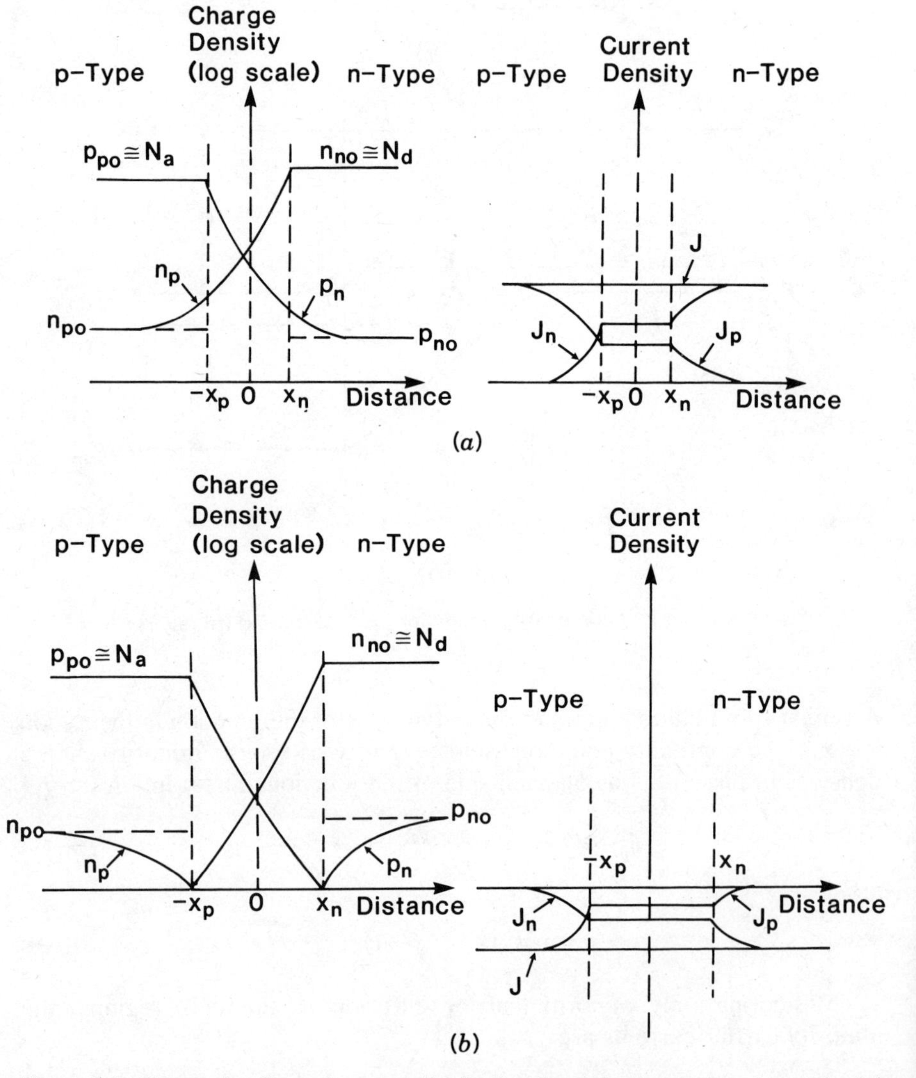

Figure 8.4 Charge density and current distribution throughout a *pn* junction under (*a*) forward and (*b*) reverse bias.

If the diode cross-sectional area is A, (8.21) is written as the ideal diode equation,

$$I = I_0(e^{qV/kT} - 1) \tag{8.22}$$

where

$$I_0 \cong qn_i^2 A\left(\frac{D_p}{N_d L_p} + \frac{D_n}{N_a L_n}\right)$$

the saturation current.

The charge and current densities in the diode under both forward and reverse bias are shown in Fig. 8.4.

8.2.2 Deviations from the Ideal Diode Equation

The derivation of the ideal diode equation considers only low injection conditions and ignores effects that, in practice, result in deviations from the $I-V$ characteristics predicted by the equation. The most significant deviations are due to

1. high-level injection conditions,
2. recombination and generation in the depletion region,
3. ohmic effects.

These effects are discussed in this section.

High-Level Injection Conditions. The largest forward bias that can be applied to a *pn* junction is the contact potential. Application of this voltage to the junction would result in a flat-band condition (i.e., no potential barrier). The ideal diode equation, however, indicates no limit to the allowed forward bias. The problem results from ignoring the majority carrier density changes at the junction. Under large forward-bias voltages the magnitude of the injected minority density can become significant compared to the background impurity density. Since space-charge neutrality conditions are maintained, the majority free charge density will redistribute and take on a profile essentially equal to that of the injected minority charge. When this effect is included in the derivation of the diode equation, a modified equation is obtained that has the form

$$I = qA\left[\frac{e^{qV/kT} - 1}{1 - e^{-2q(V_0 - V)/kT}}\right]\left[\frac{D_p p_{n0}}{L_p}\left(1 + \frac{n_i^2}{p_{p0}^2}e^{qV/kT}\right) + \frac{D_n n_{p0}}{L_n}\left(1 + \frac{n_i^2}{n_{n0}^2}e^{qV/kT}\right)\right] \tag{8.23}$$

Note that in this equation the contact potential V_0 is the limiting forward voltage for the junction.

Depletion Region Generation and Recombination. In the derivation of the ideal diode equation the generation and recombination of free charge carriers in the depletion region is ignored. Actually, a finite density of recombination centers will exist in the depletion region and there will be a net loss of charge carriers. This loss can be interpreted as a "recombination" current and can be expressed as

$$J_r \cong \frac{qWn_i}{2\tau} e^{qV/2kT} \tag{8.24}$$

where τ is the minority carrier lifetime in the depletion region. Note the factor of 2 that appears in the denominator of the exponential.

As indicated by (8.24), recombination current is most significant for

1. low bias voltages,
2. large bandgap semiconductors,
3. short-lifetime semiconductors,
4. high-impurity doping,
5. high-temperature operation.

For practical diodes, the diode equation is generally written in the form

$$J = J_0(e^{qV/nkT} - 1) \tag{8.25}$$

where $n =$ the ideality factor. An ideal diode will have $n = 1$, and diodes that are dominated by recombination currents will have $n = 2$. Generally, a good diode will have an ideality factor that approaches unity.

Under reverse bias the width of the depletion region increases and the charge generated will be swept from the depletion region and contribute to the reverse saturation current. Since the depletion region increases with reverse bias, the reverse saturation current will not saturate, but will show some sort of bias dependence. For example, in an abrupt junction diode the depletion region width is proportional to the square root of the applied voltage, and it follows that the generation current is also proportional to the square root of the voltage. That is,

$$W \propto V^{1/2}$$

and, therefore,

$$J_r \propto V^{1/2}$$

Ohmic Effects. In the derivation of the ideal diode equation, it is assumed that all the applied bias voltage appears across the junction. The finite resistivity of the bulk semiconductor regions was ignored. Actually, the finite resistivity of these regions will produce a voltage drop that will reduce the fraction of the applied voltage that actually appears across the junction. The voltage that appears across the junction is

$$V = V_{\text{appl}} - I[R_p(I) - R_n(I)] \tag{8.26}$$

where R_p and R_n are the resistances of the bulk p- and n-type regions and are functions of current, since they are dependent upon the conductivity of the material. When the ohmic effects become significant the current shows a saturation tendency.

The I–V characteristics for a *pn* junction are sketched in Fig. 8.5 where the various regions of deviation from the ideal diode equation are indicated.

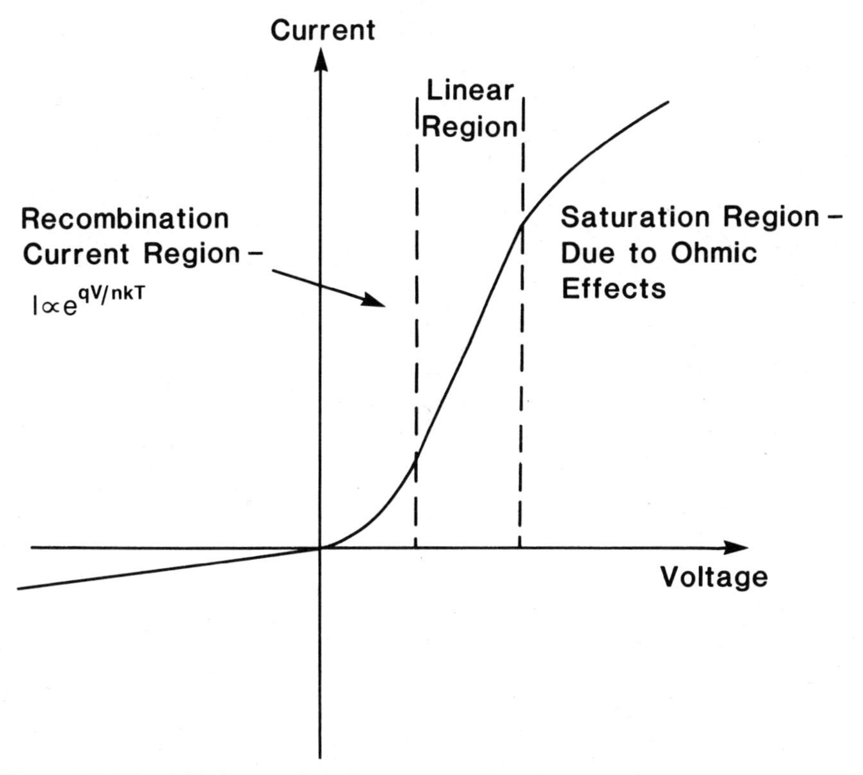

Figure 8.5 The I–V characteristic for a *pn* junction showing regions of deviation from the ideal.

8.2.3 Junction Capacitance

Capacitance is a measure of the energy storage in an electric field. It can be defined from

$$C \triangleq \frac{dq/dt}{dv/dt} = \frac{dq}{dv} \tag{8.27}$$

There are two types of capacitance associated with *pn* junctions. One is associated with the fixed space charge of the impurity atoms. This capacitance is called depletion-layer capacitance and it dominates under reverse bias. The other capacitance is associated with the charge storage of free charge carriers. This is generally termed diffusion capacitance and dominates under forward bias.

Depletion Capacitance. The depletion-layer capacitance results from the electric dipole that is created by the fixed donor and acceptor atoms of the *pn* junction. The capacitance can be calculated from the expression

$$C_j \triangleq \left|\frac{dQ}{dV}\right| \tag{8.28}$$

where Q is the stored charge on one side of the junction. For a junction of area A, $Q = AE_{\max}$. For a one-sided p^+n diode with donor density given by the expression

$$N_d(x) = ax^b \tag{8.29}$$

where a and b are constants, the depletion region width is

$$W = \left[\frac{\epsilon(b+2)(V_0 - V)}{qa}\right]^{1/(b+2)} \tag{8.30}$$

The diode capacitance is calculated to be

$$C_j = A\left(\frac{qa\epsilon^{b+1}}{b+2}\right)^{1/(b+2)}(V_0 - V)^{-1/(b+2)} \tag{8.31}$$

This is more commonly written as

$$C_j = \frac{C_0}{\left(1 - \frac{V}{V_0}\right)^{\gamma}} \tag{8.32}$$

where $\gamma = 1/(b+2)$ and C_0 is the zero-bias capacitance. For an abrupt junction $\gamma = \frac{1}{2}$, for a graded junction $\gamma = \frac{1}{3}$, and for hyperabrupt junctions $\gamma \rightarrow 1$ to 2. Hyperabrupt junctions are very important in linear tuning varactor applications.

Diffusion Capacitance. The forward-biased diffusion capacitance is due to injected minority charge. It is important for time-varying situations (i.e., transient or RF operation). By applying the continuity equation to a forward-biased junction it can be shown that the minority charge storage has two components, one due to a recombination term in which the excess charge distribution is replaced on an average of every minority carrier lifetime, and a charge buildup term where this term can indicate either increasing or decreasing charge storage. For steady state the second term is zero. An analysis (see, for example [2]) of the forward-bias condition shows that the junction can be described by the parallel combination of a conductance with magnitude

$$G_d = \frac{q}{kT} I_{dc} \tag{8.33}$$

and a capacitance with magnitude

$$C_d = \frac{q\tau}{kT} I_{dc} \tag{8.34}$$

where I_{dc} is the diode current.

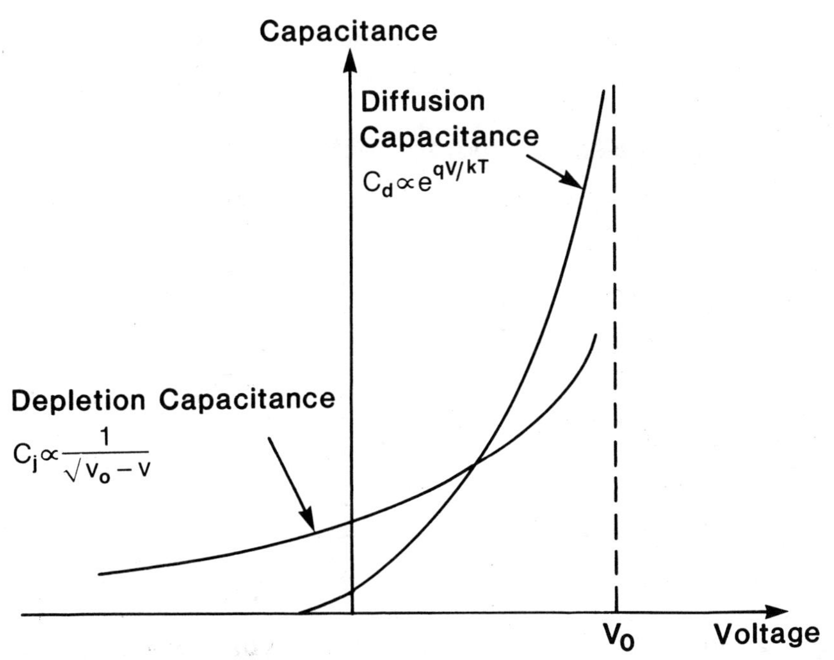

Figure 8.6 Capacitance–voltage characteristic for a *pn* junction showing comparison between depletion and diffusion capacitance.

The depletion layer and diffusion capacitances are sketched in Fig. 8.6. The diffusion capacitance can severely restrict the high-frequency and switching performance of *pn* junctions. Note that the diffusion capacitance is directly proportional to the minority carrier lifetime. Short lifetimes are desirable for minimizing charge storage effects. It will be shown that Schottky-barrier diodes are inherently faster than *pn* junctions because they do not have any minority carrier storage.

8.3 SCHOTTKY-BARRIER JUNCTIONS

Schottky-barrier junctions are formed when a metal is placed upon a semiconductor. The rectifying properties of a metal–semiconductor contact derive from the presence of an electrostatic barrier between the metal and the semiconductor. An argument for the existence of this barrier can be formulated in the following manner. Consider the energy-band diagrams for a metal and an *n*-type semiconductor as shown in Fig. 8.7. The metal has all of its allowed energy states occupied up to the Fermi energy, $\mathscr{E}_F$. The metal work function is shown in Fig. 8.7 as $q\phi_m$ and is defined as the energy required to remove an electron at the Fermi level to a reference energy

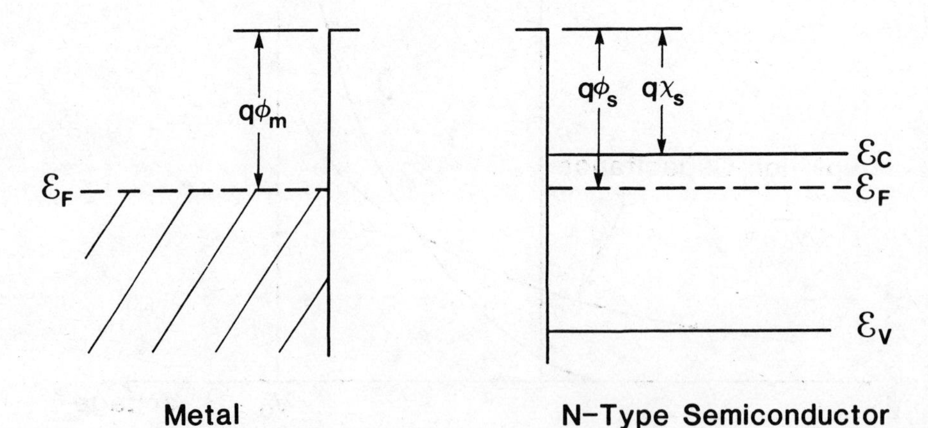

Figure 8.7 Energy-band diagrams for a metal and semiconductor.

level (assumed to be a vacuum). That is, the work function is the energy required to free an electron from the surface of the metal into a vacuum just outside the metal. Likewise, a work function can be defined for the semiconductor. In general, electrons will not exist at the Fermi energy level in the semiconductor, since it lies within the energy gap. An electron affinity (given as the symbol $q\chi_s$) is defined in a similar manner as the energy required to remove an electron at the conduction band edge to the reference level.

A rectifying barrier can be formed if the work function for the metal is greater than that for the semiconductor. That is, if

$$q\phi_m > q\phi_s \tag{8.35}$$

As the metal and semiconductor are placed in physical contact, the energy requirements dictate that at equilibrium the Fermi levels must align. This argument assumes that surface states do not exist. Electrons in the semiconductor are more energetic than those in the metal, and there is a net flow of electrons into the metal. A net positive space-charge is left in the semiconductor in the region adjacent to the metal, as illustrated in Fig. 8.8. The space charge produces a distortion (upward bending) of the energy-band edges and an electrostatic barrier is created in the semiconductor. The presence of this electrostatic barrier impedes the movement of electrons from the semiconductor into the metal. Any electron attempting to move

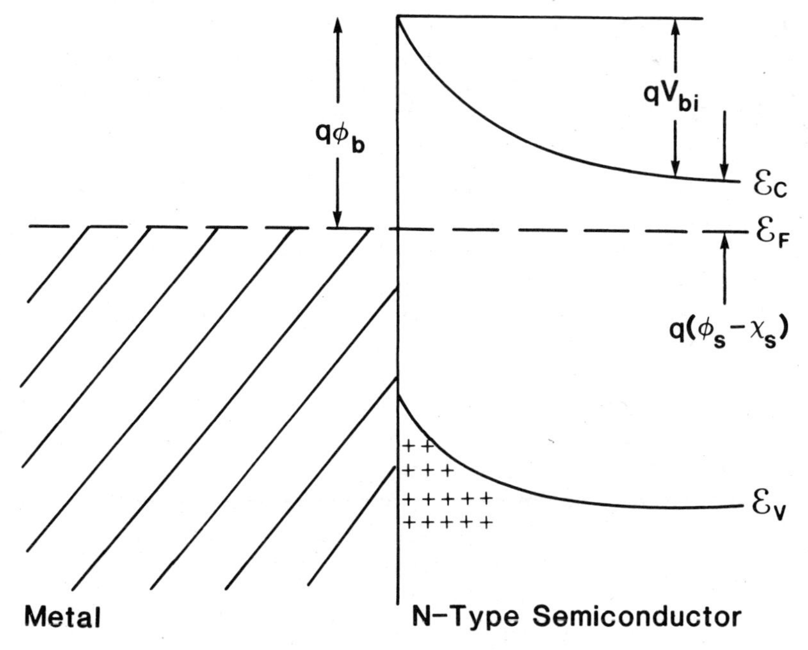

Figure 8.8 Energy-band diagram for an ideal Schottky-barrier junction.

into the metal must have energy greater than qV_0, the built-in or contact potential. For movement from the metal into the semiconductor an electron must have energy greater than the barrier height, defined as

$$q\phi_b \triangleq q(\phi_m - \chi_s) \tag{8.36}$$

This expression is called the Mott limit [3] and the barrier it defines is dependent upon both the metal and the semiconductor.

8.3.1 Surface Effects

In experiments involving the fabrication of rectifying metal–semiconductor contacts, it is often observed that when various metals are deposited upon certain semiconductors, the junctions that result do not obey the Mott formulation. The barrier height is often independent of the metal.

Bardeen [4] was the first to recognize the importance of surface states in determining the barrier height. Surface states have their origin in the discontinuous crystal structure at the crystal surface and result from unsatisfied covalent bonds. The surface atoms do not have the required four neighboring atoms to form the covalent structure, and a number of electrons are loosely bound to their atoms. The atoms with the uncompleted covalent bonds can (a) give up an electron, thereby becoming a donor, or (b) accept an electron, thereby becoming an acceptor. A distortion in the lattice structure is created that produces additional allowed energy levels within the forbidden region at the semiconductor surface.

The surface states are usually continuously distributed in energy and are characterized with a "neutral level" ϕ_0 such that the states are occupied up to ϕ_0 and empty above ϕ_0. If the Fermi level lies above ϕ_0, the surface states possess a net negative charge (i.e., they are acceptorlike) and if the Fermi level lies below ϕ_0, the states possess a net positive charge (i.e., they are donorlike). This latter situation is shown in Fig. 8.9, where it is noted that the net positive charge causes an upward bending of the semiconductor energy bands.

Some of the electric-field lines terminating on the metal are due to the surface states rather than the ionized donors in the crystal. The depletion region and the barrier height are reduced. The surface states provide a negative feedback that acts to "push" ϕ_0 down toward the Fermi level. This feedback always tends to reduce the deviation of ϕ_0 from the Fermi level.

If the number of surface states is high, the Fermi level becomes "pinned" to ϕ_0 and the barrier height becomes essentially independent of the metal. The barrier height that results can be defined if ϕ_0 is measured from the top of the valence band. The barrier height is called the Bardeen limit and is defined as

$$\phi_b \triangleq E_g - \phi_0 \tag{8.37}$$

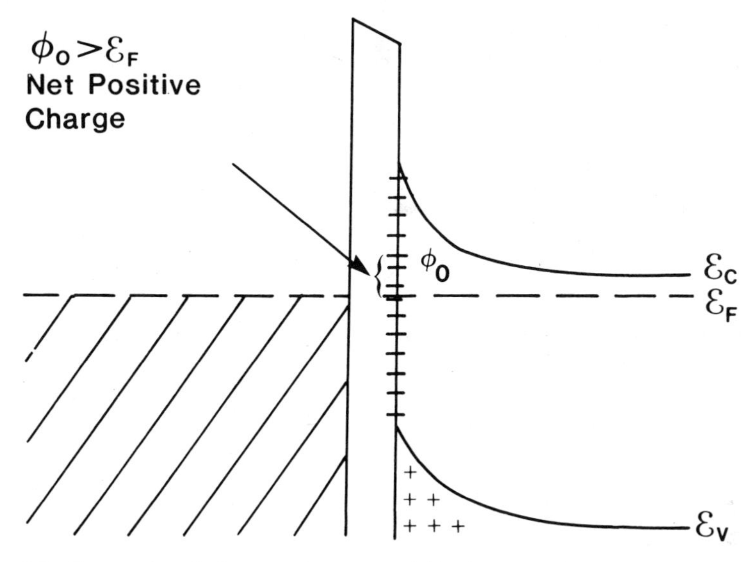

Figure 8.9 Energy-band diagram for a metal–insulator–semiconductor junction with surface states.

where E_g is the semiconductor energy gap. Note that the barrier height, as defined by the Bardeen limit, is not a function of the metal. In fact, many metal–semiconductor contacts obey the Bardeen limit. The properties of many such junctions are listed in Table 8.1.

8.3.2 Image Force Lowering

Any charged body close to a conducting plane produces an electrostatic field as if there were an equal and oppositely charged particle located at the mirror image. Since this image has the opposite charge, the total potential energy of the system is reduced. Consider the situation sketched in Fig. 8.10. An electron located at distance x in a vacuum from the metal will experience an attractive force between it and its image located at a location $-x$. The attractive force is given by the expression

$$F = \frac{-q^2}{4\pi(2x)^2\epsilon_0} = \frac{-q^2}{16\pi\epsilon_0 x^2} \tag{8.38}$$

The potential energy of the electron at distance x is calculated from the work done by an electron in moving from infinity to point x and is

$$\mathrm{PE}'(x) = \int_{\infty}^{x} F\,dx = \frac{q^2}{16\pi\epsilon_0 x} \tag{8.39}$$

TABLE 8.1 Measured Schottky-Barrier Heights (Volts at 300 K)

Semiconductor	Type	E_g (eV)	Ag	Al	Au	Cr	Cu	Hf	In	Mg	Mo	Ni	Pb	Pd	Pt	Ta	Ti	W
Diamond	*p*	5.47			1.71													
Ge	*n*	0.66	0.54	0.48	0.59		0.52		0.64			0.49	0.38					0.48
Ge	*p*		0.50		0.30				0.55									
Si	*n*	1.12	0.78	0.72	0.80	0.61	0.58	0.58		0.40	0.68	0.61		0.81	0.90		0.50	0.67
Si	*p*		0.54	0.58	0.34	0.50	0.46				0.42	0.51	0.55				0.61	0.45
SiC	*n*	3.00		2.00	1.95													
AlAs	*n*	2.16			1.20										1.00			
AlSb	*p*	1.63			0.55													
BN	*p*	7.50			3.10													
BP	*p*	6.00			0.87													
GaSb	*n*	0.67			0.60													
GaAs	*n*	1.42	0.88	0.80	0.90		0.82	0.72							0.84	0.85		0.80
GaAs	*p*		0.63		0.42			0.68										
GaP	*n*	2.24	1.20	1.07	1.30	1.06	1.20	1.84		1.04	1.13	1.27			1.45		1.12	
GaP	*p*				0.72													
InSb	*n*	0.16	0.18[a]		0.17[a]													
InAs	*p*	0.33			0.47[a]													
InP	*n*	1.29	0.54		0.52													
InP	*p*				0.76													
CdS	*n*	2.43	0.56	Ohmic	0.78		0.50					0.45	0.59	0.62	1.10		0.84	
CdSe	*n*	1.70	0.43		0.49		0.33								0.37			
CdTe	*n*		0.81	0.76	0.71										0.76			
ZnO	*n*	3.20		0.68	0.65		0.45		0.30					0.68	0.75	0.30		
ZnS	*n*	3.60	1.65	0.80	2.00		1.75		1.50	0.82				1.87	1.84	1.10		
ZnSc	*n*		1.21	0.76	1.36		1.10		0.91				1.16		1.40			
PbO	*n*		0.95						0.93			0.96	0.95					

[a] 77 K.

Source: Sze [2].

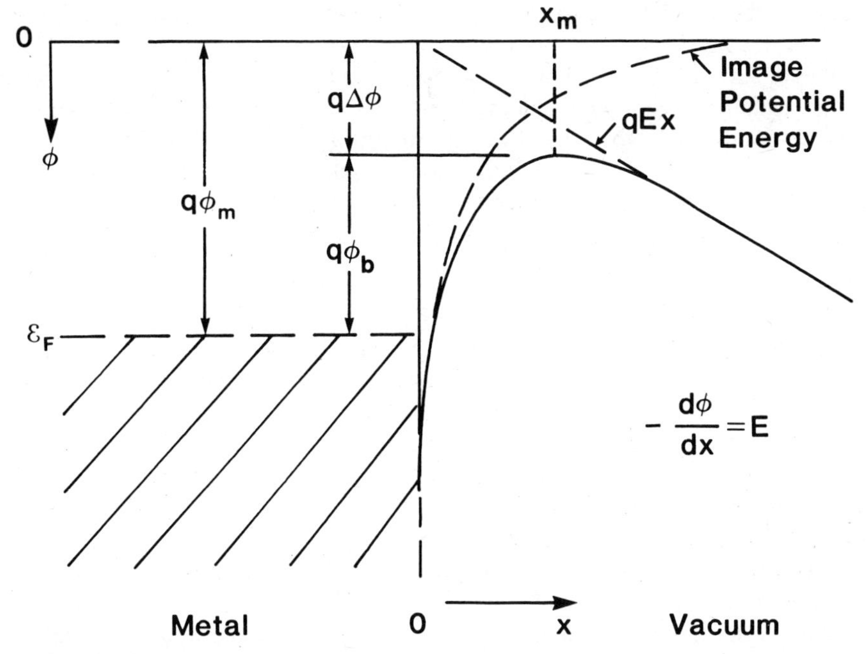

Figure 8.10 Energy-band diagram for a Schottky-barrier junction showing effects of image force lowering.

When an external electric field is also applied, the total potential energy of the system is given as

$$\mathrm{PE}(x) = \frac{q^2}{16\pi\epsilon_0 x} + qEx \tag{8.40}$$

The amount of barrier lowering due to the image force can be calculated from

$$\frac{d[\mathrm{PE}(x)]}{dx} = 0 \tag{8.41}$$

The calculation indicates that the barrier is reduced by an amount

$$\Delta\phi = \sqrt{\frac{qE}{4\pi\epsilon_0}} = 2Ex_m \tag{8.42}$$

where the distance

$$x_m = \left(\frac{q}{16\pi\epsilon_0 E}\right)^{1/2} \tag{8.43}$$

indicates the location of the potential maximum from the metal–semiconductor interface.

This result can be easily applied to situations where the electron is in a semiconductor by simply replacing the free-space permittivity with that for the semiconductor. Note that for high-magnitude electric fields the barrier height, $q\phi_b$, is significantly reduced and the effective metal work function for thermionic emission is reduced.

8.3.3 Schottky Model

Although the rectifying properties of metal-semiconductor contacts were known and studied as far back as 1874 with the work of Braun, most of the modern work on these contacts is based, at least in part, upon the models of Schottky [5] and Mott [3], which were presented in 1938 and 1939, respectively. The theoretical background presented in this early work clearly established the role of space charge as the basis for the built-in potential barrier. The Schottky model, in particular, describes how an electrostatic barrier is created when a metal is placed in contact with a semiconductor.

The model predicts the creation of a potential barrier at the metal–semiconductor interface, as shown in Fig. 8.11. If the depletion approximation is employed, the region of the semiconductor directly adjacent to the metal will be depleted of charge. This produces a charge distribution and an electric field in the interface region. By applying Poisson's equation to the junction, the maximum electric field will occur at the interface and is given by

$$E_{\max} = -\frac{qN_d W}{\epsilon} \tag{8.44}$$

The voltage across the depletion region is the built-in or contact potential and is

$$V_0 = -\frac{1}{2} E_{\max} W = \frac{1}{2} \frac{qN_d W^2}{\epsilon} \tag{8.45}$$

From these expressions the space charge in the semiconductor depletion region of junction area A can be calculated and is

$$Q = qAN_d W = A(2q\epsilon N_d V_0)^{1/2} \tag{8.46}$$

The depletion approximation is actually not valid for many applications and leads to errors. The free electron concentration will decrease from a magnitude equal to the donor density to a low value at the interface, as

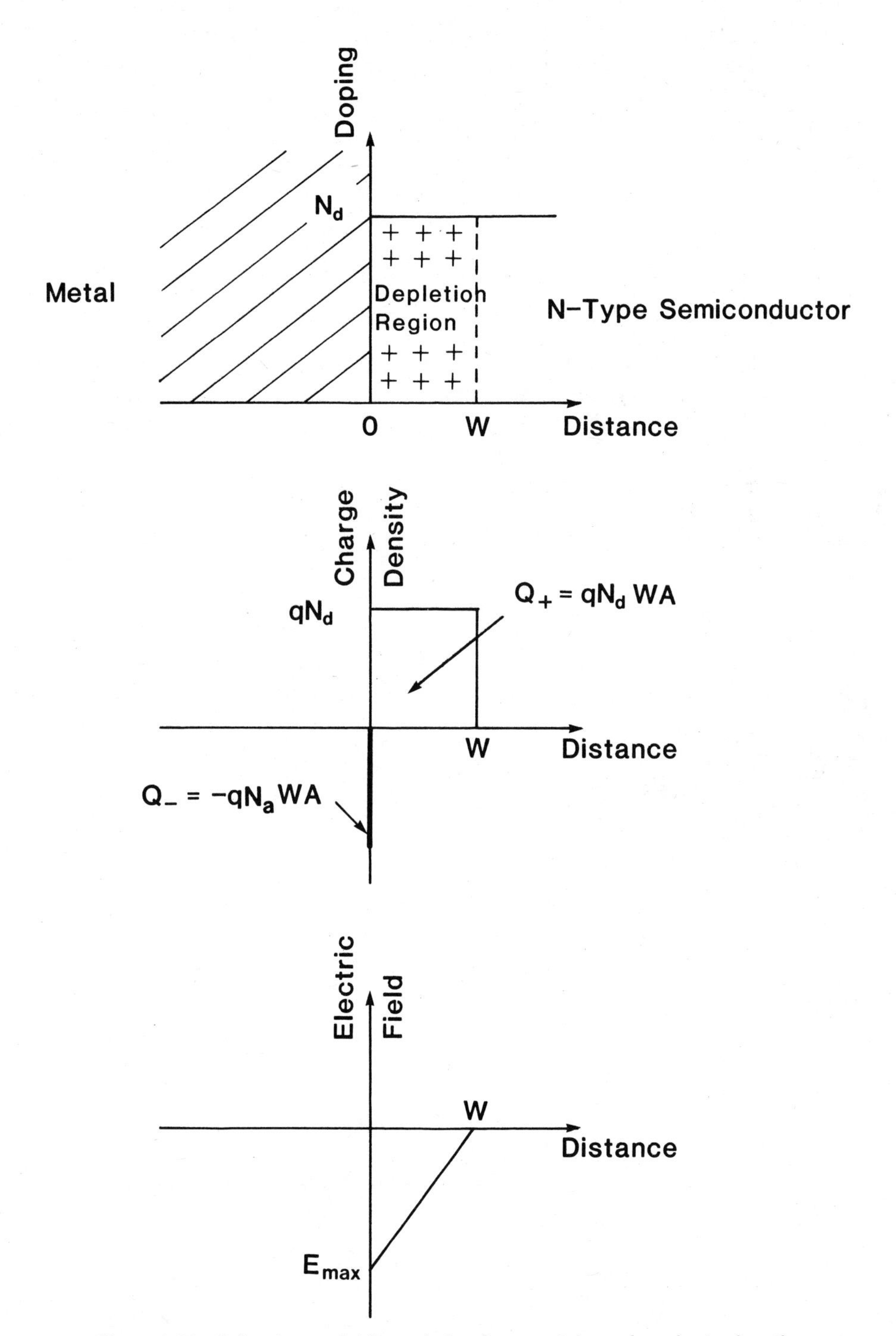

Figure 8.11 Schottky model formulation for a metal–semiconductor junction.

shown by the dotted line in Fig. 8.12. The free electron density will modify (reduce) the background space-charge. The modification required in the theory to account for the nonzero electron density results in an additional term of kT/q in (8.46), so that it is written as

$$Q = A\left[2q\epsilon N_d\left(V_0 \pm V - \frac{kT}{q}\right)\right]^{1/2} \tag{8.47}$$

where a biasing voltage V is introduced with the (−) sign indicating forward bias and the (+) sign indicating reverse bias. The bias situations are shown in Fig. 8.13.

For semiconductors with relatively high mobility (most common semiconductors) and moderate doping, the current across the junction will be dominated by thermionic emission [6]. Under these conditions the junction current is written as a function of voltage in the form

$$J = J_0(e^{qV/kT} - 1) \tag{8.48}$$

where

$$J_0 = A^* T^2 e^{-q\phi_b/kT} \tag{8.49}$$

and

$$A^* = \frac{4\pi m^* q k^2}{h^3}$$

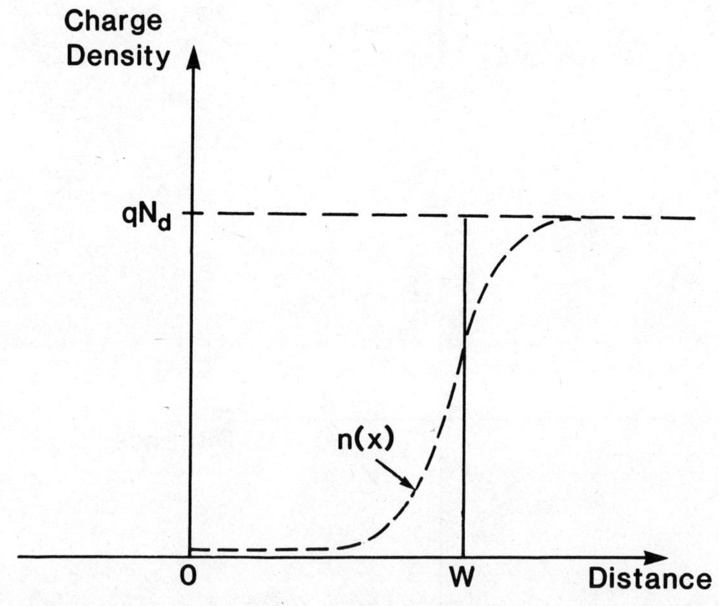

Figure 8.12 Charge distribution in a metal–semiconductor junction (dotted line indicates free charge).

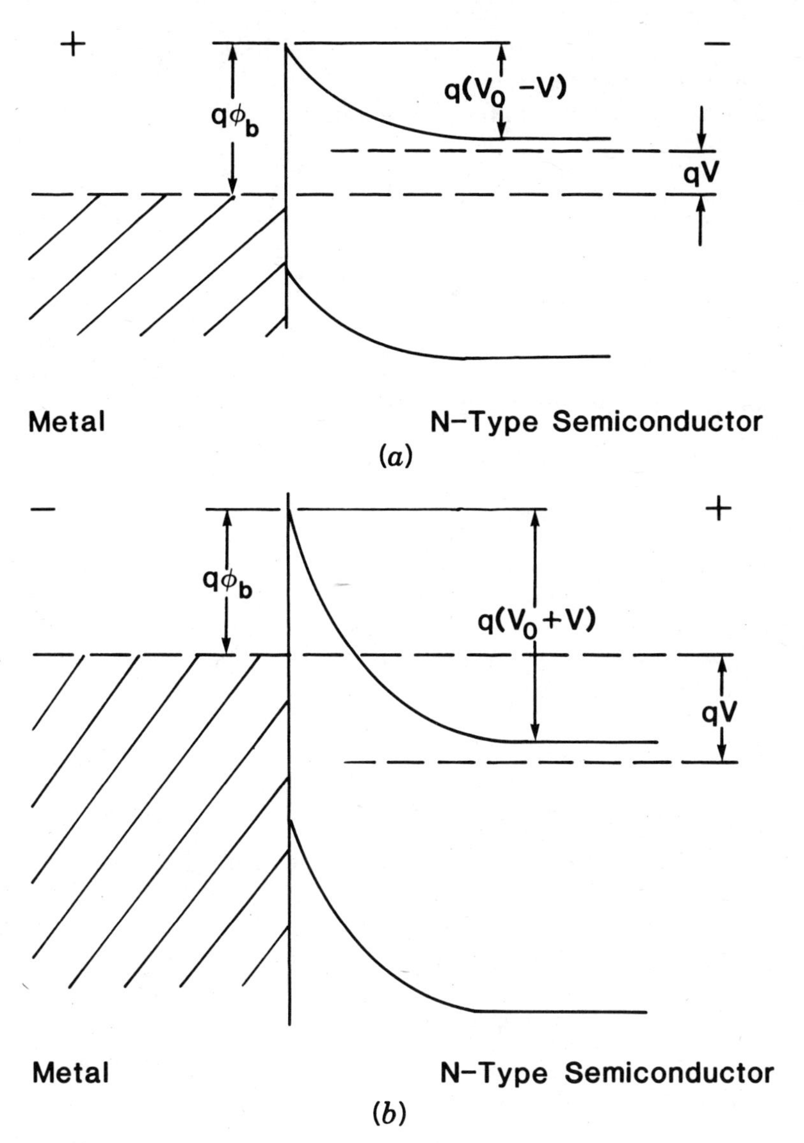

Figure 8.13 Energy-band diagrams for a Schottky-barrier junction under (*a*) forward and (*b*) reverse bias.

the effective Richardson constant. The term m^* is the electron effective mass, h is Planck's constant, and k is Boltzmann's constant.

The barrier height ϕ_b is a function of applied bias. The bias dependence of the barrier height can be modeled in a convenient manner by assuming a linear relationship,

$$\phi_b = \phi_{b0} + \beta V \tag{8.50}$$

where the coefficient β will always be positive because the barrier height always increases with forward voltage.

The current equation can be rewritten in the form

$$J = J_0 e^{qV/nkT}(1 - e^{-qV/kT}) \tag{8.51}$$

where

$$\frac{1}{n} = 1 - \beta = 1 - \frac{\partial \phi_b}{\partial V} \tag{8.52}$$

If $\partial \phi_b / \partial V$ is constant, then n is a constant. For $V > 3kT/q$ the current equation can be approximated by the commonly used form

$$J \cong J_0 e^{qV/nkT} \tag{8.53}$$

The n term is the ideality factor and its magnitude depends upon semiconductor doping and the current transport mechanism. For barrier current consisting of pure thermionic emission $n = 1$, and for current transport dominated by generation-recombination, $n = 2$. Generally, thermionic emission will dominate the current flow in good-quality diodes and a low magnitude for n is desired. Good diodes normally have n values ranging from about 1.02 to 1.2.

8.3.4 Junction Capacitance

Since Schottky-barrier junctions are majority carrier devices, minority carrier storage is not a factor in their operation. Therefore, the diffusion capacitance that dominates under forward bias in *pn* junctions is, for most RF applications of Schottky diodes, not important. When minority charge storage must be considered, as in switching applications, the minority charge storage time can be calculated from the expression

$$\tau_s = \frac{q n_i^2 \sqrt{D_p \tau_p}}{N_d J_0} \tag{8.54}$$

The elimination of the diffusion capacitance provides Schottky-barrier devices with an inherent speed advantage compared to *pn* junction diodes.

The depletion-layer capacitance can be calculated in a manner analogous to that used for *pn* junction diodes. The device can be considered to be a parallel-plate capacitor with a voltage-dependent space-charge width given by the expression

$$W = \left[\frac{2\epsilon}{qN_d}\left(V_0 - V - \frac{kT}{q}\right)\right]^{1/2} \tag{8.55}$$

The capacitance per unit surface area is determined to be

$$C_j = \frac{\epsilon}{W} = \left[\frac{q\epsilon N_d}{2[V_0 - V - (kT/q)]}\right]^{1/2} \tag{8.56}$$

The application of (8.56) to a typical microwave diode with $N_d = 2 \times 10^{17}\ \text{cm}^{-3}$, $V_0 = 0.8$ V, $\epsilon = \epsilon_r \epsilon_0 = (12.9)8.854 \times 10^{-14}$F/cm, and $q = 1.6 \times 10^{-19}$ C gives a junction capacitance of $C_j = 1.51 AfF$, where Schottky area A is in square micrometers.

8.3.5 Rectifying Contact Materials

When selecting materials for rectifying contacts it is necessary to consider a wide range of chemical, metallurgical, and physical, as well as electrical characteristics. For example, the metal films suitable for use as contacts should have the following characteristics:

1. correct barrier height,
2. good adherence to the semiconductor,
3. different etch rate from the semiconductor,
4. easy bondability,
5. resistance to metallurgical reactions,
6. resistance to diffusion,
7. resistance to oxidation and corrosion,
8. resistance to electromigration.

A variety of metals have been successfully used for Schottky junctions. Typical metals include aluminum (Al), gold (Au), tungsten (W), and titanium/tungsten (TiW). Tungsten and titanium/tungsten are refractory metals and they can be used at high temperatures. For this reason they are sometimes used for devices designed to operate at high power levels where elevated temperatures are likely to be encountered.

In order to achieve contacts with the desired characteristics, it is often necessary to use combinations of metals. For example, titanium/tungsten/gold can be used to achieve a satisfactory rectifying contact. The TiW provides for a suitable barrier height and good physical characteristics, since on Si and GaAs it is nonreactive, adheres well, and resists electromigration. The gold provides for good bonding capability and low electrical resistance. The TiW layers are very thin (in the range of hundreds of angstroms), whereas the gold layer is relatively thick (on the order of microns). The TiW layers would typically be placed upon a cleaned semiconductor surface by sputtering or evapoaration and the gold layer would be electroplated to the desired thickness.

8.3.6 Series Resistance

Series resistance of a Schottky diode [shown in Fig. 8.14(*a*)] is comprised of three parts: contact resistance (R_c), channel resistance (R_{ch}), and resistance of the Schottky junction metal (R_g). The resistance is written in the form

$$R_s = R_c + R_{\text{ch}} + R_g \tag{8.57a}$$

An approximate expression is

$$R_s = \frac{2100}{P\sqrt{a}(N_d^+)^{0.66}} + \frac{R_{\text{ch}}(R_{\text{ch}}^+ + R_{\text{ch}}')}{R_{\text{ch}} + R_{\text{ch}}^+ + R_{\text{ch}}'} + \frac{1}{3}\frac{\rho_m D}{tW} \tag{8.57b}$$

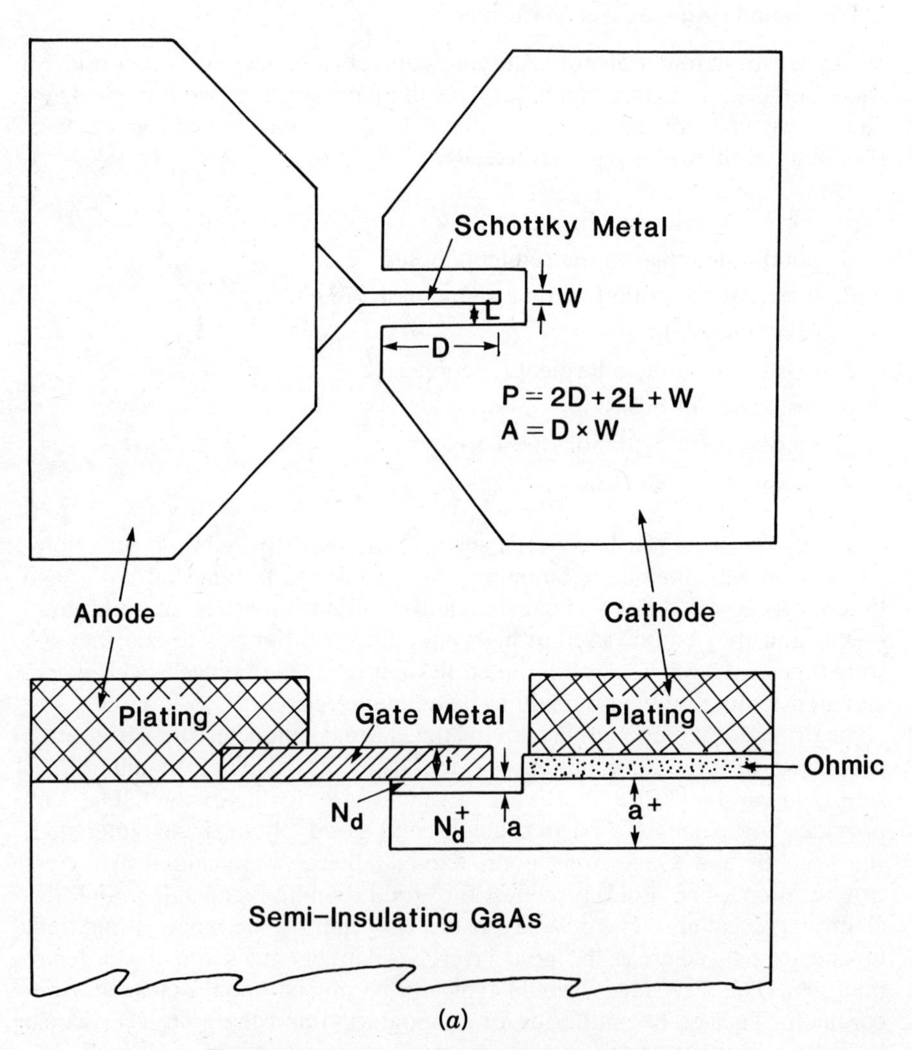

(*a*)

Figure 8.14 (*a*) Top side view of a Schottky diode.

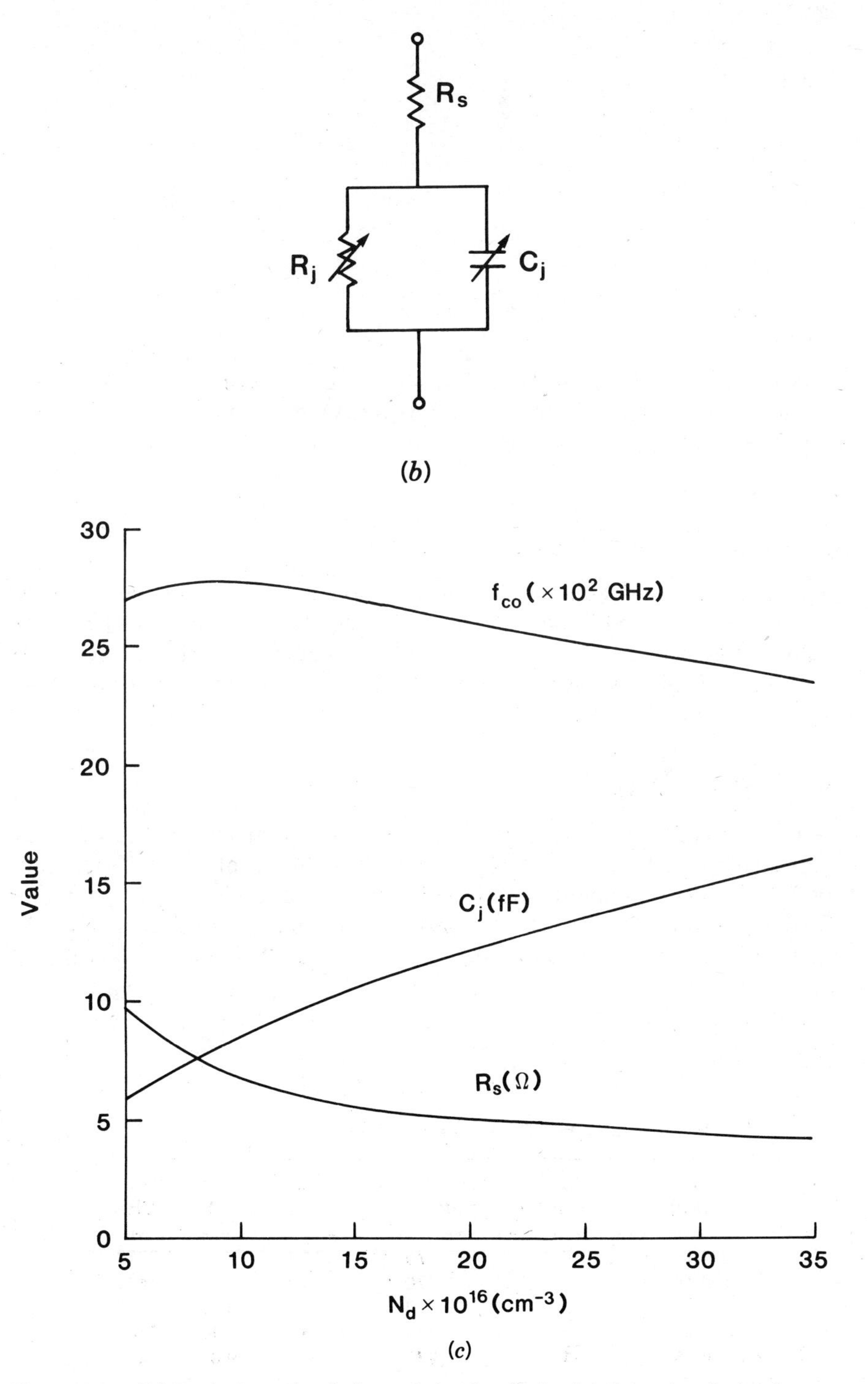

Figure 8.14 (*b*) Equivalent circuit for a Schottky diode. (*c*) Schottky diode characteristics as a function of active layer doping.

where

$$R_{\text{ch}} = \frac{1100}{aN_d^{0.82}} \frac{L}{A}$$

$$R_{\text{ch}}^{+} = \frac{1100}{a^{+}(N_d^{+})^{0.82}} \frac{L}{A}$$

$$R_{\text{ch}}' = \frac{1100}{N_d^{0.82}} \frac{a}{WD}$$

and ρ_m is the bulk specific resistivity of the gate metal ($\cong 2.5 \times 10^{-6}$ Ω-cm for gold). The terms N_d and N_d^{+} are expressed in the units of 10^{16} and the other physical dimensions are in micrometers and are defined in Fig. 8.14(*a*).

8.3.7 Equivalent Circuit

The diode equivalent circuit consists of nonlinear junction resistance R_j and capacitance C_j, and series resistance R_s as shown in Fig. 8.14(*b*). The values of R_j and C_j depend upon the dc voltage and/or RF voltage across the diode. Typical values for good Schottky diodes are $R_j = 50$ to $200\ \Omega$, $C_j \cong 0.02$ pF and $R_S = 5\ \Omega$.

8.3.8 Figure of Merit

A figure of merit is a useful predictor of how a particular device might be expected to perform in a given circuit. The figure of merit is generally useful when comparing different device types and is often used as a factor in the selection process when a particular circuit is being designed. The figure of merit for Schottky diodes is the zero cutoff frequency, f_{c0}, defined as

$$f_{c0} = \frac{1}{2\pi R_s C_{j0}} \tag{8.58}$$

TABLE 8.2 Schottky Diode Characteristics

a (μm)	N_d (10^{16} cm^{-3})	N_d^{+} (10^{16} cm^{-3})	R_s (Ω)	C_j (fF)	f_{c0} (GHz)
0.15	5	70	11.1	6	2372
0.15	5	270	7.7	6	3431
0.2	5	270	9.2	6	2853
0.15	7	270	6.4	7.1	3481
0.15	11	270	5.1	9.0	3450

Note: $t = 0.5$ μm, $D = 8$ μm, $L = 1.5$ μm, $W = 1$ μm, $a^{+} = 2$ μm, $V_0 = 0.8$ V.

where C_{j0} is the junction capacitance at zero volt bias. Table 8.2 lists f_{c0} values for various physical and electrical parameters of a beam lead diode. Figure 8.14(*c*) depicts the variation of R_s, C_j and f_{c0} as a function of N_d for $a = 0.2\ \mu\text{m}$, $a^+ = 2\ \mu\text{m}$, $N_d^+ = 2 \times 10^{18}\ \text{cm}^{-3}$, $V_0 = 0.8\ \text{V}$, $D = 8\ \mu\text{m}$, $L = 1.5\ \mu\text{m}$, and $W = 1\ \mu\text{m}$. These calculations do not include parasitic capacitances.

8.4. VARACTOR DIODES

A varactor diode is a semiconductor diode that is used as a variable reactance circuit element. The variable reactance characteristics derive from the variation of the diode depletion-layer capacitance with applied voltage (either dc or low-frequency RF). For common applications the diode is reverse biased so that the depletion-layer capacitance dominates the device characteristics. An excellent review of varactor diodes is presented in [7].

A varactor diode is a nonlinear element that can produce three fundamentally different circuit functions consisting of:

1. microwave signal tuning or modulation,
2. harmonic generation,
3. parametric amplification and/or up-conversion.

In the first two applications a bias voltage and RF signal are applied to the diode, whereas in the third application a bias voltage and multiple RF signals that may be harmonically or nonharmonically related are applied to the diode.

Varactors are commonly and extensively used in many microwave applications including voltage-controlled oscillators (VCOs), parametric amplifiers (paramps), and frequency multipliers. The VCO application is probably one of the most useful applications for varactors and these devices have been constructed using a wide range of active devices including bipolar transistors, GaAs MESFETs, transferred electron devices, and IMPATT diodes. In principle a VCO can be constructed with a varactor and any active device. Parametric amplifiers are capable of extremely low noise figure operation and, in the past, were extensively employed as front-end amplifiers in high-sensitivity receivers, such as are used in space applications. In recent years, however, the advances achieved in low-noise GaAs MESFET and HEMT technology have resulted in amplifiers constructed with these devices replacing paramps in many low-noise applications. Varactor frequency multipliers are generally used in doubler and tripler circuits. Such circuits have application for producing low phase noise sources of microwave and millimeter-wave energy. In particular, these circuits can produce RF signals at frequencies above that obtainable from a fundamental mode source, and for this reason are often used in millimeter-wave applications. Also, since varactors are capable of low phase noise performance, they are used in microwave signal synthesizers.

8.4.1 Equivalent Circuit

The equivalent circuit for a microwave varactor chip is shown in the inset in Fig. 8.15. Since the diode is generally operated in a reverse-biased mode, the equivalent circuit consists of the depletion-layer capacitance as defined in (8.32) in shunt with the junction resistance. The depletion-layer capacitance is repeated here for convenience, and is

$$C_j = \frac{C_0}{(1 - V/V_0)^{\gamma}} \tag{8.59}$$

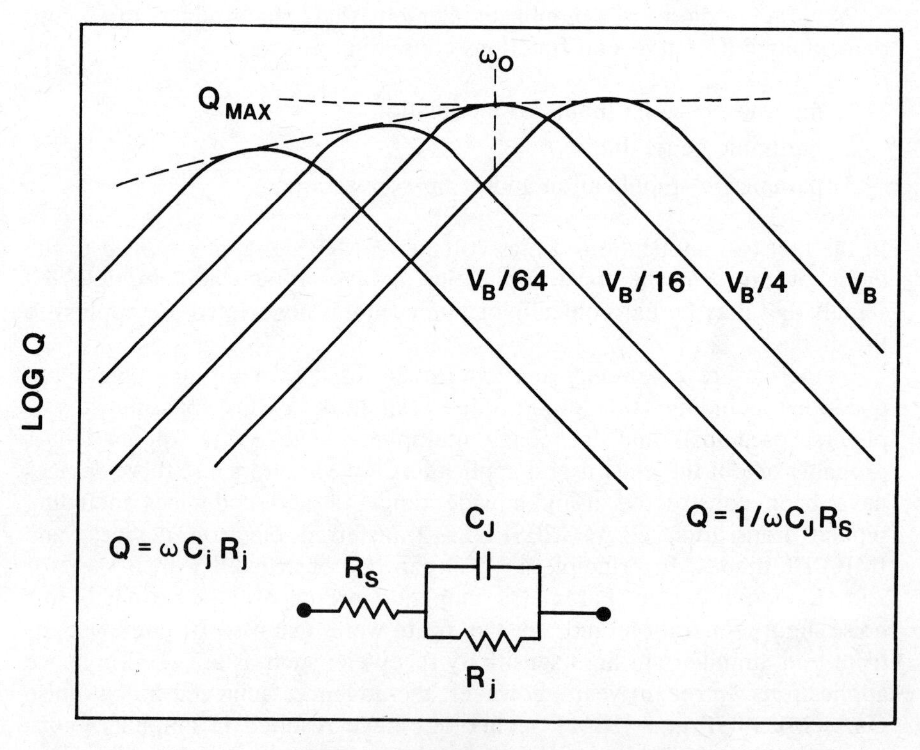

Figure 8.15 Qualitative dependence of Q vs angular frequency for various reverse-bias voltages up to the breakdown voltage V_B. (After M. H. Norwood and E. Shatz [7], 1968 IEEE, reprinted with permission.)

For reverse bias the junction resistance R_j is generally high (e.g., on the order of tens of megaohms) and can be ignored. Both the junction capacitance and resistance are functions of bias. The series resistance R_s represents the resistance of the bulk semiconductor regions extrinsic to the junction region and also the resistance due to the metal contacts. The series resistance is related to the Q for the diode, which can be defined as

$$Q \cong \frac{\omega C_j R_j}{1 + \omega^2 C_j^2 R_j R_s} \tag{8.60}$$

The diode Q is an indicator of the efficiency of the device. The variation of diode Q with frequency is shown in Fig. 8.15. The diode breakdown voltage is indicated as V_B in the figure. The Q expression can be differentiated to determine the frequency at which maximum Q occurs. The frequency is

$$\omega_0 \cong \frac{1}{C_j (R_j R_s)^{1/2}} \tag{8.61a}$$

and the maximum Q is

$$Q_{max} \cong \left(\frac{R_j}{4R_s} \right)^{1/2} \tag{8.61b}$$

Generally, GaAs varactors have higher Q values than Si devices and are preferred in many high-efficiency and low-noise applications. An exception to this is low phase noise applications. The GaAs devices suffer from high baseband $1/f$ noise due to surface states, deep levels, and other trapping effects. There is currently no known passivation that is capable of completely eliminating these effects. The Si devices, however, have the advantage of native oxide passivation that significantly reduces the low-frequency $1/f$ noise. In general, Si devices have a $1/f$ spectrum that is about 10 to 20 dB lower in magnitude compared to that from GaAs devices. The $1/f$ corner frequencies where the $1/f$ effects can no longer be seen are in the hundreds of kilohertz range for Si devices and in the hundreds of megahertz for GaAs devices.

The capacitance variation of the varactor with voltage, as indicated in (8.59), is dependent upon the magnitude of the exponent γ. The exponent γ, in turn, is directly dependent upon the device doping profile. For devices with abrupt doping profiles $\gamma = \frac{1}{2}$, for those with linearly graded profiles, $\gamma = \frac{1}{3}$, and for hyperabrupt profiles, $\gamma \rightarrow 1$ to 5. The capacitance-voltage relationships for these diodes are shown in Fig. 8.16. The hyperabrupt diodes are particularly interesting for linear tuning applications, since it is possible to obtain a capacitance variation that is inversely proportional to the magnitude of the applied bias voltage. Such devices, when used in

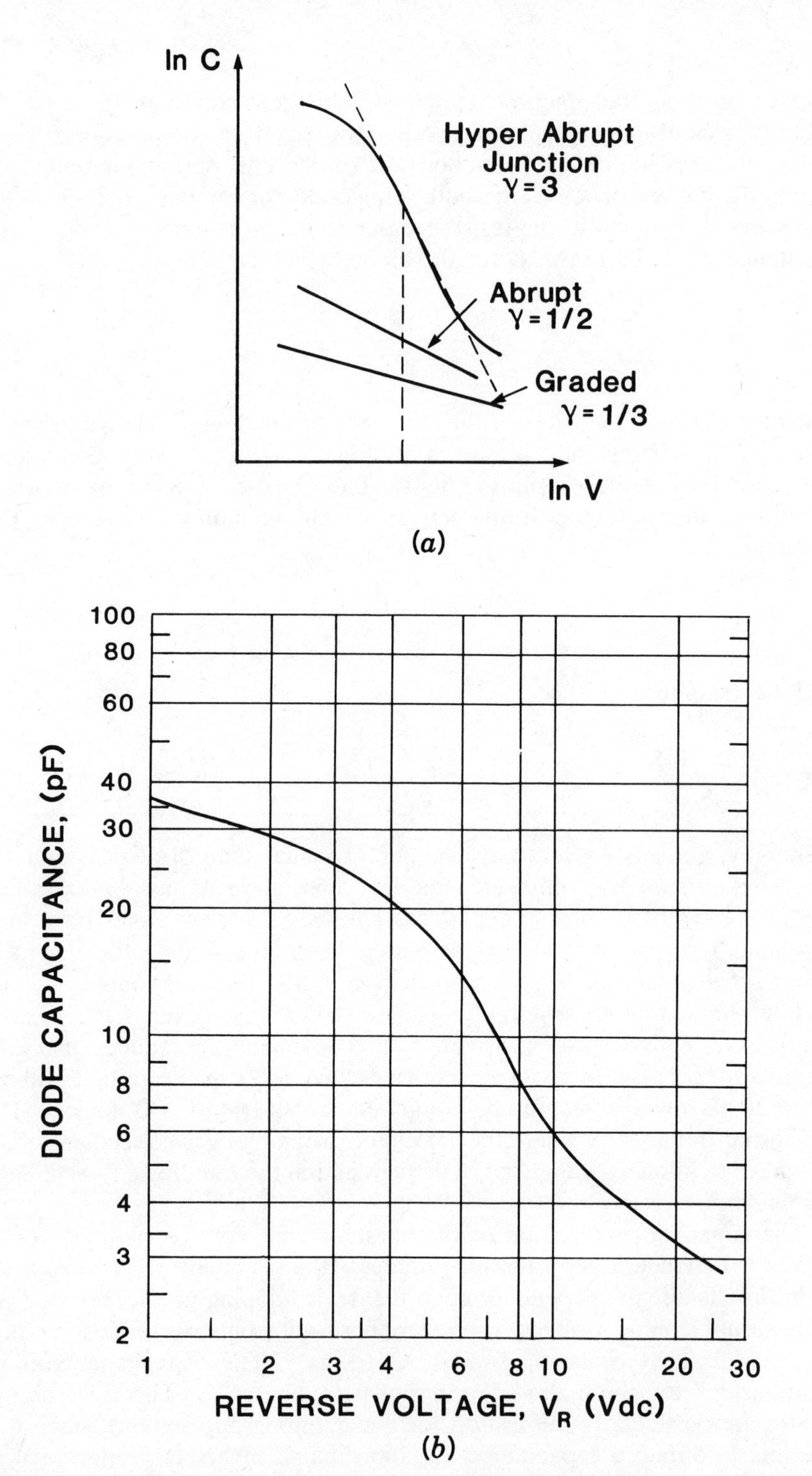

Figure 8.16 (*a*) Capacitance vs. voltage for diodes with various doping profiles. (*b*) Capacitance vs. voltage for a typical hyperabrupt diode.

VCOs, are capable of producing oscillators that tune linearly with voltage. Greater than octave frequency coverage can be obtained with these oscillators. The hyperabrupt diode, however, has a lower Q than an abrupt junction diode with the same breakdown voltage and capacitance.

The equivalent circuit shown in the inset in Fig. 8.15 includes only the elements intrinsic to the varactor chip. When the device is bonded into a circuit, either in chip form or in a package, additional parasitic elements must be added to the circuit. The equivalent circuit that results is shown in Fig. 8.17. The parallel capacitance C_f accounts for fringing capacitance due to the mounting structure, and the series inductance L_s accounts for the inductance of the bonding leads.

8.4.2 Figure of Merit

Two types of figures of merit [8], static and dynamic, have been defined for varactor diodes. The static figures of merit include:

1. The cutoff frequency defined as

$$f_{cV} \triangleq \frac{1}{2\pi R_s C_{jV}} \tag{8.62}$$

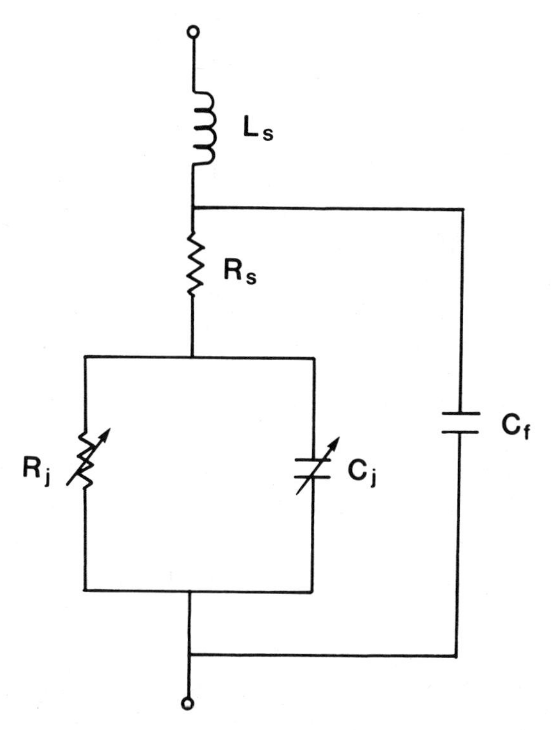

Figure 8.17 Equivalent circuit for a packaged varactor diode.

where C_{jV} is the device capacitance at a specified bias, usually zero or -4 V.

2. The quality factor or device Q where

$$Q_V \triangleq \frac{f_{cV}}{f} \tag{8.63}$$

and the Q is defined for a given bias.

The Q factor in (8.63) is defined in terms of the series resistance in (8.60).
The dynamic figures of merit include:

(a) A dynamic cutoff frequency defined as

$$f_c \triangleq \frac{[(1/C_{j,\mathrm{min}}) - (1/C_{j,\mathrm{max}})]}{2\pi R_s} \tag{8.64}$$

where the maximum and minimum capacitance values are defined as the capacitances at two bias voltages (e.g., zero and -15 V). These values are given in device data sheets.

(b) A dynamic quality factor defined as

$$\tilde{Q} \triangleq \frac{S_1}{\omega R_s} \tag{8.65}$$

where S_1 is the first Fourier component of the time-dependent elastance (i.e., the reciprocal of the capacitance). The diode elastance is written as

$$S = S_0 \left(1 - \frac{V}{V_0}\right)^{\gamma} \tag{8.66}$$

where S_0 is the zero-bias elastance.

An additional figure of merit that is useful in indicating the tuning range of the varactor, called the relative sensitivity, is defined as

$$s(V) \triangleq \frac{1}{\omega_0} \frac{d\omega}{dV} \tag{8.67a}$$

The parameter s indicates the percentage increase in frequency per volt, and a constant s indicates a linear capacitance versus bias voltage variation. In terms of the diode C–V characteristic, the relative sensitivity can be written as

$$s(V) = \frac{d}{dV}\left[\frac{C_{j0}}{C_j(V)}\right]^{1/2} \tag{8.67b}$$

where C_{j0} is the zero-bias capacitance.

The static figures of merit cannot reflect the degree of capacitance variation of the varactor. The dynamic figures of merit, since they explicitly take the device nonlinearity into account, give a more accurate indication of how a given varactor should function. Since detailed capacitance variations are ignored, however, the indication is only approximate.

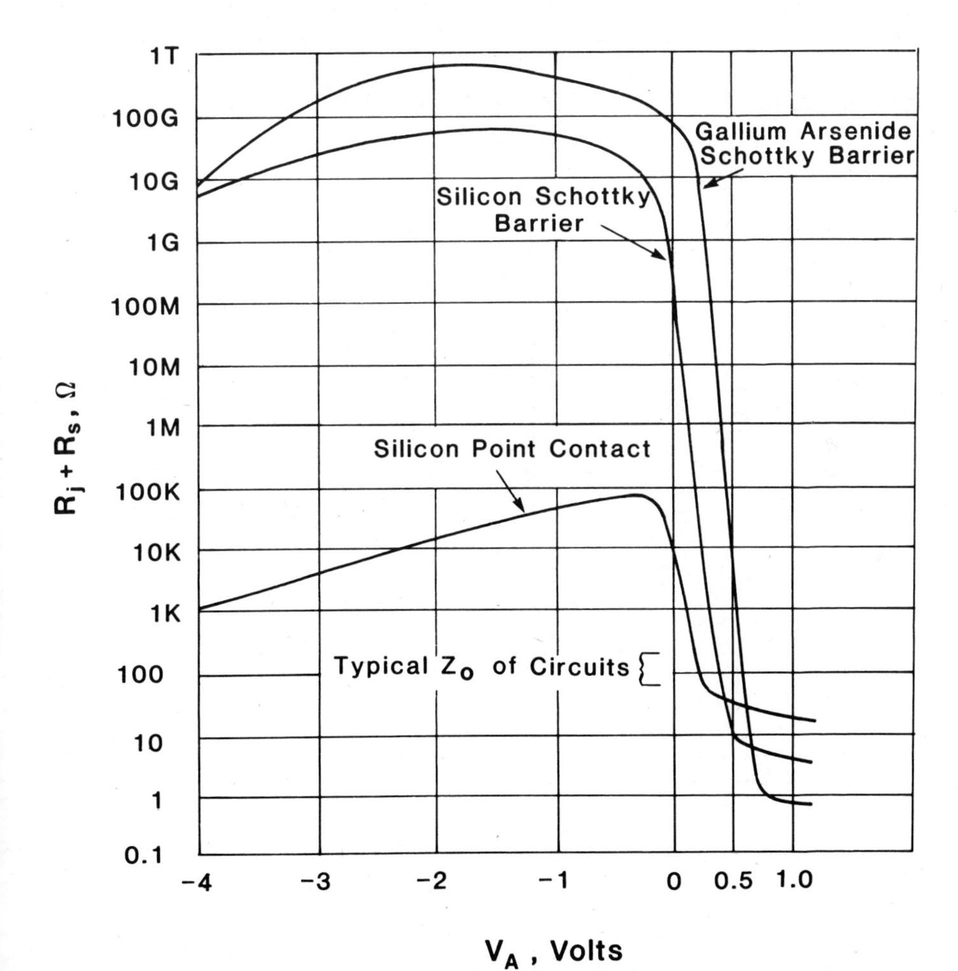

Figure 8.18 Variation of resistance with bias voltage for point-contact and Schottky-barrier varistors. (In Watson [9], reprinted with permission of McGraw Hill.)

8.5. VARISTORS

A varistor is a diode that is designed to utilize the variation of the device resistance that occurs as a function of bias voltage. The variation of resistance with bias for point-contact and Schottky-barrier varistors is shown in Fig. 8.18 [9]. For reverse bias the resistance is high and the diode is a blocking contact. For forward-bias voltages, however, a range of useful resistance values can be obtained. The diode resistance can be determined from the $I-V$ characteristic expression given for Schottky-barrier diodes in (8.53). Since varistors are normally operated in the forward-bias region, *pn* junctions cannot be used. The diffusion capacitance that dominates in these devices under forward bias prevents the device from being used as a variable resistance. The relatively large diffusion capacitance would provide a parallel path for RF energy at high frequencies, thereby tending to short out the diode resistance. The RF resistance for the diode is defined as the slope of the $I-V$ characteristic as shown in Fig. 8.19 and is given as

$$r = \frac{nkTA}{qI} \tag{8.68}$$

where n is the ideality factor previously defined.

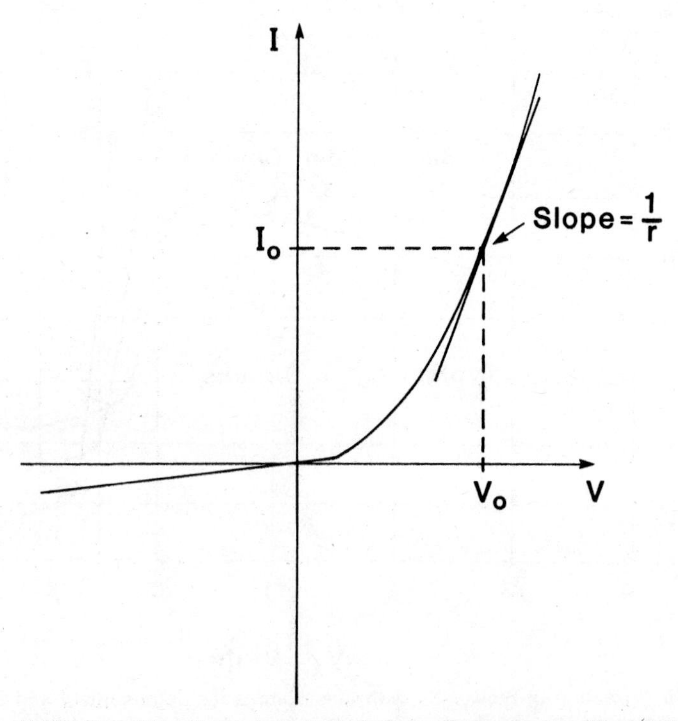

Figure 8.19 Current-voltage characteristic for a diode showing RF resistance.

8.6 *pin* DIODES

A *pin* diode is a *pn* junction device that has a very minimally doped or intrinsic region located between the *p*- and *n*-type contact regions. The addition of the intrinsic or *i*-region results in characteristics that are very desirable for certain device applications. In reverse bias the *i*-region results in very high values for the diode breakdown voltage, whereas the device capacitance is reduced by the increased separation between the *p*- and *n*-regions. The diode is very useful, therefore, for high-frequency, high-power rectifier applications. In forward bias the conductivity of the intrinsic region is controlled or modulated by the injection of charge from the end regions. The diode is a bias-current-controlled resistor with excellent linearity and low distortion. *pin* diodes are used extensively in microwave circuits for amplitude modulation, attenuation, and leveling functions. They also make excellent RF switches, phase shifters, and limiters.

8.6.1 Basic Device Physics

An analysis of *pin* diodes begins with the semiconductor equations presented in Section 7.2. Assuming the one-dimensional, abrupt junction geometry shown in Fig. 8.20, the electron continuity equation can be written in simplified form as

$$\frac{d^2n}{dx^2} + \frac{n}{L_a^2} = 0 \tag{8.69}$$

where $L_a = \sqrt{D_a \tau_a}$, the ambipolar diffusion length [10], and and the free charge density is assumed to be much larger than the thermal equilibrium density. This equation is a simple second-order differential equation and has a solution

$$n(x) = A_1 e^{x/L_a} + A_2 e^{-x/L_a} \tag{8.70}$$

The double injection condition for the diode under forward bias will result in $n(x) \cong p(x)$, and (8.70) can be considered as a general charge-density equation. It follows that a minimum will occur at $x = 0$. That is,

$$\left.\frac{dn(x)}{dx}\right|_{x=0} = 0 \tag{8.71}$$

which requires that $A_1 = A_2$, and it follows that the solution can be written as

$$n(x) = 2A_1 \cosh\left(\frac{x}{L_a}\right) \tag{8.72}$$

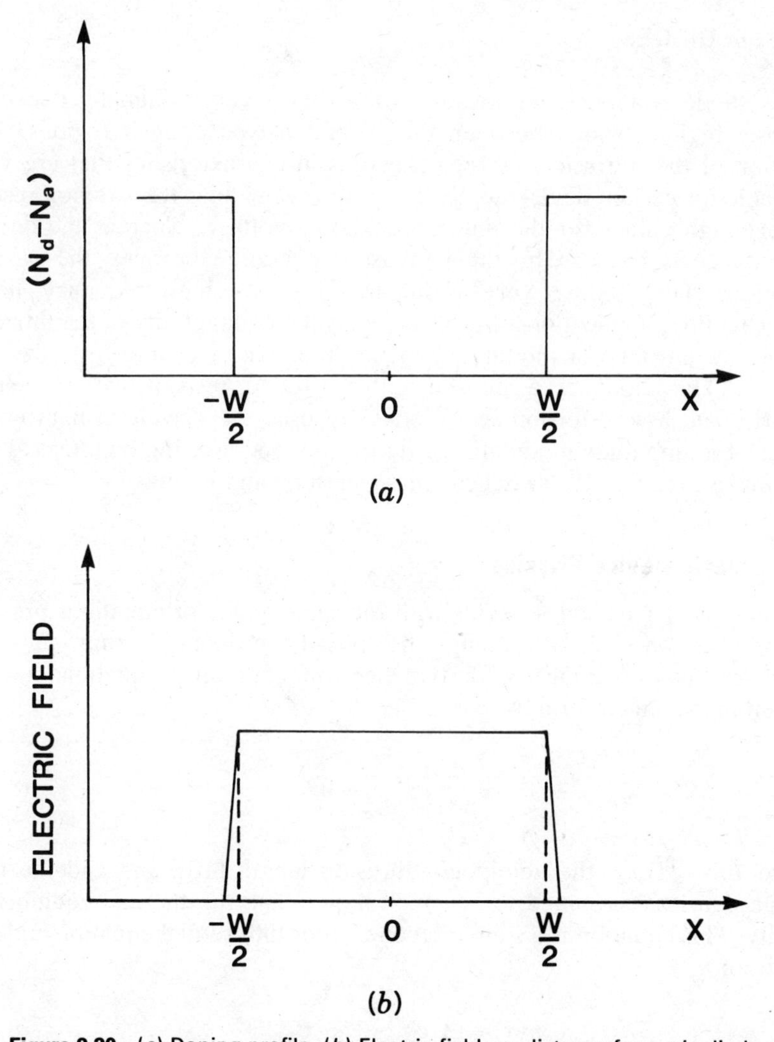

Figure 8.20 (*a*) Doping profile. (*b*) Electric-field vs. distance for a *pin* diode.

The constant A_1 is found from the boundary condition determined from the device current. That is, the current-density equations can be combined to yield the condition

$$I_0 = J_n A|_{W/2} = 2qD_a A \frac{dn}{dx} \tag{8.73}$$

where A is the cross-sectional area of the device. Applying this condition to determine the constant A_1 gives the solution

$$p(x) = n(x) = \frac{I_0 L_a \cosh\left(\frac{x}{L_a}\right)}{2qD_a A \sinh\left(\frac{W}{2L_a}\right)} \tag{8.74}$$

The i-region resistance can be calculated from the expression

$$R_i = \frac{1}{A}\int_{-W/2}^{W/2} \frac{dx}{\sigma(x)} \tag{8.75}$$

where

$$\sigma(x) = q[\mu_n n(x) + \mu_p p(x)] \cong 2q\bar{\mu}n(x)$$

and $\bar{\mu}$ is an effective mobility. Substituting and performing the integration produces an expression for the resistance that can be written for the condition that $(W/2L_a) \ll 1$ in the form

$$R_i \cong \frac{3(kT)W^2}{8qI_0L_a^2} \tag{8.76}$$

where W is the i-layer thickness (typically 10 to 100 μm). This expression shows that the resistance is proportional to $(W/L_a)^2$ and inversely proportional to the bias current. The resistance decreases with current and diffusion length (long carrier lifetimes), and increases with diode length.

The i-region is shunted by a parallel-plate capacitor determined by the p- and n-region contacts. This capacitance appears in shunt with the junction resistance and can be approximated by the expression

$$C_i = \frac{\epsilon A}{W} \tag{8.77}$$

At forward bias the junction effects must be considered. The junction capacitance and resistance are derived in the same manner as for a *pn* junction and are

$$C_j = A\left[\frac{q\epsilon N_d}{2(V_0 - V)}\right]^{1/2} \tag{8.78}$$

and

$$R_j = \frac{nkTA}{qI_0} \tag{8.79}$$

At high frequencies and at forward bias the charge storage diffusion capacitance must also be considered. It can be shown that the frequency-dependent complex admittance for the diode is

$$Y = G + jB = \frac{qA}{kT}[J_p\sqrt{1 + j\omega\tau_p} + J_n\sqrt{1 + j\omega\tau_n}] \tag{8.80}$$

For high-frequency operation where $\omega\tau \gg 1$, this expression reduces to the form

$$Y = G + jB = \frac{qI_0}{kT}\sqrt{\frac{\omega\tau}{2}}(1 + j) \tag{8.81}$$

8.6.2 Switching Speed

The switching of a *pin* diode from a low-impedance to a high-impedance state is accomplished by switching the bias current from forward to reverse bias. Under reverse bias a current will flow until the diode is depleted of charge. The switching time can be defined as the time required to remove the charge stored under forward bias less the charge that recombines in the *i*-region. The switching or reverse recovery time is composed of two components: the time required to remove most of the charge (called the delay time) from the *i*-region, and the time during which the diode is changing from a low- to a high-impedance state (called the transition time). The transition time depends upon diode geometry and doping profile, but is not sensitive to the magnitude of the forward or reverse current. The delay time is inversely proportional to the magnitude of the reverse-bias current and directly proportional to the charge-carrier lifetime. Diodes with short carrier lifetimes have short delay times (i.e., fast switching speeds), but suffer from high values of forward-bias resistance and, therefore, high insertion loss.

8.6.3 Equivalent Circuit

The equivalent circuit for a *pin* diode is shown in Fig. 8.21. The diode basically consists of two main elements in series: a diode with diffusion capacitance C_d and junction parameters R_j and C_j all in parallel; and the parallel combination of the undepleted *i*-region resistance and capacitance. The equivalent-circuit elements were defined in the previous section. The parameter L_s represents the lead inductance, C_f represents fringing capacitance from the top contact of the diode to the mounting structure, which can be neglected for first-order approximation, and C_p represents package and/or mounting structure capacitance. The parameter R_s represents the resistance of the bulk semiconductor regions plus the resistance of the contacts.

The equivalent circuit shown in Fig. 8.21 is modified under certain bias and high-frequency operating conditions. For example, under reverse bias

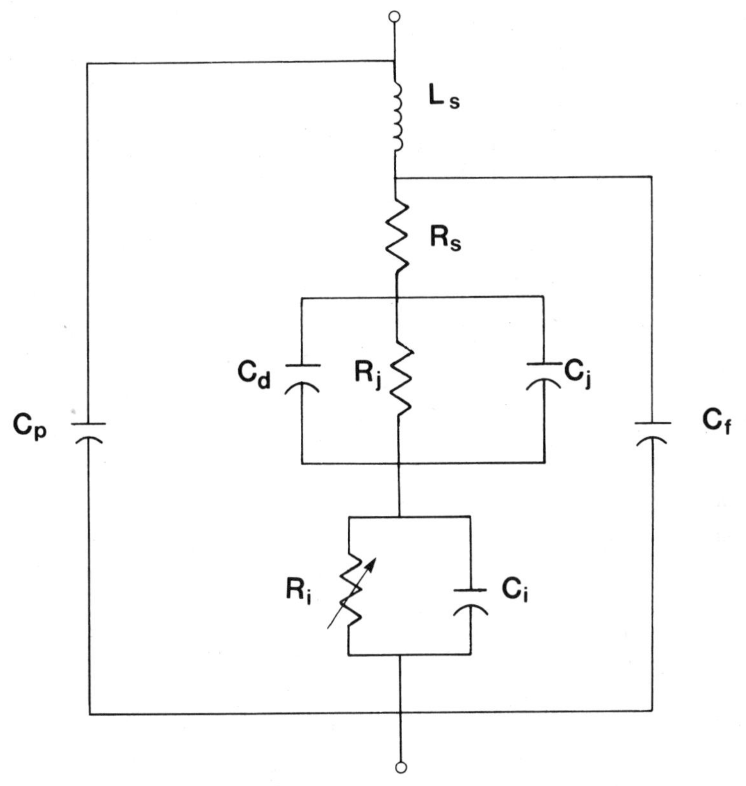

Figure 8.21 Equivalent circuit for a *pin* diode.

the diffusion capacitance C_d vanishes and the junction resistance R_j becomes very large. Only C_j remains significant in the junction network. If the reverse bias is sufficient to completely deplete the *i*-region, the R_i–C_i network will vanish. The equivalent circuit reduces to that shown in Fig. 8.22(*a*), where R_r is the total series resistance for the reverse-bias state.

Under forward-bias and high-frequency operation the diffusion capacitance C_d will be large and will short out the junction parameters. The *i*-region will be injected with charge carriers so that the R_i–C_i network is represented by R_i only. The equivalent circuit reduces to that shown in Fig. 8.22(*b*), where R_f is the total series resistance for the forward-bias state.

8.6.4 Figure of Merit

Several different methods have been used for expressing the relative merit of two-terminal switching devices exhibiting two distinct impedance states. The ratio of the resistance in the high-impedance state to that in the low-impedance state can be used as a figure of merit of the device at lower frequencies where the device reactances are insignificant. Switching diodes

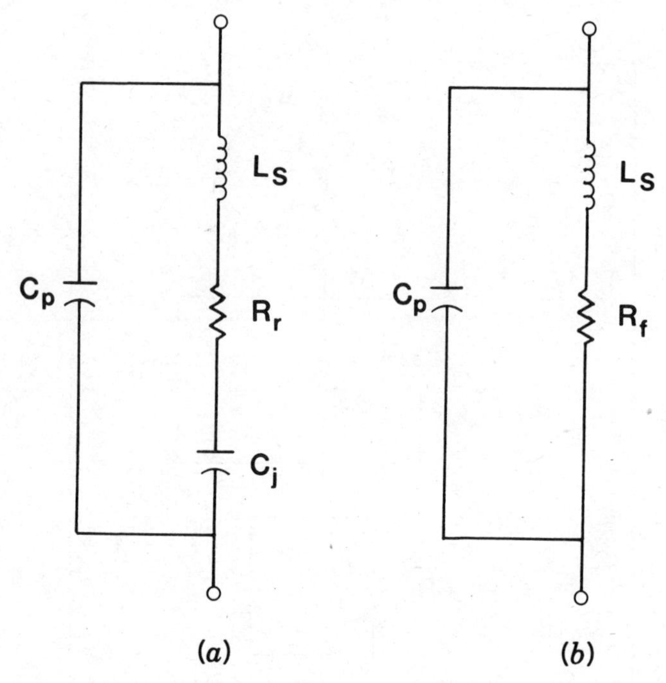

Figure 8.22 Simplified equivalent circuit for a *pin* diode under (*a*) reverse and (*b*) forward bias.

have been traditionally characterized in terms of a frequency parameter known as switching cutoff frequency f_{cs}, which is defined in terms of series resistances in two impedance states and the series capacitance in the high-impedance state. Referring to the equivalent circuit elements shown in Fig. 8.22, we have

$$f_{cs} = \frac{1}{2\pi C_j \sqrt{R_f R_r}} \tag{8.82}$$

The term switching cutoff is not intended to indicate that a particular device can be used for designing a switching circuit at frequencies as high as f_{cs}. As a matter of fact, *pin* diodes are used typically in a frequency range around one-hundredth to one-fiftieth of the switching cutoff frequency. Also it can be shown that losses in switching and phase-shifting circuits are proportional to the ratio of the switching cutoff frequency to the operating frequency.

A more general representation of the figure of merit for two-impedance-state switching devices has been proposed by Kurokawa and Schlosser [11]. When the device impedances in the two states are $z_1(=r_1 + jx_1)$ and

$z_2(= r_2 + jx_2)$, respectively, the figure of merit Q may be written as

$$Q = \frac{\sqrt{(r_1 - r_2)^2 + (x_1 - x_2)^2}}{\sqrt{r_1 r_2}} = \frac{|z_1 - z_2|}{\sqrt{r_1 r_2}} \tag{8.83}$$

For an ideal diode, the two impedances z_1 and z_2 approach zero and infinity, respectively, and the figure of merit Q tends to infinity. Also for lossless devices ($r_1 = r_2 = 0$), the factor Q tends to be infinite. An alternative expression for Q is written in terms of complex reflection coefficients Γ_1 and Γ_2 present on a lossless transmission line when terminated in the two-terminal switching device. We have [11],

$$Q' = \frac{|\Gamma_1 - \Gamma_2|^2}{(1 - |\Gamma_1|^2)(1 - |\Gamma_2|^2)} \tag{8.84}$$

This expression for Q' is more useful, as the complex reflection coefficients Γ_1 and Γ_2 can be measured conveniently (for example, by using a network analyzer). Another major advantage of using Q' as a figure of merit is that Q' is invariant to lossless transformations. That is, the measured value of Q' will remain unchanged when we interpose a lossless two-port network between the switching device and the measurement system. This lossless two-port network can represent package reactances and/or any adapters (or connectors) and/or any bias network used in the measurement setup.

For *pin* diode chips (without any bonding wire), impedances z_1 and z_2 in two bias states may be written as

$$z_1 = R_f \tag{8.85a}$$

$$z_2 = R_r + \frac{1}{j\omega C_j} \tag{8.85b}$$

and the figure of merit Q may be written as

$$Q = \frac{\sqrt{(R_f - R_r)^2 + (1/\omega C_j)^2}}{\sqrt{R_f R_r}} \simeq \frac{1}{2\pi f C_f \sqrt{R_r R_f}} \tag{8.86}$$

When $|R_f - R_r| \ll 1/\omega C_j$, (8.86) may be simplified as

$$Q \simeq \frac{f_{cs}}{f} \tag{8.87}$$

where f_{cs} is the switching cutoff frequency defined in (8.82). Values of Q at 3 GHz calculated from parameters in Table 8.3 for the MA47892 and MA47899 diodes are 118.6 and 265.3, respectively. The corresponding

TABLE 8.3 Equivalent Circuit Parameters for Two Commercially Available *pin* Diodes

Parameter	MA47892-109	MA47899-030
C_j	1 pF	0.1 pF
R_f	0.4 Ω	1 Ω
R_r	0.5 Ω	4 Ω
L_{int}	0.3 nH	0.3 nH
C_p	0.08 pF	0.18 pF
τ	5 μs	0.5 μs
f_{cs}	350 GHz	800 GHz

values for the switching cutoff frequency are 350 GHz and 800 GHz, respectively.

8.7 STEP-RECOVERY DIODES

A step-recovery diode (SRD) is a diode that has a very nonlinear conduction characteristic [12, 13]. Although in principle SRDs can be fabricated from any diode structure, practical devices are usually based upon the *pin* structure. Step-recovery diodes, which are also called snap-back diodes, are a type of charge-storage diode and are useful for frequency multiplication applications where high-efficiency harmonic generation is desirable. Due to the very nonlinear switching characteristics of these diodes, harmonic generation with an efficiency approaching $1/n$, where n indicates the harmonic number, can be obtained. This is in contrast to varactor multipliers where the harmonic generation efficiency is on the order of $1/n^2$. Step-recovery diode multipliers require no idler circuits, resulting in very simple and compact circuits. For these reasons SRDs are generally used where high-efficiency, high-order frequency multiplication is required, such as in frequency multipliers and comb generators.

8.7.1 Basic Device Physics

The operation of a SRD depends upon the charge-storage characteristics of certain semiconductors. The minority carrier lifetime in indirect bandgap materials, such as Si, can achieve magnitudes on the order of less than a microsecond to hundreds of milliseconds, depending upon the crystal quality and impurity distribution. Minority carriers in these materials can, therefore, remain in a free state without recombination for times on the order of RF cycle times at microwave frequencies. These materials are useful for fabrication of charge-storage devices. Compound semiconductors such as GaAs, for comparison, have direct bandgaps and generally have relatively short minority carrier lifetimes and are not of use for these applications. The important consideration in a SRD is that the charge extraction time, which is dependent upon the transit time of the charge carriers through the diode, be short relative to the minority carrier lifetimes. For efficient operation the charge should be removed by the terminal bias and not be lost by recombination. Minority charge recombination represents a loss mechanism that limits the harmonic generation efficiency of these devices.

The operation of a SRD can be qualitively understood with reference to the device terminal waveforms shown in Fig. 8.23. If the diode is biased with bias voltage V_b, and a sinusoidal RF voltage is placed across the diode, the diode will conduct RF current when the sum of the dc and RF voltages is greater than the forward threshold voltage, which is typically in the range

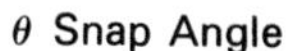

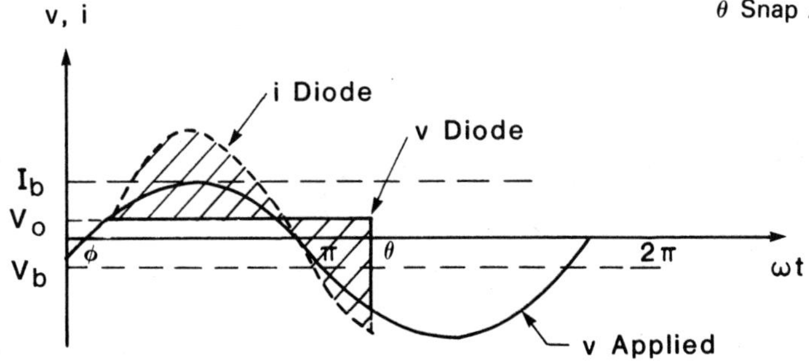

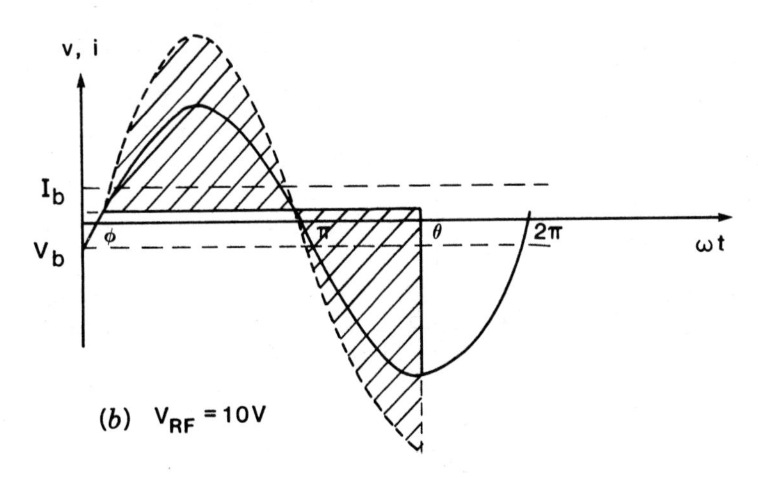

Figure 8.23 (*a,b*)

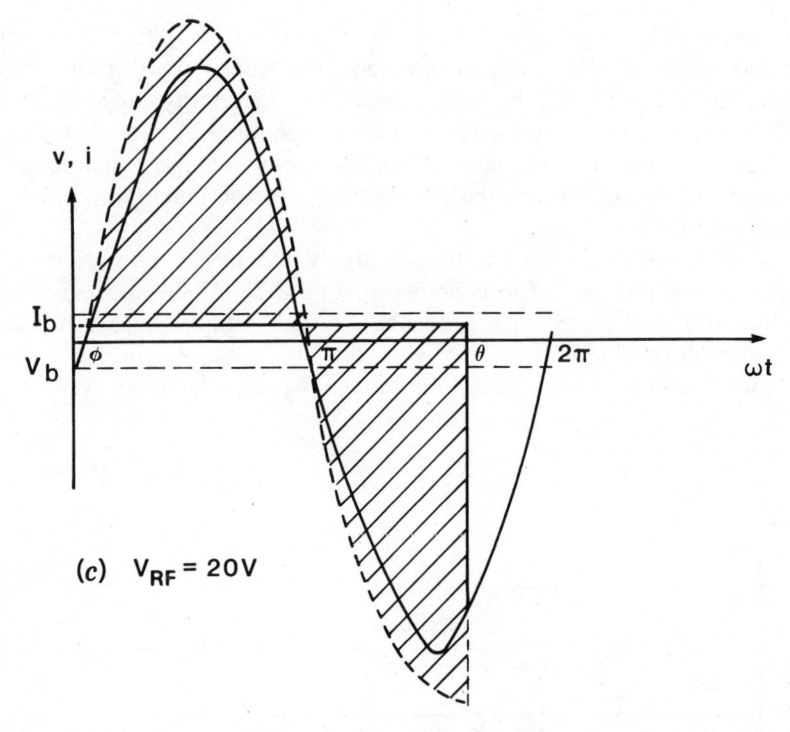

Figure 8.23 Terminal current and voltage waveforms for a step-recovery diode.

of $V_0 = 0.8\text{–}1.0$ V. The point in the RF cycle when conduction occurs is indicated as the conduction angle ϕ in Fig. 8.23. The RF current is dependent upon the RF voltage and the diode will conduct with the RF current waveform essentially following the RF voltage waveform. The charge injected from the contact regions will result in the *i*-region of the diode becoming filled with an electron-hole plasma that, due to the long carrier lifetimes, will be stored in the diode. As the diode voltage drops below threshold and begins the negative portion of the RF cycle, current conduction will continue due to the charge stored in the structure. The terminal voltage will result in charge extraction from the diode and the device terminal current will continue to follow the diode voltage. The charge extraction time is represented as the angle θ in Fig. 8.23, and it represents the time required to extract the free charges from the diode. The charge extraction time can be estimated from the switching characteristics for *pn* junction diodes. For constant amplitude waveforms, as shown in Fig. 8.24, the extraction time is approximated by the expression [14],

$$t_s = \tau \ln\left(1 + \frac{I_f}{I_r}\right) \tag{8.88}$$

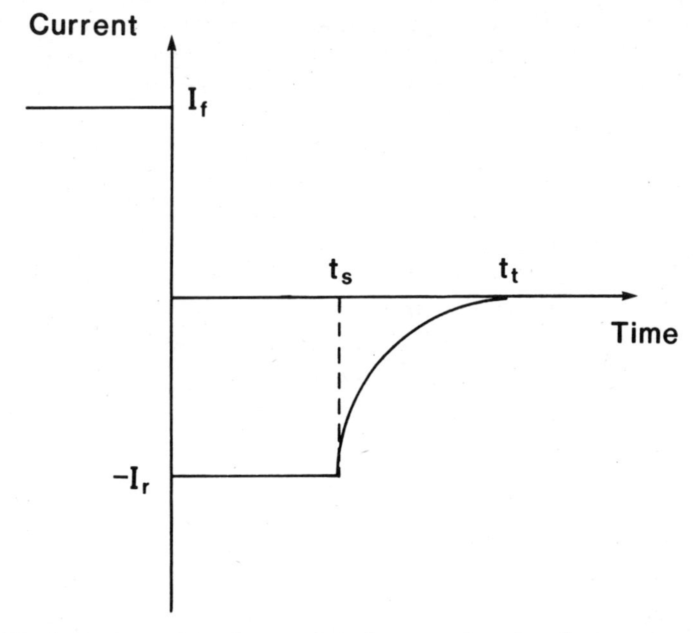

Figure 8.24 Current vs. time characteristic for a *pn* junction diode under transient, switching conditions.

where τ is the carrier lifetime, I_f is the forward current just before switching occurs, and I_r is the reverse current just after switching occurs.

Once the free charge has been removed from the diode, the current will be reduced to zero in a very short time. The current will "snap" from its reverse conduction magnitude (which can be quite substantial) to zero in a very short time. The transition time is represented by the angle θ in Fig. 8.23. The transition time for graded profile devices has been derived by Moll et al. [15] and is

$$t_t = \frac{1}{D_a}\left[\frac{Q_0/A}{4qb(N_0 b)}\right]^{2/3} \tag{8.89}$$

where D_a is the ambipolar diffusion coefficient (which is defined as $D_a = 2D_nD_p/(D_n + D_p)$, A is the diode area, Q_0 is the initial charge stored, and $(N_0 b)$ is the impurity gradient at the junction.

The rapid decrease in terminal current creates device waveforms that are very rich in harmonics. If a high-Q resonant circuit is placed across the diode, the resonant circuit will be "shocked" into a high-efficiency ringing mode. By tuning the resonant frequency of the circuit, selective harmonics can be extracted. Such a circuit should be parallel tuned so that all frequencies out of the resonators bandwidth will be terminated in a short circuit. All energy will be focused at the resonant frequency of the circuit and high-efficiency harmonic generation results. If the diode is operated in a low-Q circuit, the output signal will contain many harmonics. This type of operation is utilized in comb generators, which are useful for marker circuits.

For optimum operation, the snap angle should occur when the RF current is at a maximum. The snap angle is determined by the RF and dc voltages placed across the diode as shown in Fig. 8.25. For a given RF voltage, an increase in negative bias voltage will cause the snap angle to occur sooner in the RF cycle. In typical harmonic generator circuits, the SRD can be either self-biased through a resistor, or externally biased. Bias voltages are generally near zero and may be positive or negative. The bias voltage can be adjusted to achieve optimum harmonic generation efficiency.

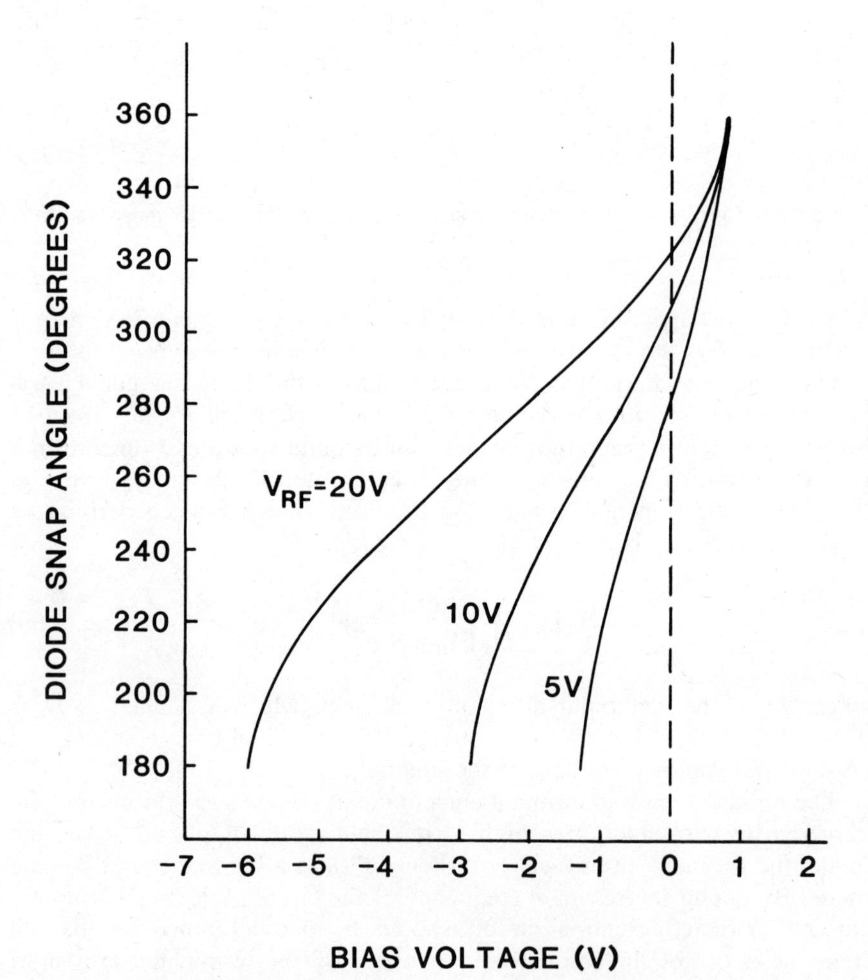

Figure 8.25 SRD snap angle vs. bias and RF voltages.

8.7.2 Frequency Limits

The operation of the SRD is dependent upon charge recombination. If the stored charge remains in the device for relatively long periods compared to the RF cycle, recombination will occur and the device efficiency will be reduced. This consideration places a lower limit upon the frequency at which a given device may be expected to operate in a satisfactory mode. The lower frequency limit is defined as the inverse of the charge lifetime as

$$f_{\text{low}} = \frac{1}{2\pi\tau} \tag{8.90}$$

The upper frequency limit results from the switching time considerations. An upper frequency at which the diode is capable of efficient harmonic generation is defined from the inverse of the transition time and is

$$f_{\text{high}} = \frac{1}{2\pi t_t} \tag{8.91}$$

8.7.3 Equivalent Circuit

The SRD is a form of electronic switch, and an equivalent circuit that is capable of describing its operation must include both forward- and reverse-bias operation. Under forward bias the equivalent circuit consists of a relatively large diffusion capacitance in shunt with a resistance. The diffusion capacitance accounts for the large charge storage in the device and the resistance accounts for the forward current that may flow. Under reverse bias very little steady state current flows. A reverse-bias equivalent circuit consists almost entirely of the depletion-layer capacitance.

A complete equivalent circuit is obtained by combining the forward and reverse bias circuits with a switch as shown in Fig. 8.26. The series

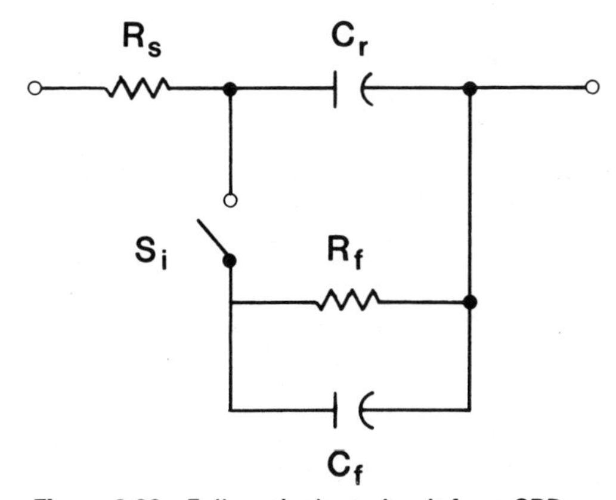

Figure 8.26 Full equivalent circuit for a SRD.

resistance is included to account for the voltage drop across the bulk material. The switch is closed under forward bias and remains closed initially during reverse bias until the time at which all the charge is extracted from the capacitor C_f. A model based upon this equivalent circuit has been presented in [16].

REFERENCES

1. Henisch, H. K., *Rectifying Semi-Conductor Contacts*, Clarendon Press, Oxford, 1957.
2. Sze, S. M., *Physics of Semiconductor Devices*, Wiley–Interscience, 1981, pp. 94–96.
3. Mott, N. F., "The Theory of Crystal Rectifiers," *Proc. Roy. Soc. (London)*, Vol. 171A, May 1939, pp. 27–38.
4. Bardeen, J., *Phys. Rev.*, Vol. 71, 1947, p. 717.
5. Schottky, W., *Naturwiss.*, Vol. 26, 1938, p. 843.
6. Howes, M. J., and D. V. Morgan, *Variable Impedance Devices*, Wiley–Interscience, 1978, pp. 51–62.
7. Norwood, M. H., and E. Shatz, "Voltage-Variable Capacitor Tuning: A Review," *Proc. IEEE*, Vol. 56, May 1968, pp. 788–798.
8. Watson, H. A., *Microwave Semiconductor Devices and Their Circuit Applications*, McGraw-Hill, 1969, Chap. 7.
9. Watson, H. A., *Microwave Semiconductor Devices and Their Circuit Applications*, McGraw-Hill, 1969, p. 353.
10. McKelvey, J. P., *Solid-State and Semiconductor Physics*, Harper & Row, 1966, Chap. 10.
11. Kurokawa, K., and W. O. Schlosser, "Quality Factor of Switching Diodes for Digital Modulation," *Proc. IEEE*, Vol. 58, Jan. 1970, pp. 180–181.
12. Boff, A. F., J. Moll, and R. Shen, "A New High-Speed Effect in Solid-State Diodes," *1960 IEEE Solid-State Circuits Conf. Digest Tech. Papers*, pp. 50–51.
13. Krakauer, S., "Harmonic Generation, Rectification and Lifetime Evaluation with the Step Recovery Diode," *Proc. IRE*, Vol. 50, July 1962, pp. 1665–1676.
14. Steetman, B. G., *Solid-State Electronic Devices*, Prentice-Hall, 1980, Chap. 5.
15. Moll, J. L., S. Krakauer, and R. Shen, "*P-N* Junction Charge-Storage Diodes," *Proc. IRE*, Vol. 50, Jan. 1962, pp. 43–53.
16. Kotzabue, K. L., "A Circuit Model of the Step Recovery Diode," *Proc. IEEE*, Dec. 1965, p. 2119.

PROBLEMS

8.1 A Schottky barrier diode is made by evaporating a dot of gold (Au) of area $A = 10^{-3}\ \text{cm}^2$ onto an n-type Si crystal doped to a level of $N_d = 10^{14}\ \text{cm}^{-3}$. For operation at room temperature:

(a) What current transport mechanism would you expect to dominate? Why?

(b) Calculate the magnitude of the "reverse saturation current," I_0, where

$$I = I_0[e^{qV/kT} - 1]$$

8.2 The electron concentration in a semiconductor can be expressed as a function of the quasi-Fermi level by the expression,

$$n = n_i e^{(E_{Fn} - E_i)/kT}$$

Show that the electron current can be expressed as a function of the spatial variation of the quasi-Fermi level, i.e.,

$$J_n \propto \frac{\partial E_{Fn}}{\partial x}$$

This calculation shows that the quasi-Fermi level must be constant with distance for zero net current flow. We used this condition in our discussion of junctions.

8.3 Two convenient devices that have rectification properties are the Si *pn* junction diode and the GaAs Schottky-barrier diode. Discuss which of these diodes you would use for

(a) high-frequency RF detection,

(b) high-speed switching, and

(c) high-power rectification.

Discuss fundamental properties and limitations. Be as complete as possible.

8.4 Consider a *pn* junction formed from uniformly (constant) doped *p*- and *n*-type semiconductor regions. Assume one dimension.

(a) Derive expressions for the minority charge densities on each side of the junctions.

(b) Sketch $p(x)$ and $n(x)$ throughout the device for forward and reverse bias.

(c) Derive expressions for the minority carrier currents on each side of the junction. State all approximations or assumptions you make.

(d) Sketch the hole, electron, and total currents that flow through the device for forward and reverse bias.

(e) Derive the ideal diode equation. State all assumptions or approximations you make.

8.5 An n^+pn^+ diode is fabricated as shown in the sketch.

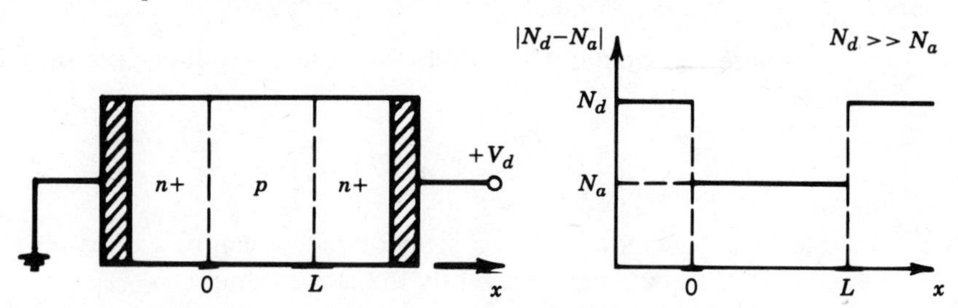

(a) Derive an expression for the punch-through voltage (V_{pt}) defined as the drain voltage required to completely deplete the p-region of mobile charge.

(b) Derive an expression for the drain voltage (V_{fb}) that would need to be applied to the device in order to completely eliminate any potential barrier internal to the device. Neglect any series resistance external to the p-region.

(c) Sketch the energy band diagrams and internal electric field versus distance for the three values of drain voltage $V_d = 0$, $V_d = V_{pt}$, and $V_d = V_{fb}$.

8.6 The limiting forward bias across a pn junction is equal to the contact potential, V_0. This effect is not predicted by the ideal diode equation. The reason for this discrepancy is the neglect of changes in the majority carrier concentrations on each side of the junction at high injection levels. For the case of high-level injection and following a procedure similar to that used in deriving the ideal diode equation, show that a general diode equation can be derived that has the form,

$$J = q\left(\frac{e^{qV/kT} - 1}{1 - e^{-2q(V_0 - V)/kT}}\right)\left[\frac{D_p p_{n0}}{L_p}\left(1 + \frac{n_i^2}{p_{p0}^2} e^{qV/kT}\right) + \frac{D_n n_{p0}}{L_n}\left(1 + \frac{n_i^2}{n_{n0}^2} e^{qV/kT}\right)\right]$$

8.7 Consider a Schottky barrier diode fabricated by placing a metal on a uniformly doped n-type semiconductor as shown in the sketch.

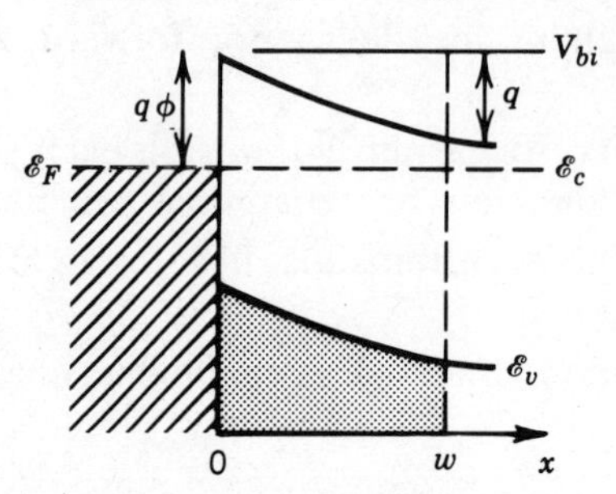

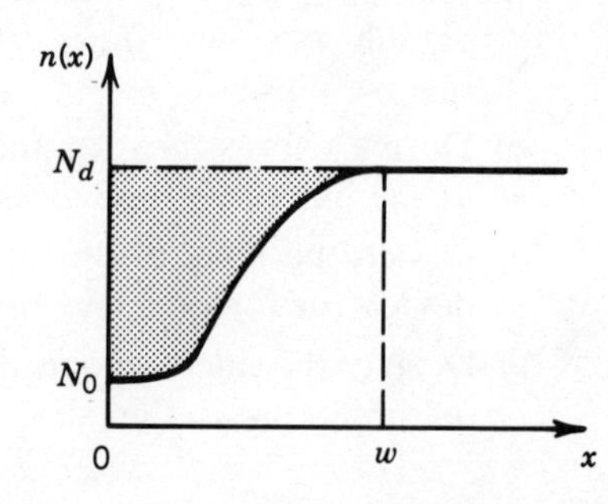

@ $x = 0$	@ $x = W$
$v = -v_{bi}$	$v \simeq 0$
$E = E_{max}$	$E \simeq 0$
$n = n_0$	$n \simeq N_d$
$\dfrac{dE}{dx} \simeq \dfrac{qN_d}{\epsilon}$	$\dfrac{dE}{dx} \simeq 0$

(a) Derive an expression for the space-charge due to the positive donor atoms within the depletion region.

(b) Use your expression from part (a) to derive an expression for the capacitance of the junction where

$$C \triangleq \left| \frac{\partial Q}{\partial v} \right|$$

(c) Use your expression from part (b) to derive an expression for the width of the depletion region W (i.e., the space-charge region).

9 OSCILLATORS

9.1 INTRODUCTION

Oscillators represent the basic microwave energy source for all microwave systems, such as radars, communications, navigation, or electronic warfare. They can be termed as dc-to-RF converters or infinite-gain amplifiers. A typical microwave oscillator essentially consists of an active device (a diode or a transistor) and a passive frequency-determining resonant element, such as a microstrip, SAW, cavity resonator, or dielectric resonator for fixed tuned oscillators and a varactor or a YIG sphere for tunable oscillators. With the rapid advancement of microwave technology, there has been an increasing need for better performance of oscillators. The emphasis has been on low noise, small size, low cost, high efficiency, high-temperature stability, and reliability for all oscillators and additionally on wider bandwidths, better tuning linearity, and reduced settling time for the tunable oscillators.

The origin of solid state microwave oscillators using Gunn and IMPATT diodes dates back to the late 1960s, before which microwave sources used to be massive klystron or magnetron tubes requiring huge power supplies. In less than two decades, solid state oscillators have come a long way. The extension of the bipolar transistor oscillator to microwave frequencies and the development of GaAs MESFET devices in the early 1970s has made available today, highly cost-effective, miniature, reliable, and low-noise sources for use right up to the millimeter-wave frequency range.

Basic active elements that can be used for microwave solid state oscillators are Gunn diodes, IMPATT diodes, and transistors. While a Gunn oscillator has the advantage of low FM noise against an IMPATT oscillator, the latter has a higher efficiency and is a higher power device as compared to the Gunn. Transistor oscillators, on the other hand, are low-noise as well as high-efficiency sources. Compared to Gunn oscillators, transistor oscillators also do not have the problems of threshold current, heat sinking, and the tendency to lock at spurious frequencies. The diodes: Gunns and IMPATTS are inherently negative-resistance devices, which means that just the application of the required dc bias is sufficient to produce negative resistance. The design of the oscillator is thus simplified to the design of the output matching circuit to deliver the desired power output. Application of a dc bias to the bipolar or the field-effect transistor, on the other hand, requires a suitable series–parallel feedback mechanism to induce the negative resistance. The frequency range over which the negative resistance is present in the diodes is determined by the physical mechanisms in the device, while in the case of transistors this frequency range is also influenced by the chosen circuit topology. The only disadvantage transistors have at the moment is the maximum oscillation frequency. While a Gunn or IMPATT oscillator is capable of delivering 100-mW power up to 100 GHz, transistor oscillators are currently limited to 10 mW at 40 GHz.

Transistor oscillators can be realized using either bipolar or GaAs FET devices. Bipolar oscillators have maximum oscillation frequency lower than that of GaAs FET oscillators, while the latter is noisier than the former. GaAs FET oscillators have been reported up to 100 GHz, while oscillators using bipolar transistors have reached 20 GHz. Typically a bipolar oscillator has 6 to 10 dB less FM noise very close to the carrier, compared to a GaAs FET oscillator. The oscillator design approach presented in this chapter is applicable to oscillators using either type of transistor, though the examples shown here are only for GaAs FETs.

Increased capability of modern technology to utilize the pure microwave signal and to measure precisely its characteristics, has necessitated the development of highly stable signal sources having minimum FM and AM noise and maximum long-term and short-term frequency stability with time and temperature. For application as a local oscillator in a communication system or radar, the frequency stability is often one of the important factors in determining the receiver IF bandwidth and for bandwidth allocations for communication systems. A narrower IF bandwidth improves the receiver performance by improving its noise figure. In Doppler radars, the short-term stability of the signal is of utmost importance for the positive detection of moving targets. Low oscillator noise in a transmitter source ensures a high S/N ratio.

Quartz crystal oscillators represent highly stable sources, but their operation is limited to a few hundred megahertz. At microwave frequencies, stable sources have commonly been realized using frequency multiplication by a factor of N, of a quartz crystal oscillator. This method increases the FM noise power by N, has very low efficiency, in addition to being very complex and expensive. Stable signals have also been generated in the past using huge metallic high-Q cavities in passive cavity stabilization systems or in complex and bulky frequency discriminator systems.

Dielectric resonators, due to their high Q, small size, and excellent integrability in MIC circuits, can be directly used as a frequency-determining element for realizing a stable MIC transistor oscillator. With the recent advent of temperature-stable materials for dielectric resonators, the transistor dielectric resonator diode (TDRO) is fast becoming an automatic choice for a vast number of applications of fixed-frequency oscillators.

Frequency-tunable oscillators find applications in instrumentation, electronic warfare systems, communication systems, and phase-locked oscillator schemes. YIG-tuned oscillators have the widest tuning bandwidths with 2- to 20-GHz tuning capability with single sphere and single transistor [1].

These oscillators are, however, slow in tuning, bulky, and much less efficient compared to varactor-tuned oscillators. YIG oscillators do present much better phase noise and tuning linearity compared to varactor-controlled oscillators (VCOs). VCOs can offer up to an octave and a half bandwidth [2] and are commonly used in applications where size, fast tuning and efficiency are the main considerations.

9.2 CONCEPT OF NEGATIVE RESISTANCE

The concept of negative resistance is as old as oscillators themselves. Contrary to a positive resistance, a negative resistance is considered as a source of electrical energy. A negative resistance implies that the device is active, while an active device does not imply a negative resistance. A two-terminal device like a Gunn or IMPATT diode inherently has negative resistance, while a three-terminal device like a transistor needs an appropriate impedance to be connected to one or more terminals to create the negative resistance. An important characteristic of the negative resistance is that it is a nonlinear function of the RF current through it. When a load R_1 is connected to the negative resistance R_n, with $R_1 < R_n$, RF current starts flowing through the circuit at a frequency at which the imaginary parts of both impedances cancel each other. This RF current causes the value of the negative resistance to change unless the oscillation condition of $-R_n = R_1$ is satisfied. The concept of negative resistance though important to an understanding of the oscillation mechanism, is not very practical to deal with at microwave frequencies. It is not possible to make any significant measurements at these frequencies in terms of R, L, and C. The use of negative resistance for oscillator analysis can be misleading unless caution is exercised to use negative conductance analysis when appropriate [3]. The most common approach is by using reflection-transmission coefficients, that is, S-parameters. Negative resistance–conductance at any terminal can, however, be converted to a reflection coefficient by using the following relations:

$$r_n = \frac{R_n}{Z_0} = \frac{1 - |\Gamma|^2}{1 - 2|\Gamma| \cos \theta + |\Gamma|^2} \tag{9.1}$$

$$g_n = \frac{G_n}{Y_0} = \frac{1 - |\Gamma|^2}{1 + 2|\Gamma| \cos \theta + |\Gamma|^2} \tag{9.2}$$

The extended Smith chart shown in Fig. 9.1 shows the negative-resistance region, that is, the area outside the normal Smith chart. The shaded area represents the normal Smith chart. This chart can be used to transform a negative impedance to a reflection coefficient and vice versa.

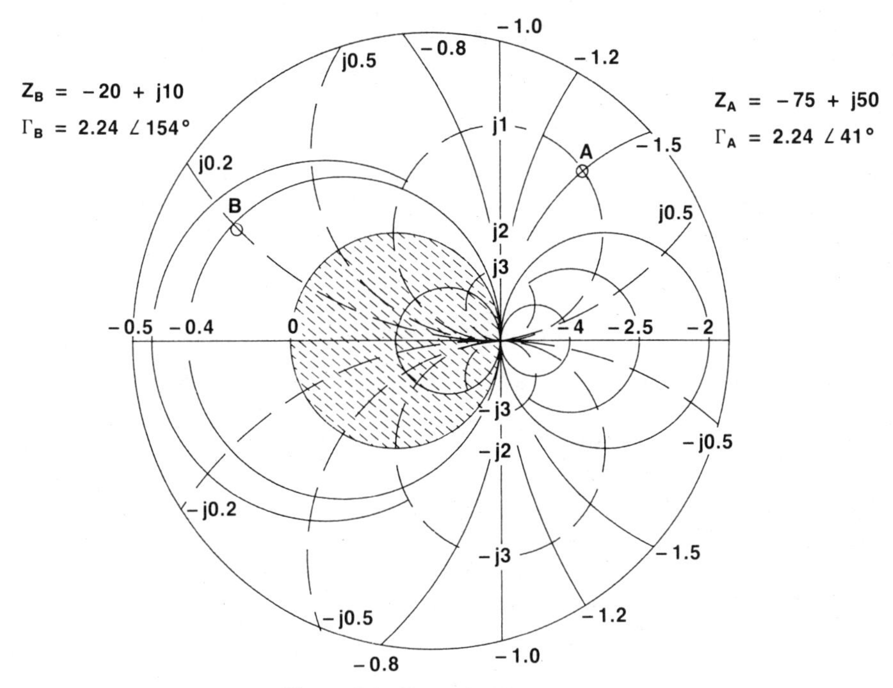

Figure 9.1 Extended Smith chart.

In view of the fact that the S-parameter approach is the most practical approach at microwave frequencies, it will be used for oscillator analysis and design here.

9.3 THREE-PORT *S*-PARAMETER CHARACTERIZATION OF TRANSISTORS

The transistor, a three-port device, is generally characterized by its two-port S-parameters with one of its ports grounded. The resulting three different configurations in the case of a GaAs FET are shown in Fig. 9.2. Each configuration has its own advantages (e.g., the common source configuration is used more often for amplifiers, common gate for wideband oscillators, and common drain for medium power oscillators). The use of three-port S-parameters, although introduced quite sometime back [4], has not often been used due to the complexity of the analysis involved. The availability of desktop computers and CAD has now made their use practical. The use of three-port S-parameters eliminates the otherwise necessary conversion to and from Z- and Y-parameters to analyze the series and parallel feedback effect as shown later in this chapter. The three-port indefinite S-matrix of the transistor holds the property of having

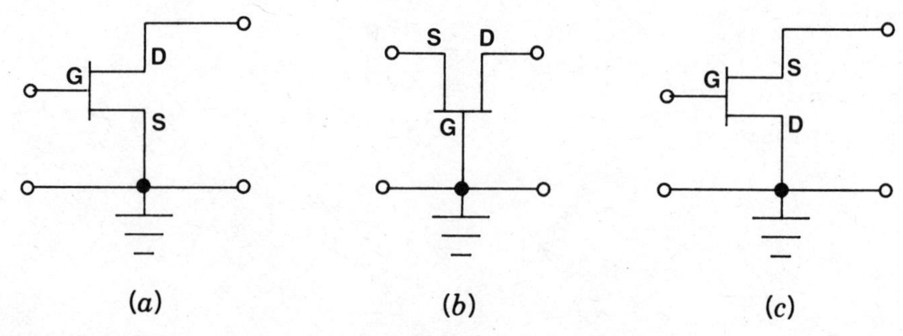

Figure 9.2 Three configurations of the transistors. (*a*) Common source. (*b*) Common gate. (*c*) Common drain.

the sums of the rows and columns to be equal to 1, which helps in determining and eventually correcting the systematic errors in the measurement or analysis. The use of the three-port *S*-parameters of the transistor sometimes become essential in the design of oscillators [5].

The transistor as a three-port device is shown in Fig. 9.3, and the *S*-matrix of the incident and reflected waves is given by:

$$\begin{bmatrix} b_1 \\ b_2 \\ b_3 \end{bmatrix} = \begin{bmatrix} S_{11} & S_{12} & S_{13} \\ S_{21} & S_{22} & S_{23} \\ S_{31} & S_{32} & S_{33} \end{bmatrix} \begin{bmatrix} a_1 \\ a_2 \\ a_3 \end{bmatrix} \tag{9.3}$$

The ports 1, 2, and 3 represent gate, drain, and source, respectively. This indefinite three-port *S*-matrix satisfies the following conditions.

$$\sum_{j=1}^{3} S_{ij} = 1 \qquad \text{for } i = 1, 2, 3 \tag{9.4}$$

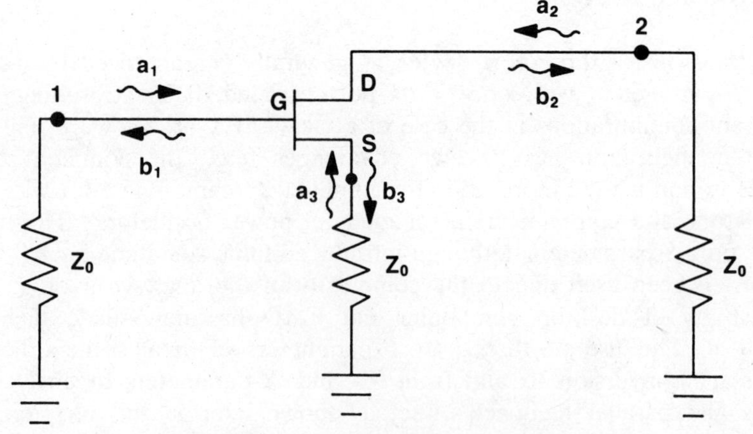

Figure 9.3 Transistor as a three-port device.

$$\sum_{i=1}^{3} S_{ij} = 1 \qquad \text{for } j = 1, 2, 3 \tag{9.5}$$

The three-port S-parameters of the transistor can be directly measured or obtained analytically from the commonly supplied two-port S-parameters using the relations given in [6]. As an example, the two-port S-parameters of a typical 0.5-μm GaAs FET at 10 GHz and the three-port S-parameters calculated using these relations are given as follows:

$$[S]_{2P} = \begin{bmatrix} 0.73\underline{/-102^\circ} & 0.1\underline{/42^\circ} \\ 2.23\underline{/96^\circ} & 0.54\underline{/-49^\circ} \end{bmatrix}$$

$$[S]_{3P} = \begin{bmatrix} 0.85\underline{/-56.6^\circ} & 0.29\underline{/66.4^\circ} & 0.61\underline{/47.3^\circ} \\ 1.26\underline{/104.7^\circ} & 0.83\underline{/-37.1^\circ} & 0.97\underline{/-47.7^\circ} \\ 0.99\underline{/-30.9^\circ} & 0.32\underline{/47.2^\circ} & 0.28\underline{/104.2^\circ} \end{bmatrix}$$

9.4 OSCILLATION AND STABILITY CONDITIONS [7]

Any oscillator can be represented in an arbitrary plane on the output line by a nonlinear impedance Z_{NL}, having a negative real part, in series with a load impedance Z_L (Fig. 9.4). We assume that the circuit has a sufficiently high-Q factor to suppress the harmonic currents. Supposing that a current

$$i(t) = I_0 \cos(\omega_0 t) \tag{9.6}$$

exists in the circuit, we can apply the Kirchhoff voltage law and write in the plane PP':

$$[Z_{NL}(I_0, \omega_0) + Z_L(\omega_0)]I_0 = 0 \tag{9.7}$$

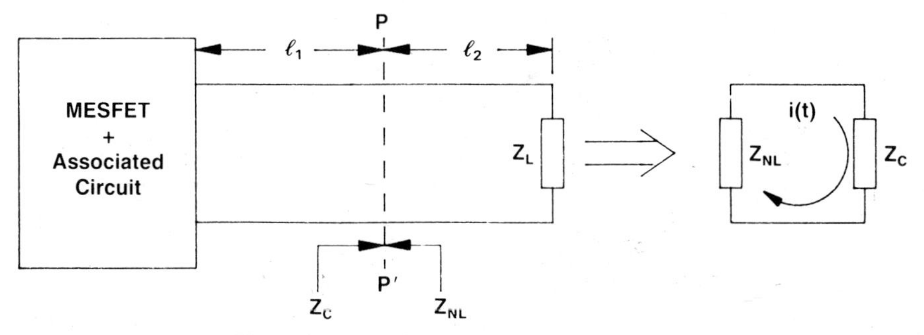

Figure 9.4 Nonlinear microwave oscillator.

Let

$$Z_{NL} + Z_L = Z_T = R_T + jX_T \tag{9.8}$$

Since I_0 is not equal to zero, (9.7) is satisfied by

$$R_T(I_0, \omega_0) = 0 \tag{9.9}$$

$$X_T(I_0, \omega_0) = 0 \tag{9.10}$$

Since $\mathrm{Re}(Z_L) > 0$, (9.9) implies that $\mathrm{Re}(Z_{NL}) < 0$. Hence, the device needs to present a negative resistance in order to be able to oscillate. The frequency of the oscillations is determined by (9.10), that is, by the requirement that the load reactance be equal and opposite to the device reactance.

Oscillators can also be represented by a nonlinear admittance Y_L. The oscillation conditions in this case can be determined in the same way to be:

$$G_T(V_0, \omega_0) = 0 \tag{9.11}$$

$$B_T(V_0, \omega_0) = 0 \tag{9.12}$$

At microwave frequencies, it is more convenient to express (9.9) to (9.12) in terms of the corresponding reflection coefficient Γ_{NL} and Γ_L as

$$|\Gamma_{NL}| \cdot |\Gamma_L| = 1 \tag{9.13}$$

$$\underline{/\Gamma_{NL}} + \underline{/\Gamma_L} = 2\pi n, \qquad n = 0, 1, 2, \ldots \tag{9.14}$$

Relation (9.13) implies that the device reflection coefficient Γ_{NL} modulus should be greater than unity.

An oscillator can be considered as a combination of an active multiport and a passive multiport (the embedding circuit), as shown in Fig. 9.5. With

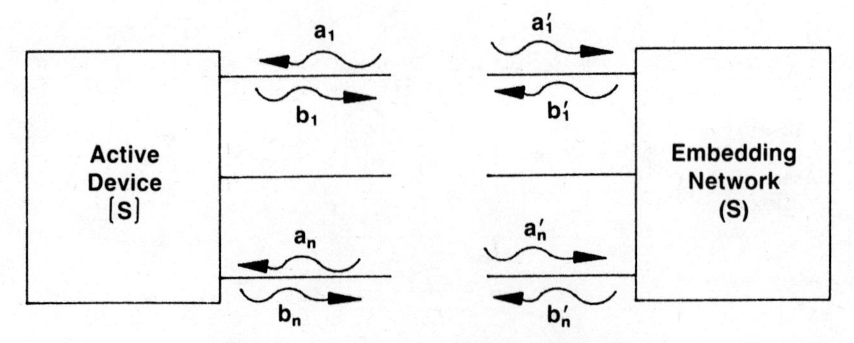

Figure 9.5 Generalized oscillator configuration.

the active device and the embedding circuit characterized by their scattering matrix, we have, for the active device [8],

$$[b]=[S][a] \tag{9.15}$$

and for the embedding circuit,

$$[b']=[S'][a'] \tag{9.16}$$

When the active device and the embedding network are connected together, we have, for the oscillation conditions,

$$[b']=[a] \tag{9.17}$$

and

$$[b]=[a'] \tag{9.18}$$

From (9.15) to (9.18) we can write

$$[a']=[S][S'][a'] \tag{9.19}$$

or

$$([S][S']-[I])[a']=0 \tag{9.20}$$

where $[I]$ is an identity matrix. Since $[a']\neq 0$, it follows that

$$[M]=[S][S']-[I] \tag{9.21}$$

is a singular matrix or

$$\det[M]=0 \tag{9.22}$$

which represents the generalized large-signal oscillation condition for an n-port oscillator.

In fact, the scattering matrix of the active device being defined at the small-signal level, the n-port oscillation condition at small signal can be represented by

$$|\det([S][S']-[I])|>0 \tag{9.23}$$

and

$$\operatorname{Arg}\det([S][S']-[I])=0 \tag{9.24}$$

The oscillations can start as soon as the preceding relations are satisfied and go on building up until the device nonlinearities cause a steady state to

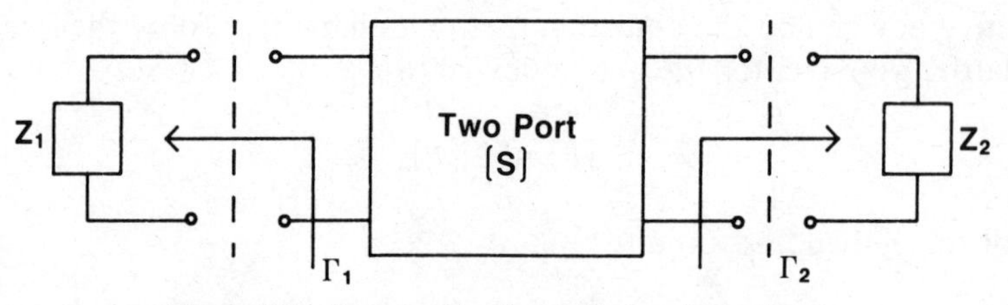

Figure 9.6 Two-port loaded with two impedances.

be reached. As an example, consider an active two-port loaded by two passive impedances, as shown in Fig. 9.6. The active device is described by the scattering matrix

$$[S] = \begin{bmatrix} S_{11} & S_{12} \\ S_{21} & S_{22} \end{bmatrix} \tag{9.25}$$

and the embedding circuit by

$$[S'] = \begin{bmatrix} \Gamma_1 & 0 \\ 0 & \Gamma_2 \end{bmatrix} \tag{9.26}$$

The oscillation condition from (9.22) is:

$$\det[M] = \det\begin{bmatrix} S_{11}\Gamma_1 - 1 & S_{12}\Gamma_2 \\ S_{21}\Gamma_1 & S_{22}\Gamma_2 - 1 \end{bmatrix} = 0 \tag{9.27}$$

which gives

$$(S_{11}\Gamma_1 - 1)(S_{22}\Gamma_2 - 1) - S_{12}S_{21}\Gamma_1\Gamma_2 = 0 \tag{9.28}$$

From the preceding, we obtain the following two well-known conditions that are simultaneously satisfied for realizing oscillations [9]:

$$S_{11} + \frac{S_{12}S_{21}\Gamma_2}{1 - S_{22}\Gamma_2} = \frac{1}{\Gamma_1} \tag{9.29}$$

$$S_{22} + \frac{S_{12}S_{21}\Gamma_1}{1 - S_{11}\Gamma_1} = \frac{1}{\Gamma_2} \tag{9.30}$$

The oscillations are considered stable if any perturbation in the RF voltage or the RF current of the oscillator circuit at any instant decays itself, bringing the oscillator back to its point of equilibrium. The oscillator stability is analyzed [10] using a quasistatic approach by applying a small perturbation to the amplitude I_0 (Fig. 9.4). The impedance Z_T defined in

(9.8), a function of I_0 and the complex frequency p, is developed in a Taylor series about I_0, ω_0. Since the perturbed current is nonzero, that is, the oscillations continue to exist after the perturbation, we should have in the plane PP'

$$Z_T(I_0, \omega_0) + \frac{\partial Z_T}{\partial p}\,\delta p + \frac{\partial Z_T}{\partial I_0}\,\delta I_0 = 0 \tag{9.31}$$

and since $Z_T(I_0, \omega_0) = 0$, we get

$$\delta p = -j\,\frac{(\partial Z_T/\partial I_0)\cdot(\partial Z_T^*/\partial \omega)}{|(\partial Z_T/\partial \omega)|^2}\cdot \delta I_0 \tag{9.32}$$

where δp can be decomposed into its real and imaginary parts

$$\delta p = \alpha + j\delta\omega$$

The oscillator will be stable if, for a positive variation of the current amplitude, the real part of the variation of the complex frequency is negative, that is, if α is negative, indicating a decreasing wave, returning to its point of equilibrium I_0.

From the previous expression of δp this condition is realized for

$$\frac{\partial R_T}{\partial I_0}\cdot\frac{\partial X_T}{\partial \omega} - \frac{\partial X_T}{\partial I_0}\cdot\frac{\partial R_T}{\partial \omega} > 0 \tag{9.33}$$

This relation represents the stability condition of an oscillator around an amplitude I_0 and angular frequency ω_0. Moreover, from (9.32), the imaginary part $\delta\omega$ vanishes if the condition

$$\frac{\partial R_T}{\partial I_0}\cdot\frac{\partial R_T}{\partial \omega} - \frac{\partial X_T}{\partial I_0}\cdot\frac{\partial X_T}{\partial \omega} > 0 \tag{9.34}$$

is fulfilled. This indicates that a variation of amplitude δI_0 will not result in a variation of the oscillator's real angular frequency ω_0. From (9.34) it can also be deduced that, for maximum stability, the device impedance $Z_T(I_0)$ and load impedance $Z_L(\omega)$ should intersect at right angles at the oscillation equilibrium point (I_0, ω_0).

9.5 FIXED-FREQUENCY OSCILLATORS

Oscillators can be mainly subdivided into two categories; fixed-frequency and frequency-tunable oscillators. While fixed-frequency oscillators find

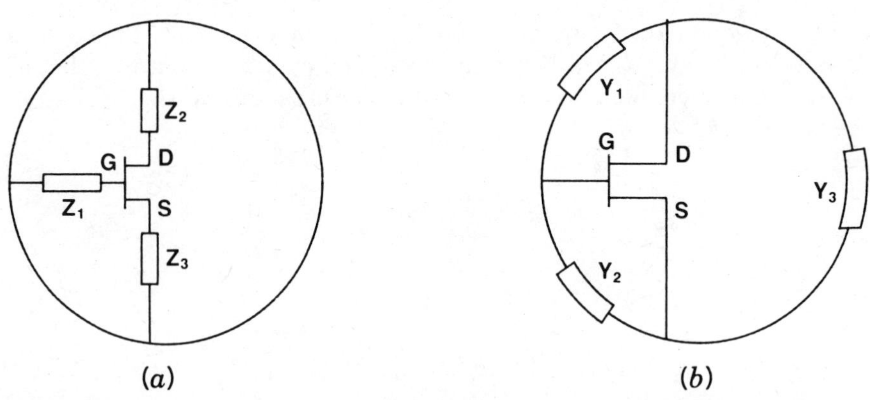

Figure 9.7 General transistor oscillator configurations. (*a*) Series type. (*b*) Parallel type.

maximum application as local oscillators in communications and radar systems, tunable and wide-band tunable oscillators are used in synthesizers, instrumentation, and EW systems. An oscillator in its most general form can be represented as a series or a parallel circuit shown in Fig. 9.7. The oscillators can be designed so that any one of the immittances shown can contain the output resistive load, while the other two immittances are normally reactive. In a fixed-frequency oscillator, all the immittances are fixed, while one or more of the immittances are made tunable by using varactor diodes or YIG resonators.

Currently, dielectric resonators have been used more commonly to realize low noise and temperature-stable fixed-frequency oscillators and are called transistor dielectric resonator oscillators (TDROs). Dielectric resonators are high-Q ceramic resonators, conveniently sized for MIC application and have been discussed in detail in Chapter 3. The DR can be used to realize one or more of the immittances shown in Fig. 9.7. The commonly used stable TDROs can be divided into two types: one using the DR as a series feedback element, and the other using the DR as a parallel feedback element. The analysis and design of both of these types [11] are discussed below.

9.5.1 Design of Oscillators

Figure 9.8 illustrates various configurations using the dielectric resonator as a series feedback element. Figure 9.8(*a*) to 9.8(*c*) use the resonator at one terminal pair, while configuration 9.8(*d*) uses the DR as a series feedback element at two terminal pairs of the transistors. As an example, we now discuss the step-by-step procedure for the realization of the configuration shown in Fig. 9.8(*a*) using the three-port S-parameters of a typical 0.5-μm GaAs FET. Comments on the design of other configurations are made where necessary.

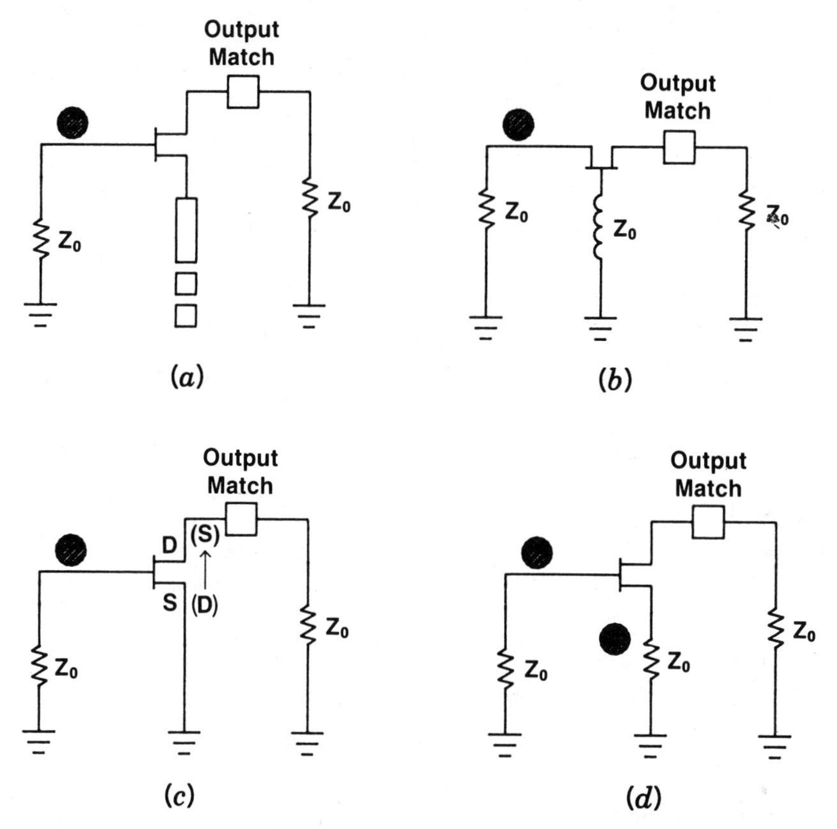

Figure 9.8 Different configurations for series feedback transistor DROs.

As a first step in the design procedure, the impedance Z_3 in Fig. 9.9 is determined. With the impedance Z_3 connected to the source terminal, the reflection coefficient Γ_3 becomes

$$\Gamma_3 = \frac{Z_3 - Z_0}{Z_3 + Z_0} \tag{9.35}$$

Substituting into (9.3), we obtain

$$b_1 = S_{11}a_1 + S_{12}a_2 + S_{13}b_3\Gamma_3 \tag{9.36}$$

$$b_2 = S_{21}a_1 + S_{22}a_2 + S_{23}b_3\Gamma_3 \tag{9.37}$$

$$b_3 = S_{31}a_1 + S_{32}a_2 + S_{33}b_3\Gamma_3 \tag{9.38}$$

Eliminating b_3 from (9.36) to (9.38), the reduced two-port S-matrix is given by

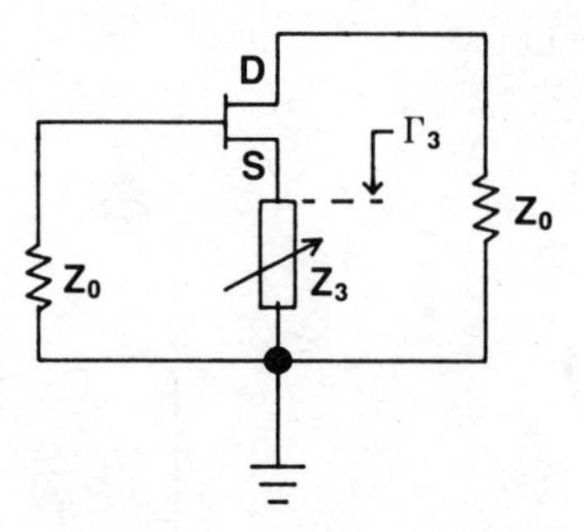

Figure 9.9 Determination of Z_3.

$$[S^T] = \begin{bmatrix} S_{11} + \dfrac{S_{31}S_{13}\Gamma_3}{1 - S_{33}\Gamma_3} & S_{12} + \dfrac{S_{13}S_{32}\Gamma_3}{1 - S_{33}\Gamma_3} \\ S_{21} + \dfrac{S_{31}S_{23}\Gamma_3}{1 - S_{33}\Gamma_3} & S_{22} + \dfrac{S_{23}S_{32}\Gamma_3}{1 - S_{33}\Gamma_3} \end{bmatrix} \tag{9.39}$$

The aim is to determine the value of the series feedback impedance Z_3 that will result in the modulus of S_{11}^T and S_{22}^T being greater than unity and, hence, create instability in the transistor. For the example considered in Fig. 9.8(a), the open-circuited microstrip line represents a purely reactive impedance. The $|\Gamma_3| = 1$ plane determines a circle when mapped in the input and output reflection coefficient planes using well-known techniques [12]. Figure 9.10 shows the $|\Gamma_3| = 1$ plane mapped into the S_{11}^T and S_{22}^T planes for the half-micron GaAs FET at 10 GHz. The shaded area represents inductive impedance and the unshaded area represents the capacitive impedance in the source.

From Fig. 9.10 it may be noted that a negative reactance greater than $-j30\ \Omega$ can be used to make both S_{11}^T and S_{22}^T greater than one. A value of $-j159\ \Omega$ is selected that can be realized by an open-circuited stub $0.048\lambda_g$ long or by a 0.1-pF capacitor. The reduced two-port S-matrix is now given by

$$[S^T] = \begin{bmatrix} 1.34\underline{/-33.5^\circ} & 0.49\underline{/70.4^\circ} \\ 0.49\underline{/156^\circ} & 1.16\underline{/-31.9^\circ} \end{bmatrix} \tag{9.40}$$

A dielectric resonator coupled to a microstrip line can also be used as impedance Z_3 in Fig. 9.9 as shown in Fig. 9.8(d). The reflection coefficient Γ_3, in this case is a function of the coupling coefficient β_3 and the distance θ_3 between the transistor plane and the resonator plane as analyzed in Chapter 3. Using the matrix coefficients from (9.39), the Γ_3 plane in terms of β_3 and θ_3 can be mapped into any S-parameter of the reduced two-port.

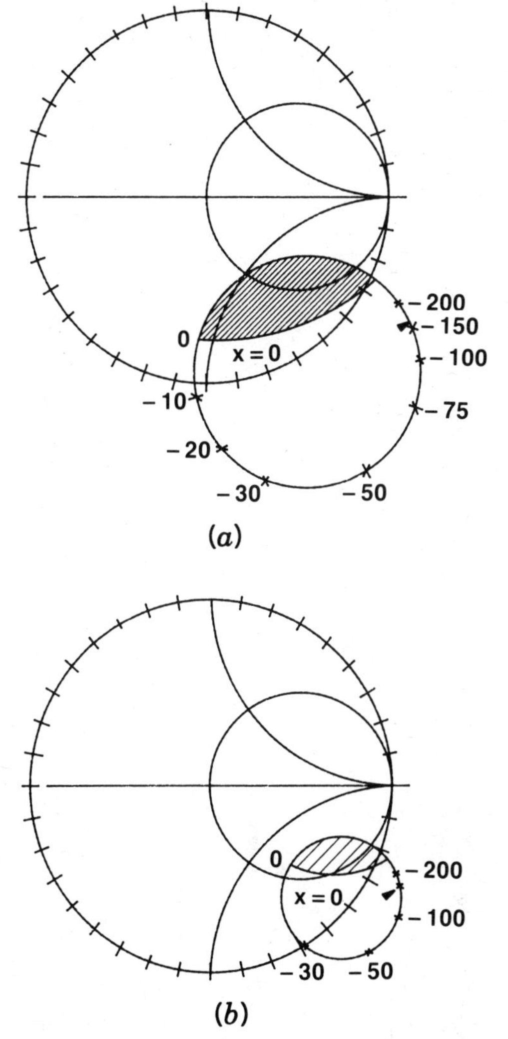

Figure 9.10 Mapping of $|\Gamma_3| = 1$ in (*a*) S_{11}^T and (*b*) S_{22}^T.

Figure 9.11 shows, for example, mapping of the DR reflection coefficient plane into all the four *S*-parameters of an *X*-band FET at 8 GHz [13].

The nonconcentric circles shown in Fig. 9.11 are the constant coupling coefficient β_3 [proportional to the $|\Gamma_3|$) circles, while the radial arcs are the constant transmission electrical line length θ_3 (proportional to arg Γ_3) arcs. The relation between Γ_3, β_3, θ_3, and frequency is given by

$$\Gamma_3 = \frac{\beta_3}{\sqrt{(1+\beta_3)^2 + \Delta^2}} \exp\left(-2j\left(\theta_3 + \tan^{-1}\frac{\Delta}{1+\beta_3}\right)\right) \tag{9.41}$$

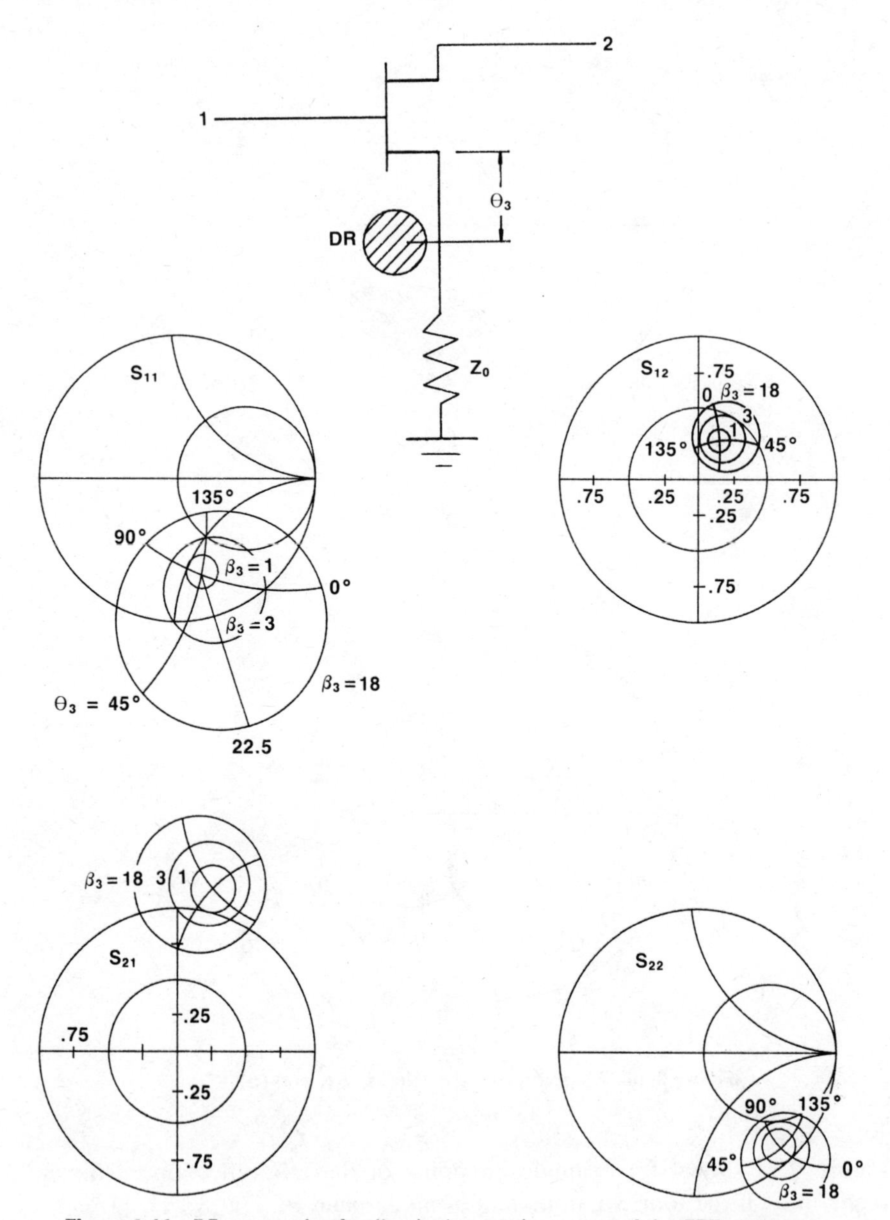

Figure 9.11 DR as a series feedback element in source of the FET oscillator.

with $\Delta = 2Q_u(f - f_0)/f_0$. Figure 9.11 can be used to determine the DR position in order to create the desired instability in the transistor.

Continuing the TDRO design example of Fig. 9.8(a), we have already determined the impedance Z_3; the resulting two-port S-parameters are given in (9.40). In the second step of the procedure, we will determine the

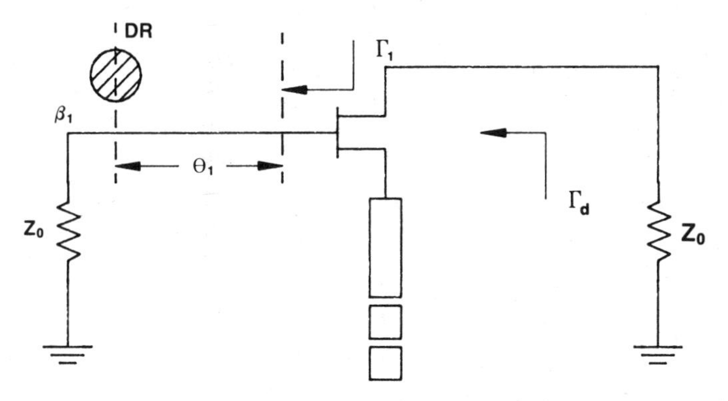

Figure 9.12 Determination of DR position in the gate.

value of reflection coefficient Γ_1, which maximizes the reflection coefficient Γ_d at the drain port (Fig. 9.12), using the following relation:

$$\Gamma_d = S_{22}^T + \frac{S_{12}^T S_{21}^T \Gamma_1}{1 - S_{11}^T \Gamma_1} \tag{9.42}$$

The reflection coefficient Γ_1 in this case is realized by a DR coupled to a microstrip line and is characterized by the coupling coefficient β_1 and the distance θ_1 between the transistor and the resonator plane.

In the present example, $|S_{11}^T|$ and $|S_{22}^T|$ being greater than unity, the mapping technique used in the first step cannot easily be used to determine the required Γ_1 and hence the position of DR. Instead, we use the constant reflection coefficient circles approach, in which case the locus of constant reflection coefficient magnitude $|\Gamma_d|$ is drawn on the reflection coefficient plane Γ_1. From (9.42) the radius R and center Ω of the constant reflection coefficient circles can be determined to be

$$R = \frac{|\Gamma_d S_{12}^T S_{21}^T|}{|\Gamma_d|^2 |S_{11}^T|^2 - |\Delta|^2} \tag{9.43}$$

$$\Omega = \frac{S_{11}^{T*}[|\Gamma_d|^2 - |S_{22}^T|]^2 + S_{22}^T S_{12}^{T*} S_{21}^{T*}}{|\Gamma_d|^2 |S_{11}^T|^2 - |\Delta|^2} \tag{9.44}$$

where

$$\Delta = S_{11}^T S_{22}^T - S_{12}^T S_{21}^T \tag{9.45}$$

Figure 9.13 shows various $|\Gamma_d| = \text{constant}$ circles on the $\Gamma_1(\beta_1, \theta_1)$ plane for the example under consideration. The value of β_1 and θ_1 can now be determined for a high value of $|\Gamma_d|$ ($\gg 1$).

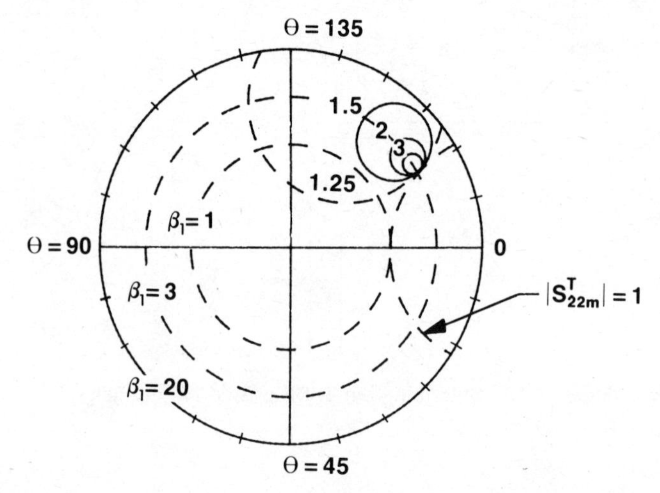

Figure 9.13 Constant $|S^T_{22m}|$ circles in the DR reflection coefficient. (β_1, θ_1) plane.

The first step and the second step used the small-signal S-parameters to determine the impedances Z_1 and Z_3 to be connected at gate and source ports, respectively, in order to achieve a high value of reflection coefficient at the drain port. A number of approaches exist to determine the load circuit impedance Z_2 in Fig. 9.7(a). The two common ones, which assume the large-signal operation, are the device-line approach and the load-pull approach.

The device-line approach [14] is based on the measurement of the inverse reflection coefficient of the one-port and the return added power for different input power levels. The load impedance for maximum oscillator output power can be determined from the device line thus obtained. This approach, however, has the limitation that requires the source resistance to be greater than the modulus of the device resistance. Otherwise, oscillation takes place and the device line cannot be measured.

The test setup for the measurement of load-pull approach [15] is shown in Fig. 9.14. The oscillator acts as the RF power source for the system. The drain port of the transistor circuit (with optimized Z_1 and Z_3 connected) is attached to the load-pull measurement system at the input port of the reflectometer through a 50-Ω line and powered up. The impedance displayed on the polar display will be the impedance presented to the output of the oscillator device. Using the output tuner, contours of constant output power can then be drawn on a Smith chart using the power readings from the output power meter and an x-y recorder connected to the polar display. A typical load-pull data is shown in Fig. 9.15. This load impedance chart can be used to design the output circuit for the transistor.

An alternate way of realizing a stable oscillator is using the DR simultaneously coupled to two microstrip lines as a parallel feedback element for

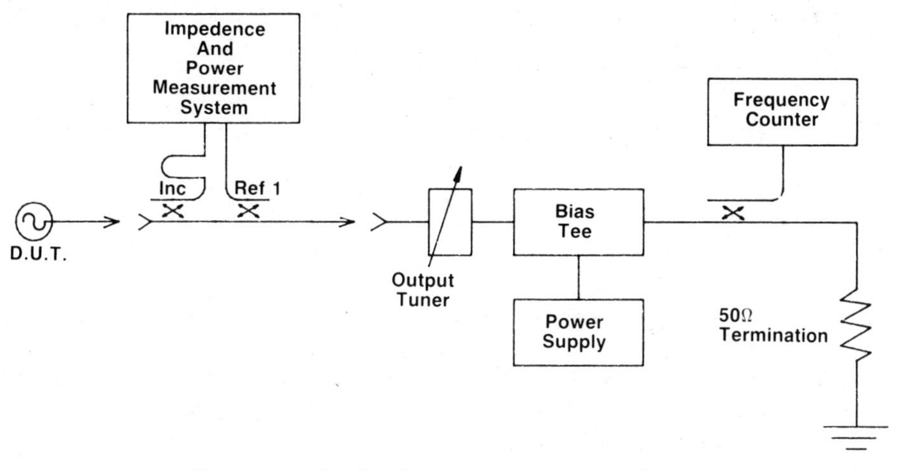

Figure 9.14 Load-pull setup for one-port oscillator.

a transistor. In this case, the transistor can be used as a two-port or a three-port device, as shown in Figs. 9.16 and 9.17, respectively.

In Fig. 9.16, the transistor is treated as a two-port [16]. In this case, the input and output matching circuits for a common source transistor are designed for a maximum transducer gain amplifier around the oscillator frequency f_0. Highly selective positive feedback between the input and the

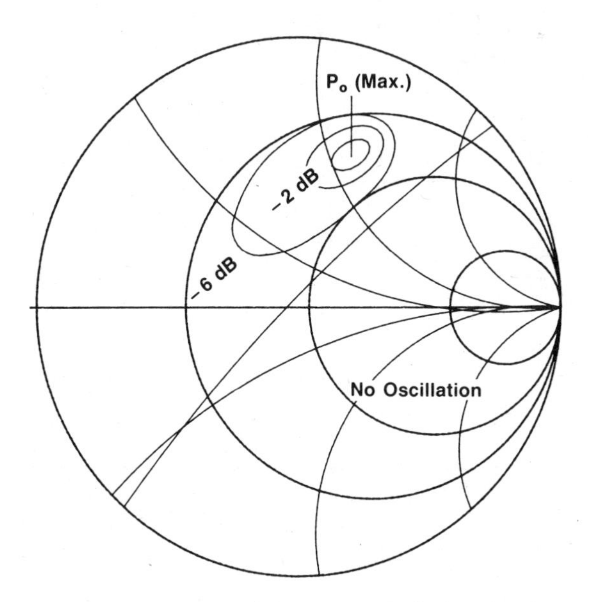

Figure 9.15 A typical load-pull data. (After D. Poulin [15], 1980 Microwaves, reprinted with permission.)

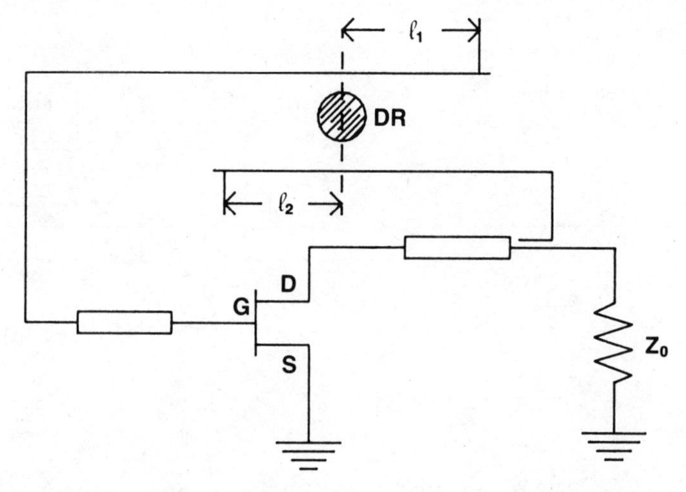

Figure 9.16 Parallel feedback transistor DRO using transistor as a two-port.

output can be used to create stable oscillations. This is achieved by feeding back a part of the output signal into the input through the dielectric resonator transmission filter. The length l_1 and l_2 are adjusted to achieve the phase shift around the loop, consisting of the amplifier and feedback circuit, equal to integer multiple of 2π radians at f_0, that is,

$$\phi_A + \phi_R + \phi_C = 2\pi k, \qquad k = 1, 2, 3 \tag{9.46}$$

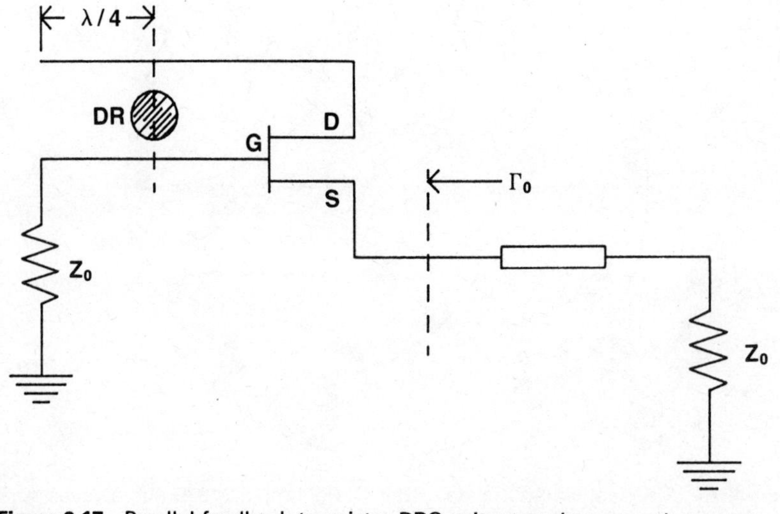

Figure 9.17 Parallel feedback transistor DRO using transistor as a three-port.

where ϕ_A, ϕ_R, and ϕ_C are respective insertion phases of the amplifier, resonator, and the remaining part of the feedback circuit at f_0. The other condition for the oscillations to take place is that the open-loop small-signal gain must exceed unity at f_0, that is,

$$G_A - L_R - L_C > 0 \text{ dB} \tag{9.47}$$

where G_A, L_R, and L_C are the amplifier gain, resonator filter loss, and loss in the other feedback components in decibels, respectively. The necessary relations for determining the resonator insertion phase ϕ_R and insertion loss L_R can be determined from Chapter 3:

$$\phi_R = \tan^{-1}\left[\frac{-j2Q_u\delta}{1+\beta_1+\beta_2}\right] \tag{9.48}$$

and

$$L_R = 10\log\frac{(1+\beta_1+\beta_2)^2}{4\beta_1\beta_2} \tag{9.49}$$

The extent of the inequality given in (9.47) and, hence, the amplifier gain compression in the steady state oscillation condition affects the output power as well as the FM noise performance of the oscillator. Excessive gain compression can adversely affect the oscillator noise due to increased amplifier noise figure and AM-to-PM conversion.

Figure 9.17 shows another configuration of a parallel feedback DRO [5, 17]. In this case, the DR transmission filter is coupled between the two terminals of the transistor and the output is taken from the third. This configuration can be analyzed as a two-port containing DR as a parallel feedback network to a three-port device, the transistor [5]. See Fig. 9.18.

Feedback parameters (i.e., the position and coupling of the dielectric resonator to the microstrip lines) can be determined to maximize the reflection gain (>1) at the output terminal. The output matching circuit can now be designed using the load-pull approach described above.

The series, as well as parallel, feedback approaches described above are both commonly used for designing TDROs. The series feedback has the advantage of ease of dielectric resonator alignment due to its coupling to microstrip on one side compared to the parallel feedback type, wherein the resonator is simultaneously coupled to both sides. This aspect also makes the series feedback configuration circuit operable over a much wider bandwidth compared to parallel feedback configuration. In the parallel feedback configuration use of a high-gain amplifier can allow significant decoupling of the resonator from the microstrip lines resulting in a higher loaded quality factor value of the DR and hence a lower phase noise oscillator. The temperature stability of the DR material, required for

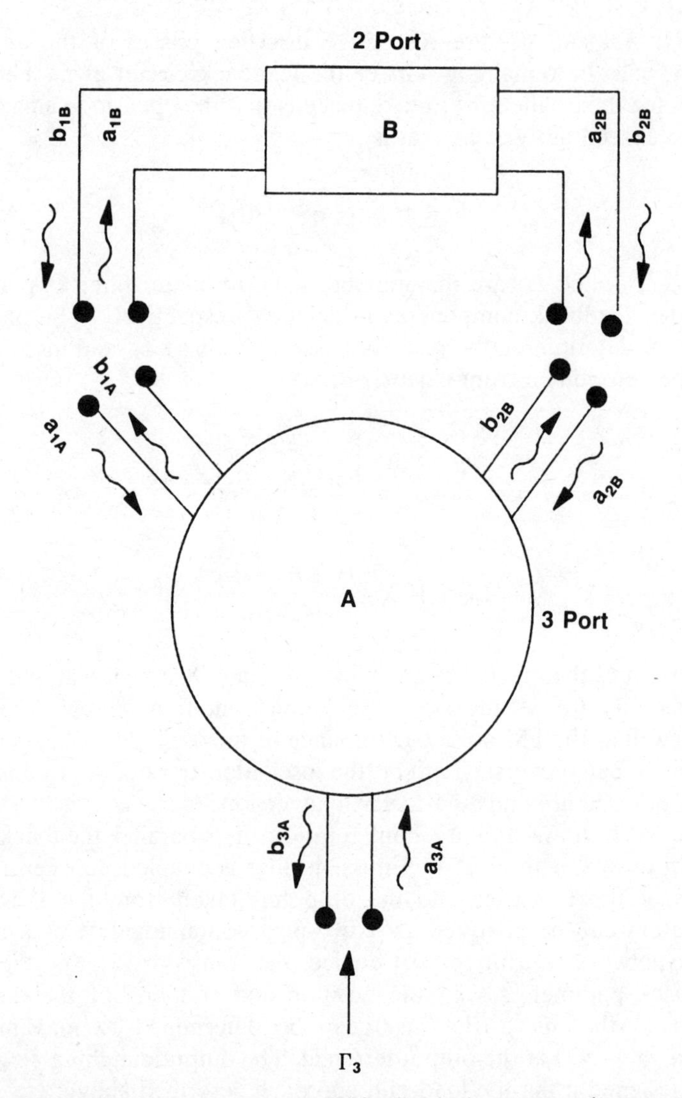

Figure 9.18 Two-port as a parallel feedback element to a three-port.

temperature compensation of the oscillator, is different in both cases due to the difference in resonator coupling configuration.

A simple design example of an oscillator circuit is described step by step in the following.

Example. The small-signal two-port S-parameters of a reverse channel 0.5-μm GaAs FET at 9 GHz are given as $S_{11} = 0.87\underline{/-155^\circ}$, $S_{12} = 0.53\underline{/-78^\circ}$, $S_{21} = 0.62\underline{/-143^\circ}$ and $S_{22} = 0.25\underline{/-100^\circ}$. (Bias: $V_{DS} = -8$ V,

$V_{GD} = -2$ V, and $I_D = 80$ mA). Design a microstrip oscillator using an open stub on the gate and a two-section (series-shunt) matching circuit on the drain to match it to a 50-Ω load. For maximum power output assume that the load impedance Z_L is related to the transistor small-signal output impedance Z_{out} by the relations $R_L = -R_{out}/3$ and $X_L = -X_{out}$. The characteristic impedance of the matching transmission lines is 50 Ω. Draw the RF circuit for a 0.63-mm-thick soft substrate with $\epsilon_r = 2.5$. Assuming a −10-V power supply, show the dc bias network.

The reverse channel GaAs FET configuration is potentially unstable. It can be verified by calculating the stabilization factor K. Using the given S-parameters, $K = 0.488$ at 9 GHz. A stability factor $K < 1$ means that the transistor can be made to oscillate by properly selecting source and load impedances, without series feedback in the source terminal. The oscillator configuration is shown in Fig. 9.19(a).

Looking into the drain of the transistor (without the matching elements l_1 and l_2):

$$\Gamma_d = S_{22} + \frac{S_{12}S_{21}}{1/\Gamma_S - S_{11}}$$

where Γ_S represents the reflection coefficient of the microstrip open-circuited stub with an electrical length of θ_g (physical length of l_g) at the gate terminal.

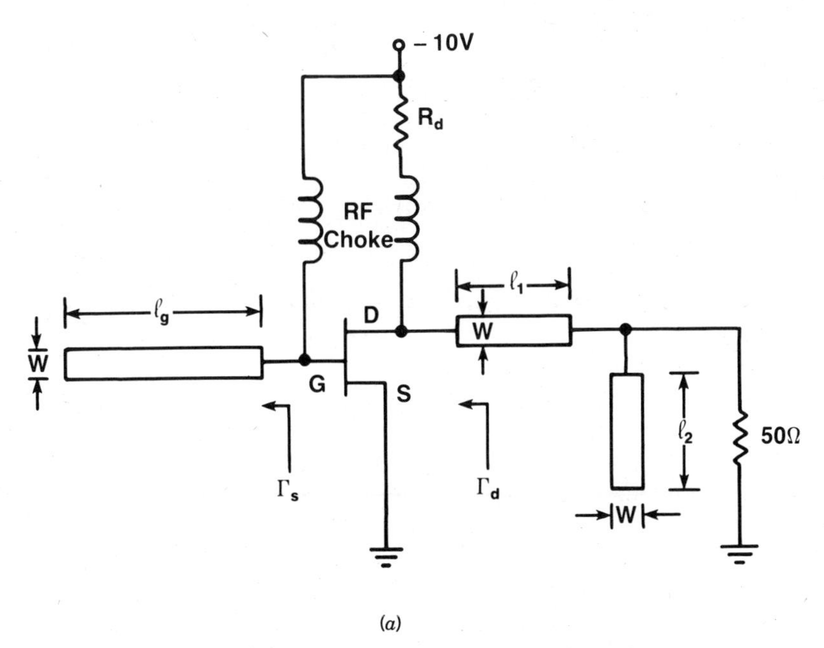

(a)

Figure 9.19 (a) A simple oscillator configuration.

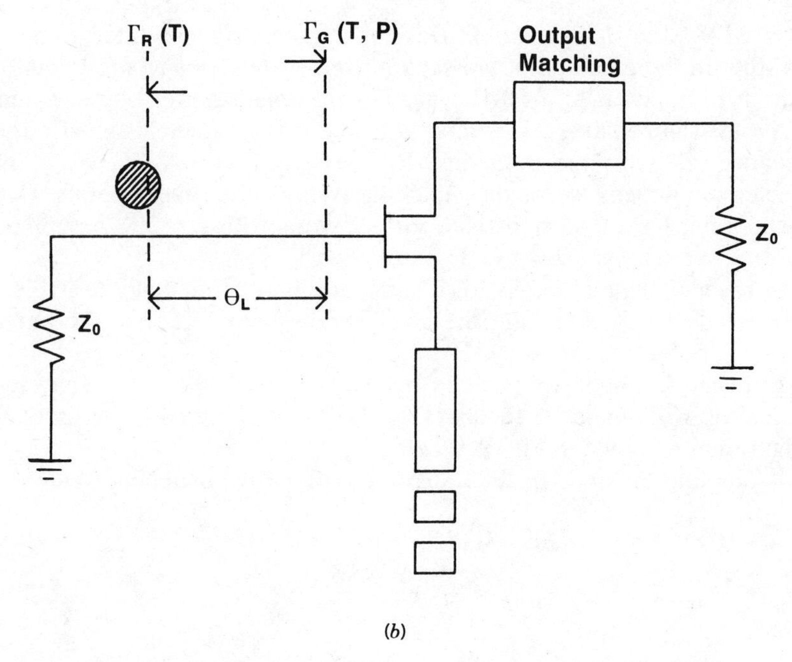

(b)

Figure 9.19 (b) Circuit used for temperature variation analysis.

Assuming a lossless open circuited stub, Γ_S may be expressed as

$$\Gamma_S = 1\underline{/\theta_S} = 1\underline{/-2\theta_g} = 1\underline{/-4\pi l_g/\lambda_g}$$

Solving for θ_S corresponding to maximum value of $|\Gamma_d|$ [10]

$$\theta_S = \theta_{22} - (\theta_{12} + \theta_{21}) + \sin^{-1}[|S_{11}|\sin\,\theta_x]$$

with

$$\theta_x = (\theta_{12} + \theta_{21}) - (\theta_{11} + \theta_{22})$$

and

$$S_{ij} = |S_{ij}|\underline{/\theta_{ij}}$$

In this example, $\theta_S = 152°$, $\theta_g = 104°$, and $l_g = 6.6$ mm ($\lambda_g = 23$ mm).

The corresponding maximum value of $\Gamma_{d\max} = 2.57\underline{/-78°}$. The small-signal output impedance of the FET with the open stub on the gate can be determined now to be

$$Z_{\text{out}} = \frac{1 + \Gamma_{d\max}}{1 - \Gamma_{d\max}} = -42.3 - j38.3\ \Omega$$

The required load impedance can be calculated as

$$R_L = -R_{out}/3 = 14.1\ \Omega \qquad \text{and} \qquad X_L = -X_{out} = 38.3\ \Omega$$

$Z_L = 14.1 + j38.3\ \Omega$ corresponds to $\Gamma_L = 0.66\underline{/105.6^\circ}$.

A two-section matching circuit shown in Fig. 9.19(a) can be designed using Smith chart or the following well-known relations:

$$l_2 = \frac{\lambda_g}{2\pi} \tan^{-1}\left(2\sqrt{\frac{|\Gamma_L|^2}{1-|\Gamma_L|^2}}\right) = 3.87\ \text{mm}$$

and

$$l_1 = \left(\frac{\theta_y - \theta_L}{720^\circ}\right)\lambda_g = 3.91\ \text{mm}$$

where

$$\theta_y = 180^\circ + \tan^{-1}\sqrt{\frac{1-|\Gamma_L|^2}{|\Gamma_L|^2}} = 228.3^\circ, \qquad \Gamma_L = |\Gamma_L|\underline{/\theta_L}$$

The width W for all the microstrip lines shown corresponds to $50\ \Omega$ and is 1.8 mm.

The resistance R_d shown in Fig. 9.19(a) can be calculated to be

$$R_d = \frac{2\ \text{V}}{80\ \text{mA}} = 25\ \Omega$$

9.5.2 Temperature Stability of DROs

Microwave sources with high-temperature stabilities have been realized in the past using Invar cavities or by phase locking the source to a frequency multiplied ovenized VHF crystal oscillator. These approaches are cumbersome as well as expensive. With the advent of high-performance dielectric resonators, TDROs now offer a miniature, elegant, and inexpensive way of achieving high stabilities.

A free-running transistor oscillator is known to have a negative temperature coefficient. A DR having a positive temperature coefficient is thus required to compensate the frequency drift with temperature. Currently, the temperature coefficient of the DR can be controlled by varying its composition. A DR with coefficients equal to anywhere between +9 to −9 ppm/°C are commercially available as discussed in Chapter 3.

The temperature stability of a TDRO can be analytically determined in terms of the coupling coefficient β, the Q-factor, and the rate of change of transistor reflection phase with temperature [18].

In Fig. 9.19(*b*), the active device including the circuitry can be presented at its gate port as a temperature and power-dependent reflection coefficient:

$$\Gamma_G(T, P) = |\Gamma_G(T, P)| e^{j\phi_G(T,P)} \tag{9.50}$$

with

$$|\Gamma_G(T, P)| > 1 \tag{9.51}$$

The stabilization circuit containing a 50-Ω microstrip line, loaded by 50-Ω damping resistor and coupled with the dielectric resonators, can be represented at the resonator plane as a temperature-dependent reflection factor:

$$\Gamma_R(T) = \frac{\beta}{1 + \beta + j2Q_u\delta(T)} \tag{9.52}$$

where

$$\delta(T) = \frac{f - f_0(T)}{f_0(T)} \tag{9.53}$$

β being the coupling coefficient between the resonator and microstrip line. Applying the oscillation condition in the plane of the gate port, we obtain

$$|\Gamma_R(T)| e^{-j2\theta_L} = \frac{1}{|\Gamma_G(T, P)|} e^{-j\phi_G(T,P)} \tag{9.54}$$

where θ_L is the inserted microstrip electrical transmission line length, in order to adjust the phase for oscillation condition. This equation, split into real and imaginary parts, leads to the following relationships for oscillation frequency and power:

$$(\beta + 1) \tan[\phi_G(T, P) - 2\theta_L] = \frac{f}{f_0(T)} - 1 \tag{9.55}$$

$$|\Gamma_G(T, P)| = \frac{1 + \beta}{\beta} \tag{9.56}$$

where $\Gamma_G(T, P)$ is known to be a monotonous decreasing function of temperature and power. A decrease of Γ_G at higher temperatures will be compensated automatically by a degradation of output power P, thus keeping $|\Gamma_G|$ constant.

For determining the frequency stability with temperature, (9.55) can be differentiated with respect to temperature and leads after certain approximations to

$$\frac{df}{f\,dT} \simeq \frac{df_0}{f_0\,dT} + \frac{\beta+1}{2Q_u}\frac{\partial \phi_G}{\partial T} \tag{9.57}$$

where $\partial \phi_G / \partial T$ represents the active circuit phase drift with temperature.

The preceding relation can be used to help design temperature-stable DR oscillators. Phase drift $\partial \phi_G / \partial T$ is known to decrease linearly with temperature, thus necessitating a positive temperature coefficient of the resonator $\partial f_0 / f_0\, dT$ varying linearly in the desired temperature range. Equation 9.57 can be written as

$$\tau_f = \tau_{f0} + \frac{1}{2Q_L}\frac{\partial \phi_G}{\partial T} \tag{9.58}$$

where τ_f, τ_{f0} are the temperature coefficients of the oscillator and resonator, respectively, and Q_L is the loaded Q of the resonator. For temperature stable operation, one requires $\tau_f = 0$, which leads to the requirement

$$\tau_{f0} = -\frac{1}{2Q_L}\frac{\partial \phi_G}{\partial T} \tag{9.59}$$

This relation tells us how to make the oscillation frequency independent of temperature. For a given device, the value of $\partial \phi_G / \partial T$ is fixed, but Q_L may be selected by adjusting the coupling coefficient between the DR and the microstrip line. Different combinations of Q_L and τ_{f0} may be tried for optimizing other properties of the DRO, like tunability or power output. Using this approach and a suitable dielectric resonator material, the temperature stabilities of 0.1 ppm/°C have been reported from −20 to +80°C [18].

A digital compensation technique to realize high-temperature stability TDRO has recently been reported [19]. In this case, a temperature sensor is mounted in the oscillator to detect the temperature changes. These data are digitized and are fed to an EPROM preprogrammed with temperature characteristics of the oscillator. The lookup table in the read-only memory provides the necessary digital temperature correction word that is converted to an analog signal (with proper synchronization between A/D and D/A) and applied to the DRO for frequency correction, as shown in the block diagram of Fig. 9.20. The correction signal can be applied to the varactor in the case of a varactor-tuned DRO [19] or to control the phase shift in the feedback loop in case of a parallel feedback DRO. Using this digital compensation technique, frequency drift of better than ±20 ppm has been achieved from −55 to +85°C.

Temperature stability of a DRO can also be enhanced significantly by inserting the oscillator package in a temperature-stabilized oven. Using a heater element, a quick response thermistor, and the associated control

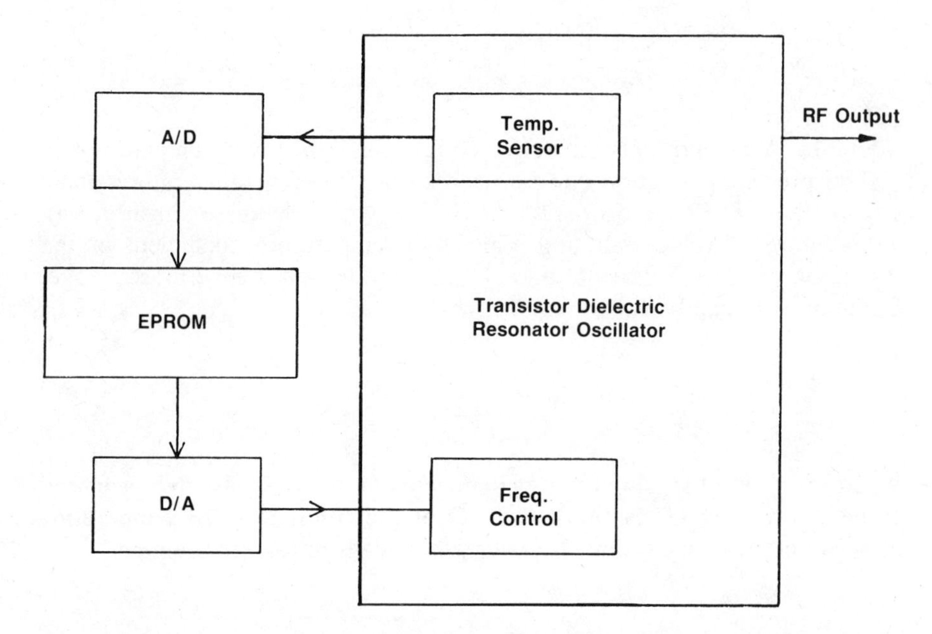

Figure 9.20 Temperature stabilization of a TDRO using digital techniques.

circuit, the package temperature can be maintained within ±5°C. The oscillator package itself is isolated from the external environment by using an outer package. The two packages are thermally connected through standoffs, which are determined taking into account the dc dissipation in the oscillator and the desired temperature of the package. The package temperature is generally controlled within 5° to 10°C above the maximum temperature at which DRO is specified. A total frequency stability of better than ±5 ppm can be obtained from −40° to 70°C. This oscillator is suitable for use as a local oscillator in radio communication and test equipment, in place of certain phase-locked oscillators.

The ovenized DRO offers better phase noise compared to the digitally controlled DRO because the oscillator does not need to incorporate electrical tuning circuitry. The digitally stabilized DRO on the other hand is smaller in size and does not need the substantial amount of heater power required by the ovenized DRO.

9.5.3 Tuning of Transistor DROs

A TDRO is basically a fixed-frequency oscillator with its frequency determined by the resonator material permittivity, resonator dimensions, and the shielding conditions, as discussed in Chapter 3. The oscillation frequency can, however, be tuned over a narrow frequency range using different approaches depending on the requirements. The frequency tuning of the

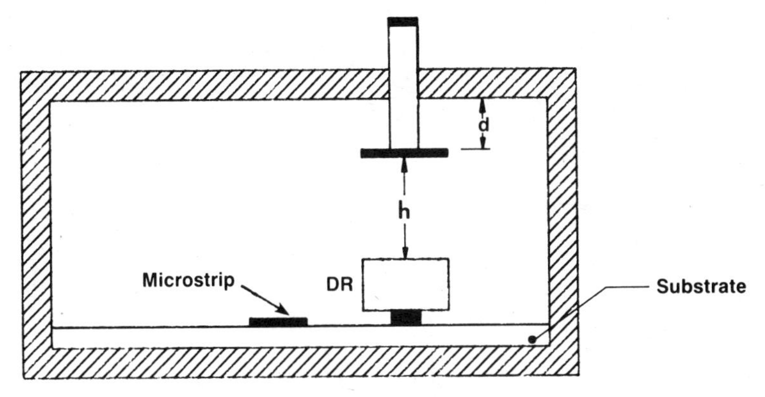

Figure 9.21 Mechanical tuning of a dielectric resonator.

TDRO can be accomplished mechanically or electrically, as discussed below.

Figure 9.21 shows the configuration used for mechanical frequency tuning of the TDRO. Use is made of the fact that the resonant frequency of the DR is highly sensitive to the shielding, that is, to the proximity of the ground plane. A tuning screw is inserted from the top cover of the package, right above the DR. The increase in the tuning screw depth d increases the resonant frequency of the DR in the commonly used mode $TE_{01\delta}$. Care should be taken to keep the distance h between the resonator and the tuning screw at least 0.5 times the resonator height in order to not degrade the DR quality factor. A mechanical frequency tuning range of $\pm 2\%$ can be obtained in a TDRO without noticeably affecting the FM noise, output power, and frequency stability over temperature.

The frequency of the TDRO can be tuned electrically by using a number of different approaches, like varactor tuning, ferrite tuning, bias tuning, and optical tuning. Electrical tuning can be affected over a very small bandwidth without significantly affecting the oscillator performance. This fast tuning can be used for different applications, like digital temperature compensation, low-deviation frequency-modulated sources, and phase locking. A brief description of the various approaches to electrically tune the TDRO follows.

In the varactor-tuned TDRO, a varactor in association with a microstrip line is made to resonate around the DR frequency. This resonant circuit is electromagnetically coupled to the dielectric resonator, forming a pair of mutually coupled resonant circuits. By varying the varactor capacitance with the bias voltage, the resonant frequency of the DR, coupled to varactor microstrip on one side and a 50-Ω microstrip line on the other, can now be tuned. Figure 9.22 shows a typical configuration for coupling the varactor and DR. Tighter coupling between the DR and varactor will result in a larger frequency control at the cost of lowered DR quality factor and,

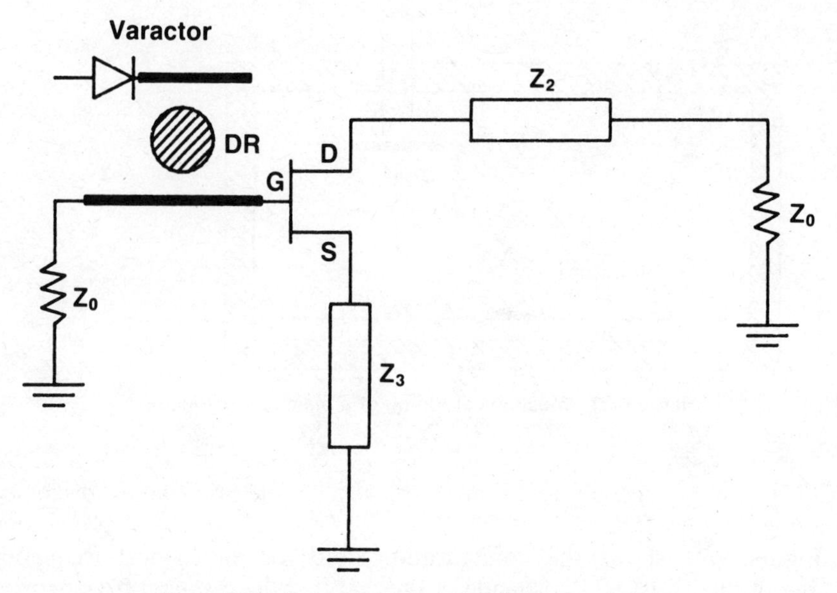

Figure 9.22 Varactor tuning of a dielectric resonator.

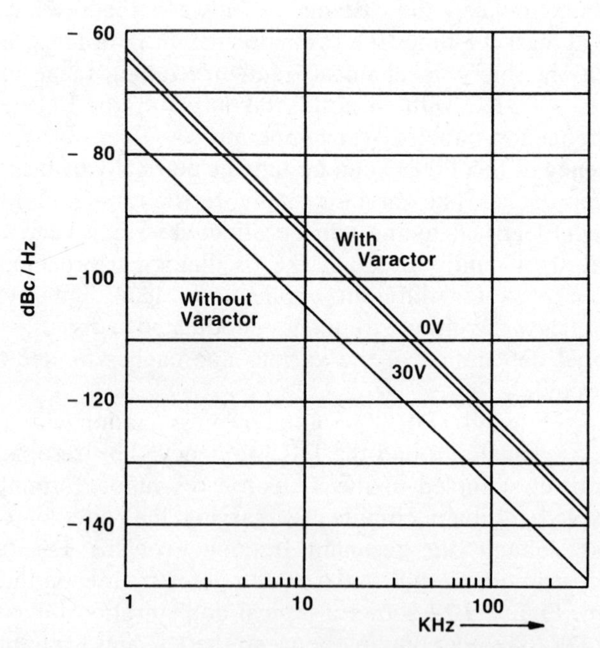

Figure 9.23 Effect of varactor tuning on DR oscillator FM noise. (After M. Camiade et al. [20], reprinted with permission of Microwave Exhibitors & Publishers Ltd.)

hence, the FM noise. Figure 9.23 shows the effect of varactor tuning on a low-noise TDRO FM noise for the varactor coupling adjusted for 0.1% frequency control at 11 GHz [20]. Using another configuration with two varactors on a quartz spacer placed directly above the resonator, a tuning bandwidth of 0.75% has been reported [21].

The DR can also be tuned by attaching a microwave ferrite on the resonator and applying a magnetic field to it. The magnetic field controls the magnetic properties of the ferrite and, hence, the field distributions in and around the DR resulting in a shift in the resonant frequency. Tuning bandwidths of the order of 0.5% [22] and 1% [21] have been reported. This method, however, cannot effectively be used in practice due to slow tuning speed, large size, and bulk of the electromagnet necessary to provide the variable magnetic field, and excessive power consumption of the driving circuit.

Another method of electronic tuning is to use the bias voltage for this purpose. The frequency of any oscillator is known to be sensitive to the bias voltage described generally by the pushing figure. Unfortunately, the change in bias voltage also affects the output power, thus making it difficult to use for frequency-modulation purposes. The bias circuit can, however, be designed in such a way as to minimize the output power variation with bias voltage. Using such a technique, a 4.5-MHz frequency-tuning range at 10 GHz with less than a 1-dB variation of oscillator power output has been reported [23]. A TDRO inherently has a high-Q factor and, hence, a low pushing figure. This limits the bias-tuned frequency range that can only be increased by intentionally reducing the oscillator Q at the cost of degradation of other oscillator characteristics like FM noise, temperature stability, etc.

Optical control of microwave devices and subsystems is a rapidly growing field of research. The resonant frequency of the DR used in the TDRO can be optically modulated and tuned as shown in Fig. 9.24. A photosensitive material like high-resistivity silicon is placed directly on the dielectric

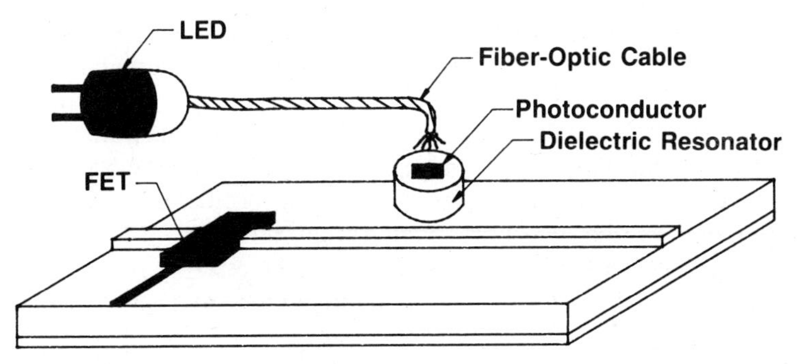

Figure 9.24 Optical tuning of the dielectric resonator. (After P. R. Hercsfeld [25], 1985 IEEE, reprinted with permission.)

resonator. Light from a laser or a LED is brought through an optical fiber to illuminate the photosensitive material, changing its conductivity and perturbing the electromagnetic field in and around the resonator. This perturbation results in a shift in the center frequency of the TDRO. Using this technique, tuning bandwidths of better than 0.1% have been achieved at X band [24, 25].

9.6 WIDE-BAND TUNABLE OSCILLATORS

The analysis and design of wide-band tunable oscillators are more complex than that of a fixed-frequency oscillator. It is important that the active element—for example, the transistor—can generate the negative resistance or the reflection gain >1 over the desired bandwidth, with the help of proper series or parallel feedback. The frequency-determining element—for example, a varactor diode or YIG sphere—should be tunable with voltage or current such that the oscillation conditions can be satisfied over the desired band [26]. The analysis using small-signal-measured S-parameters is sufficient to realize the negative resistance over the desired band. However, in order to optimize the power output it is necessary to use one of the nonlinear techniques, for example, nonlinear model, device line, or the load-pull method.

In Fig. 9.7 it can be shown that in order to obtain a maximum negative resistance in the oscillator circuit, the load impedance should be the only resistive element. In other words, the impedances connected to the terminals other than the output terminals should be low-loss reactive circuits. To realize an electronically tunable oscillator it is also necessary that one or more of the reactances connected to the transistor be electronically tunable. Though any of the transistor terminals can be used to connect the useful load, drain in FETs and collectors in bipolars are more commonly used. The electronic tuning element can be placed in either of the other two terminals. The tunable oscillator can now be represented as shown in Fig. 9.25.

The first step in the oscillator design consists of determining $X_2(\omega)$, $X_3(\omega)$, and R_2 so that the tuning element looks into an impedance having a negative real part in the desired frequency band. In wide-band tunable oscillators, the emphasis is more on the coverage of maximum frequency band rather than the power output. This calls for a design approach different than that used for fixed-frequency oscillators: Equation 9.39 already gives us the two-port S-matrix in terms of the three-port S-parameters. When a load (with reflection coefficient of Γ_2) is connected to the output terminal, the two-port can be converted to a one-port using

$$S'_{11} = S^T_{11} + \frac{S^T_{12} S^T_{21}}{1/\Gamma_2 - S^T_{22}} \tag{9.60}$$

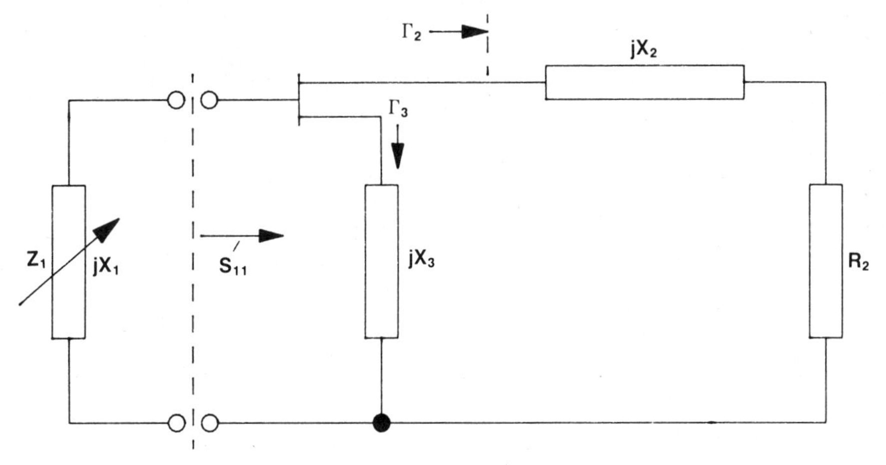

Figure 9.25 Equivalent-circuit representation of a tunable oscillator.

where S_{ij}^T are the two-port S-parameters given by (9.39). The preceding relation can be represented in terms of three-port S-parameters, Γ_2, and Γ_3 to be

$$S'_{11} = S_{11} + \nu + \Gamma_2 \frac{S_{12}S_{21}(1 - S_{33}\Gamma_3) + S_{12}S_{31}S_{23}\Gamma_3 + S_{21}S_{13}S_{32}\Gamma_3 + \nu S_{23}S_{32}\Gamma_3}{(1 - S_{33}\Gamma_3)(1 - S_{22}\Gamma_2) - S_{23}S_{32}\Gamma_2\Gamma_3} \tag{9.61}$$

where

$$\nu = S_{31}S_{13}\Gamma_3/(1 - S_{33}\Gamma_3) \tag{9.62}$$

This relation can be used to analyze the negative resistance in a given device for any combination of Γ_2 and Γ_3. The same relation can also be used to optimize the impedances Γ_2 and Γ_3 for negative resistance. Figure 9.26 shows the reflection gain of a typical 0.5-μm GaAs FET optimized in the frequency range of 6 to 18 GHz.

Two types of tuning elements are commonly used as Z_1 (Fig. 9.25) to realize wide-band oscillators: YIG spheres and varactors. The YIG sphere, which has been discussed in detail in Chapter 3, is a ferrimagnetic resonant element magneticlly tunable over a very wide bandwidth (more than a decade). The varactor is a voltage-controlled capacitance, used to accomplish wide-band tunable oscillators with about an octave bandwidth. We now briefly discuss these oscillators before comparing their performances.

9.6.1 YIG-Tuned Oscillators (YTO)

YIG-tuned oscillators are commonly used as wide-band signal sources in sweep generators, spectrum analyzers, and EW applications. YIG-tuned

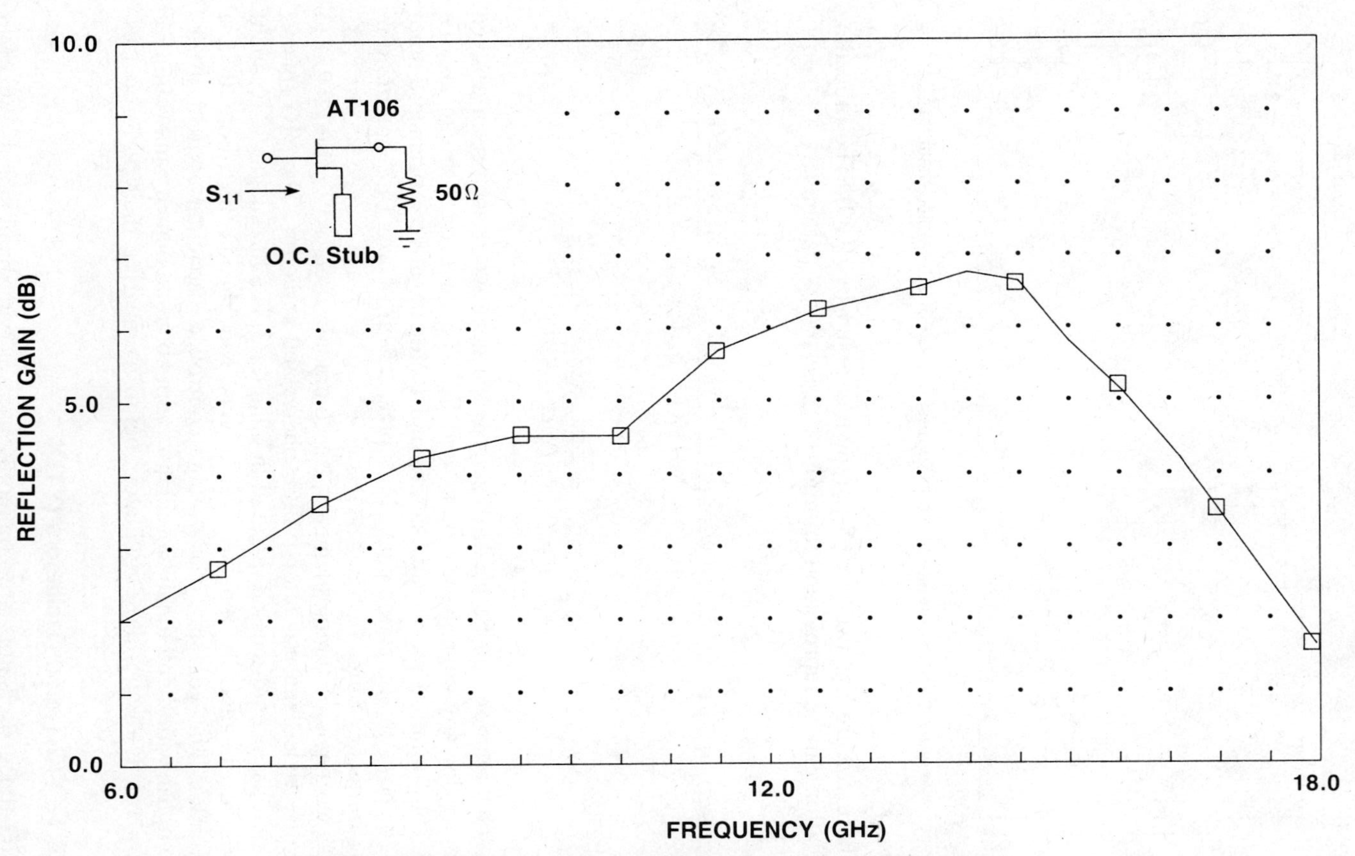

Figure 9.26 Reflection gain of a half-micron-gate-length GaAs FET.

Gunn oscillators use the early replacements of the backward wave tube oscillators. With the advent of GaAs FET and microwave bipolars, transistor YIG-tuned oscillators have been quietly taking over the field, at least up to 26.5 GHz. The main advantages offered by the transistor over Gunn diodes are better efficiency and operation at lower oscillation frequencies. While Gunns do not oscillate below about 5 GHz, the transistor YIG-tuned oscillators (YTO) are realizable down to 0.5 GHz. A YIG resonator coupled to a microstrip line can be modeled by a parallel *RLC* resonant circuit in series with a self-inductance as shown in Fig. 9.27.

The value of the elements depend on the magnetization, coupling between the sphere, and the loop and resonance line width of the YIG sphere (Chapter 3). The resonant frequency ω_0 of the circuit is a linear function of the applied magnetic field to a first-order approximation. A YTO using GaAs FET and a bipolar is represented in Fig. 9.28.

YIG sphere being a very high-Q tunable resonant circuit, the oscillation phase condition $\underline{/S'_{11}} = -\underline{/\Gamma}$ can be easily satisfied and stable oscillation achieved in the negative-resistance frequency band of the device. Under large signal conditions, the relation (9.34) should also be satisfied. Once again the high-Q characteristic or, in other words, a very high value of the reactance slope $dX/d\omega$ helps to meet this requirement.

Parasitic and Spurious Oscillations. A YTO is susceptible to parasitic oscillations due to various reasons. There exist two types of parasitic/spurious oscillations; ones that are fixed in frequency and do not tune with the resonator, and the others that do tune and originate from the YIG sphere itself. The fixed-frequency oscillations result due to the coupling loop and can be noticed even when the magnetic field is not applied. These oscillations can be minimized by the proper design of the matching circuit and loop so that the fixed-frequency oscillations fall out of the desired frequency band and parasitic oscillations should not appear at all.

The tunable parasitic oscillations due to the YIG sphere are observed due to the higher order modes of the resonator. Thus, for certain applied magnetic-field values, the oscillator can jump from the principal resonant

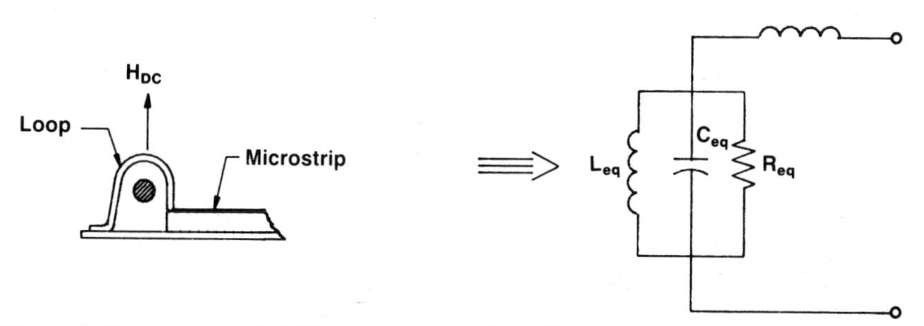

Figure 9.27 Loop-coupled YIG resonator and its *RLC* equivalent-circuit presentation.

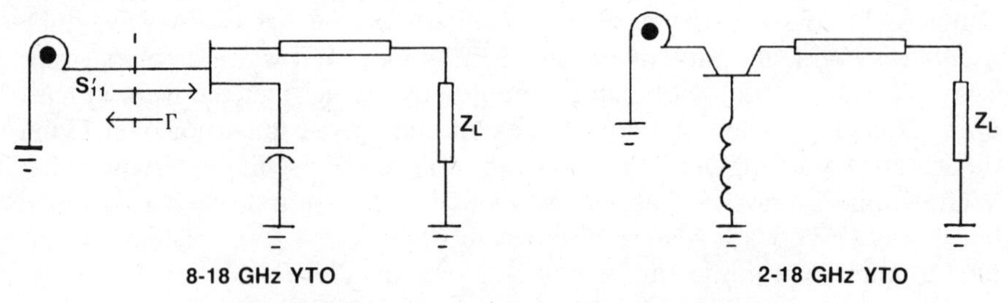

Figure 9.28 YTOs using GaAs FET and Si bipolar.

curve to a parasitic curve, and thereby provoke the phenomenon of tuning hysteresis. It is necessary to characterize the YIG sphere loop coupled to a microstrip line before using it in the oscillator circuit. The tuning hysteresis can also result from non-monotone behavior of the phase of the active circuit with frequency.

Tuning Linearity. YIG devices can be designed for excellent tuning linearity. It is important to assure that the high-permeability shell and center core pieces of the biasing electromagnet are never saturated in the desired tuning range to be covered. This unsaturated condition minimizes the tuning nonlinearity in a YTO. Another factor in YTO, which affects the

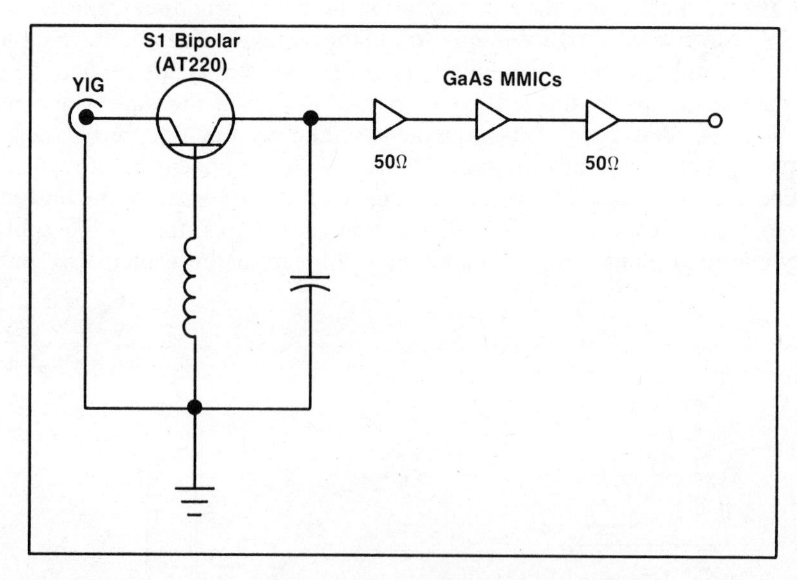

Figure 9.29 YIG oscillator constructed by buffering the oscillating output of the transistor with a series of GaAs MMIC amplifiers. (After C. C. Leung [1], 1979 IEEE, reprinted with permission.)

tuning linearity, is the matching circuits. In the oscillator design approach explained herein, it may be noted that the YIG sphere does not work exactly at its resonant frequency. In fact, this component serves as a very high-Q variable reactance that compensates the reactance presented by the transistor and the associated load and feedback circuits. Hence, deviations from linearity of the tuning curve do depend on the load circuit design. Obviously, the higher the quality factor, the higher the tuning linearity.

Performance of YTO. Initial performance results on YIG-tuned transistor oscillators were reported in 1976 [27] and 1977 [28]. Since then a number of other published reports have appeared. Oyafuso [29] presented an 8–18 GHz FET YTO in 1979, in which capacitive feedback in the source was used for the first time. Le Tron et al. [30] used a common YIG resonator coupled to both the source and gate port of the FET and obtained a tuning range of 3.5 to 14 GHz. The same group later used two separate YIG spheres in the gate and source and obtained a 2–20 GHz YIG-tuned MESFET oscillator [31]. More recently, Leung et al. [1] realized a 2–18 GHz YIG-tuned bipolar transistor oscillator using a single YIG sphere in the emitter of a bipolar, as shown in Fig. 9.29. This high-frequency performance was made possible through the use of interdigitated silicon bipolars AT220 (Avantek) transistors having 0.5-μm emitter width and 2-μm emitter–emitter pitch. This oscillator has far superior phase noise, as shown in Fig. 9.30.

Using GaAs FET, much higher frequencies have been reported. Frequency coverage of 26.5 to 40 GHz and 33 to 50 GHz in YIG FET oscillators was reported by Zensius et al. [32]. The magnetic-field saturation limitation is presently limiting the fundamental YIG transistor oscillators to a maximum of 60 GHz.

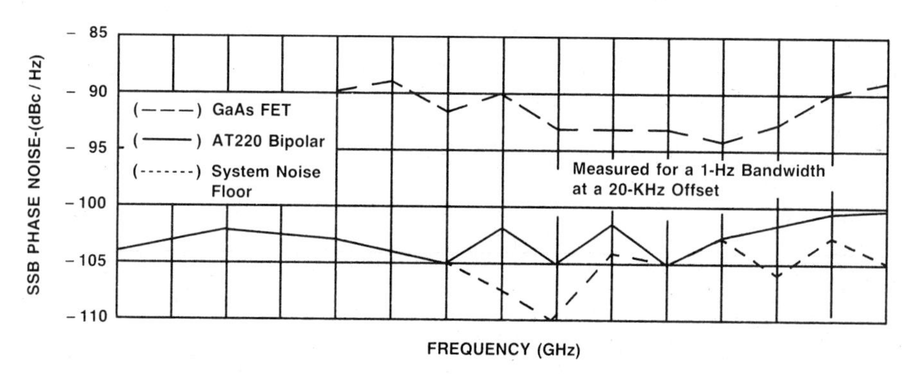

Figure 9.30 Single-sideband phase noise of the AT220 bipolar oscillator. (After C. C. Leung [1], 1979 IEEE, reprinted with permission.)

6.6.2 Voltage-Controlled Oscillators

Electronic tuning can be achieved using varactor diodes. A varactor-controlled transistor oscillator is an attractive solution for high speed and moderate tuning-range applications.

A varactor diode can be either a silicon or GaAs diode. The main difference between silicon and GaAs is that a higher Q can be obtained from GaAs devices. This is due to the lower resistivity of GaAs for a given doping level N. GaAs devices have a higher thermal resistance than silicon devices, resulting in significantly larger frequency settling times.

Figure 9.31 represents a general representation of a varactor-tuned transistor oscillator. The tuning bandwidth of a YIG-tuned oscillator is limited by the negative-resistance bandwidth, whereas it is the susceptance ratio of $C_{min}\omega_{max}/C_{max}\omega_{min}$ that limits the bandwidth of a VCO.

Let the admittance of the transistor and feedback circuit, transferred in the plane of the varactor be $Y(\omega) = G(\omega) + jB(\omega)$ with both $G(\omega)$ and $B(\omega) < 0$. Plotting $-B(\omega)$, $C_{min}\omega$, and $C_{max}\omega$ (Fig. 9.32) the electronic bandwidth is limited by the frequencies at which:

$$C_{min}\omega_{max} = -B(\omega_{max}) \tag{9.63}$$

and

$$C_{max}\omega_{min} = -B(\omega_{min}) \tag{9.64}$$

This limitation holds even if the negative-conductance bandwidth is much wider, as shown in Fig. 9.32.

To increase the electronic tuning bandwidth for a given C_{max}/C_{min} ratio, it is necessary to reduce the ratio $B(\omega_{min})/B(\omega_{max})$ and modify the variation of $B(\omega)$ to that of the susceptance of a lumped self-inductance. In order to obtain this, a self-inductance L_c is added in parallel to the transistor terminals (Fig. 9.31). In this case, the total susceptance ratio seen by the varactor is given by [33]

$$\frac{B_T(\omega_{min})}{B_T(\omega_{max})} = \frac{B(\omega_{min}) + 1/(L_c\omega_{min})}{B(\omega_{max}) + 1/(L_c\omega_{max})} \tag{9.65}$$

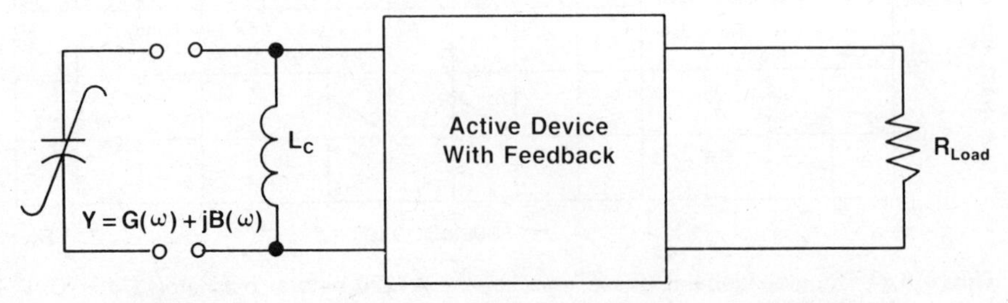

Figure 9.31 Reactance compensation in a varactor-tuned oscillator.

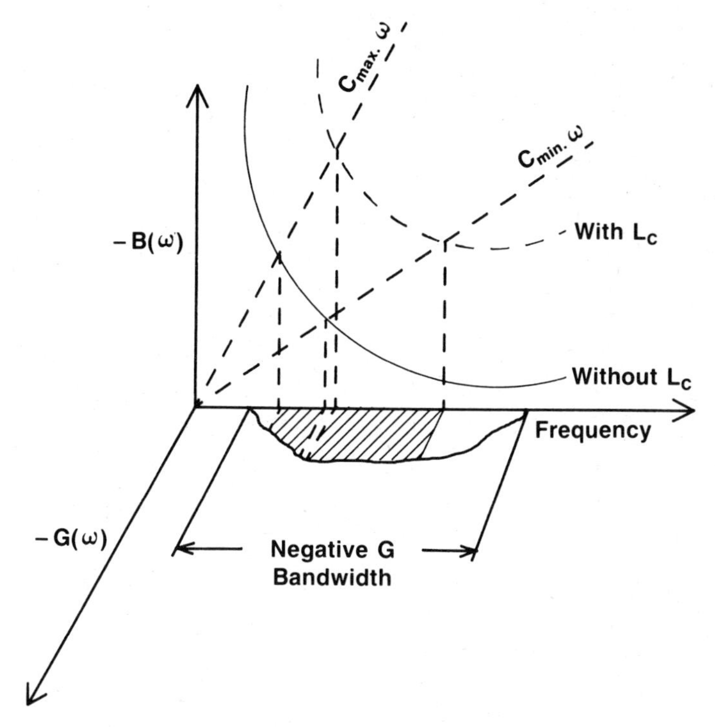

Figure 9.32 Frequency limits of varactor-tuned oscillator. (After A. P. S. Khanna [37], 1987 IEEE, reprinted with permission.)

If L_c is small, this ratio is larger and approaches $\omega_{\max}/\omega_{\min}$. Equating this ratio to that of the susceptance ratio presented by the varactor, we get

$$\frac{\omega_{\max}}{\omega_{\min}} = \sqrt{\frac{C_{\max}}{C_{\min}}} \tag{9.66}$$

The value of L_c normally has to be optimized to maximize the bandwidth, due to the parasitic inductances present in the circuit.

Figure 9.33 shows a typical VCO circuit using a bipolar transistor and a varactor diode. This configuration using NEC 567 as the bipolar transistor and a GaAs hyperabrupt varactor covers greater than an octave bandwidth from 3 GHz to 9 GHz [34]. Increased bandwidth can be realized by using two varactors, one in source and another in the gate, as shown in Fig. 9.34 for a GaAs FET VCO.

Using the preceding configuration, bandwidths greater than an octave have been realized. A low-noise VCO configuration using a bipolar transistor and a silicon abrupt diode [35] is shown in Fig. 9.35. This oscillator has an 8% tuning bandwidth at 8 GHz and has excellent phase noise of 103 dBc at 100 kHz off the carrier.

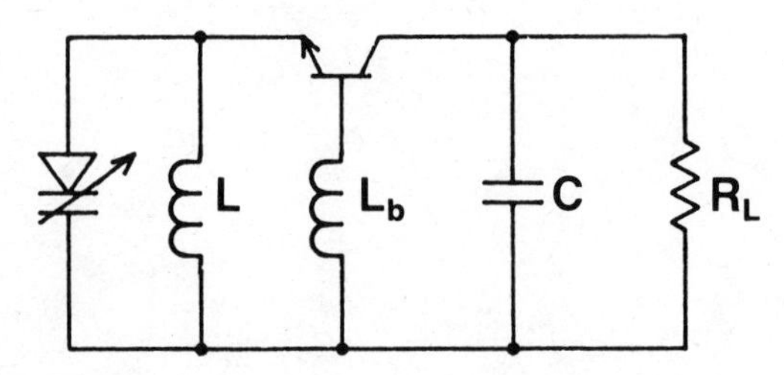

Figure 9.33 Varactor-tuned bipolar transistor oscillator.

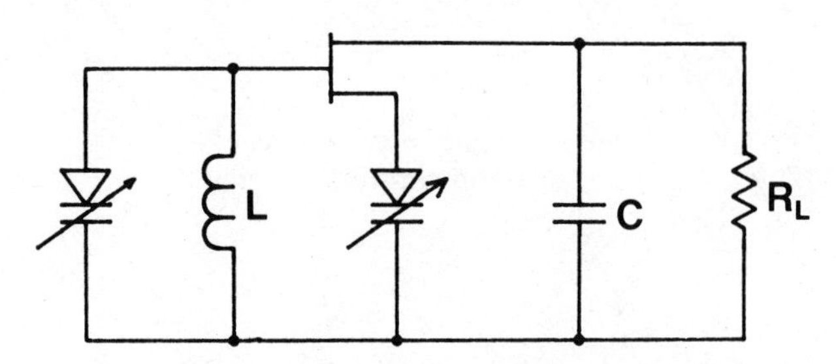

Figure 9.34 Varactor-tuned GaAs FET oscillator.

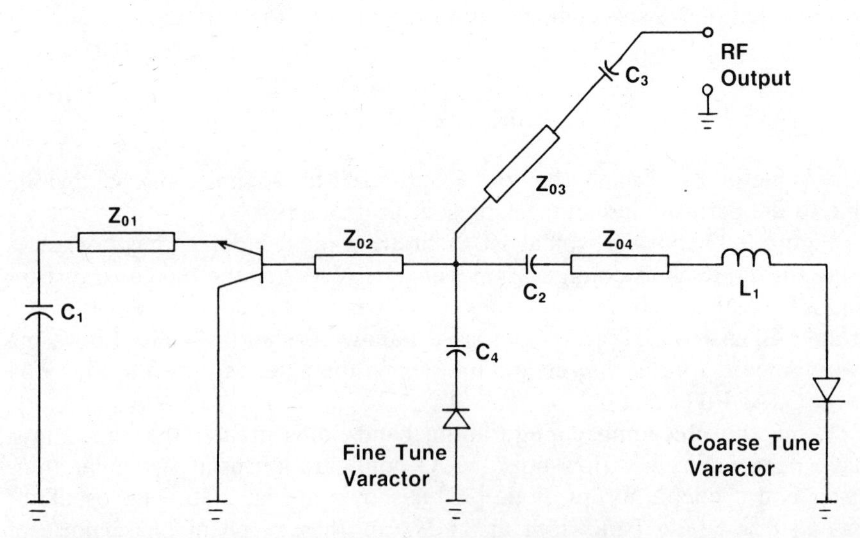

Figure 9.35 Wide-band varactor-tuned polar transistor oscillator. (After E. Niehenke and R. Hess [35], 1979 IEEE, reprinted with permission.)

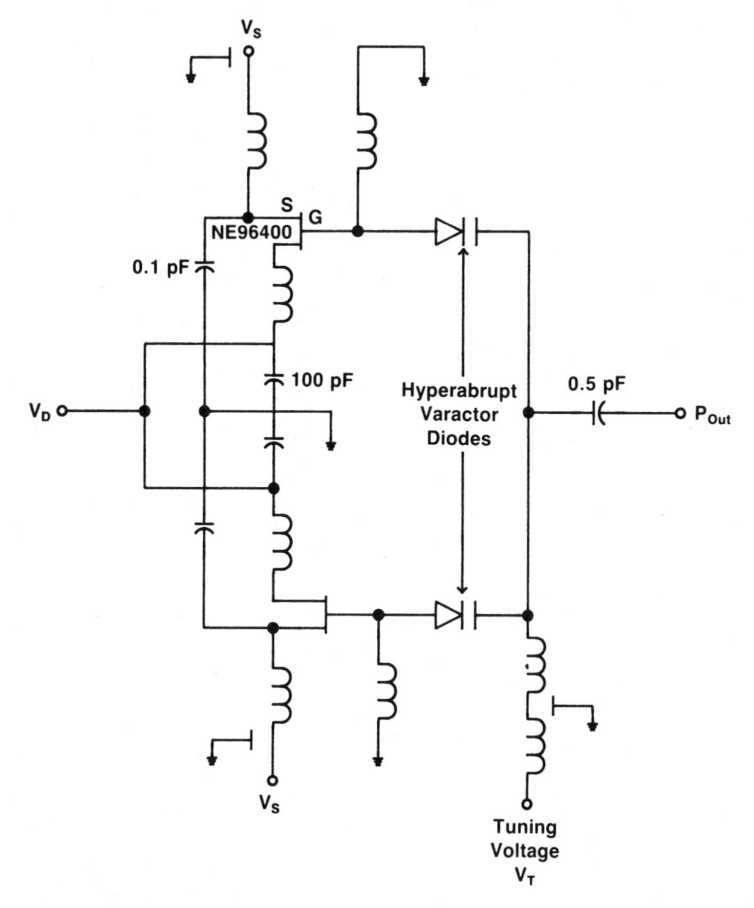

Figure 9.36 VCO circuit schematic diagram. (After R. G. Winch [36], reprinted with permission of *Electron. Lett.*)

Wideband frequency operation of a transistor VCO can be extended by using a push push type VCO configuration shown in Fig. 9.36. This circuit cancels the fundamental frequency and optimizes the second harmonic at the output point. Using hyperabrupt diodes with capacitance variation of 3.2 pF to 0.16 pF and a GaAs FET device, a VCO covering 20 to 30 GHz has been reported by Winch [36].

Settling Time and Post Tuning Drift. In a frequency-tunable oscillator, settling time is defined as the interval between the time when the input tuning drive waveform reaches its final value and the time when the VCO frequency falls within a specified tolerance of a stated final value. Figure 9.37 defines the settling time and post tuning drift. The input drive waveform reaches its final value at t_0. Settling time is the interval between t_{st}

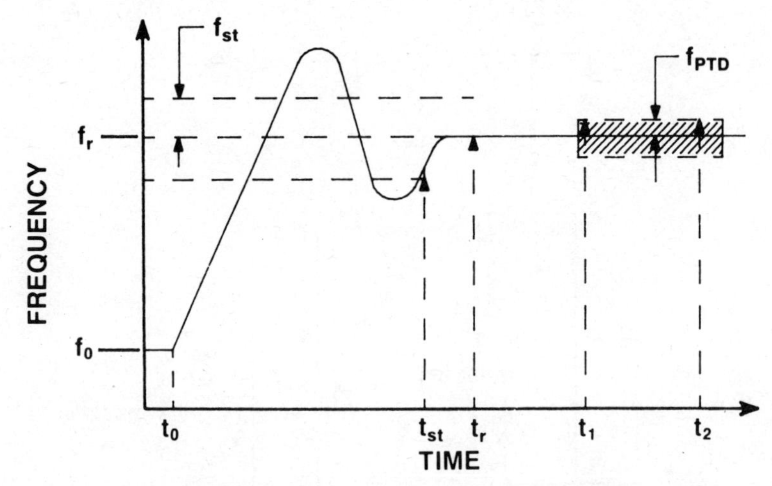

Figure 9.37 Definition of settling time and post tuning drift.

to t_0. The post tuning drift is the frequency drift that occurs between two defined times t_1 to t_2, where t_1 may be specified typically as 10 μs to 1 sec after the tuning step has been applied to the VCO. Time t_2 is generally defined in the range of 10 μsec to 1 sec after t_1.

Bias voltage drift and thermal effects are the primary contributors to post tuning drift. During the interval when a VCO is being tuned, the junction

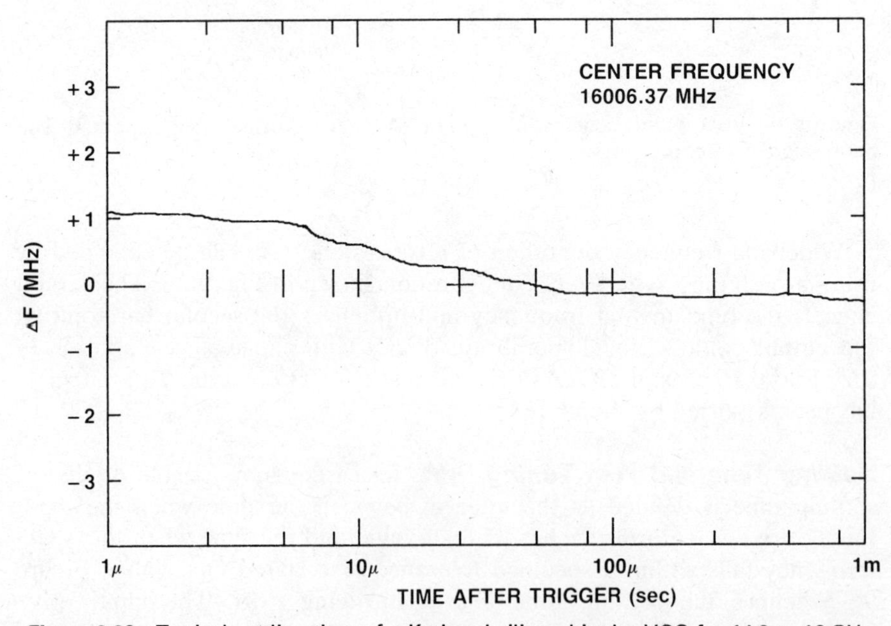

Figure 9.38 Typical settling time of a Ku-band silicon bipolar VCO for 14.3 to 16 GHz.

TABLE 9.1 Varactor-Controlled Oscillator versus YIG-Tuned Oscillator

Parameter	YIG Tuning	Varactor Tuning
Bandwidth	Wide—octave and multioctave	Narrow—octave or less
Phase noise	Excellent	Good performance in narrow bandwidth
Tuning	Current	Voltage
Linearity	±1%	±10 to 15%
Slew rate	<1 MHz/ms	1 to 10 GHz/μs
Step response time	1 to 3 ms	<0.1 μs
Post tuning drift constant	sec	μs to ms
Temperature stability	50 to 100 ppm/°C	100 to 300 ppm/°C
Frequency pushing	5 MHz/V	70 MHz/V
Oscillator *Q*	High > 1000	Low < 100
Power consumption	High—10 W maximum	Low—1 W maximum

temperatures of both the transistor and the varactor are changing due to changes in RF circuit efficiency and loading. This causes impedance changes, which result in a frequency shift. The time interval during which this happens is dependent upon the thermal impedance of the devices. Silicon abrupt and hyperabrupt diodes give excellent settling time performance. GaAs abrupt has a medium and GaAs hyperabrupt has a large settling time compared to the silicon varactor diode.

Figure 9.38 presents typical settling time of a silicon bipolar Ku-band varactor-controlled oscillator [37]. The output frequency settles within 1 MHz in less than 2 μsec.

The YIG- and varactor-tuning techniques previously discussed are significantly different from one another. Table 9.1 compares their important parameters.

9.7 OSCILLATOR MEASUREMENTS

9.7.1 Measurements Using Network Analyzer

The measurement techniques for oscillator characteristics like power, frequency, harmonics, external quality factor, and FM noise are well known. These measurements, however, require a bench loaded with expensive laboratory equipment such as power meter, frequency counter, spectrum analyzer, signal generator, circulator, variable attenuator, and sliding short. In this section we discuss the measurement of major oscillator characteristics using a commonly available network analyzer and a signal generator [38, 39].

The principle of measurements is based on injection locking the TDRO under test with the signal available at the "unknown" port of the network analyzer. An injection-locking polar diagram (ILPD) is thus obtained and displayed on the polar display. The network analyzer is used to measure the magnitude and phase of the injection gain under locked conditions. The resulting ILPD resembles ordinary impedance measurements and can be used to measure the important characteristics of the oscillator. Moreover, the ILPD can be used to optimize key oscillator operating parameters in a manner analogous to using the network analyzer in its conventional role involving filters and amplifiers.

The Hewlett-Packard HP 8410 network analyzer is used for the system described here. For low-power oscillators (up to about +10 dBm), the configuration shown in Fig. 9.39 can be used. Figure 9.40 shows a higher-power configuration that is useful up to about a 30-dBm output. To increase the gain measurement dynamic range and to protect the sensitive harmonic converter from inadvertent overload, attenuators of 30 dB and 3 dB are inserted in the test channel and reference channel, respectively. With a range of RF power of from −16 to −43 dBm at the reference channel input of the harmonic frequency converters, the use of a signal generator with a power output of from −10 to +17 dBm allows complete coverage of the range of possible power variations encountered. For example, using an oscillator having a +10-dBm output, one can measure an injection gain range of from 13 to 40 dB (Fig. 9.39).

The equipment can be calibrated for either a transmission or a reflection coefficient mode using a through-line and a short circuit. The phase calibration can be accomplished with the circuit injection-locked with very high injection gain (>30 dB). In this case, the phase difference between the locked signal and locking source at the center frequency can be assumed to be zero at the parallel tuned-circuit oscillator plane. The reference-plane extension control is adjusted to bring the polar injection-locking diagram into the desired plane. The oscillator free-running frequency can be measured using a counter to locate the θ_0 (when $\Delta\omega = 0$) point on the injection-locking diagram.

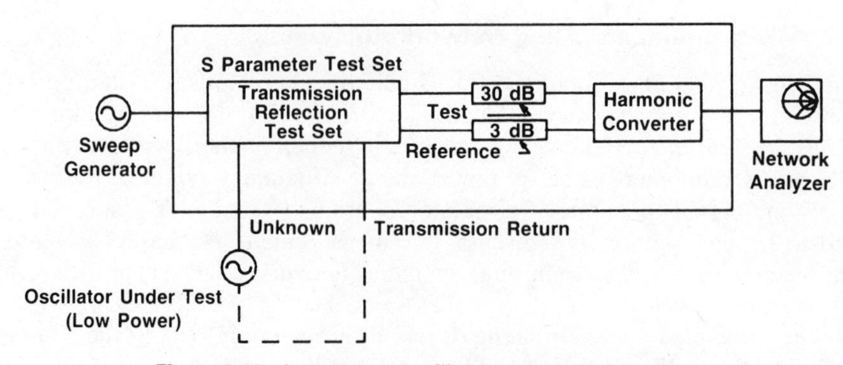

Figure 9.39 Low-power oscillator measurement setup.

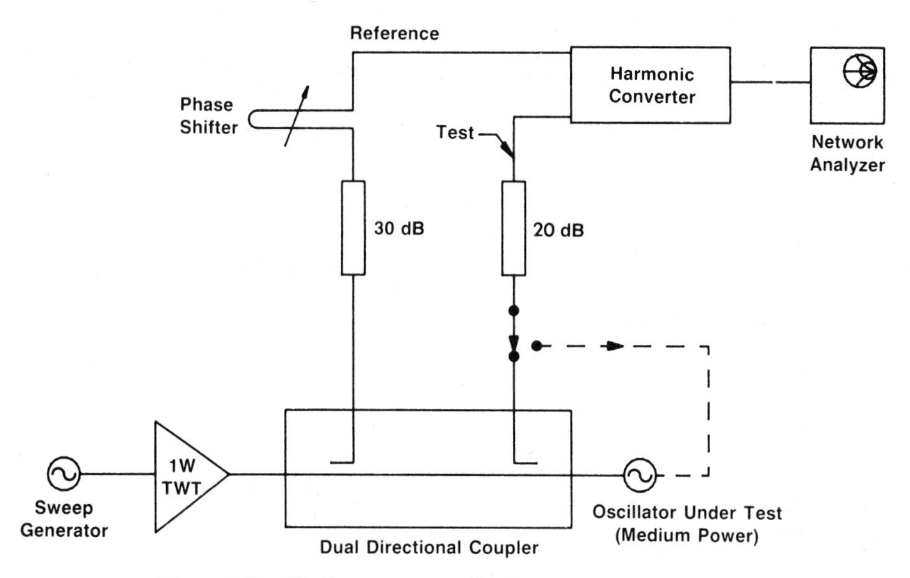

Figure 9.40 Medium-power oscillator measurement setup.

With the oscillator connected to the S-parameter test set's unknown port, and the sweep generator set to sweep across about 10% of the known oscillator's frequency, noise can be observed on both the polar and rectangular display. Next, the sweep band is gradually reduced, and the sweep generator power and the test channel gain are adjusted to obtain the desired ILPD on the polar display for various values of injection gain. The injection gain is thus directly measurable. Reducing the sweep bandwidth to a usable minimum helps eliminate the noise and improve the ILPD. Figure 9.41 shows the ILPD for different values of sweep generator power output and, hence, the injection gain. Figure 9.42 is an ILPD of a FET DRO at 9.5 GHz.

For RF power measurements, the test channel gain is set to correspond to an injection gain of 20 dB. Using the signal generator output power level control, the $G = 0$ point on the ILPD is moved to the polar display's outer edge. The oscillator power can then be determined to the accuracy of the signal generator output controls without using a power meter. For the low-power test configuration of Fig. 9.39, the oscillator output power is equal to the signal generator output power.

Having established the injection gain G on the ILPD, the remaining parameters can be easily measured. Injection bandwidth, $\Delta\omega$, is given by $|\Delta\omega_1| + |\Delta\omega_2|$ and the external quality factor, Q_{ex}, can be determined from

$$Q_{ex} = \frac{2\omega_0}{\Delta\omega} \cdot \frac{1}{\sqrt{G}} \tag{9.67}$$

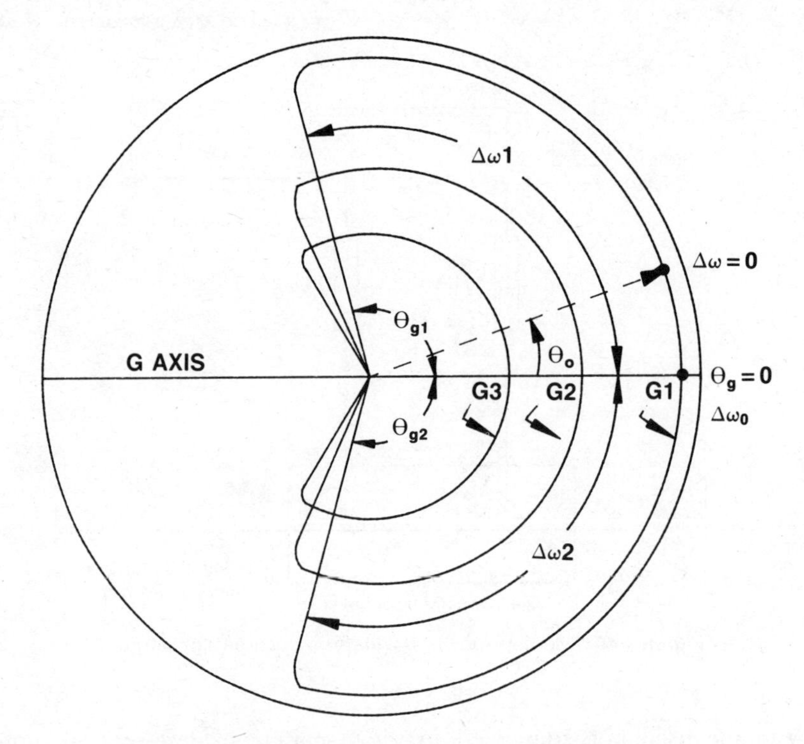

Figure 9.41 Typical ILPD and definition of various parameters. (After A. P. S. Khanna and J. Obregon [38], reprinted with permission of *Microwave* & *RF*.)

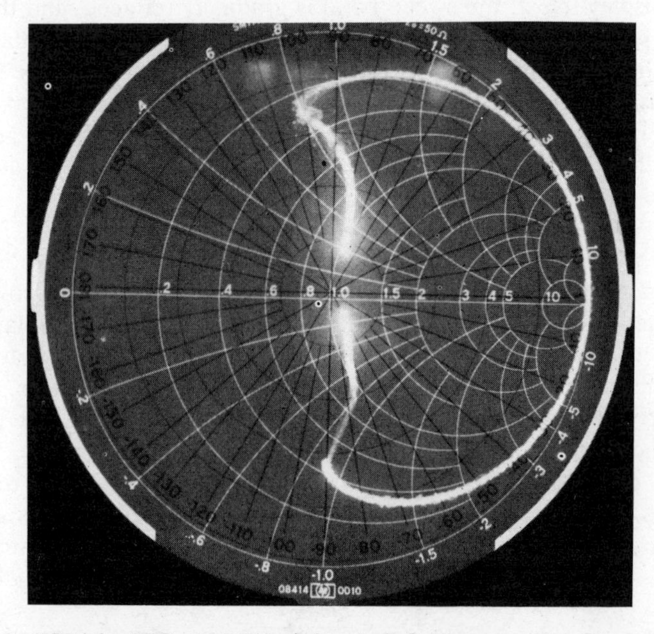

Figure 9.42 ILPD of a DRO at 9.5 GHz. (After A. P. S. Khanna and J. Obregon [39], 1983 IEEE, reprinted with permission.)

Significantly, the measurement is made without a spectrum analyzer, circulator, directional coupler, and power meter, which are usually required. Moreover, the display presented by the network analyzer is generally more informative than that obtained from a spectrum analyzer because it also gives the injection gain phase information, which is not available from a spectrum analyzer.

The nonlinear constants α and K [40] of the oscillator can be determined from the measured values of $\Delta\omega_0$, θ_0, and G using the following relations [39]

$$\alpha = \frac{G \sin \theta_0}{1 - G \cos \theta_0} \tag{9.68}$$

and

$$K = \frac{\Delta\omega_0 (G - 1)}{2\alpha} \tag{9.69}$$

This approach provides a rapid means of determining α and K and graphically depicting the effects of parameters, such as biasing voltage, on these nonlinear constants. For low values of injection gain, the elliptical power variation in the injection-locked frequency range as well as in the locking range asymmetry (which is a function of α) can be read directly from the ILPD.

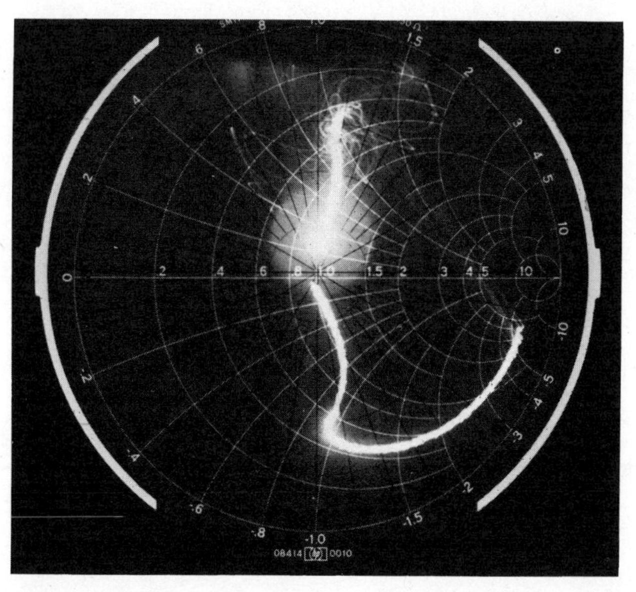

Figure 9.43 Phase jumping of TDRO. (After A. P. S. Khanna and J. Obregon [39], 1983 IEEE, reprinted with permission.)

The ILPD yields valuable insight into the oscillator output matching circuit as well. Frequency and phase jumps within the locking range are readily displayed and recognized (Fig. 9.43). The network analyzer thus provides a valuable tool for testing and aligning the output load circuit. Since the injection gain represents oscillator power output (for constant oscillator signal input), the oscillator can easily be tuned for the desired power output. The method employed is comparable to using a network analyzer for tuning a filter or amplifier. This results from the comprehensive display that immediately shows the effect of changes to load or circuit impedances.

9.7.2 Measurement of Pulling Figure

Pulling of an oscillator is defined as the total angular frequency deviation $\Delta\omega_T$ due to a known load perturbation. The load is characterized by a reflection coefficient having a constant modulus and a phase variable from 0 to 2π. Schematic of an oscillator for pulling figure derivation is shown in Fig. 9.44(*a*). Frequency pulling can be shown to be given by [40]

$$\Delta\omega_T = \frac{S^2 - 1}{S} \cdot K \cdot \sqrt{\alpha^2 + 1} \tag{9.70}$$

where

$$K = \frac{(\partial R_T / \partial I \cdot R_0)}{(\partial R_T / \partial I) \cdot (\partial X_T / \partial \omega) - (\partial R_T / \partial \omega) \cdot (\partial X_T / \partial I)} \tag{9.71}$$

$$Z_T = R_T + jX_T$$

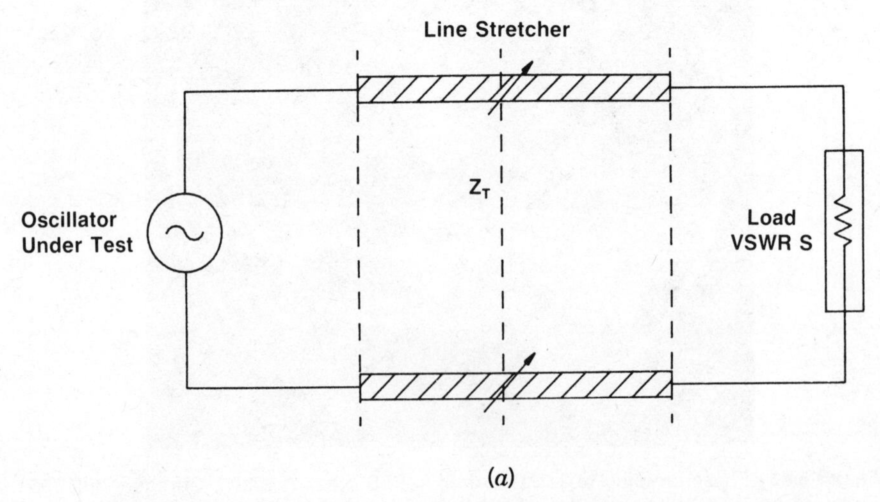

(*a*)

Figure 9.44 (*a*) Schematic of an oscillator for pulling figure derivation.

and

$$\alpha = \frac{(\partial X_T / \partial I)}{(\partial R_T / \partial I)} \tag{9.72}$$

where K and α represent the nonlinear constants of the oscillator, and S is the load VSWR.

Neglecting the nonlinearities $\partial R_T / \partial I$ and $\partial X_T / \partial I$, the simplified pulling expression can be written as

$$\Delta\omega_T = \frac{\omega_0}{2Q_{\text{ext}}} \cdot \left(S - \frac{1}{S}\right) \tag{9.73}$$

In a practical case the frequency pulling is measured on a spectrum analyzer when the oscillator is connected to a load having a constant reflection coefficient modulus and variable phase. The load is varied from 0 to 2π radians with the help of a sliding short. Figure 9.44(b) shows the experimental setup used for the measurement of pulling figure for a commonly used VSWR of 1.67 : 1.

As an example, if a typical 10 GHz DRO has a frequency pulling of 4 MHz, the external quality factor can be calculated to be

$$Q_{\text{ext}} = \frac{10{,}000}{2 \times 4}\left(1.67 - \frac{1}{1.67}\right) = 1339$$

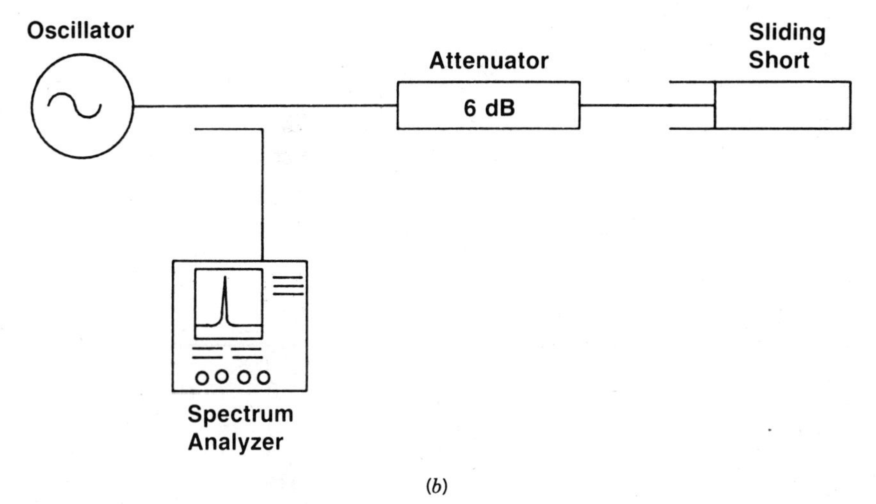

Figure 9.44 (b) Measurement setup for oscillator pulling figure.

9.7.3 Measurement of FM Noise

FM noise is an important characteristic of the oscillator. It represents the phase jitter or, in other words, the short-term stability of the oscillator. The oscillator power is not concentrated at the carrier frequency alone, but is rather distributed around it: the spectral distributions on the opposite sides of the carrier are known as noise sidebands. Noise can be discussed as a modulation phenomenon and these sidebands can be separated into amplitude- and frequency-modulation components. The amplitude-modulation components being much smaller compared to the frequency-modulation components, the discussion that follows deals mainly with FM noise. FM noise also represents the short-term stability in the time domain. Various terminologies are used to specify the FM noise of a CW (continuous wave) oscillator, causing a significant problem to the oscillator engineer. The oscillator noise performance results often reflect the measurement equipment employed, but it is often not possible without additional information to directly compare the results. FM noise can be represented in time domain or frequency domain. In the time-domain method, the rms frequency amplitude deviation from the mean-squared value of the noise is determined. A probability density function is assumed, and the time-averaged value gives little information about the spectral distribution.

In frequency domain, the FM noise is represented in various ways. The commonly used terminologies are as follows.

1. FM noise power is represented as a ratio of the power in some specified bandwidth in one sideband to the carrier power. Such power ratios are usually quoted in decibels and require specification of the bandwidth and frequency offset from the carrier at which measurement is made (Fig. 9.45). This is the most commonly used representation for FM noise.

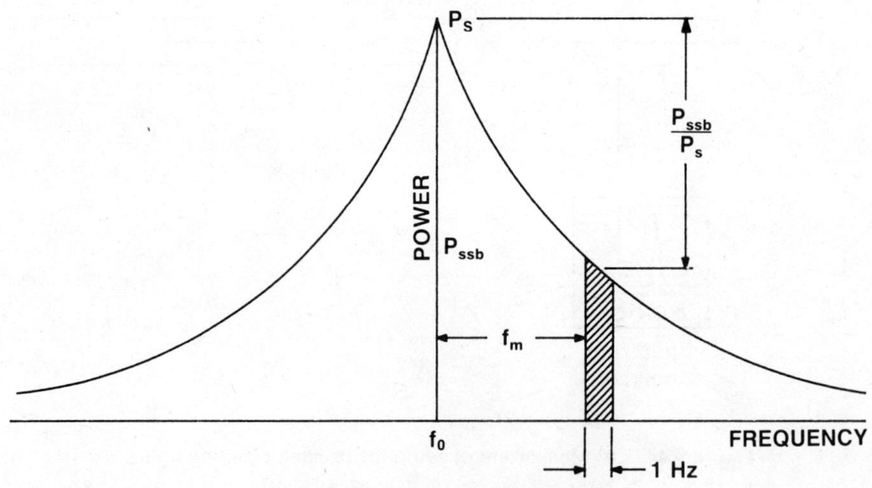

Figure 9.45 Oscillator RF spectrum.

2. FM noise can also be represented as the equivalent rms frequency deviation of a noise-free carrier modulated at a single frequency and producing the same output as the noisy carrier in a narrow band centered at the frequency of interest. This representation is related to the earlier one by the following relation and requires specification of the bandwidth and frequency offset from the carrier.

$$\text{Single Sideband dBc} = 20\log(\Delta f_{\text{rms}}/\sqrt{2}f_m) \tag{9.74}$$

Figure 9.46 represents the conversion monogram that relates the preceding two representations of the FM noise.

Theoretically an idealized spectrum analyzer with a dynamic range of greater than 150 dB and a bandwidth of 1 Hz operating at microwave frequencies would facilitate most noise measurements. Practically such performances are not possible and a variety of techniques have been developed for the measurement of oscillator FM noise. The two most common techniques are discussed below.

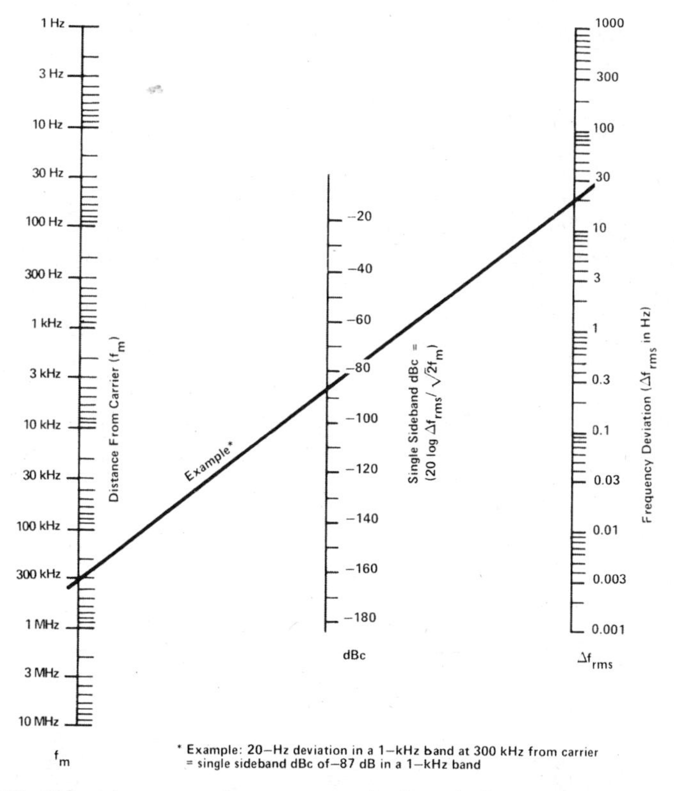

Figure 9.46 FM noise-conversion nomograph. By placing a straight edge on the frequency deviation in hertz (right column) and on the distance from the carrier at which that deviation occurs (left column), the single-sideband noise–power ratio in dBc is shown in the center column.

FM Noise Measurement Using a Cavity Discriminator **[10].** A block diagram of the simplified setup is shown in Fig. 9.47. This block diagram represents the frequency discriminator that serves to convert the frequency modulation into amplitude modulation using the frequency-response variation of a tuned circuit. The signal from the oscillator to be tested is divided into two paths: LO path and RF signal path. Adjustable phaser and attenuators are placed in the LO path, while a high-Q carrier suppression filter (a tunable cavity and a circulator) is placed in the RF path. Using the cavity in this mode, a reduced RF power corresponding only to the FM sidebands of the oscillator signal is reflected toward the balanced mixer. The phase shifter ϕ is adjusted in order to reintroduce the carrier phase advanced by 90 degrees. FM noise sidebands are thus converted to AM sidebands and are easily detected by mixer diodes, while the AM noise sidebands are converted to FM and are not detected by the mixer. The mixer output is detected by a highly sensitive, adjustable bandwidth selective voltmeter.

The system is calibrated by applying modulation to the oscillator under test. The modulation is set using the microwave spectrum analyzer at a known level (generally at modulation index of 2.405 corresponding to a carrier null) and the baseband measurement is used to determine the system calibration constants.

FM Noise Measurement Using a Delay-Line Discriminator **[41].** The block diagram of a delay line FM discriminator is shown in Fig. 9.48. This setup has less sensitivity compared to a cavity discriminator, but is wide band and easier to automate. The principle of operation of the delay-line discriminator is similar to the cavity discriminator. The length of the delay line is adjusted to place the zero crossing at the oscillator carrier frequency. Practically this means setting up to obtain zero dc output from a balanced

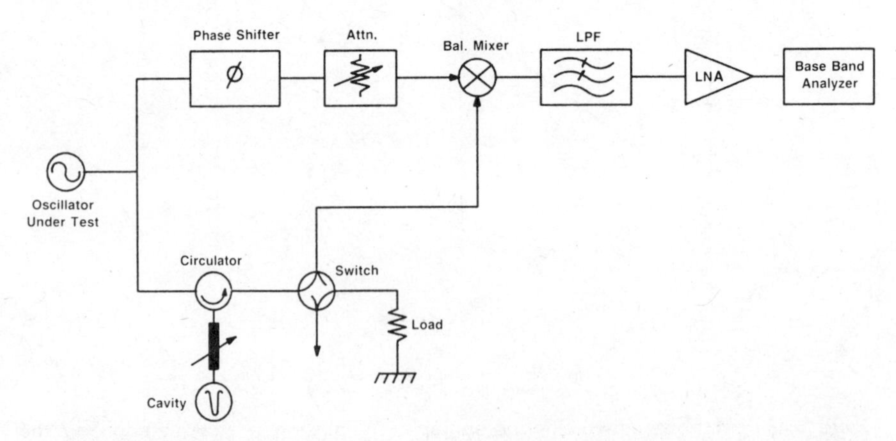

Figure 9.47 FM noise measurement setup using a cavity discriminator method.

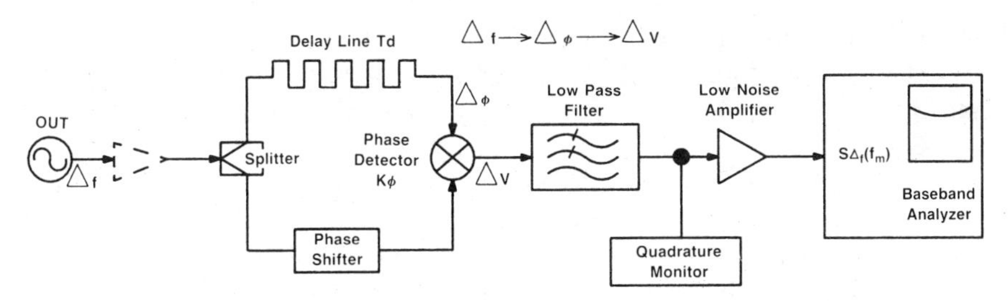

Figure 9.48 FM noise measurement setup using a delay-line discriminator method.

mixer. This minimizes the system sensitivity to the amplitude variations. The length of the delay line also controls the system sensitivity of the FM noise measurement that can be increased by increasing the delay-line length at the cost of bandwidth. The calibration procedure remains similar to that of the cavity discriminator.

Using the delay-line/mixer frequency discriminator technique, Hewlett-Packard proposes to use their carrier noise test set HP 11729C, for phase noise measurement on microwave sources from 10 MHz to 18 GHz. The block diagram is shown in Fig. 9.49. This noise test set uses an internal low-noise microwave reference signal to down convert the test signal to an IF frequency below 1.28 GHz. The IF signal is amplified and then applied to a delay-line frequency discriminator where the phase noise is demodulated and made available for analysis. The HP 11729C is GPIB controll-

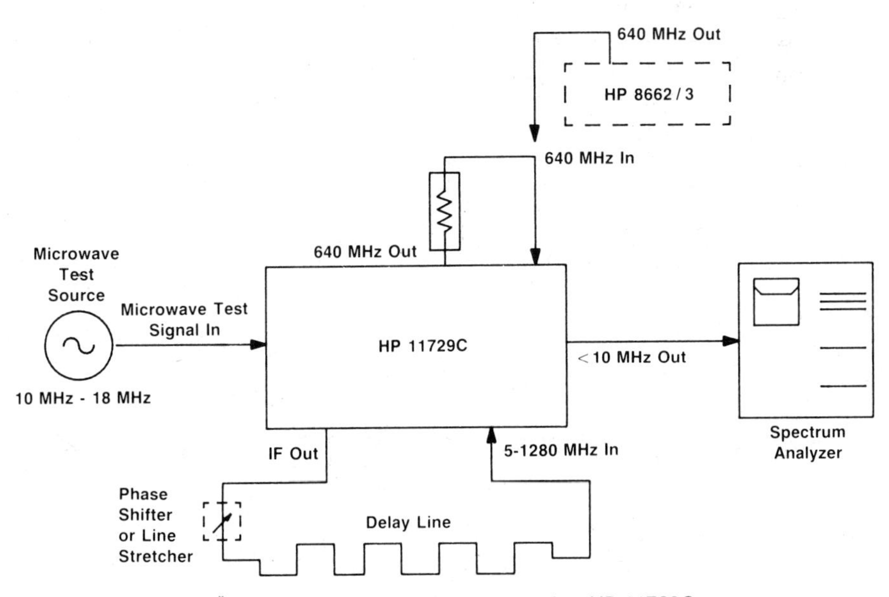

Figure 9.49 Phase noise setup using HP 11729C.

able, making automatic phase noise measurements easy. The noise floor of the test set can be significantly improved (>10 dB) at offset frequencies less than 50 kHz, by using an external lower noise 640 MHz signal from HP 8662/3.

The system sensitivity depends on both the length and attenuation of the delay line. In this system, because the delay-line discriminator works at a frequency less than 1.28 GHz, the delay line can be a common coaxial cable like RG223 providing about 1.5-ns delay per foot of the cable. The delay-line discriminator is typically used to offsets of less than $1/2\tau_d$, and the maximum delay to be used is determined by the highest frequency of interest. For example, if measurements are desired up to an offset frequency (f_m) of 1 MHz, the delay must be less than $1/2f_m = 500$ ns.

Figure 9.50 shows the phase noise of an Avantek electronically and mechanically tuned bipolar DSO at 10 GHz (DSO 4100), measured with an automated test setup using the HP 11729C. The dashed line shows the sensitivity of the system using a 100-ns delay line.

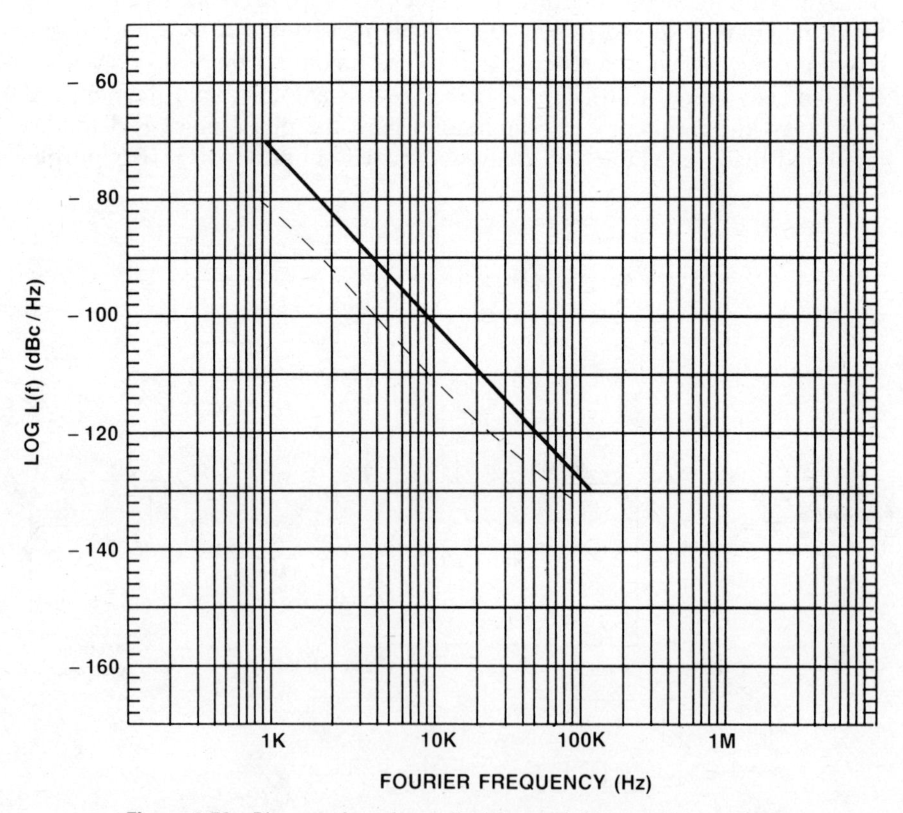

Figure 9.50 Phase noise of a typical Avantek bipolar *X*-band DRO.

REFERENCES

1. Leung, C., C. Snapp, and V. Grande, "A 0.5 μm Silicon Bipolar Transistor for Low Phase Noise Oscillator Applications Up to 20 GHz," *IEEE Int. Microwave Symp. Digest*, June 1985, pp. 383–385.
2. Camiade, M., and A. Bert, "Wide Tuning Bandwidth Ku band Varactor FET Oscillators," *Proc. 15th European Microwave Conf.*, Paris, 1985, p. 413.
3. Boyles, J. W., "The Oscillator as a Reflection Amplifier: An Initiative Approach to Oscillator Design," *Microwave J.*, Vol. 29, June 1986, pp. 83–98.
4. Bodway, G. E., "Circuit Design and Characterization of Transistor by Means of Three-Port Scattering Parameters," *Microwave J.*, Vol. 11, May 1968, pp. 55–63.
5. Khanna, A. P. S., "Parallel Feedback FETDRO Using 3-Port *S*-Parameters," *IEEE Int. Microwave Symp. Dig.*, San Francisco, 1984, pp. 181–183.
6. Khanna, A. P. S., "Three Port *S*-Parameters Ease GaAs FET Designing," *Microwave & RF*, Vol. 24, Nov. 1985, pp. 81–84.
7. Soares, R., J. Graffeuil, and J. Obregon, *Applications of GaAs MESFETs*, Artech House, Dedham, Mass., 1983.
8. Khanna, A. P. S., and J. Obregon, "Microwave Oscillator Analysis," *IEEE Trans. Microwave Theory Tech.*, Vol. MTT-29, June 1981, pp. 606–607.
9. Basawaptna, G. R., and R. B. Stancliff, "A Unified Approach to the Design of Wide-band Microwave Solid State Oscillators," *IEEE Trans. Microwave Theory Tech.*, Vol. MTT-27, May 1979, pp. 379–385.
10. Khanna, A. P. S., "Oscillateurs microondes stables integres a transistors et resonateurs dielectriques," Thesis Docteur-Ingenieur, University of Limoges, France, Sept. 1981.
11. Kajfez, D., and P. Guillon, *Dielectric Resonators*, Artech House, Dedham, Mass., 1986, Chap. 10.
12. Golio, J. M., and C. M. Krowne, "New Approach for FET Oscillator Design," *Microwave J.*, Vol. 21, Oct. 1978, pp. 59–61.
13. Khanna, A. P. S., J. Obregon, and Y. Garault, "Efficient Low Noise Three Port *X*-Band FET Oscillator Using Two Dielectric Resonators," *IEEE Int. Microwave Symp. Dig.*, Dallas, 1982, pp. 277–279.
14. Wagner, W., "Oscillator Design by Device Line Measurement," *Microwave J.*, Vol. 22, Feb. 1979, pp. 43–48.
15. Poulin, D., "Load-Pull Measurements Help You Meet Your Match," *Microwaves*, Vol. 19, Nov. 1980, pp. 61–65.
16. Galani, Z., et al., "Analysis and Design of a Single-Resonator GaAsFET Oscillator with Noise Degeneration," *IEEE Trans. Microwave Theory Tech.*, Vol. MTT-32, Dec. 1984, pp. 1556–1565.
17. Ishihara, O., et al., "A Highly Stabilized GaAsFET Oscillator Using a Dielectric Resonator Feedback Circuit in 9–14 GHz," *IEEE Trans. Microwave Theory Tech.*, Vol. MTT-28, Aug. 1980, pp. 817–824.
18. Tsironis, C., and P. Lesartre, "Temperature Stabilization of GaAsFET Oscillators with Dielectric Resonators," *Proc. 12th European Microwave Conf.*, Helsinki, 1982, pp. 181–186.

19. Lee, J., et al., "Digital and Analog Frequency-Temperature Compensation of Dielectric Resonator Oscillators," *IEEE Int. Microwave Sym. Dig.*, San Francisco, 1984, pp. 277–279.
20. Camiade, M., et al., "Low Noise Design of Dielectric Resonator FET Oscillators," *Proc. 13th European Microwave Conf.*, Nürnberg, Germany, 1983, pp. 297–302.
21. Farr, A. N., et al., "Novel Technique for Electronic Tuning of Dielectric Resonators," *Proc. 13th European Microwave Conf.*, Nürnberg, Germany, 1983, pp. 791–796.
22. Watanabe, S., et al., "Very High *Q* Dielectric Resonators Voltage Controllable Oscillator," *Proc. 8th European Microwave Conf.*, Paris, 1978, pp. 269–273.
23. Christ, S., and P. Horowitz, "Miniature *X*-band Transmitter Uses Nine-volt Battery," *Microwave System News*, Vol. 12, Dec. 1982, p. 84.
24. Hercsfeld, P. R., et al., "Optically Tuned and FM Modulated *X*-band Dielectric Resonator Oscillator," *Proc. 14th European Microwave Conf.*, Leige, 1984, pp. 268–273.
25. Hercsfeld, P. R., et al., "Optically Controlled Microwave Devices and Circuits," *IEEE MTT-S Int. Microwave Symp. Dig.*, St Louis, 1985, pp. 211–214.
26. Pengelly, R. S., *Microwave Field Effect Transistors—Theory Design and Applications*, Research Studies Press, Letchworth, Herts., England, 1982, Chap. 7.
27. Heyboer, T., and F. Emery, "YIG Tuned GaAsFET Oscillators," *IEEE Int. Microwave Symp. Dig.*, 1976, pp. 48–50.
28. Ruttan, T., "*X*-band GaAs FET YIG Tuned Oscillators," *IEEE Int. Microwave Symp. Dig.*, 1977, p. 264.
29. Oyafuso, R., "An 8–18 GHz FET YIG Tuned Oscillator," *IEEE Int. Microwave Symp. Dig.*, 1979, pp. 183–184.
30. Le Tron, et al., "Multioctave FET Oscillators Double Tuned by a Single YIG," *ISSCC Digest of Tech. Papers*, 1979, pp. 162–163.
31. Obregon, J., et al., "Decade Bandwidth FET Functions," *IEEE Int. Microwave Symp. Dig.*, 1981, pp. 141–142.
32. Zensius, D., et al., "GaAs FET YIG Oscillator Tunes from 26 to 40 GHz," *Microwaves and RF*, Vol. 22, Oct. 1983, pp. 129–139.
33. Marechal, P., and J. Obregon, "1.5 to 4.5 GHz Varactor-Tuned Transistor Oscillators," *Proc. 9th European Microwave Conf.*, Brighton, England, 1979, pp. 621–624.
34. Obregon, J., et al., "Ultrabroad Band Electronically Tunable Oscillators," *Proc. 11th Microwave Conf.*, 1981, pp. 475–479.
35. Niehenke, E., and R. Hess, "A Microstrip Low Noise *X*-band Voltage Controlled Oscillator," *IEEE Trans. Microwave Theory Tech.*, Vol. MTT-27, Dec. 1979, pp. 1075–1079.
36. Winch, R. G., "*K*-band FET Doubling Oscillator," *Electron. Lett.*, Vol. 18, Oct. 1982, pp. 946–947.
37. Khanna, A. P. S., "Fast Settling Low Noise Ku Band Fundamental Bipolar VCO," *IEEE Int. Microwave Symp. Digest*, 1987, pp. 579–581.
38. Khanna, A. P. S., and J. Obregon, "Network Analyzer Doubles as Oscillator Diagnostician," *Microwave & RF*, Vol. 23, July 1984, pp. 106–112.

39. Khanna, A. P. S., and J. Obregon, "Direct Measurement of the Non-Linear MIC Oscillator Characteristics Using Injection Locking Polar Diagram," *IEEE Int. Microwave Sym. Dig.*, Boston, 1983, pp. 501–503.
40. Obregon, J., and A. P. S. Khanna, "Exact Derivation of the Non-Linear Negative Resistance Oscillator Pulling Figure," *IEEE Trans. Microwave Theory Tech.*, Vol. MTT-30, July 1982, pp. 1109–1111.
41. Hewlett-Packard, *Phase Noise Characterization of Microwave Oscillators*, Product Note 11729C.
42. Johnson, K. M., "Large Signal GaAs MESFET Oscillator Design," *IEEE Trans. Microwave Theory Tech.*, Vol. MTT-27, Mar. 1979, pp. 217–227.

PROBLEMS

9.1 Two-port *S*-parameters of Avantek GaAs FET AT10600, in the common source configuration, are given in the table below. Calculate the two-port *S*-parameters in the common drain and common gate configurations using three-port *S*-parameters approach.

Frequency (GHz)	S_{11}		S_{21}			S_{12}			S_{22}	
	Mag	Ang	dB	Mag	Ang	dB	Mag	Ang	Mag	Ang
8	0.59	−128	8.4	2.64	91	−27.3	0.04	88	0.71	−3
9	0.56	−145	8.0	2.50	81	−26.2	0.05	88	0.70	−3
10	0.55	−164	7.5	2.37	73	−25.4	0.05	90	0.68	−5
11	0.56	178	7.1	2.27	63	−24.4	0.06	89	0.64	−8
12	0.58	161	6.5	2.11	52	−23.6	0.07	84	0.61	−12
13	0.62	150	5.7	1.92	44	−22.6	0.07	82	0.57	−18
14	0.68	142	5.4	1.86	36	−22.3	0.08	78	0.53	−24
15	0.70	133	4.7	1.71	25	−21.4	0.09	71	0.47	−37
16	0.74	128	4.4	1.65	18	−20.3	0.10	65	0.45	−50
17	0.76	117	3.7	1.52	6	−19.3	0.11	58	0.42	−65
18	0.75	105	3.1	1.43	−5	−19.0	0.11	49	0.40	−80

Bias = 4.5 V, 30.0 mA.

9.2 Using the common gate *S*-parameters at 12 GHz (determined in Problem 9.1), find the position of the DR in terms of electrical length from the GaAs FET [Fig. 9.8(*b*)] in order to maximize the reflection coefficient magnitude at the drain ($|S'_{22}|$). The DR unloaded quality factor and the coupling coefficient to the microstrip line are 3000 and 5, respectively. Plot the S'_{22} on a polar chart from 11 to 13 GHz using the interplotted *S*-parameters for the intermediate frequencies. (*Hint:* Use Equation 9.41.)

9.3 In the two-port network shown in Fig. 9.6, prove that if oscillation condition ($S'_{11}\Gamma_1 = 1$) is satisfied at the input port, it is automatically satisfied at the output port ($S'_{22}\Gamma_2 = 1$).

9.4 (a) In Fig. 9.16 the output power of an oscillator can be approximated with the equation [42]:

$$P_{out} = P_{sat}[1 - e^{-G_0 P_{in}/P_{sat}}]$$

where P_{sat} is the saturated power of the amplifier, P_{in} is the input power, and G_0 is the small-signal power gain. Show that the maximum oscillator power $[P_{osc\,(max)}]$ is given by:

$$P_{osc\,(max)} = P_{sat}\left(1 - \frac{1}{G_0} - \frac{\ln(G_0)}{G_0}\right)$$

Hint: The maximum oscillator power occurs at the point of maximum $P_{out} - P_{in}$, or where

$$\frac{\partial P_{out}}{\partial P_{in}} = 1$$

(b) The Avantek GaAs FET AT10600 has $G_0 = 9$ dB with $P_{sat} =$ 20 dBm at 12 GHz. Calculate maximum oscillator power.

9.5 (a) The frequency pulling bandwidth of a 10 dBm power output oscillator at 12 GHz is measured to be 5 MHz into a 12 dB return loss load. Calculate the external quality factor neglecting the nonlinearities.

(b) If the load is changed to 6 dB return loss, find the frequency pulling bandwidth.

(c) The frequency drift of the oscillator is measured to be 4 MHz over −30 to +70°C. Find the minimum injection locking power required to keep the oscillator locked over the temperature range.

9.6 A modulation noise measurement system is found to have an effective gain of 940 and employs a detector with a sensitivity of 100 V/W. When operated as a direct detection system connected with a source producing 1.2 mW, the detector outputs are 0.14 mV and 0.105 mV in the unbalanced and balanced configurations. What is the AM noise of the source expressed as a power ratio?

A carrier suppression filter is now inserted and the system after calibration is found to have an rms sensitivity of 1.25 μW per Hz deviation. If the output from the detector is 0.24 mV with an unbalanced mixer and 0.091 mV with a balanced mixer configuration, what is the FM noise of the source?

10 AMPLIFIERS

10.1 INTRODUCTION

For amplification of microwave signals both vacuum tubes and solid state devices are used. Tubes such as klystrons [1–3] and traveling-wave tube (TWT) [2–4] amplifiers are exclusively used for high-power applications, whereas solid state amplifiers are very suitable for low noise and medium power levels up to a couple of hundred watts. Solid state devices generally require low voltage for operation and are very compact and lightweight. These characteristics are particularly useful for space and military applications where weight and size can impose severe limitations on the choice of components and systems.

For solid state amplifiers, two types of devices are mainly used: (1) two-terminal and (2) three-terminal. The two-terminal device amplifier circuit is normally called "a reflection-type circuit," or "a negative-resistance amplifier" and generally uses diodes. The three-terminal device amplifier circuit uses transistors and is a transmission-type amplifier. Basic circuit configurations for these amplifiers are shown in Fig. 10.1. In the reflection-type amplifier, a circulator is used to isolate the input and output ports. In the transistor amplifier, input and output matching networks are required.

The two-terminal devices work on a variety of mechanisms: parametric amplification (varactor diodes) [5–8], tunneling (tunnel diodes) [7–9], transferred electron (Gunn diodes and LSA diodes) [8, 10, 11], and avalanche transit-time (IMPATT, TRAPATT, and BARITT diodes) [8, 12]. The three-terminal device amplifiers [13–16] use Si bipolar transistors and GaAs MESFETs.

Parametric amplifiers are narrow-band (<10%) and very low-noise figure devices. Tunnel-diode amplifiers are high-gain, low-noise figure, and low-power circuits. Octave bandwidth of such amplifiers is possible. Performance of Gunn-diode amplifiers is very similar to tunnel-diode amplifiers. IMPATT-diode amplifiers are high power and high efficiency. They are moderately noisy and bandwidths up to an octave are possible. The Si bipolar transistor performs very well up to about 4 GHz, with reliable performance, high power, high gain, and is low cost. The GaAs FETs perform better than the bipolar transistors above 4 GHz. They are broadband, have a wide dynamic range, are highly reliable, and are low cost. Both low-noise and medium-power amplifiers are available. They compete with uncooled parametric amplifiers as well as moderate-power IMPATTS.

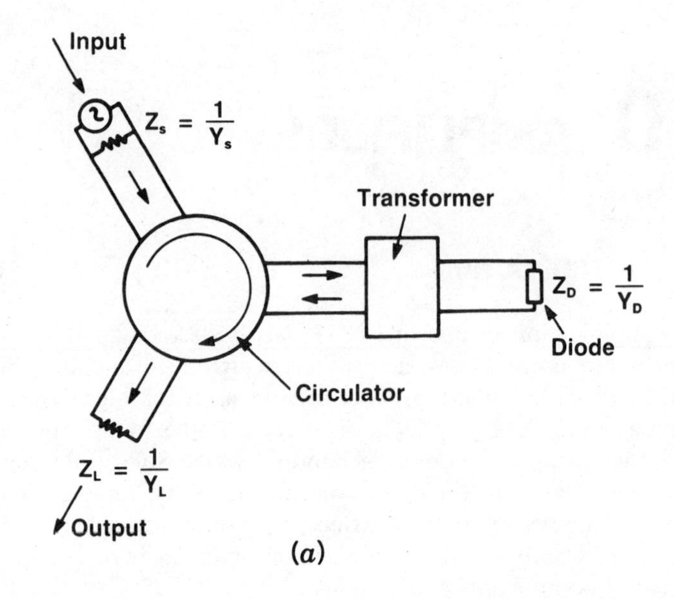

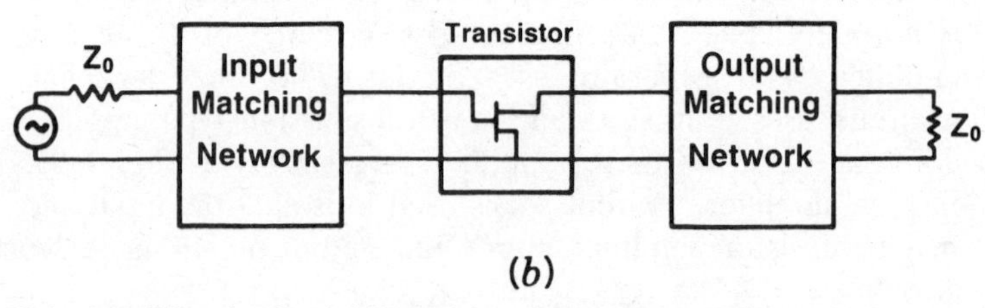

Figure 10.1 Amplifier circuits. (*a*) Negative resistance type. (*b*) Conventional transistor.

Typical noise performance of various amplifiers is compared in Fig. 10.2. Only the MASER has a noise performance that exceeds that of the best cryogenically cooled parametric amplifier. MASERs are quantum-electronic devices [17, 18] and are very bulky and expensive.

In the last several years the performance of GaAs MESFET amplifiers has developed to the point where they are having a major impact on microwave system design. These amplifiers are capable of low-noise, high-gain, and high-power operation. Up to about 20 GHz, they are replacing not only tubes but also other amplifiers using bipolar transistors, Gunn, and IMPATT diodes. They operate from a low-voltage power supply, have higher power-added efficiency, and can be highly reliable. This chapter is written for GaAs MESFETs as amplifying active devices: however, most of material is also applicable to other active devices.

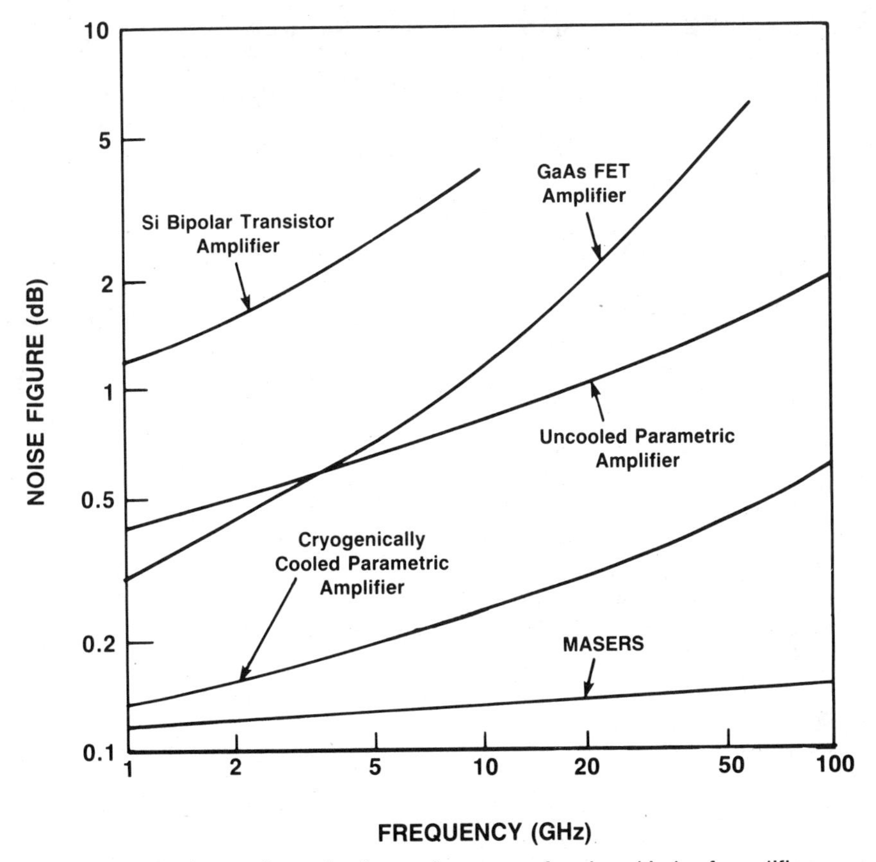

Figure 10.2 Comparison of noise performance of various kinds of amplifiers.

10.2 AMPLIFIER CHARACTERIZATION

Although many characteristics should be considered when designing an amplifier, the most important of these are the power gain, noise figure, stability, input and output VSWR, power output, 1-dB compression point, intermodulation performance, and dynamic range. These parameters are briefly described in this section. The power-added efficiency which is more important for power amplifiers is treated in Sec. 10.5.

10.2.1 Power Gain

The power gain or transducer power gain (G_T) is defined as the ratio of the power delivered to the load to the power available from the source to the network. For a two-port network (shown in Fig. 10.3),

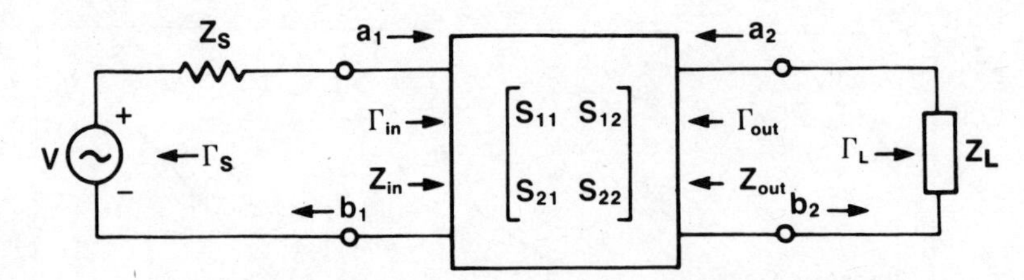

Figure 10.3 Two-port network of a transistor amplifier.

$$G_T = \frac{|S_{21}|^2(1-|\Gamma_S|^2)(1-|\Gamma_L|^2)}{|(1-S_{11}\Gamma_S)(1-S_{22}\Gamma_L)-S_{12}S_{21}\Gamma_S\Gamma_L|^2} \tag{10.1}$$

where

$$\Gamma_{L,S} = \frac{Z_{L,S}-Z_0}{Z_{L,S}+Z_0}$$

The S-parameters are measured with Z_0 as the input and output impedance, and arbitrary source impedance Z_S and load impedance Z_L are connected to the network.

For unilateral transducer power gain, the reverse power gain is set to zero (i.e., $|S_{12}|^2 = 0$), and (10.1) becomes

$$G_{TU} = \frac{|S_{21}|^2(1-|\Gamma_S|^2)(1-|\Gamma_L|^2)}{|(1-S_{11}\Gamma_S)(1-S_{22}\Gamma_L)|^2} \tag{10.2}$$

The maximum unilateral power gain is attained when $\Gamma_S = S_{11}^*$ and $\Gamma_L = S_{22}^*$, that is, when the network is conjugately matched at the input and output ports. Then the maximum unilateral power gain called the maximum available gain is given by

$$G_a = G_{TUm} = \frac{|S_{21}|^2}{(1-|S_{11}|^2)(1-|S_{22}|^2)} \tag{10.3}$$

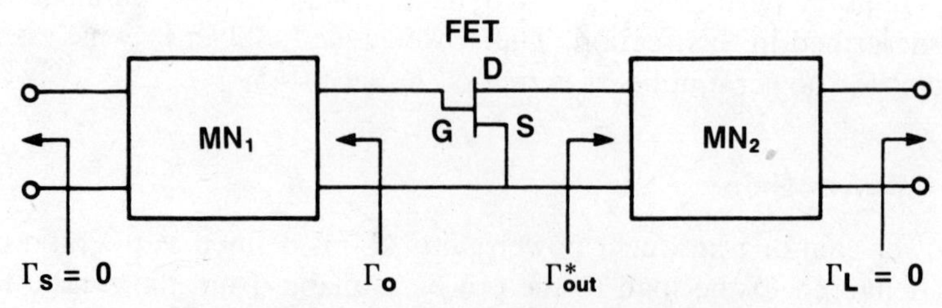

Figure 10.4 FET matched for maximum gain.

Thus the maximum available gain is the product of the transistor transducer power gain $|S_{21}|^2$ between Z_0 (usually 50-Ω) source and load impedances and the increase in gain due to matching the input port $(1-|S_{11}|^2)^{-1}$ and matching the output port $(1-|S_{22}|^2)^{-1}$ as shown in Fig. 10.4. In other words, a single-stage amplifier design consists of (1) designing an input matching network to give $\Gamma_S \simeq 0$ and stability at all frequencies and (2) designing an output matching network that simultaneously gives $\Gamma_L \simeq 0$ and stability.

10.2.2 Noise Characterization

In a microwave amplifier, a small output power can be measured, even when there is no input signal and is known as the amplifier noise power. The total output noise power consists of amplified input noise entering the amplifier plus the noise power generated in the amplifier itself.

The model of a noisy two-port microwave amplifier is shown in Fig. 10.5. The noise input power can be modeled by a noisy resistor. This noise is caused by random motion of electrons in a conductor due to thermal agitation and therefore is known as thermal or Johnson noise. The maximum available noise power, N_R from R_N is

$$N_R = kTB \tag{10.4}$$

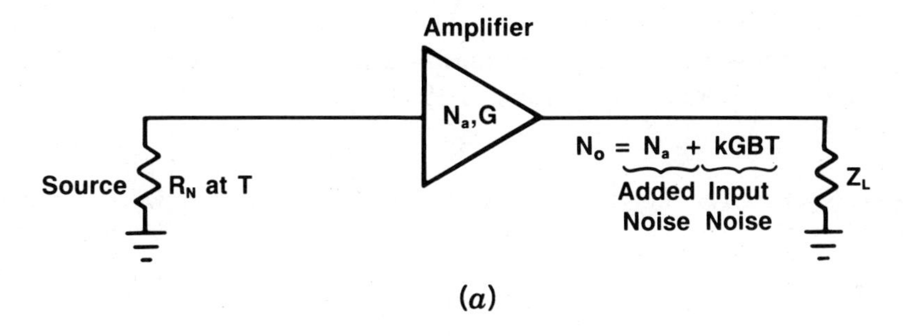

(a)

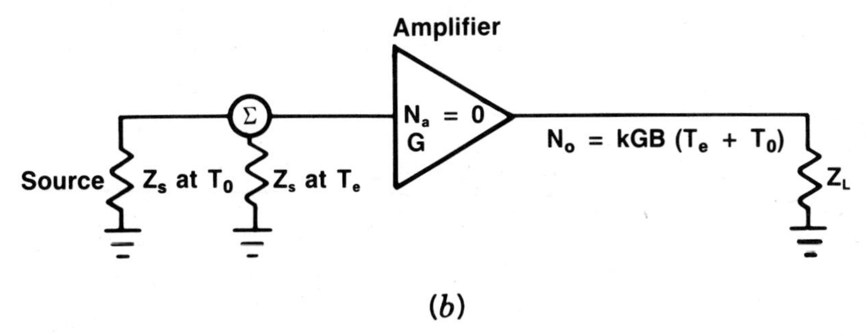

(b)

Figure 10.5 Equivalent noise power and noise temperature representation of an amplifier.

where k is Boltzmann's constant (i.e., $k = 1.38 \times 10^{-23}$ J/K), T is the resistance noise temperature expressed in K, and B is the noise bandwidth. At $B = 1$ GHz, $T = 290$ K, $kTB = -84$ dBm. Note that the available noise power is independent of the magnitude of the resistor value but, of course, the amount of noise power actually delivered to a load resistor will decrease as the ratio of the source and load resistor values varies from unity.

Equation 10.4 shows that the thermal noise power depends on the bandwidth but not on a given frequency. Such a distribution of noise is called "white noise." Obviously no true white noise sources exist because these sources would put out infinite noise power over an infinite bandwidth. Actually the formula in (10.4) breaks down at frequencies well above the millimeter-wave range, where the noise power drops. Over the microwave and millimeter-wave frequencies, most dissipative elements in electrical circuits are very well characterized as ideal kTB noise-power sources.

In addition, most resistors exhibit increased noise at low frequencies. This increase is usually referred to as the $1/f$ or flicker noise. Typically $1/f$ noise does not actually have $1/f$ frequency dependence, but it does increase with decreasing frequency. The frequency at which the $1/f$ power equals kTB is called the $1/f$ knee frequency. Flicker noise is not usually important in the design of microwave amplifiers as the knee frequency for transistors is below 100 MHz, but it is an important source of phase noise in microwave oscillators.

In microwave transistors, in addition to Johnson and $1/f$ noise, shot noise is generated by random passage of charges in the modulating channels. The shot noise is relatively constant from dc to 100 MHz. Noise in transistors is well treated in Reference 19.

Noise Figure. The noise figure of any linear two-port network can be defined as

$$F = \frac{\text{signal-to-noise ratio at input}}{\text{signal-to-noise ratio at output}}$$

$$= \frac{\text{available noise power at output}}{\text{gain} \times \text{available noise power at input}} = \frac{N_0}{GkTB} \tag{10.5}$$

where

N_0 = the available noise power at output
G = the available gain of the network over the bandwidth B

If N_a is the noise power added by the amplifier, then

$$F = \frac{GkTB + N_a}{GkTB} = 1 + \frac{N_a}{GkTB} \tag{10.6}$$

An amplifier that contributes no noise to the circuit has $F = 1$. Now consider an amplifier having n stages cascaded in series with gain values $G_1, G_2, \ldots, G_n$ and noise figure values $F_1, F_2, \ldots, F_n$, then the total noise figure of an n-stage amplifier is given by

$$F = F_1 + \frac{F_2 - 1}{G_1} + \frac{F_3 - 1}{G_1 G_2} + \cdots + \frac{F_n - 1}{G_1 G_2 \cdots G_{n-1}} \tag{10.7}$$

When $F_1 = F_2 = \cdots = F$ and $G_1 = G_2 = \cdots = G$, as $n \to \infty$ the overall noise figure obtained is called the noise measure, F_M, and is given by

$$F_M = \frac{F - 1/G}{1 - 1/G} \tag{10.8}$$

For most applications, noise measure is a much better figure of merit for an amplifier than noise figure because noise measure accounts for the noise contribution of succeeding amplifier stages. An amplifier with low gain does not guarantee good system noise performance, no matter how low the amplifier noise figure may be.

Since Johnson noise is invariant against the resistance and proportional to absolute temperature, it is used to characterize the noise power in units of temperature. Consider the amplifier shown in Fig. 10.5(b) [20]. Typically, the source and load resistors are 50-Ω, but the following definitions apply to arbitrary impedances. The output noise power is given by

$$N_o = Gk(T_0 + T_e)B \tag{10.9}$$

where T_0 is the room temperature (usually $T_0 = 290$ K) and T_e is the equivalent noise temperature of the amplifier with $N_a = 0$. From (10.5) and (10.6) when the source impedance is at T_0,

$$F = 1 + \frac{T_e}{T_0} \tag{10.10}$$

The noise figure is usually expressed in decibels as

$$\text{NF} = 10 \log(F) = 10 \log\left(1 + \frac{T_e}{T_0}\right) \tag{10.11}$$

Example. Consider an example of two amplifiers cascaded in series. If the gain and noise figures of the first amplifier are 9 dB and 2 dB, and 10 dB and 3 dB for the second amplifier, calculate the total noise figure of the cascaded amplifier. What happens to the noise figure if gain of the first amplifier is 15 dB or higher?

The overall noise figure is given by (10.7), that is,

$$F = F_1 + \frac{F_2 - 1}{G_1} \tag{10.12}$$

Here

$$NF_1 = 2\text{ dB}, \qquad F_1 = 1.58$$

$$NF_2 = 3\text{ dB}, \qquad F_2 = 2$$

$$G_1 = 9\text{ dB} = 7.94$$

$$G_2 = 10\text{ dB} = 10$$

$$F = 1.58 + \frac{2-1}{7.94} = 1.7$$

or

$$NF = 10\log(1.7) = 2.3\text{ dB}$$

when

$$G_1 = 15\text{ dB} = 31.6$$

$$F = 1.58 + \frac{2-1}{31.6} = 1.6$$

$$NF = 2.07\text{ dB}$$

Thus, as the gain of first amplifier, G_1, gets larger, the relative importance of the noise contribution for the second amplifier becomes less and less.

Example. Consider an amplifier preceded by an attenuator when they are cascaded in series. If attenuation of the attenuator is L, and the gain and noise figures of the amplifier are G_A and NF_A, and all are expressed in decibels, what is the overall noise figure of the assembly?

This example is very similar to the previous example, where

$$F_1 = 10^{L/10}$$

$$F_2 = 10^{NF_A/10}$$

$$G_1 = 10^{-L/10}$$

$$G_2 = 10^{G_A/10}$$

From (10.12)

$$F = 10^{L/10} + \frac{(10^{NF_A/10} - 1)}{10^{-L/10}}$$

$$F = 10^{L/10} \cdot 10^{\mathrm{NF_A}/10}$$

$$\mathrm{NF} = 10 \log F = L + \mathrm{NF_A}$$

In other words any lossy network at room temperature connected in front of an amplifier increases the noise figure of the amplifier by an amount equal to its loss.

Noise Bandwidth. The total noise power from an amplifier is given by

$$NT = \int_0^\infty T_A(\omega) G_A(\omega)\, d\omega \tag{10.13a}$$

where $T_A(\omega)$ and $G_A(\omega)$ are the noise temperature and power gain of the amplifier, respectively. It is often useful to treat the amplifier as having fixed T_A and G_A over a specific noise bandwidth and no gain elsewhere where the noise bandwidth (NBW) is chosen to make the total noise power correct, that is,

$$\mathrm{NBW} \cdot k \cdot T_A \cdot G_A = k \int_0^\infty T_A(\omega) G_A(\omega)\, d\omega$$

or

$$\mathrm{NBW} = \frac{\int_0^\infty T_A(\omega) G_A(\omega)\, d\omega}{T_A G_A} \tag{10.13b}$$

The range of ω is typically limited by other components in the system, or by the gain response of the amplifier.

Optimum Noise Match. In general, any noisy two-port network may be represented by a noise voltage and a noise current source connected at the input of a noiseless two-port network as shown in Fig. 10.6. If the circuit has a dominant voltage noise, using a high source impedance will minimize the transmission noise signal, but if the current noise is dominant, connecting a low source impedance will minimize the transmission of the noise

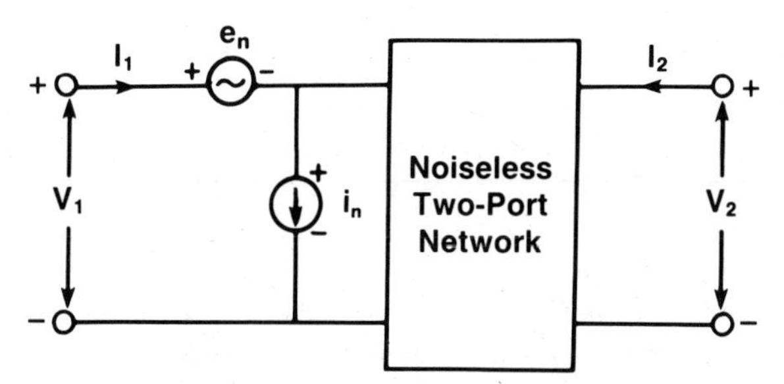

Figure 10.6 Noise equivalent circuit of a two-port network.

signal. When both noise sources are present, a minimum noise figure of the circuit results in a specific source admittance or impedance, known as the optimum source admittance. Circles of constant noise figure on the input admittance or impedance plane can be plotted using a Smith chart. How the noise figure increases from the minimum value is described by the following relation:

$$F = F_{\min} + \frac{R_n}{G_S} | Y_S - Y_O|^2$$

or

$$F = F_{\min} + \frac{R_n}{G_S} [(G_S - G_O)^2 + (B_S - B_O)^2] \tag{10.14}$$

where

F = noise figure,
$Y_S = G_S + jB_S$ = source admittance,
$F_{\min}$ = minimum noise figure,
$Y_O = G_O + jB_O$ = optimum source admittance that gives minimum noise figure,
R_n = equivalent noise resistance.

The lower the equivalent noise resistance, the less the sensitivity of noise figure increase for a nonoptimum source. The load impedance does not affect the noise figure of the stage; it does affect the gain of the stage.

We can express Y_S and Y_O in terms of the reflection coefficient Γ_S and Γ_O, and the resultant relation becomes

$$F = F_{\min} + \frac{4\bar{R}_n |\Gamma_S - \Gamma_O|^2}{(1 - |\Gamma_S|^2)|1 + \Gamma_O|^2} \tag{10.15}$$

where $\bar{R}_n = R_n / Z_0$ (Z_0 is usually 50 Ω). The quantities $F_{\min}$, R_n, and Γ_O are known as the noise parameters, and are given by the manufacturer of the transistor or can be determined experimentally.

10.2.3 Stability

Any amplifier with power gain can be made to oscillate by applying external positive feedback. At microwave frequencies unavoidable parasitics are often sufficient to cause oscillations if care is not taken in the design and fabrication of the amplifier. Any very abrupt change in the dc parameters of the amplifier, output power with no input power, circuits that are very sensitive to their surroundings, etc., are typical indications of oscillations. Oscillations may occur at frequencies that don't propagate out the amplifier because they are filtered, blocked by bias capacitors, or are below waveguide cutoff frequencies or at frequencies to which the test equipment is insensitive. It is not unusual for microwave amplifiers to oscillate anywhere between 1 MHz and 30 GHz or higher.

The best single test for oscillations involves using a combination of a broadband spectrum analyzer and sweep oscillator. By watching the broadband output spectrum as the input signal is swept, the absence of both fundamental and intermodulation oscillation signals is usually sufficient to assure stable operation. It must be emphasized that any variable change such as impedance levels, supply voltages, temperature, light intensity aging, and radiation, may quench or create oscillations. Testing often must be done over a broad range of parameters if confidence is to be obtained that the amplifier is stable under those conditions. Fortunately design techniques can be applied to build in confidence of stability.

Stability against oscillations can be examined by using two-port S-parameters. The parameters S_{12} and S_{21} form a feedback loop that, depending on the source and load impedances, may support oscillations. In an ideal amplifier, S_{12} would be zero and the amplifier would be unconditionally stable. If $S_{12} \neq 0$, input reflection coefficient Γ_{in} with arbitrary Z_L and output reflection coefficient Γ_{out} with arbitrary Z_S can be expressed as

$$\Gamma_{\text{in}} = S_{11} + \frac{S_{12}S_{21}\Gamma_L}{1 - S_{22}\Gamma_L} \qquad (10.16)$$

$$\Gamma_{\text{out}} = S_{22} + \frac{S_{12}S_{21}\Gamma_S}{1 - S_{11}\Gamma_S} \qquad (10.17)$$

If the circuit is unconditionally stable, any source or load may be connected to the input or output of the circuit without oscillations. In terms of S-parameters of the FET, unconditional stability is assured if the following inequalities are simultaneously satisfied

$$|S_{11}| < 1 \quad \text{and} \quad |S_{22}| < 1$$

$$|\Gamma_{\text{in}}| < 1 \quad \text{and} \quad |\Gamma_{\text{out}}| < 1$$

For $|\Gamma_S| < 1$ and $|\Gamma_L| < 1$, these conditions lead to the following requirement

$$K = \frac{1 + |S_{11}S_{22} - S_{12}S_{21}|^2 - |S_{11}|^2 - |S_{22}|^2}{2|S_{12}||S_{21}|} > 1 \qquad (10.18)$$

The significance of the stability factor K is that a FET is unconditionally stable for all passive source and load terminations when $K > 1$.

If we set $|\Gamma_{\text{in}}|$ and $|\Gamma_{\text{out}}|$ equal to unity, a boundary is established beyond which the device is unstable. Each condition will give a solution of a circle on a complex reflection plane whose radius (r) and center (c) are given by

$$\left.\begin{aligned} r_S &= \frac{|S_{12}S_{21}|}{||S_{11}|^2 - |D|^2|} \\ c_S &= \frac{(S_{11} - DS_{22}^*)^*}{|S_{11}|^2 - |D^2|} \end{aligned}\right\} \text{input} \tag{10.19}$$

$$\left.\begin{aligned} r_L &= \frac{|S_{12}S_{21}|}{||S_{22}|^2 - |D|^2|} \\ c_L &= \frac{(S_{22} - DS_{11}^*)^*}{|S_{22}|^2 - |D|^2} \end{aligned}\right\} \text{output} \tag{10.20}$$

where $D = S_{11}S_{22} - S_{12}S_{21}$, S and L denote source and load, and the origin of the Smith chart is at $\Gamma_{\text{in}} = \Gamma_{\text{out}} = 0$. Figure 10.7 shows typical examples of the input plane of unconditionally stable and conditionally stable networks. The shaded area represents the area of the input plane in which instability occurs. If the circuit is potentially unstable ($K < 1$), the source and load impedances should be chosen so they will not fall into the unstable region (e.g., shaded) due to device parameter changes, fabrication variation, and changes in temperature. Under such conditions, the amplifier is said to be conditionally stable and will not oscillate. Stability analysis must be carried out from dc to above the frequency where the active devices have power gain.

If the circuit is unconditionally stable ($K > 1$), the conditions required to obtain maximum power gain are $\Gamma_{\text{in}} = \Gamma_S^*$ and $\Gamma_{\text{out}} = \Gamma_L^*$. Solving for simultaneous conjugate match, the reflection coefficients to be matched are denoted by Γ_{SM} and Γ_{LM} and are given by [16, 21]

$$\Gamma_{SM} = \frac{B_1 \pm \sqrt{B_1^2 - 4|C_1|^2}}{2C_1} \tag{10.21}$$

$$\Gamma_{LM} = \frac{B_2 \pm \sqrt{B_2^2 - 4|C_2|^2}}{2C_2} \tag{10.22}$$

where

$$B_1 = 1 + |S_{11}|^2 - |S_{22}|^2 - |D|^2$$

$$B_2 = 1 + |S_{22}|^2 - |S_{11}|^2 - |D|^2$$

$$C_1 = S_{11} - DS_{22}^*$$

$$C_2 = S_{22} - DS_{11}^*$$

If $|B_1/2C_1| > 1$ and $B_1 > 0$, the solution with the minus sign produces $|\Gamma_{SM}| < 1$ and the solution with the plus sign produces $|\Gamma_{SM}| > 1$. If

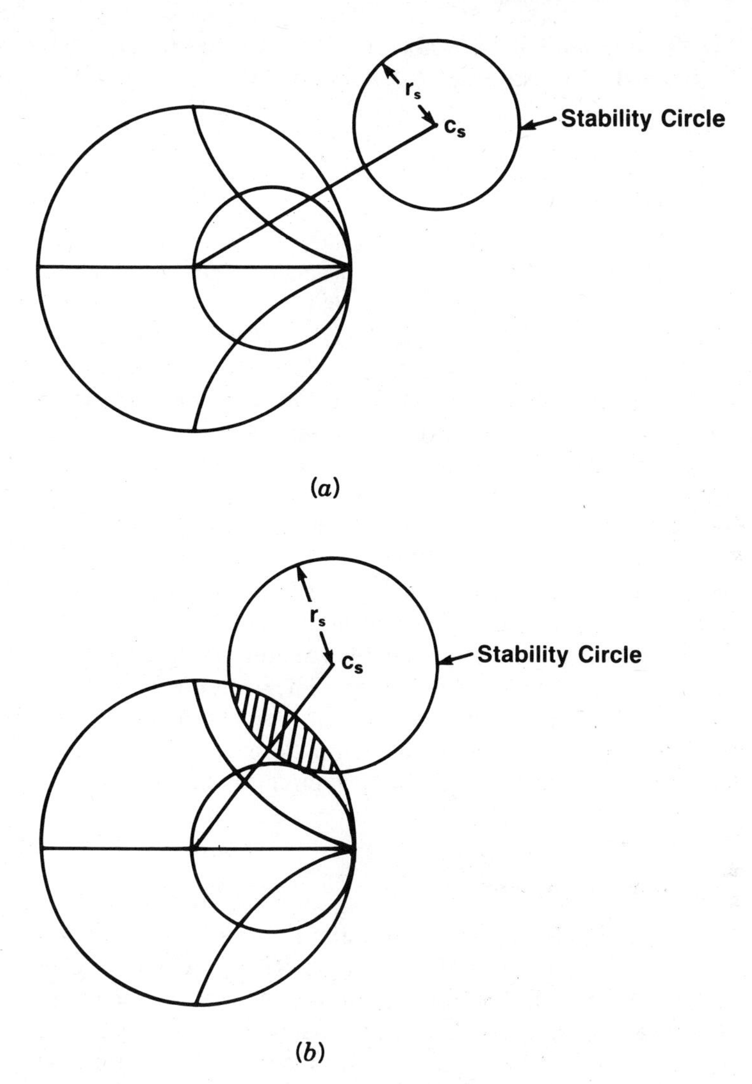

Figure 10.7 Examples of (*a*) unconditional stability and (*b*) conditional stability for $|S_{11}| < 1$.

$|B_1/2C_1| > 1$ with $B_1 < 0$, the solution with the plus sign produces $|\Gamma_{SM}| < 1$ and the solution with the minus sign produces $|\Gamma_{SM}| > 1$. Similar considerations apply to Γ_{LM}.

The value of the matched gain (MG) is

$$\mathrm{MG} = \left|\frac{S_{21}}{S_{12}}\right| (K - \sqrt{K^2 - 1}) \tag{10.23}$$

If the circuit is potentially unstable ($K < 1$), the maximum stable gain (MSG) is obtained by substituting $K = 1$ in (10.23)

$$\text{MSG} = \left|\frac{S_{21}}{S_{12}}\right| \tag{10.24}$$

Once a stability problem has been identified, it is usually not difficult to correct it, for example, by adding extra bias capacitance, putting resistances in the bias lines, or by adding inductor, capacitor, resistor, or transmission-line terminations for problem frequencies. In most microwave system applications conditional stability is usually acceptable in the frequency band where component impedances are defined, but unconditional stability may be required out of band where there are usually no system specifications.

Ferrite isolators are often used to protect amplifiers from unknown impedances. Care must be taken, since stability problems may arise far outside the isolator frequency band where the isolator is neither matched or unilateral.

The two-port stability analysis just described is applicable only to the sensitivity of the two ports to external impedances, not to internal oscillations. For example, a multistage amplifier might oscillate even though its overall K factor is greater than unity. The stability analysis must be performed on all internal two ports having active devices and on the complete circuit. A series of stable two-amplifier stages will be stable, but if additional feedback is introduced, the new circuit must also be analyzed.

10.2.4 Nonlinear Behavior

An amplifier is called linear when the output power increases linearly with the input power. The ratio of these two powers is the power gain G. As input power increases, the amplifier transfer function becomes nonlinear, that is, the output power is lower than predicted by the small-signal gain. This nonlinear behavior in amplifiers introduces distortion in the amplified signal. The output power at which the gain has dropped by 1 dB below the linear gain is called the *1-dB compression point*, P_{1dB}. Typically the gain will drop rapidly for powers above P_{1dB}. reaching a maximum or fully saturated output power within 3 to 4 decibels above P_{1dB}.

There are a number of different ways to measure the nonlinearity behavior of an amplifier. The simplest method is the measurement of the 1-dB compression power level P_{1dB}. Another method that is very popular uses two closely spaced signals 5 to 10 MHz apart. When two signals at frequencies, f_1 and f_2 are incident on an amplifier, the output of the amplifier contains these two signals, as well as intermodulation products at frequencies $mf_1 + nf_2$, where $m + n$ is known as the order of intermodulation (IM) product. Figure 10.8 illustrates the intermodulation

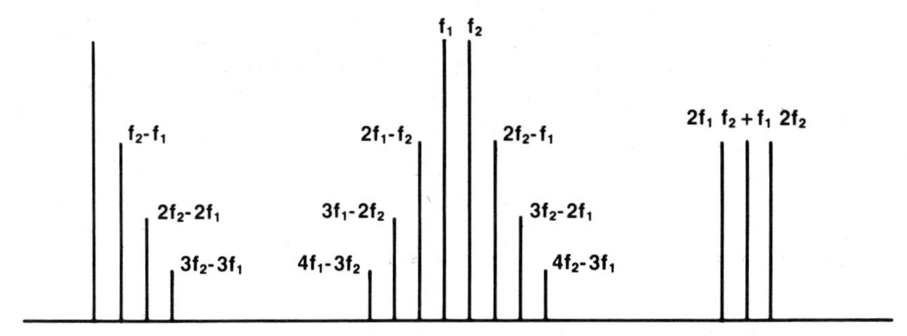

Figure 10.8 A schematic spectrum showing two signals at frequencies f_1 and f_2 and their intermodulation products.

products spectrum. The relative magnitude of the intermodulation products depends on the details of how the amplifier saturates, however, third order products which are closest in frequency to the fundamental tones usually dominate at moderate saturation levels, although second order products can be important in multioctave amplifiers.

Figure 10.9 shows the transfer characteristics of a typical solid state amplifier having a 10-dB signal gain. With proper filtering, one of the fundamental frequencies and the distortion products are measured

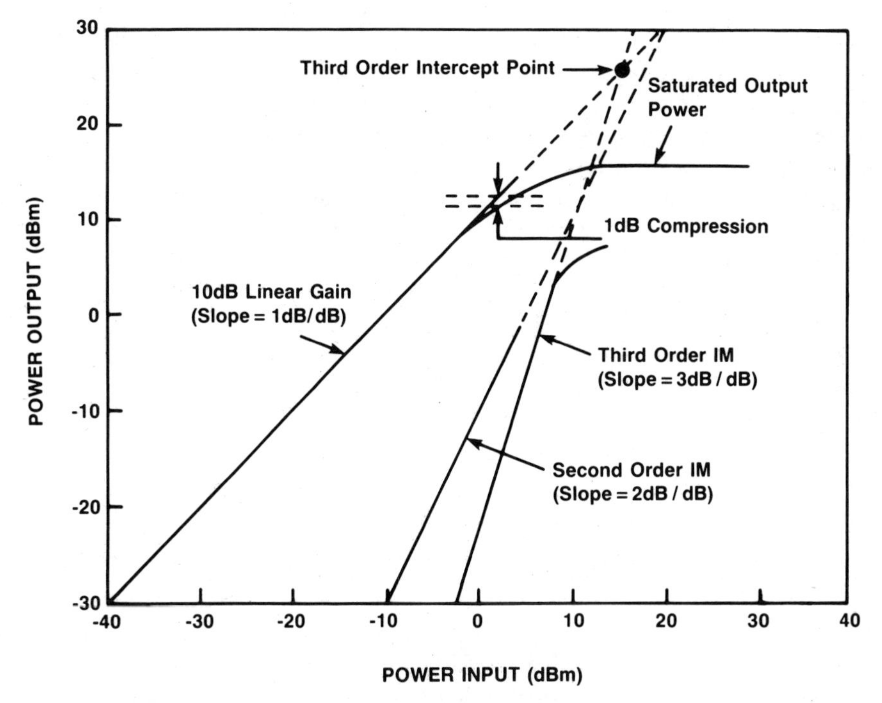

Figure 10.9 The variation of output power and intermodulation products with input power for a nonlinear amplifier.

separately. Output powers of the second- and third-order IM products are also shown in Fig. 10.9. Since second- and third-order IM products correspond to square and cubic nonlinearities, respectively, the output power for these products increases by 2 dB/dB and 3 dB/dB at low power levels. Since the third-order terms normally dominate up to the point where distortion is very severe, intermodulation distortion is often characterized by the third-order intercept, P_{3rd} as shown in Fig. 10.9. Intercept P_{3rd} is the point where the power in the third-order product and fundamental tone are equal when the amplifier is assumed to be linear. The third-order intercept power is typically 10 to 12 dB above the P_{1dB} and is a very useful parameter for calculating low level intermodulation effects.

Amplitude modulated (AM) signals are strongly attenuated by saturated amplifiers, but frequency modulated (FM) signals are often passed through power amplifiers running at the point of maximum efficiency, which is usually above P_{1dB}. Although an FM signal is less subject to distortion by gain compression than an AM signal, both are subject to distortion due to harmonic generation in nonlinear devices.

The simplest form of distortion occurs when a sinusoidal input is converted to a square-wave output by power-supply clipping. Under these conditions, the output signal will contain not only the fundamental frequency f but also the odd harmonics $(3f, 5f, \ldots)$ of the input signal. These harmonics will be more or less visible at the output, depending on the frequency response of the amplifier. If, for example, the output includes a low-pass filter that passes f but is cutoff for $3f$, the output from a single frequency input may be compressed but still be purely sinusoidal. If clipping is unsymmetrical, even as well as odd harmonics may be generated, but if filtering is applied to attenuate the $2f$ signal, the output will remain purely sinusoidal. It is the necessity to filter out harmonic distortion and intermodulation products that dictates that most microwave systems are channeled into frequency bands narrow enough to eliminate undesirable components. Such techniques work well for signal frequency inputs, but unfortunately most systems are confronted with multiple frequency signals that present more difficult design challenges to reduce or eliminate distortion.

10.2.5 Dynamic Range

The range of an input signal that can be detected by a receiver without much distortion is called the dynamic range (DR). The dynamic range of an amplifier is defined as the ratio of the 1-dB compressed power output to the amplified minimum detectable signal.

The output noise of a two-port device with noise figure F can be written from (10.5) as

$$N_o = FGkTB \tag{10.25}$$

If minimum detectable input signal is X(dB) above the noise floor, then

$$P_{\text{in}}^{\min} = N_o + X \qquad \text{(dB)} \tag{10.26}$$

$$P_{\text{out}}^{\min} = P_{\text{in}}^{\min} + G \qquad \text{(dB)} \tag{10.27}$$

Dynamic range is defined as

$$\text{DR} = P_{1\text{dB}} - P_{\text{out}}^{\min} \tag{10.28}$$

If a typical value of X is 3 dB, then from (10.25)–(10.28)

$$\text{DR} = P_{1\text{dB}} + 171 - 10\log(B) - F - G \qquad \text{(dB)} \tag{10.29}$$

where $P_{1\text{dB}}$ is in decibels above 1 mW (0 dBm).

Example. Determine the dynamic range of a low-noise amplifier with a gain of 30 dB, a noise figure of 2 dB, a 1-dB compression point of 15 dBm, and a noise bandwidth of 1 GHz.

From (10.29),

$$\text{DR} = 15 + 171 - 90 - 2 - 30 = 64\ \text{dB}$$

The spurious free dynamic range (DR_f) of an amplifier is defined as the ratio of the fundamental signal power output to the third-order intermodulation product power output, when the third-order intermodulation product is equal to the minimum detectable output signal. The spurious free dynamic range in decibels is given by [16]

$$\text{DR}_f = \tfrac{2}{3}[P_{3\text{rd}} - P_{\text{out}}^{\min}] \tag{10.30}$$

10.3 BIASING NETWORKS

Biasing of FETs is an important part of an amplifier design. The design considerations for biasing circuits are efficiency, noise, oscillation suppression, single source power supply, RF choking and impedance matching. Five practical biasing configurations are given in Table 10.1. The circuits (a), (b), and (e) require two power supplies of opposite polarity. Since in these configurations the source is grounded with minimum possible source inductance, they provide maximum gain. In these circuits, gate voltage is applied before the drain voltage. Biasing circuits (c) and (d) with source resistors are widely used for small- and medium-power applications and require only one power supply. As the supply voltage is applied, the gate is simultaneously reverse biased with respect to source by the series resistor R_s. The value of R_s is selected based on drain–source current I_{DS} and the

TABLE 10.1 Various Bias Schemes

Biasing Configuration	Typical Bias Voltages	Amplifier Characteristics	Other Comments
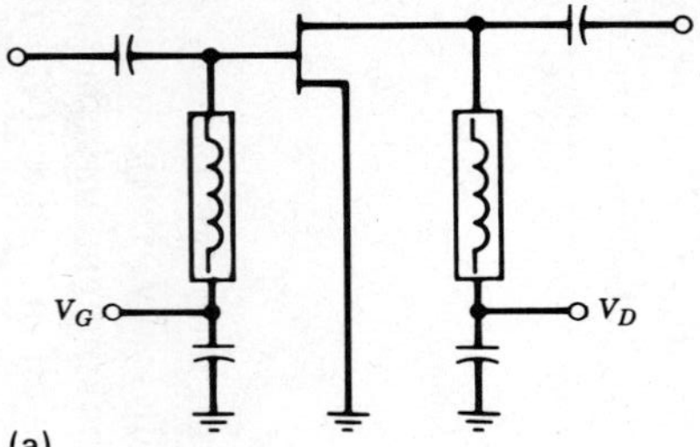 (a) V_G, V_D	$V_D = 3$ V $V_G = -1$ V	Low noise High gain High power High efficiency	Biasing networks are part of matching Insensitive to bias current
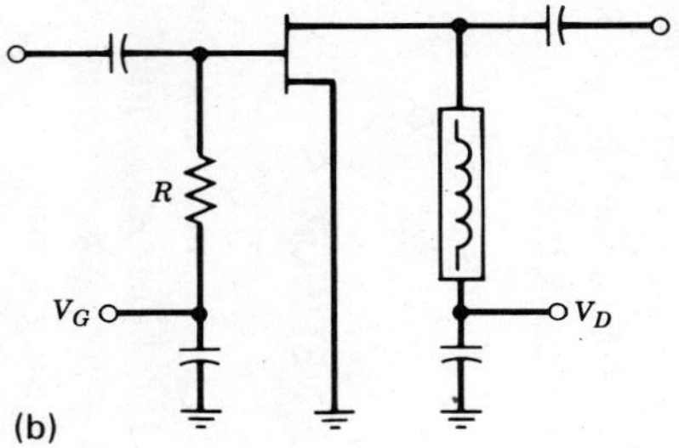 (b) R, V_G, V_D	$V_D = 3$ V $V_G = -1$ V	Moderately low noise High gain Highpower High efficiency	Biasing network is part of matching Insensitive to bias current High value of R provides higher isolation between gate and power supply

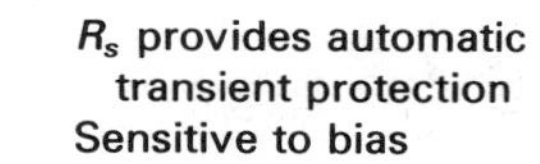

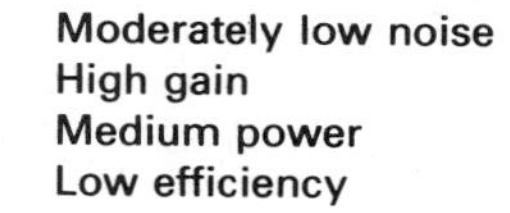
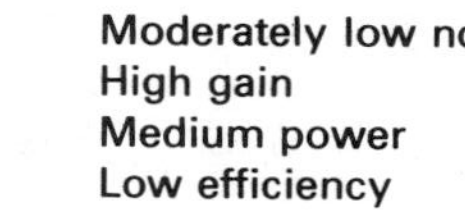

	Circuit	Bias	Characteristics	Remarks
(c)	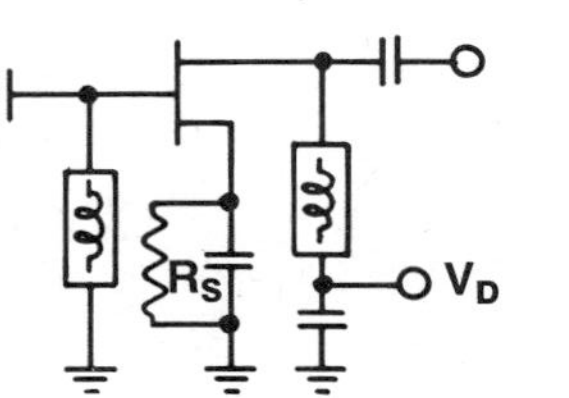	$V_D = 4$ V $I_{DS}R_S = 1$ V	Moderately low noise High gain Medium power Low efficiency	R_s provides automatic transient protection Sensitive to bias current
(d)	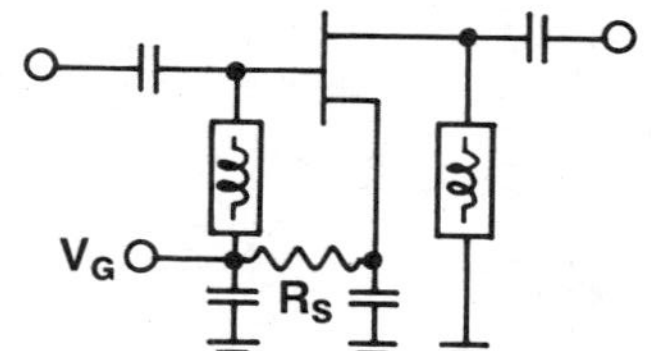	$V_G = -4$ V $I_{DS}R_S = 1$ V	Moderately low noise High gain Medium power Low efficiency	R_s provides automatic transient protection Sensitive to bias current
(e)	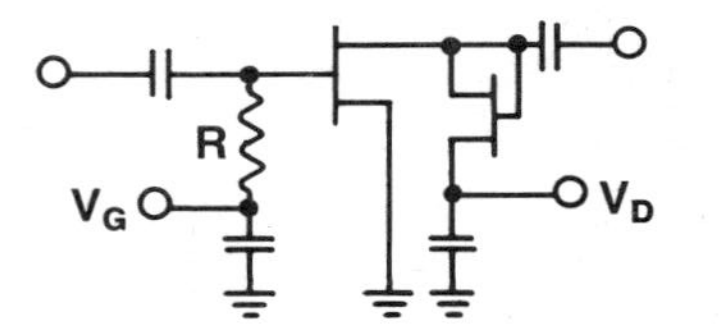	$V_D = 5$ V $V_G = -1$ V	Moderate noise High gain Medium power Low efficiency	Broadband at lower frequencies Sensitive to bias voltage

operating bias. A 50-pF bypass capacitor is usually sufficient for RF grounding and transient protection. These circuits have lower amplifier efficiency due to dc power dissipation in source resistors. Thus for power amplifiers where $I_{DS} > 500$ mA, dual polarity supplies are recommended with a sequencing circuit to bias the gate first before applying the drain bias. This sequencing is required because V_D applied at I_{DSS} (drain-source saturation current) may burn the device.

10.4 LINEAR AMPLIFIER DESIGN

In this section we discuss aspects of the design of low-noise and maximum gain GaAs FET amplifiers. Although there are many FET configurations that may be used for amplification, we concentrate on common-source amplifiers as shown in Fig. 10.10. The common source configuration has high gain, low noise, and the best amplifier stability. Low-noise amplifiers are designed to increase signal levels while introducing a minimum amount of signal-to-noise degradation.

The design of an amplifier for minimum noise figure involves several considerations. First and foremost, one must select the proper FET. For low-noise applications, the FETs are biased at relatively low current levels (typically 20% of I_{DSS}, 3-V drain–source). At these bias levels, the FET's power-handling capability is much less than that at higher bias (typically $P_{1\text{dB}} = 10$–12 dBm for a 300-μm-wide gate FET. The input and output matching networks, when designed for best noise performance, match the output (FET drain) to the load, but introduce considerable mismatch at the

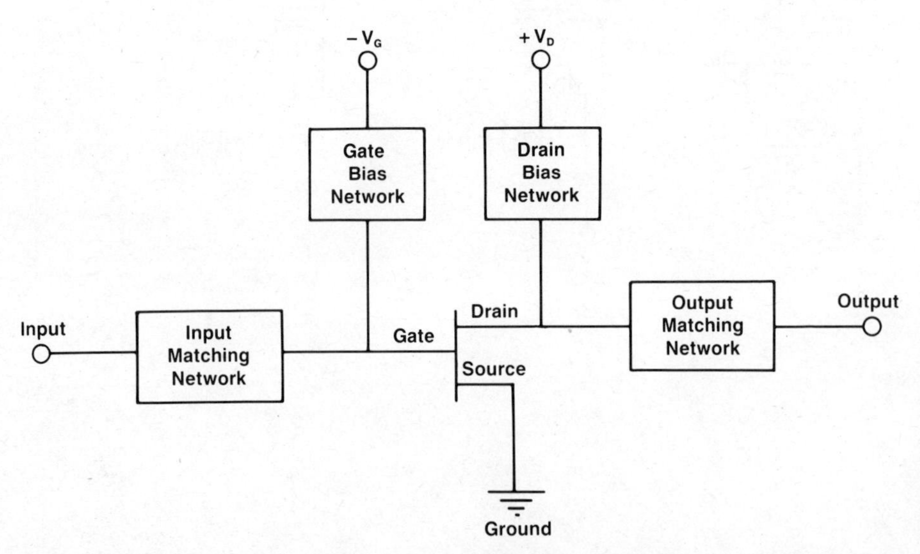

Figure 10.10 A single-stage common-source MESFET amplifier block diagram.

TABLE 10.2 Typical Low-Noise FETs

FET	Gate Length (μm)	$f = 4$			$f = 10$			$f = 18$			Package Type	Company
		NF (dB)	G_A (dB)	MSG (dB)	NF (dB)	G_A (dB)	MSG (dB)	NF (dB)	G_A (dB)	MSG (dB)		
AT8250	0.5	0.9	12.5	—	1.8	8.0	—	—	—	—	70 mil	Avantek
AT10600	0.5	1.5	14.5	—	1.7	10.5	—	2.6	6.0	7.5	Chip	Avantek
MA4F601-298	0.5	0.8	14.0	17.0	1.6	10.0	12.0	3.5	5.0	6.0	70 mil	M/A Com
NE21800	1.0	1.0	13.5	17.0	2.2	9.0	9.5	—	—	—	Chip	NEC
NE21889	1.0	1.0	13.5	16.5	2.2	9.0	9.5	—	—	—	Strip line	NEC
NE13700	0.5	0.7	14.0	17.0	1.5	10.0	12.0	2.5	7.5	8.0	Chip	NEC
NE67300	0.3	0.4	14.5	—	1.2	10.5	12.5	1.9	8.0	8.5	Chip	NEC

input (FET gate), with the result that the gain is usually considerably less than the maximum available gain. Finally for best noise (also known as optimum noise) performance, it is critical to minimize circuit loss on the input side, since loss adds directly to the amplifier noise figure.

10.4.1 FET Selection

The selection of a FET for a particular application is straightforward but sometimes tedious. One simply simulates the performance of the available FETs for the particular application and then selects the optimal FET based on the relevant trade-offs, which may include

- cost,
- reliability,
- electrical performance.

Table 10.2 lists a few of the typical FETs available commercially as of late 1986. The most important dimensions of a FET are the FET gate length and total gate periphery. Typical gate lengths are 0.3 to 1.0 μm. The FET gate is usually made up of several fingers. A typical FET might have 4 fingers, each 75 μm wide, making a total gate periphery of 300 μm, as shown in Fig. 10.11.

The electrical parameters available in vendor catalogs include S-parameters at two bias conditions (generally at 10 mA and $\frac{1}{2}I_{DSS}$), optimum reflection coefficient for minimum noise figure, equivalent noise resistance R_n, minimum noise figure, associate gain, and maximum available gain at various frequencies. Typical electrical parameters for a 300-μm-gatewidth FET are shown in Tables 10.3 and 10.4. The maximum frequency f_T (at which the FET has current gain unity) is a useful figure of merit for a FET because it defines an approximate maximum frequency $[f_T = g_m/(2\pi C_{GS})]$. The value of transconductance g_m is generally available or can be obtained from S-parameters along with C_{GS}. In addition to transconductance and C_{GS} at microwave frequencies, a FET exhibits many parasitic elements that reduce electrical performance and must be accounted for in the design. An equivalent lumped-element model for a FET as shown in Fig. 10.11 is very handy and can be obtained by computer optimization to replicate the measured S-parameters.

10.4.2 Narrow-Band Low-Noise Design

The design of a single-state narrow-band low-noise amplifier can be carried out step by step as follows:

1. Select a GaAs FET with a noise figure lower and a gain higher than the design value.

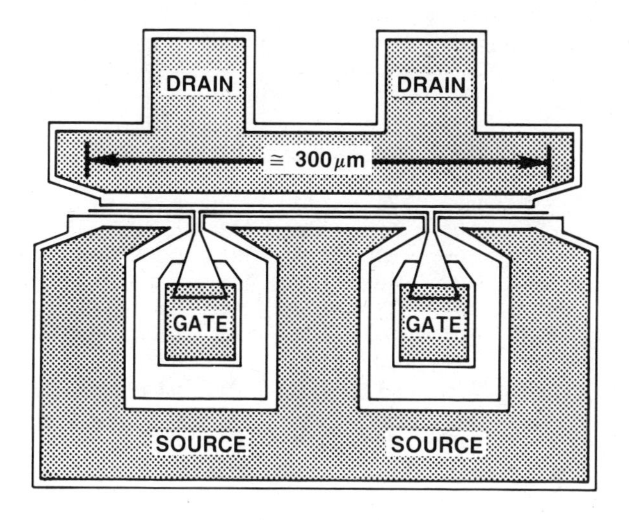

(a)

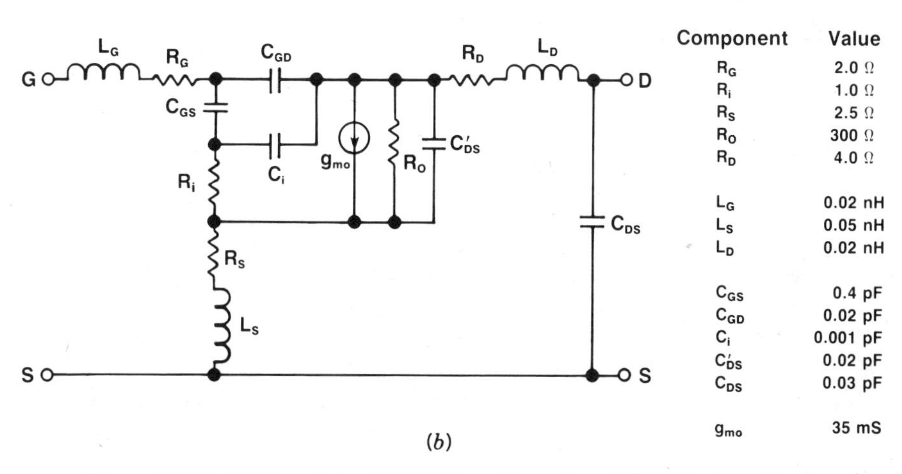

Component	Value
R_G	2.0 Ω
R_i	1.0 Ω
R_S	2.5 Ω
R_O	300 Ω
R_D	4.0 Ω
L_G	0.02 nH
L_S	0.05 nH
L_D	0.02 nH
C_{GS}	0.4 pF
C_{GD}	0.02 pF
C_i	0.001 pF
C'_{DS}	0.02 pF
C_{DS}	0.03 pF
g_{mo}	35 mS

(b)

Figure 10.11 (*a*) Four-finger MESFET configuration. (*b*) Equivalent circuit.

2. Calculate its stability factor K.
3. If $K > 1$, select suitable input and output matching networks that include biasing circuitry and complete their design.
4. If $K < 1$, plot the regions of instability on the reflection planes and select matching networks that avoid the unstable regions.
5. Calculate the amplifier performance using analytical methods or CAD tools. Check stability of the amplifier in the band as well as outside the band.
6. Work out the realization of the amplifier.

TABLE 10.3 ***S*-Parameters of a FET**

					$V_D = 3\text{ V},\ I_{DS} = 10\text{ mA}$					
Frequency (GHz)	$\lvert S_{11}\rvert$	$\angle S_{11}$	$\lvert S_{12}\rvert$	$\angle S_{12}$	$\lvert S_{21}\rvert$	$\angle S_{21}$	$\lvert S_{22}\rvert$	$\angle S_{22}$	K	MSG
2.0	0.98	−30	0.02	75	2.63	160	0.73	−7	0.18	21.2
4.0	0.93	−56	0.04	62	2.36	142	0.71	−13	0.35	17.7
6.0	0.87	−78	0.05	53	2.05	128	0.69	−19	0.53	16.1
8.0	0.83	−96	0.06	46	1.77	116	0.68	−24	0.72	14.7
10.0	0.79	−111	0.07	42	1.54	106	0.67	−28	0.91	13.4
12.0	0.77	−122	0.07	40	1.36	98	0.67	−32	1.10	11.1
14.0	0.75	−132	0.07	40	1.20	92	0.66	−37	1.29	9.2
16.0	0.74	−140	0.07	41	1.08	86	0.66	−41	1.48	7.9
18.0	0.73	−148	0.07	43	0.98	81	0.66	−45	1.67	6.8
20.0	0.72	−154	0.07	46	0.90	76	0.66	−49	1.83	5.9

Now we describe an example to illustrate the procedure. Design a low-noise single-stage amplifier using a microstrip on 0.25-mm-thick alumina substrate ($\epsilon_r = 9.9$) with the following specifications:

Frequency:	12 GHz,
Bandwidth:	5%,
Gain (min):	8 dB,
Noise figure (max):	2.5 dB,
Output VSWR (max):	1.2:1.

The FET listed in Tables 10.3 and 10.4 satisfies the requirements and $K > 1$. A simple amplifier configuration consisting of two matching elements at the input and two at the output as shown in Fig. 10.12 can be selected. At 12 GHz, the input matching circuit must transform 50 Ω to $19.1 + j26.8\ \Omega$, and the output matching circuit must transform 50 Ω to $63.9 + j107.6\ \Omega$. These matching circuit elements can be determined using

TABLE 10.4 Characteristics of a Low-Noise FET

			$V_D = 3\text{ V},\ I_{DS} = 10\text{ mA}$			
Frequency (GHz)	Minimum NF (dB)	G_A (dB)	Γ_{opt}	R_n	Z_S^* (Ω)	Z_L^* (Ω)
4	0.73	14.8	$0.66\angle 44$	13.1	$58.9 + j95.1$	$113.8 + j115.9$
8	1.44	11.1	$0.54\angle 86$	12.7	$29.0 + j44.6$	$86.0 + j114.7$
12	2.11	8.7	$0.55\angle 118$	13.1	$19.1 + j26.8$	$63.9 + j107.6$
16	2.75	6.8	$0.60\angle 140$	12.6	$14.2 + j17.1$	$47.7 + j99.0$
20	3.33	5.4	$0.64\angle 155$	13.6	$11.5 + j10.7$	$37.3 + j89.7$

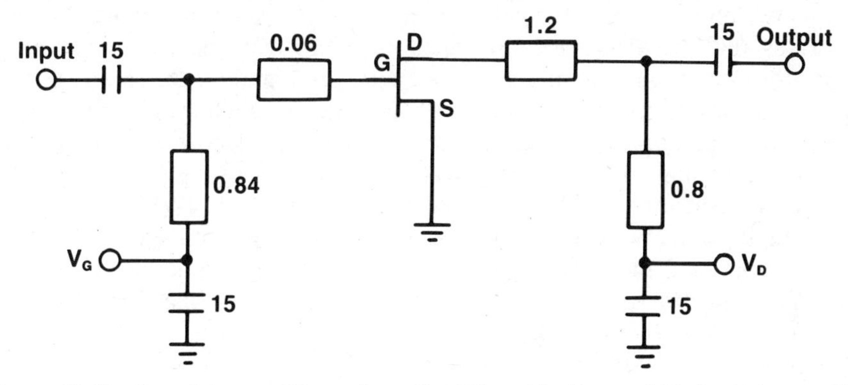

Figure 10.12 Complete amplifier schematic. Microstrip lines width is 0.11 mm. All dimensions are in millimeters. Capacitances are in picofarads.

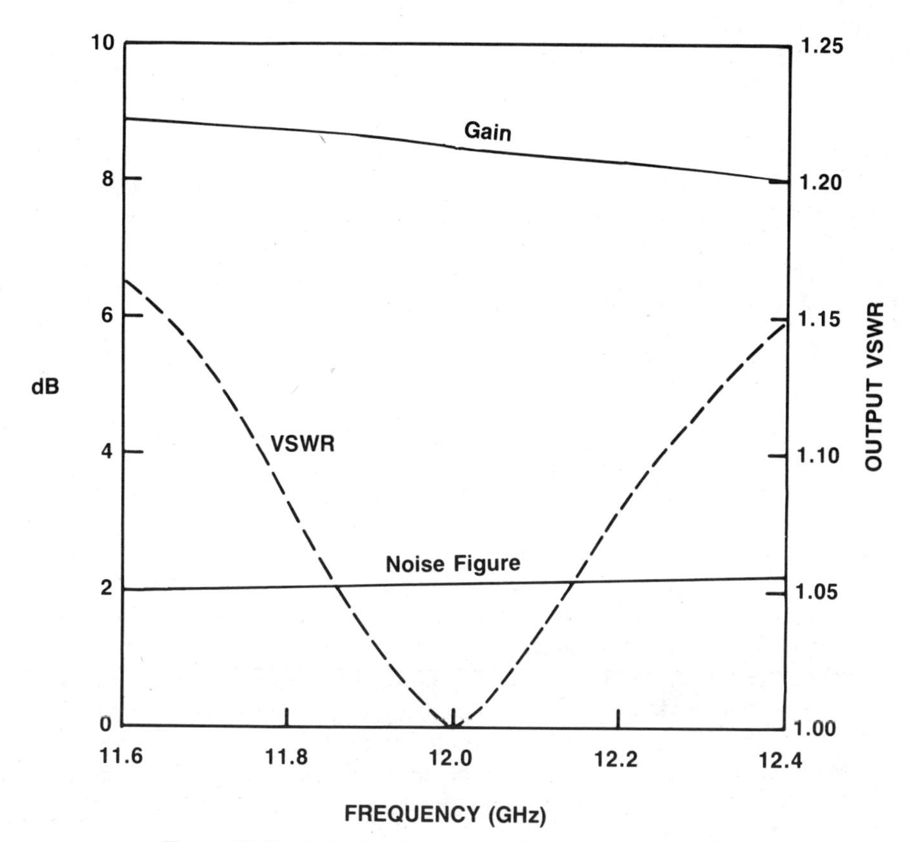

Figure 10.13 Calculated response of a low-noise amplifier.

either a Smith chart or the analytical methods described in Chapter 4. Here 70 Ω as the characteristic impedance of microstrip lines has been used. Physical dimensions for matching elements on alumina substrate ($\epsilon_r = 9.9$, $h = 0.25$ mm) are given in Fig. 10.12. In order to neglect the effect of dc blocking and bypass capacitors, their values are chosen so that their reactance values are less than 1 Ω at the lowest operating frequency. The calculated performance of this amplifier is shown in Fig. 10.13.

In multistage low-noise amplifiers, the input port is matched for optimum noise figure, and the interstages and output port are matched for maximum and flat gain. In this case, usually all FETs except in the first stage are biased approximately at half I_{DSS}.

10.4.3 Maximum Gain Amplifier Design

The design procedure for a maximum gain amplifier is the same as described for the low-noise, except that all the stages are matched for maximum gain. The following example illustrates the design of a maximum gain amplifier.

Example. Design a maximum gain amplifier at 14 GHz with a gain of 8.5 dB and VSWR better than 1.5.

The FET listed in Table 10.3 satisfies the requirements, and $K = 1.29$. Equations 10.21 and 10.22 can be used to compute the simultaneously conjugately matched source and load impedances for maximum gain as follows:

$$C_1 = 0.451\underline{/42^\circ}, \qquad C_2 = 0.325\underline{/134^\circ}$$

$$B_1 = 0.9119, \qquad B_2 = 0.6634$$

$$\text{MG} = 9.2 \text{ dB}$$

$$\Gamma_{SM} = 0.8624\underline{/136^\circ}, \qquad \Gamma_{LM} = 0.8148\underline{/46^\circ}$$

$$Z_{SM} = 4.2 + j19.1\ \Omega, \qquad Z_{LM} = 31.6 + j110.2\ \Omega$$

$$Y_{SM} = 0.0111 - j0.0499\text{ S}, \qquad Y_{LM} = 0.0024 - j0.0084\text{ S}$$

A simple amplifier configuration consisting of two matching elements at the input and two at the output as shown in Fig. 10.14 can be selected. At 14 GHz, the input matching circuit must transform 50 Ω to $4.2 + j19.1$ Ω, and the output matching circuit must transform 50 Ω to $31.6 + j111.2$ Ω. The line lengths for matching circuit elements using 50-Ω lines were determined using a Smith chart. The physical dimensions for microstrip sections on 0.25-mm-thick alumina substrate ($t = 6\ \mu$m and $\epsilon_r = 9.9$) are shown in Fig. 10.14.

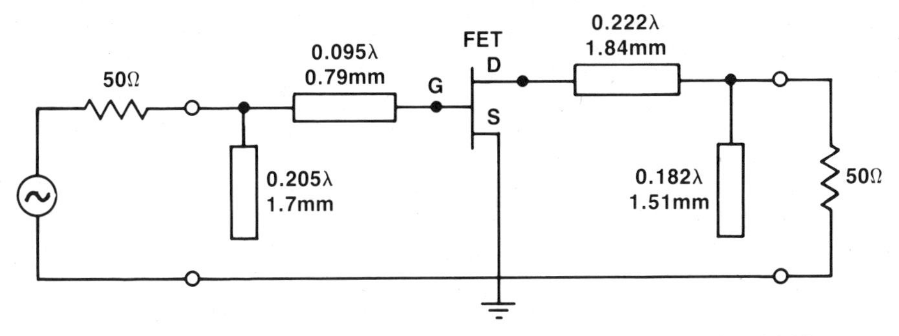

Figure 10.14 Amplifier schematic. Microstrip width is 0.24 mm and $\lambda = 8.28$ mm at 14 GHz.

10.4.4 Broadband Amplifiers

The broadband amplifier design is carried out by considering the power gain rolloff of the FET with frequency (usually 6 dB/octave), the gain-bandwidth limitations of the input and output of the FET, and the overall amplifier stability versus frequency. The design of a broadband, common-source amplifier requires a compromise between several competing requirements. The input impedance to the gate of a GaAs FET can be approximated conceptually as a series RC circuit composed of the gate resistance, the source resistance, the intrinsic resistance, and the gate capacitance. The Q of the circuit is $1/\omega RC$. For the purpose of broadband matching to the gate, the lower the Q, the easier it will be to design and realize conjugate matching circuits. There are fundamental limits to how well matched an RC circuit can be over a given bandwidth, but practical considerations are more important. If the matching circuit becomes too elaborate, it cannot be built accurately, and circuit losses mount rapidly with circuit complexity. The requirements of low FET device noise are in opposition to reduction of the input Q.

There are various techniques used to realize broadband amplifiers. Five of them are shown in Fig. 10.15. In balanced configurations [22] a single-ended reflective match amplifier (as described in the previous section) is designed, and a matched pair is used to realize a balanced configuration using two broadband Lange couplers. The single-ended amplifier is usually mismatched for flat gain, low-noise figure, and good stability. The reflections from the amplifier are terminated in 50 Ω, which usually guarantees stability. If one stage fails, the overall gain drops about 6 dB, which may provide useful fault tolerance for some applications. In feedback amplifiers [23, 24], a series RL feedback is used between the drain and gate of the FET. This configuration improves input match, output match, and stability by lowering gain at lower frequencies. The active matching method uses common-gate FET configuration at the input and common-drain FET

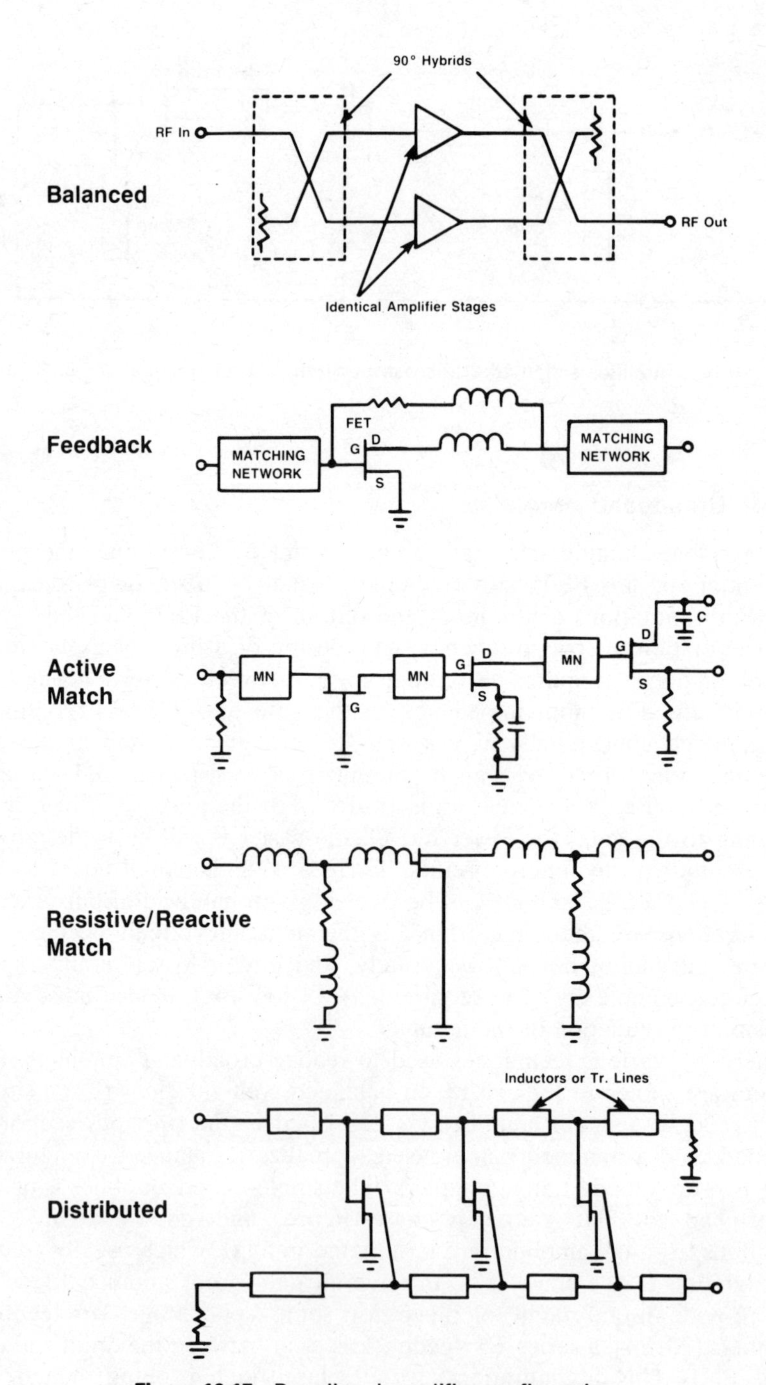

Figure 10.15 Broadband amplifier configurations.

TABLE 10.5 Comparison of Techniques used for Broadbanding of Amplifiers

Balanced Circuit	Feedback	Active Matching	Resistive Matching	Distributed Approach
Good up to a 2-octave bandwidth	Good up to a many-octave bandwidth	Good up to a many-octave bandwidth	Good up to a many-octave bandwidth	Good up to a many-octave bandwidth
Requires matched FET pairs and quadrature couplers	Requires large g_m FETs and uses negative as well as positive feedback	Requires a common-gate FET at the input and a common-drain FET at the output	Requires large g_m FETs	Requires many small FETs
Size is relatively large	Size is relatively small	Size is moderate	Large size due to greater number of stages	Size is moderate
Provides very good impedance match	Provides good impedance match	Provides good impedance match	Provides poor impedance match	Provides good impedance match
Has low-noise figure	Has high-noise figure	Has moderate noise figure	Has high-noise figure	Has moderate noise figure
Effect of fabrication tolerance is small	Effect of fabrication tolerance is moderate	Effect of fabrication tolerance is moderate	Effect of fabrication tolerance is small	Effect of fabrication tolerance is moderate
Cascading of two or more gain modules is easy	Modular approach is not easy	Modular approach is easy	Modular approach is not easy	Modular approach is easy

configuration at the output of the main FET, for matching to 50 Ω. This configuration is useful for monolithic amplifiers working up to 10 GHz [25]. The resistive matching technique uses resistors as a part of matching networks and is very similar to reflective matching methods. The distributed amplifier configuration (also commonly referred to as traveling-wave amplifier) provides the unique capability of adding device transconductance without adding device parasitic capacitance. In this configuration, FETs with series inductors behave like an effective low-pass transmission line. By terminating these lines with resistive loads, the unwanted signals are dissipated while the desired signals are added in phase at the output of the amplifier. This results in excellent gain-bandwidth product with flat gain and low VSWR. This technique has been successfully used in monolithic amplifiers [26, 27], as well as in hybrid amplifiers [28].

A qualitative comparison of techniques for broadbanding of amplifiers previously described is given in Table 10.5, and a quantitative comparison has been described by Niclas [29].

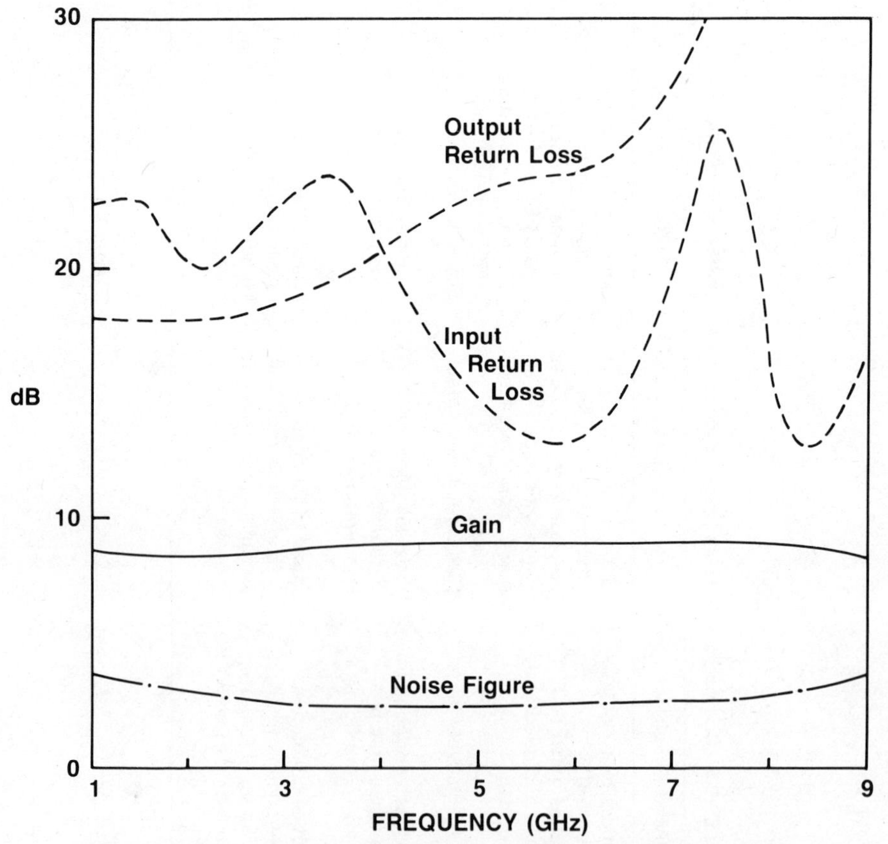

Figure 10.17 Calculated performance of a 2–9 GHz distributed amplifier.

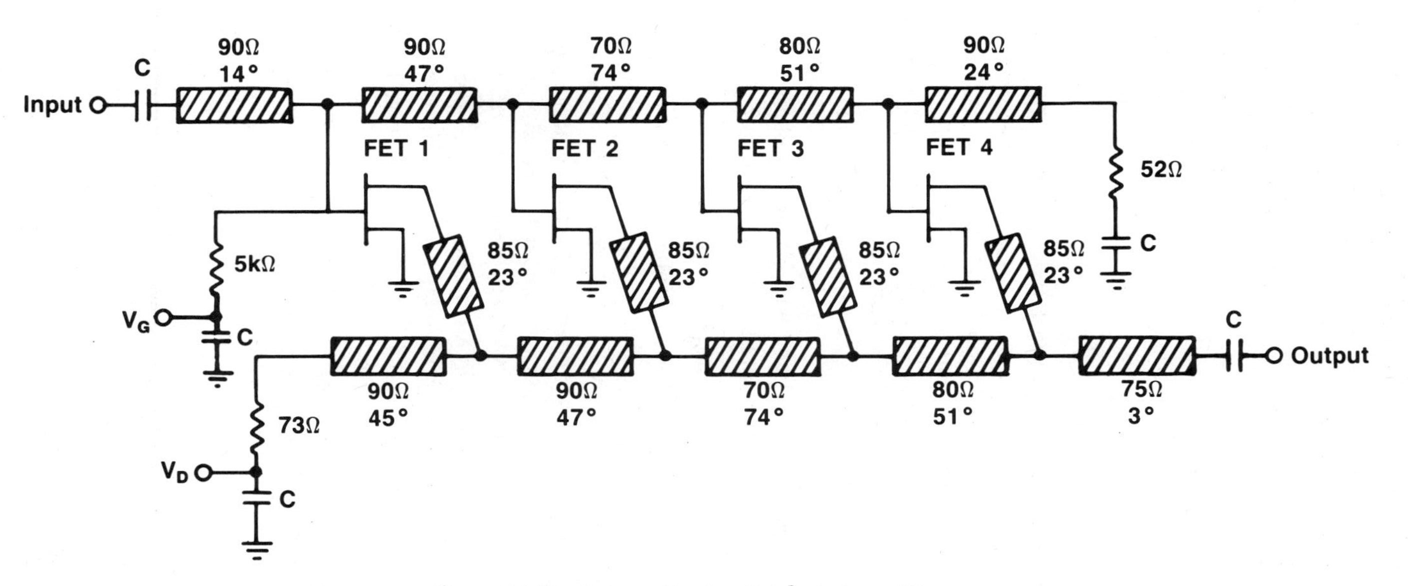

Figure 10.16 Schematic of a distributed amplifier.

Broadband amplifiers can be designed using the impedance-matching techniques described in Chapter 4 or CAD tools. CAD tools play a very important role in designing broadband amplifiers. A four-cell distributed amplifier's design is optimized based on the FETs given in Tables 10.3 and 10.4. The schematic of the amplifier is shown in Fig. 10.16 and the performance in Fig. 10.17. The amplifier has 8-dB gain, a 4-dB noise figure, and a maximum VSWR 1.6:1 over 1 to 9 GHz.

10.5 POWER AMPLIFIERS

There is a great deal of interest in finding more reliable and longer lasting substitutes for TWT amplifiers. The GaAs power amplifier is a good candidate to replace TWT amplifiers. Although GaAs FETs will not provide power levels comparative to TWT amplifiers, their use in conjunction with circuit and/or array power-combining techniques appears to offer a very viable solution. Major advancements in this area must be in the development of good yield and high-power FETs and low-loss combining techniques.

The design of narrow-band and wide-band amplifiers using power GaAs FETs requires several device and circuit considerations as given below

1. Select a suitable power device that meets the design objectives (power output and frequency range). Si bipolar transistors can deliver more RF power than GaAs FETs in the *L* and *S* band. About 20–30% margin should be included from the device P_{out} to amplifier P_{out}.

2. Power transistors with higher breakdown voltage are desirable. Use transistors that are close to industry standards. Transistors on thin substrates with via holes have low series inductance and better heat dissipation.

3. Operate the amplifier circuit within its safest operating bias. Do not exceed the maximum breakdown voltage and current ratings.

4. Junction-to-case thermal resistance should be as low as possible for better performance and reliability.

5. Load-pull measurements are essential to accurately characterize power devices for optimum amplifier design.

6. Internally matched transistors help in reducing the effect of package parasitics. They offer higher efficiency and larger bandwidth.

7. Design input matching network for maximum power transfer, while designing the output matching network for maximum power out. Matching circuits should offer minimum gain outside the desired frequencies.

8. Use lumped elements or lumped-distributed circuit elements for matching low impedance to 50 Ω in order to realize a compact circuit. Also use low-loss circuit elements at the output since the efficiency is reduced more by a given amount of loss in the output than in the input.

9. Use low-loss and 85–90% efficient power-combining techniques for high-power modules.
10. For broadband amplifiers use low-Q matching networks.

For power amplifiers, the input signal level is high, and consequently the output current is either in the cutoff or saturation region during a portion of the input signal cycle. This leads to the classification of power amplifiers into three basic modes of operation: Class A, Class B, and Class C. When the output current flows for full period of the input voltage cycle, the amplifier is operated in the Class A mode. If the output current flows for half period of the input voltage cycle, the amplifier is designated a Class B amplifier and is biased at cutoff. If the output current flows for less than half period of the input voltage cycle, the amplifier is called a Class C amplifier and is biased below cutoff. The Class A amplifier has the best linearity among these three types, whereas Class C amplifier has the best efficiency.

Efficiency of small-signal amplifiers is not as critical as for power amplifiers. For power amplifiers power-aided efficiency is defined as

$$\eta = \frac{\text{output signal power–input signal power}}{\text{dc power}} = \frac{P_{\text{out}} - P_{\text{in}}}{P_{\text{dc}}} \tag{10.31}$$

The design of narrow-band and broadband Class A power amplifiers is described in this section.

10.5.1 Selection of Power FET

Maximization of large-signal output power requires the output of the FET to have an optimum load impedance that is different from that required for a small-signal match. The optimum load impedance is calculated from the available current and voltage swing of the FET under large-signal conditions. Figure 10.18 shows the I–V characteristics of a typical $1 \times 2400\ \mu\text{m}^2$ power FET. The load line for a load resistance R_L is also shown as is V_K—a knee voltage at which current saturation is deemed to have occurred. The limiting source-drain voltage V_L is given by

$$\begin{aligned} V_L = V_{GD} &= V_p \quad \text{or} \\ &= V_{DSB} \end{aligned} \tag{10.32}$$

where V_{GD} is the gate-drain avalanche breakdown voltage for a FET biased to its pinchoff voltage V_p. Beyond V_L excess current flows between drain and source, which cannot be modulated by the RF voltage on the gate and thus is a limiting value of V_{DS} for further output power. The term V_{DSB} is the breakdown voltage between the drain and source, and typically $V_{DSB} < V_{GD}$. Assuming that the RF load and dc drain can swing to a

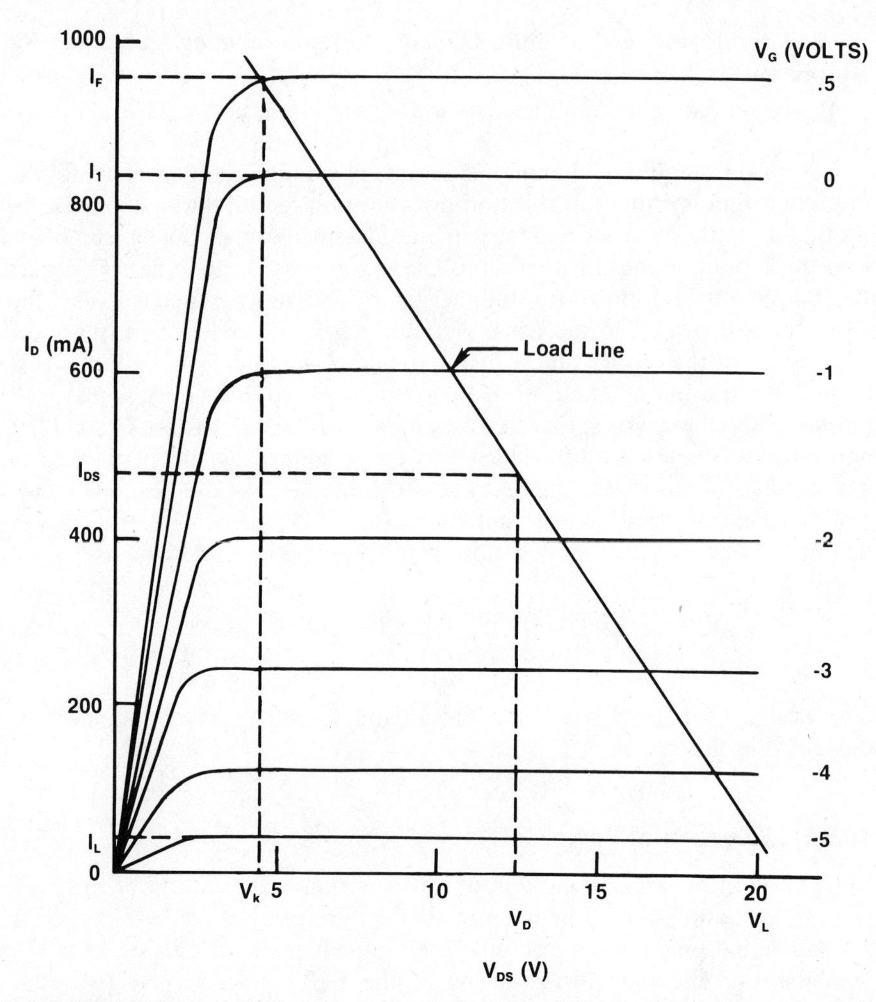

Figure 10.18 Simple representation of *I–V* and load line characteristics of a typical $1 \times 2400\ \mu m^2$ power FET.

maximum value of V_L. It is an elementary and well-known result that for maximum linear power the RF load or dynamic load must be resistive and is given by

$$R_L \cong \frac{V_L}{I_F} \tag{10.33}$$

It is possible to calculate the output power of the device from the $I-V$ characteristics. Assuming that the dc bias points are V_D and I_{DS} and that the total output power is made up of fundamental as well as of harmonics

TABLE 10.6 Typical Power FETs

Power FET	$f = 4$ GHz			$f = 8$ GHz			$f = 18$ GHz			V_D (V)	I_{DSS} (A)	Package Type	Company
	P_{1dB} (dBm)	G_{1dB} (dB)	η (%)	P_{1dB} (dBm)	G_{1dB} (dB)	η (%)	P_{1dB} (dBm)	G_{1dB} (dB)	η (%)				
AT-8141	32	9	30	31	6	21				9	1	—	Avantek
Im-7984-6				38	7	30				9	4.8	Hermetic metal/ceramic internally matched	Avantek
FLM1777-8C				39	7.0	29				10	3.6	1B/internally matched	Fujitsu
FLM1414-4C							35.5	4.5	21	10	1.8	1A/internally matched	Fujitsu
NEZ3742-3A	40	11	35							9	2.4	Hermetic flange	NEC
NEZ7984-6A				37.5	7.5	30				9	2.4	Hermetic flange	NEC

frequency, the output power available to the load is

$$P_0 \cong \frac{1}{2} \cdot \frac{1}{2} (V_L - V_k) \cdot \frac{1}{2} (I_F - I_L) = \frac{1}{8} \frac{(V_L - V_k)^2}{R_L} \tag{10.34}$$

where

$$R_L = \frac{V_L - V_k}{I_F - I_L}$$

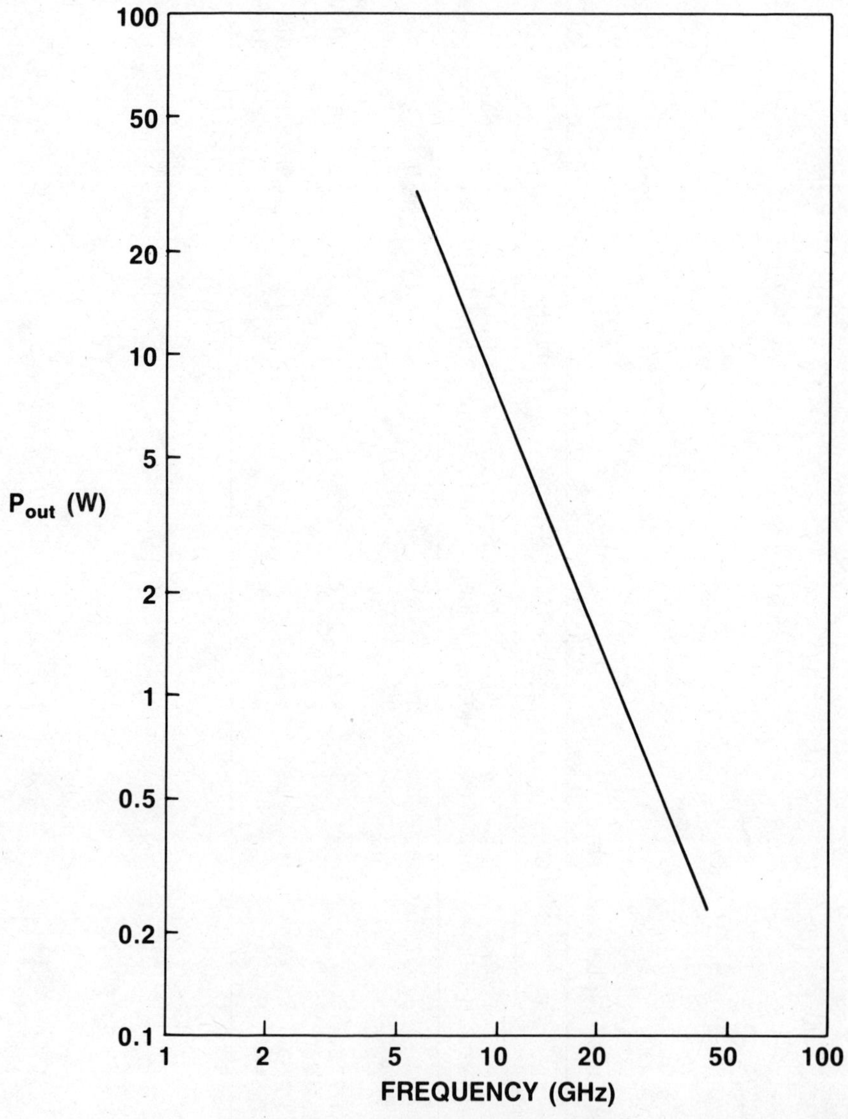

Figure 10.19 Power out of power FETs.

For example, consider a power FET with a $V_L = 20$ V, $V_k = 2$ V, $I_F = 960$ mA, and $I_L = 40$ mA. The output power as given by Equation 10.34 is 2.07 W.

The electrical parameters available in vendor catalogs include small-signal S-parameters at $\frac{1}{2}I_{DSS}$, G_{1dB}, P_{1dB}, and power-added efficiency. Usually both input and output impedance values under large-signal conditions are not available. Either one should use small-signal S-parameters with a good guess of nonlinear elements or measure input and output impedances at operating RF signal and dc bias conditions using load-pull method [30].

Table 10.6 lists a few of the typical power FETs available commercially as of late 1986 and Fig. 10.19 shows P_{1dB} versus frequency for state-of-the-art power FETs.

10.5.2 Large-Signal Characterization

To achieve the expected performance from a GaAs power FET, scattering parameters alone are insufficient. Usually S-parameters, measured with a network analyzer, are small-signal, linear measurements that are quite sufficient for many active devices and all passive circuits. Since under large RF drive, a power FET exhibits considerable nonlinear behavior in C_{GS}, g_m, C_{GD}, and R_o, a large-signal measurement must be made. Namely, the large-signal input and output impedances must be determined.

There are many methods [30] for obtaining large-signal data. Some examples include the classical load-pull technique; the active and semiautomatic load-pull, Peeling method, and large-signal S-parameters. Most of these methods suffer from limited accuracy due to two main problems. First, a disconnection and reconnection is sometimes necessary to switch a tuner to a measuring device. If the VSWR and loss of the connection and measuring path is poor, great uncertainty in the measurement results. Secondly, there usually is some physical distance between the FET and tuner. This situation prevents, in many cases, the possibility of matching the power FET at all, the consequently the true performance of the FET is unknown. Also, most of these methods provide only an output impedance (Γ_L^*), and are very time-consuming.

There is, however, a very elegant method of determining large-signal impedances. The method is an "equivalent load-pull" technique devised at NEC [31]. The equivalent load-pull technique can be explained as follows: generate a reverse traveling-wave into the output of a power FET under high RF drive. This reverse signal interacts with the forward traveling-wave generated by the FET. The ratio of these signals at the FET output is the reflection coefficient at that point. Then, by varying the magnitude and phase of the reverse wave, any terminal impedance can be simulated. This method can map the entire Smith chart, and the load contours can be generated to determine the maximum output power, which is also moni-

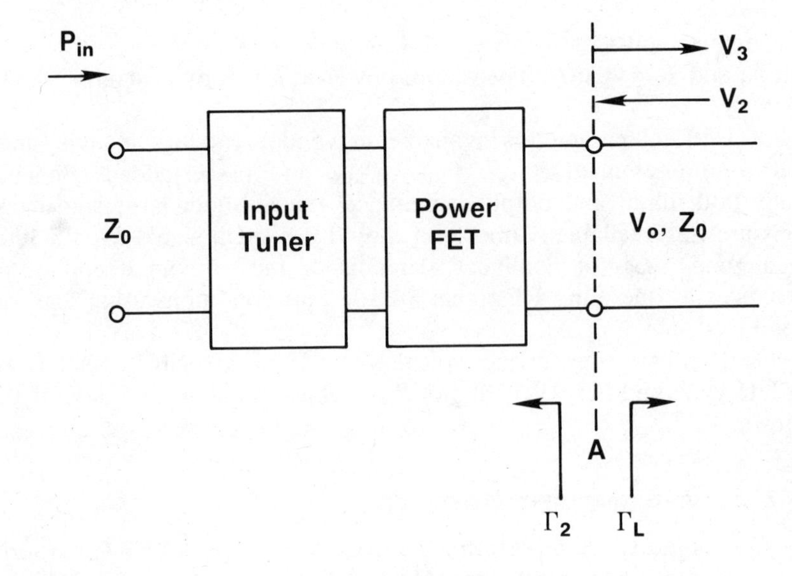

Figure 10.20 Basic network for the principal explanation of the equivalent load-pull method.

tored. This idea is illustrated in Fig. 10.20. If V_2 and V_3 are complex amplitudes of the incident and reflected voltage waves, respectively, at the output reference plane A, then

$$V_2 = \bar{V}_2 e^{j(\omega t+\theta)} \quad (10.35a)$$

$$V_3 = \bar{V}_3 e^{j(\omega t+\phi)} \quad (10.35b)$$

The resultant RF voltage at plane A is

$$V_o = V_2 + V_3 \quad (10.36)$$

The reflection coefficient looking into the device is

$$\Gamma_2 = \frac{V_3}{V_2} = \frac{1}{\Gamma_L} \quad (10.37)$$

If Y_0 is the characteristic admittance of the measurement system, then

$$V_o = \frac{2Y_0V_2}{Y_0 + Y_2} \quad (10.38)$$

where Y_2 is the output admittance looking into the device. Therefore, the

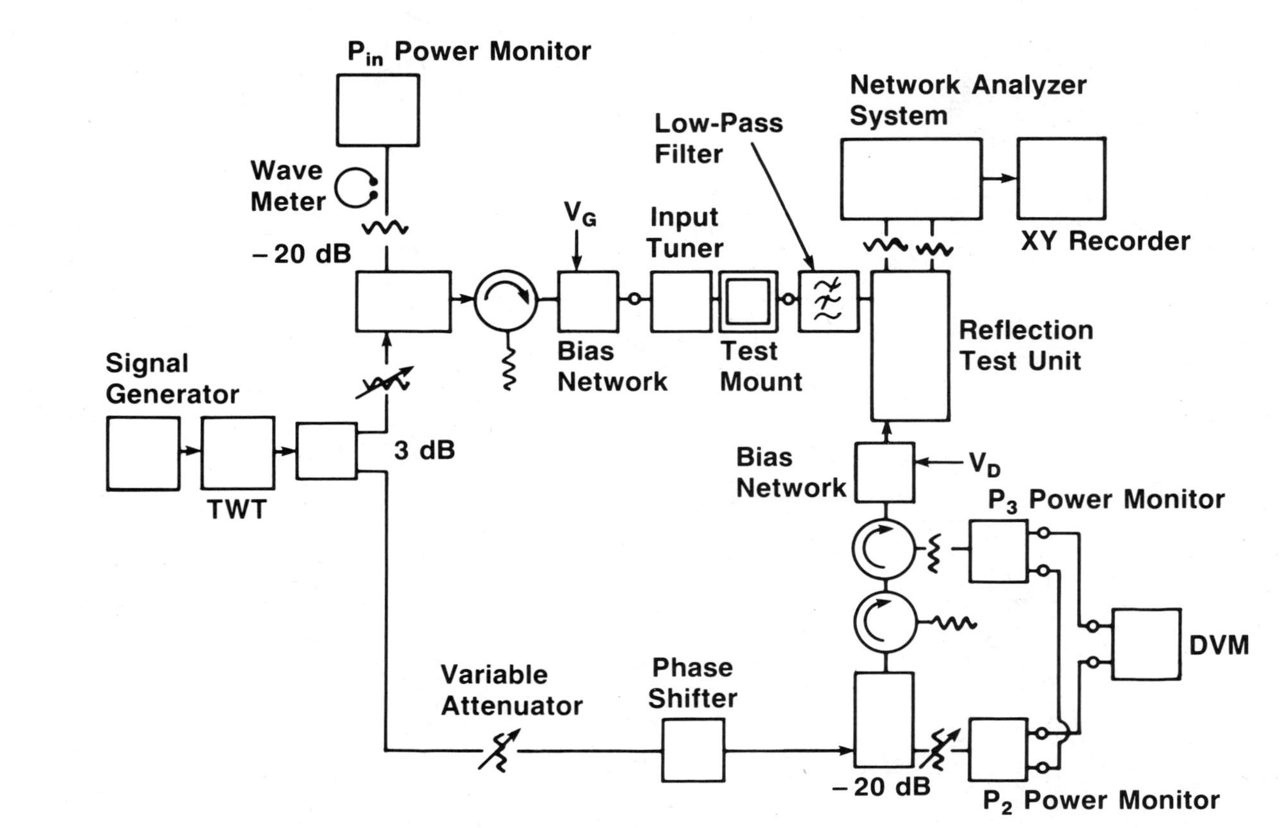

Figure 10.21 Setup for equivalent load-pull measurement. (After Y. Takayama [31], reprinted with permission of NEC Corp.)

load admittance is

$$Y_L = -Y_2 = Y_0\left(1 - 2\frac{V_2}{V_o}\right) \tag{10.39}$$

As evident from the last equation, the load admittance is a function of the reverse traveling wave V_2. Figure 10.21 shows a complete setup for the equivalent load-pull measurement method. Figure 10.22 is a Smith chart plot of load impedances realized by selecting a power level for the output of the FET and then varying the phase. This shows that the entire Smith chart can be mapped.

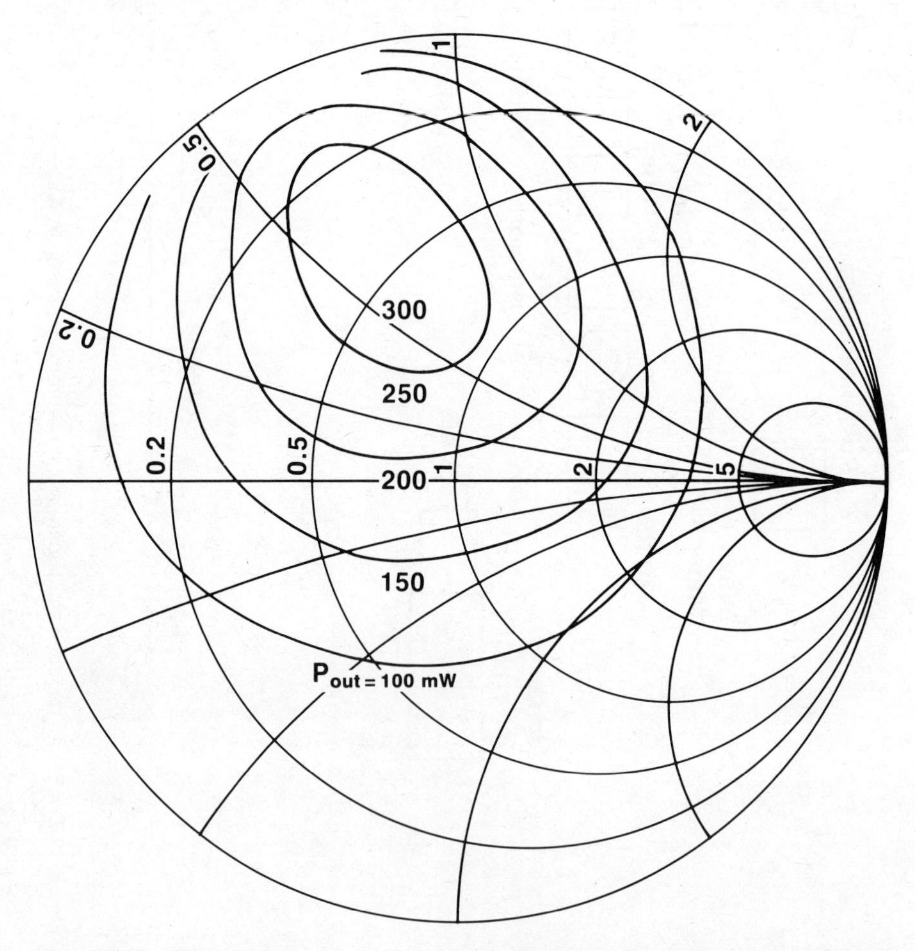

Figure 10.22 FET load impedance contours for constant power outputs at a 95.5 mW input power level and $f = 6$ GHz. (After Y. Takayama [31], reprinted with permission of NEC Corp.)

10.5.3 Power Amplifier Design

In order to transfer maximum power, low-loss matching networks must be used to match source impedance to the input of the device and to match the output impedance of the device to the load impedance. The GaAs FET's input and output impedances decrease as the power level of the device goes up (as the gatewidth becomes larger). The input impedance is the most difficult to match. Most of the time, a combination of lumped and distributed elements are used in realizing impedance matching networks.

An approximate large-signal simplified equivalent circuit of a GaAs power FET is shown in Fig. 10.23. Various circuit elements can be approximated for an X-band 1-μm gate-length FET as follows:

$$R_{\text{in}} = R_G + R_i + R_S \simeq \frac{2}{W} \quad (\Omega) \tag{10.40a}$$

$$C_{GS} \simeq W \quad (\text{pF}) \tag{10.40b}$$

$$g_m \simeq 80\,W \quad (\text{mS}) \tag{10.40c}$$

$$R_0 \simeq \frac{70}{W} \quad (\Omega) \tag{10.40d}$$

$$C_{DS} \simeq 0.12\,W \quad (\text{pF}) \tag{10.40e}$$

and

$$P_{1\text{dB}} \simeq 0.5\,W \quad (\text{Watts}) \tag{10.40f}$$

where W is the total gatewidth of the power FET in millimeters. For example, a device delivering about 2-W power out at 10 GHz has an input impedance about 0.5–j4 Ω and output impedance about 13.7–j0.41 Ω.

Two parameters that are very useful for comparing different power amplifier configurations and classes are the power output and power-added efficiency. Here Class A has been chosen for maximum linear power output. The circuit topologies are similar to those of small-signal amplifiers. The design of a narrow-band power amplifier is described step by step in the following.

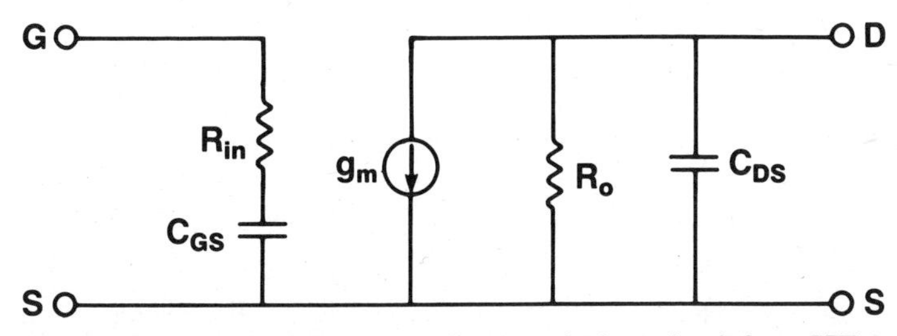

Figure 10.23 A simplified large-signal equivalent circuit for a FET.

Design a power amplifier at 5.5 GHz with 1.7 W power output, 25% power added efficiency, and 6 dB gain at 1-dB compression point. The power FET to be used has the following characteristics measured at 5.5 GHz and $P_{in} = 26$ dBm, $V_{DS} = 9$ V, $I_{DSS} = 1000$ mA.

$$\text{Gain} = 7 \text{ dB}, \qquad P_{out} = 2 \text{ W}, \qquad \eta_{add} = 35\%$$

$$S_{11} = 0.89\underline{/-157^\circ}, \quad S_{21} = 1.5\underline{/84^\circ}, \quad S_{12} = 0.049\underline{/54^\circ}, \quad S_{22} = 0.31\underline{/-145^\circ}$$

$$\Gamma_S = 0.73\underline{/150^\circ}, \qquad \Gamma_L = 0.32\underline{/160^\circ}$$

From (10.18),

$$K = 1.267$$

Therefore the transistor is unconditionally stable at 5.5 GHz. Substituting Γ_S and Γ_L into (10.1) yields $G = 8.1$ dB. Thus the device has enough gain to design a 6-dB gain power amplifier when circuit losses are incorporated.

From $\Gamma S = 0.73\underline{/150^\circ}$, $Z_S = 8.35 + j13.05\ \Omega$ and $Y_S = 0.0348 - j0.0544$ S.

A short-circuited stub has an input admittance of

$$Y = \frac{-jY_0}{\tan \beta l} = -j0.0544$$

when $Y_0 = 0.02$ and $\beta l = 20.2^\circ$. Thus the characteristic impedance of the stub is 50 Ω. To match the parallel conductance of 0.0348 S to the source admittance of 0.02 S, a quarter-wave transformer of characteristic impedance $Z_0 = \sqrt{1/(0.02 \times 0.0348)} = 37.9\ \Omega$ is used. The complete input matching network is shown in Fig. 10.24.

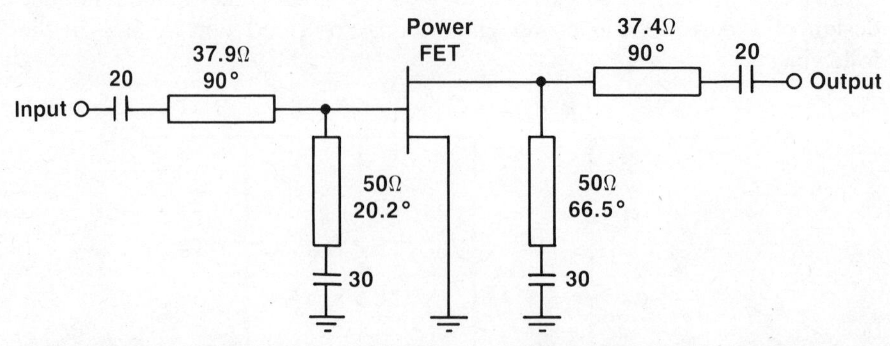

Figure 10.24 Schematic of a 1.7 W power amplifier. Electrical line lengths are at 5.5 GHz and capacitances in picofarads.

For the output matching network, $\Gamma_L = 0.32\underline{/160^\circ}$, $Z_L = 26.34 + j6.42$, $Y_L = 0.0358 - j0.00874$.

A short-circuited stub has an input admittance of $Y = -jY_0/\tan\beta l = -j0.00874$ when $Y_0 = 0.02$ and $\beta l = 66.5^\circ$. In this case also the characteristic impedance of the stub is 50 Ω. To match the parallel conductance of 0.0358 S to the load admittance of 0.02 S, a quarter-wave transformer of characteristic impedance $Z_0 = \sqrt{1/(0.02 \times 0.0358)} = 37.4\ \Omega$ is used. The complete output matching network is shown in Fig. 10.24.

10.5.4 Design of Internally Matched Power FET Amplifier

Since the total gatewidth is very large in the case of high-power GaAs FETs, the impedance of GaAs FET chips becomes so low that input and output impedances are affected by parasitic capacitance and inductance of a package. It is difficult to match the amplifier circuits out of a package, especially of high frequencies. One of the most practical methods of designing broadband power FET amplifiers is to use internal matching [32, 33] within a microwave package to deal with the low input impedance of the device.

Lumped elements and/or distributed elements for matching networks can be used. For broadband and power levels greater than 5 W, lumped elements are generally preferred for input matching of power FETs. Lumped inductors are realized by using bonding wires, and capacitors are of the metal–insulator–metal type that use high dielectric constant ceramics. Capacitors must have small parasitic inductance and resistance, sufficient thermal and mechanical strength, a small temperature coefficient, a 40-V or higher breakdown voltage, and low cost. Since the output impedance is much higher than the input impedance, the output matching networks are realized using both lumped and distributed circuit elements. Microstrip lines on ceramic substrates are generally used for distributed circuit elements.

The schematic of an internally matched power FET is shown in Fig. 10.25. A 6-GHz 15-W internally matched power FET amplifier is depicted in Fig. 10.26. This particular device has a total gate periphery of 33.6 mm, which gives a power output of about 0.45 W/mm. The real part of the input

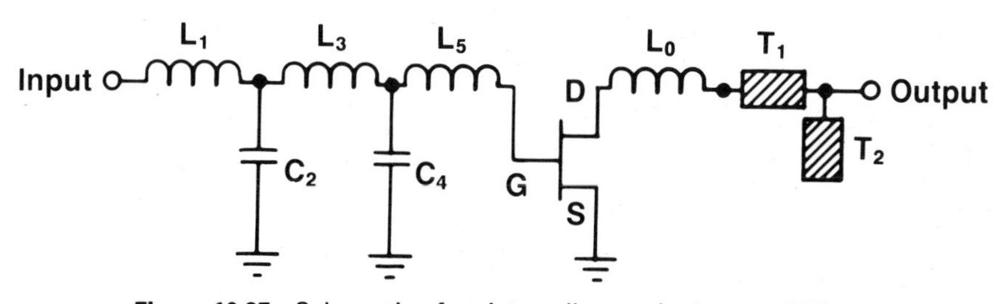

Figure 10.25 Schematic of an internally matched power FET.

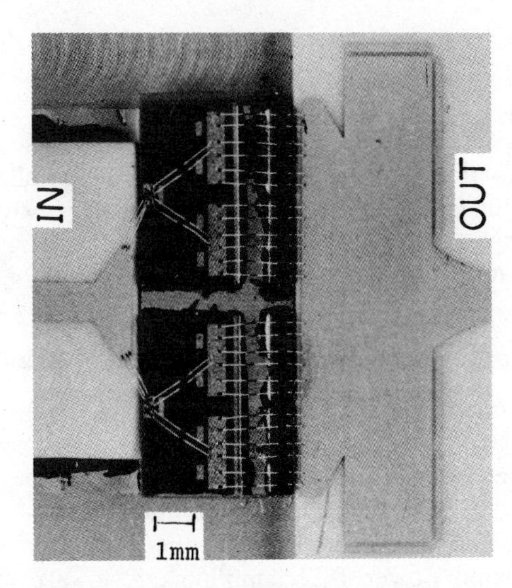

Figure 10.26 Photograph of a 6-GHz, 15-W internally matched FET amplifier. (After K. Honjo et al. [32], 1979 IEEE, reprinted with permission.)

impedance of the device is about 0.25 Ω. The design and fabrication of this circuit have been described in [32, 33].

10.5.5 Power-Combining Techniques

Solid state devices are low power, and with increasing frequency, the power output from a single solid state device decreases rapidly. In many applications RF power levels are required that far exceed the capability of any single device or amplifier. It is therefore desirable to extend the power level by combining techniques in order to take advantage of the many desirable features of solid state devices, such as small size and weight, reliability, and performance in a broader range of applications.

Although there are fundamental limitations to the power that can be generated from a single device, the achievable power levels can be significantly increased by combining a number of devices operating coherently or by accumulating the power from a number of discrete devices. This may be done in one of two ways: either by combining power at the device level or at the circuit level. Most combining techniques can provide "graceful degradation" in the case of failure of one or more devices in the combiner.

Device-Level Combining. Device-level combining is accomplished by clustering the devices in a region whose extent is small compared with a wavelength, and is generally limited to the number of devices that can be combined efficiently. Higher power is obtained by bonding several devices

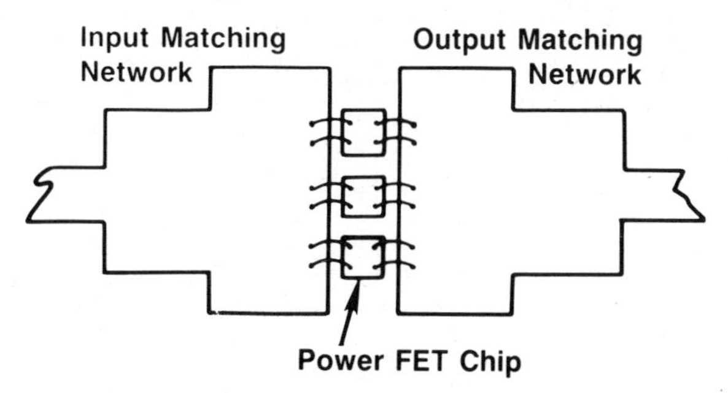

Figure 10.27 Chip combining of power FETs.

onto a heat sink or a common carrier and connecting them to input and output matching circuitry as shown in Fig. 10.27. The entire circuit may be hermetically sealed and used as a single device. Power levels of 15 W and 10 W at 6 and 8 GHz, respectively, have been demonstrated by using this technique. Monolithic power amplifiers have shown great promise as an alternative to chip combining. These amplifiers have built-in power FETs and matching networks. The advantages of monolithic power amplifiers are small size, light weight, low cost, and totally matched circuits. However, at the present time, this technology is not as fully developed and optimized as conventional hybrid circuit techniques.

Circuit-Level Power Combining. Device-level combining is generally limited due to the number of devices that can be combined and effectively matched on a small area. There are two categories of combiners: (1) those that combine the output of N devices in a single step (known as N-way combiners) and (2) tree or chain combining structures. The desirable characteristics of a combining structure are: minimum loss in the matching elements, minimum loss in the combining networks, minimum amplitude and phase imbalance, a good combiner input VSWR, even distribution of dissipated heat in the devices, and efficient removal of dissipated heat.

N-Way Combiners. N-way combining structures are the simpler of the two, and they avoid the use of several combining stages, thus making it possible to achieve high-efficiency combining. The structures can be either cavity or nonresonant combining structures. In cavity structures, cylindrical or rectangular cavities are used to combine the power output from a number of devices. These combiner–dividers have low loss ($\simeq$0.2 dB) and a combining efficiency of 85–90%. Note that, in general, combining and dividing circuits are identical. An 80-W FET amplifier using an 8-way combiner–divider working from 5.9 to 6.4 GHz has been developed [34].

Many nonresonant N-way combining techniques are also available. Essentially three types of N-way combiners are used for combining large numbers of amplifiers: Wilkinson, radial, and planar. The N-way Wilkinson

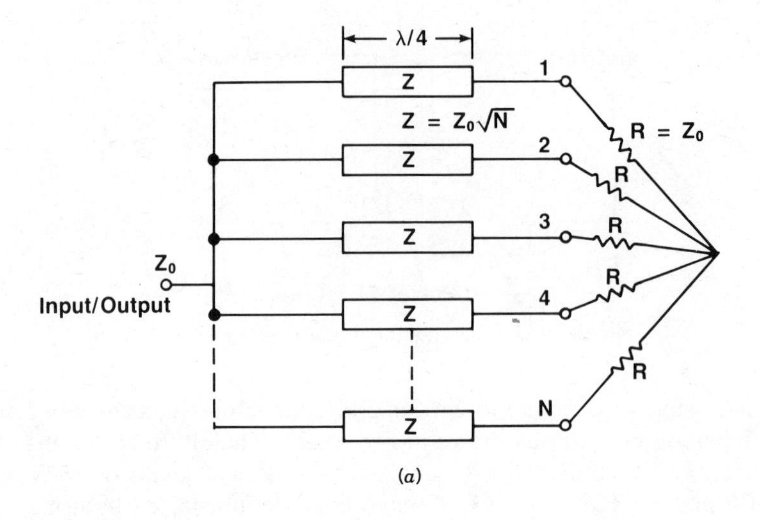

(a)

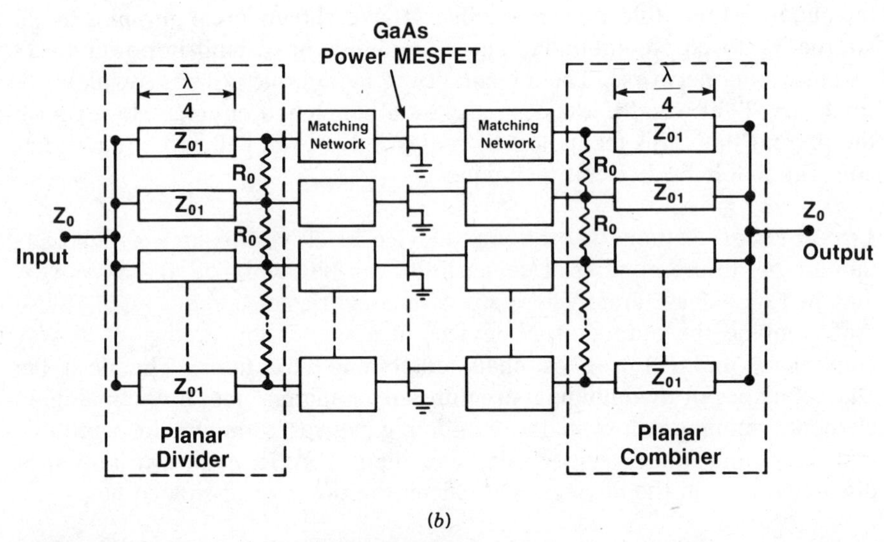

(b)

Figure 10.28 (*a*) Wilkinson *N*-way divider–combiner. (*b*) Modified *N*-way divider–combiner. Isolation resistors supress possible oscillations.

divider [35] shown in Fig. 10.28(*a*) has the advantage of low loss, moderate bandwidth, and good amplitude and phase balance. However, its major disadvantage for power applications is the "floating starpoint" isolation resistors. These resistors require a nonplanar crossover configuration, which limits the power-handling capability of the combiner. Fortunately, a simplified version as shown in Fig. 10.28(*b*) can also be used [30, 36]. This particular arrangement has a combining efficiency on the order of 90% and shows much promise for chip combining as well as for MMIC applications. Radial line and planar *N*-way combiners are shown in Fig. 10.29. The radial

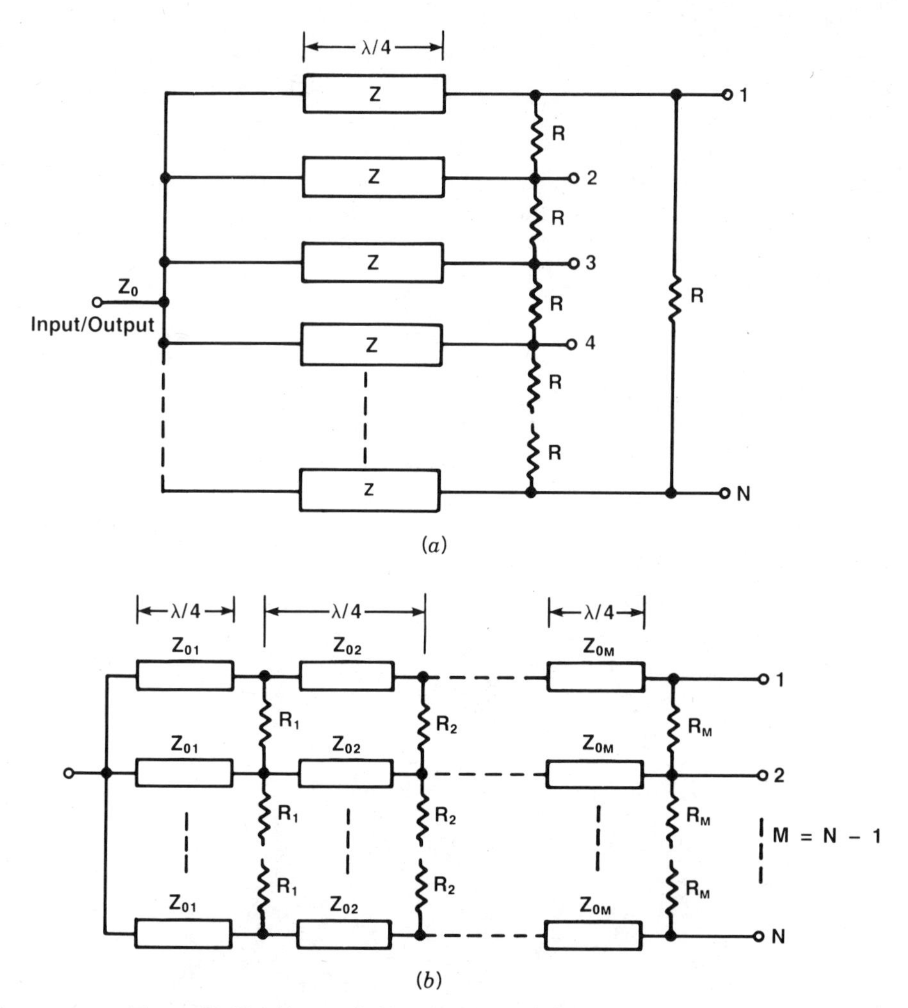

Figure 10.29 *N*-way divider–combiner. (*a*) Radial line. (*b*) planar.

line [37] combiner has low loss, inherent phase symmetry, and good isolation. Its main disadvantage is that it requires a three-dimensional structure. On the other hand, the planar N-way combiner–divider [38] requires $(N-1) \times N$ quarter-wave sections for maximum isolation and thus is very large. A compact structure using a tapered microstrip line has also been described [39]. The inherent redundancy in the N-way combiner makes it possible to obtain a graceful degradation characteristic. In an N-amplifier combiner with F failed amplifiers, the power output, relative to maximum output power, is given by $P_o/P_{max} = (N-F)/N^2$.

Example. Design a 10-W power amplifier at 5.5 GHz with 20% power added efficiency using monolithic integrated circuit amplifier chips having

50-Ω source and load impedances. The measured performance of these chips at 5.5 GHz are $P_{\text{out}} = 3$ W, Gain $= 6$ dB, and $\eta_{\text{add}} = 25\%$ at the 1-dB compression point.

In order to obtain a 10-W power output we need four such chips and two four-way divider/combiner circuits. Since the amplifier is a narrow-band circuit, a divider/combiner shown in Fig. 10.28(*b*) can be used. The parameter values for this divider/combiner on a 0.63-mm-thick alumina substrate ($\epsilon_r = 9.9$) are

$$Z_{01} = 100\ \Omega, \qquad l = 5.6\ \text{mm}, \qquad R_0 = 37\ \Omega$$

The circuit has about a 0.2-dB loss.

$$\begin{aligned}\text{Total power output} &= 4 \times \text{amplifier chip power} - \text{total loss}\\ &= 4 \times 3 \times 10^{-0.4/10} = 10.9\ \text{W}\end{aligned}$$

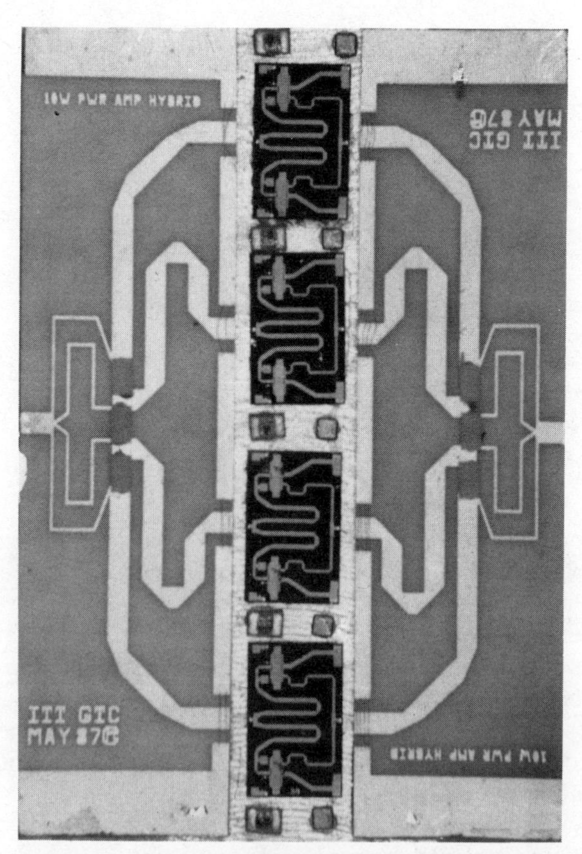

Figure 10.30 Photograph of a 10-W power amplifier using MMIC chips. (Courtesy I. J. Bahl, ITT.)

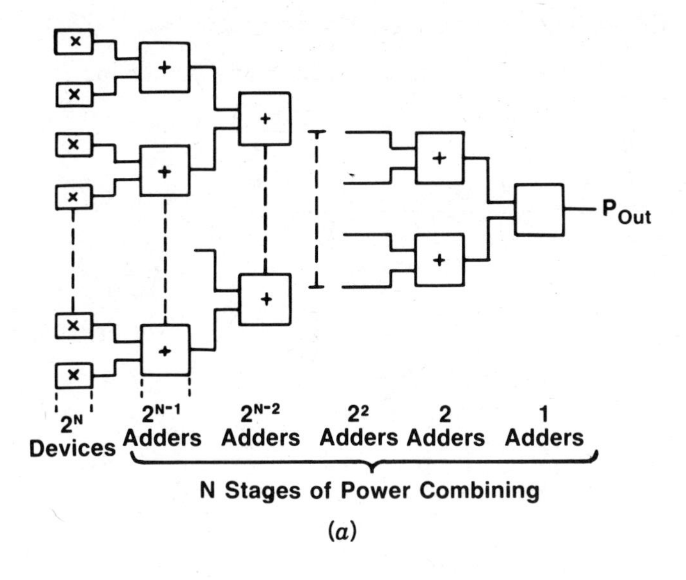

(*a*)

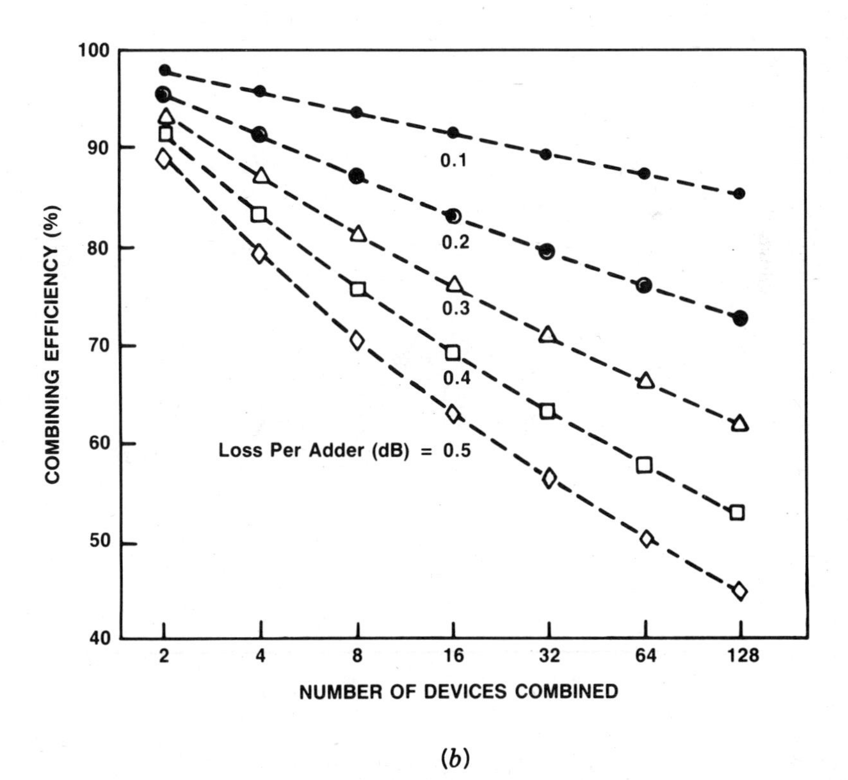

(*b*)

Figure 10.31 (*a*) Corporate combining structure. (*b*) Combining efficiency for a corporate combining structure.

$$\text{Combining efficiency} = 10^{-0.4/10} = 91.2\%$$

$$\text{Total efficiency} = \text{chip efficiency} \times \text{combining efficiency}$$
$$= 25 \times 0.912 = 22.8\%$$

Figure 10.30 shows a photograph of the 10-W power amplifier assembled using 3-W MMIC power amplifier chips.

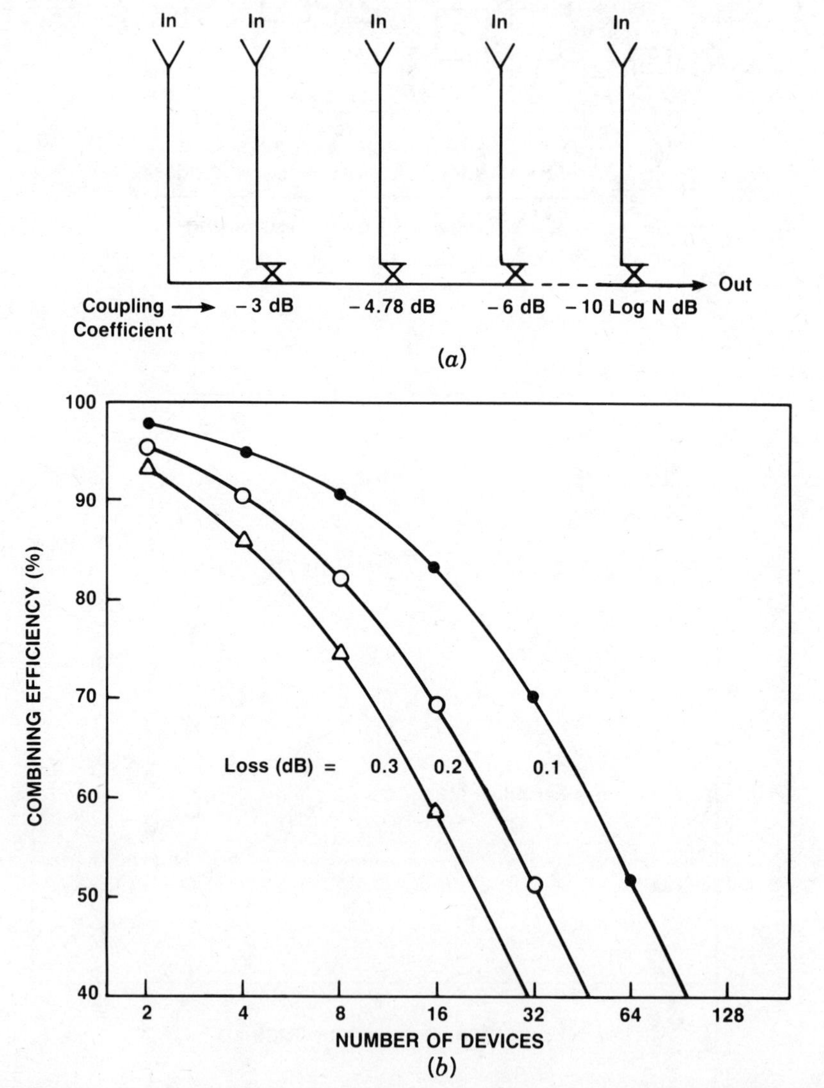

Figure 10.32 (*A*) A serial combining structure. (*b*) Combining efficiency for the chain combining structure. Loss in decibels refers to the loss in each power path in each stage's coupler.

Corporate Structures. A corporate structure (or tree) for combining power from two-way adders or combiners is shown in Fig. 10.31(*a*). The loss in adders limit the combining efficiency. Figure 10.31(*b*) illustrates [40] the combining efficiency of the corporate structure versus the number of devices for various loss per adder values. The number of devices combined in this way is binary. Examples of two-way adders are directional couplers, hybrids, and the two-way Wilkinson combiner, and are described in Chapter 5. Among the two-way adders, the Lange coupler is usually preferred because of its good isolation and wide-band properties. However, cascading these to obtain high-order combining becomes impractical beyond a four-way combiner due to the relatively high coupler loss.

Serial Structure. A serial or chain combiner is shown in Fig. 10.32(*a*). Here each successive stage of an N-way combiner adds $1/N$ of the power delivered to the output. The number of the stage determines the required coupling coefficients for that stage, as indicated in the figure. One advantage of the chain structure is that another stage can be added by simply connecting the new source to the line after the Nth stage through a coupler with $10\log(N+1)$ coupling coefficient. The roles of input and output ports are changed for the divider structure. Losses in the couplers reduce the combining efficiency and bandwidth attainable with this approach. Combining efficiency for the chain structure for each path is shown [40] in Fig. 10.32(*b*). The four-way structure on microstrip can be used to realize combining efficiency on the order of 90% over an octave or greater bandwidth.

Example. Calculate the efficiency of combining and power output of a four-way chain divider–combiner realized using couplers having 0.1-dB insertion loss. The amplifier has 3-W power output, and the coupling factors are −3, −4.78, and 6 dB.

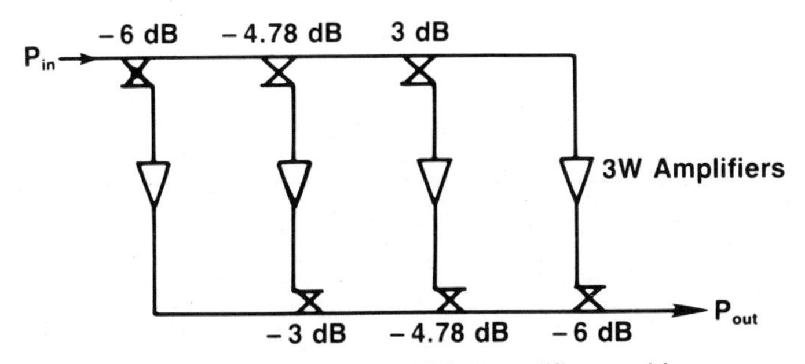

Figure 10.33 Four-way serial fed amplifier combiner.

TABLE 10.7 A Comparison of Circuit-Level Power-Combining Techniques

Combining Technique	Advantages	Disadvantages
N-way W/G cavity	Low loss High efficiency	Nonplanar Complex assembly Narrow band
N-way Wilkinson	Low loss Moderate bandwidth Good isolation High efficiency	Nonplanar Low power
N-way radial line	Low loss Good isolation	Nonplanar Complex assembly
N-way planar	Large bandwidth Good isolation Moderate loss	Large size Low efficiency
Corporate structure	Good isolation Large bandwidth	Impractical beyond Four-way due to low efficiency
Chain structure	More flexible Octave or greater bandwidth Good efficiency Good isolation	High resolution Fabrication required Complex design

$$\text{Total power out} = 4 \times \text{amplifier power} - \text{total loss} \quad (\text{W})$$

$$P_0 = 12 \times 10^{-0.6/10} = 10.5 \text{ W}$$

$$\text{Efficiency} = 100 \times \frac{10.5}{12} = 87\%$$

Figure 10.33 shows the four-way serial feed amplifier combiner.

A comparison of various circuit-level power-combining techniques appears in Table 10.7.

REFERENCES

1. Harrison, A. E., *Klystron Tubes*, McGraw-Hill, New York, 1947.
2. Liao, S. Y., *Microwave Devices and Circuits*, Prentice-Hall, Englewood Cliffs, N.J., 1980.
3. Gandhi, O. P., *Microwave Engineering and Applications*, Pergamon, New York, 1981.
4. Pierce, J. R., *Travelling Wave Tubes*, D. Van Nostrand Co., Princeton, N.J., 1950.
5. Blackwell, L. A., and K. L. Kotzebue, *Semiconductor-Diode Parametric Amplifiers*, Prentice-Hall, Englewood Cliffs, N.J., 1964.
6. Penfield, P., and R. P. Rafuse, *Varactor Applications*, MIT Press, Cambridge, Mass., 1962.

7. Chang, K. K. N., *Parametric and Tunnel Diodes*, Prentice-Hall, Englewood Cliffs, N.J., 1964.
8. Watson, H. A., *Microwave Semiconductor Devices and Their Circuit Applications*, McGraw-Hill, New York, 1969.
9. Sterzer, F., "Tunnel Diode Devices," in *Advanced in Microwaves*, Vol. 2, Leo Young (Ed.), Academic Press, New York, 1967, pp. 1–41.
10. Sterzer, F., "Transferred Electron (Gunn) Amplifiers and Oscillators for Microwave Applications," *Proc. IEEE*, Vol. 59, Aug. 1971, pp. 1155–1163.
11. Eastman, L. F., *Gallium Arsenide Microwave Bulk and Transit-Time Devices*, Artech House, Dedham, Mass., 1973.
12. Haddad, G., *Avalanche Transit-Time Devices*, Artech House, Dedham, Mass., 1973.
13. Cooke, H. F., "Microwave Transistors: Theory and Design," *Proc. IEEE*, Vol. 59, Aug. 1971, pp. 1163–1181.
14. Ha, T. T., *Solid-State Microwave Amplifier Design*, Wiley, New York, 1981.
15. Vendelin, G. D., *Design of Amplifiers and Oscillators by the S-Parameter Method*, Wiley, New York, 1982.
16. Gonzalez, G., *Microwave Transistor Amplifiers Analysis and Design*, Prentice-Hall, Englewood Cliffs, N.J., 1984.
17. Townes, C. H., et al., "The Maser—A New Type of Microwave Amplifier, Frequency Standard, and Spectrometer," *Phys. Rev.*, Vol. 99, Aug. 15, 1955, pp. 1264–1274.
18. Coleman, J. T., *Microwave Devices*, Reston Publishing Company, Inc., Reston, Va., 1982.
19. Fukui, H. (Ed.), *Low-Noise Microwave Transistors and Amplifiers*, IEEE Press, New York, 1981.
20. *Noise Figure Measurements—Principles and Applications*, Hewlett-Packard.
21. Pengelly, R. S., *Microwave Field-Effect Transistors—Theory, Design and Applications*, Wiley, New York, 1982.
22. Kurokawa, K., "Design Theory of Balanced Transistor Amplifiers," *Bell Syst. Tech. J.*, Vol. 44, Oct. 1965, pp. 1675–1698.
23. Ulrich, E., "Use of Negative Feedback to Slash Wideband VSWR," *Microwaves*, Vol. 17, Oct. 1978, pp. 66–70.
24. Niclas, K. B., et al., "The Matched Feedback Amplifier: Ultrawide-Band Microwave Amplification with GaAs MESFETs," *IEEE Trans. Microwave Theory Tech.*, Vol. MTT-28, Apr. 1980, pp. 285–294.
25. Peterson, W. C., et al., "A Monolithic GaAs 0.1 to 10 GHz Amplifier," *IEEE Int. Microwave Symp. Digest*, 1981, pp. 354–355.
26. Ayasli, Y., et al., "A Monolithic GaAs 1–13 GHz Traveling-wave Amplifier," *IEEE Trans. Microwave Theory Tech.*, Vol. MTT-30, July 1982, pp. 976–981.
27. Ayasli, Y., et al., "2–20 GHz GaAs Traveling-wave Amplifier," *IEEE Trans. Microwave Theory Tech.*, Vol. MTT-32, Jan. 1984, pp. 71–77.
28. Giamand, P., "A Complete Small Size 20 to 30 GHz Hybrid Distributed Amplifier Using a Novel Design Technique," *IEEE Int. Microwave Symp. Digest*, 1986, pp. 343–346.
29. Niclas, K. B., "Reflective Match, Lossy Match, Feedback and Distributed

Amplifiers: A comparison of Multi-octave Performance Characteristics," *IEEE Int. Microwave Symp. Digest*, 1984, pp. 215–217.

30. Soares, R., J. Graffeuil, and J. Obregon, *Applications of GaAs MESFETs*, Artech House, Dedham, Mass., 1983, Chap. 4.
31. Takayama, Y., "A New Load-Pull Characterization Method for Microwave Power Transistors," *NEC Res. Devel.*, Vol. 50, Apr. 1978, pp. 23–29.
32. Honjo, K., et al., "15-Watt Internally Matched GaAs FETs and 20-watt Amplifier Operating at 6 GHz," *IEEE Int. Microwave Symp. Digest*, 1979, pp. 289–291.
33. Honjo, K., Y. Takayama, and A. Higashisaka, "Broad-Band Internal Matching of Microwave Power GaAs MESFETs," *IEEE Trans. Microwave Theory Tech.*, Vol. MTT-27, Jan. 1979, pp. 3–8.
34. Okubo, N., et al., "A 6-GHz 80-W GaAs FET Amplifier with TM-mode Cavity Power Combiner," *IEEE Int. Microwave Symp. Digest*, 1983, pp. 276–278.
35. Wilkinson, E., "An *N*-Way Hybrid Power Divider," *IRE Trans. Microwave Theory Tech.*, Vol. MTT-8, Jan. 1960, pp. 116–118.
36. Saleh, A. A. M., "Planar Electrically Symmetric *n*-Way Hybrid Power Dividers/Combiners," *IEEE Trans. Microwave Theory Tech.*, Vol. MTT-28, June 1984, pp. 555–563.
37. Schellenberg, J. M., and M. Cohn, "A Wideband Radial Power Combiner for FET Amplifiers," *IEEE Int. Solid State Circuits Conf. Digest*, Feb. 1978.
38. Nagai, N., E. Maekawa, and K. Ono, "New *n*-way Hybrid Power Dividers," *IEEE Trans. Microwave Theory Tech.*, Vol. MTT-25, Dec. 1977, pp. 1008–1012.
39. Yau, W., and J. M. Schellenberg, "An *n*-Way Broadband Planar Power Combiner/Divider," *Microwave J.*, Vol. 29, Nov. 1986, pp. 147–151.
40. Russel, K. J., "Microwave Power Combining Techniques," *IEEE Trans. Microwave Theory Tech.*, Vol. MTT-27, May 1979, pp. 472–478.

PROBLEMS

10.1 Determine the noise figure of the receiver whose block diagram is shown below. Compare its performance with another receiver where the diode mixer is replaced by a dual-gate FET mixer having gain of 7 dB and noise figure of 10 dB.

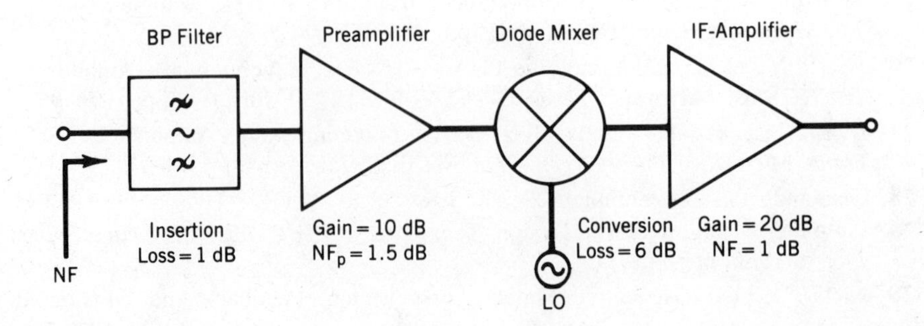

10.2 Lumped element values of a GaAs FET are given in Fig. 10.11. Determine the stability of the device and identify the element values that give rise to unstability.

10.3 The scattering parameters of a GaAs FET are given in Table 10.3, which shows that the FET is potentially unstable at 6 GHz. Draw stability circles at 6 GHz and show how one can stabilize the device.

10.4 The scattering parameters for two different transistors at 8, 10, and 12 GHz are given below. Calculate the stability factor K in each case and draw input and output stability circles on the Smith chart. Show which transistor is suitable for a 10-GHz amplifier.

Frequency (GHz)	S_{11}	S_{21}	S_{12}	S_{22}
		Device 1		
8	$0.847\angle -89°$	$2.43\angle 121°$	$0.029\angle 60°$	$0.744\angle -19°$
10	$0.810\angle -104°$	$2.15\angle 112°$	$0.030\angle 61°$	$0.734\angle -23°$
12	$0.782\angle -117°$	$1.91\angle 103°$	$0.030\angle 67°$	$0.726\angle -27°$
		Device 2		
8	$0.820\angle -89°$	$2.21\angle 118°$	$0.028\angle 66°$	$0.644\angle -24°$
10	$0.778\angle -104°$	$1.95\angle 108°$	$0.030\angle 70°$	$0.635\angle -29°$
12	$0.746\angle -117°$	$1.73\angle 99°$	$0.032\angle 78°$	$0.628\angle -34°$

10.5 If the RF output voltage, v_o, of a FET amplifier is represented by

$$v_o = C_1 v_i + C_2 v_i^2 + C_3 v_i^3$$

where the input voltage $v_i = A \cos \omega_1 t$, show that the gain of the amplifier, G, in decibels is given by

$$G = 20 \log(C_1 + 0.75 C_3 A^2)$$

10.6 In Problem 10.5 if $C_1 = 10$, $C_2 = 0$, and $C_3 = -1$, determine the 1-dB compression point for the amplifier having input and output impedances of 50 Ω.

10.7 Derive an expression for the noise figure for n identical amplifiers, with gain G and noise figure F, cascaded in series. If $G = 5$ and

$F = 2$, what is the maximum number of amplifiers that can be cascaded to obtain a noise figure less than 2.3?

10.8 A three-stage low-noise amplifier has a noise figure of 3 dB and gain of 30 dB over a 1-GHz bandwidth at room temperature. Calculate the DR and DR_f when the 1-dB compression point for the amplifier is 15 dBm.

10.9 The scattering parameters of a GaAs FET at half I_{DSS} are given in Problem 10.4 (Device 2). Design a 12-dB gain amplifier at 10 GHz with input and output VSWR better than 1.5:1 when matched to 50-Ω source and load over a 10% bandwidth.

10.10 The common-source scattering parameters of a GaAs FET at 8 GHz are given below.

$$S_{11} = 0.847\angle -89^\circ \qquad S_{21} = 2.43\angle 121^\circ$$

$$S_{12} = 0.03\angle 60^\circ \qquad S_{22} = 0.744\angle -19^\circ$$

Determine the common-gate and common-drain S-parameters and the stability factor in all the three cases.

10.11 Design a low-noise amplifier at 4 GHz using a GaAs FET whose S-parameters are given in Table 10.3 and other design parameters are given in Table 10.4. Desirable values for the noise figure and associated gain are 1.2 dB and 12 dB, respectively.

10.12 Design a two-stage amplifier with (20 ± 1)-dB gain from 8–12 GHz. The source and load impedances are 50 Ω and the minimum acceptable VSWR at output is 1.5.

10.13 Design a broadband, 10-dB flat gain MESFET amplifier using hybrid circuit matching networks in the input and output for the 6 to 8 GHz range. S-parameters are given in Table 10.3. The substrate to be used is RT/duroid ($\epsilon_r = 2.22$ and $h = 0.63$ mm). The effect of dispersion, strip thickness, and discontinuities may be ignored.

10.14 Derive an expression for the noise figure F of a balanced amplifier in terms of the individual amplifier's noise figures F_1, F_2 and gains G_1, G_2 assuming the 3-dB 90° divider/combiner is ideal.

10.15 Using Equations 10.40a through 10.40f design a 2-W power amplifier at 10 GHz assuming $V_{DSB} = 20$ V, $V_{DS} = 9$ V, and $I_{DS} =$ 0.7 A. The amplifier operates in Class A mode.

10.16 Assuming that all individual amplifiers in an N-way amplifier are perfectly matched and have the same transmission phase and 100% efficiency of the divider/combiner, prove that the P_{out} of this amplifier is given by

$$P_{out} = \frac{1}{N}\left(\sum_{n=1}^{N} \sqrt{P_n}\right)^2$$

where P_n is the output power of the nth individual amplifier.

11 DETECTORS AND MIXERS

11.1 INTRODUCTION

Detectors and mixers make use of the nonlinear characteristics of certain solid state devices to generate an output signal containing many frequency components. This is illustrated in Fig. 11.1, which shows the output current resulting from a sinusoidal input applied to the nonlinear $I-V$ characteristic of a diode. A Fourier analysis of the output signal (Fig. 11.1) shows that it contains a component at the input frequency as well as a dc and higher frequency components. The more nonlinear the diode $I-V$ curve, the more efficient the detection or mixing process, that is, the higher the percentage of power contained in frequency components other than the input frequency.

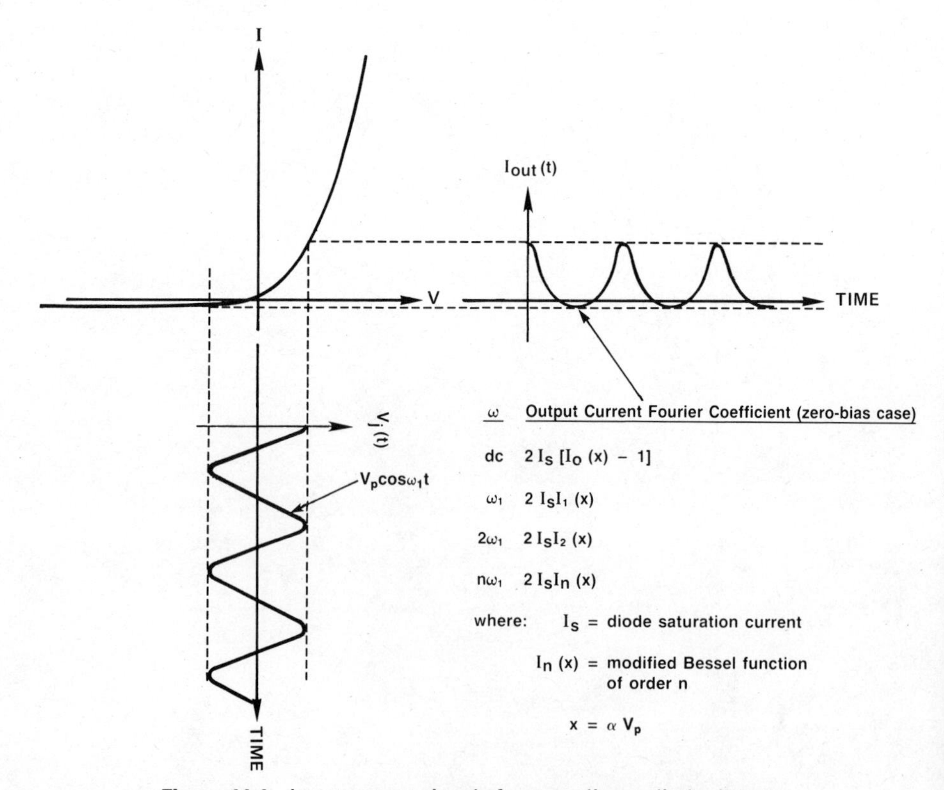

Figure 11.1 Input–output signals for a nonlinear diode detector.

The dc voltage–current relationship of a diode is given as

$$i = I_s[e^{\alpha V_j} - 1] \tag{11.1}$$

where

i = instantaneous current,
I_s = diode saturation current,
V_j = instantaneous voltage across diode junction,
$\alpha = e/nkT$,
e = electron charge = $1.6021917 \times 10^{-19}$ C,
k = Boltzmann's constant = 1.380622×10^{-23} Joule/K,
kT/e = 0.025248 V at 293 K,
n = ideality factor, typically 1 to 1.5.

For the case of a small applied ac voltage δV, (11.1) can be expressed using Taylor's expansion as

$$i = i(V_0 + \delta V) \tag{11.2a}$$

$$= i(V_0) + \delta V \left.\frac{di}{dv}\right|_{I_0} + \frac{\delta V^2}{2!} \left.\frac{d^2 i}{dv^2}\right|_{I_0} + \cdots + \frac{\delta V^n}{n!} \left.\frac{d^n i}{dv^n}\right|_{I_0} + \cdots \tag{11.2b}$$

where

V_0 = dc bias voltage,
I_0 = dc or average bias current,
δV = ac voltage across diode junction.

For $\delta V = V_p \cos(\omega_c t)$, and ignoring terms higher than second order,

$$i = i(V_0) + V_p \cos(\omega_c t) \left.\frac{di}{dv}\right|_{I_0} + \frac{V_p^2}{2} \cos^2(\omega_c t) \left.\frac{d^2 i}{dv^2}\right|_{I_0} \tag{11.3}$$

where

ω_c = carrier frequency,
V_p = peak amplitude of carrier.

The diode current consists of a dc term due to an external bias voltage, a term at ω_c, and a second-order rectified term given by

$$\Delta i = \frac{V_p^2}{4}[1 + \cos(2\omega_c t)] \left.\frac{d^2 i}{dv^2}\right|_{I_0} \tag{11.4}$$

This term consists of a dc and high-frequency component, which are proportional to the square of the voltage (i.e., power) of the input signal.

11.1.1 Basics of Detection; Video and Heterodyne

Detectors are used to provide an output signal that contains the "information" of the input signal, that is, the amplitude or amplitude variations of the signal. For the general case, consider the AM modulated input signal

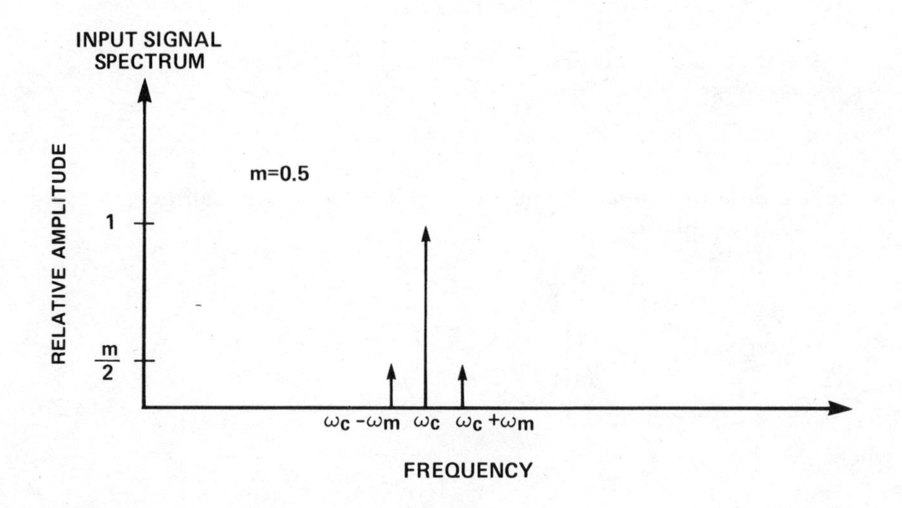

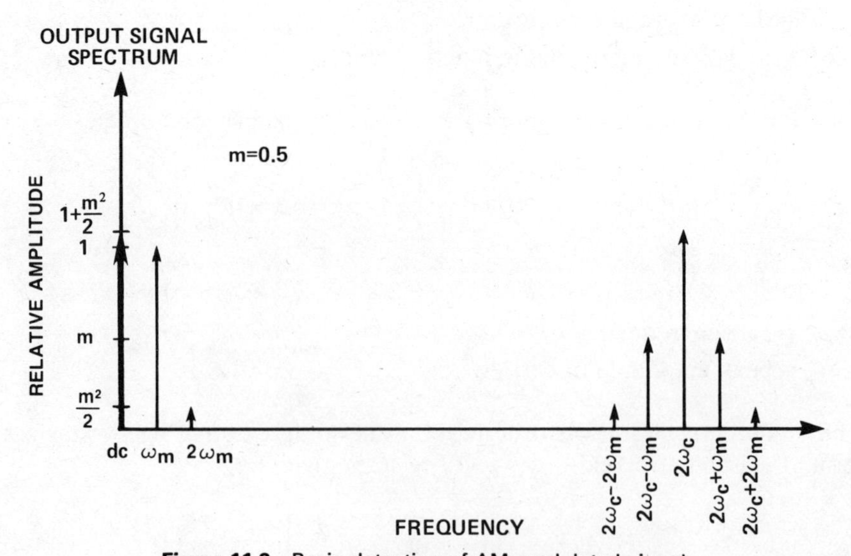

Figure 11.2 Basic detection of AM modulated signal.

given by the expression

$$v_s(t) = V_p[1 + m\sin(\omega_m t)]\sin(\omega_c t) \tag{11.5a}$$

$$= V_p\sin(\omega_c t) + \frac{mV_p}{2}[\cos(\omega_c - \omega_m)t - \cos(\omega_c + \omega_m)t] \tag{11.5b}$$

where

$v_s(t)$ = instantaneous signal voltage,
ω_c = carrier frequency,
V_p = peak amplitude of carrier,
ω_m = modulation frequency,
m = modulation index, typically 0 to 1.

The spectrum of the input signal and the square law detected output signal are shown in Fig. 11.2 for a modulation index, $m = 0.5$. The output signal components generated in the diode and relative amplitudes are listed in Table 11.1. These are obtained by substituting (11.5) into (11.2). The actual amplitudes for the output components are obtained by multiplying the relative amplitude by [1]

$$\frac{V_p^2}{4}\frac{d^2i}{dv^2}\bigg|_{I_0}$$

TABLE 11.1 Output Signal Components and Relative Amplitudes

Frequency	Relative Amplitude
dc	$1 + \frac{m^2}{2}$
ω_m	$2m$
$2\omega_m$	$\frac{m^2}{2}$
$2\omega_c$	$1 + \frac{m^2}{2}$
$2\omega_c \pm \omega_m$	m
$2\omega_c \pm 2\omega_m$	$\frac{m^2}{4}$

For $m = 0$ (no modulation), and again ignoring terms higher than second order, the diode generates output components only at dc and $2\omega_c$.

Mixers convert (heterodyne) the input frequency to a new frequency where filtering and/or gain is easier to implement. Mixing is achieved by applying both the input signal and a local oscillator to a nonlinear element. The local oscillator is a higher power signal used to pump the nonlinear device more effectively than the much smaller amplitude input signal. Again, many frequencies are generated in the nonlinear device as shown in Fig. 11.3. The output frequency components are determined from the following equation:

$$\omega_{\text{out}} = m\omega_s + n\omega_p \tag{11.6}$$

where

ω_p = local oscillator frequency,
ω_s = signal frequency,
m, n = integers, $-\infty$ to $+\infty$.

The amplitudes of the output signal components depend on the mixer design.

In a fundamental mixer, the desired output frequency is the sum or difference between ω_p and ω_s and is called the intermediate frequency (IF). The conversion loss from signal to IF is minimized by presenting the required impedances to the diode.

11.1.2 Applications

Detectors are used in a wide variety of applications in microwave systems as a source of a dc (or a low video frequency) voltage that is proportional to the amplitude of the RF signal. These applications include:

signal strength indicators,
automatic gain control (AGC),
power monitors, status indicators,
automatic leveling control (ALC).

In these applications, filtering is required to reject all but the dc component.

Detectors can also be used to detect AM modulation, since the output spectrum contains a component at the modulating frequency ω_m, as shown in Fig. 11.2. By providing filtering to reject dc and frequencies higher than ω_m, the output signal contains only the modulation signal.

Mixers are used to convert the input frequency to a frequency either higher (up-conversion) or lower (down-conversion). A common application

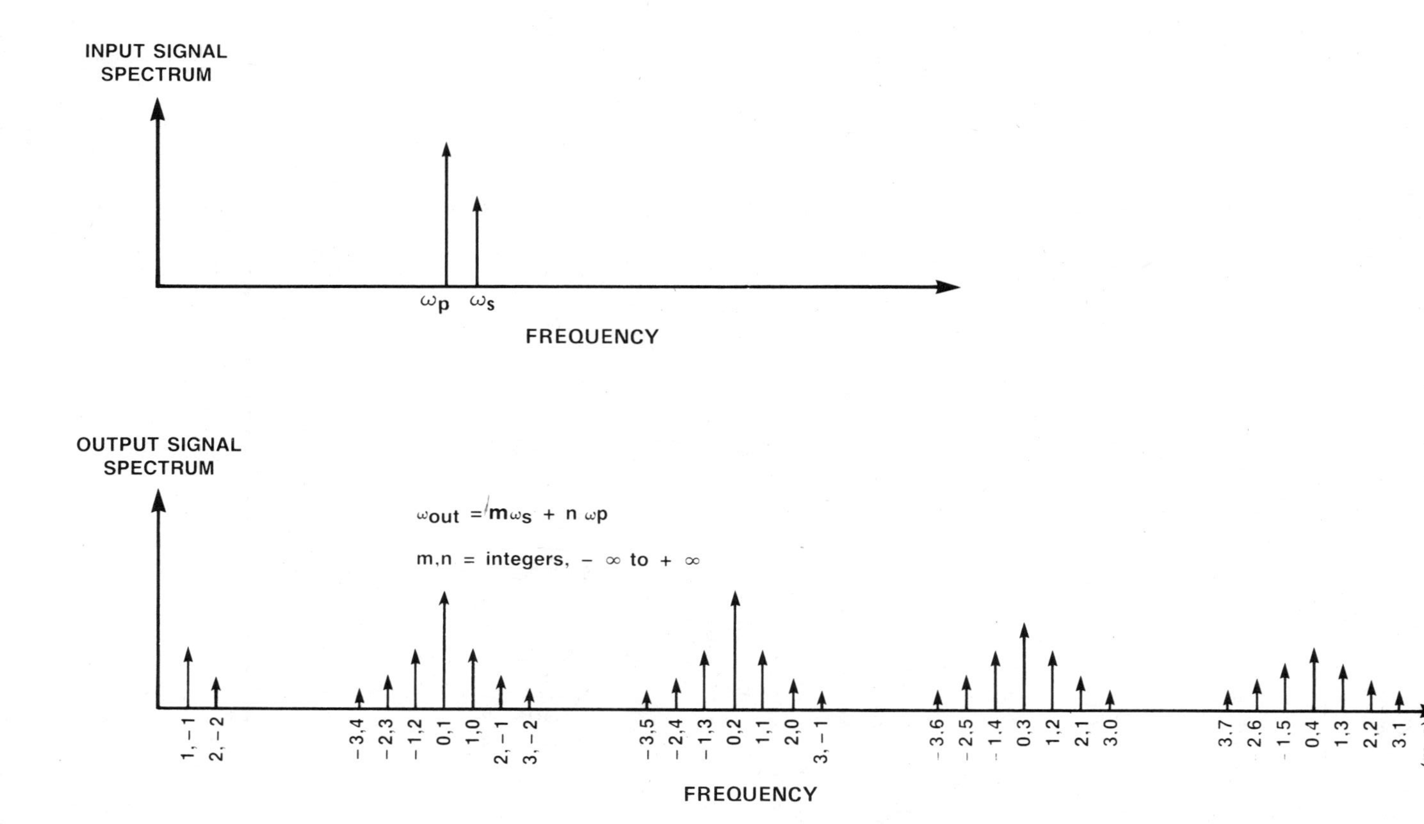

Figure 11.3 Frequency components generated in a mixer.

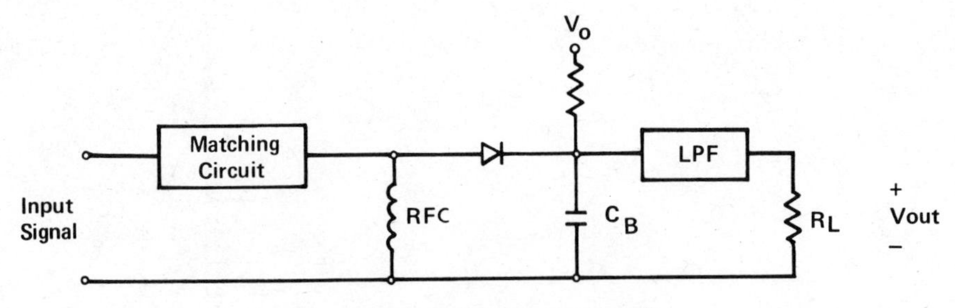

Figure 11.4 Basic detector circuit.

is in a receiver where a mixer is used to down convert the incoming frequency to a lower IF where filtering and gain are easier to implement. Mixers are used as up-converters in some transmitters where the modulation is applied at a lower frequency and then mixed up to the transmitter output frequency.

11.2 DETECTORS

As shown in Fig. 11.1, detectors make use of the nonlinear characteristics of some semiconductor devices to generate a dc and higher frequency components. The schematic for a typical detector is shown in Fig. 11.4. It utilizes a diode as the detecting device with the required RF bypass and dc return that serve to maximize the RF voltage across the diode.

11.2.1 Basic Theory

The simplified diode equivalent model shown in Fig. 11.5 consists of a bias-dependent junction resistance and capacitance R_j and C_j, as well as a series resistor R_s, due to contact, substrate, and spreading resistance. This series resistance reduces the diode sensitivity by reducing the voltage across

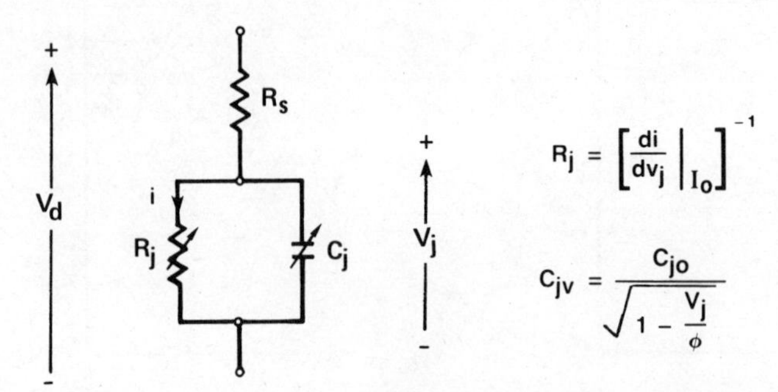

Figure 11.5 Diode equivalent model. The capacitances C_{jo} and C_{jv} are at 0 and v volt, respectively.

the diode junction. Under small-signal conditions, the junction voltage V_j, resulting from a voltage V_d, applied to a diode is given by

$$V_j = \frac{V_d}{(1 + R_s/R_j) + j\omega C_j R_s} \tag{11.7}$$

and the magnitude of the junction voltage is given by

$$|V_j| = \frac{|V_d|}{[(1 + R_s/R_j)^2 + (\omega C_j R_s)^2]^{1/2}} \tag{11.8}$$

It was shown that the rectified output current from a nonlinear junction diode is given by

$$\Delta i = \frac{\delta V^2}{2} \frac{d^2 i}{dv^2}\bigg|_{I_0} \tag{11.9}$$

where δV is the ac voltage across the junction.

For $\delta V = V_p \cos(\omega_c t)$

$$\Delta i = \frac{V_p^2}{4}[1 + \cos(2\omega_c t)]\frac{d^2 i}{dv^2} \tag{11.10}$$

where

$$\frac{di}{dv}\bigg|_{I_0} = \frac{d}{dv}[I_s(e^{\alpha V_j} - 1)] \tag{11.11a}$$

$$= \alpha I_s e^{\alpha V_j}|_{I_0} \tag{11.11b}$$

$$= \alpha(I_s + I_0) \tag{11.11c}$$

The junction resistance R_j is the inverse of the average slope of the I–V curve at the operating point. Therefore,

$$R_j = \left[\frac{di}{dv}\bigg|_{I_0}\right]^{-1} = \frac{1}{\alpha(I_s + I_0)} = \frac{nkT/e}{I_s + I_0} \tag{11.12a}$$

$$= \frac{0.025248n}{I_s + I_0} \tag{11.12b}$$

at 293 K (for I_s and I_0 expressed in amps).

The second derivative is therefore

$$\frac{di^2}{dv^2}\bigg|_{I_0} = \alpha^2(I_s + I_0) = \frac{\alpha}{R_j} \tag{11.13}$$

The rectified short circuit current is then given by

$$\Delta i = \frac{V_p^2}{4} \frac{\alpha}{R_j} \tag{11.14}$$

Adding the effect of R_s [2]

$$\Delta i = \frac{V_p^2}{4} \frac{\alpha}{R_j} \frac{R_j}{R_s + R_j} \tag{11.15}$$

The power absorbed in the diode is given by

$$P = \frac{V_d^2}{2} \mathrm{Re}[Y_d] \tag{11.16}$$

where V_d is the peak voltage across the diode. The diode admittance is given by

$$Y_d = \frac{1/R_j + j\omega C_j}{(1 + R_s/R_j) + j\omega C_j R_s} \tag{11.17}$$

and

$$\mathrm{Re}[Y_d] = \frac{(1/R_j)(1 + R_s/R_j) + (\omega C_j)^2 R_s}{(1 + R_s/R_j)^2 + (\omega C_j R_s)^2} \tag{11.18}$$

The power absorbed in the diode is therefore

$$P = \frac{V_d^2}{2} \frac{(1/R_j)(1 + R_s/R_j) + (\omega C_j)^2 R_s}{(1 + R_s/R_j)^2 + (\omega C_j R_s)^2} \tag{11.19a}$$

The peak junction voltage V_p is related to the peak diode voltage by (11.8), and the power absorbed in the diode can be expressed as

$$P = \frac{V_p^2}{2} \left[\frac{1}{R_j} \left(1 + \frac{R_s}{R_j} \right) + (\omega C_j)^2 R_s \right] \tag{11.19b}$$

The detector current sensitivity β_i is defined as the detected current divided by the power absorbed in the diode [3], that is,

$$\beta_i = \frac{\Delta i}{P} \tag{11.20a}$$

$$= \frac{(V_p^2/4)(\alpha/R_j)[R_j/(R_s + R_j)]}{(V_p^2/2)[(1/R_j)(1 + R_s/R_j) + (\omega C_j)^2 R_s]} \tag{11.20b}$$

$$= \frac{\alpha/2}{(1 + R_s/R_j)[(1 + R_s/R_j) + (\omega C_j)^2 R_s R_j]} \tag{11.20c}$$

The voltage sensitivity into an open circuit is defined as the current sensitivity times R_j, and is given by

$$\beta_v = \beta_i \left[\frac{di}{dv}\bigg|_{I_0} \right]^{-1} = \beta_i R_j = \beta_i \frac{1}{\alpha(I_s + I_0)} \tag{11.21a}$$

$$= \frac{0.5}{(I_s + I_0)(1 + R_s/R_j)[(1 + R_s/R_j) + (\omega C_j)^2 R_s R_j]} \text{ V/W} \tag{11.21b}$$

To correct for noninfinite load resistors, the sensitivity must be multiplied by $R_L/(R_v + R_L)$ where R_L is the load resistance and R_v is the diode video resistance defined as

$$R_v = R_s + R_j \tag{11.22}$$

The detector voltage sensitivity can then be expressed as

$$\beta_v = \frac{0.5 R_L}{(I_s + I_0)(R_v + R_L)(1 + R_s/R_j)[(1 + R_s/R_j) + (\omega C_j)^2 R_s R_j]} \text{ V/W} \tag{11.23}$$

If the condition $R_s \ll R_j$ is satisfied, the voltage sensitivity can be approximated by

$$\beta_v = \frac{0.0005}{(I_s + I_0)(1 + R_j/R_L)[1 + (\omega C_j)^2 R_s R_j]} \text{ mV}/\mu\text{W} \tag{11.24}$$

For a typical diode with no bias ($C_j = 0.13$ pF, $R_s = 10\ \Omega$, $I_s = 1 \times 10^{-7}$ A), the voltage sensitivity versus frequency is plotted in Fig. 11.6. It shows a strong dependence on frequency due to the change in voltage across the diode junction. For broadband applications, the drop in sensitivity can be compensated for in the RF matching circuit.

Biasing the diode also reduces the variation in voltage sensitivity, as shown in Fig. 11.6 for $I_0 = 20\ \mu$A, but at reduced sensitivity for this example diode, particularly at the lower frequencies. This is caused by the lower R_j due to the bias current I_0 and the resulting reduction in voltage across the junction. The voltage sensitivity is a parabolic-type function with I_T ($I_T = I_s + I_0$) and is a maximum value at any particular frequency at a total current given by

$$I_P = \frac{\omega C_j}{\alpha} \sqrt{\frac{R_s}{R_L}} \tag{11.25}$$

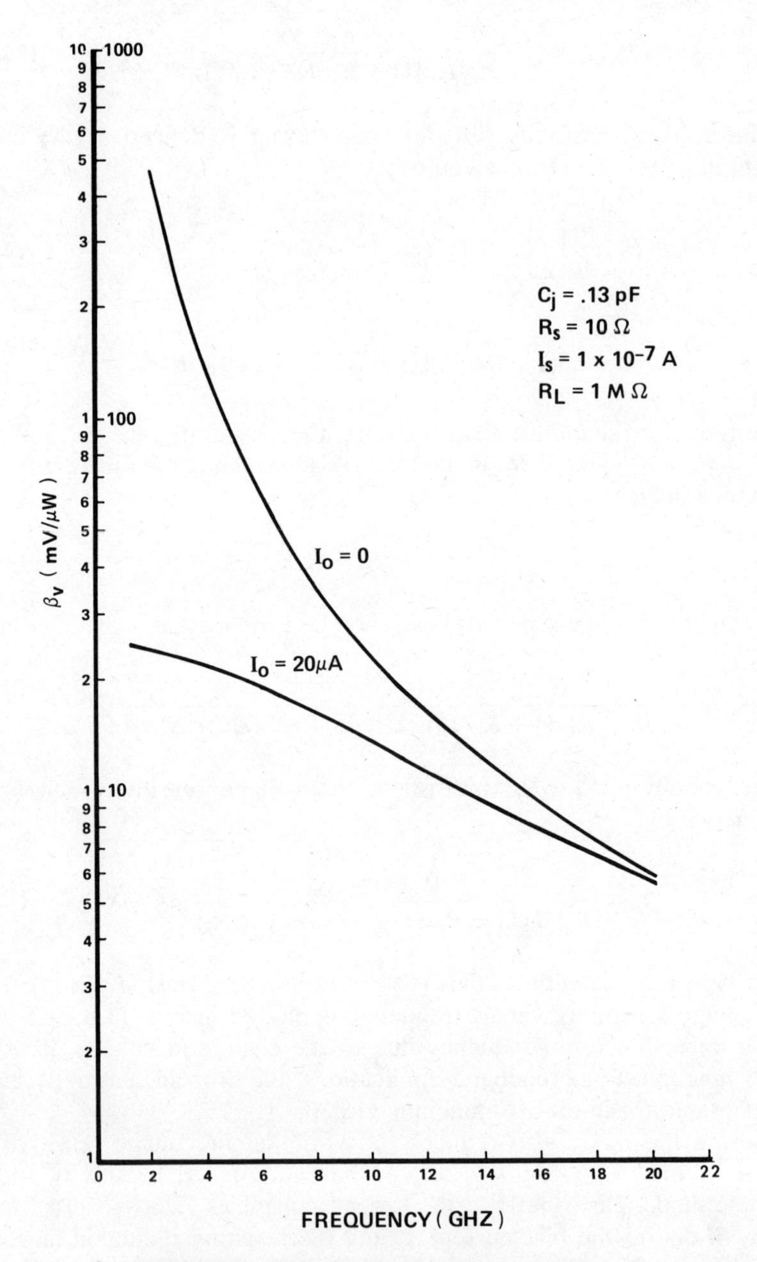

Figure 11.6 Detector diode voltage sensitivity.

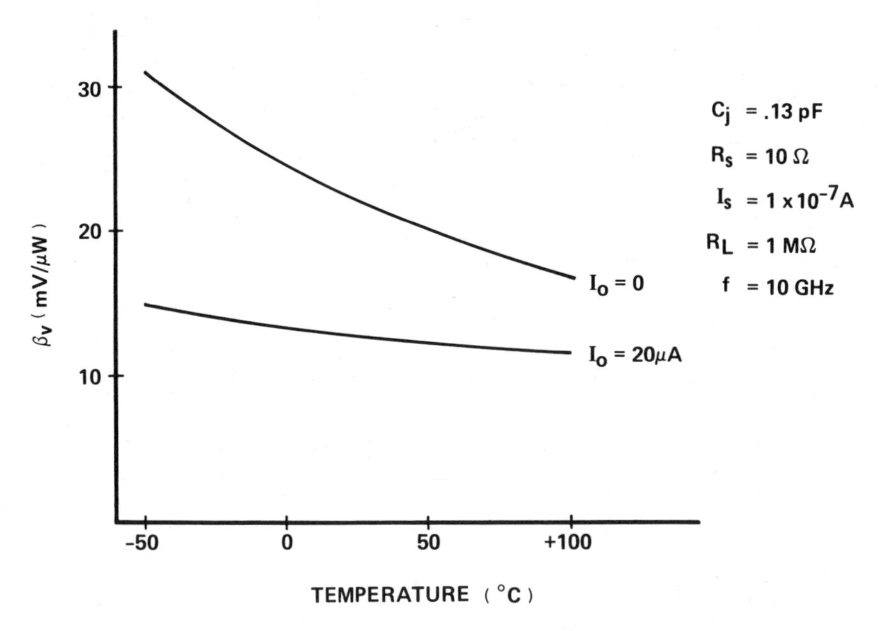

Figure 11.7 Detector voltage sensitivity variation with temperature.

For currents greater than I_P, β_v drops due to the reduced voltage across the diode junction. For currents less than I_P, R_j gets large relative to R_L and the sensitivity is reduced due to the $R_L/(R_v + R_L)$ voltage divider.

The diode sensitivity also varies with temperature since $\alpha = e/nkT$ is inversely proportional to temperature. Temperature effects can be added by letting

$$R_j = \frac{8.617 \times 10^{-5} nT}{I_s + I_0} \tag{11.26}$$

where T is in K and I_s and I_0 are expressed in A.

The variation in voltage sensitivity for the same diode (with and without bias) at 10 GHz from -50 to $+100$°C is shown in Fig. 11.7. Again, the addition of a bias current is seen to reduce the variation in voltage sensitivity.

Diode Impedance. To determine the RF matching circuit, the diode impedance including package parasitics must be calculated. The example diode equivalent model with a typical package ($C_p = 0.06$ pF, $L_p = 2.2$ nH) and the corresponding diode impedance are shown in Fig. 11.8. The impedance is also plotted for $I_0 = 20\ \mu$A and 50 μA, showing the effects of external bias. The Q of the diode impedance is defined by

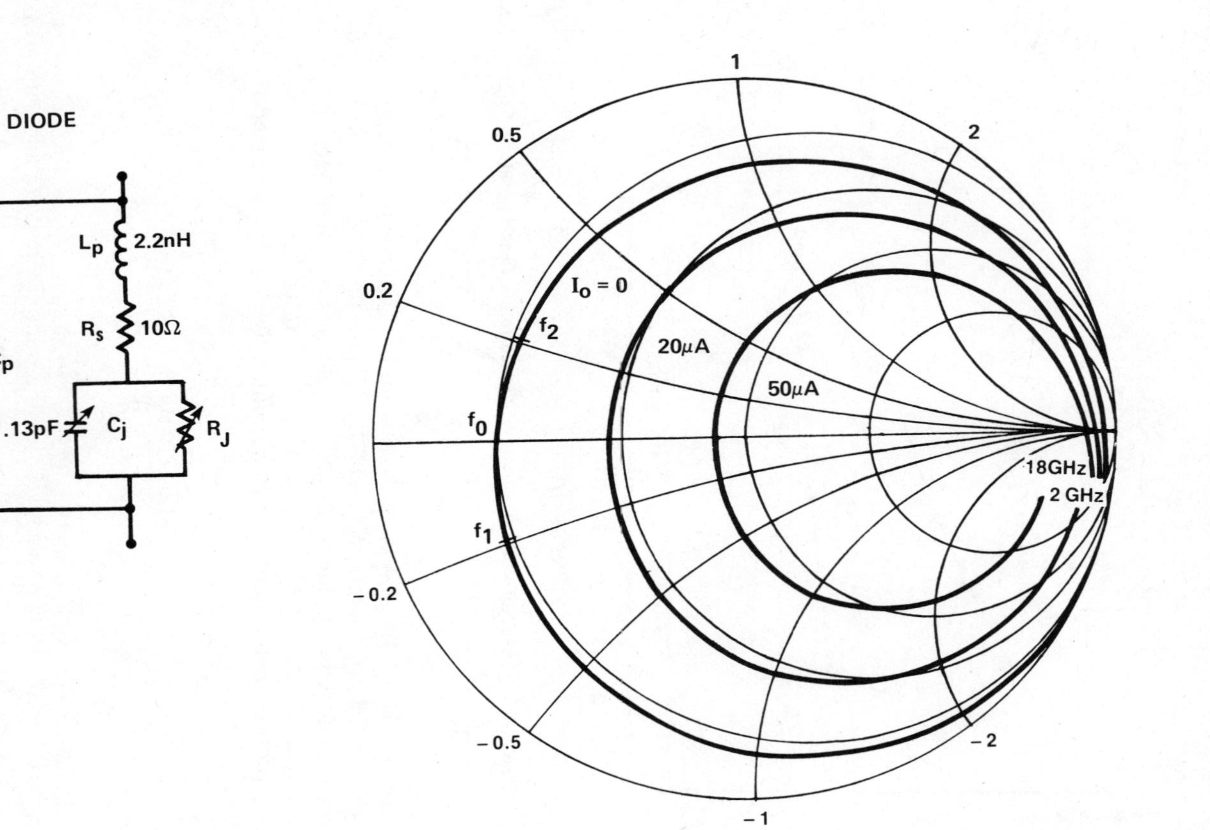

Figure 11.8 Diode impedance (2–18 GHz) with and without external bias.

$$Q = \frac{f_0}{B} \tag{11.27}$$

where

f_0 = series resonant frequency,
B = 3-dB bandwidth = $f_2 - f_1$,
f_2 and f_1 = frequencies where $|X_{\text{series}}| = R_{\text{series}}$.

For this example, the data listed in Table 11.2 were calculated.

The addition of an external bias reduces the diode Q, which reduces the complexity of the RF matching circuit.

The data presented assumed an axial lead glass package as shown in Fig. 11.9. Many other packages are available, including the pill ($C_p = 0.12$ pF, $L_p = 0.3$ nH), the double prong ($C_p = 0.15$ pF, $L_p = 1$ nH) and the beam lead packages ($C_p = 0.02$ pF, $L_p = 0.09$ nH, typical) shown in Fig. 11.9. The package selection depends upon the frequency of operation, bandwidth, and construction used in the detector.

The loss in detector sensitivity due to mismatch losses can be determined by referring to Fig. 11.10. The loss in power delivered to the diode and therefore the loss in sensitivity is given by [4],

$$\beta'_v = \beta_v(1 - |\Gamma|^2) \tag{11.28}$$

where

$$\Gamma = \frac{Z_{\text{in}} - Z_0}{Z_{\text{in}} + Z_0} \tag{11.29}$$

and Z_{in} is the diode impedance transformed through the matching circuit. A 3:1 mismatch ($|\Gamma| = 0.5$) reduces the voltage sensitivity by a factor of 0.75. Mismatch loss clearly is a dominant term in determining detector sensitivity.

Diode Matching. Any of the standard matching techniques described in Chapter 4 can be used to match the detector impedance to 50 Ω. Since the

TABLE 11.2 Diode *Q* vs. Bias Current

I_0 (μA)	f_0 (GHz)	R_{series} (Ω)	B (GHz)	Q
0	9.36	10	0.64	14
20	9.36	23	1.60	6
50	9.36	42	3.20	3

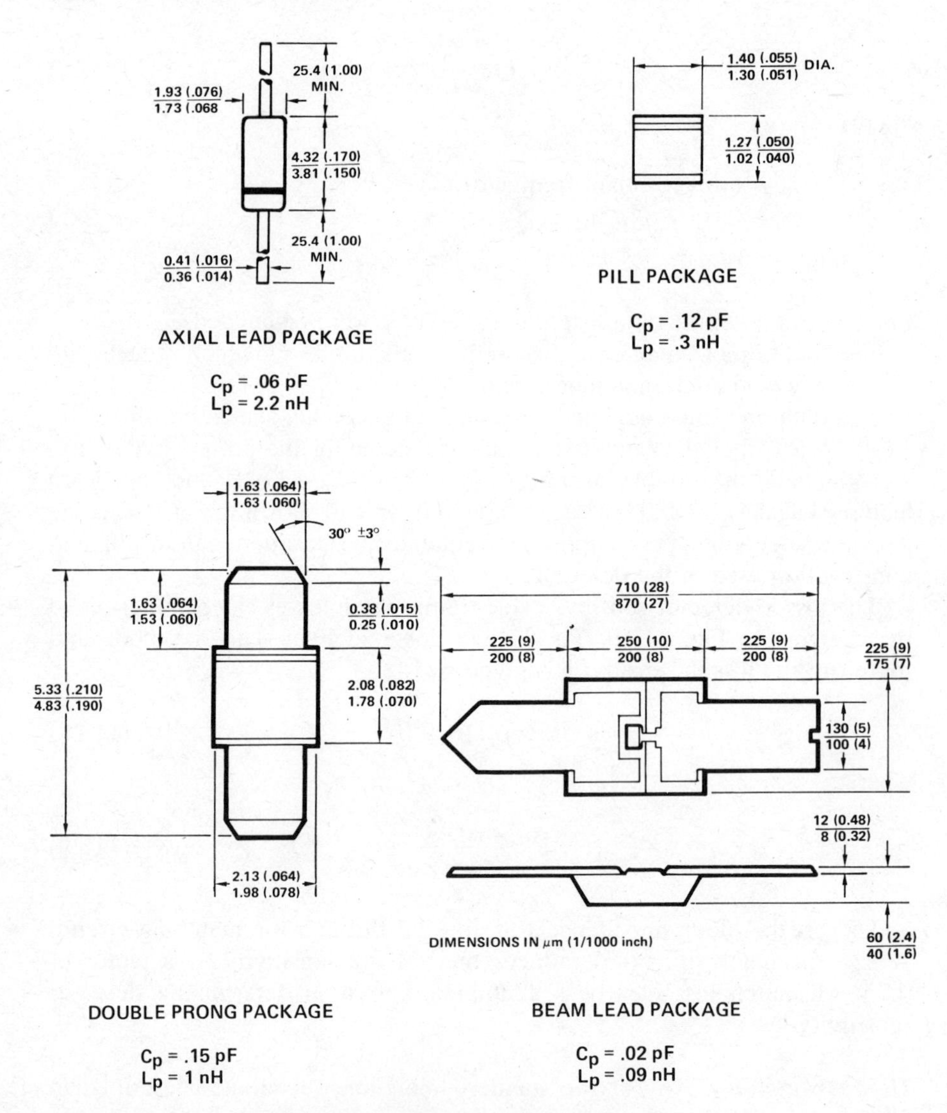

Figure 11.9 Typical diode packages.

detector is normally a small-signal device, the diode impedance is not changing with drive and conventional matching and circuit optimization techniques can be utilized. The complexity of the matching circuit depends on the bandwidth required and the desired detector VSWR. Several detector matching techniques are illustrated in the design examples presented in Section 11.2.5.

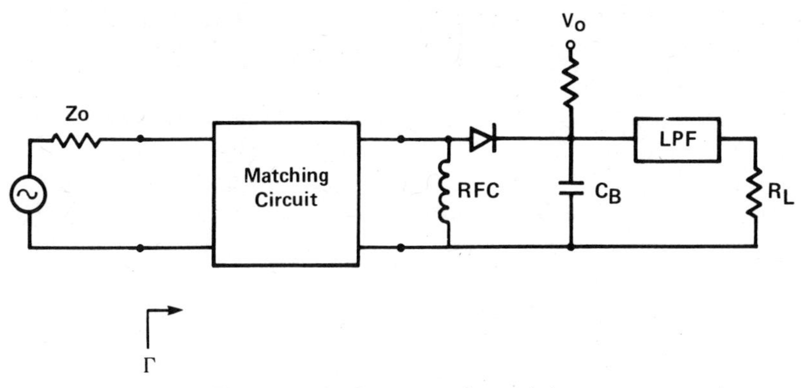

Figure 11.10 Detector mismatch loss.

Noise. Noise generated in the detector diode limits the minimum signal strength that can be detected. The noise consists of shot and thermal noise, which are independent of frequency, and flicker noise, which varies as the inverse of frequency.

Shot Noise. Van der Ziel [5] has shown that the shot noise in bandwidth B generated in a Schottky-barrier diode due to the junction resistance is given by

$$\overline{i_s^2} = 2e(I_0 + 2I_s)B \tag{11.30}$$

but

$$R_j = \frac{(nkT/e)}{I_0 + I_s} \tag{11.31}$$

therefore

$$\overline{i_s^2} = 2\frac{nkTB}{R_j}\left[\frac{I_0 + 2I_s}{I_0 + I_s}\right] \tag{11.32}$$

and

$$v_{n1}^2 = 2nkTBR_j\left[\frac{I_0 + 2I_s}{I_0 + I_s}\right] \tag{11.33}$$

Thermal Noise. The diode series resistance contributes thermal noise given by

$$v_{n2}^2 = 4kTBR_s \tag{11.34}$$

Flicker Noise. Flicker noise varies as $1/f$, and its value depends on the frequency and bandwidth of the measurement. It can be expressed as

$$v_{n3}^2 = \int_{f_L}^{f_L+B} \left(\frac{A}{f}\right) df = A \ln\left(1 + \frac{B}{f_L}\right) \tag{11.35}$$

where f_L is the lower frequency limit of the measurement bandwidth B, and A is a constant determined by the diode's flicker noise.

A corner frequency, f_c, is defined as the frequency where the flicker noise equals the shot noise generated in the junction resistance. By equating (11.35) with (11.33) at frequency f_c, f_c can be derived to be

$$f_c = \frac{A}{2nkTR_j[(I_0 + 2I_s)/(I_0 + I_s)]} \approx \frac{A}{2nkTR_j} \quad \text{for } I_0 \gg I_s \tag{11.36}$$

The flicker noise can then be approximated by

$$v_{n3}^2 = 2nkTR_j f_c \ln\left(1 + \frac{B}{f_L}\right) \tag{11.37}$$

The total noise generated in the diode is then

$$v_n^2 = 4kTB\left[R_j \frac{n}{2}\frac{I_0 + 2I_s}{I_0 + I_s} + R_s + R_j \frac{n}{2}\frac{f_c}{B} \ln\left(1 + \frac{B}{f_L}\right)\right] \tag{11.38}$$

A setup similar to the one shown in Fig. 11.11 is used to characterize the noise performance of detector diodes. It consists of the detector with sensitivity β_v and a postamplifier with gain G. The output voltage consists of signal and noise where the signal voltage $V_s = \beta_v P_{\text{in}} G$. The output noise includes the detector noise given by (11.38) multiplied by G with an additional term to account for the noise generated in the amplifier. This term is $v_n^2 = 4kTBR_a$, where R_a is the equivalent noise resistance of the amplifier, typically 1000 Ω. The total output noise from the amplifier is then

$$v_a^2 = 4kTBG\left[R_j \frac{n}{2}\frac{I_0 + 2I_s}{I_0 + I_s} + R_j \frac{n}{2}\frac{f_c}{B} \ln\left(1 + \frac{B}{f_L}\right) + R_s + R_a\right] \tag{11.39a}$$

$$= 4kTBG\left[R' + R_j \frac{n}{2}\frac{f_c}{B} \ln\left(1 + \frac{B}{f_L}\right)\right] \tag{11.39b}$$

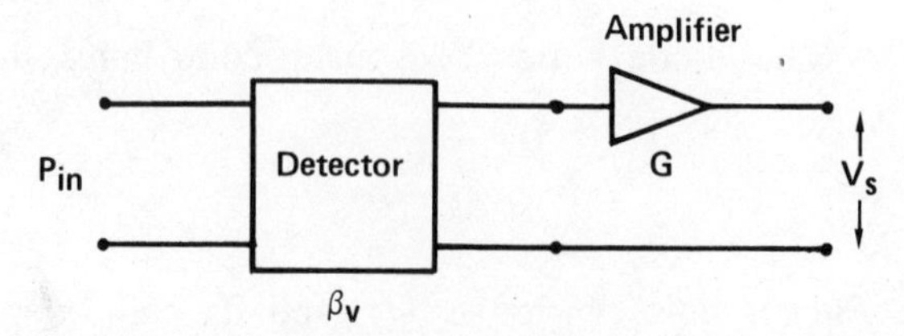

Figure 11.11 Setup to measure detector noise performance.

where

$$R' = R_j \frac{n}{2} \frac{I_0 + 2I_s}{I_0 + I_s} + R_s + R_a \tag{11.40}$$

The signal-to-noise ratio is

$$\frac{S}{N} = \frac{\beta_v P_{\text{in}}}{[4kTB(R' + R_j(n/2)(f_c/B)\ln(1 + B/f_L))]^{1/2}} \tag{11.41}$$

For diodes with low flicker noise, the signal-to-noise ratio simplifies to

$$\frac{S}{N} = \frac{\beta_v P_{\text{in}}}{\sqrt{4kTBR'}} \tag{11.42}$$

Defining a diode figure of merit M as

$$M = \frac{\beta_v}{\sqrt{R'}} \tag{11.43}$$

then

$$\frac{S}{N} = \frac{P_{\text{in}}}{\sqrt{4kTB}} M \tag{11.44}$$

The figure of merit M is dependent on the diode parameters and can be used to compare diodes by keeping R_a constant.

For the case where $I_0 = 0$ and assuming $R_s \ll R_j$ and $n = 1$, the expression for M becomes the familiar

$$M = \frac{\beta_v}{\sqrt{R_j + R_a}} \tag{11.45}$$

The minimum detectable signal can be determined by letting $S/N = 1$. The minimum input power P_{min} is therefore

$$P_{\text{min}} = \frac{\sqrt{4kTB}}{M} \text{ W} \tag{11.46a}$$

$$= 10 \log\left(\frac{\sqrt{4kTB}}{M}\right) \text{ dBW} \tag{11.46b}$$

For example, if $\beta_v = 8$ mV/μW at 300 K, $I_s = 1 \times 10^{-6}$ A, $I_0 = 0$, $n = 1.05$, $R_s = 15\ \Omega$, and $R_a = 1200\ \Omega$, then $M = 46.4$ from (11.43).

For a 3-MHz bandwidth and noting that $kT = 4.14 \times 10^{-21}$ W/Hz at

300 K, then

$$P_{\min} = \frac{\sqrt{4(4.14\times 10^{-21})(3\times 10^{6})}}{46.4}$$

$$= 4.8\times 10^{-9}\ \mathrm{W}$$

or

$$= -83.18\ \mathrm{dBW} = -23.18\ \mathrm{dB}\mu$$

The second way to characterize a detector's low-level performance is the tangential signal sensitivity (TSS) defined as the condition when the negative noise peaks riding on the detected voltage V_{det} equal the noise peaks when no signal is present. This is shown in Fig. 11.12. At the tangent condition,

$$V_{pk} = V_{\text{det}} - V_{pk} \tag{11.47}$$

then

$$V_{\text{det}} = 2\,V_{pk} \approx 2\sqrt{2}\,V_n \tag{11.48a}$$

therefore

$$V_{\text{det}} = \beta_v P_{\text{in}} \approx 2.8\,V_n \tag{11.48b}$$

For diodes with low flicker noise, the tangential power is then

$$\mathrm{TSS} = \frac{2.8\,V_n}{\beta_v} = \frac{2.8\sqrt{4kTBR'}}{\beta_v} \tag{11.49a}$$

$$= \frac{2.8\sqrt{4kTB}}{M} \tag{11.49b}$$

$$= 2.8\,P_{\min} \tag{11.49c}$$

The tangential signal sensitivity is therefore approximately 4 dB above $P_{\min}$, where $S/N = 1$. The 4 dB is only approximate because of the random

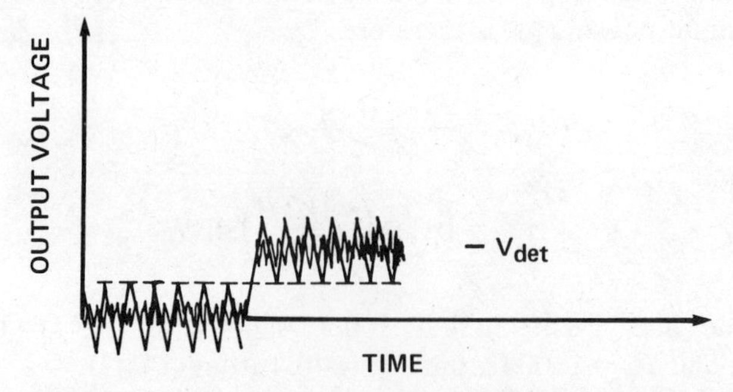

Figure 11.12 Condition for tangential signal sensitivity.

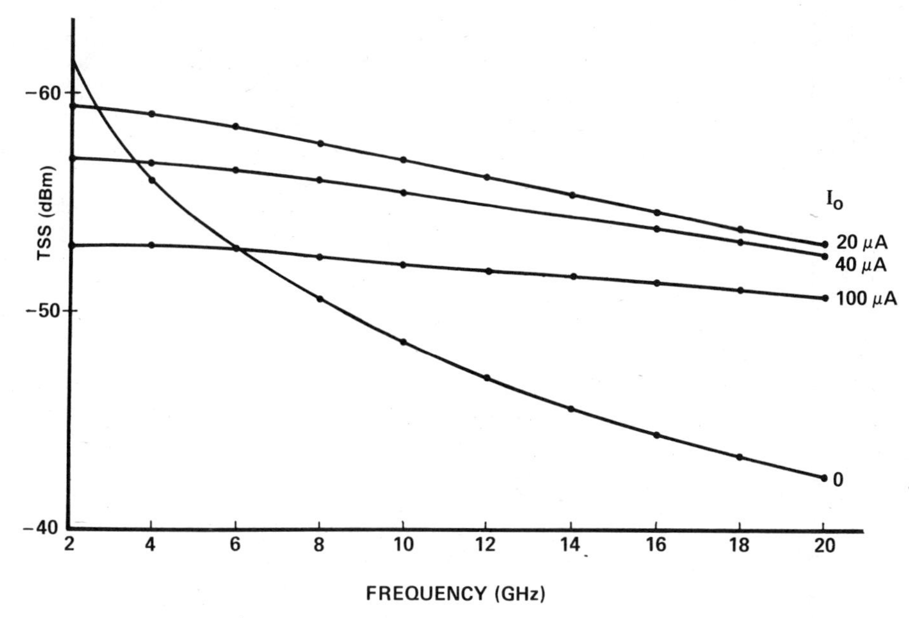

Figure 11.13 Detector TSS performance vs. frequency and bias current.

nature of the noise peaks. The measurement of tangential signal sensitivity is difficult because of the subjective definition of the tangent condition.

The detector TSS is a function of the bias current I_0 and frequency due to the variation in voltage sensitivity with frequency, as seen from (11.23) and (11.49a). A small bias current generally provides the best TSS performance, enabling the detector to work to lower input power levels. This is seen in Fig. 11.13, which is a plot of TSS versus frequency for the example diode used in Fig. 11.6. At 10 GHz, the TSS improves about 8.5 dB by biasing the diode at 20 μA of dc bias. Higher currents degrade the TSS performance.

Large-Signal Effects. For low-level input signals, the detector has been shown to behave as a square law device, that is, the output voltage is proportional to the input voltage squared, that is, the input power. As the input power increases, additional terms of (11.2) must be considered and the detector becomes linear. In this region, the change in output voltage is proportional to the change in input voltage. This is illustrated in Fig. 11.14. The break point, or compression point as it is called, is typically -20 dBm. The addition of a dc bias increases the compression point and extends the region of square law behavior. The dynamic range of the detector, defined as the range of input power between the TSS and the compression point, increases with bias current. This is because the compression point increases faster than the TSS degrades for high dc bias currents.

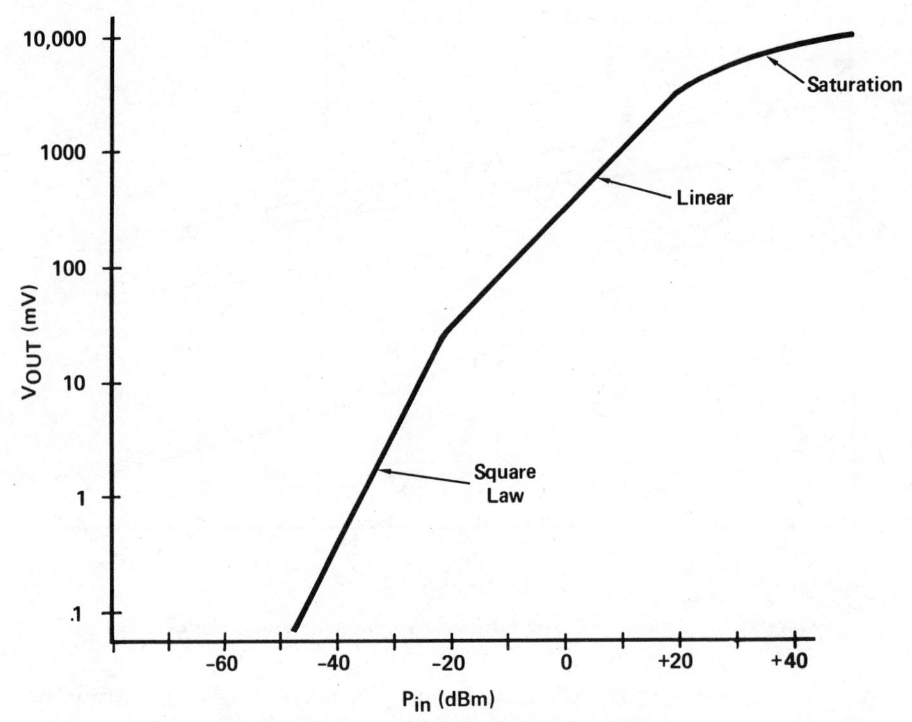

Figure 11.14 Detector transfer characteristic.

Saturation and Diode Burnout. If the input power is increased to the point where the reverse voltage on the diode reaches breakdown, reverse current flows, limiting the output of the detector. This results in a saturation region. As the input power is further increased, the junction temperature increases. Due to the small junction area (typically less than 25 μm diameter) the power density is quite high and diode failure can occur.

11.2.2 Types of Detectors

There are several types of detectors that provide different bandwidth, VSWR, and sensitivity performance characteristics. They are discussed briefly in the following paragraphs.

1. *High-Sensitivity, Narrow-band Detectors.* This type of detector uses a matching circuit to transform the diode impedance to 50 Ω and is shown in Fig. 11.15(a). High sensitivity with a low VSWR can be obtained over narrow (typically less than 10%) bandwidths.

2. *Wide-band Detectors.* To realize wide-band detectors, the matching circuit must transform the diode impedance to nominally 50 Ω over the full bandwidth. This usually results in a lower sensitivity and a higher VSWR.

The matching circuit can be designed to mismatch the diode at the lower frequency to compensate for the variation in diode sensitivity, but the resulting VSWR will be quite high. This design is sensitive to source impedance variations.

3. *Flat Detectors*. To achieve wide-band detectors with low VSWRs, the detector approach shown in Fig. 11.15(*b*) is used. The 50-Ω resistor results in a loss of sensitivity but provides a good VSWR over wide bandwidths. Higher resistor values can be used to tradeoff bandwidth and sensitivity. Typical sensitivity for this type of detector is 1 mV/μW.

4. *Temperature-Compensated Detectors*. A matched diode pair can be used to cancel the temperature variation of a detector as shown in Fig. 11.15(*c*). The change in voltage sensitivity due to the change in R_j is canceled by the same change in the matched diode.

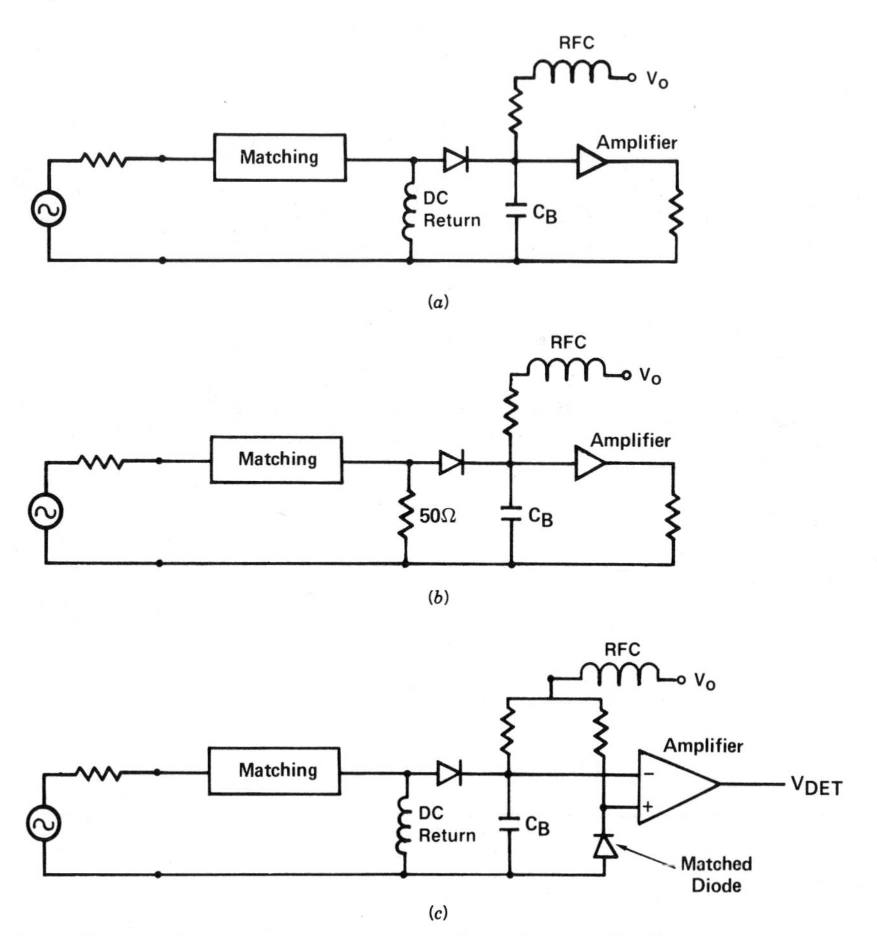

Figure 11.15 Detector types. (*a*) High-sensitivity detector. (*b*) Flat detector. (*c*) Temperature-compensated detector.

11.2.3 Detector Devices

Any nonlinear device can be used as a detector, but most microwave detectors use one of the following diodes;

Schottky-barrier diodes (conventional and zero bias),
point-contact diodes,
backward diodes.

The detector's efficiency is proportional to the degree of nonlinearity of the diode. The low-level detector performance and the bias requirements are determined by the amount of slope variation of the device's I–V curve near the origin.

The Schottky-barrier (SB) diode is formed by making a metal–semiconductor junction as discussed in Chapter 8. Typical saturation current for a SB diode is 5×10^{-10} A. The zero-bias Schottky-barrier diode has a saturation current several orders of magnitude higher than a conventional SB diode. This results in a lower voltage turn-on characteristic with video impedances ($R_v = R_j + R_s$) that are easier to work with (typically 1000–3000 Ω).

The point-contact diode uses a mechanical contact between a metal whisker and the semiconductor surface to make a rectifying junction. The whisker is made of tungsten or molybdenum and can be designed to resonate the chip capacitance to improve the detector sensitivity.

The backward diode is a type of tunnel diode that has a low peak current and the negative resistance normally associated with a tunnel diode is largely eliminated. This results in a diode with a very small forward current and a strong curvature at zero bias.

A comparison of the I–V characteristics of these diodes is shown in Fig. 11.16 [6]. The conventional Schottky diode has the highest turn-on voltage, and can be used to detect large-signal levels. To detect lower levels (<0 dBm), a dc bias current must be used to bias the diode near its nonlinear region. The zero-bias Schottky diode eliminates the need for external bias, since it has adequate nonlinearities near the origin of its I–V curve. The backward diode provides the highest curvature at low levels and makes an excellent low-level detector.

11.2.4 Design Considerations

The design of the detector is determined by tradeoffs of many parameters, including frequency of operation, bandwidth, sensitivity, dynamic range, and temperature range.

Diode Junction Parameter Selection. To obtain high voltage sensitivity β_v, the junction impedance must be high relative to the diode series

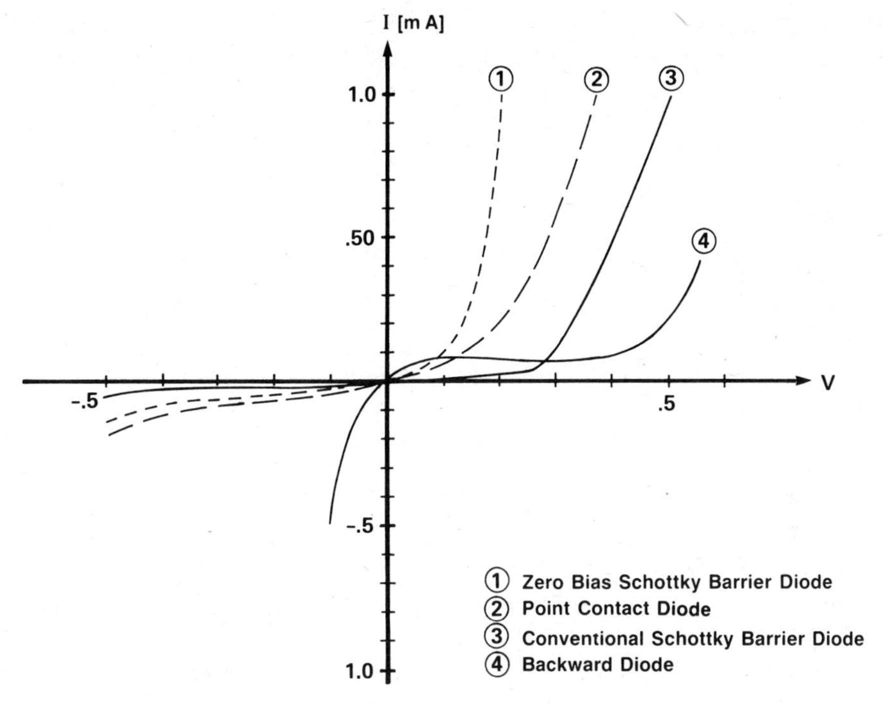

Figure 11.16 Comparison of diode types.

resistance R_s. This translates to selecting a diode with a high 0-V cutoff frequency defined as

$$f_{c0} = \frac{1}{2\pi R_s C_{j0}} \tag{11.50}$$

To maintain a junction impedance 10 times R_s at the frequency of operation f_{op},

$$R_s C_{j0} = \frac{1}{10(2\pi f_{op})} \tag{11.51}$$

or

$$f_{c0} > 10 f_{op}$$

and

$$X_c > -j150\,\Omega \text{ (for } R_s = 15\,\Omega)$$

or

$$C_{j0} < \frac{1.06}{f_{op}}$$

where C_{j0} is the 0-V junction capacitance expressed in picofarads and f_{op} is expressed in gigahertz. It was assumed that $R_j \gg 150\ \Omega$, that is $(I_0 + I_s) < 1 \times 10^{-5}$ A.

This value of C_j will give high sensitivity but may be too low for broadband applications due to the high Q of the diode impedance. Small values of junction capacitance increase sensitivity at the expense of matching circuit complexity. The losses associated with complex matching circuits limit the benefits of smaller junction capacitance. For broadband applications, diodes with 2 to 3 times the junction capacitance obtained from the preceding formulas are often used.

Package Selection. The choice of diode package is determined by the type of detector construction and the obvious desire to minimize parasitics. In some applications, the package inductance is designed to resonate the junction capacitance.

11.2.5 Detector Design Examples

The following design examples illustrate several detector design techniques.

Broadband Microstrip Detector. The following steps were used to design a 6- to 12-GHz detector implemented in microstrip.

1. *Diode Selection.* Using the preceding guidelines

$$f_{c0} > 10 f_{op} > 120 \text{ GHz}$$

$$C_{j0} < \frac{1.06}{f_{op}} < 0.09 \text{ pF}$$

$$R_s < \frac{1}{2\pi f_{c0} C_{j0}} < 15\ \Omega$$

   Because of the bandwidth required, a larger capacitance diode ($C_{j0} = 0.2$ pF) will be used. In addition, the use of an external bias will be investigated to lower the Q of the selected Schottky-barrier diode ($I_s = 1 \times 10^{-7}$ A).

2. *Package Selection.* For ease of assembly in a microstrip circuit, a beam lead package will be used.
3. *Diode Impedance.* The diode impedance with a quarter-wave open-circuited stub is shown as Z_1 in Fig. 11.17. Because of the octave bandwidth, a radial-line stub is used to obtain a low-Q short-circuit impedance behind the diode. The addition of an external bias (50 μA) is seen to increase the real part of the diode impedance (Z_2 in Fig. 11.17) and will be used to facilitate the diode matching.

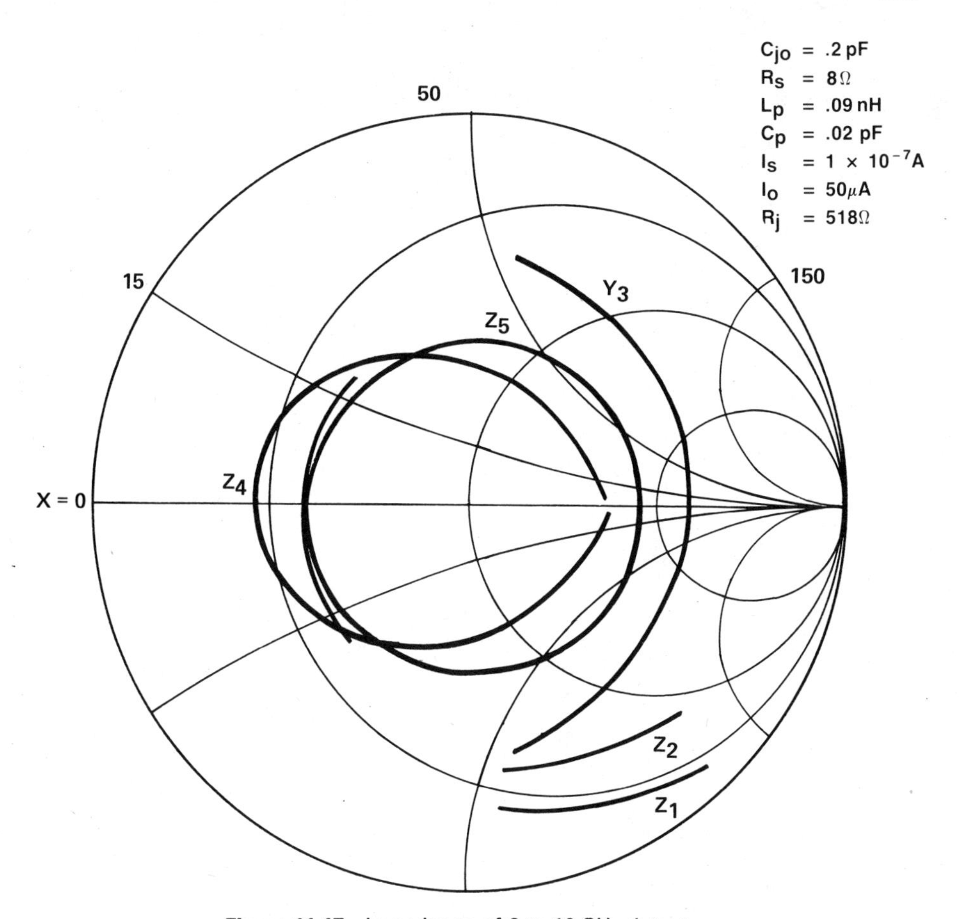

Figure 11.17 Impedance of 6 to 12 GHz detector.

4. *Matching Circuit.* One technique to match Z_2 to 50 Ω requires the following circuit components [7]:

   - A series line to rotate the diode impedance on the Smith chart until the resulting admittance has equal susceptance at the band edges (see Y_3 in Fig. 11.17). In this example, an 85-Ω line with an electrical length of 44 degrees at 9 GHz was required.
   - A short-circuited stub with the required impedance to make the net susceptance at the band edges approximately zero. The stub impedance is given by

$$Z_{sst} \approx \frac{1}{Y_{st}\tan\left(\frac{f_1}{f_0}\frac{\pi}{2}\right)} \tag{11.52}$$

where

Y_{st} = susceptance at f_1, the low end of the band,
f_0 = band center.

In this example, $Z_{sst} = 27\ \Omega$. The resulting impedance is represented by Z_4 in Fig. 11.17.

- A series line to equalize the VSWR over the band. A 43-Ω, 90-degree line (at 9 GHz) resulted in the final detector impedance Z_5, shown in Fig. 11.17. The reflection coefficient is approximately 0.44 across the band.

The final schematic is shown in Fig. 11.18. The resulting detector voltage sensitivity is listed in Table 11.3.

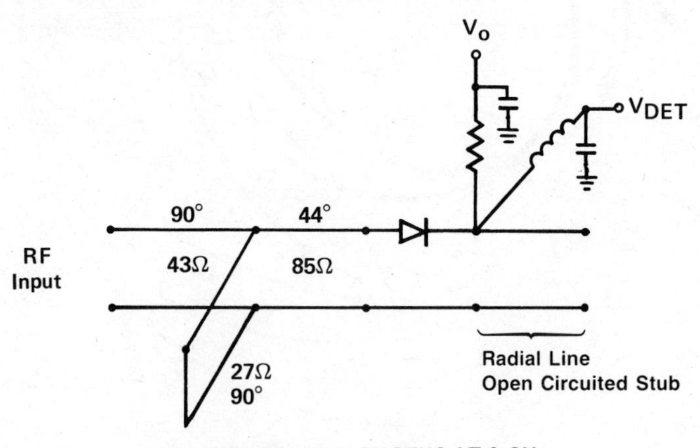

(a)

(b)

Figure 11.18 A 6 to 12-GHz detector. (*a*) Detector schematic. (*b*) Microstrip implementation.

TABLE 11.3 Microstrip Detector Performance

Frequency (GHz)	β_v (mV/μW)	$\lvert\Gamma\rvert$	β'_v (mV/μW)
6	6.6	0.44	5.3
8	5.2	0.44	4.2
10	4.3	0.44	3.5
12	3.5	0.44	2.8

The required linewidths for the particular substrate material and thickness can be obtained from Chapter 2. In addition, discontinuity models should be added to the analysis to account for the microstrip T junction and short-circuited stub impedance to ground. Computer optimization of the circuit will further improve the performance.

Coaxial Detector Design. The design a broadband coaxial detector covering 2 to 8 GHz entails the following steps:

1. *Diode Selection*

$$f_{c0} > 10 f_{op} > 80\text{ GHz}$$

$$C_{j0} < \frac{1.06}{f_{op}} < 0.133\text{ pF} \quad \text{(selected value 0.12 pF)}$$

$$R_s < 16\ \Omega \quad \text{(selected value 15 }\Omega\text{)}$$

   Again a zero-bias Schottky barrier diode ($I_s = 1 \times 10^{-7}$ A) will be used.

2. *Package Selection*. A double-prong package ($L_p = 1$ nH, $C_p = 0.15$ pF) is ideally suited for a coaxial construction.
3. *Diode Impedance*. The diode impedance Z_1 is shown in Fig. 11.19 and is seen to have a very low real part and high Q. The low impedance behind the diode is achieved with a five-section bandstop filter centered at 5 GHz, realized with a series of low- and high-impedance transmission lines.

   To reduce the diode VSWR, a 50-Ω shunt resistor is added, resulting in impedance Z_2 of Fig. 11.19.
4. *Diode Matching*. A two-section impedance transformer is used to match the diode to 50 Ω with less than a 1.5:1 VSWR over the band (Z_3 in Fig. 11.19). The final design is shown in Fig. 11.20.

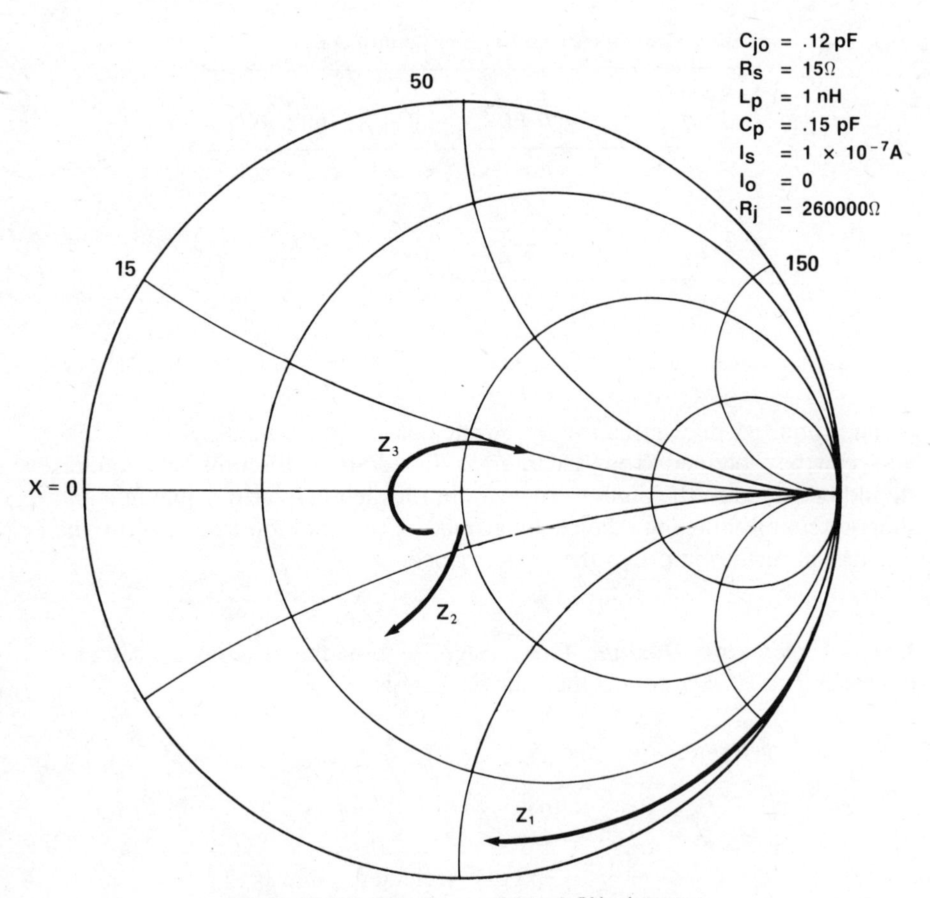

Figure 11.19 Impedance of 2 to 8 GHz detector.

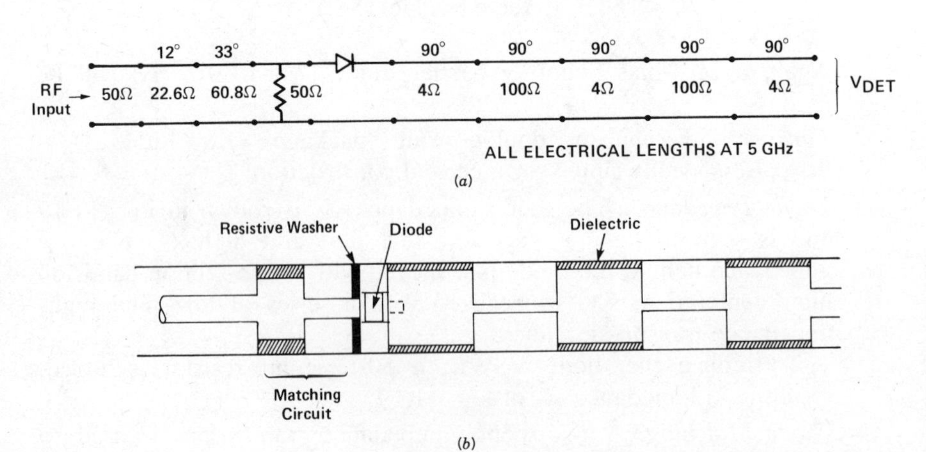

Figure 11.20 Coaxial detector design. (*a*) Detector schematic. (*b*) Coaxial implementation.

11.3 MIXERS

Mixers utilize a pumped nonlinear device to generate a spectrum of frequencies based upon the sum and difference of the harmonics of the signal and local oscillator (LO) frequencies. The diode current can be expressed as

$$i(v) = I_0 + I_s \sum_{n=1}^{\infty} \frac{\alpha^n}{n!} [V_s \sin(\omega_s t) + V_p \sin(\omega_p t)]^n \tag{11.53}$$

where

I_s = diode saturation current,
$\alpha = e/nkT$,
V_s = amplitude of signal,
ω_s = frequency of signal,
V_p = amplitude of LO,
ω_p = frequency of LO.

This results in frequencies generated at

$$f_0 = mf_s + nf_p \tag{11.54}$$

where m, n = integers, $-\infty$ to $+\infty$.

A simple block diagram of a mixer, depicted in Fig. 11.21, shows the two input signals at f_s and f_p being applied to the diode to generate the intermediate frequency (IF) output, f_0. In most applications, the IF is the difference between the signal and LO frequencies, that is, $f_0 = f_s - f_p$ or $f_0 = f_p - f_s$. Filters are used to isolate the three ports and to provide terminations to all the other frequencies generated.

In addition to the three primary mixer frequencies, the diode generates the following frequency components that must be considered in a mixer design:

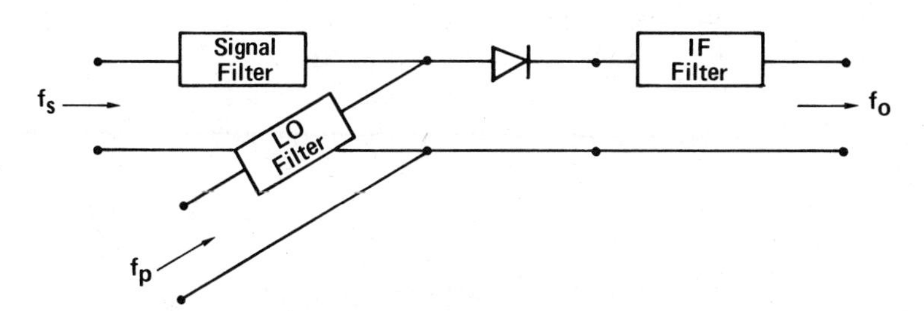

Figure 11.21 Basic mixer block diagram.

LO harmonics, nf_p, where $n = 1, 2, 3, \ldots,$
image frequency, $f_{im} = 2f_p - f_s,$
sum frequency, $f_{sum} = f_p + f_s,$
harmonic sidebands, $f_{sb} = nf_p \pm f_s.$

It is important to note that resistive terminations at the image, sum, or harmonic sidebands result in lost signal power that increases the mixer conversion loss, defined as

$$\text{Loss} = 10 \log \left[\frac{P_{out}}{P_{av}} \right] \tag{11.55}$$

where

P_{out} = power output at IF,
P_{av} = power available at signal frequency.

Reactive terminations with the correct phase have long been identified as the key to obtaining low conversion loss and are the essence of image- and sum-enhanced mixer designs [8].

The large number and range of frequencies that must be considered, plus the fact that the diode is driven into a large signal condition, make computer analysis essential to mixer design.

11.3.1 Basic Theory

The ideal mixer nonlinear element is a lossless switch configured as in Fig. 11.22. To satisfy the boundary conditions, $I_0 = I_s$ when the switch is open. Barber [9] has shown that the conversion loss is strongly affected by the duty cycle of the switch and can be expressed as

$$L = \frac{\sin[\pi(t/T)]}{\pi(t/T)} = \begin{cases} 2/\pi \quad \text{or} \quad 3.92\ \text{dB} & \text{for } t/T = 0.5 \\ 1 \quad \text{or} \quad 0\ \text{dB} & \text{for } t/T \to 0 \end{cases} \tag{11.56}$$

It can be seen in Fig. 11.22(b) that when the switch duty cycle approaches zero, a low-frequency IF current is generated with 0-dB loss (assuming $R_L = 0\ \Omega$). To work with a switch duty cycle of 50%, circuit elements that short the high-frequency components must be added as shown in Fig. 11.23 [10]. The circuit elements perform the following functions:

The signal filter provides an open circuit at image and frequencies higher than the signal.

The short-circuited stub shorts even harmonics of f_p.

The open-circuited stub shorts odd harmonics of f_p.

The IF filter shorts everything except IF.

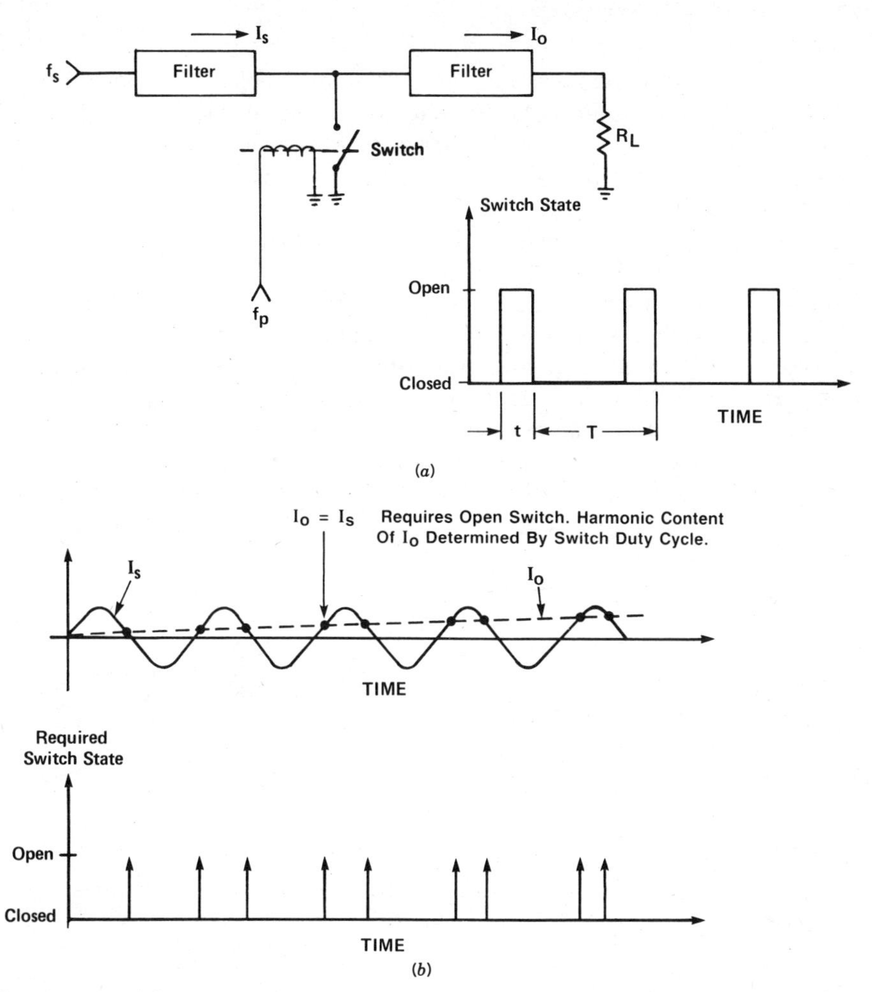

Figure 11.22 (*a*) Mixer implemented with an ideal switch. (*b*) Mixer performance as switch duty cycle approaches zero.

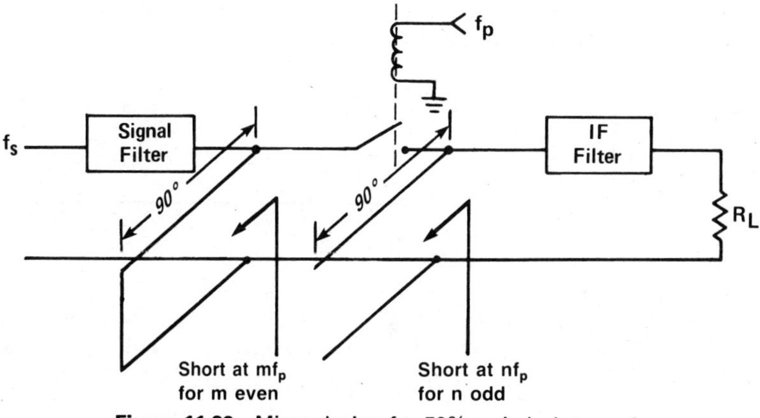

Figure 11.23 Mixer design for 50% switch duty cycle.

With this circuit configuration, a conversion loss of 0 dB is possible with a switch duty cycle of 50%. The characteristics that make a switch the ideal mixer element, shown in Fig. 11.24(*a*), are that it switches instantly between two impedance states with $|\Gamma| = 1$, separated by 180 degrees. A diode also has two impedance states when driven by a large LO voltage, Fig. 11.24(*b*), and its effectiveness in a mixer (i.e., conversion loss) can be determined by how far it degrades from the ideal conditions established in Fig. 11.24(*a*). To obtain low conversion loss, it is necessary to design a matching circuit that transforms the two diode impedance states to two impedances that are 180 degrees different in phase, equidistant from the center of the Smith chart (equal $|\Gamma|$), with $|\Gamma|$ as large as possible.

The result of the matching network transformation is shown in Fig. 11.25. The additional loss for $|\Gamma| < 1$ is $L_{add} = 1 - |\Gamma|^2$.

Hyperbolic geometry [11] provides a technique to design the matching circuit that satisfies these conditions to provide minimum loss mixing, and is explained below.

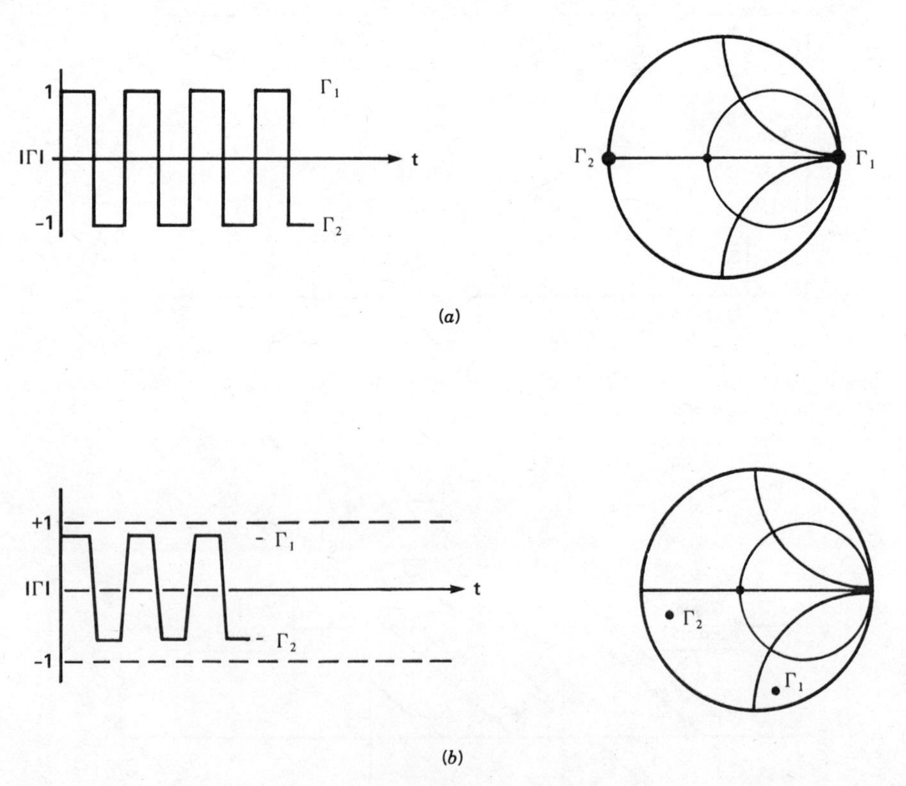

Figure 11.24 Comparison of an ideal switch and a diode for mixer applications. (*a*) Reflection coefficient of an ideal switch. (*b*) Diode reflection coefficient when driven by LO voltage.

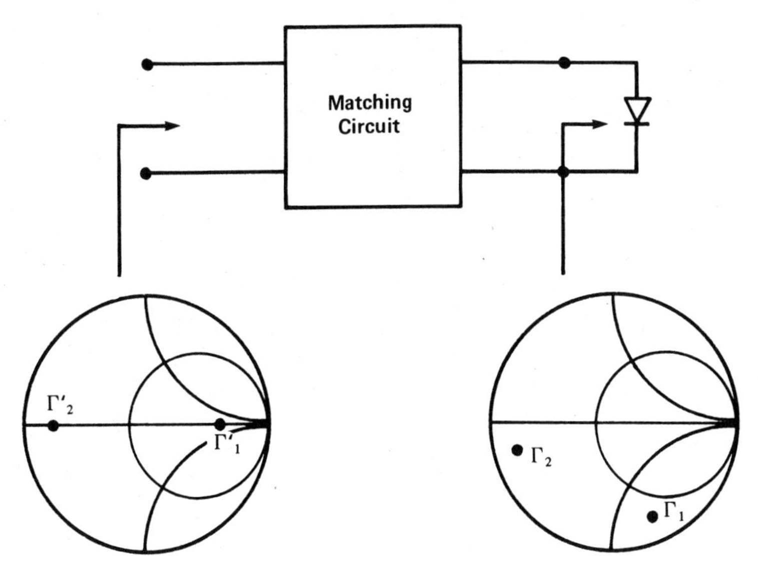

Figure 11.25 Mixer matching circuit.

Design Procedure

1. Measure the two diode impedance states at the LO frequency under large-signal conditions. As a first-order approximation, the two impedances can be calculated as outlined in Fig. 11.26. Referring to Fig. 11.27, the two states Z_{on} and Z_{off} can be used to define a dynamic Q factor as [12]

$$Q_d = \frac{\sqrt{(R_1 - R_2)^2 + (X_1 - X_2)^2}}{\sqrt{R_1 R_2}} \tag{11.57}$$

where

$$Z_{\text{on}} = R_1 + jX_1$$

$$Z_{\text{off}} = R_2 + jX_2$$

This dynamic Q is a measure of the diode's potential conversion loss (loss is inversely proportional to Q_d) and can be used to compare mixer performance with different diodes without actually fabricating the entire mixer.

2. Construct the hyperbolic circle through Z_{on} and Z_{off} (see circle C_1 in Fig. 11.27). Hyperbolic circles are those circles that are perpendicular to the outside of the Smith chart. Lines of constant reactance are hyperbolic circles.

3. Rotate Z_{on} and Z_{off} until they lie on a hyperbolic circle of constant reactance (see Z'_{on} and Z'_{off} lying on circle C_2 of Fig. 11.27). The dynamic

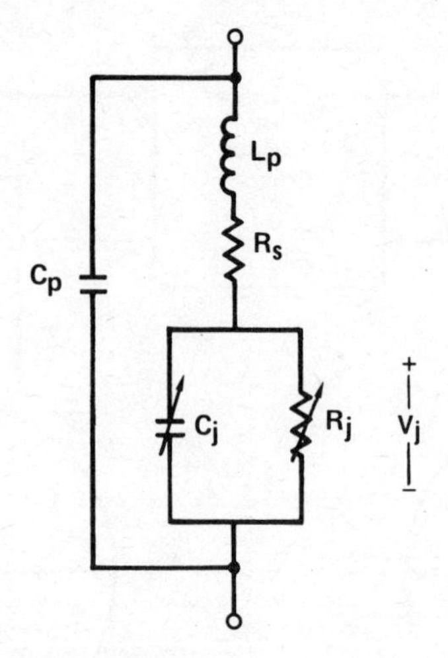

I. Forward Bias (On State)

Assume $I = .005$ A

Then $V_{j1} = .025248\,n \, \ell n\left(\frac{.005}{I_s}\right)$ n = ideality factor

$$R_j = \frac{.025248\,n}{.005} = 5.05\,n$$

$$C_j = \frac{C_{jo}}{\sqrt{1 - V_{j1}/\phi}} \qquad \text{where: } \phi = \begin{cases} 0.6\text{ V} & \text{Silicon} \\ 0.8\text{ V} & \text{GaAs} \end{cases}$$

II. Reverse Bias (Off State)

$$R_j = \frac{.025248\,n}{I_s}$$

$$C_j = \frac{C_{jo}}{\sqrt{1 + V_{j1}/\phi}}$$

Figure 11.26 Two impedance states of a diode.

Q, Q_d is unchanged by this rotation and can be expressed as

$$Q_d = \frac{\sqrt{(R_1' - R_2')^2}}{\sqrt{R_1' R_2'}} = \frac{R_1' - R_2'}{\sqrt{R_1' R_2'}} \tag{11.58}$$

where

$$Z'_{\text{on}} = R_1' + jX'$$
$$Z'_{\text{off}} = R_2' + jX'$$

The mean of Z'_{on} and Z'_{off} is given by

$$Z'_m = R' + jX' \tag{11.59}$$

where $R' = \sqrt{R_1' R_2'}$.

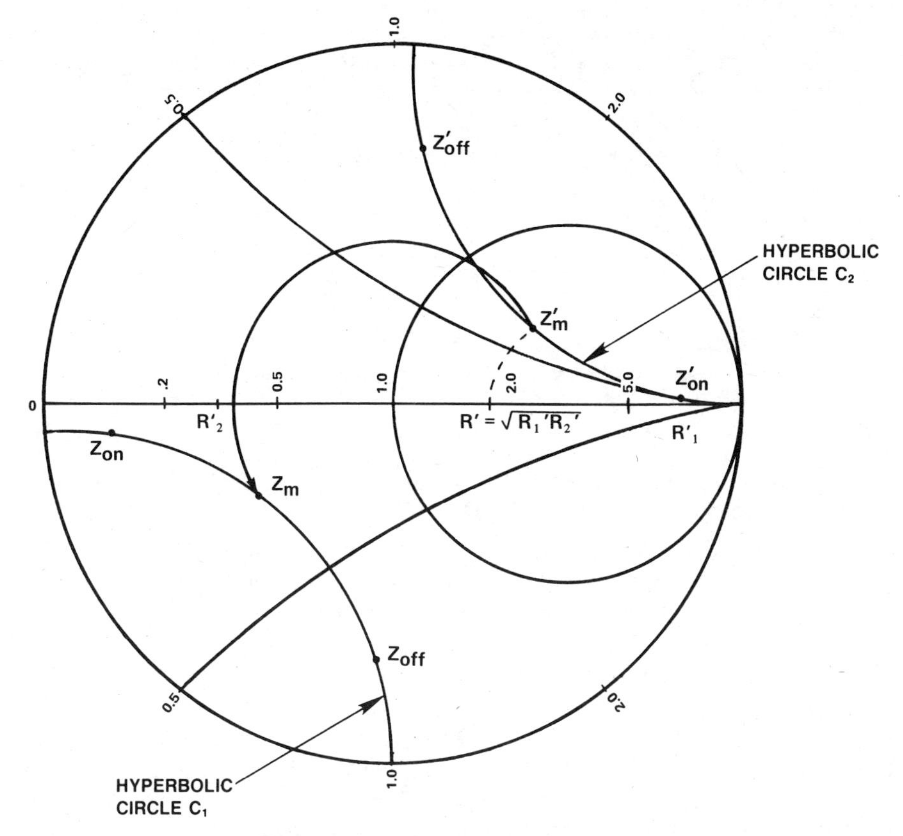

Figure 11.27 Diode impedance states.

4. Rotate Z'_m back the same angle to locate Z_m, the hyperbolic mean of Z_{on} and Z_{off}. A simplified procedure can be given as follows [13]. For

$$Z_{\text{on}} = R_1 + jX_1$$

$$Z_{\text{off}} = R_2 + jX_2$$

the hyperbolic mean is $Z_m = R_m + jX_m$, where

$$R_m = \left[R_1 R_2 \left(1 + \frac{(X_1 - X_2)^2}{(R_1 + R_2)^2} \right) \right]^{1/2} \tag{11.60}$$

and

$$X_m = X_1 + R_1 \frac{X_2 - X_1}{R_1 + R_2} \tag{11.61}$$

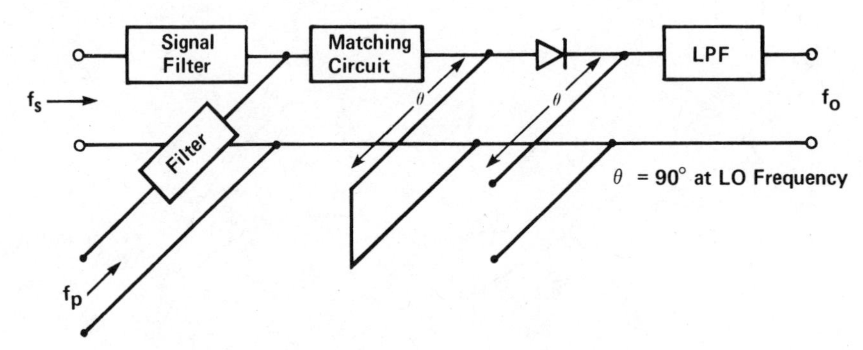

Figure 11.28 Mixer design with matching circuit.

In the example given, $Z_{on} = 5.0\text{-}j2.5\ \Omega$ and $Z_{off} = 15\text{-}j45\ \Omega$. The hyperbolic mean is therefore $Z_m = 20.3\text{-}j13.1\ \Omega$.

5. Design a matching circuit to match impedance Z_m to 50 Ω. This automatically provides maximum reflection coefficient for the given Z_{on} and Z_{off} with 180 degress between states.

For mixers required to work over a range of LO frequencies, this procedure would be repeated to yield a locus of hyperbolic means. This locus would then be matched to 50 Ω to satisfy the conditions for minimum loss mixing over the range of LO frequencies. The mixer circuit then appears as shown in Fig. 11.28.

The design of the matching circuit and the IF filter is also driven by the requirements to minimize the mismatch loss at the IF and to approximately match the diode at the signal frequency. This becomes more difficult for higher IFs because of the difference between signal and LO frequencies. The best mixer performance is generally obtained with a slight mismatch at the signal frequency.

11.3.2 Types of Mixers

There are many types of mixers that can be designed that use one diode (single ended), two diodes (single balanced or antiparallel), four diodes (double balanced) or even eight diodes (double-double balanced). The mixer design issues, however, can be addressed on the basis of a single diode and extended to any mixer configuration. Two types of single-diode mixers are illustrated in Fig. 11.29. They differ in the method of LO injection required to accommodate the range of LO frequencies. The coupler approach increases the required LO power by the coupling factor, typically 10 dB. The loop directional filter [14] is low loss at the resonance of the loop but is narrow band, so is useful only for fixed LO mixer designs. Mixers using two diodes are shown in Fig. 11.30 and include antiparallel

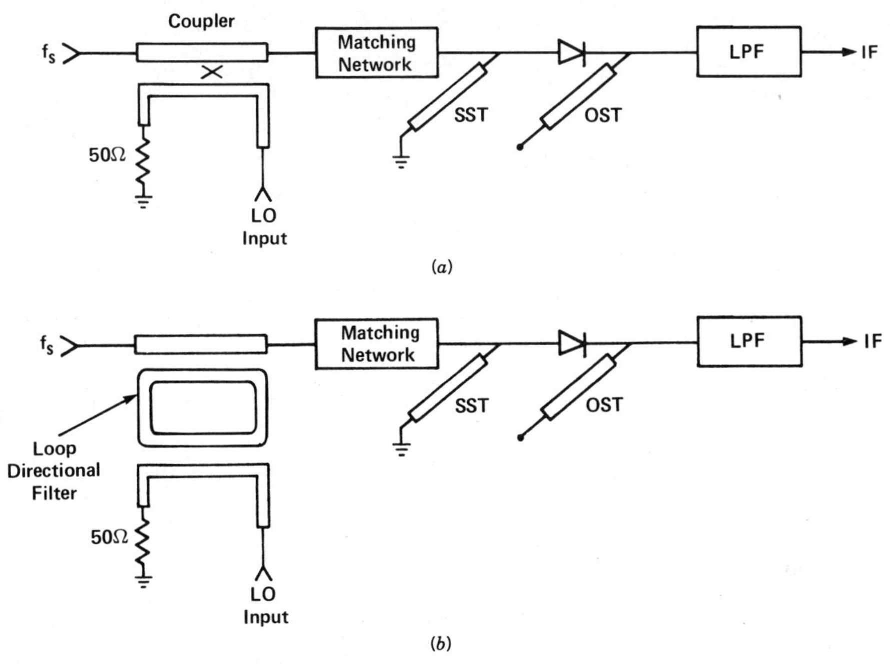

Figure 11.29 Single-ended mixers. (*a*) Broadband–LO mixer configuration. (*b*) Fixed-LO mixer configuration.

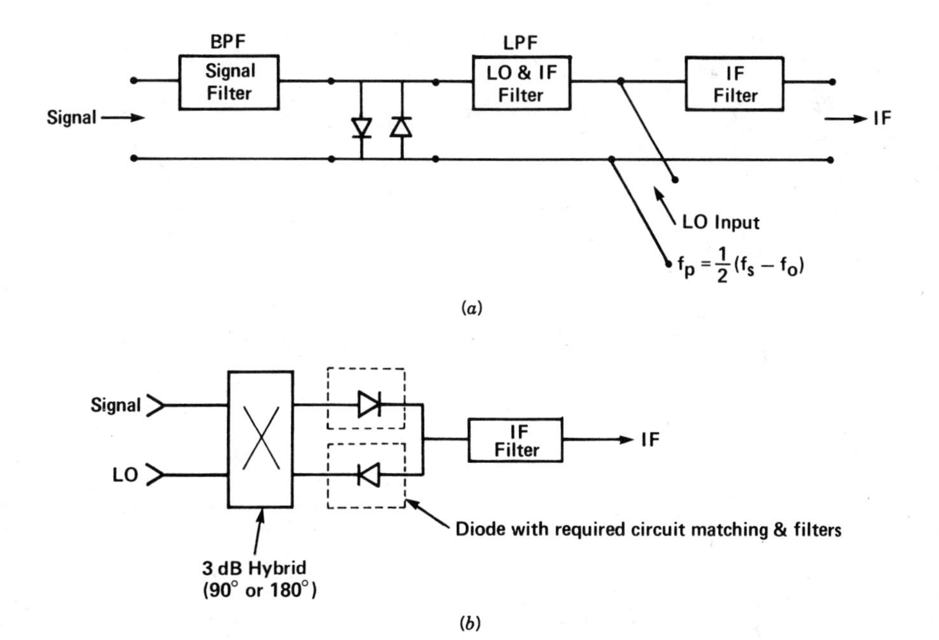

Figure 11.30 Two-diode mixer types. (*a*) Antiparallel diode mixer. (*b*) Single-balanced mixer.

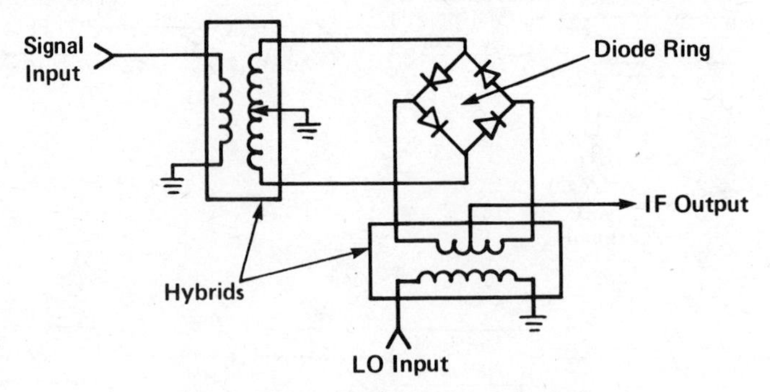

Figure 11.31 Double-balanced mixer.

and single-balanced designs. The antiparallel approach is often used for subharmonically pumped mixers due to the LO noise cancellation that can be obtained [15]. The single-balanced mixer uses a 3-dB hybrid (90- or 180-degree) to provide LO and signal power to both diodes. Double- (Fig. 11.31) and double-double-balanced mixers (Fig. 11.32) generally use diode quad rings with the required signal, LO, and IF hybrids.

The double-double-balanced mixer can be configured as an image-rejection mixer by properly phasing the IF signals as shown in Fig. 11.33. Image cancellation is accomplished by using a circuit that causes the output signal components from the image to be out of phase at the IF output port. This circuit consists of two balanced mixers, two quadrature hybrids (one at the signal and one at the IF band), and an in-phase power divider. The

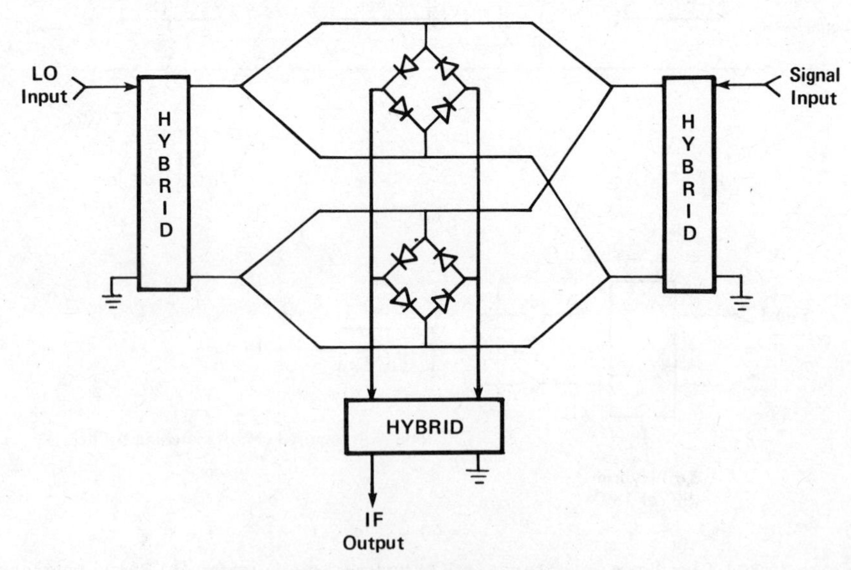

Figure 11.32 Double-double-balanced mixer.

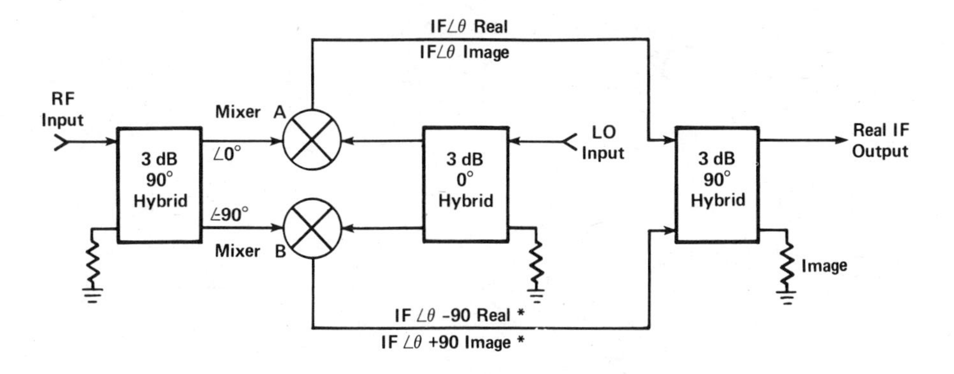

Figure 11.33 Image-rejection mixer.

signal power is split between the two mixers, with the input to mixer B lagging the input to mixer A by 90 degrees. This results in the IF output from mixer A due to both the signal and image being at a reference angle θ. The output from mixer B due to the signal will be at angle $\theta - 90°$, while the output due to the image will be at angle $\theta + 90°$. These angles assume a low-side LO frequency. The IF output signals are combined in the output hybrid. The output signals due to the signal combine in-phase at the IF port where the output signals from the image cancel. The opposite signals add and cancel at the image port, which is terminated with 50 Ω.

The key to an effective image-rejection mixer is the use of balanced mixers with well-matched amplitude and phase characteristics and high isolation. Poor hybrid phase and amplitude characteristics will also degrade the image-rejection performance. Image rejection of greater than 20 dB can be obtained with this configuration.

The image-and-sum-frequency-enhanced mixer requires four filters, as shown in Fig. 11.34 for a single-diode mixer, but offers the potential for very low conversion loss limited only by circuit and diode losses. The diode package often serves as the sum frequency filter by providing a series resonance at f_{sum}. The signal filter provides a reactive termination at the image frequency ($f_{\text{im}} = 2f_p - f_s$) that must be properly phased so that reflected power at the image enhances the mixing process. This filter can be realized as a bandpass, high-pass, or low-pass filter, depending on the signal-to-image separation, and whether the image is on the high or low side of the signal. It should be noted that both the sum and image frequency must be reactively terminated to obtain a conversion loss of less than 3.92 dB.

All of these mixers will provide a constant conversion loss for signal levels greater than 6 dB below the LO power. For higher signal levels, the signal voltage affects the bias point of the diodes, which degrades the

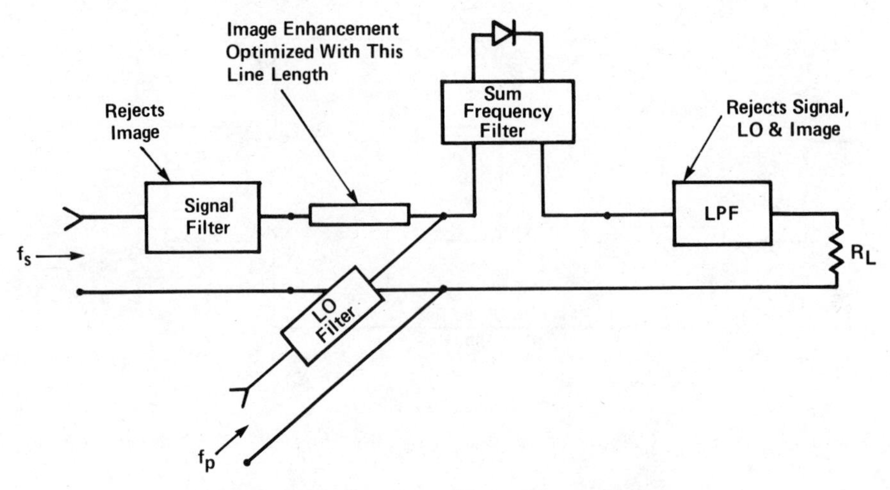

Figure 11.34 Image- and sum-enhanced mixer.

tuning and balance of the mixer. This results in higher conversion loss and lower harmonic suppression. High-level mixers are obtained by replacing a single diode by [16] two diodes in series, a resistor and a diode in series, or a parallel RC network in series with a diode. This reduces the signal voltage across any diode and increases the signal level that can be handled without degraded mixer performance. These mixers require up to 10 dB additional LO power to pump the diodes into a fully on and off state.

A comparison of some of the key tradeoff parameters for the various mixer types is shown in Table 11.4 [17].

FET Mixers. Mixers can also be realized using the nonlinearities associated with single- and dual-gate FETs. The principal advantage of FET mixers over diode mixers is that FET mixers can provide conversion gain. The FET mixer noise figure is higher than the noise figure obtainable from the same device in an amplifier due to up-conversion of low frequency noise resulting from gate capacitance modulation. The fact that the mixer has gain, however, can offset a higher mixer noise figure to provide a lower overall noise figure. This is illustrated in Table 11.5, which is a comparison of a typical diode and FET mixer followed by an IF amplifier with a 2-dB noise figure [F_{tot} from (11.81)].

The FET mixer gain has the effect of reducing the second-stage noise figure contribution to 0.2 dB versus 2.4 dB for the diode mixer. As the second-stage noise figure increases, the tradeoff between mixer gain and loss becomes more significant. For a second-stage noise figure of 3 dB, and assuming the same two mixers, the FET mixer provides a 1.2-dB lower total noise figure than the diode mixer (7.3 dB versus 8.5 dB).

TABLE 11.4 Comparison of Mixer Types

Mixer Type	Conversion Loss	VSWR LO	VSWR RF	LO/RF Isolation	LO Power Required	Spurious Rejection	Harmonic Suppression	Third-Order Intercept
Single-ended	Good	Good	Poor	Fair	+3 dBm[a]	Poor	Poor	+13 dBm
Balanced (90° hybrid)	Good	Good	Good	Poor	+5 dBm	Fair	Fair	+13 dBm
Balanced (180° hybrid)	Good	Fair	Fair	Very good	+5 dBm	Fair	Even-good Odd-fair	+13 dBm
Double balanced	Very good	Poor	Poor	Very good	+10 dBm	Good	Very good	+18 dBm
Image-reject	Good	Good	Good	Good	+7 dBm	Fair	Even-good Odd-fair	+15 dBm
Image-enhanced	Excellent	Good	Good	Very good	+7 dBm	Fair	Even-good Odd-fair	+15 dBm

[a]Does not include LO coupling factor.

TABLE 11.5 Comparison of Diode and FET Mixers

Mixer Type	Gain/Loss (dB)	NF (dB)	F_{tot} (dB)
Diode	$L = 6$	5	7.4
FET	$G = 4$	7	7.2

The model for a single-gate GaAs FET identifying the nonlinear elements is shown in Fig. 11.35(*a*). The transconductance g_m is the most important FET nonlinearity for the mixing process. A plot of the transconductance of a typical FET, shown in Fig. 11.35(*b*), indicates that the highest nonlinearity is obtained with a gate bias near pinchoff.

A single-gate FET mixer is shown in Fig. 11.36. The LO and signal are applied to the gate using a diplexer or coupler; the IF is taken from the

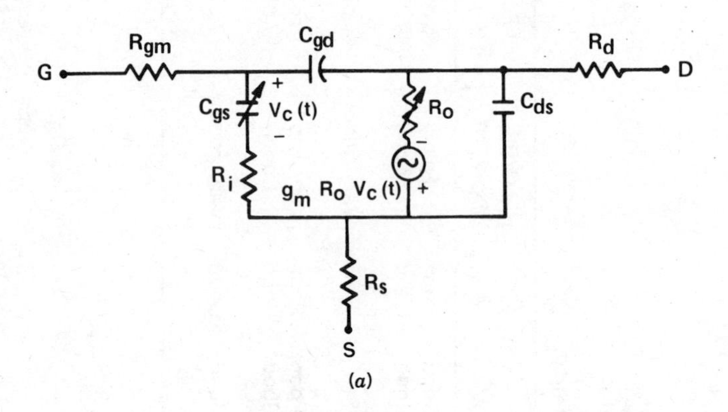

(*a*)

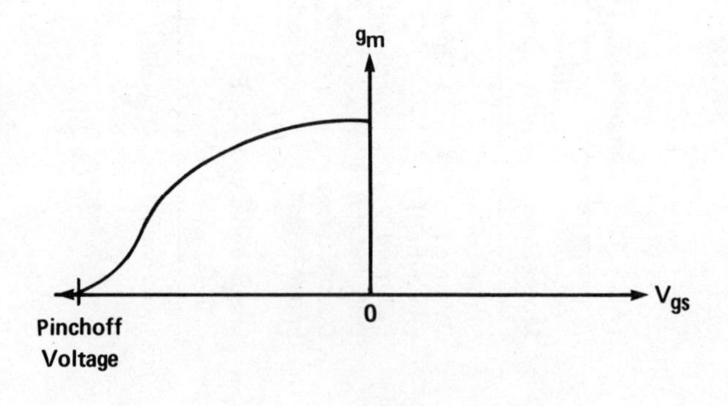

(*b*)

Figure 11.35 (*a*) Single-gate FET model. (*b*) Transconductance vs. gate voltage.

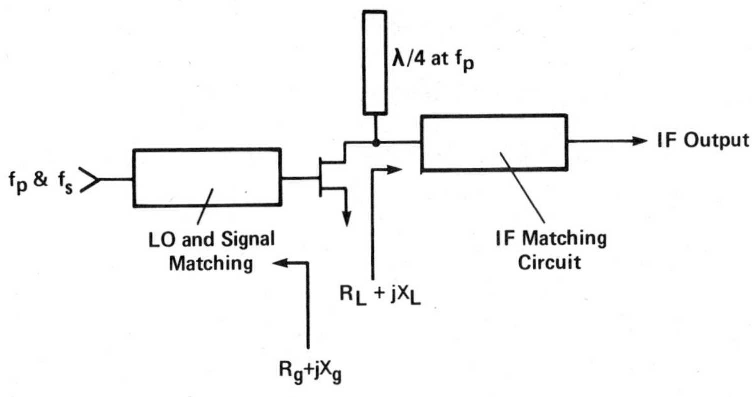

Figure 11.36 Single-gate FET mixer.

drain. The LO voltage serves to pump the FET and the input matching circuit transforms the hyperbolic mean of the FET's two impedance states to 50 Ω. As in the diode mixer, this matching network must also satisfy the matching requirements at the signal frequency. The output matching circuit minimizes the mismatch loss at the IF.

Pucel et al. [18] have presented an analysis for a single-gate mixer. The maximum available mixer conversion gain G_c is given by

$$G_c = \frac{g_1^2}{4\omega_s^2 \bar{C}_{gs}^2} \frac{\bar{R}_o}{R_{\text{in}}} \tag{11.62}$$

where

g_1 = Fourier coefficient of $g_m \approx g_m/3$ at V_{gs} near pinchoff,
$\bar{R}_o$ = average value of R_o,
ω_s = signal frequency,
$\bar{C}_{gs}$ = average value of C_{gs},
$R_{\text{in}} = R_{gm} + R_i + R_s$.

The mixer conversion gain given in (11.62) can exceed the maximum available gain for the same device in an amplifier.

The actual mixer conversion gain is reduced by circuit mismatch and, assuming that the FET output reactance approaches zero at the IF, is given by

$$G_{\text{av}} = \left(\frac{2g_1\bar{R}_o}{\omega_s \bar{C}_{gs}}\right)^2 \frac{R_g}{(R_g + R_{\text{in}})^2 + (X_g - X_c)^2} \frac{R_L}{(\bar{R}_o + R_L)^2 + X_L^2} \tag{11.63}$$

where

$$R_g + jX_g = \text{circuit source impedance}$$

$$R_L + jX_L = \text{circuit load impedance}$$

$$X_c = \frac{1}{\omega_s \bar{C}_{gs}}$$

G_{av} is a maximum given by (11.62) when

$$R_g = R_{in}$$

$$X_g = \frac{1}{\omega_s \bar{C}_{gs}}$$

$$R_L = \bar{R}_o$$

$$X_L = 0$$

Single-gate mixers can also be realized by applying the LO to the FET drain [19] and using the nonlinear drain resistance as the primary mixing element. Lower mixer noise figures can be obtained with the drain mixer due to the reduced gate capacitance modulation. The conversion gain is also lower than for the gate mixer, and the best choice depends on the value of the second-stage noise figure and the application.

Dual-gate FET mixers can provide higher conversion gain than single-gate FET mixers by utilizing the additional nonlinearities associated with the second gate. The equivalent model and the I–V characteristics of a dual-gate FET are shown in Fig. 11.37. In a mixer, the LO is applied to gate 2, the signal to gate 1, and the IF is taken from the drain. This has the advantage of separating the LO from the signal matching as seen in Fig. 11.38. As shown by Tsironis et al. [20], the FET nonlinearities are strongly dependent on the bias conditions on gates 1 and 2. The design of a dual-gate mixer requires the elements of the FET model to be determined over the range of gate 2 bias (dc + LO voltage). These can then be used to calculate the conversion gain and the mean value of the FET terminal impedances. Gate 1 is matched at the signal frequency; the drain is matched at the IF (including any required filtering); and Gate 2 is matched at the LO frequency to maximize the conversion gain. Conversion gains of 8 to 10 dB with 8- to 9-dB noise figure at 12 GHz have been obtained with dual-gate mixers [20].

A comparison of active and passive mixers is included in Table 11.6.

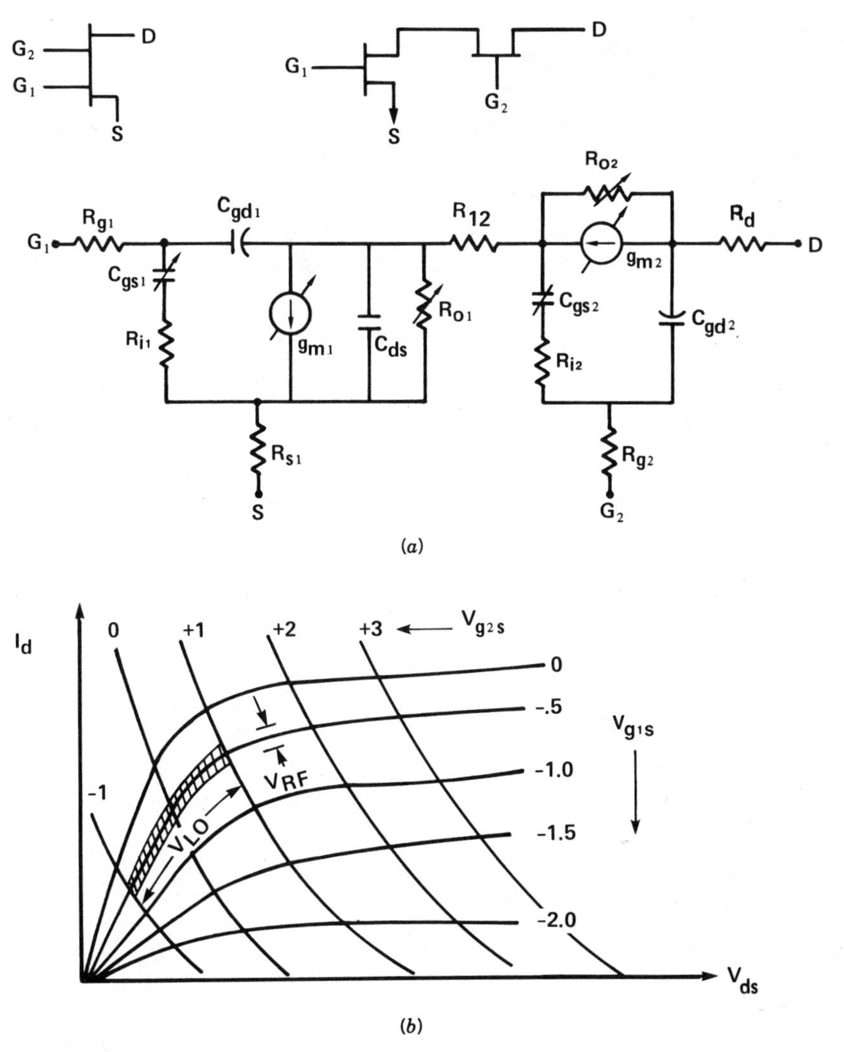

Figure 11.37 Dual-gate FET characteristics. (*a*) Equivalent model. (*b*) DC characteristics.

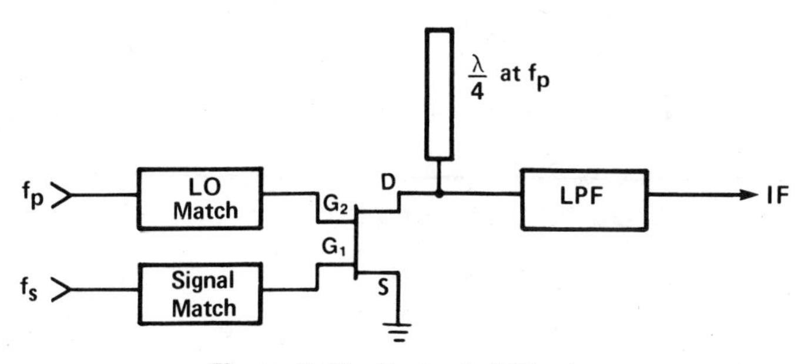

Figure 11.38 Dual-gate FET mixer.

TABLE 11.6 Comparison Between Active and Passive Mixers

Parameter	Single-Gate FET Active Mixer	Dual-Gate FET Active Mixer	Schottky-Diode Passive Mixer
Conversion factor	Low gain	Higher gain	Loss
Noise figure (typ)	6–10 dB	8–12 dB	4–6 dB
RF/LO to IF isolation	Requires filter	Requires filter	Requires filter
LO to RF isolation	Requires coupler	20 dB	Requires coupler
Bandwidth	Octave	Multioctave	Octave
Bias	Yes	Yes	No

11.3.3 Analysis Techniques

The performance of a diode mixer can be analyzed using the techniques presented by Held and Kerr [21] with computer aids such as the program developed by Siegel, Kerr, and Hwang [22]. The latter uses the diode's voltage-dependent capacitance and conductance and the impedance presented by the embedding circuit to determine the mixer performance.

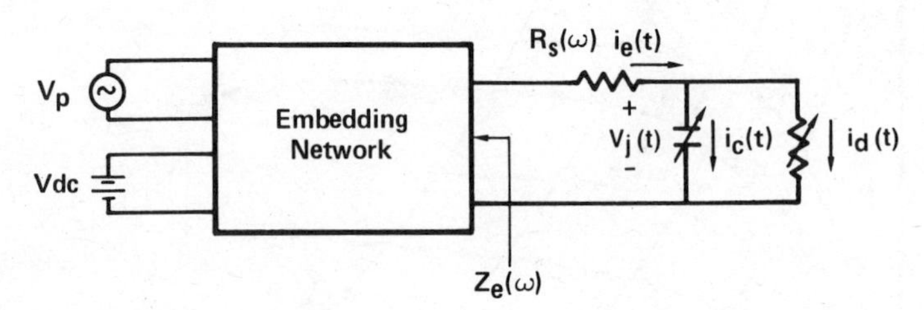

where

$$V_j(t) = \sum_{K=-\infty}^{\infty} V_K e^{jK\omega_p t} \qquad i_e(t) = \sum_{K=-\infty}^{\infty} I_{eK} e^{jK\omega_p t}$$

The diode requires the following to be satisfied:

$$i_d = I_s(e^{\alpha V_j} - 1) \qquad \text{where} \quad \alpha = \frac{e}{nKT}$$

$$i_c = C_j \frac{dV_j}{dt}$$

The embedding network requires the following steady-state conditions to be met:

$$V_0 = V_{dc} - I_{e0}[Z_e(0) + R_s(0)]$$

$$V_{\pm 1} = V_p - I_{e\pm 1}[Z_e(\pm\omega_p) + R_s(\pm\omega_p)]$$

$$\frac{V_K}{I_{eK}} = Z_D(K\omega_p) = -[Z_e(K\omega_p) + R_s(K\omega_p)] \qquad K = \pm 2, \pm 3, \ldots$$

where V_p = LO voltage.

Figure 11.39 Mixer large-signal analysis.

The circuit is characterized by the impedance it presents to the diode at the LO, harmonics of the LO, and at the sidebands. Sidebands refer to those frequencies that can mix with the LO or its harmonics to produce an IF output.

The analysis is divided into large-signal, small-signal, and noise-analysis sections. Since the LO voltage is, in general, much larger than the signal voltage, the diode current and voltage are determined by the LO. The equivalent circuit shown in Fig. 11.39 is used for the large-signal analysis. In this figure, K refers to the harmonic of the LO. The element R_s is the series resistance of the diode and is frequency dependent due to skin effects. Solutions to the boundary conditions shown in Fig. 11.39 provide the Fourier coefficients to the series expressions for diode current, conductance, and capacitance. These Fourier coefficients, together with the embedding impedance, Z_e, determine the small-signal properties of the mixer.

In the small-signal analysis, the mixer is represented as a linear multiport network, with each sideband considered as a separate port. A conversion impedance matrix, determined by the diode admittance and embedding network sideband impedance, relates a current source at sideband j to a voltage output at sideband i. This matrix serves as a basis for determining the impedance seen looking into the diode at any sideband frequency, and the conversion loss from a signal at sideband j to the output at sideband i. The difference between the diode and the circuit impedances at both the signal and IF frequencies helps to determine how to improve the mixer performance by adjusting the matching circuits. The resulting equation for the conversion loss for a conjugately matched IF, is

$$L_{ij} = \frac{1}{4|Z_{ij}|^2} \frac{|Z_{ei} + R_{si}|^2}{\mathrm{Re}[Z_{ei}]} \frac{|Z_{ej} + R_{sj}|^2}{\mathrm{Re}[Z_{ej}]} \tag{11.64}$$

where

L_{ij} = loss from sideband j to sideband i,
Z_{ei} = impedance of embedding circuit at sideband i,
R_{si} = diode series resistance at sideband i,
Z_{ij} = relates current source at sideband j to the voltage present at sideband i.

For a fundamental mixer, the conversion loss from signal to IF is given by

$$L_{01} = \frac{1}{4|Z_{01}|^2} \frac{|Z_{e0} + R_{s0}|^2}{\mathrm{Re}[Z_{e0}]} \frac{|Z_{e1} + R_{s1}|^2}{\mathrm{Re}[Z_{e1}]} \tag{11.65}$$

The actual loss is increased to the extent of the IF mismatch.

The noise analysis determines the equivalent input noise temperature from the diode series resistance, the conversion impedance matrix, and the

Fourier coefficients of the diode current. This analysis takes into account contributions from thermal, shot, and scattering noise to obtain the mixer equivalent noise temperature T_m, where

$$T_m = \frac{v_n^2}{4kB} \frac{|Z_{e1} + R_{s1}|^2}{|Z_{01}|^2 \operatorname{Re}[Z_{e1}]} \tag{11.66}$$

where v_n^2 = total noise voltage.

It is apparent from (11.65) and (11.66) that L_{01} and T_m are not minimized by the same embedding impedance and the noise figure [defined in (11.79)] can be higher or even lower than the conversion loss. This is

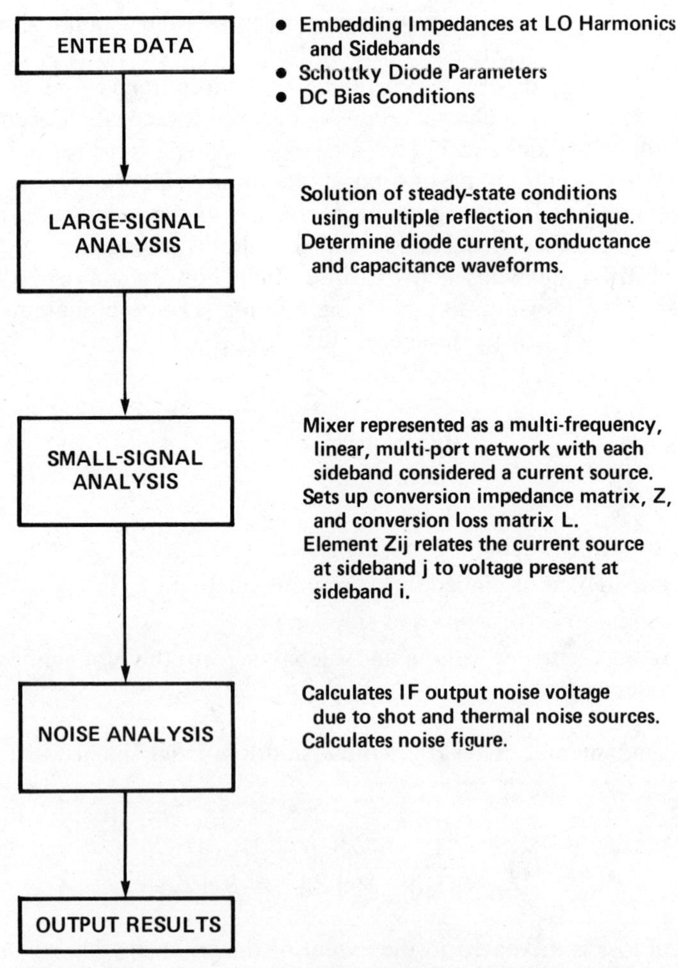

Figure 11.40 Flow chart of mixer analysis computer program.

analogous to the situation with low-noise amplifiers that require a different source impedance for minimum noise figure or maximum gain. The noise voltage of a mixer is typically minimized with an input VSWR on the order of 2:1. A flow chart of the mixer analysis is shown in Fig. 11.40.

Noise Figure. The noise figure is defined as

$$F = \frac{(S_i/N_i)}{(S_o/N_o)} = \frac{N_o}{N_i}\frac{S_i}{S_o} \tag{11.67}$$

where S_i/N_i = input signal-to-noise ratio, and S_o/N_o = output signal-to-noise ratio. The noise figure is a measure of the noise added by a circuit, where the added noise power can be expressed as

$$P_n = GkT_eB \tag{11.68}$$

in which T_e is the equivalent input noise temperature, G is the device gain, B is the bandwidth, and k is Boltzmann's constant.

For an amplifier, the noise figure can be derived from

$$N_o = GkT_0B + GkT_eB \tag{11.69}$$

$$S_o = GS_i \tag{11.70}$$

where GkT_0B is the output noise due to input noise. Therefore,

$$F = \frac{N_o}{kT_0B}\frac{S_i}{GS_i} = 1 + \frac{T_e}{T_0} \tag{11.71}$$

where $T_0 = 290$ K by convention.

For a cascade of amplifiers with gains G_i, equivalent noise temperatures T_{ei}, and noise figures F_i, the total noise figure can be derived to be

$$F_{\text{tot}} = 1 + \frac{T_{e1}}{T_0} + \frac{T_{e2}}{G_1 T_0} + \cdots \tag{11.72a}$$

$$= F_1 + \frac{F_2 - 1}{G_1} + \frac{F_3 - 1}{G_1 G_2} + \cdots \tag{11.72b}$$

For a mixer, the output signal and noise from both the signal and image responses (sidebands) must be considered. The mixer noise figure depends upon whether signal appears at both sidebands and whether the image is resistively terminated.

Case 1. Equal signal at both bands with resistive termination at all frequencies. This double-sideband (DSB) noise figure is given by

$$F_{\text{DSB}} = 1 + \frac{T_e}{T_0} \qquad (11.73)$$

where T_e is the equivalent input noise temperature.

Case 2. Signal in only the signal band with resistive terminations at all frequencies. This single-sideband (SSB) noise figure is

$$F_{\text{SSB}} = \left(1 + \frac{T_e}{T_0}\right)\left(1 + \frac{L_s}{L_i}\right) \qquad (11.74)$$

where L_s and L_i are the mixer conversion losses at the signal and image frequencies. This single-sideband noise figure can be related to the double-sideband noise figure given in (11.73) and for the case where $L_s = L_i$

$$F_{\text{SSB}} = 2\left(1 + \frac{T_e}{T_0}\right) \qquad (11.75)$$

or

$$F_{\text{SSB}} = 2F_{\text{DSB}}$$

If the noise figures are expressed in decibels, $F_{\text{SSB}} = F_{\text{DSB}} + 3$ dB.

Case 3. Signal in only the signal band with a reactive termination at the image. This "single-sideband" noise figure is given by

$$F'_{\text{SSB}} = 1 + \frac{T_e}{T_0}\left(1 + \frac{L_s}{L_i}\right) \qquad (11.76)$$

This is the noise figure obtained when a bandpass filter is placed in front of a dual-response mixer (assuming that the filter presents a reactive termination at the image frequency). For the case where $L_s = L_i$,

$$F'_{\text{SSB}} = 1 + 2\frac{T_e}{T_0} \qquad (11.77)$$

This can be related to F_{DSB} given in (11.73) to obtain

$$F'_{\text{SSB}} = 2F_{\text{DSB}} - 1 \qquad \text{for } L_s = L_i \qquad (11.78)$$

For large-noise figures, $F'_{\text{SSB}} \rightarrow F_{\text{DSB}} + 3$ dB (assuming $L_s = L_i$).

It is important to carefully define single- and double-sideband noise figures for the particular application, taking into account the mixer terminations. In addition, the comparison or conversion of single- and double-sideband noise figures must consider the mixer conversion loss at the signal and image frequencies.

In terms of the mixer noise temperature defined in (11.66), the mixer single-sideband noise figure is defined as

$$F_{\mathrm{SSB}} = 1 + \frac{T_m}{T_0} \tag{11.79}$$

The single-sideband noise figure is not necessarily equal to, and can in fact be lower than, the mixer conversion loss. The double-sided noise temperature is given by

$$T_{\mathrm{DSB}} = \frac{T_m}{1 + (L_s/L_i)} \tag{11.80}$$

The noise figure of a mixer–IF amplifier combination can be derived from (11.72) to be

$$F_{\mathrm{tot}} = F_m + L_s(F_{\mathrm{IF}} - 1) \tag{11.81}$$

where

F_{tot} = mixer/IF amplifier noise figure,
F_m = mixer single-sideband noise figure,
F_{IF} = IF amplifier noise figure,
L_s = mixer conversion loss.

where F_{tot}, F_m, L_s, and F_{IF} are expressed as power ratios.

11.3.4 Design Considerations

The mixer design and the required diode parameters are determined by the frequency of operation, bandwidth, conversion loss, VSWR, LO–RF isolation, and spurious and harmonic suppression.

Mixer Design Procedure

1. The mixer configuration is selected from the tradeoffs presented in Table 11.4.

2. The diode junction capacitance at 0 V is generally selected to provide a reactance of

$$X_{cj} = -j100\,\Omega \tag{11.82}$$

or

$$C_{j0} = \frac{1.6}{f_p} \tag{11.83}$$

where C_{j0} is the 0-V junction capacitance expressed in picofarads and f_p is the LO frequency in gigahertz. The conversion loss is increased by the diode series resistance R_s, which is minimized by selecting a diode with a high cutoff frequency. Zero-volt cutoff frequencies of 100 to 200 GHz for silicon, and 500 to 1000 GHz for GaAs diodes are available for low-loss mixer applications. Matched diodes that have similar loss and video impedance characteristics are available for balanced mixer designs.

3. The package parasitics have the effect of reducing the difference between the two diode impedance states and should be minimized. This will also serve to maximize the dynamic Q of the diode. In some applications, the package elements are selected to resonate at the sum frequency to provide a purely reactive termination to minimize losses at that frequency.

4. Calculate or measure the two impedance states of the diode under large-signal conditions.

5. Calculate the hyperbolic mean of the on and off impedances and match Z_m to 50 Ω.

6. Combine the diode and matching circuit with the open- and short-circuited stubs and low-pass filter (LPF), as shown in Fig. 11.28, to make up a baseline mixer subcircuit design. The signal and LO filter designs depend on the required signal-to-LO port isolation.

This baseline design will provide a good starting point for computer analysis. The circuit design may have to be modified to improve the impedance match at the IF and the signal frequencies to reduce the conversion loss.

11.3.5 Mixer Design Examples

The following discussion illustrates the design procedure for a mixer with the following specifications:

$$f_s = 12.0 \text{ GHz}$$

$$f_p = 10.5 \text{ GHz}$$

$$\text{Conversion loss} < 6.5 \text{ dB}$$

$$\text{Noise figure} < 6 \text{ dB}$$

$$\text{LO/RF isolation} > 20 \text{ dB}$$

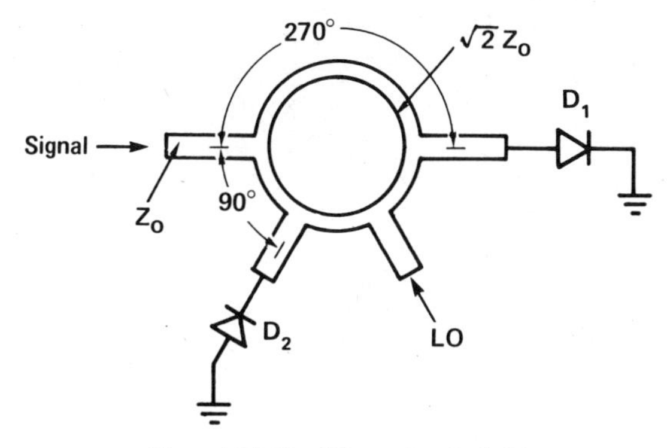

Figure 11.41 Mixer ring hybrid.

1. A single balanced mixer design using a 180-degree hybrid is selected from Table 11.4 to provide the 20-dB LO/RF isolation. A 3-dB ring hybrid shown in Fig. 11.41 is used to split the signal and LO power between the two diodes at the desired phase. The ring hybrid has the characteristic that the signal and LO are in phase at one diode and 180 degrees out of phase at the other diode. The electrical length between the signal and LO ports is 180 degrees going one way, and 360 degrees going the other way around the ring. This provides the cancellation required to achieve the 20-dB LO/RF isolation. This hybrid is easily implemented in microstrip. It will be used in the following mixer design.

2. *Diode Parameter Selection*. For a LO frequency of 10.5 GHz,

$$C_{j0} = \frac{1.6}{f_p} = 0.15\ \text{pF}$$

$$f_{c0} = 500\ \text{GHz} \quad \text{(using a GaAs Schottky-barrier diode)}$$

$$R_s = \frac{1}{2\pi f_{c0} C_{j0}} = 2.1\ \Omega$$

$$I_s = 1 \times 10^{-9}\ \text{A} \quad \text{(for the selected diode)}$$

3. *Package Selection*. The package is selected that has parasitic capacitance and inductance that are resonant approximately at the sum frequency $f_{\text{sum}} = 22.5$ GHz. The package parasitics are

$$C_p = 0.15\ \text{pF}$$

$$L_p = 0.4\ \text{nH}$$

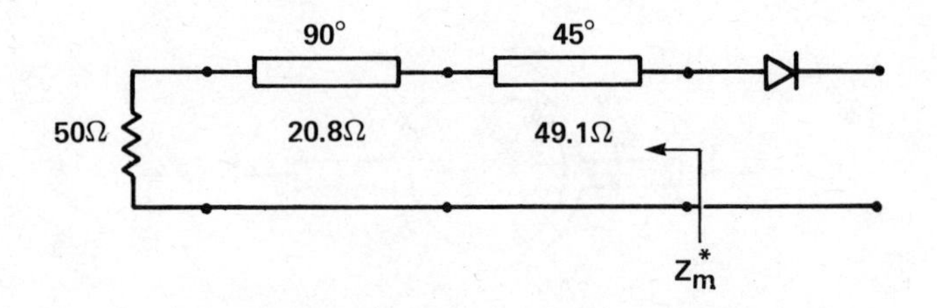

Figure 11.42 Initial diode-matching circuit.

4. *Determine the Two Diode Impedance States*. The two impedances for the diode were calculated from Fig. 11.26 to be

$$Z_{on} = 12.8 + j33.9\ \Omega$$

$$Z_{off} = 0.55 - j49.5\ \Omega$$

5. Calculate the Hyperbolic Mean from (11.60) and (11.61):

$$Z_m = 16.8 - j46.1\ \Omega$$

The two-section matching circuit shown in Fig. 11.42 transforms Z_m to 50 Ω. The resulting mixer reflection coefficients for the two diode states are

$$\Gamma_1 = 0.939\angle{-176.9°}$$

$$\Gamma_2 = 0.939\angle{+3.1°}$$

which are seen to be equal in amplitude and 180 degrees out of phase.

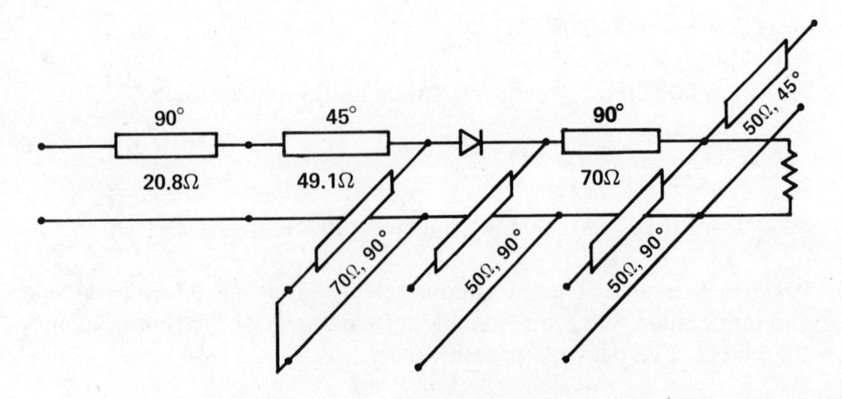

Figure 11.43 Preliminary mixer subcircuit design.

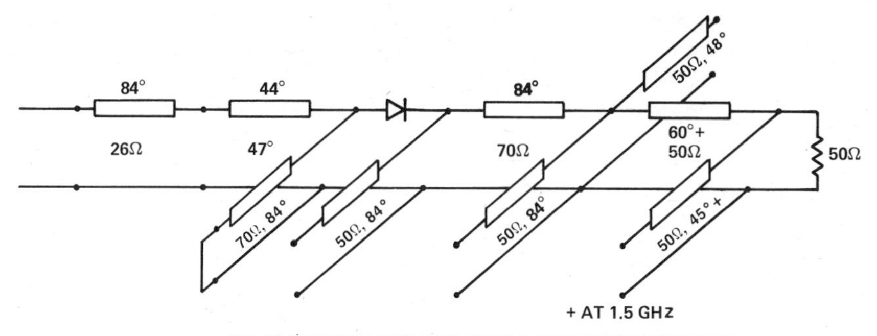

Figure 11.44 Final mixer subcircuit schematic.

6. The initial mixer subcircuit schematic (without the hybrid) including short- and open-circuited stubs, is shown in Fig. 11.43. The IF filter is realized with quarter-wavelength (at the LO frequency) open-circuited stubs separated by a quarter-wavelength transmission line. It uses the mixer open-circuited stub as the first element. Because of the periodic nature of this filter, an additional open-circuited stub (90 degrees at twice the LO frequency) is included to control the circuit impedance at the sum and second harmonic of the LO frequency. This circuit was analyzed using the Siegel, Kerr, and Hwang computer program previously described [22], and was found to have a high conversion loss of 9.3 dB due to a significant mismatch loss at the signal and IF frequencies. The signal-frequency mismatch is due to the high IF that separates the LO and signal frequencies. Based upon the computed diode impedance at these frequencies, the signal-matching circuit was modified to more closely match the diode at f_s and an IF-port matching circuit was added to reduce the IF mismatch. The modified schematic of the mixer subcircuit is shown in Fig. 11.44. The mixer subcircuit performance was again computed and is summarized in Table 11.7. This calculated performance must be modified to account for the hybrid loss that adds directly to the loss, noise figure, and required LO power. Assuming a hybrid loss of 0.5 dB, the mixer performance would be as listed in Table 11.8. A layout of the final mixer design is shown in Fig. 11.45. The linewidths and line lengths would be determined by the substrate material used. Discontinuity models for the steps, T junctions, and open-

TABLE 11.7 Mixer Subcircuit Performance

Conversion loss (L_{01})	5.7 dB
Noise temperature	296 K
Noise figure	3.1 dB
Required LO power (per diode)	1.6 mW

TABLE 11.8 Final Mixer Performance

Conversion loss (L_{01})	6.2 dB
Noise figure	3.6 dB
LO power	3.6 mW

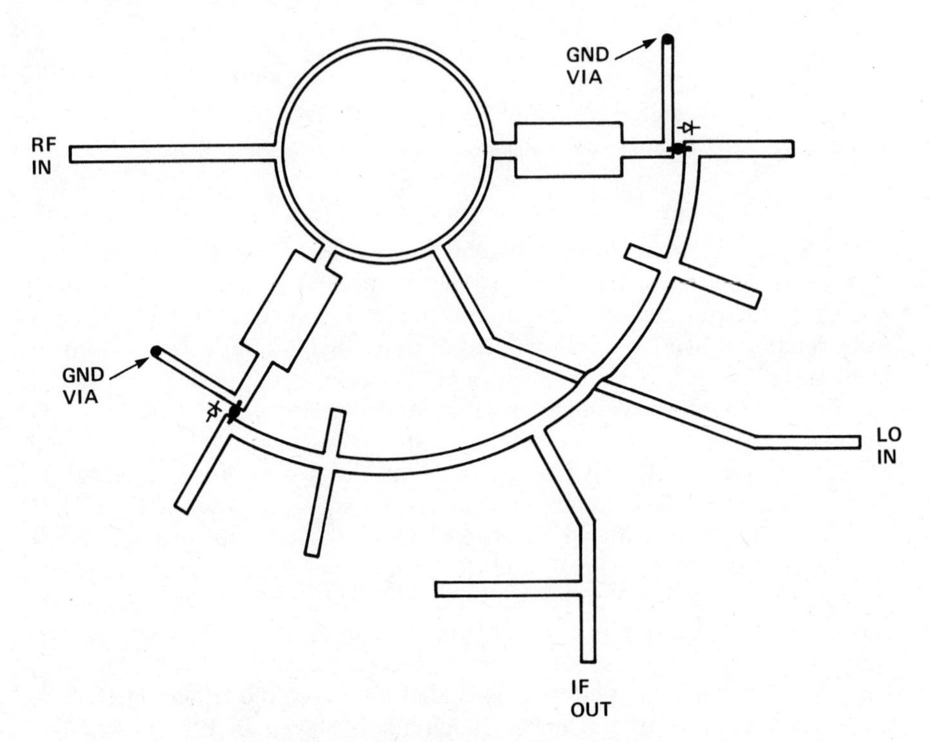

Figure 11.45 Microstrip mixer layout.

circuited stubs should be included in the analysis at this point. The desired embedding impedances are known from the analysis, and standard circuit optimization programs can be used to adjust the transmission-line lengths and impedances to compensate for the discontinuity effects.

REFERENCES

1. Bhartia, P., and I. J. Bahl, *Millimeter Wave Engineering and Applications*, Wiley, New York, 1984, pp. 427–428.
2. Torrey, H. C., and C. A. Whitmer, *Crystal Rectifiers*, MIT Radiation Lab. Series, Vol. 15, McGraw-Hill, New York, 1948, pp. 337.

3. Watson, H. A., *Microwave Semiconductor Devices and Their Circuit Applications*, McGraw-Hill, New York, 1969, pp. 379.
4. Lepoff, J., "How the New Schottkys Detect without DC Bias," *Microwaves*, Vol. 16, Feb. 1977, pp. 44–48.
5. Van der Ziel, A., "Noise in Solid State Devices and Lasers," *Proc. IEEE*, Vol. 58, Aug. 1970, pp. 1178–1206.
6. Siegal, B., and E. Pendleton, "Zero Bias Schottky Barrier Diodes as Microwave Detectors," *Microwave J.*, Vol. 18, Sept. 1975, pp. 40–43.
7. Hewlett-Packard, "Impedance Matching Techniques for Mixers and Detectors," Hewlett-Packard Application Note 963.
8. Torrey, H. C., and C. A. Whitmer, *Crystal Rectifiers*, MIT Radiation Lab. Series, Vol. 15, McGraw-Hill, New York, 1948, pp. 140–148.
9. Barber, M. R., "Noise Figure and Conversion Loss of the Schottky Barrier Mixer Diode," *IEEE Trans. Microwave Theory Tech.*, Vol. MTT-15, Nov. 1967, pp. 629–635.
10. Saleh, A. A. M., *Theory of Resistive Mixers*, MIT Press, Cambridge, Mass., 1971, pp. 64–86.
11. Deschamps, G. A., "A New Chart for the Solution of Transmission Line and Polarization Problems", *IRE Trans. Microwave Theory Tech.*, Vol. MTT-1, Mar. 1953, pp. 5–13.
12. Peterson, D. F., and D. H. Steinbrecher, "Image-optimized Frequency Scalable Mixers for Millimeter-Wave Applications," *IEEE MTT-S Int. Microwave Symp. Digest*, 1983, pp. 554–556.
13. Steinbrecher, D. H., "Circuit Aspects of Nonlinear Microwave Networks," *IEEE Proc. Int. Symp. Circuits and Systems*, 1975, pp. 477–479.
14. Matthaei, G., L. Young, and E. M. T. Jones, *Microwave Filters, Impedance Matching Networks, and Coupling Structures*, McGraw-Hill, New York, 1964, pp. 843–847.
15. Henry, P. S., B. S. Glance, and M. V. Schneider, "Local Oscillator Noise Cancellation in the Subharmonically Pumped Down Converter," *IEEE Trans. Microwave Theory Tech.*, Vol. MTT-24, May 1976, pp. 254–257.
16. Henderson, B., "Mixer Design Considerations Improve Performance," *Microwave Systems News*, Vol. 11, Oct. 1981, pp. 103–118.
17. Reynolds, J., and M. R. Rosenzweig, "Learn the Language of Mixer Specification," *Microwaves*, Vol. 17, May 1978, pp. 72–78.
18. Pucel, R. A., D. Masse, and P. Bera, "Performance of GaAs MESFET Mixers at *X* Band," *IEEE Trans. Microwave Theory Tech.*, Vol. MTT-24, June 1976, pp. 351–360.
19. Bura, P., and R. Dikshit, "FET Mixers for Communication Satellite Transponders," *IEEE MTT-S Int. Microwave Symp. Digest*, 1976, pp. 90–92.
20. Tsironis, C., R. Meierer, and R. Stahlmann, "Dual-Gate MESFET Mixers," *IEEE Trans. Microwave Theory Tech.*, Vol. MTT-32, Mar. 1984, pp. 248–255.
21. Held, D., and A. Kerr, "Conversion Loss and Noise of Microwave and Millimeter-Wave Mixers: Part I—Theory," *IEEE Trans. Microwave Theory Tech.*, Vol. MTT-26, Feb. 1978, pp. 49–55.
22. Siegel, P. H., A. K. Kerr, and W. Hwang, *Topics in the Optimization of Millimeter-Wave Mixers*, NASA Tech. Paper 2287, 1984.

PROBLEMS

11.1 What is the voltage sensitivity of the following detector diode:

$$I_s = 1 \times 10^{-7}\ \text{A}$$

$$n = 1.05$$

$$C_{j0} = 0.25\ \text{pF}$$

$$R_s = 15\ \Omega$$

$$R_L = 1\ \text{M}\Omega$$

(a) $f = 2$ GHz, $T = 23°\text{C}$, $I_0 = 0$.
(b) $f = 2$ GHz, $T = 23°\text{C}$, $I_0 = 40\ \mu\text{A}$.
(c) $f = 10$ GHz, $T = 23°\text{C}$, $I_0 = 0$.
(d) $f = 10$ GHz, $T = 100°\text{C}$, $I_0 = 0$.

11.2 What is the impedance of the following detector diode?

$$C_{j0} = 0.20\ \text{pF}$$

$$R_s = 10\ \Omega$$

$$I_s = 1 \times 10^{-8}\ \text{A}$$

$$L_p = 1\ \text{nH}$$

$$C_p = 0.25\ \text{pF}$$

$$n = 1.05$$

$$T = 20°\text{C}$$

(a) $f = 2$ GHz, $I_0 = 0$.
(b) $f = 6$ GHz, $I_0 = 0$.
(c) $f = 2$ GHz, $I_0 = 50\ \mu\text{A}$.

11.3 (a) Determine the minimum detectable signal ($S/N = 1$) at 5 GHz in a 5-MHz bandwidth if:

$$C_{j0} = 0.3\ \text{pF}$$

$$R_s = 15\ \Omega$$

$$I_s = 1 \times 10^{-8}\ \text{A}$$

$$I_0 = 20\ \mu\text{A}$$

$$n = 1.05$$

Assume $T = 27°\text{C}$, $R_a = 1200\ \Omega$, and $R_L = 1\ \text{M}\Omega$.

(b) What is the TSS?

(c) What is the TSS if $I_0 = 0$?

11.4 What is the voltage sensitivity of the detector shown below?

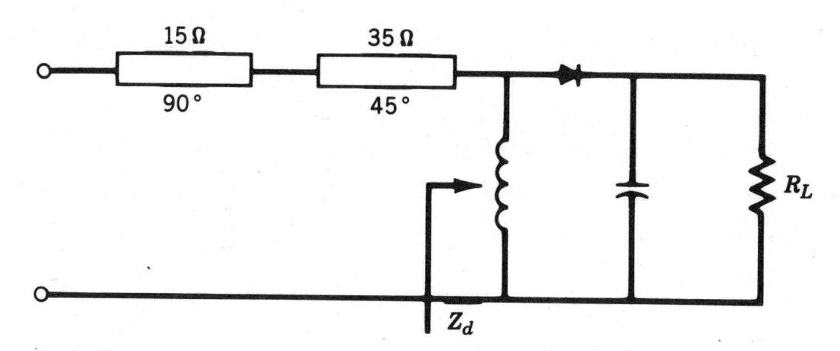

The transmission line electrical lengths are at 5 GHz. Assume the following:

$$\text{Freq} = 6\ \text{GHz}$$

$$\beta_v = 10\ \text{mV}/\mu\text{W}$$

$$Z_d = 5 - j46\ \Omega$$

11.5 (a) Design a 4-GHz microstrip detector. Assume the following:

$$T = 23°\text{C}$$

$$I_s = 1 \times 10^{-7}\ \text{A}$$

$$n = 1.1$$

$$\text{Package} = \text{beam lead}$$

Hint: Choose C_{j0} twice that calculated for high sensitivity and bias the diode at 20 μA. Use an 85-Ω series line to resonate the diode reactance and a quarter-wave transformer.

(b) Using the matching circuit shown in Fig. 11.18, broadband the design to cover 3–4 GHz.

11.6 (a) What is the dynamic Q, Q_d, of the following GaAs mixer diode?

$$f_p = 8\ \text{GHz}$$

$$C_{j0} = 0.20\ \text{pF}$$

$$R_s = 2\ \Omega$$

$$I_s = 1 \times 10^{-9}\ \text{A}$$

$$n = 1.1$$

$$L_p = 0.2\ \text{nH}$$

$$C_p = 0.12\ \text{pF}$$

(b) What is the hyperbolic mean for this pumped diode?

11.7 Design a mixer using a GaAs Schottky Barrier diode ($f_{c0} = 500$ GHz, $I_s = 1 \times 10^{-9}$ A, $n = 1.05$) that converts 4 GHz to 0.5 GHz. Assume that the package parasitics are 0.2 nH and 0.35 pF. Use a 61-Ω series line to resonate the diode reactance and a quarter-wave transformer.

11.8 (a) Calculate the conversion gain of the following 8 to 0.5 GHz FET mixer.

Source Impedance at 8 GHz $= 8 + j65\ \Omega$,
Load Impedance at 0.5 GHz $= 350 + j10\ \Omega$.

Assume the following FET parameters:

$$R_{gm} = 2\ \Omega$$

$$C_{gs} = 0.35\ \text{pF}$$

$$R_i = 1\ \Omega$$

$$C_{gd} = 0.005\ \text{pF}$$

$$R_s = 4\ \Omega$$

$$R_o = 400\ \Omega$$

$$C_{ds} = 0.02\ \text{pF}$$

$$g_m = 0.01\ \text{S}$$

(b) What is the maximum available conversion gain?

12 MICROWAVE CONTROL CIRCUITS

Microwave switches, phase shifters, and attenuators are three important "control" circuits used extensively for controlling the signal flow at microwave frequencies. A single phased array radar system, for example, may use thousands of these circuits for precise electronic control of the radiated beam. Two types of active devices used commonly in control circuits are *pin* diodes [1] and GaAs MESFETs [2]. Before looking at circuit design methods, we briefly look at the equivalent circuit representations for these devices.

12.1 DEVICES FOR MICROWAVE CONTROL CIRCUITS

12.1.1 *pin* Diodes

The physics of *pin* diodes has been described in Chapter 8. The equivalent circuit of a packaged *pin* diode may be written as shown in Fig. 12.1. In forward bias, the single-pole double-throw (SPDT) switch is in position F, and in reverse bias it is put in position R. Inductance L_p and capacitance C_p are contributed by the package. Typical values for two commercially available diodes are given in Table 12.1. See page 608.

12.1.2 GaAs MESFETs

Two different modes of operation of GaAs MESFETs are used for the design of control circuits at microwave frequencies. These are known as (1) active, and (2) passive modes. In the active mode, single-gate or dual-gate MESFETs are used as three-terminal active devices. Equivalent circuits for the active mode are identical to those used for amplifier designs and have been discussed in Chapter 7 (on solid state active devices) and Chapter 10 (on amplifiers).

In the passive mode of operation, MESFETs are used as passive two-terminal devices, with the gate terminal acting as a port for the control signal only. The RF connections are made to the drain and the source terminals, and the gate terminal looks into an open circuit for the RF signal. The RF impedance between the drain and the source terminals depends upon the dc control voltage at the gate terminal. For switching applications,

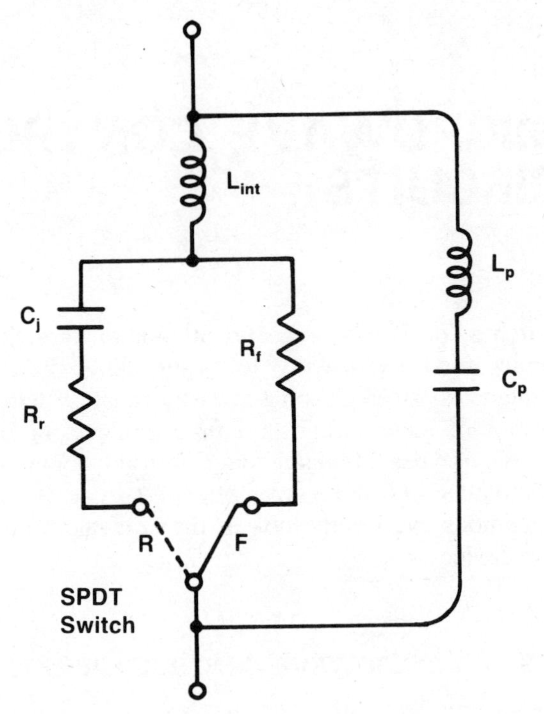

Figure 12.1 Equivalent circuit of a packaged *pin* diode.

low-impedance and high-impedance states are obtained by making the gate voltage equal to zero and by using a gate voltage greater (numerically) than the pinchoff voltage, respectively.

For equivalent circuits for these two states, refer to Fig. 12.2. In the low-impedance state, the channel presents a resistive path to the RF current between the drain and the source. The value of this resistance R_{on} can be estimated by dividing the total current path in four different types of regions as discussed by Ayasli [2]. Typical value of R_{on} for a $1 \times 1000\ \mu m^2$ gate MESFET suitable for *X*-band operation is about $2.5\ \Omega$ at around 10 GHz. Attempts at reducing the value of R_{on} using self-aligned gate technology have been reported [3].

For the high-impedance state of MESFETs operating in the passive mode, an equivalent circuit may be derived by referring to Fig. 12.2(*b*). As the channel is now pinched off, the capacitance of the depletion layer (represented by C_g) appears in series between the source and drain terminals. Also in this case, the capacitance C_{sd} (and the leakage resistance r_d) between the source and the drain terminals needs to be incorporated in the equivalent circuit. These components are present in the low-impedance state also, but need not be included in the equivalent circuit because of the small value of R_{on} appearing in parallel. A typical high-impedance-state

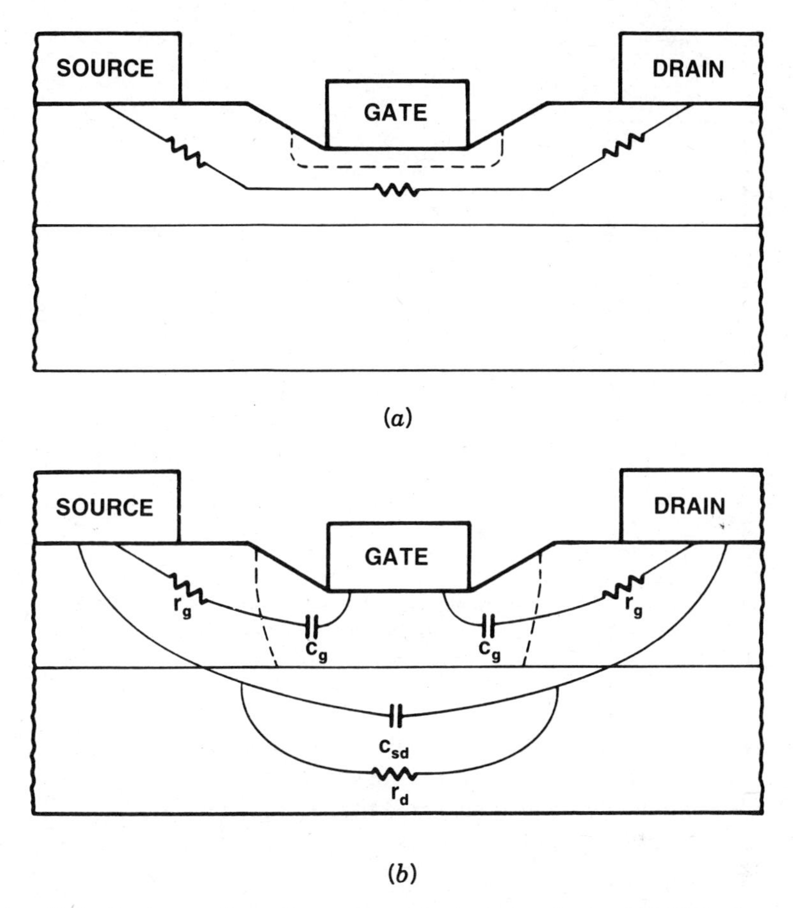

Figure 12.2 Device configurations and equivalent circuits for MESFETs in passive mode. (*a*) Low-impedance state. (*b*) High-impedance state.

equivalent circuit is shown in Fig. 12.3. For a $1 \times 1000\ \mu m^2$ gate MESFET operating at 10 GHz, typical values of the various elements in this equivalent circuit are $C_{sd} \simeq 0.25$ pF, $r_d \simeq 3$ kΩ, and $C_g \simeq 0.2$ pF. As r_g is much smaller than the reactance of C_g $[r_g \ll 1/(\omega C_g)]$, the equivalent circuit of Fig. 12.3(*a*) can be replaced by a parallel combination of R_{off} and C_{off} [shown in Fig. 12.3(*b*)], where

$$C_{\text{off}} = C_{sd} + \frac{C_g}{2} \tag{12.1}$$

and

$$R_{\text{off}} = \frac{2r_d}{2 + r_d\omega^2 C_g^2 r_g} \tag{12.2}$$

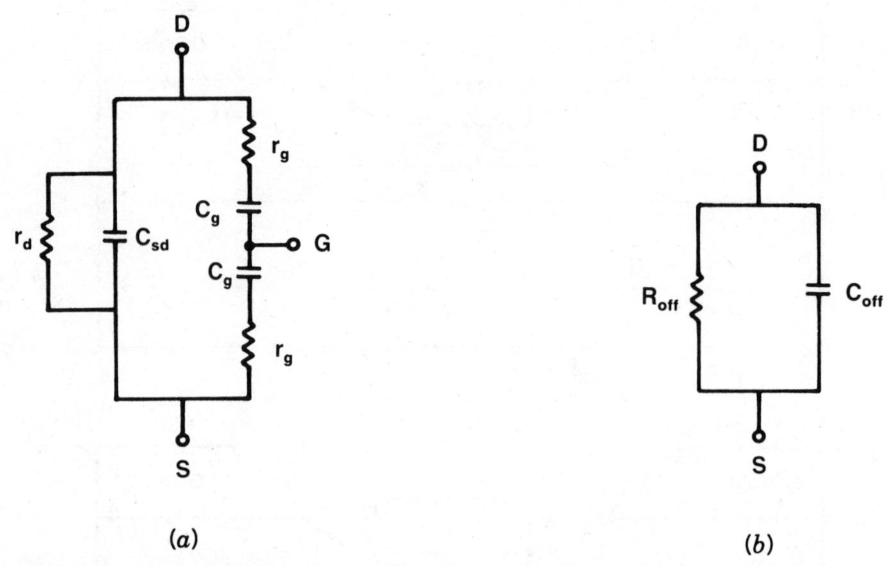

Figure 12.3 (*a*) The high-impedance-state equivalent circuit of a MESFET in the passive mode, and (*b*) Simplified circuit for the high-impedance state.

where ω is the operating frequency in radians/second. Typical values for C_{off} and R_{off} are (for $1 \times 1000\ \mu m^2$ gate devices at 10 GHz) about 0.2 pF and 2 kΩ, respectively. It may be noted that R_{off} [as indicated by (12.2)] is now a function of the operating frequency ω.

12.2 DESIGN OF SWITCHES

The design of microwave switching circuits using *pin* diodes or GaAs MESFETs is discussed in this section.

12.2.1 Basic Configurations

There are two basic configurations that may be used for a simple single-pole single-throw (SPST) switch designed to control the flow of microwave signals along a transmission line. These are shown in Figs. 12.4 and 12.5. In the first case, the switching device (SD) is mounted in series with one of the conductors of the transmission line (strip conductor of the microstrip line). In the second case (Fig. 12.5), the switching device is mounted in shunt across the two conductors of a transmission line. Inductances L in Figs. 12.4 and 12.5 are inductances of bonding ribbons, and C_i and R_f constitute the simplified equivalent circuits for the high-impedance and low-impedance states of the device chip.

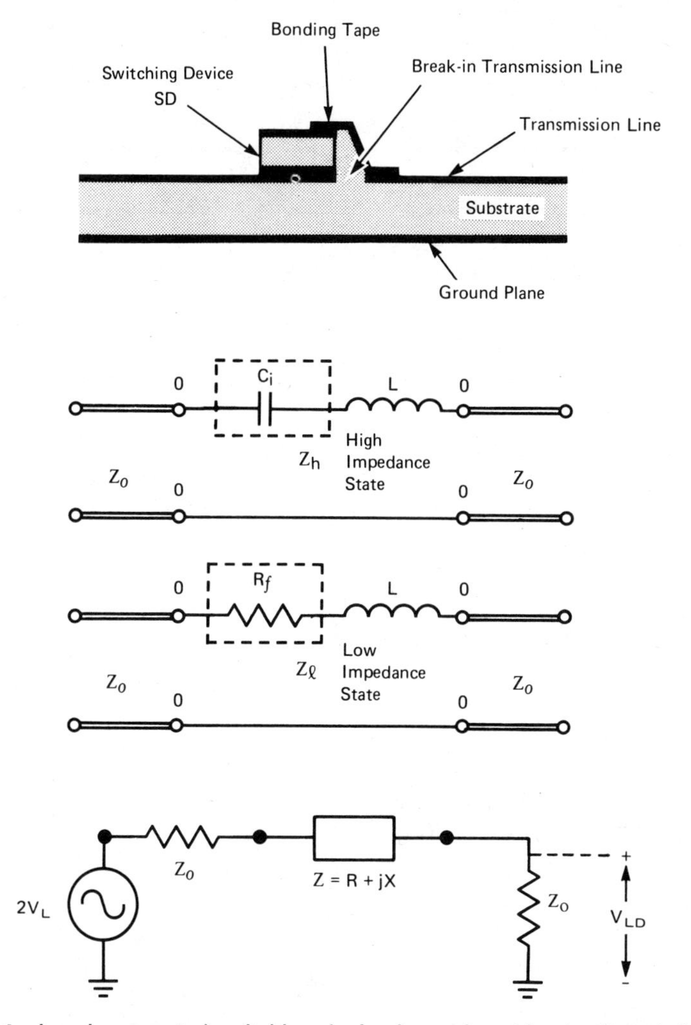

Figure 12.4 A series-mounted switching device in a microstrip circuit and simplified equivalent circuits for high- and low-impedance states. The term Z is the impedance of the switching device SD.

The two configurations are complementary in the sense that for the series configuration, the low-impedance state of the device allows the signal to propagate; whereas in the shunt-mounted configuration, the signal is delivered to the load when the device is in the high-impedance state. In the "OFF" states of both of these types of switches, the microwave power incident on the switch is mostly reflected back. However, small fractions of power get (1) dissipated in device resistances or circuit losses, and (2) get transmitted to the load (accounting for a finite isolation) because of device/circuit imperfections.

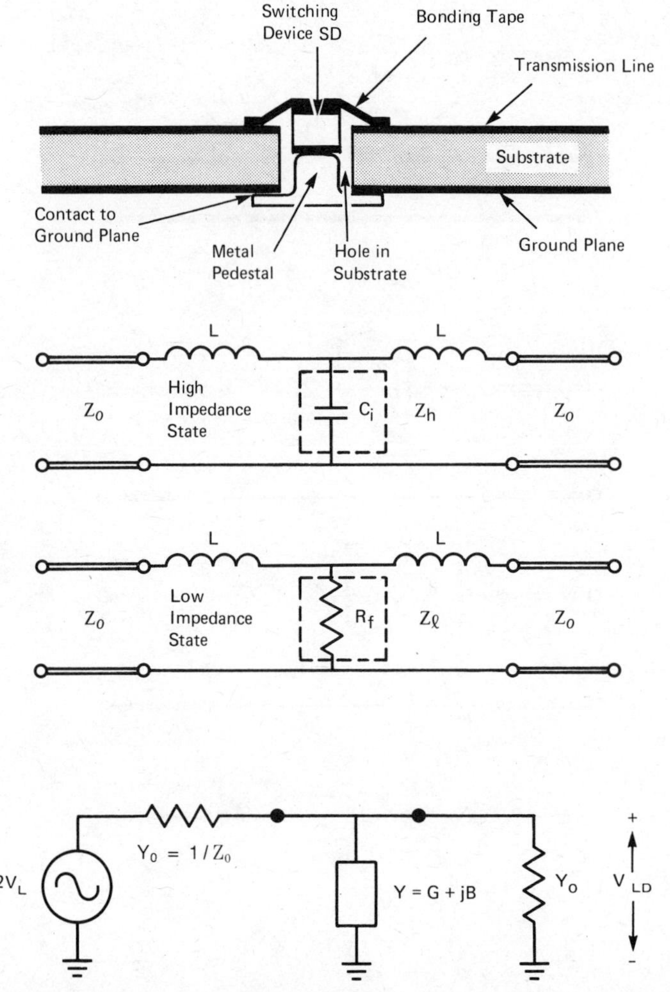

Figure 12.5 A shunt-mounted switching device in a microstrip circuit and simplified equivalent circuits for high- and low-impedance states. *Y* is the admittance of the switching device SD.

12.2.2 Insertion Loss and Isolation

Because of a finite nonzero impedance of switching devices in the low-impedance state and a definite noninfinite impedance in the high-impedance state, the switching circuits are not perfect. The performance of a practical switch can be expressed by specifying its insertion loss and isolation.

Insertion loss is defined as the ratio of the power delivered to the load in the "ON" state of the ideal switch to the actual power delivered by the practical switch (in the "ON" state). It is usually expressed in decibels. The insertion loss may be calculated by considering the equivalent circuits shown in Figs. 12.4 and 12.5. If V_L denotes the actual voltage across the load in the ideal switch, the insertion loss may be written as

$$\mathrm{IL} = \left| \frac{V_L}{V_{LD}} \right|^2 \tag{12.3}$$

For the series configuration (Fig. 12.4) we have,

$$V_{LD} = \frac{2V_L}{2 + Z/Z_0} \tag{12.4}$$

where $Z(= R + jX)$ is the impedance of the switching device. The insertion loss is given by

$$\begin{aligned}\mathrm{IL} &= \left| \frac{2 + Z/Z_0}{2} \right|^2 \\ &= 1 + \frac{R}{Z_0} + \frac{1}{4}\left(\frac{R}{Z_0}\right)^2 + \frac{1}{4}\left(\frac{X}{Z_0}\right)^2\end{aligned} \tag{12.5}$$

where R and X are the resistance and the reactance of the switching device in the low-impedance state.

For the shunt configuration (Fig. 12.5), the voltage across the load may be written by

$$V_{LD} = \frac{2V_L Y_0}{2Y_0 + Y} \tag{12.6}$$

and the insertion loss becomes

$$\begin{aligned}\mathrm{IL} &= \left| \frac{2Y_0 + Y}{2Y_0} \right|^2 = \left| 1 + \frac{G + jB}{2Y_0} \right|^2 \\ &= 1 + \frac{G}{Y_0} + \frac{1}{4}\left(\frac{G}{Y_0}\right)^2 + \frac{1}{4}\left(\frac{B}{Y_0}\right)^2\end{aligned} \tag{12.7}$$

where $Y_0 = 1/Z_0$ and G and B are the real and imaginary parts of the admittance Y of the switching device in the high-impedance state. It may be noted that the similarity of (12.5) and (12.7) originates from the dual nature of series and shunt configurations.

Isolation is a measure of the performance of the switch when it is in the OFF state. Isolation is defined as the ratio of the power delivered to the load for an ideal switch in the ON state to the actual power delivered to the load when the switch is in the OFF state. For the series configuration, the OFF state exists when the device is in the high-impedance state. The

TABLE 12.1 Equivalent-Circuit Parameters for Two Commercially Available *pin* Diodes

Parameters	MA47892-109	MA47899-030
C_j	1 pF	0.1 pF
R_f	0.4 Ω	1 Ω
R_r	0.5 Ω	4 Ω
L_{int}	0.3 nH	0.3 nH
C_p	0.08 pF	0.18 pF
τ	5 μs	0.5 μs
f_{cs} $(=1/[2\pi C_j\sqrt{R_r R_f}])$	350 GHz	800 GHz

isolation in this case is also given by (12.5) with R and X replaced by the corresponding values in the high-impedance state. Similarly, the isolation for the shunt configuration is given by (12.7) when we use the low-impedance-state values for G and B.

Example. As an example, let us calculate insertion loss and isolation for two switches using *pin* diodes with parameters listed in Table 12.1. Package capacitance is ignored and the switches are considered to operate at 3.18 GHz. For MA-47892 and for $Z_0 = 50\ \Omega$, we have $R = 0.4\ \Omega$ and $X = \omega L = 2\pi \times 3.18 \times 10^9 \times 0.3 \times 10^{-9} = 6\ \Omega$ in forward bias. The reverse-bias impedance is $(0.5 - j44)\ \Omega$. Thus for a series switch:

$$\text{IL} = 1 + \frac{0.4}{50} + 0.25\left(\frac{0.4}{50}\right)^2 + 0.25\left(\frac{6}{50}\right)^2 = 0.05\ \text{dB} \tag{12.8}$$

and

$$\text{Isolation} = 1 + \frac{0.5}{50} + 0.25\left(\frac{0.5}{50}\right)^2 + 0.25\left(\frac{44}{50}\right)^2 = 0.805\ \text{dB} \tag{12.9}$$

For this design, the value of isolation is unacceptably low because the reverse-bias impedance is comparable to the 50-Ω characteristic impedance of the transmission line. However, when we use MA-47899 with a smaller reverse-bias capacitance, we have, in forward bias $Z_f = R_f + jX_f = (1 + j6)\ \Omega$, and in reverse bias

$$Z_r = R_r + j\left(\omega L_{\text{int}} - \frac{1}{\omega C_j}\right) = 4 - j494\ \Omega$$

These yield

$$\text{IL} = 1 + \frac{1}{50} + 0.25\left(\frac{1}{50}\right)^2 + 0.25\left(\frac{6}{50}\right)^2 = 1.0237 = 0.1\ \text{dB} \tag{12.10}$$

and

$$\text{Isolation} = 1 + \left(\frac{4}{50}\right) + 0.25\left(\frac{4}{50}\right)^2 + 0.25\left(\frac{494}{50}\right)^2 = 14.06\ \text{dB} \qquad (12.11)$$

This performance is certainly more reasonable.

Shunt Switch Configuration. When MA-47892 is used in the shunt configuration, IL and isolation are given by (12.7). We have

$$Y_f = \frac{1}{Z_f} = \frac{1}{0.4 + j6} = (0.011 - j0.166)\ \text{S} \qquad (12.12)$$

which yields

$$\text{Isolation} = 1 + \frac{0.011}{0.02} + 0.25\left(\frac{0.011}{0.02}\right)^2 + 0.25\left(\frac{0.166}{0.02}\right)^2$$

$$= 18.85 = 12.75\ \text{dB} \qquad (12.13)$$

In reverse bias,

$$Y_r = \frac{1}{Z_r} = \frac{1}{0.5 - j494} = (0.00026 + j0.0272)\ \text{S} \qquad (12.14)$$

and the insertion loss is calculated to be

$$\text{IL} = 1 + \frac{0.00026}{0.02} + \frac{1}{4}\left(\frac{0.00026}{0.02}\right)^2 + \frac{1}{4}\left(\frac{0.0272}{0.02}\right)^2$$

$$= 1.336 = 1.26\ \text{dB} \qquad (12.15)$$

Corresponding calculations for MA-47899 in shunt configuration yield an isolation of 12.84 dB and an insertion loss of 0.029 dB.

12.2.3 Compensation of Device Reactances

An examination of (12.5) and (12.7), for IL and isolation of series and shunt switches, indicates that the performance of switching circuits is limited by the device reactance X or the susceptance B. Compensation of device reactances therefore provides a mechanism for improving the switch performance.

Let us consider the insertion loss of a shunt-mounted switch. The switching device is in the high impedance state. In this state, both *pin* diodes and GaAs MESFETs (passive mode) may be represented by a parallel combination of a high resistance R and a small capacitance C.

The total admittance of the high-impedance state can be reduced by connecting an inductive susceptance of an equal magnitude in parallel with the capacitance. This can be achieved either by mounting a lumped inductor or by incorporating a shorted (shorter than quarter-wave) stub. These arrangements are shown in Fig. 12.6. At the design frequency, the device capacitance C_i and externally added inductance L_c form a parallel resonant circuit, and the switching device presents only a resistive impedance R_i across the transmission line of characteristic impedance Z_0. Thus at the resonance of L_c and C_i, the insertion loss of the switch may be written as

$$\mathrm{IL} = 1 + \frac{G}{Y_0} + \frac{1}{4}\left(\frac{G}{Y_0}\right)^2 \tag{12.16}$$

If this inductive compensation is introduced in the example of the *pin* diode switch considered earlier, we find that the insertion loss of the shunt switch using MA-47892 becomes

$$\mathrm{IL} = 1 + \frac{0.00026}{0.02} + \frac{1}{4}\left(\frac{0.00026}{0.02}\right)^2 = 1.01304 = 0.0563 \text{ dB} \tag{12.17}$$

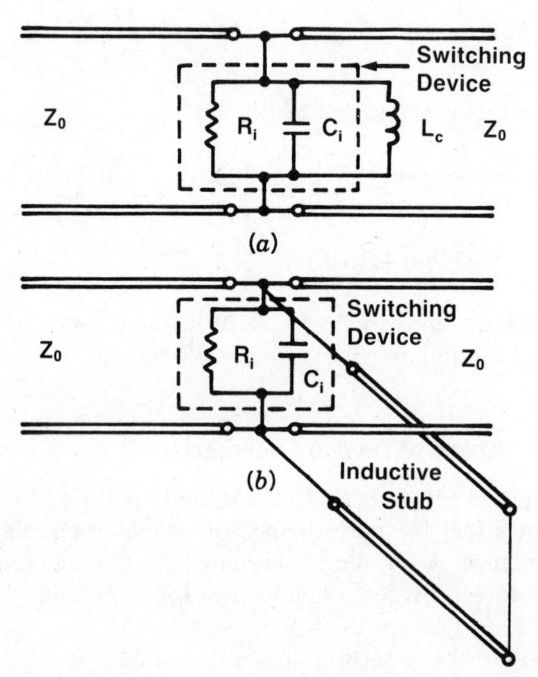

Figure 12.6 Compensation of the capacitance of a switching device in the high-impedance state by using (*a*) a lumped inductance, and (*b*) an inductive stub.

Let us consider another example of a MESFET switch with device parameters at 10 GHz being

$$R_i = R_{\text{off}} = 3\ \text{k}\Omega \qquad \text{and} \qquad C_i = C_{\text{off}} = C_{sd} + \frac{C_g}{2} = 0.25\ \text{pF}$$

Without any compensating inductance, the insertion loss for a shunt-mounted switch is found to be

$$\begin{aligned} \text{IL} &= 1 + \left\{\frac{0.00033}{0.02}\right\} + \frac{1}{4}\left\{\frac{0.00033}{0.02}\right\}^2 + \frac{1}{4}\left\{\frac{2\pi 10^{10}(0.25 \times 10^{-12})}{0.02}\right\}^2 \\ &= 1 + 0.01667 + 0.00007 + 0.15421 = 1.17095 = 0.69\ \text{dB} \qquad (12.18) \end{aligned}$$

When the capacitance is compensated by a suitable value of L_c, the insertion loss reduces to

$$\text{IL} = 1 + \left\{\frac{0.00033}{0.02}\right\} + \frac{1}{4}\left\{\frac{0.00033}{0.02}\right\}^2 = 1.01674 = 0.0721\ \text{dB} \qquad (12.19)$$

Introduction of a compensating inductance L_c in MESFET switching circuits is easier because there is no dc potential difference between the drain and the source terminals. Resonated GaAs FET devices are used extensively in microwave switching circuits [4, 5]. However, when a compensating inductance is to be introduced in a *pin* diode switching circuit, we need a dc blocking capacitor in series with L_c (lumped or stub) so that the reverse bias on the *pin* diode is not disturbed.

Isolation of shunt-mounted switches can also be improved by compensating the inductive reactance of the switching device in the low-impedance state by adding a series capacitance of suitable value.

12.2.4 Single-Pole Double-Throw Switches

Single-pole double-throw (SPDT) switches require a minimum of two switching devices. Two basic configurations for SPDT switches, namely series mounted and shunt mounted, are shown in Fig. 12.7. In the series configuration, the input signal is routed to the output 1 when the switching device SD1 is in the low-impedance state and the device SD2 is in the high-impedance state. In the shunt configuration shown in Fig. 12.7(*b*), the signal is routed to the output 1 when the device SD1 is in the high-impedance state and the device SD2 is in the low-impedance state. Thus, in either configuration, at any time, one of the devices is in the low-impedance state while the other one is in the high-impedance state.

The bandwidth of the SPDT in shunt configuration is limited because of the $\lambda/4$ line lengths required between the transmission-line junction and

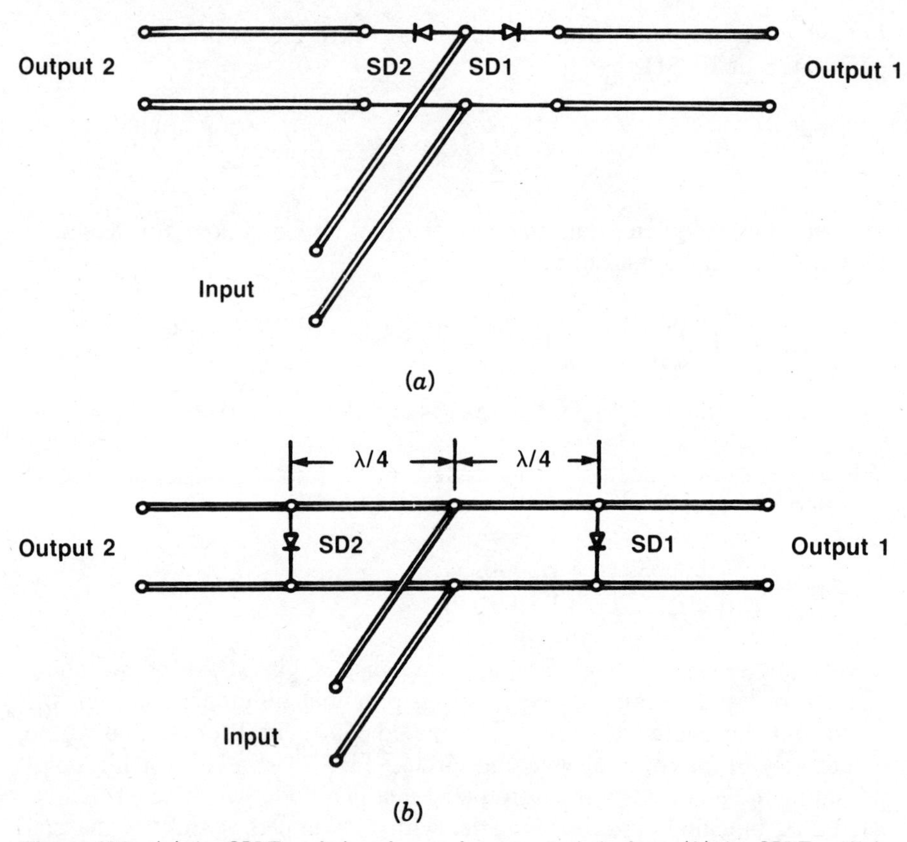

Figure 12.7 (*a*) An SPDT switch using series-mounted devices. (*b*) An SPDT switch using shunt-mounted devices.

the locations of the two switching devices. Figure 12.8 shows the performance of two SPDT configurations when the switching devices are MA-47899 *pin* diode chips. The design is centered around 3-GHz frequency. For a shunt-mounted switch, the variation of insertion loss with frequency limits the operating bandwidth. An example of a shunt-mounted SPDT switch using MESFETs is shown in Fig. 12.9. The design frequency in this case is 10 GHz, and the improvement in performance by adding two compensating inductances is also shown in the figure.

The design concept of SPDT switches may also be extended to single-pole multiple-throw switches. An SP3T switch [6] will use a minimum of three switching devices. Microstrip construction of a typical SP3T switch is shown in Fig. 12.10. In this configuration there are six devices, two associated with each of the output ports. Using a combination of series- and shunt-mounted diodes improves the switch's performance. The dc blocking capacitors and RF chokes are needed for dc biasing.

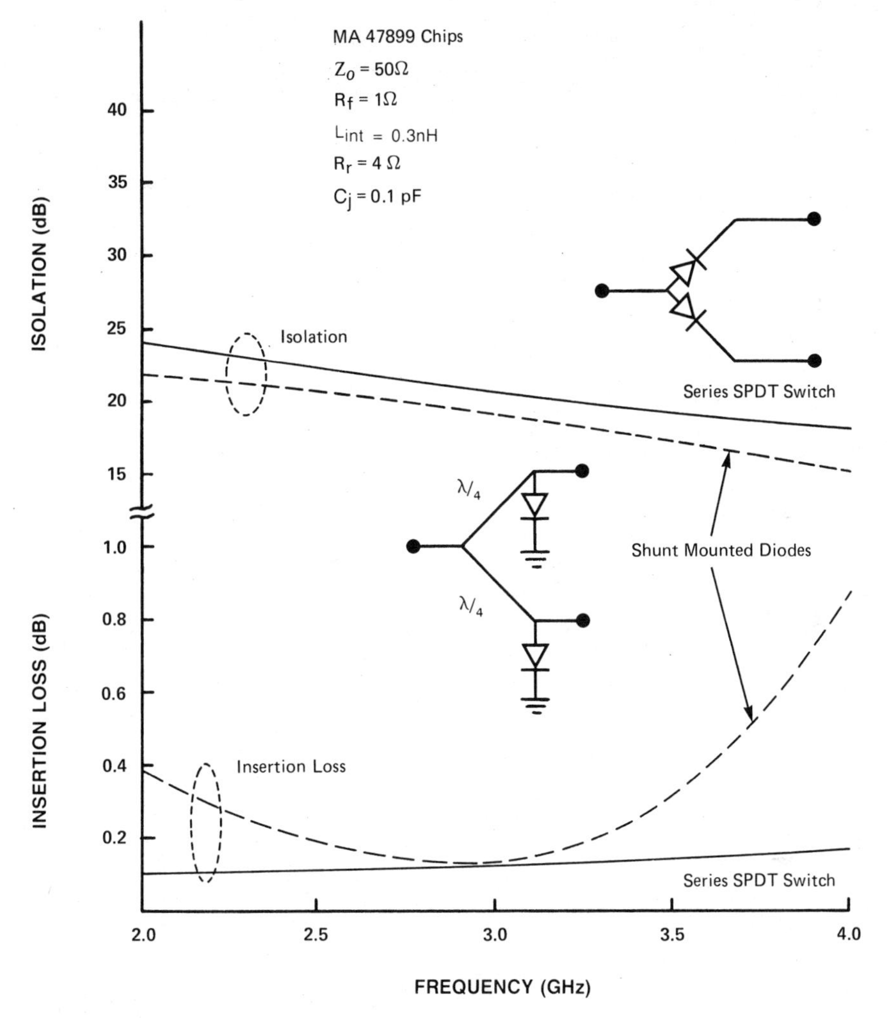

Figure 12.8 Typical insertion loss and isolation performance of SPDT switches using two *pin* diodes.

12.2.5 Series-Shunt Switching Configurations

In Section 12.2.1, we looked at the characteristics of two basic switching configurations using series-mounted and shunt-mounted switching devices. One could realize a better switching performance when both series- and shunt-mounted devices are included in a single circuit (as shown for the SP3T switch of Fig. 12.10).

The simplest series-shunt switching configuration is shown in Fig. 12.11(*a*). This switch is ON when the series device is in the low-impedance state and the shunt device is in the high-impedance state. In the OFF state

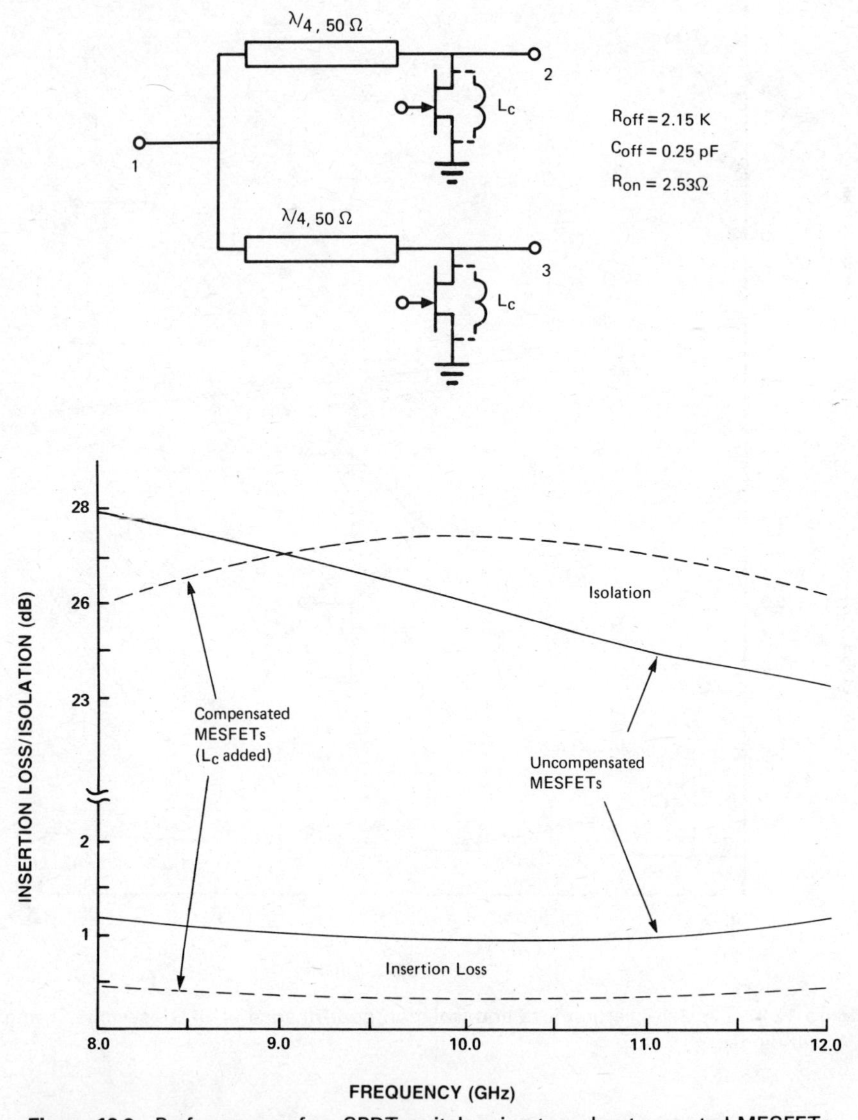

Figure 12.9 Performance of an SPDT switch using two shunt-mounted MESFETs.

of the switch, the series device is in the high-impedance state and the shunt device is in the low-impedance state. This switching circuit may be analyzed in terms of the equivalent circuit shown in Fig. 12.11(b). As shown in this figure, Z_{se} is the impedance of the series-mounted device and Z_{sh} is the impedance of the shunt-mounted device. For the ON state the device impedance Z_{se} is denoted by the low impedance Z_l, and the device impedance Z_{sh} is denoted by the high impedance Z_h. From simple circuit

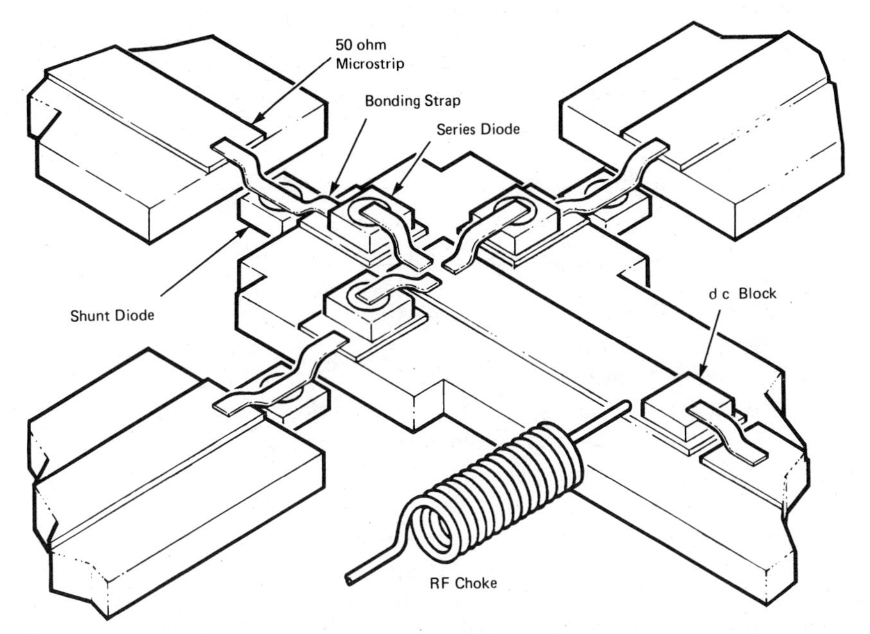

Figure 12.10 Microstrip construction of an SP3T switch using three series-mounted and three shunt-mounted devices. (After P. Chorney [6], © Microwave Jour. 1974, reprinted with permission.)

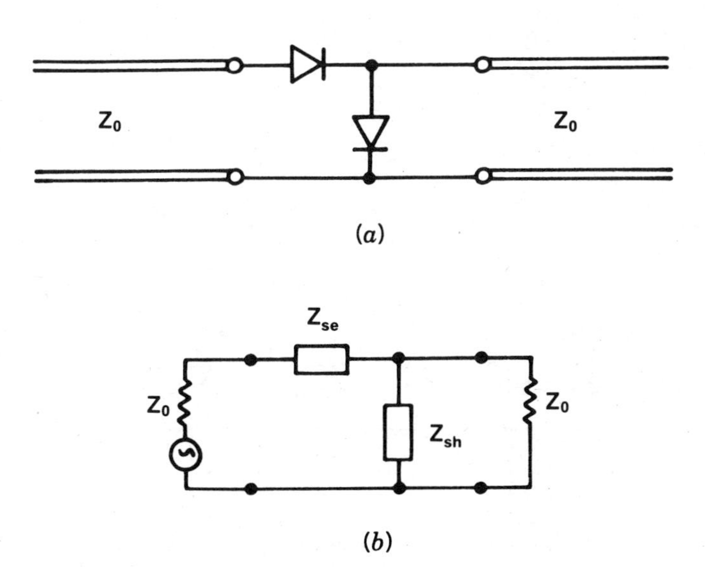

Figure 12.11 (*a*) Simplest series–shunt switching configuration. (*b*) Equivalent circuit for the series–shunt configuration shown.

analysis, the insertion loss may be written as

$$\mathrm{IL} = \left| \frac{1}{2} + \frac{(Z_0 + Z_h)(Z_0 + Z_l)}{2Z_0 Z_h} \right|^2 \tag{12.20}$$

Similarly, the isolation is written as

$$\mathrm{Isolation} = \left| \frac{1}{2} + \frac{(Z_0 + Z_l)(Z_0 + Z_h)}{2Z_0 Z_l} \right|^2 \tag{12.21}$$

Expressions (12.5) and (12.7) for series and shunt switches may be obtained as limiting cases of (12.20) for $Z_{sh}(=Z_h) \rightarrow \infty$ and for $Z_{se}(=Z_l) \rightarrow 0$, respectively. It may be noted that if nonidentical devices are used in the series and the shunt locations, values of Z_h and Z_l in (12.20) could be different from those in (12.21). As an example, let us calculate the performance of a series–shunt switch (using MA-47899 *pin* diodes in chip form and operating at 6.37 GHz), and compare the results with those for simple series and shunt switches using the same devices. This comparison is depicted in Table 12.2. We note that isolation obtained with a series–shunt configuration is much better (more than twice in decibels) than that for either the series or shunt switch. The insertion loss for the series–shunt configuration is worse than that for a shunt switch but better than that for a series switch.

At first glance it looks surprising that a switch using two (lossy) *pin* diodes can have a smaller insertion loss than that with a single diode. However, a detailed analysis shows that use of a series–shunt switch reduces the reflection loss (compared to that for a series switch) and thereby improves the insertion loss.

TABLE 12.2 Comparison of Three Switching Configurations (With MA-47899 Chip *pin* Diodes at 6.37 GHz)

Configuration	Insertion Loss	Isolation
Series Shunt (2 devices)	0.108 dB	20.17 dB
Series Switch (1 device)	0.147 dB	8.29 dB
Shunt Switch (1 device)	0.063 dB	7.52 dB

Wide-Band Series–Shunt Switch Configurations. Use of multiple switching devices in a series–shunt configuration can lead to ultra-wide-band switches. The basic concept involves the use of a ladder network structure that behaves like a low-pass filter when series devices are in the low-impedance (inductive) state and shunt devices are in the high impedance (capacitive) state. When the bias levels on the series- and shunt-switching devices are interchanged, the network behaves as a high-pass filter providing a high insertion loss below the cutoff frequency. If this cutoff frequency of the high-pass filter configuration is arranged to lie slightly above the cutoff frequency of the low-pass configuration, the network behaves as a switch with the bandwidth of the low-pass filter configuration.

Let us consider an example of three device T-structure switch shown in Fig. 12.12(*a*). In the ON state of the switch, the series devices are in the low-impedance state and shunt devices are in the high-impedance state. Wire-bonded devices exhibit a series inductance in the low-impedance state leading to an equivalent circuit of the switch as shown in Fig. 12.12(*b*). If the effect of device resistances is ignored, the circuit reduces to the low-pass filter configuration shown in Fig. 12.12(*c*). The cutoff frequency and the impedance level of this low-pass filter configuration are obtained by comparing with a three-element maximally flat low-pass design. For a prototype filter of this type, we have $L = L_s = 1$ H and $C = C_j = 2$ F (corresponding to $Z_0 = 1\ \Omega$ and cutoff frequency $\omega_c = 1$ rad/sec).

Transforming this prototype to have a cutoff frequency ω_c and an input–output impedance level of $Z_0\ \Omega$, we have

$$L = \frac{Z_0}{\omega_c}\ \text{H} \tag{12.22}$$

$$C = \frac{2}{Z_0\omega_c}\ \text{F} \tag{12.23}$$

Thus for $L = L_s$ and $C = C_j$, the cutoff frequency ω_c and the impedance level Z_0 are given by

$$\omega_c = \sqrt{\frac{2}{L_s C_j}} \tag{12.24}$$

and

$$Z_0 = \sqrt{\frac{2L_s}{C_j}} \tag{12.25}$$

Taking typical values of L_s and C_j to be 0.2 nH and 0.05 pF (typical of chip or beam-lead *pin* diodes), we get a 3-dB cutoff frequency f_c $(= \omega_c/2\pi) = 71.18$ GHz and an impedance level $Z_0 = 89.4\ \Omega$. It may be

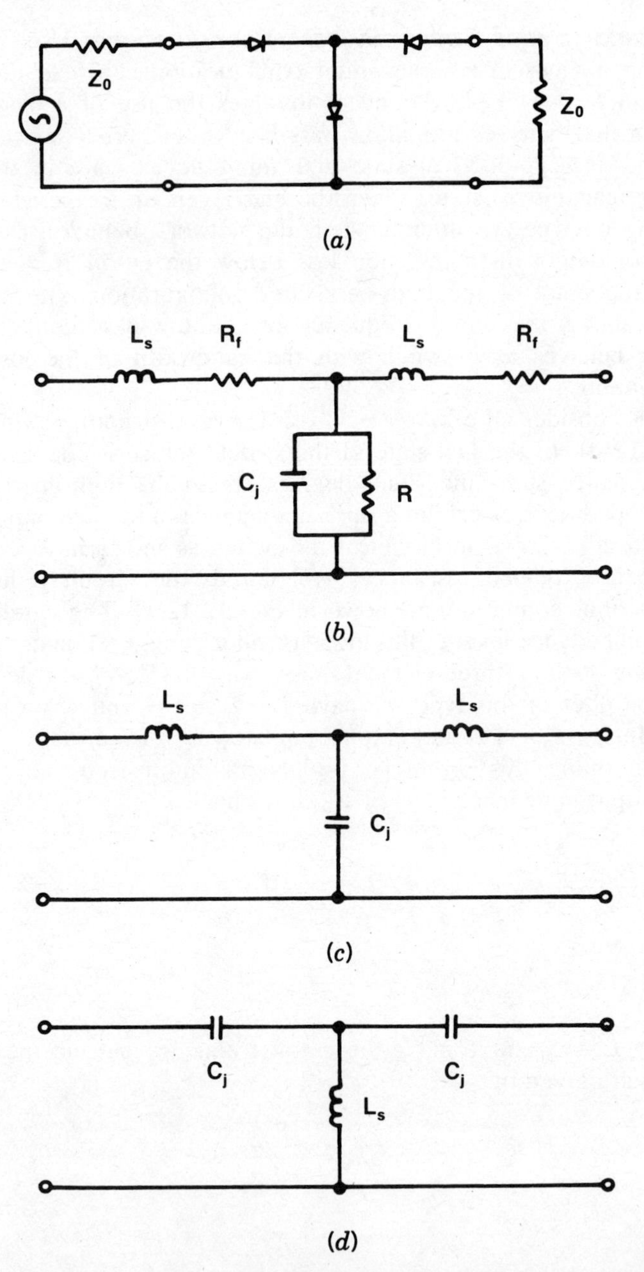

Figure 12.12 (*a*) A series-shunt switch using three switching devices. (*b*) Equivalent circuit with series devices in the low-impedance state and the shunt device in the high-impedance state (ON state for the switch). (*c*) Simplified equivalent circuit (ignoring resistances) for the ON state. (*d*) Simplified equivalent circuit for the switch in the OFF state.

noted that adjustments of ω_c and Z_0 are possible only by changing L_s and C_j. For *pin* diodes, the capacitance C_j depends on the junction area. Unless the designer has an in-house device fabrication facility, only a very limited choice of C_j can be obtained from commercially available devices. On the other hand, L_s is primarily contributed by the bonding wire inductance and can be adjusted (to some extent) by varying the length of the bonding wire.

When the series devices [Fig. 12.12(*a*)] are in the high-impedance state and the shunt device is in the low-impedance state, the switch is in the OFF state. The equivalent circuit for this state is shown in Fig. 12.12(*d*) and corresponds to a high-pass LC filter circuit. Low frequency cutoff and the impedance level of this high-pass filter may be obtained by comparing it with a prototype ($\omega_c = 1$ rad/sec and $Z_0 = 1\ \Omega$) three-element high-pass filter. For a prototype filter circuit $C = C_j = 1$ F and $L = L_s = 0.5$ H. Transforming the cutoff frequency to ω_c and the impedance level to $Z_0\ \Omega$, we get

$$C = \frac{1}{\omega_c Z_0} \text{ F} \tag{12.26}$$

and

$$L = \frac{0.5 Z_0}{\omega_c} \text{ H} \tag{12.27}$$

In terms of L_s and C_j, we obtain

$$\omega_c = \sqrt{\frac{0.5}{L_s C_j}} \tag{12.28}$$

and

$$Z_0 = \sqrt{\frac{2L_s}{C_j}} \tag{12.29}$$

Comparing the result with the corresponding results for the ON state, we find that whereas Z_0 has the same value as for the low-pass filter, the value of the cutoff frequency for the high-pass filter (OFF state) is only half that of the low-pass filter corresponding to the ON state. Since the low-frequency cutoff in the OFF state implies only 3-dB isolation, the operating range of the switch is still smaller. This operating range can be increased by choosing the shunt-mounted device such that the product $L_s C_j$ is one-fourth of that of the series-mounted device, but the ratio L_s / C_j is equal in two cases.

Considering an example with three identical switching devices (with $L_s = 0.2$ nH, $C_j = 0.1$ pF, and $R_f = 0.5\ \Omega$), the insertion loss and isolation of the series–shunt switch are plotted in Fig. 12.13 as a function of frequency up to 30 GHz. We note that the insertion loss is less than 0.1 dB throughout

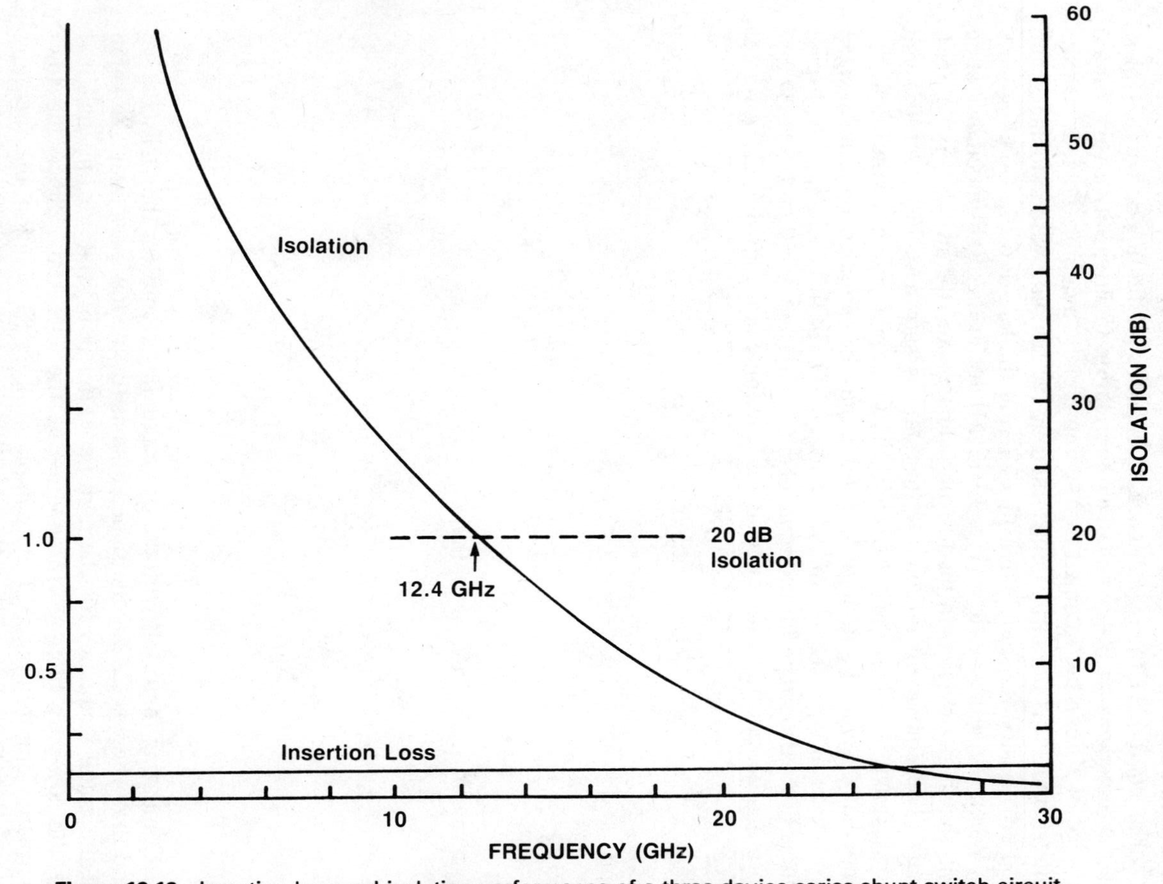

Figure 12.13 Insertion loss and isolation performance of a three-device series-shunt switch circuit.

this range, but the performance is limited by the value of isolation, which decreases monotonically. An isolation better than 20 dB is obtained for frequencies up to 12.4 GHz.

Five Device Series–Shunt Configuration. Performance obtained from the series–shunt switch configuration discussed previously can be improved by increasing the number of switching devices from 3 to 5 as shown in Fig. 12.14 (inset). When five identical devices similar to that for the three-device switch are used, the insertion loss and isolation obtained are shown in Fig. 12.14. It may be noted that the isolation is improved for frequencies below 19 GHz and is better than 20 dB for frequencies up to 16.3 GHz. In this frequency range, the insertion loss is only marginally worse ($\simeq$0.1 dB in place of $\simeq$0.08 dB for the three-device case when $R_f = 0.5\ \Omega$).

It may be noted that when five identical devices are used, the ON and the OFF state equivalent circuits do not correspond exactly to the five-element low-pass and high-pass filter designs. For realizing these filter configurations, as described in Chapter 6, three different types of switching devices (with different L_s and C_j values) are required. At times, this selection becomes difficult because of the limitations on parameter values of available devices.

12.2.6 Switching Speed Considerations

The time required to change the state of a switch from OFF to ON (or vice versa) becomes an important consideration in several switching circuits, such as those used for modulation or for ultra-high-speed digital circuits. Some of the switching speed considerations are discussed briefly in this section.

Definitions. Different terms, like turn-on delay, turn-on switching time, turn-off delay, and turn-off switching time, are used for microwave switches. These may be explained with the help of Fig. 12.15(a) and (b). Figure 12.15(a) shows an experimental setup for observing switching speed on a CRO. In an alternative experimental configuration, the crystal detector and the dual-trace oscilloscope are replaced by a sampling oscilloscope used for direct observation of microwave waveforms. In either case, the two waveforms significant for measurement of switching speed are the input from the pulse generator to the switch driver and the envelope of (or the detected) microwave signal. By comparing these waveforms on the calibrated time axis [Fig. 12.15(b)], we can define various switching speed terms. Turn-on delay is the time interval between the instants when the input signal reaches 90% of its peak value and when the detected RF envelope reaches 10% of its peak value. Turn-on switching speed is defined as the time interval between the instants when the envelope of the RF output rises from 10% of its peak value to 90% of its peak value. Turn-off delay and

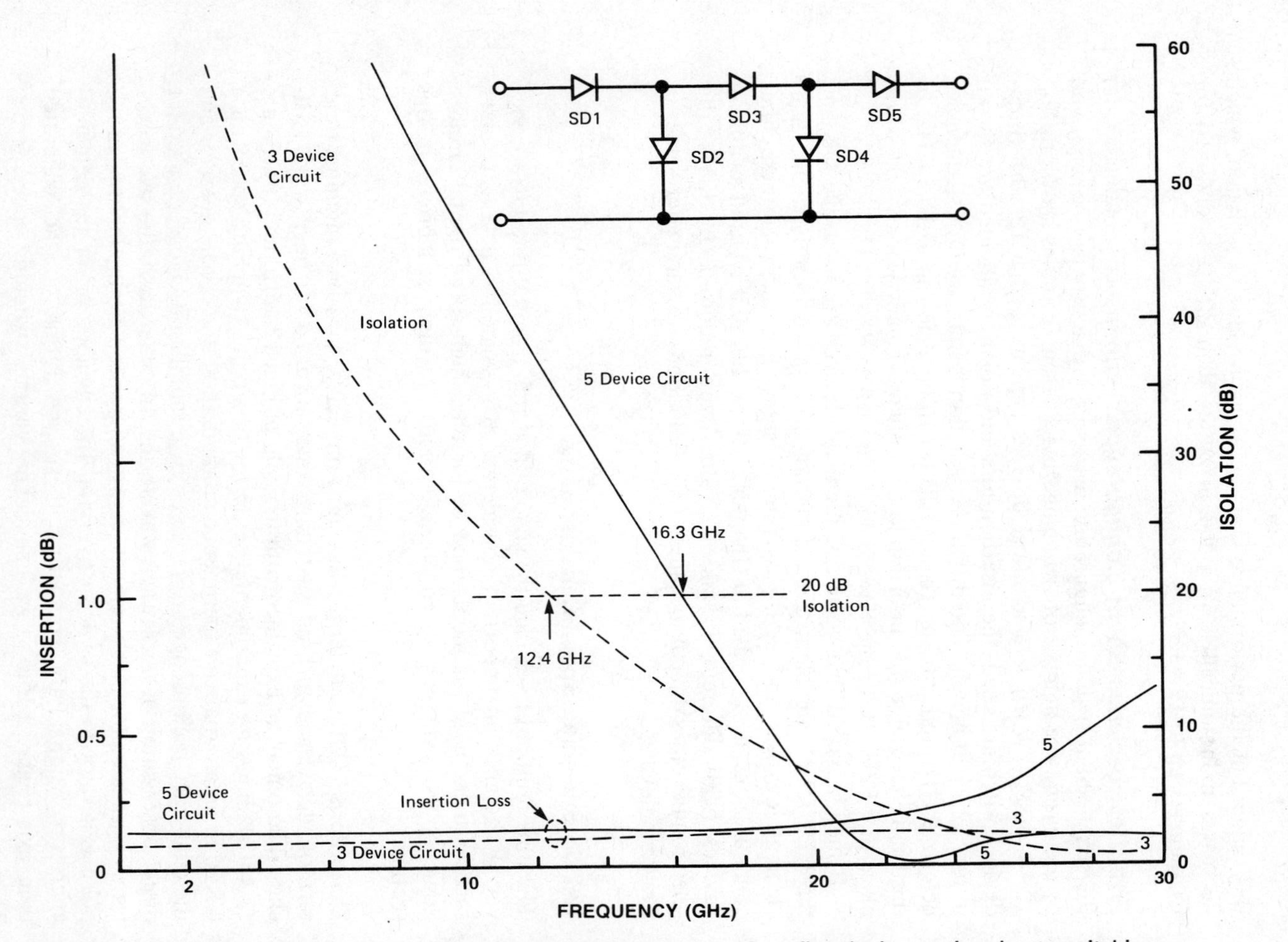

Figure 12.14 Insertion loss and isolation performance of a five-device series-shunt switching [illegible] with performance of a three-device switch (shown in dotted curves).

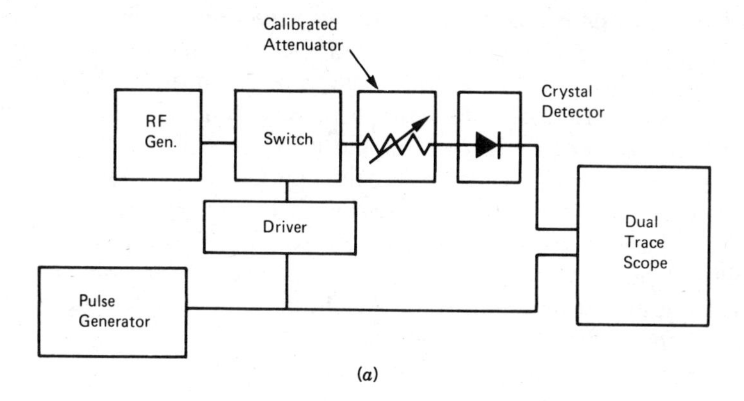

(*a*)

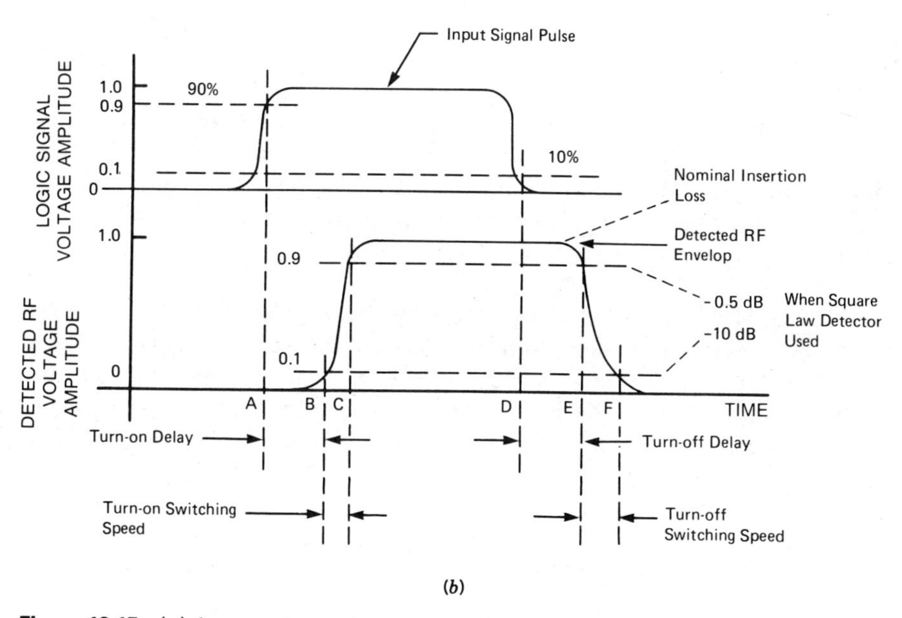

(*b*)

Figure 12.15 (*a*) An experimental setup for measuring speed of switching circuits. (*b*) Illustration of various delay terms used for microwave switches.

turn-off switching are defined in a similar manner and are shown in Fig. 12.15(*b*). Ten percent and 90% points on the detected RF envelope correspond to 10-dB and 0.5-dB points when a square law detector is used to measure the output power, but correspond to 20-dB and 1-dB points, respectively, when a sampling oscilloscope is used. Thus for interpreting switching speed results, it is necessary to know the experimental setup used.

Speed Limitations Imposed by Switching Devices. When *pin* diodes are used for microwave switching, the main factor limiting the switching speed is the time required to remove the charge from the intrinsic region when the diode bias is switched from forward to reverse. This charge removal time depends on the width (thickness) of the intrinsic layer. A decrease in the width W of the intrinsic layer makes the charge removal faster, but at the same time reduces the reverse breakdown voltage and hence the power-handling capability of the *pin* diode switches [7]. Since the power-handling capability is proportional to the square of the breakdown voltage, it is also proportional to the square of width W. On the other hand, switching time τ is proportional to width W. These simple (although crude) arguments point out a trade-off between switching time τ and the square root of the power-handling capability P. Garver [7] has shown that

$$\tau = \frac{P}{25} \quad \text{nsec} \tag{12.30}$$

when P is the power-handling capability in watts. Practical values of τ are about 25 times those given by relation (12.30).

Enhancement in switching speed of *pin* diodes can be obtained by using GaAs *pin* diodes [8] in place of the Si *pin* diodes commonly used. Electron mobility in GaAs is about four times that in silicon. This leads to faster switching times as well as lower drive current requirements for GaAs *pin* diodes.

Switching Speed Limitation Imposed by Biasing Network. In addition to the switching speed limitation imposed by the semiconductor devices used in switching circuits, switching speed may also be affected by circuit considerations. Considering the dc bias terminal(s) as a separate port, a single-pole single-throw switching circuit may be viewed as a

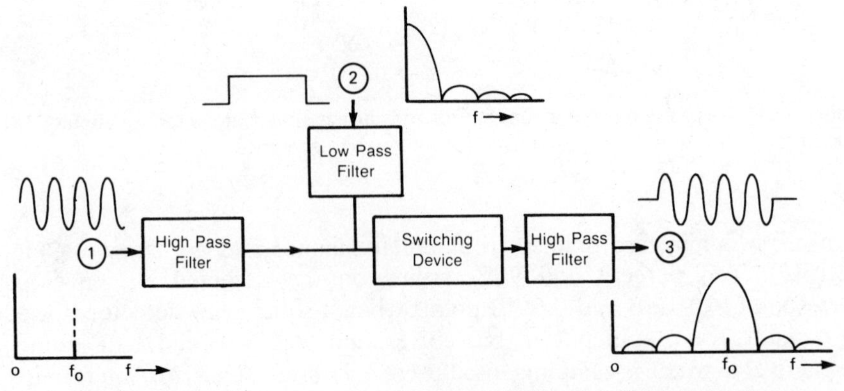

Figure 12.16 Three-port network representation of switching circuits.

three-port network [9], as shown in Fig. 12.16. A low-pass filter arrangement is needed at the bias port to ensure that the RF signal does not leak away via the bias port. High-pass filters are needed at the RF input and output ports to ensure that the dc bias (or switching pulse) does not interfere with the other parts of the circuit. In the simplest form, these high-pass filters may simply be dc blocking capacitors at either end. These filters increase the rise time of the switching pulse and thereby reduce the switching speed. Rise time τ of a pulse (10% to 90% levels) passing through a filter is related to the 3-dB bandwidth (BW) of the filter, as

$$\tau = \frac{0.44}{\mathrm{BW}} \tag{12.31}$$

Thus, in order to decrease the rise time, the 3-dB bandwidth of the filters should be as large as possible. Since we have a low-pass signal and a high-pass filter in the path of the switching pulse, optimum filter characteristics are as shown in Fig. 12.17. With this arrangement, the switching time contributed by filters is given [9],

$$\tau = \sqrt{\tau_1^2 + \tau_2^2} = \sqrt{2}\,\tau_F = \frac{1.24}{f_0} \tag{12.32}$$

where τ_1 and τ_2 are rise times for the two filters (we assume $\tau_1 = \tau_2 = \tau_F$) and f_0 is the operating frequency. Thus at an operating frequency of 1 GHz, the minimum switching delay contributed by biasing filters is about 1.24 nsec.

Switching speed limitation caused by the biasing network becomes a limiting factor for MESFET switches because the MESFET device itself is inherently much faster than *pin* diodes.

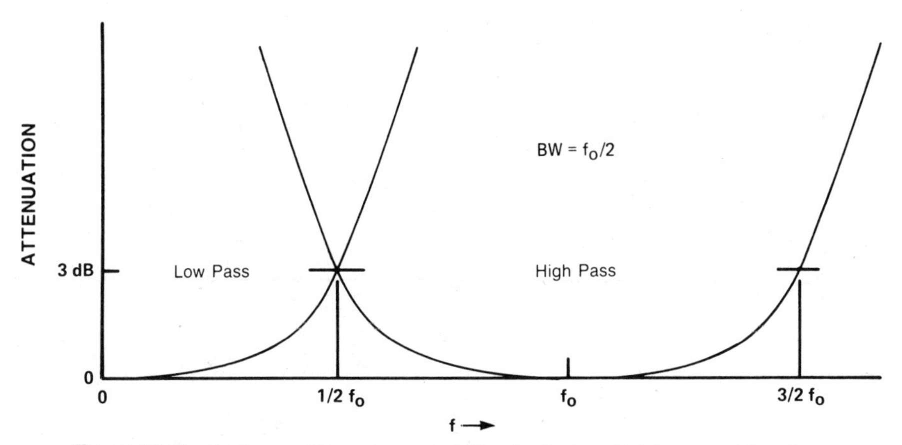

Figure 12.17 Optimum filter characteristics for fast switching considerations.

12.3 DESIGN OF PHASE SHIFTERS

12.3.1 General

A phase shifter is a two-port network with the provision that the phase difference between the output and the input signals may be controlled by a control signal (dc bias). Various phase shifter designs may be classified as shown in Fig. 12.18. Phase shifters are called "digital" when the differential phase shift can be changed by only a few predetermined discrete values, such as 180°, 90°, 45°, 22.5°, and 11.25°. On the other hand, in "analog" phase shifters, the differential phase shift can be varied in a continuous manner by a corresponding continuous variation of the control signal. Digital phase shifters find extensive applications in phased-array antenna systems. Phase control of the signals fed to the various elements of the array allows the direction of the radiated beam to be scanned electronically. Digital phase shifters are compatible with the computer control of beam scanning in phased-array systems.

There are two distinct methods for designing digital phase shifters at microwave frequencies. One is to use the properties of ferrimagnetic materials for obtaining switchable phase shift. Ferrite phase shifters have gone through substantial development during the last two decades and a fairly good description of ferrite phase shifter design is available in Skolnik's *Radar Handbook* [10]. The other major design for digital phase shifters uses semiconductor devices. These phase shifters are, in general, more compact, have lower switching times and require lower drive power when compared to ferrite phase shifters. Here, we discuss semiconductor phase shifters only.

As shown in Fig. 12.18, phase shifters using semiconductor devices can be either of the reflection type or the transmission type. In reflection-type shifters, the basic design unit is a one-port network, and it is the phase shift of the reflected signal that is changed by the control signal. These basic one-port phase shifters can be converted into useful two-port components either by using a circulator or a hybrid. Because of the ease of integration, the hybrid-coupled reflection-type phase shifters are more common. As far as the design of the basic one-port reflection phase shifter bit is concerned, one may use a switched-length–type or a switched-reactance–type design. Other designs for semiconductor phase shifters may be called transmission type. Here there are three subgroups: switched-line–type phase shifters, loaded-line phase shifters, and switched-network–type phase shifters. A special class of "transmission" phase shifters called amplifier-type use GaAs MESFETs in an active amplifier mode. This type of phase shifter is nonreciprocal and cannot be designed by using *pin* diodes. Design methods for the various types of phase shifters are discussed in this section.

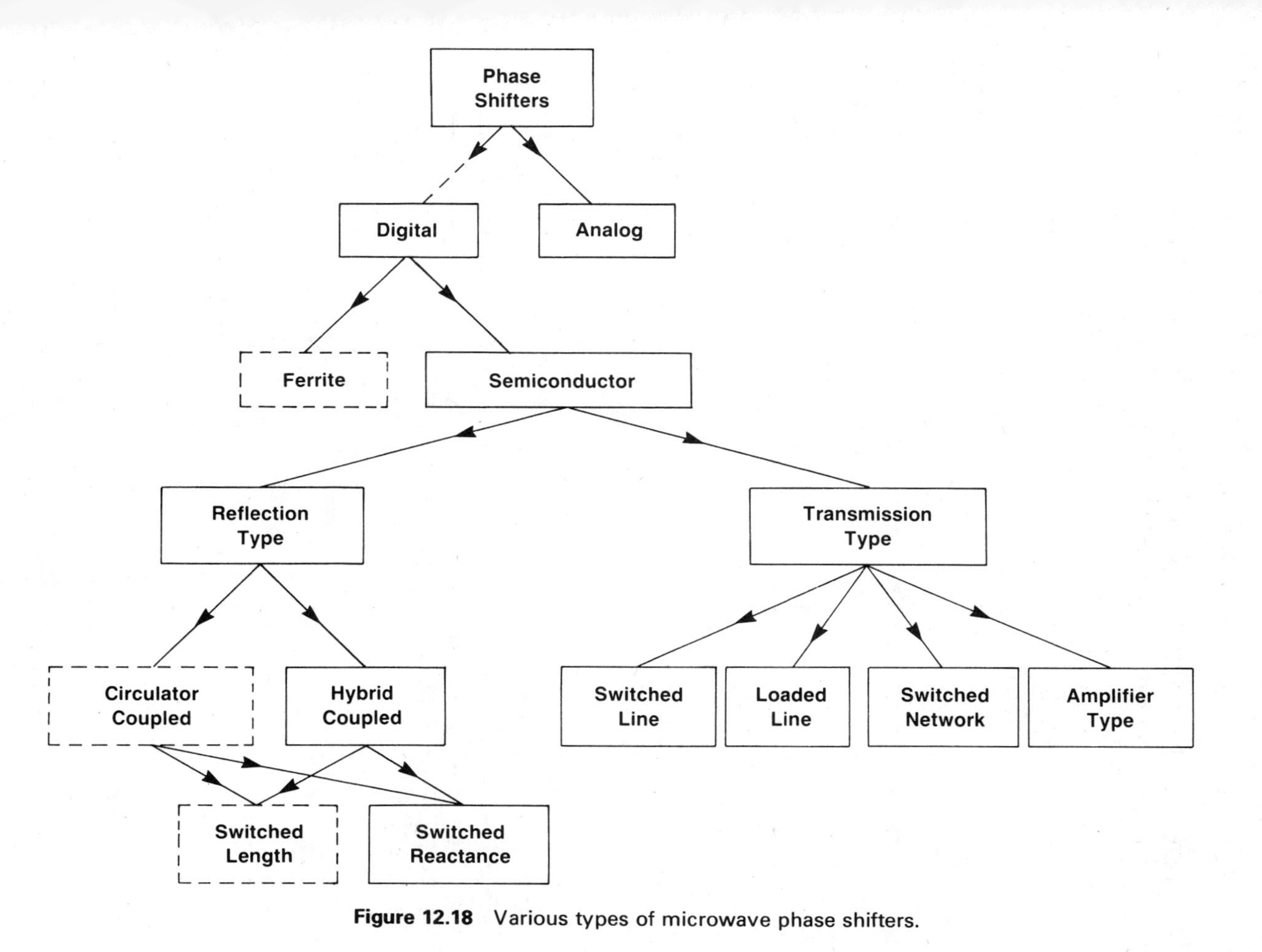

Figure 12.18 Various types of microwave phase shifters.

12.3.2 Switched-Line Phase Shifters

Design of switched-line–type phase shifters is conceptually the simplest. The basic configuration for a single-bit phase shifter is shown in Fig. 12.19. Two SPDT switches are used to route the signal via one of the two alternative transmission path lengths l_1 or l_2. When the signal passes through the longer path, it goes through an additional phase delay given by

$$\Delta\phi = \beta(l_2 - l_1) = \frac{2\pi f}{v_p}(l_2 - l_1) \tag{12.33}$$

where v_p is the phase velocity. The insertion loss of a phase shifter of this type is equal to that of the two SPDT switches (plus any line losses that may be present).

An interesting characteristic of this type of phase shifter, as can be seen from (12.33), is that the differential phase shift $\Delta\phi$ is directly proportional to frequency. Because of this feature, switched-line phase shifters are also called switchable time-delay networks. The time delay τ_d is given by

$$\tau_d = \frac{l_2 - l_1}{v_p} \tag{12.34}$$

One of the common problems in the design of the switched-line phase shifters is caused by off-path resonances. This may be illustrated by a specific example. Consider the design of a 45° phase bit for which l_1 and l_2 are chosen to be 160° and 205°, respectively, at the center frequency of 1.5 GHz. Further, the *pin* diodes used for SPDT switches can be represen-

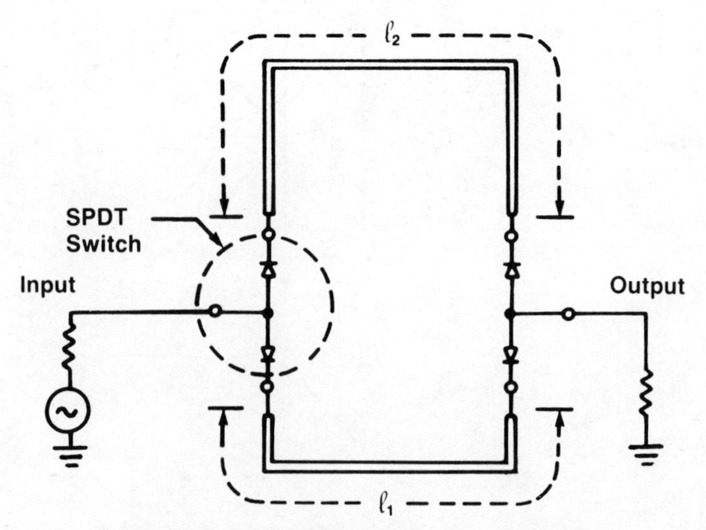

Figure 12.19 A single-bit switched-line phase shifter.

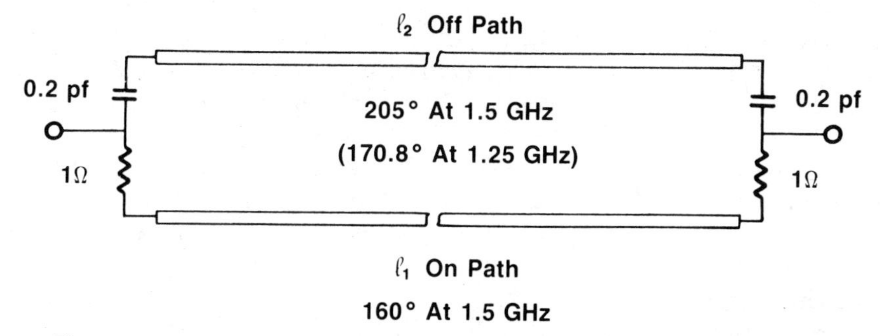

Figure 12.20 Equivalent circuit for the switched-line phase shifter example.

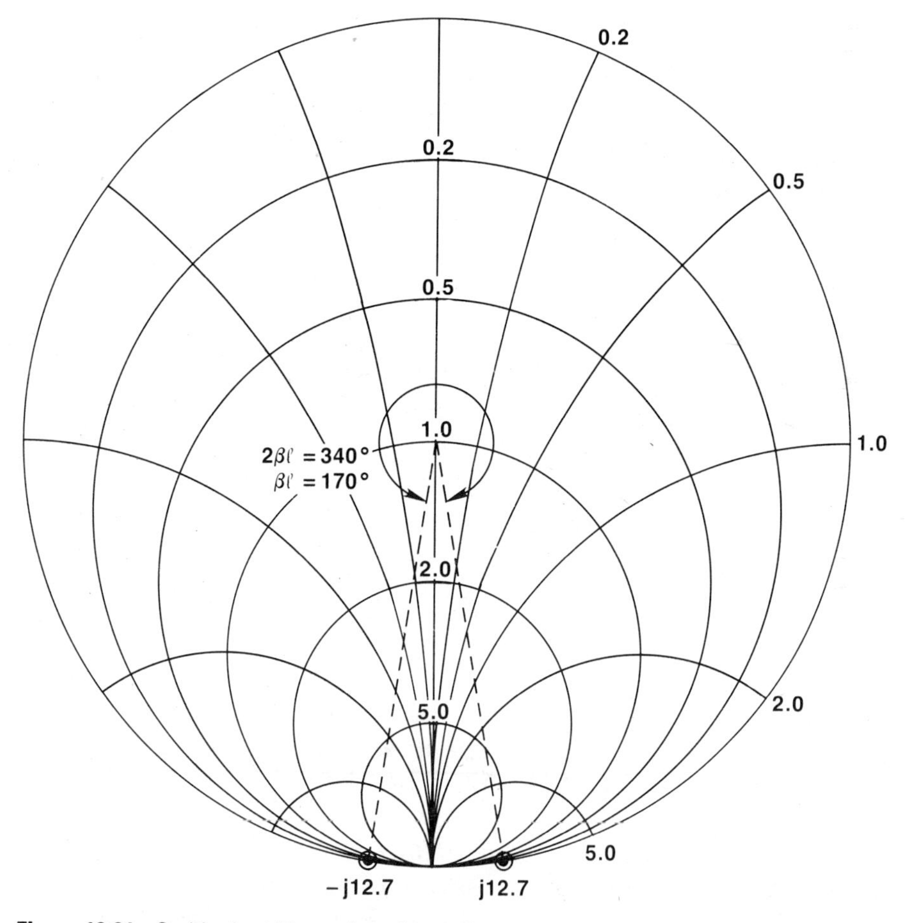

Figure 12.21 Smith chart illustration of insertion-loss resonance in switched-line phase shifters.

ted by an $R_f = 1\ \Omega$ in forward bias and a $C_j = 0.2$ pF in the reverse bias. The equivalent circuit, for l_1 in the on-path, is shown in Fig. 12.20. The design works quite well at the center frequency, but if the frequency is varied to 1.25 GHz, the length l_2 becomes 170.8° and, when associated with reactances of 0.2-pF capacitors at each end, shows a resonance. This resonance phenomenon is illustrated in Fig. 12.21. The 0.2-pF capacitors have a reactance of about $-j636\ \Omega$ each and add up to an effective line length of about 4.5° at each end. This causes a resonance that is exhibited by a peak in the insertion-loss curve as well as by a rapid phase shift variation with frequency around 1.25 GHz. Frequency variation of phase shift and insertion loss are shown in Fig. 12.22. When the signal is routed through the path l_2, the insertion-loss resonance of the length l_1 is observed near 1.58 GHz.

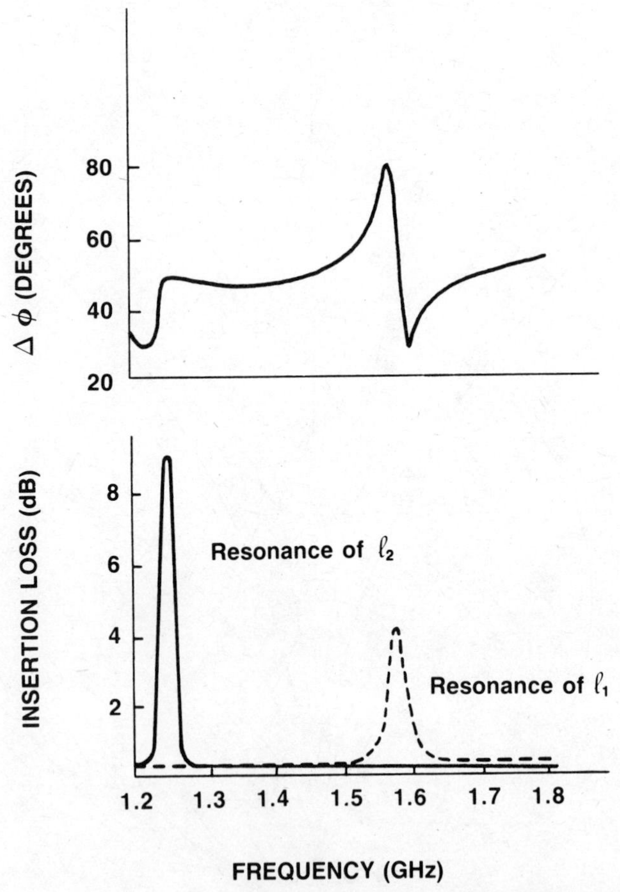

Figure 12.22 Insertion-loss resonances in switched-line phase shifter circuit shown in Figs. 12.19 and 12.20. (After J. F. White [1], © Van Nostrand 1982, reprinted with permission.)

The insertion-loss resonances may be avoided by a suitable choice of lengths l_1 and l_2. For example, in the present case, l_1, l_2 may be 50° and 95°, respectively. Another method of avoiding these resonances is by terminating the off-path by matched loads. This does, however, require additional switching devices.

12.3.3 Loaded-Line Phase Shifters

A very common design for 45° and 22.5° phase bits is known as the loaded-line phase shifter. The mechanism of phase shift in this circuit is based on the loading of a uniform transmission line by a small reactance as shown in Fig. 12.23. It can be shown that the transmitted wave undergoes a phase shift $\Delta\phi$ that depends upon the normalized susceptance $b = B/Y$. The reflection caused by b is given by

$$\Gamma = \frac{1-(1+jb)}{1+(1+jb)} = \frac{-jb}{2+jb} \tag{12.35}$$

The voltage associated with the transmitted wave is $V_T = V_I + V_R$, where V_I and V_R are voltages of the incoming wave and the reflected wave, respectively. The transmission coefficient T can therefore be written as

$$T = \frac{V_T}{V_I} = \frac{V_I + V_R}{V_I} = 1 + \Gamma = \frac{2}{2+jb} \tag{12.36}$$

$$V_T = TV_I = V_I \frac{2}{2+jb} = V_I \left(\frac{4}{4+b^2}\right)^{1/2} \exp\left\{-j\tan^{-1}\left(\frac{b}{2}\right)\right\} \tag{12.37}$$

The phase difference introduced (phase of V_I − phase of V_T) may therefore be written as

$$\Delta\phi = \tan^{-1}\left(\frac{b}{2}\right) \tag{12.38}$$

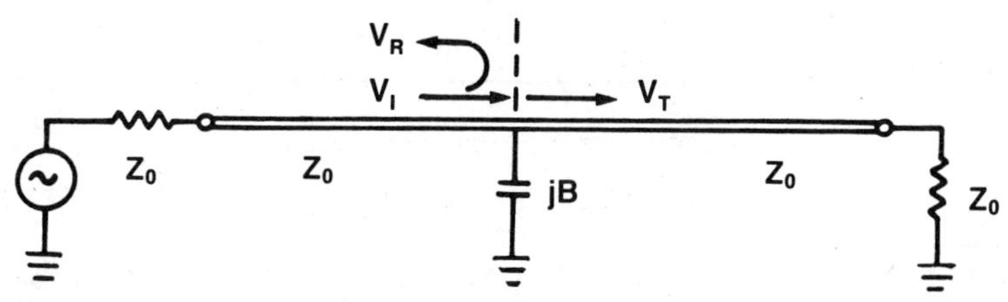

Figure 12.23 Circuit for illustrating basic mechanism of a loaded-line phase shifter.

If the normalized susceptance $b = 0.2$ (i.e., capacitive), $\Delta\phi = \tan^{-1}(0.1) = 0.1$ rad $= 5.7°$. If, on the other hand, we load the line with $b = -0.2$ (a shunt inductor) the phase delay $\Delta\phi$ is negative, that is, $\Delta\phi = \tan^{-1}(-0.1) \simeq -5.7°$. That is, the transmitted wave advances in phase as compared with the transmitted wave for $b = 0$. Another important performance parameter here is the insertion loss given by the magnitude of the ratio of V_T to V_I, which may be written as

$$\frac{V_T}{V_I} = \left(\frac{4}{4 + b^2}\right)^{1/2} \tag{12.39}$$

For $b = 0.2$, $V_T/V_I = 0.995$, which corresponds to a loss of 0.04 dB. Although, in the present case, the value of insertion loss is not significant, the presence of this loss is an unfavorable feature of the circuit shown in Fig. 12.23.

This drawback of reflection from the susceptance producing the phase shift can be overcome by using two identical susceptances separated by a quarter-wave line length. Such an arrangement is shown in Fig. 12.24. If we consider partial reflections caused by the two susceptances, these reflected waves are almost equal in magnitude and out of phase (180° phase difference) when looking into the input terminals. Thus these undesirable reflections cancel each other. A quantitative assessment of the situation may be obtained by writing a uniform transmission line equivalent as shown

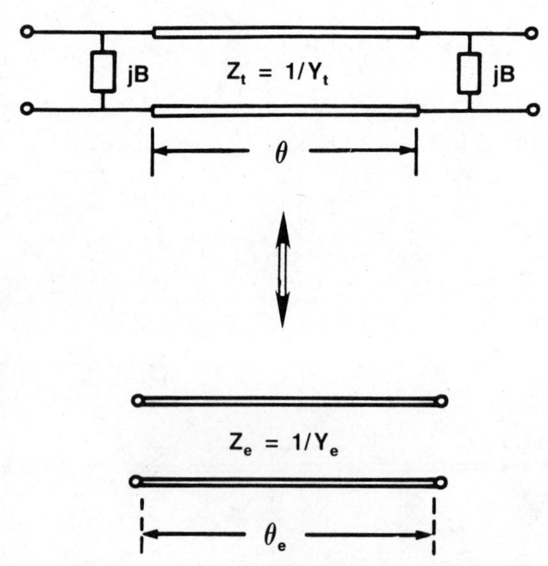

Figure 12.24 Circuit configuration for a loaded-line phase shifter and an equivalent representation.

in Fig. 12.24. Equivalent Y_e and θ_e may be obtained by comparing *ABCD* matrices of the two networks. An *ABCD* matrix of the transmission line of length θ, shunt-loaded by susceptances at either end, may be obtained by multiplying *ABCD* matrices of the shunt susceptance, of the line section, and of the second susceptance at the other end. This yields

$$\begin{bmatrix} A & B \\ C & D \end{bmatrix} = \begin{bmatrix} 1 & 0 \\ jB & 1 \end{bmatrix} \begin{bmatrix} \cos\theta & jZ_t \sin\theta \\ jY_t \sin\theta & \cos\theta \end{bmatrix} \begin{bmatrix} 1 & 0 \\ jB & 1 \end{bmatrix}$$
$$= \begin{bmatrix} (\cos\theta - BZ_t \sin\theta), & j(Z_t \sin\theta) \\ j(2B\cos\theta + Y_t \sin\theta - B^2 Z_t \sin\theta), & (\cos\theta - BZ_t \sin\theta) \end{bmatrix} \tag{12.40}$$

ABCD of the equivalent transmission line may be written as

$$\begin{bmatrix} A & B \\ C & D \end{bmatrix} = \begin{bmatrix} \cos\theta_e, & jZ_e \sin\theta_e \\ jY_e \sin\theta_e, & \cos\theta_e \end{bmatrix} \tag{12.41}$$

We can find the length of the equivalent uniform line by comparing the A terms in (12.40) and (12.41), as

$$\cos\theta_e = \cos\theta - BZ_t \sin\theta \tag{12.42}$$

The equivalent admittance Y_e is obained by comparing the ratios C/B, as

$$Y_e = Y_t\{1-(BZ_t)^2 + 2BZ_t \cot\theta\}^{1/2} \tag{12.43}$$

When $\theta = 90°$, relations (12.42) and (12.43) may be written as

$$\cos\theta_e = -BZ_t \tag{12.44}$$

or

$$\theta_e = \frac{\pi}{2} + BZ_t + \frac{(BZ_t)^3}{6} \tag{12.45}$$

$$Y_e = Y_t\{1-(BZ_t)^2\}^{1/2} \tag{12.46}$$

Thus we note that $\theta_e > \theta$ for capacitive susceptance, and $\theta_e < \theta$ for inductive susceptances. Also $Y_e < Y_t$ and its magnitude is independent of the sign of B. Thus, if only the sign of B is changed (and its magnitude kept unchanged), the circuit can remain matched in both states of the phase shifter. For example, if $b_1 = 0.2$ and $b_2 = -0.2$, we have $\theta_{e1} = 101.54°$ and $\theta_{e2} = 78.46°$, resulting in a differential phase shift $\Delta\phi = \theta_{e2} - \theta_{e1} = 23.08° \simeq 0.4$ rad. It may be noted that if $(b_1 - b_2)$ is small, $\Delta\phi$ is approximately $(b_1 - b_2)$ radians. Also in the present case, $Y_e/Y_t = 0.98$. That is, if the desired impedance level is 50 Ω, Z_t should be $0.98 \times 50 = 49$ Ω. If Z_t is retained as 50 Ω, it will result in an input VSWR of 1.02.

***Various Configurations for Loaded-Line Phase Shifters* [11].** Various designs for loaded-line phase shifters differ to the extent that the susceptances B_1 and B_2 for two states of the phase shifter are realized by different circuit configurations. We will discuss two types of circuits [11] used for this purpose. The basic design equations are obtained by combining (12.42) and (12.43) with (12.38) and may be written as

$$Y_T = Y_0 \sec\left(\frac{\Delta\phi}{2}\right) \sin\theta \tag{12.47}$$

$$B_{1,2} = Y_0 \left\{\sec\left(\frac{\Delta\phi}{2}\right) \cos\theta \pm \tan\left(\frac{\Delta\phi}{2}\right)\right\} \tag{12.48}$$

1. *Main-Line Mounted Circuit.* In this case, the switching devices are mounted directly across the main line. The idea is to use the high-impedance state capacitance C_j and the low-impedance state inductance L_s directly for B_1 and B_2, respectively. However, it becomes necessary to add an external inductance L_e in series with the device as shown in Fig. 12.25. Susceptances B_1 and B_2 are now given by

$$B_1 = \frac{\omega C_j}{1 - \omega^2 L C_j} \tag{12.49a}$$

$$B_2 = \frac{-1}{\omega L} \tag{12.49b}$$

where $L = L_s + L_e$. Using (12.48) and (12.49), we can derive the following relationship between Y_0 and θ for $\Delta\phi$ and C_j:

$$Y_0 = \omega C_j \frac{\sin\Delta\phi}{\{\sin^2(\Delta\phi/2) - \cos^2\theta\}} \tag{12.50}$$

Usually one would like to select θ so as to obtain the maximum bandwidth from the phase shifter circuit, which occurs at $\theta = 90°$. For this case

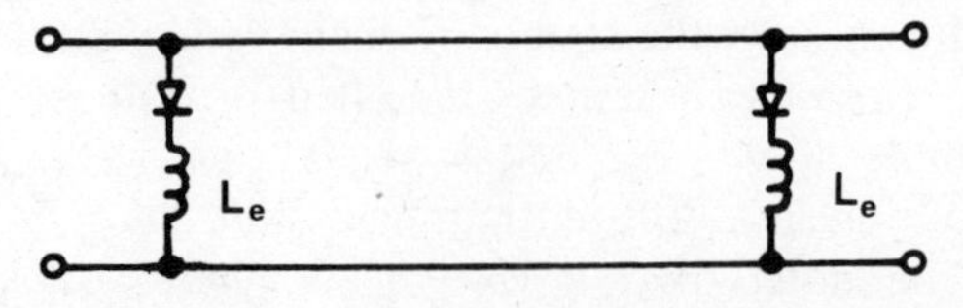

Figure 12.25 Main-line mounted-type loaded-line phase shifter circuit.

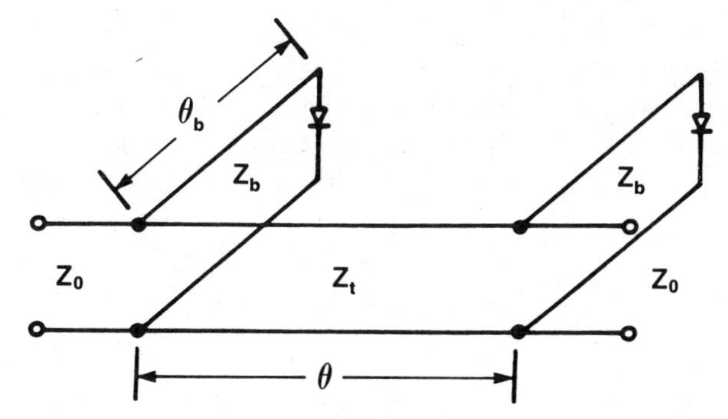

Figure 12.26 A stub-mounted-type loaded-line phase shifter circuit.

$$Y_0 = 2\omega C_j \cot\left(\frac{\Delta\phi}{2}\right) \tag{12.51}$$

This puts restrictions on the input–output line impedance unless C_j can be chosen arbitrarily, which becomes difficult because of the limited availability of switching device capacitances. Because of these difficulties, this design configuration has not become very popular.

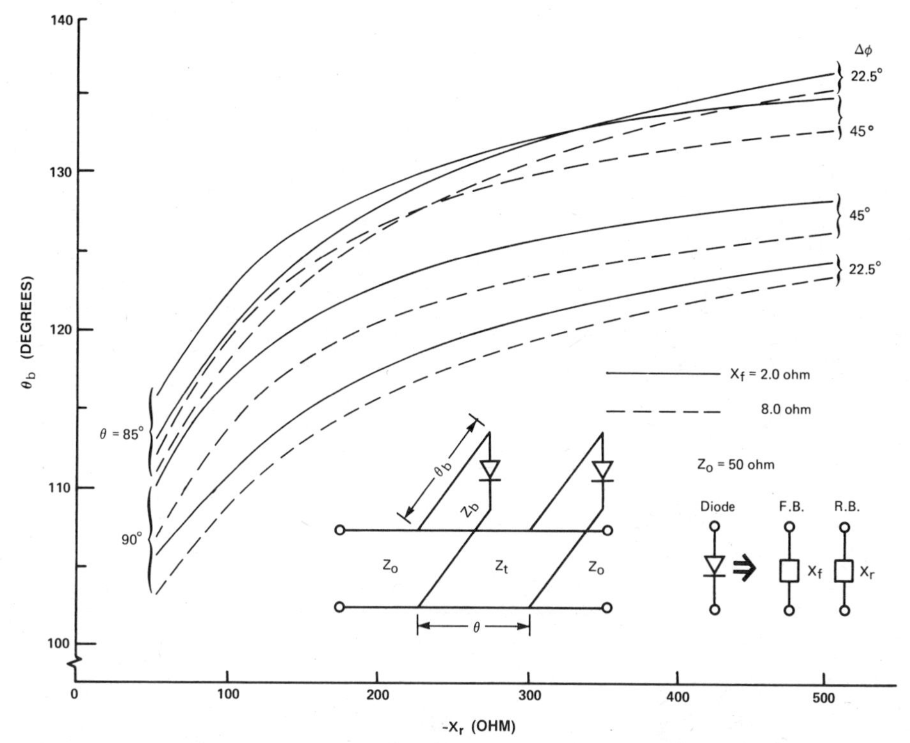

Figure 12.27 Design curves for stub length θ_B in stub-mounted loaded-line phase shifters. (After I. J. Bahl and K. C. Gupta [11], © 1980, IEEE, reprinted with permission.)

2. *Stub-Mounted-type Circuit.* In the circuit arrangement shown in Fig. 12.26, switching devices are mounted at the ends of two shunt-connected stubs separated by a line length θ. This allows greater design flexibility. Stub impedance Z_b and length θ_b are determined from device reactances X_f and X_r (in low- and high-impedance states, respectively) as follows

$$Z_b = \left\{ \frac{X_f - X_r - X_f X_r (B_1 - B_2)}{B_1 - B_2 - B_1 B_2 (X_f - X_r)} \right\}^{1/2} \tag{12.52}$$

and

$$\tan \theta_b = \frac{Z_b(1 - X_f B_1)}{X_f - B_1 Z_b^2} \tag{12.53}$$

where Z_t and θ may be selected for wide bandwidth. Figures 12.27, 12.28, and 12.29 present the design curves based on these formulas. Bandwidths shown in Fig. 12.29 are defined for phase error $<2°$ and VSWR <1.2.

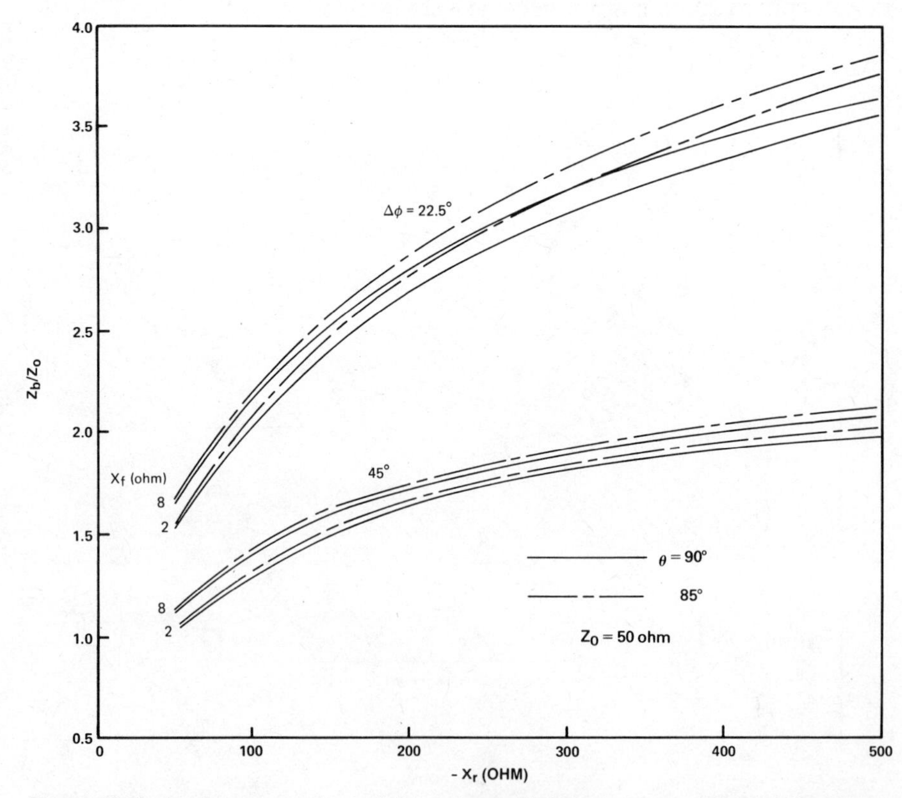

Figure 12.28 Design curves for stub impedance Z_B in stub-mounted loaded-line phase shifters. (After I. J. Bahl and K. C. Gupta [11], © 1980, IEEE, reprinted with permission.)

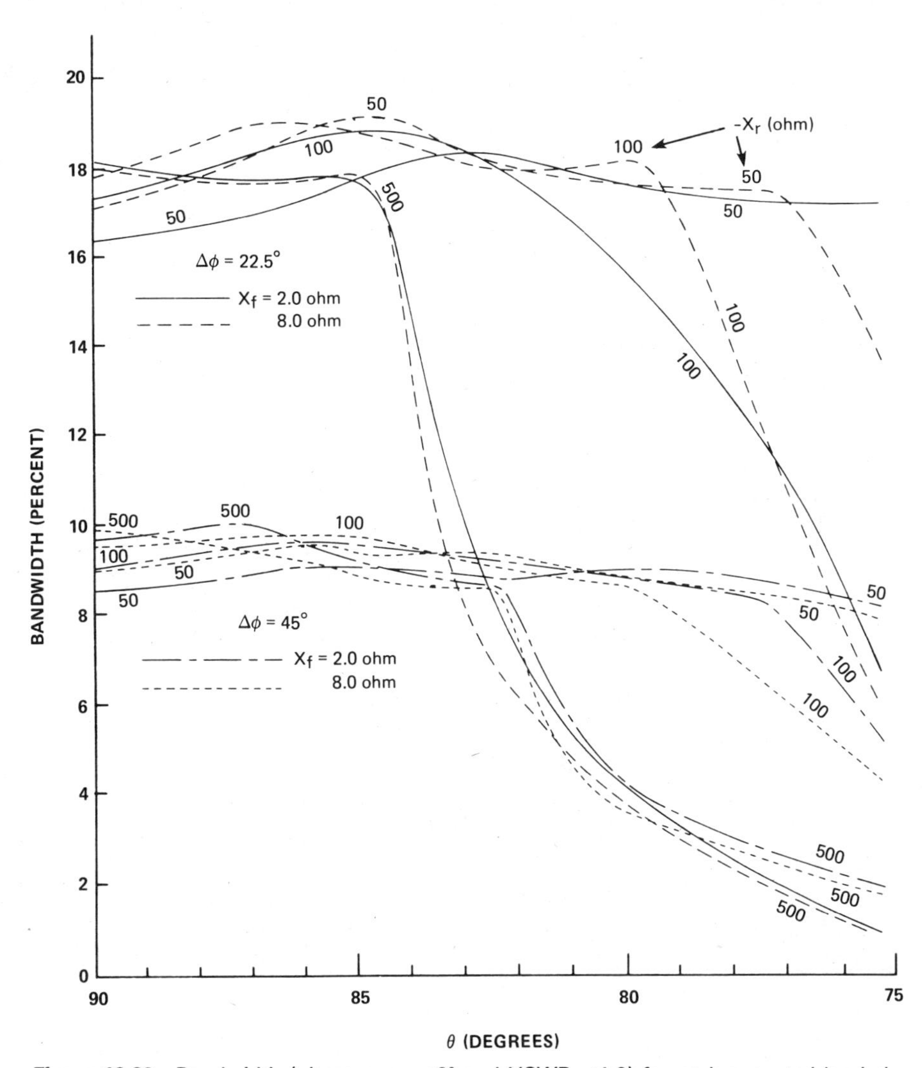

Figure 12.29 Bandwidth (phase error $<2°$ and VSWR <1.2) for stub-mounted loaded-line phase shifters. (After I. J. Bahl and K. C. Gupta [11], © 1980, IEEE, reprinted with permission.)

These design methods make use of device reactances only and assume the device resistances to be negligible. Effect of device losses on design parameters has been discussed recently [12].

12.3.4 Reflection-Type Phase Shifters

Another important class of phase shifters is constituted by reflection-type phase shifters. The basic concept is illustrated in Fig. 12.30(a). A sub-

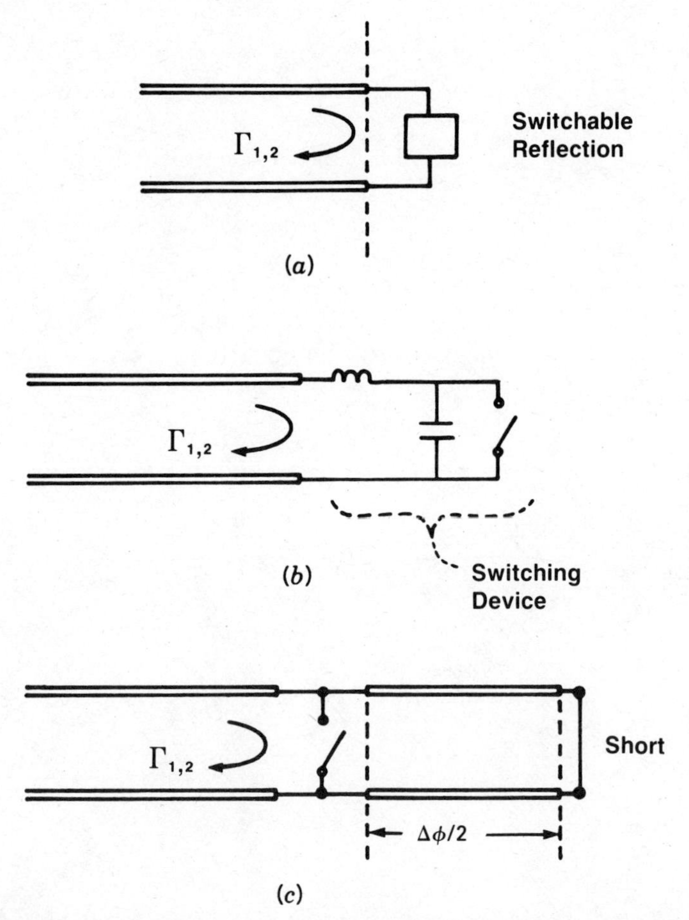

Figure 12.30 (*a*) Basic concept of reflection-type phase shifters. (*b*) Switchable-reactance-type reflection phase shifter. (*c*) Switchable-length-type reflection phase shifter.

network with switchable reflection coefficient terminates a uniform transmission line. When the reflection coefficient is switched from $\Gamma_1 = |\Gamma_1| \angle \phi_1$ to $\Gamma_2 = |\Gamma_2| \angle \phi_2$, the reflected signal undergoes a differential phase shift $\Delta\phi = \phi_1 - \phi_2$. In any state, the ratio of the reflected power to the incident power is given by $|\Gamma|^2$. Ideally $|\Gamma|$ should be unity so that there is no loss associated with the phase shifting operation.

Subnetworks providing the switchable reflection coefficients may be of two different types. In the first group, the reactance terminating the line is changed (say from inductive to capacitive) as shown in Fig. 12.30(*b*). Phase shifters using these subnetworks are called switchable reactance-type and constitute the more commonly used variety of reflection phase shifters. In the second group of these phase shifter circuits, an additional line length is added at the reflection plane by using an SPST switch. The concept is similar to that used in switched-line phase shifters. Here, as shown in Fig. 12.30(*c*), the signal travels a longer path when the switch is open. The differential phase shift, in this case, is twice the electrical length of the shorted line switched in by the SPST switch.

Transformation of a Reflection Phase Bit into a Two-Port Network. Most of the systems using phase shifters require the phase shifting circuits to be two-port transmission networks. A reflection phase bit can be converted into a two-port network either by using a circulator or by a 90° hybrid. These two arrangements are shown in Fig. 12.31(*a*) and (*b*), respectively. It may be noted that hybrid-coupled arrangements require two identical phase bits (and hence twice the number of devices). However, this circuit is preferred to the circulator-coupled phase shifters because of the following two features: (1) 90° hybrids are a lot more integrable in MICs than the circulators, and (2) the use of two switching devices in hybrid-coupled phase shifters increases the power-handling capability by a factor of 2, which is needed in several systems.

Design of Reflection-Type Phase Shifters. A hybrid-coupled reflection-type phase shifter consists of basically three elements: a hybrid, two transforming networks, and two switching devices. Design of the hybrid can be considered separately from the rest of the phase shifter circuit and will not be discussed here. One may refer to Chapter 5 or [13] for designs of hybrids in strip-line configuration. The remaining circuit consists of two identical networks connected to the two coupled ports of the hybrid. These networks may be called *phase shift networks*. An important part of the phase shift network is the switching device itself. The remaining network may be called *transforming network*, as it transforms the device impedances in two states to appropriate values at the hybrid ports, so that the desired phase shift characteristics are obtained.

Various designs for hybrid-coupled reflection phase shifters use different configurations for the two transforming networks. Let us look in more detail at the specific functions these transforming networks are intended to perform.

Design of Transforming Networks. In order to obtain a phase shift $\Delta\phi$, the reflection coefficient values (as seen at two coupled ports of the hybrid) should differ in phase by $\Delta\phi$. Ideally their magnitudes should be unity (no loss). Thus for a 180° bit, values of the reflection coefficients in two states would lie on the outer circle ($|\Gamma| = 1$) of the Smith chart and their locations would be diametrically opposite.

Although an infinite number of such combinations are possible, it can be shown that for reflection phase shifter using a two-branch branch-line hybrid, the maximum phase shifter bandwidth is obtained [14] when the two values are located at $+j$ and $-j$ points (i.e., when located symmetrically

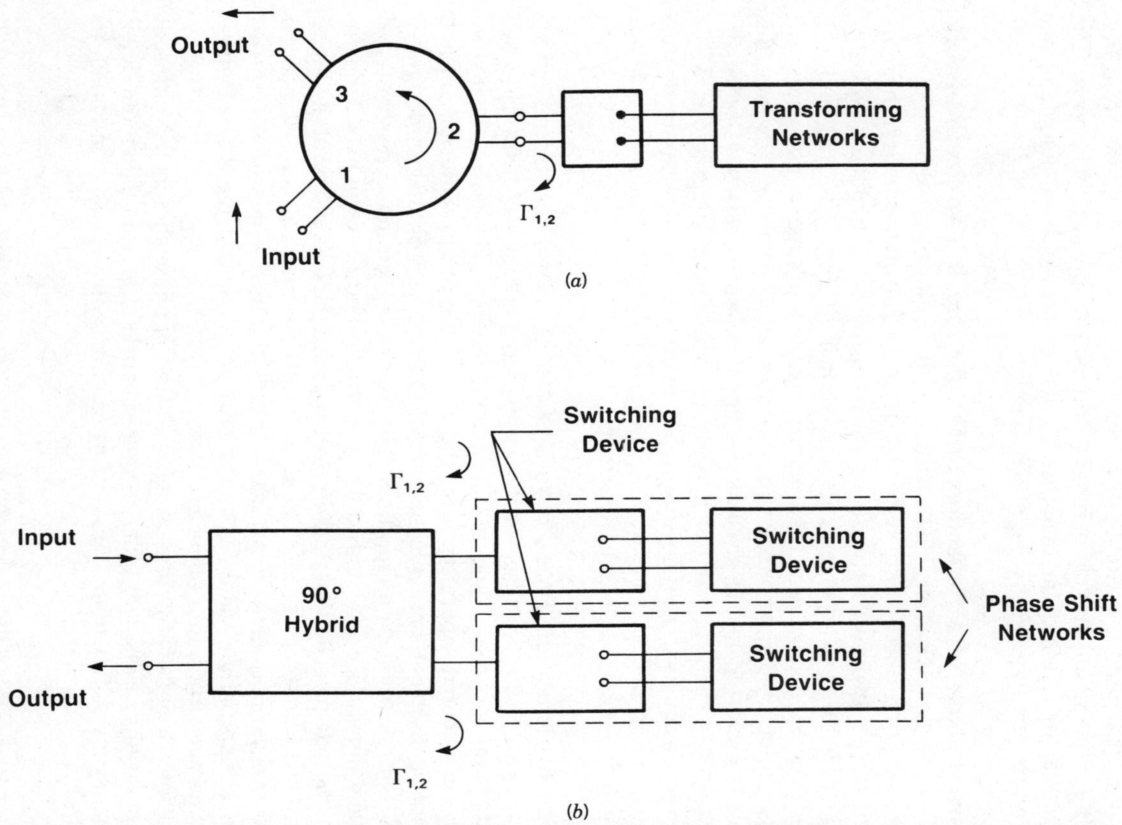

Figure 12.31 (*a*) A circulator-coupled reflection-type phase shifter. (*b*) A hybrid-coupled reflection-type phase shifter.

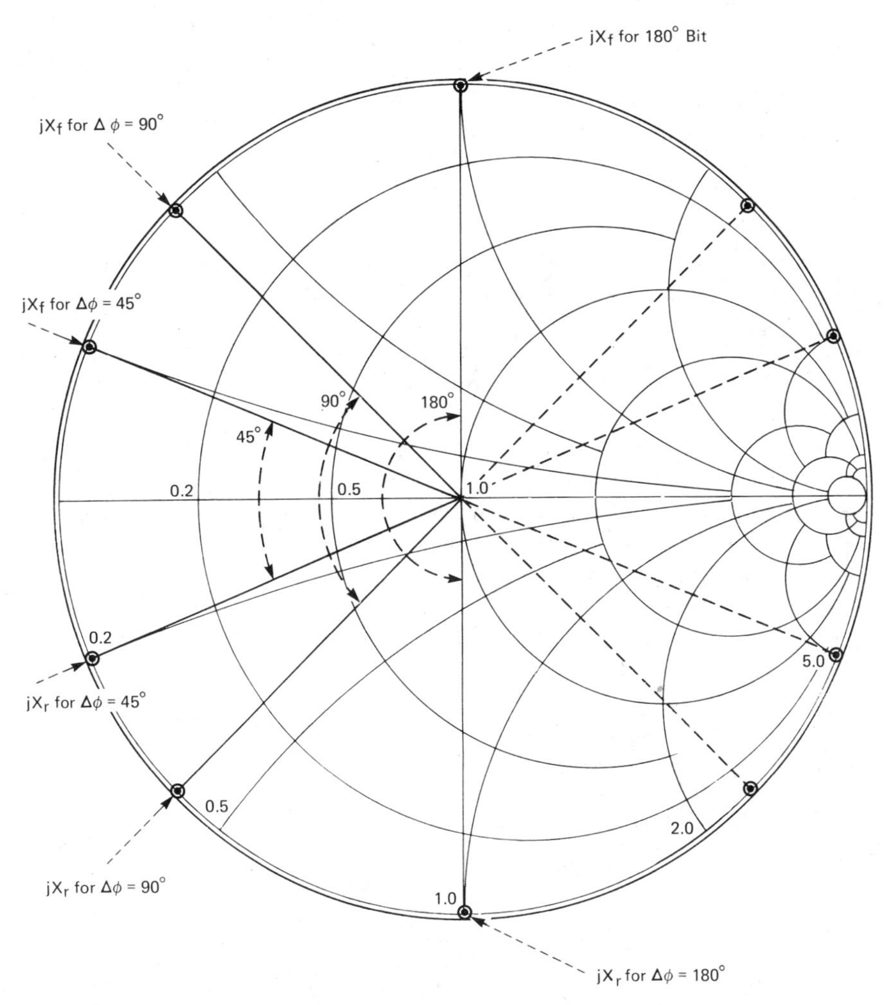

Figure 12.32 Desired location of reflection coefficients for two states of a reflection-type phase shifter using two-branch branch-line hybrids.

with respect to the $X = 0$ axis). Similarly for other values of $\Delta\phi$, Γ_1 and Γ_2 in two states should also be located symmetrically with respect to the $X = 0$ axis. These locations for $\Delta\phi = 180°$, 90°, and 45° are shown in Fig. 12.32. Note that for all values, other than $\Delta\phi = 180°$, there are two alternative sets of values for Γ_1 and Γ_2. Strictly speaking, for the positions shown in the right half of the Smith chart, the angle of Γ_f minus the angle of Γ_r would be 360° minus the corresponding values for points shown in the left half of the chart.

Based on the preceding discussion, we can divide the function of the transforming network into two parts.

1. It arranges Γ_f and Γ_r so that the desired phase shift $\Delta\phi$ is obtained, and
2. It locates Γ_f and Γ_r symmetrically about the $X = 0$ axis on the Smith chart so that the condition for maximum bandwidth is satisfied.

It may be noted that the second part of the function just mentioned is implemented by connecting a suitable line length between the transforming networks designed for (1) and the ports of the hybrid. This line length will rotate Γ_f and Γ_r points around the Smith chart without disturbing the phase relationship between them. For a given set of device impedances in two bias states (Z_f and Z_r), a desired phase shift $\Delta\phi$ can be obtained by arranging for a suitable impedance (say Z_m) to be seen when looking from the device terminals toward the rest of the network. Values of Z_m may be evaluated by considering the representation shown in Fig. 12.33.

Let a network with the equivalent Thévenin impedance Z_m terminated by an impedance Z_r (of the high-impedance state of the switching device) produce a reflection coefficient Γ, as shown in Fig. 12.33(*a*). When the same network is terminated with an impedance Z_f (of the low-impedance state of the device), we want a reflection coefficient of $\Gamma\underline{/-\Delta\phi}$ [Fig. 12.33(*b*)]. A reflection coefficient of $\Gamma\underline{/-\Delta\phi}$ will also be produced when a transmission-line section of length $\Delta\phi/2$ and impedance Z_m is connected between the network and Z_r, as shown in Fig. 12.33(*c*). Thus networks shown in Fig. 12.33(*b*) and (*c*) should be equivalent. For this equivalence to hold good, the following relationship should be satisfied:

$$Z_f = Z_m \frac{Z_r + jZ_m \tan(\Delta\phi/2)}{Z_m + jZ_r \tan(\Delta\phi/2)} \tag{12.54}$$

For low-loss switching devices, Z_f and Z_r may be approximated by jX_f and jX_r, respectively. In this case, (12.54) may be rewritten to express Z_m explicitly in terms of X_f, X_r, and $\Delta\phi$. That is, given device reactances, the desired phase shift $\Delta\phi$ may be obtained by choosing Z_m given by

$$Z_m = \frac{X_f - X_r}{2\tan(\Delta\phi/2)} \pm \sqrt{\left\{\frac{X_f - X_r}{2\tan(\Delta\phi/2)}\right\}^2 - X_f X_r} \tag{12.55}$$

For a 180° bit, Z_m becomes

$$Z_m = \sqrt{-X_f X_r} \tag{12.56}$$

and for a 90° bit, we should have

$$Z_m = \frac{X_f - X_r}{2} \pm \sqrt{\left\{\frac{X_f - X_r}{2}\right\}^2 - X_f X_r} \tag{12.57}$$

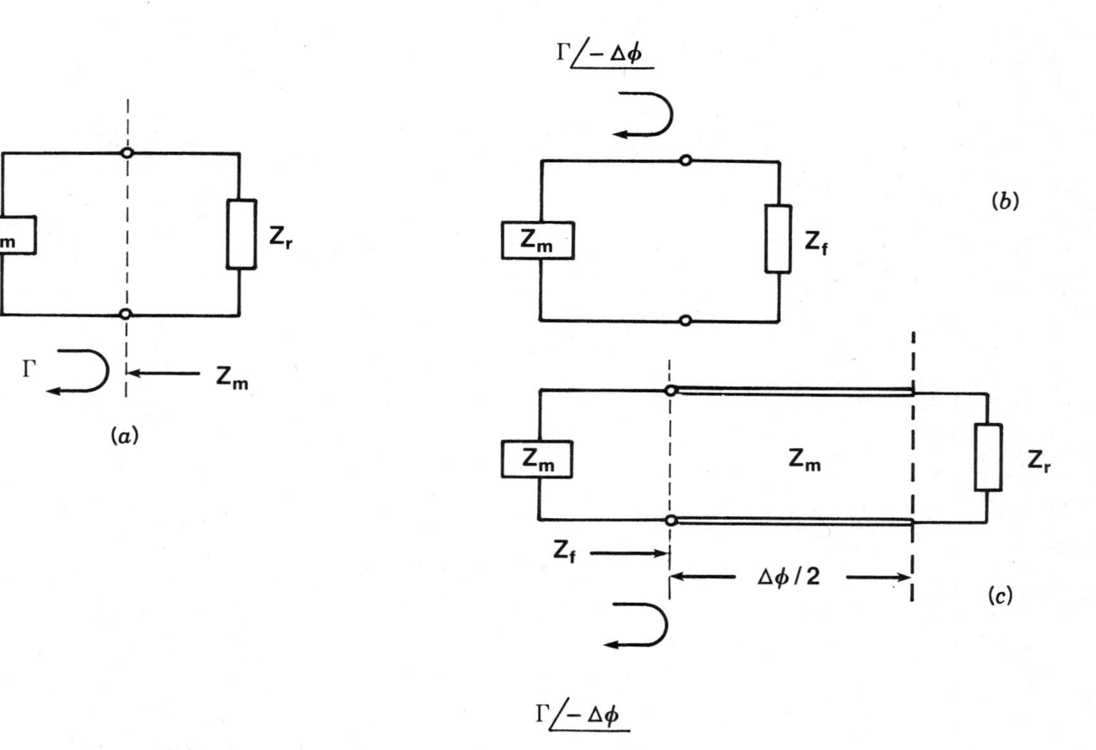

Figure 12.33 Equivalent networks for calculating Z_m (Thévenin impedance of the transforming network).

It should be remembered that the reactance X_r has a negative value (it is capacitive). Values of Z_m for various device reactances for 180° and 90° phase shifters are plotted in Figs. 12.34 and 12.35, respectively. This design approach holds good even when the switching device is lossy. However, for lossy devices Equations 12.55–12.57 get modified and Z_m may be complex.

Phase Shifter Using a λ/4 Transforming Network. The basic configuration is shown in Fig. 12.36. The design is simple. A quarter-wave line of impedance Z_t is used to transform the hybrid impedance Z_0 to a value Z_m, which the device should look into in order to provide the desired phase shift $\Delta\phi$. So

$$Z_t = \sqrt{Z_0 Z_m} \tag{12.58}$$

where Z_m is given by (12.55), (12.56), or (12.57). The line length θ_t is selected so that the impedances Z_b (in two bias states) are located symmetrically with respect to the $X=0$ axis on the Smith chart. For a 180° phase bit, Z_b in the low-impedance state should be $-jZ_0$ (since a 90° line

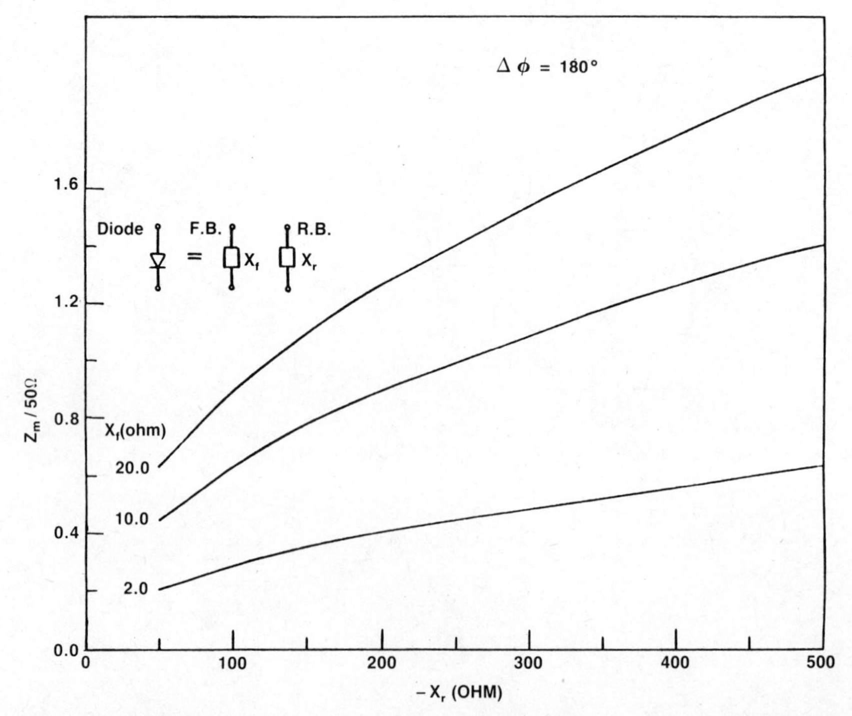

Figure 12.34 Variation of Z_m with X_r for a 180° reflection phase bit. (From [14].)

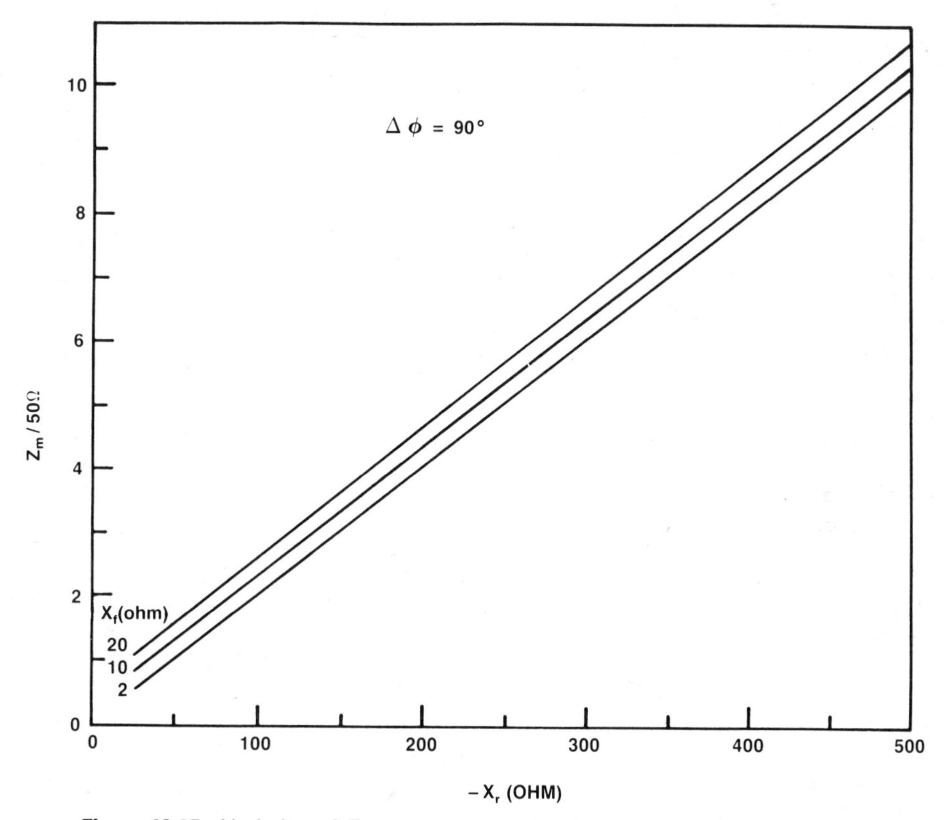

Figure 12.35 Variation of Z_m with X_r for a 90° reflection phase bit. (From [14].)

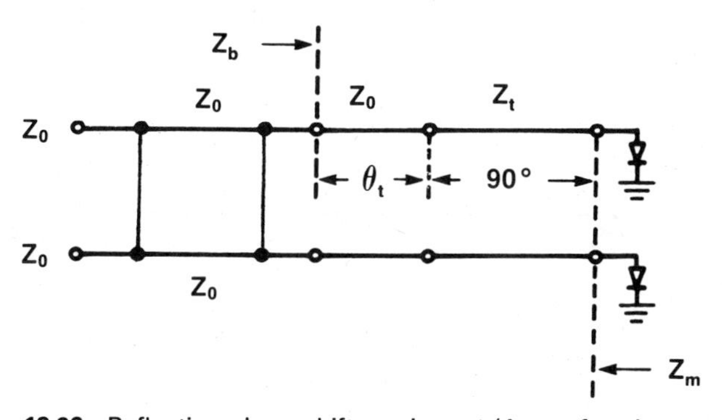

Figure 12.36 Reflection phase shifter using a λ/4 transforming network.

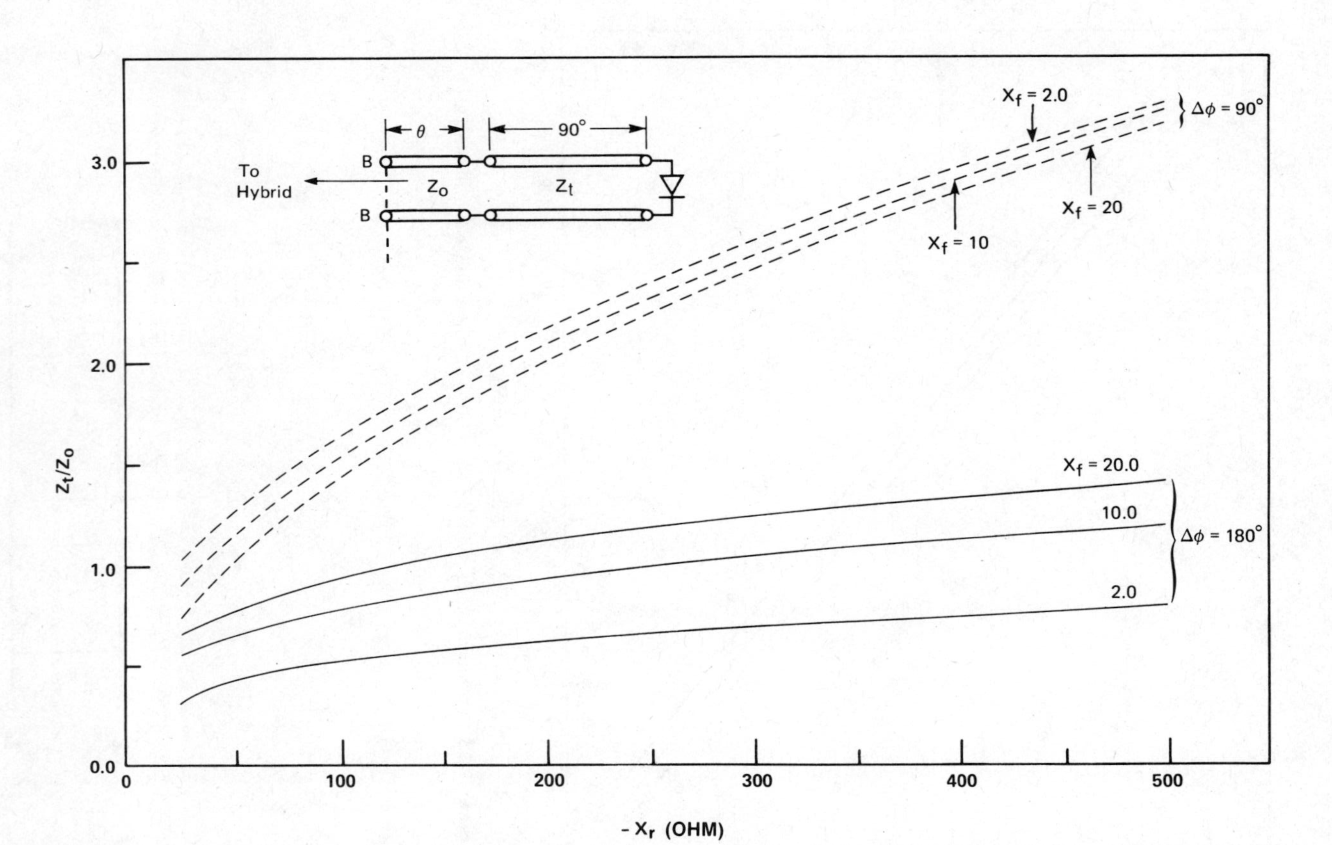

Figure 12.37 Variation of Z_t/Z_0 with X_r for 180° and 90° bits using a quarter-wave transforming network. (From [14].)

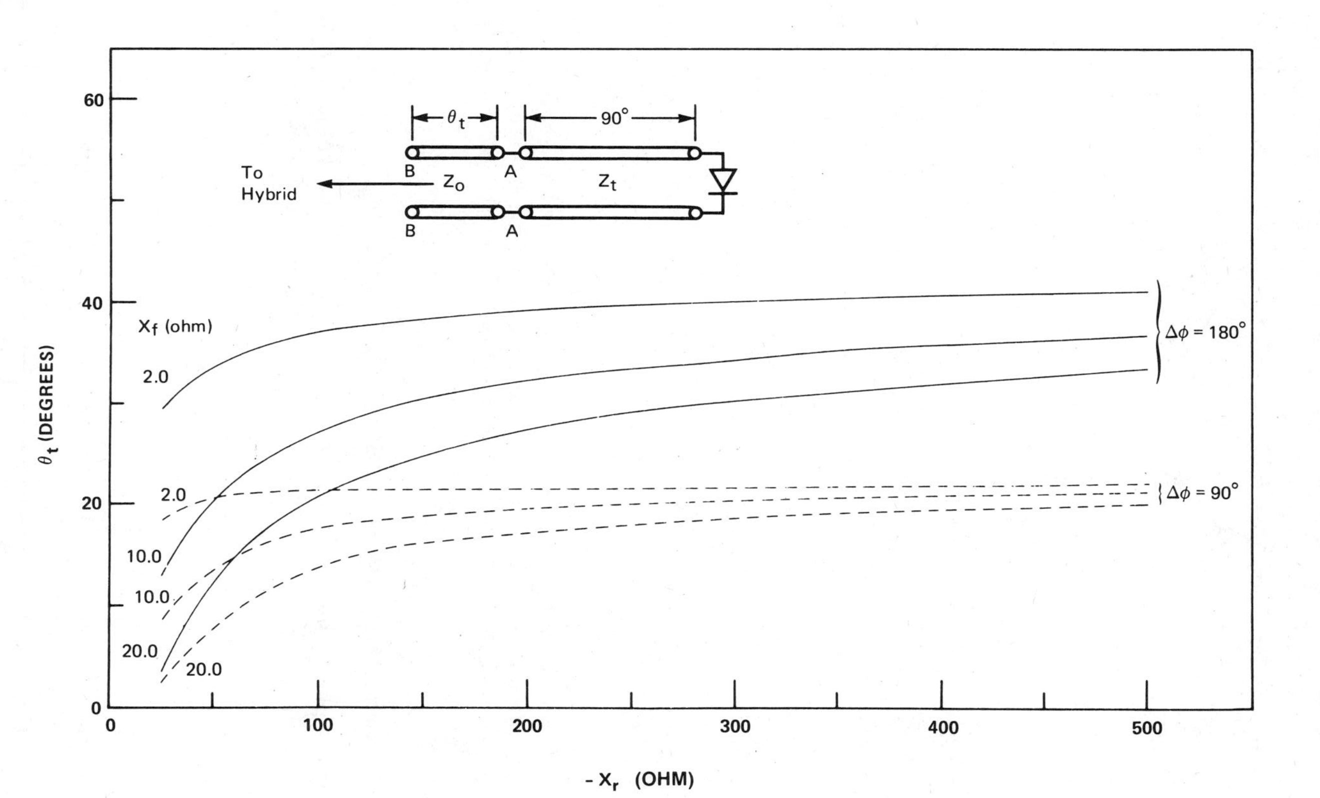

Figure 12.38 Variation of θ_T with X_r for 180° and 90° bits using a quarter-wave transforming network. (From [14].)

length has already been added). This yields

$$-jZ_0 = Z_b = jZ_0 \frac{-Z_t^2/X_f + Z_0 \tan \theta_t}{Z_0 + (Z_t^2/X_f) \tan \theta_t} \tag{12.59}$$

which gives

$$\theta_t|_{\Delta\phi=180^\circ} = \tan^{-1}\left\{\frac{Z_t^2 - Z_0 X_f}{Z_t^2 + Z_0 X_f}\right\} \tag{12.60}$$

Similarly, for a 90° bit, we should have

$$Z_b = -j2.4142 Z_0$$

which yields

$$\theta_t|_{\Delta\phi=90^\circ} = \tan^{-1}\left\{\frac{Z_t^2 - 2.4142 X_f Z_0}{Z_0 X_f + 2.4142 Z_t^2}\right\} \tag{12.61}$$

Design curves for Z_t and θ_t based on the preceding relations are presented in Figs. 12.37 and 12.38, respectively.

Phase Shifter Using Impedance-Transforming Hybrid. Rather than using a $\lambda/4$ transformer to change the impedance level from Z_0 to Z_m, this impedance transformation can be integrated within the hybrid design itself. Such a hybrid is called an impedance-transforming hybrid and has been described in [15].

12.3.5 Switched-Network Phase Shifters

This class of phase shifters may be considered to be a conceptual evolution from switched-line phase shifters. A basic block diagram of this type

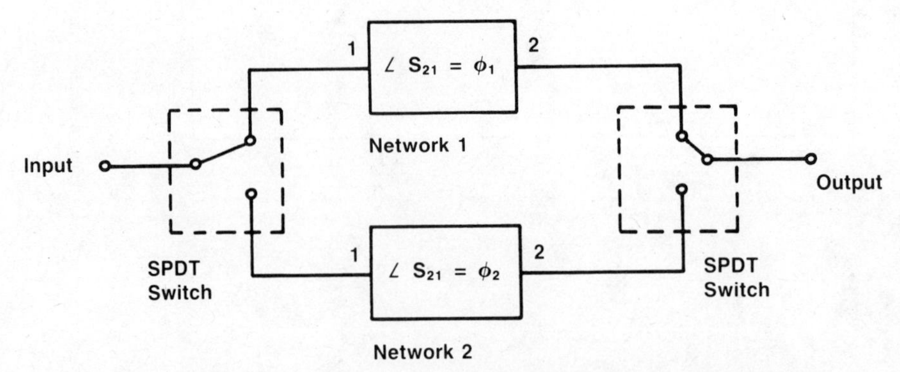

Figure 12.39 Basic block diagram of a switched-network phase shifter.

of phase shifter is shown in Fig. 12.39. When the input signal, originally passing through network 1, is switched to pass through the network 2, we get a differential phase shift ($\phi_2 - \phi_1$). The switched-line phase shifter is a special case of switched-network phase shifter with the two networks being sections of transmission lines of different lengths. The main advantage of generalizing switched-line phase shifter into a switched-network configuration is that one can design the variations of ϕ_1 and ϕ_2 with frequency appropriately and obtain a wider bandwidth or a desired frequency response of the phase shifter.

The most commonly used networks in switched-network phase shifters are the low-pass and high-pass filter configurations shown in Fig. 12.40.

The normalized *ABCD* matrix of the network in Fig. 12.40(*a*) may be written as

$$\begin{bmatrix} A & B \\ C & D \end{bmatrix}_n = \begin{bmatrix} 1 & jX_n \\ 0 & 1 \end{bmatrix} \begin{bmatrix} 1 & 0 \\ jB_n & 1 \end{bmatrix} \begin{bmatrix} 1 & jX_n \\ 0 & 1 \end{bmatrix} = \begin{bmatrix} 1 - B_nX_n & j(2X_n - B_nX_n^2) \\ jB_n & 1 - B_nX_n \end{bmatrix} \tag{12.62}$$

where X_n and B_n are the reactance and susceptance shown in Fig. 12.40 normalized with respect to transmission line impedance Z_0 and admittance Y_0, respectively. The transmission coefficient S_{21} is given, in terms of the normalized *ABCD* matrix, by

$$S_{21} = \frac{2}{A + B + C + D} = \frac{2}{2(1 - B_nX_n) + j(B_n + 2X_n - B_nX_n^2)} \tag{12.63}$$

The transmission phase ϕ is given by

$$\phi = \tan^{-1}\left\{ -\frac{B_n + 2X_n - B_nX_n^2}{2(1 - B_nX_n)} \right\} \tag{12.64}$$

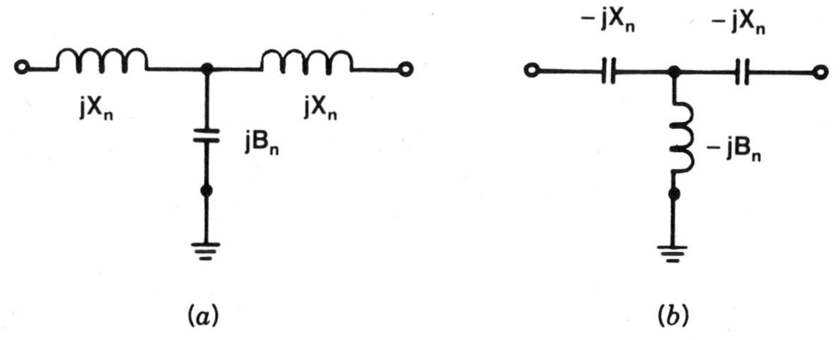

Figure 12.40 (*a*) Low-pass and (*b*) high-pass networks used in switched-network phase shifters.

When both B_n and X_n change signs [as shown in Fig. 12.40(*b*)], the phase ϕ retains the same magnitude but changes sign. Thus the phase shift $\Delta\phi$ caused by switching between low-pass and high-pass networks is given by

$$\Delta\phi = 2\tan^{-1}\left\{-\frac{B_n + 2X_n - B_n X_n^2}{2(1 - B_n X_n)}\right\} \tag{12.65}$$

For the phase shifter to be matched we need

$$|S_{11}| = 0 \tag{12.66}$$

Since we are considering a lossless case

$$|S_{11}| = \sqrt{1 - |S_{21}|^2} \tag{12.67}$$

This leads to the following relationship between B_n and X_n

$$B_n = \frac{2X_n}{X_n^2 + 1} \tag{12.68}$$

Thus the phase shift $\Delta\phi$ can be expressed in terms of X_n alone as

$$\Delta\phi = 2\tan^{-1}\left\{\frac{2X_n}{X_n^2 - 1}\right\} \tag{12.69}$$

which yields X_n in terms of $\Delta\phi$ as

$$X_n = \tan\left(\frac{\Delta\phi}{4}\right) \tag{12.70}$$

Substituting X_n in (12.68) yields

$$B_n = \sin\left(\frac{\Delta\phi}{2}\right) \tag{12.71}$$

The π-section filter shown in Figure 12.41, may also be used in place of the T-configuration shown in Fig. 12.40. For this case

$$B_n = \tan\left(\frac{\Delta\phi}{4}\right) \tag{12.72}$$

and

$$X_n = \sin\left(\frac{\Delta\phi}{2}\right) \tag{12.73}$$

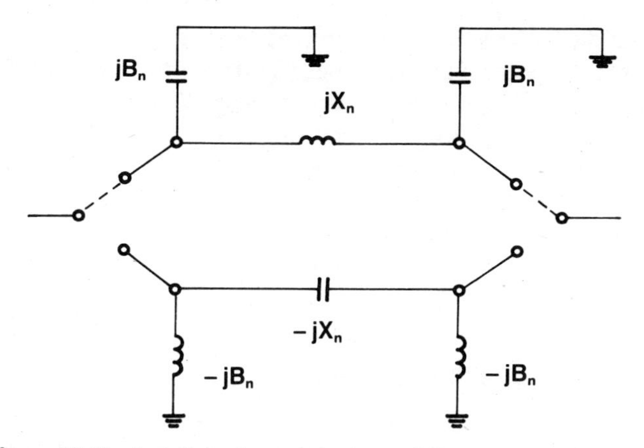

Figure 12.41 A switched-network phase shifter using a π-network.

Bandwidth of these phase shifters may be discussed by considering variations of ϕ_1 and ϕ_2 with frequency. For a T-configuration low-pass filter, phase delay ϕ_1 increases with frequency; while for a high-pass filter, phase advance ϕ_2 decreases with frequency. These two effects tend to compensate for each other and $\Delta\phi$ stays relatively constant.

A different implementation of a switched-network phase shifter using six MESFETs has been reported recently [16]. The configuration is shown in Fig. 12.42. Six MESFETs are used in two T-configurations. The MESFETs are used in the passive mode with control voltage on gate terminals used for

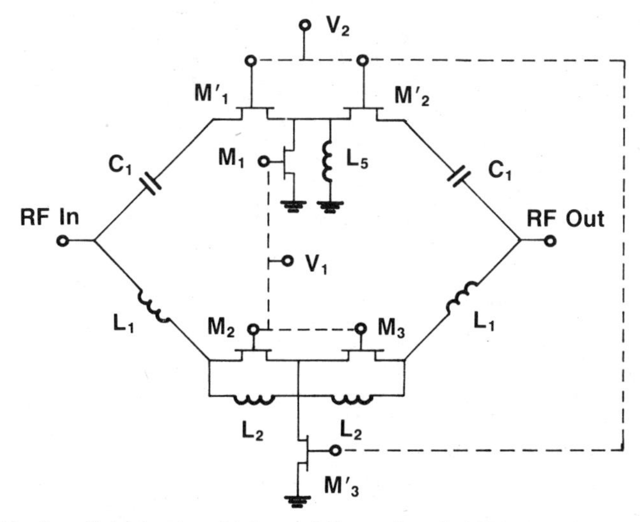

Figure 12.42 A switched-network phase shifter using six MESFETs. (After Y. Ayasli [16] © 1984, IEEE, reprinted with permission.)

switching the device from the high-impedance to the low-impedance state and vice versa. Series elements of one T and shunt element of the other T are controlled by a single gate control voltage. Thus three MESFETs M_1, M_2, and M_3 are always in the same state (high impedance or low impedance) and the other three MESFETs M_1', M_2', and M_3' are in the opposite state. External inductances and capacitances L_1 and C_1 are included to provide reactances required for low-pass and high-pass filters.

This six-MESFET design has been used for construction of a monolithic phase shifter in the 2- to 8-GHz frequency range [16].

12.3.6 Amplifier-Type Phase Shifters

Various phase shifter designs discussed in Section 12.3.2 through 12.3.5 may be realized either by using *pin* diodes or MESFETs in the passive mode. Use of MESFETs in the active amplifier mode makes several novel phase shifter designs possible. Most of the interesting active phase shifter designs have been realized using dual-gate MESFETs. Available active phase shifter designs may be grouped into three classes: (1) tuned gate dual-gate MESFET phase shifter, (2) active phase shifter using switchable SPDT amplifiers, and (3) active phase shifter using vector modulator circuits.

Active Phase Shifters Using Tuned-Gate Dual-Gate MESFET. This circuit was the first reported [17] active phase shifter circuit, but has not been very popular since. In this design, the second gate (nearer to the drain terminal) of the dual-gate MESFET is used as the signal input gate, and the first gate (nearer to the source terminal) is used as the control gate. A schematic of this type of phase shifter is shown in Fig. 12.43. As in the case of the MESFET amplifier design, matching circuits are needed both at input and output diodes. A tuning reactance (mostly inductive) is connected

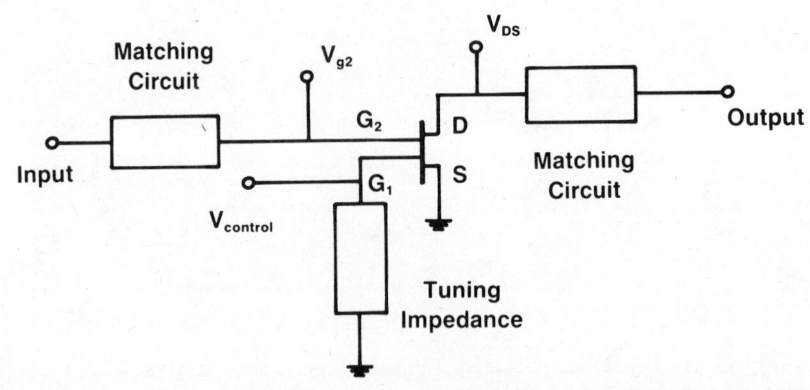

Figure 12.43 Circuit configuration for a tuned-gate dual-gate MESFET phase shifter.

between the first gate G_1 and the ground. The control voltage for controlling the phase shift between the output and input signals is applied to gate 1. Phase control is obtained because of interaction between changing device parameters (such as gate 1 to source capacitance C_{g1s}) and the externally connected tuning impedance (which could be a series inductance). A detailed analysis of the interaction is not available. Reported results point out that a phase shift of up to 100° can be obtained at 12 GHz. A continuous variation of phase shift (up to 70°) has been reported with a 3-dB gain in the 11.9- to 12.2-GHz frequency range [17]. Narrow bandwidth is one of the limitations of this type of phase shifter.

Active Phase Shifters Using Switchable SPDT Amplifiers. A block diagram of this type of active phase shifter is shown in Fig. 12.44. The input signal is switched between two identical amplifiers. At the output of one of the amplifiers, an additional line length is introduced to provide the required phase shift $\Delta\phi$. The two signal paths are combined in a power combiner circuit. A Wilkinson-type power combiner introduces a loss of 3 dB (as only one input signal is present at any time), but in view of the gain available in amplifiers this 3-dB loss is not considered serious. A convenient feature of this design is that the phase shifting part of this circuit is independent of the rest of the design, which remains invariant for different sizes of phase bits.

A detailed circuit diagram of a phase shifter of this type [18] is shown in Fig. 12.45. Signal gates and sources of two identical MESFETs are connected together in monolithic fabrication. A typical layout for this monolithic twin dual-gate MESFET is shown in Fig. 12.46. A common input matching circuit and two identical output matching circuits are included as shown in the circuit diagram (Fig. 12.45). Second gates G_2 and G_2' are used for switching the MESFETs ON and OFF. At any time only one of the MESFETs is ON and amplifies the input signal. Output of one of the MESFETs is connected to a 50-Ω transmission line whose length is selected for obtaining the required phase shift. A balanced and matched Wilkinson power divider is used to combine two signal paths.

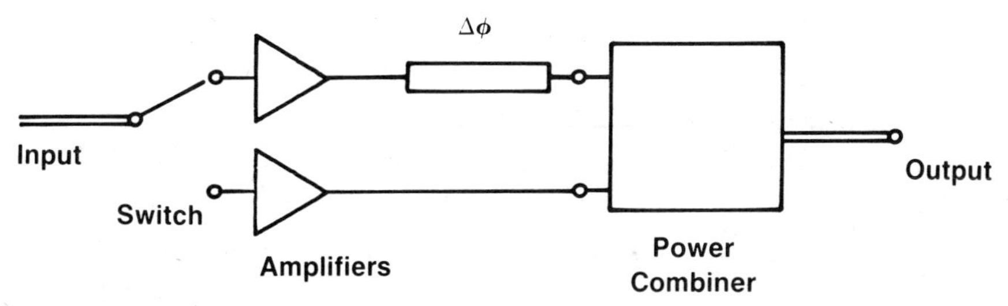

Figure 12.44 Block diagram of a phase shifter using SPDT amplifiers.

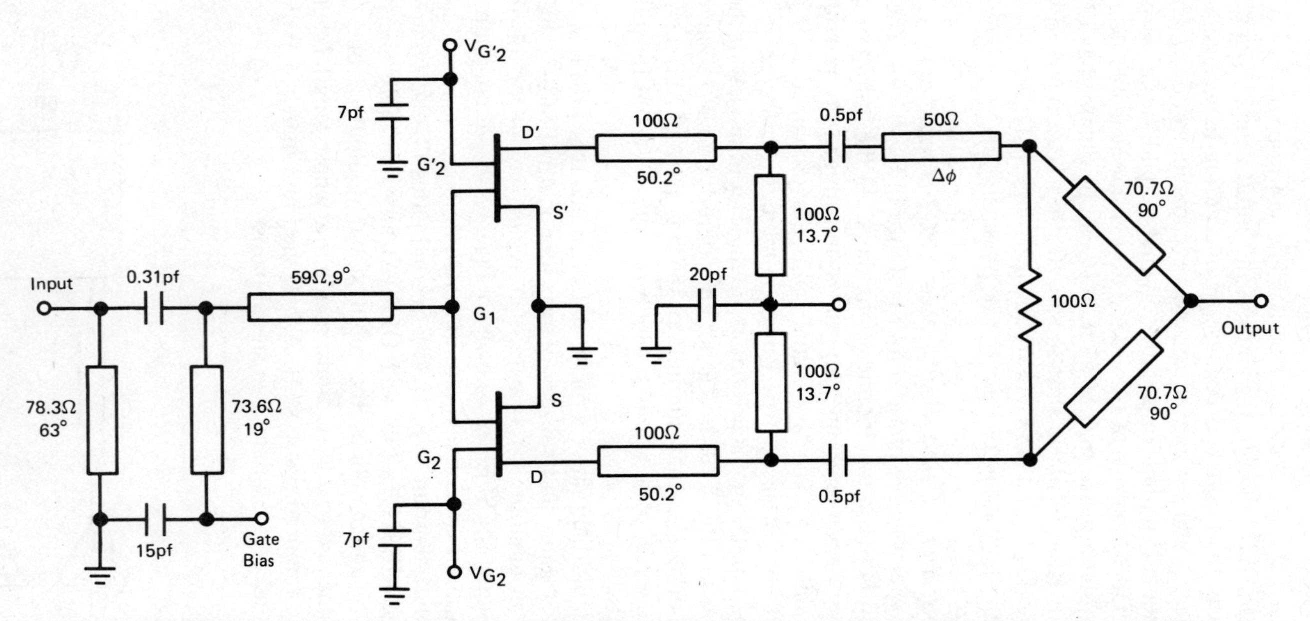

Figure 12.45 Circuit configuration for a phase shifter using switchable SPDT amplifiers. (After J. L. Varhaus et al. [18] © 1982, IEEE, reprinted with permission.)

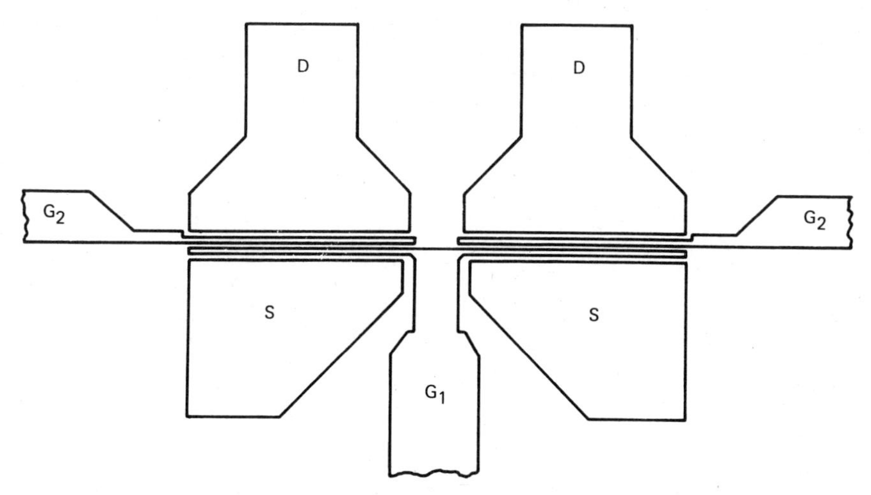

Figure 12.46 Layout of a twin dual-gate MESFET used in switchable SPDT amplifiers. (After J. L. Varhaus et al. [18] © 1982, IEEE, reprinted with permission.)

In the reported circuit [18], an overall gain of 3 dB was obtained over 10% bandwidth at X band. Dual-gate MESFETs had an on–off signal ratio of 30 dB and were capable of 9–11 dB gain at X band. The single-bit monolithic phase shifter chip size was $2.5 \times 3.0 \times 0.1$ mm. For the MESFETs used, gate length was 1 μm and separation between gates was 2 μm. Gatewidth was about 150 μm.

Active Phase Shifter Using Vector-Modulator Circuits. A vector modulator is a circuit that is capable of independently varying the amplitude and the phase of an input signal by desired amounts. The concept is shown schematically in Fig. 12.47(a). In principle, ϕ could be anywhere between 0° and 360° and A could be any reasonable factor less than or greater than 1. Two different schemes have been proposed for implementing a vector-modulator circuit at microwave frequencies. One of these uses four vectors spaced 90° apart and pointing in four different directions as shown in Fig. 12.47(b). Amplitudes of the four component vectors A_1, A_2, A_3, and A_4 are controlled independently by four different controlled-gain amplifiers. Of course, only two of these four components are nonzero at any time. Thus the vector sum could lie in any of the four quadrants and the amplitude could be adjusted by controlling the gains of two amplifiers active at that time. It may be noted that four component vectors are not needed and three vectors spaced 120°, as shown in Fig. 12.47(c), could serve the same purpose. This provides an alternative implementation for vector-modulator circuits, and its realization has been described in [19]. Clearly, a vector modulator has more versatility than that required for a

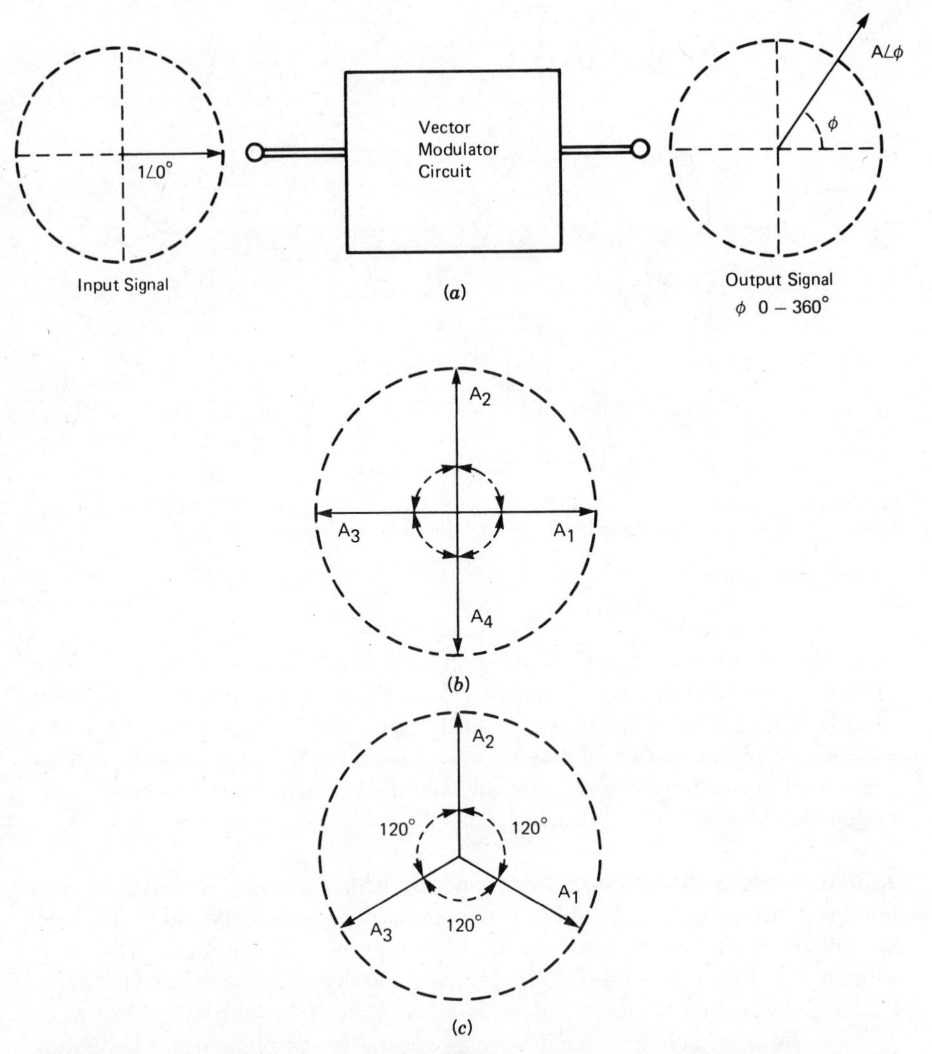

Figure 12.47 (*a*) Concept of a vector-modulator circuit. (*b*) Four vectors spaced 90° apart, used for realization of vector-modulator circuits. (*c*) Three vectors (120° apart), used for realization of vector-modulator circuits.

digital phase shifter. In fact, it is this versatility that makes the circuit very attractive for monolithic phase shifter applications.

Phase Shifter Using Segmented-Gate MESFETs. We note that in phase shifter circuits based on the vector-modulator concept, the phase shift depends upon the ratio of two gains. In order to design a digital phase shifter, we need a repeatable digital variation of the gain ratio. A monolithic device introduced recently [20] for this purpose is known as a seg-

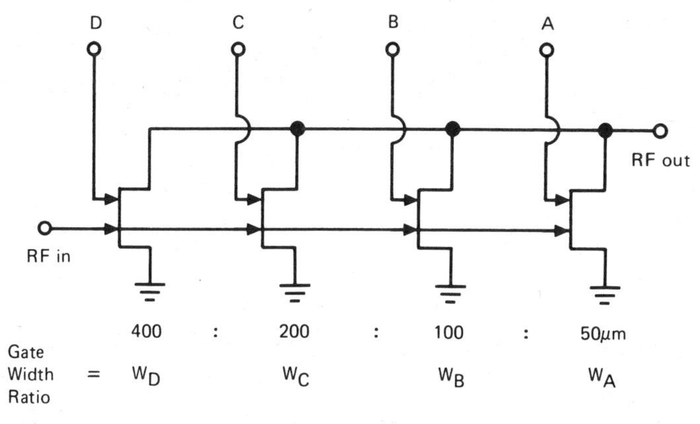

Figure 12.48 Circuit representation of a segmented dual-gate device. (After Y. Hwang et al. [20] © IEEE, 1984, reprinted with permission.)

mented dual-gate MESFET. The total gatewidth of a dual-gate MESFET is divided into several sections. A signal-gate connection is common to all of these sections, but the control gate is broken into several segments, one corresponding to each of the sections. External connections are brought out from each one of the second gate segments, so that various sections of the MESFET may be switched ON or OFF selectively. A typical schematic diagram of such a device [20] is shown in Fig. 12.48. Here, the control gate (total width 750 μm) is divided into four segments of different widths. Segments *A*, *B*, *C*, and *D* have widths of 50 μm, 100 μm, 200 μm, and 400 μm, respectively. Each one of these segments may be switched ON or OFF independently. Since the gain of the MESFET is, to a reasonable approximation, linearly proportional to the gatewidth, the gain can be varied by switching ON a particular segment or a combination of segments. If we consider the gain provided by the 50-μm segment as one unit, the gain of the device shown in Fig. 12.48 can be varied in steps of these units from a single unit (when only the 50-μm segment is on) through to fifteen units (when all four segments are turned on). This corresponds to a gain variation over a range of 23.52 dB. In the fabrication process of GaAs MESFET devices, the gatewidth is a parameter that is least sensitive to processing variables. So even if the gain of individual segments may vary from device to device, the ratios of gains exhibit an excellent repeatability. Also, these ratios are relatively independent of temperature variations.

Variable gain amplifiers may be used for the design of phase shifters. A schematic block diagram of such a phase shifter is shown in Fig. 12.49. The input signal is divided into two equal parts by using a 90° hybrid. Thus for an input of $1\underline{/0^\circ}$, the two signals at the output of block *A* are $0.707\underline{/0^\circ}$ and $0.707\underline{/-90^\circ}$, respectively. These two signals are fed to variable-gain amplifiers. If the phase shift required from the phase shifter is ϕ, the gain of

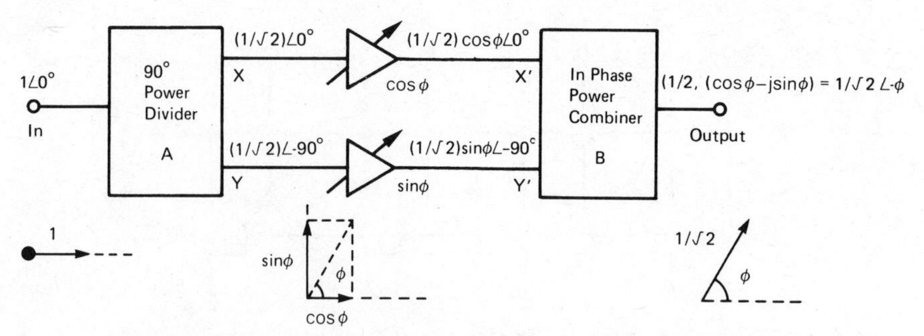

Figure 12.49 A phase shifter using two variable-gain amplifiers.

the amplifier in the upper signal path is set proportional to cos ϕ, while that of the amplifier in the lower signal path is adjusted proportionally to sin ϕ. The two signals coming out of these two amplifiers may be represented by $0.707K \cos \phi \underline{/-\theta^\circ}$ and $0.707K \sin \phi \underline{/-\theta-90^\circ}$, where K is the proportionality factor for the amplifier gains and θ is the phase common to the two channels introduced by the amplifiers and the connecting lines. The two signals are added in an in-phase power combiner circuit. The output of the combiner is $0.5K\{\cos(\theta+\phi)-j\sin(\theta+\phi)\}$, that is, a phase shift of $\theta+\phi$ compared to the signal at the input. When the amplifier in the lower path is switched OFF, the output signal lags the input by a phase angle θ. This can be considered to be the reference state. Then, the differential phase shift ϕ is obtained when gains are set proportional to cos ϕ and sin ϕ.

A digital phase shifter based on the concept shown in Fig. 12.49 can be obtained by using segmented dual-gate devices for cos ϕ and sin ϕ amplifiers. One of the possible arrangements is to use a device with three segments with widths in the ratio of 8:4:1. A fairly good approximation to cos ϕ and sin ϕ can now be obtained for $\phi=0$, 22.5°, 45°, 67.5°, and 90°. Table 12.3 shows this so-called $\frac{1}{13}$ approximation obtained by switching the

TABLE 12.3 The 1/13 Approximation for Generating sine and cosine Functions with a Segmented Gate Device*

θ degrees	sin θ	1/13 Approximation	W_8 (0-off 1-on)	W_4	W_1
0	0	0/13	0	0	0
22.5	0.0383	5/13 (0.384)	0	1	1
45.0	0.707	9/13 (0.692)	1	0	1
67.5	0.924	12/13 (0.923)	1	1	0
90.0	1.000	13/13	1	1	1

*After Y. Hwang et al. [20] © 1984, IEEE, reprinted with permission.

three segments ON or OFF. In this table, the 0s in the last three columns show that corresponding segments are OFF, and the 1s indicate the corresponding segments are switched ON.

For differential phase shifts greater than 90°, cos ϕ becomes negative. This requires a phase inversion in the upper signal path (Fig. 12.49). In this case, a possible solution is to use a 180° hybrid combiner in place of the in-phase power combiner shown in the figure. An alternative solution for obtaining ϕ greater than 90° is to cascade two stages with individual phase shifts less than 90°.

12.4 DESIGN OF LIMITER CIRCUITS

Apart from the switches and phase shifters discussed in previous sections, a "limiter" is an important control component used at microwave frequencies. An ideal limiter has no attenuation when low power is incident upon it, but has an attenuation that increases with increasing power (above a threshold level) to maintain the output power constant. Input–output characteristics for ideal limiters (solid line) and practical limiters (dotted lines) are shown in Fig. 12.50. The most common application of microwave limiters is to prevent transmitter power in a radar from reaching the receiver directly and burning its sensitive input stage. Limiters are also used to protect the receivers from other nearby radar transmitters. Other applications of limiters are for reducing the amplitude modulation of swept-frequency oscillators and for reducing amplitude modulation in phase-detection systems.

12.4.1 Various Phenomena Used for Limiting

Limiters can be compared with SPST switches in the ON state when the incident power is below the limiting threshold value, and to switches in the

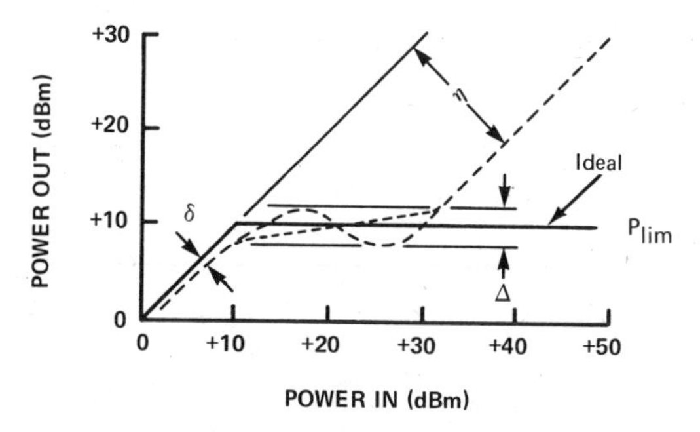

Figure 12.50 Characteristics of ideal and practical limiters. (After R. V. Garver [9] © Artech House 1976, reprinted with permission.)

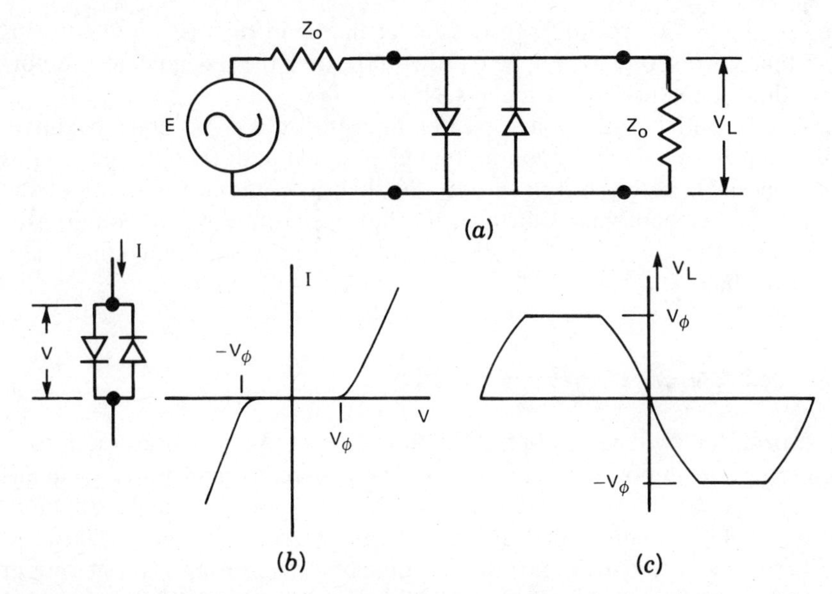

Figure 12.51 (*a*) Basic limiter circuit using two rectifying diodes. (*b*) The resulting *I*–*V* characteristic. (*c*) Output voltage waveform.

OFF state when the input exceeds the threshold. Consequently, practical limiter circuits are characterized by a finite insertion loss δ and a finite isolation η as shown in Fig. 12.50. Sometimes, another parameter, vacillation Δ about the limit power level P_{lim}, is also specified.

There are three phenomena, exhibited by microwave semiconductor diodes, that may be used for limiting. Let us discuss these very briefly.

1. *Rectification*. This is the most commonly used limiting technique at lower frequencies. A basic circuit (*a*) using two rectifying diodes, the *I*–*V* relationship (*b*), and the clipped sine wave (*c*) (obtained when the input is above the limiting threshold) are shown in Fig. 12.51. Point-contact and Schottky-barrier diodes (not *pin* diodes) can be used for rectification at microwave frequencies. The main problem in using this method is its very low power handling capability. The rectifying diodes have a very thin depletion layer (in order to have a sufficiently rapid turn-on time at microwave frequencies) and have a small junction area (in order to keep the device capacitance small). The resulting small volume of the device cannot protect it from burning out when very high-power microwave signals are present.

2. *Capacitance Variation with Voltage*. Varactor diodes have a junction capacitance that is voltage-dependent and that responds rapidly enough to change the characteristics at microwave frequency. If two diodes are set

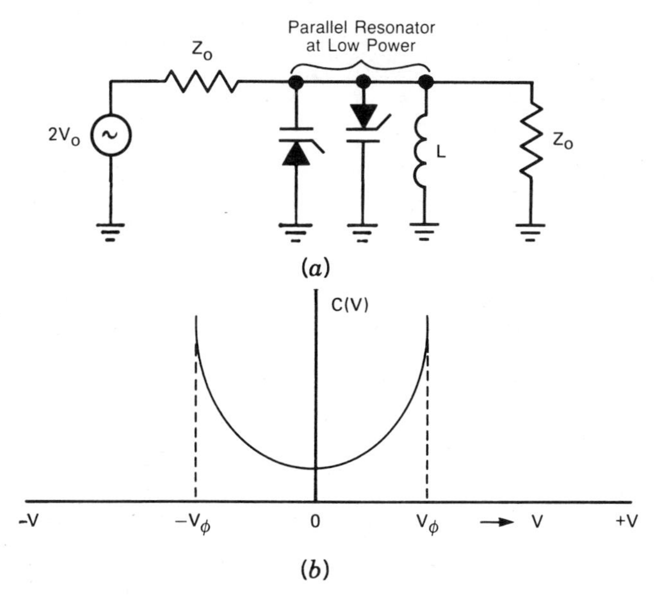

Figure 12.52 (*a*) Two varactor diodes set parallel with opposite polarity and (*b*) the resulting C–V characteristic for $|V| < |V_\phi|$.

parallel in opposite polarity (as shown in Fig. 12.52(*a*)], the resulting C–V characteristic is shown in Fig. 12.52(*b*). When these diodes are used in a parallel resonant circuit, we obtain a reflection coefficient that varies as a function of incident power. Also, some rectification occurs at high powers, and provides additional limiting. These limiters also suffer from low power-handling capability.

3. *RF Conductivity Modulation*. This phenomenon is exhibited by *pin* diodes when biased at zero voltage and subjected to high-level RF current signals. When a *pin* diode is forward biased, the free-carrier concentration in the *i*-region is not perfectly uniform but, because of the limited lifetime of the carriers, has a shape shown in Fig. 12.53(*a*). Now suppose that the dc bias is removed and replaced by a short-circuit path (dc return) as shown in Fig. 12.53(*b*). It has been shown by Leenov [21] that, when a large microwave current has been established, the distribution of the resulting carrier concentration will be as shown in Fig. 12.53(*c*). The large microwave current causes some partial carrier injection near the p^+ and n^+ boundaries during the forward-going half-cycles of the RF signal. Not all of this injected charge is withdrawn when the RF voltage reverses. The result is a trickling of electrons and holes into the *i*-region, and these get distributed as shown in Fig. 12.53(*c*). This steady state distribution is attained after a few RF cycles and remains static during the remainder of the RF pulse, serving to conductivity modulate the *i*-region. During the transient period in which the charge is built up in the *i*-region, the *pin*

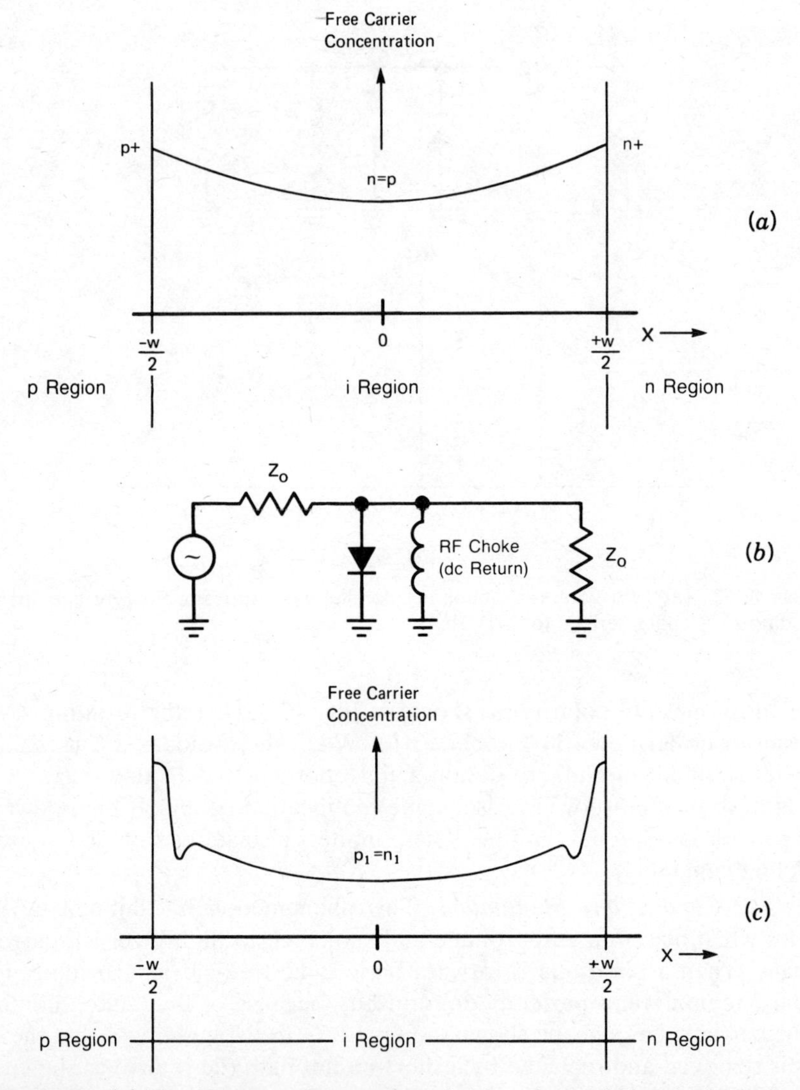

Figure 12.53 (*a*) The *i*-layer carrier concentration in a forward-biased *pin* diode. (*b*) Limiter circuit using a zero-biased *pin* diode with a dc return path. (*c*) Carrier concentration in *i*-layer *pin* diode in (*b*) for high-level RF excitation. (After J. F. White [1] © Van Nostrand 1982, reprinted with permission.)

limiter provides relatively little limiting to the high-power RF signal. The RF signal passes through the *pin* limiter with little attenuation during this spike leakage period, after which a relatively high attenuation (limiting) is achieved and the small fraction of the power that passes through is called the flat leakage. This *pin* diode limiting phenomenon is illustrated in Fig. 12.54. In order to distinguish clearly from rectification-type limiting, it may be noted that (for the limiter circuit shown in Fig. 12.54), the *pin* diode

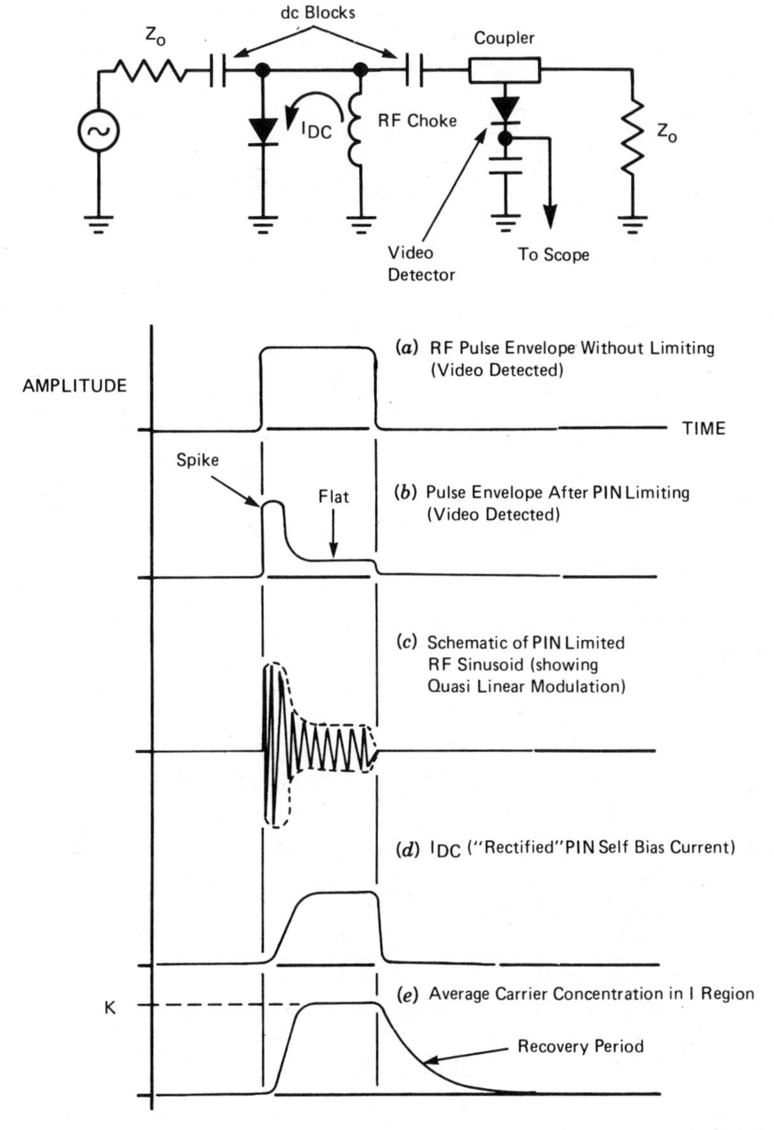

Figure 12.54 Various waveforms in the *pin* diode limiter. (After J. F. White [1] © Van Nostrand 1982, reprinted with permission.)

presents substantially the same conductivity to both forward- and reverse-going halves of the RF cycle.

12.4.2 *pin* Diode Limiters

It has been shown by Leenov [21] that the resistance of the *i*-region of a *pin* diode activated through a microwave current is given by

$$R_i = \frac{W}{\sqrt{D_{ap}/2\pi f}} \cdot \frac{1}{(e/kT)} \cdot \frac{1}{I_{rf}} \tag{12.74}$$

when the i-layer width W is much less than the carrier diffusion length L ($=43\sqrt{\tau}$ μm, where τ is the average carrier lifetime in microseconds). In the preceding relation, D_{ap} is the diffusion coefficient, W is in centimeters, and I_{rf} is in amperes rms. For Si diodes at room temperature relation (12.74) may be simplified as

$$R_i = \frac{W\sqrt{f}}{20 I_{rf}} \tag{12.75}$$

where R_i is in ohms, W is in microns, f is in gigahertz, and I_{rf} is in amperes rms.

It may be noted that the action of a high-level microwave signal is much less effective in producing conductivity modulation of the i-layer than is a dc current. For example, at 1 GHz the RF current required to reduce R_i to a value of 1 Ω [for a diode with $W = 50$ μm (2 mil)] is $I_{rf} = 50\sqrt{1}/(20 \times 1) =$ 2.5 A (rms), whereas the corresponding requirement for dc current will be only about 50 mA.

After the incident high-power microwave pulse ceases, the concentration of holes and electrons in the i-region does not disappear immediately, but decays exponentially with a time constant equal to the average carrier lifetime. During this recovery period the limiter has a relatively high insertion loss. The *recovery time* is defined as the time required for the limiter's insertion loss to return to within 3 dB of its low-level insertion loss following the cessation of the high-power pulse. In a typical radar system this period corresponds to echos from nearby targets, and a higher insertion loss in the receiver path may be tolerated.

The recovery time is related to the diode minority carrier lifetime τ. For a given τ, the recovery time is linearly proportional to the peak RF power applied up to some peak power level above which recovery time increases rapidly because of thermal heating of the diode. Measurement of the recovery time thus allows a nondestructive monitoring of the power-handling capacity of a *pin* diode limiter circuit.

12.4.3 Limiters in Microstrip Configuration

A practical implementation of a limiter circuit using two shunt-mounted diodes spaced along a microstrip line is shown in Fig. 12.55. The diodes are mounted on the ground plane (after punching out holes in the substrate dielectric material). Flexible wire (or rectangular ribbon) leads provide mechanical stress relief in the connection of the strip conductor to the diode. In addition, by proper selection of lead lengths and sizes, series

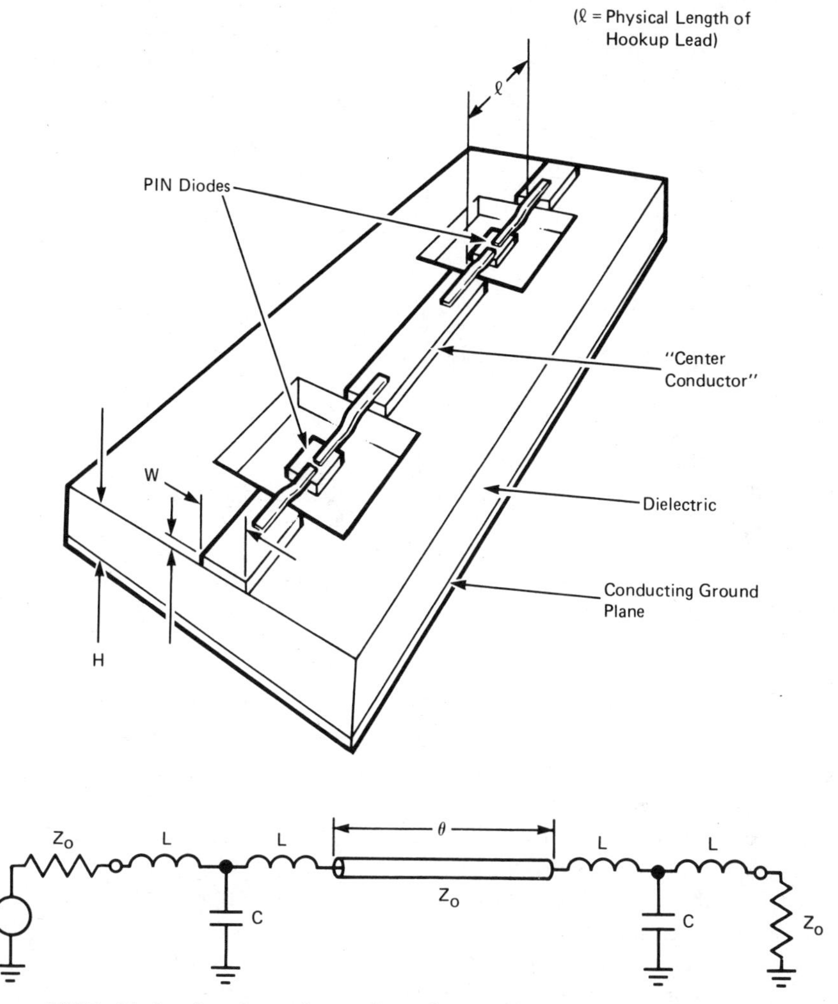

Figure 12.55 Limiter in microstrip configuration and its equivalent circuit. (After J. F. White [1] © Van Nostrand 1982, reprinted with permission.)

inductances are realized that can be used to form a matched T filter and thereby tune out the reflection that would be otherwise introduced by the diode capacitance. An equivalent circuit (with reversed-bias *pin* diodes represented by capacitances C) is also shown in Fig. 12.55. The circuit configuration shown in Fig. 12.55 is also well suited for *pin* diode switches discussed earlier.

If the total inductance, $2L$, of the two bonding straps satisfies

$$Z_0 = \sqrt{\frac{2L}{C}} \qquad (12.76)$$

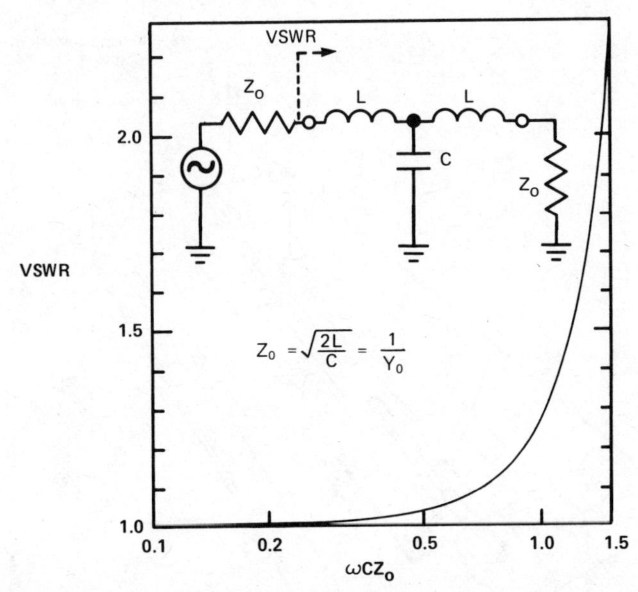

Figure 12.56 VSWR for a matched-T diode configuration. (After J. F. White [1] © Van Nostrand 1982, reprinted with permission.)

the equivalent T circuit will resemble a length of a transmission line and its transmission match will be frequency independent in the range for which $\omega L \ll Z_0$ and $\omega C \ll (1/Z_0)$. The VSWR resulting from a single T network of this type is shown in Fig. 12.56. If we consider a diode with $C = 0.32$ pF, mounted in a 50-Ω system, half of the incident power is reflected at 20 GHz (where $\omega C Z_0 = 2$), corresponding to a VSWR value of 5.8. The VSWR value is 1.28 at 10 GHz, which is a more realistic upper frequency for a limiter in this configuration. An estimate for strap inductance L is obtained by treating it as a microstrip transmission line with air dielectric and an average height above the ground plane given by means of diode chip thickness and microstrip dielectric substrate thickness used. If Z_0 is the characteristic impedance of the microstrip line, the inductance per unit length is given by (Z_0/v_p) where v_p is the phase velocity for the air medium.

12.5 DESIGN OF VARIABLE ATTENUATORS

Voltage-controlled variable attenuators are important control elements and are widely used for automatic gain control circuits. They are indispensable for temperature compensation of gain variation in broadband amplifiers. Both *pin* diodes and GaAs MESFET devices are used for design of variable attenuators. Apart from the use of MESFETs in the passive mode (described in Section 12.5.2), active MESFET amplifier circuits may also be used for variable-attenuation circuits. Dual-gate MESFET amplifiers with controlled voltage applied to the second gate are ideal for this purpose.

In this section, we discuss variable attenuators using *pin* diode and passive-mode MESFETs only.

12.5.1 *pin* Diode Attenuators

The fact that the resistance of the intrinsic layer of *pin* diodes is a strong function of the dc bias current is employed for designing current-controlled variable attenuators at microwave frequencies.

The resistance of the intrinsic layer under forward bias is given by

$$R_i = \frac{W^2}{(2\mu_{\text{ap}}\tau I_0)} \tag{12.77}$$

where W is the width of the i-layer, μ_{ap} is the ambipolar mobility ($610\,\text{cm}^2/\text{V-s}$ in silicon), τ is the lifetime of carriers, and I_0 is the dc bias current. The derivation of (12.77) is based on several simplifying assumptions. For practical diodes [22] variation with I_0 is more like $I_0^{-0.87}$ than $I_0^{-1.0}$. It is often advisable to make an experimental measurement of the variation of the R_f (which includes R_j and contact resistance) with bias current.

An important characteristic of a variable attenuator is that its input impedance should remain constant, so that the attenuator remains matched over its operating range. One way of realizing this is the π network shown in Fig. 12.57. This circuit consists of three *pin* diodes for R_1, R_2, R_2, respectively. For matching, the impedance of the network to the right of AA' (say Z_A) in parallel with R_2 should equal Z_0, that is,

$$\frac{1}{Z_0} = \frac{1}{R_2} + \frac{1}{Z_A} \tag{12.78}$$

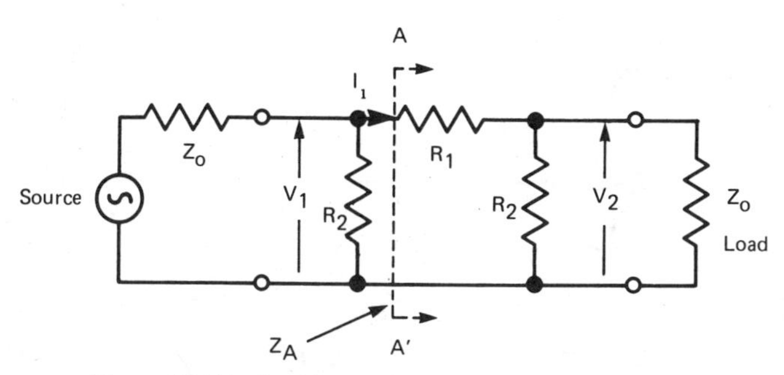

Figure 12.57 Resitive π network for *pin* attenuators.

where $Z_A = R_1 + R_2Z_0/(R_2 + Z_0)$. Since we have $V_1 = I_1Z_A$ and $V_2 = I_1R_2Z_0/(R_2 + Z_0)$, the attenuation ratio V_1/V_2 may be written as

$$\frac{V_1}{V_2} = \frac{Z_A}{R_2Z_0/(R_2 + Z_0)} \tag{12.79}$$

Substituting for Z_A we have

$$K = \frac{V_1}{V_2} = \frac{R_2 + Z_0}{R_2 - Z_0} \tag{12.80}$$

Resistance values R_2 and R_1 can be expressed in terms of the characteristic impedance Z_0 and the voltage ratio K as

$$R_2 = \frac{Z_0(K+1)}{K-1} \tag{12.81}$$

$$R_1 = \frac{Z_0(K - 1/K)}{2} \tag{12.82}$$

The values of bias currents required to produce the necessary resistance values for R_1 and R_2 are obtained by using (12.77) or by experimental measurements.

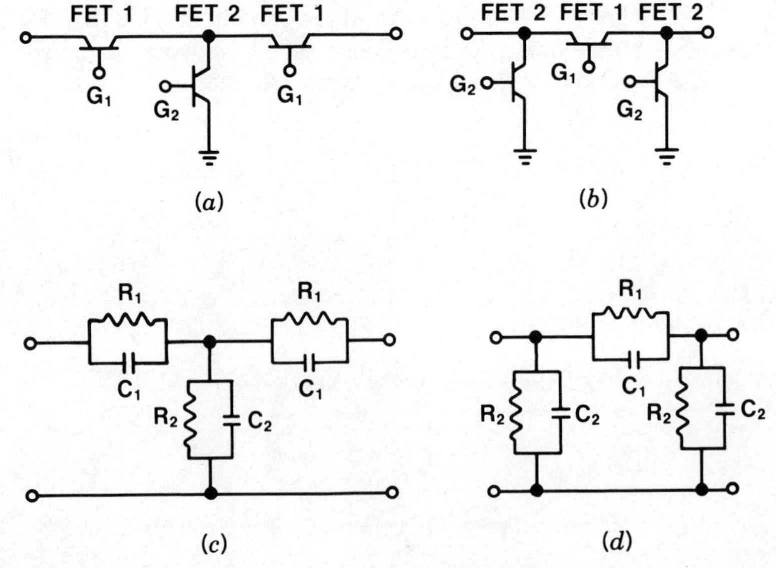

Figure 12.58 T and π attenuator networks using MESFETs and equivalent network models.

12.5.2 MESFET Attenuators

For attenuators employing MESFETs in the passive mode, the basic device mechanism used is the change in the low field resistance of a zero-biased FET controlled by the gate voltage. Three MESFETs may be connected in T or π configuration, as shown in Fig. 12.58(a) and (b), respectively. The MESFETs may be modeled by a parallel combination of R and C, as shown in Fig. 12.58(c) and (d), where the value of R varies as a function of the gate voltage. The value of R varies from zero-bias value (R_{on}) to a large value R_{off} when the gate voltage reaches the pinchoff voltage. The value of the capacitance C may be considered to be fairly constant with gate voltage.

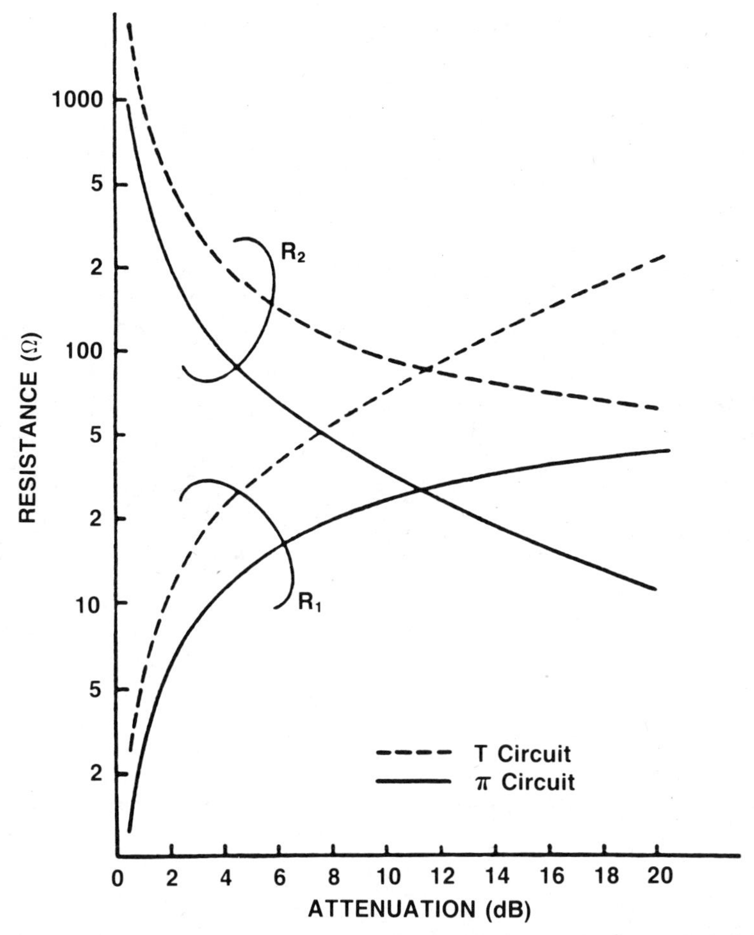

Figure 12.59 Values of resistance R_1 and R_2 needed for attenuator design. (After Y. Tajima et al. [23] © IEEE, 1982, reprinted with permission.)

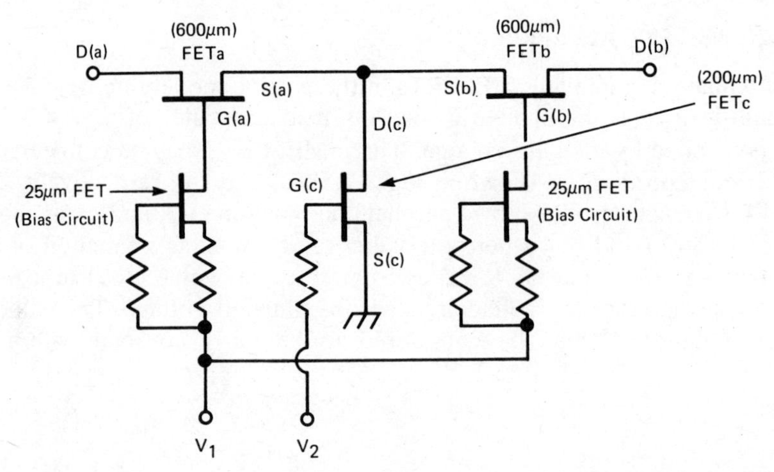

Figure 12.60 Schematic of a MESFET attenuator. (From [23] © 1982, IEEE, reprinted with permission.)

At lower microwave frequencies the effect of capacitors can be neglected. In such a case, values of R_1 and R_2 needed for various values of attenuation (for $Z_0 = 50\ \Omega$) are shown in Fig. 12.59 (see [23]). For the same R_1, the π configurations have a lower loss than the T configurations. The minimum value of R_1 $(= R_{on})$ can be reduced by increasing the number of gate fingers (although C_1 will also increase). Larger values of C_1 limit the dynamic range of attenuation at higher frequencies. From the point of view of dynamic range, the T configuration is better than the π configuration.

An attenuator with T-circuit topology (Fig. 12.60) has been reported in Reference 23. Gatewidths of 600 and 200 μm were chosen for the series and shunt MESFETs, respectively. Input and output microstrips were connected to drain contacts of MESFETs *a* and *b*, while source contacts were connected by a common air bridge that served as a connection to the drain contact of MESFET *c*. The source of MESFET *c* is grounded through a via hole. Isolation between the RF circuit and dc control circuit was achieved by thin-film resistors and FET resistors that have 25-μm-wide gates.

Experimental results [23] show this attenuator to have a 12-dB dynamic range of attenuation over the 2- to 18-GHz frequency band, and 17 dB over the 2- to 12-GHz frequency range. Return loss was better than 10 dB, and minimum insertion loss about 2 dB. The performance did not degrade up to 600-mV input signal.

REFERENCES

1. White, J. F., *Microwave Semiconductor Engineering*, Van Nostrand, Princeton, N.J., 1982, Chap. 2.
2. Ayasli, Y., "Microwave Switching with GaAs FET's," *Microwave J.*, Vol. 25, Nov. 1982, pp. 61–74.
3. Sokolov, V., et al., "A Ku-Band GaAs Monolithic Phase Shifter," *IEEE Trans. Microwave Theory Tech.*, Vol. MTT-31, 1983, pp. 1077–1082.
4. Pengelly, R. S., *Microwave Field-Effect Transistor-Theory, Design and Applications*, Research Series Press (Wiley), New York, 1982, Chap. 9.
5. McLevige, M. V., and V. Sokolov, "Resonated GaAs FET Devices for Microwave Switching," *IEEE Trans. Electron Devices*, Vol. ED-28, Feb. 1981, pp. 198–204.
6. Chorney, P., "Multi-octave, Multi-throw, *p-i-n* Diode Switches," *Microwave J.*, Vol. 17, Sept. 1974, pp. 39–42, 48.
7. Garver, R. V., "*PIN* Diode Switches: Speed vs. Power," *Microwave J.*, Vol. 21, Feb. 1978, p. 53.
8. Barratt, C., et al., "New GaAs *p-i-n* Diodes with Lower Dissipation Loss, Faster Switching Speed at Lower Drive Power," *IEEE MTT-S Int. Microwave Symp. Digest*, 1983, pp. 507–509.
9. Garver, R. V., *Microwave Diode Control Devices*, Artech House, Dedham, Mass., 1976, pp. 63–65.
10. Stark, L., et al., "Phase Shifters for Arrays," in *Radar Handbook*, M. I. Skolnik (Ed.), McGraw-Hill, New York, 1970, pp. 12.1–12.65.
11. Bahl, I. J., and K. C. Gupta, "Design of Loaded-Line *p-i-n* Diode Phase Shifter Circuits," *IEEE Trans. Microwave Theory Tech.*, Vol. MTT-28, Mar. 1980, pp. 219–224.
12. Atwater, H. A., "Circuit Designs of the Loaded-Line Phase Shifter," *IEEE Trans. Microwave Theory Tech.*, Vol. MTT-33, July 1985, pp. 626–634.
13. Howe, H., *Stripline Circuit Design*, Artech House, Dedham, Mass., 1974.
14. Trivedi, D. K., "Design, Performance and Optimization of Reflection-type *pin* Diode Phase Shifters," Ph.D. Thesis, Indian Institute of Technology, Kanpur, 1981.
15. Eking, R. B., "On the Design of Impedance Transforming Directional Couplers," *IEEE Trans. Microwave Theory Tech.*, Vol. MTT-19, April 1971.
16. Ayasli, Y., et al., "Wideband S-C Band Monolithic Phase Shifter," *IEEE Monolithic Microwave and Millimeter Wave Circuit Symp. Digest.* (San Francisco, May 1984), pp. 11–13.
17. Tsironis, C., and P. Harrop, "Dual-Gate MESFET Phase Shifter with Gain at 12 GHz," *Electron. Lett.*, Vol. 16, July 1980, pp. 553–554.
18. Vorhaus, J. L., et al., "Monolithic Dual-Gate FET Digital Phase Shifters," *IEEE Trans., Microwave Theory Tech.*, Vol. MTT-30, 1982, pp. 991–992.
19. Gazit, Y., and H. C. Johnson, "A Continuously Variable Ku-Band Phase/Amplitude Control Module," *IEEE MTT-S International Microwave Symp. Digest*, 1981, pp. 436–438.

20. Hwang, Y., Y. Chen, and R. Naster, "A Microwave Phase and Gain Controller with Segmented Dual Gate MESFETs in GaAs MMIC," *IEEE Microwave and Millimeter Wave Circuits Symp. Digest* (San Francisco, 1984), pp. 1–5.
21. Leenov, D., "The Silicon *p-i-n* Diode as a Microwave Radar Protector at Megawatt Level," *IEEE Trans. Electron Devices*, Vol. ED-11, 1964, pp. 53–61.
22. Hewlett-Packard, "An Attenuator Design Using *p-i-n* Diodes," Application Note No. 912.
23. Tajima, Y., et al., "GaAs Monolithic Wideband (2–18 GHz) Variable Attenuators," *IEEE MTT-S Int. Microwave Symp. Digest*, 1982, pp. 479–481.

PROBLEMS

12.1 An SPST switch employs two *pin* diodes in the series-shunt configuration shown below. Calculate the insertion loss and isolation at 6 GHz when two *pin* diodes have $C_j = 0.1$ pF, $R_f = 1\ \Omega$, and $L_{int} = 0.3$ nH. Package capacitances may be ignored.

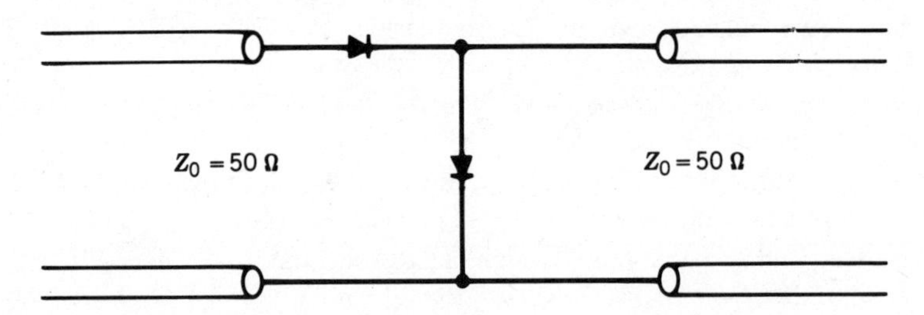

Compare the results with the switch performance when only one diode is used in (a) series configuration, and (b) shunt configuration.

12.2 An SPST switch uses two GaAs MESFETs (in passive mode) mounted in shunt across a transmission line ($Z_0 = 50\ \Omega$) and spaced a quarter wavelength apart as shown in the figure below. The switch is designed for a center frequency of 10 GHz where the MESFET equivalent circuit parameters are $R_{on} = 2.7\ \Omega$, $R_{off} = 3$ kΩ, and

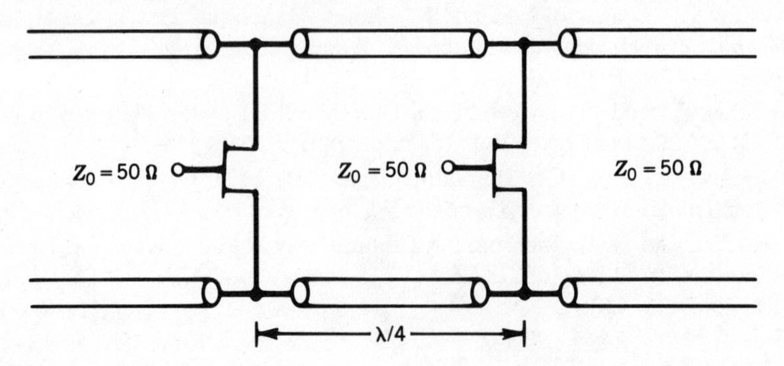

$C_{off} = 0.25$ pF. Compare the performance with shunt-mounted switch designs when (a) only one MESFET is used, and (b) two MESFETs are used at the same location (in parallel).

12.3 Evaluate and plot the response (insertion loss and isolation) of the switching circuit in Problem 12.2 as a function of frequency (from 7.5 GHz to 12.5 GHz). Assume the MESFET equivalent circuit to be valid over this frequency range. How is this performance modified when the high-impedance state capacitances of the MESFET are compensated by appropriate shunt inductances?

12.4 A *pin* diode chip is mounted in shunt across a 50-Ω microstrip line with two bonding tapes as shown in Fig. 12.5. The length of the bonding tapes and hence their inductance L can be adjusted. Find the value of the inductances needed to optimize the insertion loss at 6 GHz. In reverse bias the *pin* diode chip may be represented by a capacitance $C_j = 0.1$ pf.

12.5 An SP3T switch is designed using *pin* diode chips ($C_j = 0.1$ pF, $R_f = 1\ \Omega$, $R_r = 4\ \Omega$, and $L_{int} = 0.3$ nH). The center design frequency is 3 GHz. Calculate and plot switch performance over the frequency range 2 to 4 GHz for (a) series configuration and (b) shunt configuration.

12.6 An SP3T switch is designed using three GaAs MESFETs (passive mode) in series configuration. The MESFETs may be modeled by $R_{on} = 2.7\ \Omega$, $R_{off} = 3$ kΩ, and $C_{off} = 0.25$ pF. Calculate the insertion loss and isolation at an operating frequency of 10 GHz when all the transmission lines have a characteristic impedance of 50 Ω each.

12.7 Calculate how the performance of the SP3T switch in Prob. 12.6 is modified when:

(a) MESFET chips are used in a hybrid MIC circuit; and, in addition to the MESFET parameters specified earlier, we have a series inductance of 0.2 nH because of bonding wires.

(b) MESFET capacitance in the high impedance state is compensated by a parallel inductance of suitable value at 10 GHz. (Bonding wire inductances are not present in this case.)

12.8 A 180° single-bit switched-line phase shifter is designed using two SPDT switches employing two MESFETs each in series configurations. The MESFETs may be modeled by $R_{on} = 2.7\ \Omega$, $R_{off} = 3$ kΩ, and $C_{off} = 0.25$ pF. The design frequency is 10 GHz and the electrical length of the smaller of the switched lines is taken to be 30°.

Plot the differential phase-shift and insertion loss (for both the states) as a function of frequency from 7.5 to 12.5 GHz. Do you observe any off-path resonances? Compare the phase shift response with the performance you would expect if the SPDT switches were ideal. (Any available microwave circuit analysis program may be used for this problem.)

12.9 A main-line mounted loaded-line–type phase shifter is designed for operation at 10 GHz using *pin* diode chips. These diodes can be modeled by $R_f = 1\ \Omega$ and $L_{\text{int}} = 0.3$ nH in forward bias and by a series combination of $L_{\text{int}} = 0.3$ nH, $R_r = 4\ \Omega$, and $C_j = 0.1$ pF in reverse bias. The two diodes are spaced 90° apart and the characteristic impedance of the lines at external ports is Z_0. Find the value of extra inductance that needs to be added and the line impedance Z_0 needed to obtain a differential phase shift of 22.5°. (R_f and R_r may be ignored for this design.)

12.10 Use any available microwave circuit analysis program to plot the frequency response (VSWR, $\Delta\phi$, and insertion loss) of the phase shifter designed in Problem 12.9 over a frequency range of 8.0 to 12.0 GHz. Include the diode resistances in two bias states and recompute the phase-shifter performance in the above-mentioned frequency range in order to observe the effect of diode resistances.

12.11 A loaded-line phase shifter is designed using two stub mounted MESFETs (in passive mode). The MESFETs may be represented by $R_{\text{on}} = 2.7\ \Omega$ in the low-impedance state and $C_{\text{off}} = 0.25$ pF in the high-impedance state. Find the stub length and impedance for a 22.5° phase-shifter bit to operate at 10 GHz. The spacing between stubs is 90° and the lines at two ports have $Z_0 = 50\ \Omega$. R_{on} can be ignored.

12.12 Use any available microwave circuit analysis program to plot the frequency response (VSWR, $\Delta\phi$, and insertion loss) of the phase shifter designed in Problem 12.11 over a frequency range of 8.0 to 12.0 GHz. Repeat these computations by including R_{on} in the low-impedance state and compare the results.

12.13 A reflection-type phase shifter is designed using *pin* diode chips with $R_f = 0.4\ \Omega$, $R_r = 0.5\ \Omega$, $C_j = 1$ pF, and $L_{\text{int}} = 0.3$ nH. The design frequency is 3.18 GHz. Calculate the value of Z_m (the equivalent impedance the *pin* diodes should look into) (a) when the desired phase shift $\Delta\phi$ is 180° and (b) when $\Delta\phi$ is 90°. (Effect of diodes resistances may be ignored.)

12.14 Equation 12.55 for Z_m is valid for lossless devices only. Derive a corresponding expression for the case when the device resistances cannot be neglected; that is, in the low-impedance state $Z_f = R_f + jX_f$, and in the high-impedance state $Z_r = R_r + jX_r$.

12.15 Consider the reflection phase-shifter design of Problem 12.13. Repeat the calculations for Z_m taking the effect of diode resistances into account. (Use the expression Z_m derived in Problem 12.14.)

12.16 When reflection-type phase shifters are designed in monolithic configuration, bonding wire inductance is not present. These monolithic devices can be represented by $Z_1 = R_1$ in the low-impedance state and by $Z_h = R_h + jX_h$ in the high-impedance state. The expression for Z_m derived in Problem 12.14 can be used in this case also. Find the values of Z_m for $\Delta\phi = 180°$ and $\Delta\phi = 90°$ when $Z_1 = 0.4\ \Omega$ and $Z_h = (0.5 - j50)\ \Omega$.

12.17 Design a 180° reflection phase shifter using a quarter-wave transforming section, a two-branch hybrid, and *pin* diode chips with $C_j = 1$ pF and $L_{\text{int}} = 0.3$ nH. The center frequency for design is 3.0 GHz. Diode resistances may be ignored. Simulate the circuit using any available microwave circuit analysis program and compute $\Delta\phi$, VSWR, and insertion loss (in two states) for frequencies ranging from 2.5 to 3.5 GHz (in 0.1-GHz steps). The effect of discontinuity reactances may be ignored. Take the substrate parameters as $\epsilon_r = 2.5$, $h = 0.8$ mm, and $t = 18\ \mu$m.

12.18 Consider the case when the diodes used in the design of Problem 12.17 are lossy with $R_f = 0.4$ and $R_r = 0.5\ \Omega$. How will the performance of the phase shifter ($\Delta\phi$, VSWR, and insertion loss over 2.5 to 3.5 GHz) be modified when the design remains unchanged? Use any available microwave circuit analysis program for these computations.

12.19 Design a 180° reflection phase shifter using a two-branch impedance transforming hybrid and *pin* diodes with $C_j = 1$ pF and $L_{\text{int}} = 0.3$ nH. The design frequency is 3.18 GHz. The effect of diode resistances may be neglected. Simulate this circuit using any available microwave circuit analysis program (microstrip configuration with $\epsilon_r = 2.5$, $h = 0.8$ mm, and $t = 18\ \mu$m) and compare its performance with the design of Problem 12.17.

13 FREQUENCY MULTIPLIERS AND DIVIDERS

13.1 INTRODUCTION

This chapter concerns the realization of microwave circuits that can perform the functions of frequency multiplication and division. These functions can be used to achieve design goals unattainable by any other means and are quite different from the analog operations of frequency addition and subtraction that are performed by mixers.

13.1.1 Basics of Frequency Multiplication and Division

Several different nonlinear electrical phenomena can be used to achieve frequency multiplication or frequency division; some phenomena can be used for both purposes. In general, frequency multiplication is easily achieved, but it is usually necessary to satisfy more rigorous conditions in order to obtain frequency division. Thus whereas a given frequency divider can often also be made to operate as a multiplier, the converse is not necessarily true.

The simplest frequency-multiplying device is the nonlinear resistor, or varistor, in which there is a nonlinear static relationship between the dc current and voltage. Such behavior can be displayed at RF by *pn* junction diodes or up to a few gigahertz by low-parasitic Schottky diodes.

Microwave bipolar junction transistors (BJTs) and gallium arsenide field-effect transistors (GaAs FETs) can be operated as three-terminal nonlinear-resistance frequency-multiplying devices.

For frequency multiplication by an integer N, using a two-terminal (passive) positive nonlinear resistor, the maximum possible efficiency is $1/N^2$ [1]. In a three-terminal nonlinear-resistance frequency multiplier, however, such as a class-C BJT or GaAs FET amplifier with the output resonator tuned to the Nth harmonic of the input frequency, multiplication with gain is possible [2].

It can be shown [3] that it is impossible to obtain frequency division from a positive passive two-terminal nonlinear resistor. On the other hand, frequency division *is* possible using a two-terminal device exhibiting static negative resistance, such as a tunnel diode. Three-terminal nonlinear-resistance devices can also be made to operate as frequency dividers, even with gain, when used in conjunction with suitable feedback networks.

Two-terminal nonlinear *reactance* devices, however, can be used to realize both parametric frequency multipliers and dividers. In some instances the same circuit topology can actually be employed for both purposes. Manley and Rowe [4] have proved that both multipliers and dividers based on two-terminal nonlinear reactances have a maximum possible efficiency of 100%.

The classical nonlinear-reactance microwave device is the varactor, which is typically a Si or GaAs junction diode designed to exploit the voltage-dependent depletion-layer capacitance, which behaves as a high-Q nonlinear reactance under reverse-bias conditions. In the case of the charge storage, or step-recovery diode (SRD), the significant nonlinearity is the diffusion capacitance, which becomes important under forward-bias conditions. In all varactor devices having voltage-dependent capacitive reactances, the underlying mechanism is a nonlinear charge-voltage relationship.

Another effect that can be used to achieve either frequency multiplication or division is the frequency entrainment (or synchronization) of an oscillator by, respectively, a subharmonic or harmonic injection-locking signal. Multipliers and dividers based on this effect are referred to as injection-locked-oscillator (ILO) designs.

In addition to the varactor, SRD, and ILO types, there are at least three other kinds of frequency divider. In the regenerative, or Miller, divider the input is applied to the local oscillator (LO) port of a mixer, while the IF output is amplified and fed back to the RF port. With appropriate filtering in the feedback loop, frequency division by two can be obtained. The properties of traveling space-charge domains in transferred-electron devices (TEDs) also makes frequency division possible. Finally, both static and dynamic digital circuits can be used as wide-band microwave frequency dividers.

13.1.2 Applications

This section reviews some applications of frequency multipliers and frequency dividers.

Applications of Frequency Multipliers. An important application of frequency multipliers is in the generation of microwave power at levels above those obtainable with fundamental-frequency oscillators. Figure 13.1(a) indicates the CW and pulsed power obtainable with various solid state sources; Fig. 13.1(b) shows the corresponding powers for tube-type sources. Powers obtainable with multipliers are also indicated in Fig. 13.1(a). From this diagram it can be seen that multipliers are able to extend effective solid state source powers either under pulsed conditions at frequencies below ~5 GHz, (e.g., for weather radars [5]) or under CW conditions at frequencies above ~200 GHz. However, for those ap-

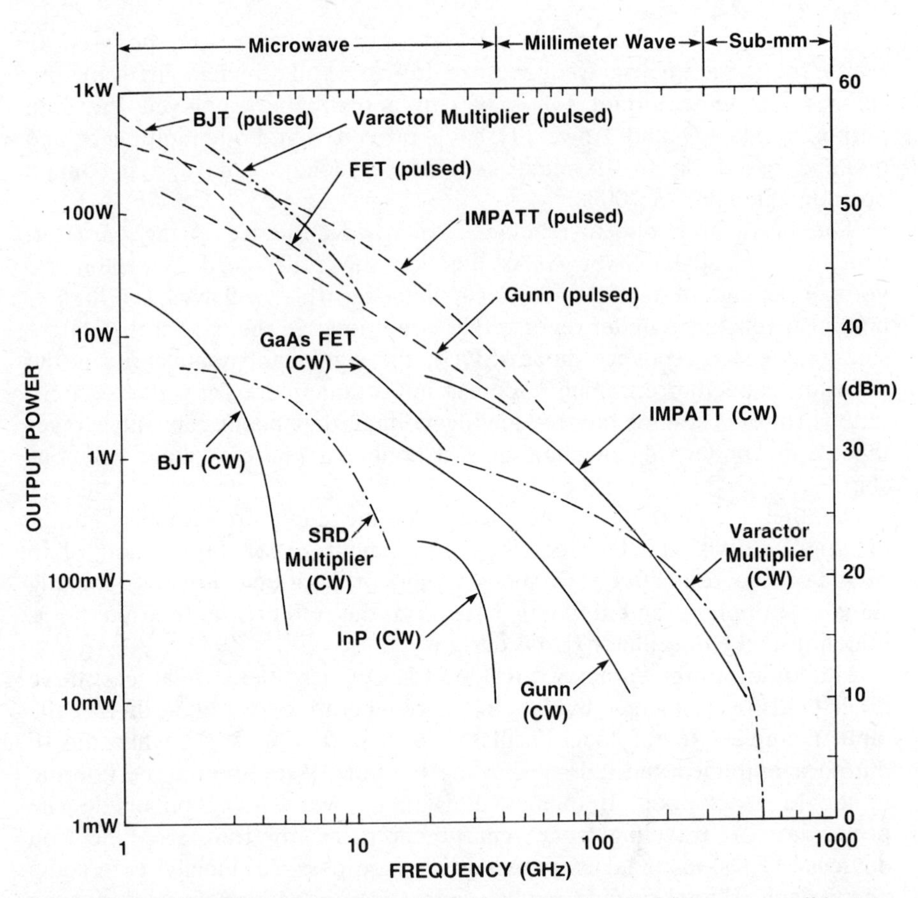

Figure 13.1 (*a*) Maximum power available from fundamental-frequency solid state sources and from frequency multipliers.

plications where the heat dissipation and/or noise of IMPATT oscillators cause problems, frequency multipliers are useful up to ~20 GHz in pulsed mode and from ~20 GHz upward in CW mode. The region between 1 and 15 GHz, formerly occupied by step-recovery-diode (SRD) multipliers, is now usurped by CW Gunn oscillators. At submillimeter frequencies efficient varactor multipliers in conjunction with lower frequency sources offer a good alternative to large and expensive primary sources, such as backward-wave oscillators (BWOs), carcinotrons, or optically pumped gas lasers (which are not tunable). Significant powers at submillimeter frequencies are needed for plasma diagnostics, spectroscopy, and particularly in local-oscillator applications in heterodyne receivers for radio astronomy.

Schottky-diode varactors have been used to generate milliwatts of power at 600 GHz [6, 7], while their potential performance as varistor-mode triplers to 900 GHz has been investigated theoretically [8].

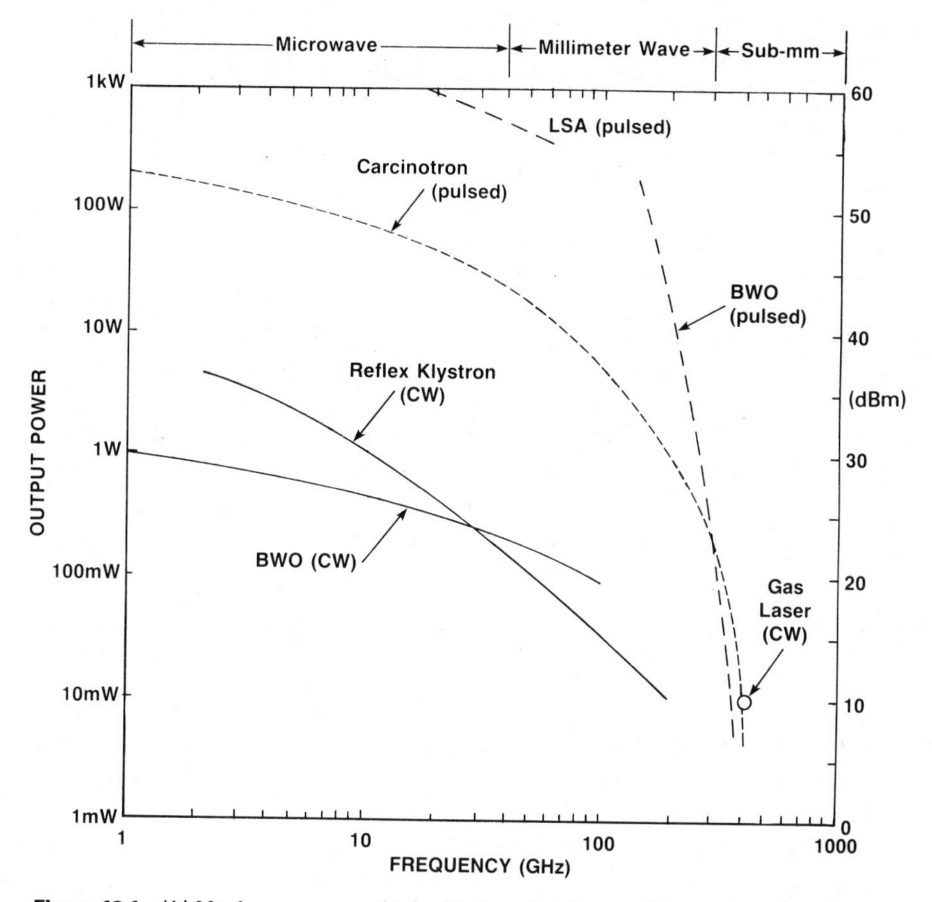

Figure 13.1 (*b*) Maximum power obtainable from fundamental-frequency tube sources.

At intermediate frequencies in the 20–60 GHz range in situations where IMPATTS cannot be used, or where small size and low cost are important in volume production, GaAs monolithic microwave-integrated-circuit (MMIC) multipliers have great potential. For example, the outputs of numerous closely similar devices can be power-combined for transmitter applications. MMIC multipliers are also effective in phased-array radars.

Applications of Frequency Dividers. Since frequency dividers are less familiar microwave components than are multipliers, some basic applications are reviewed here.

Prescalers. Divide-by-N circuits are widely used as prescalers at the front end of microwave frequency counters and digital frequency discriminators as indicated in Fig. 13.2. The maximum input frequency is limited by the capabilities of the first divide-by-two stage.

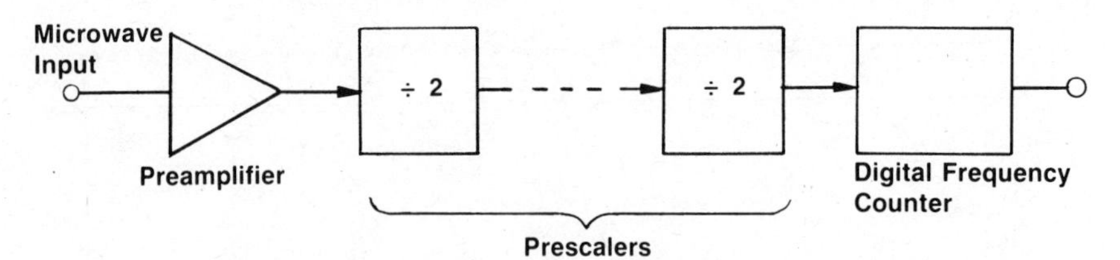

Figure 13.2 Frequency halvers applied to prescaling.

Frequency Translation. Since frequency dividers divide the input *bandwidth* as well as the input *frequency*, another application is in the translation of high-frequency (typically FM) signals to lower frequencies for digital processing. As illustrated in Fig. 13.3(a), fractional bandwidth $\Delta f/f$ is invariant under frequency division, whereas down-conversion using a mixer results in an effective *increase* in $\Delta f/f$. Digital frequency memories represent an application in which frequency division over wide bandwidths is important.

Communications Systems. Frequency dividers improve the performance of microwave communications systems in which frequency stability and low FM noise are important [9]. Conventionally, frequency stability is obtained by locking a voltage-controlled oscillator (VCO) to a stable low-frequency crystal reference using a phase-locked loop (PLL), as shown in Fig. 13.4. To achieve a 10-GHz output a frequency multiplier is needed. The external reference f_r is typically 100 MHz. A sample of the VCO output at f_1, nominally also 100 MHz, is applied to the second input of the phase comparator. If the frequencies f_r and f_1 are the same, the comparator

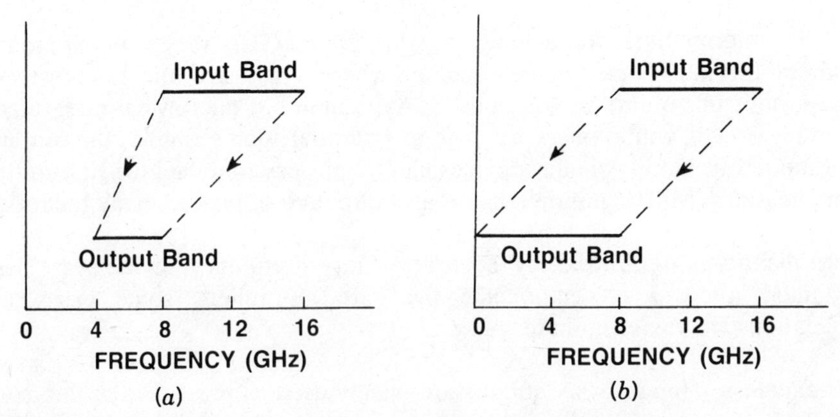

Figure 13.3 (a) Frequency translation by division. Fractional bandwidth $\Delta f/f = 66\%$ at both input and output. (b) Frequency translation by down-conversion; $\Delta f/f = 66\%$ at input; 200% at output.

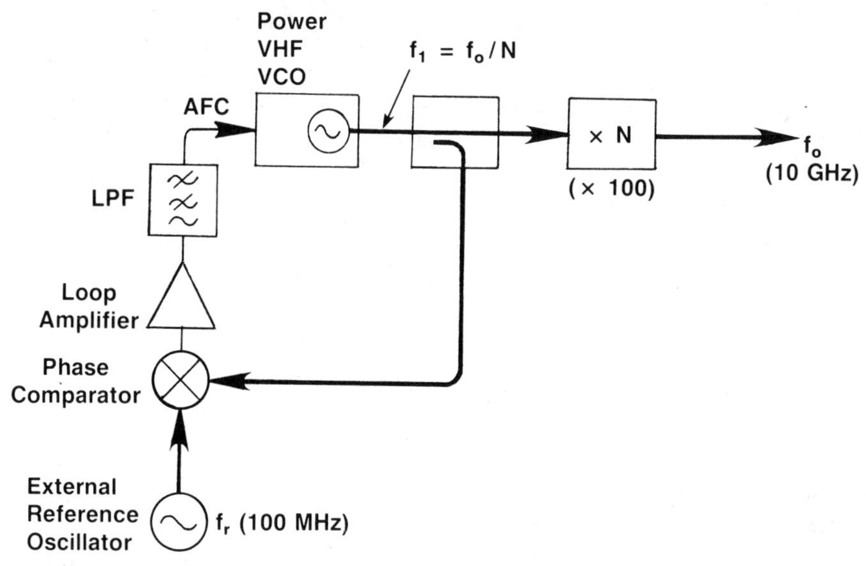

Figure 13.4 Conventional phase-locked loop (PLL) with $\times N$ multiplier to achieve microwave output.

output will be a dc voltage dependent on the phase difference between the two signals. If the frequencies are different, then the comparator output will be an ac signal at the difference frequency. In either event, upon loop lockup, the frequency f_1 will be very nearly equal to f_r. For a 10-GHz output a multiplier with $N = 100$ will be needed. Because of heterodyne mixing that will inevitably occur in the multiplier, the output will also contain upper and lower sidebands at 10.1 and 9.9 GHz. These sidebands will be difficult to filter out. Furthermore, the power in these sidebands will be *enhanced* by a factor of $20 \log N$ due to the multiplication process itself. For $N = 100$, this enhancement would be 40 dB. In addition to this, the multiplier will increase any FM noise on the signal f_1, which will track that of the reference f_r, by a factor N. Finally, the noise figure of the multiplier can contribute to the overall noise figure for frequencies greater than about 1 MHz from the carrier.

An alternative approach is to use a *microwave* VCO and to *frequency divide* a sample of its output f_0 to permit phase comparison with a much lower crystal reference frequency f_r, as indicated in Fig. 13.5. Here a microwave VCO is controlled directly by the reference f_r. A sample of the output frequency f_0, for example, 16 GHz, is frequency divided by a factor $N_m = 2^K$, using a sequence of K microwave frequency halvers. In this instance, $K = 3$, so $N_m = 8$ and the input to the digital frequency divider is 2 GHz, well within the capability of contemporary, commercially available units. Digital frequency division then reduces the frequency by a factor N_d to permit phase-comparison with the reference f_r. Here $N_d = 32$, giving a

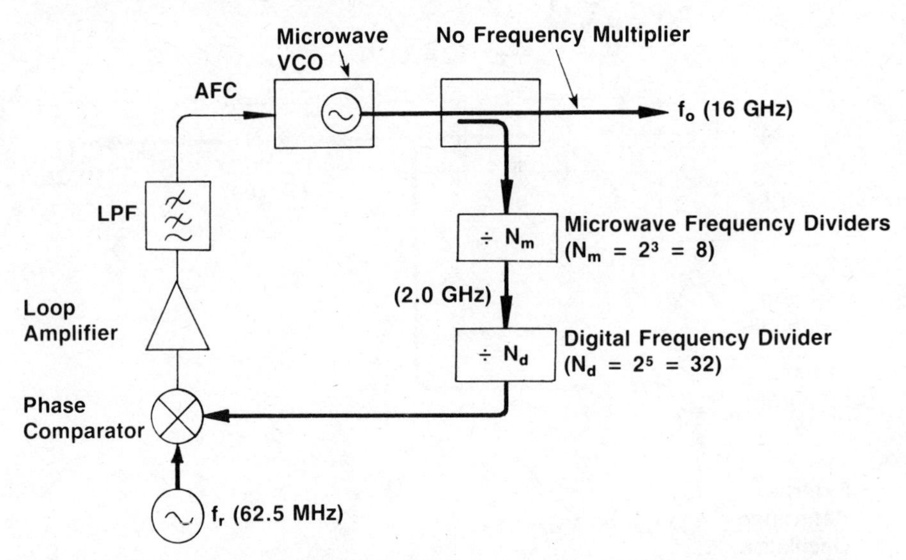

Figure 13.5 Phase-locked loop with direct crystal control of microwave VCO.

total division ratio $N_m N_d = 256$, so that a crystal reference frequency $f_r = 62.5$ MHz would be chosen. The advantage of this approach is that the FM noise characteristics of the output signal f_0 are those of the crystal reference oscillator itself. There is no degradation due to frequency multipliers, since none are used.

Frequency-Modulated PLL. In a conventional phase-locked loop, direct FM modulation of the VCO is possible, as shown in Fig. 13.6(*a*). However, the modulating frequency f_m must be greater than about 1.5 times the 3-dB loop bandwidth [10, 11]; otherwise, the PLL will treat the modulation as an error signal and cancel it out. At the same time, the maximum deviation Δf must not exceed f_m, otherwise the loop can lock to a modulation sideband. Thus the modulation index must be less than 1.0, severely limiting the utility of the conventional system. On the other hand, if FM modulation is applied to a PLL with a divide-by-N frequency divider in the loop [see Fig. 13.6(*b*)], the deviation Δf as well as the output f_0 will be divided by N. This means that the reference frequency can also be divided by N and that the phase comparison can now be done at f_0/N rather than f_0. Consequently the system can now operate correctly when $\Delta f/(Nf_m) < 1.0$, that is, the effective modulation index of the system has been increased by a factor N.

Frequency Counters. It is often necessary to measure the frequency of signals that may consist of short RF pulses and/or may have large FM components. Conventional heterodyne or transfer-oscillator techniques depend on mixing the unknown signal with the Nth harmonic of a local

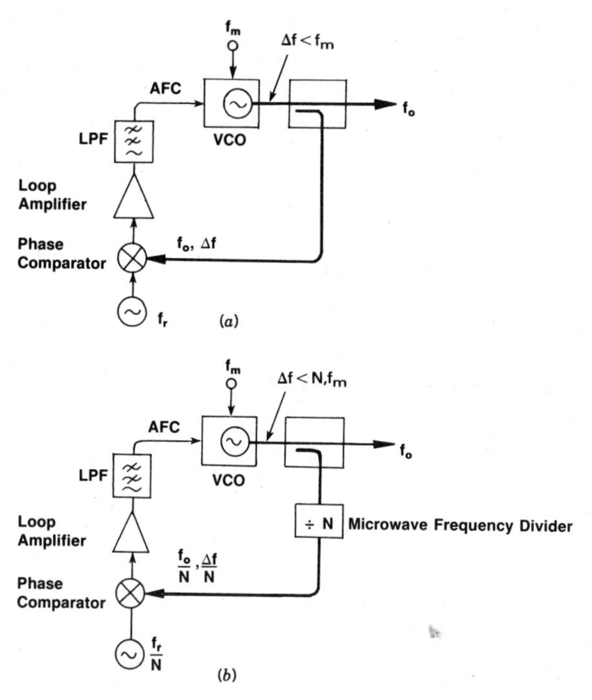

Figure 13.6 Frequency modulation applied to (*a*) conventional phase-locked loop (PLL), and (*b*) PLL with microwave frequency divider.

oscillator. The heterodyne counter measures the resulting difference frequency, which must be sufficiently low, using a direct digital counting circuit. Since the LO frequency and the harmonic number N are both known, the system can calculate and display the input frequency.

In the transfer oscillator approach, the LO frequency is tuned until the frequency difference between the input and the Nth harmonic of the LO is zero. A digital counter then measures the LO frequency. This information, together with knowledge of the harmonic number N, makes it possible to calculate and display the original input frequency.

Heterodyne oscillators are limited in their ability to measure the frequency of pulsed microwave signals. Transfer oscillator techniques, on the other hand, can measure the fundamental frequency of pulsed microwave signals, but have little ability to handle signals with large FM components.

A third alternative is to apply the unknown input signal to a sequence of n divide-by-two stages and to count the resulting output digitally. Since the number n is known, it is possible to calculate and display a frequency equal to 2^n times the value counted, that is, the original input frequency. If microwave frequency halvers with good RF pulse response and FM capability are used, for example, the varactor halvers to be discussed in Section 13.3.2, the disadvantages of both the heterodyne and transfer-oscillator techniques can be overcome.

Phase-Noise Reduction. In principle, the phase noise of a microwave reference source can be reduced by a factor of 20 log N by following it by a divide-by-N stage. In practice there will be some phase-noise degradation at very small offsets from the reference frequency.

The phase noise incurred in translating a reference frequency f_r down to a lower frequency f_r/N will be compared for the cases of down-conversion by mixing and by frequency division. For down-conversion by mixing, as in Fig. 13.7(a), it is assumed that the reference oscillator voltage is

$$v_r(t) = V_r \sin(\omega_r t + \psi \sin \omega_n t) \tag{13.1}$$

where

V_r = peak amplitude of reference oscillator signal,
$\omega_r = 2\pi f_r$ = reference-oscillator angular frequency,
ψ = peak amplitude of phase noise,
ω_n = angular frequency of the phase noise that frequency modulates ω_r.

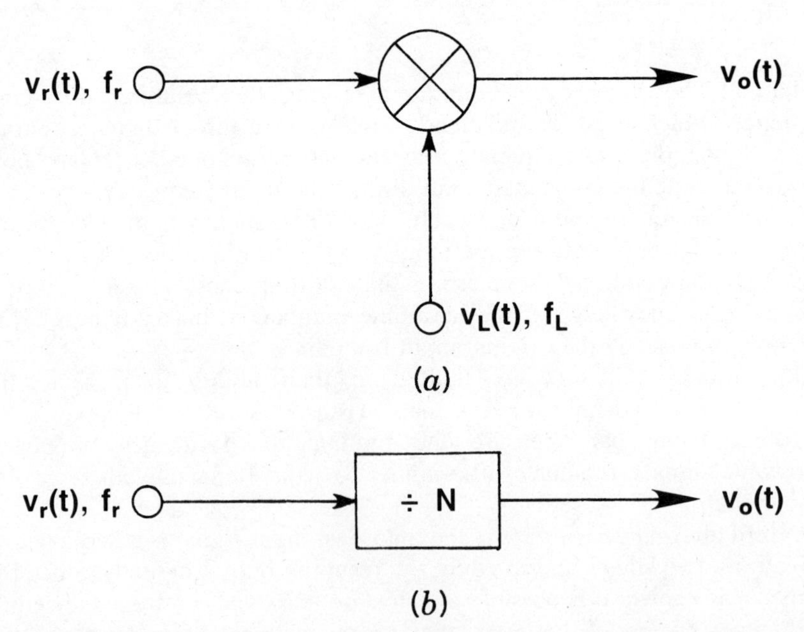

Figure 13.7 (a) Down-conversion with ideal balanced mixer. (b) Down-conversion with frequency divider.

If it is assumed that the LO signal is

$$v_L(t) = V_L \sin \omega_L t \tag{13.2}$$

where

V_L = peak LO amplitude,
ω_L = LO angular frequency,

and that the output signal of interest is given by

$$v_o(t) = v_r(t) \cdot v_L(t) \tag{13.3}$$

then

$$v_o(t) = \frac{1}{2} V_r V_L [\cos\{(\omega_r - \omega_L)t + \psi \sin \omega_n t\} - \cos\{(\omega_r + \omega_L)t + \psi \sin \omega_n t\}] \tag{13.4}$$

To achieve the design objective, the LO frequency is chosen to be $\omega_r(1 - 1/N)$, which makes the lower sideband frequency

$$\omega_r - \omega_L = \frac{\omega_r}{N} \tag{13.5}$$

If all the phase noise is represented by $\psi \sin \omega_n t$, then the lower sideband signal

$$v_{oL}(t) = \frac{1}{2} V_r V_L \cos\left\{\left(\frac{\omega_r}{N}\right) t + \psi \sin \omega_n t\right\} \tag{13.6}$$

has a worst case timing error of $N\psi/\omega_r$. On the other hand, when the down-conversion is done by frequency division [see Fig. 13.6(*b*)], with the input signal still as given in Equation 13.1, the output signal becomes

$$v_o(t) = V_o \sin\left\{\left(\frac{\omega_r}{N}\right) t + \left(\frac{\psi}{N}\right) \sin \omega_n t\right\} \tag{13.7}$$

where a possible fixed phase shift has been ignored. In this case the peak timing error is only ψ/ω_r, that is, N times less.

An Application Involving Simultaneous Use of Multipliers and Dividers. To demodulate suppressed-carrier phase-shift-keyed (PSK) signals, an artificial carrier must be synthesized from the PSK signals themselves. One scheme for doing this in the case of a biphase signal is shown in

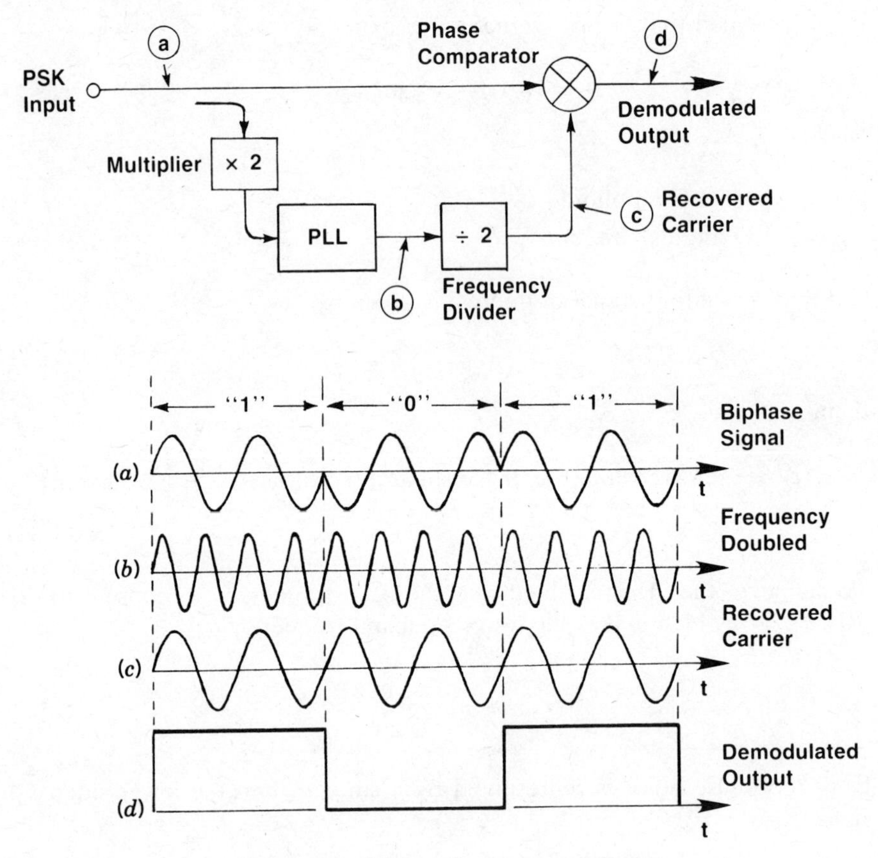

Figure 13.8 PSK carrier-recovery scheme.

Fig. 13.8. A sample of the PSK signal (a) is multiplied by 2 to yield a frequency-doubled signal in which the phase coding has been eliminated. To remove sidebands due to the phase switching, this signal is passed through a tracking PLL that acts as a very narrow-band filter and produces the sinusoidal signal (b). Next, it is frequency divided by 2 to yield a *recovered carrier* (c) at the original input frequency. Finally, the input PSK signal and the recovered carrier are fed to the phase comparator. This produces an output dc level (d) that depends on whether the input phases are equal or opposite, that is, it produces a demodulated output. Similar schemes are possible for 4-, 8-, 16-phase, etc., PSK signals. In such cases multiplication and division by 4, 8, 16, etc., would be used. Note that in conventional systems, demodulation is done at an IF frequency in the 10–70-MHz range. By using appropriate frequency division and multiplication techniques microwave PSK signals can be demodulated directly.

13.2 FREQUENCY MULTIPLICATION

13.2.1 Types of Multipliers

Frequency multiplication can be achieved in at least six different ways:

1. harmonic generation due to the static nonlinear V–I relationship of varistor diodes,
2. parametric multiplication using the nonlinear reactance of varactor diodes,
3. harmonic generation using step-recovery diodes (SRDs),
4. simultaneous harmonic generation and gain in bipolar junction transistors (BJTs), typically class-C amplifiers,
5. multiplication with gain in GaAs FETs,
6. oscillators injection locked at a submultiple of the natural frequency.

Item (3) formerly received considerable attention [12–20], as did item (4) [21–26]. However, Fig. 13.1 shows that SRD and BJT multipliers are now less important than they were, and so they are not discussed here.

13.2.2 Passive Multipliers Using Diodes

Theory of Nonlinear-Resistance Multipliers. Pantell [27], Page [28], and Clay [29] show that for a positive nonlinear resistor (varistor), such that the voltage v is a single-valued function of i and such that $\partial i/\partial v \geq 0$, for all v, the power P_N generated at the Nth harmonic of an input frequency f_{in} is related to the power P_1 absorbed at f_{in} by

$$\frac{P_N}{P_1} \leq \frac{1}{N^2} \tag{13.8}$$

Thus resistive doublers and triplers have maximum possible efficiencies of 25% and 11.1%, respectively. This is a fundamental limitation on *all* multipliers using positive nonlinear resistors, including the ideal exponential diode:

$$i = I_s\left[\exp\left(\frac{ev}{nKT}\right) - 1\right] \tag{13.9}$$

Page [1] demonstrates that the preceding result even applies to the ideal lossless rectifier of Fig. 13.9, which is often (erroneously) assumed to provide 100% multiplier efficiency.

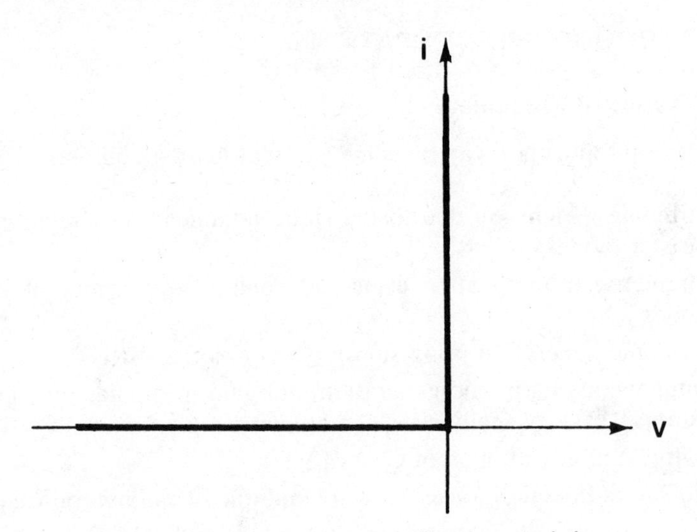

Figure 13.9 Ideal lossless rectifier characteristic.

A harmonic generator using the exponential diode of Equation 13.9 with a series resistance R_s has been analyzed [30]. The result is shown in Fig. 13.10. For $N > 4$ the conversion loss increases with N much faster than is predicted by Equation 13.8.

Clay [29] shows that for odd-order harmonic generation the optimum nonlinear resistor, which gives an efficiency of $1/N^2$, has the V–I charac-

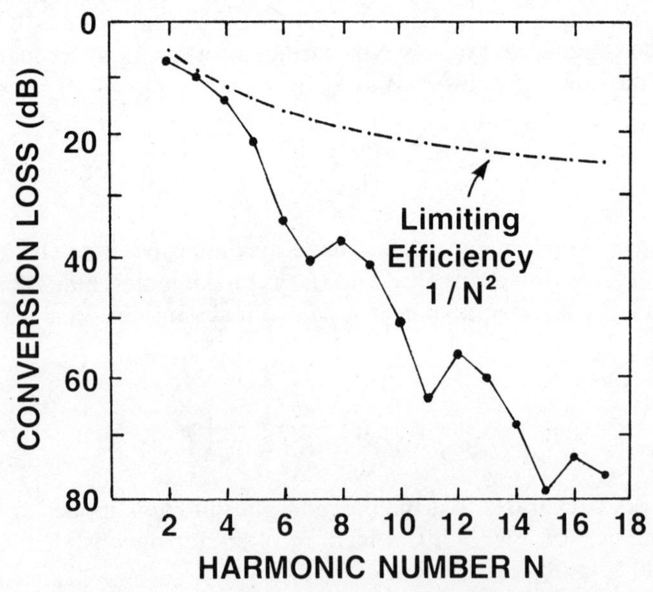

Figure 13.10 Efficiency of multiplier vs. harmonic number, exponential-law analysis. The dots are connected for clarity. Parameters: $nkT/e = 25$ mV, $I_s = 1$ μA, $R_s = 100$ Ω. (After Benson and Winder [30], reprinted with permission of IEE.)

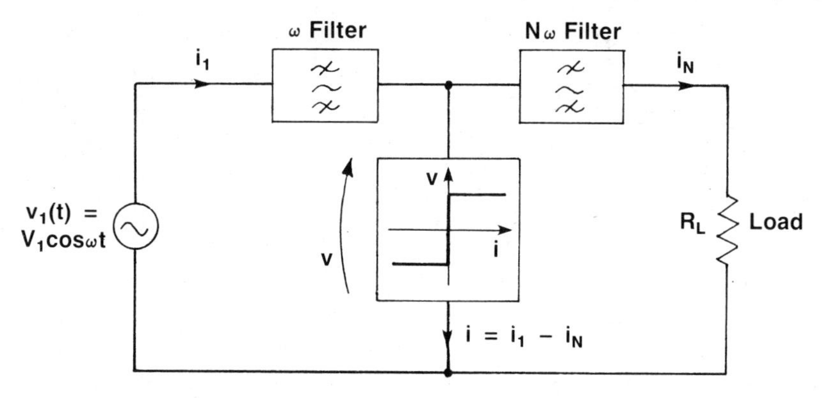

Figure 13.11 Odd-order harmonic generation using maximum-efficiency nonlinear resistor. (After R. Clay [29], reprinted with permission of Wiley-Interscience.)

teristic of Fig. 13.11, that is, it is an ideal limiter. As will be seen in Section 13.2.4, this conclusion is relevant to the operation of GaAs FET multipliers.

To estimate the ultimate performance of varistor multipliers Benson and Frerking [8] have investigated the theoretical efficiency of GaAs Schottky-diode triplers for $f_{in} = 300$ GHz, with the result shown in Fig. 13.12. To

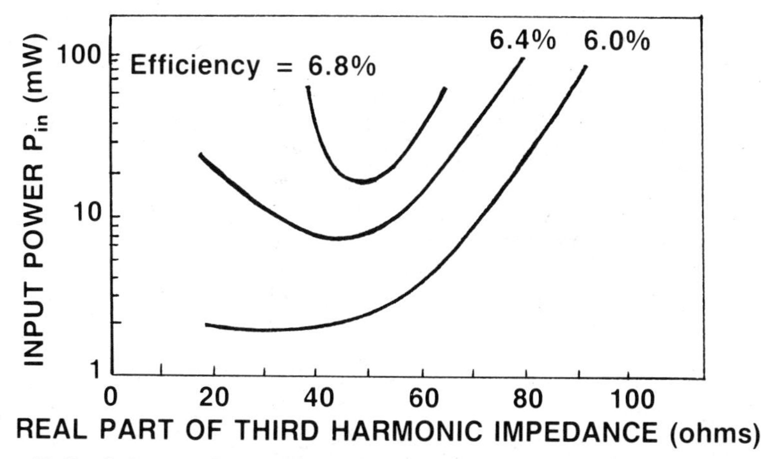

Figure 13.12 Contours of constant computed tripler efficiency as a function of input power P_{in} and the real part of the third-harmonic embedding impedance. The bias current I_0 is chosen at each point to optimize efficiency. The assumed Schottky diode parameters are:

$$R_S = 10\ \Omega, \quad I_s = 10^{-6}\ \text{A}, \quad C_j = 1.0\ \text{fF}$$

$$\phi_0 = 1\ \text{V}, \quad \gamma = 0, \quad n = 1.2$$

(After K. Benson and M. A. Frerking [8], 1985 IEEE reprinted with permission.)

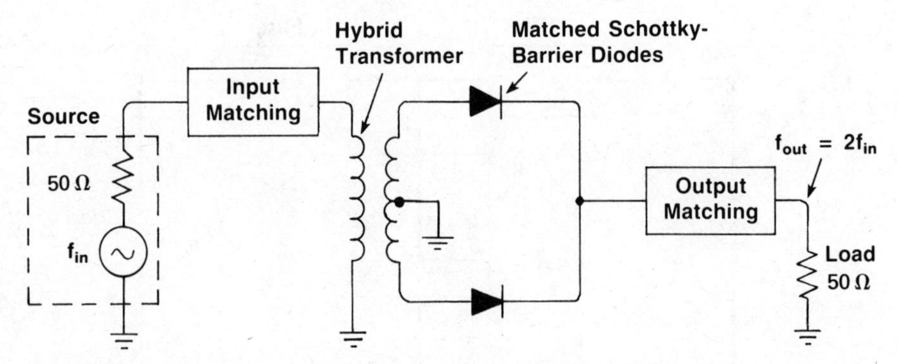

Figure 13.13 Equivalent circuit of wide-band resistive frequency doubler using Schottky-barrier diodes in a balanced configuration.

suppress voltage-dependent depletion capacitance effects due to

$$C_j(v) = C_j(0)\left(1 - \frac{v}{\phi_0}\right)^{-\gamma} \tag{13.10}$$

the exponent γ is set to zero, so that the device resembles a Mott diode. The maximum efficiency is about 7%, compared with 11.1% for an ideal tripler.

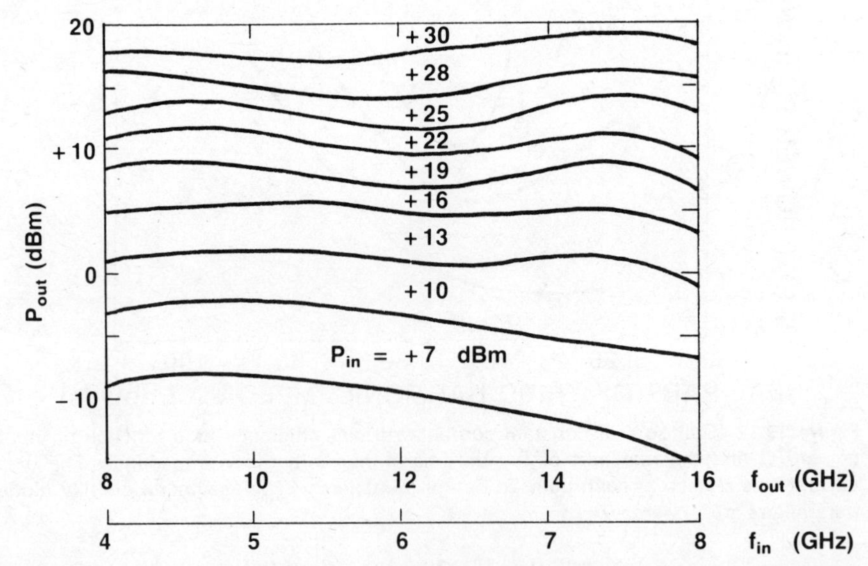

Figure 13.14 Typical frequency response of Schottky-diode doubler (type RX16000) for a range of values of P_{in}. (Courtesy of TRW Microwave Inc.)

Practical Nonlinear-Resistance Multipliers. Figure 13.13 is the schematic of a balanced Schottky-diode resistive doubler that suppresses odd harmonics. The 180° hybrid could be a wirewound transformer for input frequencies $f_{in} < 500$ MHz, or a four-port magic-T at microwave frequencies. Frequency-response curves for an 8–16 GHz unit are given in Fig. 13.14. The best midband conversion efficiency is 9%, (−10.5 dB), compared with the theoretical maximum of 25% (−6 dB) given by Equation 13.8.

13.2.3 Parametric Multipliers Using Varactors

Theory of Nonlinear-Reactance Multipliers. The Manley–Rowe equations [4] show that for parametric nonlinear-reactance multipliers, the power P_N generated at the Nth harmonic of f_{in} is related to the power P_1 absorbed at f_{in} by

$$\frac{P_N}{P_1} \leq 1 \tag{13.11}$$

that is, the maximum possible efficiency is 100%.

At microwave frequencies, practical nonlinear reactances are varactor diodes, the physical details of which have been given in Chapter 8. The simplified nonlinear equivalent circuit of Fig. 13.15 can model both *pn* junction and Schottky varactors as well as SRDs and is similar to the diode model in the SPICE program [31]. Here $C_j(v)$ represents the depletion capacitance, dominant for $v < 0$, as given by Equation 13.10, while $C_d(v)$ represents the diffusion capacitance, dominant under forward bias $v > 0$. The large-signal behaviors of $C_j(v)$ and $C_d(v)$ are best represented by the junction charge

$$q_j(v) = \frac{C_j(0)\phi_0}{\gamma - 1}(\phi_0 - v)^{1-\gamma} \tag{13.12}$$

and the diffusion charge

$$q_d(v) = \tau_L I_s \left[\exp\left(\frac{ev}{nkT}\right) - 1\right] \tag{13.13}$$

respectively, where

ϕ_0 = built-in potential barrier,
γ = capacitance-law exponent,
$C_j(0)$ = depletion capacitance at zero bias,

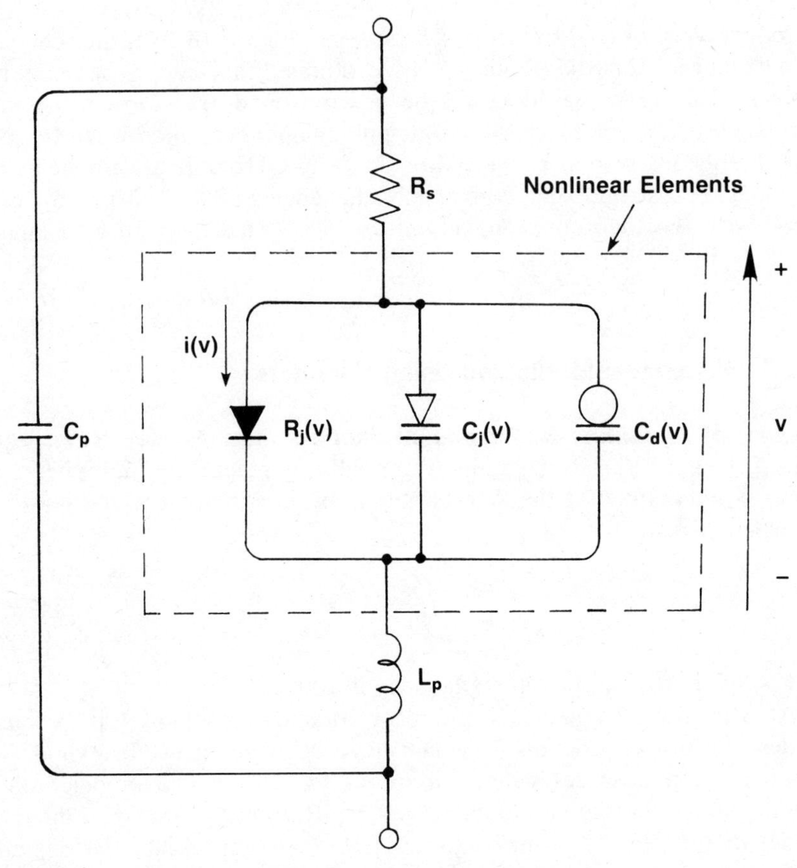

Figure 13.15 Nonlinear equivalent circuit applicable to *pn* junction, Schottky-barrier, and step-recovery varactors. Breakdown effects are not included.

τ_L = average lifetime of injected minority carriers,
I_s = diode reverse saturation current,
e = charge on an electron (1.602×10^{-19} C),
k = Boltzmann's constant,
T = absolute temperature in degrees Kelvin,
n = diode ideality factor.

Conduction current is represented by $R_j(v)$ and is given by Equation 13.9.

For Schottky and abrupt *pn* junction varactors $\gamma \simeq \frac{1}{2}$. The $q_j(v)$ relation of Equation 13.12 can then be expressed in a simple inverse form:

$$v = \phi_0 \left\{ 1 - \left[\frac{q_j(v)}{q_j(0)} \right]^2 \right\} \tag{13.14}$$

where

$$q_j(0) = -2C_j(0)\phi_0 \tag{13.15}$$

is the zero-bias depletion charge. The quadratic $v(q_j)$ relation of Equation 13.14 greatly simplifies the analysis of parametric varactor multipliers and dividers. It is depicted in Fig. 13.16(*a*), while Fig. 13.16(*b*) shows $C_j(v)$.

A simple expression for the diffusion charge is found by comparing

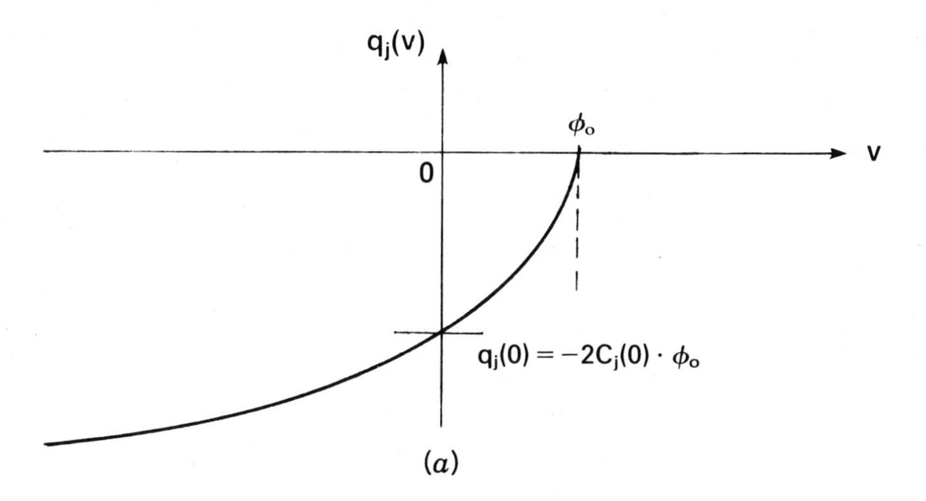

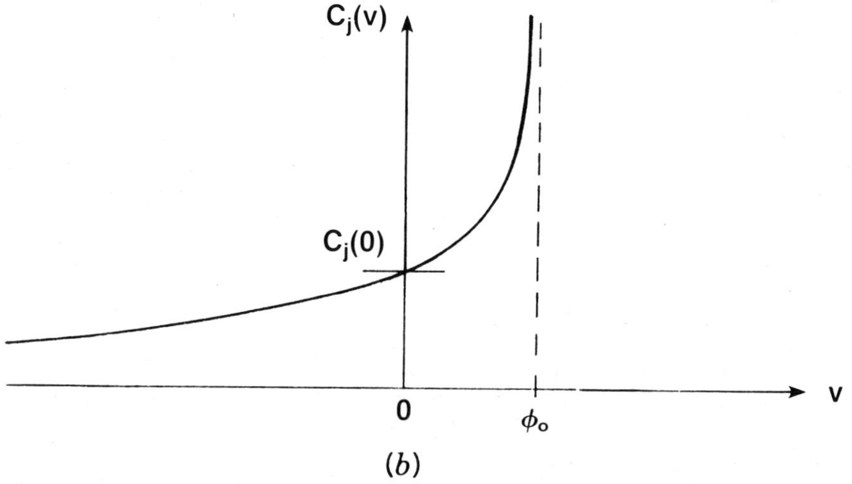

Figure 13.16 (*a*) Depletion-layer charge q_j vs. junction voltage v for an ideal Schottky-barrier or abrupt *pn* junction varactor with $\gamma = \frac{1}{2}$. (*b*) Derived depletion-layer capacitance $C_j(v) = \partial q_j / \partial v$.

Equations 13.9 and 13.13:

$$q_d(i) = \tau_L \cdot i \tag{13.16}$$

Under *fully pumped* conditions, it is appropriate to define a *large-signal* varactor cutoff frequency

$$\mathbf{F}_c \triangleq \frac{1}{2\pi R_s}\left[\frac{1}{C_j(V_B)} - \frac{1}{C_j(0)}\right] \tag{13.17}$$

where V_B is the reverse breakdown voltage. This is different from the usual small-signal cutoff frequency quoted in data sheets:

$$f_c(V_0) = \frac{1}{2\pi R_s C_j(V_0)} \tag{13.18}$$

where V_0 is the bias voltage (e.g., 0 V or −6 V). Using Equations 13.10, 13.17, and 13.18, the large-signal cutoff frequency can be written

$$\mathbf{F}_c = f_c(0)\left[\left(1 + \frac{V_B}{\phi_0}\right)^{\gamma} - 1\right] \tag{13.19}$$

A comprehensive analytical treatment of *fully pumped* depletion-varactor multipliers with $\gamma = \frac{1}{2}$ has been given by Penfield and Rafuse [3], who included in their varactor model only C_j and R_s, and assumed circuits that limit varactor currents, and hence time-varying charges, to

1. the input frequency f_{in},
2. the output frequency Nf_{in},
3. necessary idler frequencies when $N > 2$.

Tang [32] obtained elegant closed-form solutions for fully pumped doublers with $\gamma = \frac{1}{2}$, under slightly different conditions. The theory was extended by Sard [33] to include not only skin effect, which increases R_s at high frequencies, but also partial varactor pumping, which is necessary for high spectral purity applications where the noisy breakdown region must be avoided. Burckhardt [34] used computer simulations to predict the behavior of depletion multipliers with arbitrary varactor capacitance laws and drive levels.

Figure 13.17 compares the maximum efficiencies η_c of fully pumped doublers (with $\gamma = \frac{1}{2}$) for the maximum-efficiency/maximum-power-output case (Penfield and Rafuse) with the minimum-dissipation/fixed-power-input

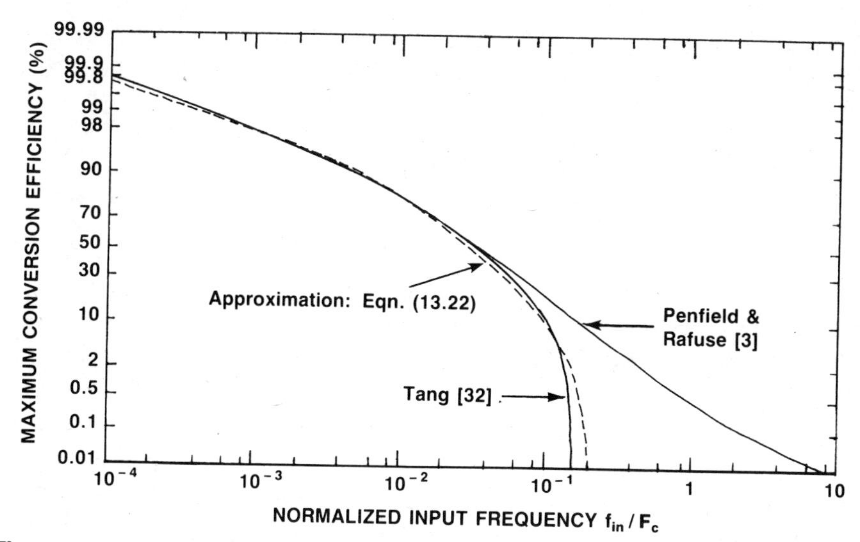

Figure 13.17 Maximum efficiency of fully pumped depletion-varactor frequency doublers, with $\gamma = \frac{1}{2}$, as a function of input frequency f_{in}.

case (Tang), both as functions of $f_{in}/\mathbf{F}_c$. Tang's expression is [32]

$$\eta_c = \frac{1 - \sqrt{2} \cdot k(f_{in}/\mathbf{F}_c)}{1 + 2\sqrt{2} \cdot k(f_{in}/\mathbf{F}_c)} \tag{13.20}$$

where

$$k = 2\sqrt{(\sqrt{5}+1)} + \sqrt{(\sqrt{5}-1)} \simeq 4.71 \tag{13.21}$$

Also shown in Fig. 13.17 is the low-frequency approximation [34, 35]

$$\eta_c \simeq e^{-2\alpha f_{in}/\mathbf{F}_c} \tag{13.22}$$

in which the coefficient α is $\simeq 11.1$ for a doubler with $\gamma = \frac{1}{2}$. Related design curves for input, output and dissipated powers, optimum source and load terminations, and bias voltages are given in [3, pp. 328–412] for multipliers of orders 2, 3, 4, 5, 6, and 8.

Unlike nonlinear-resistance multipliers based on varistors or GaAs FETs, or those employing SRDs, depletion-varactor multipliers require both input and output tuning, as well as idlers if $N > 2$, and are consequently narrow-band circuits. For high conversion efficiency it is also necessary to ensure that the input, output, and idler circuits are coupled only via the nonlinear reactance and that the input and output circuits are conjugately matched to the dynamic varactor impedance at the respective input and output frequencies. A further requirement is that, ideally, varactor currents should flow only at the input, output, and idler frequencies [35].

Analytic treatments of AM–AM and AM–PM conversion, additive noise, and hysteresis phenomena in "$\gamma = \frac{1}{2}$" varactor multipliers have been given by Bava et al. [36, 37].

Practical Nonlinear-Reactance Multipliers. This section reviews some planar multiplier realizations at microwave and millimeter wave frequencies.

Microwave Varactor Multipliers. Designs for doublers and triplers with outputs up to ~40 GHz have been given by several authors, for example, [38–42]. For high output powers, good heat sinking mandates a shunt-

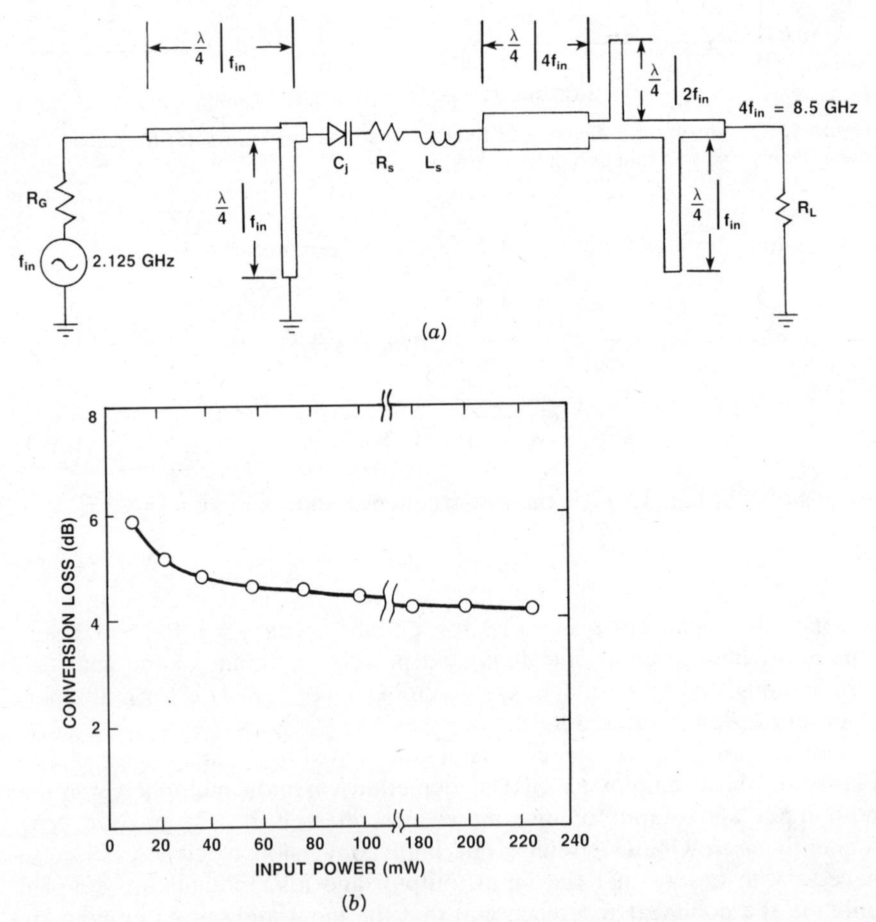

Figure 13.18 (*a*) Schematic of 2.125–8.500 GHz microstrip quadrupler. Parameters: $\gamma = 0.45$, $V_B = 26$ V, $f_c(-6\text{ V}) = 200$ GHz. (*b*) Conversion loss vs. input power for $f_{in} =$ 2.125 GHz. Bias is optimized at each point. (After J. B. Horton [40], 1967 IEEE, reprinted with permission.)

varactor configuration. Swan [38] describes such a tripler for 12-GHz output: its performance conforms closely to the Penfield–Rafuse theory. For planar circuits, particularly where power dissipation is not a problem, series-varactor configurations are topologically convenient. Moreover, as pointed out by Roulston and Boothroyd [39], such circuits permit reasonably efficient high-order multiplication *without* the use of idlers. Figure 13.18(*a*) depicts the layout of a thin-film series-mode quadrupler due to Horton [40]. It uses a beam-lead varactor, and is designed according to the theory of Burckhardt [34]. As seen in Fig. 13.18(*b*), conversion loss is ~4.5 dB ($\eta_c \simeq 35\%$) over a wide range of input powers.

For frequencies in the 20–60 GHz range and for power outputs up to ~100 mW, an alternative to varactor multipliers is offered by the rapidly increasing performance of GaAs FET multipliers, as discussed in Section 13.2.4. However, at the shorter millimeter wavelengths, that is, up to 300 GHz, and into the submillimeter region, the sources of choice remain depletion-varactor multipliers in conjunction with lower frequency oscillators.

Millimeter-Wavelength Varactor Multipliers. Monolithic microwave-integrated-circuit (MMIC) realizations of depletion-varactor multipliers have potential at millimeter wavelengths.

Chu et al. [43, 44] describe GaAs MMIC varactor doublers with outputs of 300 mW at 25 GHz and 150 mW at 37 GHz. Figure 13.19 shows an example of a two-diode series-mode doubler for output frequencies of ~37 GHz. Since the power capability of varactor multipliers is proportional to breakdown voltage V_B, connecting n diodes in series increases the overall breakdown voltage to nV_B and the output power by a factor of n^2. To maintain overall values of capacitance and series resistance, the active area of each diode has to be increased by a factor n. In this design the diameter of each anode is ~38 μm, the overall V_B is 30 V, and the total zero-bias capacitance is 0.3 pF. The input and output matching circuits are $\lambda/4$-transformers with open-circuited 60° radial line stubs. These stubs are resonant at the input and output frequencies; radial lines are used to maximize the resonance bandwidths. Their locations are chosen to reflect the appropriate impedances at the position of the varactors. The important steps in fabricating these multipliers are given in References 43 and 44.

Test results are given in Fig. 13.20. Maximum chip-level power output is 150 mW with a conversion efficiency of −6.2 dB (24%). The peak efficiency of −4.6 dB (35%) occurs at $P_{out} = 95$ mW.

For the higher frequency part of the millimeter wavelength region (i.e., ~100–260 GHz), Archer [45] has described mechanically tunable doublers with conversion efficiencies of about −10 dB (10%). His technique, which involves the interconnection of crossed input and output waveguides by means of a suspended quartz-substrate strip-line circuit, resembles the approach of Takada et al. [6, 46]. A somewhat similar crossed-waveguide

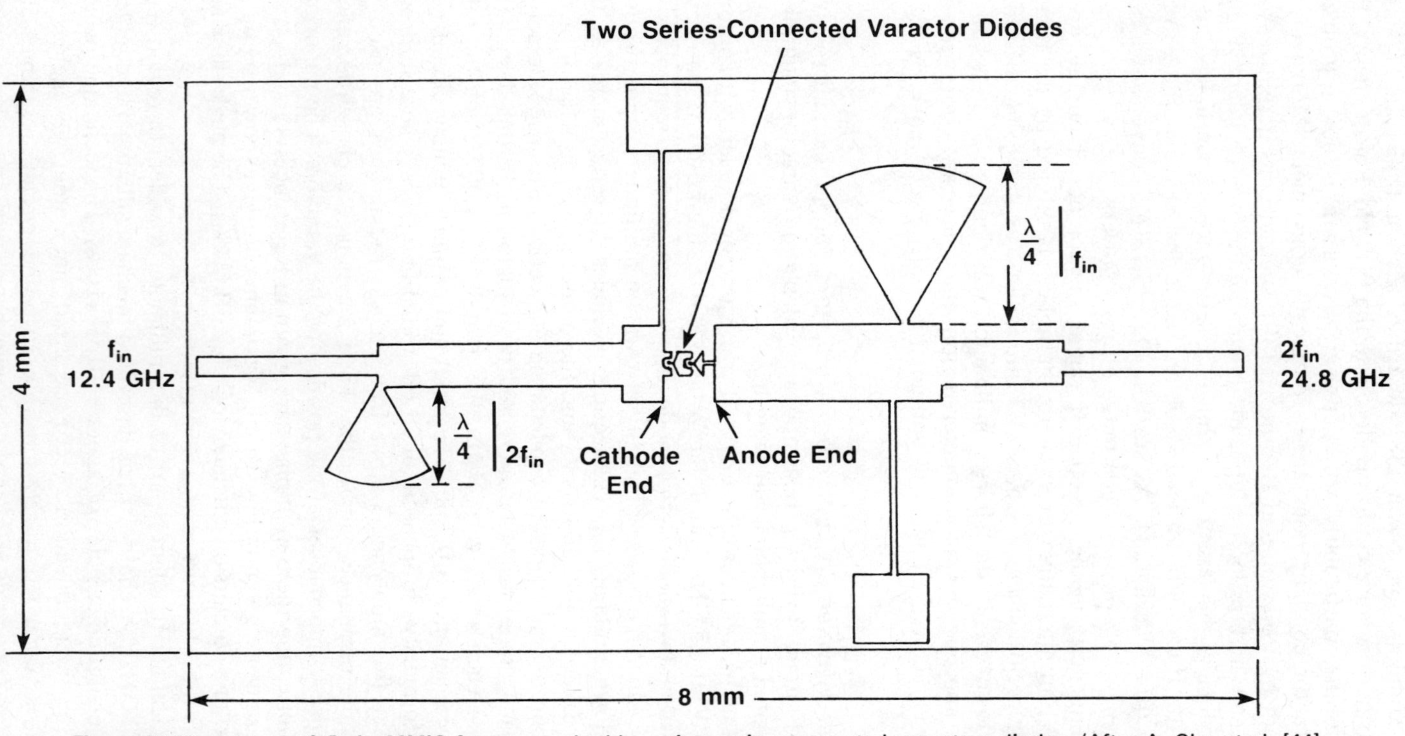

Figure 13.19 Layout of GaAs MMIC frequency doubler using series-connected varactors diodes. (After A. Chu et al. [44], 1984 IEEE, reprinted with permission.)

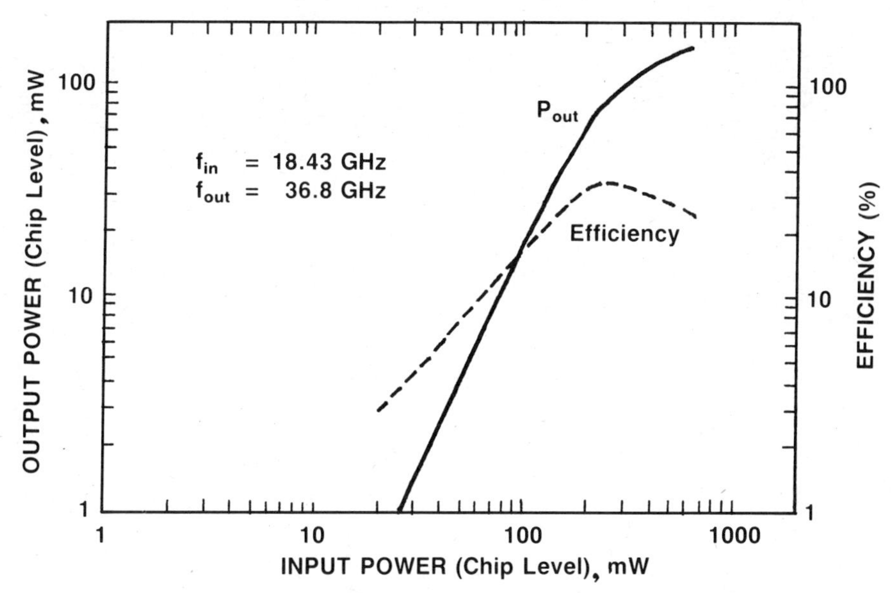

Figure 13.20 Chip-level output power P_{out} and conversion gain of MMIC frequency doubler. Bias = −11 V. (After A. Chu et al. [44], 1984 IEEE, reprinted with permission.)

multiplier, with a doubling efficiency of −11.5 dB (7%) at 268 GHz, has been reported by Erickson [47].

13.2.4 Active Multipliers Using GaAs FETs

Compared with varistor and varactor multipliers, active multipliers using GaAs FETs offer both higher efficiency (even gain) and broad operational bandwidth for outputs up to at least 60 GHz. Since these multipliers do not need idlers, wide bandwidth of operation is possible. Additionally, conversion gains of ~10 dB have been obtained and input requirements are low, being of the order of 10 mW.

In addition to outlining single-gate and dual-gate multipliers, this section describes self-oscillating multipliers, which, when dielectric-resonator stabilized, can replace Gunn and IMPATT oscillators in the millimeter wavelength region [48].

Theory of GaAs FET Multipliers. Figure 13.21(*a*) shows the major lumped linear and nonlinear elements of a GaAs FET operating in its active region. Figure 13.21(*b*) is the corresponding equivalent circuit. The significant nonlinear-resistance elements for multiplication are as follows

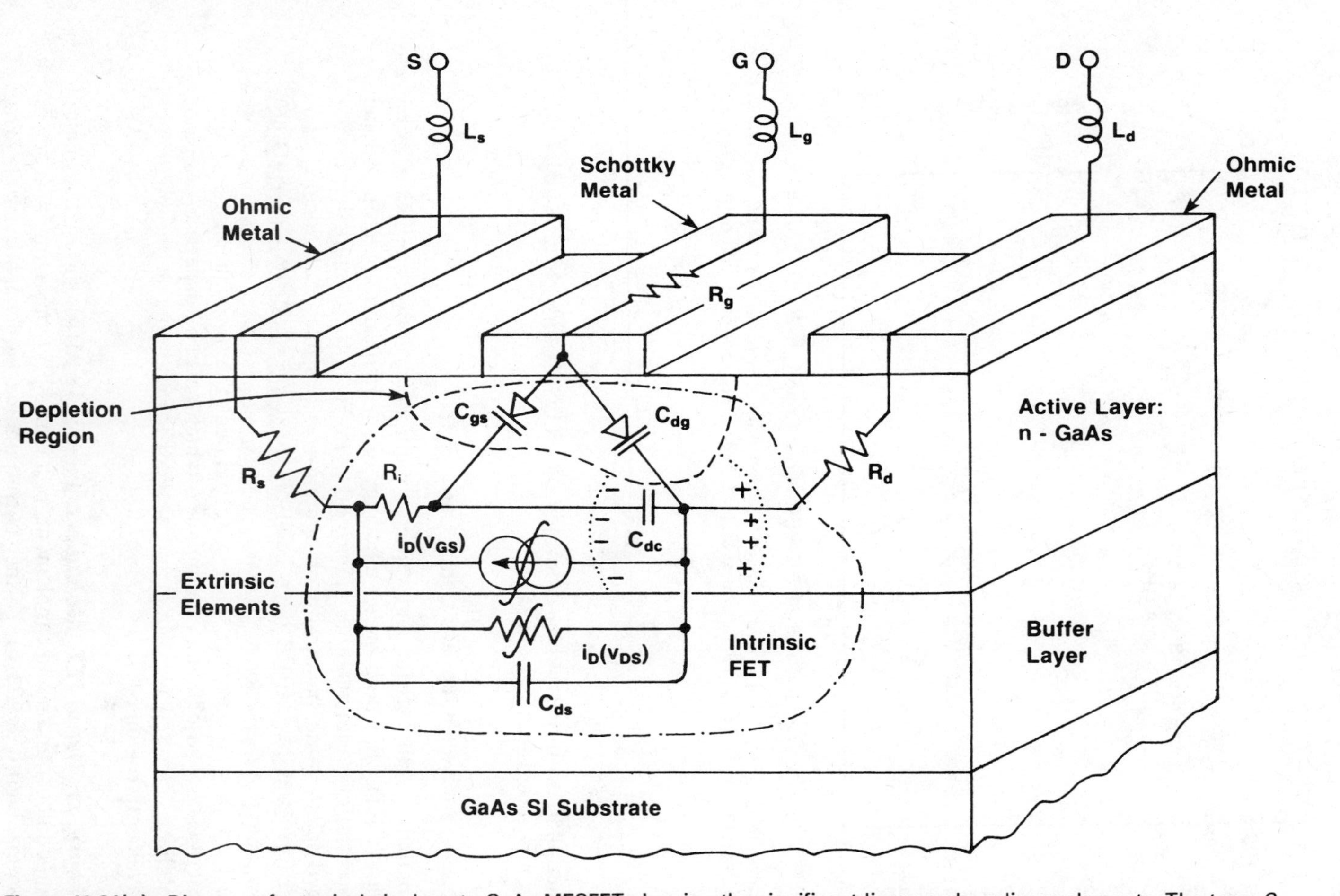

Figure 13.21(*a*) Diagram of a typical single-gate GaAs MESFET showing the significant linear and nonlinear elements. The term C_{dc} represents the space-charge dipole-layer capacitance.

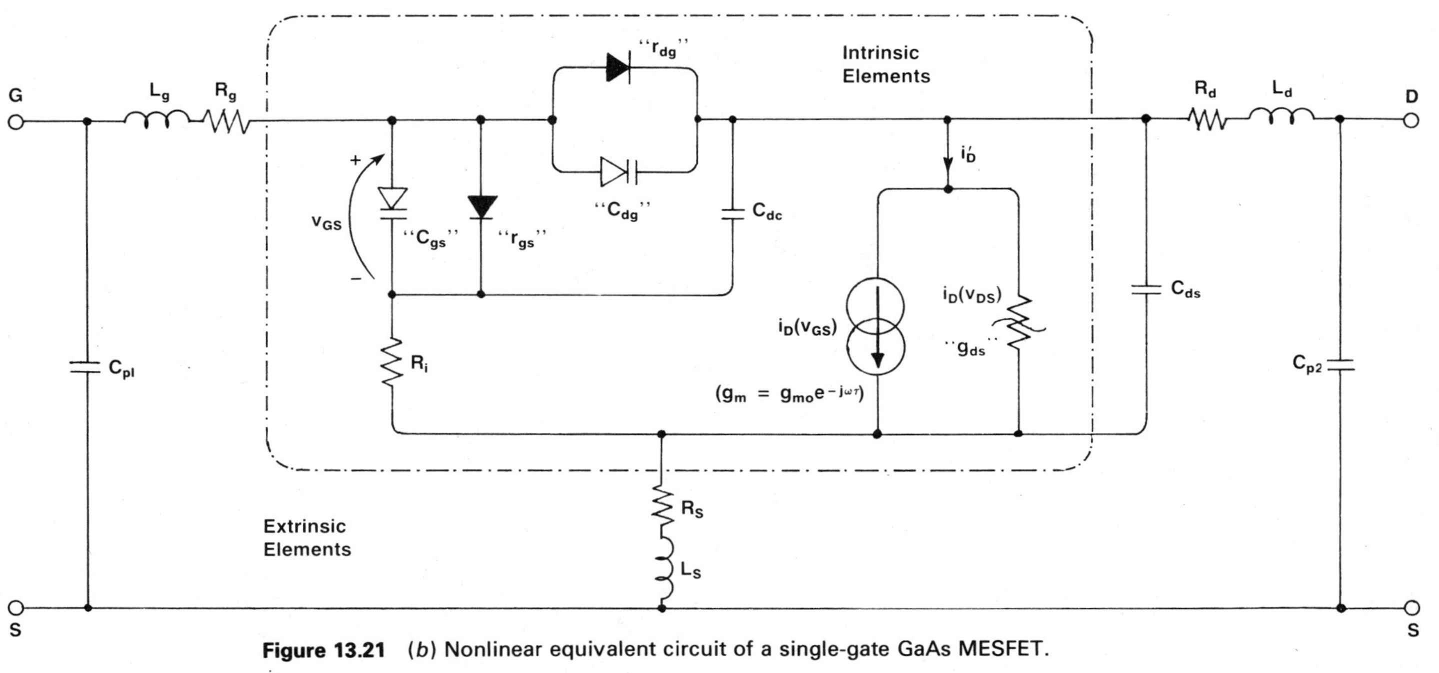

Figure 13.21 (*b*) Nonlinear equivalent circuit of a single-gate GaAs MESFET.

1. The transconductance (g_m) represented by

$$i_D(v_{DS}) = \begin{cases} I_{DSS}\left(1 - \dfrac{v_{GS}}{V_P}\right)^2, & V_p < v_{GS} < 0 \\ 0, & v_{GS} < V_P \end{cases} \tag{13.23}$$

2. The output conductance (g_{ds}) of the pinched-down channel, which can be included by modifying the preceding equation to read [49]

$$i_D(v_{GS}, v_{DS}) = \left(1 - \frac{v_{GS}}{V_P}\right)^2 (I_{DSS} + G_{DS}v_{DS}) \tag{13.24}$$

3. The rectifying property of the Schottky gate (r_{gs}), with conduction current

$$i_G(v_{GS}) = I_{G0}\left[\exp\left(\frac{ev_{GS}}{nkT}\right) - 1\right] \tag{13.25}$$

where the leakage current I_{G0} is extremely small. The reactive nonlinearities due to the gate-channel depletion region are represented by the voltage-dependent "varactor" capacitances

$$C_{gs}(v_{GS}) = C_{gs}(0)\left(1 - \frac{v_{GS}}{\phi_0}\right)^{-1/2} \tag{13.26}$$

and

$$C_{dg}(v_{DG}) = C_{dg}(0)\left(1 - \frac{v_{DG}}{\phi_0}\right)^{-1/2} \tag{13.27}$$

In principle, all five of these nonlinearities can contribute to harmonic generation. However, the effects of C_{gs} and C_{dg} are small since the large resistances R_s and R_d of the bulk semiconductor and R_i of the undepleted channel cause low cutoff frequencies

$$f_{cgs}(0) = \frac{1}{2\pi(R_s + R_i)C_{gs}(0)} \tag{13.28}$$

and

$$f_{cds}(0) = \frac{1}{2\pi R_d C_{dg}(0)} \tag{13.29}$$

Computer simulations [49, 50] show that for doubler operation the principal nonlinearity is the i_D clipping effect due to v_{GS} swinging either positively, causing gate conduction, as in Equation 13.25, or below the pinchoff voltage V_p, see Equation 13.24. For class-A biasing at $V_{GS} = 0$ V [see Fig. 13.22(*a*)], the v_{GS} waveform gets clipped due to gate conduction, and has a half-wave rectified form. If the negative swing does not exceed the pinchoff voltage V_p, this waveform is transferred to i_D via the transconductance characteristic. Thus $i_D(t)$ contains harmonics of all orders. If

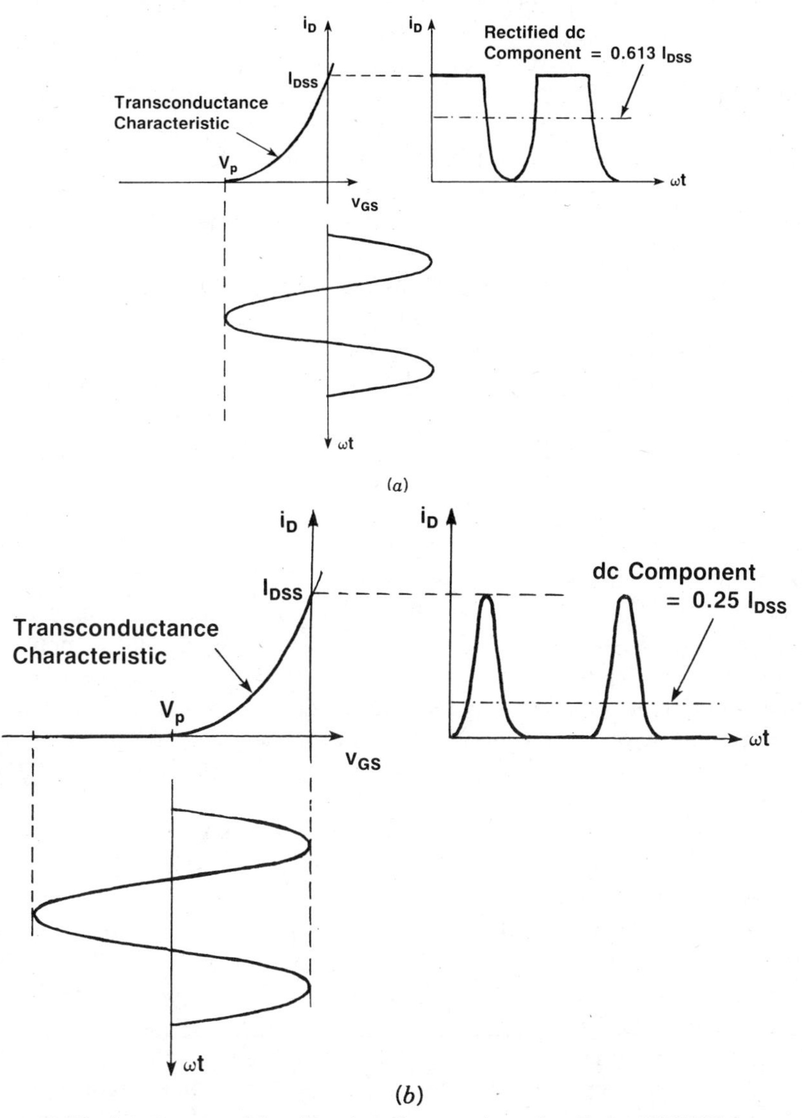

Figure 13.22 (*a*) Bias condition for class-A operation of a GaAs MESFET frequency multiplier. In this case the device is biased at $V_{GS} = 0$. (*b*) Bias condition for class-B operation of GaAs FET frequency multiplier. The device is biased at $V_{GS} = V_p$. (After G. S. Dow and L. S. Rosenheck [51], reprinted with permission of *Microwave J.*)

the negative swing of v_{GS} just reaches V_P, and if the quadratic conductance characteristic of Equation 13.24 is assumed, then the second harmonic component of i_D is theoretically 10.4 dB below the fundamental, favoring doubling.

Figure 13.22(b), for class-B biasing with $V_{GS} = V_p$, illustrates the pinchoff-clipping effect. If the positive swing of v_{GS} just reaches 0 V, then the second-harmonic component of i_D is theoretically only 4.6 dB below* the fundamental, again favoring doubling.

Although the class-A mode appears to offer a 1–2 dB frequency conversion-efficiency advantage [50], it requires a larger dc drain current than the class-B mode, leading to lower dc-to-RF conversion efficiency and more risk of FET failure due to gate-current spikes [50].

If the gate is biased between $v_{GS} = 0$ and $v_{GS} = V_P$, and if the input swing is sufficient to cause clipping at both extremes, the $i_D(t)$ will approximate a square wave. For a symmetrical square wave the second-harmonic component will be small, favoring frequency tripling.

The underlying relationships between doubler performance and the principal design parameters have been elucidated by Rauscher [50] in a series of simulations in which the following factors were taken into account:

1. device-circuit interactions at harmonic frequencies,
2. FET matching at *both* input and output ports at *both* f_{in} and $2f_{in}$,
3. device-external feedback options.

The device model used is essentially that of Fig. 13.21(b), and class-B biasing is used. For a computationally reasonable simulation, using a variant of the well-known harmonic balance technique, the nonlinear elements C_{gs}, R_i and C_{dg} are replaced by their large-signal time-averaged values, while diode conduction is modeled by limiting the diode forward voltage to 0.7 V. The remaining dominant nonlinearities, $i_D(v_{GS})$ and $i_D(v_{DS})$, are both represented by a composite controlled-current source [i'_D in Fig. 13.21(b)] that incorporates the transit delay τ.

Figure 13.23 shows the doubler configuration. Optimum matching conditions are found by an iterative search procedure. The calculated conversion gain G_c versus $P_{out}(2f_{in})$ is depicted in Fig. 13.24, in which the fundamental frequency load impedance $Z_{TL}(f_{in})$ is a parameter. For these results, it is assumed that the input and output ports are conjugately matched and that there is no feedback *external* to the device. The strong dependence of G_c on $Z_{TL}(f_{in})$ is a linear effect due to parasitic feedback *internal* to the FET. This internal feedback, which is at $2f_{in}$, causes the second-harmonic input termination $Z_{TI}(2f_{in})$ to become involved in the

*The values of 8.7 dB and 2.1 dB for the class-A and class-B cases, respectively, as given in [51], are incorrect.

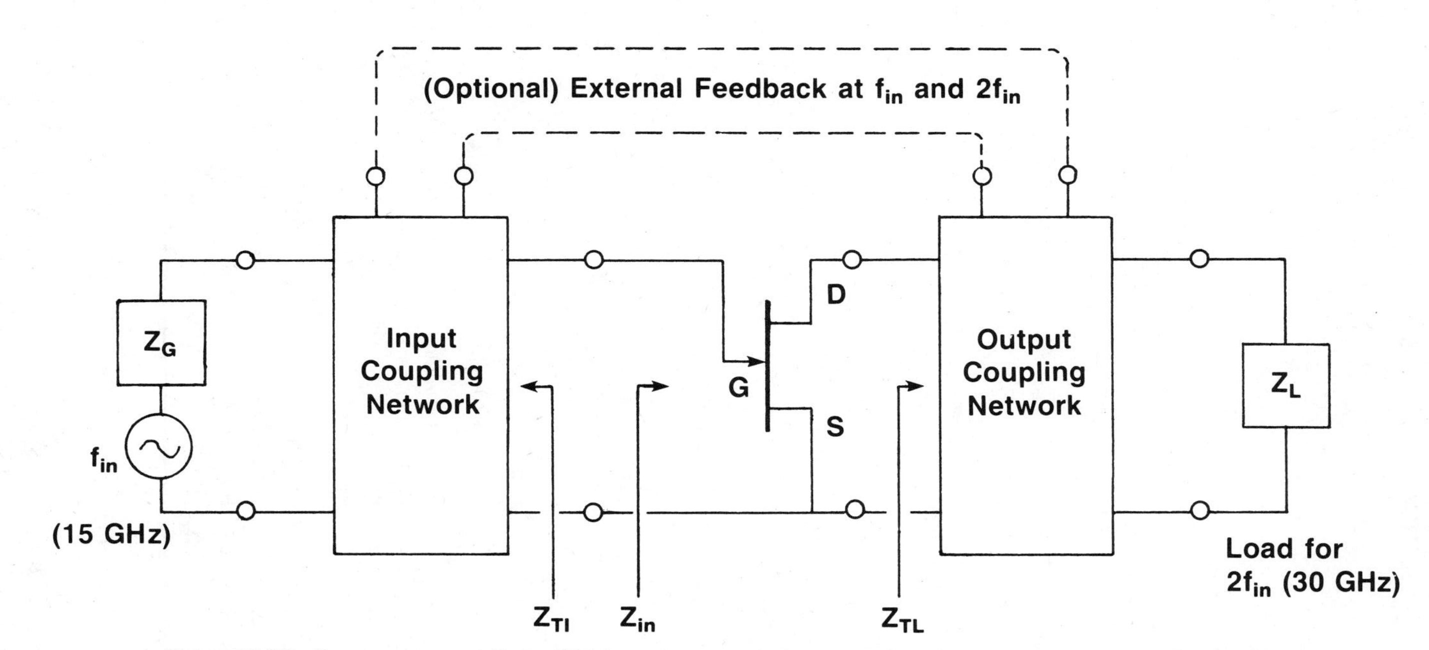

Figure 13.23 Common-source GaAs FET frequency-doubler configuration. (After C. Rauscher [50], 1983 IEEE, reprinted with permission.)

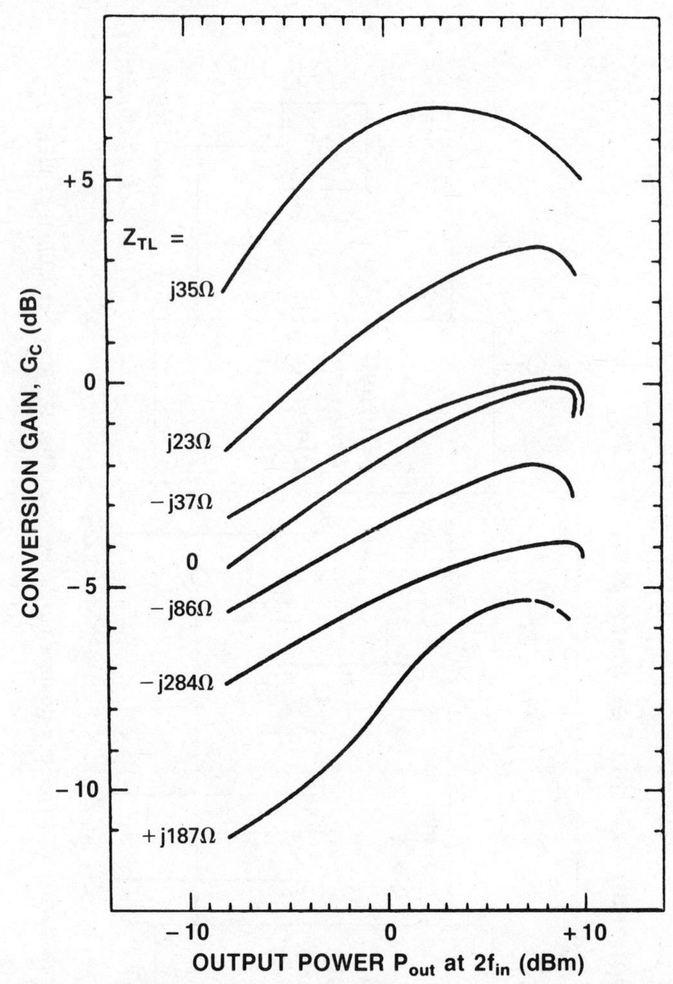

Figure 13.24 Simulated large-signal conversion gain G_C as a function of fundamental-frequency load reactance $Z_{TL}(f_{in})$ and second-harmonic power $P_{out}(2f_{in})$. (After C. Rauscher [50], 1983 IEEE, reprinted with permission.)

doubler performance. At low operating frequencies, $Z_{TI}(2f_{in})$ can cause a destructive-interference effect, but its role diminishes at higher frequencies.

The simulations also show that G_c is essentially independent of $Z_{TL}(f_{in})$. This indicates that the device-circuit interactions at f_{in} and $2f_{in}$ are separable phenomena, a conclusion useful in doubler design.

Practical GaAs FET Frequency Multipliers

Single-Gate Multipliers. Figure 13.25 shows the schematic of a 15–30 GHz doubler [50], the design of which is based on the simulation of the

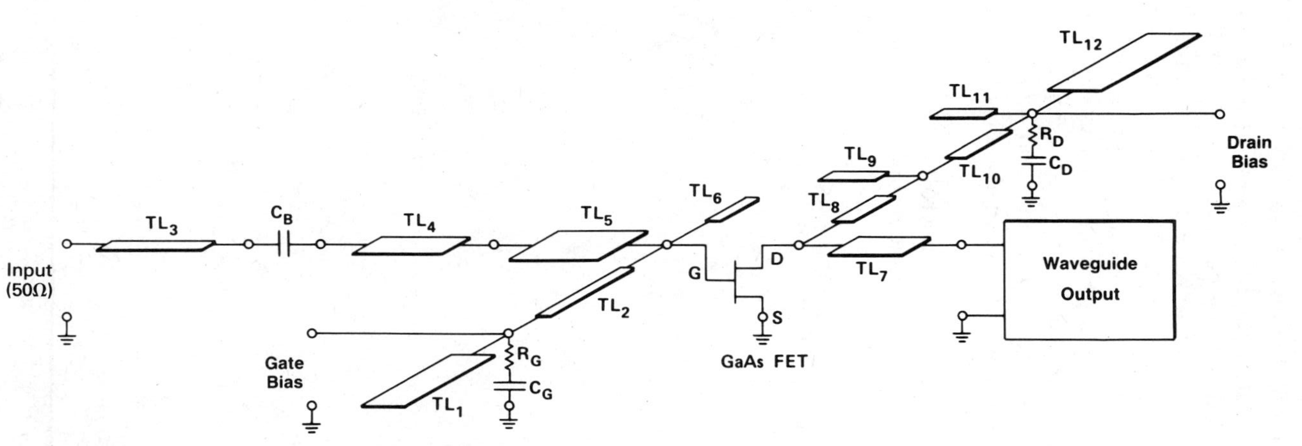

Figure 13.25 Schematic of single-gate GaAs FET 15–30 GHz frequency doubler. The *TL* elements are microstrip lines. (After C. Rauscher [50], 1983 IEEE, reprinted with permission.)

previous section. The gate bias circuit consists of R_G, C_G, TL_1, and TL_2, while the input network includes the transmission lines TL_3, TL_4, and TL_5, a dc block capacitor C_B and an open-circuited stub TL_6, which is $\lambda/4$ long at $2f_{\text{in}} = 30$ GHz. These elements provide a conjugate match at the input while simultaneously blocking $2f_{\text{in}}$ by short circuiting the gate at 30 GHz. Output power at $2f_{\text{in}}$ is probe coupled via line TL_7 to a backshorted WR-28 waveguide. At f_{in}, the probe, with its stray capacitances, acts as an open-circuited stub at f_{in}, thereby establishing a reactive load $Z_{TL}(f_{\text{in}}) = -j50\ \Omega$. The drain bias circuit, consisting of R_D, C_D, and TL_8 to TL_{12} inclusive, rejects both f_{in} and $2f_{\text{in}}$. No deliberate external feedback is applied at either f_{in} or $2f_{\text{in}}$.

This design has been implemented in microstrip, as shown in Fig. 13.26, using a GaAs FET chip chosen to resemble the device in the simulation

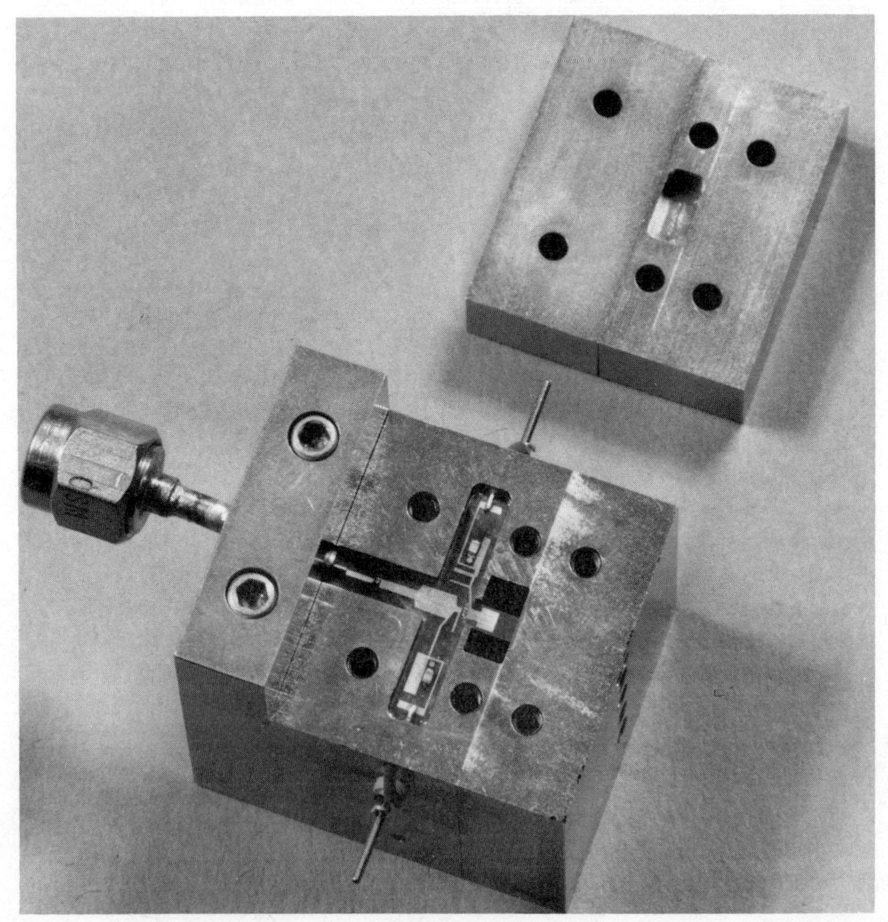

Figure 13.26 Single-transistor 15–30 GHz frequency doubler. Input is coaxial (SMA); output is waveguide (WR-28). (After C. Rauscher [50], 1983 IEEE, reprinted with permission.)

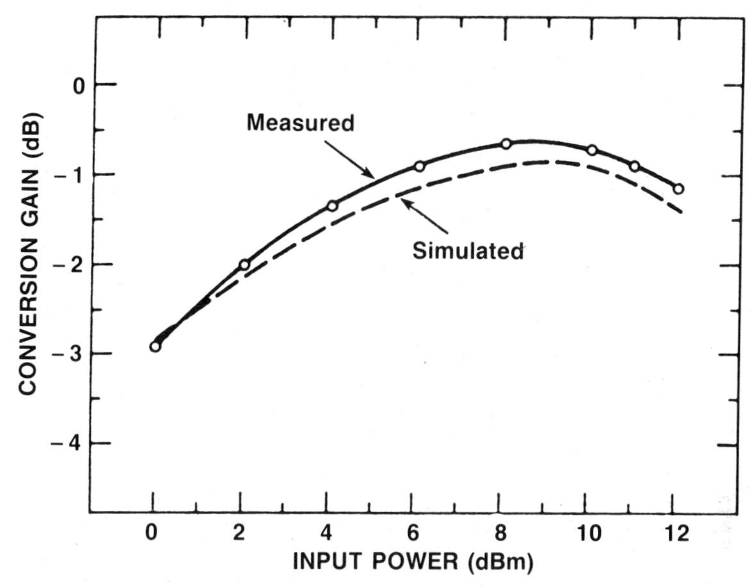

Figure 13.27 Measured and predicted performance of 15–30 GHz frequency doubler. (After C. Rauscher [50], 1983 IEEE, reprinted with permission.)

previously discussed. Figure 13.27 shows a good agreement between simulated and measured results: this is due in part to an accurate treatment of device and probe parasitics at 30 GHz.

Figure 13.28 is a block diagram of a novel octave-bandwidth GaAs FET doubler due to Gilmore [52] and also designed using the harmonic balance technique.* Rejection of f_{in} is particularly difficult to achieve in octave-bandwidth doublers using conventional approaches, since high-Q matching circuits are needed for rejection, while low-Q circuits are required for bandwidth. Here f_{in} rejection is obtained in two ways.

1. By antisymmetric coupling of two FET half-wave rectifiers using two Lange couplers. These couplers provide a total phase difference of 180° and provide good wide-band matching. The antisymmetry cancels out all odd-order frequencies.
2. By providing an output band-reject amplifier to achieve further cancellation: the Lange couplers here are designed for the output band and to reject the input band.

Measured results are given in Fig. 13.29. Highest conversion efficiency is obtained for $P_{in} = +5$ dBm, and ranges from -5 dB (32%) at $f_{out} = 8$ GHz to -1 dB (10%) at $f_{out} = 16$ GHz. The corresponding rejections of f_{in} are 15 dB and 20 dB, respectively.

*In this case implemented on an IBM PC AT.

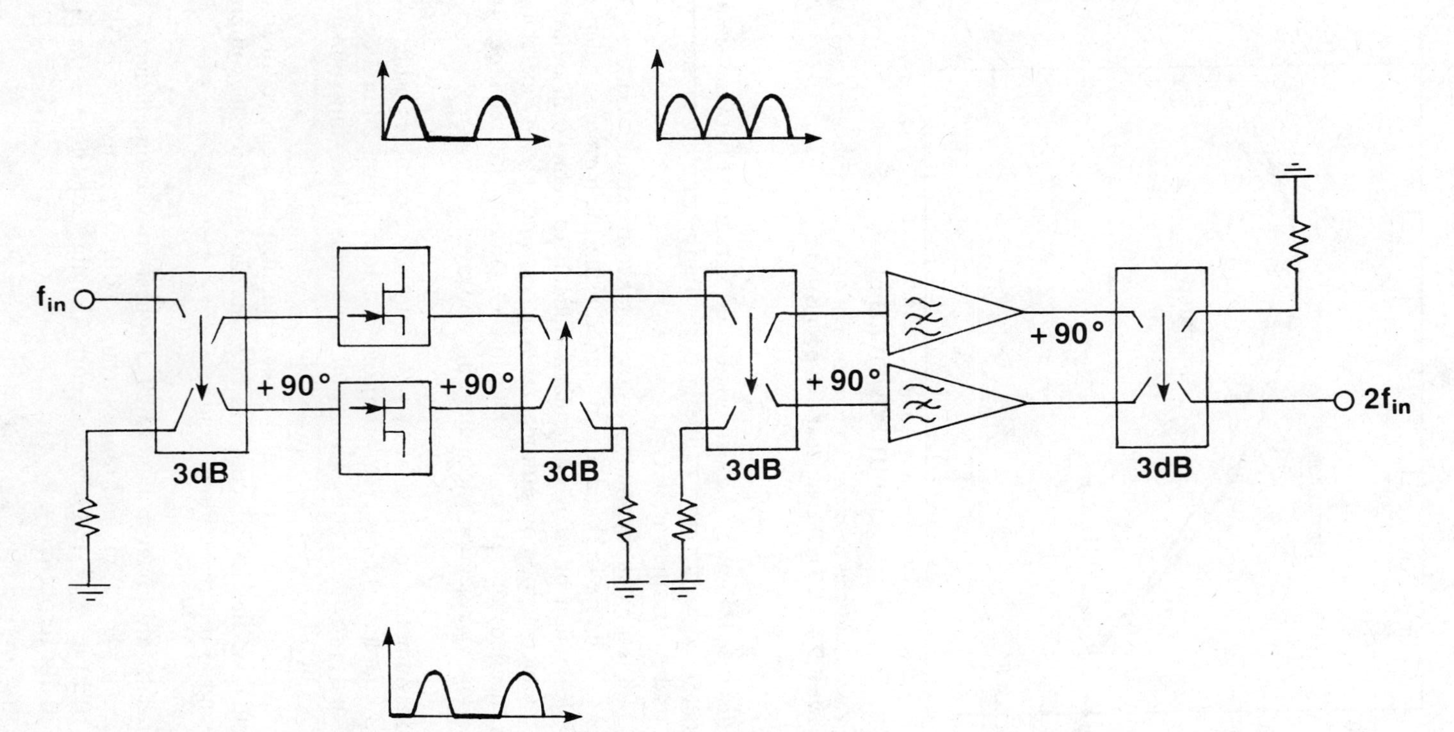

Figure 13.28 Block diagram of octave-bandwidth GaAs FET doubler. (After R. Gilmore [52], 1986 IEEE, reprinted with permission.)

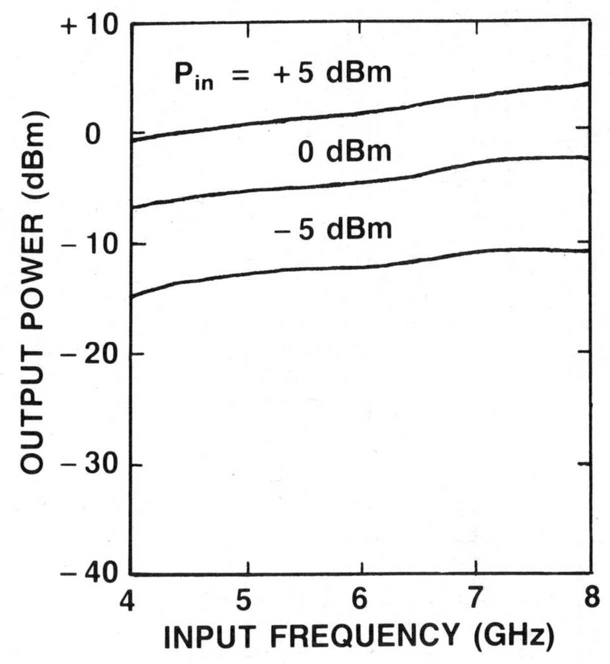

Figure 13.29 Experimental output power of octave-bandwidth GaAs FET doubler at $2f_{in}$ as a function of input frequency f_{in}. (After R. Gilmore [52], 1986 IEEE, reprinted with permission.)

Dual-Gate Multipliers. Dual-gate GaAs FET doublers and triplers have been described by Chen et al. [53, 54] and by Stancliff [55]. Compared with the single-gate device, the dual-gate GaAs FET offers larger small-signal gain, better isolation, and more nonlinear characteristics. Figure 13.30 illustrates the principle of operation. The dual-gate device is modeled as a cascade of two single-gate devices, with the drain of the first feeding the source of the second. The input signal at the first gate (G_1) modulates the g_m of the first device. After amplification, it is fed to the second FET, in which it is clipped by forward conduction in the second gate (G_2), thereby generating harmonics. Amplified harmonics are extracted from the second drain (D_2).

In an experimental circuit the input frequency f_{in} (6–9 GHz) was applied to the first gate (G_1) via a 9.2-GHz low-pass filter; a tuner at D_2 maximized P_{out} at $2f_{in}$ or $3f_{in}$. The high-pass output filter was essential to suppress the input frequency. As in dual-gate GaAs FET mixers [56] the optimum impedance at G_2 is a pure reactance. This was realized here as a sliding short, and was adjusted for optimum conversion gain. Figure 13.31(a) and (b) show measured doubler and tripler gains versus output frequency as functions of input power $P_{in}(f_{in})$. The values obtained exceed those obtained with single-gate FETs or with BJT multipliers operated at lower

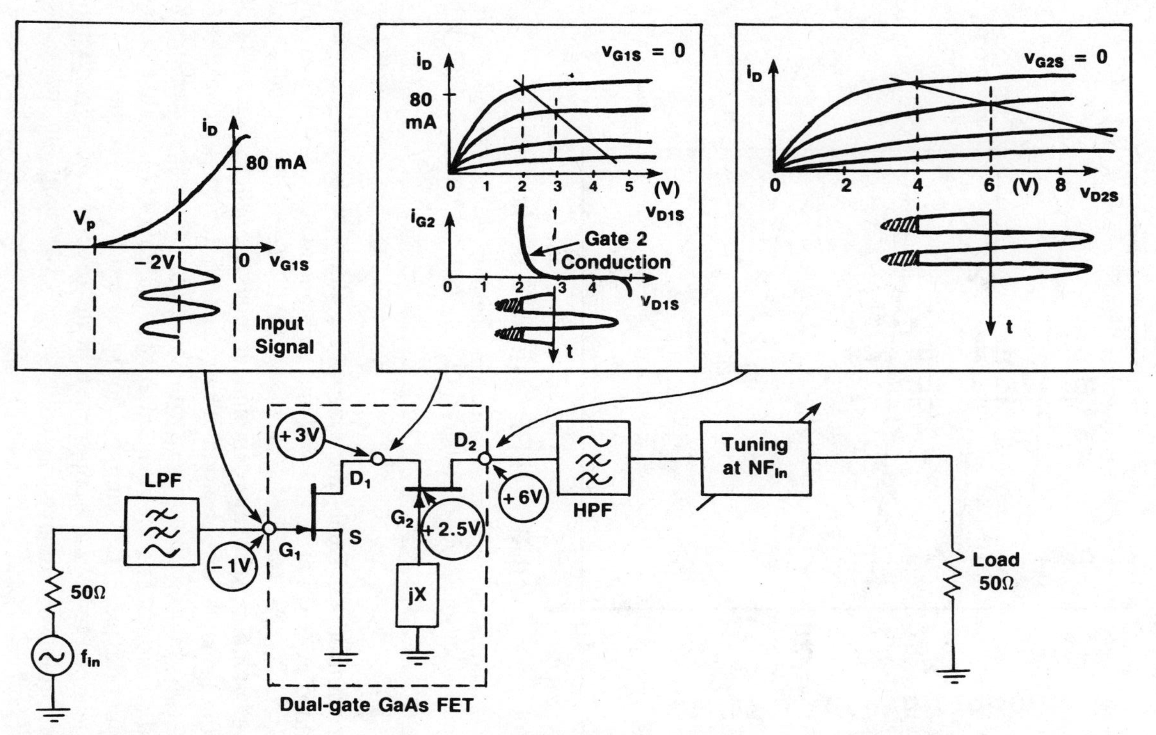

Figure 13.30 Basis of dual-gate MESFET frequency multiplier. Here the dual-gate FET is represented as a cascade of two single-gate devices. (After P. T. Chen et al. [54], 1979 IEEE, reprinted with permission.)

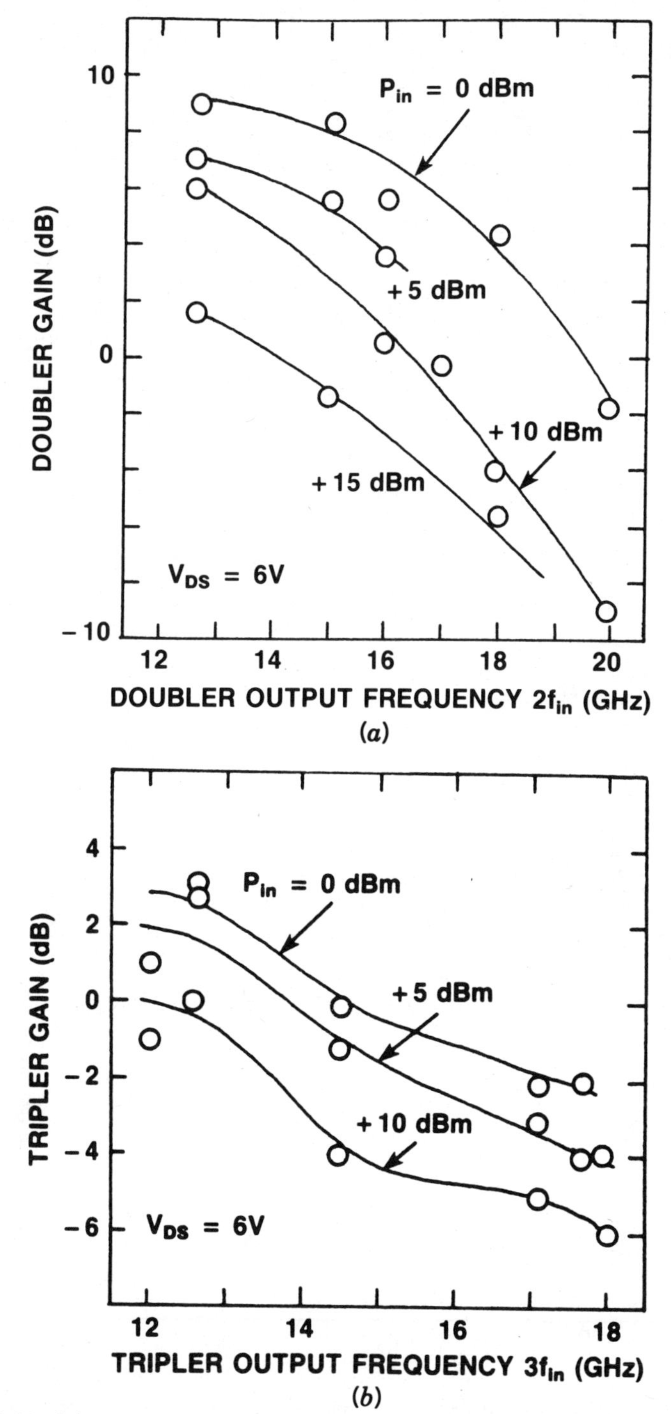

Figure 13.31 (*a*) Conversion gain of dual-gate GaAs FET doubler vs. output frequency as a function of input power. Both gate biases are adjusted at each data point. (After P. T. Chen et al. [53, 54], 1979 IEEE, reprinted with permission.) (*b*) As in part (*a*), but for tripler operation. (After P. T. Chen et al. [53, 54], 1979 IEEE, reprinted with permission.)

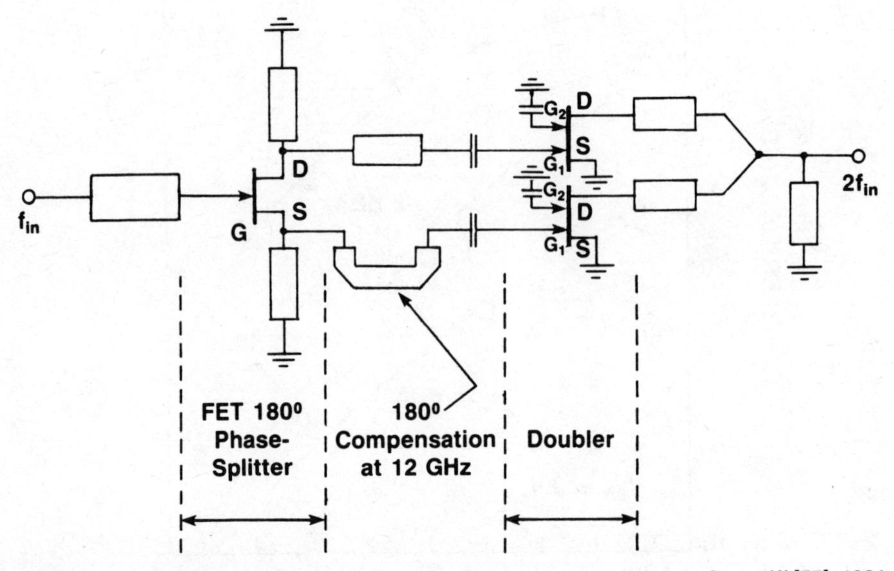

Figure 13.32 Balanced active phase-splitter/doubler circuit. (After R. Stancliff [55], 1981 IEEE, reprinted with permission.)

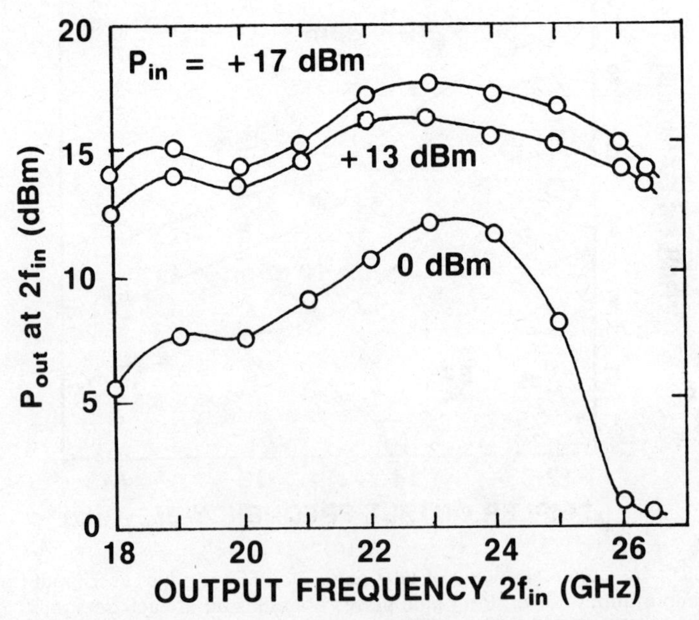

Figure 13.33 Output power of balanced dual-gate GaAs FET doubler vs. f_{in} and $P_{in}(f_{in})$. (After R. Stancliffe [55], 1981 IEEE, reprinted with permission.)

frequencies [21]. The decline of gain with P_{in} is due to self-biasing of both gates.

Even higher gains are reported by Stancliff [55], using the balanced dual-gate structure of Fig. 13.32. Experimental results for an 18–26.5-GHz doubler, shown in Fig. 13.33, indicate 12 dB of conversion gain at midband for an input of $P_{in} = 0$ dBm.

Self-Oscillating Doublers. A single GaAs FET can be used as an oscillator at a fundamental frequency f_0 and simultaneously as a doubler, delivering power at $2f_0$ to a load. Figure 13.34 is the schematic of a self-oscillator doubler using the external feedback option indicated in Fig. 13.23. Parallel coupled lines (CL) form the output/feedback network, while stubs TL_2, TL_3, and TL_4 ensure proper oscillation conditions at the fundamental f_0 as well as optimum feedback at $2f_0$. Line elements TL_5 to TL_8 and the two RC elements form the gate- and drain-bias circuits. This circuit, employing an Avantek M106 chip, was designed to provide $P_{out}(2f_0) = 9.1$ dBm at 30 GHz with a dc-to-RF conversion efficiency of 9.1%. The measured performance, using bias voltages of $V_{GS} = -1.2$ V and $V_{DS} = +3.0$ V, agrees well with predictions, P_{out} being +8.8 dBm at 29.34 GHz with an efficiency of 8.9%. These results are relatively insensitive to gate-bias variations in the range $-1.2\text{ V} \leq V_{GS} \leq 0$ V.

Dual-gate GaAs FET self-oscillating doublers have also been reported. Chu et al. [57] describe varactor-tuned circuits with outputs of ~+16 dBm at 12 GHz and ~+9 dBm at 18 GHz, while Saito et al. [48] report a self-oscillating dual-gate doubler that produces ~+10 dBm at 45 GHz. As in other dual-gate circuits, a capacitive termination at the second gate maximizes efficiency.

Advantages and Disadvantages of GaAs FET Multipliers. Advantages of single-gate MESFET frequency multipliers include conversion gain, broadband operation, low noise figure, low input power, high dynamic range, and some isolation between input and output. As the frequency of operation is increased, these desirable characteristics are degraded.

Dual-gate MESFET multipliers offer both increased conversion gain and very low input power requirements. The fact that excellent performance is obtained with 1 mW of input power is worth emphasizing when compared with the typical 100-mW input power requirement of a BJT multiplier or the 300 mW typically needed by a varactor or SRD multiplier. However, dual-gate multipliers tend to have larger noise figures.

Since GaAs FET multipliers are essentially nonlinear-resistance frequency converters, no idlers are needed. This means that AM-to-PM conversion is low and that broadband operation is possible.

The problem of a high-level fundamental signal appearing at the output port can be alleviated through the use of balanced multiplier configurations.

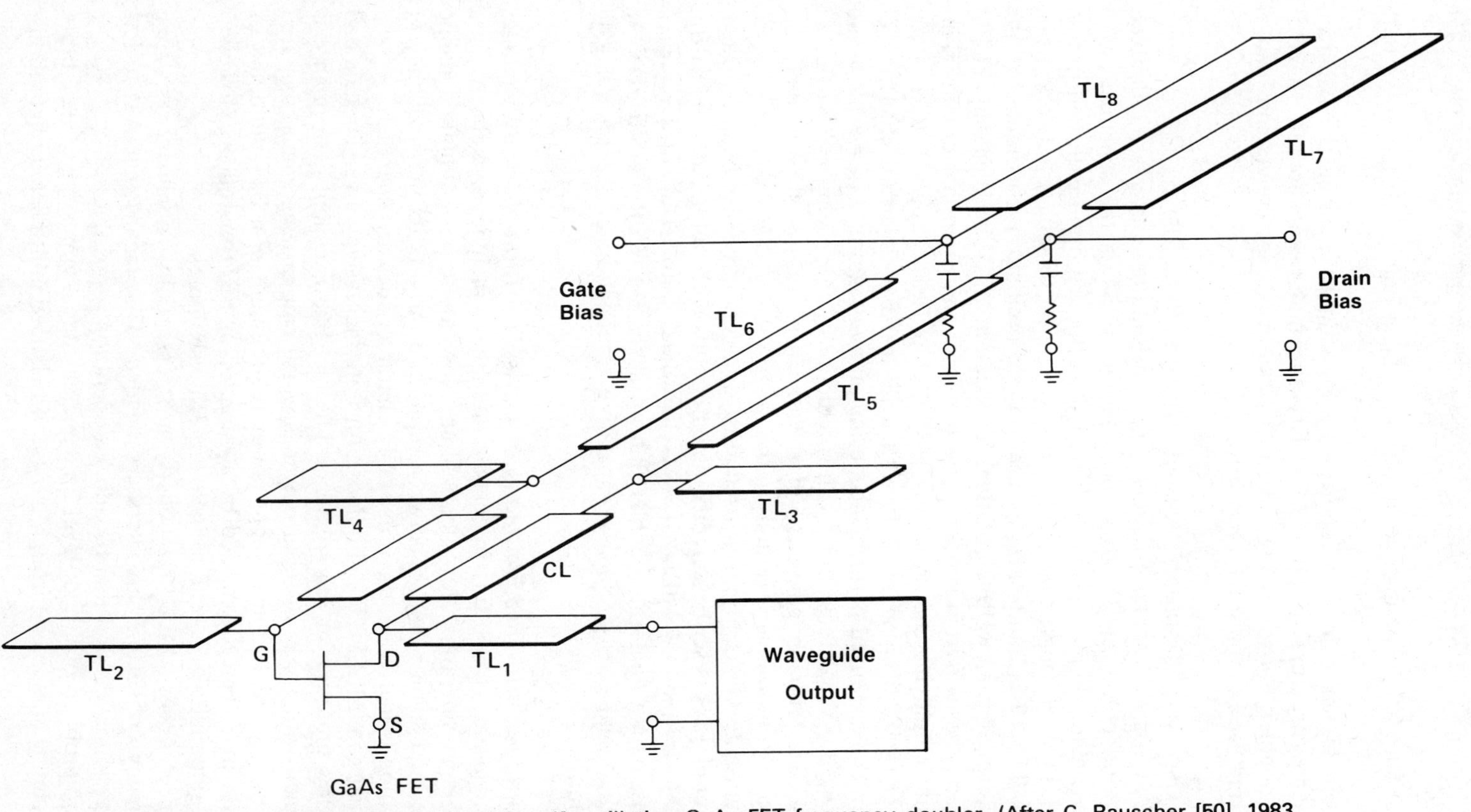

Figure 13.34 Schematic of 30-GHz self-oscillating GaAs FET frequency doubler. (After C. Rauscher [50], 1983 IEEE, reprinted with permission.)

A possible disadvantage for some applications is that although GaAs FETs have low intrinsic thermal noise, they do exhibit a $1/f$-like flicker noise spectrum near the carrier frequency.

13.3 FREQUENCY DIVISION

13.3.1 Types of Dividers

Frequency division can be obtained by the following methods.

1. Parametric subharmonic generation using varactors,
2. Utilization of charge-storage phenomena in step-recovery diodes (SRDs),
3. Mixer-with-feedback, or regenerative, dividers,
4. Effects in transferred-electron devices (TEDs),
5. Digital frequency division,
6. Use of oscillators injection-locked at a multiple of the natural frequency,
7. Phase-locked loops (PLLs) with multipliers in the feedback path.

Of the preceding, methods 1, 3, and 5 are of considerable current interest and are discussed here. Although item 2 is closely related to item 1, it has received but little theoretical attention [58].

Method 4, which involves TEDs, was initially felt to offer much promise [59, 60]; indeed division by all integers up to 34 (with $f_{\text{in}} = 38.4$ GHz, $f_{\text{out}} = 1.13$ GHz) was demonstrated [61]. However, not only do TED dividers exhibit narrow operational bandwidths of $\sim 10\%$ but they also suffer from large power dissipation due to the requirement for high electric fields over many micrometers of device length. As well as being unsuitable for integration with GaAs FETs, they also have problems with temperature compensation [62] and have been neglected since about 1977.

Method 7 is a straightforward PLL application and is not considered suitable for inclusion in the present text.

13.3.2 Parametric Frequency Dividers Using Varactors

Theory of Parametric Dividers. The Manley–Rowe equations [4] predict that for ideal nonlinear-reactance frequency dividers, the power $P_{1/N}$ generated at the Nth subharmonic of f_{in} is related to the input power P_1 by

$$\frac{P_{1/N}}{P_1} \leq 1 \qquad (13.30)$$

that is, the maximum possible efficiency is 100%. Frequency division cannot be obtained from positive nonlinear resistors.

Practical parametric frequency dividers use varactor diodes, which have been described in Chapter 8. A simplified nonlinear varactor equivalent circuit was given in Fig. 13.15. Under reverse-bias conditions, the voltage-dependent depletion charge $q_j(v)$ (see Equation 13.12), is the significant nonlinearity. Wide-band dividers often operate under forward bias, in which case the conduction current $i(v)$, and (in the case of *pn*-junction varactors) the diffusion charge $q_d(v)$, given by Equations 13.9 and 13.13, respectively, also become important.

For abrupt-junction and Schottky varactors, for which $\gamma = \frac{1}{2}$, the $v(q_j)$ relationship assumes the particularly simple quadratic form of Equation 13.14. This is the basis of the approximate large-signal theoretical treatment of balanced varactor halvers given by Harrison [63, 64]. This analysis assumes that the halver can be represented by the lumped-element equivalent circuit of Fig. 13.35, which includes the effect of varactor cutoff frequency. Defining the normalized *sum* of the varactor charges as

$$u = \frac{q_j(v_1) + q_j(v_2)}{\phi_0} \tag{13.31}$$

and the normalized *difference* between them as

$$z = \frac{q_j(v_1) - q_j(v_2)}{\phi_0} \tag{13.32}$$

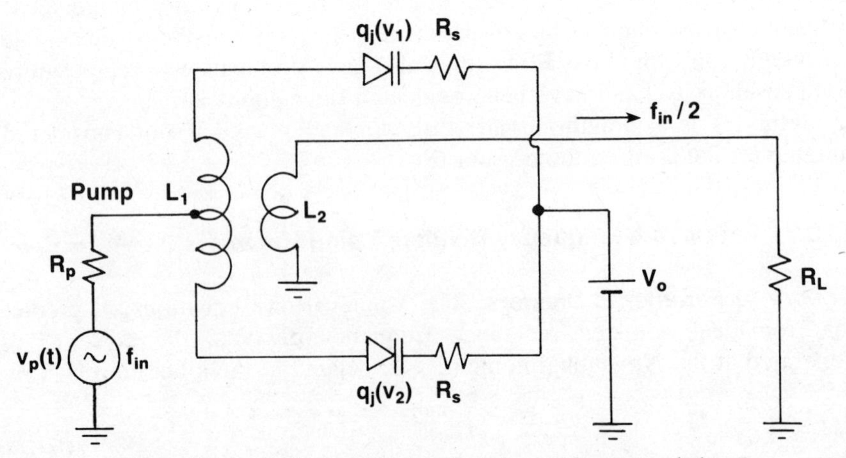

Figure 13.35 Equivalent circuit of a balanced frequency halver.

the differential equation (DE) for the even-mode excitation is

$$(\xi_s + 2\xi_p)\frac{du}{d\tau} + 1 - \frac{1}{4}(u^2 + z^2) = x(\tau) + X_0 \tag{13.33}$$

while the DE for the odd-mode resonance is, assuming R_L to be large,

$$\frac{d^2z}{d\tau^2} + \xi_s z = \frac{1}{2}uz \tag{13.34}$$

where

$\tau = \omega_0 t$, the normalized time;
$\omega_r = [\frac{1}{2}L_1 C_j(0)]^{-1/2}$, the small-signal resonant frequency at zero bias;
$\xi_s = \omega_r/(2\pi f_c(0)) = \omega_r C_j(0) R_s$, the ratio between ω_r and the zero-bias varactor cutoff frequency;
$\xi_p = \omega_r C_j(0) R_p$, pump-resistance parameter;
$x = (V_p/\phi_0)\cos 2\omega t$, normalized pump voltage at frequency 2ω;
$X_0 = V_0/\phi_0$, normalized bias voltage; and
ϕ_0 = varactor built-in potential.

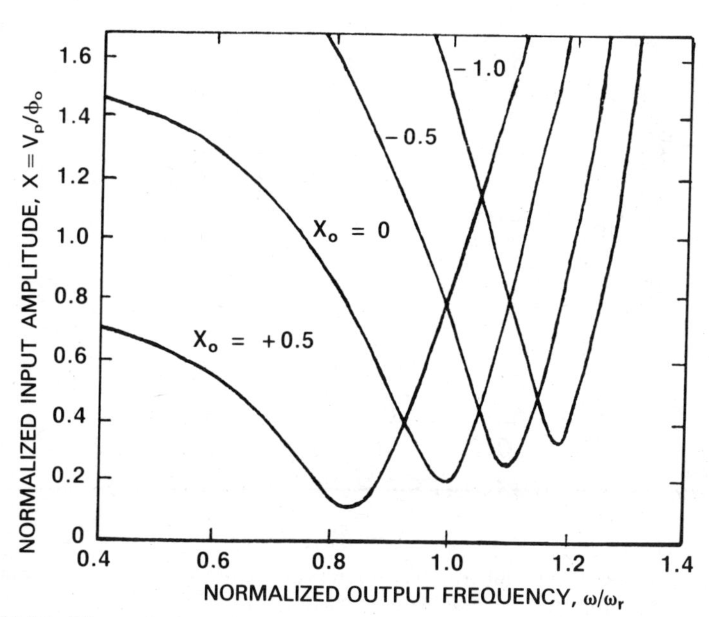

Figure 13.36 Theoretical region of frequency halving for the case $\xi_s = 0.1$, $\xi_p = 0$. The parameter is the normalized bias voltage $X_0 = V_0/\phi_0$.

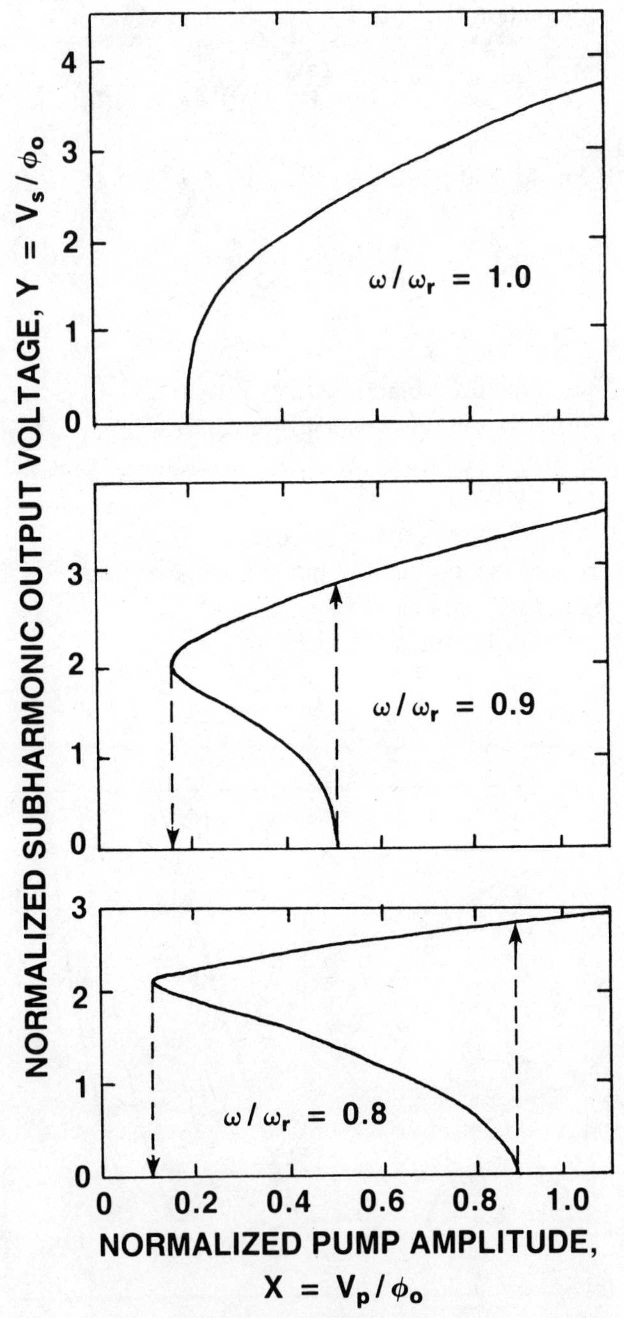

Figure 13.37 Theoretical subharmonic output voltage V_s as a function of pump amplitude V_p and output frequency ω. Here $\xi_s = 0.1$, $\xi_p = 0$.

Solving Equations 13.33 and 13.34 using Hayashi's approximate method [65] yields the steady state solution

$$\left[\frac{3}{2}(\nu^4 - 1 + X_0) + \xi_s\left(\frac{9}{2}\xi_s + 8\xi_p\right)\nu^2 + \frac{5}{64}\frac{Y^2}{\nu^4}\right]^2 + [2\xi_s\nu^3]^2 = X^2 \quad (13.35)$$

where

$X = V_p/\phi_0$,
$\nu = \omega/\omega_r$ is the normalized frequency, and
$Y = V_s/\phi_0$ is the normalized subharmonic amplitude.

Figure 13.36 shows the resulting theoretical region of division of 2, for biases between −1 and +0.5. The effect of bias on both resonant frequency and required pump threshold is apparent. The half-frequency response as a function of pumping level is given in Fig. 13.37 for several values of detuning. This illustrates the typical amplitude hysteresis for frequencies below resonance. Figure 13.38 is a sketch of the halving response as a function of both frequency and amplitude.

For significantly forward-biased operation, the $q_d(v)$ and $i(v)$ varactor nonlinearities become important and the full device model of Fig. 13.15 must then be used. This means that algebraic solutions must give way to

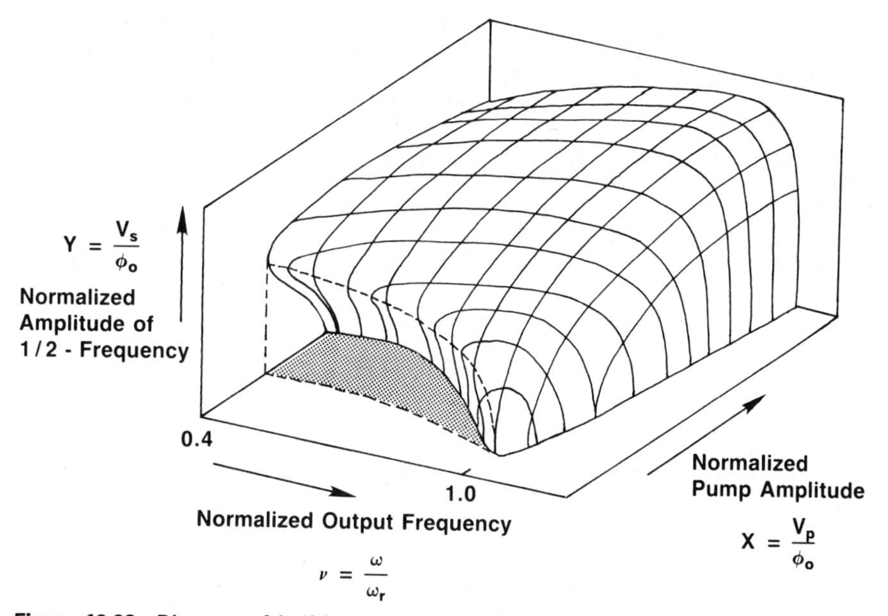

Figure 13.38 Diagram of half-frequency response surface as a function of frequency and pump amplitude, assuming zero bias ($X_0 = 1$), $\xi_6 = 0.1$, $\xi_p \cong 0$.

numerical simulations, typically using the well-known harmonic balance method. An example for a narrow-band single-varactor halver has been given by Lipparini et al. [66, 67] and is discussed in the next subsection.

Practical Parametric Frequency Dividers. The earliest true microwave parametric divider was a single-varactor parametric phase-locked oscillator intended as a phase-switching computer element [68, 69] (see Fig. 13.39). With a 4-GHz input and a 2-GHz output the best conversion efficiency was ~ -18 dB (1.58%).

Figure 13.40 is the schematic of a much more recent single-varactor halver [66, 67, 70]. The input matching network, consisting of sections 1 to 6, also acts as a filter to isolate the half-frequency from the generator. The low-pass output filter and matching network, sections 7 to 13, minimizes feedthrough of the input frequency to the load R_L. The microstrip dimensions are listed in Table 13.1. The behavior of the entire circuit is analyzed—assuming the varactor model of Fig. 13.15—using a version of the well-known harmonic-balance method that simultaneously optimizes the elements of the entire network. Figure 13.41(a) compares the predicted and measured conversion efficiencies versus detuning for a narrow-band halver designed for a center frequency of 2.375 GHz. Figure 13.41(b) depicts the computed voltage waveform at the varactor junction: the limiting of the forward voltage due to conduction is apparent.

Balanced designs [71, 72], eliminate the need for RF filters and improve the transient response. Figure 13.42(a) depicts a basic microstrip/varactor subharmonic resonator circuit. An input signal at f_{in} pumps the two varactors in phase, that is, in an *even* mode, while the subharmonic resonance at $f_{in}/2$ occurs in an *odd* mode. In a practical wide-band frequency halver embodying this principle [see Fig. 13.42(b) and (c)], two GaAs varactors are used. A coplanar waveguide (CPW) balun coupled to the microstrip resonator provides an unbalanced output. Figure 13.43 shows the performance of this design for $f_{in} \sim 4$–8 GHz. The presence of the CPW balun leads to out-of-band asymmetry, causing leakage of f_{in} to the output. This problem is avoided by using the completely symmetrical microstrip/slotline

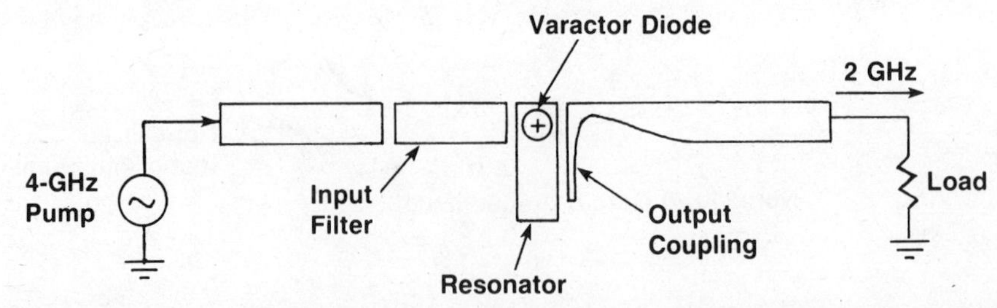

Figure 13.39 Early (1959) microstrip parametric frequency divider. (After F. Sterzer [68], 1959, IRE, reprinted with permission.)

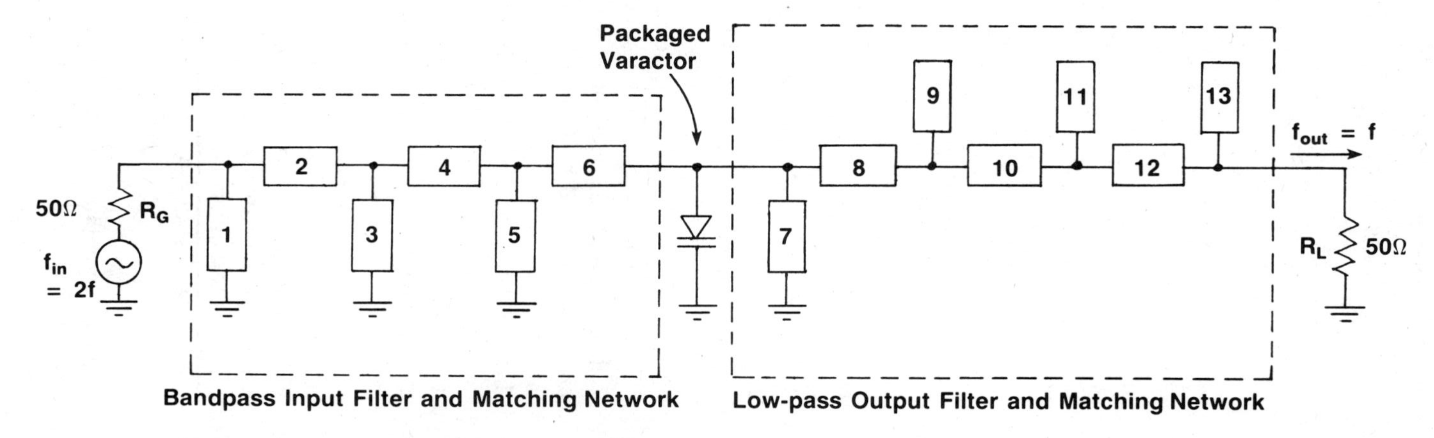

Figure 13.40 Schematic of single-varactor parametric frequency divider. (After V. Rizzoli et al. [70], 1985 IEEE, reprinted with permission.)

TABLE 13.1 Microstrip Dimensions for Fig. 13.40

Line Number	Width (mm)	Length (mm)
1	8.9	25.4
2	8.4	18.4
3	6.6	34.1
4	6.8	23.6
5	8.1	19.9
6	7.7	22.0
7	8.9	14.0
8	1.1	13.1
9	6.1	20.7
10	5.1	22.5
11	4.6	14.3
12	3.2	20.6
13	1.3	24.4

Substrate: 1.58-mm Duroid

resonator of Fig. 13.44 [73]. This consists of a substrate of thickness h carrying on one side an input microstrip line of width W and on the other a ground plane incorporating a collinear slotline of width S. The voltage-dependent reactances of the two varactors, together with the overlapped slot-line section of length l, form the parametric subharmonic resonator. An input–output coupling (at $2f$ and its harmonics) is minimized by the balanced phase. Because of the nonlinear coupling mechanism between this mode and the odd-mode resonance, as expressed by Equations 13.33 and 13.35, energy is transferred from $2f$ to f, causing subharmonic currents to flow through the closed path indicated in Fig. 13.44 in broken line. Unwanted input–output coupling (at $2f$ and its harmonics) is minimized by the balanced circuit configuration.

Figure 13.45 shows an implementation of this structure using discrete components and a simple slot-line–microstrip output transition [74]. Here the 50-Ω impedance Z_0 of the input microstrip line is matched to the *pumped* impedance of the two varactors in parallel by the line sections of impedance Z_1 and Z_2. The output transition converts the balanced subharmonic signal developed across the slot line to an unbalanced signal on the output microstrip line. Such a circuit is suitable for use with inputs up to ~4 GHz. The performance of a design for the 2–4 GHz input band is given in Fig. 13.46.

At higher frequencies, the design of Fig. 13.47, which incorporates a fourth-order Marchand balun as a combined transition and output matching structure, is more appropriate [73]. Such a network can be designed as described in [75].

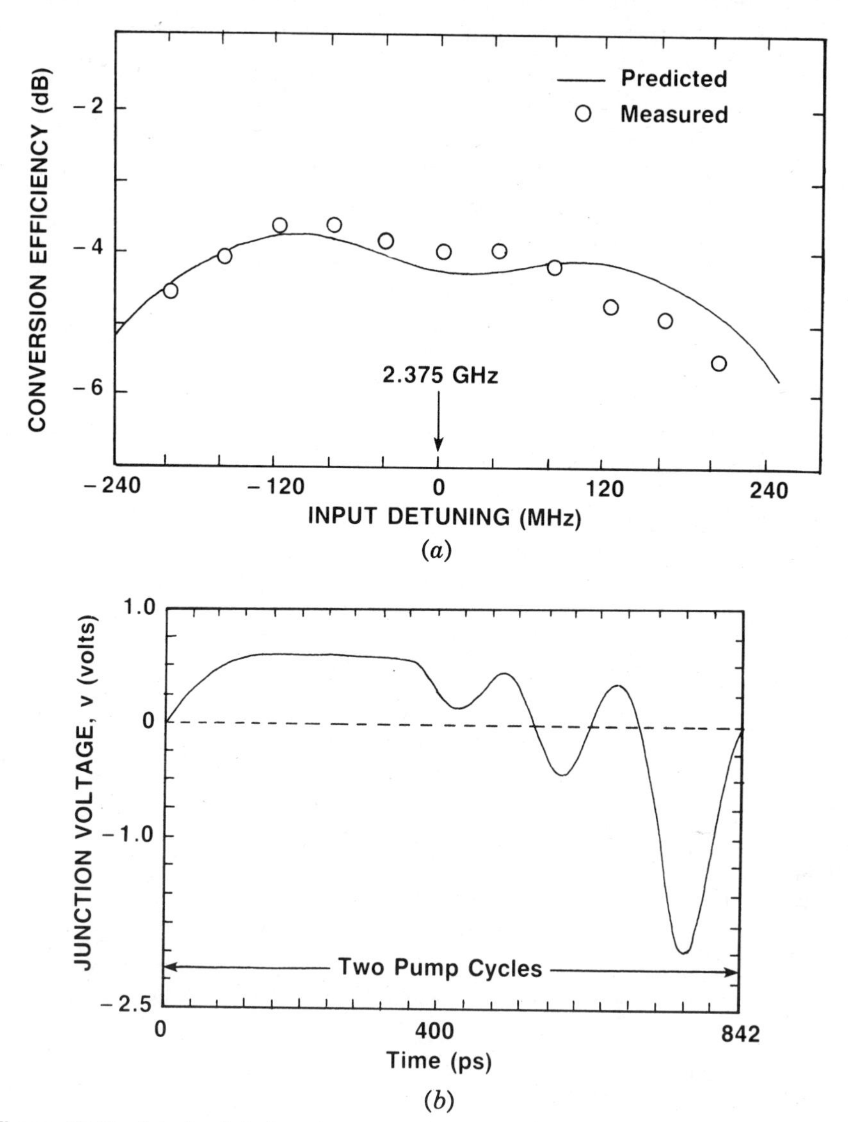

Figure 13.41 (*a*) Predicted and measured conversion efficiencies of narrow-band varactor halver. (*b*) Computed varactor junction voltage at band center. (After A. Lipparini et al. [67], 1982 IEEE, reprinted with permission.)

For octave-band operation conversion loss is fairly high, but can be reduced in optimized designs to ~6 dB for narrower operational bandwidths, with threshold powers as low as −8 dBm [76].

Advantages and Disadvantages of Parametric Dividers. Parametric varactor frequency halvers have the following practical advantages [76].

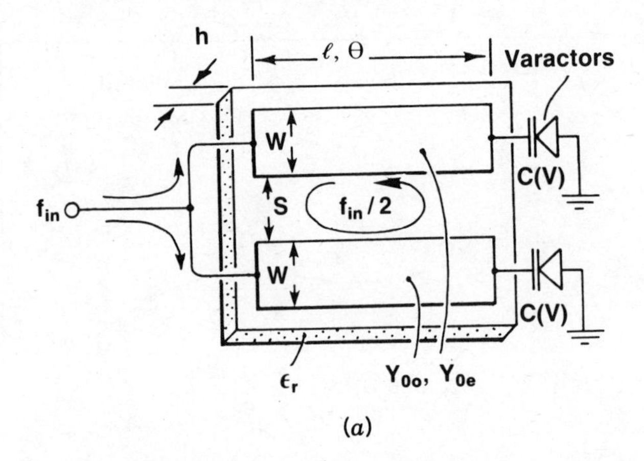

(*a*)

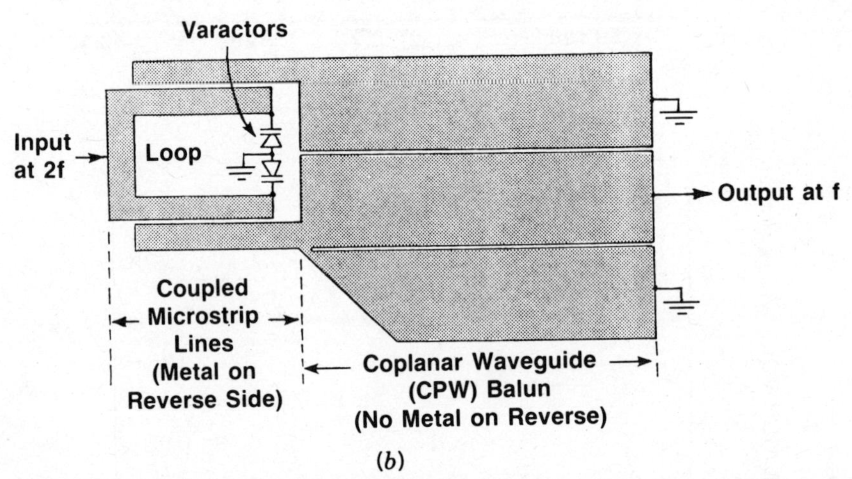

(*b*)

Figure 13.42 (*a*) Basic balanced subharmonic resonator circuit. Y_{0o} and Y_{0e} are the odd- and even-mode impedances of the coupled microstrip lines of length ℓ. (*b*) Layout of a microstrip/CPW balanced varactor frequency divider. (After R. G. Harrison [72], 1977 IEEE, reprinted with permission.)

1. As seen in Fig. 13.48, they have very good phase-noise characteristics. Digital dividers, on the other hand, have poor phase-noise properties.
2. As Fig. 13.49 indicates, extreme temperature variations have a negligible effect on the operational bandwidth and only a small effect on the conversion loss.
3. Compared with the digital frequency dividers to be discussed in Section 13.3.4, varactor dividers are much easier to use since there are no heat-dissipation problems.

A disadvantage of varactor halvers is that pre- and/or postamplifiers are

(*c*)

Figure 13.42 (*c*) Photograph of a broadband balanced varactor frequency halver for the 4–8 GHz input band. Substrate dimensions are 14 × 28 mm. (After R. G. Harrison [72], 1984 IEEE, reprinted with permission.)

needed to overcome the insertion loss. Although FM signals can be divided successfully, spurious-signal generation can occur under multifrequency input conditions. For two input signals at different frequencies and power levels, the larger signal will generally be divided by two, while the smaller signal will mix with the divided output signal and cause heterodyne sidebands. Similar behavior occurs for the regenerative dividers described in the next subsection.

13.3.3 Mixer-with-Feedback Dividers

Regenerative frequency dividers were simultaneously and independently described in 1939 by Fortescue [77] and Miller [78]; a brief description was also given by Kharkevich [79] in 1962.

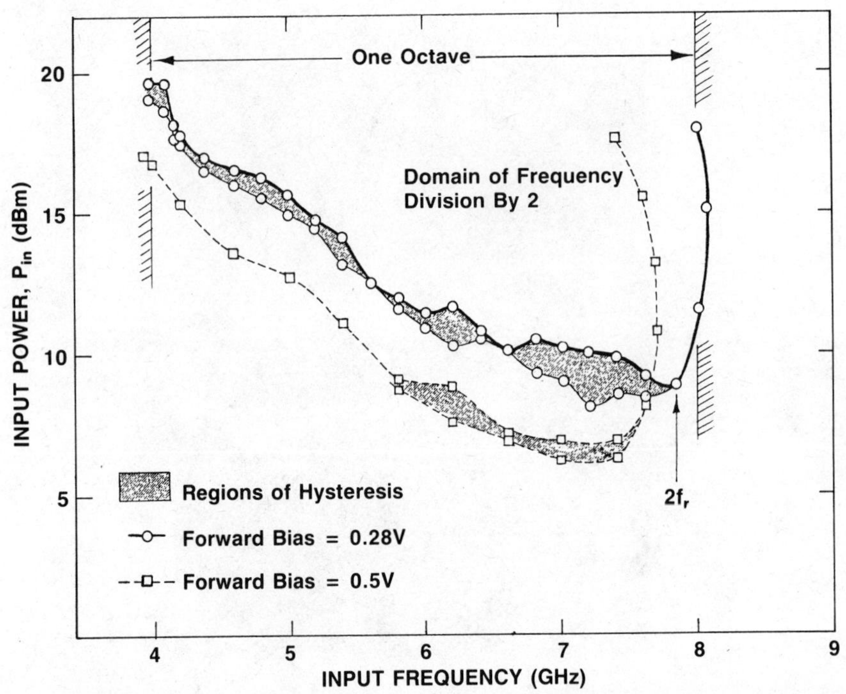

Figure 13.43 Domain of frequency division for the halver shown in Fig. 13.42(*c*), for two different values of bias V_0. (After R. G. Harrison [72], 1984 IEEE, reprinted with permission.)

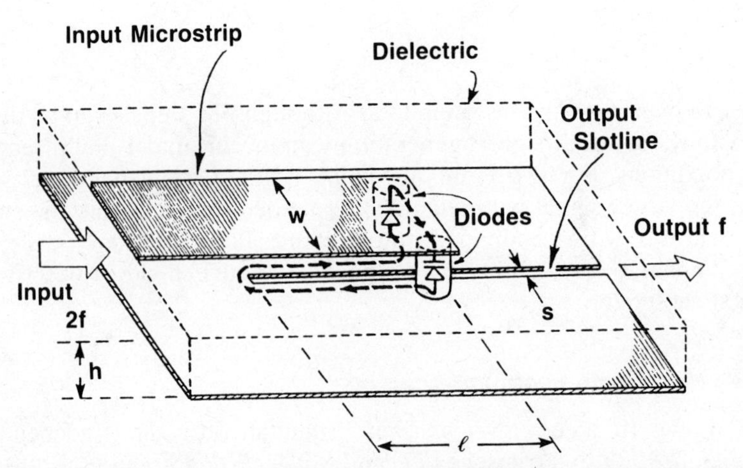

Figure 13.44 Basic slot-line–microstrip symmetrical frequency-halver structure. (After G. A. Kalivas and R. G. Harrison [73], 1985 IEEE, reprinted with permission.)

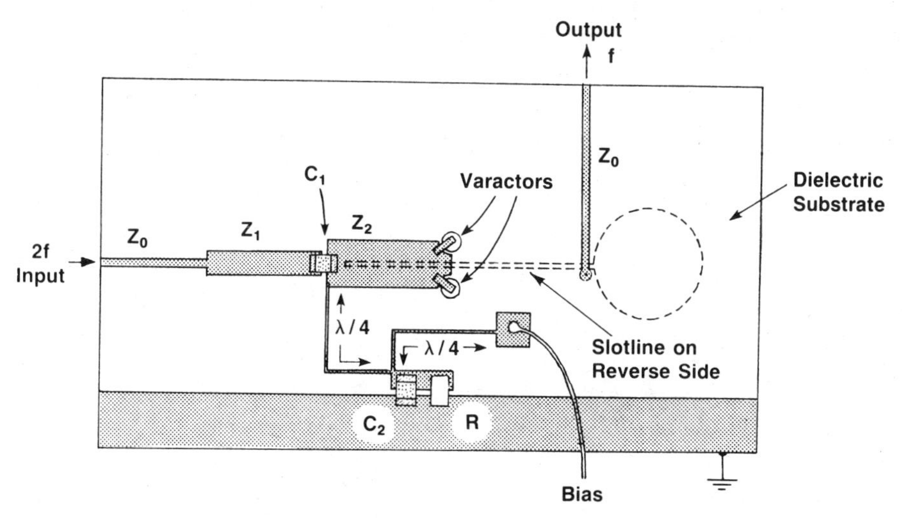

Figure 13.45 A practical implementation of the slot-line–microstrip frequency halver for the 2–4 GHz input band. The varactors are mounted in holes in the substrate. C_1 is a chip blocking capacitor, while the bias circuit consists of chip capacitor C_2, resistor R, and two high-impedance quarter-wave lines.

Theory of Regenerative Frequency Dividers. A generalized scheme for regenerative frequency division by N is given in Fig. 13.50. For regeneration to occur, (1) a finite signal at the output frequency f must be present in the loop initially, and (2) the loop gain must exceed unity. However, to avoid spurious oscillations, the loop gain must be less than unity in the absence of an input signal [80]. For regeneration, the input frequency Nf (N an integer) is mixed with a frequency $(N-1)f$ to give the dominant sidebands

$$Nf \pm (N-1)f = (2N-1)f, f \tag{13.36}$$

The frequency $(N-1)f$ is obtained from f by means of a multiplier-by-$(N-1)$. If the multiplier is accurate, and if the sideband f is selected by filtering, then accurate division by N is obtained. Mixer loss is compensated for by amplification in the loop. In a degenerate scheme, the multiplier is absent and division by 2 occurs. For stable microwave performance, the loop delay due to filters and other dispersive elements must be much less than the FM modulation period. The band of stable division is generally bounded by asynchronous and submultiple-division modes. A detailed theoretical treatment of stability and response to wide-band FM signals has been given by Immovili and Mantovani [81].

Practical Regenerative Frequency Dividers. Microwave regenerative dividers for ~4 GHz [81–83] and ~12 GHz [84] inputs have been constructed with discrete components. Loop delays are consequently long,

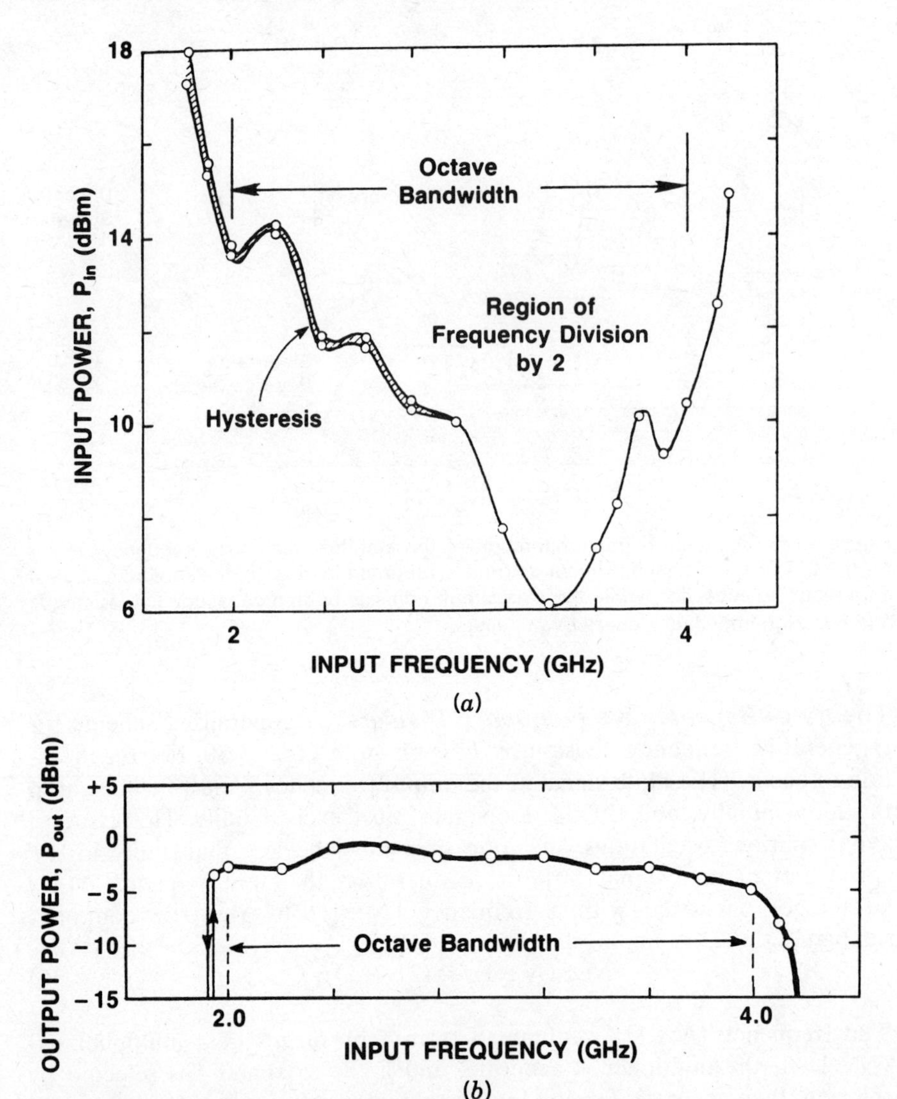

Figure 13.46 Octave bandwidth microstrip–slot-line frequency-halver response. (*a*) Region of frequency division. (*b*) Power output P_{out} at $f_{in}/2$ for $P_{in} = +15$ dBm. Bias is $V_0 = 0.87$ V.

leading to narrow fractional bandwidths in the range 1% to 4%. As theoretically predicted [81], a series of frequency bands of stable and unstable operation is often obtained. Additionally, transient responses are slow [84]. Conversion gains up to +6 dB have been reported [83].

Wider bandwidth and faster transient response can be achieved by reducing the loop size through circuit integration. Rauscher [80] has undertaken a comprehensive investigation of an MIC regenerative halver

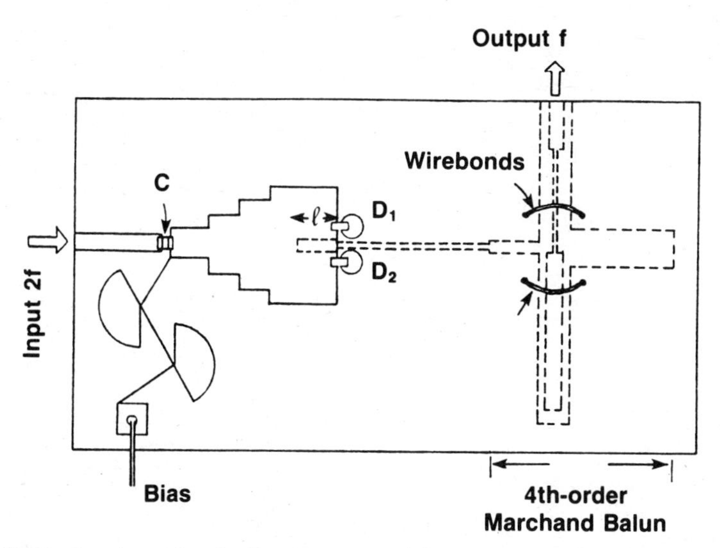

Figure 13.47 A microstrip–slot-line frequency halver using a four-section input matching transformer, radial lines in the bias decoupling circuit, and a fourth-order Marchand balun for output matching. (After G. A. Kalivas and R. G. Harrison [73], 1985 IEEE, reprinted with permission.)

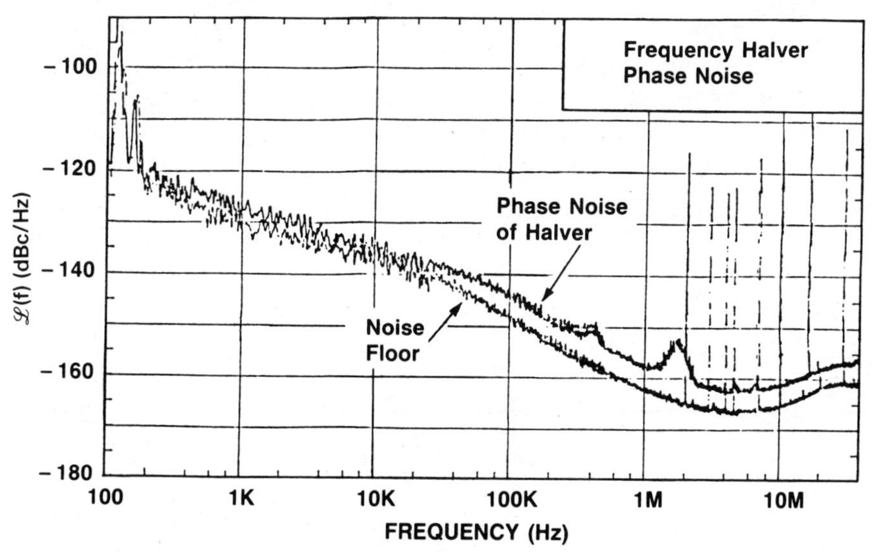

Figure 13.48 Phase noise of a typical varactor frequency halver. (Courtesy of Telemus Electronic Systems, Inc.)

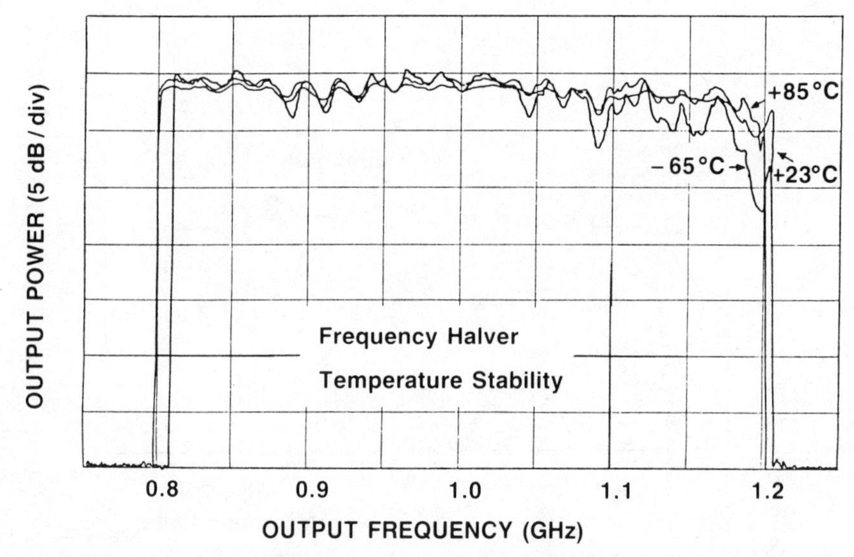

Figure 13.49 Temperature stability of output response of a typical varactor frequency halver. (Courtesy of Telemus Electronic Systems, Inc.)

for inputs in the 15.5–16.5 GHz region. Small size is achieved by using a single GaAs FET chip to perform as both mixer and amplifier, as indicated in the schematic of Fig. 13.51(a). The critical subharmonic feedback and filtering functions are performed by the 300-μm-diameter air-coil inductor L_1 and SiN blocking capacitor C_3, connected in series between drain and gate. Characteristic impedances and electrical lengths of the microstrip input and output coupling networks are listed in Table 13.2. Figure 13.51(b) is a photograph of the assembled unit; the SMA connectors indicate its dimensions. As can be seen from Fig. 13.51(c), the small loop size increases the fractional bandwidth to ~6%. A conversion efficiency of

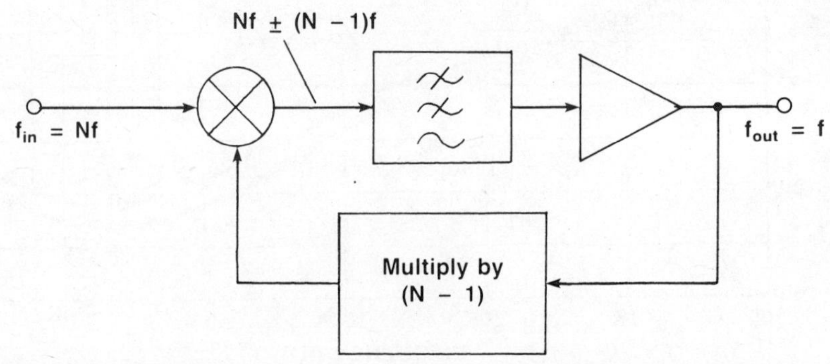

Figure 13.50 Generalized scheme for regenerative division by N. For division by 2, the multiplier is not required.

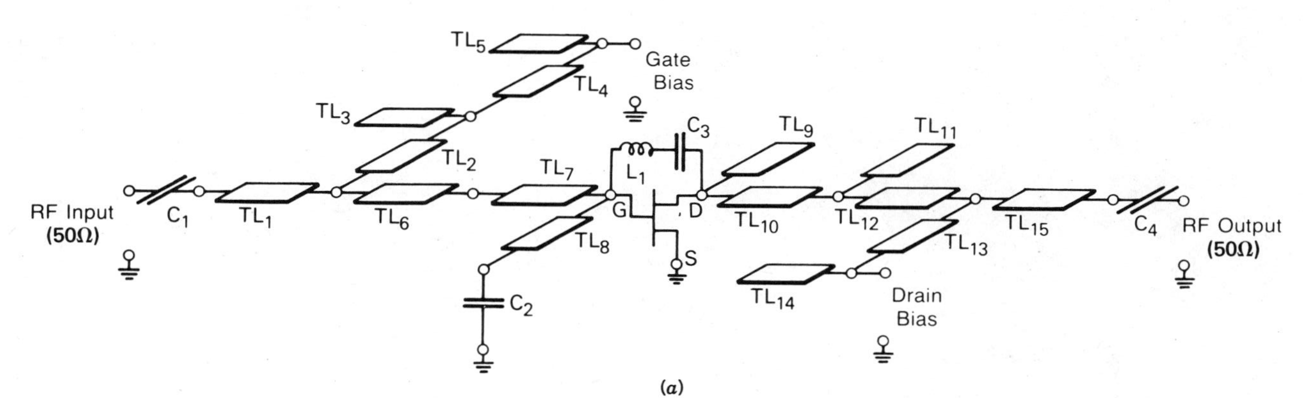

(*a*)

Figure 13.51(*a*) Schematic of a 16–18 GHz regenerative frequency divider using a single GaAs FET. The subharmonic feedback path consists of L_1 (5 nH) and C_3 (5 pF). (After C. Rauscher [80], 1984 IEEE, reprinted with permission.)

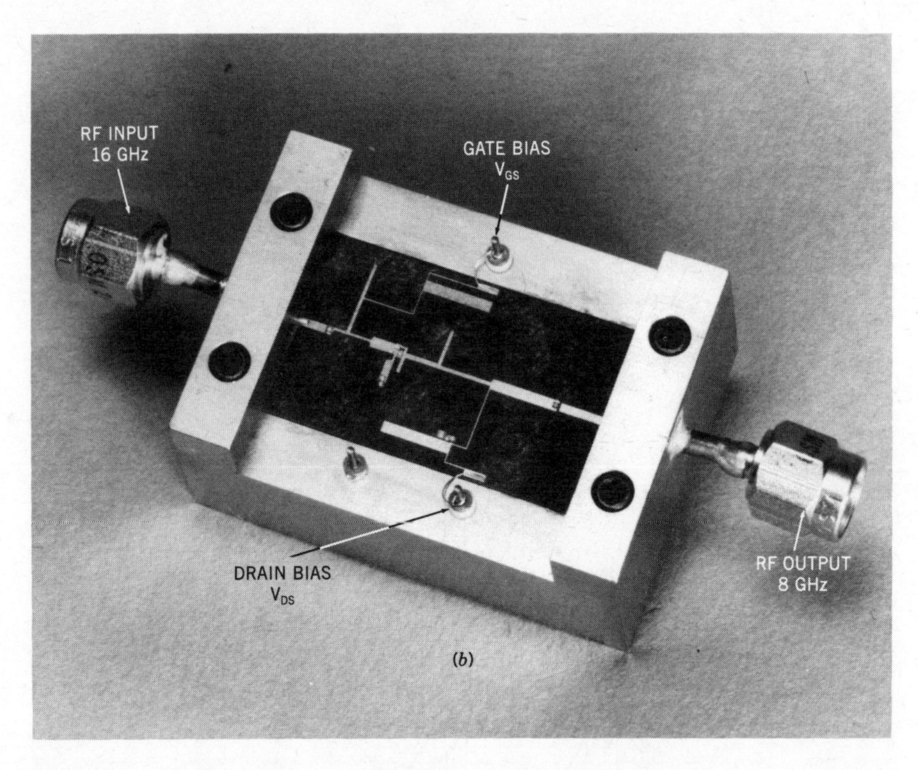

Figure 13.51 (*b*) A 16–18 GHz regenerative frequency divider employing a single GaAs FET. (After C. Rauscher [80], 1984 IEEE, reprinted with permission.)

-3 dB with $P_{out} = +12$ dBm is obtained for $f_{in} = 16$ GHz, while the pulse rise time is of the order of 20 ns (160 cycles of $f_{in}/2$), decreasing with increasing P_{in}.

Further reductions in loop size have been achieved by Derksen et al. [85, 86] who employ a standard Si bipolar technology to fabricate a wide-band regenerative divider IC operating with inputs from 1.4–5.3 GHz, power dissipation being approximately 135 mW. The circuit is shown in Fig. 13.52, in which T_1–T_6 act as a double-balanced mixer that suppresses even harmonics of $f_{in}/2$. Normally, f_{in} is applied to node I, while $\bar{I}$ is blocked with an external capacitor. The feedback voltage v_f is applied to T_4 and T_5, while the transimpedance amplifier enhances both gain and bandwidth. Use of the regenerative principle allows approximately twice the frequency of operation possible using a conventional master–slave D-type flip-flop (MS-D-FF) in the same technology. Here f_{in}^{max} is as high as the transit frequency of the transistors in the loop (5.3 GHz at $V_{CE} = 1$ V), whereas in the MS-D-FF, f_{in}^{max} is limited by two gate delays in the loop. On the other hand, the regenerative divider has a higher f_{in}^{min} because the $3f_{in}/2$ mixer product may not be suppresssed if $f_{in} < f_{in}^{max}/3$. Figures 13.53 and 13.54 show,

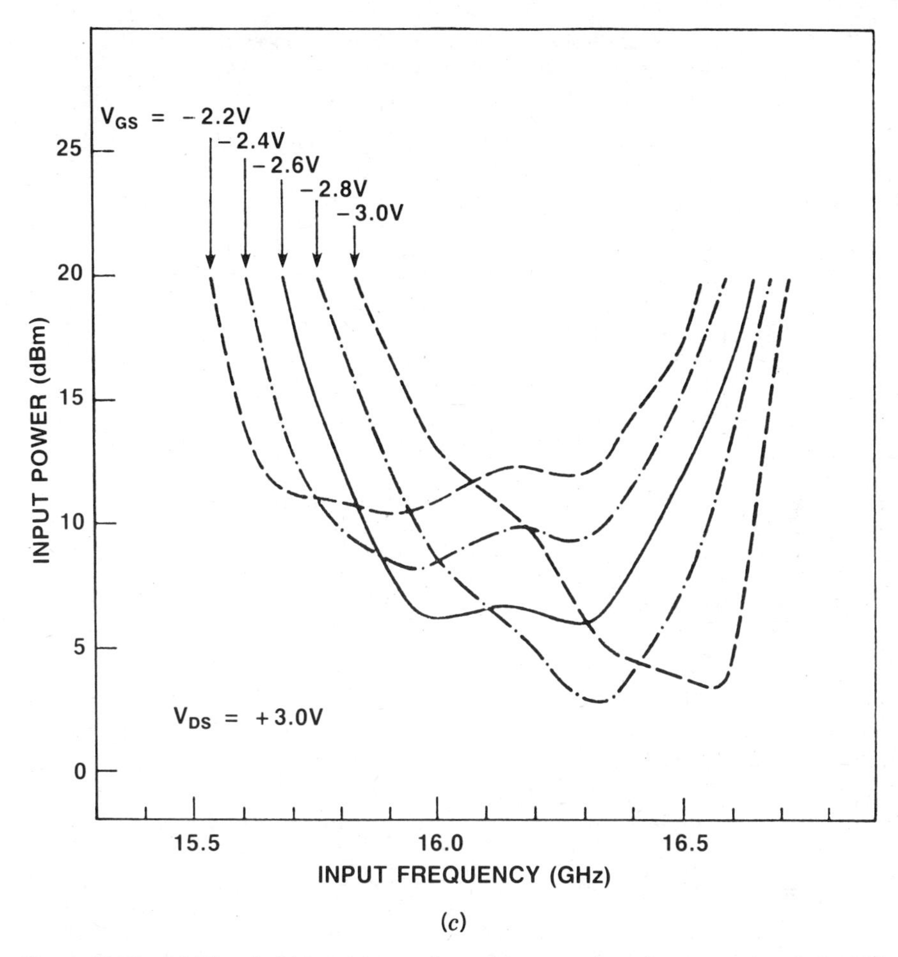

(*c*)

Figure 13.51 (*c*) Threshold input power for stable operation of regenerative GaAs FET divider as a function of input frequency and bias voltage V_{GS}. [After C. Rauscher [80], 1984 IEEE, reprinted with permission.)

respectively, part of the actual chip layout and the region of frequency division. Improved circuits of this type [87] divide by 2 and 8 at frequencies up to 9 GHz with reduced power consumption.

Preliminary results obtained by Kaminsky et al. [88] indicate that the dual-gate GaAs FET is also a candidate for MMIC regenerative divider realizations. A schematic is given in Fig. 13.55. Here f_{in} is applied to the first gate G_1, while the $f_{in}/2$ feedback signal passes through a low-pass filter to the second gate G_2. A conversion gain of +4 dB was obtained at $f_{in} = 6.65$ GHz. The fractional bandwidth was about 14% at $P_{in} = 10$ dBm, increasing to approximately 23% at $P_{in} = 15$ dBm.

TABLE 13.2 Values of Characteristic Impedances Z_0 and Electrical Lengths θ at 16 GHz for the Transmission-Line Elements of Fig. 13.51(*a*)

TL_1: $Z_0 = 95\ \Omega$, $\theta = 33.5°$	TL_9: $Z_0 = 80\ \Omega$, $\theta = 43°$
TL_2: $Z_0 = 95\ \Omega$, $\theta = 90°$	TL_{10}: $Z_0 = 70\ \Omega$, $\theta = 90°$
TL_3: $Z_0 = 95\ \Omega$, $\theta = 90°$	TL_{11}: $Z_0 = 80\ \Omega$, $\theta = 90°$
TL_4: $Z_0 = 95\ \Omega$, $\theta = 180°$	TL_{12}: $Z_0 = 80\ \Omega$, $\theta = 130°$
TL_5: $Z_0 = 40\ \Omega$, $\theta = 180°$	TL_{13}: $Z_0 = 95\ \Omega$, $\theta = 180°$
TL_6: $Z_0 = 95\ \Omega$, $\theta = 56.5°$	TL_{14}: $Z_0 = 40\ \Omega$, $\theta = 180°$
TL_7: $Z_0 = 42\ \Omega$, $\theta = 75°$	TL_{15}: $Z_0 = 42\ \Omega$, $\theta = 180°$
TL_8: $Z_0 = 51\ \Omega$, $\theta = 71°$	

The Values of the Lumped Elements

$C_1 = C_2 = C_4 = 20$ pF
$C_3 = 5$ pF (SiN capacitor)
$L_1 = 5$ nH (300 μm diameter air coil)

Source: C. Rauscher, "Regenerative Frequency Division with a GaAs FET," *IEEE Trans. Microwave Theory Tech.*, Vol. MTT-32, Nov. 1984.

Advantages and Disadvantages of Regenerative Frequency Dividers. These circuits have potential as moderate-bandwidth dividers that can be realized in a variety of technologies. They can be designed for zero output in the absence of an input signal [80]. This is not the case with PLL dividers, for example. However, they do have several disadvantages.

1. The dynamic range is small. This, together with limiting action, causes partial loss of amplitude information.
2. For pulsed signals [80, 84], leading-edge delays can be long.
3. Like ILO dividers, regenerative dividers can exhibit very noisy behavior for multiple input signals closely spaced in frequency.

13.3.4 Digital Frequency Dividers

At present, analog GaAs MMICs can operate with fundamental frequencies up to at least 60 GHz. In digital GaAs MMICs, however, where "square" waveforms are required, harmonics up to at least the third must be present, thereby limiting the fundamental frequency to about 20 GHz. In general, for a given technology, digital dividers will operate at maximum frequencies considerably lower than will their analog counterparts.

Theory of Digital Dividers. Digital frequency dividers (DFD) operate on the well-known flip-flop principle, typically using master–slave divide-by-two circuits as described, for example, in [92]. They are either of the

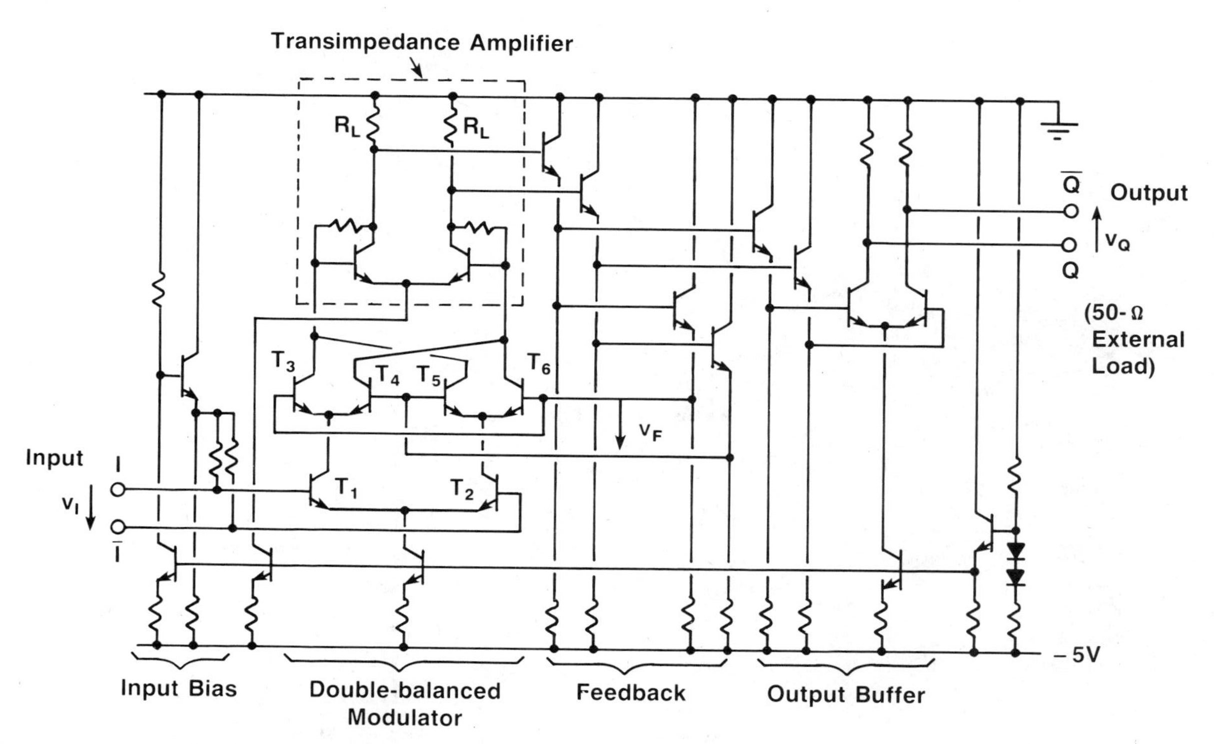

Figure 13.52 Circuit diagram of monolithic regenerative frequency divider, with output buffer, using a bipolar Si technology. (After R. H. Derksen et al. [85, 86], reprinted with permission of Electronics Letters.)

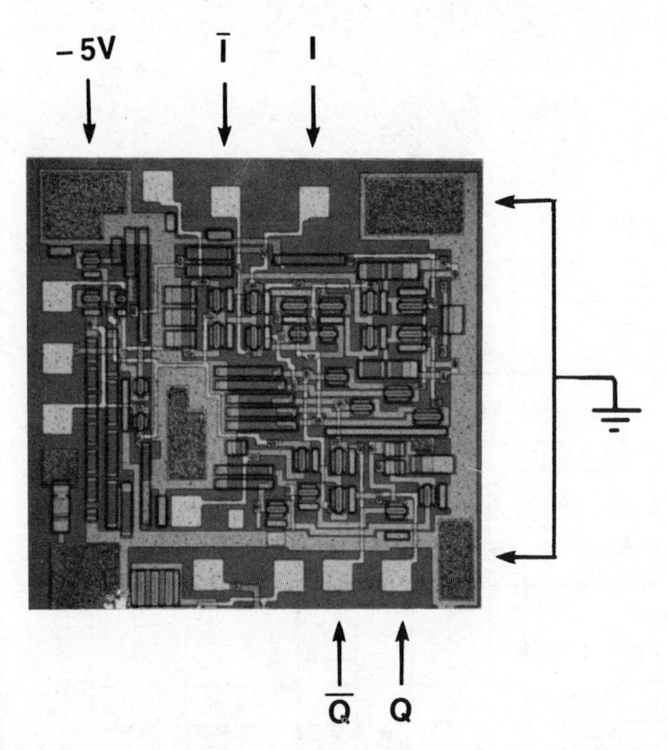

Figure 13.53 Part of chip layout of the circuit shown in Fig. 13.52. (After R. H. Derksen et al. [85, 86], reprinted with permission of Eletronics Letters.)

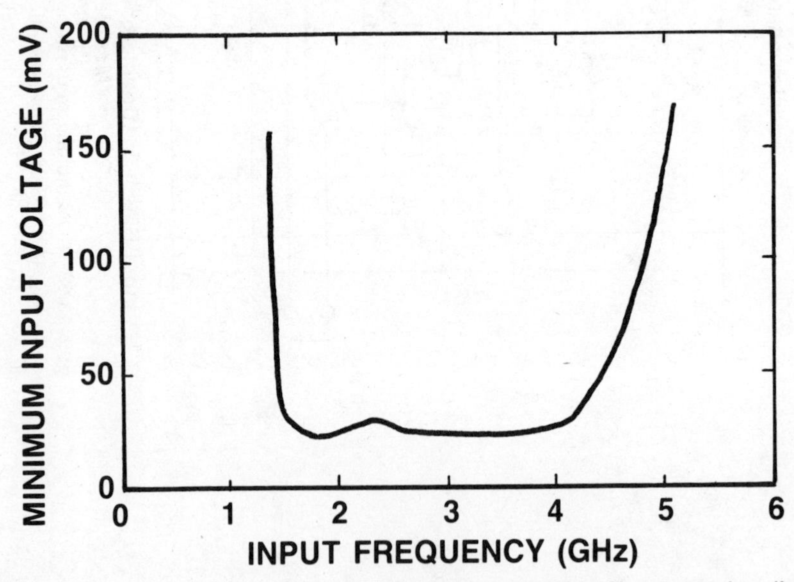

Figure 13.54 Region of frequency of division for monolithic Si regenerative divider. The input voltage is measured without circuit loading. (After R. H. Derksen et al. [85, 86], reprinted with permission of Electronics Letters.)

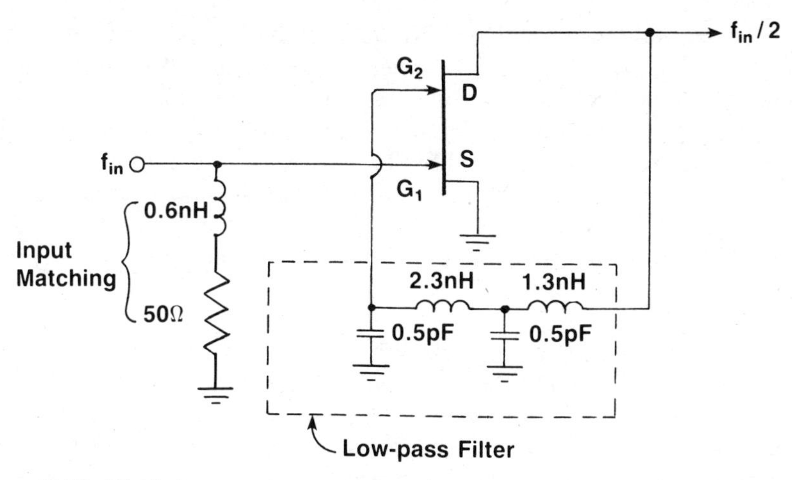

Figure 13.55 Divide-by-two circuit using dual-gate GaAs FET with regenerative feedback. (After D. Kaminsky et al. [88], 1983 IEEE, reprinted with permission.)

dynamic type, or of the *static* type. *Dynamic* DFDs [89–92] use astable flip-flops with *ac* cross couplings, and therefore tend to produce an output in the absence of an input, but cannot operate down to dc. *Static* DFDs, on the other hand, have *dc* cross couplings and do operate down to dc. Since they do not self-oscillate, there is no output in the absence of an input.

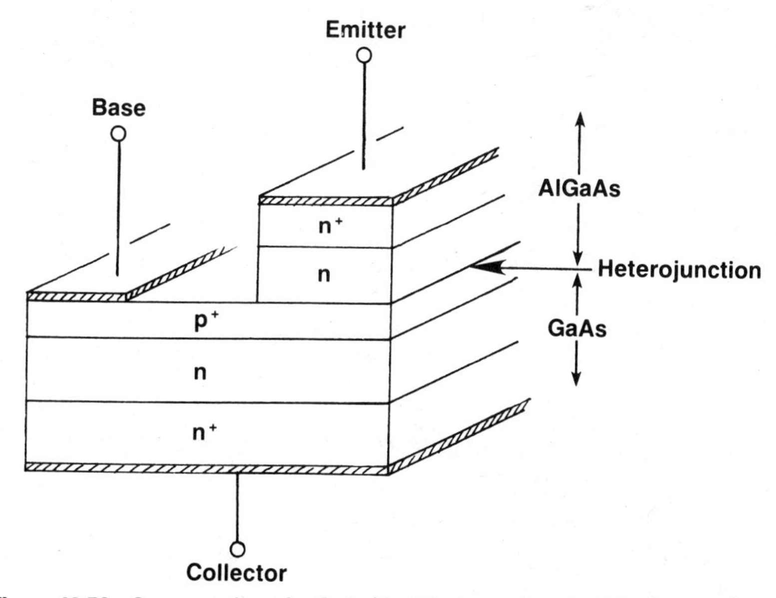

Figure 13.56 Cross section of a GaAs/GaAlAs heterojunction bipolar transistor.

Practical Digital Dividers. Digital frequency dividers operating at ~4 GHz were reported in 1977 in both Europe, by Noordanus [89] (astable divider using discrete Si bipolar technology), and in the United States, by Van Tuyl et al. [93, 94] (static divider using monolithic GaAs FET technology). Since that time progress has been intense. The following subsections review four static digital divider technologies utilizing: (1) Si bipolar transistors, (2) heterojunction bipolar transistors (HBT), (3) GaAs FETs, and (4) GaAs/GaAlAs (and other III–IV material) high-electron-mobility transistors (HEMT).

Si Bipolar Digital Dividers. Currently, oxide-isolated Si bipolar master–slave flip-flop (MS-FF) dividers achieve $f_{in} = 2.6$ GHz with $P_{diss} = 110$ mW, with 3.5 GHz projected [95]; whereas Derksen et al. [85, 86] use the same technology to get $f_{in} > 5$ GHz using the regenerative divider principle (see Section 13.3.3). With super-self-aligned technologies, reduced size and hence higher speed is possible: a 5.5-GHz divide-by-eight circuit with $P_{diss} = 800$ mW has been realized [95] using devices with 0.7-μm emitter width. Further reduction of emitter width to 0.35 μm has resulted in a 9-GHz divider circuit [96]. It can be anticipated that emitter-width reductions to ~0.2 μm will make possible digital dividers for input frequencies in the region of 16 GHz.

Heterojunction Bipolar Transistor Digital Dividers. The performance of bipolar frequency dividers can be increased by using heterojunction bipolar transistor (HBT) technology. Figure 13.56 is a cross section of a GaAs/GaAlAs HBT. Because of the wide-gap emitter, the base doping can be increased without reducing emitter efficiency, yielding low base resistance and emitter capacitance [95]. Although initial HBT results were disappointing [97], Asbeck et al. [98] have now obtained $f_{in}^{max} = 8.6$ GHz, for power dissipation $P_{diss} = 210$ mW at room temperature, using transistors with 1.2-μm emitter width and cutoff frequencies as high as 40 GHz in current-mode logic (CML) circuitry.

GaAs FET Digital Dividers. Reviews of progress in GaAs-FET divider technology have been given by Bosch [99, 100] and by Eden et al. [101]. Compared with Si ICs, GaAs FETs offer much reduced dissipation-times-delay products. The basic GaAs IC approaches are as follows.

1. Direct-coupled FET logic (DCFL), using enhancement-mode FETs as switches, and depletion-mode devices as active loads;
2. Schottky-diode FET logic (SDFL), using diodes for logic and depletion FETs for other functions;
3. Buffered FET logic (BFL), a very fast but high-P_{diss} technology employing depletion FETs.

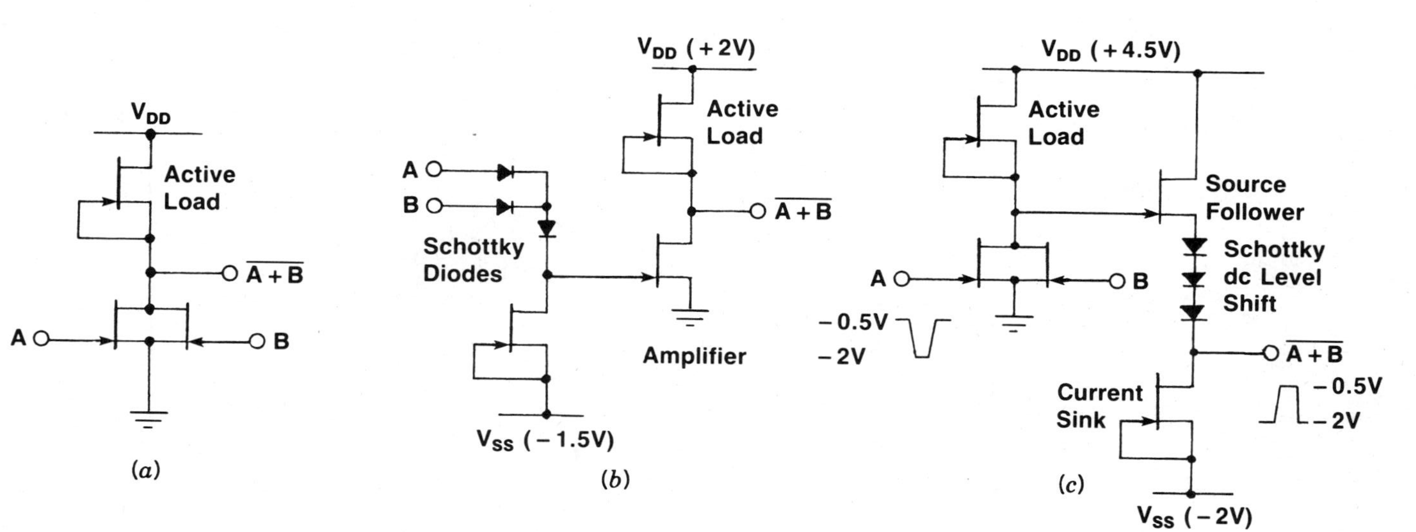

Figure 13.57 Examples of basic GaAs FET (NOR) gates from which flip-flop frequency dividers can be constructed. (*a*) Direct-coupled FET logic (DCFL). (*b*) Schottky-diode FET logic (SDFL). (*c*) Buffered FET logic (BFL).

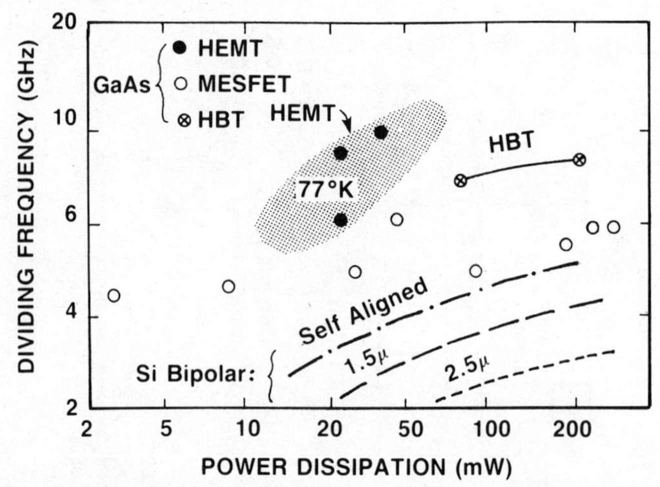

Figure 13.58 Comparison of results obtained from dividers fabricated in various Si and GaAs technologies. (After K. Wörner et al. [95], 1986 IEEE, reprinted with permission.)

Typical (NOR) gates implemented in these technologies are shown in Fig. 13.57.

Of these digital approaches, the best frequency divider results are: $f_{\text{in}}^{\max}$ values of 1.9 GHz ($P_{\text{diss}} = 15$ mW) for SDFL [102], 3.8 GHz ($P_{\text{diss}} = 24$ mW)

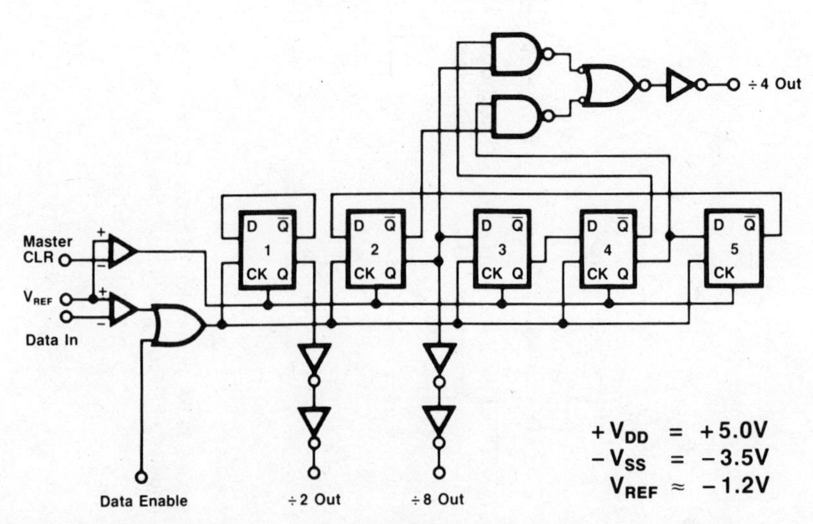

Figure 13.59 (*a*) Logic diagram of type HMD-11016-1 2/4/8 binary counter. (Courtesy of Harris Microwave Semiconductor, Inc.)

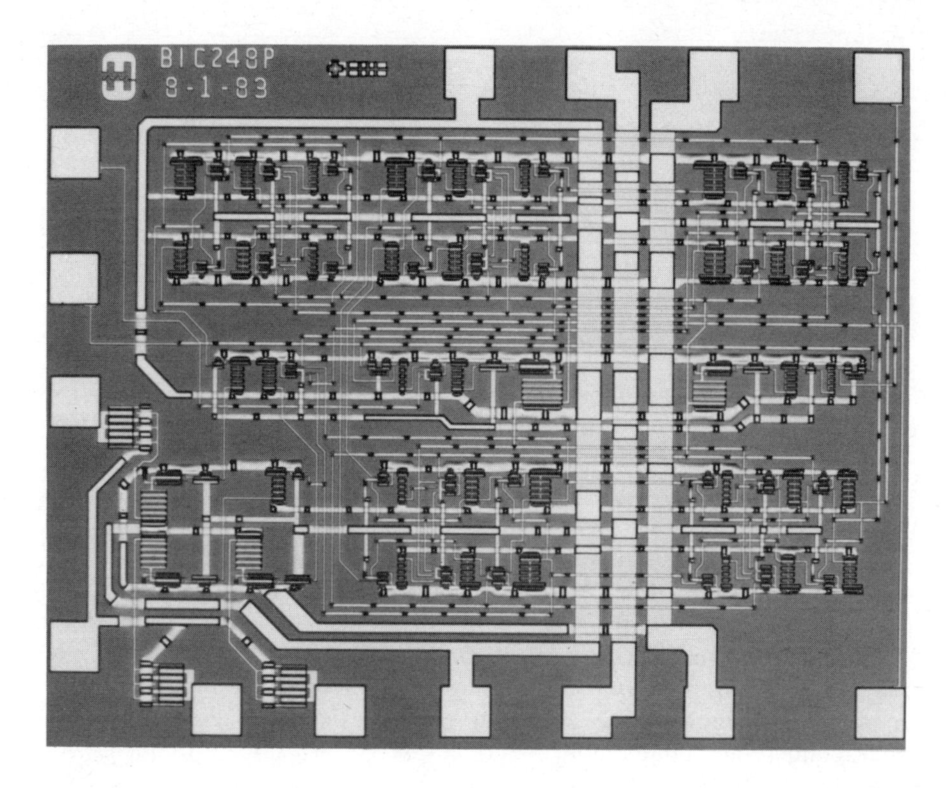

Figure 13.59 (*b*) Photomicrograph of a 2-GHz GaAs divide-by-2/4/8 binary counter, type HMD-11016-1. (Courtesy of Harris Microwave Semiconductor, Inc.)

for DCFL [95], and 5.7 GHz for BFL [103] using 0.8-μm-gate FETs. Interestingly, a BFL master–slave flip-flop (MS-FF) operating at 4.5 GHz ($P_{diss} = 160$ mW) with 1-μm gates, and reported by Van Tuyl et al. [93, 94] in 1977, was one of the first digital dividers to exceed 1 GHz. For comparison, other BFL results include 4.1 GHz for 1.2-μm gates [104] and 5.5 GHz for 0.6-μm gates [105].

HEMT Digital Dividers. The high-electron-mobility FET [also called modulation-doped FET (MODFET), selectively-doped heterostructure transistor (SDHT), or TEGFET] is a heterostructure fabricated using a GaAs/GaAlAs, GaAs/InGaAs, or InAlAs/GaInAs interface. Of these, the first is the more advanced technology. Geometrically, it resembles a GaAs FET. Recent MS-FF dividers have yielded maximum input frequencies f_{in}^{max} of 8.9 GHz (Abe et al., 1983 [106]) and 10.1 GHz (Hendel et al., 1984 [107]) at 77 K. At room temperature, however, f_{in}^{max} is only 5.5 GHz in each case.

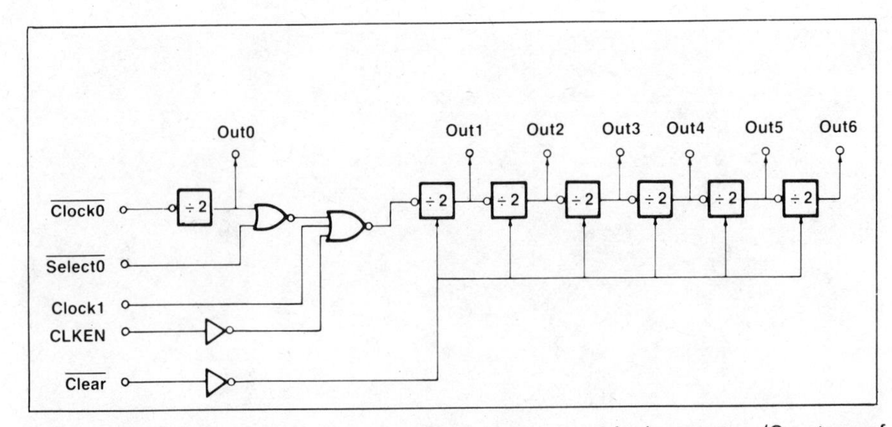

Figure 13.60 Block diagram of type 10G065 seven-stage ripple counter. (Courtesy of Gigabit Logic.)

Summary. Frequency-divider results obtained using the Si and GaAs technologies described previously are summarized in Fig. 13.58.

Examples of Practical GaAs MMIC Dividers. Figure 13.59(a) is the logic diagram of a 2/4/8 binary counter using D-flip-flops, while Fig. 13.59(b) is a photomicrograph of the chip that employs the BFL technology with ECL-compatible input–output interfaces [108]. Maximum frequency of operation is 2 GHz.

Another example is a 7-stage 2.5-GHz ripple counter with the block diagram shown in Fig. 13.60. This unit uses depletion-mode GaAs FETs in capacitor-diode FET logic (CDFL), wherein reverse-biased Schottky diodes act as capacitors [109]. All outputs up to $f_{in}/128$ are available.

REFERENCES

1. Page, C. H., "Harmonic Generation with Ideal Rectifiers," *Proc. IRE*, Vol. 46, Oct. 1958, pp. 1738–1740.
2. Harrison, R. G., "A Nonlinear Theory of Class C Transistor Amplifiers and Frequency Multipliers," *IEEE J. Solid-State Circuits*, Vol. SC-2, Sept. 1967, pp. 93–102.
3. Penfield, P., and R. P. Rafuse, *Varactor Applications*, MIT Press, Cambridge, Mass., 1962.
4. Manley, J. M., and H. E. Rowe, "Some General Properties of Nonlinear Elements, Part I, General Energy Relations," *Proc. IRE*, Vol. 44, July 1956, pp. 904–913.
5. Brobst, J. D., and R. J. Weber, "*L-X* Band 25 μs Coherent Power Sources," *IEEE Int. Solid-State Circuits Conf. Digest*, Feb. 1981, pp. 134–135.

6. Takada, T., and M. Ohmori, "Frequency Triplers and Quadruplers with GaAs Schottky-Barrier Diodes at 450 and 600 GHz," *IEEE Trans. Microwave Theory Tech.*, Vol. MTT-27, May 1979, pp. 519–523.
7. Hirayama, M., T. Takada, T. Ishibashi, and M. Ohmori, "Submillimeter Wave Frequency Multipliers and IMPATT Oscillators," *IEEE MTT-S Int. Microwave Symp. Digest*, June 1978, pp. 435-437.
8. Benson, K., and M. A. Frerking, "Theoretical Efficiency for Triplers Using Real Varister Diodes at Submillimeter Wavelengths," *IEEE-MTT-S Int. Microwave Symp. Digest*, June 1985, pp. 315–318.
9. Harrison, R. G., and T. W. Tucker, "Frequency Division Solves Systems Problems," *Microwave Systems News*, Vol. 8, Oct. 1978, pp. 97–101.
10. Payne, J. B., "Recent Advances in Solid-State, Phase Locked Microwave Signal Sources: Part I," *Microwave Systems News*, Feb./Mar. 1976, pp. 79–87.
11. Payne, J. B., "Recent Advances in Solid-State, Phase Locked Microwave Signal Sources: Part II," *Microwave Systems News*, Apr./May 1976, pp. 118–129.
12. Boctor, S., and D. J. Roulston, "High Efficiency Conditions for an Ideal Nonlinear Capacitor Frequency Multiplier," *Proc. IEEE*, Vol. 57, Apr. 1969, pp. 688–689.
13. Johnston, R. H., "Optimized Frequency Multipliers Using an Ideal Step-Recovery Diode Without Idlers," *Electron. Lett.*, Vol. 5, Apr. 1969, pp. 136–137.
14. Moll, J. L., and S. A. Hamilton, "Physical Modeling of the Step Recovery Diode for Pulse and Harmonic Generation Circuits," *Proc. IEEE*, Vol. 57, July 1969, pp. 1250–1259.
15. Kotzebue, K. L., and G. L. Matthaei, "The Design of Broadband Frequency Doublers Using Charge Storage Diodes," *IEEE Trans. Microwave Theory Tech.*, Vol. MTT-17, Dec. 1969, pp. 1077–1086.
16. Winkelman, R. E., and E. G. Cristal, "Filter Matches to SRD Impulse," *Microwaves*, Vol. 9, Sept. 1970, pp. 34–38.
17. Kodali, V. P., "Narrow Band and Broad Band Step Recovery Diode Frequency Multipliers for Microwave Power Generation," *J. Inst. Telecommun. Eng. (New Delhi)*, Vol. 18, 1972, pp. 457–463.
18. Hewlett-Packard, "Step-Recovery Diode Multipliers," Application Note No. 913.
19. Hewlett-Packard, "Harmonic Generation Using SRD and SRD Modules," Application Note No. 920.
20. Hewlett-Packard, "*Ku* Band SRD Multipliers," Application Note No. 928.
21. Caulton, M., H. Sobol, and R. L. Ernst, "Generation of Microwave Power by Parametric Frequency Multiplication in a Single Transistor," *RCA Rev.*, Vol. 26, June 1965, pp. 286–311.
22. Minton, R., and H. C. Lee, "Designing Transistor Multipliers," *Microwaves*, Vol. 4, Nov. 1965, pp. 18–28.
23. Hyde, F. J., "Parametric Action in Transistors: Theory," *Proc. IEE (London)*, Vol. 113, Feb. 1966, pp. 209–213.

24. Gök, I. and F. J. Hyde, "Parametric Action in Transistors: Experiment," *Proc. IEE* (*London*), Vol. 113, Feb. 1966, pp. 214–218.
25. Anderson, A. P., "Circuit Aspects of Transistor Parametric Frequency Doublers," *IEEE Trans. Electron Devices*, Vol. ED-14, Feb. 1967, pp. 86–89.
26. Harrison, R. G., "A Nonlinear Theory of Class C Transistor Amplifiers and Frequency Multipliers," *IEEE J. Solid-State Circuits*, Vol. SC-2, Sept. 1967, pp. 93–102.
27. Pantell, R. H., "General Power Relationships for Positive and Negative Nonlinear Resistive Elements," *Proc. IRE*, Vol. 46, Dec. 1958, pp. 1910–1913.
28. Page, C. H., "Frequency Conversion with Positive Nonlinear Resistance," *J. Res. Nat. Bur. Stand.*, Vol. 56, Apr. 1956, pp. 179–182.
29. Clay, R., *Nonlinear Networks and Systems*, Wiley–Interscience, New York, 1971.
30. Benson, F. A., and F. Winder, "Nonlinear-Resistance Harmonic Generators," *Electron. Lett.*, Vol. 3, Dec. 1967, pp. 534–535.
31. Nagel, L. W., "SPICE 2: A Computer Program to Simulate Semiconductor Circuits," Memor. No. ERL-M520, Electronics Research Lab., University of California, Berkeley, May 1975.
32. Tang, C. C. H., "An Exact Analysis of Varactor Frequency Multipliers," *IEEE Trans. Microwave Theory Tech.*, Vol. MTT-14, Apr. 1966, pp. 210–212.
33. Sard, E. W., "Closed-Form Analysis of the Abrupt-Junction Varactor Doubler," *IEEE Trans. Microwave Theory Tech.*, Vol. MTT-27, June 1979, pp. 604–611.
34. Burckhardt, C. B., "Analysis of Varactor Frequency Multipliers for Arbitrary Capacitance Variation and Drive Level," *Bell Syst. Tech. J.*, Vol. 44, April 1965, pp. 675–692.
35. Rafuse, R. P., "Recent Developments in Parametric Multipliers," *Proc. Nat. Electronics Conf.*, October 1963, pp. 461–470.
36. Bava, E., G. P. Bava, A. Godone, and G. Rietto, "Analysis of Varactor Frequency Multipliers: Nonlinear Behavior and Hysteresis Phenomena," *IEEE Trans. Microwave Theory Tech.*, Vol. MTT-27, February 1979, pp. 141–147.
37. Bava, E., G. P. Bava, A. Godone, and G. Rietto, "Transfer Functions of Amplitude and Phase Fluctuations and Additive Noise in Varactor Doublers," *IEEE Trans. Microwave Theory Tech.*, Vol. MTT-27, Aug. 1979, pp. 753–757.
38. Swan, C. B., "Design and Evaluation of a Microwave Varactor Tripler," *IEEE Int. Solid-State Circuits Conf. Digest*, Feb. 1965, pp. 106–107.
39. Roulston, D. J., and A. R. Boothroyd, "A Large-Signal Analysis and Design Approach for Frequency Multipliers using Varactor Diodes," *IEEE Trans. Circuit Theory*, Vol. CT-12, June 1965, pp. 194–205.
40. Horton, J. B., "A Thin-Film *X*-Band Varactor Quadrupler," *IEEE Trans. Microwave Theory Tech.*, Vol. MTT-15, Dec. 1967, pp. 752-754.
41. Johnson, K. M., "A High-Performance Microwave Integrated-Circuit

Frequency Quadrupler," *IEEE Int. Solid State Circuits Conf. Digest*, Feb. 1968, pp. 84–85.

42. Staecker, P. W., and R. W. Chick, "10 to 40 GHz Doubler-Doubler Chain for Satcom Applications," *Microwave J.*, Vol. 28, Dec. 1985, pp. 87–108.
43. Chu, A., W. E. Courtney, L. J. Mahoney, R. A. Murphy, and R. W. McLelland, "Monolithic Frequency Doublers," *IEEE Microwave and Millimeter-Wave Monolithic Circuits Symp. Digest* (Boston, Mass., 1983), pp. 45–49.
44. Chu, A., W. E. Courtney, L. J. Mahoney and R. W. McLelland, "GaAs Monolithic Frequency Doublers with Series Connected Varactor Diodes," *IEEE Microwave and Millimeter-Wave Monolithic Circuits Symp. Digest* (San Francisco, Calif., 1984), pp. 74–77.
45. Archer, J. W., "Millimeter Wavelength Frequency Multipliers," *IEEE Trans. Microwave Theory Tech.*, Vol. MTT-29, June 1981, pp. 552–557.
46. Takada, T., M. Takashi, and M. Ohmori, "Hybrid Integrated Frequency Doublers and Triplers to 300 and 450 GHz," *IEEE Trans. Microwave Theory Tech.*, Vol. MTT-28, Sept. 1980, pp. 966–973.
47. Erikson, N. R., "A 200–300 GHz Heterodyne Receiver," *IEEE Trans. Microwave Theory Tech.*, Vol. MTT-29, June 1981, pp. 557–561.
48. Saito, T., M. Iwakuni, T. Sakane, and Y. Tokomitsu, "A 45 GHz GaAs FET Oscillator-Doubler," *IEEE MTT-Int. Microwave Symp. Digest*, 1982, pp. 283–285.
49. Gopinath, A., and J. B. Rankin, "Single-Gate MESFET Frequency Doublers," *IEEE Trans. Microwave Theory Tech.*, Vol. MTT-30, June 1982, pp. 869–875.
50. Rauscher, C., "High Frequency Doubler Operation of GaAs Field Effect Transistors," *IEEE Trans. Microwave Theory Tech.*, Vol. MTT-31, June 1983, pp. 462–473.
51. Dow, G. S., and L. S. Rosenheck, "A New Approach for mm-Wave Generation," *Microwave J.*, Vol. 26, Sept. 1983, pp. 147–162.
52. Gilmore, R., "Design of a Novel FET Frequency Doubler Using a Harmonic Balance Algorithm," *IEEE MTT-S Int. Microwave Symp. Digest*, June 1986, pp. 585–588.
53. Chen, P. T., C.-T. Li, and P. H. Wang, "Dual-Gate GaAs FET Multiplier as a Frequency Multiplier at *Ku*-Band," *IEEE MTT-S Int. Microwave Symp. Digest*, June 1978, pp. 309–311.
54. Chen, P. T., C.-T. Li, and P. H. Wang, "Performance of a Dual-Gate GaAs MESFET as a Frequency Multiplier at *Ku*-Band," *IEEE Trans. Microwave Theory Tech.*, Vol. MTT-27, May 1979, pp. 411–415.
55. Stancliff, R., "Balanced Dual Gate GaAs FET Frequency Doublers," *IEEE MTT-S Int. Microwave Symposium Digest*, June 1981, pp. 143–145.
56. Tsironis, C., R. Meierer, and R. Stahlmann, "Dual Gate MESFET Mixers," *IEEE Trans. Microwave Theory Tech.*, Vol. MTT-32, Mar. 1984, pp. 248–255.
57. Chu, A. S., and P. T. Chen, "An Osciplier up to *K*-Band Using Dual-Gate

GaAs MESFET," *IEEE MTT-S Int. Microwave Symp. Digest*, June 1980, pp. 383–386.

58. Ryan, W. D., and H. B. Williams, "The Carrier-Storage Frequency Divider: A Steady-State Analysis," *IEEE Trans. Circuit Theory*, Vol. CT-11, Sept. 1964, pp. 396–403.
59. Upadhyayula, L. C., and S. Y. Narayan, "Microwave Frequency Division Using Transferred Electron Devices," *Electron. Lett.*, Vol. 9, Feb. 1973, pp. 85–86.
60. Huang, C. J., J. Tsui, R. T. Kemerly, and G. L. McCoy, "Dynamic Microwave Frequency Division Characteristics of Coplanar Transferred-Electron Devices," *IEEE Trans. Microwave Theory Tech.*, Vol. MTT-24, Jan. 1976, pp. 61–63.
61. Claxton, D., T. Mills, and L. Yuan, "High Ratio Frequency Division With Three-Terminal Gunn Effect Devices," *Proc. 5th Biennial Cornell Electrical Engineering Conf.* 1975, pp. 195–203.
62. Dodson, D., T. Mills, and L. Tichauer, "TED Applied to Phase Locked Loop Circuit in a Frequency Divider," *Proc. 6th Biennial Cornell Electrical Engineering Conf.*, Aug. 1977, pp. 359–367.
63. Harrison, R. G., "Theory of Varactor Frequency Halver," *IEEE MTT-S Int. Microwave Symp. Digest*, June 1983, pp. 203–205.
64. Harrison, R. G., "Theory of Broadband Abrupt-Junction and Schottky-Barrier Frequency Halvers," Tech. Rep. CERL-85-3, Carleton Univ., Ottawa, Ont. Canada, Mar. 1985.
65. Hayashi, C., Y. Nishikawa, and M. Abe, "Subharmonic Oscillations of Order 1/2," *IRE Trans. Circuit Theory*, Vol. CT-7, June 1960, pp. 102–111.
66. Lipparini, A., E. Marazzi, and V. Rizzoli, "Computer-Aided Design of Microwave Parametric Frequency Dividers," *IEEE MTT-S Int. Microwave Symp. Digest*, 1981, pp. 229–231.
67. Lipparini, A., E. Marazzi, and V. Rizzoli, "A New Approach to the Computer-Aided Design of Nonlinear Networks and its Application to Microwave Parametric Frequency Dividers," *IEEE Trans. Microwave Theory Tech.*, Vol. MTT-30, July 1982, pp. 1050–1058.
68. Sterzer, F., "Microwave Parametric Subharmonic Oscillators for Digital Computing," *Proc. IRE*, Vol. 47, Aug. 1959, pp. 1317–1324.
69. Onyshkevych, L. S., W. F. Kosonocky, and A. W. Lo, "Parametric Phase-Locked Oscillator—Characteristics and Applications to Digital Systems," *IRE Trans. Electronic Computers*, Vol. EC-8, Sept. 1959, pp. 277–286.
70. Rizzoli, V., and A. Lipparini, "General Stability Analysis of Periodic Steady-State Regimes in Nonlinear Microwave Circuits," *IEEE Trans. Microwave Theory Tech.*, Vol. MTT-33, Jan. 1985, pp. 30–37.
71. Abeyta, I., F. Borgini, and D. R. Crosby, "A Computer Subsystem Using Kilomegacycle Oscillators," *Proc. IRE*, Vol. 49, Jan. 1961, pp. 128–135.
72. Harrison, R. G., "A Broad-Band Frequency Divider Using Microwave Varactors," *IEEE Trans. Microwave Theory Tech.*, Vol. MTT-25, Dec. 1977, pp. 1055–1059.
73. Kalivas, G. A., and R. G. Harrison, "A New Slotline-Microstrip Frequency

Halver," *IEEE MTT-S Int. Microwave Symp. Digest*, June 1985, pp. 683–686.

74. Schiek, B., and J. Köhler, "An Improved Microstrip-to-Microslot Transition," *IEEE Trans. Microwave Theory Tech.*, Vol. MTT-24, Apr. 1976, pp. 231–233.
75. Harrison, R. G., and G. A. Kalivas, "Impedance Measurement in Frequency-Halving Networks Using a Two-Frequency Synthetic Loading Technique," *IEEE Trans. Microwave Theory Tech.*, Vol. MTT-32, Dec. 1984, pp. 1591–1597.
76. Cornish, W. D., Telemus Electronic Systems Inc., Nepean, Ont., Canada, private communication, Aug. 1985.
77. Fortescue, R. L., "Quasi-Stable Frequency-Dividing Circuits," *J. IEE*, Vol. 84, 1939, pp. 693–698.
78. Miller, R. L., "Fractional-Frequency Generators Utilizing Regenerative Modulation," *Proc. IRE*, Vol. 27, July 1939, pp. 446–457.
79. Kharkevich, A. A., *Nonlinear and Parametric Phenomena in Radio Engineering*, Rider, New York, 1962.
80. Rauscher, C., "Regenerative Frequency Division with a GaAs FET," *IEEE Trans. Microwave Theory Tech.*, Vol. MTT-32, Nov. 1984, pp. 1461–1468.
81. Immovili, G., and G. Mantovani, "Analysis of Miller Frequency Divider by Two in View of Applications to Wideband FM Signals," *Alta Freq.*, Vol. 17, Nov. 1973, pp. 313–323.
82. Ahamed, S. V., J. C. Irvine, and H. Seidel, "Study and Fabrication of a Frequency Divider-Multiplier Scheme for High-Efficiency Microwave Power," *IEEE Trans. Commun. Technol.*, Vol. COM-24, Feb. 1976, pp. 243–249.
83. Jacob, A., and M. Jenett, "Performance of Microwave FM Signal Frequency Division Circuits," *IEEE MTT-S Int. Microwave Symp. Digest*, June 1985, pp. 207–210.
84. Harrison, R. G., "Development and Evaluation of Microwave Frequency Dividers," unpublished report, Oct. 1977.
85. Derksen, R. H., and H.-M. Rein, "A Monolithic 5 GHz Frequency Divider IC Fabricated with a Standard Bipolar Technology," *ESSCIRC 1985 Digest* (Toulouse, France, Sept. 1985), pp. 396–400.
86. Derksen, R. H., H.-M. Rein, and K. Wörner, "Monolithic Integration of a 5.3 GHz Regenerative Frequency Divider Using a Standard Bipolar Technology," *Electron. Lett.*, Vol. 21, Oct. 1985, pp. 1037–1039.
87. Derksen, R. H., and H.-M. Rein, "Bipolar Monolithic Integrated Frequency Dividers Operating up to 6.7 GHz," private communication, Nov. 1985.
88. Kaminsky, D., P. Goussu, R. Funck, and A. G. Bert, "A Dual-Gate GaAs FET Analog Frequency Divider," *IEEE MTT-S Int. Microwave Symp. Digest*, 1983, pp. 352–354.
89. Noordanus, J., "Hybrid MIC Digital Frequency Dividers at 4.5 and 9 GHz," in *Proc. 7th European Microwave Conf.*, Sept., 1977, pp. 198–202.
90. Kasperkovitz, D., and D. Grenier, "Travelling-wave Dividers: A New Concept in Frequency Division," *Microelectron. Rel.*, Vol. 16, 1977, pp. 127–134.

91. Noordanus, J., and G. Meiling, "Hybrid MIC's for a 6 GHz Phase-Locked Divider Loop," *Proc. 8th European Microwave Conf.*, 1978, pp. 613–617.
92. Kasperkovitz, W. D., "Frequency Dividers for Ultra-High Frequencies," *Philips Tech., Rev.*, Vol. 38, 1978/79, pp. 54–68.
93. Van Tuyl, R., C. A. Liechti, R. E. Lee, and E. Gowan, "GaAs MESFET Logic with 4-GHz Clock Rate," *IEEE J. Solid-State Circuits*, Vol. SC-12, Oct. 1977, pp. 485–496.
94. Van Tuyl, R., C. Liechti, R. Lee, and E. Gowan, "4 GHz Frequency Division with GaAs MESFET IC's," *IEEE Int. Solid-State Circuits Conf. Digest*, Feb. 1977, pp. 198–199.
95. Wörner, K., and A. Colquhoun, "High-Speed Si-Bipolar and GaAs Technologies," *IEEE J. Selected Areas in Commun.*, Vol. SAC-4, Jan. 1986, pp. 24–31.
96. Konaka et al., "A 30-ps Si Bipolar IC Using Super Self-Aligned Process Technology," *ICSSDM Digest of Tech. Papers* (Kobe, Japan, 1984), pp. 209–212.
97. Asbeck, P. M., D. L. Miller, R. J. Anderson, R. N. Deming, L. D. Hou, C. A. Liechti, and F. H. Eisen, "4.5 GHz Frequency Dividers Using GaAs/(GaAl)As Heterojunction Bipolar Transistors," *IEEE Int. Solid-State Circuits Conf. Digest*, Feb. 1984, pp. 50–51.
98. Asbeck, P. M., D. L. Miller, R. J. Anderson, R. N. Deming, R. T. Chen, C. A. Liechti, and F. H. Eisen, "Applications of Heterojunction Bipolar Transistors to High-Speed, Small Scale Digital Integrated Circuits," *IEEE Gallium Arsenide Integrated Circuit Symp. Tech. Digest*, Oct. 1984, pp. 133–136.
99. Bosch, B. G., "Gigabit Electronics—A Review," *Proc. IEEE*, Vol. 67, Mar. 1979, pp. 340–379.
100. Bosch, B. G., "Device and Circuit Trends in Gigabit Logic," *Proc. IEE (London)*, Part I, Vol. 127, Oct. 1980, pp. 254–265.
101. Eden, R. C., B. M. Welch, R. Zucca, and S. I. Long, "The Prospects for Ultra High-Speed VLSI GaAs Digital Logic," *IEEE Trans. Electron Devices*, Vol. ED-26, April 1979, pp. 299–317.
102. Long, S. I., F. S. Lee, R. Zucca, B. M. Welch, and R. C. Eden, "MSI High-Speed Low-Power GaAs Integrated Circuits Using Schottky Diode FET Logic," *IEEE Trans. Microwave Theory Tech.*, Vol. MTT-28, May 1980, pp. 466–472.
103. Gloanec, M., J. Jarry, and G. Nuzillat, "GaAs Digital Integrated Circuits for Very High-Speed Frequency Division," *Electron. Lett.*, Vol. 17, Oct. 1981, pp. 763–765.
104. Yamamoto, R., and A. Higashisaka, "High Speed GaAs Digital Integrated Circuit with Clock Frequency of 4.1 GHz," *Electron. Lett.*, Vol. 17, Apr. 1981, pp. 291–292.
105. Cathelin, M., G. Durand, M. Gavant, and M. Rocchi, "5 GHz Binary Frequency Division on GaAs," *Electron. Lett.*, Vol. 16, July 1980, pp. 535–536.
106. Abe, M., T. Mimura, K. Nisiuchi, A. Shibatomi, and K. Kobayashi, "HEMT LSI Technology for High Speed Computers," *Gallium Arsenide Integrated Circuit Symp. Tech. Digest*, Oct. 1983, pp. 158–161.

107. Hendel, R. H., S. S. Pei, R. A. Kiehl, C. W. Tu, M. D. Feuer, and R. Dingle, "A 10-GHz Frequency Divider Using Selectively Doped Heterostructure Transistors," *IEEE Electron Device Lett.*, Vol. EDL-5, Oct. 1984, pp. 406–408.

108. Barrera, J., B. Hoffmann, and D. Wilson, "Digital GaAs ICs Hit Gigaherz Speed Mark," *Microwaves and RF*, Vol. 23, Feb. 1984, pp. 133–138.

109. Lee, F., and R. Miller, "4-GHz Counters Bring Synthesizers up to Speed," *Microwaves and RF*, Vol. 23, June 1984, pp. 113–118.

PROBLEMS

13.1 The bridge-rectifier doubler of Fig. 13.61 uses ideal rectifiers defined by:

$$v = 0, \qquad i \geq 0$$

$$i = 0, \qquad v \leq 0$$

$$vi = 0$$

$$\langle vi \rangle = 0$$

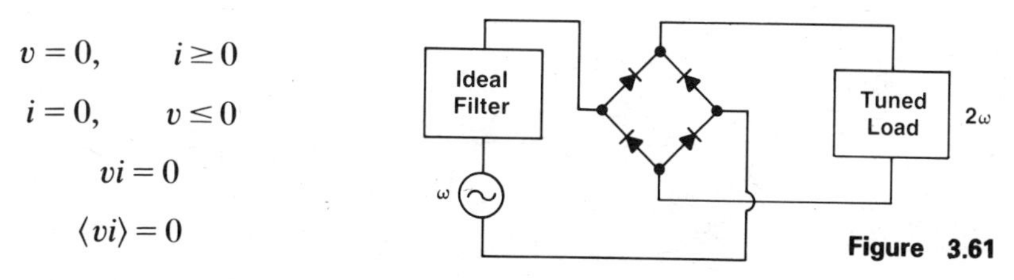

Figure 3.61

where $\langle \cdot \rangle$ means time-average. The source supplies power at the fundamental frequency (ω) only, while the ideal filter and tuned load eliminate all harmonics above the second (2ω) from the load current. Show that the maximum possible conversion efficiency is 25%.

13.2 A tuned frequency doubler has the equivalent circuit of Fig. 13.62, in which the ideal filters pass currents at the indicated frequencies only. The abrupt-junction varactor has series resistance R_s, is always operated under reverse bias and is characterized by the voltage-charge relationship of Equation 13.14. Assuming the varactor to be fully driven between reverse breakdown ($v = -V_B$) and the onset of forward conduction ($v = \phi_0$), derive the following design formulas under the constraints of given input power and maximum efficiency:

(a) Fixed charge on varactor:

$$Q_0 = -C_j(0)\sqrt{\phi_0}\sqrt{V_B + \phi_0}$$

(b) Current amplitude at the input frequency:

$$I_1 = -\frac{8\pi}{k} f_{\text{in}}\, C_j(0)\sqrt{\phi_0}\sqrt{V_B + \phi_0}$$

where, as in Equation 13.21, $k = 2\sqrt{(\sqrt{5}+1)} + \sqrt{(\sqrt{5}-1)}$.

(c) Efficiency, as given by Equation 13.21:

$$\eta = \frac{1-\sqrt{2}k(f_{\text{in}}/\mathbf{F}_c)}{1+2\sqrt{2}k(f_{\text{in}}/\mathbf{F}_c)}$$

(d) Optimum load resistance:

$$R_L = \frac{1}{2\sqrt{2}\pi k f_{\text{in}} C_j(-V_B)} - R_s$$

(e) Optimum generator resistance:

$$R_G = \frac{R_L}{2} + \frac{3}{2} R_s$$

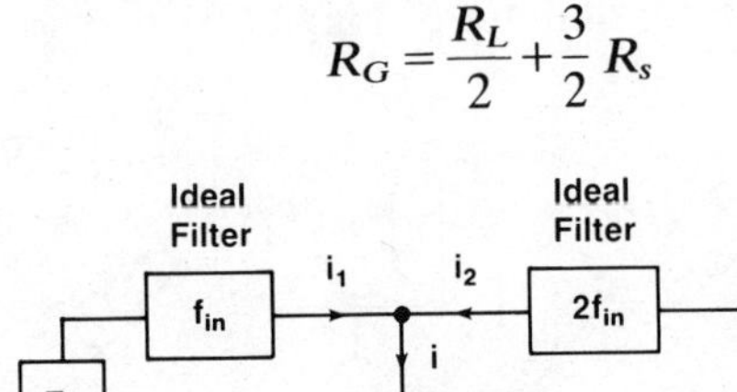

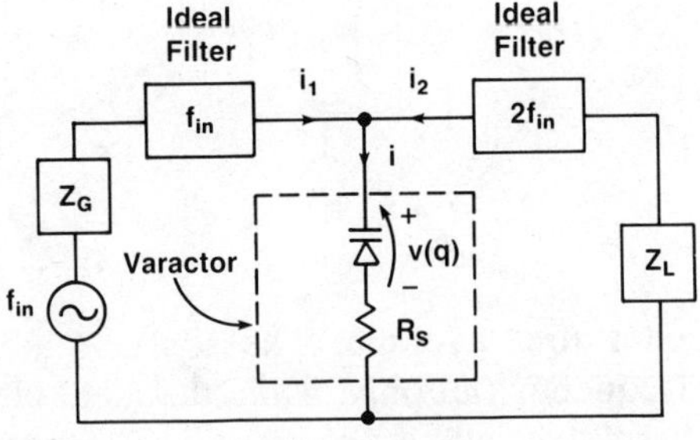

Figure 13.62

13.3 Given that the tuned frequency doubler of Problem 13.2 uses a varactor with the following parameters:

$$F_c(0) = 50 \text{ GHz}, \ C_j(0) = 1.0 \text{ pF}, \ \gamma = 0.5, \ V_B = 30 \text{ V}, \ \phi = 0.7 \text{ V},$$

determine (a) the maximum possible efficiency, (b) the optimum load resistance R_L, and (c) the optimum generator resistance R_G.

13.4 A GaAs FET is used in a harmonic-generator mode, as described in Reference 51. The important device nonlinearity is the transconductance characteristic, which is assumed to be approximated by:

$$i_D = \begin{cases} I_{DSS}\left(1-\dfrac{V_{GS}}{V_p}\right)^2, & V_P < V_{GS} < 0 \\ I_{DSS}, & V_{GS} > 0 \end{cases}$$

(a) Show that in the class-A mode of Fig. 13.22(*a*), in which the gate is biased to $V_{GS} = 0$ and where the peak gate voltage is equal to $|V_p|$, the second-harmonic output power is 10.4 dB below the fundamental.

(b) Show that in the class-B mode of Fig. 13.22(*b*), in which the gate is biased to V_P and where the peak gate voltage is again equal $|V_p|$, the second-harmonic output power is 4.6 dB below the fundamental.

13.5 Explain why idler circuits are theoretically unnecessary in frequency doublers or in frequency halvers which use abrupt *pn*-junction or Schottky-barrier varactors, but are needed for all other orders of multiplication or division when such varactors are used.

13.6 A balanced parametric frequency halver employs a matched pair of Si varactors in a configuration which has the approximate equivalent circuit of Figure 13.35. The varactor parameters are:

$$C_j(0) = 1.0 \text{ pF}$$

$$f_c(0) = 50 \text{ GHz}$$

$$\gamma = 0.5$$

$$\phi_0 = 0.8 \text{ V}.$$

Assume the following circuit parameters:

$$L_1 = 2.026 \text{ nH}$$

$$R_p = \text{very small}$$

$$R_L = \text{very large}$$

(a) Estimate the subharmonic voltage appearing across L_1 under the following conditions: a 10-GHz input signal of peak amplitude 0.5 V, bias voltage $V_0 = -0.2$ V.

(b) As above, but with $V_0 = 0$ V.

(c) Repeat with an 8-GHz input signal of peak amplitude 0.32 and bias $V_0 = 0$ V.

(d) Comment on the meaning of the multiple-valued solution in (b).

(e) Investigate the effect of finite R_p in (a).

14 COMPUTER-AIDED DESIGN

Design procedures for various types of microwave circuits have been discussed in previous chapters. In order to implement these designs to meet the given circuit or subsystem specifications successfully, a computer-aided design (CAD) approach becomes necessary. When the CAD approach is used, experimental modifications of the circuit (which are unavoidable in the conventional design procedure) are replaced by a computer-based simulation and optimization of the initial design. CAD procedures become still more essential when integrated-circuit techniques (discussed in Chapter 15) are used for circuit fabrication. For monolithic microwave integrated circuits, any experimental iteration of a design becomes prohibitively costly. This chapter summarizes basic aspects of the CAD procedure as applied to microwave circuit design.

14.1 BASIC ASPECTS OF CAD

A flow chart for a typical CAD process is shown in Fig. 14.1. This flow chart is written specifically for circuit design but holds good for antenna design (or any other design) also. One starts with the given circuit specifications. Synthesis methods and available design data (at times pre-stored in computer memory) help to arrive at the initial design. The performance of this initial design is evaluated by a computer-aided circuit-analysis package. Numerical models for various components used in the initial design are needed for the analysis. These are usually called from the library of subroutines developed for this purpose. Performance characteristics obtained, as a result of the analysis, are compared with the given specifications. If the results fail to satisfy the desired specifications, the designable parameters of the design are altered in a systematic manner. This constitutes the key step in the optimization. Several optimization strategies include sensitivity analysis of the design for calculating changes in the designable parameters. The sequence of design analysis, comparison with the desired performance, and parameter modification are performed iteratively until the specifications are met or the optimum performance of the design (within the given constraints) is achieved. The circuit designed is now fabricated and experimental measurements are carried out. As indicated in the lower part of Fig. 14.1 (portion inside the dashed rectangle), some modifications may still be required if the modeling and/or analysis has

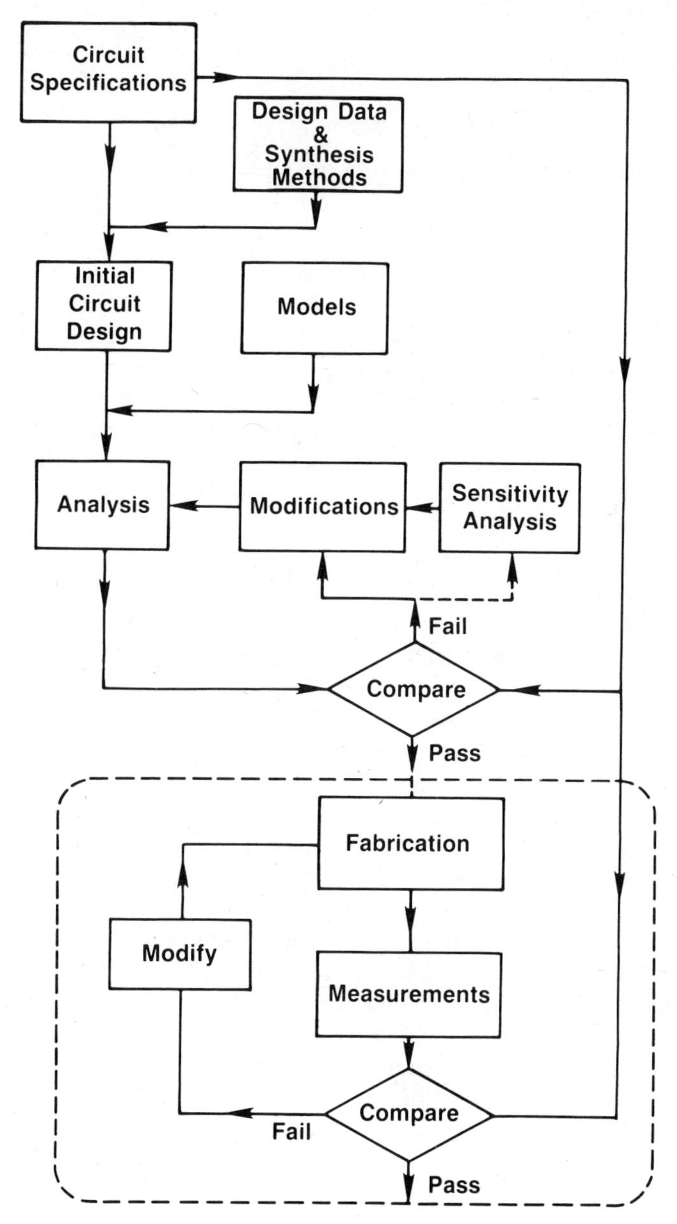

Figure 14.1 Typical flow chart for a CAD procedure.

not been accurate enough. However, these modifications, hopefully, should be small and the aim of the CAD is to minimize the experimental iterations as far as practicable.

The process of CAD, as previously outlined, consists of three important segments, namely: (1) modeling; (2) analysis, and (3) optimization. These three aspects of CAD as applied to microwave circuit design have been discussed in detail in [1]. In the following sections, we review the important features of these three aspects.

14.2 MODELING OF CIRCUIT COMPONENTS

An accurate and reliable characterization of microwave circuit components is one of the basic prerequisites of successful CAD. The degree of accuracy to which the performance of microwave integrated circuits can be predicted depends on the accuracy of characterization and modeling of components. Kinds of elements and passive devices that need to be characterized depend upon the transmission medium used for circuit design. Most research efforts aimed at characterization for CAD purposes have been reported for microstrip lines [2, 3]. Some results are available for slot lines and fin lines also [2, 4]. However, for coplanar lines and suspended substrate transmission structures, CAD techniques are still in a state of infancy. Characterization and modeling of various types of transmission structures have been discussed in Chapter 2.

In addition to the transmission media, implementation of CAD requires characterization of various junctions and discontinuities in transmission structures. Effects of these junctions and discontinuities become more and more significant as one moves from the microwave frequency range to millimeter waves. At higher frequencies, radiation associated with discontinuities needs to be considered. Variations of effects of two typical discontinuities with frequency are illustrated in Fig. 14.2. Figure 14.2(*a*) shows the input VSWR introduced by right-angled bends in 25-Ω and 75-Ω microstrip lines (on a 0.079-cm-thick substrate with $\epsilon_r = 2.5$). We note that the input VSWR caused by the 90° bend increases monotonically with frequency and is larger for low impedance (wider) lines, the specific values being 3.45 for a 25-Ω line at 10 GHz as compared to 1.30 for 75 Ω at the same frequency. Figure 14.2(*b*) shows the effect of discontinuity reactance on the behavior of a step (change-in-width) discontinuity. The reflection coefficient increases from its nominal value. At 12 GHz, this increase is about 1% for a 40-Ω to 60-Ω step, and about 3 percent for a 30-Ω to 70-Ω impedance change. These results are based on quasistatic characterization of these discontinuities [1] and are included to show that the discontinuity reactances should not be ignored for design at *X*-band or higher frequencies.

Quasistatic results for microstrip discontinuities characterization have been available for some years [1, 2, 5] and are used in some of the commercially available CAD packages [6, 7]. A more accurate analysis of bends, T junctions, and crossings (based on field matching using equivalent parallel-plate waveguide models of discontinuity configurations) became available later [8]. Computer programs based on these techniques are commercially available [9], but have not been integrated in CAD packages.

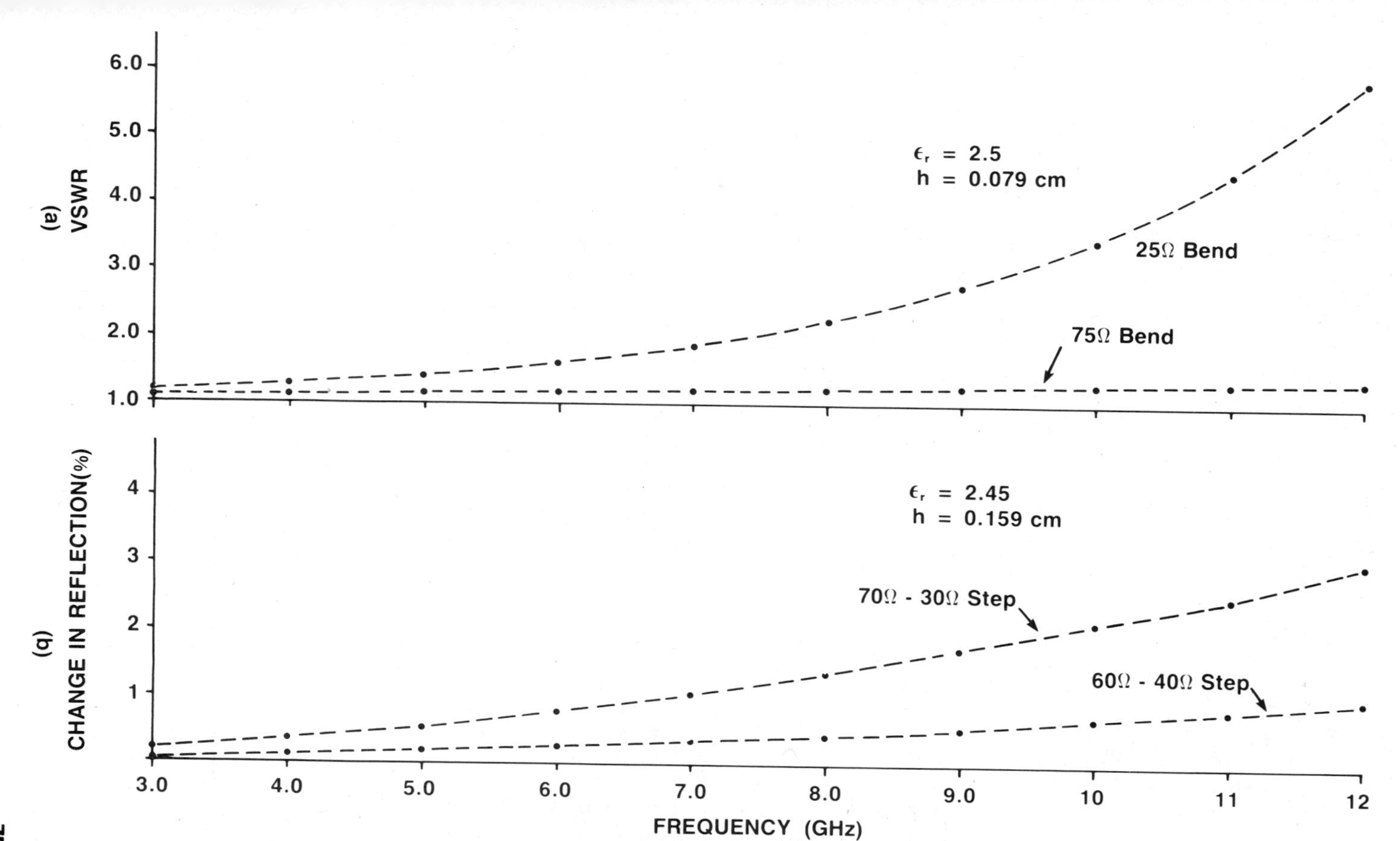

Figure 14.2 (*a*) Frequency variations of VSWR introduced by right-angled bends in 25-Ω and 75-Ω microstrip lines. (*b*) Frequency variations of the percentage changes in reflection coefficients introduced by two step-discontinuities in microwave circuits.

Rigorous hybrid-mode frequency-dependent characterizations of microstrip open ends and gaps were reported around 1981–1982 [10, 11, 12]. These results are based on Galerkin's method in spectral transform domain. This technique has been extended to other types of discontinuities also [13], and results for a step discontinuity have been reported recently [14]. Other rigorous approaches (which could account for discontinuity induced radiation also) have been attempted recently and are being followed up by several groups. One of these is based on the moment method using full-wave Green's function for a grounded dielectric slab [15, 16, 17]. Characterization of microstrip open ends and gaps based on this method has been reported only very recently (October 1985 [15, 16]). Similar results are currently not available for other discontinuities.

Even when accurate numerical results are available, efficient transfer of these results for CAD is not straightforward. Development of a suitable mechanism for interfacing discontinuity analysis programs with CAD software is an equally important problem that needs to be addressed. As discussed in Chapter 2, closed-form expressions (derived by curve fitting of the numerical results) have been used as discontinuity models. Such closed-form expressions suffer from limited validity with respect to various parameters (impedance, frequency, geometry, etc.) and limit the accuracy of CAD. It appears necessary to build up (and interface with the CAD software) a mechanism based on multidimensional interpolation of data files backed by a generalized rigorous full-wave analysis program for characterization of discontinuities. The concept of such an adaptive (continually updated) data base for discontinuity models is shown in Fig. 14.3.

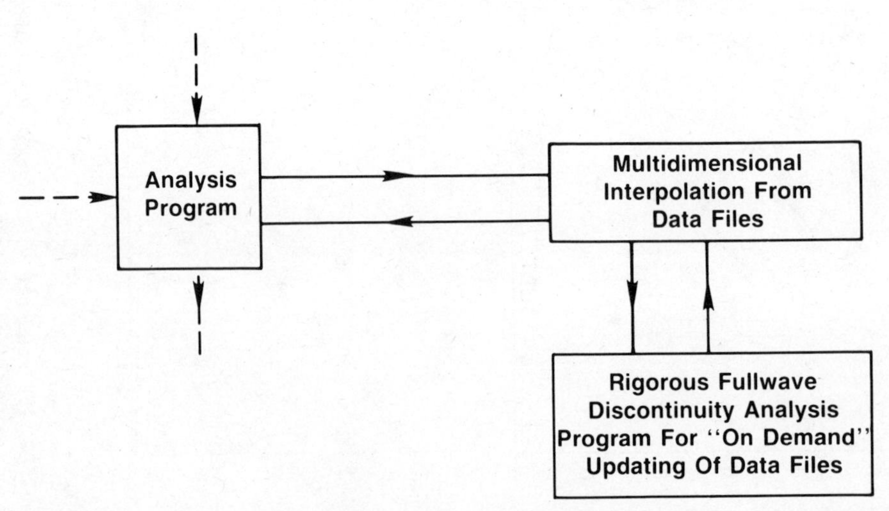

Figure 14.3 Concept of "adaptive" (continually updated) data base for discontinuity models.

In addition to the modeling of transmission lines and their discontinuities, implementation of the CAD procedure requires accurate models for various active devices like GaAs MESFETs, varactors, and *pin* diodes. These equivalent-circuit models for active devices have been discussed in Chapters 7 and 8.

14.3 COMPUTER-AIDED ANALYSIS TECHNIQUES

As shown in Figs. 14.1 and 14.3, computer-aided analysis constitutes the key step in the CAD procedure. Since the analysis forms a part of the optimization loop, the analysis subprogram is executed again and again (typically 100 times or more) for a specific circuit design. For this reason, an efficient analysis algorithm constitutes the backbone of any CAD package.

Any general microwave circuit discussed in previous chapters can be viewed as an arbitrarily connected ensemble of multiport components as illustrated in Fig. 14.4. The circuit shown here consists of eight components (A, B, C, D, E, F, G, H), of which G is a one-port component; A, E, and C are two-port components; D, F, and H are three-port components; and B is a four-port component. Different ports of these components are connected together. There are two external ports (1 and 2). The analysis problem in this case may be stated as follows: characterization of components A through H being known (say in terms of individual S-matrices), find the S-matrix of the overall combination with reference to the two external ports 1 and 2. The outcome of this analysis will be a 2×2 S-matrix. The process of evaluating the circuit performance from the known characterizations of its constituents is termed *circuit analysis*. It requires two inputs: components characterizations and the topology of interconnections. The analysis process is depicted symbolically in Fig. 14.5.

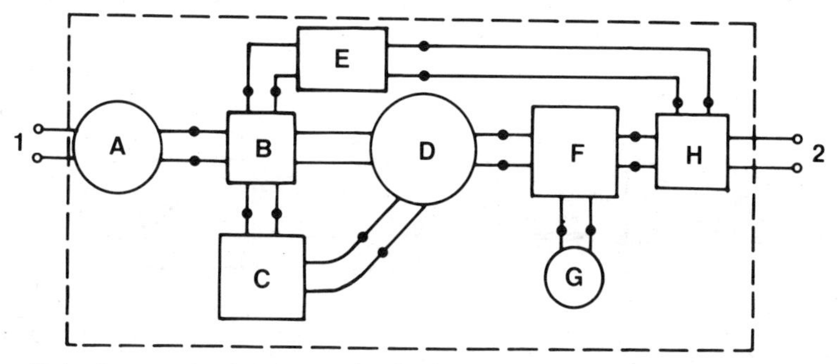

Figure 14.4 A general microwave circuit viewed as an ensemble of multiport components.

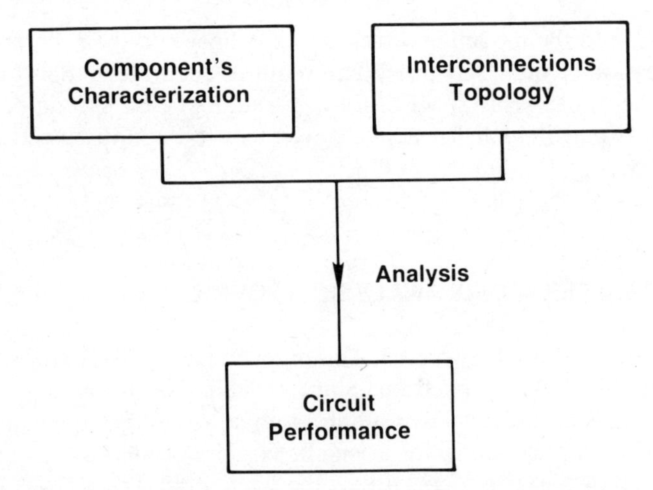

Figure 14.5 Block diagram representation of the circuit analysis process.

14.3.1 General Scattering-Matrix Analysis

General scattering-matrix analysis is applicable to any general microwave circuit configuration when all the circuit components are modeled in terms of their scattering parameters. In this method of analysis [1, 18], the circuit diagram is configured such that there are no unconnected ports. That is, sources (along with their source impedances) are connected to the input ports, and all output ports are terminated in respective loads. For circuit analysis, it is adequate to consider matched sources and matched loads. Consider the example of an impedance matching circuit and its equivalent representation shown in Fig. 14.6(a) and (b). For implementing this

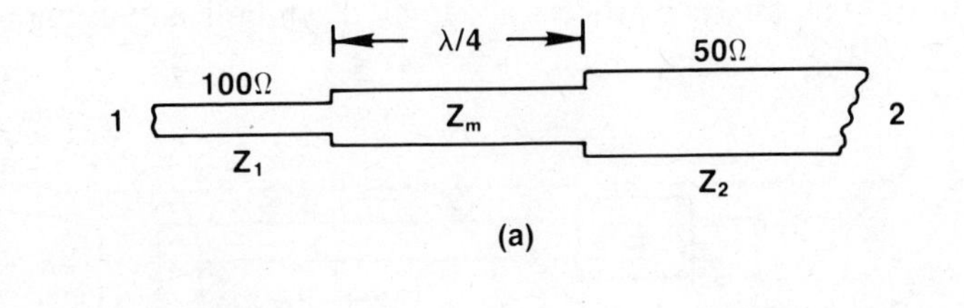

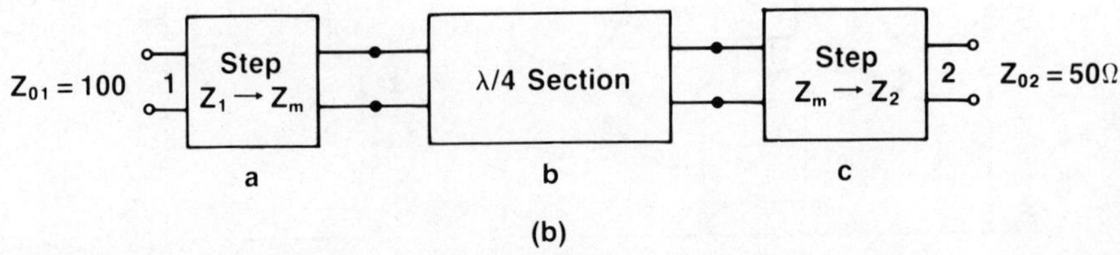

Figure 14.6 (a) A quarter-wave matching network. (b) Interconnected multiport representation of the circuit in part (a).

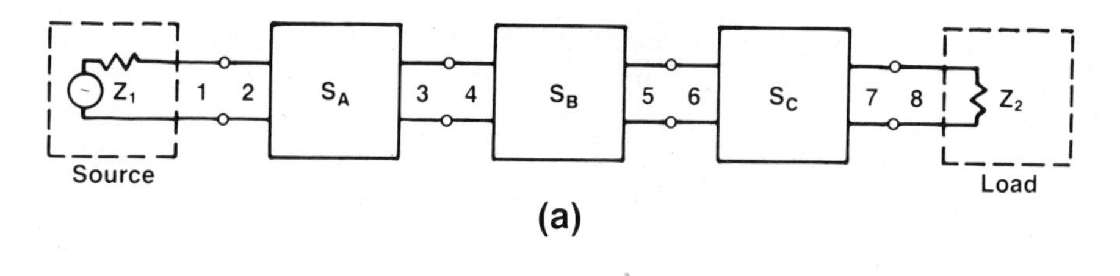

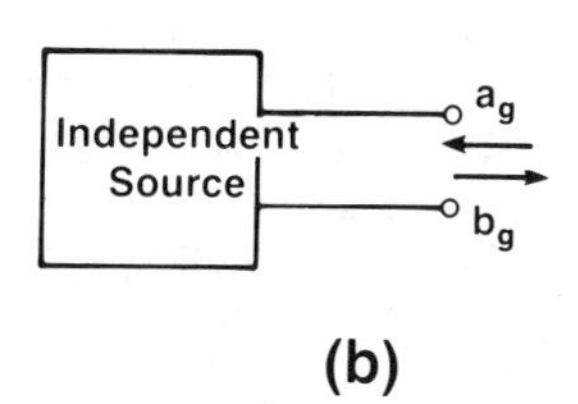

Figure 14.7 (*a*) Network of Fig. 14.6(*b*) rewritten for implementing the generalized scattering-matrix analysis procedure. (*b*) Independent source introduced for scattering-matrix analysis.

method, this circuit may be depicted as shown in Fig. 14.7(*a*). Independent sources [Fig. 14.7(*b*)] in such a representation may be described by the relation

$$b_g = s_g a_g + c_g \tag{14.1}$$

where c_g is the wave impressed by the generator. For sources that are matched (or isolated) $s_g = 0$ and $b_g = c_g$. All other components in the circuit are described by

$$\mathbf{b}_i = S_i \mathbf{a}_i \tag{14.2}$$

where $\mathbf{a}_i$ and $\mathbf{b}_i$ are incoming and outgoing wave variables, respectively, for the *i*th component with n_i ports and S_i is the scattering matrix.

The governing relation for all components (say the total number is *m*) in the circuit can be put together as

$$\mathbf{b} = S\mathbf{a} + \mathbf{c} \tag{14.3}$$

where

$$\mathbf{b} = \begin{bmatrix} \mathbf{b}_1 \\ \mathbf{b}_2 \\ \vdots \\ \mathbf{b}_m \end{bmatrix}, \quad \mathbf{a} = \begin{bmatrix} \mathbf{a}_1 \\ \mathbf{a}_2 \\ \vdots \\ \mathbf{a}_m \end{bmatrix}, \quad \text{and} \quad \mathbf{c} = \begin{bmatrix} \mathbf{c}_1 \\ \mathbf{c}_2 \\ \vdots \\ \mathbf{c}_m \end{bmatrix} \tag{14.4}$$

Here $\mathbf{b}_1, \mathbf{a}_1, \ldots,$ are themselves vectors with the number of elements equal to the number of ports of that particular component. The size of the vectors **b**, **a**, or **c** is equal to the sum total of all the ports of all the components in the modified network representation. For the network shown in Fig. 14.7(a), this number is 8; counting one port for the source, two ports each for the three components A, B, and C; and one port for the load. The vector $\mathbf{c}$ will have nonzero values only for the output ports of independent sources in the network. For the network of Fig. 14.7(a),

$$\mathbf{c} = [c_1, 0, 0, 0, 0, 0, 0, 0]^t \tag{14.5}$$

The output port of the source has been numbered 1, and the superscript t indicates transpose of a vector or matrix.

The matrix S is a block diagonal matrix whose submatrices along the diagonal are the scattering matrices of various components. In the general case, the matrix S may be written as

$$S = \begin{bmatrix} (S_1) & 0 & — & — & 0 \\ 0 & (S_2) & — & — & 0 \\ — & — & (—) & — & — \\ — & — & — & (—) & — \\ 0 & 0 & — & — & (S_m) \end{bmatrix} \tag{14.6}$$

where 0 represents null matrices. For the network of Fig. 14.7(a), the S-matrix will look like

$$S = \begin{bmatrix} [S_{11}^S] & 0 & 0 & 0 & 0 \\ 0 & \begin{bmatrix} S_{22}^A & S_{23}^A \\ S_{32}^A & S_{33}^A \end{bmatrix} & 0 & 0 & 0 \\ 0 & 0 & \begin{bmatrix} S_{44}^B & S_{45}^B \\ S_{54}^B & S_{55}^B \end{bmatrix} & 0 & 0 \\ 0 & 0 & 0 & \begin{bmatrix} S_{66}^C & S_{67}^C \\ S_{76}^C & S_{77}^C \end{bmatrix} & 0 \\ 0 & 0 & 0 & 0 & [S_{88}^L] \end{bmatrix} \tag{14.7}$$

where S_{11}^S and S_{88}^L are one-port S-parameter characterizations for the source and the load, respectively. When both the source and the load are matched, $S_{11}^S = S_{88}^L = 0$. The other three (2×2) S-matrices characterize components A, B, and C, respectively.

Equation 14.3 contains the characterizations of individual components but does not take into account the constraints imposed by interconnections. For a pair of connected ports, the outgoing wave variable at one port must equal the incoming wave variable at the other (assuming that wave variables at two ports are normalized with respect to the same impedance level). For example, if the port j of one component is connected to port k of the other component, as shown in Fig. 14.8, the incoming and outgoing wave

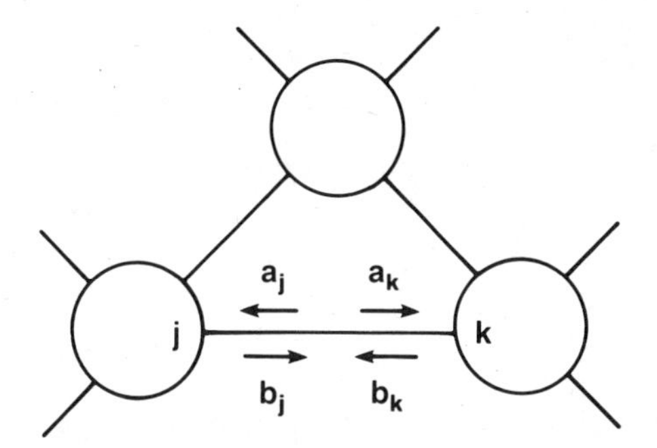

Figure 14.8 Interconnection of ports *j* and *k*.

variables are related as

$$a_j = b_k \qquad \text{and} \qquad a_k = b_j \tag{14.8}$$

or

$$\begin{bmatrix} b_j \\ b_k \end{bmatrix} = \begin{bmatrix} 0 & 1 \\ 1 & 0 \end{bmatrix} \begin{bmatrix} a_j \\ a_k \end{bmatrix} = [\Gamma]_{jk} \begin{bmatrix} a_j \\ a_k \end{bmatrix} \tag{14.9}$$

We can extend $[\Gamma]_{jk}$ to write an interconnection matrix describing all the connections in the circuit. We express,

$$\mathbf{b} = \mathbf{\Gamma a} \tag{14.10}$$

where Γ is a square matrix of the same size as that of S. The size of the Γ-matrix is given by the sum total of the ports in all the components in the circuit. The size is 8×8 in the example of Fig. 14.7(a). For the example of Fig. 14.7(a), the Γ matrix may be written as

$$\begin{bmatrix} b_1 \\ b_2 \\ b_3 \\ b_4 \\ b_5 \\ b_6 \\ b_7 \\ b_8 \end{bmatrix} = \begin{bmatrix} 0 & 1 & 0 & 0 & 0 & 0 & 0 & 0 \\ 1 & 0 & 0 & 0 & 0 & 0 & 0 & 0 \\ 0 & 0 & 0 & 1 & 0 & 0 & 0 & 0 \\ 0 & 0 & 1 & 0 & 0 & 0 & 0 & 0 \\ 0 & 0 & 0 & 0 & 0 & 1 & 0 & 0 \\ 0 & 0 & 0 & 0 & 1 & 0 & 0 & 0 \\ 0 & 0 & 0 & 0 & 0 & 0 & 0 & 1 \\ 0 & 0 & 0 & 0 & 0 & 0 & 1 & 0 \end{bmatrix} \begin{bmatrix} a_1 \\ a_2 \\ a_3 \\ a_4 \\ a_5 \\ a_6 \\ a_7 \\ a_8 \end{bmatrix} \tag{14.11}$$

Note that the matrix is symmetrical and there is only one 1 in any row or any column. The latter signifies that any port is connected only to one other port. This excludes the possibility of three ports being connected at one single point. If such a junction exists in a circuit, it must be considered a three-port component, as shown in Fig. 14.9. The ports a', b', and c' of this

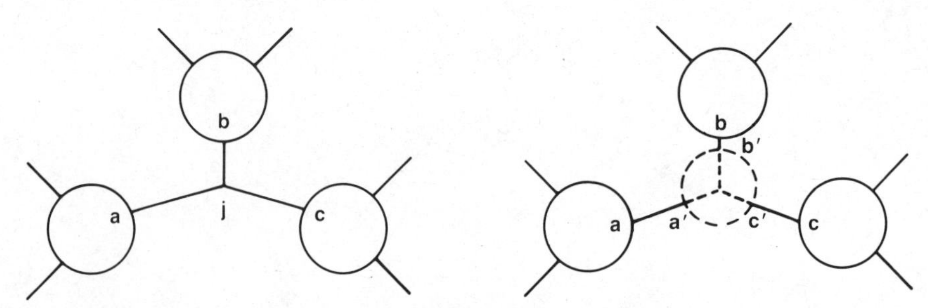

Figure 14.9 A three-way connection at j treated as a three-port component.

new component depicting the junction j are connected to the parts a, b, and c, respectively, in a one-to-one manner and the nature of the Γ-matrix is similar to the one as indicated in (14.11).

Equations 14.3 and 14.10 describe the circuit completely and may be combined into a single equation as

$$\Gamma \mathbf{a} = S\mathbf{a} + \mathbf{c} \tag{14.12}$$

or $(\Gamma - S)\mathbf{a} = \mathbf{c}$.

Setting $(\Gamma - S) = W$, we have

$$\mathbf{a} = W^{-1}\mathbf{c} = M\mathbf{c} \tag{14.13}$$

W is called the connection-scattering matrix. For our example of Fig. 14.7(a), W may be written as

$$\begin{array}{c|cccccccc} & 1 & 2 & 3 & 4 & 5 & 6 & 7 & 8 \\ \hline 1 & -S_{11}^{S} & 1 & 0 & 0 & 0 & 0 & 0 & 0 \\ 2 & 1 & -S_{11}^{A} & -S_{12}^{A} & 0 & 0 & 0 & 0 & 0 \\ 3 & 0 & -S_{21}^{A} & -S_{22}^{A} & 1 & 0 & 0 & 0 & 0 \\ 4 & 0 & 0 & 1 & -S_{11}^{B} & -S_{12}^{B} & 0 & 0 & 0 \\ 5 & 0 & 0 & 0 & -S_{21}^{B} & -S_{22}^{B} & 1 & 0 & 0 \\ 6 & 0 & 0 & 0 & 0 & 1 & -S_{11}^{C} & -S_{12}^{C} & 0 \\ 7 & 0 & 0 & 0 & 0 & 0 & -S_{21}^{C} & -S_{22}^{C} & 1 \\ 8 & 0 & 0 & 0 & 0 & 0 & 0 & 1 & -S_{11}^{L} \end{array} \tag{14.14}$$

We note that the main diagonal elements in W are the negative of the reflection coefficients at the various component ports. The other (non-diagonal) elements of W are negative of the transmission coefficients between different ports of the individual components. All other elements are zero except for those corresponding to the two ports connected together (the Γ-matrix elements) which are 1s. The zero/nonzero pattern in the W-matrix depends only on the topology of the circuit and does not change with component characterizations or frequency of operation. The tri-diagonal pattern of (14.14) is a characteristic of a chain of two-port components cascaded together.

The circuit analysis involves solution of the matrix equation 14.13 to find components of the vector **a**. Variable a's corresponding to loads at various external ports are found by choosing the vector **c** to consist of a unity at one of external ports and zeros elsewhere. This leads to the evaluation of a column of the S-matrix of the circuit. Say, for $c_j = 1$ and all other $c_{i,i \neq j} = 0$, we find values of vector **a**. Then

$$S_{lj} = \frac{b_l}{a_j} = a_{l'} \qquad \left(a_i = \begin{cases} 1, & \text{for } i = j \\ 0, & \text{otherwise} \end{cases} \right) \tag{14.15}$$

where l' is the load port corresponding to the lth port of the circuit and j is the other circuit port (for which the impressed wave variable is unity). Equation 14.15 yields the jth column of the scattering matrix. Considering the example of Fig. 14.7(a), if we make $c_1 = 1$, other c's being zero, then the first column of the matrix for the circuit in Fig. 14.6 (i.e., cascade of three components A, B, and C) is given by

$$S_{11} = a_1 = M_{11}, \qquad S_{21} = a_8 = M_{81} \tag{14.16}$$

Here subscripts of S refer to circuit ports of Fig. 14.6 and subscripts of a correspond to component ports of Fig. 14.7(a). For obtaining the second column of the S-matrix, we take $c_8 = 1$ and other c's to be zero and find a_1 and a_8 again to yield

$$S_{21} = a_1 = M_{81} \qquad \text{and} \qquad S_{22} = a_8 = M_{88} \tag{14.17}$$

It may be noted that (14.13) yields much more information (namely wave variables at all the internal (connected) ports also) than what is needed for calculation of the S-matrix of the circuit. Knowledge of the wave variables at internal ports is needed for carrying out a sensitivity analysis (discussed in Section 14.4) of the circuit using the adjoint network method. For this reason, the present method is suitable for CAD purposes.

14.4 CIRCUIT OPTIMIZATION

Optimization is an important step in the CAD process and, as depicted in Fig. 14.10, converts an initial (and quite often unacceptable) design into an optimized final design meeting the given specifications. Optimization procedures involve iterative modifications of the initial design, followed by circuit analysis and comparison with the specified performance.

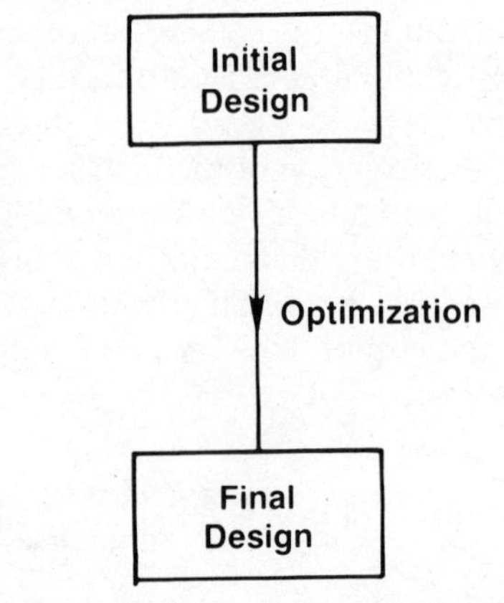

Figure 14.10 Role of the optimization process.

There are two different ways of carrying out the modification of designable parameters in an optimization process. These are known as gradient methods and direct search methods of optimization. Gradient methods use information about the derivatives of the performance functions (with respect to designable parameters) for arriving at the modified set of parameters. This information is obtained from the sensitivity analysis. On the other hand, the direct search methods do not use gradient information, and parameter modifications are carried out by searching for the optimum

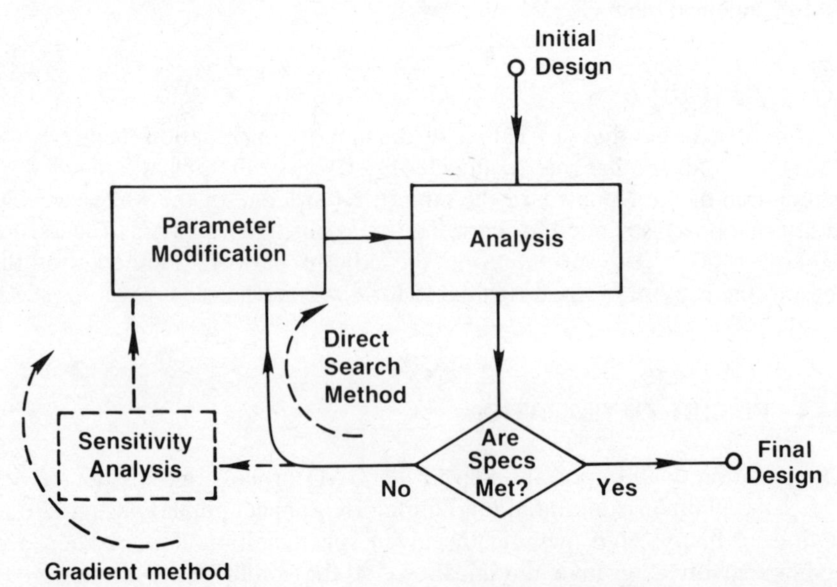

Figure 14.11 Flow charts for direct search and gradient methods of optimization.

in a systematic manner. A flow chart for the optimization process is shown in Fig. 14.11.

The optimization methods used for microwave circuit design are identical to those used in other disciplines and are well documented in the literature [1]. In all of these methods, the optimization problem is formulated as the minimization of a scalar objective function $U(\phi)$, where $U(\phi)$ is an error function that represents the difference between the performance achieved at any stage and the desired specifications. For example, in the case of a microwave transistor amplifier, the formulation of $U(\phi)$ will involve specified and achieved values of the gain, input and output VSWR, the noise figure, etc., over the required frequency range.

In this section, we review two commonly used optimization techniques; namely, the pattern search method and the conjugate gradient method.

14.4.1 Pattern Search Optimization Method

Pattern search is one of the most popular search methods, and has been used for CAD of microwave circuits [19]. As the name suggests, the algorithm searches for the minimum of the objective function by carrying out a sequence of one-dimensional searches in what is called "pattern direction." The pattern direction is established by a set of "exploratory moves" around a base point $\boldsymbol{\phi}_1$. These exploratory moves may be defined by

$$\phi_{2,i} = \begin{cases} \phi_{1,i} + \Delta\phi_i, & \text{if } U(\phi_1 + \Delta\phi_i \mathbf{u}_i) < U(\phi_1) \\ \phi_{1,i} - \Delta\phi_i, & \text{if } U(\phi_1 - \Delta\phi_i \mathbf{u}_i) < U(\phi_1) \\ \phi_{1,i}, & \text{if } U(\phi_1) < \min\{U(\boldsymbol{\phi}_1 + \Delta\phi_i \mathbf{u}_i),\, U(\boldsymbol{\phi}_1 - \Delta\phi_i \mathbf{u}_i)\} \end{cases} \tag{14.18}$$

where $\phi_{1,i}$ is the ith component of ϕ_1 and denotes a designable parameter, $U(\boldsymbol{\phi}_1)$ is the value of the scalar objective function U at $\boldsymbol{\phi}_1$, $\mathbf{u}_i$ is a unit vector in the ith direction, and $\boldsymbol{\phi}_2$ is the new base point. The pattern direction $\mathbf{s}_1$ is established by

$$\mathbf{s}_1 = \boldsymbol{\phi}_2 - \boldsymbol{\phi}_1 \tag{14.19}$$

A one-dimensional search is carried out along the direction $\mathbf{s}_1$ and the position of the minimum along s_1 yields the new base point $\boldsymbol{\phi}_3$

$$\boldsymbol{\phi}_3 = \boldsymbol{\phi}_2 + \lambda \mathbf{s}_1 \tag{14.20}$$

where λ is the move along the pattern direction $\mathbf{s}_1$. Exploratory moves are now repeated with ϕ_3 as the base point and the process continues. If there is no progress in a particular direction $\mathbf{u}_i$, the step length $\Delta\phi_i$ is reduced (say by a factor of 2) and the exploratory moves continue. The process is assumed to have converged whenever no progress is made in a particular set of exploratory moves and step lengths fall below a small quantity ϵ, that is, when

$$\max_i\{\Delta\phi_i\} < \epsilon \qquad \text{and} \qquad s_j = 0 \tag{14.21}$$

The procedure is summarized in the flow chart shown in Fig. 14.12.

14.4.2 Conjugate Gradient Method

A critical part of the gradient method of optimization is the evaluation of gradients of the objective function $U(\boldsymbol{\phi})$ with respect to various designable parameters ϕ_i. Gradient is defined as

$$\boldsymbol{\nabla} U \triangleq \left\{\frac{\partial U}{\partial \phi_1}, \frac{\partial U}{\partial \phi_2}, \ldots, \frac{\partial U}{\partial \phi_k}\right\}^{t} \tag{14.22}$$

Various derivatives $\partial U/\partial \phi_i$ are obtained by sensitivity analysis. Second-order partial derivatives of U are used to define a Hessian matrix $\mathbf{H}$ as

$$\mathbf{H} \triangleq \begin{bmatrix} \dfrac{\partial^2 U}{\partial \phi_1^2}, & \dfrac{\partial^2 U}{\partial \phi_1 \partial \phi_2}, \ldots, & \dfrac{\partial^2 U}{\partial \phi_1 \partial \phi_k} \\ \dfrac{\partial^2 U}{\partial \phi_2 \partial \phi_1}, & \dfrac{\partial^2 U}{\partial \phi_2^2}, \ldots, & \dfrac{\partial^2 U}{\partial \phi_2 \partial \phi_k} \\ \cdots & \cdots & \cdots \\ \dfrac{\partial^2 U}{\partial \phi_k \partial \phi_1}, & \dfrac{\partial^2 U}{\partial \phi_k \partial \phi_2}, \ldots, & \dfrac{\partial^2 U}{\partial \phi_k^2} \end{bmatrix} \tag{14.23}$$

Gradient methods make use of the property that at the optimum point $\boldsymbol{\phi}_{\min}$,

$$\boldsymbol{\nabla} U(\boldsymbol{\phi}_{\min}) = 0 \tag{14.24}$$

In the vicinity of this optimum point $\boldsymbol{\phi}_{\min}$ (where $\phi_{\min} = \phi + \Delta\phi$), we can use a multidimensional Taylor series expansion to write

$$U(\boldsymbol{\phi} + \boldsymbol{\Delta\phi}) = U(\boldsymbol{\phi}) + \nabla \mathbf{U}^{t} \cdot \boldsymbol{\Delta\phi} + \tfrac{1}{2}\boldsymbol{\Delta\phi}^{t} H(\phi)\boldsymbol{\Delta\phi} + \cdots \tag{14.25}$$

Differentiating and using (14.24),

$$0 \approx \boldsymbol{\nabla} U(\boldsymbol{\phi}) + H(\phi)\Delta\phi \tag{14.26}$$

When $H(\boldsymbol{\phi})$ is a nonsingular matrix, the set of linear equations given by

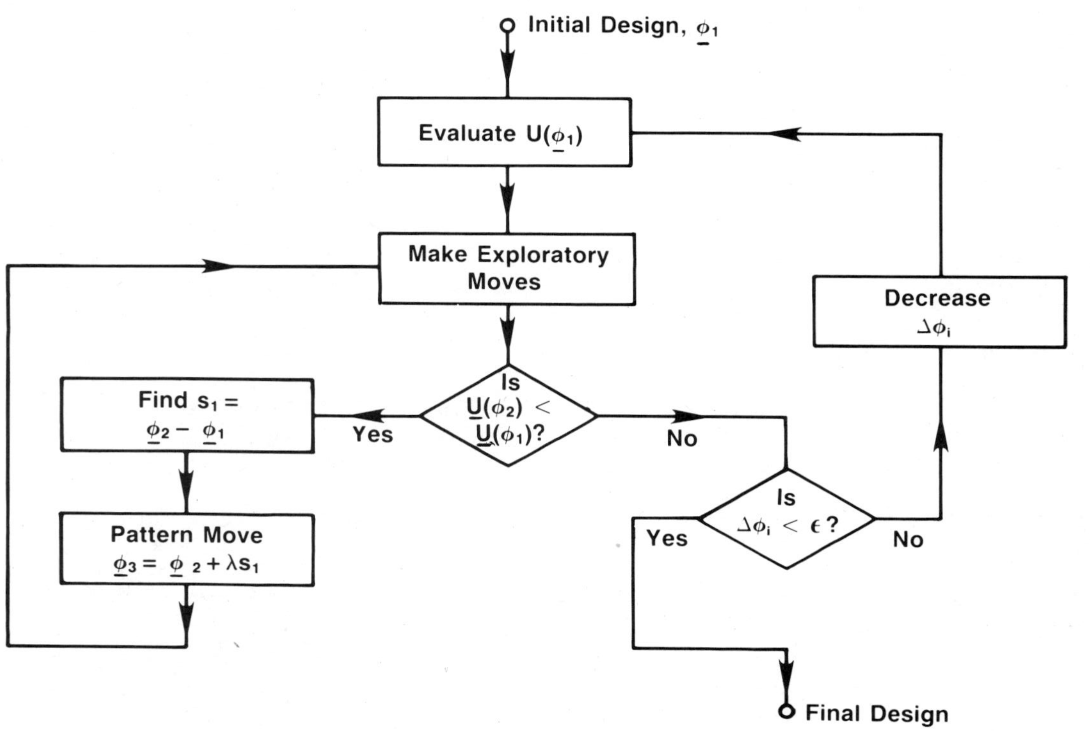

Figure 14.12 Flow chart for pattern search optimization method.

(14.26) can, in principle, be solved for $\Delta\phi$ as

$$\boldsymbol{\Delta\phi} \approx -H^{-1}(\phi)\nabla U(\phi) \tag{14.27}$$

where $H^{-1}(\phi)$ is the inverse of the Hessian matrix at $\boldsymbol{\phi}$. The increment $\boldsymbol{\Delta\phi}$ allows one to move from $\boldsymbol{\phi}$ to the optimum point $\boldsymbol{\phi}_{\text{min}}$. However, the main difficulty in implementing the preceding procedure is the evaluation of H and its inverse at each stage of iteration.

The Davidon–Fletcher–Powell method [20] for conjugate gradient optimization overcomes this difficulty by using an approximation to the inverse of Hessian. This approximation is gradually improved in the successive iterations. Thus, at the $(i+1)$th iteration, an approximation to the inverse of H (say G_i) is used to define the direction of search by

$$\mathbf{s}_i = -G_{i-1}\nabla U(\boldsymbol{\phi}_i) \tag{14.28}$$

Successively improved approximations to G are obtained by

$$G_i = G_{i-1} + \frac{\lambda_i \mathbf{s}_i \mathbf{s}_i^{\text{t}}}{\nabla U(\boldsymbol{\phi}_i)^{\text{t}} G_{i-1} \nabla U(\boldsymbol{\phi}_i)} - \frac{G_{i-1}\mathbf{Y}_i \mathbf{Y}_i^{\text{t}} G_{i-1}}{\mathbf{Y}_i^{\text{t}} G_{i-1} \mathbf{Y}_i} \tag{14.29}$$

where $\mathbf{Y}_i$ is the difference in the values of the gradients in two successive iterations given by

$$\mathbf{Y}_i = \nabla U(\boldsymbol{\phi}_{i+1}) - \nabla U(\boldsymbol{\phi}_i) \tag{14.30}$$

To start the iterations, G_0 may be taken as an identity matrix I. To ensure convergence, G needs to be positive definite. Choosing $G_0 = I$ ensures the positive definiteness of G_0. It can be shown that if G_i is positive definite, G_{i+1} will also be positive definite provided λ_i (i.e., the move in the direction $\mathbf{s}_i$) is computed accurately. However, it is advisable to check for positive definiteness of G as iterations proceed and, if necessary, to reset G to an identity matrix. The optimization procedure is summarized in the flowchart shown in Fig. 14.13.

Sensitivity Analysis. As pointed out earlier, evaluation of gradients is the key step in the implementation of the gradient method of optimization. As the response of a microwave circuit may be expressed in terms of outgoing wave variables (b's) at the various external ports of the circuit, we need to evaluate changes in these wave variables as a result of incremental changes in various designable parameters (ϕ_i) of the circuit. General scattering matrix analysis, discussed in Section 14.3, can be extended to yield the sensitivity information also.

The key step used in sensitivity calculations is the concept of "adjoint network." An adjoint network has (by definition) the same topology as the original network being considered. The number of components, the number

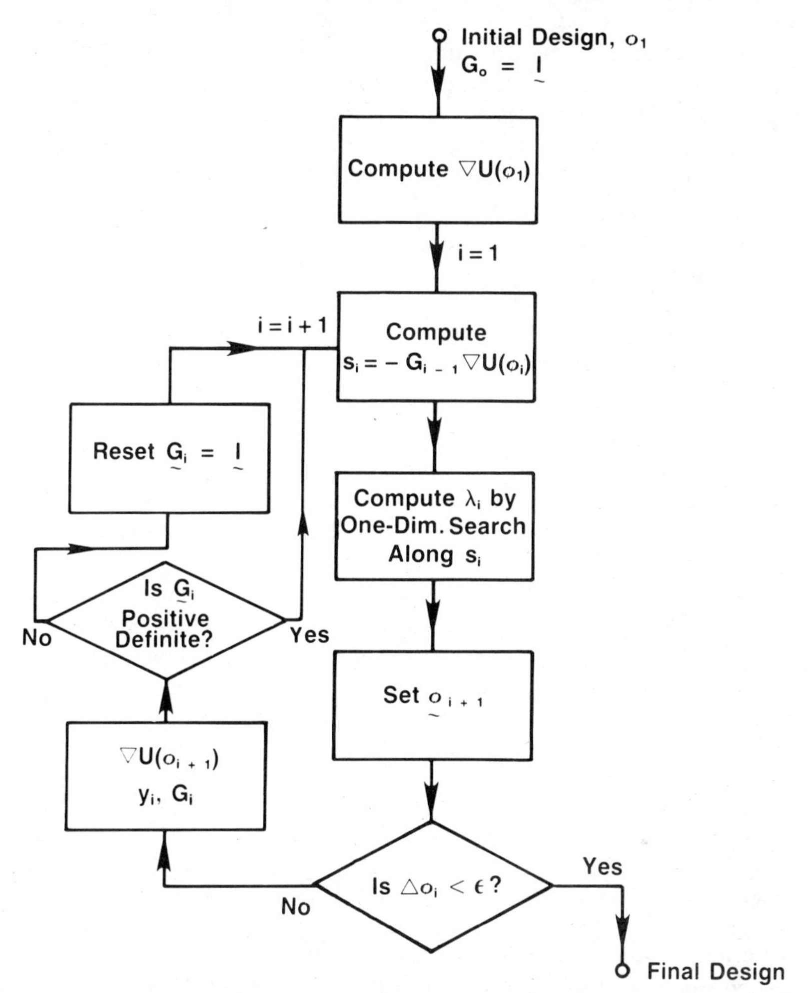

Figure 14.13 Davidon–Fletcher–Powell method for conjugate gradient optimization.

of ports in each of these components, and interconnections among these ports are identical in the two cases. In addition, the scattering matrix of each component in the adjoint network is the transpose of the scattering matrix of the corresponding component in the original network. Thus, if all the components in a circuit are reciprocal, the adjoint network is identical to the original network. However, if a circuit includes nonreciprocal components (like transistors), the adjoint network will be different from the original network.

The method of sensitivity calculations using the adjoint network [1, 20] involves an analysis of the original network and, in addition, an analysis of the adjoint network, also. These two analyses should be carried out to yield the values of wave variables not only at the external ports but also at internal connected ports. The connection-scattering-matrix analysis method discussed in Section 14.3 is well-suited for this purpose. As a matter of fact, the vector **a** evaluated by (14.13) does include the values of wave variables at all the connected ports also. Corresponding b's are written by using connection matrix Γ as in (14.10) without any additional computations.

The procedure for evaluation of sensitivities is obtained by applying Tellegen's theorem [1] to the original and the adjoint networks. If **a** and **b** denote the wave variables for the original network, and $\boldsymbol{\alpha}$ and $\boldsymbol{\beta}$ denote the corresponding wave variables for the adjoint network, the following relationship may be derived [1] by using Tellegen's theorem (in wave variables),

$$\frac{\partial \mathbf{b}_p^t}{\partial \phi_k} \boldsymbol{\alpha}_p - \frac{\partial \mathbf{a}_p^t}{\partial \phi_k} \boldsymbol{\beta}_p = \mathbf{a}^t \frac{\partial S^t}{\partial \phi_k} \boldsymbol{\alpha} \tag{14.31}$$

In this equation, the subscript p refers to the external ports of the circuit. Thus $\mathbf{b}_p$ denotes the outgoing wave variables at the external ports of the circuit. Also, ϕ_k represents a circuit parameter. We need to find out the sensitivities of S-parameters of the circuit with respect to the designable parameters like ϕ_k. These are given by the derivatives $\partial \mathbf{b}_p^t / \partial \phi_k$. The superscript t denotes the transpose of a vector or matrix. For example

$$S_{ij} = \frac{b_i}{a_j} \tag{14.32a}$$

When $a_j = 1$,

$$\frac{\partial S_{ij}}{\partial \phi_k} = \frac{\partial b_i}{\partial \phi_k} \tag{14.32b}$$

The vector $\mathbf{a}_p$ denotes the incoming wave variables (excitations) at the various external ports of the original network. As $\mathbf{a}_p$ is independent of ϕ_k, we have

$$\frac{\partial \mathbf{a}_p^t}{\partial \phi_k} = 0 \tag{14.33}$$

and Equation 14.31 reduces to

$$\frac{\partial \mathbf{b}_p^t}{\partial \phi_k} \boldsymbol{\alpha}_p = \boldsymbol{a}^t \frac{\partial S^t}{\partial \phi_k} \boldsymbol{\alpha} \tag{14.34}$$

where the $\boldsymbol{\alpha}_p$ denote the excitations of the adjoint network. The $\boldsymbol{\alpha}_p$ are chosen appropriately in conjunction with the excitations $\mathbf{a}_p$ for the original network. For example, if $\partial S_{ij}/\partial \phi_k$ is to be evaluated, we choose $a_j = 1$ (all other a's = 0) and $\alpha_i = 1$ (all other α's = 0). For this set of choices, the left-side of (14.34) becomes

$$\frac{\partial \mathbf{b}_p^{\mathrm{t}}}{\partial \phi_k} \boldsymbol{\alpha}_p = \frac{\partial S_{ij}}{\partial \phi_k} \tag{14.35}$$

On the right side of (14.34), $\mathbf{a}^{\mathrm{t}}$ is the transpose of the wave variables $\mathbf{a}$ obtained as the response of the original network. Similarly, $\boldsymbol{\alpha}$ denotes the response of the adjoint network. The matrix S is a matrix similar to S in (14.3) and in (14.7). It has a block diagonal form, submatrices along the diagonal being the individual S-matrices of various components in the circuit. The S-matrix satisfies the relations

$$b = S\mathbf{a} \tag{14.36}$$

and

$$\boldsymbol{\beta} = S^{\mathrm{t}}\boldsymbol{\alpha} \tag{15.37}$$

Differential scattering matrix $(\partial S^{\mathrm{t}}/\partial \phi_k)$ will have only a few nonzero elements. These are the elements that are effected by changes in the parameter ϕ_k. For example, if ϕ_k is a parameter of a two-port component in the circuit, we could possibly have only four nonzero elements in $\partial S/\partial \phi_k$. In such a case, evaluation of $\partial S_{ij}/\partial \phi_k$ needs values of only those a and α wave variables that appear at the ports of this two-port component involving ϕ_k. Differential scattering matrices $\partial S/\partial \phi_k$ are obtained from S-parameter characterizations of individual components in the circuit.

It may be pointed out that for evaluating other sensitivities of the circuit, say $\partial S_{ij}/\partial \phi_l$, it is not necessary to reevaluate responses $\mathbf{a}$ and $\boldsymbol{\alpha}$. Only the differential scattering matrix to be used on the right side of (14.34) now becomes $\partial S^{\mathrm{t}}/\partial \phi_l$. Thus (14.34) provides an efficient method of evaluating all the circuit sensitivities with only two circuit analyses (one for the original network and one for the adjoint network).

These sensitivities $\partial S_{ij}/\partial \phi$ are used for writing $\boldsymbol{\nabla} U(\boldsymbol{\phi})$ used in gradient optimization methods (Equations 14.28 to 14.30) and also for carrying out tolerance analysis of the circuit [1, chap. 13].

14.5 CAD OF NONLINEAR CIRCUITS

The computer-aided circuit analysis techniques discussed in Section 14.3 are applicable to linear circuits only. For design of nonlinear circuits, such as oscillators, frequency multipliers, and mixers, it becomes necessary to modify the analysis technique. A general CAD procedure for nonlinear microwave circuits is discussed in this section.

14.5.1 Linear and Nonlinear Subnetworks

In most of the nonlinear microwave circuits, the nonlinearity is restricted to a single device (a MESFET or Schottky diode, for example) operating in the nonlinear region. Since the rest of the circuit is linear, it is desirable to separate the circuit into linear and nonlinear subnetworks that may be treated separately. Nonlinear and linear subnetworks are shown schematically in Figure 14.14. It may be noted that the linear part of the device model and the package equivalent circuit are also included in the linear subnetwork.

The nonlinear subnetwork consists of nonlinear circuit elements in the device model (nonlinear controlled current sources, diodes, voltage-dependent capacitors, current-dependent resistors, etc.). These nonlinear components (and hence the resulting nonlinear subnetwork) are usually best simulated in terms of time-domain voltage and current vectors $\mathbf{v}(t)$ and $\mathbf{i}(t)$. A general time-domain representation of the nonlinear subnetwork could be of the form [21, 22],

$$\mathbf{i}(t) = \mathbf{f}\left\{\mathbf{i}(t), \frac{d\mathbf{i}}{dt}, \mathbf{v}(t), \frac{d\mathbf{v}}{dt}\right\} \tag{14.38}$$

where vector $\mathbf{f}$ is a nonlinear function of various currents, voltages, and their time derivatives. Higher order derivatives may also appear. For the present discussion, it is assumed that the vector function $\mathbf{f}$ is known (hopefully analytically) from the nonlinear modeling of the device.

The linear subnetwork, on the other hand, can be easily analyzed by a frequency-domain circuit analysis program and characterized in terms of an admittance matrix as

$$\mathbf{I}(\omega) = Y(\omega)\mathbf{V}(\omega) + \mathbf{J}(\omega) \tag{14.39}$$

where $\mathbf{V}, \mathbf{I}$ are the vectors of voltage and current phasors at the subnetwork ports, $\mathbf{Y}$ represents its admittance matrix, and $\mathbf{J}$ is a vector of Norton equivalent current sources.

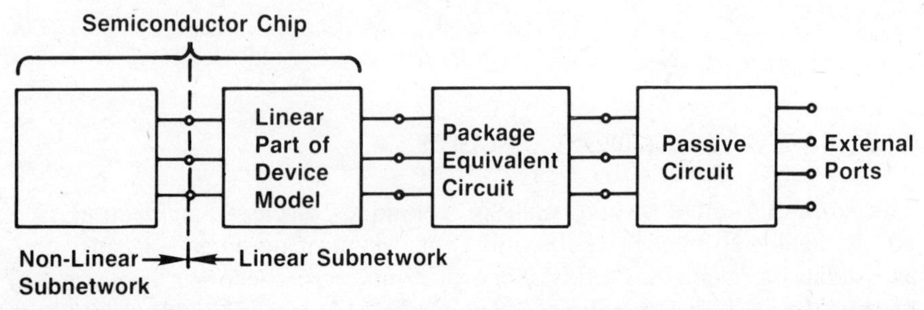

Figure 14.14 Separation of a nonlinear microwave circuit in linear and nonlinear subnetworks.

Analysis of the overall nonlinear circuit (linear subnetwork plus nonlinear subnetwork) involves continuity of currents given by (14.38) and (14.39) at the interface between linear and nonlinear subnetworks.

14.5.2 Harmonic Balance Method

In principle, the solution to the nonlinear circuit problem can always be found by integrating the differential equations that describe the system. However, in most microwave nonlinear circuits we are interested in the steady state response with periodic excitations and periodic responses with a limited number of significant harmonics. In such a situation, the "harmonic balance technique" is much more efficient.

In the conventional harmonic balance technique [23, 24], each state variable in the total nonlinear network is represented by a Fourier series that satisfies the requirement of periodicity. An optimization algorithm is then used to adjust the coefficients of the Fourier series such that the system equations are satisfied with least error. Although this method avoids the computationally expensive process of numerically integrating the dynamic equations, its main disadvantage is the large number of variables that must be optimized. A modification of this harmonic balance method is the piecewise harmonic balance technique proposed by Nakhla and Vlach [25]. In this modified technique, the nonlinear network is decomposed into a minimum possible number of linear and nonlinear subnetworks as discussed in Section 14.5.1. The frequency-domain analysis of the linear subnetwork is carried out at a frequency ω_0 and its harmonics. The characterization in (14.39) is thus generalized as:

$$I_k(k\omega_0) = Y(k\omega_0)\, V_k(k\omega_0) + J_k(k\omega_0) \tag{14.40}$$

where $k = 1, \ldots, N$, N being the number of significant harmonics considered. The nonlinear subnetwork is analyzed in the time domain and the response obtained is in the form of (14.38). A Fourier expansion of the currents yield

$$\mathbf{i}(t) = \left\{ \mathrm{Re} \sum_{k=0}^{N} \mathbf{F}_k(k\omega_0) \exp(jk\omega_0 t) \right\} \tag{14.41}$$

where coefficients $\mathbf{F}_k$ are obtained by a FFT algorithm. The piecewise harmonic-balance technique involves a comparison of (14.40) and (14.41) to yield a system of equations as

$$\mathbf{F}_k(k\omega_0) - Y(k\omega_0)\mathbf{V}_k(k\omega_0) - J_k(k\omega_0) = 0, \qquad k = 0, 1, \ldots, N \tag{14.42}$$

The solution of this system of equations yields the response of the circuit in terms of voltage harmonics V_k. Numerically, the solution of (14.42) is obtained by minimizing the harmonic balance error

$$\Delta\epsilon_b(\mathbf{V}) = \left\{ \sum_{k=0}^{N} |\boldsymbol{F}_k(k\omega_0) - Y(k\omega_0)\mathbf{V}_k(\omega k_0) - \mathbf{J}_k(k\omega_0)|^2 \right\}^{1/2} \tag{14.43}$$

Optimization techniques discussed in Section 14.4 may be used for this purpose.

14.5.3 Optimization of Nonlinear Circuits

As in the case of linear circuits, for nonlinear circuits also, it becomes necessary to adjust the various designable parameters to optimize the circuit performance. As most of the designable parameters are in the linear subnetwork part of the circuits, it becomes necessary to obtain repeated solutions for (14.40) and (14.43). Minimization of (14.43) provides an objective function for optimization of the circuit performance with respect to designable parameters. This leads to a solution scheme consisting of two nested optimization loops, the inner one providing the objective function for the outer. Obviously, this demands cónsiderable computational effort.

Recently, Rizzoli et al. [26] have proposed a nonlinear circuit design procedure that combines the previously mentioned two optimization loops into a single iterative loop. Let a vector $\boldsymbol{\phi}_i$ represent a set of designable parameters of the circuit; $\boldsymbol{\phi}_0$ being the values in this set for the initial design, $\boldsymbol{\phi}_1$ after the first iteration in the optimization loop, etc. As the linear subnetwork is analyzed for various values of $\boldsymbol{\phi}$, we can rewrite (14.40) as

$$\mathbf{I}_k = \mathbf{Y}(k\omega_0, \boldsymbol{\phi})\mathbf{V}_k(k\omega_0) + \mathbf{J}_k(k\omega_0, \boldsymbol{\phi}) \tag{14.44}$$

and the harmonic balance error may be written as

$$\Delta\epsilon_b(\mathbf{V}) = \left\{ \sum_{k=0}^{N} |\mathbf{F}_k(k\omega_0) - Y(k\omega_0, \boldsymbol{\phi})\mathbf{V}_k(k\omega_0) - J_k(k\omega_0, \boldsymbol{\phi})|^2 \right\}^{1/2} \tag{14.45}$$

Rather than trying to minimize (14.45), we select an objective function such that the circuit performance is optimized at the same time that the harmonic balance error is being minimized. This does imply that, in the intermediate iterations, the circuit performance is being evaluated for a configuration for which $\Delta\epsilon_b \neq 0$. Such intermediate configurations will be devoid of physical meaning. But after a successful optimization the harmonic balance will be restored, and a physically meaningful optimized network solution is obtained. The formulation of objective functions for the nonlinear circuit design using this procedure is discussed in [26].

14.6 USE OF SUPERCOMPUTERS

In order to make microwave circuit analysis and optimization more efficient, it is desirable to look into applications of now easily accessible vector processors like Cyber 205 or Cray X-MP for microwave CAD. The key advantage of using such machines lies in the fact that similarly structured computational steps in analysis and optimization algorithms can be

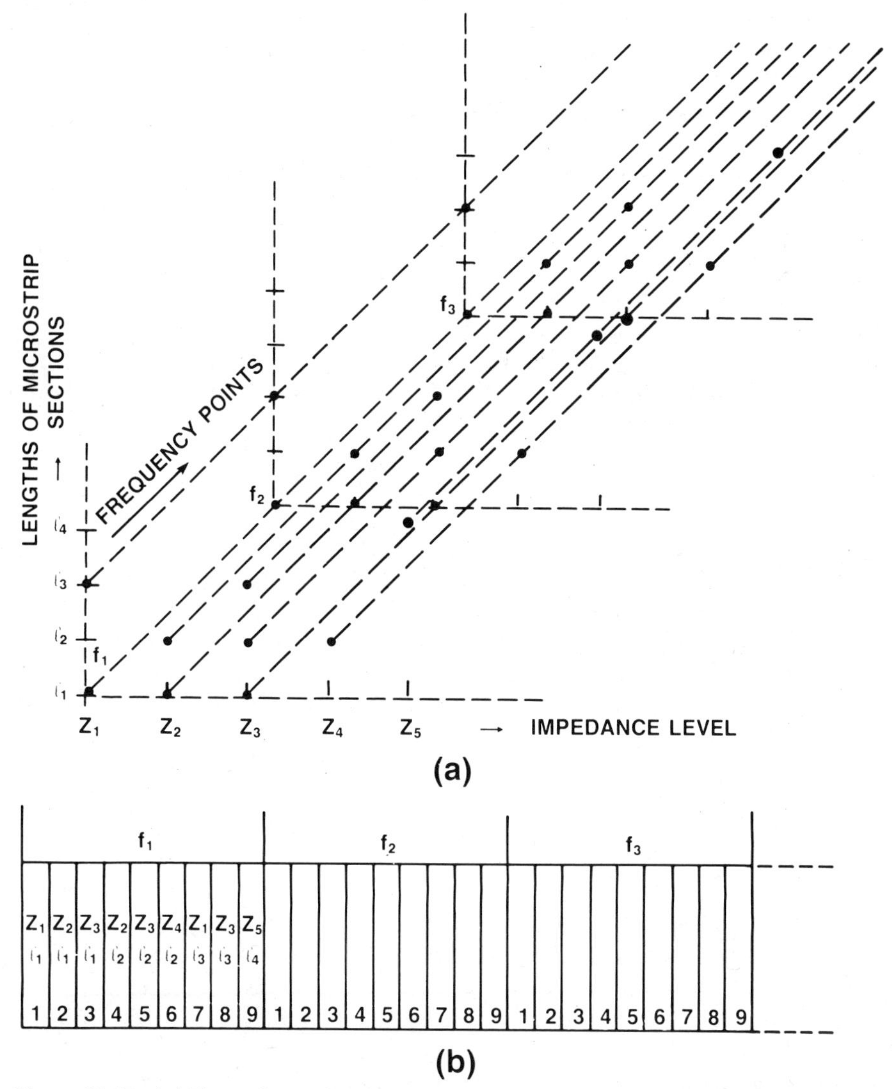

Figure 14.15 (*a*) Three-dimensional set of parameters for various microstrip sections in a typical circuit. (*b*) A single "super component" for all the nine microstrip sections in the circuit analyzed at three frequency points.

grouped together and carried out in parallel (in the pipeline sense [27]) by the specialized vector hardware. Such algorithms appear to be very well-suited for microwave CAD for the following two reasons. First, typical microwave circuits contain only a few types of components repeating themselves in different parts of the circuit with slightly different parameters. For example, a circuit will contain a large number of microstrip sections with different impedances and lengths located at various parts of the circuit. Secondly, in most of the circuits we need the analysis to be carried out at a large number of frequency points. When a vector processor is being used for analysis of such a circuit, all these microstrip sections may be combined into a multidimensional "super component" having ($m \times n \times p$) dimensions, where m is the number of impedances, n is the number of different lengths of the microstrip sections, and p is the number of frequency points where the circuit performance is to be evaluated. This "super component" concept is illustrated in Fig. 14.15. For the example considered, the three-dimensional set of parameters for various microstrip sections in the circuit is shown in Fig. 14.15(a), and the corresponding single "super component" is shown in Fig. 14.15(b). It has been pointed out [28] that such "semantic" vectorization of algorithms could save considerable computer time and would result in a much lower computational cost as compared to similar algorithms on present-day scalar processors. Thus there is a need for investigations leading to the optimum selection of processor architecture and algorithms suitable for microwave CAD.

REFERENCES

1. Gupta, K. C., R. Garg, and R. Chadha, *Computer Aided Design of Microwave Circuits*, Artech House, Norwood, Mass., 1981.
2. Gupta, K. C., R. Garg, and I. J. Bahl, *Microstrip Lines and Slotlines*, Artech House, Norwood, Mass., 1979.
3. Edwards, T. C., *Foundations for Microstrip Circuit Design*, Wiley, New York, 1981.
4. Bhartia, P., and P. Pramanick, "Fin-line Characteristics and Circuits," in *Infrared and Millimeter Waves*, Vol. 17, Academic Press, New York, 1988.
5. Hammerstad, E. O., and F. Bekkadal, "Microstrip Handbook," ELAB Report, STF44 A74169, University of Trondheim, Norway, 1975.
6. *Super-Compact Users Manual*, Version 1.6, Compact Software, Inc., Palo Alto, Calif., July 1982.
7. *Touchstone 002 Users Manual*, Version 1.4, EEsof, Inc., Westlake Village, Calif., Dec. 1985.
8. Mehran, R., "Frequency Dependent Equivalent Circuits for Microstrip Right Angle Bends, T-junctions and Crossings," *Arch. Elek. Übertragug*, Vol. 30, 1976, pp. 80–82.
9. "MCAD Software and Design," brochure on microstrip computer programs, Aachen, West Germany, 1984.

10. Jansen, R. H., and N. H. L. Koster, "A Unified CAD Basis for Frequency Dependent Characterization of Strip, Slot and Coplanar MIC Components," *Proc. 11th European Microwave Conf.* (Amsterdam, the Netherlands, 1981), pp. 682–687.
11. Jansen, R. H., "Hybrid Mode Analysis of End Effects of Planar Microwave and Millimeter-wave Transmission Lines," *Proc. IEE*, Part H, Vol. 128, 1981, pp. 77–86.
12. Koster, N. H. L., and R. H. Jansen, "The Equivalent Circuits of the Asymmetrical Series Gap in Microstrip and Suspended Substrate Lines," *IEEE Trans. Microwave Theory Tech.*, Vol. MTT-30, Aug. 1982, pp. 1273–1279.
13. Jansen, R. H., "The Spectral Domain Approach for Microwave Integrated Circuits," *IEEE Trans. Microwave Theory Tech.*, Vol. MTT-33, Oct. 1985, pp. 1043–1056.
14. Koster, N. H. L., and R. H. Jansen, "The Microstrip Step Discontinuity: A Revised Description," *IEEE Trans. Microwave Theory Tech.*, Feb. 1986, pp. 213–223.
15. Katehi, P. B., and N. G. Alexopoulos, "Frequency Dependent Characteristics of Microstrip Discontinuities in Millimeter-wave Integrated Circuits," *IEEE Trans. Microwave Theory Tech.*, Vol. MTT-33, Oct. 1985, pp. 1029–1035.
16. Jackson, R. W., and D. M. Pozar, "Full-wave Analysis of Microstrip Open-end and Gap Discontinuities," *IEEE Trans. Microwave Theory Tech.*, Oct. 1985, pp. 1036–1042.
17. Mosig, J. R., and F. E. Gardiol, "Analytical and Numerical Techniques in the Green's Function Treatment of Microstrip Antennas and Scatterers," *Proc. IEE*, Part H, Vol. 130, 1983, pp. 175–182.
18. Monaco, V. A., and P. Tiberio, "Computer-Aided Analysis of Microwave Circuits," *IEEE Trans. Microwave Theory Tech.*, Vol. MTT-22, Mar. 1974, pp. 249–263.
19. Gelnovatch, V. G., and I. L. Chase, "DEMON—An Optimal Seeking Computer Program for the Design of Microwave Circuits," *IEEE Trans. Solid-State Circuits*, Vol. 5, Dec. 1970, pp. 303–309.
20. Bandler, J. W., and R. E. Seviora, "Wave Sensitivities of Networks," *IEEE Trans. Microwave Theory Tech.*, Vol. MTT-20, Feb. 1972, pp. 138–147.
21. Chua, L. O., *Introduction to Non-Linear Network Theory*, McGraw-Hill, New York, 1969.
22. Chua, L. O., and P. M. Lin, *Computer-Aided Analysis of Electronic Circuits*, Chapts. 5, 7, 10, and 11, Prentice-Hall, Englewood Cliffs, N.J., 1975.
23. Baily, E. M., "Steady State Harmonic Analysis of Non-Linear Networks," Ph.D. dissertation, Stanford University, Stanford, Calif., 1968.
24. Lindenlaub, J. C., "An Approach for Finding the Sinusoidal Steady State Response of Non-Linear Systems," *Proc. 7th Annu. Allerton Conf. Circuit and System Theory*, University of Illinois, Chicago, 1969.
25. Nakhla, M. S., and J. Vlach, "A Piecewise Harmonic Balance Technique for Determination of Periodic Response of Nonlinear Systems," *IEEE Trans. Circuits Syst.*, Vol. CAS-23, 1976, pp. 85–91.
26. Rizzoli, V., et al., "A General-Purpose Program for Nonlinear Microwave Circuit Design," *IEEE Trans. Microwave Theory Tech.*, Vol. MTT-31, Sept. 1983, pp. 762–770.

27. Hwang, K., et al., "Vector Computer Architecture and Processing Techniques," in *Advances in Computers*, Vol. 20, Academic Press, New York, 1981, pp. 115–197.
28. V. Rizzoli, et al., "Vectorized Program Architectures for Supercomputer-Aided Circuit Design," *IEEE Trans. Microwave Theory Tech.*, Vol. MTT-34, Jan. 1986, pp. 135–141.

15 MICROWAVE INTEGRATED CIRCUITS

15.1 INTRODUCTION

The current trend in microwave technology is toward miniaturization and integration. Component size and weight are prime factors in the design of electronic systems for satellite communications, phased-array radar systems, electronic warfare, and other airborne applications. Microwave integrated circuits (MICs) fulfill this requirement. A MIC consists of an assembly that combines different circuit functions formed by strip or microstrip transmission lines and incorporates planar semiconductor devices and passive distributed or lumped circuit elements. MICs promise higher reliability, reproducibility, better performance, smaller size, and lower cost than conventional waveguide and coaxial microwave circuits.

Microwave integrated circuits can be divided into two categories, namely, hybrid MICs and monolithic MICs (MMICs). In the hybrid technology, solid state devices and passive circuit elements are added to a dielectric substrate. The passive circuits either use distributed elements or lumped elements or a combination of both. The distributed and lumped elements are fabricated by using thin- or thick-film technology. The distributed circuit elements are generally single-level metallization. The lumped elements are either fabricated by using multilevel deposition and plating techniques or attached to the substrate in chip form. In the case of monolithic microwave integrated circuits, all active and passive circuit elements or components and interconnections are formed into bulk, or onto the surface of a semi-insulating substrate by some deposition scheme such as epitaxy, ion implantation, sputtering, evaporation, diffusion, or a combination of these processes.

Over the past two decades, MICs have been upgraded from laboratory curiosities to qualified military and commercial hardware in production. A major factor in the success of MICs has been the advancement in the development of microwave solid state devices as described in Chapters 7 and 8. In fact, MICs have brought about a revolution in the microwave industry because so many microwave functions can be packaged into a given space. The concept of implementing a "system on a chip," which was pioneered in low-frequency silicon integrated-circuit technology (microprocessors, speech synthesizers, AM/FM broadcast receivers, and broadband transmitters), is now becoming a reality in monolithic microwave technology. MMICs provide low cost, improved reliability, reproducibility, small size, low weight, broadband performance, circuit design flexibility, and multifunction performance on a single chip.

Three general types of circuit components can be utilized for MICs: distributed transmission lines (microstrip, coplanar, and strip line, etc.), lumped elements (resistors, inductors, and capacitors), and solid state devices (field-effect transistors, Schottky-barrier diode, etc.). The theory and design of these circuit components have already been discussed in earlier chapters. In this chapter, the choice of materials, mask layouts and their fabrication, circuit design considerations, and some key features of fabrication technology are described. Some examples of the different types of MICs (hybrid, miniature, and monolithic) are also given.

15.2 MATERIALS

The basic materials for fabricating MICs, in general, are divided into four categories:

1. Substrate materials—sapphire, alumina, beryllia, ferrite/garnet, silicon, RT/duroid, quartz, GaAs, InP, etc.,
2. Conductor materials—copper, gold, silver, aluminum, etc.,
3. Dielectric films—SiO, SiO_2, Si_3N_4, Ta_2O_5, etc.,
4. Resistive films—NiCr, Ta, Ti, TaN, Cermet, GaAs, etc.

15.2.1 Substrate Materials

The substrate selected for MICs must have the following general characteristics [1–3]:

1. The cost of the substrate must be justifiable for the application.
2. The choice of thickness and permittivity determines the achievable impedance range as well as the usable frequency range.
3. There should be low loss tangent for negligible dielectric loss.
4. The substrate surface finish should be good (~0.05–0.1 μm finish), with relative freedom from voids, to keep conductor loss low and yet maintain good metal–film adhesion.
5. There should be good mechanical strength and thermal conductivity.
6. No deformation should occur during processing of circuit.
7. A substrate with sufficient area for the particular application and complexity should be available.

The dielectric constant of the substrate should be high while meeting the other criteria to keep circuit size small. A variety of substrate materials are listed in Table 15.1 with their properties.

Use of a high dielectric constant substrate $\epsilon_r \simeq 10$ is highly desirable. However, the substrate thickness is limited by modal problems. High-impedance lines on thin substrates require very narrow conductors, which become lossy and the definition of these narrow conductors can be difficult. The temperature dependence of the dielectric constant of substrates (such as rutile) can lead to problems in certain applications, where the temperature variations are large.

For low frequencies up to about 4–6 GHz for circuits and up to and beyond 20 GHz for array antennas, plastic substrates ($\epsilon_r \simeq 2$–4) are often used. Alumina (Al_2O_3) is one of the most suitable substrate material for use up to 20 GHz. The grade of the Al_2O_3 used depends upon the fabrication technology employed: thin or thick film. Alumina with 85% purity has high dielectric losses and poor reproducibility and is normally not used. The dielectric constant of alumina may be high for millimeter-wave circuits because the high impedance lines with required tolerances are difficult to fabricate and are lossy. Quartz with a dielectric constant of 4 is more suitable and widely used for high-frequency (>20 GHz) microwave and millimeter-wave integrated circuits.

Beryllia is a good conductor of thermal energy and is suitable for power applications where heat dissipation from active devices is large and a low thermal resistance substrate is required. GaAs is one of the most suitable substrates for MMICs, since most of the active devices, such as low-noise MESFETs, power MESFETs, and Schottky diodes, are fabricated on a GaAs substrate, and the material has semi-insulating properties.

15.2.2 Conductor Materials

The conductor material used for microwave integrated circuits should have high conductivity, a low temperature coefficient of resistance, low RF resistance, good adhesion to the substrate, good etchability and solderability, and be easy to deposit or electroplate [1, 3]. The resistance is determined by the RF surface resistivity and skin depth: the skin depth determines the thickness required. The conductor thickness should be at least three to four times skin depth, to include 98% of the current density.

Table 15.2 shows the properties of some widely used conductor materials for MICs. These materials have good conductivity and can be deposited by a number of methods. Some conductors have good electrical conductivity but poor substrate adhesion, whereas others have poor electrical conductivity and good substrate adhesion. To obtain good adhesion with high conductivity material, a very thin layer (100–500 Å) of a poor conductor is deposited between the substrate and the good conductor. Some examples of typical conductor combinations to obtain good adhesion and good conductivity material are Cr/Au, Pd/Au, and Ta/Au for hybrid MICs, and Cr/Au, Ti/Pt/Au, and Ti/Pd/Au for MMICs. The selection of the conductors is

TABLE 15.1 Properties of Substrates for MICs

Material	Surface Roughness (μm)	Loss Tangent ($\tan \delta$) at 10 GHz (10^{-4})	Relative Dielectric Constant (ϵ_r)	Thermal Conductivity K(W/cm-°C)	Dielectric Strength (kV/cm)	MIC Applications
Alumina						
99.5%	2–8	1–2	10	0.3	4×10^3	Microstrip, suspended substrate
96%	20	6	9	0.28	4×10^3	
85%	50	15	8	0.2	4×10^3	
Sapphire	1	1	9.3–11.7	0.4	4×10^3	Microstrip, lumped element
Glass	1	20	5	0.01	—	Lumped element
Beryllia (BeO)	2–50	1	6.6	2.5	—	Compound substrate
Rutile	10–100	4	100	0.02	—	Microstrip
Ferrite/Garnet	10	2	13–16	0.03	4×10^3	Microstrip, coplanar
GaAs (high resistivity)	1	6	12.9	0.46	350	High-Frequency, microstrip, monolithic MIC
Si (high resistivity)	1	10–100	12	1.5	300	MMICs
Quartz	1	1	3.8	0.01	10×10^3	Microstrip, high frequency
Polyolefin	1	1	2.3	0.001	~300	
InP	—	—	14	0.68	—	MMICs

TABLE 15.2 Properties of Conductors for MICs

Material	Surface Resistivity ($\Omega/sq \times 10^{-7}\sqrt{f}$)	Skin Depth δ at 2 GHz (μm)	Coefficient of Thermal Expansion ($\alpha_t/°C \times 10^6$)	Adherence to Dielectrics	Deposition Technique
Ag	2.5	1.4	21	Poor	Evaporation
Cu	2.6	1.5	18	Poor	Evaporation plating
Au	3.0	1.7	15	Poor	Evaporation plating
Al	3.3	1.9	26	Poor	Evaporation
Cr	4.7	2.7	9.0	Good	Evaporation
Ta	7.2	4.0	6.6	Good	Electron-beam sputtering
Ti	—	—	—	Good	Evaporation sputtering
Mo	4.7	2.7	6	Fair	Electron-beam sputtering evaporation
W	4.7	2.6	4.6	Fair	Sputtering, vapor phase, electron-beam, evaporation
Pt	—	3.6	9	—	Sputtering, electron beam
Pd	—	3.6	11	—	Evaporation, sputtering, electroplating

determined by compatibility with other materials required in the circuit and the process required. A typical adhesion layer may have a surface resistivity ranging from 500 to 1000 Ω/square, but does not contribute to any loss.

15.2.3 Dielectric Materials

Dielectric films in microwave integrated circuits are used as insulators for capacitors, protective layers for active devices, and insulating layers for passive circuits. The desirable properties of these dielectric materials are reproducibility, high breakdown voltage, low-loss tangent, and the ability to undergo processing without developing pin holes [3]. Table 15.3 shows some of the properties of commonly used dielectric films in MICs. SiO is not very stable and can be used in noncritical applications, such as by-pass and dc blocking capacitors. A quality factor Q of more than 100 can be obtained for capacitors using SiO_2, Si_3N_4, and Ta_2O_5 materials. These materials can be deposited by sputtering and plasma-enhanced chemical vapor deposition (CVD). For high-power applications, high breakdown voltage in excess of 200 V is required. Such capacitors can be obtained with fairly thick dielectric films ($\sim 1\ \mu$m) with low probability of pin holes.

15.2.4 Resistive Films

Resistive films in microwave integrated circuits are required for fabricating resistors for terminations, attenuators, and for bias networks. The properties

TABLE 15.3 Properties of Dielectric Films for MICs

Material	Method of Deposition	Relative Dielectric Constant (ϵ_r)	Dielectric Strength (V/cm)	Microwave Q
SiO	Evaporation	6–8	4×10^5	30
SiO_2	Deposition	4	10^7	100–1000
Si_3N_4	Vapor-phase sputtering	7.6	10^7	
Al_2O_3	Anodization evaporation	7–10	4×10^6	
Ta_2O_5	Anodization sputtering	22–25	6×10^6	100

required for a resistive material are: good stability, low temperature coefficient of resistance (TCR), and sheet resistivities in the range of 10 to 2000 Ω/square [3, 4]. Table 15.4 lists some of the thin-film resistive materials used in MICs. Evaporated Nichrome and tantalum nitride are the most commonly used materials.

15.3 MASK LAYOUTS AND MASK FABRICATION

Any MIC design starts with a schematic diagram for the circuit. After the circuit is finalized, a rough layout is drawn. The next step is to obtain an accurate mask layout for producing a single mask layer for hybrid MICs or a set of masks for miniature MICs and MMICs. Finally, hybrid MIC substrates are etched using these masks for the required pattern, and for miniature and monolithic MICs various photolithic steps are carried out using a set of masks.

15.3.1 Mask Layouts

For MICs the layout is carefully prepared keeping in mind the chip or packaged devices (active and passive), crosstalk considerations, microstrip and layout discontinuities, and tuning capability. There are several techniques that have been used to produce accurate layouts for MICs.

TABLE 15.4 Properties of Resistive Films for MMICs

Material	Method of Deposition	Resistivity (Ω/square)	TCR (%/°C)	Stability
Cr	Evaporation	10–1000	−0.100–+0.10	Poor
NiCr	Evaporation	40–400	+0.001–+0.10	Good
Ta	Sputtering	5–100	−0.010–+0.01	Excellent
Cr-SiO	Evaporation or cement	up to 600	−0.005–−0.02	Fair
Ti	Evaporation	5–2000	−0.100–+0.10	Fair

In addition to manually prepared printed circuit taping and rubylith methods [5, 6], the digitally controlled methods being used are (1) automated rubylith, (2) MiCAD™ or Autoart™, and (3) GDS II–CALMA.

Automated Rubylith. Rubylith consists of a Mylar base sheet covered with a peelable layer of a soft transparent plastic (red or orange) material. The Mylar sheet is about 50 to 100 μm thick, while the thickness of the soft red material is about 25 to 50 μm. The required pattern is cut using an instrument called a coordinatograph and by peeling it away from the clear base layer. A coordinatograph consists of a lighted table, a digitally controlled motorized cutting tool head, and a small computer. The whole system is basically a precision drafting instrument that allows a high degree of accuracy, repeatability, and resolution in cutting rubylith and artworks. The complete circuit in terms of the x and y coordinates and other dimensions is entered into the computer, then the computer control blade cuts the required pattern on the rubylith sheet placed on the lighted table. The size of the artwork could be 5×, 10×, etc., depending on the accuracy required. The final step still involves manually removing the undesired rubylith film. This layout is now photographically reduced to produce the final size 1× mask or mask set.

MiCAD™ or Autoart™. Both MiCAD and Autoart are interactive, two-dimensional drafting programs for microwave circuits that translate the circuit descriptions into mask layouts. The operation of Autoart is similar to MiCAD. MiCAD software gives the user the ability to create the camera-ready artwork for MICs. This software works with Touchstone CAD software. The user can edit or tune the Touchstone circuit file to modify the autoprocessed picture on the screen. MiCAD takes the abstract model of the Touchstone circuit file and builds a graphical data base from it and displays a picture of the circuit on the color graphics screen. Then it automatically converts the circuit into a physical layout.

GDS II–CALMA. GDS II is a turnkey interactive graphic CAD system introduced by CALMA (General Electric Company) for integrated-circuit layout. The basic function of the system is to serve as an aid for the layout phase of IC design, that is, to start with the circuit diagram or hand-drawn mask layout and obtain its finished mask layout (single layer for hybrid MICs or multilayer for monolithic MICs) in the form of a coordinate printout, pen plot of the circuit, and the complete circuit on a magnetic tape that can be given to a mask manufacturer.

A GDS II System consists of a number of graphic editing stations, the central processing unit (CPU), a tape reader, and a printer. The software package that accompanies it provides graphics and other functions. Each editing station as shown in Fig. 15.1 comprises a keyboard, a color graphic monitor to show the layout, an alphanumeric monitor to show the numerical information entered, and a tablet with an electronic pen for entering data points.

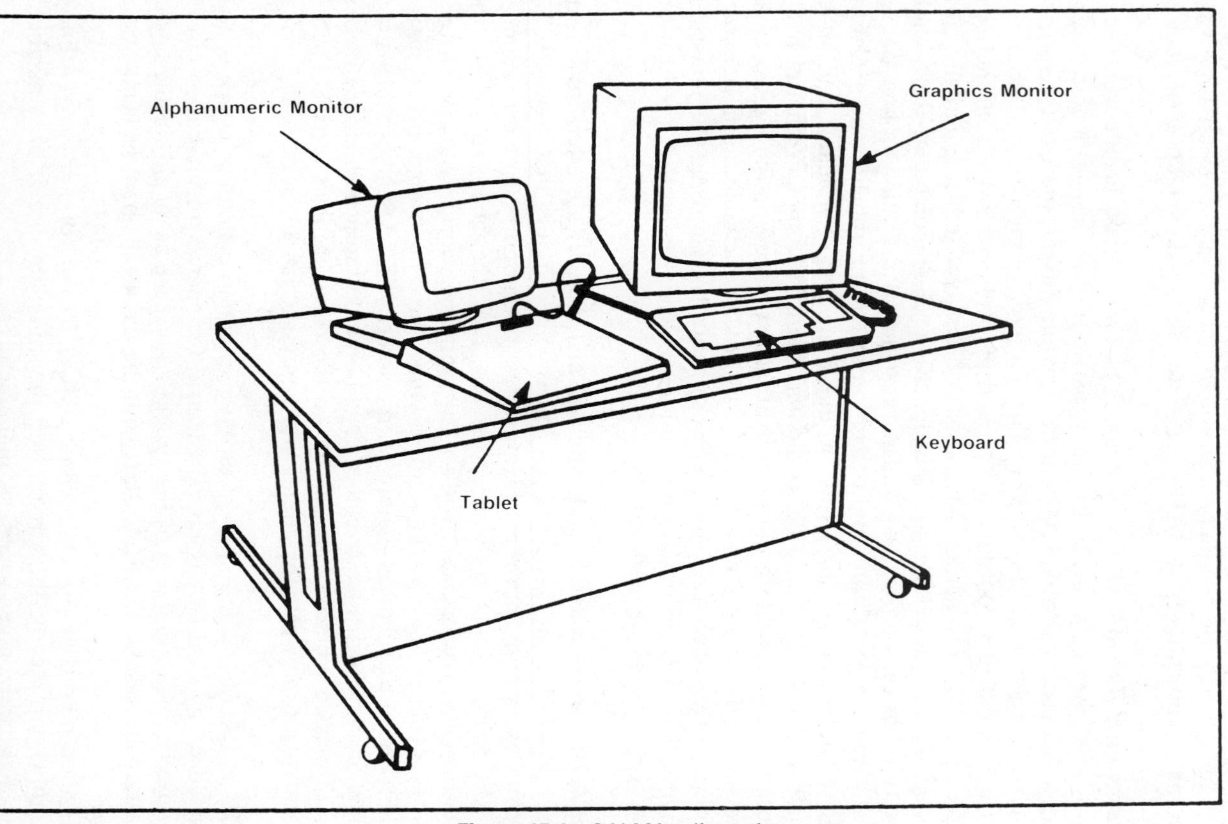

Figure 15.1 CALMA edit station.

Figure 15.2(*a*) shows a composite CALMA layout of a Lange coupler (see Section 5.3 for details) and Fig. 15.2(*b*) shows a composite CALMA layout of a two-stage low-noise amplifier. Both circuits are designed using the monolithic approach.

15.3.2 Mask Fabrication

Optical masks are usually used for hybrid MICs. However, in MMICs, new lithography techniques (considered very important for good process yield and fast turnaround) are headed in the direction of beam writing, including electron beam, focused ion beam, and laser beam. However, except for a small percentage of direct writing on the wafers (only critical geometries), optical masks are widely used. These masks are usually generated using optical techniques or electron-beam lithography.

Masks consist of sheets of glass or quartz (also called blanks) with the desired pattern defined on them in thin-film materials such as photoemulsion (silver halide based), chromium, or iron oxide. Emulsion mask coatings are still the most widely used for hybrid MICs and for noncritical working plates where geometries above 3 μm are common. Silver halide based emulsions have numerous advantages such as low cost, high photosensitivity, good image resolution and contrast, and reversal processing. Their major disadvantages are scratch sensitivity and higher image defect density. Shiny chrome is the most popular hard-surface coating for glass blanks and has been proven successful for high-resolution work when used with positive optical photoresists. The main difficulty with chromium is its high reflectivity, which is solved by using an antireflection layer of chromium oxide. Iron oxide is another hard-surface coating material that has very low reflectivity and is used commonly to make "see-through" masks. Iron oxide is transparent at longer wavelengths, allowing the operator to "see through" the entire mask when aligning it to the pattern on the wafer. High-frequency light, at which the photoresist is sensitive and the iron oxide mask is opaque, is then used to make the exposure.

There are many different ways available for transferring digital pattern data onto wafers. Figure 15.3 shows [7] several different process options available to arrive at a mask or a means of patterning the wafer. The trend in mask imaging has generally been toward a shorter process between the pattern digital data and the wafer imaging operation. The magnetic tape on which the pattern data are stored is loaded into the console, and a light-field emulsion reticle, typically at 10×, is obtained through computer control of the exposure shapes and placement. This reticle is then contact printed to yield a dark-field emulsion reticle. The next step is to make a 10× reticle on a hard-surface blank and step-and-repeat it into 1× emulsion master masks for the complete die. Finally these emulsion masters are contact printed to make hard-surface working plates.

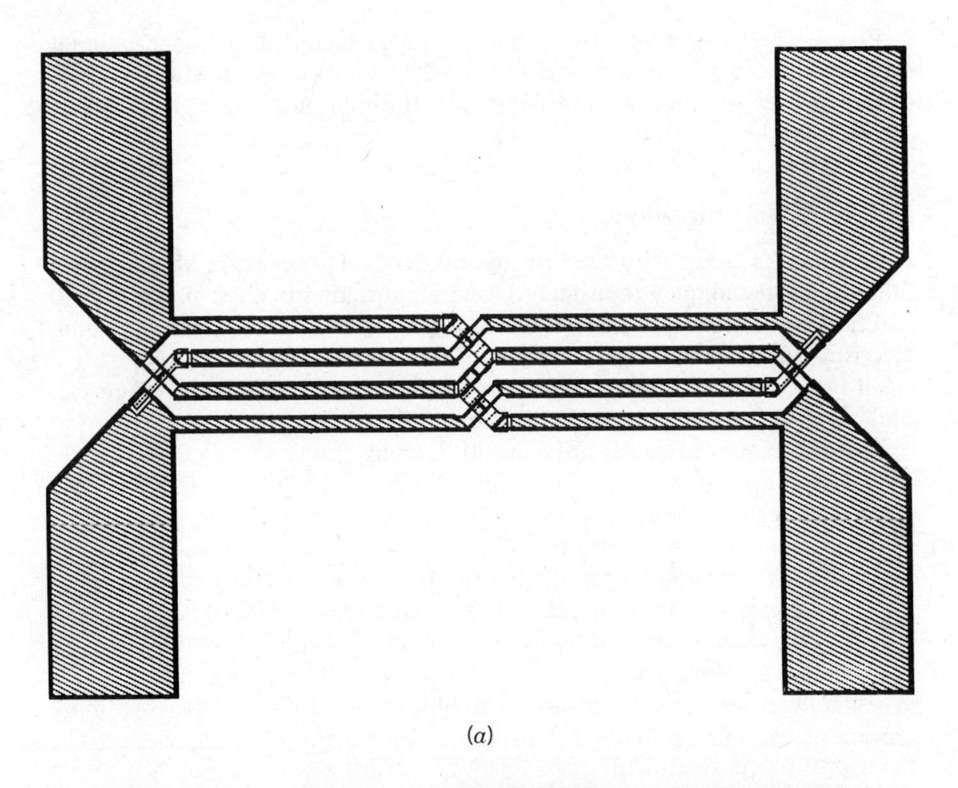

(*a*)

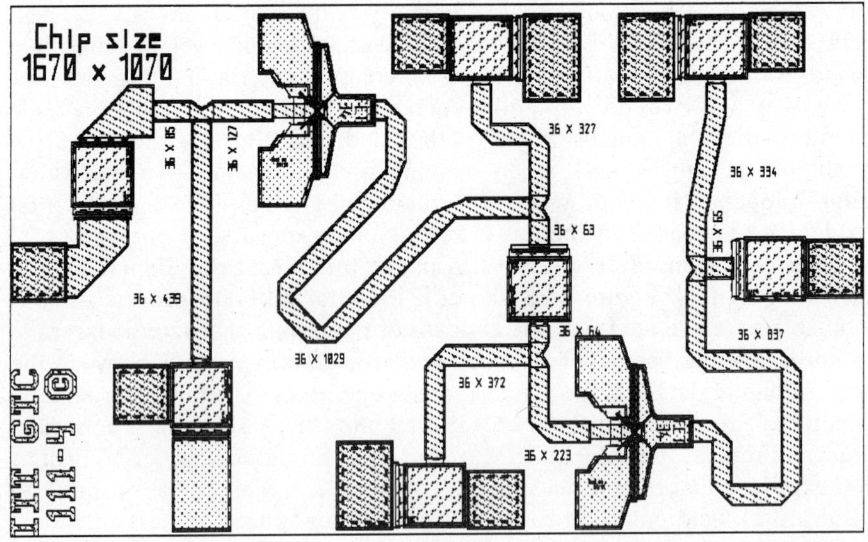

(*b*)

Figure 15.2 Composite CALMA layouts. (*a*) Lange coupler. (*b*) Two-stage low-noise amplifier.

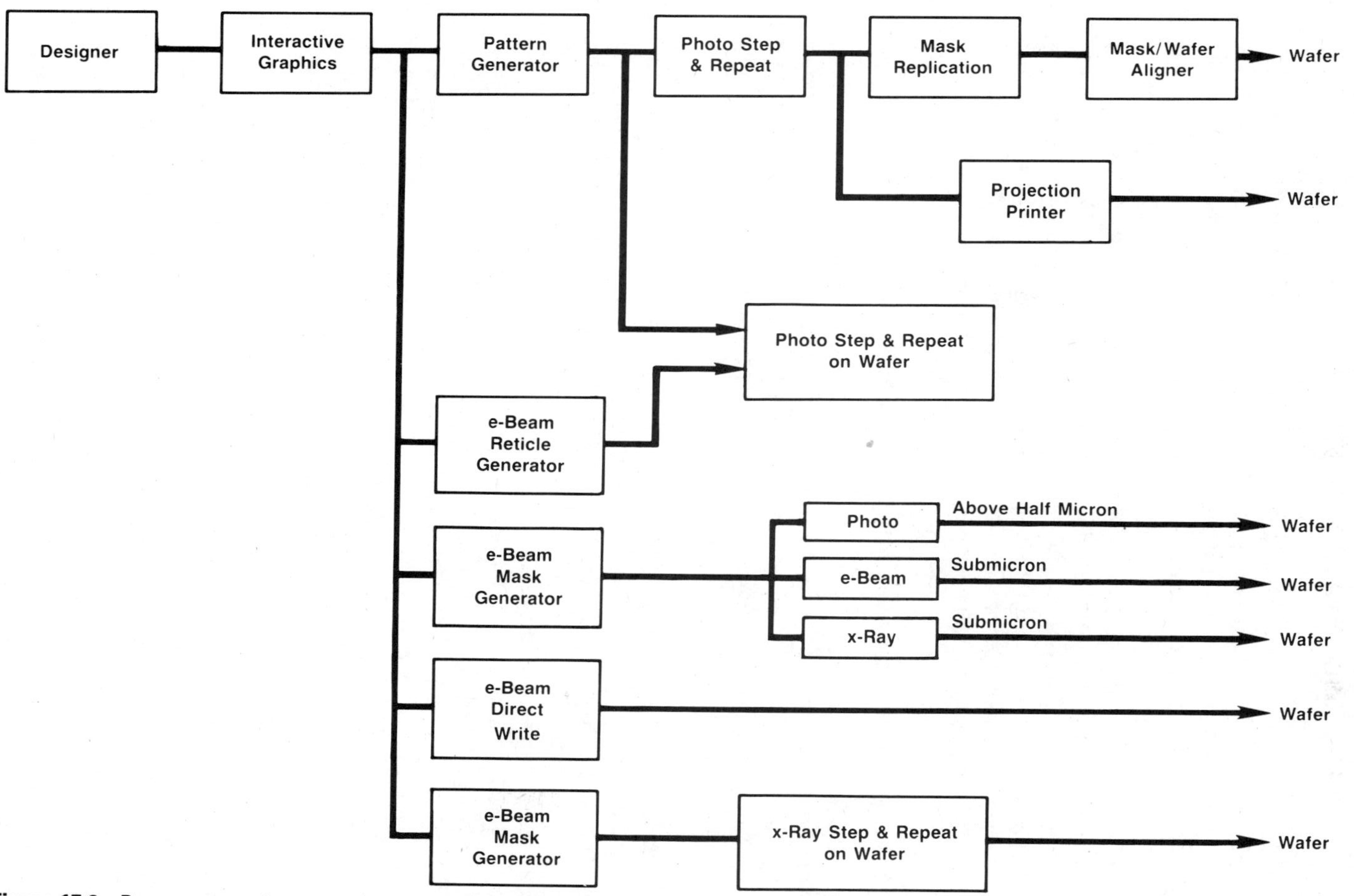

Figure 15.3 Process steps for various mask making techniques. (After D. J. Elliott [7], reprinted with the permission of McGraw-Hill.)

15.4 HYBRID MICROWAVE INTEGRATED CIRCUITS

Hybrid MICs can be divided into two categories: (1) hybrid MICs and (2) miniature hybrid MICs. Hybrid MICs use the distributed circuit elements that are fabricated on a substrate using a single-level metallization technique. Other circuit elements, such as inductors, capacitors, resistors, and solid state devices, are added to the substrate. Miniature hybrid MICs use multilevel elements, such as inductors, capacitors, resistors, and distributed circuit elements, deposited on the substrate and the solid state devices are attached to the substrate. The circuit fabricated using this technology is smaller in size than with hybrid MICs but larger than MMICs. In the following sections, techniques of fabrication, design considerations, and some examples of these two types of MICs are presented.

15.4.1 Hybrid Microwave Integrated Circuits

Hybrid microwave integrated circuits have been used almost exclusively in the frequency range 1 to 20 GHz for space and military applications, because they meet the requirements for shock, temperature conditions, and severe vibration. This section is intended to provide a brief introduction of this technology.

Fabrication Technology. Fabrication technology used for microwave integrated circuits is constantly developing to meet the requirements of increasing frequency of operation, higher yield, and reduced costs. This can be achieved by a carefully controlled thin-film manufacturing process in a clean room environment that is accurate and repeatable.

The first step in the fabrication process is the deposition of a first layer (seed layer) of metal film on the substrate. The selection of the film is made based on the criteria discussed in Section 15.2.2. Good adhesion to the substrate is one of the important factors in selection of conductor material for the first layer of metal film.

Some precautions particular to MIC conductors should be mentioned with regard to the deposition techniques. At RF frequencies the electromagnetic fields are confined to several skin depths of the conductors. In order to achieve low loss, the layer of high-resistance material such as chromium must be extremely thin. The main conductor must have a low bulk dc resistivity for low-loss propagation. Improper processing techniques can result in high RF loss for a low sheet resistance material made up of a thin chromium and a thicker gold structure. In particular, as a result of very high substrate temperatures ($>300°C$) sometimes encountered during sputtering, this thin sputtered chromium layer will diffuse into an overlaying gold film. This results in a high RF loss, even though the sheet resistance may be low with a thick gold layer. Therefore, techniques such as sputtering must be used with care for MIC materials.

The metal films are deposited on the substrates by the following three methods.

1. Vacuum evaporation,
2. Electron-beam evaporation,
3. Sputtering.

As discussed in Section 15.2 on materials, a proper combination of metal layers is used to obtain good adherence to the substrate, low loss, and stability. A typical metal combination for alumina substrate is Cr/Cu/Au or NiCr/Ni/Au. A very thin seed layer of suitable metal is deposited on the substrate by one of the preceding techniques and then the bulk conductor metal is deposited by electroplating techniques. The seed layers of metal provide mechanical and electrical foundation layers on which to electroplate good-quality bulk conductor metal.

The circuit definition can be accomplished by a plate-through technique or by an etchback technique. The techniques that are used to define patterns in metal layers can influence the deposition choice. Figure 15.4 illustrates the fabrication techniques [8].

The plate-through technique begins with a substrate coated with a thin layer of evaporated metal. This is followed by an application of a thick photoresist, as shown in Fig. 15.4(*a*). The thickness of this photoresist is similar to the thickness of the final metal film required. After defining a pattern in the photoresist, the second metal layer is plated up to the desired thickness with precise definition, and only in the areas where metal is required. The photoresist layer is then washed away and the thin seed metal is etched with very little undercut from the undesired areas. This technique is also suitable for fabricating lines that are 25 to 50 μm wide and/or when the separation between them is 25 to 50 μm.

The second technique is the etchback technique. This technique, as illustrated in Fig. 15.4(*b*), utilizes a thick metal layer obtained either completely by evaporation or by a combination of a thin evaporated layer and a thicker plated layer. A thin photoresist layer is used as a mask to define the circuit pattern. The undesired areas of metal are then removed by etching. This technique results in undercutting the metal film by about twice the line conductor thickness. The plate-through technique not only permits better definition for thick conductors, but also saves on cost in that the required material is deposited.

Design Considerations for Hybrid MICs. Hybrid microwave integrated-circuit technology using a single-layer metallization technique to form the RF components is a mature technology. In this section, some of the design considerations of hybrid MICs are described.

For a hybrid MIC, the first design consideration starts with the selection of the substrate material. Some of the details of the substrate materials were

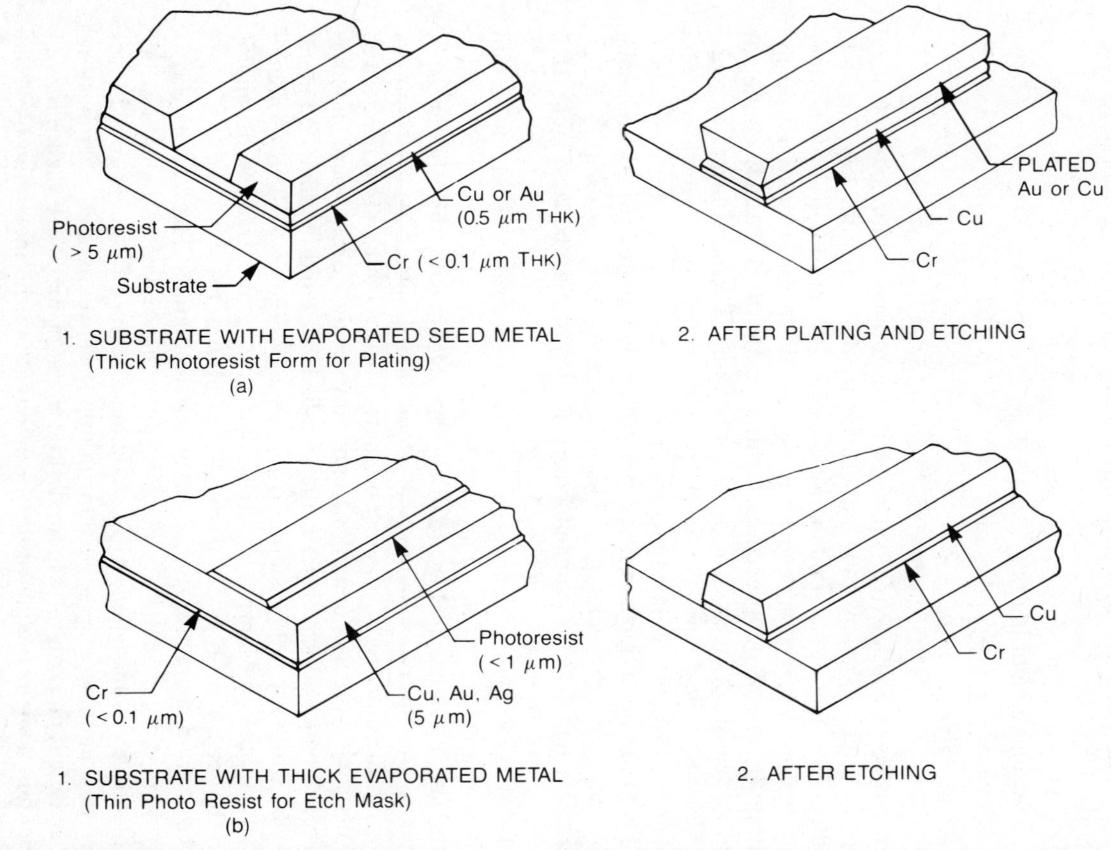

Figure 15.4 Techniques for defining conductor pattern in hybrid MICs. (*a*) Fabrication by plating and etching. (*b*) Fabrication by etching thin plated metal.

described in Section 15.2, and a list of substrate materials with their properties is given in Table 15.1. The choice of substrate material depends on the frequency of operation, application, etc. High dielectric constant substrates result in small-size circuits; however, a trade-off with size versus performance has to be made. The choice of thickness of the substrate depends on the frequency of operation. Thinner substrates are required as the frequency becomes higher. For example, 0.635-mm-, 0.380-mm-, and 0.254-mm-thick substrates are suitable for 1–8 GHz, 8–12 GHz, and 12–20 GHz, respectively. In the millimeter-wave frequency range, 0.1- to 0.2-mm-thick substrates are used. A discussion on the advantages and disadvantages of substrate thickness is presented in Chapter 2.

The circuit components can be integrated with close proximity and packaged in a small size. However, the coupling between two adjacent circuit components should be minimal, typically less than 30 dB. As a rule of thumb, two adjacent microstrip lines should be separated by three substrate thicknesses.

Examples of Hybrid MICs. This technology is widely used in various applications, for example, electronic systems, instruments, satellite communications, phased-array radar systems, electronic warfare, and other airborne applications. Passive as well as active circuits have been used in these applications. Amplifiers (low noise, broadband, and power), mixers, transmit-receive switches, phase shifters and oscillators, employing hybrid MIC technology are being used in the previously mentioned systems. Many other examples of hybrid MICs have been described in the preceding chapters.

Amplifiers have been the most important application for hybrid MICs. Figure 15.5 shows a 12-W GaAs FET amplifier [9] operating over 3.7–4.2 GHz. This amplifier has a gain of 58 dB and power-added efficiency of 26%. The amplifier shown in Fig. 15.5 is space qualified and is currently being used in communication satellites [9] as a replacement for a traveling-wave tube amplifier. A 0–360° continuously variable phase shifter that operates over the 4–8 GHz band is shown in Fig. 15.6. The phase shifter uses dual-gate FETs [10], and is capable of performing multiple modulations.

15.4.2 Miniature Hybrid Microwave Integrated Circuits

In miniature hybrid MICs, passive circuits are batch fabricated on the substrate and the solid state devices are attached to these circuits. The circuit components that are fabricated on the substrate include transmission lines, inductors, capacitors, resistors, air bridges, and distributed components [11, 12]. The solid state devices, GaAs FETs in pellet form or flip-chip [13], are mounted on the substrate surface. A low parasitic ground connection for the FET source contacts is provided by a metal septum that

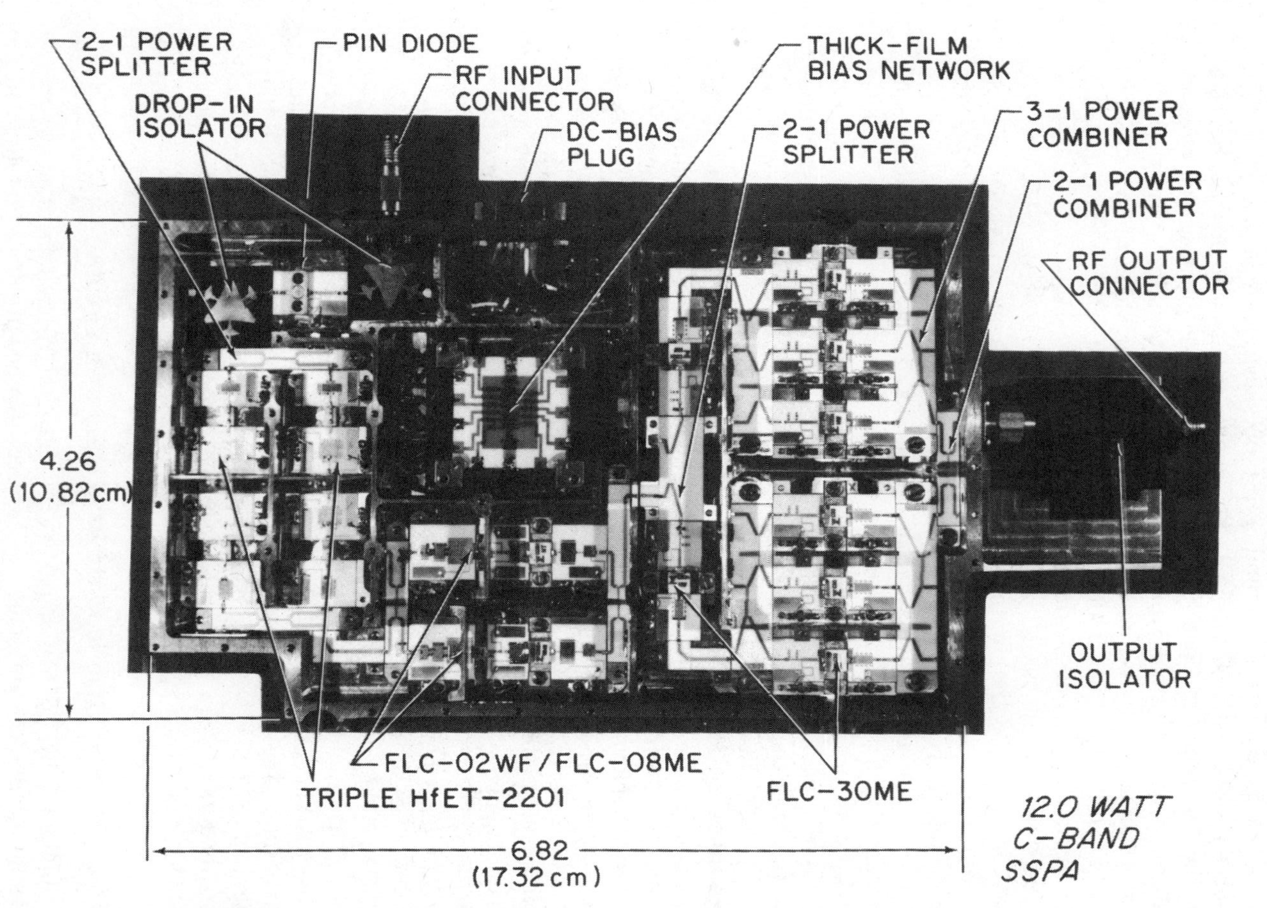

Figure 15.5 Photograph of a 12-W *C*-band solid state power amplifier. (After B. Dornan, et al. [9], reprinted with the permission of RCA Corp.)

Figure 15.6 Photograph of a 0–360° continuously variable dual-gate FET phase shifter. (After M. Kumar et al. [10], 1984 IEEE, reprinted with permission.)

is fabricated as an integral part of the substrate. Low inductance ground connections are also provided on the substrate through the septums whenever the circuit technology requires it. The advantages of such circuit technology are small size, light weight, excellent heat dissipation, and low loss. Batch fabrication reduces the assembly cost when compared to the hybrid circuits previously described. Also tuning of circuits can be done as in hybrid MICs by providing the tuning elements on the circuit and connecting them by wire bonds to get the best performance. The device selection is made independently of the circuit, thus giving the flexibility to obtain the best results.

The thermal expansion of BeO and GaAs are about equal, and thus the mounting of the devices can be achieved with very low thermal stress. For low-noise circuits, for example, low-noise amplifiers and mixers, the heat

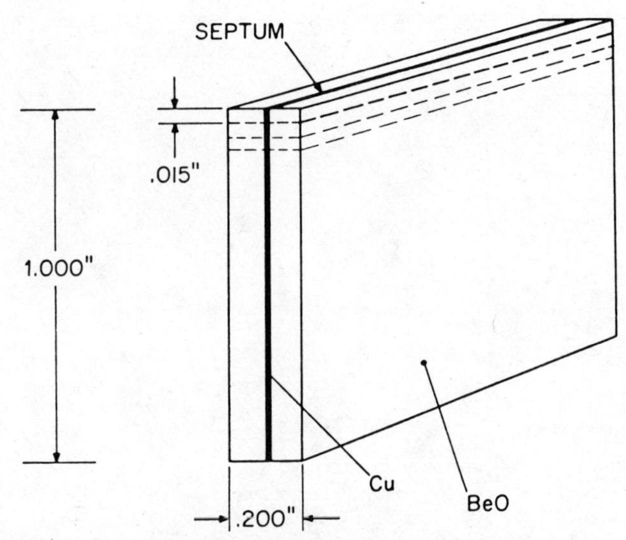

Figure 15.7 Fabrication of ceramic substrates with septum. (After F. Sechi, et al. [14], reprinted with the permission of *Microwave J.*)

dissipation through the substrate is not very important. Other substrates like alumina (Al_2O_3) are used for such applications.

Fabrication Technology [14]. The fabrication process starts with the preparation of a special ceramic substrate that includes a copper septum for a low-inductance ground return. Beryllium oxide or alumina ceramic plates—two for a single septum structure—are copper metallized on one side. The two plates are stacked with two metal layers in contact as shown in Fig. 15.7, pressed in a fixture, and heated at a high temperature in a furnace. The solid diffusion between the copper layers (no liquid phase is involved) results in a very strong thermocompression bond. The sandwich, as shown in Fig. 15.7, is then sliced to the required substrate thickness, typically 0.25 to 0.38 mm. The substrates, with copper septum embedded in the ceramic, are then polished on both sides.

The fabrication process of the circuits is illustrated in Fig. 15.8. The substrate strip is covered on the top surface with a glaze. This is required because the surface quality of ceramic, even when polished, is too rough and is not suitable for the fabrication of thin-film capacitors. The glazing is done selectively and the areas such as the metal septum and the FET bonding area are left uncovered. For glazing the substrate, a glass frit is selectively deposited on the substrate using silk-screening techniques and is fired in an oven at a high temperature. The glass particles of the frit melt form a very smooth film of glass over selected areas of the substrate. Thin films of metal and dielectric are deposited and photo defined using multilevel masks to form inductors, thin-film capacitors, air bridges, trans-

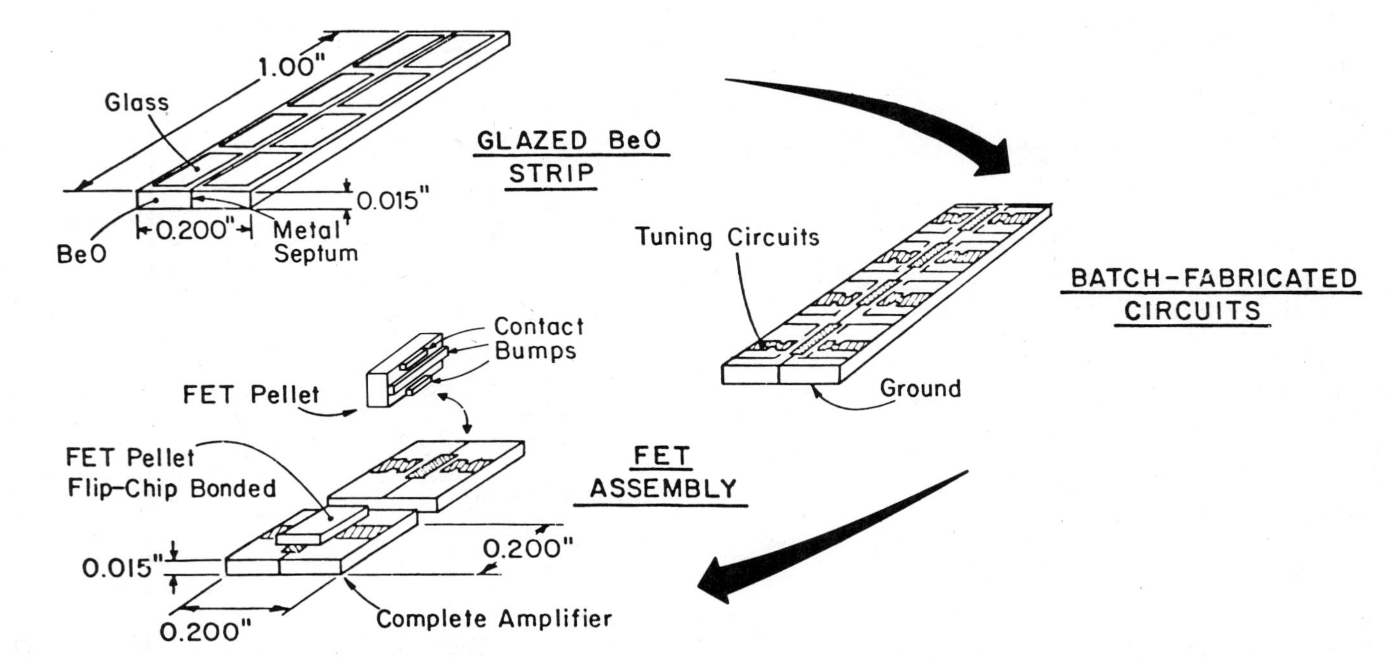

Figure 15.8 Miniature ceramic circuit fabrication process. (After F. Sechi, et al. [14], reprinted with the permission of *Microwave J.*)

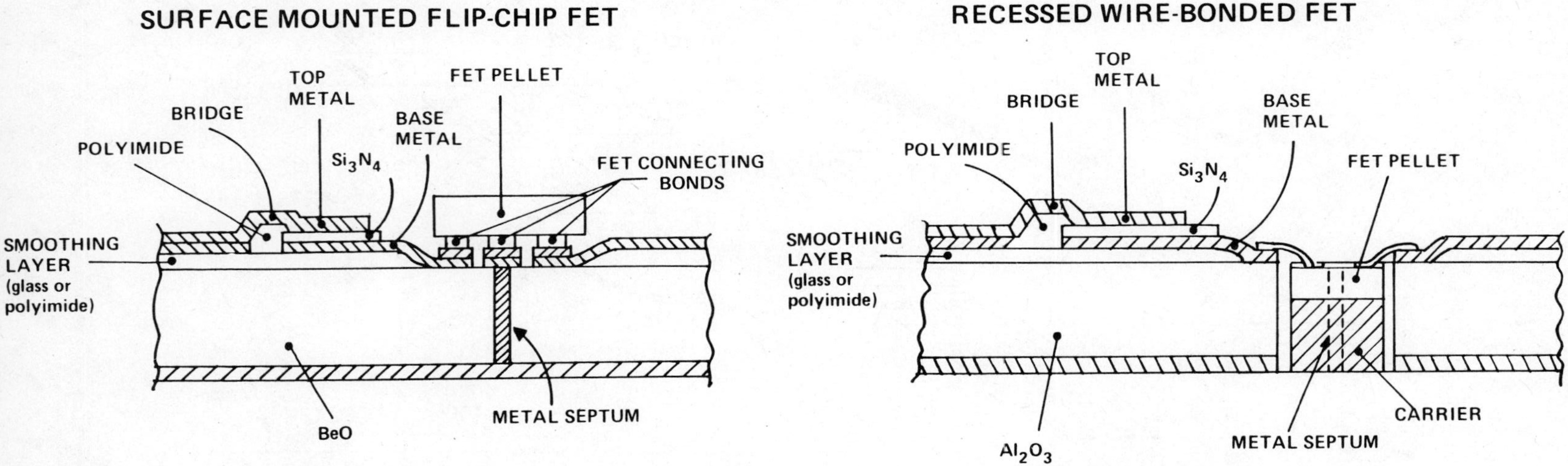

Figure 15.9 Cross section of miniature circuit. (After F. Sechi, et al. [14], reprinted with the permission of *Microwave J.*)

mission lines, resistors, etc. The circuits are batch fabricated and have small size—a combination that leads to low fabrication cost. The processed substrates are diced into separate circuits, and FETs, in chip form, are bonded to the circuit.

Figure 15.9 shows in detail a cross section of the miniature ceramic circuit structure, specifically the individual metal and dielectric layers and the high thermal conductivity beryllium oxide substrates as used for high-power applications. The thin-film base conductor forms the ground contacts, the bottom electrodes of the capacitors, inductive lines, and the first-level of interconnections. A thin film of silicon nitride is the dielectric for the capacitors. Other dielectric films, tantalum oxide (Ta_2O_5), and silicon dioxide (SiO_2), can also be used. Another metal layer forms the second layer of interconnections, the top electrodes of the capacitors and air bridges supported by polyimide. In this circuit technology, high-Q circuit components such as inductors are fabricated by using thick substrates (0.25–0.38 mm thick).

Design Considerations. For a hybrid miniature microwave integrated circuit, the design considerations for the passive circuit are similar to those of monolithic microwave integrated circuits (MMICs), which is described in Section 15.5.3. In miniature circuits, the passive circuits and solid state devices are fabricated separately and then solid state devices are attached to the substrate having passive circuits. The selection of substrate depends upon the specific application wanted. For example, for power amplifiers, a good heat dissipation is required through the substrate. Beryllium oxide has excellent heat dissipation properties and high Q. In miniature hybrid MICs, the substrate thickness is chosen independent of the device thermal characteristics. The other design considerations for designing the circuits follow similar principles described in Section 15.5.3.

Some Circuit Examples. Several circuits have been fabricated using this technology, for example, a high-efficiency X-band amplifier, a 1-W Ku-band amplifier, and a T/R module. Figure 15.10 shows the photograph of a 1-W Ku-band amplifier [13]. The amplifier exhibited 4.5-dB gain over a 400-MHz bandwidth at 16 GHz, an output power of 1 W and efficiency of 14%. The FET pellet is flip-chip bonded at the center of the substrate.

Another example of the circuit is a T/R module [14]. It consists of a Ku-band transmitter with electronic tuning and biphase modulation capability and a receiver with a low-noise amplifier and phase detector. The photograph of the T/R module is shown in Fig. 15.11. The size of this T/R module is $7.0 \times 9.0 \times 1.6$ cm. The module operates over a 16.0–16.5 GHz frequency range, with 3.9–4.4 W peak power, has a tuning voltage range of +2.5 to +2.9 V, and a noise figure of 5 dB.

Low-noise amplifiers have been fabricated on alumina substrate [15]. An example of such a circuit is a two-stage low-noise amplifier shown in Fig.

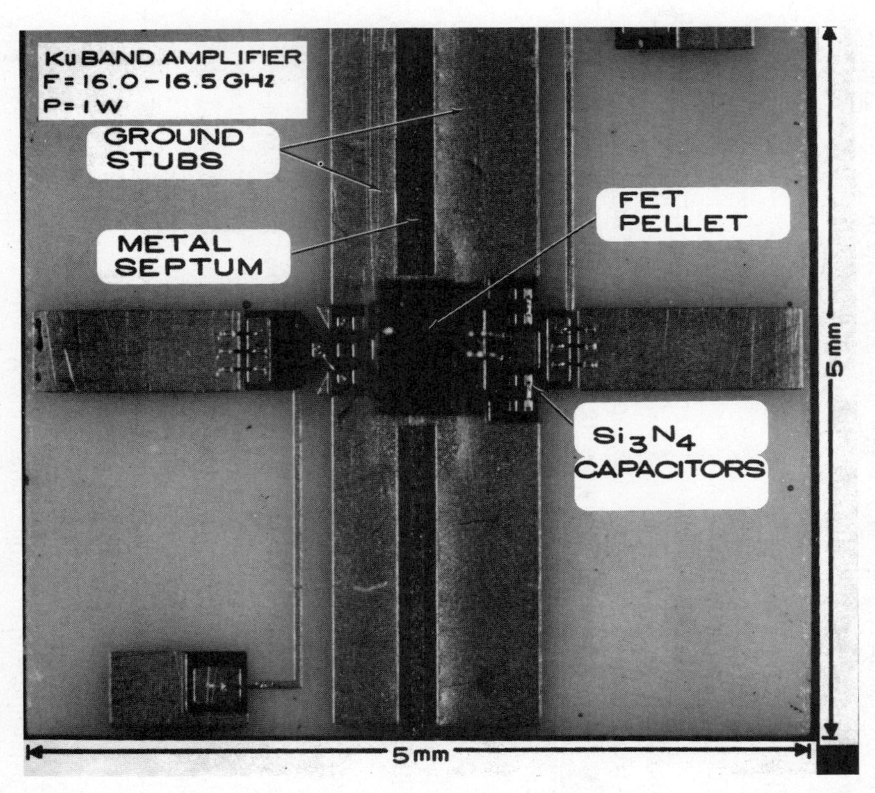

Figure 15.10 Photograph of a 1-W *Ku*-band amplifier. (After F. Sechi, et al. [13], 1983 IEEE, reprinted with permission.)

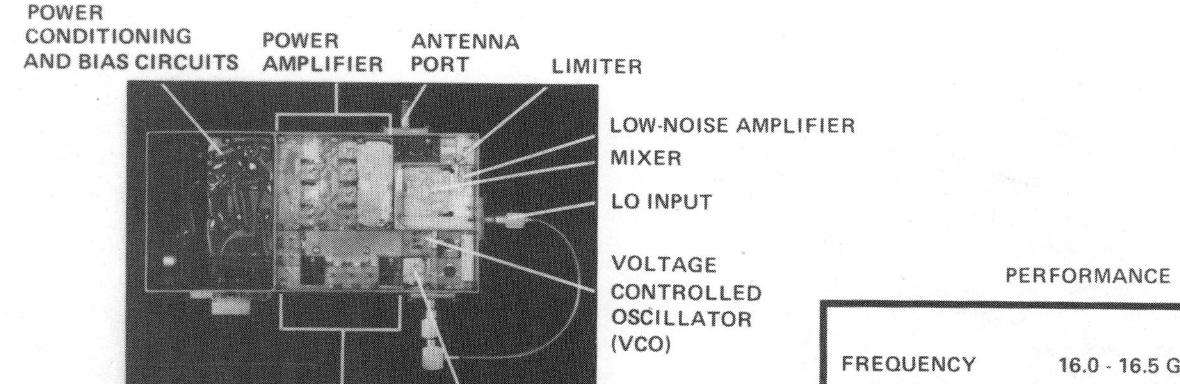

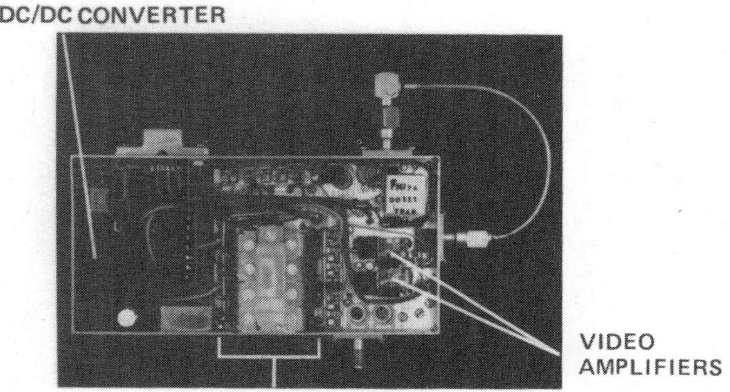

PERFORMANCE

FREQUENCY	16.0 - 16.5 GHz
TUNING VOLTAGE	+2.5 to +29V
POWER OUTPUT	3.9 - 4.4 W PEAK
PULSE LENGTH	0.2 - 10 μs
DUTY CYCLE	up to 50%
NOISE FIGURE	5 dB (LNA = 3 DB)
DC POWER	14.5 V @ 1.1A (1.0 μs P.L. & 10% D. C.)
SIZE	7.0 x 9.0 x 1.6 cm
WEIGHT	104 gr

Figure 15.11 Photograph of a T/R module. (After F. Sechi, et al. [14], reprinted with the permission of *Microwave J.*)

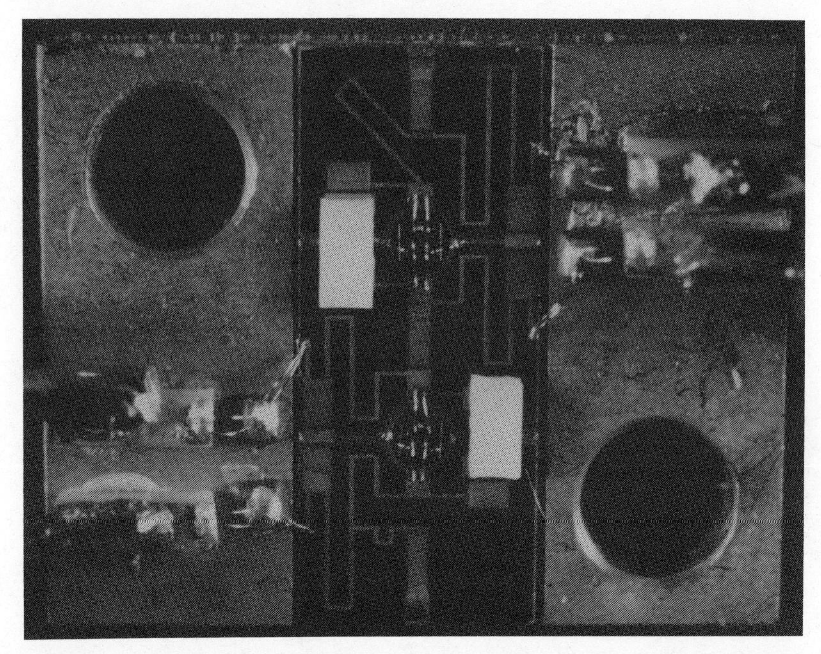

Figure 15.12 Photograph of a dual-stage 4-GHz amplifier. (After B. Dornan, et al. [15], 1985, IEEE, reprinted with permission.)

15.12. The amplifier operates over a frequency range of 3.7–4.2 GHz. The key feature of this circuit is that the FETs in chip form were placed into a 0.625-mm-diameter hole drilled into the 10-mil-thick alumina substrate. The circuit size is 0.5×0.25 cm.

The circuit examples shown here are representative of the current status of miniature circuit technology. This technology, which can be termed as quasi-monolithic, is at present being developed at the laboratory level.

15.5 MONOLITHIC MICROWAVE INTEGRATED CIRCUITS

The trend in advanced electronic systems is toward increasing integration, reliability, radiation hardness, and lower cost in large volume production. Hybrid microwave integrated circuits have been established in the microwave industry since the early 1970s. However, recent new developments in microwave system designs in the military, commercial, and consumer markets demand a new approach for mass production and for multi-octave bandwidth circuits. In addition, a new upsurge in millimeter-wave applications has created an interest in a new technology where the effect of bond-wire parasitics can be minimized and the use of discrete elements can be avoided. An answer to all these problems lies in the monolithic approach.

15.5.1 Brief History

Monolithic is a multilevel process approach comprising all active and passive circuit elements and interconnections formed into the bulk or onto the surface of a semi-insulating substrate. The concept of MMICs was derived from low-frequency ICs. The first MMIC results for T/R modules using silicon technology were reported in 1964. The results were not promising because of the low resistivity of silicon substrates [16]. One of the main reasons why monolithic technology did not take off until the mid-1970s is due to the fact that microwave solid state circuits require a large variety of solid state devices, such as transistors, mixer diodes, *pin* diodes, varactors, impact avalanche transit-time diodes (IMPATTs), and Gunn diodes, and it is not easy to standardize the process specifications so that several kinds of devices can be produced simultaneously in an optimum fashion. In 1968 Mehal and Wacker [17] revived the approach and attempted to fabricate a 94-GHz receiver front-end by using Schottky-barrier diodes and Gunn diodes on a semi-insulating GaAs substrate. The results were disappointing due to the absence of the high-temperature processing required for GaAs. It was not until 1976 when Pengelly and Turner [18] applied this approach to an *X* band amplifier based on the GaAs MESFET that results were very encouraging. By 1980 many MMIC results using MESFETs for various circuits had been reported. The reasons for the recent increase in MMIC research and development can be summarized as follows:

1. Rapid development of GaAs material technology, namely epitaxy and ion implantation.
2. Rapid development of low-noise MESFETs up to 60 GHz and power MESFETs up to 30 GHz. The MESFET is a very versatile component in the designer's arsenal of tricks.
3. MESFETs, dual-gate MESFETs, Schottky-barrier diodes, and switching MESFETs can be fabricated simultaneously using the same process and almost any microwave solid state circuit can be realized using these devices.
4. Excellent microwave properties of semi-insulating GaAs substrates ($\epsilon_r = 12.9$ and $\tan \delta = 0.0005$).
5. Availability of CAD tools for reasonably accurate modeling and optimizing of microwave circuits.

15.5.2 Why GaAs for MMICs

Any assessment of MMIC technology options available to the microwave designer will generally be in terms of its size, weight, cost, reliability, reproducibility, and maximum frequency of operation. Various substrate materials used for MMICs are bulk silicon, silicon-on-sapphire

TABLE 15.5 Comparison of Monolithic Integrated-Circuit Substrates

Property	Silicon	Silicon-on-Sapphire	GaAs	InP
Semi-insulating	No	Yes	Yes	Yes
Resistivity (Ω-cm)	10^3–10^5	$>10^{14}$	10^7–10^9	$\sim 10^7$
Dielectric constant	11.7	11.6	12.9	14
Electrical mobility[a] (cm²/V-sec)	700	700	4300	3000
Saturation electrical velocity (cm/sec)	9×10^6	9×10^6	1.3×10^7	1.9×10^7
Radiation hardness	Poor	Poor	Very good	Good
Density (g/cm³)	2.3	3.9	5.3	4.8
Thermal conductivity (W/cm-°C)	1.45	0.46	0.46	0.68
Operating temperature (°C)	250	250	350	300
Handling	Very good	Excellent	Good	Poor
Cost (2 in. diameter)	$15	$50 (base sapphire)	$100	$300

[a] A 10^{17} cm^{-3} doping level.

(SOS), GaAs, and InP. Their electrical and physical properties are compared in Table 15.5. The semi-insulating property of the base material is crucial to providing higher device isolation and lower dielectric loss for MMICs. Bulk Si MMIC technology is based on the bipolar transistor and is usable to *S* band, although the maximum operating frequency for the bipolar transistor is about 10 GHz. The basic active device in the SOS MIC technology is a silicon MESFET. The maximum operating frequency for a 1-μm gate-length MESFET is about 6 GHz [19], while a similar geometry GaAs MESFET has a maximum operating frequency of about 15 GHz. The electron mobility in GaAs is five to six times higher than in silicon and thus shorter gate-length FET devices are capable of operating as high as 100 GHz. FETs up to 60 GHz have already been demonstrated. It has also been established that GaAs semi-insulating substrates are suitable for MMICs up to 100 GHz. The GaAs FET as a single discrete transistor is already being widely used in amplifiers (low noise, broadband, medium power, high power), mixers, multipliers, switching circuits, and gain control circuits. A GaAs FET mixer, however, produces its own gain. The great popularity of GaAs FETs can be attributed to their higher frequency of operation and versatility. The versatility of GaAs FETs is a major reason for the choice of GaAs substrates for MMICs. InP has been used [20] for millimeter-wave monolithic integrated circuits, but very little work has been done on InP MMICs. The low Schottky-barrier height of metals on *n*-type InP is a chronic impediment to the development of an InP MESFET technology of equivalent performance to that of GaAs.

15.5.3 Design Considerations

Since the first reported GaAs MMIC, the MESFET has been the workhorse for all analog ICs (viz., amplifiers, oscillators, mixers, detectors, phase shifters, switches, modulators, and multipliers) used in receiver or transmitter applications. As the MMIC technology is progressing toward maturity, greater emphasis is being placed on analog ICs that operate at higher and higher frequencies. It is projected that by the end of the 1980s, the maximum frequency of analog circuits will approach 100 GHz.

The MMIC designer is presented with a number of considerations at the outset of each design. Five of the most important of these are listed as follows [21].

1. The design must be consistent with manufacturing capabilities.
2. The design should be tolerant to process variations.
3. The MMIC must meet the electrical requirements.
4. The MMIC should be as small as possible for best manufacturing yield and lower cost.
5. The MMIC thermal design must permit dissipation of any heat generated on the chip, taking into account the package thermal impedances and the environmental conditions. Acceptable operating temperatures must be chosen based on a trade between reliability and temperature. GaAs MMICs will operate for a while at 300°C, but high reliability applications generally require maximum temperatures to be below 150°C.

The relative importance of the preceding five considerations vary with the MMIC application. Obviously, any design must be manufacturable so item (1) is always important. A major portion of the MMIC design effort is directed toward the choice and accurate modeling of the active devices and the matching networks. MESFETs and Schottky-barrier diodes are the most important circuit devices in a GaAs IC. These devices (treated in Chapters 7 and 8) must be compatible with the manufacturing processes. Microstrip transmission lines are the most commonly used distributed matching elements for ICs. Some of the more important properties of microstrip lines are summarized in Chapter 2. The basic idea is that wide lines have low impedance and low loss. The range of transmission-line characteristic impedances that can be realized goes from about 30 to 100 Ω. Any deviation from the straight uniform-width transmission line causes the introduction of discontinuity parasitics that must also be modeled. Some of the other elements available to the designer are lumped elements: capacitor, resistors, and inductors (refer to Chapter 2 for details). Although distributed elements occupy more space compared to lumped elements, they have lower loss and offer the best manufacturing control.

The designed monolithic circuit must meet the electrical specifications. Out of three possible FET configurations, an optimum topology by properly choosing FETs must be worked out to meet the electrical specifications. Also the same FET can be capable of very different power and noise operation depending on the impedances presented to the FET gate–source and drain–source.

Information on the devices used in the design, describing parameter variations as a function of manufacturing tolerances must also be available. The design must provide acceptable performance over as wide a range of parameter values as possible to maximize IC yield. In this respect, a successful design technique is to select matching networks that are more tolerant to process variations. As ratios are less sensitive to process variations than absolute values, a good design should depend on ratios of resistances, inductances, capacitances, or other parameters rather than on their absolute values. The effect of fabrication tolerance on distributed element circuits is significantly less than in lumped circuits. Thus a MMIC design based on microstrip rather than lumped elements is less sensitive to process variations.

The size of the circuit is a constant concern and is greatly affected by the choice of topology. The circuit chip should be as small as possible for maximum yield and lowest cost. Small chip size and higher density give rise to potential problems of undesirable cross talk or coupling within the circuit elements. In order to avoid cross talk between two adjacent lines, line spacings on the order of two to three substrate thicknesses are required in most cases. However, the circuit function, location and the length of the coupled microstrip lines, etc., determine the spacing between the lines for best location usage on the chip.

Small-signal MMICs rarely generate enough heat to present thermal problems, but power MMICs may require a major effort to distribute and dissipate the heat generated.

15.5.4 Design Procedure

The evolution of a typical MMIC design generally follows the flow diagram depicted in Fig. 15.13. The design starts with the customer specifications, which depend on the system requirements. System requirements also dictate the circuit topology along with the types of active elements to be used, for example, single or dual gate FETs, and low-noise or power FETs. In systems, the extent to which integration can be accomplished is limited to yield and cost.

Since it is impossible to tune GaAs ICs, an accurate and comprehensive modeling of each device and circuit component is required to save expensive and time-consuming iteration of mask fabrication and evaluation. The MESFET is electrically characterized by measuring RF and dc parameters. The S-parameters are generally measured over the frequency band of

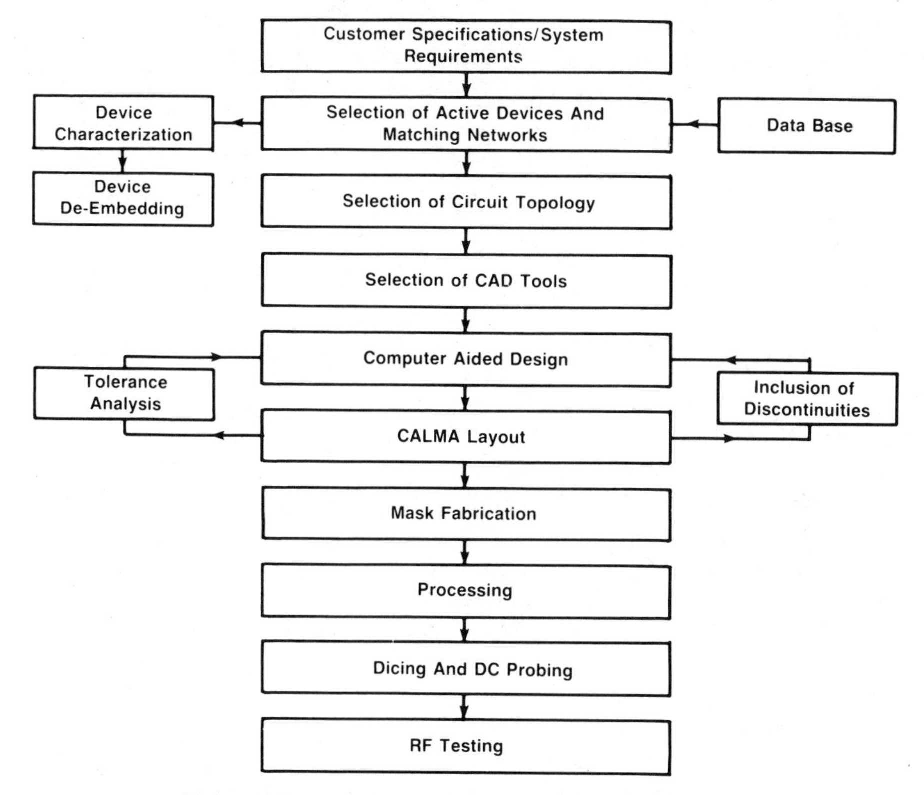

Figure 15.13 Flow diagram for a MMIC design procedure.

2–26 GHz at the required bias conditions. A lumped-element model is obtained by computer optimization to replicate the measured S-parameters. Similarly, other active devices and circuit elements are characterized and lumped-element models are obtained. In MMIC design, lumped-element models are preferred for the following reasons.

1. One can check how good the device is.
2. It allows extrapolation and interpolation of the device performance at frequencies other than the one measured.
3. It simplifies tolerance analysis.

The design of a MMIC requires proper CAD software. A brief description of CAD tools commercially available for main frame computers is given by Pucel [22]. Super Compact and Touchstone are available for personal computers (PC). The MMIC design is optimized and completed with the performance of tolerance studies and stability analysis (for amplifiers and oscillators).

All MMICs have built-in bias circuitry and dc blocking capacitors, making them truly self-contained circuits. After designing the complete circuit, a layout can be drawn using a GDS II–CALMA system. Once the basic layout has been determined, design iteration is necessary to account for discontinuities and other differences between the layout and the initial design. Once the final CALMA layout is complete, a pattern-generation tape is prepared and sent to a photolithographic mask manufacturer. The processed wafers are diced and good chips are picked up by dc probing. Finally, MMICs are electrically tested by packaging them or mounting them along with 50-Ω input and output microstrip lines typically on a 0.25-mm-thick alumina substrate on a gold-plated carrier.

15.5.5 MMIC Fabrication

There are many ways to fabricate MMICs. In order to give the reader an understanding of the relative complexity of GaAs MMIC manufacturing, a typical "generic" process flow chart is given in Fig. 15.14. This flow chart describes the main process steps while the details of each step, in-process electrical characterization, and optical inspection have not been included. An example of the small-signal IC process is shown schematically in Fig. 15.15. The process includes the fabrication of active devices, resistors, capacitors, inductors, air bridges, and via holes for ground connections through the substrate. Each step requires four to six steps to complete its function.

Each pattern is defined using standard photoresist and lift-off techniques. In the case of photolithography, diffraction effects limit the resolution to dimensions on the order of a wavelength of light used for exposure. Therefore, such a technique is usually employed to define gate lengths up to half a micron. Below half-micron dimensions, electron-beam lithography (taking advantage of the much smaller wavelength) is used. Photolithographic techniques are suitable for higher throughput purposes, while direct electron-beam lithography is required for fast turnaround time. A brief description of a MMIC fabrication process is given below [23–30].

Active Layers. The MMIC process starts with the formation of an active layer on/into qualified semi-insulating GaAs substrate. There are basically two methods of forming an n-type active layer: ion implantation, Fig. 15.15(a), and epitaxy. In the ion-implantation technique, the dopant atoms are bombarded on the GaAs substrate. Typical energies are 30 to 400 keV and doses are 10^{12} to 10^{14} atoms/cm^2. A suitable combination of energy and dose is used for the particular application in mind. For small-signal applications, a suitable silicon implant is 5×10^{12}/cm^2 at 60 keV. During ion implantation, the crystal lattice of GaAs is greatly damaged and the implanted atoms come to rest at random locations with the material. A high-temperature (850°C) annealing step is performed to anneal out the

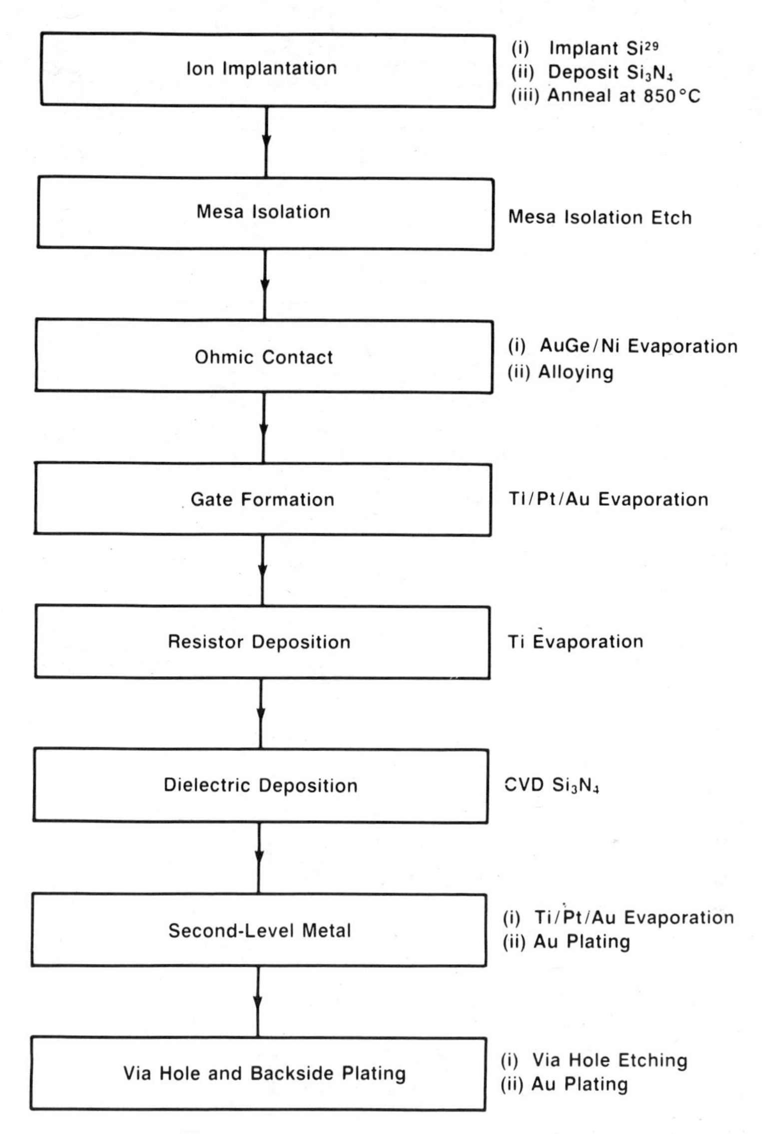

Figure 15.14 MMIC process flow chart.

lattice damage and allow the implanted atoms to move onto lattice sites. This technique has the advantage of selective ion implantation and is capable of fabricating ion-implanted substrates in large quantities for MMIC production.

In the epitaxial technique, an additional GaAs material layer is grown on the surface of the GaAs substrate in a manner that preserves the crystal structure. Ga and As atoms are brought in contact with the crystal surface under suitable temperatures, concentrations, and other conditions that

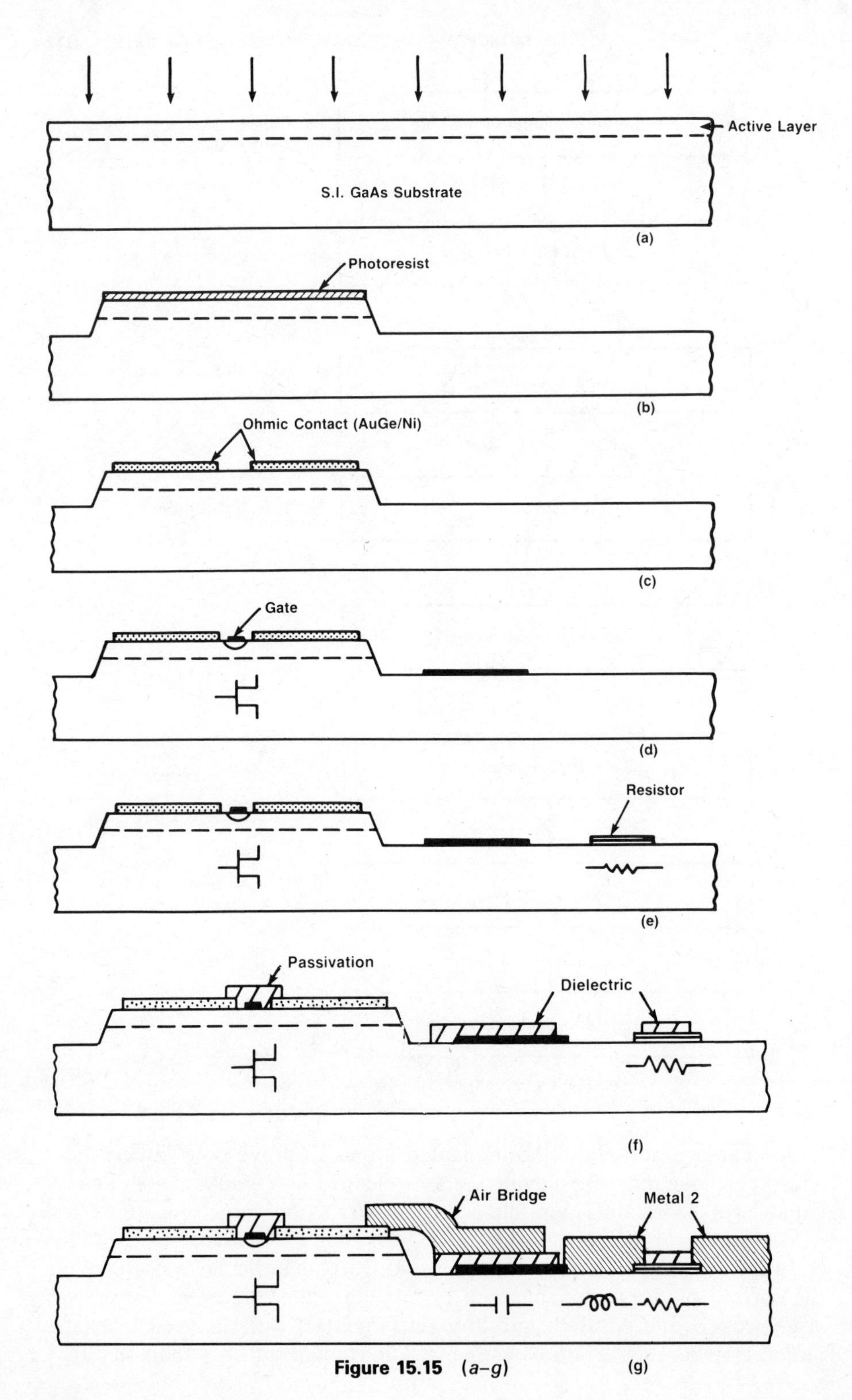

Figure 15.15 (*a–g*)

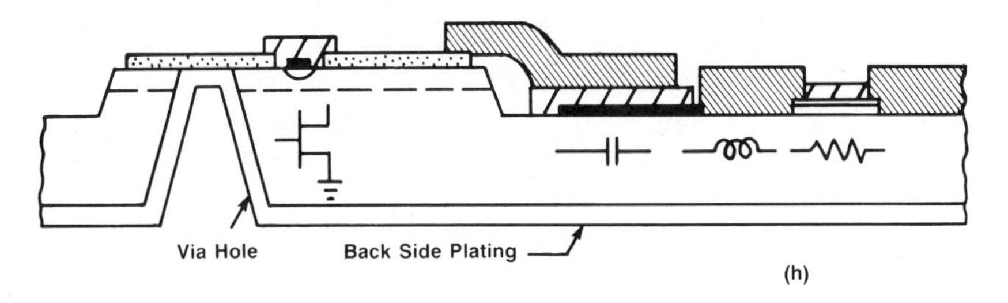

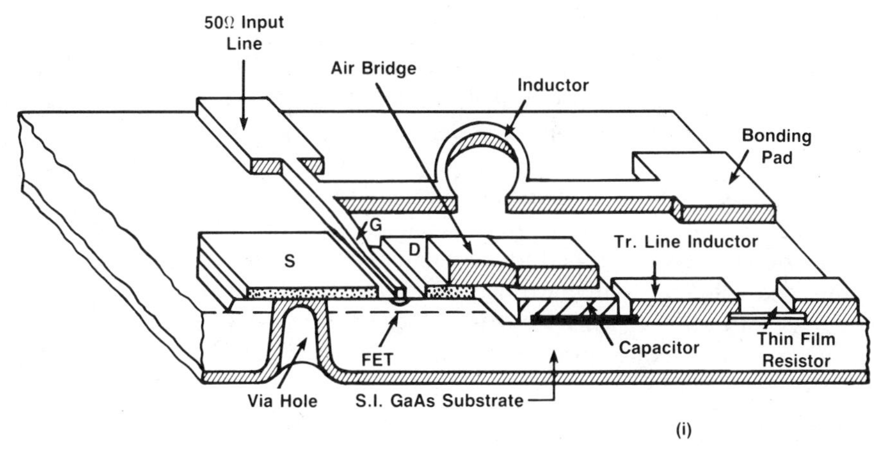

Figure 15.15 Small-signal microwave ICs process levels. (*a*) Ion implantation. (*b*) Mesa isolation. (*c*) Ohmic contact. (*d*) Gate formation. (*e*) Resistor evaporation. (*f*) Dielectric deposition. (*g*) Metal 2. (*h*) Via-hole etching and backside plating. (*i*) Completed MMIC.

result in crystalline growth. There are three basic types of epitaxy that have been used for GaAs: liquid-phase epitaxy (LPE), vapor-phase epitaxy (VPE), and molecular-beam epitaxy (MBE). LPE is the oldest technique used to grow epitaxial layers on GaAs crystals, while MBE is the most recent and powerful. MBE's advantages are that it can produce almost any epitaxial layer composition, layer thickness, and doping with the highest possible accuracy and uniformity across a wafer. However, VPE is the most advanced and the most widely used in the field.

Isolation. The actual process starts with the isolation pattern. The term "isolation" means restricting the active areas to specific parts so that electrical current is restricted from flowing to other areas. In active devices, it allows the current flow only to the desired portions and also isolates the devices from each other. In passive circuits, it reduces the parasitic capacitance and conductor loss in transmission lines. High-Q microstrip lines are fabricated on good dielectric materials.

Basically there are two techniques used by the GaAs MMIC industry for isolation: mesa etching [Fig. 15.15(*b*)] and ion implantation. Isolation by etching is done by etching away the whole surface, leaving mesas of the active layer in the desired locations. The thickness of material removed is on the order of 0.8 μm. The etched surface remains in either the buffer (if one was present) or semi-insulating GaAs substrate. Mesa etching is the simplest means of providing isolation and for this reason it is widely used.

In ion implantation, the unmasked areas of the active layer are bombarded with ions (hydrogen, oxygen, or boron), which cause considerable damage to the crystal lattice. This damage creates many electron trapping centers and renders the bombarded areas insulating. The masking substances prevent the ions from damaging the protective areas. Implanted-ion isolation is a planar process and provides better isolation as assessed by leakage current measurements and backgating than does mesa etching.

Ohmic Contact. Device ohmic contacts are made next [Fig. 15.15(*c*)]. The purpose of an ohmic contact on a semiconductor material is to provide a good contact between a pad and a semiconductor surface. Since the noise figure and gain are particularly sensitive to such resistances, ohmic contacts are very important to GaAs MMICs. If the active layer of the GaAs has a very high carrier concentration on the order of 10^{19} cm^{-3}, almost any metal placed in close contact with the surface will result in an ohmic contact. However, the most common approach in the industry to fabricate ohmic contacts on GaAs is by alloying a suitable metal at suitable temperature. During alloying and the cooling period, a component of the metal enters into the GaAs and highly dopes the active layer. The commonly used alloy is gold germanium (88% Au and 12% Ge by weight), whose melting point is 360°C. First of all a thin layer of AuGe eutectic alloy, followed *in situ* by a thin layer of Ni is deposited by evaporation. The total layer thickness is about 2000 Å. Then the lift-off technique is used to define ohmic contact pads, followed by alloying at 400°C in an H_2 ambient. After completing this step the ohmic metal semiconductor contacts and sheet resistance are checked.

Schottky or Gate Formation. The Schottky-barrier gate is another one of the more important elements of GaAs devices. The quality and placement of the gate metal is critical to FET performance in both low-noise and power FETs. Any metal placed on the active layer will form the Schottky barrier [Fig. 15.15(*d*)]. The choice of the gate metal is generally based on good adhesion to GaAs, electrical conductivity, and thermal stability. Metals meeting all these requirements are aluminum (Al), titanium (Ti), chromium (Cr), and molybdenum (Mo). Aluminum has the best electrical conductivity, but is not highly reliable. Other metals (Ti, Cr, Mo) are more resistive and require an overlay metal (generally gold) to enhance conductivity. Another barrier metal (Pt or Pd or W) is used between the Schottky metal (Ti, Cr) and gold, as gold diffuses in these metals. From the reliability point of view, Cr is not acceptable as it diffuses in GaAs at higher temperatures. Out of all gate metal combinations, a Ti/Pt/Au composition has proved highly reliable and is being used for GaAs MMICs throughout the industry. Typical thicknesses of these metals are on the order of 1000 Å, 1000 Å, and 4000 Å, respectively.

The dc current of the MESFET is adjusted just prior to any gate metal deposition. The part of the channel left exposed by the gate resist opening is slowly etched (commonly called recessed) until the saturated current between the drain and source pads drops to a specified value. When the gate metals are deposited, the saturated channel current will be lower due to the depletion region under the gate. This recessing process minimizes the parasitic channel resistances between the gate and the ohmic contacts by leaving a thicker GaAs channel outside of the channel. At this point (based on in-process dc and low-frequency tests), the wafers on which the FETs do not meet, process control specifications are rejected, thereby saving subsequent processing costs.

First-Level Metal. First-level metal refers to the overlay metallization applied for increased conductivity, formation of lower plate for overlay capacitors, and formation of inductors and transmission lines. In many processes, the Schottky gate metal and first-level metal are the same and are deposited at the same time [Fig. 15.15(*d*)]. Thus the preferred first-level metal compositions are Cr/Pt/Au, Mo/Au, Al, and Ti/Pt/Au, and the thickness is typically 6000 Å.

Resistor Deposition. Resistors are used in MMICs for several purposes including FET biasing, termination, feedback, isolation (power dividers and combiners), voltage dividers in bias networks, and attenuation. Both GaAs material (using mesa or selective ion-implanted techniques) and resistive films (Cr, Ni, Ti, NiCr, TaN) have been used in the fabrication of resistors. GaAs resistors are defined between two ohmic pads and have several problems: current saturation, Gunn domain formation, and temperature coefficient. In the case of thin-film resistors [Fig. 15.15(*e*)], very tight control of resistivity and thickness is required. Ni, Cr, Ti, etc., thin films are deposited by evaporation or sputtering, while TaN resistors are usually produced by sputtering. Resistors other than GaAs generally require at least one additional mask level to define the pattern, while GaAs resistors can be fabricated using the existing isolation and ohmic contact masks. In the case of GaAs resistors, an additional trimming or tuning mask is usually used for better accuracy.

Dielectric Deposition. Dielectric films are used in GaAs MMICs for passivation of active areas of FETs, diodes, and resistors, for metal insulator metal (MIM) capacitor, and for crossover insulation [Fig. 15.15(*f*)]. Besides polyimide, the dielectric materials usually used are silicon nitride (Si_3N_4) or silicon dioxide (SiO_2). These two materials are easily deposited either by plasma-assisted chemical-vapor deposition or sputtering. SiO_2 films have a lower dielectric constant than Si_3N_4 films, making them preferable for insulation; on the other hand, Si_3N_4 is suitable for capacitor formation. Furthermore, Si_3N_4 is less permeable to ions and therefore makes the superior encapsulant. The thickness of the dielectric film determines the capacitance per unit area of the MMIC capacitor. The thickness is usually between 1000 and 3000 Å, and is optimized to have minimum *pin* holes, large breakdown voltage, and maximum possible capacitance. Typically, a 1500-Å-thick layer of Si_3N_4 provides a 300-pF/mm^2 MIM capacitor with a breakdown voltage of over 30 V.

Second-Level Metal. Interconnection of components, air bridges, and the top plate of MIM capacitors is accomplished with the second-level layer Ti/Pt/Au metal system. In order to realize low-resistance connections, gold plating (3 to 5 μm thick) is used. Figure 15.15(*g*) shows this step along with an air dielectric crossover or air bridge. The gold-plated air bridge has a maximum current capability of 6 mA/μm of width and a sheet resistance of less than 25 mΩ/square [24].

Backside Polishing and Via-Hole Connection. Backside processing (polishing, via hole, and plating) are even more important than the frontside processing. In a production environment, a significant investment has been made in the wafer by the time the frontside processing is completed. Also, several of the backside operations critically affect the circuit function and the yield as a whole.

After the frontside process, the wafer is thinned to the required thickness. The starting thickness is usually 400 to 600 μm, while for low-noise circuits the final thickness is 100 to 200 μm. MMICs require accurate control of the polished substrate thickness to realize the designed transmission-line impedances. The GaAs substrate is usually thinned by the lapping technique, which is nothing but precision grinding. After lapping, the backside is well cleaned and smoothed.

State-of-the-art MMICs require via holes for low-inductance ground connections and better thermal dissipation paths. In via-hole technology, holes are etched (using a dry or wet process) through the GaAs substrate to provide a good connection from the frontside metallization to the backside ground plane. The backside of the polished wafer is patterned using photolithographic techniques (similar to other steps) to open holes (usually 50 to 100 μm in diameter) at the desired ground connections. The exposed GaAs is etched away until a hole is etched completely through the substrate. Then the backside is metallized. Sputtering is the best technique to ensure that the metal enters the holes and covers the slopes as shown in Fig. 15.15(*h*). Figure 15.15(*i*) depicts a completed MMIC.

Dicing. After all processing operations are completed successfully, it remains to cut and select the good chips. The first check for a good circuit is an automatic dc wafer probe. This test is performed on each chip. The microwave test's yield, which may be measured on a wafer with microwave probes [30, 31], is usually measured on a sample basis. After dc screening and on wafer RF probing, if the wafer has an acceptable yield, it is diced into separate chips. This can be done in several ways: scribing (diamond or laser), sawing, or by etching along the defined lines.

15.5.6 Examples of MMICs

Since this technology is growing rapidly, new examples of its application are evolving. No attempt to include an exhaustive sampling is made here, as many MMICs have already been described in the previous chapters. Instead, a selection of circuits that have been developed for various applications is described so that the diversity of the technology may be exposed. Circuits chosen for exposition include several amplifiers (low noise, broadband, and power), balanced mixer, transimpedance amplifier, switch matrix, phase shifter, switch, voltage-controlled oscillator, and voltage-controlled amplifier.

From the preceding selection of circuits, it is obvious that MMIC technology is very diverse in its capability to perform a wide range of required circuit functions [32]. In addition to single-function chips, multifunction chips, such as receiver front-ends and T/R modules, have also been developed. Progress in the area of increased integration level is being paced by achievable yields versus performance.

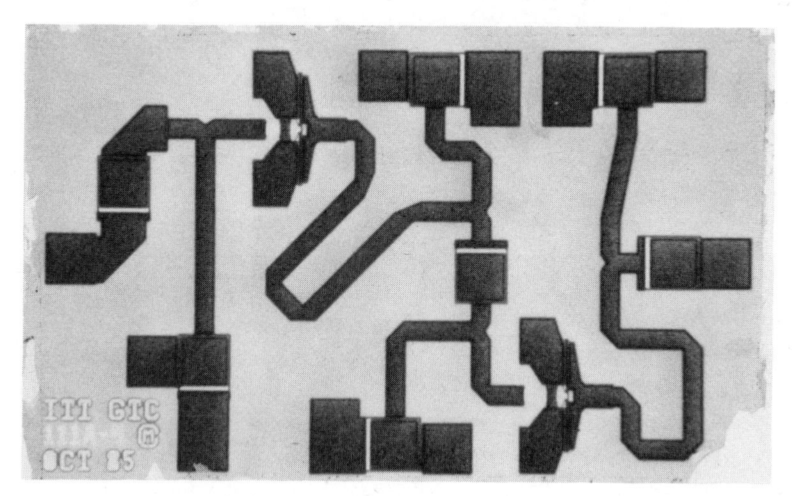

Figure 15.16 Photograph of a two-stage low-noise *Ku*-band amplifier. Chip size is $1.67 \times 1.07 \times 0.125$ mm. (Courtesy, I. J. Bahl, ITT.)

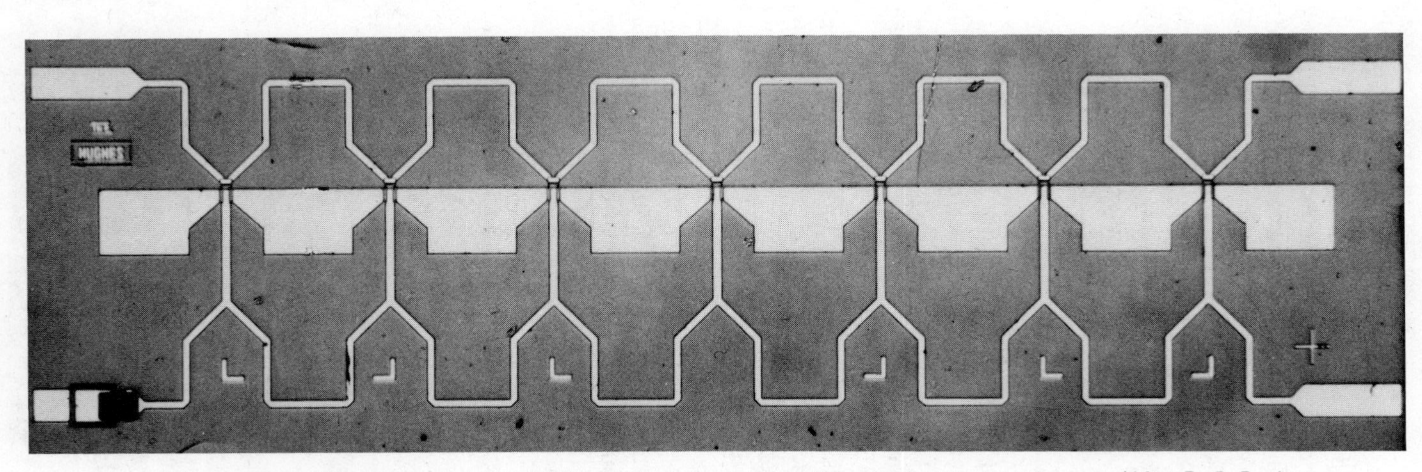

Figure 15.17 Photograph of a 2–40 GHz distributed amplifier. Chip size is $3.2 \times 1.1 \times 0.1$ mm. (After R. G. Pauley, et al. [33], 1985 IEEE, reprinted with permission.)

To date, amplifiers have been the most important application for MMICs. The variety of amplifiers include single- and multistage low-noise, broadband, and power amplifiers. The photograph of a two-stage low-noise amplifier using 0.5-μm gate-length FETs is shown in Fig. 15.16. This circuit exhibited a noise figure of 3.5 ± 0.5 dB and associated gain of 16 ± 0.5 dB in the 14.5–15.5 GHz band. There are many techniques used for broadband amplifiers. An attractive approach for multioctave bandwidths is the distributed amplifier shown in the photograph of Fig. 15.17. The seven-section circuit has a 476-μm total gate periphery and uses 0.25-μm gate-length FETs. The amplifier has a gain of 4 dB over the entire 2–40 GHz band with a noise figure of less than 9 dB in *K* band and 7 dB at 10 GHz [33]. An alternative approach for obtaining wider bandwidths is the resistive feedback technique. Figure 15.18 is the photograph of a three-stage *RLC*-feedback amplifier. The total gate periphery is 2100 μm, with a nominal gate length of 1 μm. The measured gain for this amplifier is 16 ± 1 dB over 2–9 GHz. A 6–18 GHz two-stage conjugately matched amplifier for ECM applications is shown in Fig. 15.19. The total gate width is 650 μm with gate length of 0.5 μm. The amplifier exhibited 10 ± 1 dB gain with 20-dBm output power over 6–18 GHz. A single-stage power amplifier with lower gain but with increased power output is shown in Fig. 15.20. A saturated output power of 3 W over a 1-GHz band centered at 5.5 GHz was obtained for this circuit. Figure 15.21 represents a completely different design philosophy to those considered so far. Shown is a photograph of a transimpedance amplifier designed for very high bit rate optical communciation receivers [34]. The amplifier exhibited a 3-dB bandwidth of 1.2 GHz and a sensitivity of -30 dBm at a 10^{-9} bit error rate.

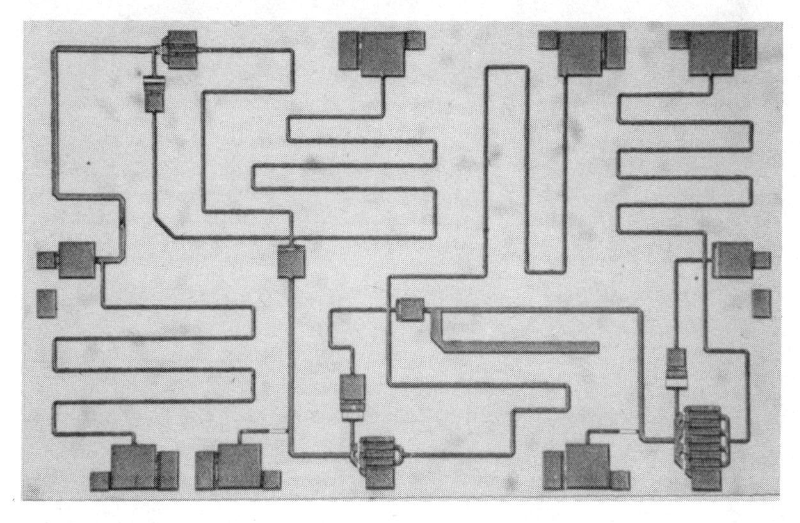

Figure 15.18 Photograph of a broadband feedback amplifier. Chip size is $4.0 \times 2.5 \times 0.125$ mm. (Courtesy, I. J. Bahl, ITT.)

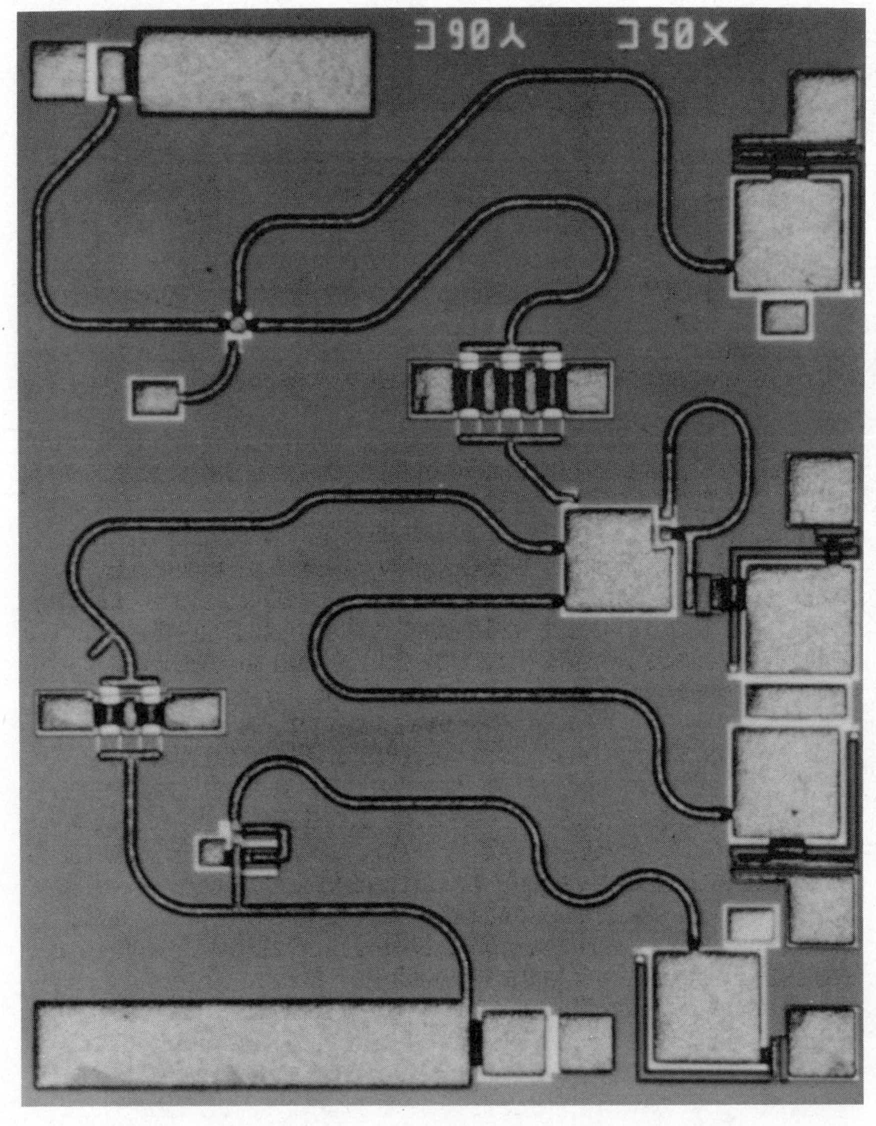

Figure 15.19 Photograph of a 6–18 GHz amplifier. Chip size is 1.7 × 1.4 × 0.1 mm. (Courtesy, M. Kumar, MSC.)

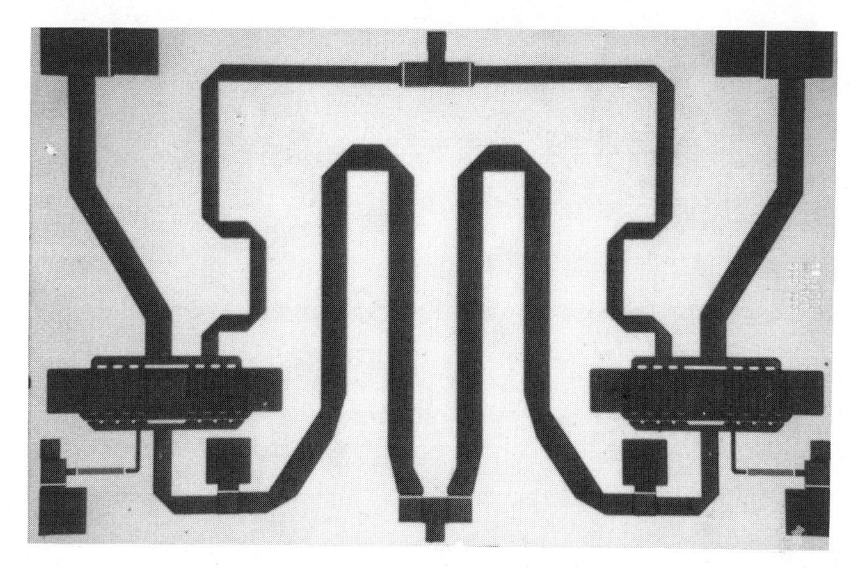

Figure 15.20 Photograph of a *C*-band power amplifier. Power output is 3 W. Chip size is 2.9 × 4.6 × 0.12 mm. (Courtesy, I. J. Bahl, ITT.)

Other standard components developed in MMIC form include mixers, oscillators, switches, phase shifters, and modulators. Figure 15.22 shows a photograph of a balanced mixer using Schottky diodes designed for direct broadcast satellite (DBS) applications. Conversion loss was less than 6.5 dB over the 11.5–12.3 GHz range. A voltage-controlled oscillator covering the 11–14 GHz band is shown in Fig. 15.23. The output power obtained was about 6 dBm [35]. Oscillators with 20-dBm power output have also

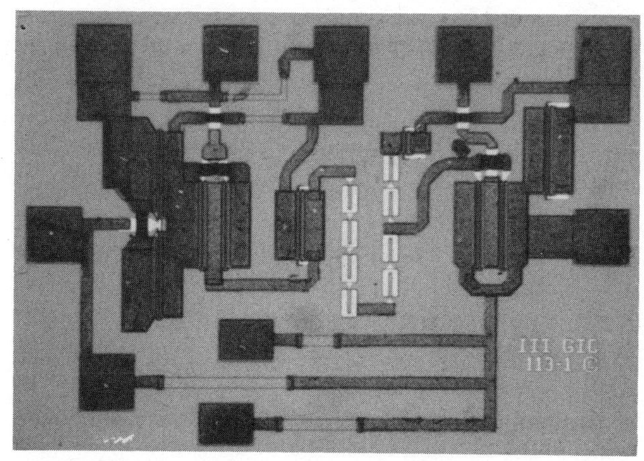

Figure 15.21 Photograph of a transimpedance amplifier. Chip size is 1.25 × 0.9 × 0.2 mm. (Courtesy, I. J. Bahl, ITT.)

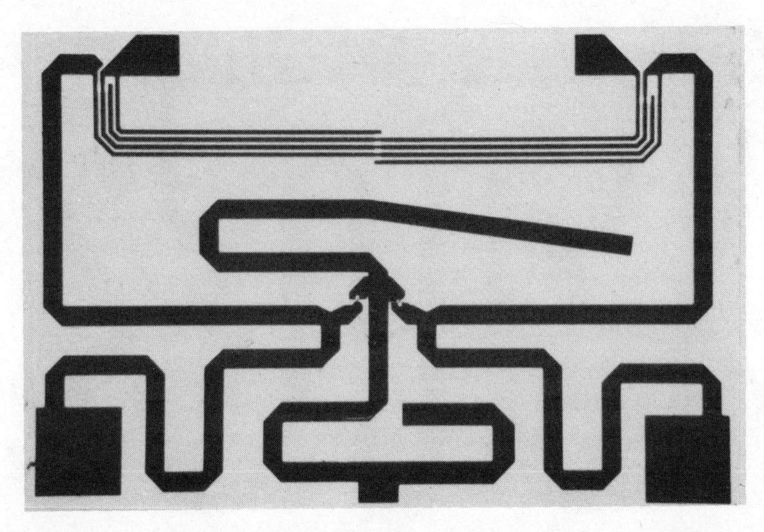

Figure 15.22 Photograph of a balanced mixer using Schottky diodes. Chip size is $2.6 \times 1.8 \times 0.2$ mm. (Courtesy, I. J. Bahl, ITT.)

been reported [36]. Figure 15.24 shows a photograph of a 1×1 switch module, which is the building block of the $m \times n$ microwave switch matrix operating over 3.7–4.2 GHz used in the SS–TDMA system for spot beam communication. The switch module uses a two-stage dual-gate FET amplifier switch that provides a 20-dB gain in the "on" state and a 50-dB loss in the "off" state. An important element in the T/R module is the phase shifter used in both the transmit and receive modes. A 6-bit phase shifter is shown in Fig. 15.25. A phase variation of $\pm 1°$ was achieved for the 5.6°,

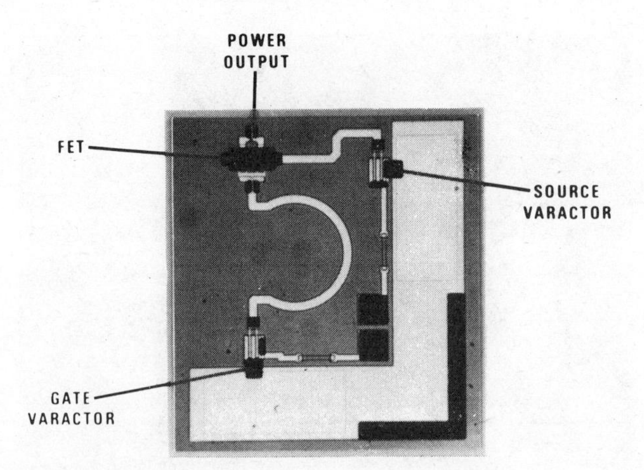

Figure 15.23 Photograph of a voltage-controlled oscillator. Chip size is $1.1 \times 1.2 \times 0.2$ mm. (After B. N. Scott and G. E. Brehm [35], 1982 IEEE, reprinted with permission.)

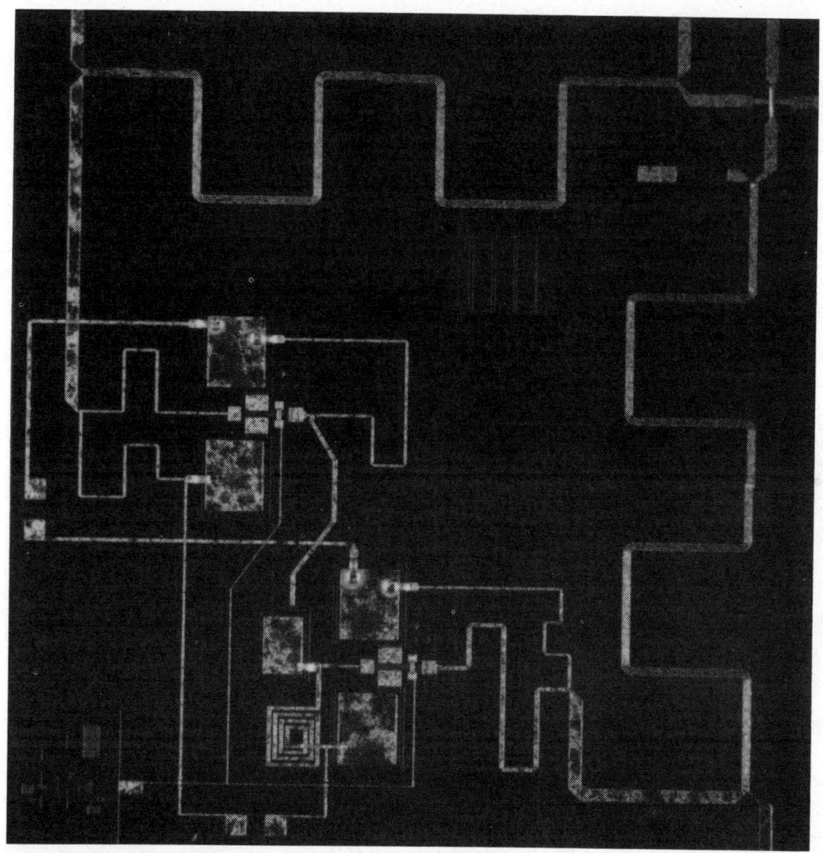

Figure 15.24 Photograph of a 1 × 1 microwave switch matrix. Chip size is 5 × 5 × 0.1 mm. (Courtesy, M. Kumar.)

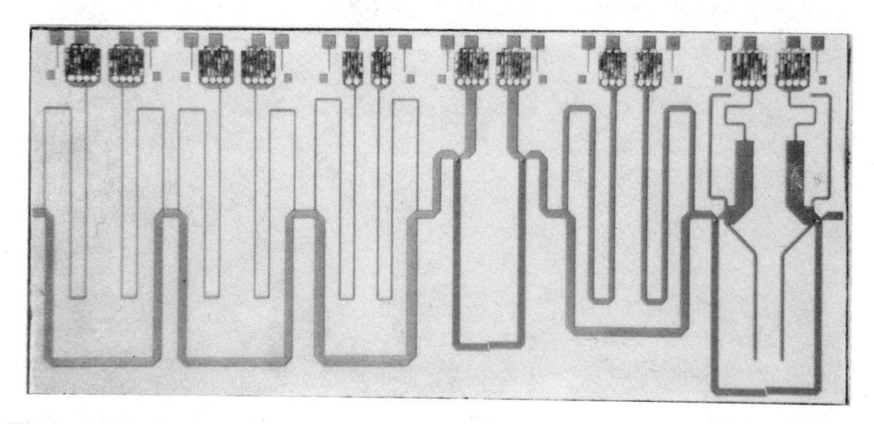

Figure 15.25 Photograph of a 6-bit phase shifter. Chip size is 9.43 × 4.2 × 0.125 mm. (Courtesy, I. J. Bahl, ITT.)

11.25°, 22.5°, and 45° loaded line bits, while for the 90° and 180° bits, which are of the reflection type, the phase variation was ±6° over 5 to 6 GHz [37]. Figure 15.26 shows a continuously variable phase shifter using dual-gate FETs [38, 39]. This phase shifter is capable of rotating the phase from 0° to 360° over 4–8 GHz in a continuous fashion by varying the bias voltage on the second gate of the dual-gate FETs. This phase shifter can also be used as an n-bit phase shifter where each bit can be fine-tuned for obtaining accurate phase shift.

Up until now, we have described single-function monolithic circuits. The next two circuits represent MMIC integration on a function level. The first

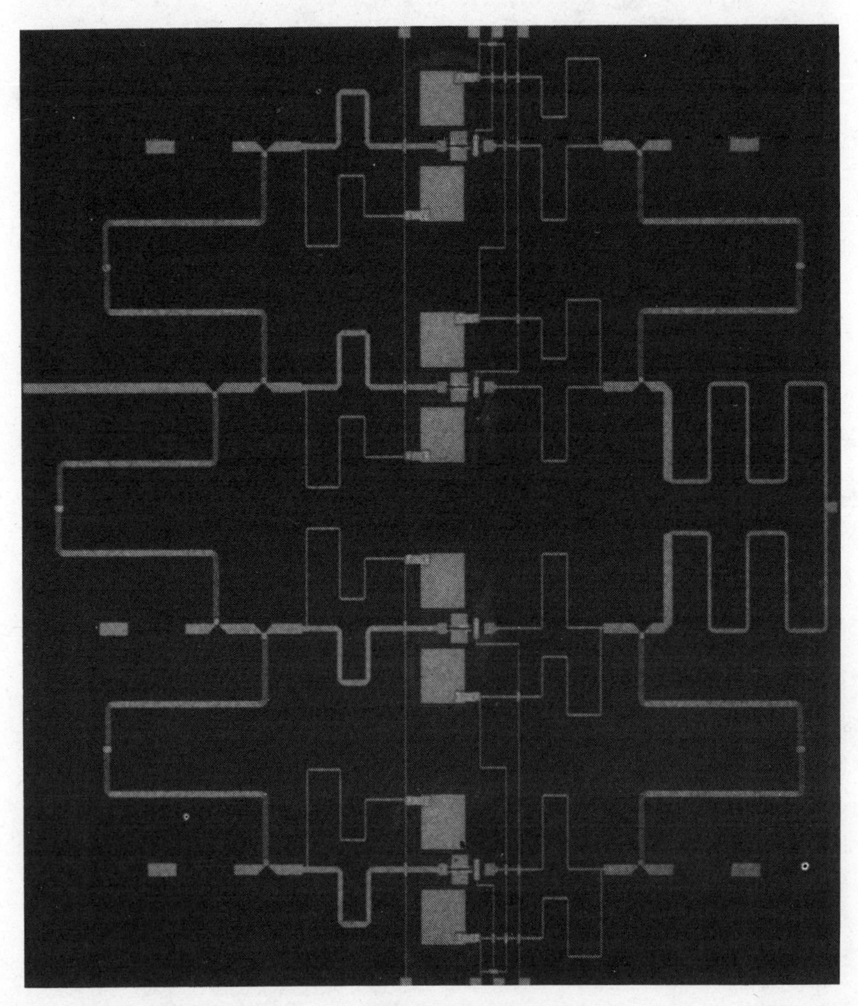

Figure 15.26 Photograph of a 0°–360° continuously variable dual-gate FET phase shifter. Chip size is 8 × 7 × 0.1 mm. (Courtesy M. Kumar.)

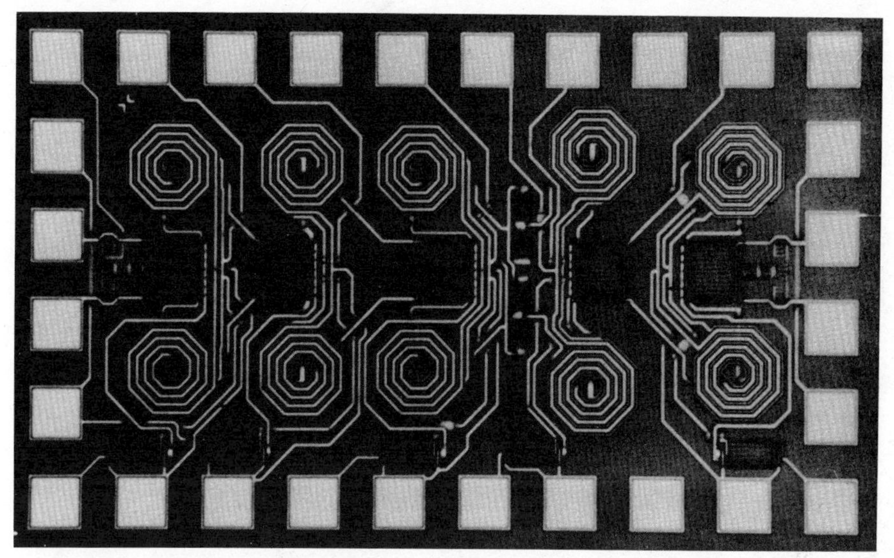

Figure 15.27 Photograph of a *C*-band monolithic receiver. Chip size is 1×1.65 mm. (After J. Culp, et al. [40], reprinted with permission of *Microwave Systems News*.)

example of a functional block is the *C*-band receiver designed for avionics applications [40]. The receiver as shown in Fig. 15.27 consists of limiters, three preamplifier stages, a double-balanced mixer, and two local oscillator buffer stages. This circuit utilizes coupled planar spiral inductors as transformers—a first for MMICs. The second example is of a transmit module operating over the 17.7–20.2 GHz band. The module as shown in Fig. 15.28 consists of a 5-bit phase shifter, five gain stages, and digital interface circuitry [41].

The present production yield ($\leq 10\%$) for functional blocks (two to three functions) is not cost effective. However, it is feasible to develop MMICs at the component level with acceptable yields (20–30%) and then brassboard the chips to obtain the complete functional module. The next three examples describe this kind of integration of MMIC chips. Figure 15.29 shows the complete assembly of an *S*-band image-rejection receiver module [42]. The receiver system incorporates the following MMIC chips: low-noise amplifier, in-phase signal splitter, quadrature local oscillator splitter, and dual-channel mixer. A photograph of a brassboard *X*-band T/R module consisting of eleven MMIC chips is shown in Fig. 15.30. This 9- to 10-GHz module consists of a 4-bit phase shifter, a two-stage low-noise amplifier, four SPDT FET switches, and a cascade pair of two-stage power amplifiers [43]. A DRFM analog subsystem comprising 10 GaAs MMIC chips (six limiting amplifiers, one quadrature IF down converter, two IF amplifiers, and one quadrature up converter) has been developed and integrated into an 11.8 cm^3 (0.72 in^3) module. The input signal dynamic range for the

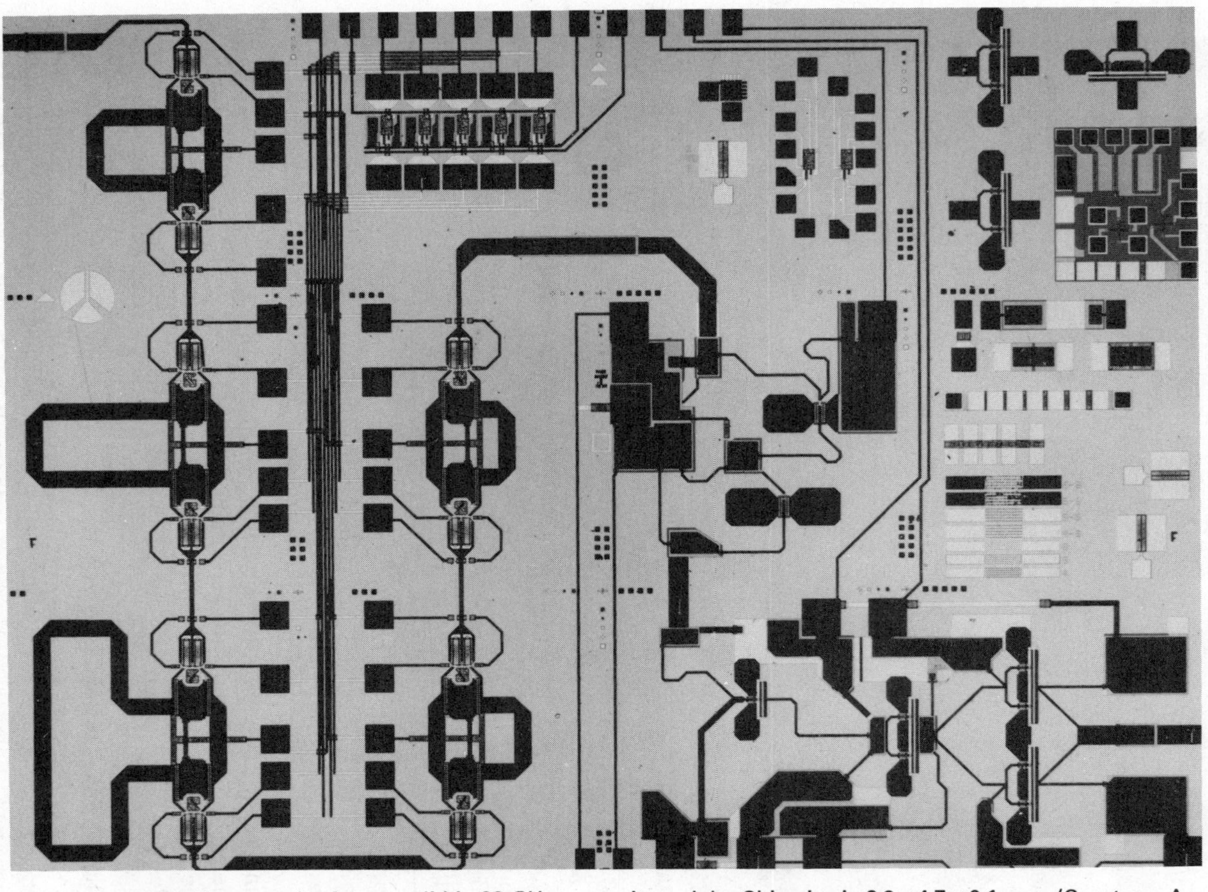

Figure 15.28 Photograph of a monolithic 20-GHz transmit module. Chip size is 6.3 × 4.7 × 0.1 mm. (Courtesy, A. Gupta, Rockwell International.)

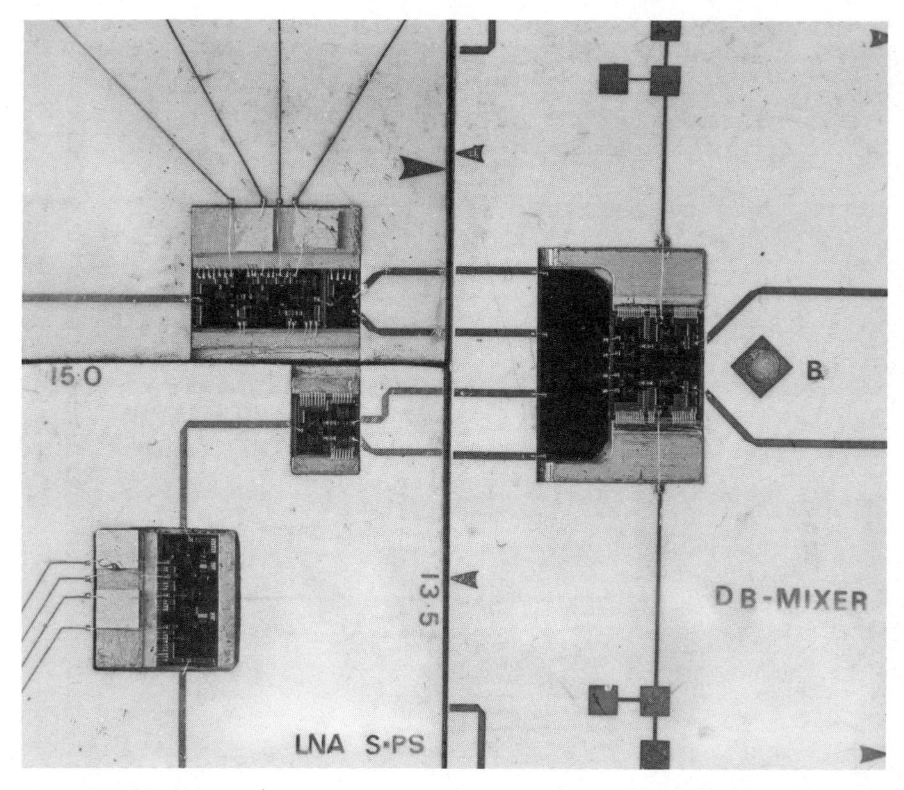

Figure 15.29 Photograph of an assembled *S*-band image-rejection receiver. (Courtesy, G. Beech, Plessey.)

module is 50 dB from 2 to 6 GHz, the image sideband rejection is 20 dB typical, and the LO to RF isolation is 27 dB minimum. A photograph of the module assembly is shown in Fig. 15.31. In all these modules, MMIC chips are interconnected by appropriate microstrip lines on alumina substrates that are coplanar with the monolithic GaAs chips.

15.6 HYBRID VERSUS MONOLITHIC MICs

Hybrid MICs use a large number of techniques such as microstrip, suspended microstrip, slot line, coplanar lines, fin lines, as well as lumped elements, and can realize optimum circuit functions on high dielectric-constant substrates (alumina, sapphire, epsilam, etc.), or on low dielectric-constant materials (RT/duroid, quartz, etc.), or on good thermally conducting material (beryllia). On the other hand, most monolithic MICs use microstrip on GaAs, as well as lumped elements. Having discussed hybrid and monolithic MICs briefly in the last two sections, the relative merits of these

Figure 15.30 Photograph of an assembled *X*-band T/R module. (Courtesy, R. A. Pucel, Raytheon.)

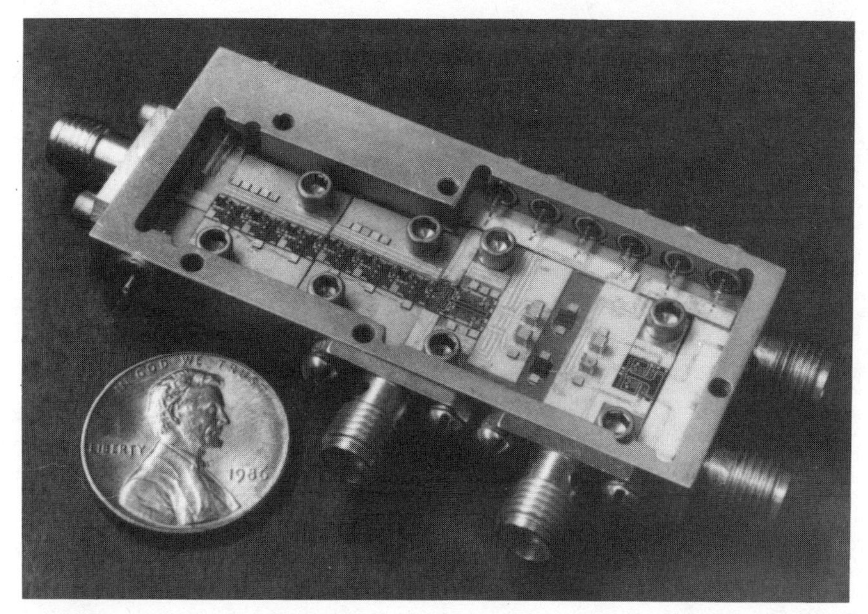

Figure 15.31 Photograph of the DRFM analog subsystem. (After G. K. Lewis et al., 1987 IEEE, reprinted with permission.)

two technologies [45–48] are highlighted under the following headings:

- Cost
- Size and weight
- Design flexibility
- Circuit tweaking
- Broadband performance
- Reproducibility
- Reliability

15.6.1 Cost

In the modern technology manufacturing world, cost is often the overriding consideration in the selection of a process or method for very large-scale manufacture. The manufacture of hybrid circuits tends to be very labor-intensive (unless automated), involving individual-circuit components, whereas monolithic circuits are fabricated on wafers in batches, and hundreds and thousands can be manufactured at the same time. Clearly, monolithic circuits have a great advantage in terms of the manufacturing cost per unit when produced in large quantities, with an acceptable yield, say, of approximately 40–50%. However, MMICs have a cost drawback relating to the areas of the expensive semiconductor wafer wasted in relation to the area actually occupied by the active circuit elements. For example, in MMIC balanced amplifiers, the active devices take less than 1% of the area with matching networks, and 3-dB hybrids occupy the rest of the GaAs area, which is very expensive as compared to alumina. Assuming acceptable fabrication and packaging yields, semiautomatic microwave testing, and available packaging facilities, it has been estimated that in a few years, a MMIC chip having 1 mm^2 area will cost about $10–25 when produced in the thousands.

15.6.2 Size and Weight

Many of today's MICs find application in spaceborne systems, where size and weight are at a premium. Here monolithic MICs clearly have an edge over bulky hybrid MICs. Table 15.6 attempts [45] to show the difference in characteristics of amplifiers fabricated using the two technologies. It is worth mentioning that the weight of a chip resistor or a chip capacitor is more than that of an average MMIC chip.

TABLE 15.6 Comparison of Hybrid and Monolithic High-Gain Amplifiers

Parameter	Details
Hybrid	
Frequency	8–18 GHz
Noise figure	4.5 dB typical
Gain	40 ± 2 dB
Number of wire bonds	400
Number of FETs	16
Number of substrates	16
Number of carriers	8
Number of chip resistors (dc and RF)	32
Number of chip capacitors	40
Size	Approximately 64 × 8 × 4 mm (unpackaged)
Monolithic (using via technology and alumina Lange couplers)	
Frequency	7.5–18.5 GHz
Noise figure	5.2 dB typical
Gain	57 ± 1.5 dB
Number of wire bonds	14
Number of FETs	16
Number of GaAs chips	5
Number of carriers	1
Size	19.5 × 9.5 × 5 mm (packaged)

15.6.3 Design Flexibility

Perhaps the most interesting aspect of MMIC development to date has been the way in which the GaAs MESFETs have been used to perform a wide variety of functions: amplification, oscillation, mixing, multiplication, division, switching, and phase shifting. Significantly in many cases, the FET has outperformed its predecessors in these particular circuit areas. MMICs have advantages over hybrid circuits in terms of using FET size and its configuration, for example, common source, common gate, and common drain, in order to realize the best possible matching structure. However, there are certain cases (examples of low-noise amplifiers and narrow-band filters) where the superior Q factors of hybrid circuit techniques are advantageous.

15.6.4 Circuit Tweaking

Hybrid circuits have the advantage over monolithic circuits in the ability to perform postmanufacture tuning or "tweaking" to obtain the optimum performance from the circuit or make adjustments in the circuit at the development stage. This is very difficult to perform on monolithic circuits, but in hybrid circuits it is commonly done by varying lumped-element values or using extra microstrip patches, which may be bonded in as required. This means that in MMICs, very accurate modeling of each component used in the circuit is required and also the circuit performance must not be very sensitive to manufacturing tolerances.

15.6.5 Broadband Performance

Performance of single-ended, balanced, low-noise, and multistage monolithic GaAs amplifiers is comparable to hybrid amplifiers. However, multioctave bandwidth performance, for example, 2–40 GHz, is not possible with hybrid MIC technology. The 2–40 GHz distributed amplifiers with 4-dB gain demonstrates how well one can design and fabricate broadband monolithic amplifiers.

The broadband performance of a circuit is most dependent on the parasitics associated with the lumped elements, active devices, and connecting bond wires or ribbons. In MMICs where the number of bond wires is reduced to those at the periphery of the circuit (with via-hole technology only input and output wire connections are required) and where the active devices and passive circuits are embedded into/onto the substrate, these parasitics are minimized or accounted for, and so these circuits tend to give superior broadband performance when compared with hybrid circuits.

15.6.6 Reproducibility

One of the most important advantages for the MMIC is that of reproducibility of performance from chip to chip. Reproducibility of MMICs mainly results from the excellent definition and repeatability of passive components, as well as from the choice of an optimum circuit for the function needed. Well-controlled processes or good designs that can accommodate small variations in process (e.g., doping concentration, channel depth, and GaAs substrate thickness) from batch to batch result in reproducible MMIC chips.

Since MMIC chips are small in size, the phase-angle repeatability of performance (very important in phased-array radars) is often superior to hybrid MICs. Improvements have also been made in hybrid circuit repeatability in the last few years with the increased use of semiautomation [49] (e.g., semiautomatic wire bonders, chip placement, and laser-scribing die eutectic).

15.6.7 Reliability

Both hybrid and monolithic MIC technologies are considered reliable. However, a well-qualified IC process can be highly reliable, particularly in

TABLE 15.7 Comparison Between Monolithic and Hybrid MICs

Feature	Monolithic	Hybrid
Substrate	Semi-insulator	Insulator
Interconnections	Deposited	Wire-bonded/ deposited
Distributed elements	Microstrip	Microstrip and coplanar lines
Lumped elements	Deposited	Discrete/deposited
Solid state devices	Deposited	Discrete
Controlled parasitics	Yes	No
Labor intensive	No	Yes
Repairability	No	Yes
Equipment costs	High	Low
Mass production	Yes	No
Debugging	Difficult	Easy
Integration with digital and electrooptic ICs	Possible	Impossible
Cost	Low	High
Size and weight	Small	Large
Design flexibility	Very good	Good
Circuit tweaking	Impractical	Practical
Broadband performance	Relatively good	Limited
Reproducibility	Excellent	Good
Reliability	Excellent	Good

the areas (namely, lumped-element bonding, thin-film resistors, and wire bonding) where hybrid circuits have difficulties. On the other hand, the hybrid circuit producer can easily rework the circuits if failure occurs.

Table 15.7 summarizes the comparison between these two technologies.

REFERENCES

1. Kesiter, F. Z., "An Evaluation of Materials and Processes for Integrated Microwave Circuits," *IEEE Trans. Microwave Theory Tech.*, Vol. MTT-16, July 1978, pp. 469–475.
2. Caulton, M., and H. Sobol, "Microwave Integrated Circuit Technology—A Survey," *IEEE J. Solid-State Circuits*, Vol. SC-5, Dec. 1970, pp. 292–303.
3. Sobol, H., "Applications of Integrated Circuit Technology to Microwave Frequencies," *Proc. IEEE*, Vol. 59, Aug. 1971, pp. 1200–1211.
4. Sobol, H., "Technology and Design of Hybrid Integrated Circuits," *Solid State Technol.*, Vol. 13, Feb. 1970, pp. 49–57.
5. Laverghetta, T. S., *Microwave Materials and Fabrication Techniques*, Artech House, Dedham, Mass., 1984.
6. Jones, R. D., *Hybrid Circuit Design and Manufacture*, Dekker, New York, 1982.
7. Elliott, D. J., *Integrated Circuit Mask Technology*, McGraw-Hill, New York, 1985.
8. Young, L., and H. Sobol (Eds.), *Advances in Microwaves*, Vol. 8, Academic Press, New York, 1974.
9. Dornan, B., et al., "Advances in the Design of Solid-State Power Amplifiers for Satellite Communications," *RCA Rev.*, Vol. 45, Dec. 1984, pp. 619–630.
10. Kumar, M., et al., "Broad-Band Active Phase Shifter using Dual-Gate MESFET," *IEEE Trans. Microwave Theory Tech.*, Vol. MTT-29, Oct. 1981, pp. 1198–1202.
11. F. Sechi, et al., "Miniature Beryllia Circuits—A New Technology for Microwave Power Circuits," *RCA Rev.*, Vol. 43, 1982, p. 363.
12. F. Sechi, et al., "*Ku*-band VCO using Miniature Beryllia Circuit Technology," *Proc. 13th European Microwave Conf.*, Sept. 1983, pp. 208–212.
13. F. Sechi, et al., "Miniature Beryllia Circuits for *Ku*-band Power Amplifiers," *IEEE MTT-S Int. Microwave Symp. Digest* (Boston, Mass.), June 1983, pp. 530–532.
14. F. Sechi, et al., "New Miniature Beryllia Circuit Technology," *Microwave J.*, Vol. 28, Mar. 1985, pp. 117–124.
15. B. Dornan, et al., "Miniature Ceramic Circuit Components for *Ku* band Receivers," *IEEE MTT-S Int. Microwave Symp. Digest* (St. Louis, Mo.), June 4–6, 1985), pp. 597–600.
16. Hyltin, T. M., "Microstrip Transmission on Semiconductor Substrates," *IEEE Trans. Microwave Theory Tech.*, Vol. MTT-13, Nov. 1965, pp. 777–781.
17. Mehal, E., and R. W. Wacker, "GaAs Integrated Microwave Circuits," *IEEE Trans. Microwave Theory Tech.*, Vol. MTT-16, July 1968, pp. 451–454.

18. Pengelly, R. S., and J. S. Turner, "Monolithic Broadband GaAs FET Amplifiers," *Electron. Lett.*, Vol. 12, May 13, 1976, pp. 251–252.
19. Laighton, D., J. Sasonoff, and J. Selin, "Silicon-on-Sapphire (SOS) Monolithic Transceiver Module Components for *L*- and *S*-band," *IEEE MTT-S Int. Microwave Symp. Digest*, 1981, pp. 37–39.
20. Sleger, K. J., et al., "InP Monolithic Integrated Circuits for mm-Wave Applications," *Microwave J.*, Vol. 27, May 1984, pp. 175–189.
21. Bahl, I. J., "GaAs Monolithic Integrated Circuit Amplifiers," presented at the Int. Symp. Microwave Technology in Industrial Development, Campinas, São Paulo, Brazil, July 22–25, 1985.
22. Pucel, R. A., "On the Design of MMICs," *IEEE GaAs IC Symp. Digest*, 1984, pp. 173–176.
23. Williams, R. E., *Gallium Arsenide Processing Techniques*, Artech House, Dedham, Mass., 1984.
24. Andrade, T., "Manufacturing Technology for GaAs Monolithic Microwave Integrated Circuits," *Solid State Technol.*, Feb. 1985, pp. 199–205.
25. Dilorenzo, J. V., and D. D. Khandelwal (Eds.), *GaAs FET Principles and Technology*, Artech House Dedham, Mass., 1982.
26. Pengally, R. S., *Microwave Field-Effect Transistors-Theory, Design and Applications*, Wiley, New York, 1982.
27. Soares, R., J. Graffeuil, and J. Obregon (Eds.), *Applications of GaAs MESFETs*, Artech House, Dedham, Mass., 1983.
28. Pucel, R. A. (Ed.), *Monolithic Microwave Integrated Circuits*, (A Reprint Volume), IEEE Press, Piscataway, N.J., 1985.
29. Ferry, D. K. (Ed.), *Gallium Arsenide Technology*, Howards Sams, Indianapolis, Ind., 1985.
30. Van Tuyl, R. L., et al., "A Manufacturing Process for Analog and Digital Gallium Arsenide Integrated Circuits," *IEEE Trans. Microwave Theory Tech.*, Vol. MTT-30, July 1982, pp. 935–942.
31. Carlton, D. E., K. R. Gleason, and E. W. Strid, "Microwave Wafer Probing," *Microwave J.*, Vol. 28, Jan. 1985, pp. 121–129.
32. Pucel, R. A., "Design Considerations for Monolithic Microwave Circuits," *IEEE Trans. Microwave Theory Tech.*, Vol. MTT-29, June 1981, pp. 513–534.
33. Pauley, R. G., et al., "A 2 to 40 GHz Monolithic Distributed Amplifier," *IEEE GaAs IC Symp. Digest*, 1985, pp. 15–17.
34. Bahl, I. J., et al., "A High Speed GaAs Monolithic Transimpedance Amplifier," *IEEE Microwave and Millimeter-Wave Monolithic Circuit Symp. Digest*, 1986, pp. 35–37.
35. Scott, B. N. and G. E. Brehm, "Monolithic Voltage Controlled Oscillator for *X*- and *Ku*-bands," *IEEE Trans. Microwave Theory Tech.*, Vol. MTT-30, Dec. 1982, pp. 2172–2177.
36. Scott, B. N., M. Wurtele, and B. B. Cregger, "A Family of Four Monolithic VCO MIC's Covering 2–18 GHz," *IEEE Microwave and Millimeter-Wave Monolithic Circuits Symp. Digest*, 1984, pp. 58–61.
37. Andricos, C., I. J. Bahl, and E. L. Griffin, "*C*-band 6-bit GaAs Monolithic Phase Shifter," *IEEE Trans. Microwave Theory Tech.*, Vol. MTT-33, Dec. 1985, pp. 1591–1596.

38. M. Kumar, et al., "Monolithic GaAs Dual-Gate FET Phase Shifter," *Proc. 1984 Government Microcircuit Applications Conf.*, (Las Vegas, Nev.), Nov. 1984, pp. 241–244.
39. L. C. Upadhyayula, M. Kumar, and H. Huang, "GaAs MMICs Could Carry the Waves of the High Volume Future," *Microwave Syst. News*, Vol. 13, July 1983, pp. 58–82.
40. Culp, J., et al., "Integration is Paramount in Gallium-Arsenide Receiver Design," *Microwave Syst. News*, Vol. 13, Apr. 1983, pp. 90–98.
41. Gupta, A., et al., "A 20-GHz 5-bit Phase Shift Transmit Module with 16 dB Gain," *IEEE GaAs IC Symp. Digest*, 1984, pp. 197–200.
42. Beech, G., C. W. Suckling, and J. R. Suffolk, "An *S*-band Image Rejection Receiver Front-End Incorporating GaAs MMICs," *Focus on GaAs MMICs*, Plessey Research Caswell Limited, Sept. 1985, pp. 21–23.
43. Pucel, R. A., et al., "A Multi-Chip GaAs Monolithic Transmit-Receive Module for *X*-band," *IEEE Int. Microwave Symp. Digest*, 1982, pp. 489–492.
44. Lewis, G. K., et al., "GaAs MMICs for Digital Radio Frequency Memory (DRFM) Subsystems," *IEEE Microwave and Millimeter-Wave Monolithic Circuit Symp. Digest*, 1987, pp. 53–56.
45. Pengelly, R. S., "Hybrids vs. Monolithic Microwave Circuits—A Matter of Cost," *Microwave Syst. News*, Vol. 13, Jan. 1983, pp. 77–114.
46. Shillady, R. W., "Microwave Integrated Circuits Offer Alternatives to the Radar System Designer," *Microwave Syst. News*, Vol. 13, Aug. 1983, pp. 100–119.
47. Decker, D. R., "Are MMICs a Fad or a Fact," *Microwave Syst. News*, Vol. 13, July 1983, pp. 84–92.
48. Howes, M. J., and P. Shepherd, "Passive Components Essential to Microwave Integrated Circuit Design," *Microwave Syst. News*, Vol. 14, 1984, pp. 113–122.
49. Yamaskai, H., and D. Maki, "Hybrid vs. Monolithic—Is More Monolithic Better," *Microwave J.*, Vol. 25, Nov. 1982, pp. 95–100.

PROBLEMS

15.1 Describe the advantages of microwave integrated circuits over conventional circuits.

15.2 Describe the advantages of monolithic millimeter wave monolithic integrated circuits over hybrid integrated circuits.

15.3 (a) List the basic properties required for an ideal substrate material, conductor material, resistive film, and capacitor dielectric film used in integrated circuits.

(b) Compare the advantages and disadvantages of thin-film versus active-layer resistors.

16 MICROWAVE OPTIC, ACOUSTIC, AND MAGNETOSTATIC CIRCUITS

16.1 INTRODUCTION

Advances in semiconductor laser technology, optical fibers, and high-speed modulation of light have demonstrated the potential for enhancement of capabilities in the state-of-the-art microwave systems. RF transmission over optical fibers can serve as a direct replacement for short coaxial lines or long multigigabit trunk lines, intrasatellite guided-wave switches, antenna remoting, and feedlines for phased-array antenna elements. Fiber-optic transmission of RF signals offers several advantages commonly associated with optical circuits, such as immunity from electromagnetic interference, ground loops, cross talk, and interception. In addition, optical fibers have low propagation loss, light weight, and large instantaneous bandwidth. Figure 16.1 compares the attenuation of several types of microwave transmission lines. Coaxial cables are extremely lossy for signals in the gigahertz range, free-space transmission is restricted to line-of-sight systems, and metallic waveguides are bulky. Low propagation loss of silica fibers (less than 0.5 dB/km at 1.3-μm wavelength) is orders of magnitude lower than the loss in RG-122/U coaxial line ($\simeq$500 dB/km at 1 GHz) and the loss in 0.141 semirigid coaxial line ($\simeq$700 dB/km at 5 GHz). Thus, microwave fiber-optic links using modulated light sources as transmitters and photodetectors as demodulators offer an attractive alternative to coaxial cable transmission of microwave and millimeter-wave signals. Microwave delay lines can use optical fibers as a delay medium. Optical delay lines and transmission links using optical fibers are discussed in Section 16.3.1.

The word LASER is an acronym for light amplification by stimulated emission of radiation. It is essentially an optical oscillator that consists of an amplifying medium placed inside a suitable optical resonator or a cavity. An optical cavity consists of two mirrors, curved or plane, one or both of the mirrors partially transmitting. Semiconductor diode lasers offer high dc-to-light conversion efficiency and several milliwatts of optical power in a small size, making them highly attractive for many airborne applications and other microwave systems.

Acoustooptic technology has become a leading candidate for the next generation radar warning receivers (RWR), which require fast frequency

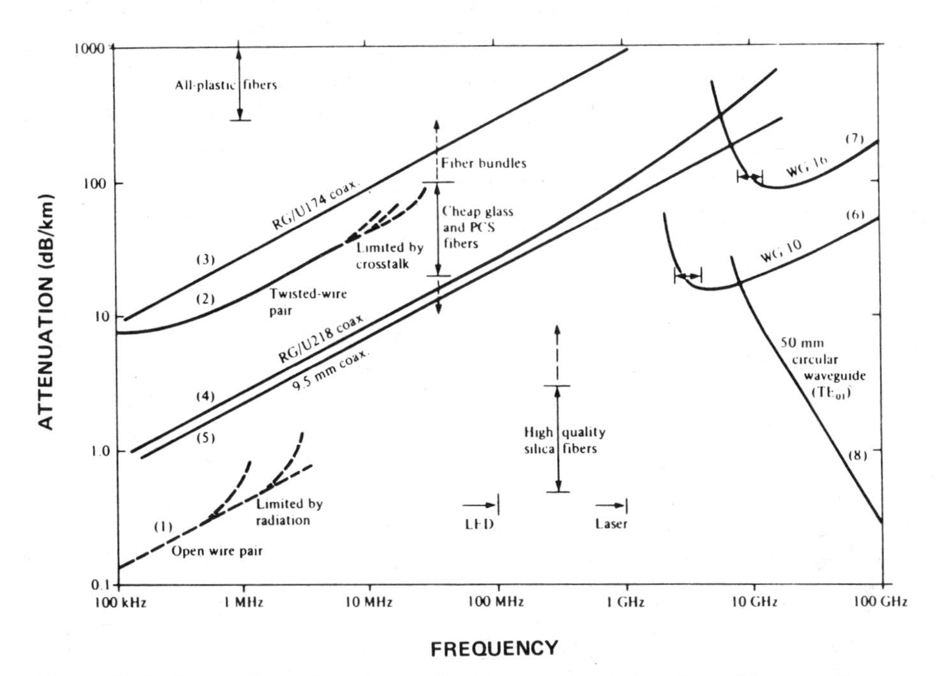

Figure 16.1 Comparison of various microwave transmission lines. (From J. Gowar, *Optical Communication Systems*, Prentice Hall, 1984, reprinted with permission.)

channelization, ability to handle simultaneous signals in a dense electronic-warfare (EW) environment, large instantaneous bandwidth, and moderate frequency resolution.

Direct dc-to-RF conversion utilizing picosecond optoelectronic switching has the potential of generating microwave frequency waveforms that have high output power and high switchout efficiency for applications that include high-resolution radar, microwave signal generation, and time-domain meteorology.

The material in this chapter covers RF, microwave, and millimeter wave applications of optoelectromagnetics. The discussion is covered under four separate topics, which are as follows.

1. Microwave modulation of optical sources for optical RF links and delay lines.
2. Acoustooptic, electrooptic, and magnetooptic interaction for modulators, switches, and spectrum analysis.
3. Optoelectronic control of solid state microwave devices for intensity and phase modulation.
4. Optical fabrication technologies as applied to millimeter-wave component development.

The material in this chapter deals with the application of optical technology for microwave system needs more than the optical components and devices themselves. The reader is assumed to have some basic understanding of optical sources (LASERS), photodetectors, optical fibers, and acoustooptic interaction. Some references [1, 2, 3] are included at the end of this chapter for additional information about these devices, since it is beyond the scope of the book to cover each one of them separately.

16.2 MICROWAVE MODULATION OF OPTICAL SOURCES

The transmission of microwave signals over an optical fiber is very similar to the transmission of an amplitude-modulated RF signal over a coaxial cable or a rectangular waveguide. The microwave signal to be transmitted modulates an optical carrier, which is sent over the optical fiber to be recovered at the other end using a photodetector. Thus, high-frequency modulation of optical sources is essential for transmitting analog or digital microwave signals over optical fibers, and/or for fabricating microwave delay lines using fibers as a delay medium. An optical source may be modulated either directly by modulating the bias current of the source, such as a light-emitting semiconductor diode, or by means of an external modulator following a CW light source, such as a He–Ne laser. Direct modulation of a laser diode is relatively simple to implement and it is much more efficient than external modulation, but the modulation bandwidth is limited by the diode characteristics. External modulation of light has the potential for larger operational RF bandwidths, but it has higher loss and requires additional expense for the modulator. External modulation of a light source allows CW operation of the laser with no variation in its peak emission wavelength, while direct modulation results in varying the wavelength of the light output due to the internal heating. The modulation speed in a bias modulation scheme is proportional to the drive level.

16.2.1 Direct Modulation

Figure 16.2 shows the direct bias current modulation of a laser diode. The modulation bandwidth of the device is limited by (1) the relaxation resonance frequency, a fundamental characteristic of all laser diodes, (2) circuit parasitic elements associated with the diode contacts and the diode package, and (3) by the poor impedance match between the low impedance of the device and the 50-Ω input impedance of the RF drive circuit. The modulation bandwidth can be increased by optimizing the laser and the device package. Relaxation oscillation of the intensity has been observed in most types of lasers [3a]. It is present because of an interplay between the oscillation field in the cavity and the atomic inversion [3]. The dynamics of the coupling between the drive power and the stimulated emission limits efficient modulation frequencies below the relaxation resonance frequency.

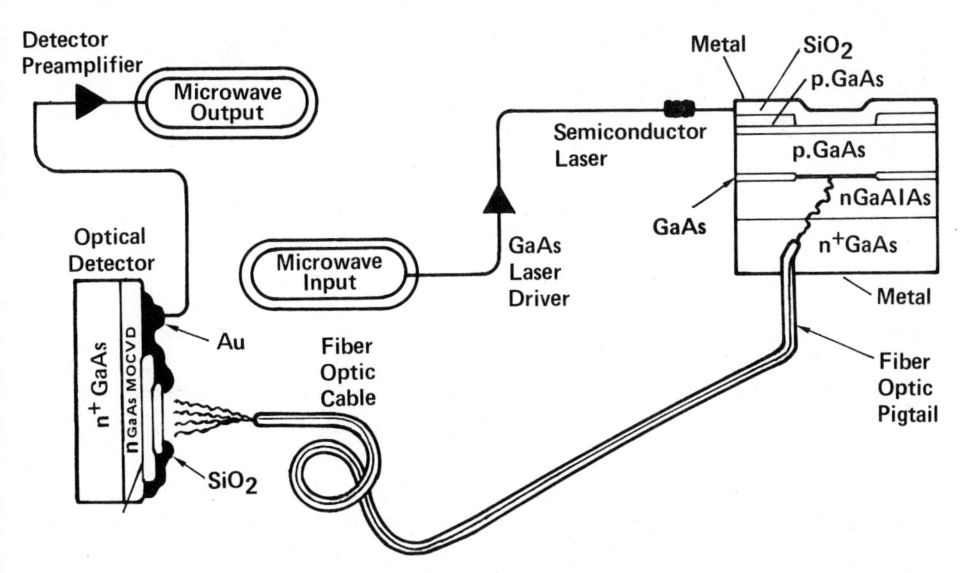

Figure 16.2 Direct modulation of a laser diode.

The relaxation resonance frequency is calculated from the rate equations for the electron and the photon densities in the diode lasers. That is, [4]

$$\frac{dn}{dt} = \frac{I}{eV} - r(n) - v_{gr}\, g(n) p \tag{16.1}$$

and

$$\frac{dp}{dt} = v_{gr}\, g(n) p \Gamma - \frac{p}{\tau_p} + \beta r(n) \Gamma \tag{16.2}$$

where n and p are, respectively, the electron and the photon densities, e is the charge of an electron, V is the active layer volume, I is the dc current, v_{gr} is the optical mode group velocity, $g(n)$ is the optical power gain of the medium, $r(n)$ is the spontaneous recombination rate, β is a fraction of spontaneous emission coupled into the laser mode, and Γ is the fraction of optical mode confined within the active layer. The photon lifetime τ_p is given by

$$\tau_p = \frac{1}{v_{gr}[\alpha + (1/L)\ln(1/R)]} \tag{16.3}$$

where α is the distributed internal loss in the cavity, L is the length of the cavity, and R is the mirror reflectivity. A standard small-signal analysis of the rate equations gives the relaxation resonance frequency to be [4]

$$\begin{aligned} \nu_{\text{rel}} &= \frac{1}{2\pi}\left(\frac{AP_o}{\tau_p}\right)^{1/2} \\ &\simeq \frac{1}{2\pi}\left(\frac{I}{I_{\text{th}}} - 1\right)^{1/2} \end{aligned} \tag{16.4}$$

where P_o is the steady state photon density in the active region, A is the optical gain coefficient, and I_{th} is the laser threshold current. Quartnery lasers have higher photon density, thus they have larger modulation bandwidths. From (16.4), it is clear that the direct modulation bandwidth of a diode laser can be increased by increasing the optical gain coefficient A and the steady state photon density P_o, or by reducing the photon lifetime τ_p. The photon lifetime may be reduced by shortening the laser cavity length. The minimum cavity length is limited by the thermal effects, since reducing the laser cavity length is equivalent to increasing the driving current density in the device which, in turn, results in excessive heating. The gain coefficient A can be increased by simply cooling the device. It is possible to improve the gain coefficient by a factor of 5 when the device is cooled from room temperature to about 77 K. Modulation bandwidth can be further improved by increasing the dc bias current to the diode as given in (16.4). Unfortunately, this is associated with higher optical output power densities that can cause mirror damage in the laser cavity. The optical output power density of a laser diode is given as

$$P_{\text{out}} = \frac{1}{2} P_o h\omega v \ln \frac{1}{R} \tag{16.5}$$

where h is Planck's constant, ω is the angular frequency, v is the group velocity of the optical beam, and R is the mirror reflectivity. Typically, for a laser with mirror reflectivity of 0.3, catastrophic mirror damage can occur for power densities as low as 1 MW/cm^2. The modulation bandwidths and power densities of a typical laser are plotted as a function of cavity length and pump current density in Fig. 16.3 at two different temperatures (300 K and 77 K). It becomes obvious that a short cavity length laser offers a larger modulation bandwidth for the same output power density. This is associated with an increase in the current density requirement of the device. A class of lasers known as "window" lasers can be operated at much higher current densities without any catastrophic facet damage and any appreciable degradation of reliability. The three commonly used configurations for window lasers are the window stripe, window buried structure, and the Crank TJS [5, 6, 7].

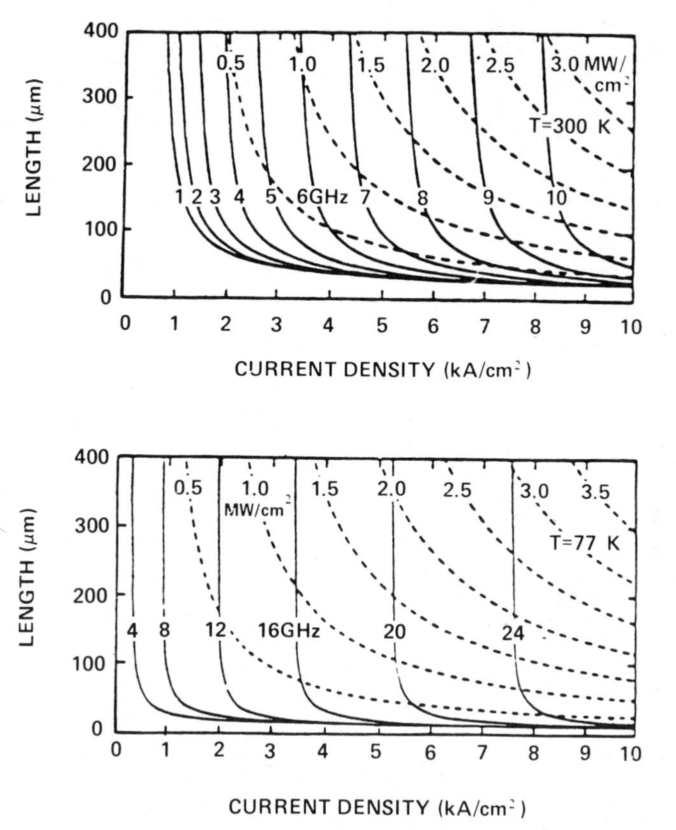

Figure 16.3 Modulation bandwidths and power densities of a typical laser at 300 K and 77 K. (After K. Y. Lau [4], reprinted with permission of *Applied Physics Letters.*)

Effects of parasitic circuit elements associated with the laser contacts and the device package are minimized by fabricating the device on a semi-insulating substrate. Such substrates are usually boat grown with chromium or oxygen doping [8]. Typically, the resistivity of a semi-insulating substrate may be of the order of 10^7 Ω-cm. Speed and noise performance of communication systems using devices fabricated on semi-insulating substrate improve significantly compared with the performance of conventional lasers grown on highly conductive, heavily doped substrates. In addition, it is easier to integrate optical as well as electronic devices on the same substrate when semi-insulating substrates are used.

It is desirable to have a flat light output response of a directly modulated laser for most system applications. This demands that the low-impedance laser diode be well matched to the 50-Ω driving circuit over a wide usable modulation bandwidth. For example, in the development of a FET driver one has to compensate for the variations in the FET gain with frequency and the light output of the device, for a flat response. It is often necessary

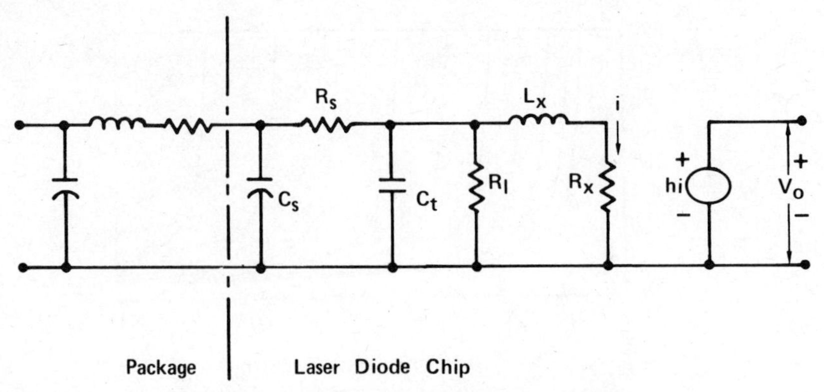

Figure 16.4 Laser diode equivalent circuit.

to determine the dynamic response of a direct modulated laser to a given RF input signal for this purpose. One can use an analytical approach based upon the circuit model of the device. Alternatively, one can measure experimentally, the laser S-parameters in a 50-Ω system at different power levels [9, 10]. Depending upon the complexity of the equivalent circuit, it is possible to include the effects of various parameters, such as package parasitics, heterojunction I–V characteristics, carrier diffusion characteristics, and the space-charge capacitance. Figure 16.4 shows an equivalent circuit of a laser diode based on the single-mode rate equations [10, 11]. The resistance R_1 includes the junction resistance of the diode, C_s is the parasitic capacitance, and the capacitance C_t represents a sum of the active layer diffusion capacitance C_d and the space charge capacitance C_{sc} of the diode. The transimpedance h of the current-controlled voltage source is independent of the frequency. The term R_x approaches infinity and C_d goes to zero at zero bias. The resistor R_s represents the parasitic resistance associated with the contact and the bulk semiconductor chip. Typical values for the circuit elements for two different types of laser diodes are given in Table 16.1 [10]. The broad stripe laser is a double heterojunction laser diode, LCW-10, from Laser Diode Labs, and the buried heterostructure is an HLP-3400 diode manufactured by Hitachi. Figure 16.5 shows an experimental arrangement to measure the reflection coefficient and frequency response of a laser diode. To assume minimum

TABLE 16.1 Typical Values for Circuit Elements in Fig. 16.4 for Different Types of Laser Structures

Laser Structure	C_d (pF)	C_{sc} (pF)	R_1 (Ω)	R_x (mΩ)	L_x (pH)
Broad stripe	8500	140	0.4	2–20	2–0
Buried heterostructure	285	10	0.5–1.0	4–25	3–9

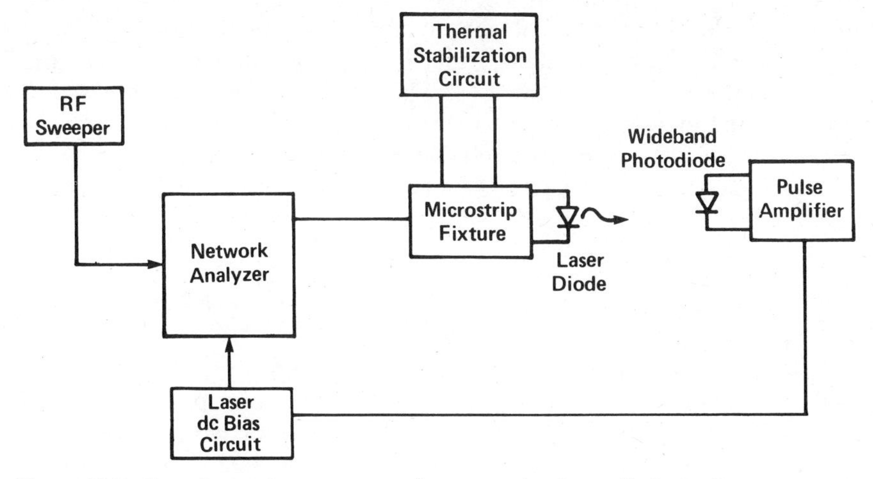

Figure 16.5 Experimental arrangement for measuring laser-diode device parameters.

drift in the bias threshold current and the modulation response characteristics, the text fixture is temperature stabilized. Parameters of the device are obtained using computer-aided fitting of the modeled response to the measured data. The equivalent circuit of the laser diode at the RF bias port can be incorporated in a microwave circuit analysis program to match the microwave driver/amplifier with the low dynamic impedance of the device. Figure 16.6 shows the block diagram of a FET driver-laser circuit.

Incorporating many of the techniques previously described, laser modulation bandwidths can be extended beyond 20 GHz for peak emission wavelengths between 800 and 870 nm. A much higher modulation bandwidth is possible for higher emission wavelength (1300 nm and 1550 nm) laser diodes [12]. Microwave swept response of a laser diode/driving amplifier circuit can be obtained using a network analyzer setup. The observed modulation response of the laser diode should be normalized to the photodiode response at each frequency [13].

16.2.2 Indirect Modulation

External modulators are required when the bandwidth requirements of a system exceed the relaxation resonance frequency of the laser diode. An

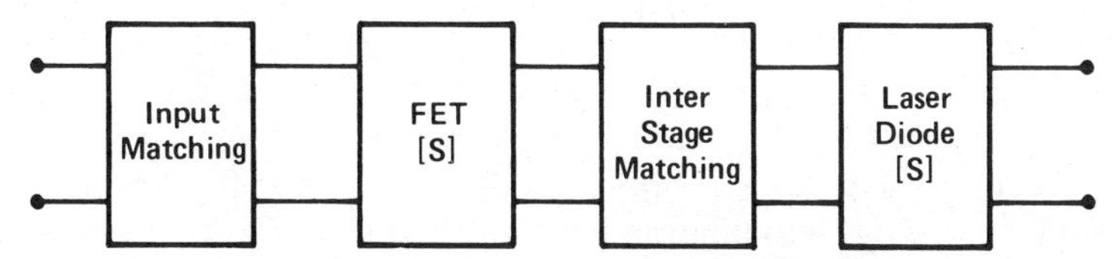

Figure 16.6 Laser driver/amplifier circuit block diagram.

anisotropic dielectric material can support two linearly polarized modes ("ordinary" and "extraordinary") that have unique directions of polarization and different phase velocities. The linear electrooptic effect is the change in the indices of refraction for each mode that is caused by and is proportional to the externally applied electric or RF field. The propagation characteristics in crystals can be described by means of the index ellipsoid equation [3].

$$\frac{1}{n_1^2}x^2+\frac{1}{n_2^2}y^2+\frac{1}{n_3^2}z^2+\frac{2}{n_4^2}yz+\frac{2}{n_5^2}xz+\frac{2}{n_6^2}xy=1 \tag{16.6}$$

In the absence of an electric field n_1, n_2, and n_3 correspond to the three principal dielectric axes directions x, y, and z, respectively, and $n_4=n_5=n_6=0$. The change in coefficients when an electric field is applied is given as

$$\Delta\left(\frac{1}{n_i^2}\right)=\sum_{j=1}^{3} r_{ij}E_j \tag{16.7}$$

where E_j is the applied electric field. The 6×3 matrix with elements r_{ij} is known as the electrooptic tensor. Obviously, crystals that are centrosymmetric ($r_{ij}=0$) do not give any change in the indices of the ordinary and extraordinary modes, and thus they are not suitable for electrooptic modulation. Linear electrooptic coefficients for some commonly used crystals, for modulator applications, such as CdTe, GaP, KD_2PO_4, $LiNbO_3$, and $LiTaO_3$, are available in several acoustic and optics books [3, 14]. The RF signal can be impressed electrooptically as an amplitude modulation on the laser beam and subsequently be recovered by a photodetector. The RF electric field can be applied along the direction of light propagation (longitudinal modulation mode) or it can be applied normal to the direction of propagation (transverse-mode modulation).

High-frequency operation of electrooptic modulators is limited by the parasitics associated with the crystal, the power needed to obtain peak retardation (phase difference at the output plane between the two mutually orthogonal transverse components of the optical field) and the time it takes for light to go through the crystal (transit time τ_D).

The transit time limits the operation of an electrooptic modulator to a few gigahertz. This can be overcome if the modulation signal is applied in the form of a traveling wave [15]. Traveling-wave modulation can be applied in the transverse mode because the RF field in most propagating structures (microstrip, strip line) is predominantly transverse. A coherent light traveling-wave electrooptic modulator can offer multigigahertz bandwidth [16]. Wide modulation bandwidth at low power can be obtained, in a traveling-wave modulator, by matching the phase velocities of the optical signal in the electrooptic substrate and that of the RF modulation signal. The modulation index may be increased either by increasing the incident RF power (which is limited by crystal damage considerations) or by lengthening the signal interaction path in the modulator. Velocity compensation techniques to match the phase velocities require either to decrease the phase velocity of the optical signal or to increase the velocity of the RF signal. The former design approach presents some fabrication problems and the later results in reduced modulation power, due to the partial distribution of the modulating field in a nonactive material. In addition, the modulator impedance has to be well matched to the impedance of the RF drive circuit over a broad frequency range. Attenuation of the RF signal in a modulator results in frequency dispersion of the output waveform.

Electrooptic modulators have been fabricated using bulk crystals and using thin-film deposited waveguides. Thin-film guided-wave modulators require less power per unit bandwidth because they do not encounter beam spreading normal to the plane of propagation (a common problem in bulk modulators). The optical field is confined to dimensions comparable to optical wavelengths, which makes it possible to achieve the large electric fields required for modulation with relatively small RF signals. The absence of diffraction in a guided beam makes thin films attractive for traveling-wave modulators, because it is possible to have longer interaction lengths. Figure 16.7 shows various types of guided-wave intensity modulators having coplanar electrodes on an electrooptic substrate. The electrode configurations for transverse applied fields and for normal applied fields are shown in Fig. 16.8. The phase-modulation index is given in terms of a refractive index change in the device. That is,

$$\Delta n = \tfrac{1}{2}\eta n^3 \gamma E \tag{16.8}$$

where n is the refractive index of the material, γ is the electrooptic coefficient, E is the effective modulating field, and η is an empirically determined efficiency factor that represents nonuniformities of electrical and optical fields and, the fact, that both fields do not overlap completely in the device.

A $LiNbO_3$ (lithium niobate) substrate material is preferred for electrooptic modulators because it has a large ($n^3\gamma$) factor value, almost five (5) times larger than the value for a GaAs substrate ($\simeq 59 \times 10^{-12}$ m/V). To date, the largest modulation bandwidth of 17 GHz has been reported using a Mach-Zehnder interferometer configuration for the modulator (Fig. 16.9). The modulator has been fabricated by diffusing the photolithographically defined titanium (Ti) pattern into a Z-cut $LiNbO_3$ substrate. The electrodes were electroplated gold strip lines with one of the strips connected to a RF ground [16]. Low resistive loss strip lines and minimum impedance mismatch have been used for wide-band operation. Figure 16.10 shows the response of this traveling-wave electrooptic modulator. These modulators have, however, limited power-handling capability (typically, 100 μW) due

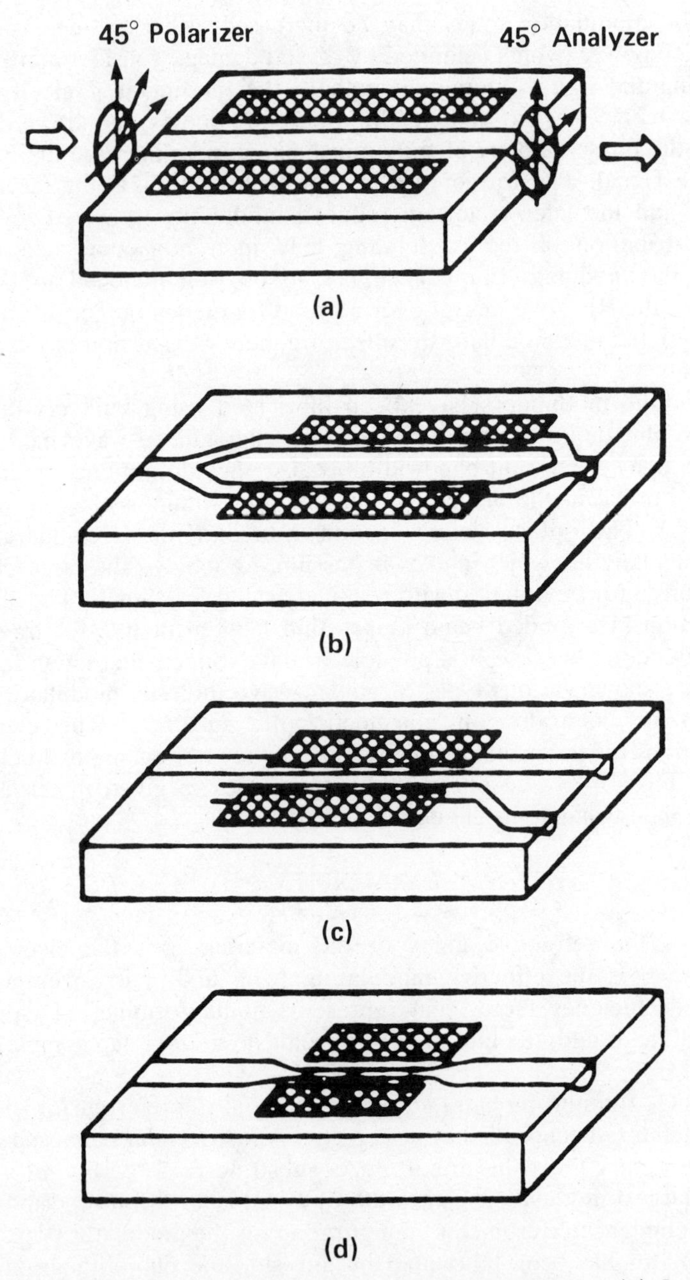

Figure 16.7 Various types of guided-wave intensity modulators. (*a*) Crossed-Nicol type. (*b*) Interferometer type. (*c*) Directional coupler type. (*d*) Cutoff type.

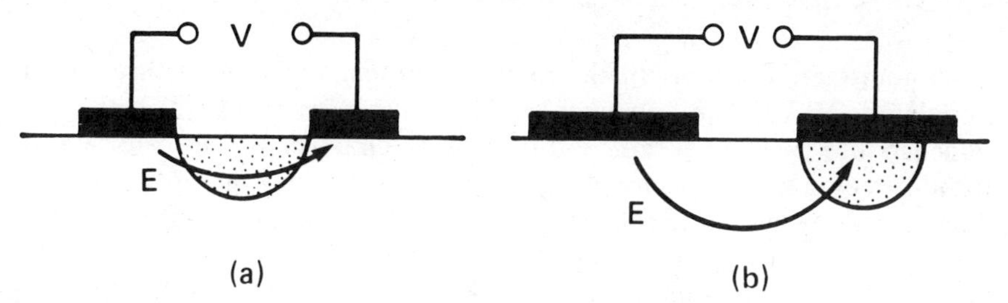

Figure 16.8 Electrooptic phase-modulator electrode configurations for (*a*) transverse applied fields, and (*b*) normal applied fields.

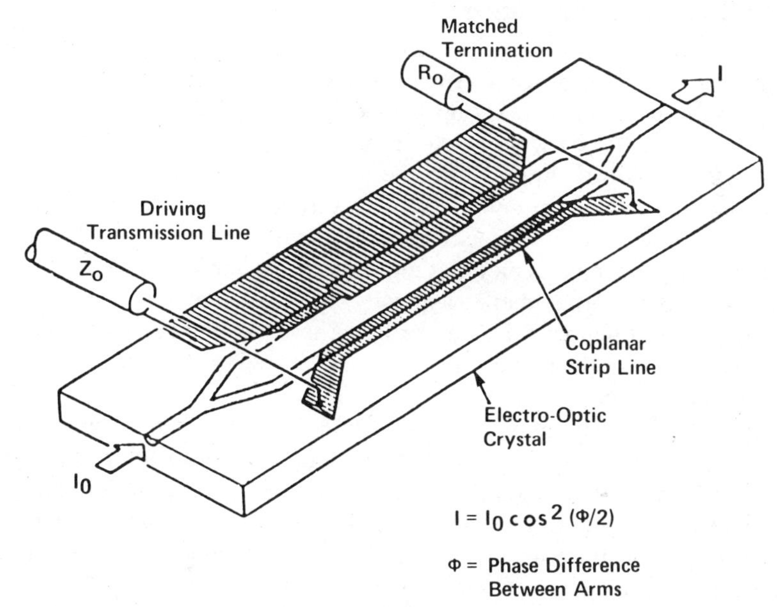

Figure 16.9 Traveling-wave electrooptic modulator. (After C. M. Gee and G. D. Thurmond [16], reprinted with permission of SPIE.)

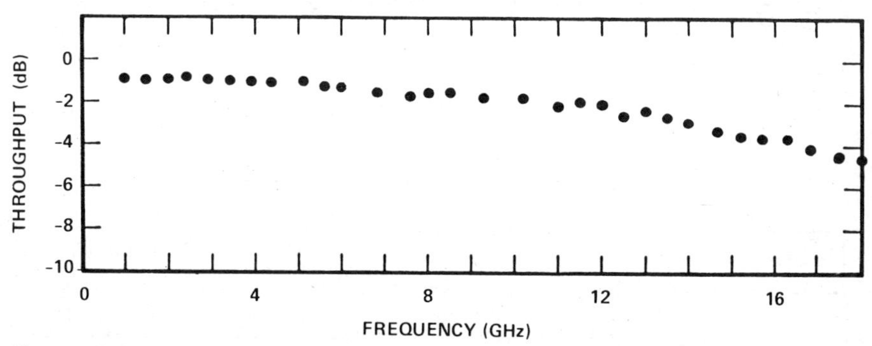

Figure 16.10 Microwave throughput vs. frequency for an electrooptic modulator. (After C. M. Gee and G. D. Thurmond [16], reprinted with permission of SPIE.)

to photorefractive effects in the material. Propagation losses through the modulator and input/output fiber coupling losses result in still less power launched into the optical link, and thereby much lower signal-to-noise ratio at the detector than in a link using direct modulation of the optical source. Thus, a microwave system designer would prefer using the direct optical modulation scheme whenever possible. An improvement in power-handling capability of the modulator is possible when a source of higher peak emission wavelength is used [16].

16.3 FIBER-OPTIC RF LINKS

Modulation of optical sources at microwave and millimeter-wave frequencies has made it possible to set up fiber-optic RF links as an alternative to conventional coaxial or waveguide links. In addition to low loss and wide bandwidth, fiber-optic links offer the usual advantages associated with optical fibers, such as immunity from electromagnetic interference, light weight, and low cost. There is, however, some loss associated with the optical modulation and demodulation. But this is easily offset by the low propagation loss in the fibers. Typically, a fiber-optic RF link less than 150 ft long operating at 5 GHz can be less lossy than a coaxial link. The loss advantage in an optical fiber link becomes much more apparent at still higher frequencies.

Figure 16.11 shows a typical fiber-optic RF link. The principal components of the link are an impedance-matched optical source, an optical fiber, an impedance-matched detector, and a low-noise amplifier. The laser diode microwave matching network can be fabricated photolithographically on a dielectric substrate using standard microwave integrated circuit (MIC) techniques. The important characteristics of a RF link are:

1. the frequency response,
2. the noise characteristics, and
3. the group delay, distortion, and linearity.

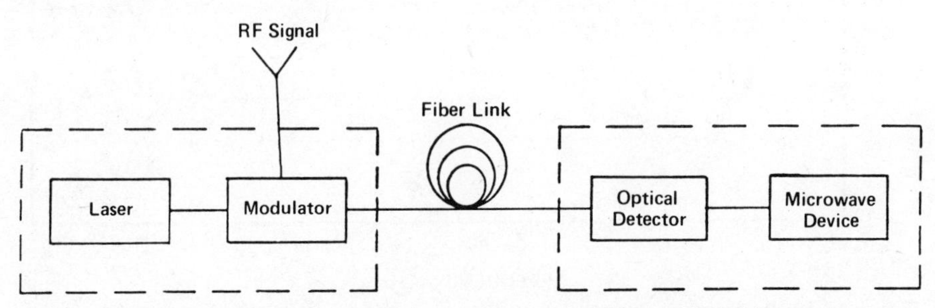

Figure 16.11 A generalized schematic for a fiber-optic RF link.

The frequency response of the link is primarily governed by the responses of the laser diode and the detector. For greater accuracy, however, one can take into account effects of chromatic dispersion, microbending and internal imperfections in the fiber, and/or the frequency response of the amplifier. In most cases, one can assume the fiber and amplifier responses to be nearly flat over the RF band, particularly compared with the response of the source and the photodetector. The frequency response of a laser diode, excluding parasitic effects, is of the form

$$|S_{21}|^2 = \frac{1}{(\nu^2 - \nu_{\text{rel}}^2) + \gamma^2 \nu^2} \tag{16.9}$$

where ν_{rel} is the relaxation resonance frequency given by (16.4). At the receiving end, the frequency response of the photodetector is determined by the transit time of electron hole pairs through the high-field drift region, the RC time due to the shunting effect of the junction capacitance plus any other parasitic capacitance, and by the diffusion of minority carriers generated outside the high-field region. High-speed operation of a photodiode requires all of the incident light to be absorbed within the depletion layer, thus eliminating slow diffusion of carriers, generated outside the depletion layer, into the depletion layer. A GaAs Schottky photodiode having a planar mesa configuration on a semi-insulating substrate provided a 3-dB bandwidth in excess of 100 GHz [13]. Figure 16.12 shows a simple equivalent circuit to evaluate the frequency and transit-time response of the photodiode. Typical parameters for a 100-GHz device are: the junction capacitance, $C_j \simeq 20$ fF, the parasitic capacitance, $C_p \simeq 15$ fF, the bond wire inductance, $L_p \simeq 0.08$ nH, the depletion layer resistance, $R_d \simeq 100$ MΩ, and the parasitic series resistance, $R_s \simeq 2$ Ω measured in a test system having source resistance equal to 50 Ω and a load resistance of 25 Ω.

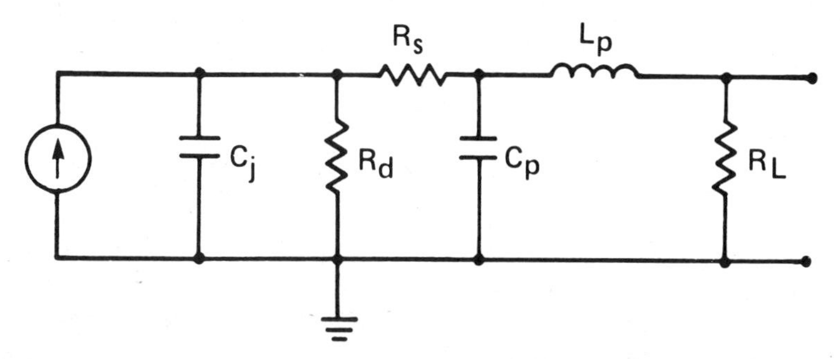

Figure 16.12 A simplified photodiode equivalent circuit to evaluate the frequency and transit time response of the diode: C_j: junction capacitance; R_d: the depletion resistance; C_p: parasitic capacitance; R_s: parasitic series resistance; R_L: load resistor; L_p: the inductance associated with the bond wire.

The overall signal power of the fiber-optic RF link may be given as

$$S = KI_D^2 G_A |S_{21}|^2 H_D^2(\nu) \tag{16.10}$$

where K is a proportionality constant, I_D is the photodiode current, G_A is the amplifier gain, $|S_{21}|^2$ and H_D^2 are the frequency responses of the source and the detector, respectively.

The amplifier thermal noise power N_A, the photodetector shot noise N_D, and the intrinsic laser intensity noise N_L, represent three primary sources of noise in a multigigahertz fiber-optic RF link. The noise distribution per unit bandwidth is given as a sum of these three noise sources.

$$N_T = [(F-1)kT + K_1 I_D H_D^2(\nu) + K_1 I_D^2 N^2(\nu) \cdot H_D^2(\nu)]^{1/2} G_A \tag{16.11}$$

where F is the amplifier noise figure, K_1 is a parameter of the detector, and $N^2(\nu)$ gives the excess noise distribution of the laser diode that peaks at the resonance frequency of the device. From (16.10) and (16.11) it can be interpreted that the signal-to-noise ratio (SNR) of the link is independent of the photodetector current if laser noise is dominant. The laser noise places an upper limit on SNR in the case of direct-modulated lasers, near relaxation resonance frequency. But that limit can be on the order of 50 dB at 5 GHz, which is sufficient for an FM format television signal of 5-MHz bandwidth.

Fiber-Optic RF Delay Lines. Microwave delay lines, transversal filters, correlators, and adaptive array distribution networks are fundamental building blocks for many state-of-the-art signal processing and communication systems. High-frequency modulation of optical sources and ultrafast photodetectors have made it possible to realize delay lines having low loss, long time delay, and large bandwidth. A silica fiber has a propagation delay of approximately 5 μs/km, while the propagation loss may be even less than 1 dB/km. Thus, it is possible to obtain 50-μs delay for attenuation values of less than 10 dB at C or X band. A practical fiber-optic delay line, however, has additional losses due to source-to-fiber coupling and bending losses when a long fiber is wound into a coil. Bending loss is attributed to higher order modes that escape out of the fiber when a fiber is bent. Bending loss increases with decreasing bend radius of curvature. Source-to-fiber coupling loss is present due to the limited numerical aperture of the fiber and Fresnel reflection at the fiber end. Grin (graded index) rod microlenses have been used to reduce this type of loss. Design requirements of a fiber-optic delay line are very similar to those for a fiber-optic RF link, discussed in Section 16.2. In addition, one has to worry about the temperature coefficient of the time delay and the time bandwidth product of the delay line. The thermal characteristics of a delay line depend on the temperature coefficient of the fiber's dielectric material ($\Delta\epsilon/\epsilon$) and the thermal expansion of the delay line ($\Delta l/l$). The thermal expansion coefficient of fused quartz can vary from 0.5 ppm/°C to about 10 ppm/°C, depending upon its alkali content. The changes in refractive index with temperature are, typically, a few parts per million per degree Celsius positive. Special fibers optimized for delay-line applications are possible in the near future.

The concept of a simple fiber-optic delay line can be easily extended to tapped delay lines and recirculating lines. Tapped delay lines may be fabricated using access couplers along a long fiber or by using separate fibers for each time delay required. For exceedingly large bandwidths (a few gigahertz) and long time delays (of the order of milliseconds) required in some applications, such as moving target detection, it becomes necessary to use single-mode (low-dispersion) fibers and extra amplification along the lossy propagation path. A recirculating delay line shown in Fig. 16.13 minimizes the requirements for the hardware and the fiber that can run into hundreds of kilometers for delays of the order of milliseconds. In a recirculating delay line, the SNR after N circulations, $(\mathrm{SNR})_N$, can be expressed in terms of the SNR after the first pass, $(\mathrm{SNR})_1$ by [17]

$$(\mathrm{SNR})_N = (\mathrm{SNR})_1 \frac{\beta^{-2} - 1}{\beta^{-2N} - 1} \tag{16.12}$$

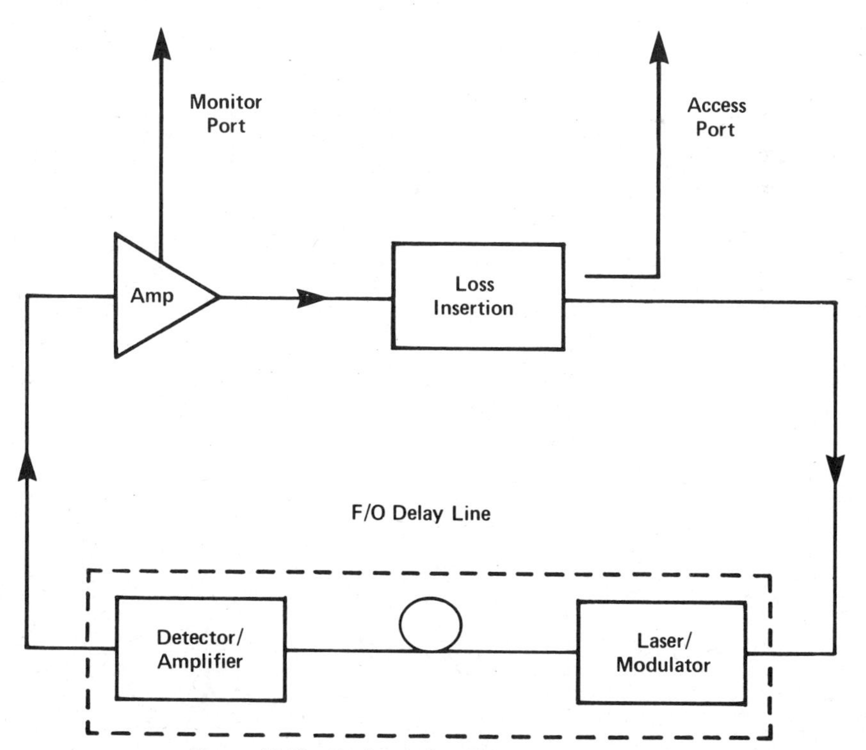

Figure 16.13 Recirculating fiber-optic delay line.

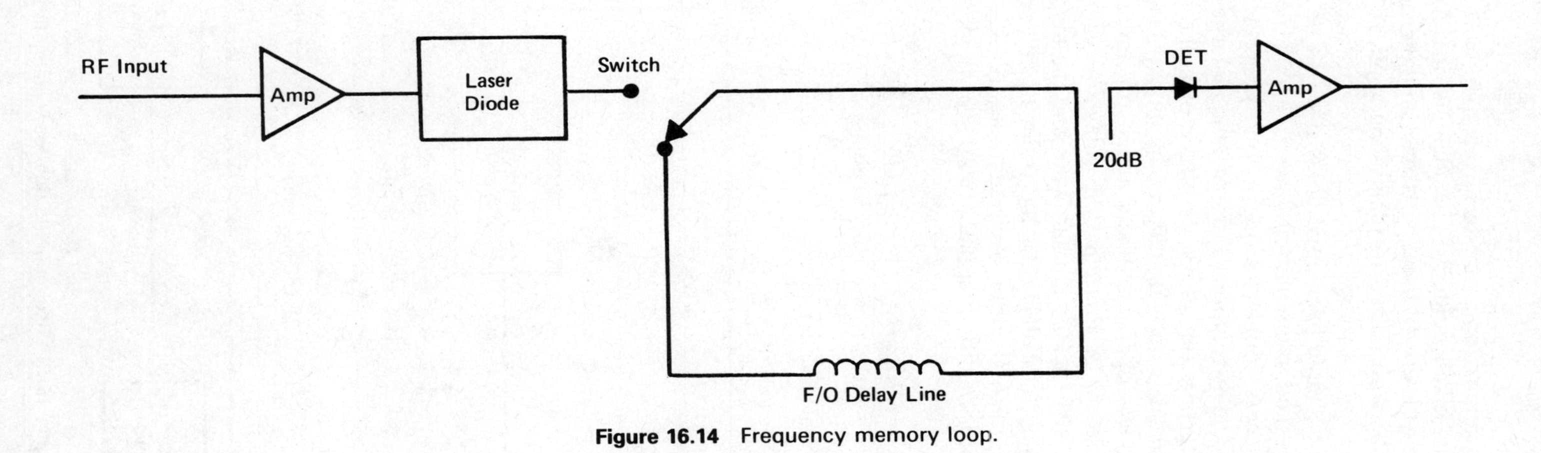

Figure 16.14 Frequency memory loop.

where β is the recirculating loop gain that is kept at less than one. Two important design considerations are: (1) the input signal pulsewidth has to be less than the delay time per pass to avoid pulse overlapping, and (2) the loop loss has to have enough margin to allow for variation in the amplifier gain and the optical device transfer characteristics. A higher loop loss may also be desirable to reduce distortion in some applications.

The loss and dispersion requirements may require the use of single-mode fibers and optical sources having peak emission wavelength around 1.3 μm. Preliminary results for a single-mode fiber 1-GHz–1-ms recirculating delay line in small size have been promising [17]. Other EW subsystems where delay lines and RF links play a key role include frequency memory loops (FML), antenna remoting, and beam forming in phased arrays. A frequency memory loop "memorizes" the radar's incident pulse frequency and retransmits it at a carefully controlled time. It is an electronic countermeasure designed to mislead the threat radar. The process employed is called range gate pulloff (RGPO). A FML consists of a recirculating delay line, an input RF switch, a loop switch, and an amplifier, as shown in Fig. 16.14. The RGPO action is limited by the time for which a FML is able to store the incident radar pulse. The maximum storage time is limited by the losses in the delay line and noise performance of the amplifier in the loop. Low-loss fiber-optic delay lines allow the signal to go around the loop many more times without amplification, and thus promise an interesting solution to this requirement.

Fiber-optic delay lines may be used to configure a beam-forming network to feed individually the correct amplitude and phase to all elements of a multiple antenna or a phased array. Replacement of conventional coaxial delay lines by fiber-optic delay lines in the network promises higher operating frequencies, wider bandwidths, and better phase accuracy. A typical multielement beam-forming network is shown in Fig. 16.15. An acceptable level of intermodulation limits the optical source power, since a laser is the dominant source of intermodulation. The quality of construction of fiber nodes affects the power division and combination characteristics and, thus, determines the amplitude accuracy in the network. Integrated-optics techniques may be employed to construct the input or output nodes. Use of a fiber-optic distribution system in a phased array is more important for operation at frequencies in the millimeter-wave band because of the small component size. At high frequencies, the interelement spacing is such that physical connector sizes will not only influence the interface design, but may preclude some distribution techniques. A typical active element of a phased-array antenna has to interface with various interfaces as shown in Fig. 16.16. All signals in and out of the antenna array element except for power supplies can be distributed optically. Further, some or all of these signals will require connections to other parts of the system. Many of these signals may be transmitted on the same fiber using wave-division multiplexing techniques [18]. Interestingly, the requirement to transmit diverse signal

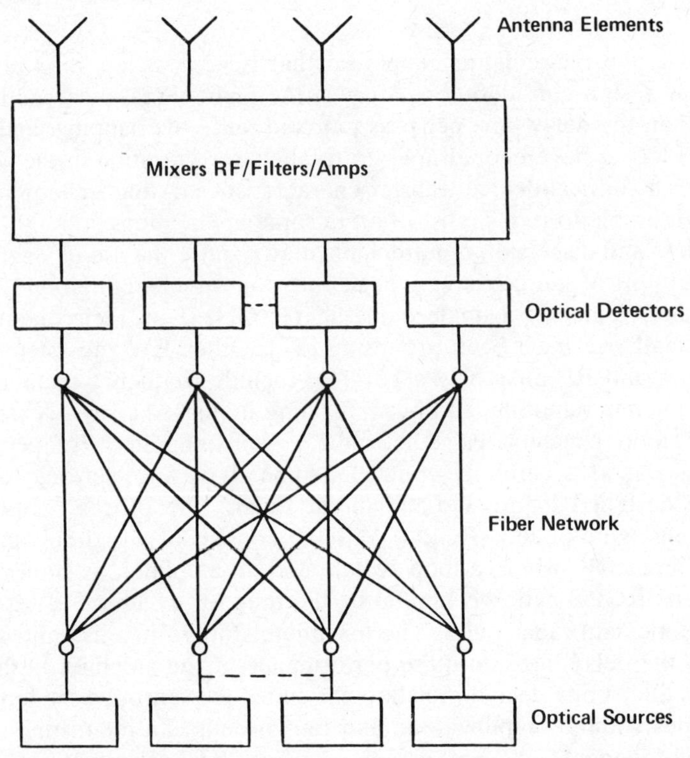

Figure 16.15 Optical fiber IF beam-former for a multielement phased-array network.

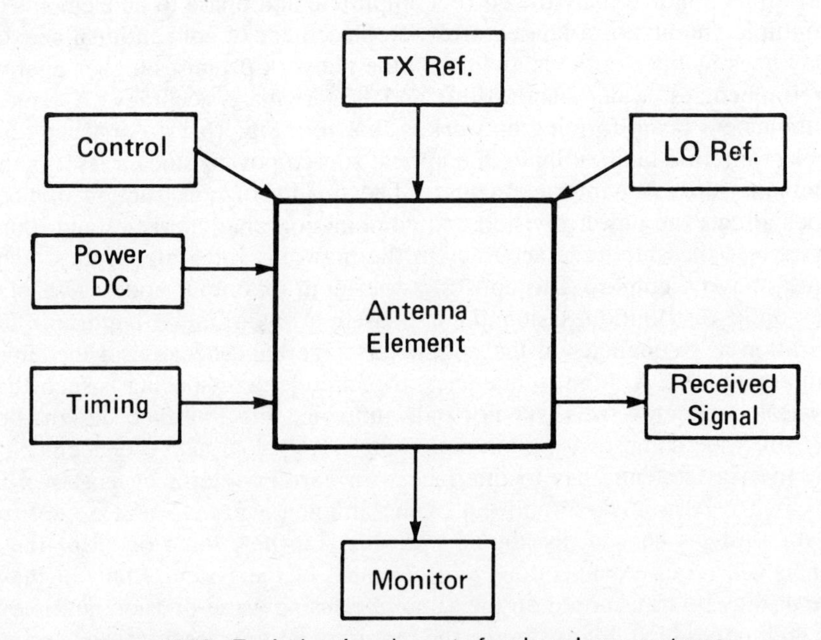

Figure 16.16 Typical active element of a phased-array antenna.

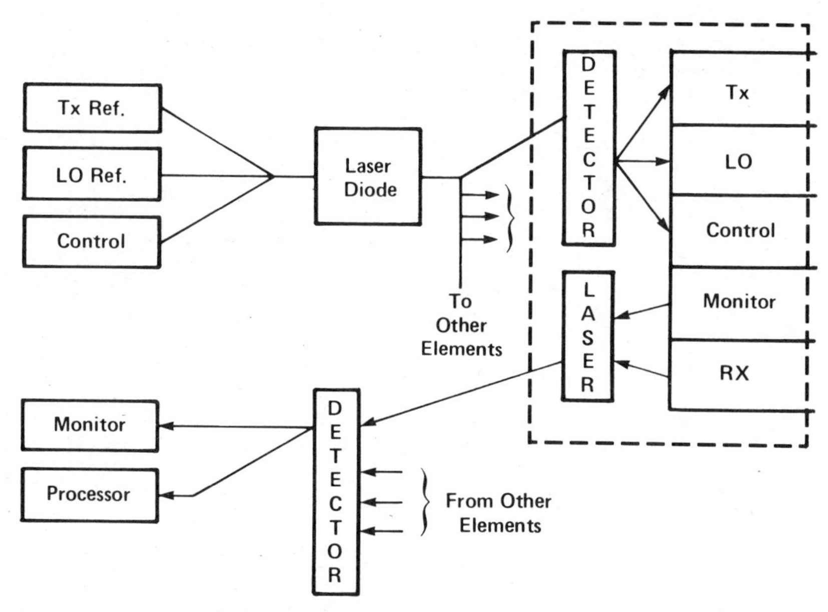

Figure 16.17 Two-way network distribution system.

types going in and out of an element (i.e., control and timing signals in digital form, transmitter reference signal in microwave and millimeter-wave bands, IF signals in microwave, VHF or UHF band) could be satisfied easily by the use of optical-fiber communication techniques. Figure 16.17 shows a possible two-way network distribution system for the transit/receive (T/R) element in a phased-array antenna. Only two fibers are required to interface to the array element and the control circuit, which is an important feature when interelement spacing decreases for high-frequency operation.

It should be pointed out here that considerable care is needed in the assembly of the distribution network, because the phase (delay) inequalities in the network result in increased sidelobes in the transmitter radiation pattern. The effect of random errors decreases with an increasing number of elements in an array, for a given sidelobe level. However, the distribution networks become more complex and it becomes necessary to allow for errors between multiple-source amplifiers and other components in the phase-error budget. Effect of temperature variation over a large array (generally, a nonrandom process) should also be compensated for in determining the overall antenna radiation pattern.

16.4 RF/OPTICAL INTERACTION

Proper interception and classification of parameters, of a RF signal, such as frequency, modulation, time of arrival (TOA), pulsewidth (PW),

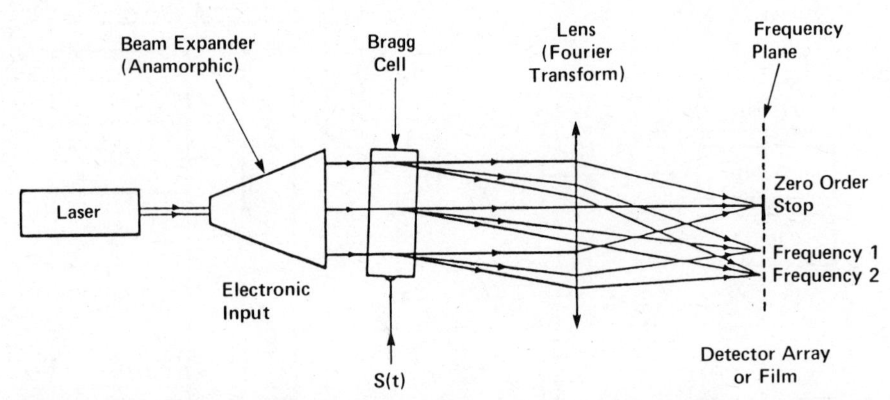

Figure 16.18 Basic acoustic spectrum analyzer. (From D. L. Hecht, "Acoustooptic Signal Processing Device Performance," SPIE Proceedings, Vol. 180, April 1979, reprinted with permission of SPIE.)

amplitude, and direction of arrival (DOA) are essential for effective power management and jamming in a dense EW environment. Acousto-optic (A/O) techniques have the potential of a high probability of intercept even in the presence of simultaneous signals in such an environment. Additional constraints that make A/O technology attractive for radar warning receivers (RWR) are size, weight, processing complexity, and cost [19]. In addition, analog technologies such as optical signal processing hold the promise of satisfying projected signal requirement for applications in areas such as radar, communication, and signal processing.

In an acousto-optic Bragg-cell receiver, the RF channelizing function is performed by the interaction of a collimated laser beam with a traveling acoustic wave as shown in Fig. 16.18. Two conditions must be met for Bragg diffraction. The first condition requires that an acoustic beam be wide enough to allow the development of interference effects in the diffracted beam, and the second condition requires that the acoustic beam and the optical beam interact at the Bragg angle. Both conditions can be mathematically expressed as

$$Q = \frac{2\pi\lambda L}{n\lambda_a^2} \gg 10 \tag{16.13}$$

and

$$2\lambda_a \sin\theta_B = \frac{\lambda}{n} \tag{16.14}$$

where Q is the Bragg region parameter, λ is the optical wavelength, n is the refractive index of the substrate material, λ_a is the acoustic wavelength, L is the width of the acoustic beam, and θ_B is the Bragg incident angle. An

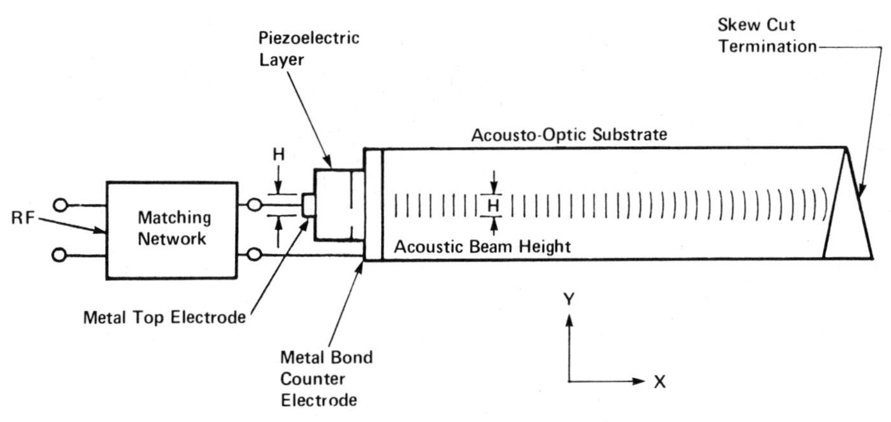

Figure 16.19 Acoustic device configuration. (From D. L. Hecht, "Acoustooptic Signal Processing Device Performance," SPIE Proceedings, Vol. 180, April 1979, reprinted with permission of SPIE.)

input transducer converts the RF signal into a traveling acoustic wave in the crystal (Fig. 16.19). It is a sinusoidal perturbation in the density of the substrate material, or it can be considered as a strain that causes a change in the material index of refraction. The laser beam incident at the Bragg angle gets deflected by the acoustic beam to a new direction that depends upon the acoustic wavelength. The presence of N signals, simultaneous or otherwise, results in N deflected spots that can be detected by placing a photodetector array at the output plane. Each element in the detector array corresponds to a known frequency interval, and since most photodetectors have square law response, the output signal from any element is linearly proportional to the RF power at that frequency. Postprocessing requirements are simplified by reading the output from several detectors in parallel. A complete analysis of A/O interaction is beyond the scope of this book and the reader should refer to any of the many published articles in this field [14]. Key aspects of the device performance are diffraction efficiency, dynamic range, and bandwidth. Other important parameters include operating wavelength, frequency resolution, optical scatter, intermodulation products, and power-handling capability. Many of these parameters are interrelated requiring trade-offs based on the overall system requirements. It is, therefore, important to point out here that irrespective of the developments in lasers, bulk-wave acoustics or SAW technologies, and high-speed light detectors it is necessary that in A/O receiver design one has to optimize the individual components based on the detailed analysis of their effect on the entire system. For example, the Bragg diffraction bandwidth and the diffraction efficiency of a cell are given by [19].

$$\mathrm{BW} = 1.8nV^2/fL\lambda \tag{16.15}$$

TABLE 16.2 Bulk Bragg-Cell Materials

Material	Density (10^3 kg/m^3)	Velocity (km/sec)	Index	Attenuation (dB/μsec at GHz)	Figure of Merit (10^{-15} sec^2/kg) M_2
$LiTaO_3$	7.45	6.19	2.18	0.062	1.37
$LiNbO_3$	4.64	6.57	2.20	0.098	7.00
TiO_2	4.23	8.03	2.584	0.566	3.93
TeO_2	6.00	4.20	2.26	6.30	34.5
GaP	4.13	6.32	3.31	3.80	44.6
SbN	5.40	5.50	2.299	2.20	38.6

$$\eta(\%/\mathrm{W}) = \frac{\pi^2}{2\lambda^3 f^{3/2} \tau^{1/2}} \cdot M_2 \cdot \frac{L}{H} \cdot \eta_e \tag{16.16}$$

where n is the reflective index of the substrate material, η_e is the transducer coupling efficiency, f is the RF frequency, τ is the aperture time, V is the acoustic velocity, M_2 is the acoustooptic figure of merit, L/H is the ratio of normalized interaction length and acoustic beam height. From (16.15) and (16.16), it is apparent that short optical wavelengths are desirable for wide-band operation and high efficiency of the cell. But it is incompatible with minibench or integrated optic technologies, since typical small-size semiconductor laser diodes have peak emission wavelengths greater than 800 nm. Again, a high-velocity mode and a small interaction length may be attractive for increasing the bandwidth but both factors compromise the efficiency. Table 16.2 and Fig. 16.20 compare several materials suitable for acousto-optic devices. A high-index material such as Gallium phosphide (GaP) is promising for a high-frequency wide-band device having reasonable diffraction efficiency.

The single-tone dynamic range of a Bragg-cell receiver is defined as the difference in maximum deflected optical power from the Bragg cell and the tangential sensitivity of a photodiode. It is limited by the amplifier noise figure, photodiode noise floor, laser power, and the efficiency of the Bragg cell. Low Bragg-cell efficiency, particularly for higher peak emission wavelength semiconductor laser diodes, results in less diffracted light for a given RF power. Increasing the RF power puts significant strain on microwave amplifier noise characteristics and increases nonlinear acoustic effects. Compatibility problems arise between the diffracted light intensity and the pnotodiode noise floor. For example, assume that the sensitivity of a junction silicon photodiode extends to only 100 nW and the laser-diode output power is 10 mW. Then for 3% diffracted light from the Bragg cell, the system dynamic range is about 35 dB for a 3-MHz video bandwidth. Improvements can be achieved by using more sensitive photodiodes (such

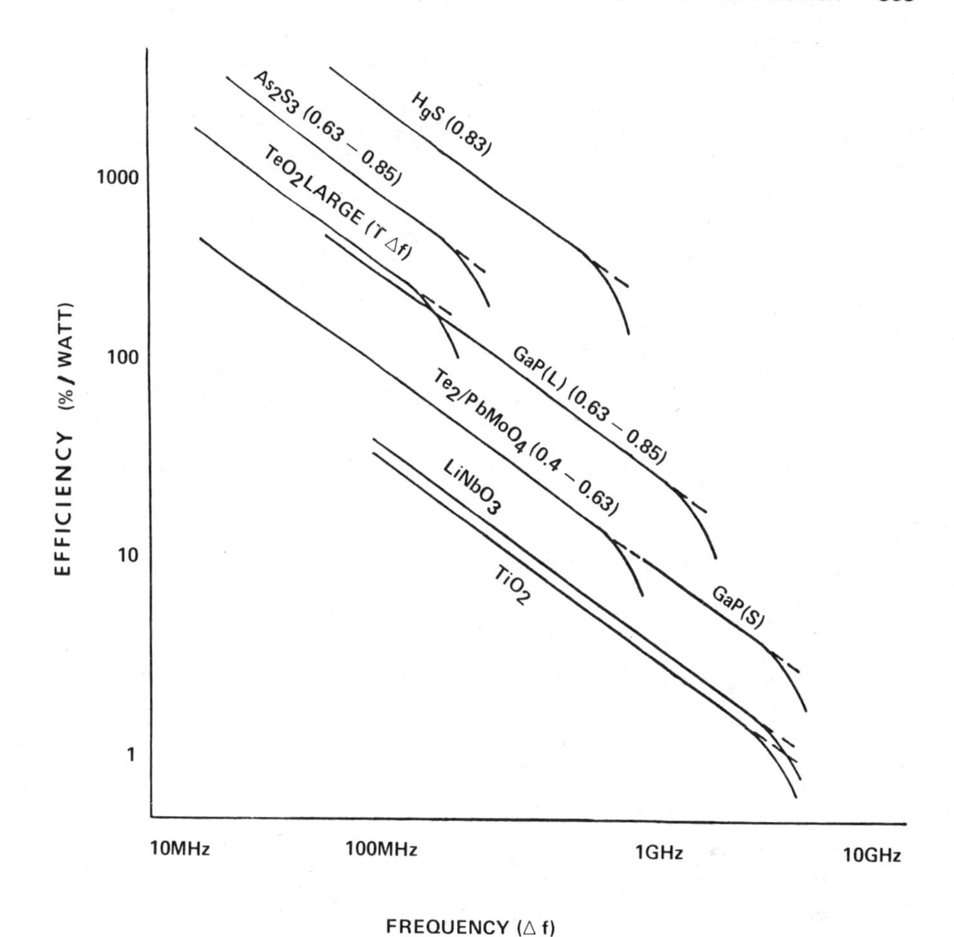

Figure 16.20 Bragg-cell material choices. (After S. K. Yao [22], reprinted with permission of SPIE.)

as avalanche photodiodes), higher laser power, or a narrower video bandwidth.

Maintenance of linear response in the presence of strong signals and avoidance of spurious signals in the presence of multiple RF signals limit the maximum allowable diffracted light intensity (typically, 1–5%) by any one signal. The two-tone dynamic range is limited by the third-order intermodulation products. The maximum diffracted light intensity should not exceed about 3% of the incident laser power for a two-tone dynamic range ≥ 50 dB. Since the laser power is constant, an incident RF pulse will cause depletion of a portion of the undiffracted light leaving less intensity for a second signal, and thus causing a nonlinear response.

Acousto-optic spectrum analyzers have been fabricated using integrated optics as well as bulk-optic techniques. Some minibench models have used

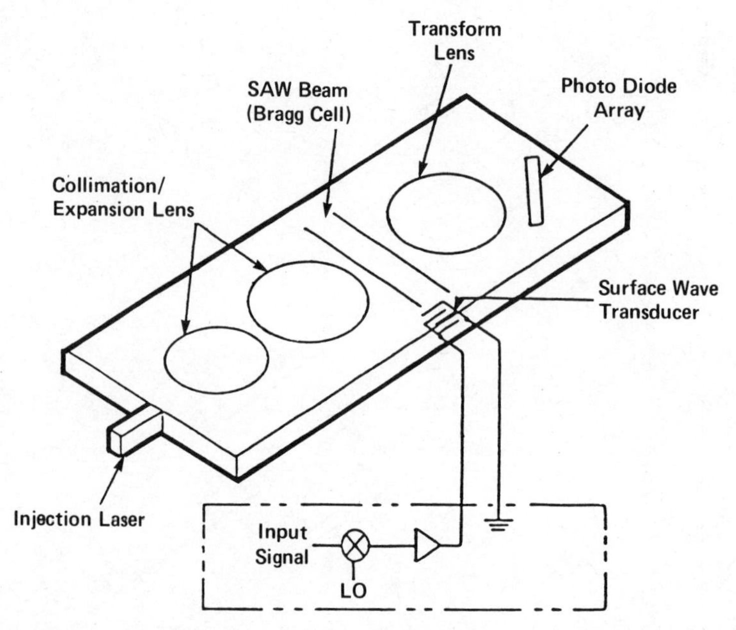

Figure 16.21 Integrated-optic spectrum analyzer. (From C. B. Hofmann and A. R. Baron, Wideband ESM Receiving Systems: Part I", Microwave J., Sept. 1980, reprinted with permission.)

bulk-optic components of classical three-dimensional design and bulk Bragg cells. Integrated optic spectrum analyzers (IOSA) have been implemented using surface acoustic-wave transduction techniques. The light beams in an IOSA are constrained in a thin guiding layer grown on a substrate. The integrated optics substrate contains the waveguide, the beam-forming optics, the acoustic-wave transducers, and the transform lens (Fig. 16.21). IOSA are inherently small in size, rugged, reproducible, and light weight. These characteristics make IOSA best suited for airborne ESM receivers. Even with multiple tilted transducers, the bandwidth of an IOSA is relatively low (around 500 MHz at 1 GHz) compared with bulk devices. The IOSA bandwidth is primarily limited by the high attenuation of high–frequency surface waves in $LiNbO_3$. Performance of IOSA is further limited by the fabrication techniques for near-perfect geodesic lenses. It is relatively unlikely that the current IOSA results will be significantly improved in the near future. If, however, fabrication technologies which can achieve narrow diffraction spots on anisotropic substrates such as $LiNbO_3$ are simplified, then the IOSA may yet prove to be a useful and relatively rugged and inexpensive receiver with a high probability of intercept.

In bulk Bragg cells, a laser interacts with a bulk acoustic wave propagating in the volume of a single-crystal material. Bulk Bragg cells for microwave application are fabricated either by using piezoelectric crystal platelet transducers that are welded to the crystal or by depositing piezoelectric ZnO thin-film transducers on the crystal itself (Fig. 16.19). While both approaches have potential for large bandwidths, platelets are not easily usable beyond 1 GHz because multigigahertz operation of the device requires very thin transducers (typically, a few microns). But platelet transducers have better coupling efficiency, and they can be used for longitudinal as well as shear-wave transduction. Multilayer ZnO films have been used to improve the coupling efficiency of the deposited electrodes. The main disadvantage of a bulk-wave acousto-optic spectrum analyzer (AOSA) is the loss of mechanical stability and ruggedness inherent in the assembly of many discrete components. Several bulk-wave AOSAs have been reported using He–Ne and semiconductor diode lasers [19, 20]. He–Ne laser-based devices use much simpler optics due to a high degree of collimation of the light source. Until the early 1980s, a bulk-wave "standard" Bragg cell used a $\langle x \rangle$ propagating longitudinal acoustic wave in $LiNbO_3$ crystal with light entering in the YZ-plane at an angle of 36° from the Y axis. Preliminary design calculations for the "standard" cell (using 16.15) show that the Bragg bandwidth centered at 1.8 GHz could reach almost 1 GHz with an interaction length L of 127–178 μm. This assumes the acoustic-wave velocity to be approximately 6.6 km/sec. But the propagation mode used in this design has relatively poor efficiency (approximately 1%/W at 6328 Å). The low efficiency places significant strain on the RF amplifier and the photodetector when a high dynamic range is required. The Bragg cell efficiencies are still lower when semiconductor laser diodes (peak emission wavelengths around 8300 nm) are used for minibench design. This follows from (16.16), which shows the inverse wavelength dependence of the cell efficiency due to the negative dispersion-photoelastic constants and index of refraction for $LiNbO_3$.

Improvement in two key parameters, the instantaneous bandwidth and the efficiency of a Bragg-cell device, are possible with at least one of the following techniques.

1. Acoustic anisotropy,
2. Birefringent Bragg diffraction, and
3. Phased-array transducer design.

The anisotropic interaction and the interaction of light with acoustic shear waves instead of the longitudinal waves result in greater efficiency. This is primarily because of a favorable phase-matching condition that enables the use of longer interaction length using shear waves than with longitudinal waves and because of the higher figure of merit for acoustic shear waves when an appropriate direction of propagation is chosen. The operating frequency of a shear-wave acousto-optic device is a strong function of the direction of propagation and is given by [21]

$$f_s = \frac{V}{\lambda} \sqrt{n_1^2 - n_2^2} \tag{16.17}$$

where V is the acoustic velocity, λ is the light wavelength, and n_1 and n_2 are indices of the ordinary and extraordinary waves, respectively. This adds another constraint to the device because the anisotropic Bragg cell has to be designed so that its transducer bandshape coincides with the inherent acousto-optic bandshape. An optimization procedure involves rotating the crystal in three-dimensional-space to find phase-matching conditions that reduce f_s in (16.17) while maximizing the figure of merit [14]. A factor of 3 or better improvement in figure of merit is possible for shear-wave propagation versus x-propagating longitudinal waves in $LiNbO_3$.

Use of shear waves involves diagonal components of the photoelastic tensor; these components flip the polarization of an incident laser beam. If this flip causes the polarization to change from an ordinary to an extraordinary wave (or vice versa), birefringent diffraction occurs. The birefringence phenomenon occurs only in certain crystallographic phases of anisotropic materials. A change in the index of refraction between the incident and refracted light loosens the Bragg diffraction constraint, which allows significant improvement in the device bandwidth, and (16.15) does not apply anymore. The center frequency of operation of the Bragg-cell device is given by (16.17), where n_1 and n_2 are the indices of incident and diffracted light, respectively. The birefringence mode of diffraction is characterized by the presence of a broad range of frequencies, around f_s, in which angle of incidence is nearly independent of the RF frequency, while the angle of the diffracted beam continues to be linear with frequency. In addition, polarization flip aids in reducing optical noise resulting from scattering of the incident beam at the device and the optical interfaces. This can result in a typical improvement of 10–15 dB in dynamic range of the receiver.

Use of phased-array transducers provides lower spurious response, multiple gigahertz bandwidth, and improved efficiency bandwidth product. The improvement in spurious response is due to the significantly reduced power density (acoustic energy spread over a much larger area) and, thus, lower acoustic nonlinearities. This effect is material-dependent, being particularly serious in TeO_2 (even though paratellurite is a semiconductor) and much less serious in cubic GaP. Figure 16.22 shows the schematic of a phased-array transducer for a Bragg device, and Fig. 16.23 shows the diffraction efficiency response of an eighty-four-element phased-array transducer on $LiNbO_3$ [22]. An approximate midband for the acoustic beam-steering Bragg cell is given by [23]

$$f_p = \sqrt{\frac{n_d V^2}{S\lambda}} \tag{16.18}$$

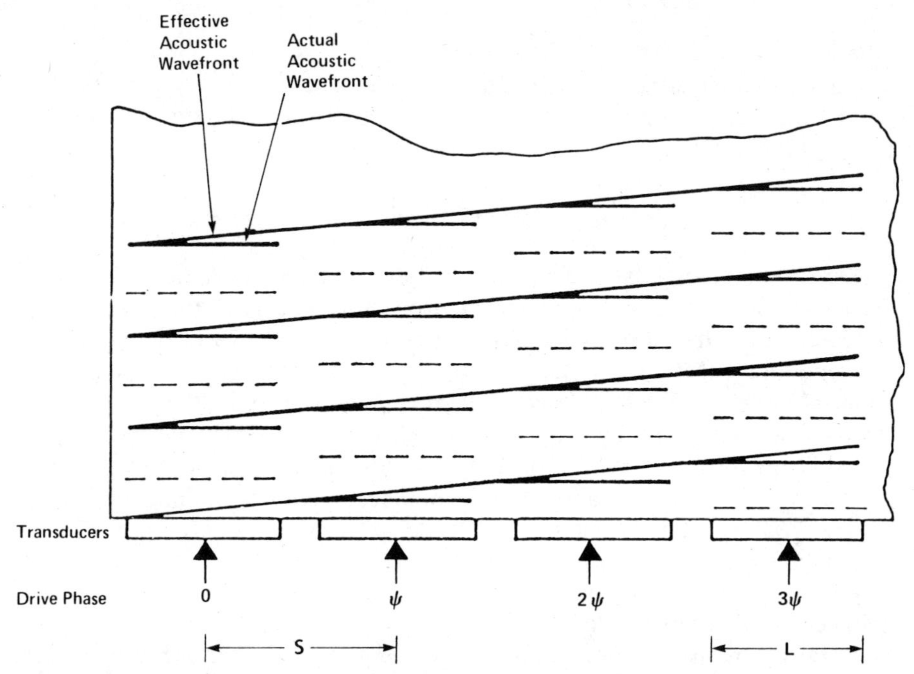

Figure 16.22 Phased-array transducers for acousto-optic applications. The term L is the interaction length. (After S. K. Yao [22], reprinted with permission of SPIE.)

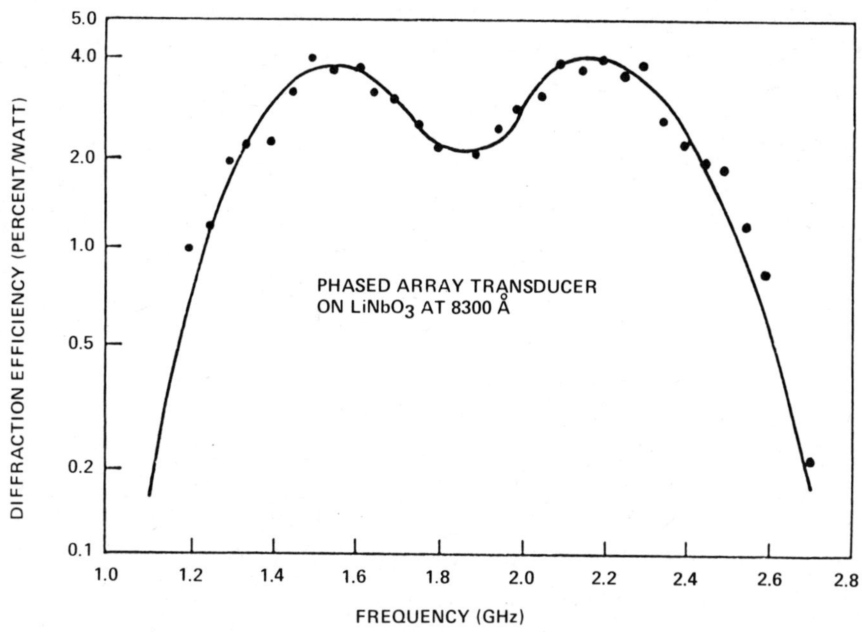

Figure 16.23 Diffraction efficiency vs. frequency of an eighty-four element phase array. (After S. K. Yao [20], reprinted with permission of SPIE.)

where n_d is the index of the diffracted light and S is the center-to-center spacing between adjacent elements. The midband frequency for a wide-band Bragg cell using beam-steering birefringence is approximated from f_s and f_p, that is, (16.17) and (16.18). It is

$$f_0 = [f_s^2 + f_p^2]^{1/2} \tag{16.19}$$

Detection cross talk and optical scatter limit the dynamic range performance of an acousto-optic signal processor. For A/O processing, the detectors are either a one-dimensional photodetector array of metal–oxide-semiconductor (MOS) sensors, discrete Si photodiodes, or avalanche photodiodes (APDs). Individual detector elements are accessed either by random access switching or by charge-coupled-device (CCD) registers. The crosstalk problem between adjacent detector elements in an array is addressed by sophisticated fabrication techniques, including etching or milling of grooves between the elements. Where space is not at a premium, it is usually advantageous to use discrete photodiodes. Each detector element is followed by its own amplifier, so that the cross talk between channels is reduced. Each photodiode may be fed by an optical fiber from the Bragg-cell transform plane (Fig. 16.24). Alignment of the fiber array with the detector array is critical.

Substantial improvement in dynamic range is possible using the heterodyne-detection scheme. Theoretically, shot-noise-limited detection can be achieved by a common laser diode and an off-the-shelf photodiode. Heterodyne detection makes use of the fact that the diffracted wave carries the RF modulation information upshifted in frequency by the laser.

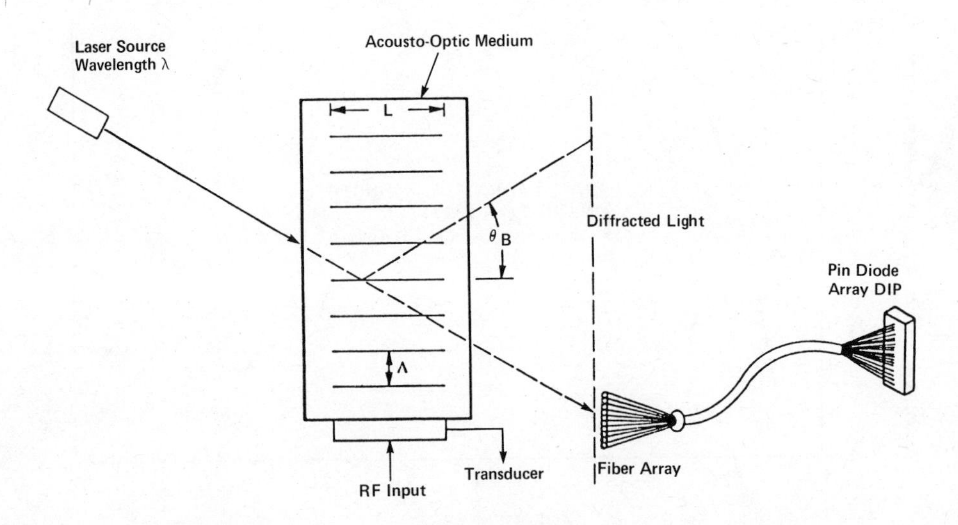

Figure 16.24 Fiber-optic arrangement for a *pin* detector pickup array.

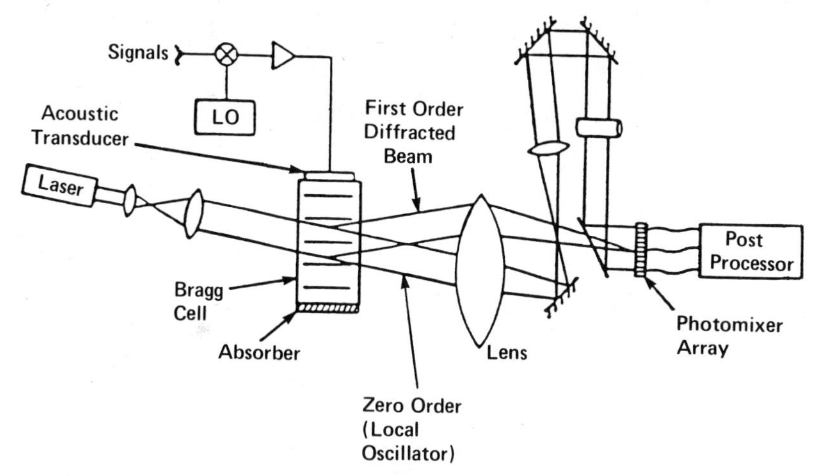

Figure 16.25 Bragg-cell optical heterodyne receiver. (After F. Freye, et al. [24], reprinted with permission of SPIE.)

Configuring the receiver as a Mach-Zender interferometer and using the undiffracted light as a local oscillator allows coherent detection in the photodetector array (Fig. 16.25). The increased dynamic range is possible because the photosensor output is proportional to the light amplitude rather than the light intensity. Dynamic-range improvement up to 20–30 dB is possible using this scheme [24]. In an alternative detection scheme two Bragg cells are used, one driven by a known RF signal and other by the signal to be detected, as shown in Fig. 16.26. Using this scheme, a minimum detectable power of 1 pW (corresponding to 78-dB dynamic range) has been observed [25]. However, this scheme is much more complex to implement.

The acousto-optic spectrum analyzer is one of the various methods for RF channelization using optical and RF interaction. An alternative scheme uses Bragg diffraction of light by magnetostatic waves (MSWs) [26]. MSWs propagate in thin-film ferrite media, such as yitrium iron garnet (YIG) grown by liquid-phase epitaxy techniques on a highly polished gadolinium gallium garnet (GGG) substrate [27]. Bragg diffraction of light by MSWs is also capable of providing a wide-band frequency channelizer. The frequency of operation is determined by the external magnetic bias field and can be extended beyond the 20-GHz range, limited primarily by the insertion loss. Figure 16.27 shows the schematic of a MSW-optical frequency channelizer. It is possible to transduce the high-frequency RF signal into the YIG film using a simple strip-line transducer, the width of the transducer being inversely proportional to its bandwidth. Light is coupled into the YIG film using any of the integrated-optic techniques, such as lenses, prisms, or tapered film couplers. The light transparency window of YIG limits the use of optical signals to the near-infrared region

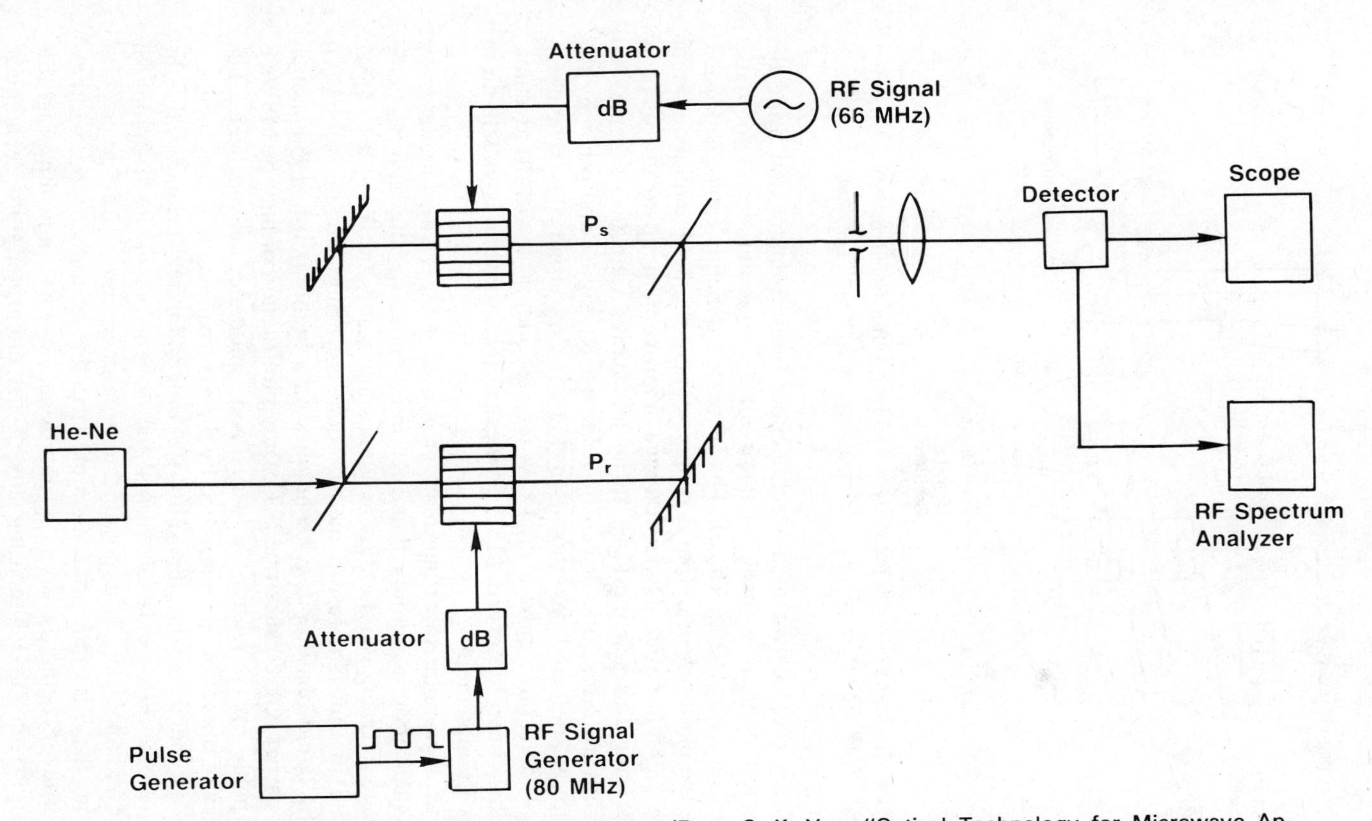

Figure 16.26 Two Bragg-cell detection scheme. (From S. K. Yao, "Optical Technology for Microwave Applications, Part II: Microwave Signal Processing," SPIE's Technical Symp. East '85.)

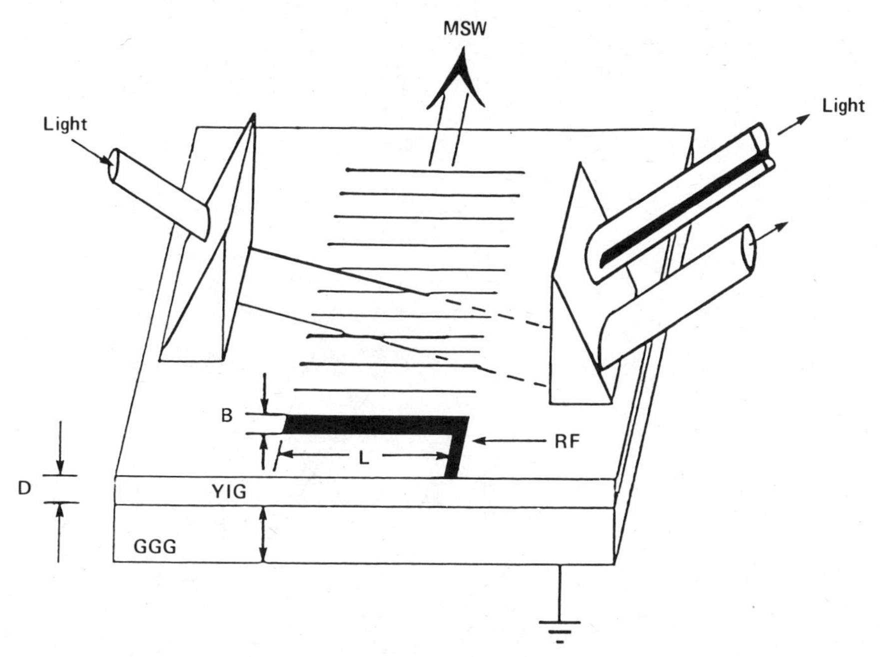

Figure 16.27 Basic configuration of a magnetostatic-wave and optical-wave interaction device. (From J. N. Lee and A. D. Fisher," Optical-Magnetostatic Wave Interaction for Spectrum Analysis," SPIE Proceedings, Vol. 477, May 1984, reprinted with permission.)

(1.1–5 μm). The detection can be performed using Ge detectors at room temperature with or without fiber pigtails for convenient fanout structure. Heterodyne detection may be used for improved dynamic-range performance.

In addition to spectrum analysis applications, acousto-optic interaction will have an impact in real-time signal processing of RF signals. For example, Fig. 16.28 shows a Bragg-cell receiver to obtain frequency and direction of arrival information instantly with almost 100% probability of intercept [28].

16.5 OPTICAL CONTROL OF MICROWAVE DEVICES

Recent advances in understanding the behavior of microwave devices under optical illumination have led to increased interest in the control of microwave signals using optoelectromagnetic techniques. Optical techniques promise ultimate speed and effective control of high-power highly nonlinear oscillators, eliminating currently used complex and expensive means of stabilization. Additional advantages include miniaturization, immunity of control from other electromagnetic devices, and capability of

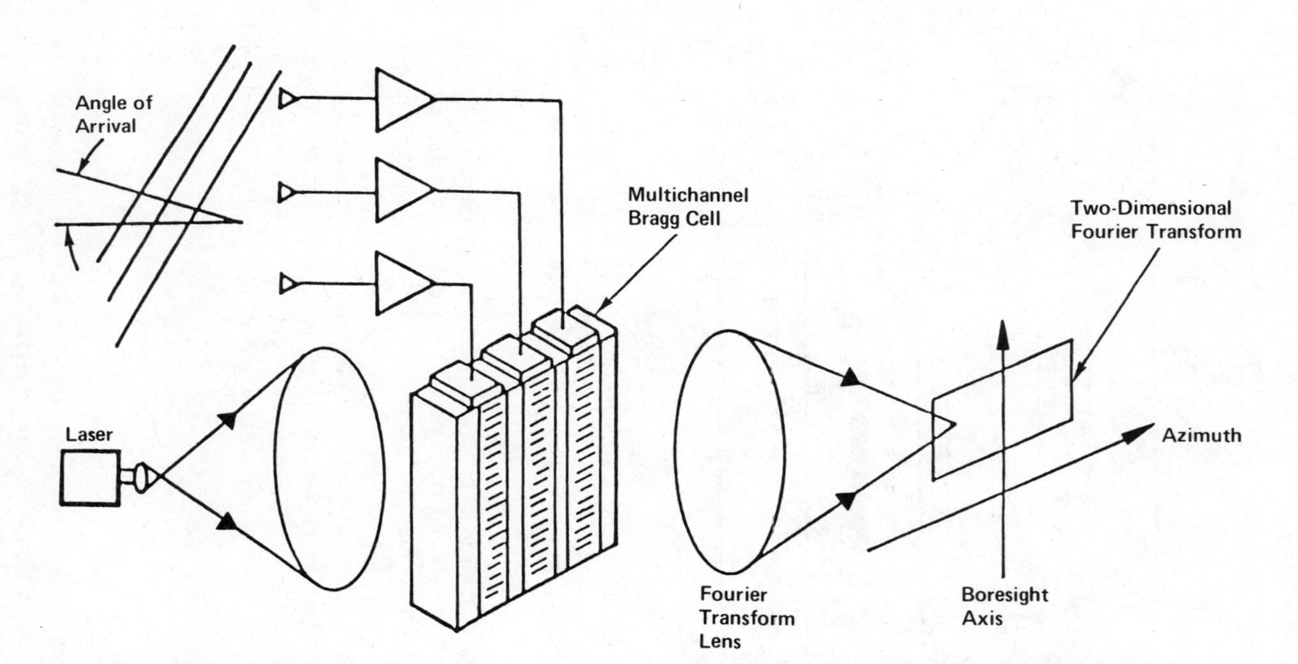

Figure 16.28 Phase interferometric four-channel Bragg-cell system for target frequency and DOA determination. (From S. K. Yao, "Optical Technology for Microwave Applications, Part II: Microwave Signal Processing," SPIE's Technical Symp. East '85.)

phase shift through optical delay. The injection of light is like introducing an extra terminal that has inherent optical isolation and does not require any unwanted decoupling structure. Light has been used to phase lock and control microwave generation devices, such as avalanche diodes, dielectric resonators, and injection locking of MESFET oscillators.

16.5.1 Switching Applications

Two-Terminal Diodes. Optical beams have been used to both phase lock and switch IMPATT diodes, with relatively low power required to control the device [29, 30]. The laser controls the operation of an IMPATT diode or any other microwave generation device, by generating photoinduced carriers in the active region. Both the density and the distribution of these carriers affect the device operation. Figure 16.29 shows the schematic of an optically coupled microwave power generation diode. It is important that most of the opotical power is coupled into the active region of the diode. Burnstein shift, interband transition, and tuning of E around E_g (bandgap of the semiconductor from which the microwave diode is made) have been used to enhance the light-coupling efficiency [29]. Quantum efficiencies of these devices vary with the change in optical wavelength and the bias voltage, as shown in Fig. 16.30. Quantum efficiencies as high as 50% have been observed at $\lambda = 0.875\ \mu$m. Further improvement is possible if an antireflection coating is used. However, the quantum efficiency spectra peaks at the same optical wavelength at different bias voltages [29].

Pulsewidth control, fast on/off switching, and intrapulse frequency shift (chirp) have been achieved by laser illumination of a pulsed millimeter-wave IMPATT-diode oscillator operating at 70 GHz [30].

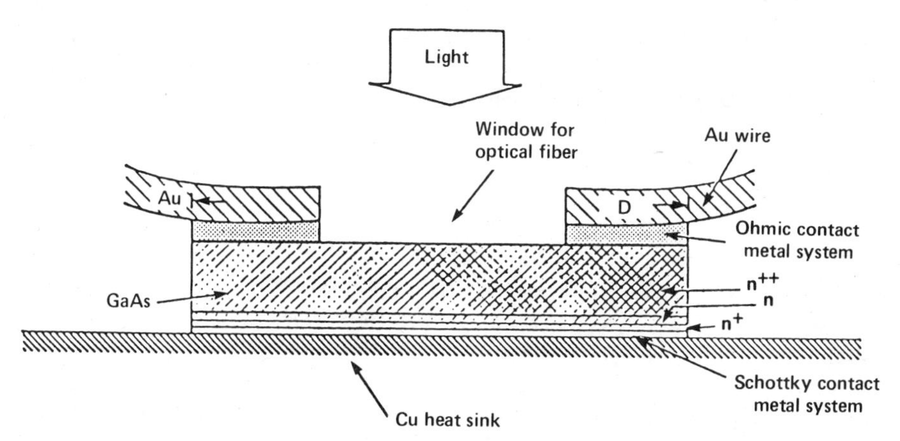

Figure 16.29 Optical control of IMPATT diodes. (After W. Chen [29], reprinted with permission of SPIE.)

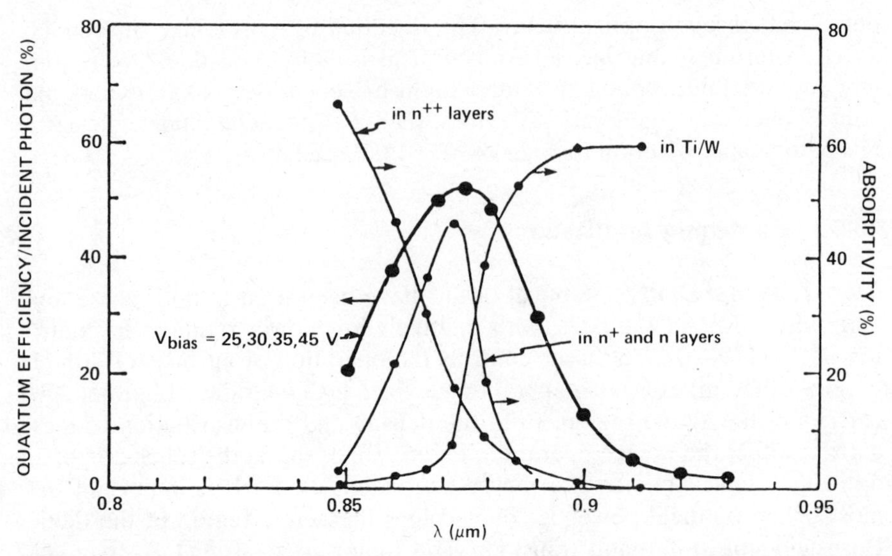

Figure 16.30 Experimental results for optical control of IMPATTs. (After W. Chen [29], reprinted with permission of SPIE).

Three-Terminal Devices. The performance of GaAs field-effect transistors under optical illumination is of interest for varying a FET's dc characteristics: to switch a FET from a low-impedance to a high-impedance state, to adjust gain of a FET amplifier, to tune FET oscillators using very low optical powers, and for high-speed optical detection using FETs. Optical illumination of the MESFET produces free carriers within the semiconductor when light of photon energy greater than the semiconductor bandgap is absorbed. The FET active region, buffer layer, and the substrate absorb light through the gaps between the gate and the drain regions. The optical power density decreases exponentially as it penetrates the material with an absorption coefficient α, which is typically $10^4\ \text{cm}^{-1}$ for n-type GaAs, used for MESFETs. That is

$$S = S_0 e^{-\alpha d} \tag{16.20}$$

where S_0 is the optical power density on the surface and d is depth of penetration. Absorption of light in the device results in a change of carrier density contributing to the drain current flow (photoconductive effect) and change of voltage across the depletion regions (photovoltaic effect). Photoconductive and photovoltaic effects can be represented by variable elements in the device equivalent circuit [32]. Figure 16.31(a) shows a simplified MESFET under optical illumination, together with its equivalent circuit [Fig. 16.31(b)]. Major variation in the transfer parameter S_{21} of the FET with illumination has been observed when the metal–gate

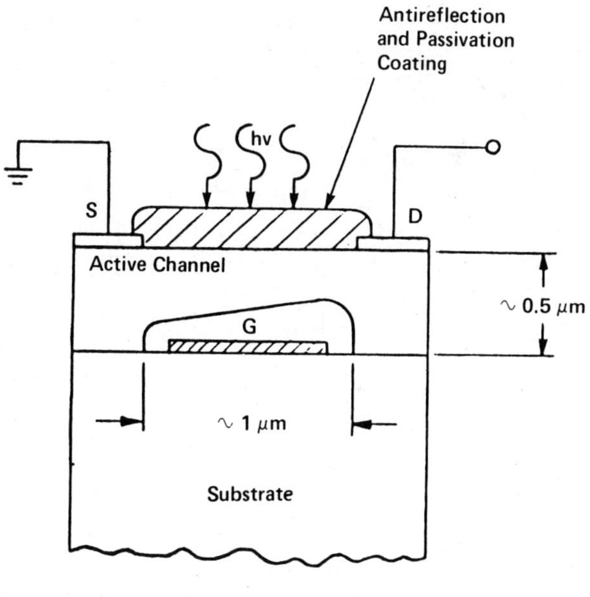

BURIED GATE MESFET

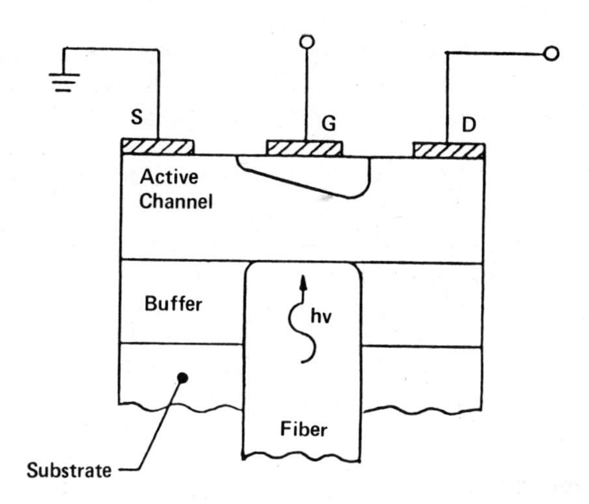

BOTTOM ILLUMINATION

(*a*)

Figure 16.31 (*a*) Optical illumination of MESFETs.

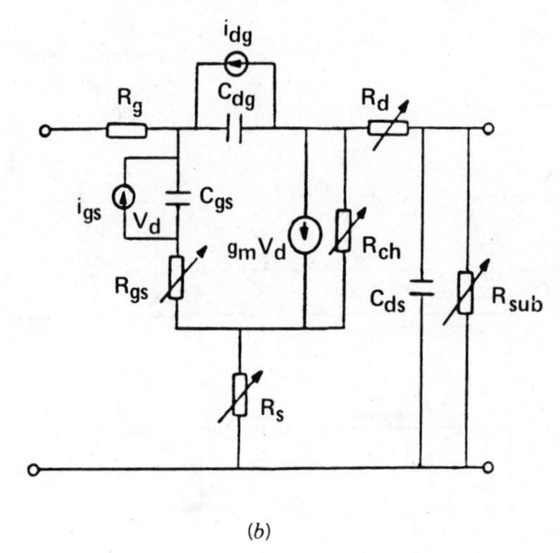

(b)

Figure 16.31 (*b*) Equivalent circuit of an optically illuminated MESFET.

resistance [R_g in Fig. 16.31(*b*)] is high. This is due to the significant influence of the photovoltage developed in the gate junction. The effect has been shown by externally connecting a large resistance (~1 MΩ) at the gate junction [32] to the gate circuit. Both the device transconductance g_m and the gate-to-source capacitance C_{gs} increase significantly (10–20%) under these conditions. The S_{11}, S_{12}, and S_{22} parameters do not change significantly. External microwave matching networks can be optimized to exploit the change in device parameters, depending upon the application. For example, a change in transconductance of the MESFET can be used to vary the gain of an amplifier. The gain variation and the time constant associated with it depend upon the device parameters (C_{gs}, R_{gs}, etc.), the matching circuit, and the bias condition. The largest gain variation occurs when the FET gate is biased close to the pinchoff ($V_{gs} \simeq V_p$). A few microwatts of optical power from an inexpensive LED can be used to vary gain up to 20 dB in a typical FET. Since the input and output impedance of the device change very little with illumination, it is possible to design matching circuits using conventional *S*-parameters of the device.

16.5.2 Oscillator Tuning

FET oscillators can be tuned by varying the gate-to-source capacitance of the device using optical illumination. A few nanowatts of coupled optical power is sufficient to tune an *X*-band oscillator over a 10% bandwidth [32]. The source series feedback configuration is best suited for this application because the gate circuit, being the frequency-discriminating element, benefits the most from a change in the gate-to-source capacitance.

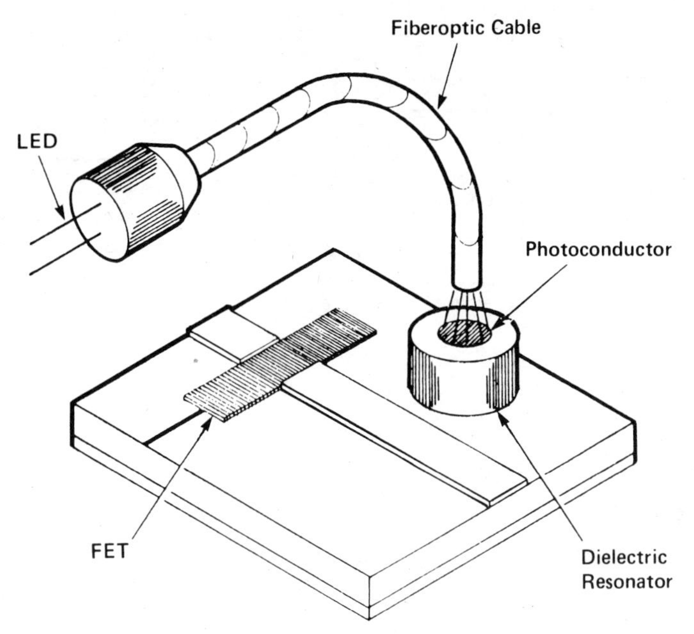

Figure 16.32 Optically controlled DRO.

However, the tuning of an oscillator using optical illumination is associated with a change in oscillator power output, too. This can be compensated in the design of matching networks. Optical illumination of a FET also improves the FM noise of the oscillator, probably as a result of the trapping and detrapping process in different regions of the device.

An alternate scheme to tune and FM modulate an oscillator optically alters the microwave network by incorporating a photosensitive material into the circuit and controlling it with light. For example, Fig. 16.32 shows the implementation for tuning and modulating a dielectric resonator oscillator [33]. Here a photosensitive material is deposited on top of the dielectric resonator so that it can be illuminated from the top. The incident optical signal changes the conductivity of the sample dielectric resonator, perturbing its electromagnetic fields, and thus resulting in a shift of the center frequency of the DRO.

16.5.3 Injection Locking

Optical injection locking of multiple free-running oscillators is useful in active phased arrays and other ESM, ECM, and ECCM systems that require coherent detection. A free-running MESFET oscillator is injection locked when the light incident on the active region of the device is modulated at a frequency close to the frequency of the free-running

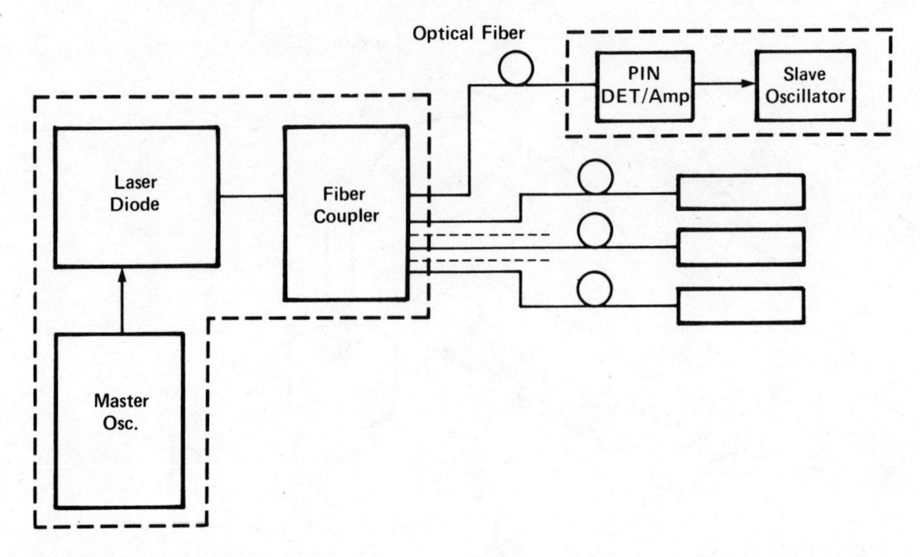

Figure 16.33 Injection locking of multiple oscillators using indirect optical techniques.

oscillator. The locking range of the oscillator depends strongly on the optical power absorbed in the active region of the device. It is given by [32]

$$2\Delta\omega = \frac{\omega}{Q}\frac{g}{C_{gs}}\frac{|IL|}{(2G_L P_{out})^{1/2}} \tag{16.21}$$

where $|IL|$ is the optically injected current, Q is quality factor of the gate circuit, G_L is load conductance presented to the FET, ω is the frequency of operation, and P_{out} is the power output from the oscillator. For a typical FET this is of the order of a few megahertz, but it can be increased by more efficient coupling of the RF-modulated light into the device.

Direct injection locking of a FET oscillator, by illuminating the active device with a modulated optical signal, is limited up to the relaxation oscillation frequency of the optical source. Subharmonic modulation of the light signal to lock an oscillator at a frequency higher than the relaxation frequency of the laser has met with limited success.

An alternate indirect-optical injection-locking scheme shown in Fig. 16.33 has been used to lock oscillators at X band and above [34]. This technique can extend the injection-locking frequency range well into the millimeter-wave range. The locking signal is harmonically generated, from the RF-modulating signal, due to nonlinearities that exist in an optical fiber link (in laser diode, fiber, and the detector). The laser diode bias is optimized to generate the maximum amplitude of the desired harmonic, depending upon the frequency of the modulating signal and the frequency of the oscillator to be locked. This technique can be easily extended to lock

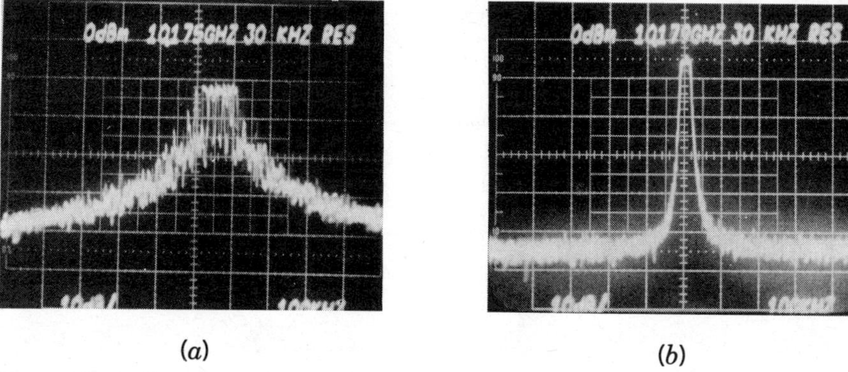

(*a*) (*b*)

Figure 16.34 Spectrum of (*a*) a free-running and (*b*) an injection-locked oscillator. (After P. Wahi et al. [34], 1986 IEEE, reprinted with permission.)

multiple free-running oscillators simultaneously from the same master oscillator. The optically modulated signal is divided into the desired number of parts (equal to the number of free-running oscillators to be locked) using optical power dividers and transmitted via separate fibers to individual oscillators. The injection-locking bandwidth decreases as the power is divided to lock more oscillators, since the locking range is proportional to the incident optical power. Figure 16.34 shows the results for two oscillators locked at around 10 GHz using a master oscillator that modulated the laser diode at around 3.3 GHz [34]. The indirect injection-locking technique has been used to lock multiple oscillators up to 35 GHz [35].

16.6 OPTICAL TECHNIQUES FOR MILLIMETER-WAVE CIRCUITS

Several microscopic optical components currently developed using integrated optical techniques are functionally and conceptually similar to microwave components. Integrated optics has miniaturized many of the optical components to the point that they can be integrated on a single substrate just like microwave supercomponents. The bulk of the components are developed using dielectric waveguide technology. At high millimeter- and submillimeter-wave frequencies conventional microwave integrated-circuit techniques become less applicable because they become unbearably small and very lossy and difficult to manufacture. Dielectric waveguide techniques developed for integrated optical circuits offer an attractive alternative for millimeter- and submillimeter-wave components [36, 37]. Although the principles of propagation remain unchanged regardless of the frequency of operation, some modifications of the structures may be needed for dielectric waveguides at millimeter- and submillimeter-wave frequencies. The circuit structure is important to minimize radiation losses

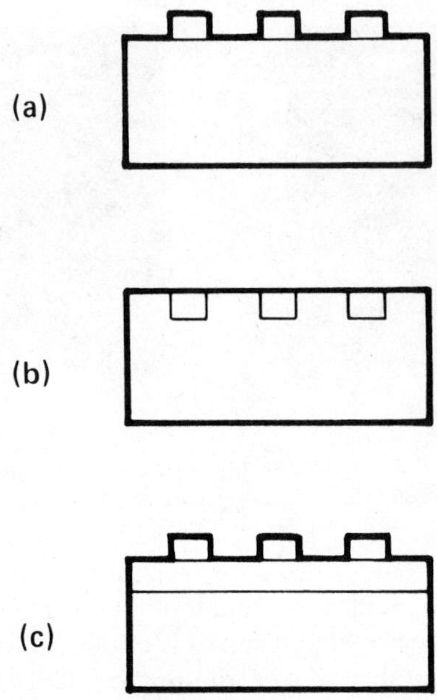

Figure 16.35 Channel waveguide cross sections. (*a*) Raised ridge. (*b*) Embedded channel. (*c*) Optical strip line.

and to maximize circuit performance. While two-dimensional thin films may be useful for integrated optical circuits, channel waveguides that function like microwave rectangular guides are really needed in millimeter-wave circuits. Raised ridge, embedded channel, and optical strip-line structures shown in Fig. 16.35 represent three different fabrication techniques. A raised ridge is made by photomasking and etching a thin-film waveguide, an embedded channel guide is fabricated by diffusion and implantation, and the optical strip line is an analog of dielectric microstrip line in which the energy is trapped beneath the strip rather than in the defined channel. Dielectric waveguides for millimeter-wave applications are made from materials having dielectric constants ranging from 2.5 to 15. This is different from waveguides used for optical integrated circuits, where the difference in dielectric constant for material used for the core region and the cladding (surrounding) region is only a few percent at most. The high dielectric-constant material used for millimeter-wave applications ensures concentration of the energy in the rod. Table 16.3 lists the dielectric constant and loss tangent for some of the materials suitable for millimeter- and submillimeter-wave dielectric waveguides [38]. For a material having a loss tangent of around 10^{-4} and a relative dielectric constant of 2, the attenuation constant for plane waves in the bulk material is less than 1.3 dB/m at 100 GHz, which is better than 3 dB/m for conventional metallic waveguides.

TABLE 16.3 Properties of Dielectric Materials for Millimeter and Submillimeter Waveguide Applications [38]

Material	Relative Dielectric Constant	Test Frequency (GHz)	Loss Tangent
Alumina	9.7	10	2×10^{-4}
KRS-5	30.5	94.75	1.9×10^{-2}
KRS-6	28.5	94.75	2.3×10^{-3}
$LiNbO_3$	6.7	94.75	8×10^{-3}
Polyethylene	2.3	50	10^{-4}

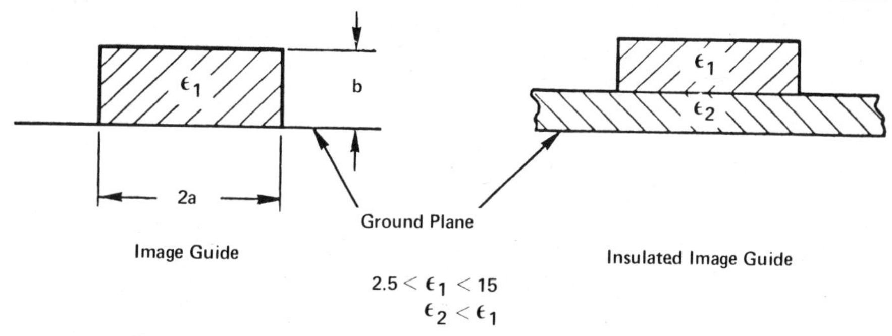

Figure 16.36 Dielectric waveguide for millimeter-wave circuits.

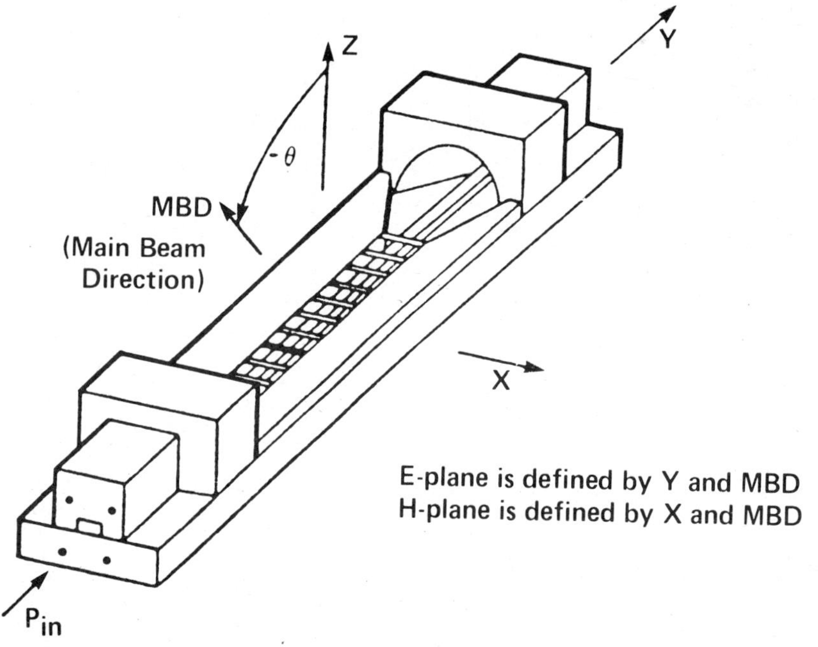

Figure 16.37 Leaky wave antenna made of a trapped image guide. (After T. Itoh [39], reprinted with permission of SPIE.)

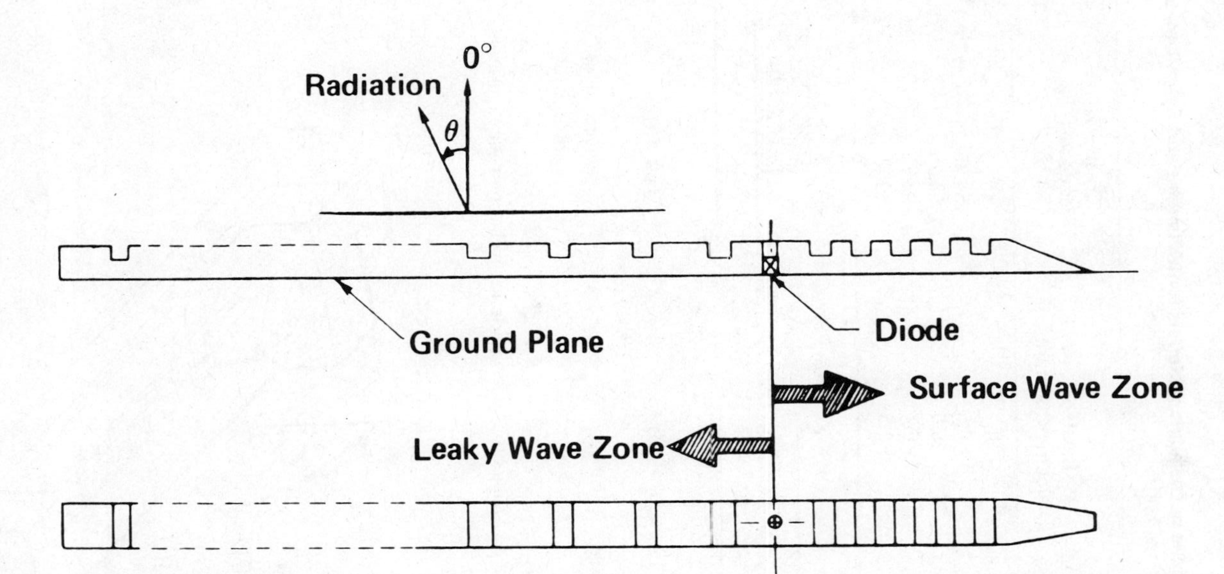

Figure 16.38 Distributed Bragg reflector Gunn oscillator. (After T. Itoh [39], reprinted with permission of SPIE.)

Image guides, and insulated image guide shown in Fig. 16.36, have also been used for millimeter-wave circuits. Both passive and active millimeter-wave components have been fabricated using these dielectric guide configurations. The most common of all circuits is a directional coupler using proximity for coupling energy between two parallel guides. Filters and leaky wave antennas have borrowed grating structure techniques from optical circuits [39]. Figure 16.37 shows a leaky wave antenna fabricated using trapped image guide. In this configuration, one can change the direction of the main beam by changing the frequency. Distributed Bragg reflected techniques have provided an alternative to conventional millimeter-wave generation techniques using two-terminal diodes. Figure 16.38 shows a schematic of a distributed Bragg reflector Gunn oscillator [39]. The band-reject nature of the grating placed on both sides of the device provides the frequency-sensitive feedback. Efforts have also been made in developing isolators using dielectric waveguide technology at millimeter-wave frequencies. A comprehensive treatment of millimeter-wave circuits and components is given in Reference 40.

REFERENCES

1. Cherin, A. H., *An Introduction to Optical Fibers*, McGraw-Hill, New York, 1982.
2. Kressel, H., "Semiconductor Devices for Optical Communication," in *Topics in Applied Physics*, Vol. 39, Springer-Verlag, New York, 1982.
3. Yariv, A., *Optical Electronics*, 3rd ed., Holt, Rinehart & Winston, New York, 1985.

3a. Brinbaum, G., *Optical Masers*, Academic Press, New York, 1964, p. 91.

4. Lau, K. Y., et al., "Direct Amplitude Modulation of Short Cavity GaAs Lasers Up to *X*-band Frequencies," *Appl. Phys. Lett.*, Vol. 43, July 1983, pp. 1–3.
5. Takamiya, S., et al., "AlGaAs TJS Lasers with Very Low Failure Rate of 1 Percent at Estimated Operating Time of 400000 Hours at 25 Degrees Centigrade," *Proc. 6th European Conf. Optical Communication*, York, England, September 16–19, 1980, pp. 187–190.
6. Yonezu, H. O., et al., "AlGaAs Window Structure Laser," *IEEE J. Quantum Electronics*, Vol. QE-15, 1979, pp. 775–781.
7. Blauvelt, H., et al., "Large Optical Cavity AlGaAs Buried Heterostructure Window Laser," *Appl. Phys. Lett.*, Vol. 40, 1982, p. 1029.
8. Bar-chaim, N., and A. Hilbert, "Semi-insulating Semiconductor Materials," *Lasers Appl.*, Nov. 1983, pp. 51–54.
9. Angelov, I., et al., "Investigation of Microwave Properties of GaAlAs Injection Laser," *Proc. 13th European Microwave Symp.* (Nurenberg, W. Germany), Sept. 1983, pp. 494–497.
10. Tucker, R. S., and D. J. Pope, "Microwave Circuit Models of Semiconductor Injection Lasers," *IEEE Trans. Microwave Theory Tech.*, Vol. MTT-31, Mar. 1983, pp. 289–294.

11. Tucker, R. S., "Large Signal Circuit Model for Simulation of Injection-Laser Modulation Dynamics," *Proc. Inst. Elec. Eng.*, Part 1, Vol. 128, Oct. 1981, pp. 180–184.
12. Su, C. B., et al., "17 GHz Direct Modulation Bandwidth and Impedance Characteristics of Vapor Phase Regrown 1.3 μm LnGaAsρ Buried Heterostructure Lasers," *Proc. SPIE*, Vol. 545 (*Optical Technology for Microwave Applications II*, S. K. Yao, Ed.), Virginia, Apr. 9–10, 1985.
13. Wang, S. Y., "Ultra High Speed Photodiode," *Laser Focus/Electro Optics*, Dec. 1983, pp. 99–106.
14. Sapriel, J., *Acousto Optics*, Wiley, New York, 1984.
15. Peters, L. C., "Gigacycle Bandwidth Coherent Light Traveling Wave Phase Modulators," *Proc. IEEE*, Vol. 51, 1943.
16. Gee, C. M., and G. D. Thurmond, "Wideband Travelling-Wave Electro-Optic Modulator," *Proc. SPIE*, Vol. 477 (*Optical Technology for Microwave Applications*, S. K. Yao, Ed.), Virginia, May 1984, pp. 17–22.
17. Muoi, T. V., "Single Mode Fiber Recirculating Delay Line," *Proc. SPIE*, Vol. 477 (*Optical Technology for Microwave Applications*, S. K. Yao, Ed.), Virginia, May 1984, pp. 57–64.
18. Wallington, J. R., and J. M. Griffin, "Optical Techniques for Signal Distribution in Phased Arrays," *GEC J. Res.*, Vol. 2, 1984, pp. 62–75.
19. Chang, I. C., and L. S. Lee, "Acousto-Optic Bragg Cells for EW Applications," *Proc. SPIE*, Vol. 477 (*Optical Technology for Microwave Applications*, S. K. Yao, Ed.), Virginia, May, 1984, pp. 23–28.
20. Yao, S. K., et al., "Wideband ZnO Phased Array Techniques," *Proc. SPIE*, Vol. 477 (*Optical Technology for Microwave Applications*, S. K. Yao, Ed.), Virginia, May 1984, pp. 29–35.
21. Dixon, R., "Acoustic Diffraction of Light in Anisotropic Media," *IEEE J. Quantum Electron.*, Vol. QE-3, Jan. 1967, pp. 85–93.
22. Yao, S. K., "Wideband Bragg Cell Efficiency Enhancement Techniques," *Proc. SPIE*, Vol. 545 (*Optical Technology for Microwave Applications*, S. K. Yao, Ed.), Virginia, Apr. 1985, pp. 72–79.
23. Lee, L. S., and I. C. Chang, "Acousto Optic Bragg Cell Techniques," *Proc. SPIE*, Vol. 545 (*Optical Technology for Microwave Applications*, S. K. Yao, Ed.), Virginia, Apr. 1985, pp. 68–71.
24. Freye, F., et al., "A Bragg Cell-Optical Heterodyne Channelized Rf Receiver," *Proc. SPIE*, Vol. 330 (*Optical Technology for Microwave Applications*, S. K. Yao, Ed.), Jan. 1982.
25. Chen, T. S., et al., "Investigation of Large Dynamic Range Heterodyne Acousto Optic Spectrum Analyzer," *Proc. SPIE*, Vol. 477 (*Optical Technology for Microwave Applications*, S. K. Yao, Ed.), Virginia, May 1984, pp. 144–149.
26. Craig, A. E., et al., "Magnetostatic Wave Optical Bragg Cell Devices," *SPIE*, Vol. 545 (*Optical Technology for Microwave Applications*, S. K. Yao, Ed.), Virginia, Apr. 1985.
27. Adam, J. D., and J. H. Collins, "Microwave Magnetostatic Delay Devices Based on Epitaxial Yittrium Iron Garnet," *Proc. IEEE*, Vol. 64, May 1976, pp. 794–800.

28. Gatenby, P. V., and R. J. Sadler, "Acousto-Optic Signal Processing," *GEC J. Res.*, Vol. 2, 1984, pp. 88–95.
29. Chen, W., et al., "Optically Controlled IMPATT Diodes," *Proc. SPIE*, Vol. 477 (*Optical Technology for Microwave Applications*, S. K. Yao, Ed.), Virginia, May 1984, pp. 105–108.
30. Vaucher, A. M., et al., "Optically Controlled Millimeter-Wave Devices," *Proc. SPIE*, Vol. 477 (*Optical Technology for Microwave Applications*, S. K. Yao, Ed.), Virginia, May 1984, pp. 109–113.
31. Chang, C. S., et al., "Direct dc to rf Conversion by Picosecond Optoelectronic Switching," *Proc. SPIE*, Vol. 477 (*Optical Technology for Microwave Applications*, S. K. Yao, Ed.), May 1984, pp. 101–104.
32. DeSalles, A. A. A., "Optical Control of GaAs MESFETs," *IEEE Trans. Microwave Theory Tech.*, Vol. MTT-31, Oct. 1983, pp. 812–820.
33. Herczfeld, P. R., et al., "Optically Tuned and FM Modulated *X*-Band Dielectric Resonator Oscillator," *Proc. 14th European Microwave Conf.* (Leige, Belgium), Sept. 1984.
34. Wahi, P., et al., "Comparison of Indirect Optical Injection-Locking Techniques of Multiple *X*-Band Oscillators," *IEEE, MTT-S Int. Microwave Symp. Digest* (Baltimore, Md.), June 1986, pp. 615–618.
35. Daryoush, A., "Optical Beam Control of Millimeter Wave Phased Array Antennas for Communications," *Microwave J.*, Vol. 30, Mar. 1987, pp. 97–104.
36. Rudokas, R., and T. Itoh, "Passive Millimeter Wave IC Components Made of Inverted Strip Dielectric Waveguides," *IEEE Trans. Microwave Theory Tech.*, Vol. MTT-29, Dec. 1976, pp. 978–981.
37. Yamamoto, K., "A Novel Low Loss Dielectric Waveguide for Millimeter and Submillimeter Wavelength," *IEEE Trans. Microwave Theory Tech.*, Vol. MTT-28, 1980, pp. 580–584.
38. Yeh, C., and F. Shimabukura, "Design of New Millimeter Waveguide Using Optical Concepts", *Proc. SPIE*, Vol. 545 (*Optical Technology for Microwave Applications*, S. K. Yao, Ed.), Apr. 1985, pp. 45–49.
39. Itoh, T., "Recent Developments of Dielectric Waveguide Technology," *Proc. SPIE*, Vol. 477 (*Optical Technology for Microwave Applications*, S. K. Yao, Ed.), May 1985, pp. 90–93.
40. Bhartia, P., and I. J. Bahl, "Millimeter Wave Engineering and Applications," Wiley, New York, 1984.

PROBLEMS

16.1 Determine the photon lifetime of a laser having a 150-μm long cavity. The distributed internal loss in the cavity is $10\,\text{cm}^{-1}$ and the velocity is 5×10^{10} cm/sec. The end mirror reflectivity is equal to 0.3.

16.2 Calculate the relaxation resonance frequency of the laser having a photon density of 1.2×10^{10} photons/cm^3, optical gain coefficient equal to 2×10^{-6} cm^3/sec, and photon lifetime of 1 psec.

16.3 Calculate the density of light emitted at 8500 Å from a laser having photon density of 1.2×10^{15} photons/cm^3 and group velocity of 5×10^{10} cm/sec. The end mirror reflectivity is equal to 0.3.

16.4 Determine the degradation in signal-to-noise ratio of a recirculating delay line, after five circulations, if the loop gain is 0.95.

16.5 Determine the Bragg angle for 8300-Å light interacting with a 200-MHz RF signal in lithium niobate (refractive index = 2.20). Calculate the Bragg diffraction bandwidth for an interaction length of 1 cm.

17 FUTURE TRENDS IN MICROWAVE CIRCUITS

17.1 MONOLITHIC MICROWAVE INTEGRATED CIRCUITS

During the past two decades, remarkable advances have been made in the integration of microwave circuits using hybrid and monolithic technologies. After hybrid microwave integrated circuits (MICs), monolithic MICs (MMICs) have emerged as another exciting technology that is expected to exert a profound impact on future microwave systems in the military and commercial markets. During the 1980s, excellent progress has been made by this technology in the 1–30 GHz frequency range. One of the driving forces behind such new advances is the requirement for millions of identical circuits with good amplitude and phase tracking for transmit/receive (T/R) modules, which are the key elements in active phased-array radars. A low-cost 12-GHz direct broadcast satellite (DBS) home receiver is another potential example of the need for high volume production of MMICs.

Multipurpose broadband MMICs have received special attention recently. A market appears to be developing for a set of generic MMICs to serve such functions as amplification, mixing, and attenuation. Distributed amplifiers are capable of providing 4–5 dB gain over the 2–40 GHz range. There is every expectation that in the near future the upper frequency limit of such amplifiers will be extended beyond 50 GHz.

In the future, the use of a combination of monolithic and hybrid circuits in an optimal fashion is expected in systems. Some systems will be completely monolithic or completely hybrid, but a large percentage of these will consist of MMICs in a MIC hybrid environment. With the combined strength of existing hybrid MIC technology, MMIC technology will move rapidly in four major areas: (1) newly developing high-volume microwave markets, such as DBS, collision avoidance radar, global positioning system (GPS) receivers and transmitters, expendable electronic countermeasure (ECM) decoys, phased-array radar T/R modules (low cost and high reliability are the key factors in this area); (2) special-purpose microwave markets, such as satellite, aircraft, and ground-mobile applications (here small size, low weight, and improved performance are the essential features); (3) commercial instrument market, such as in complex test instrumentation and telecommunications where combining analog and digital functions yield a significant performance (cost) advantage; and (4) the emerging millimeter-wave market, above 20 GHz where MMICs reduce circuit interconnection parasitic problems compared with conventional MIC technology (low cost and improved performance are the primary benefit for this area).

17.1.1 MMIC Subsystems

Microwave subsystem designs during the last forty years have evolved in their construction from waveguide, coaxial, strip line, and microstrip. The concept of implementing subsystems on a single chip in the microwave area is becoming a reality now. A few examples of such systems are T/R modules, digital RF memories, DBS, and optical *pin*-FET receivers.

A phased-array radar T/R module consists of a low-noise amplifier, phase shifter, attenuator, power amplifier, and low- and high-power T/R switches. The T/R module is one of the most important military applications for MMICs.

Recent advances in ultrahigh-speed digital ICs have made it possible to develop a new component known as the digital RF memory (DRFM) for ECM applications. In this device, the incoming RF signal is down-converted with a fixed local oscillator, sampled at the baseband frequency by an analog-to-digital (A/D) converter at approximately gigahertz rates, and stored in a high-speed digital semiconductor memory from which it can be read at more leisurely rates by a digital computer. Because of the gigahertz sampling rate, this process makes possible the generation, storage, and reproduction of RF signals with wide bandwidths. The availability of both high-frequency and high-speed logic GaAs ICs affords lower cost and better performing DRFMs owing to very high levels of possible integration.

DBS receivers are one of the most important commercial applications of MMICs. A DBS receiver comprises four microwave circuits: a front-end preamplifier, a mixer, an IF amplifier, and a local oscillator. At 12 GHz, total system noise figure of less than 3 dB and gain above 35 dB are achievable from a 2×2-mm MMIC chip.

In addition to monolithically integrated microwave functions, a significant amount of development work is taking place to monolithically integrate optical and electrical functions. An example of such a component is a short-wavelength (<0.89-μm) optical receiver that includes a photodetector, a microwave preamplifier, a postamplifier, and a digital clock restoration circuit. Optoelectronic IC technology will open up a whole range of new applications.

17.1.2 Millimeter-Wave Monolithic ICs

Millimeter-wave technology is an important area whose growth is dependent on the monolithic approach. A major trend in this technology will be large-scale integration, that is, a complete subsystem on a single chip. Cost, size, and performance factors are driving this trend. At present most millimeter-wave monolithic IC efforts have been in the integration of Schottky-diode mixers and low-noise IF amplifiers on a single GaAs chip. Balanced mixers using Schottky diodes have been demonstrated up to 94 GHz with acceptable performance. Low-noise amplifiers using 0.2–0.3 μm gate-length FETs have been demonstrated up to 60 GHz. Extension of these components up to 100 GHz awaits the development of suitable three-terminal devices. Above 100 GHz, GaAs will most probably give way to InP technology and quasi-optical dielectric transmission media.

The general future trends are as follows.

- Development of millimeter-wave monolithic integrated receivers,
- Development of millimeter-wave monolithic integrated transmitters with GaAs power FET amplifiers replacing Gunn and IMPATT amplifiers,
- Development of new three-terminal devices.

17.2 OPTICS FOR MICROWAVE APPLICATIONS

Advances in laser technology, high-speed modulation of light, and optical interaction with solid state microwave devices have created the potential for the integration of optics in microwave circuits for many applications. Current developments in the general area of optoelectromagnetics include: microwave fiber optics and optical control of microwave devices.

17.2.1 Microwave Fiber Optics

Fiber-optic assemblies are finding their way into an ever-increasing range of applications that have traditionally been the domain of microwave technology. Typical applications include:

1. fiber-optics communication links,
2. phased-array scanning,
3. lightweight connection of antennas to receivers,
4. missile guidance,
5. spectrum analyzers,
6. signal processing,
7. ultrawide-band communications.

Major factors in the rapidly growing popularity of fiber optics for microwave applications are: very low loss (3 dB/km at 0.85-μm and less than 0.5 dB/km at 1.3-μm wavelength, independent of microwave modulation frequency), light weight, immunity to RF interference, electromagnetic pulse (EMP) and lower vulnerability to radiation, very low electrical noise, nonconductive (no coupling with other electrical circuitry), high flexibility, and higher operating temperatures.

Improvement in modulation capability of laser diodes is pushing fiber-optic technology toward millimeter-wave frequencies. Modulation frequencies up to 20 GHz will be possible for direct modulation of the laser diode, and over 40 GHz for indirect, evanescent coupled modulation. Optical fibers can transmit microwave signals up to several hundred gigahertz, and 100-GHz detector capability has already been demonstrated.

17.2.2 Optical Control of Microwave Devices

Optical control of microwave solid state devices can be utilized in a variety of ways. Several RF control functions such as switching, limiting, phase shifting, phase locking, oscillator tuning and the gain control of amplifiers, have already been demonstrated by illuminating microwave semiconductor devices with focused optical beams. A number of experiments have been carried out with devices such as IMPATTs, TRAPATTs, bipolar transistors, and MESFETs, but most interest now centers on MESFETs because of their wide future applications. Changes in MESFET gate–source capacitance may be used to alter oscillation frequency. Changes in channel resistance may be used to provide switching action or gain control in amplifiers. Optically generated currents in MESFETs may be used to provide an injection-locking signal. Injection locking of oscillators by optical means is expected to be superior to the conventional complex and expensive method of frequency stabilization. The injection of light in effect introduces an extra terminal to the device. An optically controlled MESFET oscillator working at 10.174 GHz has a phase noise of about −78 dBc/Hz at 100 kHz from the carrier.

Optical tuning of MMICs is an important future topic as optical devices, optical guiding structures, and MMICs can all be fabricated on the same GaAs substrate.

17.3 MICROWAVE ACOUSTIC TECHNOLOGY

Advances in microwave acoustic technology and its usage in commercial and military electronic systems have been steadily accelerating for the past two decades. Progress has been remarkable not only in the surface-acoustic-wave (SAW) area but in bulk-acoustic-wave (BAW) technology as well. This progress has resulted in higher frequencies of operation, and more efficient and compact devices for more demanding systems applications, primarily as a result of submicron photolithography. Filters, dispersive filters, delay lines, dispersive delay lines, resonators, and convolvers have been realized using SAW devices. When combined with hybrid and integrated-circuit technologies, they have been configured into programmable filters, programmable tapped delay lines, oscillators, memory correlators, frequency synthesizers, and acousto-optic devices.

SAW filters working at 4.4 GHz have been developed and the upper frequency limit for SAW/BAW bandpass filters is projected to be about 5 GHz. SAW delay lines could be realized up to 2 GHz, while BAW delay lines can be operated up to 20 GHz. SAW resonators represent another area of SAW technology, which is gaining popularity in L- and S-band oscillators. These resonators have been demonstrated with Q-values in excess of 50,000 at 2 GHz. SAW oscillators can be constructed using either SAW resonators or delay lines. The resonator approach achieves high Q, while the delay-line approach provides a wider tuning range. Significant recent improvements are in higher frequency of operation, better long-term aging and improved temperature stability. SAW oscillators have been reported above 1 GHz and BAW oscillators up to 5.2 GHz.

Emerging acousto-optic technology will greatly increase the processing capability of equipment having electronic warfare applications. Wide bandwidth and linear processing characteristics are features of acousto-optics that make this technology attractive for signal processing in dense electromagnetic environments. Small size, light weight, reduced cost, and lower fabrication complexity are the other attractions of acousto-optics technology.

17.4 MAGNETOSTATIC-WAVE TECHNOLOGY

Magnetostatic-wave (MSW) technology has opened a new door to high-frequency signal-processing devices to supplement the SAW technology. Filters, transversal filters, delay lines, resonators, directional couplers, tunable oscillators, and correlators have been realized using MSW technology.

MSWs propagate in three different modes depending on the direction of the magnetic field; the forward volume MSW, the backward volume MSW, and the magnetostatic surface wave. The frequency of operation is controlled by the magnitude of the magnetic field and therefore these devices can be magnetically tuned. A magnetostatic surface-wave tunable bandpass filter has achieved usable tunability range from 2 to 3.5 GHz. A 3.4–7 GHz FET tunable oscillator using a magnetostatic volume wave resonator with a low FM noise of -90 dBc/Hz at 10 kHz from the carrier has been reported. MSW has the advantage of having higher loaded Q than the YIG sphere technology ($\simeq 1000$ as compared to 200). Oscillators using planar technology are possible replacements for YIG-tuned oscillators.

To date, much of the work on MSW devices has been performed in the 1–10 GHz range, although oscillators and filters can be built to operate at frequencies in excess of 20 GHz. Since there is little commercial interest in MSW devices above 20 GHz, millimeter-wave MSW devices are not expected to be developed in the near future.

APPENDIX A
UNITS AND SYMBOLS

A.1 SI UNITS AND THEIR SYMBOLS

In 1960 the International System of Units was established as a result of a long series of international discussions. This modernized metric system, called SI, from the French name, Le Systéme International d'Unités, is now, as a general world trend, to replace all former systems of measurement, including former versions of the metric system.

In the SI system, four physical quantities are classified as fundamental: length, mass, time, and charge. For practical purposes, temperature is included here as a basic unit. In Table A.1 the first five are basic quantities

TABLE A.1 SI Units and Their Symbols

	SI		
Quantity	Unit	Symbol	Dimensions
Length	meter	m	basic
Mass	kilogram	kg	basic
Charge	coulomb	C	basic
Time	second	s	basic
Temperature	kelvin	K	basic
Frequency	hertz	Hz	1/s
Energy	joule	J	$kg \times m^2/s^2$
Force	newton	N	$kg \times m/s^2$
Power	watt	W	J/s
Pressure	pascal	Pa	N/m^2
Electric-current	ampere	A	C/s
Electric-potential (voltage)	volt	V	J/C
Electric field	volts/meter	V/m	J-m/C
Resistance	ohms	Ω	V/A
Resistivity	ohms-meter	Ω-m	V-m/A
Conductance	siemens	S	A/V
Capacitance	farad	F	C/V
Permittivity	farads/meter	F/m	F/m
Magnetic field	amperes/meter	A/m	A/m
Inductance	henry	H	$V \times s/A$
Permeability	henrys/meter	H/m	H/m

TABLE A.2 SI Prefixes

Prefix	Symbol	Factor by Which the Unit is Multiplied
exa	E	10^{18}
peta	P	10^{15}
tera	T	10^{12}
giga	G	10^{9}
mega	M	10^{6}
kilo	k	10^{3}
hecto	h	10^{2}
deca	da	10^{1}
		10^{0}
deci	d	10^{-1}
centi	c	10^{-2}
milli	m	10^{-3}
micro	μ	10^{-6}
nano	n	10^{-9}
pico	p	10^{-12}
femto	f	10^{-15}
atto	a	10^{-18}

and the rest are derived quantities, that is, their dimensions can be expressed as a combination of the first five.

A.2 METRIC PREFIXES

The nomenclature in this decimal structure is derived from a system of prefixes, which are attached to units of all sorts. For example, the prefix "kilo" means 1000, hence kilometer, kilogram, and kilowatt mean 1000 meters, 1000 grams, and 1000 watts, respectively. Most of our everyday experiences with metric units will involve some of the prefixes listed in Table A.2.

A.3 DECIBEL UNITS

The ratio of signals between the output and input ports of a network is expressed in decibels and its absolute powers are measured in dBm or dBW.

The Decibel (dB). The decibel is a logarithmic unit of power ratio, although it is commonly also used for current ratio and voltage ratio. If the input power P_i and the output power P_o of a network are expressed in the same units, then the network insertion gain or loss is

$$G = 10 \log \frac{P_o}{P_i} \text{ dB} \tag{A.1}$$

For example, if $P_i = 5$ W and $P_o = 20$ W, then $G = 10 \log 4 = 6$ dB, that is, power gain of 6 dB. If $P_i = 5$ W and $P_o = 2.5$ W, then $G = 10 \log 0.5 = -3$ dB, and the network is said to have a power loss of 3 dB.

The dBm and dBW. The absolute power levels of a network are expressed in dBm, which is defined as the power level P in reference to 1 mW, that is,

$$P(\text{dBm}) = 10 \log \frac{P(\text{mW})}{1 \text{ mW}} \tag{A.2}$$

Thus $P = 1 \text{ mW} = 0 \text{ dBm}$, $P = 100 \text{ mW} = 20 \text{ dBm}$, and $P = 0.5 \text{ mW} = -3 \text{ dBm}$. If power unit reference is 1 W, the decibels are expressed in dBW.

APPENDIX B PHYSICAL CONSTANTS AND OTHER DATA

Permittivity of vacuum, $\epsilon_0 = 8.854 \times 10^{-12} \simeq (1/36\pi) \times 10^{-9}$ F/m
Permeability of vacuum, $\mu_0 = 4\pi \times 10^{-7}$ H/m
Impedance of free space, $\eta_0 = 376.7 \simeq 120\pi\,\Omega$
Velocity of light, $c = 2.998 \times 10^8$ m/sec
Charge of electron, $e = 1.602 \times 10^{-19}$ C
Mass of electron, $m = 9.107 \times 10^{-31}$ kg
$\eta = e/m = 1.76 \times 10^{11}$ C/kg
Mass of proton, $M = 1.67 \times 10^{-27}$ kg
Boltzmann's constant, $k = 1.380 \times 10^{-23}$ J/K
Planck's constant, $h = 6.547 \times 10^{-34}$ J-sec
10^7 ergs = 1 J
1 joule = 0.6285×10^{19} eV
1 electron volt = energy gained by an electron in accelerating through a potential of 1 V
Energy of 1 electron volt = equivalent electron temperature of 1.15×10^4 K
Electron plasma frequency

$$f_p = \frac{e}{2\pi}\left(\frac{N}{m\epsilon_0}\right)^{1/2} = 8.97\sqrt{N}\ \text{Hz}$$

where N is the number of electrons per cubic meter
Electron cyclotron frequency, $f_c = eB/2\pi m = 28.000B$ MHz for B in webers per square meter; $f_c = 2.8B$ MHz for B in gauss
10^4 gauss = 1 Wb/m^2
Conductivity of copper, $\sigma = 5.8 \times 10^7$ S/m
Conductivity of gold, $\sigma = 4.1 \times 10^7$ S/m

APPENDIX C
ABCD AND *S*-PARAMETERS

At microwave frequencies, the generalized circuit constants matrix ($ABCD$) and scattering matrix (S) methods of circuit analysis are exclusively used. Calculations are most easily made using $ABCD$ parameters because (1) lumped elements and transmission-line elements are related to the matrix elements by simple expressions, and (2) elements are cascaded simply by multiplying their matrices. On the other hand, scattering matrix formulation is a more general method of representing microwave networks and can handle three or more ports.

C.1 *ABCD* PARAMETERS

$ABCD$ parameters for a two-port network, such as shown in Fig. C.1, are defined

$$\begin{bmatrix} V_1 \\ I_1 \end{bmatrix} = \begin{bmatrix} A & B \\ C & D \end{bmatrix} \begin{bmatrix} V_2 \\ I_2 \end{bmatrix} \tag{C.1}$$

Note that I_2 is shown to flow outward and becomes I_1 of the next two-port network in a cascaded chain. The $ABCD$ matrices for commonly used microwave circuit elements are listed in Table C.1. The operations for combining $ABCD$ matrices in series and in parallel are given in Fig. C.2. $ABCD$ matrices exhibit the following characteristics.

1. For reciprocal networks,

$$AD - BC = 1$$

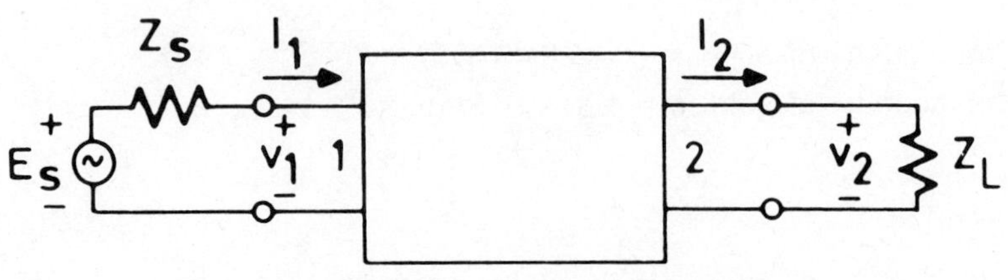

Figure C.1 A two-port network.

TABLE C.1 ***ABCD*** **Matrices of Commonly Used Two-port Ladder Type Networks**

Network	ABCD Matrix
1. A transmission line section 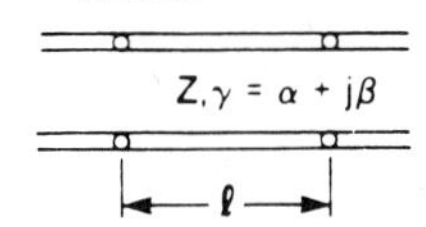	$\begin{bmatrix} \cosh \gamma \ell & Z \sinh \gamma \ell \\ \dfrac{\sinh \gamma \ell}{Z} & \cosh \gamma \ell \end{bmatrix}$
2. A series impedance 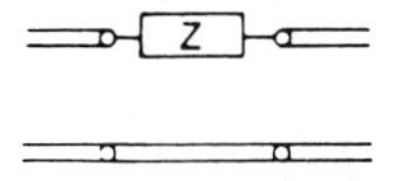	$\begin{bmatrix} 1 & Z \\ 0 & 1 \end{bmatrix}$
3. A shunt admittance 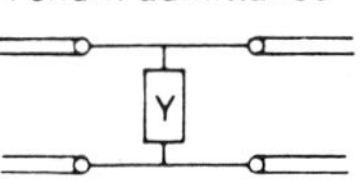	$\begin{bmatrix} 1 & 0 \\ Y & 1 \end{bmatrix}$
4. An ideal transformer 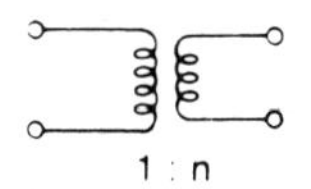	$\begin{bmatrix} \dfrac{1}{n} & 0 \\ 0 & n \end{bmatrix}$
5. π-network 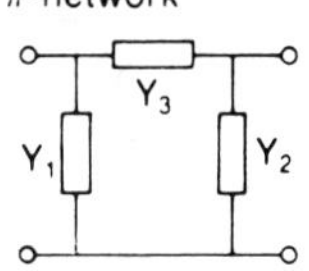	$\begin{bmatrix} 1 + \dfrac{Y_2}{Y_3} & \dfrac{1}{Y_3} \\ Y_1 + Y_2 + \dfrac{Y_1 Y_2}{Y_3} & 1 + \dfrac{Y_1}{Y_3} \end{bmatrix}$
6. T-network 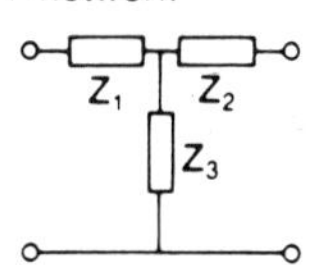	$\begin{bmatrix} 1 + \dfrac{Z_1}{Z_3} & Z_1 + Z_2 + \dfrac{Z_1 Z_2}{Z_3} \\ \dfrac{1}{Z_3} & 1 + \dfrac{Z_2}{Z_3} \end{bmatrix}$

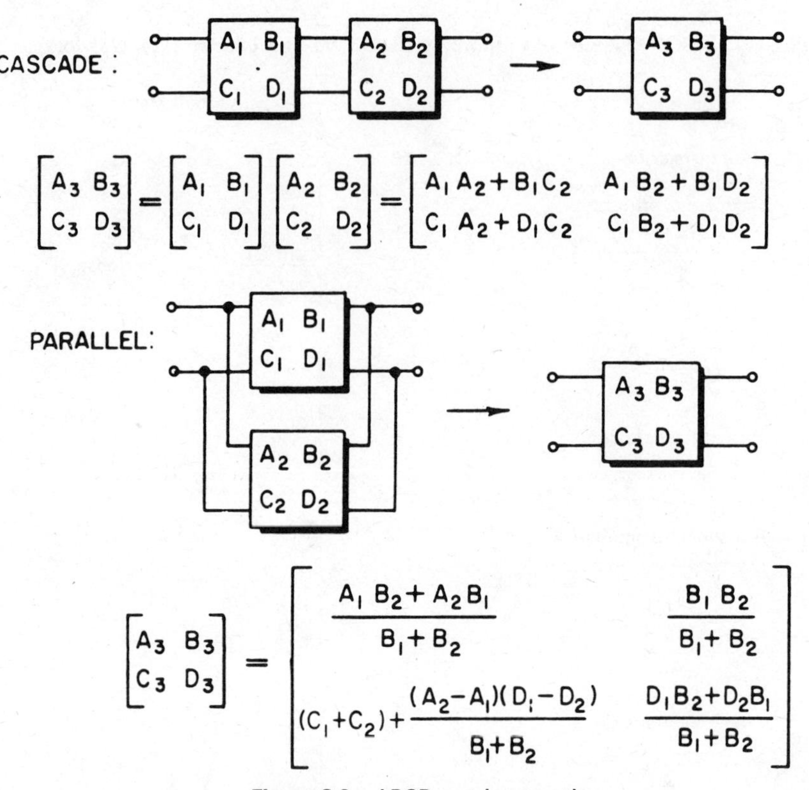

CASCADE :
A_1 B_1
C_1 D_1
A_2 B_2
C_2 D_2
A_3 B_3
C_3 D_3
$\begin{bmatrix} A_3 & B_3 \\ C_3 & D_3 \end{bmatrix} = \begin{bmatrix} A_1 & B_1 \\ C_1 & D_1 \end{bmatrix} \begin{bmatrix} A_2 & B_2 \\ C_2 & D_2 \end{bmatrix} = \begin{bmatrix} A_1A_2+B_1C_2 & A_1B_2+B_1D_2 \\ C_1A_2+D_1C_2 & C_1B_2+D_1D_2 \end{bmatrix}$
PARALLEL:
A_1 B_1
C_1 D_1
A_2 B_2
C_2 D_2
A_3 B_3
C_3 D_3
$\begin{bmatrix} A_3 & B_3 \\ C_3 & D_3 \end{bmatrix} = \begin{bmatrix} \dfrac{A_1B_2+A_2B_1}{B_1+B_2} & \dfrac{B_1B_2}{B_1+B_2} \\ (C_1+C_2)+\dfrac{(A_2-A_1)(D_1-D_2)}{B_1+B_2} & \dfrac{D_1B_2+D_2B_1}{B_1+B_2} \end{bmatrix}$

Figure C.2 *ABCD* matrix operations.

2. For symmetrical networks (which remain unaltered when the two ports are interchanged), we have $A = D$.

C.2 *S*-PARAMETERS

The use of *ABCD* parameters at microwave frequencies is not very convenient from the measurements point of view. Also, its main advantage of cascading network components does not hold when the network consists of components with three or more ports and when the topology is different.

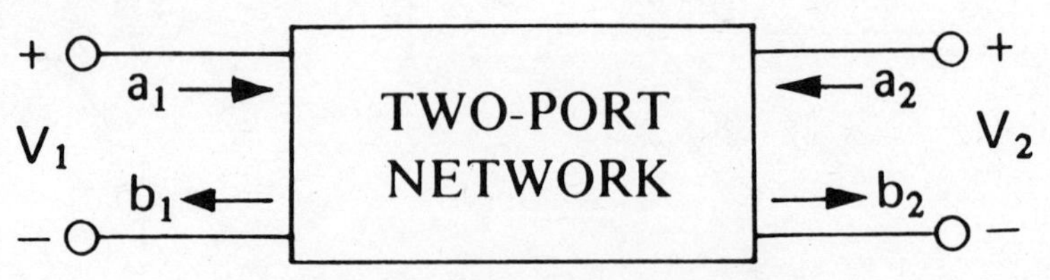

Figure C.3 *S* parameter representation.

Scattering matrix formulation is a more general method for representing microwave networks.

A scattering matrix represents the relationship between variables a_n (proportional to the incoming wave at the nth port) and variables b_n (proportional to the outgoing wave at the nth port) defined in the following manner

$$a_n = \frac{v_n^+}{\sqrt{Z_{0n}}} \tag{C.2}$$

$$b_n = \frac{v_n^-}{\sqrt{Z_{0n}}} \tag{C.3}$$

where v_n^+ and v_n^- represent voltages corresponding to the incoming and the outgoing waves in the transmission line (or the waveguide) connected to the nth port, and Z_{0n} is the characteristic impedance of the line (or waveguide). Knowledge of v_n^+ and v_n^- is not required to evaluate coefficients of the scattering matrix. Relationships between b_n and a_n for the two-port network shown in Fig. C.3 may be written as

$$b_1 = S_{11}a_1 + S_{12}a_2 \tag{C.4}$$

and

$$b_2 = S_{21}a_1 + S_{22}a_2 \tag{C.5}$$

In general, for an n-port network, we have

$$[b] = [S][a]$$

Some of the important characteristics of S-matrices are listed below.

1. For a reciprocal network the S-matrix is symmetrical, that is,

   $$S = S^{\mathrm{t}}$$

   where the superscript t indicates the transpose of a matrix.
2. For a lossless passive network

   $$\sum_{n=1}^{N} |S_{ni}|^2 = \sum_{n=1}^{N} S_{ni}S_{ni}^* = 1 \tag{C.6}$$

   for all $i = 1, 2, \ldots, N$.

3. Again for lossless passive networks, the power conservation condition yields an orthogonality constraint given by

$$\sum_{n=1}^{N} S_{ns} S_{nr}^{*} = 0 \tag{C.7}$$

for all s, $r = 1, 2, \ldots, N$, $s \neq r$.

The relationships between *ABCD* parameters and *S*-parameters are given in the following.

From S-Matrix to ABCD Matrix

$$A = \frac{(1 + S_{11} - S_{22} - \Delta S)\sqrt{Z_{01}/Z_{02}}}{(2S_{21})} \tag{C.8}$$

$$B = \frac{(1 + S_{11} + S_{22} + \Delta S)\sqrt{Z_{01}Z_{02}}}{(2S_{21})} \tag{C.9}$$

$$C = (1 - S_{11} - S_{22} + \Delta S)(2S_{21}\sqrt{Z_{01}Z_{02}}) \tag{C.10}$$

$$D = \frac{(1 - S_{11} + S_{22} - \Delta S)\sqrt{Z_{02}/Z_{01}}}{(2S_{21})} \tag{C.11}$$

where Z_{01} and Z_{02} are normalizing impedances for *S*-parameters at ports 1 and 2, respectively, and

$$\Delta S = (S_{11}S_{22} - S_{21}S_{12}) \tag{C.12}$$

From ABCD Matrix to S-Matrix

$$S_{11} = \frac{AZ_{02} + B - CZ_{01}Z_{02} - DZ_{01}}{AZ_{02} + B + CZ_{01}Z_{02} + DZ_{01}} \tag{C.13}$$

$$S_{12} = \frac{2(AD - BC)\sqrt{Z_{01}Z_{02}}}{AZ_{02} + B + CZ_{01}Z_{02} + DZ_{01}} \tag{C.14}$$

$$S_{21} = \frac{2\sqrt{Z_{01}Z_{02}}}{AZ_{02} + B + CZ_{01}Z_{02} + DZ_{01}} \tag{C.15}$$

and

$$S_{22} = \frac{-AZ_{02} + B - CZ_{01}Z_{02} + DZ_{01}}{AZ_{02} + B + CZ_{01}Z_{02} + DZ_{01}} \tag{C.16}$$

APPENDIX D TRANSFER FUNCTION RESPONSES

This appendix presents two popular and practical responses used in matching networks, couplers, filters, namely, the Butterworth and Chebyshev responses. Both responses can be expressed mathematically as

$$H(j\omega) = \frac{K_n}{[1 + R_n^2(\omega)]^{1/2}}, \qquad 0 \le K_n \le 1 \tag{D.1}$$

where $R_n(\omega) = \omega^n$ for Butterworth and $R_n(\omega) = \epsilon T_n(\omega)$ for Chebyshev. Here ϵ is the equiripple amplitude and $T_n(\omega)$ is the Chebyshev polynomial of order n.

D.1 BUTTERWORTH RESPONSE

A simple normalized Butterworth response is given by

$$H(j\omega) = \frac{1}{[1 + \omega^{2n}]^{1/2}} \tag{D.2}$$

Near $\omega = 0$,

$$[1 + \omega^{2n}]^{1/2} = 1 + \tfrac{1}{2}\omega^{2n} - \tfrac{1}{8}\omega^{4n} + \tfrac{1}{16}\omega^{6n} + \cdots \tag{D.3}$$

This expansion shows that the first $2n - 1$ derivatives are zero at $\omega = 0$. Since $R_n(\omega)$ is an nth-order polynomial, $2n - 1$ are the maximum number of derivatives that can be made zero. Thus the slope is as flat as possible near $\omega = 0$. For this reason the Butterworth response is also known as the maximally flat response.

The poles of Butterworth response $H(s)$ are the zero of the polynomial

$$1 + (-1)^n s^{2n} = 0 \tag{D.4}$$

which are given by

$$s_k = \exp \frac{j\pi(2k-1+n)}{2n}, \qquad k = 1, 2, \ldots, 2n \tag{D.5}$$

These poles are located on the unit circle in the s-plane.

D.2 CHEBYSHEV RESPONSE

Instead of a maximally flat characteristic, an equiripple response is obtained by making the reflection in the network vary according to a Chebyshev polynomial; hence the name Chebyshev response. A Chebyshev polynomial of degree n is defined as

$$T_n(\omega) = \cos(n \cos^{-1} \omega), \quad \omega < 1 \tag{D.6a}$$

$$= \cosh(n \cosh^{-1} \omega), \quad \omega > 1 \tag{D.6b}$$

The first seven Chebyshev polynomials may be written as

$$\begin{aligned}
T_0(\omega) &= 1 \\
T_1(\omega) &= \omega \\
T_2(\omega) &= 2\omega^2 - 1 \\
T_3(\omega) &= 4\omega^3 - 3\omega \\
T_4(\omega) &= 8\omega^4 - 8\omega^2 + 1 \\
T_5(\omega) &= 16\omega^5 - 20\omega^3 + 5\omega \\
T_6(\omega) &= 32\omega^6 - 48\omega^4 + 18\omega^2 - 1
\end{aligned} \tag{D.7}$$

A recurrence relation useful for finding the nth-degree polynomial from $(n-1)$th- and $(n-2)$th-degree polynomials is

$$T_n(\omega) = 2\omega T_{n-1} - T_{n-2} \tag{D.8}$$

Also

$$T_{-n}(\omega) = T_n(\omega) \tag{D.9}$$

For $\omega < 1$, we note that $-1 \leq T_n(\omega) \leq 1$ and increase in magnitude monotonically for $\omega > 1$.

INDEX

A

ABCD matrix:
  filters, 272
  parameters, 892
  phase shifter, 649
  from S matrix, 896
  transmission line, 633
Acoustic anisotropy, 861
Acoustic-Optic Spectrum Analyzers (AOSA), 861
Acoustooptics, 836
  bragg-cell receiver, 856
  figure of merit, 858
  spectrum analyzers, 859
Active multipliers using GaAs FETs, 699
Admittance inverters, 262
AGC, 544
ALC, 544
Amplifier, 483
  biasing networks, 499
  broadband, 509
  cascading, 489
  characterization, 485
  design of, 502
  distributed, 512
  dynamic range, 498
  gain, 485
  intermodulation effects, 497
  matched gain, 495
  maximum gain design, 508
  noise, 484, 487
  non-linearity, 496
  parametric, 483
  power, 514
  stability, 492
  stability factor, 493
  three-terminal, 483
  two-terminal, 483
Amplifier type, 652
Aperture coupling, 187
  analysis, 188
  coupling coefficients, 191
  design, 192
  between different lines, 188
  effect of aperture, 191
  multiple aperture, 195
  single aperture couplers, 187
Asymmetric coupler, 209
Atmospheric attenuation, 2
Attenuators, 666
  MESFET, 669
  pin diode, 667
Autoart™, 787

B

Backward diodes, 562
Band designation chart, 3
Bandpass filter, 244
  design, 247
  printed circuit, 281
  transformation from low-pass, 260
Bandstop filters, 244
  design, 247
  transformation from low-pass, 261
Bipolar transistors, 311
  alpha current gain, 317
  base transport factor, 317
  current gain, 316
  early effect, 319
  emitter-injection efficiency, 317
  I-V characteristics, 314

Bipolar transistors (*Continued*)
Kirk effect, 319
microwave transistor, 320
operation, 312
second order effects, 318
Birefringent bragg diffraction, 861
Boltzman's constant, 375, 395, 488, 498
Bragg angle, 856
Bragg Cell AO receiver, 856
Bragg diffraction, 862
by magnetostatic waves, 865
Broadband amplifiers, 509
comparison of, 511
configurations, 510
design of, 509
Broadband LO mixer, 577
Broadband microstrip detector, 564
Broadside coupled Strip lines, 27, 29
attenuation, 29, 31
characteristic impedance, 29, 31
coupling coefficients, 32
inhomogeneous media, 32
multilayer, 32
Bulk-wave acoustics, 857, 886
Butterworth low-pass filter, 248
Butterworth response, 897

C

CAD, 754
analysis techniques, 756
flowchart, 755
MMIC design, 809
modeling, 756
of non linear circuitry, 773
optimization, 765
using supercomputers, 777
CAD tools, 170, 754
Canonic networks, 144
Capacitors, lumped, 52
interdigital, 52
microstrip, 52
MIM, 52, 815
Schottkey, 52
Cavity resonators, 69
Characteristic impedance, 7, 9
coaxial lines, 11
coplanar waveguide, 24
coupled lines, 28, 29
inverted microstrip, 19
microstrip, 18
slot line, 21, 22
strip line, 14, 15
suspended microstrip, 19
waveguides, 12
Chebyshev filter nomogram, 250
Chebyshev polynomials, 137, 195
Chebyshev response, 898
Chebyshev transformer, 137
Chip combining, 526, 528
Circuit component modeling, 756
Circuit-level combining, 527
Circuit-optimization, 765
Circular waveguide resonators, 73
Coaxial detector design, 567
Coaxial lines, 10
characteristics, 11
discontinuities, 34
equations, 11
resonators, 69
Coaxial resonators, 69
Combiners, 214, 531
chain, 524
device level, 526
N-way, 527
planar, 529
power, 526
radial line, 529
structures, 523
COMINT, 5
Comparison of active and passive mixers, 586
Comparison of FET and diode mixers, 582
Comparison of hybrid and MMIC technology, 827

broadband performance, 831
circuit breaking, 831
cost, 829
design flexibility, 831
reliability, 832
reproducibility, 832
size and weight, 830
Comparison of MIC transmission lines, 24, 26
Conductor materials, 782, 784
Conjugate gradient method, 768
Contact potential, 375
Continuity equations, 306
Coplanar strips, 24
Coplanar waveguides, 10, 13, 23
attenuation, 25
characteristic impedance, 24
effective dielectric constant, 24
Coupled lines, 26
broadside coupled, 198
characteristics, 28
edge coupled, 198
effect of conductor thickness, 31
equivalent current, 157
fin lines, 212
image lines, 212
quasi-TEM, 202
strip lines, 28
structures, 27
TEM, 201
Couplers, 173
aperture-coupled, 177
applications, 177
asymmetric coupler, 206
branch-line, 175
coupled lines, 26, 185
coupling coefficient, 174
direct coupled, 176
directivity, 174
directivity improvement, 231
isolation, 174
multisection, 179
parallel-coupled line, 176
single-section, 175
symmetric coupler, 205
Crank TJS laser, 840
Current-density equations, 305

D

De Ronde coupler, 223
Decibel unit, 889
Design of capacitors, 52
Design of inductors, 46
equivalent circuit, 48
expressions, 48
inductance, 47
loop, 48
resistance, 47
spiral, 48
strip, 48
unloaded Q, 47
Design of resistors, 55
Design of transforming networks, 639
Detectors, 540, 546
applications, 544
broadband microstrip, 564
coaxial design, 567
current sensitivity, 548
design, 562
devices, 562
efficiency, 562
heterodyne, 542
large signal effects, 559
minimum detectable signal, 557
narrowband, 560
noise, 555
noise figure, 557
tangential signal sensitivity, 558
temperature compensated, 561
theory of, 546
types of, 560
video, 542
voltage sensitivity, 549
wideband, 560
Device-level combining, 526
Dielectric films, 782, 785

Dielectric resonators, 98, 436
  coupling, 103
  filters, 284
  frequency tuning, 108
  insulated, 101
  materials, 99
  MIC configuration, 101
  properties, 100, 291
  resonant frequency, 100
  spurious modes, 107
Diffusion coefficient, 305
Digital frequency dividers, 736
  HEMT, 743
  theory, 736
  types, 740
Diodes, 373
  dipletion capacitance, 384
  diffusion capacitance, 385
  equation, 378
  GUNN, 426
  IMPATT, 426
  I-V characteristic, 378, 540, 569
  junction capacitance, 384
  pin, 409
  Schottky-barrier, 386
  step-recovery, 416
  varactor, 401
  varistors, 408
Diplexers, 298
  design, 298
Direct broadcast satellite, 883
Directional couplers, 173. *See also* Couplers
  aperture coupled, 187
  asymmetric, 206
  coupled-line, 185
  De Ronde, 222
  design, 192
  directivity, 135
  distributed type, 211
  interdigital, 209
  multiconductor, 209, 211
  multisection, 204
  quasi-TEM, 202
  real compensated, 230
  real uncompensated, 229
  symmetric, 205
  Tandem, 222
  TEM-line, 198, 201
  Wilkinson, 214
Direct search method for optimization, 766
Directivity, 195, 231
Directorate of arrival (DOA), 856
Discontinuities, 33
  coaxial line, 34
  compensation, 45
  cross junction, 33
  effects, 33
  microstrip, 39
  open circuits, 33
  rectangular waveguide, 37
  short circuits, 33
  step changes, 33
  strip line, 38
  T-junctions, 33
Distributed amplifier, 512
Distributed matching networks, 153
Distributed-type couplers, 211
Dividers, *see* Frequency dividers
Doping charge densities, 375
Double balanced mixers, 576
Double-double balanced mixers, 576
DRFM subsystem in MMIC, 825
Dual gate multipliers, 711

E

Early effect in bipolars, 319
Ebers-Moll model, 323
Effective dielectric constant:
  coplanar waveguides, 24
  inverted microstrip lines, 19
  microstrip lines, 18
  suspended microstrip, 19
Electromagnetic spectrum, 2
Electron bean evaporation, 793
Electronic warfare, 5

Electrooptics, 836
  modulators, 844
  tensor, 844
ELINT, 5
Elliptic waveguide resonators, 76
Etchback technique, 793
Expendable decoys, 883
Experimental filter design, 265

F

Fabrication of MMICs, 810
  active layers, 810
  backside polishing, 816
  dielectric deposition, 815
  isolation, 813
  ohmic contact, 814
  resistor deposition, 815
  Schottky gate formation, 814
  via-hole connection, 816
FANO's constraint relations, 168
Faraday's Law, 307
FET amplifier design, 502
  FET selection, 504
  narrow-band, low noise design, 504
FET mixers, 580
Fibre-optic RF link, 848
  delay lines, 850
  microwave applications, 886
Field effect transistors, 4, 329
  biasing networks, 499
  equivalent current, 342
  I-V characteristics, 331
  low noise, 506
  mixers, 580
  model, 337
  noise analysis, 350
  operation, 330
  parameters, 342
  power, 360
  power combining, 361
  power devices, 517
  S-parameters, 506
Figure of merit:
  acoustooptic, 858
  bipolars, 325
  FETs, 348
  GaAs MESFETS, 346
  pin diodes, 413
  Schottky barrier diodes, 400
  varactor diodes, 405
Filters, 237
  applications, 243
  bandwidth, 288
  combine, 278, 292
  definition of terms, 237, 238
  design from low-pass synthesis, 246
  dielectric resonator, 284
  electric tuning, 293
  experimental design, 265
  finite Q, 288
  group delay, 293
  insertion loss, 240
  interdigital bandpass, 283
  measurements, 245
  modeling, 271
  numerical techniques, 275
  parameters, 239
  power handling, 290
  printed circuit, 277
  realization, 276
  return loss, 240
  size, weight and cost, 287
  synthesis, 246
  temperature effect, 290
  transformation, 256
  types, 243
Filter modeling, 271
  ABCD-matrix, 272
  analysis, 271
  Kirchhoff's equations, 272
  narrow band approximation, 271
Filter responses:
  Chebyshev, 250
  elliptic function, 254
  generalized Chebyshev, 255
  maximally flat, 248

Finline, 10
  coupled, 211
  resonators, 94
Finline resonators, 94
Fixed-frequency oscillators, 435
  design of, 436
  series feedback TDROs, 437
Fixed LO mixer, 577
Flat detectors, 561
FM noise conversion nomograph, 475
FM noise measurement with cavity discriminator, 476
FM noise measurement with delay line discriminator, 476
Frequency counters, 682
Frequency dividers, 676, 717
  applications, 679
  digital, 736
  regenerative, 727
  types of, 717
  using varactors, 717
Frequency halves, 726
Frequency memory loops, 852
Frequency modulated PLL, 682
Frequency multipliers, 676, 687
  applications, 677
  basics, 676
  nonlinear-resistance, 696
  theory of, 687
  types of, 687
  using diodes, 687
  using GaAs FETs, 699
  using varactors, 691
Frequency translation, 680

G

GaAs MESFETS, 304, 307
  amplifiers, 484
  attenuators, 669
  bipolar, comparison with, 366
  for control circuits, 601
  deep levels, 356
  doping profile model, 356
  equivalent circuit, 342, 602
  figure of merit, 396
  frequency multipliers, 676, 699
  high impedance state, 602
  I-V characteristics, 331
  large signal operation, 363
  low impedance state, 602
  maximum operation frequency, 347
  MMICS, 806
  model, 337
  noise analysis, 350
  operation, 329, 334
  oscillators, 426
  parameters, 342
  phase shifter using, 651
  power gain, 347
  small signal model, 341
  S-parameters, 427, 429
GaAs varactor, 403
Gain of amplifier, 485
Gallium arsenide, 4
  FETs, 4
  MMIC, 814
GDS II-CALMA, 787
Global positioning system, 883
Gradient method for optimization, 766

H

Heterodyne detectors, 542
High-pass filters, 244
  design, 247
  transformation from low-pass, 258
Hurwitz polynomials, 143
Hybrid MIC, 792, 883
  comparison with MMIC, 827
  design factors, 793
  examples, 796
  fabrication, 792
  miniature, 796
Hybrids, 173

branch line, 181
design of, 178
losses in, 231
matched *T*, 181
90° hybrids, 178
ring form, 181, 182

I

Ideal diode equation, 378
deviation factors, 381
Image mixer, 578
Impedance inverters, 262
Impedance matching circuits, 153
impedance transformers, 157
Impedance matching networks, 131
applications, 131
gain-bandwidth limitations, 159
lossless, 138
one-port/two-port networks, 132
quarter wavelength transformer, 136
real frequency approach, 162
transmission line, 133
Inductors:
design of, 46
loop, 48
spiral, 48
strip, 48
unloaded *Q*, 49
wire, 51
Insertion loss, 240
measurement, 245
for switches, 606
Integral Optic Spectrum Analyzers (IOSA), 860
Interacting resonant structures, 92
Interdigital capacitors, 52
Inverted microstrip lines, 10, 13, 19
characteristic impedance, 19
effective dielectric constant, 21
Ion implantation, 814

J

J-Inverter circuits, 264
Johnson noise, 489

K

K-Inverter circuits, 264
Kirk effect in bipolars, 319
Kuroda's identities, 153, 156

L

Lasers, 836
board stripe, 842
crank TSS, 840
diode, 838
modulation BW, 843
window buried, 840
window stripe, 840
Limiter circuits, 601
design of, 659
methods for limiting, 659
in microstrip, 664
pin-diode, 663
Liquid Phase Epitaxy, (LPE), 813
Loaded line phase shifters, 631
Loop inductors, 48
Lossless matching networks, 138
approximate solution, 150
insertion loss, 151
ladder networks, 145
network theory, 141
synthesis, 142
transfer function, 138
Low-pass filters, 244
Butterworth prototype, 248
design, 247
microstrip, 277
transformation to desired band, 257
Lumped elements, 45
capacitors, 52
design, 46, 52

Lumped elements (*Continued*)
inductors, 46
resistors, 55

M

Magnetostatic wave technology, 887
MASER, 484
Mask fabrication, 790
Mask layout for MMIC, 786
Mason's gain, 323
Matched gain of amplifier, 495
Matched hybrid *T*, 181
S matrix, 182
Matching networks, 131
Measurement of oscillator parameters, 467
FM noise, 474
load pull setup, 443
low power setup, 468
medium power setup, 469
paramcters, 467
pulling figure, 472
Mesa etching, 814
Metric prefixes, 889
MiCad™, 787
Microstrip lines, 10, 13
broadside coupled, 198
capacitors, 52
characteristics, 16, 18
closed-form expressions, 17
coupled lines, 29
discontinuities, 39
edge-coupled, 198
effective dielectric constant, 17, 18
filters, 278
impedance, 16, 18
limiters, 664
maximum frequency of operation, 17
phase constant, 16, 18
radiation *Q*, 19
resonators, 77
Microstrip resonators, 77
circular disk, 79
elliptic disk, 90
hexagonal, 89
rectangular, 79
ring, 82
triangular, 84
Microwave acoustics, 886
Microwave bipolar transistor, 319
equivalent circuit, 323, 326
figure of merit, 325
frequency multiplexer, 676
maximum oscillation frequency, 324
noise figure, 326
operation, 320
unilateral gain, 323
Microwave control circuits, 601
GaAs MESFETS, 601
limiters, 659
phase shifters, 626
switches, 604
variable attenuators, 666
Microwave modulation of optical sources, 838
direct modulation, 838
indirect modulation, 843
Microwave planar circuits:
applications, 4
history, 3, 4
Microwaves:
band designation, 2
characteristics, 2
spectrum, 2
MICs, 1, 4, 781. *See also* MMIC
dielectric resonator, 101
hybrid, 792
mask fabrication, 790
mask layout, 786
materials, 782
millimeter wave, 884
substrates, 782
Millimeter waves:
band designation, 2
characteristics, 2

MMICs, 884
spectrum, 2
MIM capacitors, 52, 815
Miniature hybrid MIC, 756
design factors, 801
examples, 801
fabrication, 758
Minimum detectable signal, 557
Mixers, 540, 569
active/passive comparison, 586
analysis, 583, 586
block diagram, 569
comparison of, 581, 582
conversion gain, 583
design, 573, 591
noise analysis, 587
noise figure, 589
theory, 570
types of, 576
MMIC, 4, 781, 804, 883
amplifier, 530
capacitors, 52
comparison hybrids, 827
design factors, 807
design procedure, 808
discontinuity effects, 34
dividers, 736, 744
examples of, 817
fabrication, 810
GaAs, 805
history, 805
inductors, 51
multipliers, 679, 698
optical tuning, 886
resistors, 55
substrates, 806
subsystems, 884
Molecular Beam Epitaxy (MBE), 813
Multiconductor couplers, 209
design of, 210
Multiplexers, 193
common junction, 295
contiguous, 296
noncontiguous, 296
realization, 299
shunted manifold, 295
techniques for, 194
Multiplier, *see* Frequency multipliers
Multisection couplers, 179, 204
Multisection Wilkinson coupler, 215

N

Narrowband detectors, 560
Network synthesis, 163
design, 163
insertion loss, 163, 164
microwave realization, 169
parasitic element compensation, 166
topology selection, 166
Noise, 555
analysis for mixers, 587
flicker, 555
shot noise, 555
thermal, 555
Noise figure, 488
amplifiers, 484, 487, 498
bipolar/FET comparison, 366
detectors, 555
FETs, 350
GaAs MESFETs, 350
microwave bipolar, 326
mixers, 589
Non-linear circuits, 773
CAD of, 773
harmonic balance design, 775
optimization, 776
Non-linear resistance multipliers, 696
Norton's transformations, 158
N-way cobiners, 527

O

Optical control of devices, 455, 867, 886
advantages, 867

Optical control of devices (*Continued*)
injection locking, 873
oscillator tuning, 872, 886
switching applications, 869
Optical fibre, 836, 838
Optical techniques for millimeter waves, 875
Optimization, 765
conjugate gradient method, 768
direct-search method, 766
gradient method, 766
pattern search, 767
sensitivity analysis, 770
Oscillators, 426
diode, 426
fixed-frequency, 435
frequency tunable, 427
measurements, 467
negative-resistance, 428
optical tuning, 872
oscillation condition, 434
quartz crystal, 427
stability conditions, 431
transistor, 427
voltage controlled, 462
wide band tunable, 456
YIG-tuned, 457

P

Parallel resonant circuit, 66
fractional bandwidth, 67
quality factor, 66
resonant frequency, 66
Parametric amplifiers, 483
Parametric division using varactors, 717
Parametric multipliers using varactors, 691
Passive multipliers using diodes, 687
Pattern search method, 767
Phase shifters, 601
active, using SPDT amplifiers, 653
amplifier-type, 652
design of, 626
loaded line, 631
reflection type, 637, 639
switched line, 628
switched network, 648
types of, 627
using $\lambda/4$ transforming network, 644
using segmented gate MESFET, 656
using vector modulator circuits, 655
Phase velocity, 7
Phased array transducer design, 861
Physical constants, 891
Pin diode, 409, 501
attenuators, 667
device physics, 409
equivalent circuit, 412, 602
figure of merit, 413
junction capacitance, 411
limiters, 663
parameters, 416, 608
resistance, 411
switching speed, 412
transition time, 412
Pi-section matching networks, 154
Planar combiners, 529
Planar transmission lines, 13
characteristics, 13
resonators, 77
Planck's constant, 840
Plate-through technique, 793
P-n junctions, 374
Point contact diodes, 562
Poissons equation, 306, 375, 392
Power FETs, 358
Prescalers, 679
Power amplifiers, 514
design of, 514, 523
efficiency, 515
FET selection, 515
internally matched power FET, 525
large signal analysis, 519
load-pull measurement, 521

Power combining techniques, 526
  circuit-level, 527
  device-level, 526
  N-way, 527
  structures, 531
Power dividers, 214
  Wilkinson-unequal split, 221

R

Radar Warning Receiver (RWR), 836, 856
Radial line combiners, 529
Range Gate Pull-Off (RGPO), 853
Rate equations, 839
Reactance function, 143
Real compensated coupler, 230
Real uncompensated coupler, 229
Rectangular waveguide resonators, 72
Rectifying junctions, 373
Reentrant coaxial resonators, 71
Reflection type phase shifters, 637
Regenerative dividers, 727
  advantages, 736
  theory, 729
Relaxation resonance frequency, 839
Resistive films, 782, 785
Resistors, 55
  characteristics, 56
  design of, 55
  thin film, 55
RF/optical interaction, 855
Richardson constant, 395
Richard's transform, 169
Resonator measurements, 120
  single-port resonator, 121
  two-port resonator, 124
Resonators, 65
  cavity, 69
  coaxial, 69
  coupling of, 68
  damping factor, 67
  dielectric, 98
  finline, 94
  fractional BW, 67
  interaction, 92
  loaded quality factor, 67
  measurements, 120
  microstrip, 77
  parameters, 65
  quality factor, 66
  rectangle waveguide, 72
  reentrant coaxial, 71
  resonant frequency, 65
  SAW, 426
  YIG, 109
Return loss, 240
  measurement, 245
Rubylith, 787

S

SAW cavity resonators, 426
SAW devices, 857, 887
Schottky barrier:
  capacitors, 52
  current equation, 396
  detectors, 562
  diodes, 386, 562, 678
  equivalent circuit, 400
  figure of merit, 400
  image-force lowering, 389
  junction capacitance, 396
  junction current, 394
  materials, 397
  model, 392
  series resistance, 398
  surface effects, 388
Self-oscillating doublers, 715
Semi-conductor equations, 304
Semiconductor materials, 307
  parameters, 308
Sensitivity analysis, 773
Series resonant circuit, 66
  fractional bandwidth, 67
  quality factor, 66
  resonant frequency, 66
Shear waves, 862
SIGINT, 5
Single balanced mixers, 576

Single ended mixers, 576
Single gate multipliers, 706
Single section coupler, 175
  frequency response, 203
SI units, 888
Slot line, 10, 13, 21
  characteristic impedance, 21, 22
  line wavelength, 21, 22
  propagation mode, 21
S-matrix analysis, 760
S-parameters, 246
  in ABCD terms, 894
  amplifier, 486
  GaAs FET, 431, 506
  oscillators, 428
  transistor characterization, 429
SPDT switch, 611
Special cases of De Ronde couplers, 228
Spiral inductors, 48
  typical parameters, 51
SPST switch, 604
SP3T switch, 615
Sputtering, 793
Stability of amplifier, 492
Stability factor for amplifier, 493
Step-recovery diodes, 416
  equivalent-circuit, 421
  extraction time, 418
  frequency divider, 677
  frequency limits, 421
  physics, 416
  snap angle, 420
  transition time, 419
Strip inductors, 47, 48
  design, 49
Strip line, 10, 13
  broadside coupled, 29
  characteristics, 14
  coupling, 28
  discontinuities, 38
  equations, 14, 15
  equivalent circuits for resonators, 279
Substrate materials, 782
Sum-enhanced mixer, 580
Suspended microstrip lines, 10, 13, 19
  characteristic impedance, 19
  effective dielectric constant, 20
Switchable length-type reflection phase shifter, 638
Switchable reactance-type reflection phase shifter, 638
Switched line phase shifters, 628
Switched network phase shifters, 648
Switches:
  design of, 604, 612
  device reactance compensation, 609
  insertion loss, 606, 609, 616
  isolation, 607, 616
  series configuration, 605, 612
  series-shunt configurations, 613
  shunt configuration, 606, 609, 612
  SPDT, 611
  speed, 621
  SPST, 604
  SP3T, 615, 616
  wideband, 617
Symmetric coupler, 205

T

Tandem couplers, 222
Tangential signal sensitivity, 558
Tapped delay lines, 851
Temperature compensated detectors, 561
Three section maximally flat coupler, 205
Time of Arrival (TOA), 855
Transistor dielectric resonator, 427
  backward, 562
  equivalent circuit, 546
  frequency multiplication, 687
  impedance, 551
  junction resistance, 547
  laser, 838

matching, 553
packaging, 554, 564
point-contact, 562
Transistor dielectric resonator diode, 427
design of, 438
digital compensation, 451
electronic tuning, 455
parallel feedback, 444
temperature stability, 449
tuning of, 452
varactor tuned, 453
Transistor oscillators, 427
configurations, 430
S-parameters, 429
Transmission line matching networks, 133
double stub matching, 134
quarter wavelength-transformer, 136
series single section, 133
single stub matching, 134
tapered lines, 138
Transmission lines, 7
attenuation, 8
basics, 7
characteristics, 101
dielectric rods, 7
general formulas, 9
lumped parameters, 8,
parameters, 8
power handling, 8
types of, 10
waveguides, 7
Transmission matrix, 201
T/R Modules, 5, 801, 883
T-section matching networks, 155

V

Vacuum evaporation, 793
Vapor-phase epitaxy (VPE), 813
Varactor diodes, 401
capacitance variation, 403
cutoff frequency, 405
equivalent circuit, 402
figure of merit, 405
frequency dividers, 717
frequency halver, 726
frequency multipliers, 691, 696
properties, 401
resistance variation with bias, 407
tuned DRO, 453
Variable attenuators, 601, 666
design of, 666
MESFET, 669
pin diode, 667
Varistors, 408, 678
I-V characteristics, 408
resistance, 408
VHSIC, 1
Video detectors, 542
Voltage Controlled oscillators, 462
comparison with YIG, 467
electronic tuning, 462
post-drift tuning, 466
settling time, 465
typical circuit, 464

W

Waveguides, 10
characteristic impedance, 12
circular, 10, 11
discontinuities, 37
equations, 12
fundamental mode, 13
properties, 12
rectangular, 10, 11
resonators, 72, 73, 76
Wideband detectors, 560
Wide-band tunable oscillators, 456
analysis, 456
design, 456
voltage controlled, 462
YIG-tuned, 457
Wilkinson couplers, 214
Window buried laser, 840
Window stripe laser, 840
Wire inductors, 51

Y

YIG resonators, 109
  equivalent circuitry, 115
  frequency of operation, 112
  magnetic tuning, 119
  material characteristics, 110
  quality factor, 112
  resonant frequency, 111
  spurious modes, 118
YIG-tuned oscillators, 457
  applications, 459
  comparison with VCO, 467
  parasitic oscillations, 459
  performance, 461
  tuning linearity, 460
Y-parameters, 429

Z

Z-parameters, 429